The Hvac/r Professional's Field Guide to
ICE MACHINE SERVICE

BNP BUSINESS NEWS PUBLISHING COMPANY
TWENTY YEARS OF TECHNICAL BOOK PUBLISHING

The Hvac/r Professional's Field Guide to
ICE MACHINE SERVICE

Richard Jazwin

BNP BUSINESS NEWS PUBLISHING COMPANY
Troy • Michigan

Copyright © 1990
Business News Publishing Company

All rights reserved. No part of this book may be reproduced or transmitted in any form or by any means—electronic or mechanical, including photocopying, recording or by any information storage and retrieval system—without written permission from the publisher, Business News Publishing Company.

Library of Congress Cataloging-in-Publication Data
Jazwin, Richard.
The Hvac/r professional's field guide to ice machine service/
Richard Jazwin.
p. cm.
1. Ice—Manufacture. I. Title.
TP492.7.J39 1990
621.5'8—dc20 90-2103
CIP

Printed in the United States of America
7 6 5 4 3 2 1

The Hvac/r Professional's Field Guide to
Ice Machine Service

Table of Contents

DISCLAIMER

Neither the author or the publisher assume any liability for any information contained in this book. The service manuals are reprinted unedited and the author and publisher are not responsible for their content. The text portions of this book are a generic guide to ice machine operation and troubleshooting and are intended to offer only a basic understanding of ice machines. The author and publisher have neither liability or responsibility for the use, misuse, misunderstanding or misapplication of any of the information contained herein. Damage of any kind, including but not limited to personal or material alleged to be caused by information contained herein, is not the liability or responsibility of the author or publisher.

DEDICATION

To my wife Jan, who loves commas.
I am thankful that she loves me more.

Introduction

Ice machine repair is based on a principle common to all types of mechanical service: in order to fix a particular machine, the repair technician must first *understand* it. Information is the key element to successful service.

In the automotive industry, manuals covering service procedures for many different models of cars have been available for years. These books have made the mechanic's job easier by providing a variety of service data in one convenient source. The need for such a collection of information also exists in the field of ice machine service. Few technicians can master the enormous variety of units in operation today without some reference. It is my hope that this book will begin to fill the need for information on ice machines.

This information is more important than ever because the ice machine industry is rapidly expanding. The biggest contributor to this growth is the extensive construction of convenience stores, fast food outlets and restaurants. As a result, the number of ice machines in use has increased dramatically. Industry sources estimate that the average convenience store uses over 650 lbs. of ice daily. A single ice machine operating in a convenience store produces over 237,000 lbs. of ice annually. Considering the fact that ice machines operate 24 hours a day, 7 days a week, there is little doubt that they will require service by a qualified technician.

Today's ice machine is an excellent piece of equipment. The fact that a substantial number of ice machines receive little or no regular service and yet continue to operate is a credit to the technology of the manufacturers. However, while all manufacturers agree that ice machines of the future will be easier to service, understanding the ice-making process will always be a prerequisite for successful machine service.

All ice-making systems have two things in common: all are based on the same laws of refrigeration, and the raw material to make ice—namely, water—is a common element to all units. However, there the similarity ends. Each manufacturer's machine uses a different technique to produce ice. The technique employed is called the unit's *sequence of operation.* Because the operating sequence varies from manufacturer to manufacturer, unit-specific information is quite valuable. The manufacturer's service manuals included here supply the technician with the operating sequences of many different units. These manuals are reprinted unedited, and contain operating sequences for both cube and flake ice machines.

For various reasons, not all service manuals for ice machines manufactured today are included here. However, the exclusion of any brand from this book does not imply that there is anything wrong with that brand. Some of the manufacturers chose not to participate in this project. Others manufacture ice machines that are too specialized, or produce quantities of ice that are too large, for inclusion in this book. I attempted to limit coverage of machines to those brands that will be encountered in light commercial service, producing between 200 to 800 lbs. of ice per day.

In many cases the service manual provided will not exactly match the model being serviced. This is due to the fact that as sales of a particular model increase, the manufacturer makes refinements to controls and components. Remember, upgrading is the rule of the marketplace.

It should be noted that factory service seminars are the best source of unit-specific information. Most manufacturers schedule these seminars regularly. While it is not possible for not every technician to attend every seminar, it is recommended that he attend as many as possible. This book is not a replacement for attendance at a factory seminar.

In closing, I wish to thank all the manufacturers who have contributed service manuals. Providing the field technician with readily available, unit-specific information will certainly result in more successful ice-maker installation, service and maintenance.

Richard Jazwin
Chief Instructor, Universal Technical Institute, Phoenix, Arizona

The Ice-Making Cycle

There are two basic types of ice makers: flakers and cubers. *Flakers* produce flaked ice, which is used primarily for product packing. Flaked ice is also often used in hospitals, restaurants, display cases and even soft drink service, but it is chiefly used to package perishable products en route to market. *Cubers* create ice cubes in distinctive shapes and sizes. Ice cubes are generally formed into squares, rectangles and circles, but may take any shape. Ice cubes are used to cool products, such as beverages, without rapidly watering them down or otherwise altering their quality.

An ice cube's shape is designed to allow maximum surface area to contact the product requiring cooling. The more contact the cube surface exposes to the product, the faster the heat transfer occurs. This is important because rapid heat transfer results in faster product cooling.

Cube shape is also regarded as a trademark of the unit's manufacturer. Each manufacturer produces a unique cube shape. An experienced technician can usually look at an ice cube and tell what make and model machine produced that particular cube.

The ice machine itself is called the *head* and is typically placed on top of the ice storage bin. The *bin,* underneath the ice machine, is a separate component sized for the owner's particular ice production requirements.

Each manufacturer uses a different operating sequence to accomplish the ice-making process. However, while manufacturers may add or delete controls or components to improve performance, the four basic steps of producing ice are the same for all units. These basic steps are:

1. Supply water to the machine.
2. Freeze the water into ice.
3. Remove the ice from the evaporator once the cube reaches the proper size (*harvest* process).
4. Start the machine on the next ice-making cycle when harvest is complete, or turn the machine off if the storage bin is full.

Supplying Water to the Unit

In supplying water to the ice machine, some method of controlling inlet water quantity is necessary. As ice forms, the water level in the machine drops. As a result, additional water must be supplied. The water control system must always allow sufficient water to enter the machine so ice production is not hampered.

Water pressure poses another important consideration. Water entering an ice machine should have a pressure of at least 20 psig but not more than 80 psig.

Controlling Water Quantity

Primary water control is accomplished by using either a float valve or a water solenoid. When a *float valve* is used, water constantly flows into the machine. The formation of ice causes the water level in a float chamber or reservoir to gradually decrease. As a result, the float drops and admits more water through an orifice valve connected to the float assembly.

In a unit using a *water solenoid*, the solenoid valve admits water for a fixed time period. During this interval, an electrically operated solenoid remains open. The length of this interval determines how much water is introduced.

One manufacturer's unit employs a unique alternate method utilizing a solenoid valve. Rather than relying on a set time period, the manufacturer calculates the weight for the proper quantity of water. When the proper weight is attained, a small tank drops and de-energizes the open water solenoid, stopping water flow.

In addition to the water necessary to make ice, some units require additional water to aid in harvest. The additional water is dispersed over the finished cube to actually help "melt" the cube off the evaporator. While hot gas defrost is the primary method used in harvesting ice, some units assist the hot gas defrost with water spray. If water is used to assist harvest, sufficient water for both harvest and ice production is essential for proper operation.

Flaker ice makers generally utilize a float that allows water to enter the evaporator as ice is produced. The float is mounted so the water level in the float chamber maintains a level that is consistent with the water level in the evaporator. The evaporator water level is regulated so water is always in contact with liquid refrigerant. As ice is produced, the water level in the evaporator drops, resulting in a lower float chamber water level. The lower level in the float chamber allows more water to enter, replacing the water that has turned into ice.

Cubers sometimes use a float or a solenoid. A float maintains a constant level in a reservoir. The water is "picked up" from the reservoir by a water pump and flows across the cold evaporator to make ice. As ice is produced, the float allows make-up water to enter the reservoir. If a cuber employs a water solenoid, rather than a float, the solenoid opens for a set duration allowing a measured amount of water to enter the machine and produce a batch of ice. The manufacturer's design determines whether a float or water solenoid is used as a primary water control.

Controlling Water Level

Water level is a fundamental consideration for all ice machines. If the water level is kept too low, production decreases and serious problems may result. This is demonstrated by comparison to a refrigeration system. In any refrigeration system, insufficient load results in poor operation, valve failure and possible compressor burn-out due to liquid refrigerant slugging. In an ice-maker, water is the load, and problems are apt to result if the load is insufficient.

Conversely, if the water level is too high, decreased production is also possible since the load is too great. In flakers, excessive water level places the water above the refrigerant-filled coils, retarding the freezing process. In some machines, this condition forces the ice through unfrozen water, making the ice extremely wet.

Freezing the Water

For water to change into its solid state, it must freeze. To accomplish this state change, the water temperature must drop as quickly as possible to 32° F or less. Speed is advantageous to the freezing process because the longer it takes to freeze water, the more prolonged the ice-making cycle. Moreover, longer cycles result in lower production. In order to freeze water, an evaporator must operate at a suction pressure equivalent to a refrigerant temperature below 32° F. The evaporator must always be colder than the water to maintain proper heat transfer.

Suction Pressure

Flakers do not switch between ice-making and harvesting modes. As a result, flakers operate at a constant suction pressure. When a flaker produces ice, the ice thickness on the evaporator is constant. Using R-12 refrigerant, a flaker maintains a suction pressure of 9-12 psig with an equivalent evaporating temperature of 0 to 5° F. Flaker ice-makers typically employ an automatic expansion valve or capillary tube as a metering device.

Unlike flakers, however, most cubers operate at constantly changing suction pressures and temperatures. When water initially comes in contact with the evaporator, the water is warm, creating a high heat load. Water continually cascades over the evaporator, causing the water temperature to decrease. The decrease in water temperature decreases the load on the evaporator. The decreased load causes the evaporator pressure and temperature to drop. Ice formation also "insulates" the water from the refrigerant, and further causes the evaporator pressure and temperature to decrease. Cubers use a thermostatic expansion valve (TEV) or capillary tube as a metering device. A cap tube or TEV can usually compensate for fluctuating suction pressures.

In a cuber, decreasing pressure and temperature indicates that the ice cube is growing larger. Some machines use this decline in pressure and temperature to signal the start of harvest cycle. When the cube reaches the required size, the lowered evaporator pressure indicates full cube size and harvest begins.

Evaporators

Evaporator types vary from machine to machine. Flakers typically utilize a metal cylinder called a cylindrical steel evaporator. Embedded within the cylinder are the refrigerant coils. The height of the refrigerant coils is such that the liquid refrigerant is maintained at a level equal to the water level in the evaporator. Maintaining the liquid refrigerant at the same level as the water ensures even heat transfer and maximum cooling.

Cubers use many different styles of evaporators. A common style is the mold type, which produces individual cubes similar to those made in household freezers. If the molds are spaced far apart, harvest of individual cubes is possible. If the molds are close together, the cubes tend to grow past the outer limits of each mold and bridge (freeze together), forming a slab of cubes. Bridging accelerates harvest because the slab weighs more than the individual cubes, and the added weight allows the cubes to clear the evaporator sooner.

Some cubers utilize a channel-type evaporator. This design has paths (channels) that the water flows through. The channels contain cold high spots, and the refrigerant coils are located behind these areas. When the water comes in contact with the cold high spot, it begins forming an ice cube.

Other types of evaporators include the inverted mold or flat plate. In the end, however, the construction and design of the evaporator ultimately depends on the manufacturer.

Harvest

Once an ice cube is formed properly, it is harvested. Harvest must occur as quickly as possible so a new freeze cycle can begin. Ice machines generally use hot gas defrost to accomplish harvest. Hot gas defrost involves routing discharge gas from the high side of the system directly to the evaporator. The presence of hot gas in the evaporator halts the freezing process and initiates separation of the cubes from the evaporator. After a given interval the cubes fall, individually or in a bridged slab.

Melting some ice in harvest is normal since the ice must be freed from the evaporator surface. However, if a unit remains in harvest too long, some of the newly formed cubes begin to melt beyond design limits. The smaller cubes that result weigh less than they should. In addition, the excess time spent in harvest creates a longer cycle, resulting in decreased overall ice production.

The first requirement for successful harvest is to initiate harvest when the cube is sized properly. If the machine continues to freeze ice past the time necessary to size the cube, the result is a prolonged freeze cycle. This creates larger cubes but fewer total cubes in a 24-hour period. Clearly, proper harvest initiation requires two important capabilities: that the machine be able to sense when the cubes are properly sized and start the harvest immediately.

Harvest Initiation: Cuber

Various methods are used to initiate harvest, depending on the unit's specifications. The most common methods of initiating harvest on a cuber include:

- time
- refrigerant pressure
- refrigerant temperature
- water weight
- cube size
- float level

Time: In a time-initiated cycle, a timing device is set to allow a certain amount of time for freezing and sizing cubes. After the cube is sized, a timed cycle is established for hot gas to harvest the cube.

Pressure Defrost: The process of water changing into ice affects the refrigerant pressure and temperature. The refrigerant pressure decreases as ice cubes grow larger because the ice insulates the refrigerant in the coil from the water flowing over the evaporator. The large cube and colder water have less heat to transfer to the refrigerant, and the lower rate of heat transfer causes the refrigerant pressure to drop measurably. The decreased refrigerant pressure indicates that the cube has grown to finished size.

A pressure control is installed in the hot gas circuit. The lower evaporator pressure causes contacts in the control to close. The low pressure control is reverse acting. The contacts in a reverse acting low pressure control (RALPC) close on a pressure drop. When the contacts actually close, a circuit is made to the hot gas solenoid and harvest begins.

If the bridge size needs to be increased after the low pressure control has closed, a solid state device called a potentiometer may be employed. The closing of the RALPC establishes a circuit to the potentiometer which allows additional freeze time to finish the bridge to proper size. At that point, harvest begins.

A cuber may start the freeze cycle at 27 psig (no cube). As the cube grows, the pressure drops. Once the RALPC reaches the setting of 11 psig (cube approaching size for harvest), the RALPC closes. The closed RALPC sets up a series of events that causes hot gas to melt the formed cube away from the evaporator.

Refrigerant Temperature: As the ice cube becomes larger, the suction line temperature drops. A bulb or sensor mounted on the suction line measures suction line temperature. When a preset temperature is sensed, the bulb or sensor closes contacts and harvest begins.

Weight of Water: To utilize water weight as an indicator for harvest, one manufacturer's unit starts out with a tank holding a measured amount of water. As ice is forms on the evaporator, this tank becomes lighter. At a certain point the tank overcomes spring pressure and rises, triggering a switch that initiates hot gas defrost.

Cube Size: When cubes reach finished size, they come in contact with metal probes mounted in front of the evaporator. The cube then functions as a conductor and completes the circuit initiating hot gas defrost.

Float level: As the water becomes ice, the float level drops. This acts as a signal to the machine that the water is frozen. Inside the float, a coil allows increased current flow to a solid state board. The solid state board sets up the events necessary to begin harvest.

Keep in mind these two important points concerning the harvest cycle in a cuber:

1. When the ice is ready, a signal must be issued to initiate hot gas defrost.
2. The harvest cycle is very hard on an ice machine. Any unnecessary time spent in harvest shortens machine life and lowers production.

Harvest Initiation: Flaker

Initiating harvest in a flaker is actually much easier than in a cuber. Since a flaker operates at a constant suction pressure, ice is always present on the evaporator surface. Harvest is usually performed by means of an auger (cutting blade) which scrapes flakes of ice off the evaporator. The auger turns constantly and moves the flaked ice up and out of the freezing chamber. Because ice production and harvest are an ongoing process in a flaker, no signal or mode switch is necessary to send the machine into a separate harvest cycle.

Next Ice Cycle *or* Turning Off the Machine

When harvest is complete, either the next ice cycle must start, or (if the ice storage bin is full) the machine must shut off until more ice is needed. Either action is performed by a bin switch, which can be thermostatic (temperature-operated) or mechanical.

If the bin switch is thermostatic, a bulb or thermistor comes in contact with the ice in the bin and issues a signal indicating the bin is full. The bin switch is wired into line power; when the switch receives bin full signal, line power is broken.

If the bin switch is mechanical, depressing a button on the switch or breaking a light beam causes the machine to turn off. Mechanical bin switches are also used when solid state components are involved in the harvest process. When ice falls through a trap door, the door depresses the mechanical bin switch and causes a momentary electrical outage. This momentary outage causes a solid state board to reset its switches and returns the machine to its freeze mode. If the machine has been turned off because the bin is full, removing ice from the bin causes the machine to turn on again and resume the ice-making process.

Solid State Controls

Solid state boards with electromechanical controls are used increasingly more on ice-makers to:

- Finish sizing the cube.
- Control bridge thickness.
- Begin the harvest cycle.
- Terminate the harvest cycle.
- Re-enter freeze mode.
- Stop the machine when the bin is full.

Solid state boards generally use timer circuits designed to accomplish specific tasks in the ice-making process. Replacement boards are readily available from the ice machine distributors.

When troubleshooting a solid state board, the primary questions to ask are:

1. Is power available to the board when the proper time in the sequence occurs?
2. Is sufficient power issued from the board when a specific component requires energizing?

Production

Essentially, an ice machine serves one function: to make ice. Consequently, it is the service technician's responsibility to ensure that the machine produces ice at its rated capacity. Unfortunately, several factors work against efficient operation. Ice machines operate 24 hours a day, and they are often located in areas where their production is actually inhibited. For example, water supply piped into the same line as the dishwasher burdens the ice maker with the disposal of an excessive heat load. The larger the heat load on the machine, the longer the ice-making process takes. In addition, scale, rust and bacterial growth all hinder ice production.

Entering water and air temperatures greatly affect the rated production of an ice machine. As air and water temperatures increase, production decreases. For example, a cuber rated at 400 pounds per day produces that much ice *only* if provided with correct air and water temperatures. Manufacturers furnish production charts in their service manuals. These charts adjust production based on the actual ambient air temperature and entering water temperature. Final production checks should always be compared to manufacturer's specifications using these charts.

Ice machines are rated according to the amount of ice produced in a twenty-four hour period (1440 minutes). However, the approach to figuring production for a given ice-maker differs between cubers and flakers. The procedures for each type are listed on the opposite page.

Calculating Production for a Cuber:

The amount of a time a cuber requires to make ice varies with each unit. To figure production for a cuber, follow these steps:

1. Time a full cycle. A cycle is defined from start of freeze to the actual drop into the bin. Use a stopwatch and be sure your time is accurate.
2. Divide the total cycle time into 1440. The resulting number indicates how many times per day the machine drops a load of ice. For example, say the machine took 30 minutes to drop a full load; 30 divided into 1440 equals 48 which means the machine drops ice 48 times in each 24 hour period.
3. Determine how much a load weighs. Weigh a single load of ice, and do not use a start-up load. At least let the machine run through one full cycle before attempting to weigh a load.
4. Multiply the load weight (step 3) by the number of loads per day (step 2). For example, if the load weighed 5 pounds, total production would be 240 pounds per day (5 x 48).

Calculating Production for a Flaker:

Since ice production in a flaker is constant, it is much easier to measure:

1. Select an interval and time a cycle for that duration. For example, choose a 20 minute cycle and divide this number into 1440. In this case, 1440 divided by 20 = 72. This translates into the fact that the flaker drops ice 72 times in a 24-hour period.
2. Multiply the load weight by the number of loads per day (step 1). In this example, if each ice drop weighs 10 pounds, the flaker produces 720 pounds of ice daily. ♦

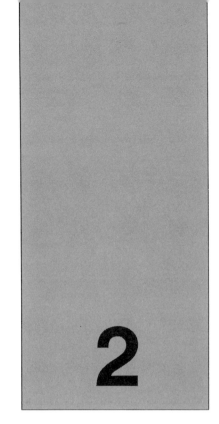

Troubleshooting Ice Machines

This section covers general troubleshooting information for basic failures common to all ice machines. (Specific troubleshooting information is provided in the manufacturers' service manuals reproduced in the appendices). By using both the information in this section and the service manuals, the technician should be able to identify the cause of most ice machine failures.

This section also identifies areas of service that are unique to ice machines. Ice machines, like any other equipment, occasionally malfunctions. However, there are two major differences between troubleshooting an ice machine and troubleshooting most refrigeration equipment. First, since different models use different sequences to make ice, the technician must understand the sequence for the particular machine requiring service for successful repair. In order to understand what caused the failure, the technician must first know what the machine should be doing at each step in the ice-making cycle so, at any given point in the sequence, the failing component becomes identifiable.

Second, when an ice machine fails, the interval between actual failure and the service call is often prolonged because most owners only become aware that their ice machine has failed when the bin level is low or empty. The service call may be even further delayed because a malfunctioning ice machine may continue to produce some ice. As a result, the lower production may not be evident for a considerably long time. This type of delay usually works against both the technician and the longevity of the machine.

Owners of ice makers often know more about their machine than they are aware or care to admit, so ask questions before attempting to repair the unit. You may discover that the machine actually does not need repair. The following questions are designed to identify areas that may cause problems:

- *When did the user first notice a problem?* Customers tend to postpone service calls until absolutely necessary. Finding out when the problem initially occurred may save valuable troubleshooting time.
- *Has the machine made any strange noises?* Customers are aware of how their equipment normally sounds. Strange noises generally indicate mechanical problems.
- *Has the water and power supply been constant, or has the location experienced a power failure?* Water turned off for new construction, and power brown-outs or outages due to construction or storms can severely affect overall production.
- *Did the ice level in the bin drop over the past few weeks or last few days?* If the drop occurred over a long period of time, business conditions may be the cause. For example, if the owner has run a special promotion, the machine may not be able to keep up with the increased demand.
- *Has the owner installed any new equipment recently?* Perhaps a new hot water heater was installed and the ice machine piped to the hot water supply rather than the cold water. Also ask if the heat load in the ice machine area may have increased since a new oven or dishwasher was installed.
- *Did the machine actually ever meet the owner's requirements?* An ice maker sold in the winter may not be sized to meet the increased demand of summer soft drink service.

Troubleshooting Tips

Bin Level

A low bin level does not necessarily indicate mechanical problems. Many problems unrelated to machine failure can cause a low bin level. For example, new construction in the area may require the water to be turned off for a short time. This causes the flaker's evaporator to freeze up, and in response, the machine shuts down on low pressure safety. By the time the technician arrives, however, the water flow is fine and the unit is running great because the evaporator defrosted and the machine returned to normal operation. But because the bin is low on ice, the owner assumes the machine is inoperative. In this case, the lower bin level merely indicates that the machine fell behind and needs time to catch up.

Consequently, if a problem is not evident, run a production check upon arrival to determine if the machine actually failed.

Production

Always check production before trying to fix a machine, especially if the cause of the failure is not clear cut or even if a known malfunction exists. The machine may not need repair. A key factor in troubleshooting ice machines is that *production varies according to entering air and water temperatures*. The temperature of supply water varies from season to season. And since water is the load in an ice machine, when the load temperature varies, the production also varies. For example, a given machine may take 19 minutes to freeze a batch in January, but the same machine may take 29 minutes in July due to increased air and water temperatures. The extra 10 minutes results in decreased production overall but does not indicate a malfunction. It is merely the machine's response to ambient conditions. Remember, while production is the final measure of whether or not a machine is working properly and efficiently, it is not the only one. Decreased production does not always indicate a malfunction.

On the other hand, the fact that a machine is running does not mean that it is working properly. To demonstrate, a car rated at 20 miles to the gallon but only getting 10 is running, but not performing properly. Manufacturers publish charts that furnish production for a given machine, and the technician should check this data against actual production to determine if a problem exists.

Seasonal Effects

Seasons greatly influence ice machine production, especially if the unit is sized during a season of low demand. A unit operating in the winter can produce sufficient ice to meet the user's needs, yet still not be making anywhere near the unit's capacity. When warm weather arrives, ice demand increases while ice production simultaneously decreases. If the unit was sized to meet winter demand, it may be unable to keep up with summer demand.

The failure can also be as simple as undercharge on the unit, resulting from a winter service call that occurred only a few months earlier. The previous service technician may have removed hoses and blown off too much charge. When operating in a cool ambient, the unit ran fine, but the hotter operating conditions of summer may prove too much for the unit to handle.

Unit Location

The machine's physical location can cause problems. Ice machines are generally installed in areas that are too hot for efficient operation. In a hot location, an air-cooled machine produces less ice and takes longer to produce each batch–and no amount of repair can change that.

Cube Shape & Size

Every cuber ice machine manufacturer makes a unit that produces a cube of unique design. Consequently, cubes that are poorly formed or irregular in size provide evidence that the unit is not functioning properly.

The technician should learn to identify ice machines by the shape of the cube produced. Hollow spaces in one brand are normal, in another brand abnormal. Clearly, knowing what an ice cube for a particular unit should look like is essential if the diagnosis is to be accurate.

If the ice cubes are perfectly formed and all of the cubes from a single load appear identical, the technician can assume that the machine's evaporator is full of liquid refrigerant. But if the cubes are too small or irregularly shaped, refrigerant undercharge may be suspected. If an ice machine is undercharged, the top rows of the evaporator may not have sufficient liquid refrigerant to freeze the cubes properly. In this instance, heat transfer cannot be uniform over the evaporator surface. This causes the cubes to be thinner on the top rows of the evaporator and thicker on the bottom rows. Because liquid refrigerant generally enters the evaporator at the bottom and exits at the top, undercharge leaves the top rows of the evaporator short of refrigerant. In this manner abnormal cube size or shape can indicate problems.

Unit Failure

Ice machine failure results from breakdown in three basic areas:

1. *Water*—Industry sources feel that most ice machine failures result primarily from water-related problems.
2. *Electrical*—Electrical failures are more complex to troubleshoot. Cube sizing and harvest controls require an understanding of the machine as well as basic electrical theory.
3. *Mechanical*—Undercharging, wear and tear, and the constant presence of water all take their toll on the mechanical system of an ice machine.

Water-Related Problems

Clarity

An ice cube is only as good as the water from which it comes. Ice produced by a cuber is more pure than the water supplied to the machine because the constant flow of water over the evaporator "cleans out" many of the water impurities. As a result, cubers tend to produce clear cubes. On the other hand, flakers produce cloudy ice since there is no water flowing over the evaporator.

Purity

The flowing water in a cuber does not remove all impurities. Some impurities will remain in the water and affect both operation and production if not treated.

The mineral content of water produces *scale*. Scale is formed from the calcium and magnesium salts that naturally occur in water, and this varies according to locale. Scale is measured in terms of hardness, and creates problems for the ice machine. When scale forms a deposit on the evaporator surface, production is affected in the following ways:

- Scale insulates the refrigerant from the water affecting production. The insulating effect of scale slows heat transfer.
- Scale can clog openings in water distributors, lines and pumps. This clogging restricts water flow and eventually production.
- Scale increases friction between moving parts and causes premature failure.
- During harvest, scale causes the ice to remain on the evaporator longer than necessary. The scale creates a rough service which slows down the ice leaving the evaporator. This slow down causes ice to remain on the evaporator longer during harvest resulting in smaller cubes. During the freeze process, time and energy are invested in properly sizing the cube which then shrinks during the lengthened harvest cycle.

When the mineral content of water becomes too high, water treatment becomes necessary. Common water treatment employed today involves using a water softener to combat hardness. However, using water softeners does not work with an ice machine since most softeners use salt which will eventually pit the evaporator surface and creates additional problems for the ice machine.

Caution: Do not use a water softener to treat ice machine water.

Cleaning & Filtration

Fungus, bacterial growth and rust are ever-present invaders of the water supply. These organisms enter the machine with the supply water. The growth of these organisms is visible as rust-colored residue in the sump, and black or green coloring in the plastic water piping.

In some cases, the growths are not visible to the eye. However, these growths must be treated. Ice is essentially a food product, and as such, care must be taken in its production.

One method for treating hard water in ice machines is a polyphosphate feeder. The polyphosphates in the feeder encase the impurities in the water, and as a result, the impurities are less able to adhere to the ice machine surfaces. If a polyphosphate feeder is used, a dump cycle or some type of bleed must be employed to periodically clean the sump of the machine.

Household bleach can be an effective cleaner to fight bacterial problems, although repeated treatments may be necessary. Water-borne organisms can also be treated by filtration. Many filter systems are available today to treat the supply water. These systems can often be tailored to combat the problem present in a specific machine. Filters that combat problems such as bacterial growth, high or low ph and excessive mineral content are readily available. Some of these filter systems also contain polyphosphate feeders that dispense pre-measured amounts of this substance into the supply water at given intervals.

Another enemy of ice machines is sediment. To control sediment, special filters trap and hold particles that would otherwise clog up the small openings in an ice machine. Sediment filters are particularly beneficial in areas where new construction exists because water line breaks are common and the breaks can contaminate supply lines with sediment. Even in an area where the water lines are old, a sediment filter may save far more than its cost in preventing repeat service calls caused by plugged-up orifices in the unit.

Organic growth in an ice machine often results from airborne sources. For example, the spores present in the air in a pizza parlor or the smoke-laden air of a bar are possible sources of contamination. The most effective method of treating airborne invasion is an electronic air cleaner. The air cleaner should be installed in area that allows it to scrub the air around the ice machine. In the event the technician is having a constant growth problem and water filtration is not curing the problem, consider the air as the source of contaminants.

In some cases it may be desirable to install a charcoal filter, especially when it comes to soft drinks. Soft drinks, a major source of profit for any restaurant, are mixed in precise blends using water as the main ingredient. Today's water supplies contain purification agents. These agents may end up in the ice and affect the taste of the drink. When a customer complains that a drink tastes "funny," consider installing a charcoal filter system. Charcoal filter systems are the primary means of correcting peculiar tasting water.

Mechanical Failure

Mechanical failure results from component wear, component failure or refrigerant problems. Most mechanical failures are evident—the water pump is not running, or the plastic piping is kinked and cracked.

Refrigerant Charge

One manufacturer of ice machines maintains that the majority of service calls are caused by incorrect refrigerant charge. Ice machines are critically charged units, and a slight over- or undercharge results in low ice production. The loss of just a few ounces leaves the machine without sufficient refrigerant. Because correct charge in an ice machine is so critical, the charge should always be weighed-in. A small shortage of refrigerant can easily result in a 100 pound loss of ice production.

Undercharge typically results from leaks in the system, but can also result when installing or removing manifold gauges. Consequently, allow for loss of refrigerant when checking a machine. When possible, use valves that cut down on loss of refrigerant during manifold gauge removal.

Check production before installing any manifold gauges. Remember, manifold gauges do not have to be installed on every service call. If the unit is meeting its rated production, there is no reason for attaching manifold gauges. Refraining from attaching unnecessary gauges eliminates the potential of creating undercharge conditions.

Mineral Build-up

When the water pump fails electrically, the technician has a normal call and replaces the pump. However, if the pump fails mechanically due to scale build-up, the technician must clean the unit in addition to replacing the pump. Scale contributes to and is also the source for many mechanical problems. Plugged-up distributors restrict proper water flow and must be cleaned out. Scale formation between moving parts, such as a water pump impeller, creates premature wear. The surface of the evaporator deteriorates due to continual usage and mineral build-up. Over a period of time, the evaporator's surface begins to change deteriorate due to pitting caused by scale. Some evaporators even lose their plating finish, and the rough surface holds the ice on the evaporator longer than necessary during harvest. And longer harvest, as discussed, decreases cube size and production.

Although difficult, scale removal is possible with a good solvent and proper cleaning (instructions for cleaning an ice machine are included in Part Two). However, when the evaporator surface has deteriorated to the point where production is significantly retarded, consider replacement. Most technicians agree that it is better to replace than re-coat the surface of an evaporator. And once replaced, a filtration system installation is recommended to combat future scale build-up and corrosion.

When components fail, consider using only original equipment replacements. Components such as TEVs and water pumps are manufactured for specific applications on specific machines. Replacing a TEV with an over-the-counter model may get the unit working, but the replacement will probably not meet the exact superheat setting specifications required by the manufacturer. All water pump replacements should have National Sanitation Foundation approval in order to comply with local health codes.

Cleaning & Sanitizing

Cleaning is a controversial area in ice machine service. Cleaning for the sake of cleaning is not recommended since most ice machine cleaners are acid-based and harm the component surfaces in an ice machine. To avoid damage, use only ice machine cleaners recommended by the manufacturers. Some manufacturers specify that only a certain brand of cleaner be used on their machine, and use of any other brand can even void the machine warranty. Above all, clean an ice machine only when necessary, not as part of every service call.

If the ice machine has organic growth problems, it must be sanitized. Household bleach is one of the best agents for sanitizing an ice machine. However, keep in mind that the first treatment may not kill all of the agents causing the growths. If the growth reappears after the first treatment, a new application of bleach is necessary. Remember, during the sanitizing process, empty the bin and dispose of any and all ice.

Electrical Problems

Electrical problems may affect operation in two ways: complete breakdown, or malfunction of a particular component.

Complete Unit Failure

If this occurs, check the following:

- *Is power available to the machine, and is the power adequate?* Is the neighborhood experiencing brown outs? Check the compressor for voltage and ohm out the compressor if voltage is present and the compressor is not running.

- ***Has a safety control opened up?*** Safety controls such as overload, high pressure switches and low pressure switches can open, causing the machine to shut down. One safety control unique to some flakers is a protector inside the auger motor. This protector (torque control) shuts the flaker down in the event the auger cannot cut through the ice. The protector is wired so that the contactor energizing the compressor cannot be energized unless the auger motor is turning.
- ***Are the double-pole double-throw switch and the bin switch working?*** The DPDT switch may be faulty, or the bin switch may have failed so the machine is not receiving power. Many cubers use a double-pole double-throw switch to energize the machine. This switch, in conjunction with the bin switch or bin thermostat, allows the machine to run and still allow a separate wash cycle for cleaning purposes. Note, a flaker uses a single-pole single-throw switch since it does not have a wash requirement.

Component Failure

This type of electrical failure is difficult to troubleshoot because diagnosis involves finding a specific malfunctioning component at a given time that is creating production problems.

If the ice machine is operating but the ice production or quality is not up to par, it may be staying in harvest too long. Different units accomplish harvest with different control approaches, but all share one common goal: to get the cube off the evaporator immediately after the cube has been formed. If the machine is making the proper size cube in the freeze cycle but staying in harvest too long, the cube will be small. If harvest is initiated too early, the machine "cooks" the cube in harvest before releasing it to the bin. And when harvest begins, the hot gas makes the small cube even smaller.

Electrically, harvest is generally initiated in one of three ways:

- single control
- solid state board
- combination of a control and a solid state board

Single Control Systems

A single control, such as a time clock or reverse-acting low pressure control, initiates harvest. The control can measure cube size, suction line temperature, elapsed time or refrigerant pressure. When the control signals for harvest, it directs harvest to begin. In most units using a control, the action of the control closing determines the energizing path for the hot gas solenoid. If the harvest control fails to close when it receives the proper signal, the unit bypasses harvest.

When troubleshooting a single control system, consider the following possible areas of malfunction:

- Has the control itself failed?
- Has the hot gas solenoid failed?
- Is the control out of adjustment? For instance, too low a pressure setting on the reverse-acting low pressure control causes the unit to stay in freeze cycle too long.
- Is the control not sensing that harvest is complete and ordering termination? The control is responsible for ordering harvest to commence and terminate. The control also establishes the length of the harvest cycle. If the unit is in harvest too long, cubes melt. Be sure to check that the machine is coming out of harvest as soon as the batch of cubes has dropped. Any additional time spent in harvest is hard on the machine and the cubes.

Solid State Boards

A thermistor acts as a sensor. The sensor can measure cube size, suction line temperature, and send a signal to create a time circuit. The sensor sends a signal to a solid state board to finish the cube bridge and start harvest. When harvest is finished, the sensor advises the machine to re-enter freeze cycle.

If a solid state board is the ultimate harvest control, note the following:

1. ***Check the board*** for power-in and power-out at the appropriate time. When harvest is ordered, the board has power at a given terminal that energizes a component. If no power is present at the correct terminal, assume board failure and replace the board. If output power is present at the board, then check the hot gas solenoid.
2. ***Check the thermistor.*** Solid state boards rely on thermistors to gather the data necessary for operation. When you suspect a defective thermistor, ohm it out. Thermistors have an ohm value relative to the temperature at which the ohm check is being made. If the ohm value in incorrect, replace the thermistor.
3. ***Check thermistor placement.*** Thermistors cannot operate correctly if they are mounted improperly, so make sure that the thermistor is not loose or hanging in the air.
4. ***Check the potentiometer.*** The potentiometer on a solid state board is a timer. If the ice bridge is too thin, adjusting the potentiometer lengthens the freeze cycle. However, always check the whole cycle before attempting any adjustment on a potentiometer.

5. *Check the bin switch.* In most solid state systems the bin switch is what causes the solid state board to reset the switches in the board. If the bin switch is not operating, the machine cannot return to freeze cycle.

Combination of Control Board & Solid State Board

In this combination, the control senses the condition necessary to commence harvest and sends a signal to the solid state board to initiate defrost. When a reverse-acting low pressure control senses that the suction pressure is low enough, its contact closes. The closed contact creates a path to the solid state board, which finishes freeze and orders harvest.

Remember that in a combination system, two variables exist. Both the control and the solid state board have adjustable settings. These adjustments control overall cycle time. Adjustment controls exist so the technician can customize the machine to a specific environment. For example, reverse-acting low pressure controls can be adjusted to trigger harvest at different times, and potentiometers can be adjusted to change cube size or harvest time.

However, *do not adjust these controls unless you are certain that the control needs adjustment.* If the machine has been running well for years, adjustment of these controls will probably not solve the problem. Similarly, if the evaporator is covered with lime scale, adjusting a control to increase freeze time does not solve the problem. Remember, any control adjustment affects overall cycle time and production. A good percentage of ice machine service requests result from improper timing of the freeze or harvest cycle.

After the Call

After making the repair, inform the customer that the bin will not fill up immediately. Many owners call back for service the day after a service call because they expect the bin to be full of ice once the machine is fixed. However, this is impossible—if the machine produces 400 lbs per day and the user is using 350 lbs per day, it will take 8 days for the bin to refill. Customers often do not realize this, so tell the customer how much the machine is producing and how long it should take to refill the bin. Taking a few minutes to give the customer this information may eliminate an unnecessary call-back.♦

Appendix:
Manufacturers' Service Data

This appendix contains service and installation manuals for many popular models of ice machines currently utilized in the field. These manuals have been furnished by the manufacturers and are reproduced herein as originally published by those manufacturers. The manuals are arranged in no particular order, and no order of preference is implied or intended by the author or publisher.

The author and publisher wish to thank the manufacturers for their participation and cooperation.

Crystal Tips

AUTOMATIC ICE CUBER SERVICE MANUAL

MODELS

200, 300, 400

CRYSTAL TIPS INC.
JUNCTION OF U.S. HIGHWAYS 9 & 71
SPIRIT LAKE, IOWA 51360
(712) 336-2350

TABLE OF CONTENT

FORWARD

THIS SERVICE MANUAL IS PREPARED TO ASSIST YOU DURING THE INSTALLATION, OPERATION AND SERVICING OF YOUR CRYSTAL TIPS ICE MAKER. IT CONTAINS INFORMATION IN REGARDS TO UNPACKING, INSTALLING AND MAINTENANCE OF THE ICE MAKER. IT ALSO INCLUDES THE VARIOUS SPECIFICATIONS AT WHICH THE MAKER IS DESIGNED TO OPERATE.

WE ENCOURAGE YOU TO THOROUGHLY READ THIS MANUAL SO YOU CAN FOLLOW RECOMMENDED PROCEDURES IN ORDER TO OBTAIN MAXIMUM PERFORMANCE FROM YOUR ICE MAKER.

IF YOU ENCOUNTER ANY SITUATION NOT COVERED IN THIS SERVICE MANUAL, FEEL FREE TO CONTACT OUR SERVICE DEPARTMENT. WE WILL BE HAPPY TO ASSIST YOU IN ANY MANNER WHICH IS NECESSARY.

KEEP THIS SERVICE MANUAL HANDY FOR FUTURE REFERENCE. REFER TO PAGE 42 BEFORE CONTACTING THE SERVICE DEPARTMENT FOR ASSISTANCE.

FOR SERVICE ASSISTANCE, CONTACT:

CRYSTAL TIPS ICE PRODUCTS
Junction Highways 9 and 71
Spirit Lake, Iowa 51360
(712) 336-2350

TO ORDER PARTS, CONTACT:

CRYSTAL TIPS ICE PRODUCTS
655 Glenwood Avenue
Smyrna, Delaware 19977
(302) 653-3015

SPECIFICATIONS

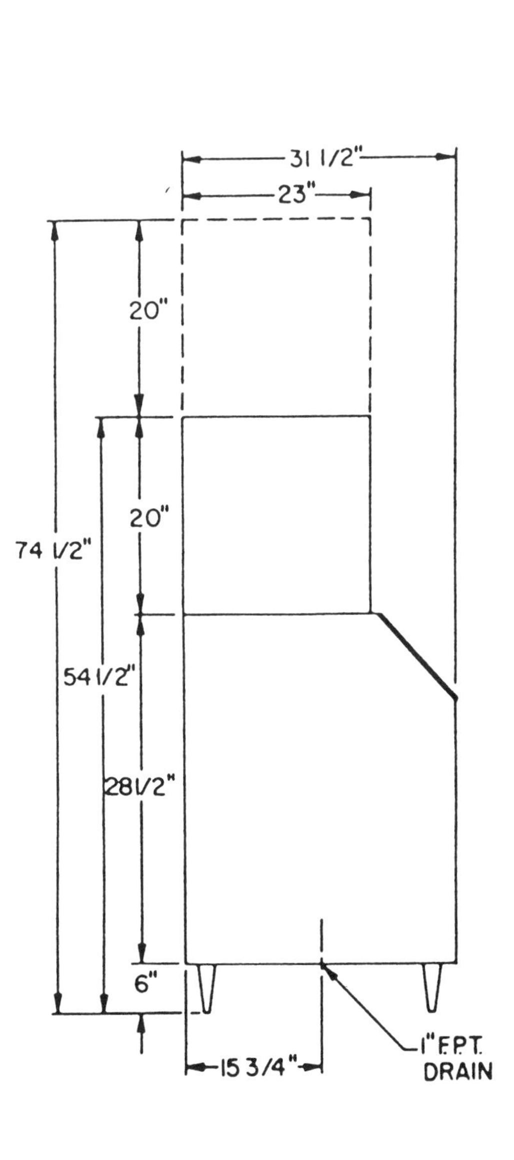

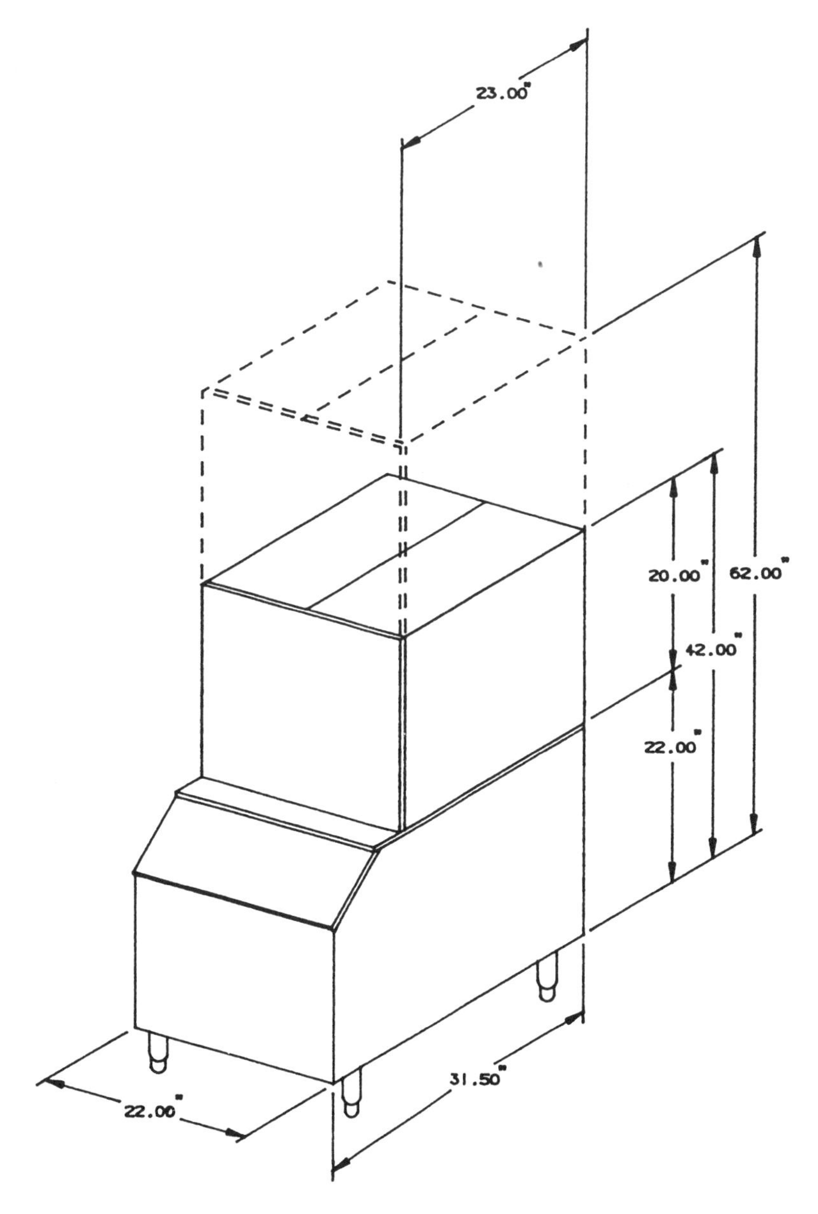

MODEL NUMBER IDENTIFICATION

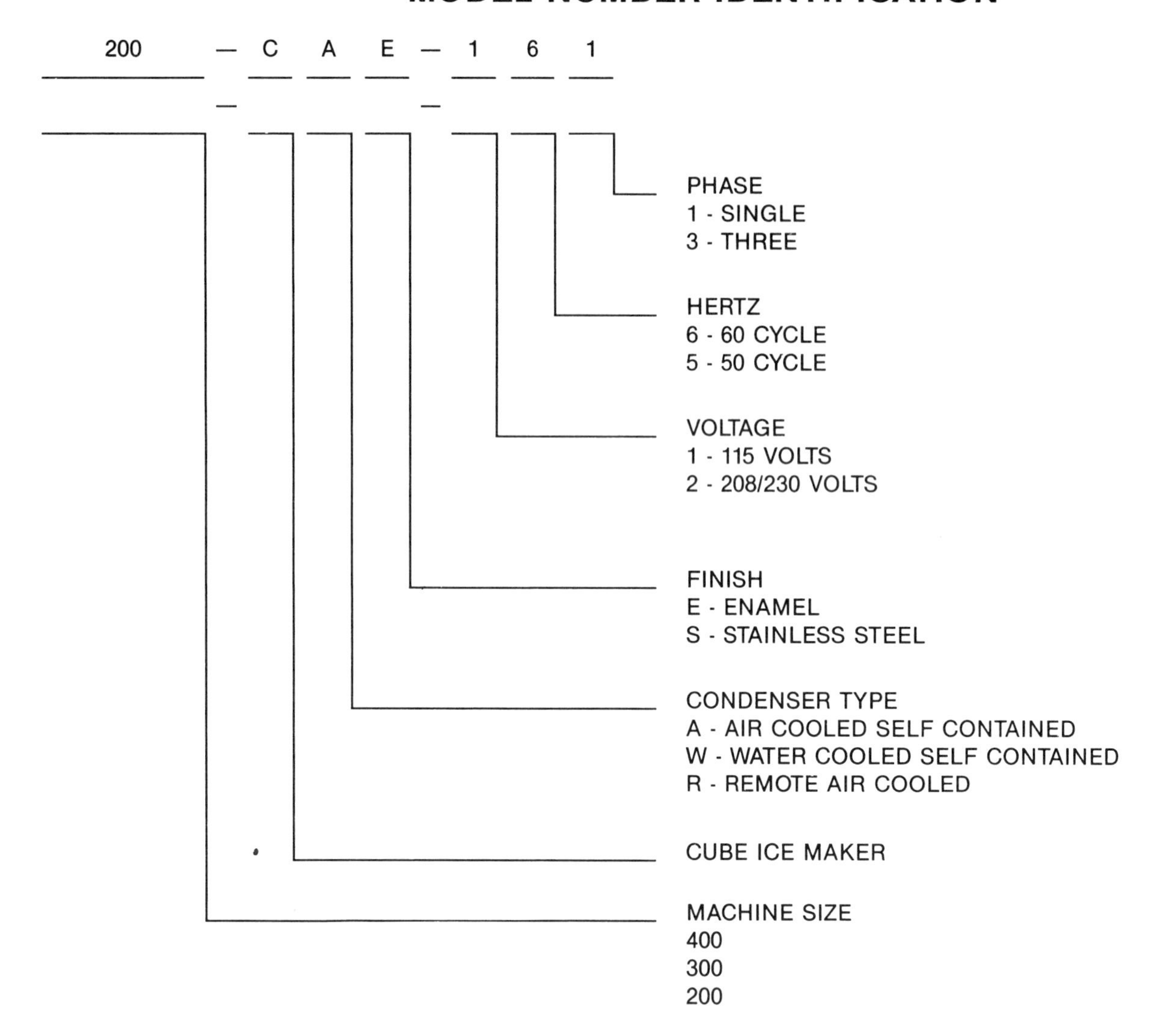

GENERAL SPECIFICATIONS

MPORTANT: We reserve the right to make product improvements at any time. Specifications and design could hange at any time without notice and without consequences.

All of the following information was arrived at with the ambient temperature of 90 F and water temperature of 70 F and is approximate information, your ice maker may vary slightly.

200 CA*-161

LECTRICAL REQUIREMENTS:	115 VOLT, 60 HERTZ, 1 PHASE
MPERAGE:	10.9 FULL LOAD AMPS
AXIMUM FUSE SIZE:	25.0 AMPS
INIMUM CIRCUIT AMPACITY:	16.5 AMPS
LECTRIC CONSUMPTION PER 100 LBS OF ICE PRODUCED:	13.17 K.W.H.
OTABLE WATER CONSUMPTION PER 100 LBS. OF ICE PRODUCED:	16.1 GALLONS
EFRIGERANT CHARGE:	17 OZ/R502
IMENSIONS OF ICE MACHINE:	22" WIDE, 23" DEEP, 20" HIGH
EIGHT OF MACHINE:	137 LBS.
E PRODUCTION:	143 LBS PER 24 HOURS:

200 CA*-251

LECTRICAL REQUIREMENTS:	230 VOLTS, 50 HERTZ, 1 PHASE
MPERAGE:	6.01 FULL LOAD AMPS
AXIMUM FUSE SIZE:	10 AMPS
INIMUM CIRCUIT AMPACITY:	7.3 AMPS
LECTRIC CONSUMPTION PER 100 LBS OF ICE PRODUCED:	12.5 K.W.H.
OTABLE WATER CONSUMPTION PER 100 LBS OF ICE PRODUCED:	12.7 GALLONS
EFRIGERANT CHARGE:	17 OZ R502
IMENSIONS:	22" WIDE, 23" DEEP, 20" HIGH
EIGHT OF ICE MACHINE:	137 LBS.
E PRODUCTION:	143 LBS. PER 24 HOURS

200 CW*-251

LECTRICAL REQUIREMENTS:	230 VOLTS, 50 HERTZ, 1 PHASE
MPERAGE:	5.66 FULL LOAD AMPS
AXIMUM FUSE SIZE:	10 AMPS
INIMUM CIRCUIT AMPACITY:	6.9 AMPS
LECTRIC CONSUMPTION PER 100 LBS OF ICE PRODUCED:	7.1 K.W.H.
OTABLE WATER CONSUMPTION PER 100 LBS OF ICE PRODUCED:	14.3 GALLONS
ONDENSER WATER CONSUMPTION PER 100 LBS OF ICE PRODUCED:	130 GALLONS
EFRIGERANT CHARGE:	14 OZ R502
IMENSIONS:	22" WIDE, 23" DEEP, 20" HIGH
VEIGHT OF ICE MACHINE:	137 LBS.
CE PRODUCTION:	145 LBS. PER 24 HOURS

300 CA*-161

LECTRICAL REQUIREMENTS:	115 VOLTS, 60 HERTZ, 1 PHASE
MPERAGE:	9.3 FULL LOAD AMPS
AXIMUM FUSE SIZE:	35.0 AMPS
INIMUM CIRCUIT AMPACITY:	19.2 AMPS
LECTRIC CONSUMPTION PER 100 LBS OF ICE PRODUCED:	11.57 K.W.H.
OTABLE WATER CONSUMPTION PER 100 LBS OF ICE PRODUCED:	13.06 GALLONS
EFRIGERANT CHARGE:	16 OZ R502
IMENSIONS:	22" WIDE, 23" DEEP, 20" HIGH
VEIGHT OF ICE MACHINE:	143 LBS.
CE PRODUCTION:	240 LBS. PER 24 HOURS

300-CW*-161

ELECTRICAL REQUIREMENTS:	115 VOLTS, 60 HERTZ, 1 PHASE
AMPERAGE:	7.8 FULL LOAD AMPS
MAXIMUM FUSE SIZE:	30.0 AMPS
MINIMUM CIRCUIT AMPACITY:	17.7 AMPS
ELECTRIC CONSUMPTION PER 100 LBS OF ICE PRODUCED:	9.48 K.W.H.
POTABLE WATER CONSUMPTION PER 100 LBS OF ICE PRODUCED:	12.2 GALLONS
CONDENSER WATER CONSUMPTION PER 100 LBS OF ICE PRODUCED:	140 GALLONS
REFRIGERANT CHARGE:	17 OZ R502
DIMENSIONS:	22" WIDE, 23" DEEP, 20" HIGH
WEIGHT OF ICE MACHINE:	143 LBS.
ICE PRODUCTION:	240 LBS. PER 24 HOURS

300 CA*-261

ELECTRICAL REQUIREMENTS:	NOT AVAILABLE
AMPERAGE:	5.3 FULL LOAD AMPS
MAXIMUM FUSE SIZE:	15 AMPS
MINIMUM CIRCUIT AMPACITY:	7.6 AMPS
ELECTRIC CONSUMPTION PER 100 LBS OF ICE PRODUCED:	11.25 K.W.H.
POTABLE WATER CONSUMPTION PER 100 LBS OF ICE PRODUCED:	15.39 GALLONS
REFRIGERANT CHARGE:	NOT AVAILABLE
DIMENSIONS:	NOT AVAILABLE
WEIGHT OF ICE MACHINE:	NOT AVAILABLE
ICE PRODUCTION:	NOT AVAILABLE

300-CA*-251

ELECTRICAL REQUIREMENTS:	230 VOLTS, 50 HERTZ, 1 PHASE
AMPERAGE:	6.5 FULL LOAD AMPS
MAXIMUM FUSE SIZE:	10 AMPS
MINIMUM CIRCUIT AMPACITY:	7.9 AMPS
ELECTRIC CONSUMPTION PER 100 LBS OF ICE PRODUCED:	11.4 K.W.H.
POTABLE WATER CONSUMPTION PER 100 LBS OF ICE PRODUCED:	12.15 GALLONS
REFRIGERANT CHARGE:	16 OZ R502
DIMENSIONS:	22" WIDE, 23" DEEP, 20" HIGH
WEIGHT OF ICE MACHINE:	143 LBS.
ICE PRODUCTION:	240 LBS. PER 24 HOURS

300-CW*-251

ELECTRICAL REQUIREMENTS:	230 VOLTS, 50 HERTZ, 1 PHASE
AMPERAGE:	6.2 FULL LOAD AMPS
MAXIMUM FUSE SIZE:	10.0 AMPS
MINIMUM CIRCUIT AMPACITY:	7.3 AMPS
ELECTRIC CONSUMPTION PER 100 LBS OF ICE PRODUCED:	8.77 K.W.H.
POTABLE WATER CONSUMPTION PER 100 LBS OF ICE PRODUCED:	14.42 GALLONS
CONDENSER WATER CONSUMPTION PER 100 LBS OF ICE PRODUCED:	155 GALLONS
REFRIGERANT CHARGE:	17 OZ R502
DIMENSIONS:	22" WIDE, 23" DEEP, 20" HIGH
WEIGHT OF ICE MACHINE:	143 LBS.
ICE PRODUCTION:	240 LBS. PER 24 HOURS

400-CA*-161

ELECTRICAL REQUIREMENTS:	115 VOLTS, 60 HERTZ, 1 PHASE
AMPERAGE:	11.1 FULL LOAD AMPS
MAXIMUM FUSE SIZE:	40 AMPS
MINIMUM CIRCUIT AMPACITY:	24.4 AMPS
ELECTRIC CONSUMPTION PER 100 LBS OF ICE PRODUCED:	8.83 K.W.H.
POTABLE WATER CONSUMPTION PER 100 LBS OF ICE PRODUCED:	14.5 GALLONS
REFRIGERANT CHARGE:	27 OZ R502
DIMENSIONS:	22" WIDE, 23" DEEP, 20" HIGH
WEIGHT OF ICE MACHINE:	167 LBS.
ICE PRODUCTION:	350 LBS. PER 24 HOURS

400-CW*-161

ELECTRICAL REQUIREMENTS:	115 VOLTS, 60 HERTZ, 1 PHASE
AMPERAGE:	10.0 FULL LOAD AMPS
MAXIMUM FUSE SIZE:	40 AMPS
MINIMUM CIRCUIT AMPACITY:	22.9 AMPS
ELECTRIC CONSUMPTION PER 100 LBS OF ICE PRODUCED:	8.16 K.W.H.
POTABLE WATER CONSUMPTION PER 100 LBS OF ICE PRODUCED:	18.9 GALLONS
CONDENSER WATER CONSUMPTION PER 100 LBS OF ICE PRODUCED:	158.1 GALLONS
REFRIGERANT CHARGE:	24 OZ R502
DIMENSIONS:	22" WIDE, 23" DEEP, 20" HIGH
WEIGHT OF ICE MACHINE:	167 LBS.
ICE PRODUCTION:	338 LBS. PER 24 HOURS

400-CA*-251

ELECTRICAL REQUIREMENTS:	230 VOLTS, 50 HERTZ, 1 PHASE
AMPERAGE:	8.6 FULL LOAD AMPS
MAXIMUM FUSE SIZE:	15 AMPS
MINIMUM CIRCUIT AMPACITY:	10.5 AMPS
ELECTRIC CONSUMPTION PER 100 LBS OF ICE PRODUCED:	7.94 K.W.H.
POTABLE WATER CONSUMPTION PER 100 LBS OF ICE PRODUCED:	13.47 GALLONS
REFRIGERANT CHARGE:	27 OZ R502
DIMENSIONS:	22" WIDE, 23" DEEP, 20" HIGH
WEIGHT OF ICE MACHINE:	167 LBS.
ICE PRODUCTION:	350 LBS. PER 24 HOURS

400-CW*-251

ELECTRICAL REQUIREMENTS:	230 VOLTS, 50 HERTZ, 1 PHASE
AMPERAGE:	8.3 FULL LOAD AMPS
MAXIMUM FUSE SIZE:	15 AMPS
MINIMUM CIRCUIT AMPACITY:	10.2 AMPS
ELECTRIC CONSUMPTION PER 100 LBS OF ICE PRODUCED:	6.58 K.W.H.
POTABLE WATER CONSUMPTION PER 100 LBS OF ICE PRODUCED:	13.87 GALLONS
CONDENSER WATER CONSUMPTION PER 100 LBS OF ICE PRODUCED:	165.6 GALLONS
REFRIGERANT CHARGE:	24 OZ R502
DIMENSIONS:	22" WIDE, 23" DEEP, 20" HIGH
WEIGHT OF ICE MACHINE:	167 LBS.
ICE PRODUCTION:	338 LBS. PER 24 HOURS

TEST PRESSURES

All ice makers have been tested on the High Pressure Side to 440 psig. (31 Kg/cm2) and on the low side to 230 psig (16 Kg/cm2).

ICE MAKING METHOD

A water pump is used to move the water from the water tank to a distributor pan, which allows the water to flow over the vertical surface of the evaporator. Any water which has not frozen on the evaporator as the water flows over it drains back into the water tank where the water pump can recirculate it.

COMPRESSOR SPECIFICATIONS

MACHINE MODEL NUMBER	200-C**-161
COMPRESSOR MANUFACTURE	COPELAND
COMPRESSOR MODEL NUMBER	JRE1-0033-IAA
RESISTANCE VALUE	
COMMON TO START	5.7
COMMON TO RUN	1.05
START TO RUN	6.75
TYPE OF COMPRESSOR OIL	SUN 150 3GS
AMOUNT OF COMPRESSOR OIL	18 OZ
BTU CAPACITY	4100 BTU/HR.
START CAPACITOR	233/280MFD/110V
RUN CAPACITOR	NONE

MACHINE MODEL NUMBER	200-C**-251
COMPRESSOR MANUFACTURE	TECUMSEH
COMPRESSOR MODEL NUMBER	AK9458J
RESISTANCE VALUE	
COMMON TO START	7.1
COMMON TO RUN	2.7
START TO RUN	9.8
TYPE OF COMPRESSOR OIL	SUN 150 3GS
AMOUNT OF COMPRESSOR OIL	15 FL OZ
BTU CAPACITY	4950 BTU/HR
START CAPACITOR	72-88MFD/330V
RUN CAPACITOR	15 MFD/370 V

1.	MACHINE MODEL NUMBER	300-C**-161	300-C**-261
2.	COMPRESSOR MANUFACTURE	TECUMSEH	TECUMSEH
3.	COMPRESSOR MODEL NUMBER	AK9466J	AK9466J
4.	RESISTANCE VALUE		
5.	COMMON TO START	5.0	10.3
6.	COMMON TO RUN	.7	1.8
7.	START TO RUN	5.7	12.1
8.	TYPE OF COMPRESSOR OIL	SUN 150 3GS	SUN 150 3GS
9.	AMOUNT OF COMPRESSOR OIL	15 FL OZ	15 FL OZ
10.	BTU CAPACITY	6600 BTU/HR.	6600 BTU/HR.
11.	START CAPACITOR	72-88 MFD/250V	72-88/330
12.	RUN CAPACITOR	15 MFD/330V	15-370

1.	MACHINE MODEL NUMBER	300-C**-251
2.	COMPRESSOR MANUFACTURE	TECUMSEH
3.	COMPRESSOR MODEL NUMBER	AK9474J
4.	RESISTANCE VALUE	
5.	COMMON TO START	9.5
6.	COMMON TO RUN	2.2
7.	START TO RUN	11.7
8.	TYPE OF COMPRESSOR OIL	SUN 150 3GS
9.	AMOUNT OF COMPRESSOR OIL	15 FL OZ
10.	BTU CAPACITY	6100 BTU/HR
11.	START CAPACITOR	72-88MFD/330V
12.	RUN CAPACITOR	15 MFD/370 V

1.	MACHINE MODEL NUMBER	400-C**-161	400-C**-251
2.	COMPRESSOR MANUFACTURE	COPELAND	COPELAND
3.	COMPRESSOR MODEL NUMBER	RSN2-0075-PAA	RSH2-0101-PAG
4.	RESISTANCE VALUE		
5.	COMMON TO START	4.5	8.8
6.	COMMON TO RUN	.6	2.0
7.	START TO RUN	5.1	10.8
8.	TYPE OF COMPRESSOR OIL	SUN 150 3GS	NOT AVAILABLE
9.	AMOUNT OF COMPRESSOR OIL	24 FL. OZ	24 FL OZ
10.	BTU CAPACITY	7000 BTU/HR.	7000 BTU/HR.
11.	START CAPACITOR	180-130MFD/220V	
12.	RUN CAPACITOR	25 MFD/370 V.	20-370

CONDENSER CONSTRUCTION

AIR COOLED — Fin and Tube Type Blow Through

WATER COOLED — Tube Within a Tube Type, Opposed Flow.

HEAD PRESSURE CONTROL

AIR COOLED: N/A

WATER COOLED: Automatic Water Regulating Valve set at 255 Head Pressure.

EVAPORATOR CONSTRUCTION

Stainless Steel Evaporator plate containing tin plated copper buttons.

REFRIGERANT CONTROL

External equalized thermostatic expansion valve.

ICE SIZE CONTROL

Single pole double throw thermostat which initiates the defrost mode when the capillary tube senses 38° F and switches its contacts back to a freeze mode when it reaches 43° F.

DEFROST MODE CONTROL

Evaporator defrost is accomplished when the hot gas solenoid valve is opened by the ice size control thermostat.

BIN CONTROL

Single pole, single throw thermostat.

WATER LEVEL CONTROL

Inlet Water is controlled by a float valve. The water purge is controlled by a siphon tube.

POTABLE WATER SUPPLY FOR ICE

200 - 1/4" F.P.T. female fitting
300 - 1/4" F.P.T. female fitting
400 - 1/4" F.P.T. female fitting

WATER SUPPLY FOR WATER COOLED CONDENSER

3/8" F.P.T. female fitting

DRAIN SYSTEM

On air cooled systems a 3/4" F.P.T. female fitting is used. On water cooled ice makers, two 3/4" F.P.T. female fittings are used, and requires two independent drains to be used. Follow all applicable local plumbing codes.

ELECTRICAL SYSTEM

Connect wires inside the control box. Follow all applicable local electrical codes.

OPERATING CONDITIONS

Ambient temperature range:	60 F - 100 F
Water temperature range:	34 F - 90 F
Water pressure range:	20 psig - 60 psig
Voltage range on:	230 volt – +/– 10%
	208 volt – – 5%
	115 volt – +/–10%

TRANSPORTATION

WARNING

During transporting of the ice maker to its installation point, do not tip the ice maker more than 45 degrees. Tipping allows the lubricant in the compressor to travel through the refrigeration lines to other refrigerant components. If the ice maker has been tipped, allow it to stand upright for at least 1 hour. This will permit most of the oil to drain back into the compressor.

Carry or transport the ice maker carefully. Do not drop, handle roughly or allow the ice maker to be impacted. Damage can easily result not only to the exterior cabinet, but also the interior components.

Do not step on the carton as this could damage the exterior panels of the ice maker. The carton has been constructed with a corner post in each corner, thus making the corners the strongest part of the carton. This allows the ice maker to be stacked 5 units high in the warehouse.

CAUTION

Do not allow the cartons to become wet. Moisture will greatly reduce the strength of the carton and increases the chances of damage to the ice maker. Stacked ice makers may tip over due to loss of corner support strength.

UNPACKING AND INSPECTION

Visually inspect the exterior of the shipping carton. Any damage noticed should be reported to the delivering carrier as soon as possible.

Then remove the carton from the ice maker being careful not to destroy the carton until the installation has been completed.

Remove the front panel from the ice maker by turning the two (2) quarter turn fasteners and lifting the front panel up off its top hanger. Then inspect the interior of the ice maker for any concealed damage. If damage is noticed, file a freight claim immediately.

IMPORTANT

If a container or its contents show damage STOP. Report the damage to the delivering carrier immediately. Consignee is responsible for the filing of the freight damage claim. Crystal Tips responsibility ends at our dock!

ICE MAKER LOCATION

Select a location for the ice maker and bin which is level and firm enough to support the ice maker and the bin when full of ice. Failure to locate the equipment on a good surface may result in the ice maker tipping over, drain problems, noise and vibration, and improper water flow over the evaporator.

Avoid locating the ice maker in an area exposed to direct sunlight, rain, heat or splashing water.

CAUTION

THIS ICE MAKER IS DESIGNED FOR INDOOR USE ONLY.

Exposure to direct sunlight or heat could cause a large reduction in ice production and cause an excessively warm bin, which in turn will cause the ice in the bin to melt at a more rapid rate.

If the electrical circuit of the ice maker becomes wet from rain or splashing water, the controls may become shorted out electrically. This will also increase the possibility of electric shock and/or death.

The ice maker must be installed where the ambient temperature does not exceed a low of 60 F or a high of 100 F. If the ice maker is located in an area where the ambient can drop below 60 F, it may not be able to generate enough heat to properly defrost the evaporator during the harvest cycle. The low ambient may also cause the bin thermostat or the ice size thermostat to operate in an erratic manner. (If ice maker is shut off during the winter months, please refer to winterizing procedures listed on page 18 of this manual.

During periods of high humidity, condensation may form on the exterior panels of the ice maker and on the outer walls of the bin. Install the ice maker where some drippage from these conditions will not cause any damage to floors or surroundings.

Place the ice maker in an area near a water supply and a drain system. Make sure the drain in the building is lower than the bin drain. This will prevent any drainage back flow to the bin which would contaminate the ice, making it unusable.

Make sure there is adequate room left between the ice maker and the walls, at least 6" is required. This applies to all ice maker models, not only air cooled. This 6" spacing will allow proper air to flow around the ice maker and also provides room for the water, drain and electrical connections to the ice maker.

On air cooled ice makers, if there isn't 6" of space, the ability of the ice maker to operate efficiently will be greatly reduced.

Finally select an area to install the ice maker which is clean and well ventilated and accessible for both the person who removes ice from the bin and the service person.

A well ventilated room allows the ice maker to operate as efficiently as possible even in high ambient conditions.

Make sure the electrical supplied to the area where you intend to install the ice maker corresponds to the electrical specifications on the ice makers nameplate.

INSTALLATION OF ICE MAKER AND BIN

After the location of the ice maker has been selected. Remove the carton from the bin. Flatten the carton, and then lay the back of the bin on the flattened carton. This will prevent the back of the bin from becoming scratched. Before laying the bin on its back, remove the carton from inside the bin, which contains the (4) bin legs. Remove the (2) bolts which hold the shipping skid to the bin bottom. Then thread the (4) legs into the base of the bin and secure tightly.

Before uprighting the bin, make sure the adjustable levelers on each leg are turned clockwise until they are finger tight. The bin may now be tipped upright and set as close as possible to its final location.

CAUTION

Avoid sliding the bin across the floor as much as possible. The sliding action could cause the leg bolts to become bent which would cause the bin to become unstable after the ice maker is installed on the bin.

Allow enough room so the ice maker can be placed on top of the bin easily. After the ice maker is installed on top of the bin, the entire unit can be moved to its final position.

NOTE

If you are using an oversize bin, you will need an adapter plate kit to decrease the bin top opening to accommodate the ice maker. Please refer to the ice maker/storage bin combination selection chart on page 13 or in your Crystal Tips Ice Products Suggested Retail Price List.

BIN MODEL SERIES	BR*150	BR*260	BR*400	BR*500	BR*675
200	X	X	1	1	2
300	X	X	1	1	2
400	X	X	1	1	2

BIN ADAPTER GUIDE

COMPATIBILITY CHART NO.	PART NUMBER
X = EXACT FIT	NO ADAPTER REQUIRED
1 = BIN ADAPTER REQUIRED	80317901
2 = BIN ADAPTER REQUIRED	80317801

Before installing the ice maker on the bin, locate the insulating tape that is packed inside the bin. It will serve as a gasket between the ice maker and the bin. The gasket should be positioned on the bin so that its dimensions equal that of the perimeter of the ice maker. The ice maker may now be set on top of the bin in it's proper position.

CAUTION

Ice makers are heavy. Use a mechancial lifting device or have adequate personnel available to safely lift the ice maker in place. Do not attempt to slide the ice maker on the sealing gasket. Carefully lift and lower the ice maker until you have achieved the proper alignment between the ice maker and the bin.

The entire ice maker/storage bin unit may now be located in its permanent location. Remember to provide the 6" of clearance all the way around the ice maker. Failure to do so may result in damage to the ice maker due to the lack of proper air flow around the ice maker. This may cause overheating and reduce the efficiency of the ice maker.

Check to make sure that the entire unit is level from front to back and from left to right. This may be accomplished by turning the leveler portion of the bin legs. This step of leveling the entire unit will permit adequate and even water flow over the evaporator and allow for proper water purge during the defrost cycle. Also, this will prevent water from accumulating in the bin and will insure proper bin drainage.

WATER SUPPLY AND DRAIN CONNECTIONS

All plumbing should be done by a licensed plumber and should meet all local plumbing codes.

WATER SUPPLY CONNECTIONS

Water supply pressure should be between 20 psig and 60 psig. If the water pressure exceeds 60 psig, use a pressure reducing valve.

The quality of the water supplied to the ice maker has the most adverse affect on the quality of the ice produced and the overall performance of the ice maker than any other factor. Therefore, the use of external water filters or treatment may be required, depending on the water quality at the ice makers location. Any questions concerning water quality should be directed to your Crystal Tips Distributor or refer to water problems in this manual, page 20.

Flush out all water lines before connecting them to the ice maker. This action will remove many of the foreign particles in the line which could clog the float valve in the ice maker.

The water supply inlet connection is a 1/4" F.P.T. female fitting located on the left side of the rear panel of the ice maker, when viewed from the front of the ice maker. It is recommended that a shut off valve be located in the supply line and which provides easy access for the service person. This connection is providing potable water to the ice maker which the ice will be made from and should be of good quality water.

Water cooled units require a second water inlet identified on the rear of the ice maker. This is a 3/8" F.P.T. female fitting and provides cooling water to the condenser. This water supply does not need to be potable water.

DRAIN CONNECTIONS

Located on the rear of the ice maker is a 3/4" F.P.T. female fitting which is identified as a drain. You will need to supply a 3/4" drain line to this fitting. This drain line should lead to an open, trapped or vented drain and must comply to all local codes. This drain line must be dedicated solely to this one drain fitting.

Water cooled ice makers have a second 3/4" FPT female drain fitting located on the rear and should be marked condenser drain. A second 3/4" drain line should be connected to this fitting and must be dedicated solely to the condenser drain. This drain will contain warm condenser water and should never be connected to the ice maker drain or the bin drain. Interconnection of this drain to the ice maker drain could allow warm water to enter the ice maker causing the stored ice to be melted.

Each bin contains a drain which may be located at either the left rear base corner or the bin bottom center. If the drain is at the left rear base corner, it will be a 3/4" M.P.T. male fitting. If the drain is in the bottom center of the bin, it will be a 1" F.P.T. female fitting. This drain allows the melted ice water to drain from the storage area. This line should have a 1/4" drop per foot of horizontal run which will allow the water to drain to an open spill drain.

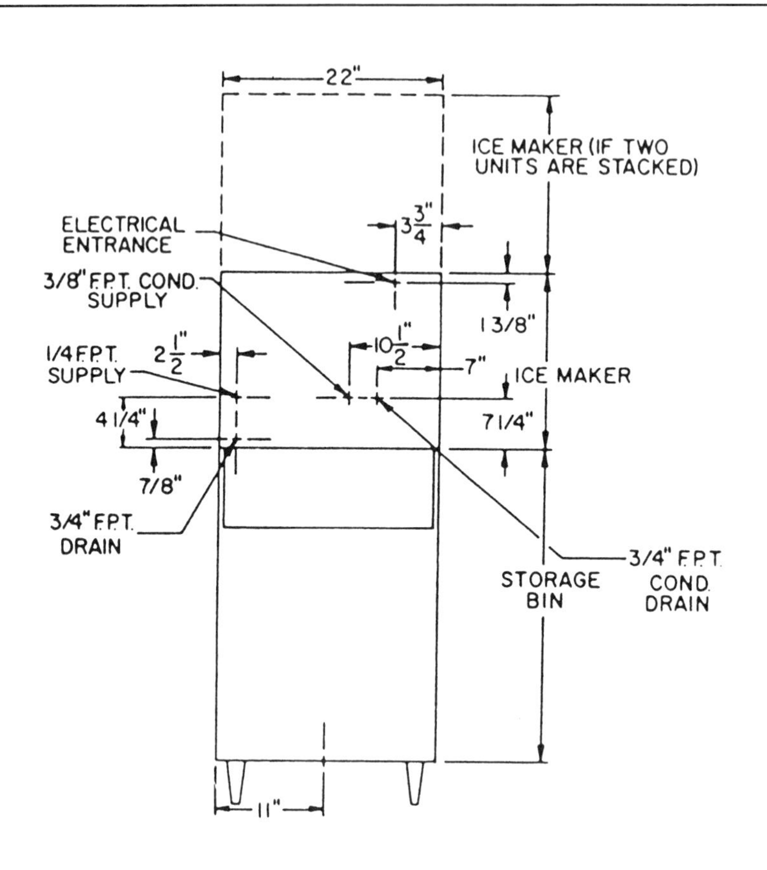

CAUTION

All drains must be installed according to all local codes. Always have one drain pipe for each function. Improper drain installation can cause drain water to back up inside the storage bin or ice maker and will in turn contaminate these areas.

NOTE

In areas of high humidity, all pipes should be covered with plumbing insulating materials to prevent condensation.

ELECTRICAL CONNECTIONS

All electrical connections should be done by a licensed electrician and should conform to all National, State and local electrical codes. On self contained ice makers, the electrical supply should be routed into the rear of the control box and the leads should be connected to the appropriate terminals marked A and B on the terminal board. Connect the ground wire to the green ground screw located in the control box.

CAUTION

1. Use electrical wire which is rated equal or greater than the minimum circuit ampacity rating specified on the ice makers name plate. Failure to do so could cause the wires to overheat and lead to a fire. Always follow applicable electrical codes.
2. The ice maker must be on a separate fused circuit and fuse size must not exceed the maximum fuse size specified on the name plate.
3. The voltage supplied to the ice maker must not fluctuate more than 10% of the ice makers rated name plate voltage.
4. Always make sure the ice maker is properly grounded.

CAUTION

DO NOT START THE ICE MAKER AT THIS POINT

FINAL CHECKS BEFORE STARTING

1. All packing materials should be removed from the evaporator section In order to accomplish this, the water pump must be removed from the water tank by removing the two wing type thumbscrews located on the compressor compartment side of the ice maker. The supply hose from the pump to the distributor pan must be removed from the pump. Then the pump may be lifted out of the water tank and can be set over on the compressor compartment side so that it is out of the way. The ice rack, located under the evaporator, may now be lifted out.

   Once the ice rack has been removed, all shipping debris including the cardboard cover over the ice chute may be removed. The entire water tank should be wiped clean with a clean wet cloth. Then replace the ice rack, making sure that it is resting on its supports in the water tank. The water pump may now be replaced in the water tank. Make sure that the supply hose is reconnected to the water pump.

2. Make sure the siphon tube assembly is on the drain opening and is in the vertical position in the water tank. Turn on the water supply to the ice maker and allow the water tank to fill until the float stops the incoming water flow. The water flow should stop when it reaches a level of 1/8" below the bottom of the inverted U-shaped siphon tube. If the water flow fails to stop at the proper level, turn off the water supply to the ice maker and refer to the adjustment section on page 59-60, figure 24 of this manual.

3. Install the bin thermostat mounting bracket to the underside of the ice maker base inside the bin. It will be located to the right of the ice chute and should be fastened to the base with the two (2) thumbscrews packed with the mounting bracket. After the mounting bracket has been secured, locate the capillary tube of the bin thermostat which is coiled up outside the electrical control box. Carefully uncoil the capillary tube, making sure that it does not become kinked. Then route the capillary tube through the black plastic sleeve which is located in the base next to the compartment divider. This allows the capillary tube to enter the bin close to the mounting bracket where it can be routed through the eyelets of the mounting bracket as shown in the figure 1.

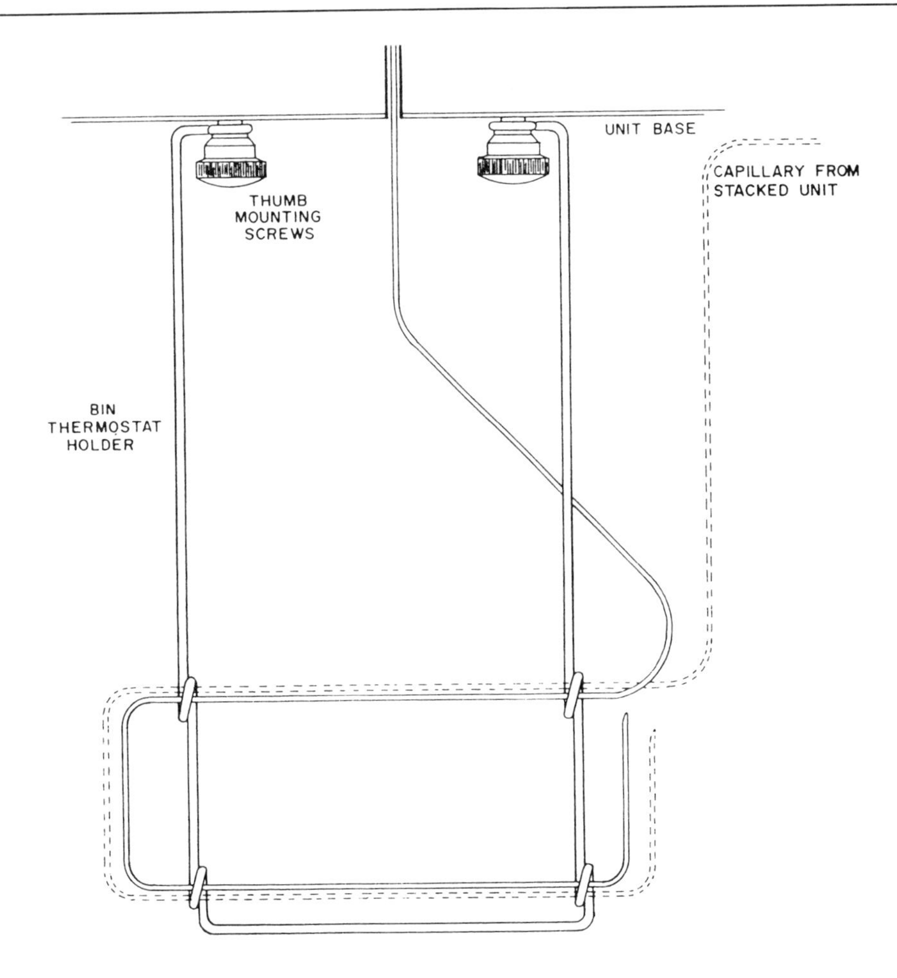

FIGURE 1

4. Using clean water, wash and rinse out the interior of the bin to remove all dust and dirt which may have collected in the bin during transportation and installation.

NOTE

Do not use any soap or cleansers to wash the bin, residue from these cleansers may cause bad tasting ice.

5. Check the electrical supply voltage to insure that it complies with the ice makers name plate rating.
6. Check all water and drain line connections and fittings for any water leaks.
7. Check the ice maker's refrigeration lines to make sure that the tubing is not touching one another. During operation, vibration will produce a rubbing action which would wear a hole in the tubing and cause the refrigerant to be lost and then cause the ice maker to fail.

STACKING ICE MAKERS

Do not stack more than two (2) ice makers on one (1) bin. Stacking of more than two (2) ice makers will cause the installation to become top heavy and could cause the entire installation to be very unstable.

When two (2) ice makers are to be stacked, the top panel over the evaporator section of the bottom ice maker must be removed. Place foam insulation around the evaporator opening of this unit. This will form a seal between the two evaporator sections of both ice makers and will prevent contaminants from entering the ice making compartments.

Then, locate the knock-out in the compressor compartment top cover of the lower ice maker. It is approximately 3/4 of the way back and on the left side of this panel. Remove this knock-out. This will allow the bin thermostat capillary tube from the upper ice maker to pass through the hole on downward through the lower ice maker and into the bin.

The ice makers may now be stacked.

CAUTION

Ice makers are heavy. Use a mechanical lifting device or have adequate personnel available to safely lift the machine in place.

When placing the ice maker on top of the lower ice maker, lift the top ice maker into place. Do not slide the top ice maker across the top of the lower ice maker. Make sure both ice makers are square with each other. Uncoil the bin thermostat capillary tube in the top ice maker, route the tube through the plastic collar on the left side of the compressor section of the ice maker. The plastic collar should be lined up with the knock-out in the top compressor compartment cover of the lower ice maker. Both capillary tubes from the two ice makers will be routed through the bin thermostat holder in the same manner as a single ice maker installation.

The water, drain, and electrical connections are to be preformed the same as a single ice maker installation.

IMPORTANT

Each ice maker must have its own water supply and a separately fused electrical circuit.

CAUTION

If you are installing a Crystal Tips ice maker on a storage bin other than Crystal Tips, the ice maker must be joined to the bin in such a way as to be free of any openings between the ice makers bottom and the storage bins top.

After completing the installation of the ice maker and bin, please fill out the warranty registration card and place it in the mail. Failure to do so may lead to a delay in processing any warranty claims made at a later date.

MACHINE START UP

1. Turn the water supply on to the ice maker and energize the electrical supply to the ice maker.
2. Locate the ice, off, wash switch on the front of the control box. First, switch to the wash position. This will energize only the water pump which will cause the water level to drop in the water tank. This will cause the float valve to open and fill the water tank to its original level.

   Check the water flow at the bottom of the evaporator, it should appear to be even and consistent across the entire evaporator bottom. If water flow appears abnormal, refer to adjustments page 59-60 of this manual.
3. Move the switch to the ice position. This will energize the fan motor on air cooled models and the compressor on all models and the water pump should remain operating. The ice maker should start making ice on the evaporator.
4. Check for any abnormal or peculiar sounds or noises. Locate the source of the noise and determine what should be done to correct the problem.

OPERATION CHECK

1. Check for normal compressor operating pressures.

200-CA*-***		
Room Temperature	Pressure Suction	PSIG Head
60 F	22	146
70 F	24	175
80 F	26	205
90 F	28	228
100 F	29	260

300-CA*-***		
Room Temperature	Pressure Suction	PSIG Head
60 F	17	155
70 F	20	182
80 F	22	212
90 F	24	242
100 F	26	270

400-CA*-***		
Room Temperature	Pressure Suction	PSIG Head
60 F	21	175
70 F	23	196
80 F	25	224
90 F	27	250
100 F	29	290

ALL WATER COOLED		
Room Temperature	Pressure Suction	PSIG Head
60 F	22	250
70 F	23	250
80 F	23	250
90 F	25	250
100 F	26	250

2. Check for air bubbles in the water supply tube from the water pump to the water distributor pan. Any air bubbles in this line indicates a low water level in the water tank which is causing the water pump to cavitate. Improper adjustments of the float valve (see page 34 of this manual for adjustments.) or lack of water to the ice maker may cause this condition.

3. Is the ice maker producing ice fast enough? The following chart is to be used only as a reference. Ambient temperatures, water temperatures, siphon purge during each harvest cycle, cleanliness of the evaporator and condenser all have a decisive part in determining how fast ice is produced and how fast ice is harvested.

AIR TEMPERATURE	FREEZE CYCLE	HARVEST CYCLE	TOTAL CYCLE
90 F	13.5 MINUTES	1.7 MINUTES	15.2 MINUTES
70 F	10.5 MINUTES	2.3 MINUTES	12.8 MINUTES

4. Does the ice maker go into harvest when a full batch of ice is made? When the machine goes into harvest, does the condenser fan motor, and water pump stop operating? Does the hot gas valve become energized? If this is not the sequence of events, please refer to trouble shooting, page 64-66 of this manual.

5. Does the hot gas valve open and close properly? To determine if the valve is functioning, locate the outlet side of the valve, which will be the side going to the evaporator. During harvest the tube on this side of the valve should become very warm, and when the harvest cycle terminates, this tube should become cool. If the hot gas valve fails to operate properly, please refer to trouble shooting, page 51 of this manual.

6. Are the ice cubes the proper size for the end user of the ice maker. If not, please refer to adjustments page 52-53 of this manual.

NOTE

When the water cooled model goes into the harvest cycle, there should not be any water flowing through the condenser. If there is water flow through the condenser, please refer to trouble shooting, page 46 of this manual.

7. Once the ice maker has gone through a complete cycle, gather some ice in the ice scoop and place the ice around the bin thermostat capillary tube and bracket. This should cause the ice maker to shut down within one (1) minute. If the ice maker does not stop, locate the bin thermostat in the electrical control box and turn the adjustment knob counter clockwise until the ice maker stops. Once the ice maker stops, remove the ice from the capillary tube. The ice maker should start within 2 minutes. If further adjustments are needed, refer to adjustments, page 57 of this manual.

WINTERIZING PROCEDURES

If the ambient temperature surrounding the ice maker is subject to freezing or below, the ice maker must be winterized. This will prevent damage to the water supply system and on water cooled models will prevent damage to the water cooled condenser and its refrigerant circuit.

A. Winterizing an air cooled model.

1. Turn off water supply at the valve and disconnect the line from the ice maker.
2. Place the Ice-off-Wash switch located in the electrical control box in the off position (center position). Then make sure the circuit breaker or the disconnect switch of the electrical supply is placed in the off position.
3. Remove the siphon tube assembly from the water tank drain (NOTE: DO NOT REINSTALL THE SIPHON TUBE ASSEMBLY)
4. Blow compressed air through the ice makers water line to remove any water which may be left in the line.

CAUTION

Water cooled ice makers will stop operating when the water supply is shut off for a short period of time. When there is no water supplied to the condenser, the high pressure control will open because of the high pressure developed in the refrigerant circuit. The high pressure control will reclose when the high pressure is reduced.

B. Winterizing a water cooled model.

1. Turn off the water supply at the valve to both the water tank and the condenser and disconnect the lines at the connection to the ice maker.
2. Using compressed air, blow through the water inlet to the condenser. When the ice maker shuts off on the high pressure cutout, the water regulating valve will be open and will allow for free passage of air through the condenser. This will remove any remaining water from the condenser.
3. Place the Ice-Off-Wash switch located on the electrical control box in the off position (center position). Then make sure the circuit breaker or disconnect switch that is suppling electricity to the ice maker is in the off position.
4. Remove the siphon tube assembly from the water tank drain. (DO NOT REINSTALL THE SIPHON TUBE ASSEMBLY).
5. Blow compressed air through the ice maker water line to remove any water left in the line.

START UP AFTER WINTERIZATION

1. Reconnect all water lines and open valves.
2. Reinstall siphon tube assembly to the drain in the water tank.
3. Check water level and make sure that it is 1/8" below the inverted U-shaped siphon tube. If unable to make adjustment, please refer to adjustments page 59-60 of this manual.
4. On air cooled models, check condenser for cleanliness and oil the condenser fan motor with 3 drops of oil.
5. On water cooled models, check the water condenser and its water circuit for any water leaks.
6. Turn on electrical supply to the ice maker at the circuit breaker or disconnect switch.
7. Place the Ice-Off-Wash switch on the electrical control box in the wash position to allow the water to be circulated over the evaporator.
8. Then place the Ice-Off-Wash switch in the ice position, which will allow the ice maker to start making ice.
9. Check operating pressures.

MAINTENANCE AND CLEANING INSTRUCTIONS

DANGER

Whenever working on or cleaning any electrical parts on the ice maker, always disconnect the electrical supply to the ice maker. This will prevent any possible electric shock and or death.

EXTERIOR

The exterior of your Crystal Tips ice maker will be of a special epoxy paint or of stainless steel which have been selected because of their durability. However, the exterior could rust if it is not properly maintained from time to time, dust accumulations should be removed with a soft clean cloth. If the exterior is to be washed, use only warm water and a neutral cleaning agent.

NOTICE

Do not use a cleaning agent which contains chlorine or one that contains an abrasive. These cleansers can permanently damage the finish of the ice maker.

INTERIOR

Clean and sanitize the ice maker and bin every 90 days or sooner if the ice maker becomes dirty due to bad water conditions. If more frequent cleaning is required, it is recommended that Crystal Tips Water Filtration be used.

CLEANING PROCEDURES

1. Remove all ice from the ice bin.

CAUTION

Remember, ice is a FOOD! Failure to remove ice from the bin during the cleaning procedure could allow the ice to become contaminated and make it unfit for human consumption. Ice machine cleaners are an acid.

2. Remove the front panel and top evaporator cover from the ice maker.
3. Place the ice, off, wash switch to the off position.
4. Turn off water supply to ice maker and then remove siphon tube assembly to allow the water tank to drain.
5. Remove water supply tube from the water distributor pan and remove the pan from the top of the evaporator.
6. In a clean container mix a solution of water and ice maker cleaner, following the directions on the ice maker cleaner container. Mix enough solution to make 2 gallons.
7. Place in the solution the water distributor pan and clean thoroughly.
8. Install the distributor pan over the evaporator and connect the water supply tube.
9. Install the siphon tube assembly to the drain and pour half of the cleaning solution into the water tank.
10. Switch the ice, off, wash switch to the wash position. The water pump will now circulate the cleaning solution. Add the remainder of the cleaning solution and allow it to circulate for 15 minutes.
11. Place the ice, off, wash switch in the off position and remove the distributor pan from above the evaporator and the ice rack from below the evaporator.
12. Locate the brush that was supplied with the ice maker and use it to scrub the evaporator surfaces to remove any remaining foreign materials.
13. Replace the distributor pan and connect the supply tube and also replace the ice rack.
14. Again, recirculate the cleaning solution for 5 minutes.
15. Stop the water pump and then with a clean cloth soaked in the cleaning solution from the water tank, wipe down the entire inside of the storage bin.
16. You may now drain the cleaner solution from the water tank.
17. Restore the water supply to the ice maker and allow the water tank to fill with fresh water.
18. Place the switch in the wash position to circulate water over the evaporator and rinse it of the cleaning solution. Also rinse the inside of the bin with fresh water.
19. Place the switch in the off position and drain the rinse water from the water tank. Then, refill with fresh water.
20. Place the ice, off, wash switch in the ice position. The ice maker will now begin to make ice. Catch the ice that is made and check the quality of the ice. Discard this first batch of ice.
21. The ice making portion and the storage bin are now cleaned. Restore the ice maker to its normal operation. Be sure to replace all panels and parts that may have been removed.

ROUTINE MAINTENANCE

The condenser on air cooled ice makers should be cleaned periodically to insure adequate heat exchange through the condenser. This may be accomplished with the use of a fin brush and a vacuum cleaner or air pressure. If you use air pressure, direct the air stream directly at the condenser fins to avoid bending them. If the fins are bent, straighten them with the correct size fin comb. The condenser on the 200 and 300 have 14 fins per inch. The 400 condenser has 15 fins per inch.

Remember, a clean condenser is a condenser you can shine a light thru.

Take a clean dry cloth and wipe off the fan blades and the fan motor. Then add about 3 drops of oil to the oil hole on each end of the fan motor.

BIN LINER RUSTING

All commercial grades of stainless steel will corrode or rust when certain chemicals come into contact with it. One chemical which will attack stainless steel is chlorine and most compounds of chlorine, such as hydrochloric acid. The speed with which this corrosion takes place, depends on the concentration of the chlorine and the length of time it is in contact with the stainless steel.

In many bin applications, a rust stain or brown deposit will appear at the top of the side and far walls of the bin liner and any other stainless steel parts inside the bin, which do not normally get covered by ice. The lower part of the liner walls usually stay clean if the bin is used regularly due to the washing action of the ice and meltage water draining down these walls.

This staining or rusting can come from two sources.

1. Foreign Materials:
   If the ice makers panels are painted, it is possible that the unpainted edges are exposed where the ice maker sits on the bin. These edges could rust. The rust stain could drip down the bin liner.
2. Material expelled during the freeze cycle:
   The Crystal Tips ice maker produces a clear ice by freezing out the impurities, chlorine gas is an impurity, which is expelled during the ice making cycle. This material, being heavier than air, will drop into the ice bin. In the bin, the chlorine gas will combine with water vapor and form a mild hydrochloric acid which will condense on the bin walls above the level of the ice. This area of the bin is not scrubbed by the action of the ice. If the bin walls are not scrubbed, the hydrochloric acid will form a brown stain.

CLEANING OF THE BIN LINER

Because the brown staining or rusting is due to the materials expelled during the ice making cycle, the bin liner must be cleaned to prevent this staining from causing pitting of the stainless steel. The time between cleanings will depend on the water conditions.

When the staining is slight, it can be removed by washing with a non-chlorine cleanser, such as Bon-Ami or Copper-Glo and water. After cleaning, rinse thoroughly with clean water.

NOTE: Do not use a chlorine cleanser to clean this staining. The chlorine is what caused the stain.

It may be necessary to use a stainless steel wool to remove some of the bad stains.

DO NOT USE plain steel wool, because some of the steel fibers will get imbedded in the liner which will cause more serious rusting.

WATER FILTRATION

When installing an ice maker, the quality of the water supplied to the ice maker should be determined before the unit is installed. A complete water analysis is available from Crystal Tips Service Department to determine what type of filtration, if any is necessary. The size of the ice maker determines the water usage and the size of the water filter that is required.

The three most common water related problems associated with ice makers are lime scale build up on the evaporator, water distributor holes plugging up from suspended solids, and chlorine treated water supply.

Lime scale is caused by calcium and magnesium forming deposits on the evaporator. These deposits reduce the evaporators ability to transfer heat. It plugs the holes in the distributor pan, which reduces the machines capacity and could cause an evaporator freeze up.

Suspended solids are usually fine particles of sand, rust or dirt. These materials can accumulate in the water tank of the ice maker which in turn would be picked up and circulated by the water pump, causing the holes in the distributor pan to become plugged.

Chlorine is added to municipal water supplies to control bacteria. Chlorine will kill most of these organisms, but chlorine will cause rusting of stainless steel and leave an objectional taste and odor in the ice.

If you have these concerns about the water quality supplied to the ice maker, see your Crystal Tips Dealer or Distributor. They will be able to assist you.

PRINCIPAL OF OPERATION

The evaporator section in a Crystal Tips ice maker is made up of several stainless steel evaporator plates. These plates have tin plated copper buttons pressed into them.

The buttons are soldered to a copper refrigerant line on the back side. Water flows over the surface of the stainless steel plates and freezes on the buttons. The water will continue to freeze on the buttons until it reaches the size determined by the ice size thermostat. The ice size thermostat then changes its contacts to stop the water pump and to energize the hot gas solenoid valve. The hot gas solenoid valve will then allow hot discharge gas to enter the evaporator through the copper line that is soldered to the buttons. This will release the ice from the evaporator buttons. The ice will fall off the evaporator and onto the ice rack, which directs the harvested ice through the ice chute built into the ice maker cabinet and down into the ice storage bin.

When the evaporator surfaces warm, the ice size thermostat will switch back to its original, or ice making position, setting up the next freeze cycle. As the suction line warms to 55 F the harvest termination switch will open putting the ice maker back into the freeze cycle.

WIRING DIAGRAMS

Electricity is supplied to the ice maker at the terminal board located in the electrical box, with connections being made to the posts marked A and B and the ground wire connected to the grounding screw.

The electrical circuit varies slightly between air cooled and water cooled and also between compressors on the three different models. The compressor on the 200 series requires a start capacitor while the compressors of the 300 and 400 require both a start and run capacitor.

When the electrical circuit is energized, electricity will flow from terminal A of the terminal board to the following:

1. The wash side of the ice, off, wash switch (S1)
2. The high pressure control (HP1) onto the bin thermostat (TC1) of the water cooled or directly to the bin thermostat (TC1) of the air cooled and then to the ice side of the ice, off, wash switch (S1)
3. The line side (L1) of the compressor contactor (K2) Please refer to the figure below for the electrical flow with the switch in the OFF position.

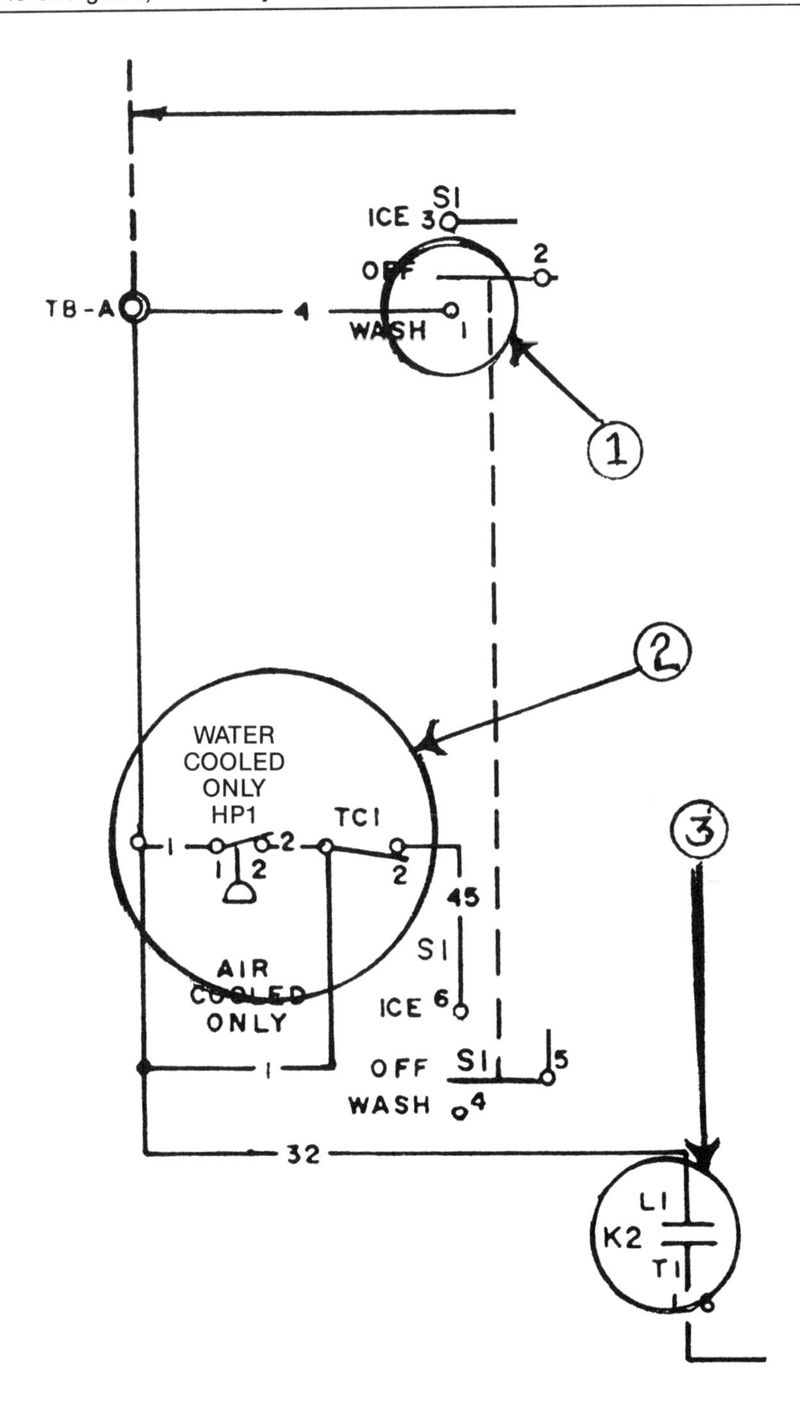

FIGURE 2

The electrical flow stops at the ice, off, wash switch (S1) when the switch is in the off position. To understand the electrical flow from here, lets first look at the switch itself.

The ice, off, wash switch is a double pole-double throw switch, or more simply, two switches in one. One side of the switch is marked with terminals 1, 2 and 3, the other side 4, 5 and 6. When the switch is placed in the wash position, the contacts at 1 and 2 and 4 and 5 are closed. This allows electrical flow through 1 and 2 and onto the water pump and from there to B of the terminal board thus completing the electrical circuit. Although there is electrical flow through 4 and 5, there is no electrical circuit beyond 4. The water pump should be the only component operating with the switch in this position. Please refer to the figure 3 below for the electrical flow with the switch in this position.

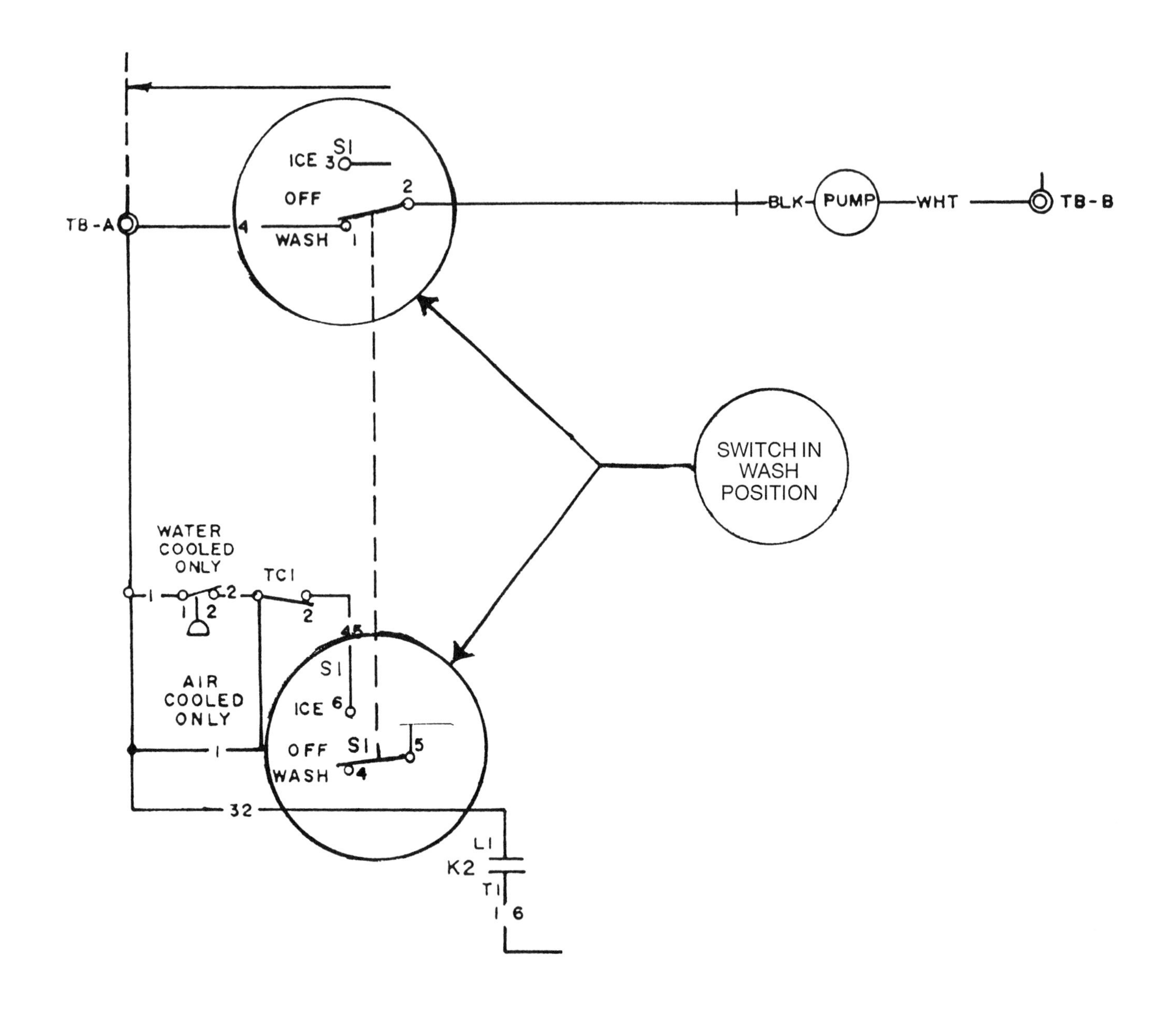

FIGURE 3

When the ice, off, wash switch is placed in the ice position, the contacts at 2 and 3 and 5 and 6 are closed. This allows electrical flow through 5 and 6 and on to the terminal board terminal C which provides electricity to the following:

1. The coil of the compressor contactor (K2) which in turn closes the contacts of L1 and T1 of the compressor contactor (K2). This provides electrical flow through the starting components of the compressor and onto the compressor and then onto terminal B of the terminal board thus completing its circuit.
2. The electric ice size heater which is connected to terminal board B and C, is energized.
3. The contacts of the defrost thermostat (TC3) which is connected to the defrost relay (R2) at terminal 6. Please refer to defrost cycle, air cooled, page 25 or defrost cycle water cooled, page 27 of this manual for the operation of this relay.
4. The contacts of the ice size thermostat (TC2) and then onto the contacts of 7 and 1 of the defrost relay (R2). From there electricity flows to terminal E of the terminal board which allows electrical flow to the fan motor (air cooled only) and onto terminal B of the terminal board, thus completing its circuit. It also allows electrical flow to the ice, off, wash switch (S1), through its contacts 3 and 2 which provide electrical flow to the water pump and then flows to terminal B of the terminal board thus completing its circuit. Please refer to figure 4 below to interpret the electrical flow of the ice maker.

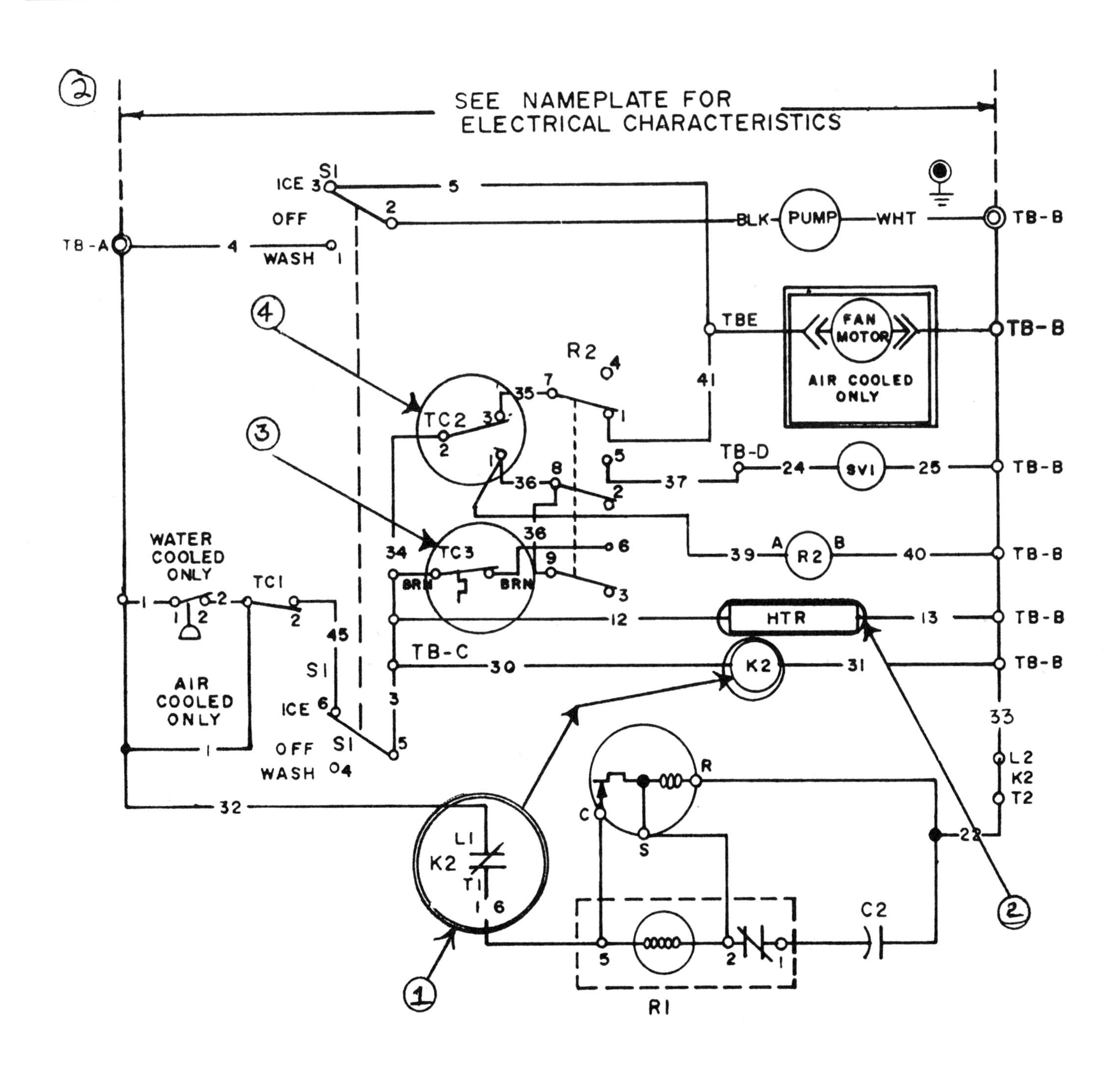

FIGURE 4

ELECTRICAL CIRCUIT OF AIR COOLED DURING FREEZE CYCLE

Electricity flows from terminal A of the terminal board through the bin thermostat (TC1), whose contacts are normally closed and open only when ice touches its capillary tube located in the bin, and then to the ice, off, wash switch (S1). With this switch in the ice position, electricity is supplied to the ice size heater. (HTR). This heater provides a small amount of heat to the ice size sheath which is located in the evaporator section. Electricity has also energized the compressor contactor (K2) which closes its contacts. This supplies electricity to the compressor which pumps the refrigerant through the refrigeration circuit. Electricity is also supplied through the ice size thermostat (TC2) to the condenser fan motor that provides the air flow through the condenser. Electricity is also supplied back to the ice, off, wash switch (S1) which in turn supplies electricity to the water pump. The water pump supplies water over the evaporator to make ice. Refer to figure 5 for the electrical circuit during the freeze cycle.

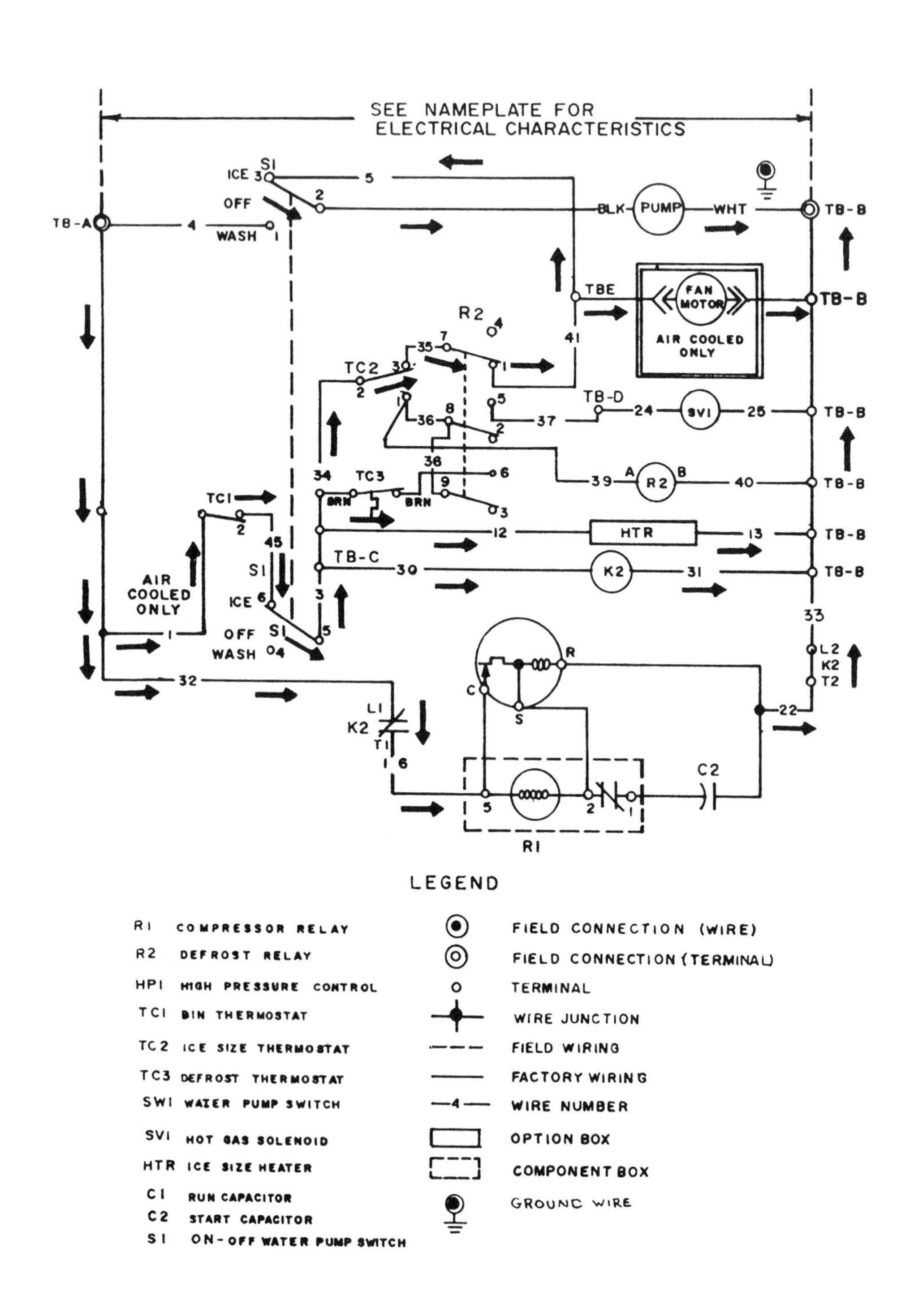

FIGURE 5

ELECTRICAL CIRCUIT OF AIR COOLED DURING HARVEST CYCLE

When the ice size thermostat (TC2) senses ice in the evaporator, the contact in the thermostat changes from 2 and 3 to 2 and 1, which changes the electrical flow and causes the condenser fan motor and the water pump to de-energize. The electrical flow now energizes the coil of the defrost relay (R2) which causes the defrost relay (R2) to change its contacts. The defrost relay has 3 sets of contacts or more simply, 3 single pole-single throw switches controlled by the relay coil. When in harvest, the contacts are closed between 6 and 9 and 8 and 5 and 7 and 4. This allows electricity supplied through the contacts of the defrost thermostat (TC3) to flow to 6 and 9 contacts and on the 8 and 5 contacts which in turn energizes the hot gas valve (SV1). When the hot gas valve opens, hot discharge gas is allowed to enter the evaporator and release the ice from the surface of the evaporator. The harvest will not terminate until the harvest termination switch (TC3) has sensed a temperature of 55 F at its location on the suction line. At that temperature the contacts open and thus opens the contacts of 6 and 9 and 8 and 5 of the defrost relay (R2) which de-energizes the defrost relay coil and the hot gas valve (SV1). This action places the ice maker back to the freeze cycle. Please refer to figure 6 for the electrical circuit during harvest.

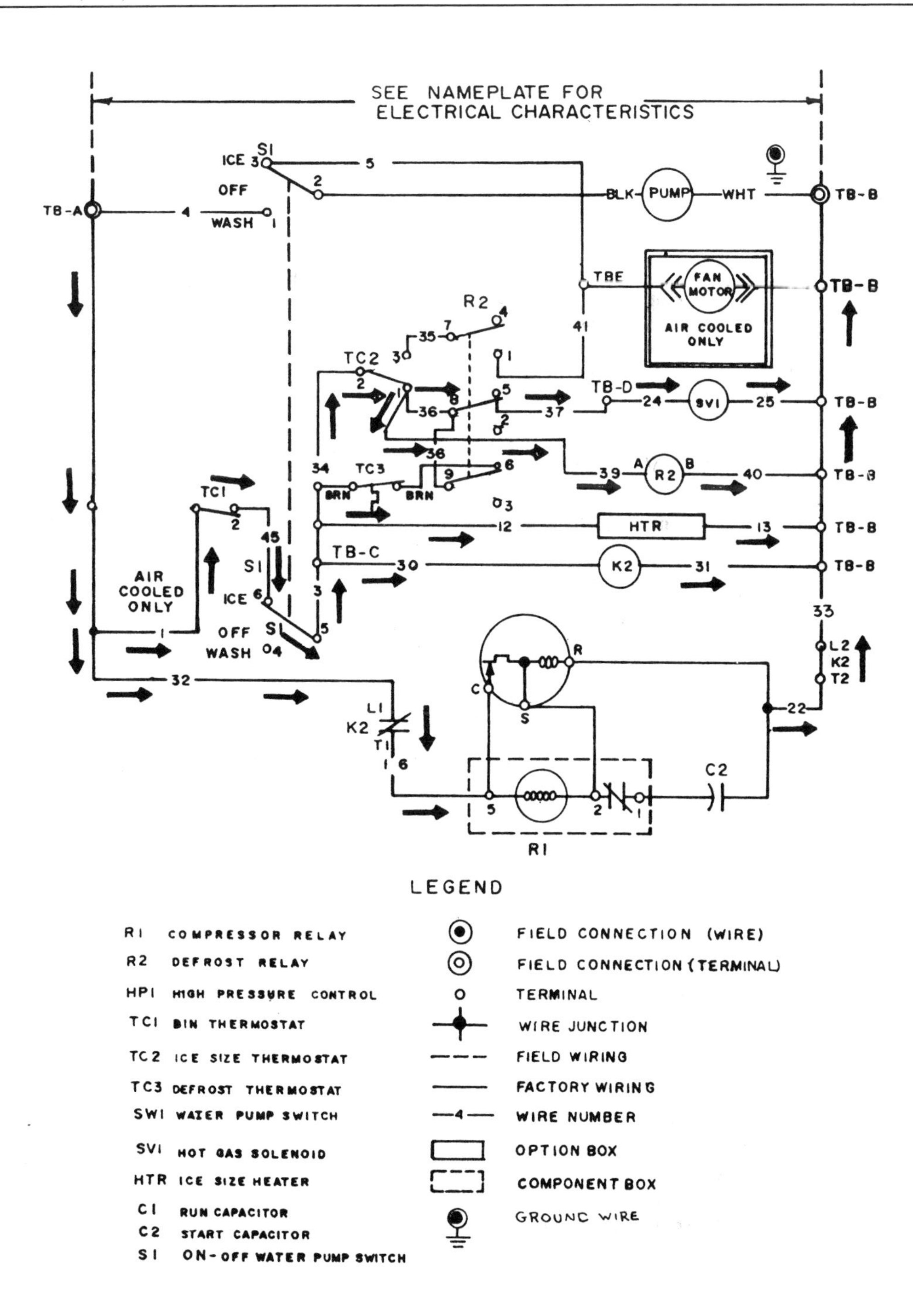

FIGURE 6

ELECTRICAL CIRCUIT OF WATER COOLED DURING FREEZE CYCLE

Electricity flows from terminal A of the junction board through the bin thermostat (TC1), whose contacts are normally closed and open only when ice touches its capillary tube located in the bin, and then to the ice, off, wash switch (S1). With this switch in the ice position, electricity is supplied to the ice size heater (HTR). This heater provides a small amount of heat to the ice size sheath which is located in the evaporator section. Electricity has also energized the compressor contactor (K2) which closes its contacts. This supplies electricity to the com pressor which moves the refrigerant through the refrigerant ci cuit. Electricity is supplied back to the ice, off, wash switc (S1), which in turn supplies electricity to the water pump. Th water pump supplies water over the evaporator to make ic Refer to figure 7 for the electrical circuit during the freeze cycl

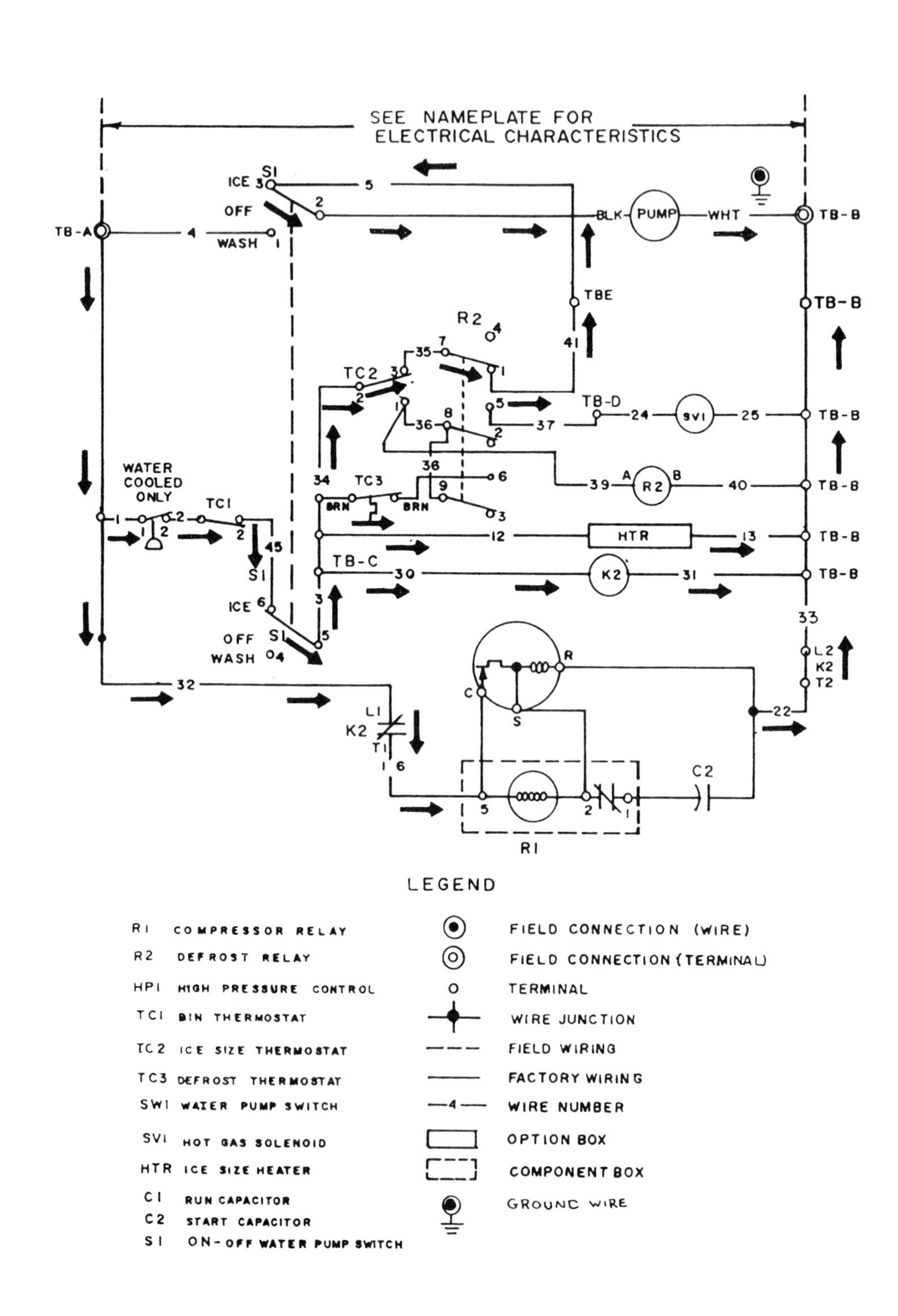

FIGURE 7

ELECTRICAL CIRCUIT OF WATER COOLED DURING HARVEST CYCLE

Vhen the ice size thermostat (TC2) senses ice in the evaporator, he contact in the thermostat changes from 2 and 3 to 2 and , which changes the electrical flow and causes the water ɔump to de-energize. The electrical flow now energizes the coil ɔf the defrost relay (R2) which causes the defrost relay (R2) o change its contacts. The defrost relay has 3 sets of contacts ɔr more simply, 3 single pole single throw switches controlled ɔy the relay coil. When in harvest, the contacts are closed between 6 and 9 and 8 and 5. This allows electricity supplied hrough the contacts of the defrost thermostat (TC3) to flow o 6 and 9 contacts and onto the 8 and 5 contacts which in turn energizes the hot gas valve (SV1). When the hot gas valve opens, hot discharge gas is allowed to enter the evaporator and release the ice from the surface of the evaporator. The harvest will not terminate until the harvest termination switch (TC3) has sensed a temperature of 55 F at its location on the suction line. At that temperature the contacts open and thus opens the contacts of 6 and 9 and 8 and 5 of the defrost relay (R2) which de-energizes the defrost relay coil and the hot gas valve (SV1). This action switches the ice maker back to the freeze cycle. Please refer to figure 8 for the electrical circuit during harvest.

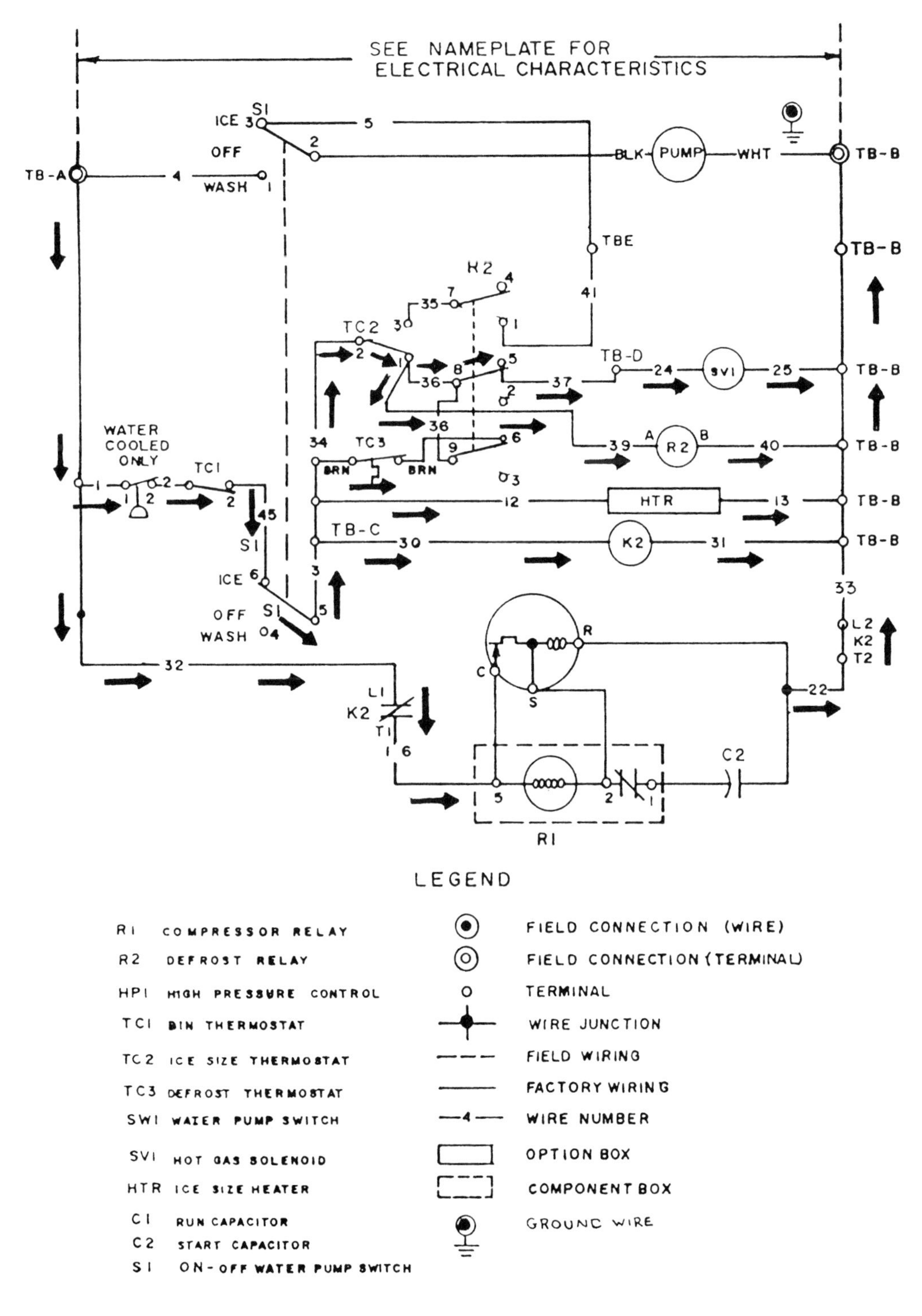

FIGURE 8

ICE STORAGE BIN FULL

When the ice storage bin is full of ice, the bin thermostat (TC1) will open which de-energizes the entire control circuit of the ice maker. This stops the water pump, condenser fan and de-energizes the compressor contactor which opens the contacts of the contactor and opens the circuit to the compressor.

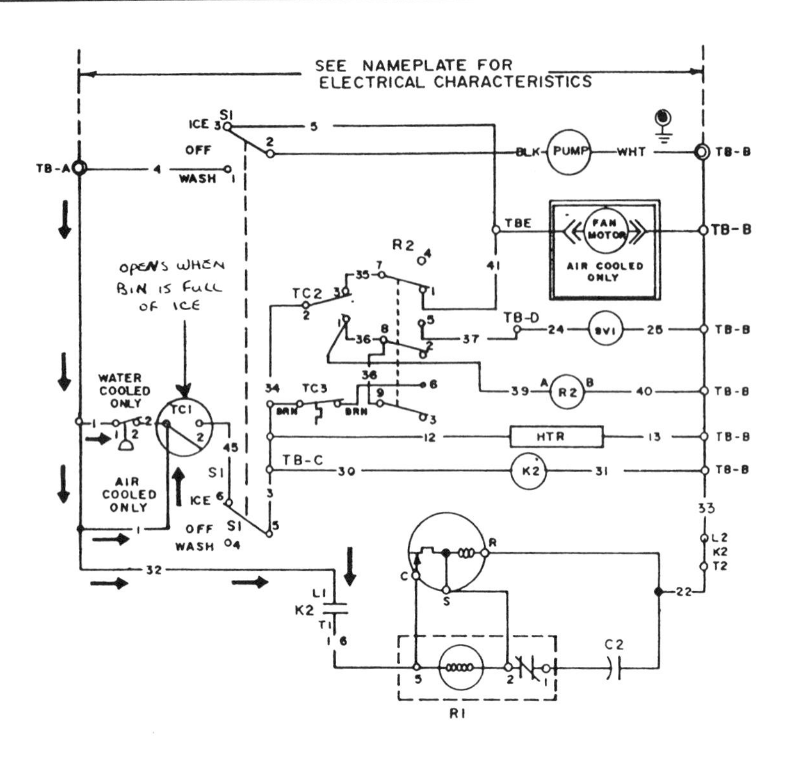

FIGURE 9

ICE-OFF-WASH IN WASH POSITION

When the ice maker is to be cleaned, the refrigeration system must be stopped, but you want the water pump operational. This is done by switching the ice-off-wash switch (S1) to the wash position. This energizes only the water pump and allows it to circulate the ice machine cleaner.

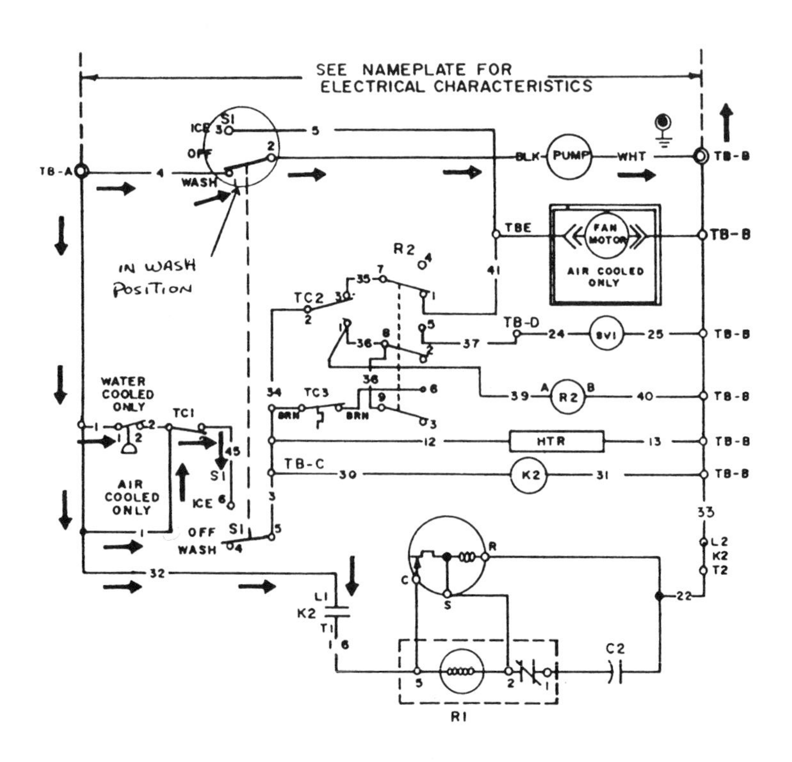

FIGURE 10

REFRIGERATION SYSTEM

The Crystal Tips ice maker uses a compressor, condenser and evaporator that are properly sized for each other to get the maximum efficiency from the system.

AIR COOLED IN FREEZE CYCLE

The refrigerant leaves the compressor as a high pressure vapor which goes to the air cooled condenser. As the high pressure vapor flows through the condenser, the heat from the high pressure vapor is removed. The high pressure vapor becomes a high pressure liquid. The high pressure liquid then flows through the liquid line which is soldered to the suction line and accumulator where the high pressure liquid gives up some more heat and adds this heat to the return gas in the suction line which reduces the chance of liquid returning to the compressor. The high pressure liquid is then passed through the filter drier which is designed to keep the refrigerant clean and dry. After the filter drier, the high pressure liquid then flows to the expansion valve which reduces the pressure of the liquid refrigerant. The refrigerant becomes a low pressure liquid which as it passes through the evaporator will absorb heat from the water passing over it. The absorbtion of heat into the low pressure liquid will cause the liquid to become a vapor which is then drawn back to the compressor through the suction line, accumulator and heat exchanger. At the compressor, the low pressure vapor is compressed back into a high pressure vapor and sent on to the condenser.

FREEZE CYCLE

AIR COOLED CONDENSER

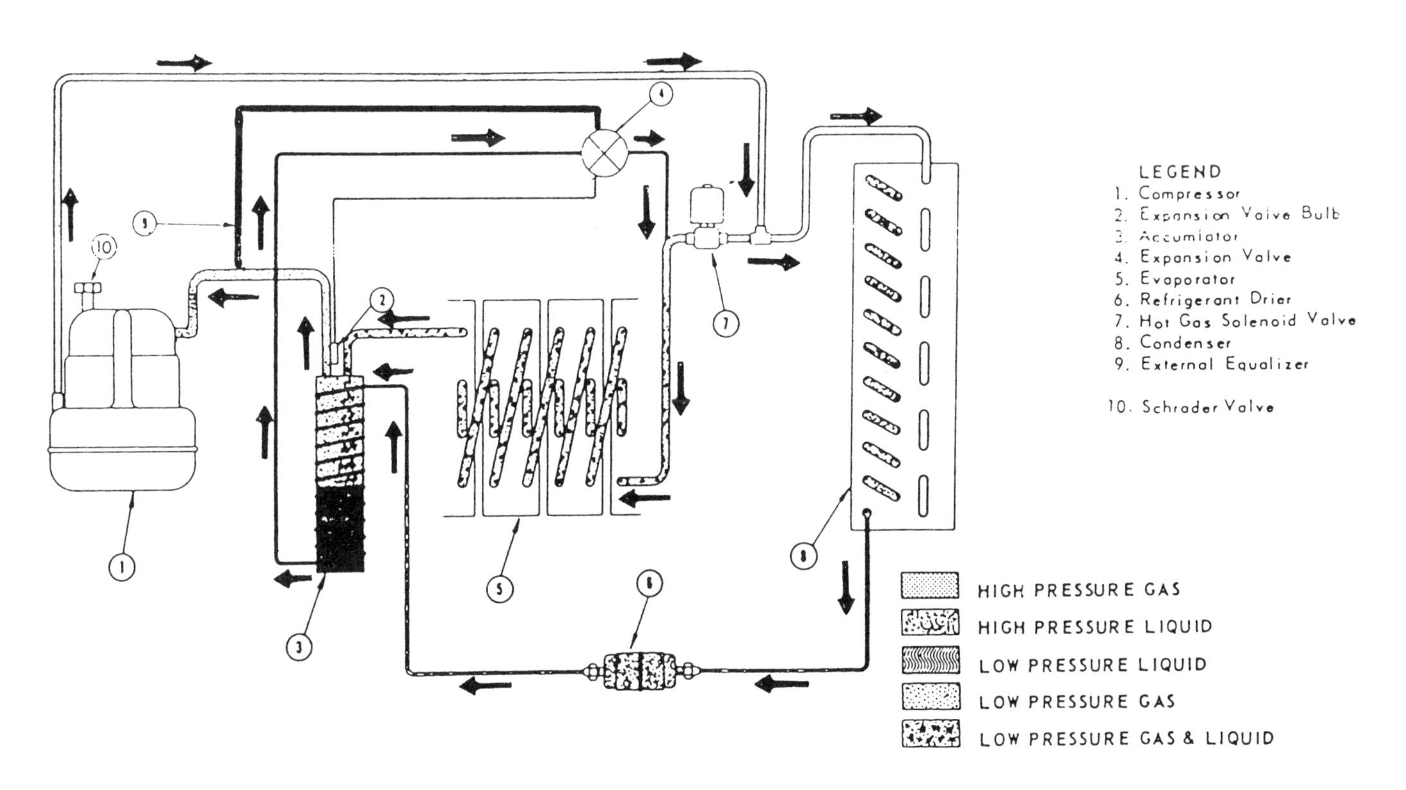

FIGURE 11

AIR COOLED IN DEFROST CYCLE

During the defrost cycle the hot gas solenoid valve opens allowing hot discharge vapor to enter the evaporator. The discharge vapor will take the path of least resistance. Therefore, it will not go to the condenser. The hot discharge vapor being released into the evaporator will cause the ice to be released from the evaporator surface. The pressure of the hot discharge gas against the outlet of the expansion valve will over come the sensing bulb pressure causing the expansion valve to close. The suction accumulator acts as a storage vessel for this liquid so it is not returned to the compressor. When the defrost thermostat terminates the defrost cycle, the hot gas solenoid valve is de-energized which causes the solenoid valve to close.

HARVEST CYCLE

AIR COOLED CONDENSER

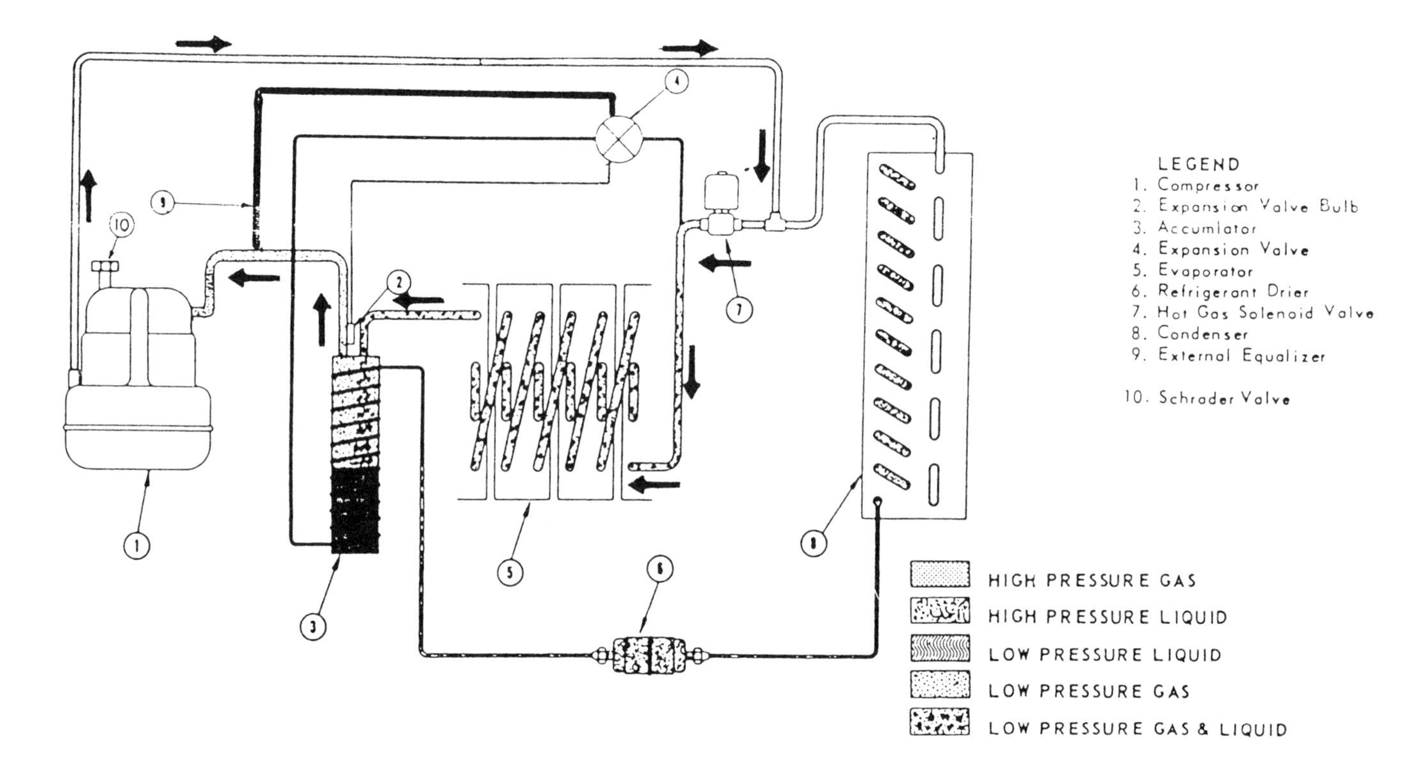

FIGURE 12

WATER COOLED IN FREEZE CYCLE

A water cooled ice maker operates in much the same way as an air cooled unit does. There is one exception however, the high pressure vapor enters the water cooled condenser where it gives up its heat to the water flowing through the condenser. The amount of water that is required to condense the high pressure vapor into a high pressure liquid is controlled by the water regulating valve. The water regulating valve is set to maintain a refrigerant pressure on the high side of 250 psig.

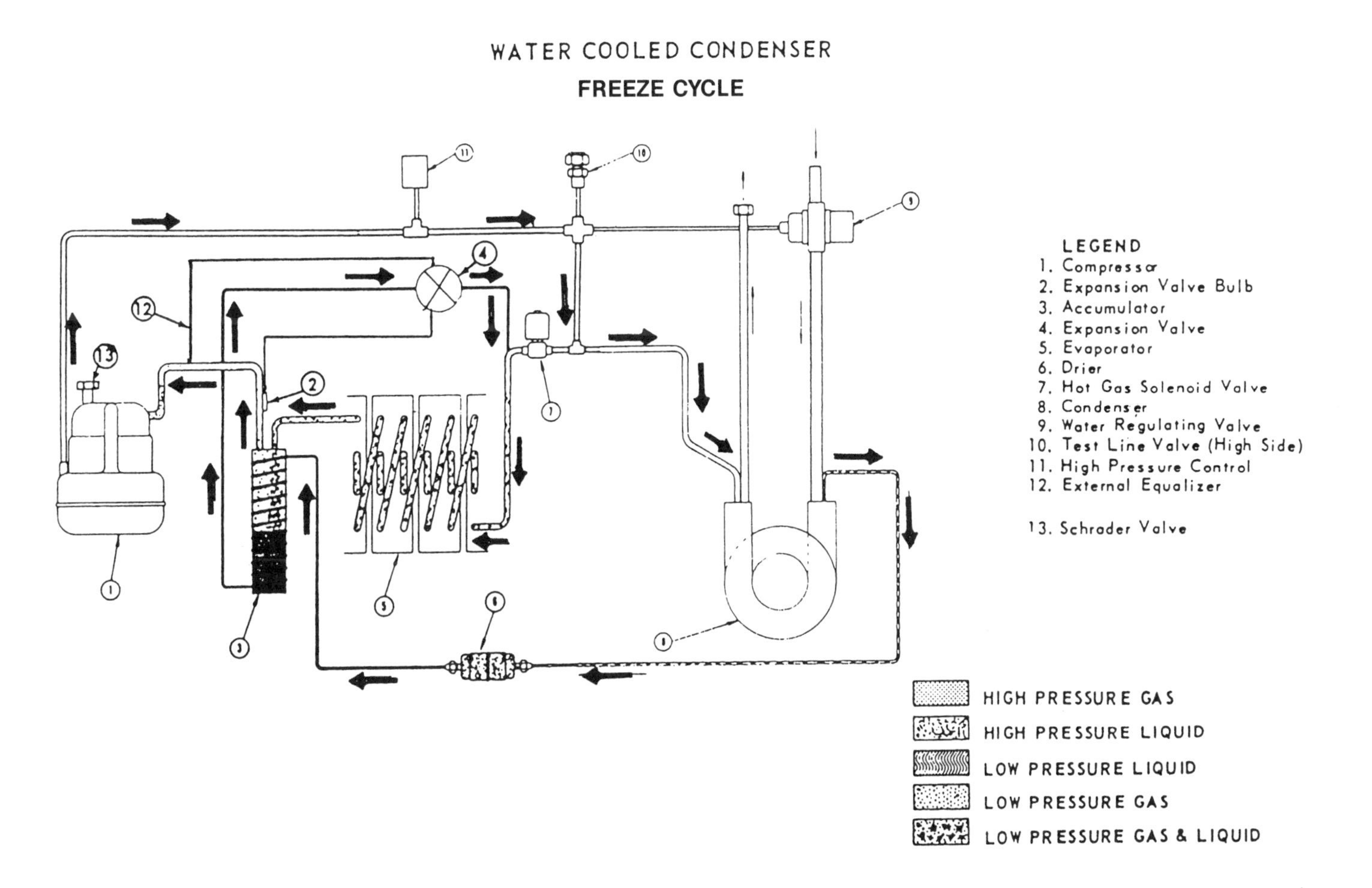

FIGURE 13

WATER COOLED DEFROST CYCLE

The water cooled ice maker operates the same as an air cooled ice maker during the defrost cycle. When the hot gas solenoid valve opens, the high pressure of the vapor is reduced to the water regulating valve. This pressure reduction causes the water regulating valve to close and will remain closed all through the defrost cycle. When the defrost cycle terminates, the hot gas solenoid valve will close causing the high side pressure to raise which will reopen the water regulating valve.

WATER COOLED CONDENSER

HARVEST CYCLE

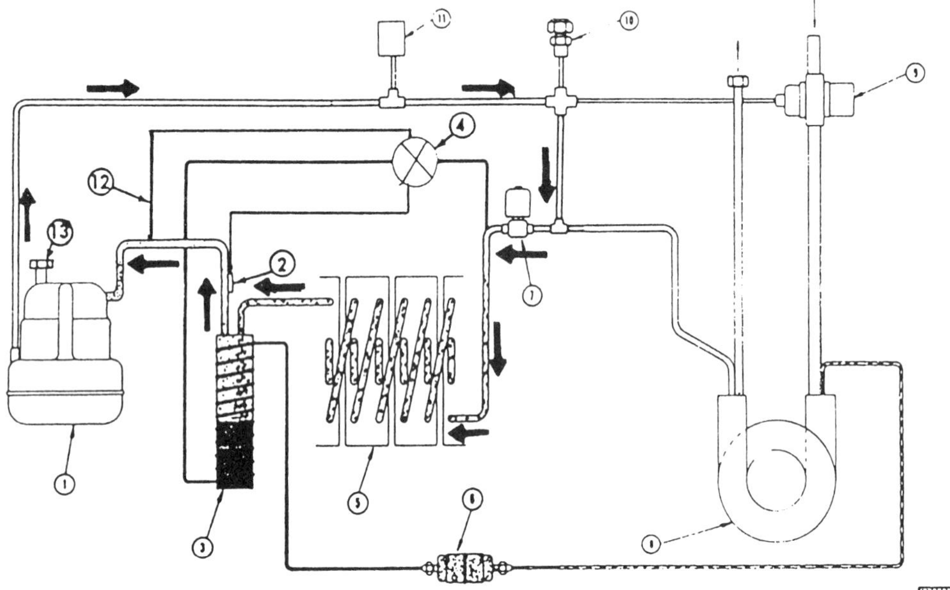

LEGEND

1. Compressor
2. Expansion Valve Bulb
3. Accumulator
4. Expansion Valve
5. Evaporator
6. Drier
7. Hot Gas Solenoid Valve
8. Condenser
9. Water Regulating Valve
10. Test Line Valve (High Side)
11. High Pressure Control
12. External Equalizer

13. Schrader Valve

HIGH PRESSURE GAS

HIGH PRESSURE LIQUID

LOW PRESSURE LIQUID

LOW PRESSURE GAS

LOW PRESSURE GAS & LIQUID

FIGURE 14

WATER DISTRIBUTION

The water enters the ice maker through the float assembly, which controls the amount of water in the water tank. The water level should be 1/8" below the bottom of the inverted U-shape of the siphon tube when the ice maker has been shut off and the water pump is not operating. When the water pump starts, it will pump the water from the water tank up through the water line to the distributor pan. The distributor pan has a series of small holes which allows the water to be distributed over the entire area of the evaporator. The water draining through the holes in the distributor pan is deflected over the ice making surface of the evaporator by the angle of the evaporator plate. The water forms a sheet which wets the entire evaporator plate. As the water flows over the evaporator plate, some of it passes over the evaporator buttons which is where the ice is frozen. All water which has not frozen on the evaporator buttons drains from the evaporator section through the ice rack and returns to the water tank, where the water pump recirculates it until the ice size thermostat puts the ice maker into a defrost cycle.

When the ice maker enters the defrost cycle, the water pump stops. This action allows all the water that is in circulation to return to the water tank. This will cause the water level to rise which will allow the water to overflow the siphon tube which draws the impurities from the bottom of the water tank. When the defrost cycle is terminated the water pump restarts causing the water level to drop and the float valve to add more fresh water to the system.

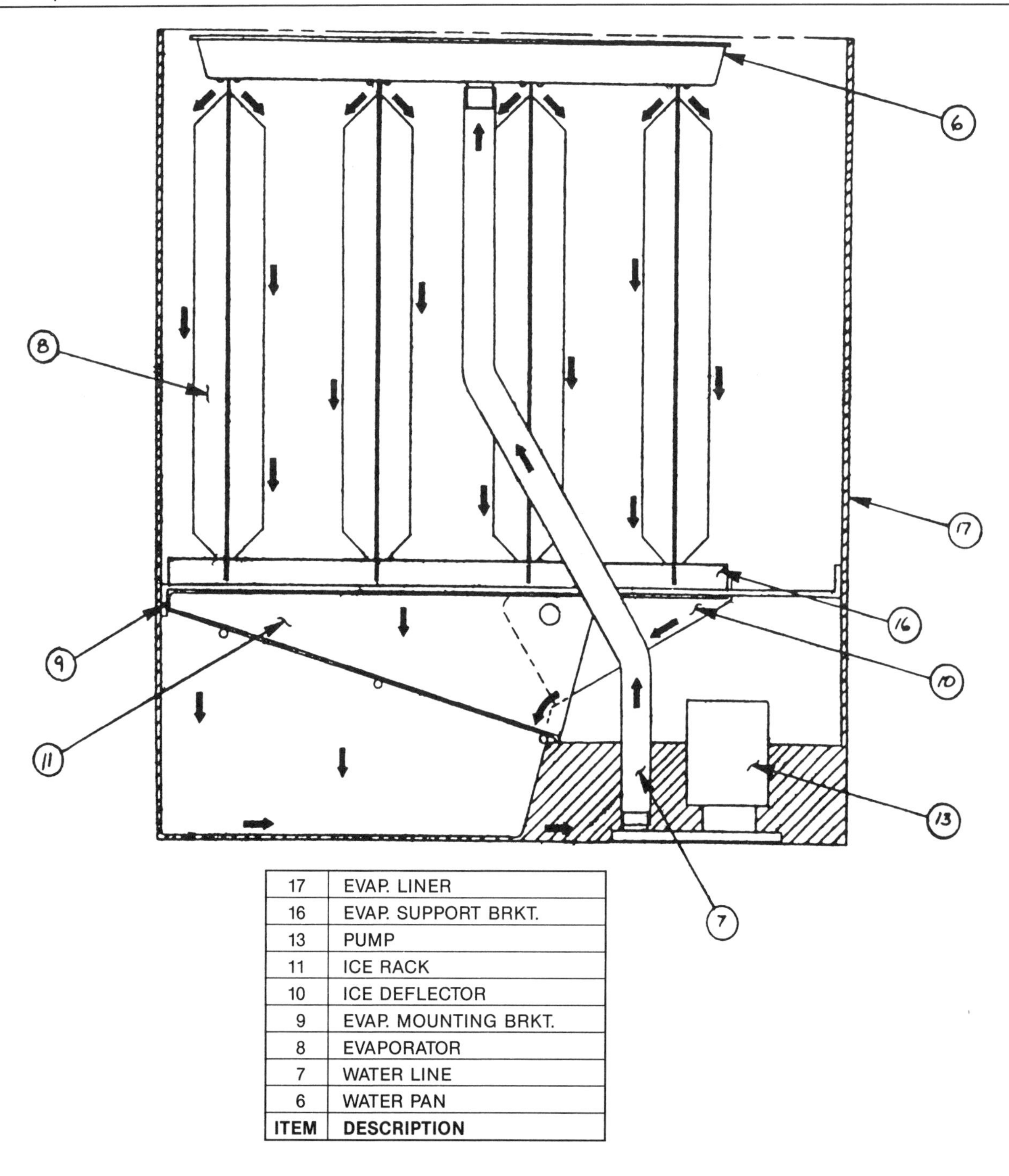

ITEM	DESCRIPTION
17	EVAP. LINER
16	EVAP. SUPPORT BRKT.
13	PUMP
11	ICE RACK
10	ICE DEFLECTOR
9	EVAP. MOUNTING BRKT.
8	EVAPORATOR
7	WATER LINE
6	WATER PAN

FIGURE 15

ADJUSTMENTS

FLOAT VALVE ASSEMBLY

The amount of water which enters the ice maker is controlled by a float valve. If the valve allows too much water to enter the water tank, there will be a constant draining of water from the water tank through the siphon drain. The water in the water tank will not get cold enough to make a consistent size ice cube over the entire surface of the evaporator.

If the float valve doesn't allow enough water to enter the water tank, the water pump may cavitate during the entire freeze cycle, which will cause air to be frozen into the ice giving the ice cubes a cloudy or milky appearance.

NOTE: It is normal for the water pump to cavitate for 2 to 3 minutes after the water pump has been started, depending on water pressure supplied to the ice maker. It may require this amount of time for the float valve to replenish the water level in the water tank.

The water level in the water tank should be adjusted while the water pump is not operating, and there is no water draining from the ice maker.

The water level is correctly set when the water is 1/8" below the bottom of the inverted U-shape siphon tube.

The best method for adjusting the water level in the water tank is by bending the float arm between the float valve and the float ball. The upper portion of the arm should be held stationary while the float bulb is gently moved either up or down to raise or lower the water level.

NOTE: WE DO NOT RECOMMEND FREQUENT BENDING OF THE FLOAT ARM. Frequent bending action may cause the arm to break, or may break the seal between the float arm and the float ball which will allow water to fill the bulb.

The float valve is made up of a plastic body and has a stainless steel valve seat. The water inlet of the valve is a 1/4" compression nut with a plastic ferrule. The water supply line is connected to this fitting. The float arm, which attaches to the valve body and the float bulb, is made of stainless steel and is the portion which must be bent to make the float adjustments.

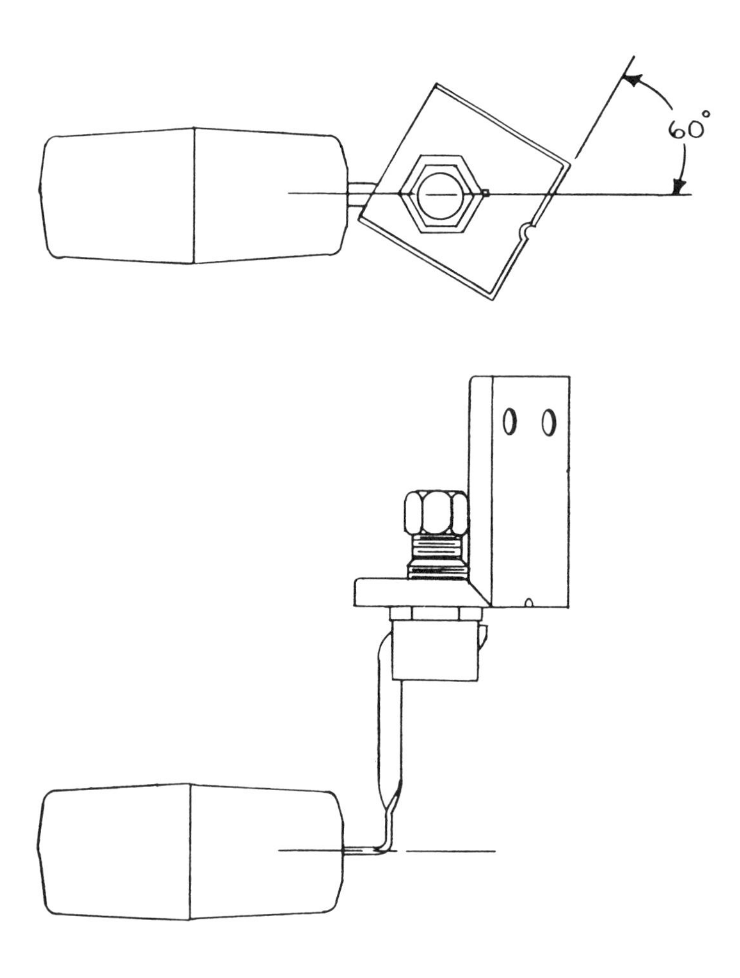

FIGURE 16

BIN THERMOSTAT

The bin control thermostat has a capillary tube which is to be woven through the thermostat holder at the time of installation. See Fig. 1, page 15. When ice fills the bin to the level of the capillary tube, some ice will come in contact with the tube. When this happens, the ice maker should stop within one (1) minute. When the ice is removed from the capillary tube, the ice maker should restart in two (2) minutes.

The bin thermostat has a maximum cold setting of 35 F +/− 2 F which is when the adjusting drive or shaft is turned clockwise as far as it will turn.

The maximum warm setting (shaft turned all the way counter clockwise) is 50 F +/− 4 F.

If the bin is not full and the bin thermostat needs adjustments, hold a few ice cubes in your hand and hold them against the capillary tube, which is woven through the holder in the ice bin.

IMPORTANT

Make sure your hand does not come into contact with the capillary tube. Make contact only with ice cubes.

If the ice maker does not stop within one (1) minute, rotate the bin thermostat shaft counter clockwise until the ice maker does stop. Then remove the ice from the capillary tube, the ice maker should restart in two (2) minutes. If it does not start, rotate the adjusting shaft clockwise until it does restart.

When the thermostat is adjusted properly, the differential between the cut out and the cut in will be 4 1/2 F +/− 1 F.

IMPORTANT

The Crystal Tips Ice Maker must be installed in an area where the ambient temperature does not drop below 60 F. If the temperature should become too cold in the area of the ice maker, the ice maker may shut off on the bin thermostat.

WARNING

If the bin thermostat is not properly adjusted or is taken out of the electrical circuit, the ice maker will not stop making ice when the bin is full. This will cause the evaporator to freeze up which in turn will cause severe damage to the refrigeration system.

If these adjustments cannot be made, the bin thermostat should be replaced.

ICE SIZE THERMOSTAT

The ice size thermostat has a capillary tube which is inserted into the ice size sheath, which is located in the evaporator section of the ice maker.

As the ice maker operates in the freeze cycle, the ice on the buttons of the evaporator continues to get larger. When they become large enough to force the water flowing over the evaporator to come into contact with the ice size sheath, the ice size thermostat will sense the temperature drop. This will cause the contacts in the thermostat to switch. This action causes the ice maker to come out of freeze cycle and go into a harvest cycle. Once the ice maker is in the harvest cycle, the hot gas valve opens, allowing hot discharge gas to warm the evaporator. When the evaporator section warms sufficiently the ice size thermostat will switch its contacts back to the freeze position. The ice maker will remain in harvest until the harvest termination switch opens its contacts which in turn switches the contacts of the defrost relay back to the freeze position. The ice maker will return to the freeze cycle at this point.

The thermostat will normally switch from the freeze cycle to the harvest cycle when the capillary tube senses 38 F +/− 2 F. When the capillary tube senses an evaporator temperature of 43 F, the ice size thermostat sets up the ice maker for the next cycle.

If the thermostat is not functioning in this manner, there are no adjustments to make on the control, but there are a number of other components to check. (See ice maker does not harvest, page 65 of service manual).

HARVEST TERMINATION SWITCH

The harvest termination switch has a tube mounting cup which fits on to the outlet line of the evaporator and is located on the machine compartment side of the ice maker. It has a spring steel clamp that holds the mounting cup firmly to the suction line. When the ice maker is in the freeze cycle, the thermo disc of the thermostat is closed, because the suction line is cold. When the ice maker goes into the harvest cycle, the hot discharge gas warms the evaporator and also the suction line. The defrost thermostat opens when the suction line reaches 55 F +/− 5 F. This causes the defrost relay to switch its contacts which brings the ice maker back into the freeze cycle. This cools the suction line which allows the thermo disc of the harvest termination switch to reclose at 40 F +/− 5 F.

The harvest termination switch allows the evaporator to warm sufficiently to release all ice from the evaporator.

DEFROST RELAY

The defrost relay is a 3 pole - double throw switch, whose switching action is controlled by its electric coil. More simply, this relay is actually 3 switches in one body. During the freeze cycle, one set of switch contacts are closed which are 7 and 1 and provides electricity to the fan motor on air cooled models and the water pump on all models, while the other two sets of switch contacts are open, 8 and 2, and 9 and 3. When the ice size thermostat senses ice and switches its contacts, the contact at 7 is now with 4 which opens the circuit of the water pump and the fan motor. The contact of 8 is now with 5 and 9 is with 6 which means that the contacts of 9 and 6 are now keeping the coil of the defrost relay energized and that the contacts of 8 and 5 will energize the hot gas valve, which will harvest the ice from the evaporator.

The harvest cycle will not terminate until the harvest termination switch senses a suction line temperature of 55 F +/– 5 F. This causes the contacts of the thermostat to open which in turn de-energizes the coil of the defrost relay thus returning the switches of the defrost relay back to their original position. This places the ice maker back into the freeze cycle.

ICE SIZE SHEATH

The ice size sheath is a tube arrangement which holds the ice size thermostat capillary tube in the proper position in the evaporator section of the ice maker. The sheath is assembled into a bracket with the ice size heater and the ice size selector knob. When the entire assembly is located between the evaporator plates by the two holes in the bracket, the sheath will be positioned at the correct clearance from the evaporator plate. The sheath will be closer to one evaporator plate than the other plate. This is the proper position. It is important that the clearance between the sheath and evaporator plate is the same along its entire length.

If the clearance is not the same, the sheath must be checked for straightness.

The ice size adjustment bracket assembly must be removed from the evaporator section in order to remove the sheath. Remove the two (2) screws holding the bracket assembly to the evaporator plates. With the assembly out of the evaporator section, the sheath assembly may now be removed from the bracket assembly. Completely loosen the brass nut holding the sheath to the bracket. The sheath should now slide out of the ice selector knob, the plastic sleeve, the ice size heater, the brass nut and the bracket.

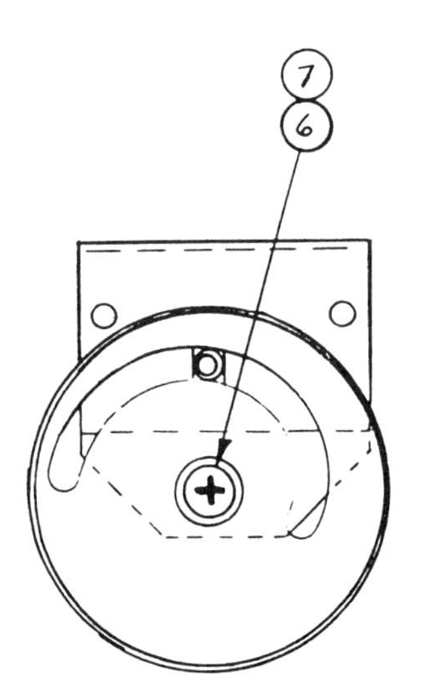

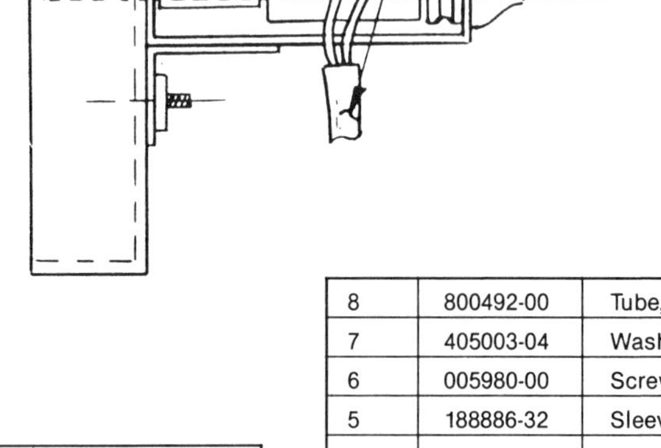

ITEM	PART NO.	DESCRIPTION	QTY.
8	800492-00	Tube, Plastic 3/8 I.D. x .75 Lb.	1
7	405003-04	Washer #10	1
6	005980-00	Screw #10.32 S.S.	1
5	188886-32	Sleeve, Insulation	1
4	800488-00	Capillary Sheath & Nut	1
3	801219C-01	Knob-Ice Size	1
2	See Tab.	Bracket-Cube Size F.A.	1
1	See Tab.	Resistor Ass'y	1

PART NO.	ITEM #1	ITEM #2	MODEL
802904B-01	118005 -03	802905A-01	800,1000,1200
802904B-02	118005 -01	802909A-01	200,300,400

FIGURE 17

Once the sheath has been removed, slide the brass connector and the rubber boot assembly off the sheath. It must be removed from the plugged end of the sheath. Then lay the sheath on a flat surface. While watching the center portion of the sheath, slowly roll the sheath back and forth until the high point of the sheath from the flat surface is located.

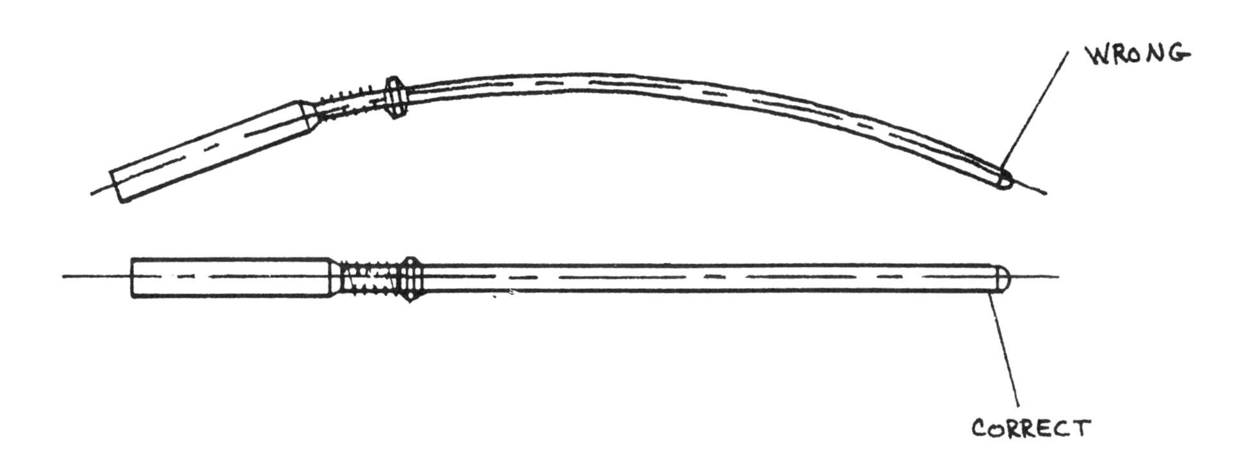

FIGURE 18

At that high point, gently tap on the sheath with a soft-faced hammer. Repeat this step until the sheath rolls true to center. The sheath may now be reassembled in the reverse order and then may be installed back into the evaporator section.

The ice size adjust knob, identified as the ice size selector knob, has a series of numbers on it. The smaller the number, the smaller the ice cube and the opposite for the larger the number. Normal cube size should be at position 5. Smaller or larger cubes may be made depending on customer preference.

WATER REGULATING CONTROL VALVE (WATER COOLED UNITS ONLY)

The water regulating valve is connected to the water inlet supply line to the condenser. The valve controls the amount of water that flows through the condenser. The valve reacts to the ice makers high side pressure, as the pressure rises, the valve will allow more water to pass through the condenser. As the high side pressure drops, the valve will restrict the amount of water passing through the condenser.

The water regulating valve must be set to maintain a high side pressure of 250 psig during the freeze cycle. If the high side pressure is not being maintained at 250 psig, the water regulating valve requires adjustments. On the top of the valve there is an adjusting screw. Turning the adjusting screw counter clockwise will decrease the flow of cooling water and increase the high side pressure. Turning the adjusting screw clockwise increases the cooling water flow which will decrease the high side pressure.

To determine the high side pressure of the ice maker, connect the high side service gauge to the service valve identified as discharge port.

If the water regulating valve does not adjust the high side pressure, please refer to Main Parts Function and Replacement Procedure in the service manual, page 38.

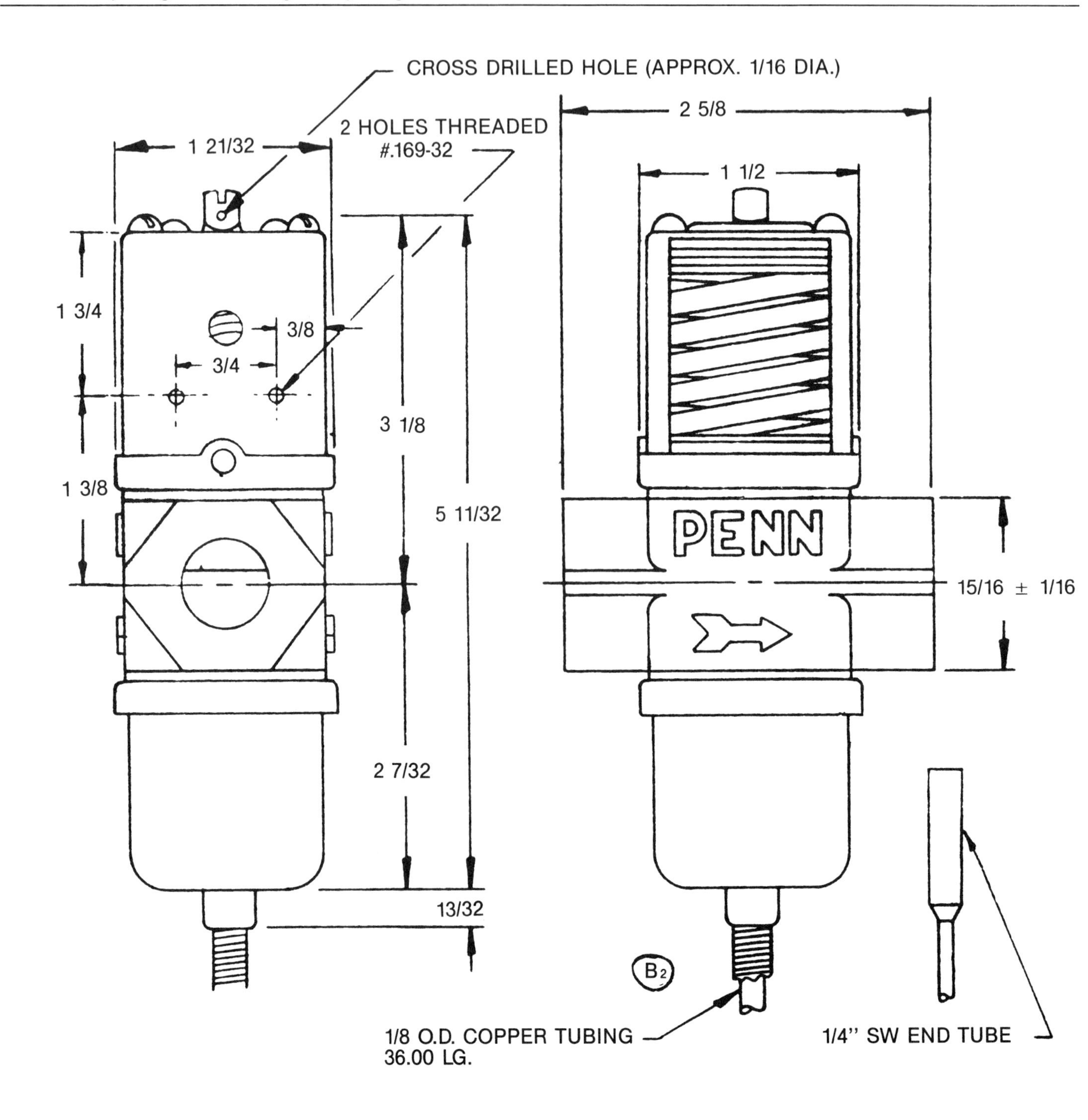

FIGURE 19

PARTS FUNCTION AND REPLACEMENT PROCEDURE

DANGER

Always disconnect the electrical and water supply to the ice maker when providing service. Failure to do this may result in electrical shock to yourself or water damage to the surroundings.

PART IN THE REFRIGERATION CIRCUIT

1. REFRIGERANT

Refrigerant in a high pressure liquid form is fed to the expansion valve where the refrigerant is reduced to a low pressure liquid. Under this low pressure, the liquid will absorb heat from the evaporator causing the liquid to change to a vapor. This vapor is then drawn into the compressor where the temperature and pressure of the vapor are increased. The high temperature, high pressure vapor flows to the condenser where the heat is removed causing the vapor to return to the liquid form, making the refrigerant ready to flow back to the evaporator to pick up more heat.

The Crystal Tips ice maker uses R-502 refrigerant. Cylinders containing refrigerant R-502 are color coded in orchid (pale purple).

R-502 has all the qualities of other halogenated refrigerants. It is nontoxic, nonflamable, nonirritating, stable and noncorrosive. Leaks can be detected with electronic leak detectors, a halide torch or soap solution.

When a leak in the refrigerant system developes or one of the refrigeration parts requires changing, make sure the room is thoroughly ventilated before starting to work on the ice maker. Using a set of service gauges, check the pressure with in the refrigeration system. Always check the replacement refrigerants R-number before charging to avoid mixing refrigerants.

WARNING

Make sure there are no lighted flames near a system which is suspected of having a leak. Refrigerant contact with an open flame will create a corrosive and toxic gas.

In a confined area with no ventilation, refrigerant can cause intoxication due to overinhalation. This could cause suffocation because the refrigerant will replace the oxygen supply in the room.

Wear goggles and gloves whenever you are charging or discharging the refrigerant. These will protect your eyes and skin. Liquid refrigerant on the skin may freeze the skin surface. If this should happen, wash the affected area immediately with water.

Any accident involving refrigerant should be reported to a doctor.

Never attempt to solder a refrigeration line which contains refrigerant. The increase in temperature will cause a rapid rise in the systems pressure which could cause the tubing to explode.

IMPORTANT

When discharging the refrigerant from the ice maker, capture as much of the refrigerant as possible with a reclaim devise. This will prevent the refrigerant from entering the atmosphere.

METHOD OF DISCHARGING REFRIGERANT

1. Connect the service gauges to the low side and the high side service ports on the ice maker. The valve opener inside the hoses will depress the schrader valve core.
2. Open the high side valve of the set of service gauges slowly so you gradually discharge the ice maker.

When the refrigerant stops flowing from the high side, open the low side valve on the service gauge to allow the refrigerant on the low side to discharge.

WARNING

Allowing the refrigerant to discharge with a completely open valve on the high side may cause the condenser on a water cooled unit to freeze and rupture. Rapid discharge of the refrigerant may also cause the oil in the compressor to be discharged.

METHOD OF CHARGING REFRIGERANT

1. In order to achieve a properly charged refrigeration system, the system must be completely evacuated.

In order to achieve a complete evacuation you will need a service gauge manifold set with properly maintained hoses and a vacuum pump capable of pulling a 50 micron vacuum. This will require a two stage pump.

2. Connect the service gauge manifold to the high and low side service ports and to the vacuum pump. Make sure the valves on the gauge manifold are closed, then start the vacuum pump.

NOTE: Do not use a refrigeration compressor as a vacuum pump. Compressors are able to pull only 50,000 micron.

After the vacuum pump has been started, open the valves on the gauge manifold. This will allow the refrigeration system to start being evacuated.

If there has not been an excessive amount of moisture in the system allow the vacuum pump to pull the system down to about 200 microns or 29.9 hg or less. Once this has been achieved, allow the vacuum pump to operate for another 30 minutes. Then, close the valves on the gauge manifold and stop the vacuum pump. Then, watch your gauges rise to 500 microns in three (3) minutes or more. This indicates a dry system under a good vacuum.

If your gauge registers a more rapid rise, the system either has moisture remaining or there is a leak in the system, requiring a check for the leak and repair and another complete evacuation.

NOTE: Seal the ends of the gauge mainfold hose and pull them into a deep vacuum to determine if the leak is not in the hoses. The gauge manifold should be able to hold the vacuum for three (3) minutes.

If the refrigeration system is extremely wet, use radiant heat to raise the temperature of the system. This action will cause the moisture to vaporize at less of a vacuum.

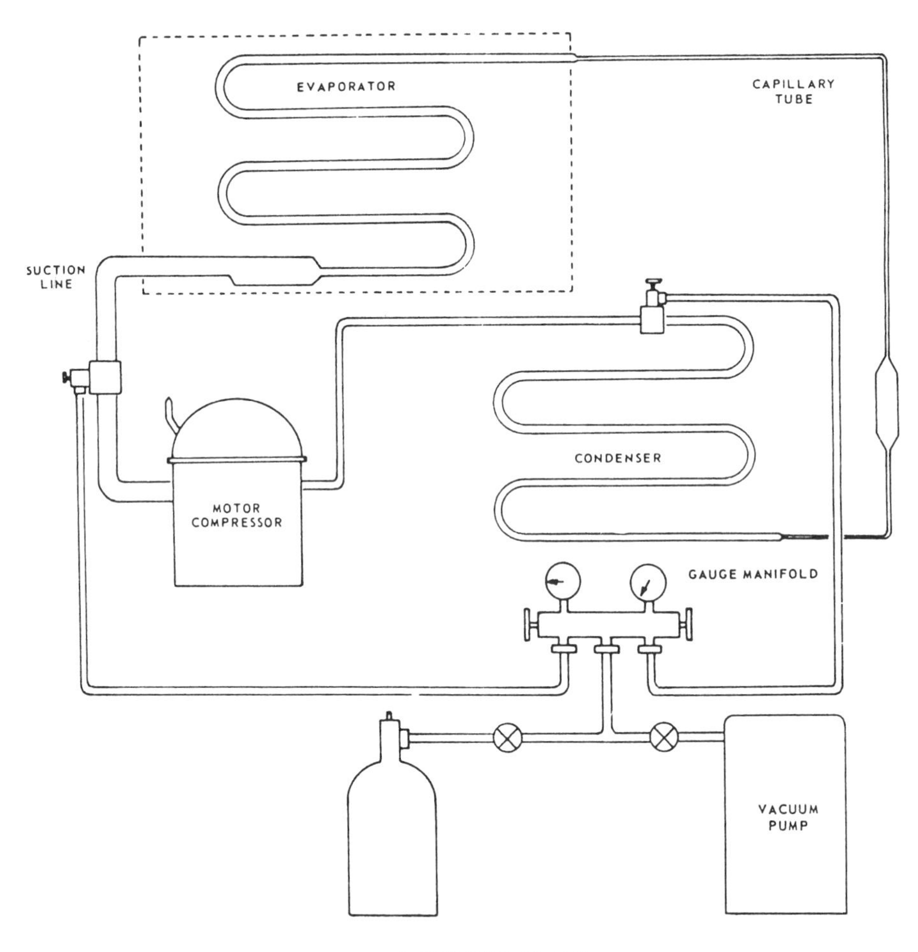

FIGURE 20

Figure 20 above is a typical service connection. The use of two (2) hoses, one between the vacuum pump and gauge manifold and the other hose between the refrigerant cylinder and the gauge manifold allows you to evacuate and charge the system with out disconnecting any hoses. If the hoses were disconnected, air or moisture will have the opportunity to enter the hoses and then the system.

A properly charged ice maker is a service persons greatest ally.

Proper charging will allow any concern with the ice maker to be accurately diagnosed.

The refrigerant charge must be weighed into the ice maker either by a charging scale or with a dial-a-charge.

The amount of R-502 refrigerant required for the ice maker is printed on the serial tag plate attached to the ice maker and is listed under ice maker specifications in the manual.

WARNING

Use only vapor refrigerant to charge a self contained air cooled or water cooled ice maker. On an ice maker with a remote condenser, liquid refrigerant can only be charged into the outlet port on the receiver.

Charging the ice maker in any other manner may cause severe damage to the compressor.

Never attempt to start the compressor while the system is in a vacuum.

In some cases the complete charge may not enter the refrigeration system. In those instances, close the gauge manifold high side valve and disconnect the manifold hose from the high side port. Then, start the compressor and allow the low side to draw in the remainder of the refrigerant vapor.

When the ice maker is completely charged, secure the caps to the service ports and check to make sure the ports are not leaking refrigerant.

SERVICE PORTS

The Crystal Tips ice maker is equipped with high side and low side service (schrader valves). These are provided to give easy access for the service person to check the operating pressure, and to evacuate and recharge the ice maker.

These service valves have a core which must be depressed with the valve opener inside the gauge manifold hoses. There is a sealing type valve cap which closes or caps the valve when it is not being used. These caps provide the primary seal to prevent the valve from leaking.

If the valve becomes damaged, it must be replaced. To replace a valve, follow the method of discharging refrigerant listed in this manual. Then, using a torch, remove the damaged valve and resolder in the new one. Then, follow the refrigerant charging method described in this manual on page 40.

CAUTION

When soldering on the ice maker, be careful not to let the torch flame come in contact with the electrical wires or the insulation.

IMPORTANT

Any time the refrigeration system has been opened, proper refrigeration procedure requires the liquid line drier to be replaced.

COMPRESSOR

The compressor draws the low pressure refrigerant vapor from the evaporator and compresses this vapor into a high pressure, high temperature vapor. The compressor is the main component in the refrigeration system.

There are two basic ways a compressor can become inoperative. Mechanical breakdown or electrical breakdown.

Mechanical breakdowns are experienced when the oil has been washed out of the compressor crankcase, or liquid refrigerant has entered the compressor causing valve damage or other compressor components to become damaged.

Electrical breakdowns are caused by the presence of moisture in the system or improper electrical supply to the compressor.

To properly diagnose the compressor failure, you must determine the type of failure and then determine the cause. To simply replace a defective compressor is not enough! Without determining the reason for the failure and correcting it, the replacement compressor may not operate in a satisfactory manner.

If the compressor is running but is not producing any ice, or very little ice, the operating pressures should be checked. But first, make certain the hot gas solenoid valve is closed and the ice maker is not stuck in a harvest cycle. If it is stuck in the harvest cycle (the evaporator is warm), you need to check the solenoid valve or ice size thermostat.

Check the temperature of the evaporator. The temperature should be approximately the same over the entire evaporator surface and the suction pressure should comply with those pressures listed on page 17 of this manual.

If the temperature of the evaporator is uneven and the suction pressure is low, this may indicate a lack of refrigerant, a restriction in a line, or a defective expansion valve.

If the ice maker is low on refrigerant, follow the method of charging on page 40 of this manual.

CAUTION

Do not add any refrigerant to the system if the ice maker has a known full charge or the actual amount of the charge is unknown.

A restriction in the liquid line or drier will be indicated by a temperature drop across the restriction.

If the expansion valve is suspected, check the valve according to the procedures listed on page 49 of this manual.

An abnormally high discharge (head) pressure will also cause a loss of capacity and may be caused by a dirty condenser, a defective fan blade or motor or possibly a noncondensable in the system.

If the suction pressure is high and all components seem to be functioning properly, check the compressor amperage draw. An amperage draw near or above the compressors name plate rating indicates a normal operation. However, it is possible the compressor may have damaged valves.

An amperage draw below the name plate rating indicates a broken valve or connecting rod in the compressor. The compressor must be replaced.

IMPORTANT

The compressor used in Crystal Tips Ice Makers have an internal overload protector. If the compressor is warm and there is no continuity through the windings, the overload protector may be open. Wait until the compressor has cooled normally, then check for continuity again.

If there is no continuity through the windings, the compressor must be replaced.

If you suspect the compressor is grounded, do a continuity check between the common terminal and a spot on the compressor body which has had the paint scratched off. If you get continuity, the compressor is grounded and must be replaced.

REPLACEMENT PROCEDURES

1. Disconnect the power supply to the ice maker and turn off the water supply.

WARNING

When the Ice-Off-Wash switch is in the off or center position, the electricity is NOT disconnected from the ice maker. Always disconnect the electrical supply to the ice maker when working on the ice maker. Failure to disconnect the electrical supply to the ice maker may result in electrical shock or death.

2. Remove the ice makers exterior panels for easy access to the compressor compartment.
3. Remove the lead wires from the compressor terminals.
4. Discharge the refrigerant from the system following the procedure listed under method of discharge page 39 of this manual.
5. Unsolder and disconnect the service port, suction line and discharge line from the compressor.

CAUTION

When soldering on the ice maker, be careful not to let the torch flame contact the electrical wires or the insulation.

6. Remove the bolts which secure the compressor to the ice maker. Then take the compressor out of the unit.

WARNING

The compressor is extremely heavy, be careful when lifting or moving the compressor. Always make sure the floor in the work area is dry and clean and there are no slip or trip hazards which could cause injury to yourself or others nearby. Always lift properly to prevent back injuries.

7. Remove the rubber mounting grommets and sleeves from the defective compressors mounting feet and insert them in the new compressor mounting feet.
8. Before removing the new compressors port plugs, clean the solder connection area of the ports.
9. Carefully lift and place the replacement as it was originally installed. Then secure the compressor to the ice maker base.
10. Remove the port plugs and connect the refrigerant tubing to the compressor.
11. Replace the filter drier.

IMPORTANT

Always replace the drier whenever the refrigeration system has been opened for any repair.

12. Connect the vacuum pump and evacuate the refrigeration system according to the procedures outlined on page 40 of this manual.
13. While the system is being evacuated, connect the electrical supply wires to the proper terminals of the compressor. REMEMBER! The compressor requires three (3) wire connections to operate. There are six (6) ways to wire, but only one (1) is correct. Check your wiring diagram for the proper sequence.
14. Clean the condenser, fan motor and fan blade while you are waiting for the complete evacuation.
15. Recharge the ice maker with the correct amount of refrigerant according to the method of charging described on page 40 of this manual.
16. Check the entire operation of the ice maker, replace the panels.

COMPRESSOR RELAY, START AND RUN CAPACITOR

On single phase compressors, there may be a start and run capacitor and a compressor relay in the circuit. If the compressor attempts to start but does not or if the compressor hums, one possible cause is a defective relay.

Remove the wires from the relay and capacitors, and using a high voltage ohmmeter, check the continuity through the relay coil terminals number 2 and 5 on the relay. If there is no continuity, replace the relay.

Then using the ohmmeter, check across the relay contacts terminals 1 and 2. When the relay is not energized, those contacts should be closed. If the contacts are open, replace the relay.

To replace the relay, remove the two (2) screws which secure it to the control box of the ice maker. When installing the relay, make sure the identified side is in the proper position. (this side up). If it is in an incorrect position, the new relay will not operate as it is designed to.

Any start or run capacitor which is bulging, leaking or damaged in any other way must be replaced.

CAUTION

Always make sure any capacitor has been discharged before checking or handling. A capacitor is an electrical device, capable of storing electrical energy. This energy can remain in the capacitor even though the electrical supply has been disconnected.

Using an ohmmeter, check for continuity between each terminal on the capacitor and the capacitors case. Any continuity indicates the capacitor is shorted and must be replaced.

The simplest way to test a capacitor in the field is to substitute the suspected capacitor for a capacitor that is known to be good. If the compressor operates properly, replace the capacitor.

Capacitors can be tested with an ohmmeter for open or shorted circuits. Set the ohmmeter on its highest resistance scale and connect its leads to the capacitor terminals.

1. If the ohmmeter first moves to zero then gradually increases to infinity, the capacitor is good.
2. If there is no movement of the ohmmeter, an open circuit is indicated.
3. If the ohmmeter moves to zero and stays there, showing a low resistance reading, the capacitor is shorted.

To replace a defective capacitor, remove the capacitor strap. The capacitor is now free of the electrical control box. When replacing the capacitor and securing it with the strap be careful not to crack or damage the capacitor housing in any way.

TROUBLE SHOOTING GUIDE FOR COMPRESSORS

1. Compressor will not start	a. Incorrect wiring b. Control contacts are open c. Control circuit open d. Compressor defective
2. Compressor will not start; Compressor hums;, then cycles off on overload	a. Low voltage b. Incorrect wiring c. Compressor relay contacts open or relay defective d. Start capacitor defective e. Wrong start capacitor f. Run capacitor defective g. Wrong run capacitor h. Defective compressor i. Excessive head pressure.
3. Compressor starts, but start winding stays in the circuit	a. Low voltage b. Incorrect wiring c. Defective compressor relay d. Defective start capacitor e. Defective run capacitor f. Defective compressor g. Excessive head pressure
4. Compressor starts and runs, but cycles on overload.	a. Low voltage b. 3 phase supply out of balance on 3 phase system c. Defective run capacitor d. Defective compressor e. Excessive head pressure f. Dirty condenser or defective fan motor.
5. Compressor starts after several attempts	a. Low voltage b. Defecive relay c. Defective start capacitor d. Restricted liquid or discharge line e. Non-condensable in the refrigeration system.
6. Compressor starts but immediately stops on overload	a. Low voltage b. Defective relay c. Defective start capacitor d. Defective run capacitor e. Defective compressor
7. Run capacitor continually burns out.	a. Low voltage b. Defective relay
8. Compressor short cycles	a. Overcharge b. Undercharge c. Defective control d. Defective expansion valve e. Defective solenoid valve f. Defective compressor
9. Start capacitor continually burns out.	a. Low voltage b. Defective relay c. Defective start capacitor d. See compressor short cycles (8)
10. Compressor relay burns out.	a. High voltage b. Low voltage c. Incorrect wiring d. Relay defective e. Relay installed in wrong position. f. Defective start capacitor g. Defective run capacitor h. Defective compressor

CONDENSER

The condenser in the refrigeration system removes the heat from the refrigerant vapor, which was picked up in the evaporator. On a 200, 300, 400 model Crystal Tips ice maker, there are two types of condensers used.

1. Self contained Air Cooled
2. Self contained Water Cooled.

SELF CONTAINED AIR COOLED

Under normal operating conditions, the condenser is extremely reliable, but on occasion, the condenser refrigerant line may develop a leak due to something rubbing against the line or from abuse.

Since moving air is used to condense the high pressure, high temperature refrigerant, the air cooled condenser can become plugged and may not allow any air to flow through it. This makes it necessary to clean the fin area of the condenser.

AIR COOLED CONDENSER CLEANING

1. Disconnect the electrical supply and turn off the water supply to the ice maker.

WARNING

When the Ice-Off-Wash switch is in the off or center position, the electricity is NOT disconnected from the ice maker. Always disconnect the electrical supply to the ice maker when working on the ice maker. Failure to disconnect the electrical supply to the ice maker may result in electrical shock or death.

2. Remove the front panel, top panel over the compressor compartment and the right side panel.
3. Separate the male and female electrical connections to the fan motor.
4. Remove the screws along the right side of the condenser, which secures the fan shroud to the condenser.
5. Slide the fan motor shroud assembly to the right to release the assembly from the securing clips on the left side and place the fan shroud assembly where it won't become damaged.
6. Use a vacuum cleaner or compressed air to clean the condenser. Remember, the condenser is clean when you can see a light through it.
7. Before reinstalling the shroud assembly, wipe the fan motor and fan blade off, making sure these are clean also.
8. Install the fan shroud in the ice maker. Before connecting the electrical supply, make sure the fan turns freely.

FAN MOTOR AND FAN BLADE

The fan motor and fan blade are the devices which move the air through the air cooled condenser. The fan motor will operate only when the ice maker is in the freeze cycle.

The fan motor is sensitive to the voltage supplied, as the voltage drops, so does the speed of the motor.

The fan motor does require lubrication and should be oiled every six months with SAE 20 oil. Use only 1 drop of oil.

If the fan motor becomes defective, remove the entire fan shroud assembly from the ice maker. When this has been removed, loosen the set screw which secures the fan blade to the motor and slide the blade off the motor shaft. Then, loosen the motor band which secures the motor to the fan cage.

IMPORTANT

Note the location of the fan blade on the fan motor shaft and the location of the motor in the band. This should be done so the new motor and fan assembly are properly aligned. This will prevent the fan blade from hitting either the shroud or the condenser.

Slide the fan motor out of the band, then install the new motor, making sure it is properly aligned and the motor band is tight and secure.

Inspect the fan blade, if it is bent or damaged in any way, the blade must be replaced. Before reinstalling the blade make sure it is clean. Then, reinstall the fan blade making sure it is tight and secure to the motor shaft. Then, replace the entire fan shroud assembly into the ice maker as described under Air Cooled Condenser Cleaning, page 19 of this manual.

WATER COOLED CONDENSER

Crystal Tips water condensers are a tube-within a tube and are used because of their reliability. On this type of condenser, water passes through the inside tube cooling the refrigerant in the outside tube. The outside tube is exposed to the room temperature, which provides double cooling which improves the condensers efficiency.

Water enters the condenser at the point where the refrigerant leaves the condenser and water leaves the condenser where the hot vapor from the compressor enters the condenser. This is called counterflow design. The warmest water leaving is adjacent to the warmest refrigerant entering and the coolest refrigerant leaving is adjacent to the coolest water entering.

WATER COOLED CONDENSER CLEANING

In areas where the quality of water being supplied to the condenser is in question, it may become necessary to clean the condenser periodically. A condenser which has become partially plugged with mineral deposits, will cause an increase in head pressure and an increase in water consumption.

The need for cleaning can be determined by an increase in head pressure for no apparent reason and an increase in the amount of water which is flowing through the condenser at a lower than normal temperature at the water exit.

Since the condenser can not be opened to clean, it must be cleaned with a scale dissolving acid such as the ice maker cleaner you use to clean the evaporator area. The ice maker cleaner should be circulated through the condenser. A small recirculating pump will work fine for this.

Mix a weak solution of ice maker cleaner and water in a container. Then, connect a hose from the recirculating pump which is submerged in the ice maker cleaner solution to the inlet water supply to the condenser. Then open the water pressure control valve by turning the adjusting stem clockwise until it stops. Connect a hose from the condenser outlet to the container of cleaning solution. Start the recirculating pump to provide a flow of cleaning solution through the condenser. Allow the weak solution to circulate for approximately 10 minutes. Then gradually add more ice maker cleaner to increase the strength of the cleaning solution until you reach the concentration of the ice maker cleaner recommended by the manufacturer. Allow this solution to circulate for 30 minutes to 1 hour.

CAUTION

Disconnect the electrical supply to the ice maker when working on the condenser, always start cleaning the water cooled condenser with a weak solution, then gradually increase its strength.

A strong solution may cause large chunks of scale to come off. These chunks can block the flow through the condenser which will slow the cleaning process. A weak solution will gradually dissolve the scale and carry it out of the condenser.

After the condenser has been cleaned, reconnect the incoming water supply and the water outlet line. Connect a set of service gauges to the high and low side service ports. Then, reconnect the electrical supply to the ice maker and start it. Once it is started, adjust the water pressure control valve so it will maintain a high side pressure of 250 psig.

WARNING

Ice maker cleaners are acids and must be handled with care and should be disposed of properly. Always wear goggles and rubber gloves when handling or working with ice maker cleaners.

Dispose of ice maker cleaner according to all applicable local laws and do not contaminate any food products.

WINTERIZING OF WATER COOLED CONDENSERS

If the ice maker is to be taken out of operation during the winter months, it is necessary to make certain there is no water left in the condenser.

To remove the water from the condenser, shut off the water supply to the ice maker and disconnect the water line from the condenser water inlet. Then start the ice maker and let it cut out on the high pressure cut out control. This causes the pressure control water valve to open. Using a minimum of 80 psi air pressure, blow through the water inlet to remove all water from the condenser.

If the pressure operated valve closes during this operation, restart the ice maker to build head pressure so the valve will reopen.

PRESSURE OPERATED WATER VALVE

To maintain the proper head pressure on an ice maker with a water cooled condenser, a pressure control valve is used which opens or closes depending on the operating head pressure of the ice maker. As the head pressure rises, the pressure is exerted against the bellows of the valve causing the valve to allow more water to flow through the water cooled condenser. The valve will continue to increase the water flow as long as the ice makers head pressure continues to increase.

As the ice makers head pressure starts to fall, the pressure on the bellows in the valve will decrease the amount of water flowing through the condenser.

This valve will maintain a constant head pressure. Incoming water temperatures over 70 F may cause a slightly elevated head pressure. Water should flow through the valve only during the freeze cycle. The valve must be set to maintain a 250 psig head pressure.

No water should flow through the valve and into the condenser during the defrost or harvest cycle. Any water flowing at this time will remove heat from the refrigeration system which is needed to harvest the ice.

If the valve does not control the proper amount of head pressure during the freeze cycle, an adjustment may be required.

ADJUSTMENT OF PRESSURE OPERATED WATER VALVE

Attach a set of service gauges to the high side and low side service ports to check the operating pressures of the ice maker.

The pressure operated water valve is located near the water cooled condenser and is secured to the back panel of the compressor compartment. On the top of the valve is the adjusting stem. With the use of a screwdriver or refrigeration wrench, the adjusting stem can be turned counter clockwise and the head pressure will rise. This rise will be shown on your service gauges. By turning the stem clockwise, the head pressure will decrease. If the pressure does not respond to the adjustments, it may be necessary to replace the valve.

REPLACEMENT OF PRESSURE OPERATED WATER VALVE

1. Disconnect the electrical supply and turn off the water supply to the ice maker. Then drain the water from the supply and disconnect the lines from the back of the ice maker.

WARNING

When the Ice-Off-Wash switch is in the off or center position, the electricity is NOT disconnected from the ice maker. Always disconnect the electrical supply to the ice maker when working on the ice maker. Failure to disconnect the electrical supply to the ice maker may result in electrical shock or death.

2. Remove the panels from the refrigeration compartment (right side).
3. Discharge the refrigerant from the ice makers refrigeration system (see refrigerant page 40).
4. Disconnect the water line from the outlet of the valve.
5. Using a torch, unsweat the valves capillary tube from the compressor discharge line.

WARNING

The R 502 refrigerant is not flammable or explosive by itself. It is not poisonous nor will it cause skin problems or affect food. However, there are some precautions which must be followed while handling it.

1. Torch flame or excessive heat applied to a system not completely discharged will cause the system pressure to increase excessively, which could cause an explosion.
2. Contact with open flame creates a corrosive toxic gas.
3. In a closed non-ventilated room, there is a danger of suffocation, if refrigerant is allowed to replace the oxygen in the room.
4. In its liquid state, refrigerant can cause frostbite.

Therefore when discharging the system, make sure the area is well ventilated and there is no open flame or fire. In the liquid state, do not allow it to come in direct contact with skin. Always wear goggles to protect your eyes and always discharge the refrigerant from the system before unsweating any part from the system.

6. From the back of the ice maker, remove the two (2) screws which attach the valve to the back panel.
7. The valve is now free of the ice maker and can be removed.
8. To install the replacement valve reverse the removal procedures.

CAUTION

1. Coat the flare contact surface with refrigeration oil to prevent the flare surface from becoming deformed which may cause a water leak.
2. When opening a refrigeration system always replace the drier. Failure to do so may cause a refrigeration system failure.

9. Evacuate and recharge following procedure listed under the heading "Refrigerant" in this manual.
10. After reconnection of water and electricity to the ice maker, start the ice maker and check for water and refrigerant leaks.
11. Adjust the valve to maintain 250 psig head pressure.
12. Remove the service gauges, replace the service valve caps and check for refrigerant leaks around the service ports.
13. Check the entire ice maker for proper operation.

ACCUMULATOR

The accumulator is a safety device designed to prevent liquid refrigerant from flowing to the compressor. If liquid refrigerant were to enter the compressor, damage to the internal components can be expected.

Any liquid refrigerant that enters the accumulator will be evaporated and vapor will be supplied to the compressor. Under normal usage the accumulator will function properly for many years.

However, if a failure should occur, it most likely will be a refrigerant leak. Any leak which may develop in the accumulator can be repaired. Remove the insulation from the accumulator to locate the leak. Once the leak has been found remove all of the glue from the area. Then, follow these steps to make the repair:

ACCUMULATOR REPAIR

1. Disconnect the electrical supply to the ice maker.

WARNING

When the Ice-Off-Wash switch is in the off or center position, the electricity is NOT disconnected from the ice maker. Always disconnect the electrical supply to the ice maker when working on the ice maker. Failure to disconnect the electrical supply to the ice maker may result in electrical shock or death.

2. Discharge the refrigerant from the ice maker.

WARNING

The R 502 refrigerant is not flammable or explosive by itself. It is not poisonous nor will it cause skin problems or affect food. However there are some precautions which must be followed while handling it.

1. Torch flame or excessive heat applied to a system not completely discharged will cause the system pressure to increase excessively which could cause an explosion.
2. Contact with open flame creates a corrosive toxic gas.
3. In a closed non-ventilated room, there is a danger of suffocation, if refrigerant is allowed to replace the oxygen in the room.
4. In its liquid state refrigerant can cause frostbite.

Therefore when discharging the system make sure the area is well ventilated and there is no open flame or fire. In the liquid state do not allow it to come in direct contact with skin. Always wear goggles to protect your eyes and always discharge the refrigerant from the system before unsoldering any part from the system.

3. Solder the leak with solder suitable for the refrigeration system.
4. Evacuate and recharge the ice maker following the procedures listed on page ??? of this manual.
5. Reconnect the electrical supply and start the ice maker.
6. Check the repaired accumulator for refrigerant leaks.
7. Reinsulate the accumulator.
8. Check the entire ice maker for proper operation.

FILTER - DRIER

Filter driers are designed to remove harmful elements from the refrigerant circulating in the system, before they can damage the system. All refrigeration systems have contaminates. It doesn't matter if its a new system or a system which has been opened for repairs. These contaminates can be in the form of metal filings, airborne dirt, wax, acid or moisture. Any of these contaminates can cause a system failure. Because of this contamination, good refrigeration practice dictates that any time a system is opened, the filter drier must be replaced.

REPLACEMENT OF THE FILTER DRIER

The filter drier can be replaced when other components are changed. The precautions which need to be exercised are:

1. Make sure the refrigerant is flowing in the direction indicated by the arrow stamped on the drier body.
2. Protect the drier from excessive heat during soldering, by wrapping the drier with a wet cloth.
3. Wrap any other area which may become overheated while soldering the filter drier.

EXPANSION VALVE

The thermostatic expansion valve is designed to regulate the rate of refrigerant liquid flow into the evaporator in the exact proportion to the rate of evaporation.

There are three (3) forces which determine the expansion valves operation.

1. The remote bulb pressure and power element.
2. The evaporator pressure.
3. The pressure exerted by the valve spring.

Because of the length of the suction line on the evaporator and the use of the refrigerant distributor after the expansion valve, together, both cause a considerable pressure drop in the refrigeration system. The Crystal Tips 400 Series Icemaker uses an external equalized thermostatic expansion valve. An external equalized valve allows for the valve to sense the outlet-pressure of the evaporator. This in turn allows the expansion valve to maintain the proper amount of refrigerant in the evaporator. The Crystal Tips 300, 200 Series Icemakers use an internally equalized expansion valve since the refrigerant distributor is not used on these units.

A vapor is said to be superheated whenever its temperature is higher than the saturation temperature corresponding to its pressure. The function of the thermostatic expansion valve is to control the superheat of the suction gas leaving the evaporator. The superheat setting of the expansion valves used on Crystal Tips ice makers is 3 to 6. When checking the operation of the expansion valve, it is important that the superheat be measured as accurately as possible.

SUPER HEAT MEASUREMENT

When measuring superheat, the recommended practice is to install a pressure gauge you know is accurate to the suction service port on the compressor. To check the temperature of the suction gas at the bulb location, use either an accurate pocket thermometer with a bulb clamp or an electronic thermometer with thermocouples. Tape the thermocouples to the suction line at the remote bulb location and then insulate the area so you are not picking up any ambient temperatures. Using a temperature pressure conversion chart, compare the two temperatures. The difference is the superheat. The temperature reading from the thermometer should always be 3 to 6 degrees warmer than the temperature from the conversion chart.

The bulb must be located on a clean portion of the suction line, so the entire length of the bulb is in contact with the line. The bulb must be secured to the suction line with two (2) worm gear clamps.

IMPORTANT

Do not use a copper strap. Because of the temperature changes during the freeze and defrost cycle, the strap may stretch which will result in a loose bulb on the suction line.

IMPORTANT

Do not use a copper strap. Because of the temperature changes during the freeze and defrost cycle, the strap may stretch which will result in a loose bulb on the suction line.

The expansion valve bulb and suction line must be completely insulated to prevent the ambient air from affecting the valves operation.

Checking the superheat is the first step in checking the thermostatic expansion valves operation.

If there is not enough refrigerant being fed into the evaporator, the superheat will be high.

If there is too much refrigerant being fed into the evaporator, the superheat will be low.

The expansion valve used on the Crystal Tips ice maker has a permanent bleed port. The function of the bleed port is to equalize the high and low side pressures during the off cycle. This will reduce the required starting torque of the compressor. If the system does not start to equalize after the system has been shut off, it is possible the bleed port has become restricted by dirt or other foreign material. If this is the case, the valve should be replaced.

An expansion valve performs only one function. It keeps the evaporator supplied with enough refrigerant to satisfy all load conditions.

EXPANSION VALVE TROUBLE SHOOTING GUIDE

PROBLEM	CAUSE
Valve does not feed enough refrigerant. Superheat too high	1. Moisture frozen in the valves port. System must be evacuated, change filter drier and recharge. 2. System has a low charge, evacuate, change drier and recharge. 3. Leaking hot gas solenoid valve. Check for operation page 51 of this manual. 4. On remote system check for proper operation of head pressure control valve. 5. Defective thermostatic expansion valve.
Expansion valve feeds too much refrigerant. Superheat too low.	1. Moisture frozen in the valves port. System must be evacuated change filter drier and recharge. 2. Dirt or foreign materials lodged in valve port, evacuate system, change drier and expansion valve and recharge. 3. Incorrect bulb installation. Make sure bulb is tight to suction line, properly clamped and insulated. 4. Inefficient compressor 5. Defective expansion valve.
Expansion valve feeds too much refrigerant at start up only No superheat	1. External equalizer tube plugged or restricted. Replace the equalizer tube.
Valve doesn't feed properly. Superheat normal or low	1. Unequal heat load over the evaporator. Check to make sure water distributor pan holes are all open. 2. Damaged or kinked liquid lines from distributor to the evaporator section, replace the damaged line. 3. Lack of heat load over the surface of the evaporator. Check the water flow and entire water supply system.
Expansion valve hunts, Superheat flucuates	1. Expansion valve bulb not tight to suction line. 2. Damaged or kinked liquid lines from distributor to the evaporator section, replace the damaged line. 3. Lack of heat load over the surface of the evaporator. Check the water flow and entire water supply system.
Expansion valve won't feed properly	1. The thermostatic expansion valve has been physically abused. This is the usual result of a mistaken analysis. It is assumed that if a valve does not feed properly, it is stuck. Beating the valve body with a hammer will only distort the body and make it impossible for the valve to work once the real cause has been determined.

EXPANSION VALVE REPLACEMENT PROCEDURES

1. Disconnect the electrical supply to the ice maker.

WARNING

When the Ice-Off-Wash switch is in the off or center position, the electricity is NOT disconnected from the ice maker. Always disconnect the electrical supply to the ice maker when working on the ice maker. Failure to disconnect the electrical supply to the ice maker may result in electrical shock or death.

2. Remove the panels from the compressor compartment.
3. Discharge the systems refrigerant.
4. Carefully remove the insulation which covers the expansion valve bulb. It must be reused to cover the bulb of the replacement valve. If damaged or rolled up in any way, replace the insulation.
5. Take a note of the expansion valve bulb location then loosen the two (2) worm gear clamps and remove the bulb from the suction line. The bulb should be located a 3 or 9 o'clock position if it is installed on a horizontal tube.

WARNING

R 502 refrigerant is not flammable or explosive by itself. It is not poisonous nor will it cause skin problems or affect food. However, there are some precautions which must be followed while handling it.

1. Torch flame or excessive heat applied to a system not completely discharged, will cause the system pressure to in crease excessively which could cause an explosion.
2. Contact with open flame creates a corrosive toxic gas.
3. In a closed non-ventilated room, there is danger of suffocation, if refrigerant is allowed to replace the oxygen in the room.
4. In its liquid state, refrigerant can cause frostbite.

Therefore, when discharging the system, make sure the area is well ventilated and there is no open flame or fire. In the liquid state, do not allow it to come in direct contact with the skin. Always wear goggles to protect your eyes and always discharge the refrigerant from the system before unsoldering any part of the system.

6. Remove the insulating tape from the expansion valve body and the 3 soldered connections, save the tape to be used to insulate the replacement valve.
7. Unsolder the defective valve.
8. Clean tubing ends to insure good solder joints on the replacement valve.
9. Solder in the replacement valve.

CAUTION

It is not necessary to disassemble the valve when soldering it to the connecting lines. Any of the commonly used refrigeration solders are satisfactory. It is important, regardless of the solder used, to direct the torch flame away from the valve body and wrap a wet cloth around the body and element during the soldering operation. This will avoid putting excessive heat on the valves diaphragm.

10. Replace the filter drier.
11. Evacuate and recharge the system according to procedures listed on page 40 of this manual.
12. While the system is being evacuated, reinstall the expansion valve bulb on the suction line at either the 3:00 o'clock or 9:00 o'clock position. Tighten the bulb in place by tightening the 2 worm gear clamps. Make sure the bulb is tight but not so tight that the clamps dent or distort the bulb. Then, reinsulate the bulb and suction line so the ambient air will not affect the valves operation. Also, reinsulate the body to prevent condensation from dripping into the compressor compartment.
13. Reconnect the electrical supply and restart the ice maker.
14. Check the system for refrigeration leaks and for proper operation.

IMPORTANT

The expansion valve used on the Crystal Tips ice maker is a special designed valve which is NOT available at most refrigeration supply stores. The valve has been preset at the factory and does not need any further adjustments.

HOT GAS SOLENOID VALVE

Crystal Tips ice makers utilize hot discharge gas to harvest the ice from the evaporator. During a hot gas defrost cycle, hot refrigerant vapor is pumped through the evaporator tubing. The system has a refrigerant line running from the compressor discharge line directly to the evaporator. This line is opened and closed by a solenoid valve. When the ice size thermostat determines there is ice in the evaporator, the hot gas solenoid will open to allow the hot vapor to enter the evaporator.

Three things must be considered to assure a properly operating valve.

1. The valve must be designed for the refrigerant in the system. Crystal Tips ice makers use R 502 refrigerant.
2. The valve must be properly sized so adequate hot gas is allowed in the evaporator. Valve size is determined by the orifice not the line size.
3. The coils electrical requirements must be matched to those of the entire ice maker.

HOT GAS SOLENOID VALVE TROUBLE SHOOTING GUIDE

SOLENOID VALVE WILL NOT OPEN

CAUSE	REPAIR REQUIRED
Movement of the plunger restricted 1. Corrosion 2. Foreign materials 3. Dented enclosure tube 4. Warped body caused by: a. Improper brazing b. Crushed in a vise 5. Oil above plunger	Clean or replace affected parts. Find and correct the source of the corrosion or foreign materials
Coil burnt	See coil burnout below
Improper wiring	Check electrical circuit for loose or broken wire connections.
Faulty ice size thermostat	Check or replace ice size thermostat
Electrical supply incorrect a. Low voltage b. High voltage c. Incorrect hertz	Check the voltage and hertz requirements as stamped on the coil assembly. If different, change the coil. a. If low voltage, locate voltage drop or install a transformer to boost voltage b. If high voltage, install a transformer to reduce the voltage.

SOLENOID VALVE WILL NOT CLOSE

CAUSE	REPAIR REQUIRED
The plunger is restricted due to: 1. Corrosion 2. Foreign material 3. Dented enclosure tube 4. Warped valve body due to: a. Improper brazing b. Improper handling	Clean affected parts or replace. Correct the cause of foreign material or corrosion in the system.
Defective ice size thermostat which keeps the coil energized	1. Check ambient ice maker is operating in. 2. Check primary resistor 3. Check operation of ice size thermostat.

SOLENOID VALVE CLOSES BUT FLOW CONTINUES

CAUSE	REPAIR REQUIRED
Foreign material under valve	Clean and remove foreign material
Valve seat warped or broken	Replace valve
Valve not properly assembled	Take apart and reassemble

SOLENOID VALVE COIL BURNS OUT

CAUSE	REPAIR REQUIRED
Supply voltage to coil too low, below 85% of rated voltage	Locate the cause of the low voltage such as a loose wire or incorrect wire size. Correct the cause.
Supply voltage to coil too high, more than 10% above rated voltage	Install a transformer to reduce voltage.
Plunger restricted due to: a. Corroded parts b. Foreign material c. Dent enclosure tube d. Warped valve body	Clean affected parts or replace. Correct the cause of corrosion of the source of foreign material in system.
Coil exposed to excessive moisture	Determine where excessive moisture is coming from and stop it.
Coil energized while removed from the valve	Never energize a coil when it is removed from the valve.

SOLENOID VALVE REPLACEMENT

1. Disconnect the electrical supply to the ice maker.

WARNING

When the Ice-Off-Wash switch is in the off or center position, the electricity is NOT disconnected from the ice maker. Always disconnect the electrical supply to the ice maker when working on the ice maker. Failure to disconnect the electrical supply to the ice maker may result in electrical shock or death.

2. Discharge the refrigerant from the system, only if there is an internal problem with the valve. If only the coil is defective, the refrigerant charge does not need to be discharged.

WARNING

R502 refrigerant is not flammable or explosive by itself. It is not poisonous nor will it cause skin problems or affect food. However, there are some precautions which must be followed while handling it.

1. Torch flame or excessive heat applied to a system not completely discharged, will cause the system pressure to increase excessively which could cause an explosion.
2. Contact with open flame creates a corrosive toxic gas.
3. In a closed non-ventilated room, there is danger of suffocation, if refrigerant is allowed to replace the oxygen in the room.
4. In its liquid state, refrigerant can cause frostbite.

Therefore, when discharging the system, make sure the area is well ventilated and there is no open flame or fire. In the liquid state, do not allow it to come in direct contact with the skin. Always wear goggles to protect your eyes and always discharge the refrigerant from the system before unsoldering any part of the system.

3. Unplug the electrical leads from the valve coil.
4. Remove the fastener which secures the coil to the valve stem and lift the coil off the valve body. If only the coil is defective, replace the coil reversing the above procedure
5. Unsolder the defective valve body from the system.
6. Replace valve body.

IMPORTANT

When installing the solenoid valve, be sure the arrow points in the direction of the refrigerant flow. The Crystal Tips replacement valves have extended tubing eliminating the need to disassemblu the valve before soldering. When soldering, do not use too hot of a torch flame and point the flame away from the valve body. Wrapping the valve body with a wet cloth will keep it cool and decreased the chance of damage from overheating.

7. Replace the drier.
8. Evacuate and recharge following the procedures listed on page 40 of this manual.
9. Reinstall coil and secure it to the valve body. Then reconnect the electrical leads to the coil.
10. Reconnect electrical supply and restart the maker.
11. Check the system for refrigerant leaks and the entire ice maker for proper operation.

EVAPORATORS

The evaporator is the part of the ice maker where the ice is made. The evaporator is a non-movable part of the ice maker. It is constructed of stainless steel for easy maintenance and designed for long life. The ice is frozen on circular disc which are called buttons which have been pressed into the stainless steel plates. The number of plates which make up a complete evaporator assembly varies according to the size of the ice maker. There is a copper serpentine soldered to the buttons which provides a method of getting the refrigerant to each button in the evaporator.

Most times, when a problem occurs, the evaporator is suspected. However, in most cases something other than the evaporator has caused the problem.

The following trouble shooting guide is intended to provide some guide lines in determining what can be causing evaporator problems.

ICE MAKER MAKING SMALL CUBED ICE

CAUSE	REMEDY
1. Ice size sheath out of adjustment	1. Readjust according to procedures listed on page 56 of this manual.
2. Faulty primary resistor	2. Check according to procedures listed on page 56-57 of this manual.
3. Ice size thermostat not properly adjusted	3. See adjusting procedure listed on page 55 of this manual.
4. Evaporator partially froze up.	4. a. Check water flow over the evaporator b. Check for dirty evaporator c. Check for dirty or plugged distributor pan.

EVAPORATOR FREEZE UP COMPLETELY

CAUSE	REMEDY
1. Water distributor pan holes plugged	1. Clean the pan with ice maker cleaner.
2. Ice rack out of position	2. Reposition ice rack.
3. Defective bin thermostat	3. Replace bin thermostat, see procedures on page 35 of this manual.
4. Defective ice size thermostat	4. Check out thermostat following procedures listed on page 57 of this manual.
5. Hot gas solenoid valve not opening	5. Check valve following procedures listed on page 51 of this manual.
6. System low on refrigerant	6. Locate refrigerant leak and repair.

CLOUDY SOFT ICE

CAUSE	REMEDY
1. Poor water conditions	1. a. Increase the amount of water purged from the system during each harvest cycle. b. Install a Crystal Tips water filter.
2. Air being trapped in the ice	2. a. Adjust water level b. Check water tube for holes where air may enter the system.

INCOMPLETE HARVEST CYCLE

CAUSE	REMEDY
1. Dirty evaporator	1. Clean evaporator with ice machine cleaner.
2. Ice maker is low on refrigerant	3. Locate and repair leak, then recharge.
3. Defective ice size thermostat or thermo disc.	3. Check control and replace if defective.
4. Restricted hot gas supply	4. Check for restriction and remove it.
5. Air temperature is too low.	5. Warm the area around the ice maker.

EVAPORATOR REPLACEMENT PROCEDURES

1. Disconnect the electrical supply to the ice maker.

WARNING

When the Ice-Off-Wash switch is in the off or center position, the electricity is NOT disconnect from the ice maker. Always disconnect the electrical supply to the ice maker when working on the ice maker. Failure to disconnect the electrical supply to the ice maker may result in electrical shock or death.

2. Turn off the water supply to the ice maker.
3. Remove the front panel and the top left panel from the ice maker.
4. Discharge the refrigerant from the ice maker.

WARNING

R 502 refrigerant is not flammable or explosive by itself. It is not poisonous nor will it cause skin problems or effect food. However, there are some precautions which must be followed while handling it.

1. Torch flame or excessive heat applied to a system not completely discharged, will cause the system pressure to increase excessively which could cause an explosion.
2. Contact with open flame creates a corrosive toxic gas.
3. In a closed non-ventilated room, there is danger of suffocation, if refrigerant is allowed to replace the oxygen in the room.
4. In its liquid state, refrigerant can cause frostbite.

Therefore, when discharging the system, make sure the area is well ventilated and there is no open flame or fire. In the liquid state, do not allow it to come in direct contact with the skin. Always wear goggles to protect your eyes and always discharge the refrigerant from the system before unsoldering any part of the system.

5. Remove the water distributor pan, ice rack and ice deflector from the evaporator section of the ice maker.
6. Using either a tubing cutter or a torch, disconnect the inlet and outlet refrigeration lines as close to the edge of the evaporator as possible. If a torch is used to separate the lines, use a wet cloth as a heat sink.
7. With a utility knife, cut the sealant which seals the evaporator front and back edge to the frame of the evaporator support.
8. Pull the damaged evaporator out of the ice maker through the top opening. If the evaporator appears to be stuck, recut the sealant and gently rock the evaporator plate front to back.
9. Clean the remaining sealant from the front and back evaporator support.
10. Clean the tubing ends to provide a clean solder connection and to remove the protective coating on the refrigeration lines.
11. Insert the replacement evaporator copper connectors on the refrigerant lines.
12. Insert the replacement evaporator into the evaporator frame.
13. Replace the water distributor pan in the ice maker and use the water holes as an alignment fixture for the top of the evaporator.
14. Slide the copper connectors over the ends of the tubing to be connected and solder the tubing together.

CAUTION

Use a wet cloth to prevent the evaporator section from becoming overheated.

15. Replace the filter drier.
16. Evacuate and recharge the system following the procedures listed on page 40 of this manual.
17. Apply sealant where the evaporator sits in the front and back support frame.
18. Coat the refrigeration lines to the evaporator with a sealant material Crystal Tips part number 80364804, to prevent the line from oxidizing.
19. Reconnect the electrical and water supply to the ice maker and start it.
20. Check the system for refrigeration leaks and the entire ice maker for proper operation.

IMPORTANT

If one of the evaporator plates which is used to mount the ice size control needs replacement, remove the two (2) screws which secures the ice size control to the evaporator and remove the entire ice size control assembly, before proceeding to STEP #6 above.

Use these mounting holes as the template for determining the location of the hole in the new evaporator plate.

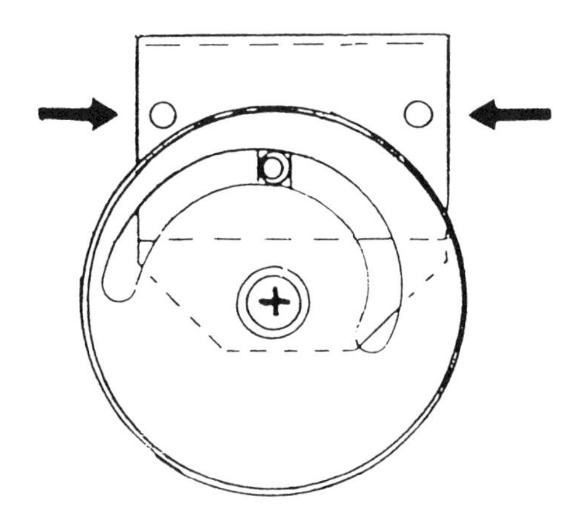

FIGURE 21

ICE SIZE CONTROL REPLACEMENT

Before proceeding with STEP #13 above, replace the ice size control in the evaporator section by locating it with one screw which will thread into the one evaporator with the screw hole. When it is secured with the one screw, using the second hole in the mounting bracket as a template, drill a small hole in the replacement evaporator to accept the second screw to secure the ice size adjustment mounting plate. Then readjust using the procedures listed on page 56 of this manual.

ICE SIZE THERMOSTAT

The ice size thermostat determines when the ice maker should go into a defrost cycle. This thermostat senses the temperature of the evaporator section of the ice maker. As the temperature of the evaporator reaches 38 F +− 2 F, the single pole double throw switch will change position, opening the circuit between terminal numbers 2 and 3 and close the contacts between terminals number 2 and 1. This energizes the hot gas solenoid valve.

On ice makers equipped with a defrost termination switch, the ice size control puts the ice maker into the defrost cycle. When it senses an evaporator temperature of 38 F +− 2 F, the thermostat opens its contacts between terminals 2 and 3 and closes the contacts between 1 and 2 which energizes the defrost relay, which in turn opens the hot gas solenoid valve.

ICE SIZE THERMOSTAT REPLACEMENT PROCEDURES

1. Disconnect the electrical supply to the ice maker.

WARNING

When the Ice-Off-Wash switch is in the off or center position, the electricity is NOT disconnected from the ice maker. Always disconnect the electrical supply to the ice maker when working on the ice maker. Failure to disconnect the electrical supply to the ice maker may result in electrical shock or death.

2. Remove the ice size thermostat capillary tube from the sheath assembly located in the evaporator section.
3. On the 200, 300 and 400, remove the two (2) screws which secures the thermostat to the electrical control box.
4. Remove the wires from the ice size thermostat terminals.
5. Gently pull the capillary tube through the dividers and remove the thermostat from the ice maker.
6. Reverse the procedures to install the replacement thermostat.

CAUTION

Be careful not to cause the capillary tube to kink during the installation procedures.

7. Connect wires numbers 39 and 36 to terminal 1, wire number 34 to terminal 2 and wire number 35 to terminal 3.
8. Check the operation of the entire ice maker.

ICE-OFF-WASH TOGGLE SWITCH

The ice-off-wash switch provides a method of turning off the entire ice maker or stopping just the refrigeration system for cleaning.

WARNING

When the Ice-off-Wash switch is in the off or center position, the electricity is NOT disconnected from the ice maker. Always disconnect the electrical supply to the ice maker when working on the ice maker. Failure to disconnect the electrical supply to the ice maker may result in electrical shock or death.

When the switch is in the ice position or up, the ice maker is in its ice mode. When the switch is in the wash position, only the water pump operates to provide ice maker cleaner circulation.

If the switch fails, it will only fail in one of three ways. Either the ice maker will not stop when switched to the off position, or will not start when switched to the ice position, or the pump will not operate when switched to the wash position.

ICE-OFF-WASH SWITCH REPLACEMENT PROCEDURE

1. Disconnect the electrical supply to the ice maker.

WARNING

When the Ice-off-Wash switch is in the off or center position, the electricity is NOT disconnected from the ice maker. Always disconnect the electrical supply to the ice maker when working on the ice maker. Failure to disconnect the electrical supply to the ice maker may result in electrical shock or death.

2. Remove the Hex trim nut which secures the switch to the electrical control box.
3. Disconnect the wires from the switch terminals.
4. Connect the wires to the terminals of the replacement switch. The following is a guide to assist you with the proper placement of the wires.
   a. Hold the switch so the terminals are toward you with the mounting slot in the shank of the switch pointed up.
   b. By holding in this position, terminal 1 will be on the left side the top terminal. Terminal 2 is left center, terminal 3 is left bottom, terminal 4, top right, terminal 5, right center and terminal 6, bottom right.
   c. Keeping step B in mind connect the wires in the following manner.
      Terminal 1 - wire number 4
      Terminal 2 - wire number 11
      Terminal 3 - wire number 44
      Terminal 4 - is not used
      Terminal 5 - wire number 5
      Terminal 6 - wire number 45
5. Mount switch in the ice maker and secure with the trim nut.
6. Reconnect the electrical supply to the ice maker and restart it.
7. Check the entire ice maker for proper operation.

SHEATH ASSEMBLY

The ice size sheath is a tube which contains the ice size thermostat capillary tube. The sheath is sealed on the end which extends into the evaporator.

The sheath must be located in the center of the evaporator envelope on the older ice makers. On the new ice makers, the sheath is not located in the center of the spacing between the evaporator.

No matter which style ice maker you have, the sheath must always be straight, free of mineral deposits and the sealed end extending into the evaporator section must be completely sealed. If water is allowed to go into the sheath, it will cause the ice size thermostat to operate in an erratic manner.

SHEATH ASSEMBLY REPLACEMENT

1. Disconnect electrical supply to the ice maker.

WARNING

When the Ice-off-Wash switch is in the off or center position, the electricity is NOT disconnected from the ice maker. Always disconnect the electrical supply to the ice maker when working on the ice maker. Failure to disconnect the electrical supply to the ice maker may result in electrical shock or death.

2. Remove the ice size thermostat capillary tube from the sheath from the knob end.
3. Remove the two (2) screws which secure the plate or bracket to the evaporator.
4. Loosen the brass thumb nut from the sheath.
5. Slide the sheath out of its mounting bracket.
6. To replace the sheath, slide the tube through the mounting bracket hole. Then, slide the brass thumb nut over the sheath. Then slide the sheath through the primary resistor, then the plastic sleeve and then tighten the brass thumb nut in place.
7. Then secure the entire assembly in place.
8. Replace the knob, the larger the number setting on the knob the larger the ice cube.
9. Reconnect the electrical supply to the ice maker and check the entire ice maker for proper operation.

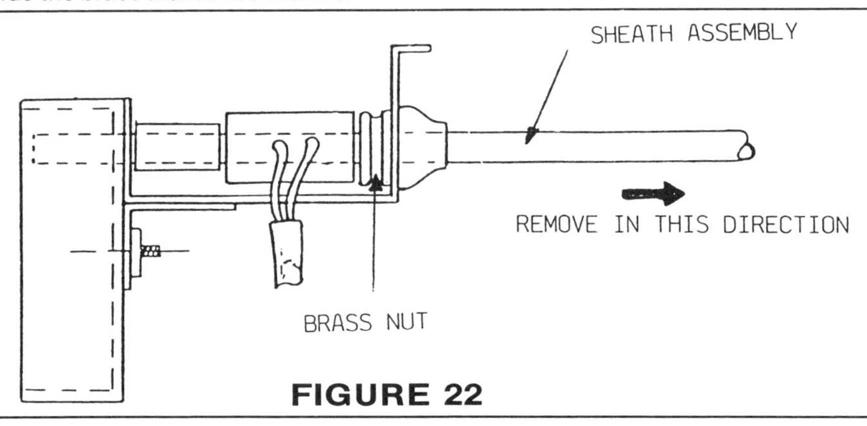

FIGURE 22

PRIMARY RESISTOR (ICE SIZE HEATER)

The primary resistor is a small heater which provides a small amount of heat to the ice size sheath. The heat is added because the air and water temperature in the evaporator area is below 38 F. This additional heat will be sufficient to over come the lower air temperatures and any water which may splash on the sheath and cause the ice maker to go into a premature defrost.

When the ice builds and gets close to the sheath more water will come into contact, removing the heat from the sheath allowing the ice size thermostat to put the ice maker into a defrost cycle.

If the primary resistor is not working properly, it could cause the ice maker to go into a premature harvest cycle.

WARNING

Before checking the resistance, make sure the electrical supply has been disconnected to the ice maker. The red or yellow lead from the resistor should be removed from terminal number 4 on the terminal board.

Failure to disconnect the electrical supply to the ice maker may cause electrical shock or death.

Failure to disconnect the leads from the terminal board will not allow you to get an accurate reading of the resistance value.

The primary resistor may be checked with an ohmmeter. The following readings should be detected across the leads:

a. If the ice maker is 115 volt, the resistor leads should be yellow and white and should read 6,000 ohms +/−5%.

b. If the ice maker is 230 volt, the resistor leads should be red and white and should read 24,000 ohms +/− 5%.

Also, the resistor may be discharging to ground. Place the resistor leads together with one ohmmeter lead and place the other ohmmeter lead to an unpainted surface to provide a good grounding path. An ohm reading indicates that the resistor must be changed.

PRIMARY RESISTOR REPLACEMENT PROCEDURE

1. Disconnect the electrical supply to the ice maker.

WARNING

When the Ice-off-Wash switch is in the off or center position, the electricity is NOT disconnected from the ice maker. Always disconnect the electrical supply to the ice maker when working on the ice maker. Failure to disconnect the electrical supply to the ice maker may result in electrical shock or death.

2. Disconnect the primary resistors red wire or yellow from terminal #4 and the white wire from terminal #1 on the terminal board. Then, pull the two wires through the refrigeration divider.
3. Remove the ice size thermostat capillary tube from the sheath.
4. Loosen the brass thumb nut from the sheath and slide the sheath into the evaporator section.
5. Remove the plastic sleeve and heater from the sheath.
6. To replace the primary resistor, reverse the removal procedures.
7. Reconnect the electrical supply and check the entire ice maker for proper operation.

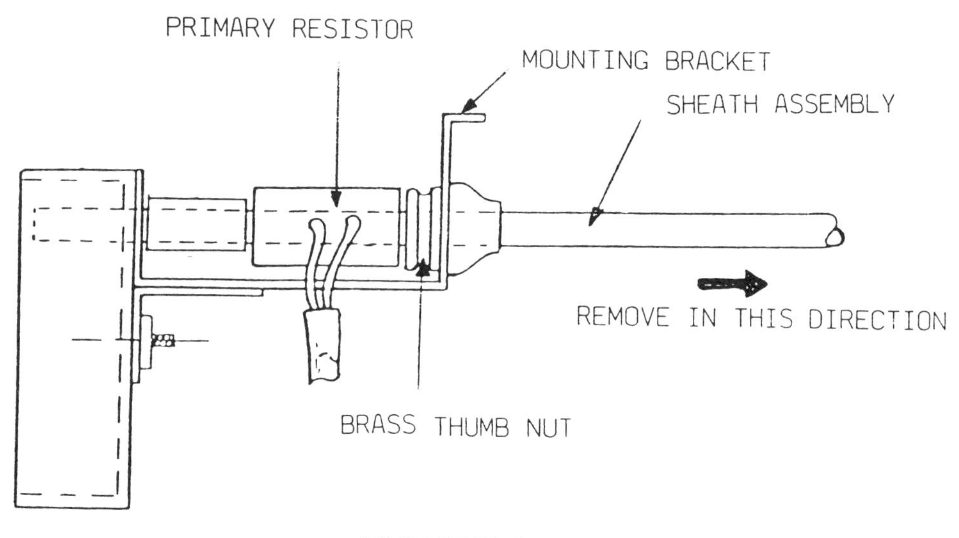

FIGURE 23

BIN THERMOSTAT

The bin thermostat provides a method of controlling the amount of ice that the ice maker produces and stores in the bin. When the ice bin fills with ice, the ice will be in contact with the capillary tube mounted in the thermostat bracket extending into the bin from the bottom of the ice maker. This contact causes the ice maker to stop making ice until some ice has been removed from the capillary tube. At that time the ice maker will restart, filling the bin until the ice comes in contact with the capillary tube, causing the ice maker to stop again.

The bin thermostat is an adjustable control, which has a cut out setting of 35 +− 2 F, a differential of 4 1/2 +− 1 F and a cut in of 50 1/2 +− 4 F.

The thermostat must be checked for proper operation periodically. To do this, place some ice on the capillary tube in the bin. The ice maker should shut off within 1 minute. If the ice maker doesn't stop within the first minute, rotate the adjusting stem counter clockwise until it does stop.

Then, remove the ice from the capillary tube and the ice maker should start within 2 minutes. If the ice maker does not start, rotate the adjusting stem clockwise until it does restart.

IMPORTANT

If the temperature in the area of the ice maker is below 60 F, the bin thermostat may not get warm enough to start the ice maker. Any adjustments made or attempted will not allow the thermostat to control the ice maker properly.

If after the adjustments, the bin thermostat still doesn't operate properly, it should be replaced.

BIN THERMOSTAT REPLACEMENT

1. Disconnect the electrical supply to the ice maker.

WARNING

When the Ice-off-wash switch is in the off or center position, the electricity is NOT disconnected from the ice maker. Always disconnect the electrical supply to the ice maker when working on the ice maker. Failure to disconnect the electrical supply to the ice maker may result in electrical shock or death.

2. Remove the bin thermostat capillary tube from the mounting bracket inside the ice bin.
3. Pull the capillary tube up into the compressor compartment through the sleeve in the ice makers base.
4. Disconnect the two (2) wires which connect to the thermostat terminals.
5. Remove the two (2) screws which secure the thermostat in the electrical control box.
6. To install the replacement thermostat, reverse the removal procedures.

NOTE

Wire number 1 connects to thermostat terminal 1, wire number 45 connects to thermostat terminal 2.

IMPORTANT

When reinstalling the thermostat, be careful so the capillary tube does not become kinked or broken.

7. When the thermostat is installed, reconnect the electrical supply and adjust the thermostat for proper operation. Also, check the operation of the entire ice maker.

HIGH PRESSURE CUT OUT

The water cooled models use an automatic reset control.

On the water cooled models if the head pressure should exceed a preset pressure, the reset will open, stopping the ice maker. When the head pressure drops below the differential of the control, the control will cut back in, restarting the ice maker. The cut-out for the 300 series water cooled is 475 psig. And when the pressure drops below 410 psig, the control will cut-in. The cut-out for the 400 series water cooled is 380 psig, with a cut-in pressure of 315 psig.

The following trouble shooting guide provides some of the reasons you may be experiencing excessive head pressures

CAUSE	REMEDY
1. Refrigerant overcharge	1. Evacuate and recharge according to procedures listed on page 40 of this manual.
2. Non condensable in the refrigeration system.	2. Change drier and evacuate and recharge according to procedures listed on page 40 of this manual.
3. Dirty condenser	3. Clean condenser coil
4. Defective head pressure control	4. Replace head pressure control
5. Defective head pressure control water valve	5. Replace the pressure control valve
6. Insufficient water supply to condenser.	6. Increase water supply

If the control develops a refrigerant leak or the electrical contacts become defective, the control must be replaced.

HIGH PRESSURE CUT-OUT REPLACEMENT

1. Disconnect the electrical supply to the ice maker.

WARNING

When the Ice-off-wash switch is in the off or center position, the electricity is NOT disconnected from the ice maker. Always disconnect the electrical supply to the ice maker when working on the ice maker. Failure to disconnect the electrical supply to the ice maker may result in electrical shock or death.

2. Discharge the refrigerant from the ice maker.

WARNING

R 502 refrigerant is not flammable or explosive by itself. It is not poisonous nor will it cause skin problems or effect food. However, there are some precautions which must be followed while handling it.

1. Torch flame or excessive heat applied to a system not completely discharged, will cause the system pressure to increase excessively which could cause an explosion.
2. Contact with open flame creates a corrosive toxic gas.
3. In a closed non-ventilated room, there is danger of suffocation, if refrigerant is allowed to replace the oxygen in the room.
4. In its liquid state, refrigerant can cause frostbite.

Therefore, when discharging the system, make sure the area is well ventilated and there is no open flame or fire. In the liquid state, do not allow it to come in direct contact with the skin. Always wear goggles to protect your eyes and always discharge the refrigerant from the system first before unsweating any part of the system.

3. Unsolder the capillary bulb from the compressor discharge line.
4. Disconnect the two (2) wires from the pressure control terminals.
5. Remove the two (2) screws, which secures the control to the electrical control box. Then, remove the control from the ice maker.
6. Replace the filter drier.
7. Solder the capillary bulb to the discharge line.
8. Evacuate and recharge, following the procedures listed on page 40 of this manual.
9. While the system is being evacuated, secure the control in place and reconnect the wires. Recharge the system.
10. Reconnect the electrical supply and restart the ice maker. Check for refrigerant leaks and check the entire ice maker for proper operation.

WATER PUMP

The water pump recirculates water in the ice maker. The pump picks up the water from the water tank and forces that water through a water tube up to the distributor pan. In the distributor pan, the water is allowed to drain through a series of holes over the evaporator surface where some of the water is frozen. The remainder drains back into the water tank to be picked up by the water pump to be recirculated.

The water pump must be capable of suppling an adequate amount of water to the distributor pan so it in turn will provide an even flow of water over the entire surface of the evaporator.

If the water pump can not supply enough water to the distributor pan, the ice may become cloudy and irregular shaped or may cause a freeze up condition to develop in the evaporator area.

If the water supply appears to be inadequate to provide a good flow of water, check the following:

1. Restricted water tube.
2. Mineral deposits in distributor pan.
3. Mineral deposits on water pump.
4. Broken or cracked water tube.
5. Broken distributor pan.
6. Is there an adequate water level in the water tank?

If all of the above are correct and not causing a water flow problem, the water pump should be replaced.

WATER PUMP REPLACEMENT

1. Disconnect the electrical supply to the ice maker.

WARNING

When the Ice-off-wash switch is in the off or center position, the electricity is NOT disconnected from the ice maker. Always disconnect the electrical supply to the ice maker when working on the ice maker. Failure to disconnect the electrical supply to the ice maker may result in electrical shock or death.

2. Disconnect water pump electrical leads from the terminal board. The black lead from terminal 2 of S1 and the white lead from terminal TB-B. Then, pull the wires out of the electrical box.
3. Remove the two (2) screws which secures the cover channel to the refrigeration divider.
4. Remove the two (2) thumb screws which secures the water pump to the refrigeration divider.
5. Remove the water tube from the outlet of the water pump.
6. Remove the water pump from the ice maker.
7. Lift the cover off the pump motor.
8. To install the replacement pump, reverse the above procedures.
9. Reconnect the electrical supply, restart the ice maker and check the entire ice maker for proper operation.

WATER DISTRIBUTOR PAN

The distributor pan provides an even supply of water over the entire evaporator surface. The water flows through a series of small holes in the bottom of the distributor pan, which are directly over the evaporator plates.

The distributor pan must be kept free of mineral and dirt build up to provide adequate water flow over the evaporator surface. If the distributor pan is broken, it must be replaced.

WATER DISTRIBUTOR PAN REPLACEMENT

1. Turn the ice-off-wash switch to the off position.
2. Disconnect the water tube from the distributor pan.
3. Slide the distributor pan out through the front of the evaporator section of the ice maker.
4. Reverse the above procedures to install the new distributor pan, making sure the holes in the pan line up over the evaporator sections and make sure it is level.
5. Restart the ice maker and check the entire ice maker for proper operation.

FLOAT ASSEMBLY

The float assembly maintains the water level in the water tank at all times to make sure the ice maker has an adequate water supply. The amount of water in the water tank is determined by the siphon drain in the water tank. See siphon drain page 59-60 of this manual, refer to figure below. If the float does not maintain the water level, one of the following may be causing the problem.

1. Foreign material stuck in valve orifice.
2. The float bulb may be leaking and filling with water.
3. Orifice in valve body may be damaged.

If the problem is dirt in the valve, clean it. If it is some other concern the valve assembly must be replaced.

WATER FLOAT VALVE REPLACEMENT

1. Turn off water supply to the ice maker and switch the ice-off-wash switch to the off position.
2. Disconnect the compression nut on the water line from the brass valve body.
3. Remove the brass nut which secures the valve body to the bracket.
4. Remove valve body from the bracket.
5. Install the replacement float by reversing the above procedures.
6. Turn the water supply on and restart the ice maker. Then check the entire ice maker for proper operation.

SIPHON DRAIN

As ice is formed on the evaporator, pure water is frozen first. The water flow action washes the impurities from the evaporator surface and carries them back to the water tank where they become more and more concentrated. These impurities become increasingly unstable, finally precepitating out of the water, settling on the bottom of the water tank as a fine talc type material. This talc material is made up of dissolved minerals which could not attach themselves to the evaporator or other areas. The only way to remove these impurities from the water tank is to siphon them out of the system during the defrost cycle, using the siphon drain desiged into the ice maker.

The siphon drain is an adjustable device which lets you determine the amount of water to be removed from the system. In extremely bad water conditions, the siphon should be adjusted for a maximum purge. Stop the water pump and make sure there is no water entering the water tank from the float valve. When the water level has stabilized, adjust the 180 return bend so the small hole on the inside of the bend does not touch the water. Then slide the water inlet tube either up or down so its inlet end is 1/8" to 1/4" from the bottom of the water tank.

To reduce the amount of water purged from the ice maker each cycle, raise the 180 return bend so the distance between the water and small hole becomes greater. Always adjust the siphon purge to keep the bottom of the water tank clean. The inlet tube must be adjusted so it is 1/8" to 1/4" from the bottom of the water tank.

The small hole on the inside of the 180 return bend will allow air to enter the siphon tube. The air entering the tube will stop the siphoning action and set the ice maker up for the next freeze cycle.

Should the siphon not be completed by the time the ice maker returns to the next freeze cycle, the drain line is restricted either by dirt and foreign material or the drain line has an air trap somewhere between the back of the ice maker and the floor drain. To prevent an air trap, it may require that the drain line be vented vertically at the rear of the ice maker.

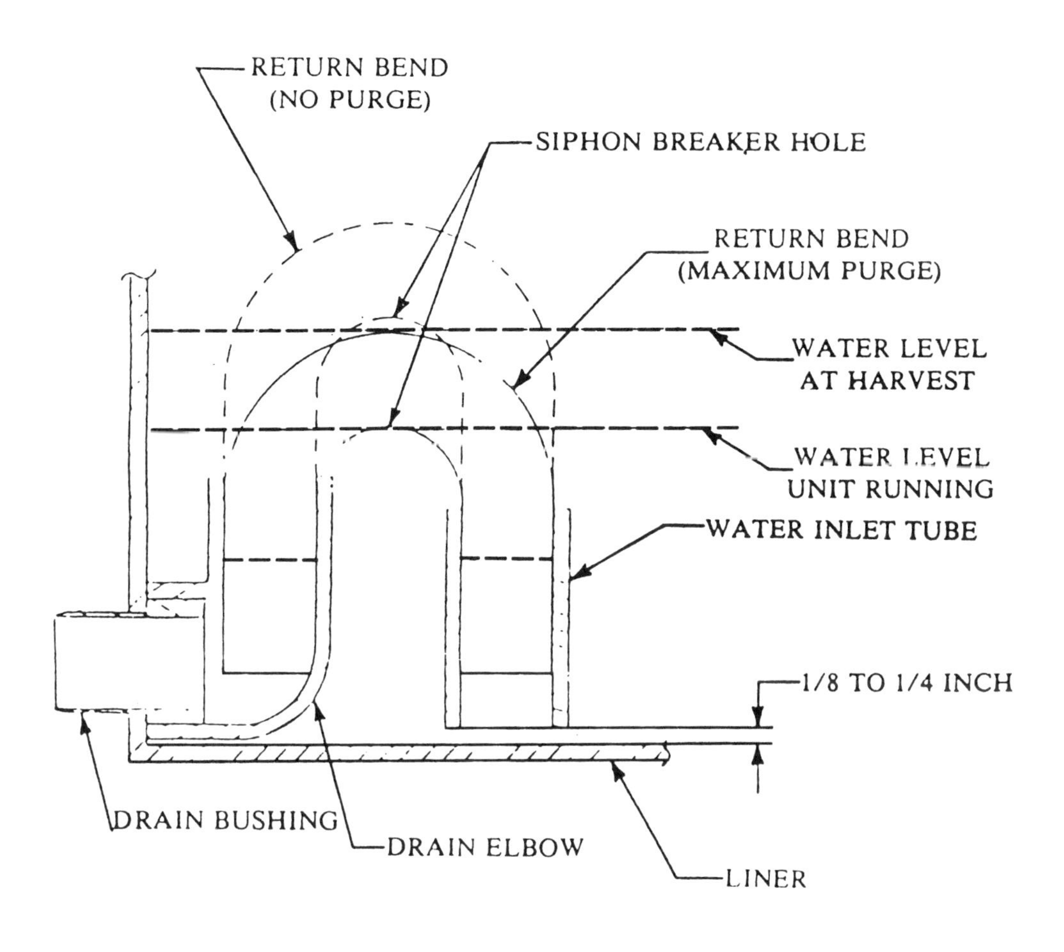

FIGURE 24

HARVEST TERMINATION SWITCH AND RELAY

The defrost or harvest cycle is terminated by a thermo disc attached to the suction line. After the ice size thermostat puts the ice maker into the harvest cycle, the hot gas solenoid valve opens causing the evaporator and suction line to warm. The thermo disc will remain closed until the suction line warms to 55 +/− 5. At this point the thermo disc will open electrically, thus returning the ice maker to the freeze cycle.

When the ice size thermostat puts the ice maker into the harvest or defrost mode, it also energizes the defrost relay through wire number 39 which is connected to terminal number 1 of the ice size thermostat and terminal identified as A on the defrost relay. Terminal B of the defrost relay is connected to terminal letter B on the terminal board by wire identified as number 40. This causes the three contacts in the relay to shift, opening the contacts between terminal number 7 and 1 which de-energizes the fan motor and water pump and closes the contacts between 7 and 4. Terminal number 4 does not control anything.

The third set of contacts open between terminals 9 and 3 and closes the contacts between terminals 9 and 6. This action now supplies electricity through the brown wires of the thermal disc which connect to terminal number 6. From terminal number 9, the electrical supply passes through wire number 36 to terminal number 8 of the relay and to terminal number 1 of the ice size thermostat, which then supplies electricity back to the defrost relay coil and provides electricity to the relay during the time after the ice size thermostat has changed its contacts back to the freeze position.

However, the ice maker remains in the harvest mode until the defrost termination switch opens, which interrupts the electrical supply to the defrost relay coil. This causes the contacts of the defrost relay to switch back to their original position. This action returns the ice maker back to the freeze mode.

If the ice maker fails to come out of the defrost cycle, check to make sure the suction line is above 55 +− 5 and the defrost termination switch is open.

If the defrost termination switch is not open, it should be replaced. If the defrost termination switch is open, either the ice size thermostat has not switched back to the freeze position or the hot gas solenoid valve is stuck open. If the hot gas solenoid valve is stuck open, hot gas will be flowing through the evaporator and the water pump will not operate.

If the ice maker goes into a defrost cycle but returns to the freeze cycle before the ice has released from the evaporator or before the suction line reaches 55 +− 5, the defrost relay is defective and should be replaced.

DEFROST TERMINATION SWITCH REPLACEMENT

1. Disconnect the electrical supply to the ice maker.

WARNING

When the Ice-off-wash switch is in the off or center position, the electricity is NOT disconnected from the ice maker. Always disconnect the electrical supply to the ice maker when working on the ice maker. Failure to disconnect the electrical supply to the ice maker may result in electrical shock or death.

2. Disconnect one (1) brown lead from terminal letter C on the terminal board, the other brown lead from terminal number 6 on the defrost relay. Then pull the two (2) brown wires from the electrical control box.
3. Locate the defrost termination switch on the suction line.
4. Unclip the switch from the suction line and remove it from the ice maker.
5. Install the replacement defrost termination switch by reversing the above procedures.

IMPORTANT

Install the replacement defrost termination switch in the same location the original switch was located in.

6. Reconnect the electrical supply and check the entire ice maker for proper operation.

DEFROST RELAY REPLACEMENT

1. Disconnect the electrical supply to the ice maker.

WARNING

When the Ice-off-wash switch is in the off or center position, the electricity is NOT disconnected from the ice maker. Always disconnect the electrical supply to the ice maker when working on the ice maker. Failure to disconnect the electrical supply to the ice maker may result in electrical shock or death.

2. The relay is attached to the electrical control box by two (2) screws, loosen the screw on the end of the relay where the wires are attached and remove the screw on the opposite end.
3. Remove the relay from the electrical box and remove the wires from the relay terminals. The easiest method to do this, is to remove a wire from the defective relay and place it on the corresponding terminal of the replacement relay.
4. To install the replacement relay, reverse the removal procedures, reconnect the electrical wires before securing the relay to the electrical control box.

   There are 11 terminals on the relay, the following is a guide to reconnect the wires to the proper terminal.

   terminal number 1 - wire number 41
   terminal number 2 - is not used
   terminal number 3 - is not used
   terminal number 4 - is not used
   terminal number 5 - wire number 37
   terminal number 6 - brown wire from defrost termination switch
   terminal number 7 - wire number 35
   terminal number 8 - wire number 36
   terminal number 9 - wire number 36
   terminal A - wire number 39
   terminal B - wire number 40
5. After the relay has been installed, reconnect the electrical supply to the ice maker and check the entire ice maker for proper operation.

COMPRESSOR CONTACTOR

The contactor is designed to carry the electrical load of the compressor. The use of this contactor allows a method to supply electricity to the compressor without the compressors full load going through the control circuit of the ice maker. When the high pressure cut out is closed, the bin thermostat is closed and the ice-off-wash switch is in the ice position, the contactor coil is energized which will close the contacts of the contactor, starting the compressor. If the contacts should open for any reason, the contactor coil will become de-energized which will cause the contacts to open stopping the compressor.

If electricity is supplied to the contact or coil but the contacts do not close or if the contacts appear to close but there is no electrical flow through the contacts, the contactor must be replaced.

COMPRESSOR CONTACTOR REPLACEMENT

1. Disconnect the electrical supply to the ice maker.

WARNING

When the Ice-off-wash switch is in the off or center position, the electricity is NOT disconnected from the ice maker. Always disconnect the electrical supply to the ice maker when working on the ice maker. Failure to disconnect the electrical supply to the ice maker may result in electrical shock or death.

2. Remove the four (4) screws which secures the contactor to the electrical control box.
3. Disconnect all electrical wires from the contactor and reconnect the wires to the replacement contactor. The easiest method to do this, is to remove a wire from the defective contactor and place it on the corresponding terminal of the replacement contactor.
4. When all the wires are in place, secure the replacement contactor to the electrical control box.
5. Reconnect the electrical supply to the ice maker and check the entire ice maker for proper operation.

TERMINAL BOARD

The terminal board provides a place to tie all electrical components together. It is made of a special insulating material secured to the electrical control box of the ice maker.

If a terminal should break off, it is not necessary to replace the board. By using a duplex connector on the good terminals, you will be able to keep the ice maker operating.

ICE RACK

The ice rack is placed under the evaporator to prevent the ice cubes from falling into the water tank during the harvest cycle, but lets the water run through it during the freeze cycle.

The ice rack can be removed from the ice maker by removing the water pump from the water tank and then sliding the ice rack toward you out of the ice maker. Follow water pump replacement procedures on page 58 of this manual, steps 1, 3, 4, 5, 6.

This rack must be kept clean, free of mineral deposits, and rust deposits, to enable the ice to slide off it and into the bin.

When placing the ice rack in the ice maker, it must be positioned on the shoulder in the ice makers plastic liner. Then, install the water pump back into the water tank.

Failure to do this can result in excessive amounts of water dripping into the ice bin and the angle may not be steep enough for the ice to slide off during the harvest cycle.

ICE DEFLECTOR

The ice deflector is a stainless steel panel secured to the bottom right edge of the evaporator, which spans the entire front to back area of the right side of the evaporator.

This deflector is to prevent ice and water from splashing into the bin during the ice makers operation.

The deflector is held in place by two (2) thumb screws. These screws allow you to remove the deflector for cleaning.

SPLASH SHIELD

When the front panel is removed from the ice maker, you will notice a white plastic panel which prevents the water splashing in the water tank from getting on the front panel, which, then would run out of the ice maker and onto the floor.

To provide proper operation without water splashing, this shield must be in place.

TROUBLE SHOOTING GUIDE

WARNING

FOR SAFETY REASONS ALWAYS DISCONNECT THE ELECTRICAL SUPPLY AND SHUT OFF THE WATER WHENEVER TROUBLE SHOOTING.

A. NO ICE PRODUCTION:

TROUBLE: 1. Unit will not start

POSSIBLE CAUSE	REMEDY
1. Open fuse or disconnect	1. Check circuit for possible short. Replace fuse.
2. Ice on bin thermostat capillary tube	2. Remove ice from capillary tube.
3. High pressure switch is in open position (water cooled)	3. Check switch, if it is defective, replace
4. Faulty compressor or start components	4. Replace if defective
5. Low voltage	5. Use only correct voltage, not less than 10% or more than 10% of rated voltage.
6. Faulty bin thermostat	6. Replace or adjust thermostat, ambient conditions are below 60F.
7. Electrical power wires (loose connections or wrong connections)	7. Connect correctly
8. Ice-Off-Wash is in off position or wash position	8. Turn Ice-Off-Wash switch to "Ice"
9. Shorted or grounded compressor	9. Replace compressor
10. Unequalized pressure.	10. Let unit stabilize

TROUBLE: 2. Compressor won't start, but everything else works

POSSIBLE CAUSE	REMEDY
1. Ice off wash switch in wash position/defective	1. Switch to ice/replace
2. Compressor circuit has loose wires	2. Repair or replace
3. Bad start component or wrong start component	3. Replace per manufacturers part number involved
4. Compressor locked or shorted out	4. Replace
5. Voltage too low or too high	5. Should be +− 10% of rated voltage on serial plate
6. Overcharged with refrigerant	6. Drop charge, evacuate, change liquid drier and recharge per serial plate.
7. Dirty condenser	7. Check to see if the high side pressure control is open and clean condenser.
8. Water supply shut off or below 20# (especially water cooled models)	8. Open supply valve check to see if there is a minimum of 20# of pressure and no more than 60#
9. Ambient and water temperature too high	9. Cool off room and water supply.
10. Compressor internal motor protector open	10. Find cause and or replace compressor
11. Burnt motor windings open circuit	11. Replace compressor
12. Improperly wired	12. Refer to schematic
13. Internal compressor mechanical damage	13. Replace compressor

B. NORMAL APPEARANCE OF ICE BUT LOW PRODUCTION

TROUBLE: 1. Water distributor pan plugged.

POSSIBLE CAUSE	REMEDY
1. Dirty or plugged	1. Clean pan

TROUBLE: 2. Water level too high.

POSSIBLE CAUSE	REMEDY
1. Water running over the siphon drain	1. Lower water level

TROUBLE: 3. Partial freeze

POSSIBLE CAUSE	REMEDY
1. Dirty, plugged distributor pans or dirty evaporator plates.	1. Clean and sanitize
2. Faulty ice size thermostat, bin thermostat or primary resistor	2. Replace
3. Ice size thermostat capillary is not bottomed out in sheath	3. Place in all the way
4. Ambient conditions are too cold (any thing below 60°F)	4. Warm up above 60°F
5. Head pressure too low air cooled and water cooled	5. Check pressure, repair any refrigerant leaks, maintain 250 psi on water cooled units
6. Ice rack out of position	6. Install correctly
7. Hot gas valve not opening	7. Replace or repair
8. Ice built up on evaporator tubing due to water leak	8. Repair leak
9. Setting too cold on bin thermostat	9. Readjustment needed
10. Ice size sheath out of adjustment	10. Readjustment needed
11. Not enough of a water purge	11. Adjustment needed

C. IRREGULAR SHAPED ICE AND POOR PRODUCTION PERFORMANCE

TROUBLE: 1. White irregular shape ice

POSSIBLE CAUSE	REMEDY
1. High concentrated impurities in recirculating water	1. Adjust the drain system, clean and sanitize
2. Impurity level too high in water supply	2. Install correct filtration
3. Incorrect water level	3. Adjust water level
4. Low charge	4. Evacuate and recharge
5. Evaporator plates or distributor pan dirty	5. Clean and sanitize

TROUBLE: 2. Low capacity

POSSIBLE CAUSE	REMEDY
1. Excessive ambients	1. Cool room off
2. Water over flowing siphon drain	2. Adjust water level
3. Hot gas solenoid leaking	3. Repair or replace
4. Under charged	4. Repair leak and evacuate unit and recharge

TROUBLE: 3. Incomplete harvest cycle

POSSIBLE CAUSE	REMEDY
1. Under charged	1. Repair leaks and evacuate
2. Faulty ice size thermostat or primary resistor	2. Replace
3. Restriction in hot gas solenoid valve	3. Repair or replace
4. Dirty evaporator or water distributor pan is dirty	4. Clean and sanitize
5. Incorrect ice size thermostat position	5. Adjustment needed
6. Ambients below 60	6. Warm room up

TROUBLE: 4. Compressor cuts out during harvest

POSSIBLE CAUSE	REMEDY
1. Loose wires on compressor	1. Repair
2. Faulty run capacitor	2. Replace
3. Faulty overload in the compressor	3. Replace compressor

TROUBLE: 5. Constant water draining from the water tank.

POSSIBLE CAUSE	REMEDY
1. Drain siphon not in place or incorrectly position	1. Replace correctly
2. Excessive high water pressure	2. Not to exceed 60#
3. Water level too high	3. Adjust float arm
4. Defective or partial plugged float valve	4. Repair or replace

TROUBLE: 6. Only half of plate has ice.

POSSIBLE CAUSE	REMEDY
1. Faulty expansion valve	1. Repair or replace
2. Low refrigerant charge	2. Repair leak,evacuate, and recharge

D. OTHER PROBLEMS

TROUBLE: 1. Ice maker won't stop with full bin of ice.

POSSIBLE CAUSE	REMEDY
1. Faulty bin control	1. Adjust or replace

TROUBLE: 2. Abnormal noise

POSSIBLE CAUSE	REMEDY
1. Pump motor	1. Replace or repair
2. Fan motor	2. Replace motor, or check fan blade for accumulated dirt or bent blades
3. Compressor bearings worn or valve broken	3. Replace compressor
4. Too high of voltage	4. Inspect and have it corrected
5. Excessive head pressure	5. Determine cause, such as dirty condenser, non-condensables, high ambients
6. Tubing vibrating against panels or other tubing	6. Bend away from cause

TROUBLE: 3. Ice in storage bin melts

POSSIBLE CAUSE	REMEDY
1. High ambients	1. Lower room temperatures
2. Plugged drain line	2. Clean bin drain

TROUBLE: 4. Unit short cycles

POSSIBLE CAUSE	REMEDY
1. Shorted out primary resistor	1. Replace
2. Dirty distributor pan	2. Clean and sanitize
3. Faulty ice size thermostat	3. Adjust or replace
4. Setting too warm on bin thermostat	4. Adjust or replace
5. Unit is not level	5. Level unit
6. Partial evaporator freeze up	6. Refer to B-3

TROUBLE: 5. Water pump does not operate

POSSIBLE CAUSE	REMEDY
1. Pump windings burnt out	1. Replace pump
2. Impellor loose or broken	2. Tighten or replace

SERVICE QUESTIONNAIRE

When you have a concern about a Crystal Tips Ice Machine and factory assistance is requested, please have all of the following information available when contacting the factory.

MODEL NUMBER: **SERIAL NUMBER:**
AIR TEMPERATURE WHERE ICE MAKER IS LOCATED:
AIR TEMPERATURE TO CONDENSER ON A REMOTE SYSTEM:
WATER TEMPERATURE TO ICE MAKER:
FREEZE CYCLE TIME:
DEFROST CYCLE TIME:
WHAT IS ACTUAL VOLTAGE BEING SUPPLIED TO ICE MAKER:
WHAT IS THE SUCTION PRESSURE AT THE BEGINNING OF FREEZE CYCLE:
WHAT IS THE SUCTION PRESSURE AT 8 MINUTES IN FREEZE CYCLE:
WHAT IS THE SUCTION PRESSURE AT THE END OF THE FREEZE CYCLE:
WHAT IS THE SUCTION PRESSURE AT THE BEGINNING OF HARVEST CYCLE:
WHAT IS THE SUCTION PRESSURE AT THE END OF THE HARVEST CYCLE:
WHAT IS HIGH SIDE PRESSURE DURING FREEZE CYCLE:
WHAT IS HIGH SIDE PRESSURE DURING HARVEST CYCLE:
WHEN WAS ICE MAKER LAST CLEANED:
DOES BIN CONTROL START AND STOP THE ICE MAKER PROPERLY:
WHAT IS THE BIN MODEL NUMBER:

WE MUST HAVE THE ACTUAL TEST READINGS

DO NOT ESTIMATE

By providing correct information, we will be able to assist you with your concerns in a much more expedient manner.

Crystal Tips Service Department can be reached during normal work hours, Monday thru Friday, except Holidays.

Crystal Tips Ice Products
Junction of U.S. Highways 9 and 71
Spirit Lake, Iowa 51360
(712) 336-2350

SAFETY REVIEW

The service person must always be alert to service situations which could:

1. cause serious injury to the service person
2. cause the ice maker to fail in the future.

Normal service work is not hazardous. There are recommended procedures which must be followed to assure that service is performed under the safest possible conditions. If the service person understands the construction and the purpose of the various components in the ice maker, it will help in the safe and timely servicing of the machine.

A knowledge of the hazards which can develop because of unsafe practice should make the service person aware of safe procedures and allow the service person to wear protective equipment, such as goggles.

The refrigeration service person should keep their tools clean and in proper repair as well as practice good houskeeping, while on the job. Clean tools and clean parts assist in keeping the refrigeration system clean.

It is the little things that count most in servicing the ice machine. Care in tightening a flare connection or make sure a solder connection is clean before soldering it or a tight wire connection, will determine if the system will operate properly after the service is completed.

A slight amount of moisture or dirt from an improperly cared for set of service gauges or a vacuum pump which has not had the oil changed recently, can cause the expansion valve to freeze up or cause sludge to form in the compressor.

These ice makers have been engineered to use specific parts, off the shelf parts are not a direct replacement part. Always use Crystal Tips replacement parts to assure you of an efficient operating ice machine for many years to come.

Ice-O-Matic

Service & Parts Manual

DIAGNOSTIC CUBER

C-61,81,121-C

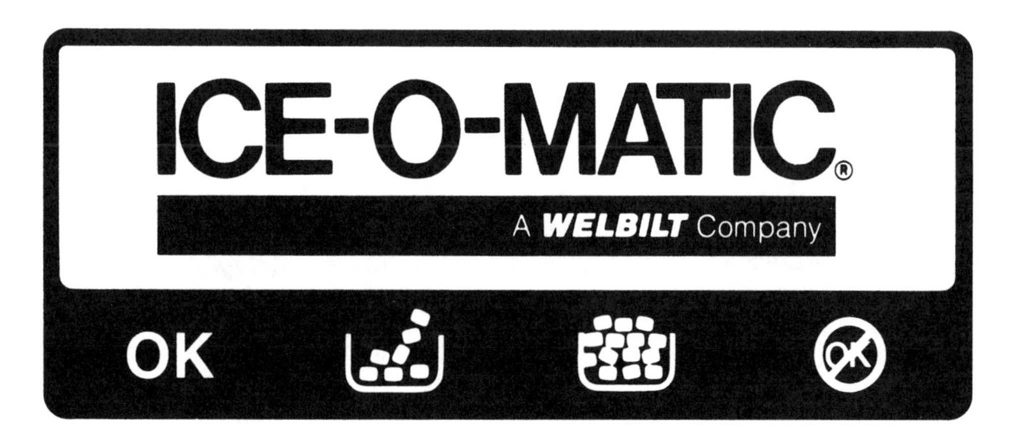

11100 E. 45TH AVE.
DENVER, COLORADO 80239
(303) 371-3737

© MILE HIGH EQUIPMENT COMPANY 1987

P/N 9081067–01

Revised February 1988

MILE HIGH EQUIPMENT COMPANY
ICE MACHINE TWO-YEAR WARRANTY
To the original owner of an ICE-O-MATIC MACHINE.

We warranty every Ice-O-Matic Ice Machine to be free from defects in material and factory workmanship if properly installed, cared for and operated under normal conditions with competent supervision. **WITHIN TWO YEARS FROM DATE OF ORIGINAL INSTALLATION WHEN REPORTED ON A WARRANTY CARD RETURNED TO THE FACTORY** we will replace, F.O.B. our plant, Denver, CO, or F.O.B. authorized Mile High Equipment Co. Parts Depots, without cost to the customer, that part of any such machine that becomes defective, BUT no part or assembly which has been subject to any alteration or misuse, accident, or is from a machine on which the serial number has been altered or removed. We will also honor warranty for **27 MONTHS (THIS TWO YEAR WARRANTY SHALL APPLY TO MACHINES SHIPPED ON OR AFTER JANUARY 1, 1978)** from date of shipment whichever period is last to expire, provided in either such case installation of the equipment is reported on warranty card and returned to the factory at the time of installation. After thorough examination, the decision of our Service Department shall be final. Any defective parts to be repaired or replaced must be returned to us, transportation charges prepaid, and they must be properly sealed and tagged. The serial and model number of the machine, and the date of original installation of such machine must be given. We will not, however, assume any responsibility for any expenses (including labor) incurred in the field incidental to the repair or replacement of equipment covered by this warranty. Our obligation hereunder to repair or replace a defective part is the exclusive remedy for breach of this warranty; and we will not be liable for any other damages or claims, including consequential damages.

No representation, dealer, distributor or any other person is authorized or permitted to make any other warranty or obligate MILE HIGH EQUIPMENT COMPANY, to any liability not strictly in accordance with this policy. THIS WARRANTY IS IN LIEU OF ALL OTHER WARRANTIES EXPRESSED OR IMPLIED, INCLUDING ANY WARRANTY OF MERCHANTABILITY, AND OF ALL OTHER OBLIGATIONS OR LIABILITIES ON OUR PART.

MILE HIGH EQUIPMENT COMPANY
ICE MACHINE FIVE-YEAR WARRANTY
For Motor Compressor

On the said ICE-O-MATIC Ice Machine, we warrant to the original purchaser for a period of five years following the date of installation, the repair or replacement of the motor compressor only if it is shown to our satisfaction that the motor-compressor is not operative due to defects in factory workmanship or material as originally supplied and that normal use and reasonable care have been exercised. We will not, however, assume any responsibility for any expenses "including labor" incurred in the field incidental to the repair or replacement of compressor covered by this warranty. Our obligation hereunder to repair or replace the motor compressor is the exclusive remedy for breach of this warranty; and we will not be liable for any other damages or claims, including consequential damages.

This warranty does not apply to destruction or damage caused by alterations by unauthorized service, using other than identical replacements, risk of transportation, accidents, misuse, abuse, damage by fire, flood or acts of God. After thorough examination, the decision of our Service Department shall be final.

No representative, dealer, distributor or any other person is authorized or permitted to make any other warranty or obligate MILE HIGH EQUIPMENT CO., to any liability not strictly in accordance with this policy. THIS WARRANTY IS IN LIEU OF ALL OTHER WARRANTIES EXPRESSED OR IMPLIED, INCLUDING ANY WARRANTY OF MERCHANTABILITY, AND OF ALL OTHER OBLIGATIONS OR LIABILITIES ON OUR PART.

ICE-O-MATIC
C SERIES DIAGNOSTIC CUBERS

TABLE OF CONTENTS

SECTION A: GENERAL INFORMATION

SECTION B: INSTALLATION

SECTION C: OPERATION

SECTION D: SERVICE

SECTION E: REMOTE CONDENSERS

SECTION F: PARTS

ICE-O-MATIC® | MODEL NUMBERING SYSTEM

The Model Number fully describes the unit. A letter code identifies the model series. A two or three digit number identifies the basic family size and electrical characteristics. A second letter code identifies the type of condenser. Cabinet size is identified by a two digit number. Finish is identified by a letter code.

The following examples illustrate how Model Numbers are developed:

CUBERS

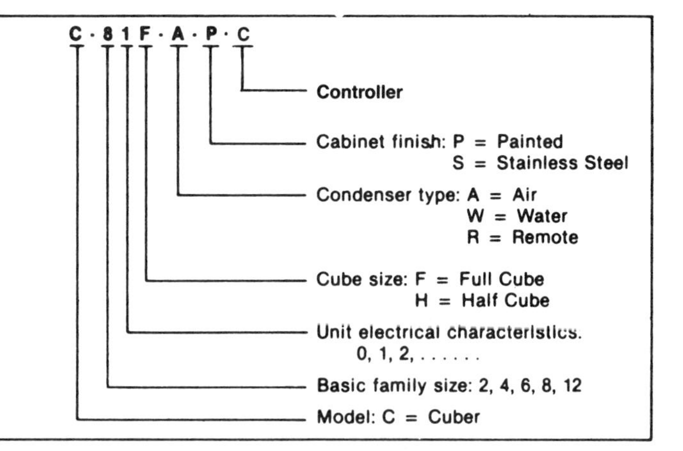

FLAKERS

F · 7 5 6 · A · 3 8 P

- Cabinet finish: P = Painted, S = Stainless Steel
- Cabinet size: 32, 38, or 48
- Condenser type: A = Air, W = Water, R = Remote
- Unit electrical characteristics: 0, 1, 2, ...
- Basic family size: 25, 35, 60, 75, 100
- Model: F = Flaker, FP = Flaker with Cold Plate, MF = Modular Flaker, FC = Flaker Chassis only

REMOTE CONDENSERS

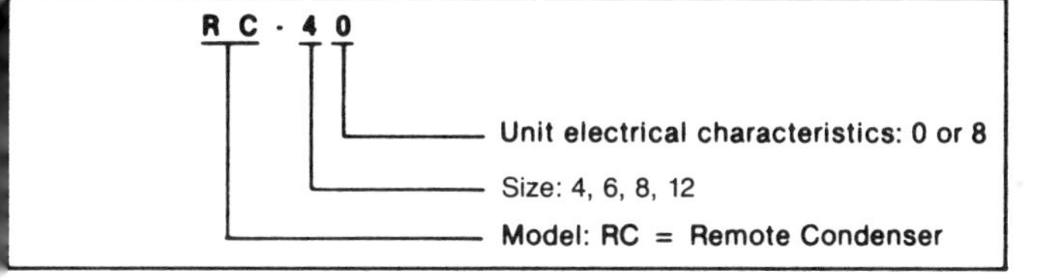

BINS

B · 55 · P

- Cabinet finish: P = Painted, S = Stainless Steel
- Size: 550 lbs. capacity
- Model: B = Bin

FLAKER/DRINK DISPENSERS

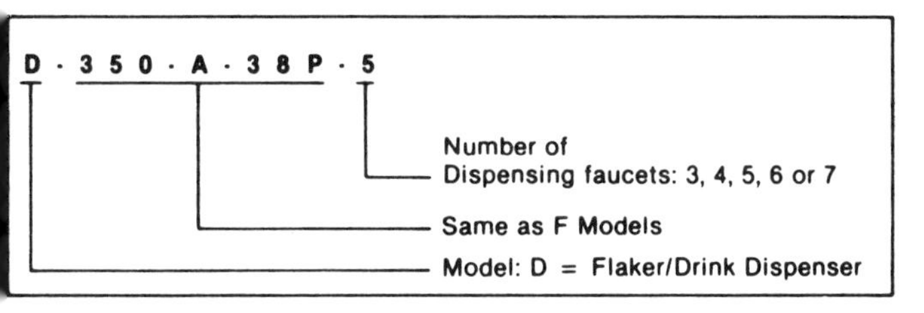

DISPENSERS

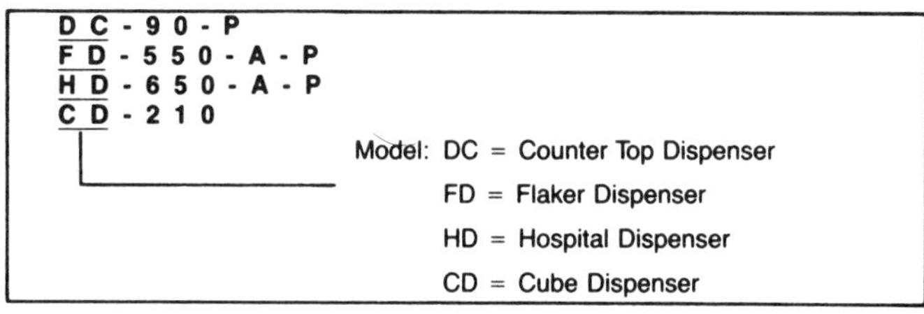

Other model nomenclature same as Flaker or Cuber nomenclature.

Electrical characteristics of all units are coded in the Model Number by a one digit number appearing after the basic family size. The code is as follows:

Code Number	*Volts	**Cycles	Phase	No. of Service Wires Required
0	115	60	1	2
1	208-230/115	60	1	3
2	208	60	1	2
3	230	60	1	2
4	230/115	60	3	4
5	230	50	1	2
6	208-230	60	1	2
7	208-230	60	3	3
8	208-230	50-60	1	2

*Two voltage values separated by a hyphen (-) means that the unit is capable of operating over this voltage range.
Two voltage values separated by a slant line(/) means that both voltages must be supplied at the installation. This is usually done by a neutral wire.
**Two cycle values separated by a hyphen (-) means that the unit is capable of operating at either cycle value.

SERIAL NUMBER FORMAT

Month	Year
A - January	
B - February	3 - 1983
C - March	4 - 1984
D - April	5 - 1985
E - May	6 - 1986
F - June	7 - 1987
G - July	8 - 1988
H - August	9 - 1989
I - September	0 - 1990
J - October	
K - November	
L - December	

Model	Code	Model	Code
ALL BINS	-00	MF-400	-30
		MF-600	-31
		MF-750	-32
		MF-1000	-33
RC40	-01	MF-2500	-34
RC60	-02	MF-5000	-35
RC80	-03	MF-700	-36
RC120	-04		
C-20	-11	HD-350	-41
C-40	-12	HD-650	-42
C-60	-13	HD-750	-43
C-81	-14		
C-121	-15	FD-550	-51
C-10	-16		
F-250	-21	D-250	-61
F-350	-22	D-350	-62
F-600	-23	D-600	-63
F-750	-24	D-750	-64
F-1000	-25	D-1000	-65
F-400	-26	D-400	-66
F-700	-27	D-700	-67
		DX (used)	-99

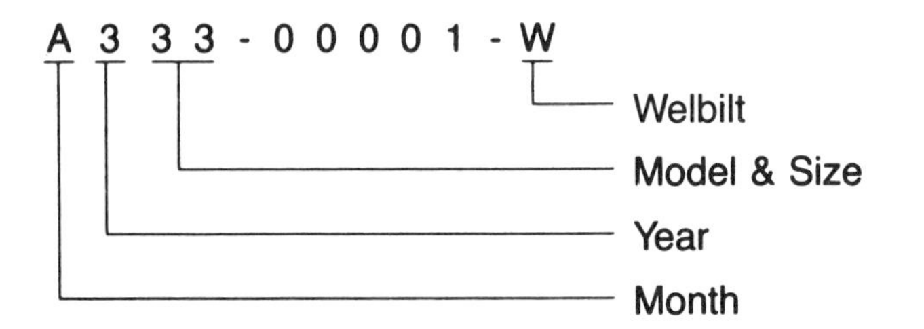

FOREWORD

We, ICE-O-MATIC, present this service manual to aid the service men and users in the installation, operation, and maintenance of your equipment.

If, at any time, you encounter conditions that are not answered in this manual, write or call the service department of ICE-O-MATIC explaining the conditions in detail, giving THE MODEL NUMBER AND SERIAL NUMBER of the unit, and we will give your questions our immediate attention and reply.

ICE-O-MATIC
11100 E. 45th Avenue
Denver, Colorado 80239
(303) 371-3737

ICE-O-MATIC®

MODEL C-61-C

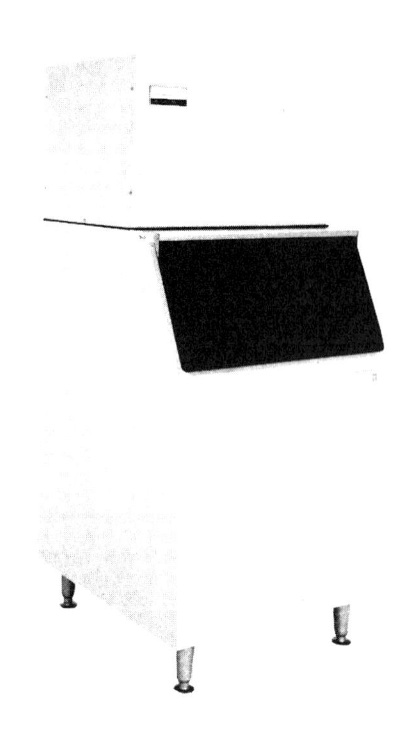

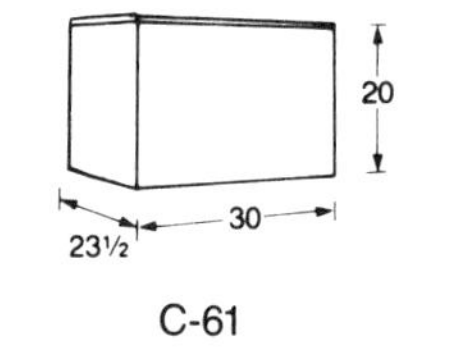

C-61
DIMENSIONS (Inches)

STORAGE BINS

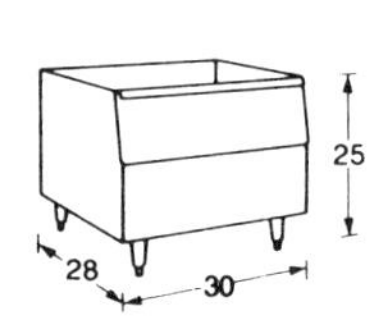

B-20
Storage Capacity 170 lb.

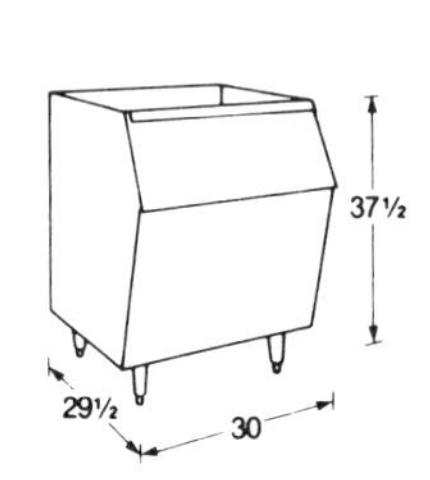

B-40
Storage Capacity 365 lb.

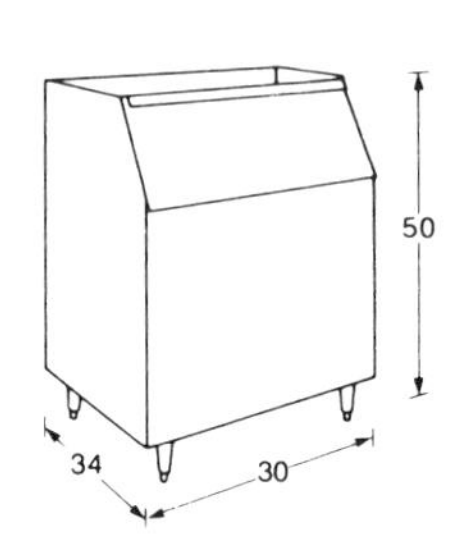

B-55
Storage Capacity 550 lb.

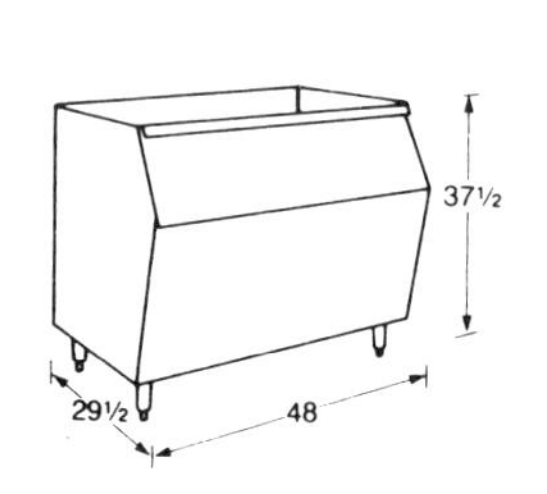

B-60
Storage Capacity 580 lb.)
(KBT-5 Bin Top Required)

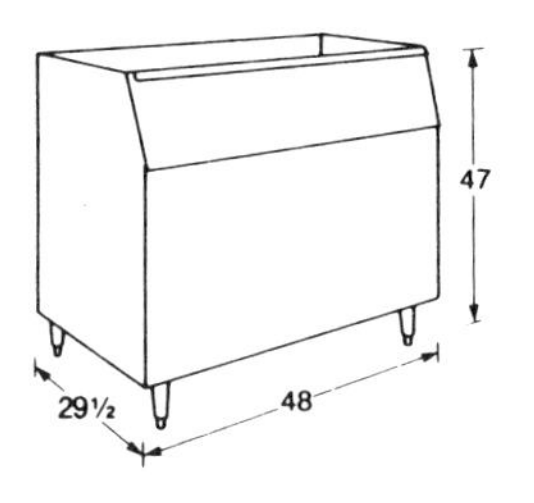

B-80
Storage Capacity 800 lb.
(KBT-5 Bin Top Required

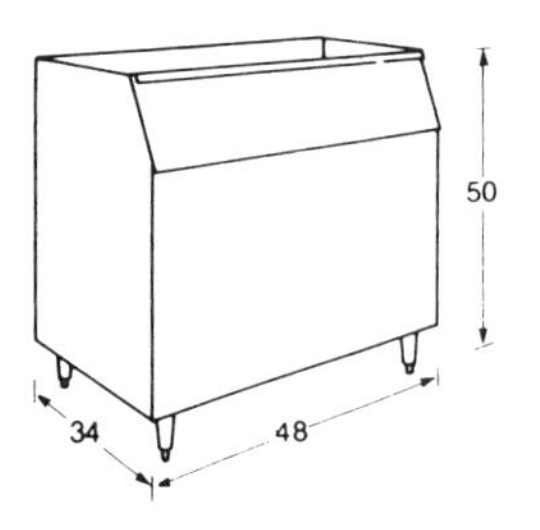

B-100
Storage Capacity 1000 lb.
(KBT-5 Bin Top Required)

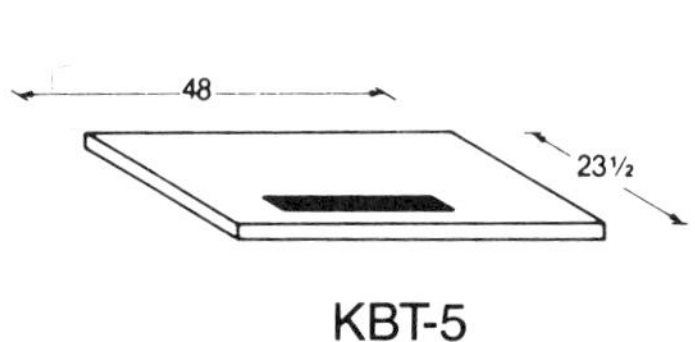

KBT-5

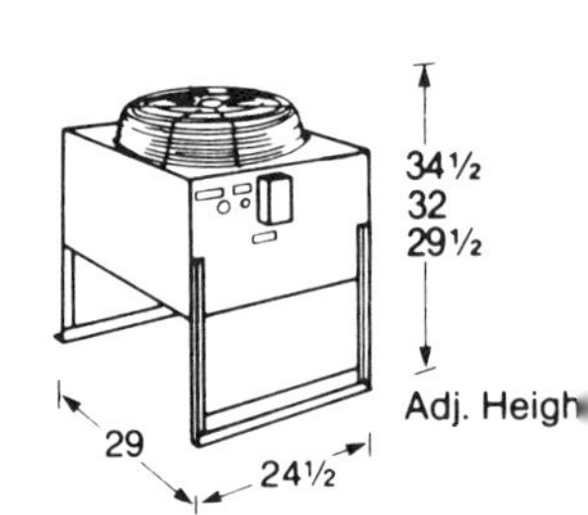

RC-60-2
Remote Condenser

ICE-O-MATIC®

MODEL C-81/121-C

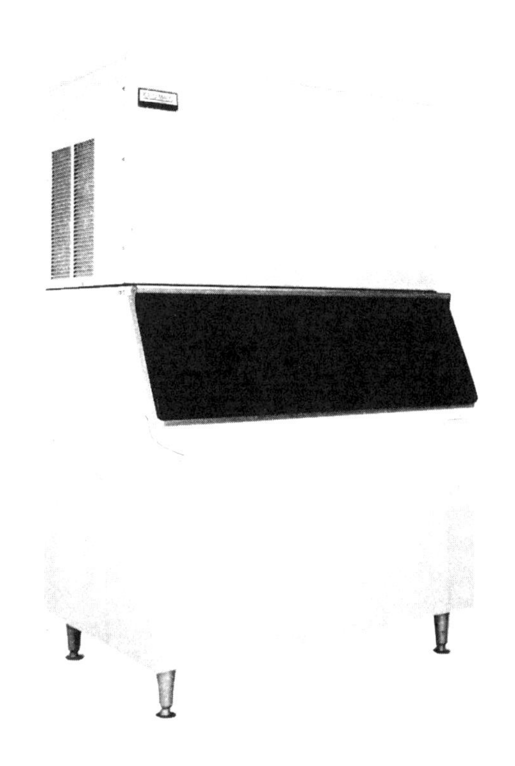

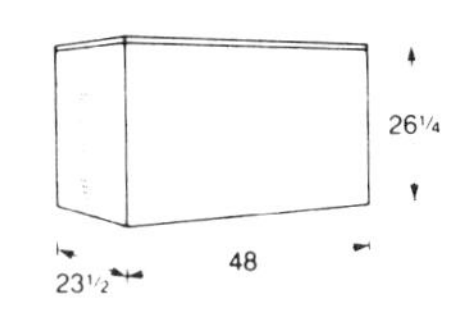

C-81/121
DIMENSIONS (Inches)

STORAGE BINS

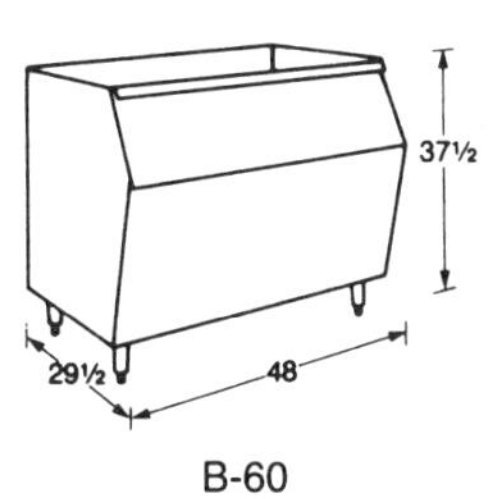

B-60
Storage Capacity 580 lb.

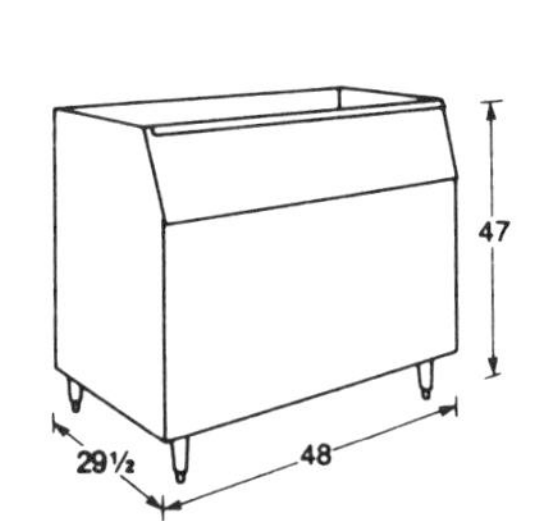

B-80
Storage Capacity 800 lb.

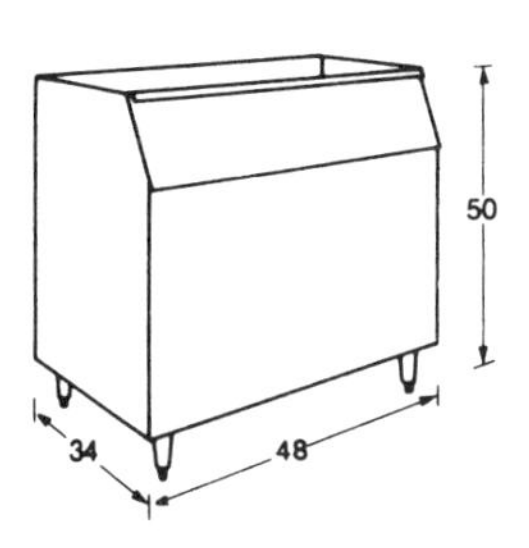

B-100
Storage Capacity 1000 lb.

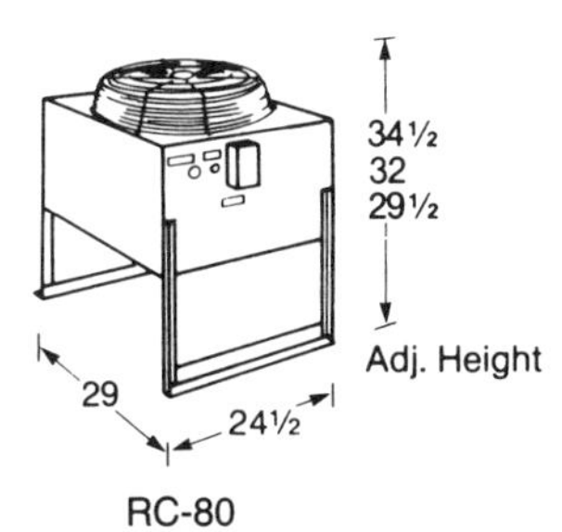

RC-80

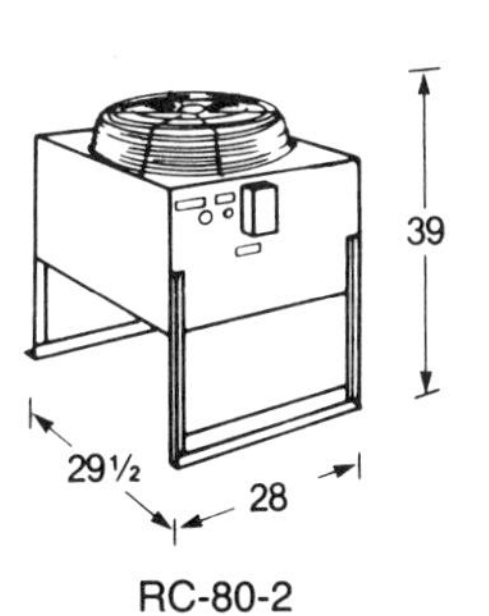

RC-80-2

Remote Condensers

RC-120

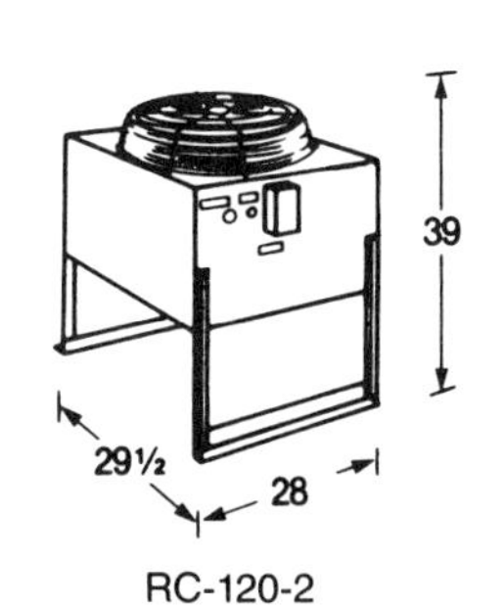

RC-120-2

Remote Condensers

ELECTRICAL AND MECHANICAL SPECIFICATIONS
SERIES C CUBERS

MODEL NUMBER	CONDENSING UNIT	COMP. H.P.	VOLTAGE REQUIREMENTS	NO. OF WIRES	OPERATING AMPS @ RATED VOLTAGE	MAX. AMP FUSE SIZE	REFRIGERANT TYPE	REFRIGERANT CHARGE (oz.)	APPROX. SHIPPING WEIGHT (lbs.)
C-61-A-C	AIR	¾	208-230/115-60/1	3	8.9	15	R-502	28	180
C-61-W-C	WATER	¾	208-230/115/60/1	3	8.2	15	R-502	18	188
C-61-R-C	REMOTE AIR	¾	208-230/115/60/1	3	12.7	20	R-502	144	186
C-81-A-C	AIR	1	208-230/115/60/1	3	12.2	20	R-12	44	292
C-81-W-C	WATER	1	208-230/115/60/1	3	10.5	20	R-12	30	286
C-81-R-C	REMOTE AIR	1	208-230/115/60/1	3	14.0	20	R-12	208	295
C-84-A-C	AIR	1	208-230/115/60/3	4	8.2	15	R-12	44	292
C-84-W-C	WATER	1	208-230/115/60/3	4	6.6	15	R-12	30	286
C-84-R-C	REMOTE AIR	1	208-230/115/60/3	4	9.2	15	R-12	208	295
C-121-A-C	AIR	1½	208-230/115/60/1	3	15.1	20	R-502	50	297
C-121-W-C	WATER	1½	208-230/115/60/1	3	13.5	20	R-502	32	293
C-121-R-C	REMOTE AIR	1½	208-230/115/60/1	3	16.1	20	R-502	208	302
C-124-A-C	AIR	1½	208-230/115/60/3	4	10.4	15	R-502	50	297
C-124-W-C	WATER	1½	208-230/115/60/3	4	8.8	15	R-502	32	293
C-124-R-C	REMOTE AIR	1½	208-230/115/60/3	4	13.4	15	R-502	208	302

ICE PRODUCTION CHARTS
POUNDS PRODUCED PER 24 HOURS

CONDENSING UNIT	AIR TEMP. (°F)	WATER TEMP. (°F)			AIR TEMP. (°F)	WATER TEMP. (°F)			AIR TEMP. °F)	WATER TEMP. (°F)		
		50°	70°	80°		50°	70°	80		50°	70°	80°
AIR	70°	636	550	537	70°	815	785	715	70°	1160	1040	1005
	80°	587	503	490	80°	732	698	630	80°	1070	950	920
	90°	541	455	443	90°	670	621	585	90°	990	855	825
WATER	70°	640	566	540	70°	733	700	683	70°	1275	1137	1040
	80°	635	561	534	80°	693	651	638	80°	1268	1123	1026
	90°	630	556	527	90°	654	697	598	90°	1261	1109	1012
REMOTE AIR	0°	598			0°	802			0°	1155		
	30°	596			30°	790			30°	1153		
	50°	594	555		50°	780	700		50°	1152	1090	
	70°	593	550		70°	775	690		70°	1135	1056	
	90°	570	510	473	90°	710	662	567	90°	1005	895	841
	110°	526	473	433	110°	605	555	480	110°	871	782	700
MODEL NUMBER		**C-61-C**				**C-81-C**				**C-121-C**		

ELECTRONIC CONTROLLER GLOSSARY OF TERMS

Crystalline — A substance composed of crystals whose atoms are arranged with order and regularity.

Error Code — Occurs if or when there is a malfunction with a component that is being monitored by the controller or a malfunction of the controller itself. The error code is visually indicated on the digital LED display on the electronic controller. Designated EC-1, -2, etc. throughout manual.

Light Emitting Diode — Abbreviated LED. A semiconductor diode that converts electrical energy into visible light. Also known as a solid-state lamp.

Loop — A closed path or circuit over which a signal can circulate and cycle back to its point of origin.

Microcomputer — A single silicon chip on which the arithmetic and logic functions of a computer are placed.

Momentary Switch — A push-type switch that completes a circuit only while it is depressed.

Random-Access Memory — Abbreviated RAM. A computer memory in which the data can be retrieved at a speed which is independent of its location in the memory.

Read-Only Memory — Abbreviated ROM. A computer memory in which the data is pre-programmed and permanent. It is used primarily for rapid information retrieval applications.

Semiconductor — A group of solid crystalline materials having electrical conductivity properties between those of conductors and insulators.

Solid-State — Relating to any device, usually crystalline semiconductors, that can control electrical current without moving parts or vacuum tubes.

Static Electricity — The transfer of an electric charge from one object to another by means of a spark that bridges an air gap between the objects.

Thermistor — A solid-state semiconducting device whose electrical resistance varies with temperature.

Time Delay Switch — A rocker-type, dual in-line pin (DIP) switch that is soldered directly onto a circuit board. Moving the individual switch segments to the 'ON' position increases the time (or delay) that the switch remains active in the circuit.

Triac — A three-terminal semiconductor switch used for a-c power control.

HOW THE ICE-O-MATIC DIAGNOSTIC CUBERS WORK

Description of Controller Operational and Diagnostic Functions:

The Ice-O-Matic diagnostic cubers contain an electronic controller which is microcomputer based. It is designed to control the motors, compressor, solenoid valves, and pumps used for ice-making. The automatic control decisions throughout the ice-making process are made based on information continuously received from temperature sensors, pressure controls and switches, time, and the control program itself.

The electronic controller is connected to four machine-status lights (LEDs) which are displayed on the machine front panel. The status lights and their symbols indicate the current operational status of the ice machine. The colors, symbols, and meanings of the indicators are shown below:

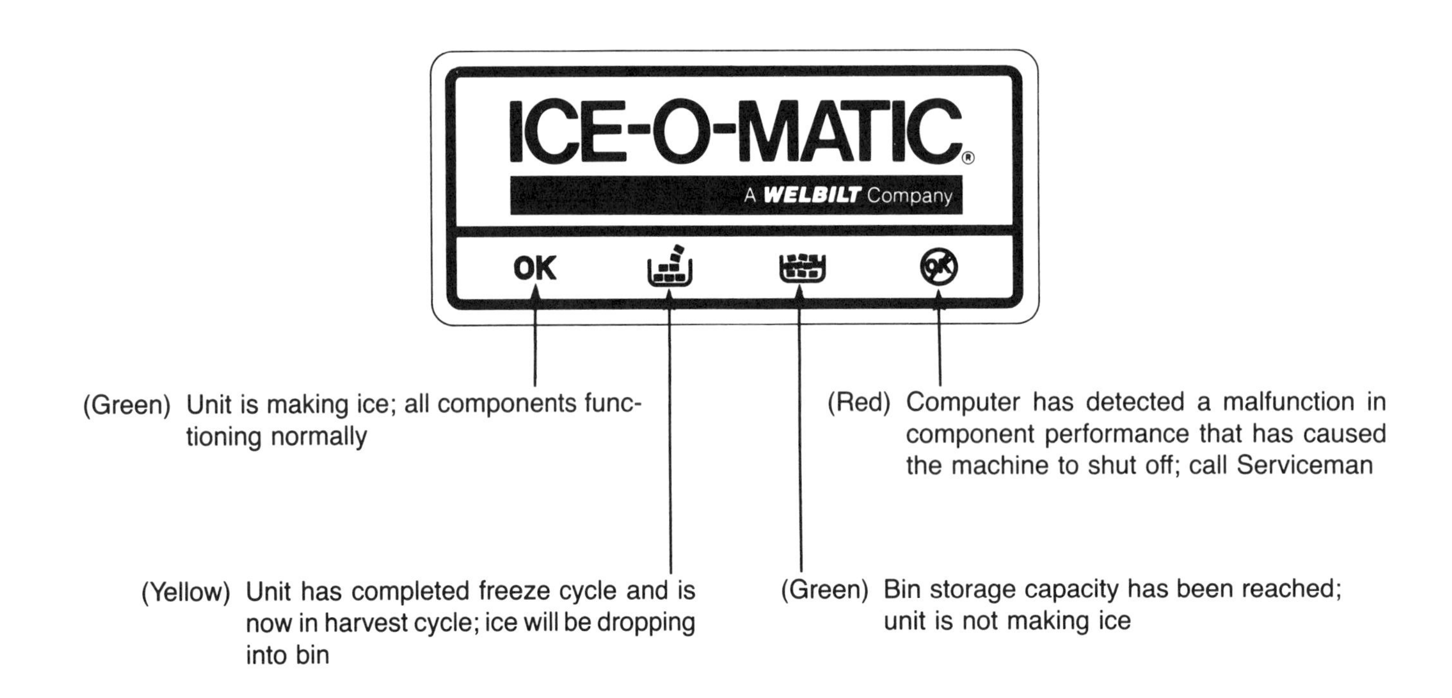
ICE-O-MATIC®
A WELBILT Company
OK
(Green) Unit is making ice; all components functioning normally
(Red) Computer has detected a malfunction in component performance that has caused the machine to shut off; call Serviceman
(Yellow) Unit has completed freeze cycle and is now in harvest cycle; ice will be dropping into bin
(Green) Bin storage capacity has been reached; unit is not making ice

HOW THE ICE-O-MATIC DIAGNOSTIC CUBERS WORK
(Continued)

The electronic controller also has several integral self-diagnostic capabilities which operate continuously. The controller monitors and stores various types of operational data, such as number of harvest cycles and any malfunction since initial start-up or last reset. When malfunctions are detected, in either the controller's operation or in the operation of the machine being controlled, the machine may discontinue operation and a specific error code description will be stored in the controller's memory until the machine is serviced. In addition, the red 'Call Serviceman' LED will remain lit until the controller is reset by a Serviceman. The LED display is located on the controller board behind the machine front panel in the electrical control box area.

The electronic controller consists of a microcomputer with two types of internal memory that perform distinct functions in the ice machine:

1. Read-Only Memory (ROM) — contains the control program and controls the sequence of operations of the ice machine based on input signals received from devices and components external to the controller.
2. Random-Access Memory (RAM) — stores the system's operational history data and logs the number of harvests and any error codes.

The computer program and its process that controls the ice machine can be thought of as a 'Loop' or cycle in which a predetermined sequence of events are repeated indefinitely. At initial power-up or upon a reset condition, the microcomputer will execute the program from the top (or beginning) of the loop which initiates the freeze cycle. During normal operation, when the bottom of the loop (harvest cycle) is reached and concluded, the program will automatically return itself and the ice machine to the top of the loop and begin another freeze cycle.

DANGER: Electrical shock and/or injury from moving parts inside this machine can cause serious injury. Disconnect electrical supply voltage to machine prior to performing any adjustments or repairs.

NOTE: If above cleaning procedures are not performed every six (6) months, harvest problems may occur.

INSTALLATION

A. Uncrating & Inspection Procedures

1. Inspect exterior of shipping carton and skid for any signs of shipping damage.

2. Using a pry bar or large screwdriver, remove all staples from bottom of carton. Do not pry against carton.

3. Lift carton straight up, entirely off machine. Do not discard carton until satisfactory inspection of machine is complete.

4. Inspect machine cabinet for any signs of shipping damage.

5. Remove top, front, left and right hand side panels for access to skid attachment bolts. With panels removed, inspect machine for damaged refrigeration lines and any other signs of shipping damage. On air cooled machines, make certain that fan blade(s) turn freely. Should any damage be noted, follow Freight Claim Procedures on page B2.

6. Remove the four skid bolts. Lift machine off skid and place on bin.

7. Securely attach machine to bin using the supplied attachment straps and hardware located on the rear of the machine and bin.

8. Remove tape from splash curtain(s).

9. Fill out and return Installation and Warranty card to factory.

FREIGHT CLAIM PROCEDURES

IMPORTANT

THIS MERCHANDISE HAS BEEN CAREFULLY INSPECTED AND PACKED IN ACCORDANCE WITH THE CARRIER'S PACKING SPECIFICATIONS. RESPONSIBILITY FOR SAFE DELIVERY HAS BEEN ASSUMED BY THE CARRIER. IF LOSS OR DAMAGE OCCURS, YOU AS CONSIGNEE MUST FILE A CLAIM WITH THE CARRIER, AND HOLD CONTAINER FOR CARRIER'S INSPECTION.

CONCEALED LOSS OR DAMAGE

IF LOSS OR DAMAGE DOES NOT APPEAR UNTIL MERCHANDISE HAS BEEN UNPACKED MAKE A WRITTEN REQUEST FOR INSPECTION BY THE CARRIER WITHIN 15 DAYS OF THE DELIVERY DATE. THEN FILE A CLAIM WITH THE CARRIER.

DO NOT RETURN DAMAGED MERCHANDISE TO US WITHOUT WRITTEN PERMISSION. INSPECT PROMPTLY AND FILE CLAIM WITHOUT DELAY.

VISIBLE LOSS OR DAMAGE

ANY EXTERNAL EVIDENCE OF LOSS OR DAMAGE MUST BE FULLY DESCRIBED AND NOTED ON YOUR FREIGHT BILL OR EXPRESS RECEIPT AND SIGNED BY THE CARRIER'S AGENT. CLAIM SHOULD THEN BE FILED ON A FORM AVAILABLE FROM THE CARRIER ON REQUEST.

DO NOT RETURN DAMAGED MERCHANDISE TO ICE-O-MATIC

FILE YOUR CLAIM AS ABOVE

ICE-O-MATIC

MODEL C-61-C

PLUMBING/ELECTRICAL CONNECTIONS DIAGRAM:

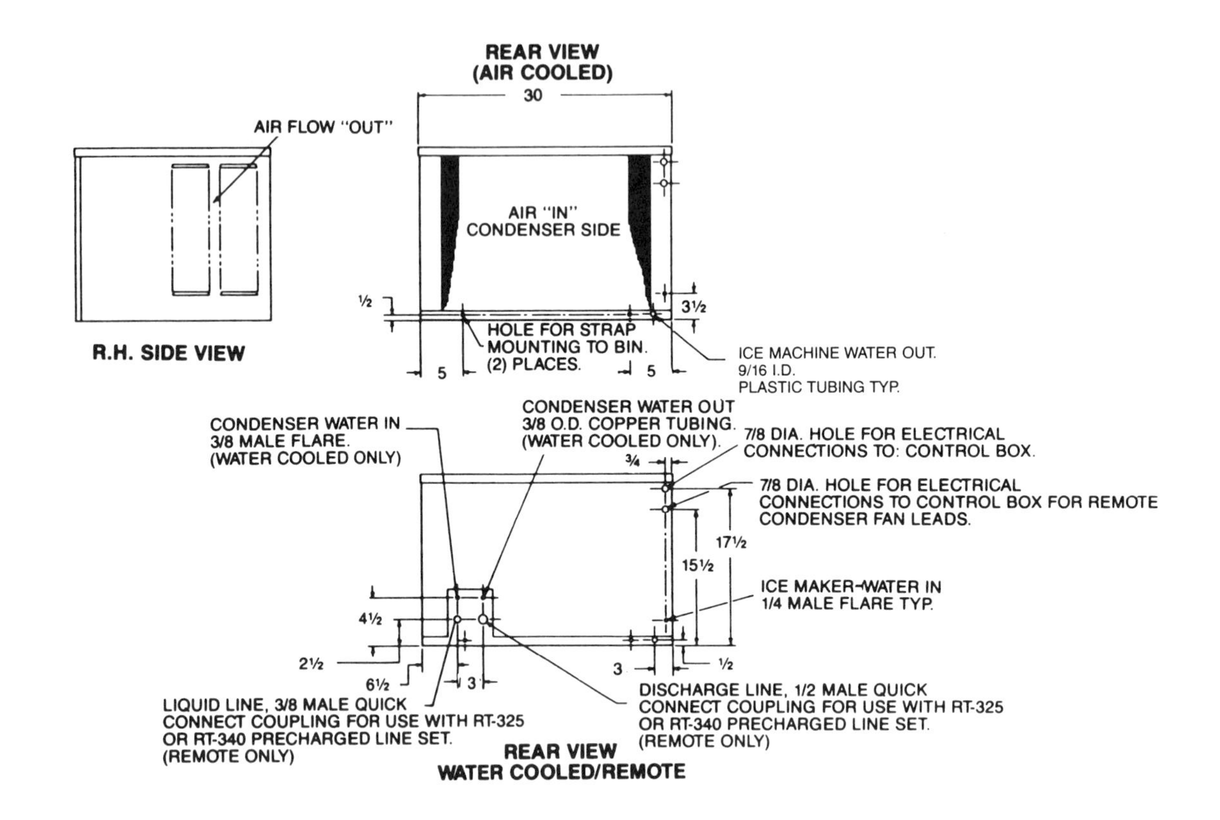

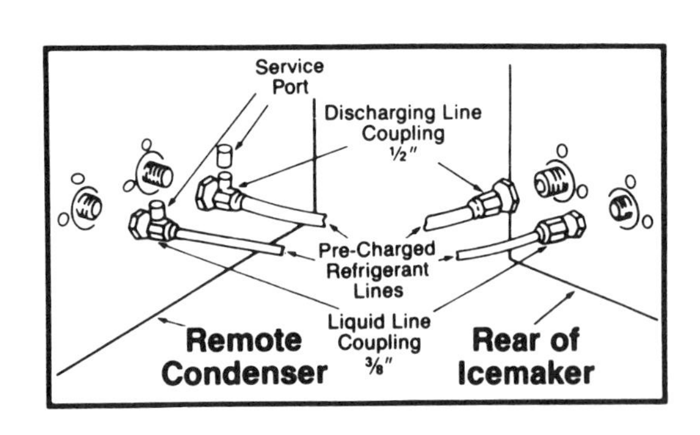

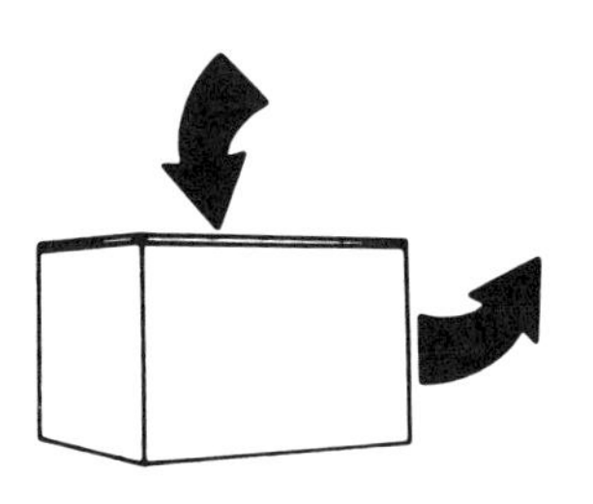

Air Circulation Pattern
(Air Cooled Models)

ICE-O-MATIC®

MODEL C-81/121-C

PLUMBING/ELECTRICAL CONNECTIONS DIAGRAM:

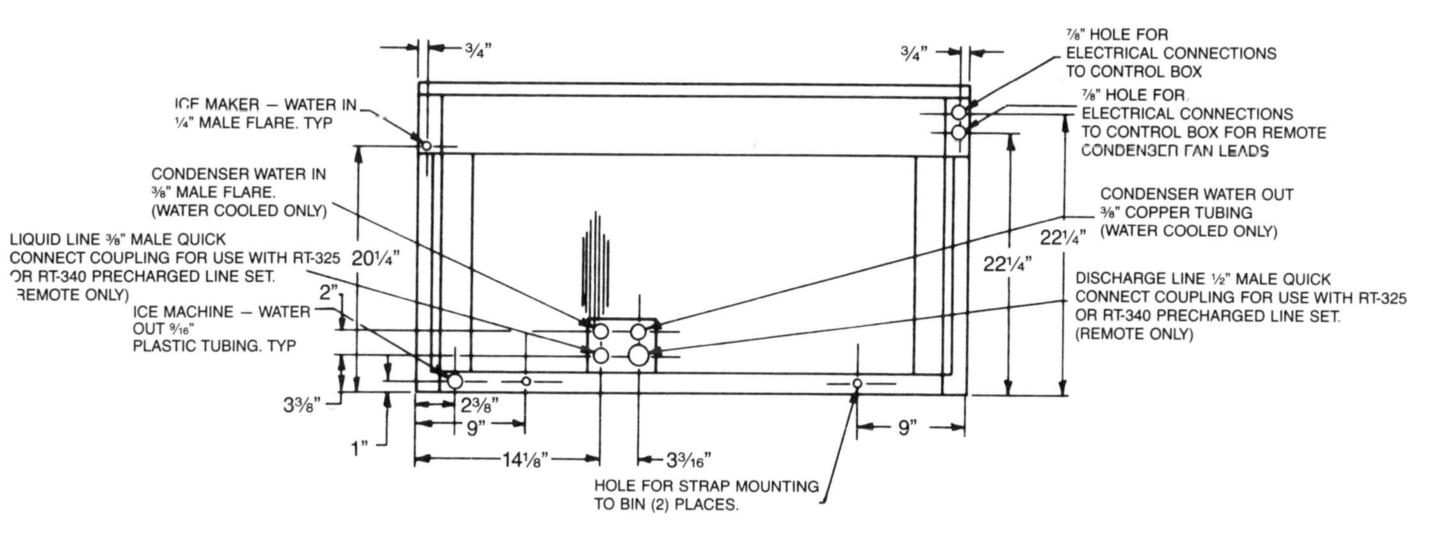

REAR

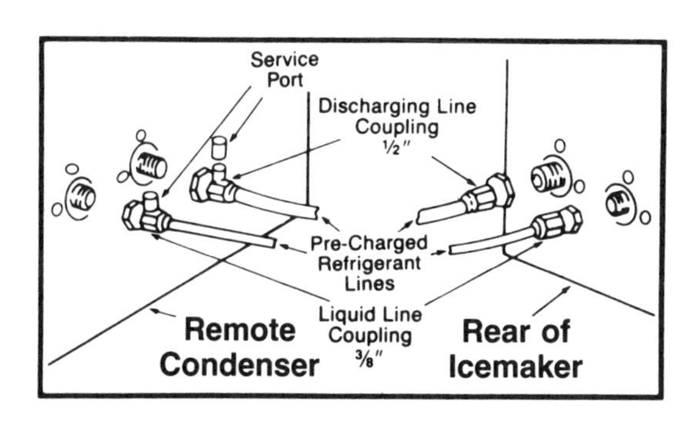

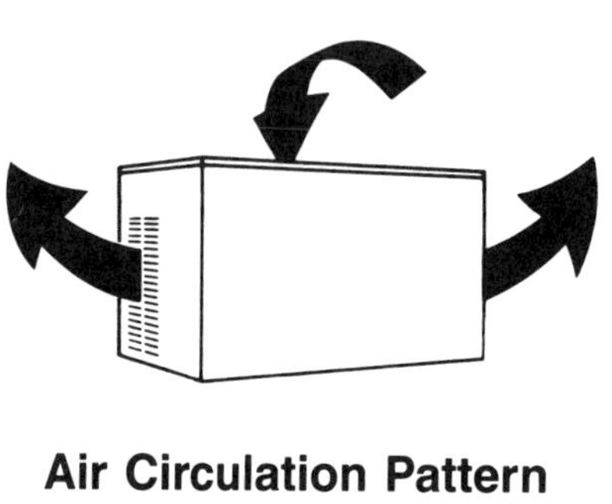

Air Circulation Pattern
(Air Cooled Models)

INSTALLATION

DANGER: Electrical shock and/or injury from moving parts inside this machine can cause serious injury. Disconnect electrical supply voltage to machine prior to performing any adjustments or repairs.

B. **Set-up Procedure**

Note: All installations must conform with local plumbing and electrical codes.

Caution: Make certain all power to equipment is off during installation!

1. Check for proper utility connections for installation. Refer to diagrams on pages B3 and B4.

   ELECTRICAL: Check for proper voltage, current and fusing size according to requirements on model/serial number plate on rear of machine.

Note: On single voltage machines (208 or 230), voltage must be ±10% of machine voltage requirements. On dual voltage machines (208-230), voltage must be +10% and −5% of machine voltage requirements.

Note: A separate electrical circuit must be provided specifically for the ice machine.

   WATER: Inlet water pressure must range from 20 to 60 p.s.i.g.; ¼" male flare water supply inlet connection is standard.

   Wate.· cooled machine connections:
   Condenser in — ⅜" male flare
   Condenser out — ⅜" copper tube
   Connections must be run separately

   DRAIN: Connections
   Bin Drain — ¾" F.P.T.
   Purge Drain — 9/16" I.D. Plastic Tube

2. Connect the purge, bin and condenser drains (water cooled only) to well-pitched drain lines. Run all drain lines *separately* to an open or trapped drain.

3. Electrical connections are made in the electrical control box with the left side panel removed. Route lines from a standard electrical circuit (as required and indicated on serial plate) through side of cabinet to terminal block. Terminal block connections are made with #10 fork terminals.

Note: All wiring must conform with local codes.

4. Place machine/bin unit in its permanent location while maintaining a minimum clearance of 5" between the back, left and right hand sides and any wall. This clearance is necessary to allow for adequate machine air ventilation.

5. After ice machine is in permanent location, make certain that unit is level side-to-side and front-to-back. Accurate leveling is essential for proper machine operation.

INSTALLATION
(Continued)

6. Install ice deflector per instructions below.
   a. Locate the four threaded inserts, two on the L.H. front, lower side panel and two on the R.H. front, lower side panel. Refer to figure 1.

   b. Install screws into each lower hole. (Note: leave these screws loose)

   c. Install the ice deflector over these screws as per figure 2. Install the two screws in the top mounting holes and securely tighten all four screws.

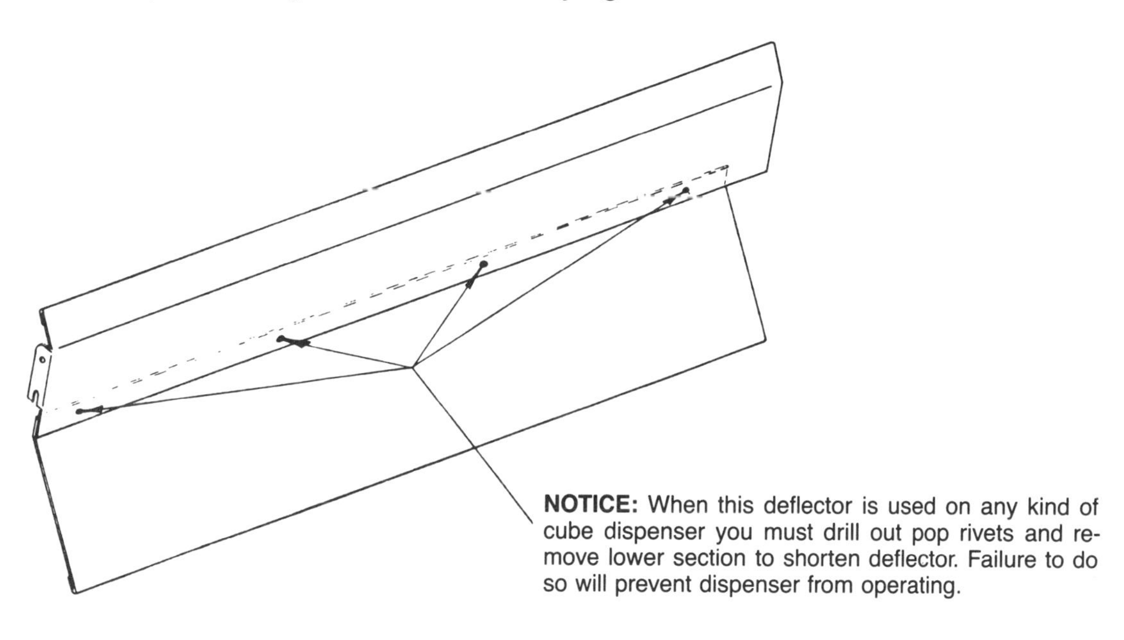

NOTICE: When this deflector is used on any kind of cube dispenser you must drill out pop rivets and remove lower section to shorten deflector. Failure to do so will prevent dispenser from operating.

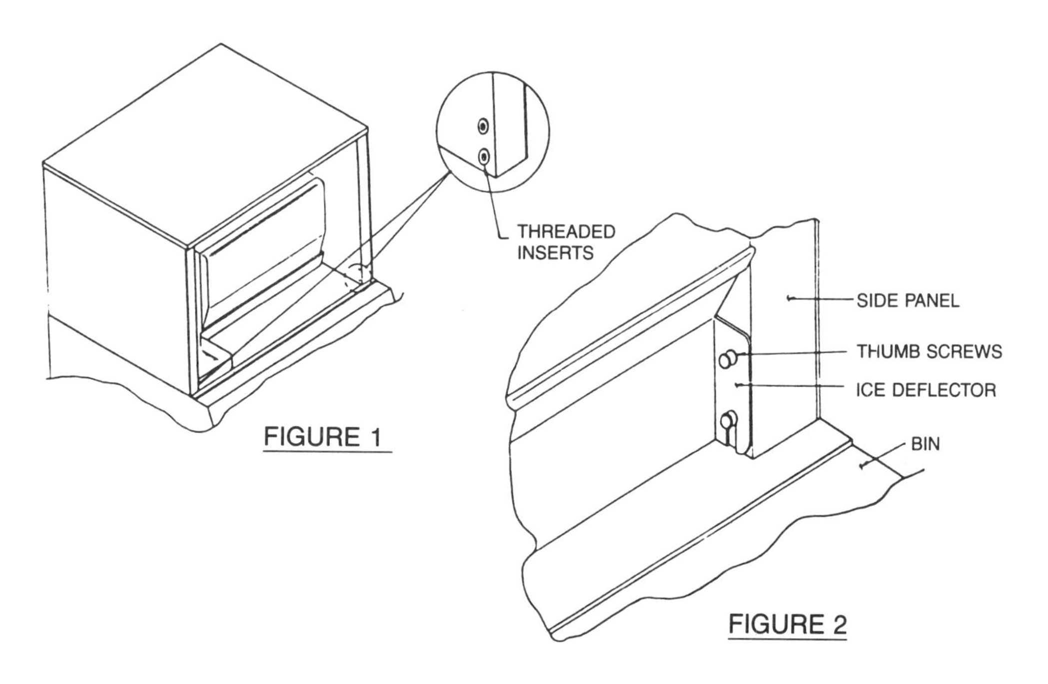

FIGURE 1

FIGURE 2

INSTALLATION
(Continued)

7. Check probe alignment per instructions below.

   It is possible for the evaporator to shift in transit, causing misalignment between the evaporator probe guide and the probe assist motor assembly. It is recommended to check alignment of the probe(s) at the time of installation. This should be done as follows:

   a. Make sure evaporator mounts are resting in the small part of the key hole slot. If not, loosen mounting screws, re-position and re-tighten mounting screws.

      At this time remove allen screw from the arm part of the probe shaft. Rotate the arm by hand in direction of arrow following the clutch as if it were in normal operation.

   If there is any binding, re-adjust by loosening mounting screws on the assembly bracket, once shaft is free (there should be no drag at all) re-tighten mounting screws.

   NOTE: When aligning probes, check probe lock nut (indicated by arrow marked ①) is tight and that tip of probe shaft does not extend out of the evaporator bushing into the freezing surface of the evaporator. When re-tightening probe lock nut always secure with a non-hardening Loc-tite.

   Also check probe clutch allen set screw in stainless steel portion of clutch (indicated by arrow marked ②) to insure it's tight against the flat portion of the harvest motor drive shaft.

Audibly check cam switch. Switch should change contacts when moving in and out of cam depression. Adjustment should be made by bending the switch arm.

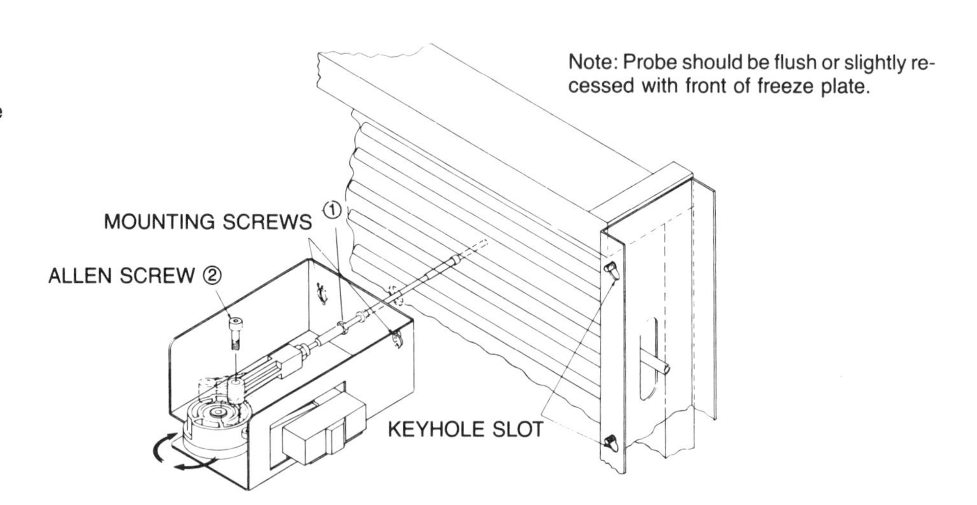

8. Turn water supply on. Check water flow through float assembly. Check water level in water trough; level should be ½" above water pump impeller housing (base).

OPERATION

DANGER: Electrical shock and/or injury from moving parts inside this machine can cause serious injury. Disconnect electrical supply voltage to machine prior to performing any adjustments or repairs.

A. Start-up Descriptions and Procedure

1. Check position of ICE/OFF/WASH selector switch and test purge function (Step C).

   a. ICE Position: The bin (curtain) switch(es) will be tested. If the bin is not full, all ice-making components will turn on, the 'OK' LED will illuminate, and the machine will begin a freeze cycle from the top of the loop.

      If the bin is full, and the bin (curtain) switch(es) are actuated, all controlled components will be turned off except the 'BIN FULL' LED, which will be illuminated. The controlled components will remain off until the bin switch(es) indicate that the bin is no longer full, then the unit will begin a freeze cycle, deactivate the 'BIN FULL' LED, and illuminate the 'OK' LED. Ice making cycles will continue until the bin again becomes full or the selector switch is moved to either the OFF or WASH position.

   b. OFF Position: Entire machine and all components will be off when selector switch is in this position, except the LED display on the electronic controller which will still display evaporator and condenser temperatures.

   c. WASH Position: All controlled machine components will be off except the water pump. The water pump will remain on until the selector switch is moved to either the OFF or ICE position.
      At this time the purge function can be tested by depressing and holding down the purge switch. The purge valve will open allowing all water in the water trough to be removed by the water pump.

2. Move selector switch to ICE position for initial self-test and freeze cycle.

   When a machine is powered-up for the first time, upon a manual reset, or upon a memory loss condition (as would occur in a severe partial or total power loss), and the bin is not full, the controller will run a self-test of itself and check the status of the single/dual evaporator switch which determines the number of evaporators being controlled. This switch is preset at the factory and should not be changed. Refer to page D-2 for further details on the single/dual evaporator switch.

   Once the initial self-test process is completed and the bin is not full, the freeze cycle will begin. Water is now flowing over the evaporator and freezing and the 'OK' LED is illuminated.

OPERATION (Continued)

3. Check LED display on the controller board for monitoring of evaporator and condenser temperatures.

   The controller constantly monitors the evaporator outlet and condenser temperatures. When the machine is operating properly the temperatures are displayed in a sequence (detailed below) on the seven-segment LED found on the controller board.

   The LED displays the evaporator outlet temperature as 'E' and the condenser temperature as 'C' in the following sequence:

   - An 'E' will display for ½ of a second (indicating evaporator temperature to follow).

     The readout will go blank for ¼ of a second.

     The first digit of the temperature will display for ½ of a second.

     The readout will go blank for ¼ of a second.

     The second digit of the temperature will display for ½ of a second (if necessary).

     The readout will go blank for ¼ of a second.

     The third digit of the temperature will display for ½ of a second (if necessary).

     The readout will go blank for ½ of a second.

   - A 'C' will display for ½ of a second (indicating condenser temperature to follow).

     The readout will go blank for ¼ of a second.

     The first digit of the temperature will display for ½ of a second.

     The readout will go blank for ¼ of a second.

     The second digit of the temperature will display for ½ of a second (if necessary).

     The readout will go blank for ¼ of a second.

     The third digit of the temperature will display for ½ of a second (if necessary).

     The readout will go blank for 1 second.

     The sequence repeats itself.

     Note: The readout will not display '0' preceding any temperature reading. i.e., 5°F evaporator temperature will display as E5.

     All temperatures are displayed in degrees Fahrenheit.

OPERATION (Continued)

4. Observe first slabs of ice for proper ⅛" bridge thickness.

   If adjustments are necessary, follow procedure below.

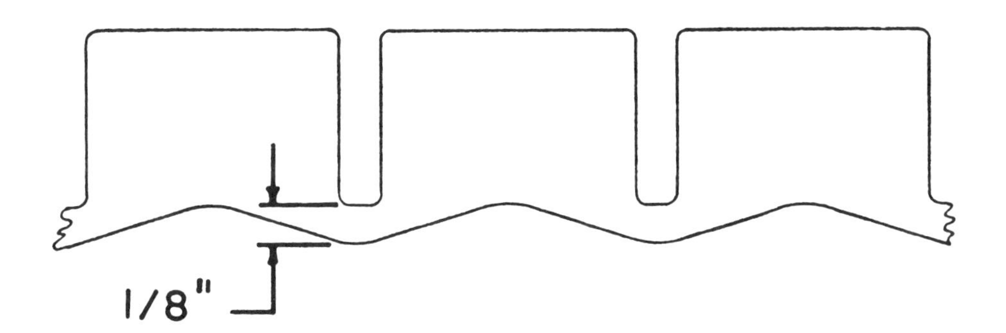

Proper Bridge thickness

CHANGING ICE BRIDGE THICKNESS

The ice bridge thickness is controlled by the time delay switch. To increase the bridge thickness, increase the timer setting. To decrease the bridge thickness, decrease the timer setting.

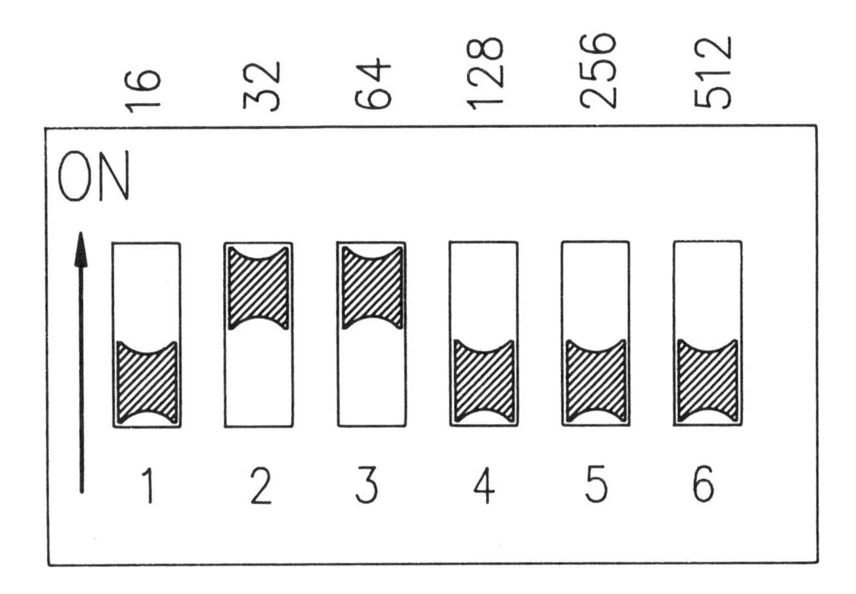

TIMER ADJUSTMENT

To change timer settings — Move the switch next to the number to an 'ON' position (note the ON marking on the timer) to activate that number. The numbers indicate seconds. To determine total time delay add up the numbers moved to the ON position. Timer shown above has a 96 second setting.

Note: System will not purge during cycle if there is no time set on timer.

OPERATION (Continued)

5. Check bin control for proper operation.

   The bin control is a mechanical switch that is actuated by the opening of the splash curtain. When the curtain is opened for more than 10 seconds at the end of the freeze cycle, the switch is actuated, all machine components except the controller should turn off and stay off until the curtain can close to its original position. Machine should now resume normal operation.

   If adjustments are necessary, follow procedure below.

   a. Make sure power is disconnected.

   b. Disconnect one lead from bin switch.

   c. Connect ohm meter to bin switch terminals as shown in figure 1 to read through bin switch.

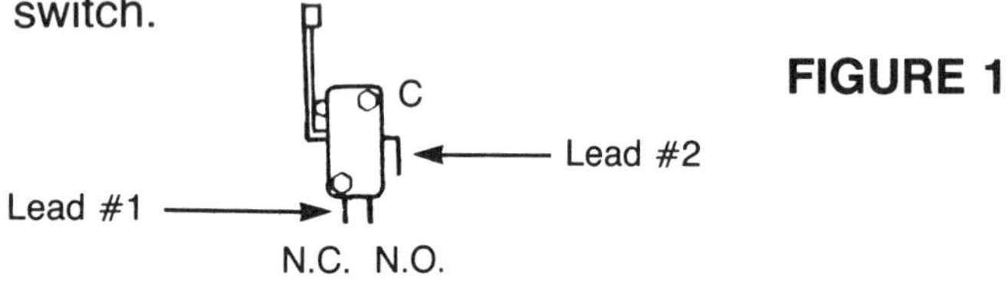

FIGURE 1

   d. Loosen the adjustment screw on the bin switch bracket (ref. to figure 2) while holding the evaporator splash curtain even with the edge of water trough (ref. to figure 3) connect a continuity test across terminals N.C. and C. terminals, you should have no continuity. Once the splash curtain is beyond the outside edge of water trough, you should have continuity.

   e. Check the curtain and switch movement per figure 3, and re-adjust as necessary.

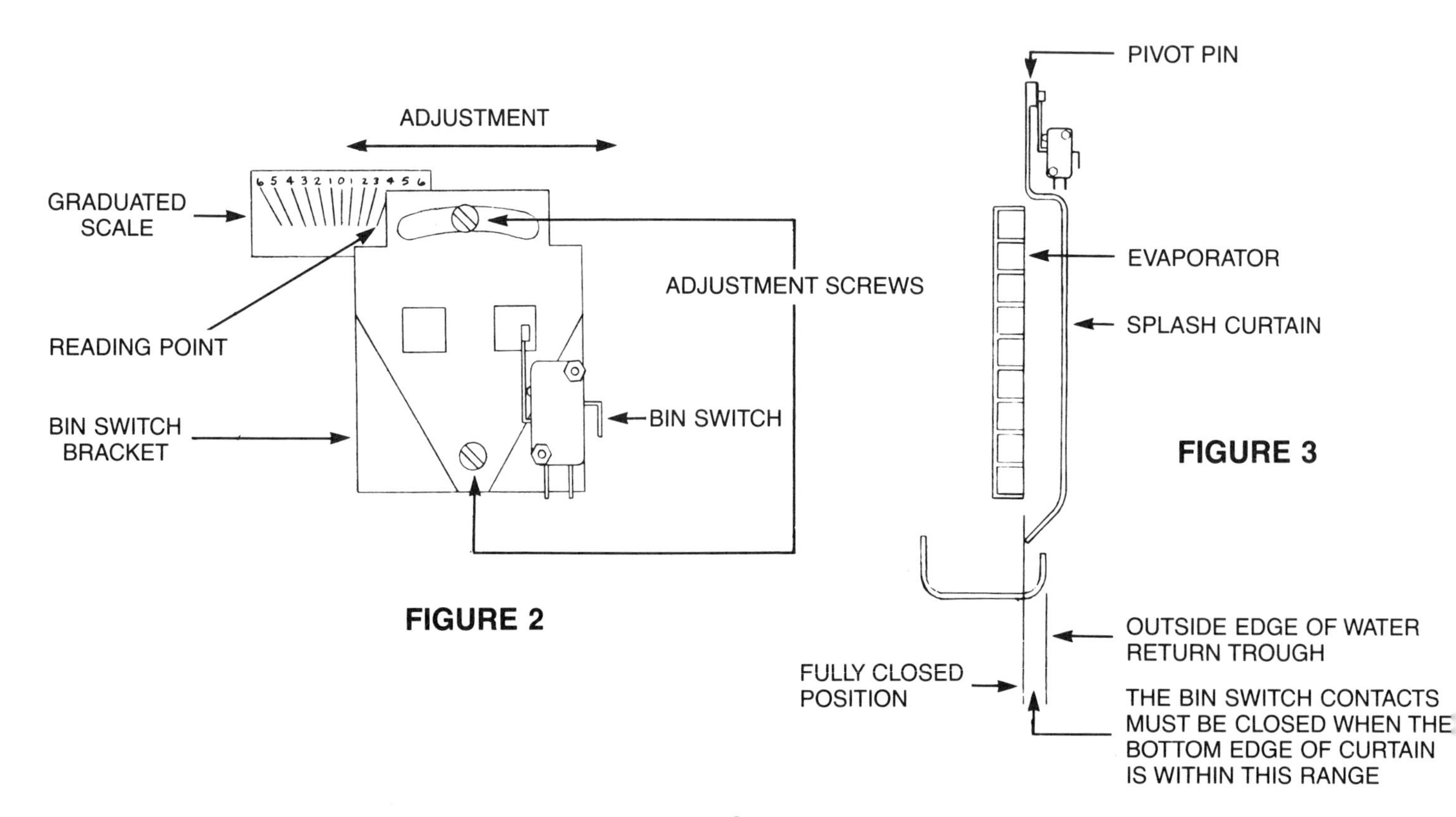

FIGURE 2

FIGURE 3

Description of Devices Connected to the Electronic Controller

Refer to diagram on page C-8.

External Input Devices

The following are devices external to the controller which have inputs to the controller board. The input signals that the external devices provide to the controller board influence and determine the sequence and timing of the ice making cycle.

Note: The external input devices will be referred to hereafter by the term shown in parentheses.

1. Evaporator Outlet Temperature Thermistor (Evap)
2. Condenser Temperature Thermistor (Cond)
3. Ice-Off-Wash Toggle Selector Switch (Ice/Wash)
4. Momentary Purge Switch (Purge Sw)
5. Door Switch #1 (Door 1)
6. Door Switch #2 (Door 2)
7. Time Delay Switch (Six Segment Time Delay Switch)
8. Momentary Reset Switch (Summary/Reset Switch)
9. Cam Switch #1 (Cam 1) (Probe 1)
10. Cam Switch #2 (Cam 2) — only on C-81/121 (Probe 2)
11. Stacking Connection (Stack 1 and Stack 2)

Note: Fan Controls are present on machines with electronic controllers that do not have single/ dual evaporator switches. Fan operation is regulated by the electronic controller if it has a single/dual evaporator switch.

External Controlled Devices

The following are devices external to the controller which are directly regulated by the controller board. The controlled devices are activated only when they receive power based on decisions made by the electronic controller.

Note: The external controlled devices will be referred to hereafter by the term shown in parentheses.

1. Hot Gas Solenoid (Hot Gas)
2. Water Pump (Pump)
3. Compressor (Comp)
4. Purge Solenoid (Purge)
5. Harvest Motor #1 (Motor 1)
6. Harvest Motor #2 — only on C-81/121 (Motor 2)
7. Machine 'OK' green LED (Ok LED)
8. Machine 'Harvest' yellow LED (Har LED)
9. 'Bin Full' green LED (Full LED)
10. Machine malfunction 'Not OK' red LED (Service LED)
11. LED Digital Readout (Readout)

DESCRIPTION OF DEVICES
(Continued)

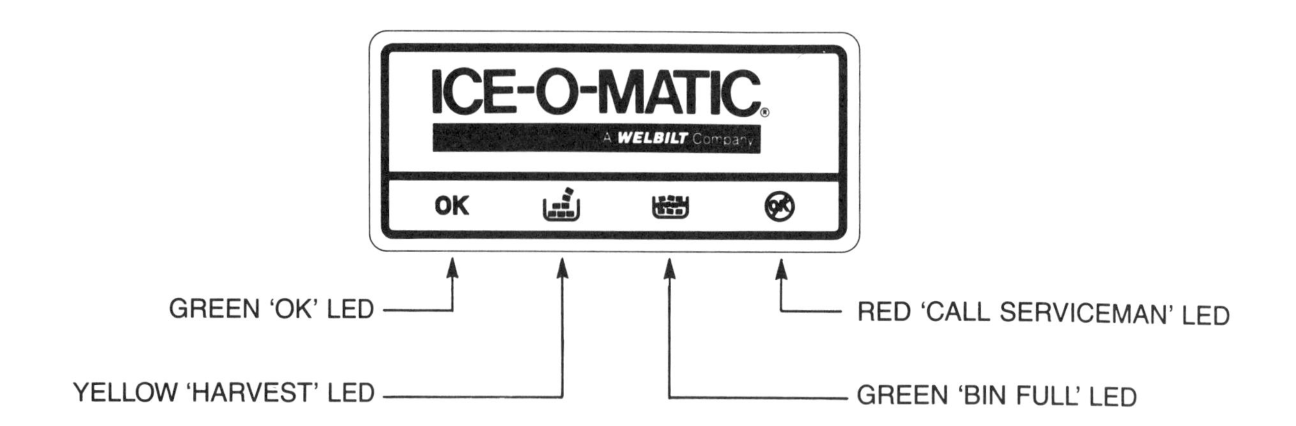

STATUS INDICATOR LEDS

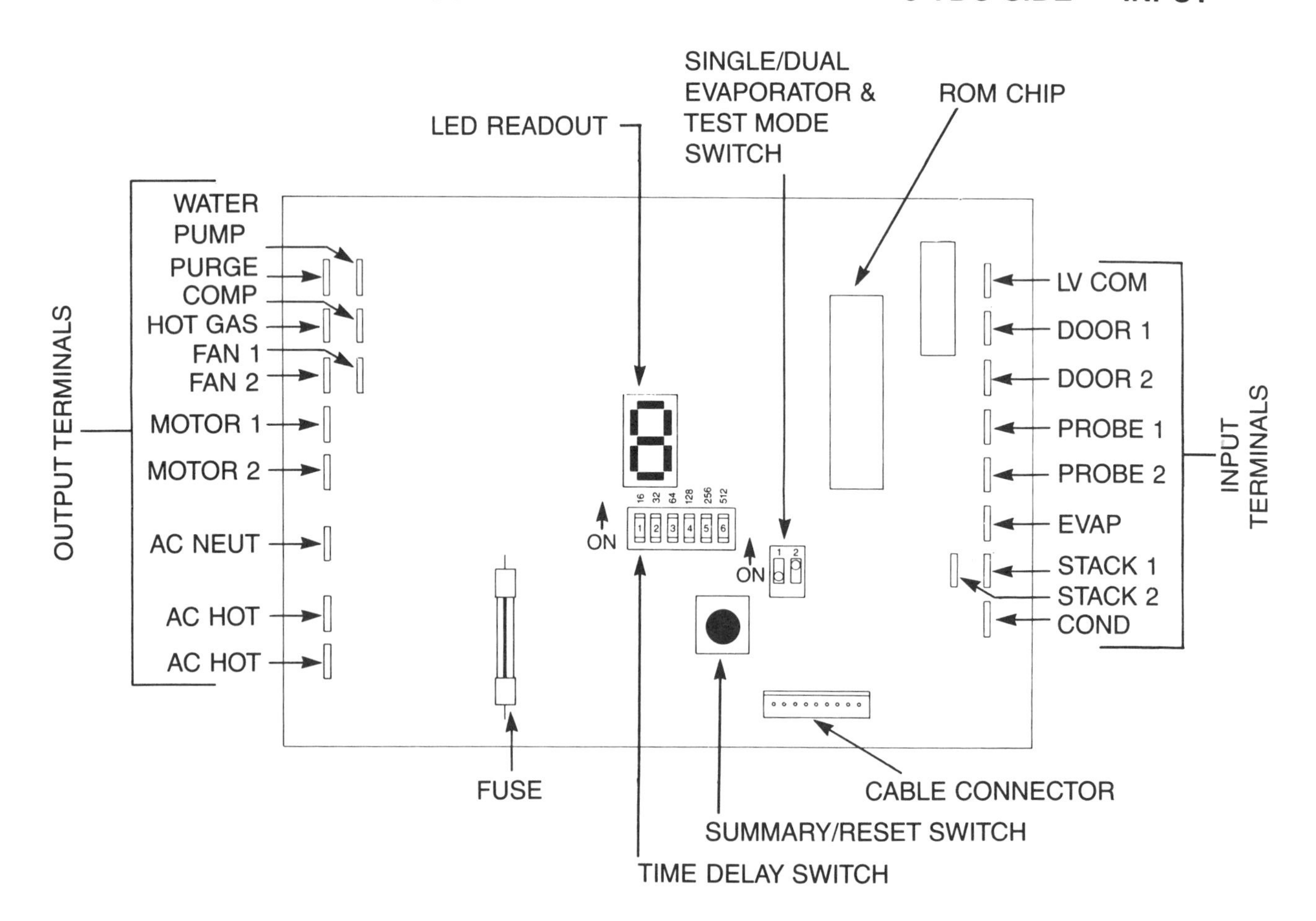

ELECTRONIC CONTROLLER CIRCUIT BOARD

MAINTENANCE INFORMATION

ROUTINE MAINTENANCE PROCEDURES

To insure economical, trouble-free operation of your machine, it is recommended that the following procedures be followed every 3 months.

1. Check the Cam Switch Setting.
2. Check to see that the T.X. Valve Bulb is securely fastened.
3. Check for leaks of any kind, water refrigerant, oil, etc.
4. Check the Bin Control, it should shut the machine off.
5. Check all electrical connections.
6. Clean the condenser to insure air passage across the fins.
7. Clean the system out with Ice Machine Cleaner. (See Cleaning Procedure.) NOTE: It may be necessary to clean the machine more often due to local water conditions.

CLEANING PROCEDURES

1. Turn main electrical service switch to the ice maker "OFF".
2. Remove front panel, and turn the ON/OFF/WASH switch to the "OFF" position.
3. Remove all ice from the ice bin.
4. Depress float valve and allow water trough to fill.
5. Add four ounces of a regular liquid ice machine cleaner to water trough (not nickel safe).
6. Turn ON/OFF/WASH switch to the "WASH" position and re-install panel (safety precaution).
7. Allow the solution to circulate (15 minutes).
8. Push purge valve switch with on/off wash switch in wash position. This will flush ice machine cleaner out through drain. Allow water trough to fill with fresh water.
9. Repeat Step 1 and see that all visible deposits are removed from the evaporator. If not, repeat Steps 4 through 7.
10. If necessary, remove the water distribution tube assembly. Use a brush to remove any scale buildup inside either the inner or outer tube. Refer to page D-22 for water distribution tube assembly and disassembly instructions.
11. Repeat Step 1 and wipe out any scale deposits in the water tank.
12. Re-install the water distributor tube and wipe all areas of the ice machine evaporator and water trough areas with a sanitizing solution.
13. Wipe out the ice storage bin with warm, soapy water; rinse and sanitize.
14. Turn ON/OFF/WASH switch to the "ON" position and replace front panel.
15. Discard the first two batches of cubes.

NOTE: If above cleaning procedures are not performed every six (6) months, harvest problems may occur.

MAINTENANCE INFORMATION

WINTERIZING AND STORAGE PROCEDURE

1. Shut the water off to the machine.
2. Turn the On-Off Switch on the Control box to the “OFF” position.
3. Let the machine stand for one hour (or as long as necessary to melt all the ice out of the evaporator freeze plate.
4. Disconnect the tubing between the pump discharge and the water distributor manifold.
5. Drain complete system. Do not replace the tubes.
6. On water cooled machines, blow condenser completely dry.
7. Wipe out the storage bin.

SERVICE

CONTROLLER SELF-DIAGNOSTIC FUNCTION

The self-diagnostic function of the electronic controller will assist the Serviceman in troubleshooting and diagnosis of machine problems. The controller automatically and continuously monitors and tests itself as well as key controlled components in the ice machine. If the controller detects a malfunction or fault, an error code is identified and logged into the random access memory (RAM) of the controller. Certain malfunctions and corresponding error codes (detailed on pages D-5 through D-6) will cause the machine to shut down and the 'SERVICE' LED to illuminate. The specific malfunction that caused the machine to shut down can be determined by observing the illuminated error code that appears on the seven-segment LED on the controller board. Other less serious malfunctions may occur that do not shut down the machine, but their corresponding error codes will still be logged into the controller's memory. Regardless of whether or not the machine has been shut down by the controller, the nature of error codes can be displayed at any time. Also, the total number of ice harvests since power-up, power loss, or last reset (up to 65,000) can be displayed at any time.

DISPLAYING SUMMARY TEST MODE DATA: The error codes and/or number of harvests can be displayed at any time by momentarlly pressing the Summary/Reset switch on the controller board. This can be done regardless of the position of the selector switch and will not shut the machine down.

The summary test mode is entered by momentarily pressing the Summary/Reset switch. When the Summary/Reset switch is pressed, the LED display on the controller board will display error codes that have an accumulated total greater than zero and the number of harvests. This information is displayed as follows:

1. An 'E' will display for ½ a second.
2. There will be a ¼ second pause (LED display is blank).
3. A 'C' will display for ½ a second (indicating error codes to follow).
4. The error code number (1-8) will display for ½ a second.
5. There will be a 1 second pause (LED display is blank).
6. The controller will then go on to the next error code or the harvest count.
7. The harvest count will follow all error codes. An 'H' will display for ½ a second.
8. A '—' (dash) will display for ½ a second.
9. The number of accumulated harvests will display, one number at a time, each for ½ a second, with a ¼ second delay between digits. (Harvests will count to 65,000 and then hold until reset to '0' [zero].)
10. The readout will go blank for 1 second.
11. The readout will start showing 'evap temp' and 'cond temp'.

To reset the count on error codes, you must hold the Summary/Reset switch for a minimum of 5 seconds. At this point all accumulated error codes will reset to zero. To reset harvest counts the Summary/Reset switch must be held in for 10 seconds.

Power Failure Considerations

In the event of a power failure, voltage drops (more than 10%), or the ice machine is unplugged, all accumulated data in the RAM will be lost. Machine will remain off until power is resumed, with the controller resetting itself.

Test Mode Operation

The serviceman can put the electronic controller into a test mode to check the 115 volt outputs from the controller (purge, hot gas, etc.). To access the test mode, the procedure below must be followed:

1. With power being supplied to the ice machine, move the ICE/OFF/WASH selector switch to the OFF position. The LED display on the controller board will still display evaporator and condenser temperatures.
2. Remove all time from the time delay switch by moving all switches to the OFF position.
3. Depress and hold the Summary/Reset switch through Step 4.
4. Refer to figure below while observing positions of single/dual evaporator and test mode switches.

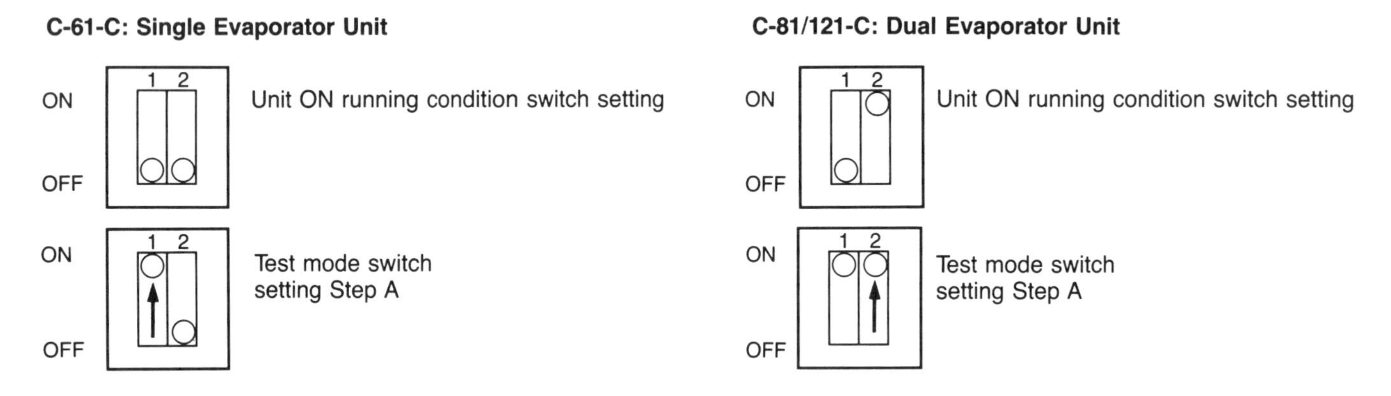

As shown above, with either single or dual evaporator machines, move switch 1 to ON position (Step A).

5. Release the Summary/Reset switch. The LED display on the electronic controller will stop displaying evaporator and condenser temperatures. The electronic controller is now in the test mode.

   Note: If the LED display continues to display evaporator and condenser temperatures, start again at Step 1.

6. Move the ICE/OFF/WASH selector switch to the ON position.

   The 115 volt outputs can now be checked individually by moving to ON each of the six segments on the time delay switch one at a time. The six switch segments control the output in the test mode as follows from left to right:
   1. Purge Solenoid
   2. Hot Gas Valve Solenoid
   3. Water Pump
   4. Fans (Air-cooled only)
   5. Compressor
   6. Harvest Motor (Right hand motor as viewed from front on C-81/121-C)

7. Connect either a multimeter or 115 volt test lamp from the 115 volt (AC) side component terminal being tested to the AC HOT terminal on the electronic controller. If component is to be energized, you will read 115 volts on your meter. (See Figure 1 below.)

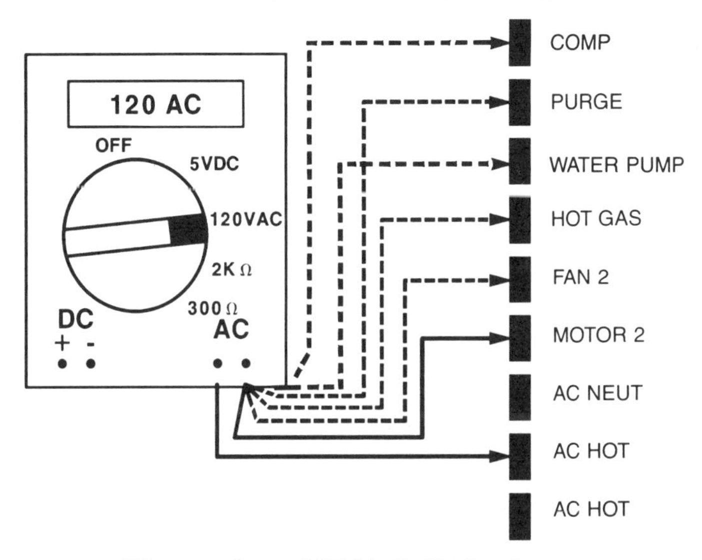

Figure 1 — 115 Volt Output

8. To bring the ice machine out of the test mode and resume normal operation, the following procedure must be followed:

   1. Move the ICE/OFF/WASH selector switch to the OFF position.

   2. Move switch 1 on Test Mode switch to OFF position (Step B).

   3. Depress the Summary/Reset switch momentarily.

   4. Reset the time delay switch to the time that was on it before entering the test mode.

   5. Move ICE/OFF/WASH selector switch to ICE position to resume normal ice-making operation.

   6. The ice machine is now out of the test mode and the LED display on the controller board will again display evaporator and condenser temperatures.

Error Code Definitions, Possible Causes and Possible Solutions

Note: All error codes are designated EC-1, EC-2, etc.

IMPORTANT: The error codes listed *assist* the serviceman in diagnosing and troubleshooting the ice machine. The error codes do *not* diagnose every problem that could occur. Refer to the complete Diagnosis and Troubleshooting Guide beginning on page D-7 for problems not addressed by the electronic controller error codes.

If LED is displaying appropriate data, microcomputer on electronic controller is functioning correctly.

Resetting Error Code and Harvest Count Memory

To clear and reset the type of Error Code, press the Summary/Reset switch on the electronic controller for a minimum of 5 seconds. All previous error codes will be erased and the controller will reset all accumulated data to zero.

To clear and reset the Harvest Count, press the Summary/Reset switch on the electronic controller for a minimum of 10 seconds. All previously accumulated data on harvest counts will be erased and reset to zero.

Once the controller memory has been cleared and reset, move the selector switch to ICE position, reinstall front panel, and resume the ice-making process. The icemaking process will now start at the top of the loop.

IMPORTANT NOTICE

ALL SOLID STATE ELECTRONIC CONTROLLER COMPONENTS ARE, IN GENERAL, SUSCEPTIBLE TO DAMAGE FROM STATIC ELECTRICITY, ELECTRICAL, MAGNETIC FIELDS AND POWER SURGES. WE AT THE FACTORY, TO PROTECT THE CONTROLLER, DESIGNED IN AS MANY SAFETY DEVICES AS POSSIBLE. TO REDUCE THE POSSIBILITY OF DAMAGE TO THE CONTROLLER, PLEASE SEE THE FOLLOWING.

A. BEFORE WORKING ON THIS UNIT, A PERSON SHOULD GROUND THEMSELVES, TO DISCHARGE STATIC ELECTRICITY (I.E. TOUCH THE UNIT'S GROUND SCREW, OR WEAR A U.L. APPROVED WRIST OR ANKLE GROUND STRAP).

B. NYLON OR OTHER STATIC GENERATING MATERIAL SHOULD NOT COME IN CONTACT WITH THE CONTROLLER OR ITS COMPONENTS.

C. IT IS STRONGLY RECOMMENDED THAT ALL ELECTRICAL EQUIPMENT BE KEPT AWAY FROM ANY PORTION OF THE CONTROL BOARD. (I.E. FAN MOTORS, HAND DRILLS, ETC.)

D. IT IS STRONGLY RECOMMENDED THAT ALL MAGNETIC DEVICES BE KEPT AWAY FROM THE CONTROL BOARD AND ITS COMPONENTS. (I.E. SCREWDRIVER, HAND WRENCHES, ETC. WITH MAGNETIC HEADS.)

E. WHEN HANDLING THE CONTROL BOARD, PLEASE DO SO BY THE CORNERS. IF YOU MUST HANDLE A COMPONENT, IT IS STRONGLY RECOMMENDED THAT YOU GROUND YOURSELF AS PER ITEM "A" ABOVE.

If replacing a controller board, make certain it remains in its protective plastic bag to prevent damage to it.

ERROR CODES & DIAGNOSIS

EC-1 — Freeze cycle exceeds 50 minutes, or the evaporator(s) fail to reach 40°F in 6 minutes. Machine does not shut down. 'SERVICE' LED does not illuminate. EC-1 will display only when the summary switch is depressed.

Possible Cause	Possible Solution
— Low on refrigerant charge	— Check for leaks, evacuate & recharge unit
— Defective compressor	— Check and/or replace compressor
— Dirty air condenser	— Clean air condenser
— Blocked air flow	— Check for and remove obstruction
— Defective evaporator thermistor	— Check and/or replace thermistor

EC-2 — Both cam switches do not activate and deactivate within 45 minutes of harvest initiate. Machine is shut down and red 'SERVICE' LED is illuminated.

Possible Cause	Possible Solution
— Stuck or bent harvest probe	— Adjust harvest probe assembly or replace harvest probe
— Defective harvest assist motor	— Check and/or replace harvest assist motor
— Cam switch out of adjustment or defective	— Check and/or replace cam switch

EC-3 — Splash curtain(s) do not open within 5 minutes after harvest initiate. Machine does not shut down. 'SERVICE' LED does not illuminate. EC-3 will display only when the summary switch is depressed.

Possible Cause	Possible Solution
— Ice slab hung-up in evaporator	— Free ice slab and check adjustment of harvest probe assembly
— Poor harvest probe alignment or bent probe	— Adjust harvest probe assembly to proper alignment and/or replace harvest probes
— Defective harvest assist motor	— Check and/or replace harvest assist motor
— Defective hot gas valve	— Check and/or replace hot gas valve
— Curtain switch shorted	— Check and/or replace curtain switch
— Curtain switch open	— Check and/or replace curtain switch
— Poor curtain switch alignment	— Adjust curtain switch to proper alignment
— No ice slab is made	— Check that water pump is functioning properly
— Thin ice slab, probe produces holes in or breaks slab	— Check for proper water circulation over evaporator, check that machine is level, add time to time delay switch.

EC-4 — All of controller read-only memory (ROM) lost. Machine is shut down and red 'SERVICE' LED is illuminated.

Possible Cause	Possible Solution
— Memory loss	— Replace controller board

EC-5 — Condenser thermistor malfunction. Machine is shut down and 'SERVICE' LED is illuminated.

Possible Cause	Possible Solution
— Defective condenser thermistor	— Check and/or replace thermistor

EC-6 — Evaporator thermistor malfunction. Machine is shut down and 'SERVICE' LED is illuminated.

Possible Cause	Possible Solution
— Defective evaporator thermistor	— Check and/or replace thermistor

EC-7 — Air or water condenser temperature exceeds 150°F. Machine will shut down and red 'SERVICE' LED will be illuminated.

Possible Cause	Possible Solution
— Stuck or defective fan (air cooled unit)	— Check that fan turns freely and/or replace fan
— Defect condenser water valve (water cooled unit)	— Check and/or replace condenser water valve
— Surrounding (ambient) air temperature exceeds 120°F	— Wait until ambient air temperature decreases to start machine
— Dirty air condenser	— Clean air condenser
— Defective condenser thermistor	— Check and/or replace thermistor

EC-8 — Freeze Cycle exceeds 80 minutes. Machine is shut down and red 'SERVICE' LED is illuminated.

Possible Cause	Possible Solution
— Low on refrigerant charge	— Check for leaks, evacuate and recharge unit
— Defective compressor	— Check and/or replace compressor
— Defective evaporator thermistor	— Check and/or replace thermistor

GENERAL SERVICE INFORMATION

SERVICE DIAGNOSIS

Condition	Possible Cause or Remedy
1. Unit runs but no ice production.	a. Water shut off. b. Hot gas valve defective. c. Unit out of gas. d. Motor compressor not pumping e. ON/OFF/ON switch on "WASH" position. f. Refrigerant drier plugged. g. TX valve defective. h. Defective probe assist motor. i. Cam switch defective or not tripping. j. Water pump defective k. Thermistor defective. l. Defective electronic controller. m. Low line voltage.
2. Low ice production	a. Dirty condenser b. Valves in compressor not functioning properly c. Refrigeration system under or over charged d. Leak in water circulating system. e. Hot gas valve not seating properly. f. Evaporator in need of cleaning.
3. Dimples in ice too large.	a. Time not properly set b. Problem with refrigerant feed. (Low on charge, T.X. valve defective, dirty condenser, etc.
4. Cloudy ice.	a. Water system in need of cleaning. b. Water pump not pumping properly. c. Purge drain plugged up. d. Purge valve not opening.
5. Machine fails to shut off when bin is full.	a. Bin switch defective or out of adjustment.
6. Machine is noisy.	a. Defective compressor b. Loose components c. Fan hitting shroud. d. Defective purge valve.

GENERAL SERVICE INFORMATION

SERVICE DIAGNOSIS

(CONTINUED)

Condition	Possible Cause or Remedy
7. Probe assist motor runs continuously.	a. Cam switch defective or needs adjustment.
8. Machine will not go into harvest cycle.	a. Defective hot gas solenoid. b. Defective probe assist motor. c. Defective thermistor. d. Defective electronic controller.
9. Machine in defrost too long.	a. System low on refrigerant b. Defective probe assist motor. c. Head pressure setting too low. d. Probe assist assembly needs adjustment.
10. Compressor cuts out on overload.	a. Low line voltage. It should be within 10% of rated voltage. b. High head pressure. Dirty condenser. c. Defective compressor unit, starting capacitor relay or overload device. d. Loose electrical connection, probably in compressor junction box.
11. Motor compressor runs but condenser fan does not.	a. Loose electrical connections. b. Fan blade cannot turn due to obstruction. c. Motor burned out. d. Fan control inoperative.
12. Machine not operating at all.	a. Check voltage supply. b. Check ON/OFF switch. c. Check bin control with continuity tester. d. Check for loose connections. e. Defective wiring harness. f. Defective contactor. g. Out of refrigerant. h. Water supply shut off (water cooled).

GENERAL SERVICE INFORMATION

SERVICE DIAGNOSIS

(CONTINUED)

Condition	Possible Cause or Remedy
13. High head pressure.	a. Dirty condenser. b. System is overcharged. c. Air in refrigeration system. d. Defective fan motor.
14. Low suction pressure.	a. Restricted flow through filter-drier. b. System is low on refrigerant. c. Moisture in refrigeration system. d. Water distribution restricted. e. Defective T.X.V.
15. Ice slabs break when moving out of evaporator.	a. Bridging across front of ice slab too thin. Must be a minimum of 1/8" thick for proper harvest. b. Probe shaft is loose and has worked out into evaporator. c. Dirty evaporator (see page C-9 for cleaning instructions).
16. Water dripping into ice storage area from evaporator plate.	a. Cam switch out of adjustment. b. Dirty evaporator (see page C-9 for cleaning instructions) c. Bin control switch out of adjustment (see bin control switch adjustment page C-6)

STEP-BY-STEP WIRING DIAGRAM DESCRIPTION

The following machine conditions are assumed:

1. Ice/Off/Wash selector switch is in Ice position.
2. Bin control switches are in closed position.

Freeze Cycle — Step 1.

The cycle begins when the electronic controller signals that the machine is at the top of the loop. Throughout Step 1 the following components are energized through the electronic controller board: water pump, fan motor(s) and contactor coil. The compressor is energized through the contactor. The green 'OK' LED on the machine front panel is illuminated.

Freeze Cycle — Step 2.

When the suction line temperature drops to 14°F, the evaporator thermistor energizes the time delay on the electronic controller board. The machine remains in the freeze cycle for the amount of time set on the time delay. Throughout Step 2 the water pump, fan motor(s) contactor coil and compressor remain energized. The green 'OK' LED is illuminated.

Freeze Cycle — Step 3.

When the time remaining on the time delay reaches 12 seconds on a C-61 and 20 seconds on a C-81/121, the electronic controller energizes the water purge valve solenoid. The machine will now purge (empty) the water from the water trough. Throughout this process the machine is still in the freeze mode. Throughout Step 3 the water pump, fan motor(s), contactor and compressor remain energized. The green 'OK' LED is illuminated.

Harvest Cycle — Step 4.

When the harvest cycle begins, the water pump, fan motor(s) and purge valve are shut off by the electronic controller. The contactor coil, compressor, hot gas solenoid and harvest assist motor(s) are energized and the yellow 'Harvest' LED is illuminated.

As the ice is pushed off the evaporator plate(s) by the harvest assist motor(s) the bin switch(es) will momentarily open. The open bin switch(es) will momentarily illuminate the green 'Bin Full' LED.

After the harvest assist motor(s) have made a complete rotation and the ice has cleared the evaporator plate(s), the cam switch(es) return to the N.C. position and terminate the harvest cycle.

The machine will begin another freeze cycle from the top of the loop and the green 'OK' LED will be illuminated. Should the bin switch(es) remain open for more than 10 seconds, the machine will shut off and the 'Bin Full' LED will remain illuminated.

NOTE: On C-61-C only — Machine has one fan motor, one harvest assist motor, one bin switch and one cam switch.

STEP-BY-STEP WIRING DIAGRAMS
C-61-C

Freeze Cycle — Step 1

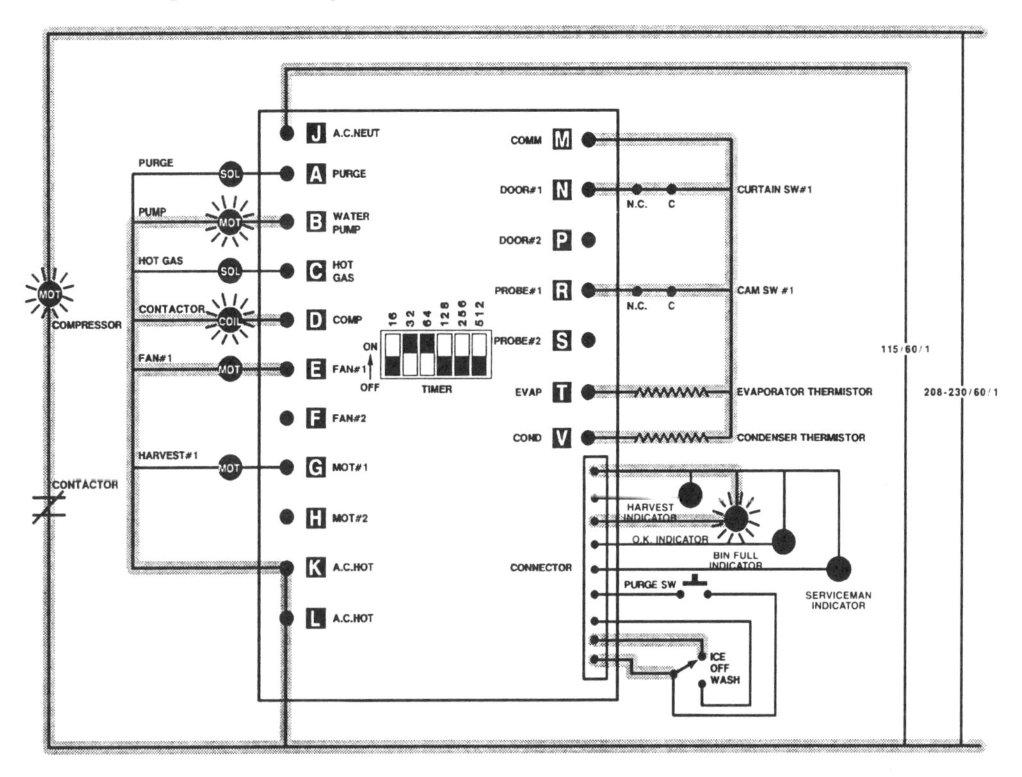

Freeze Cycle — Step 2

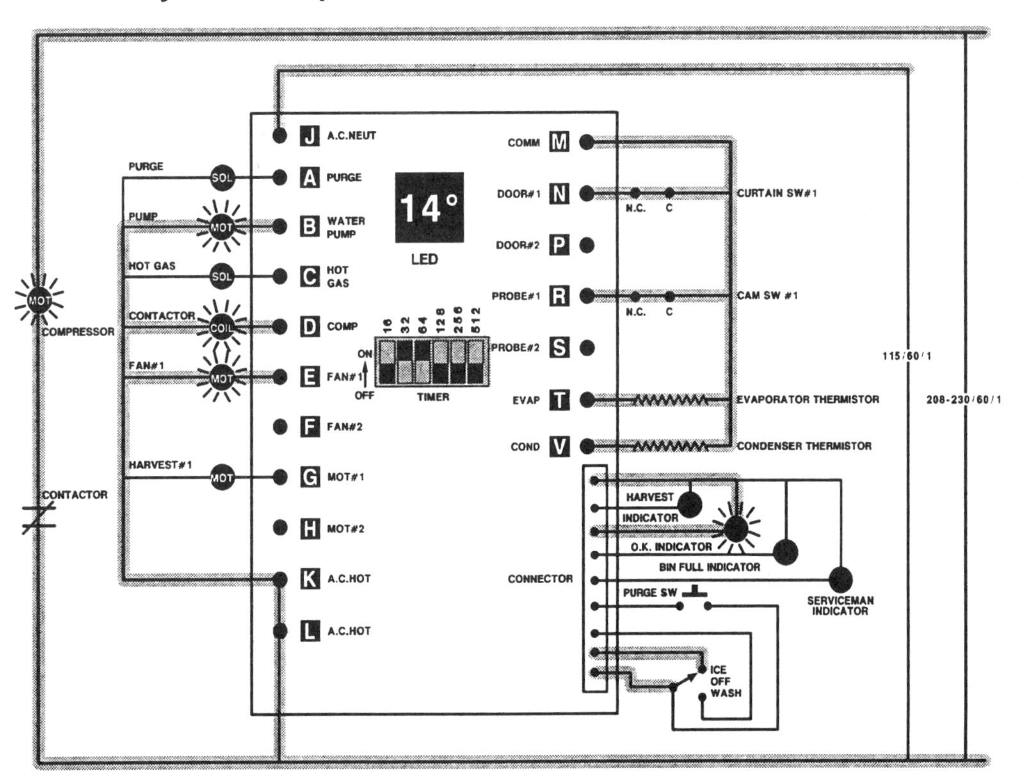

STEP-BY-STEP WIRING DIAGRAMS
C-61-C

Freeze Cycle — Step 3

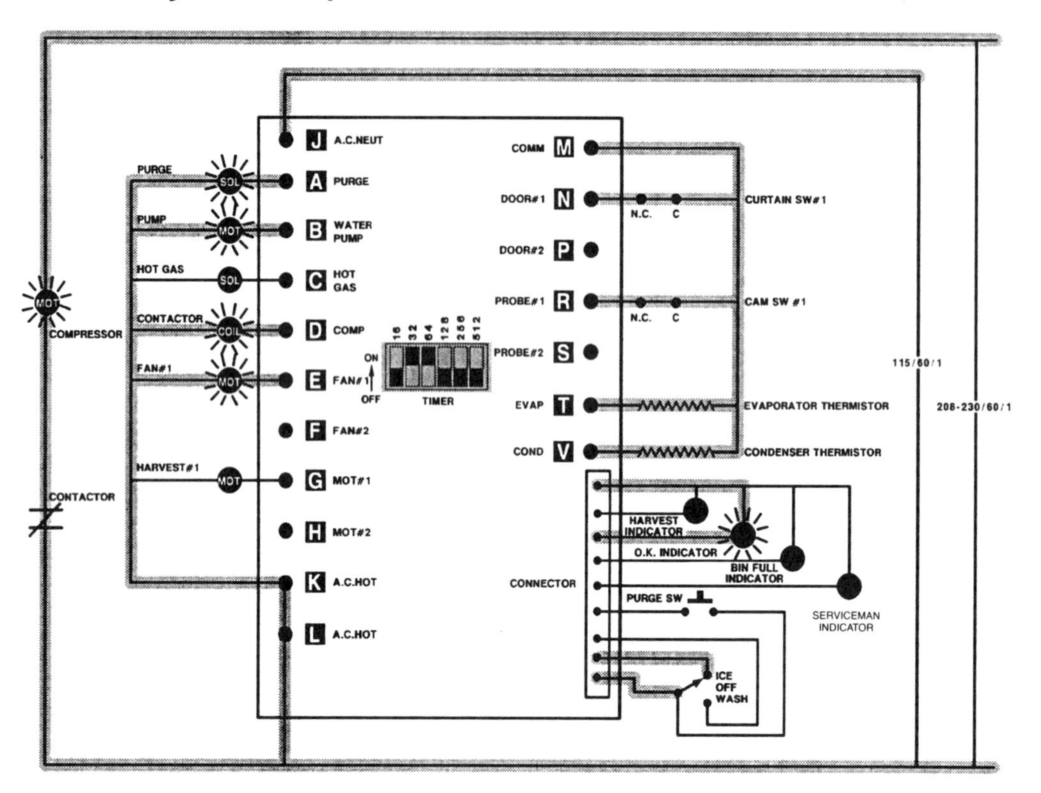

Harvest Cycle — Step 4

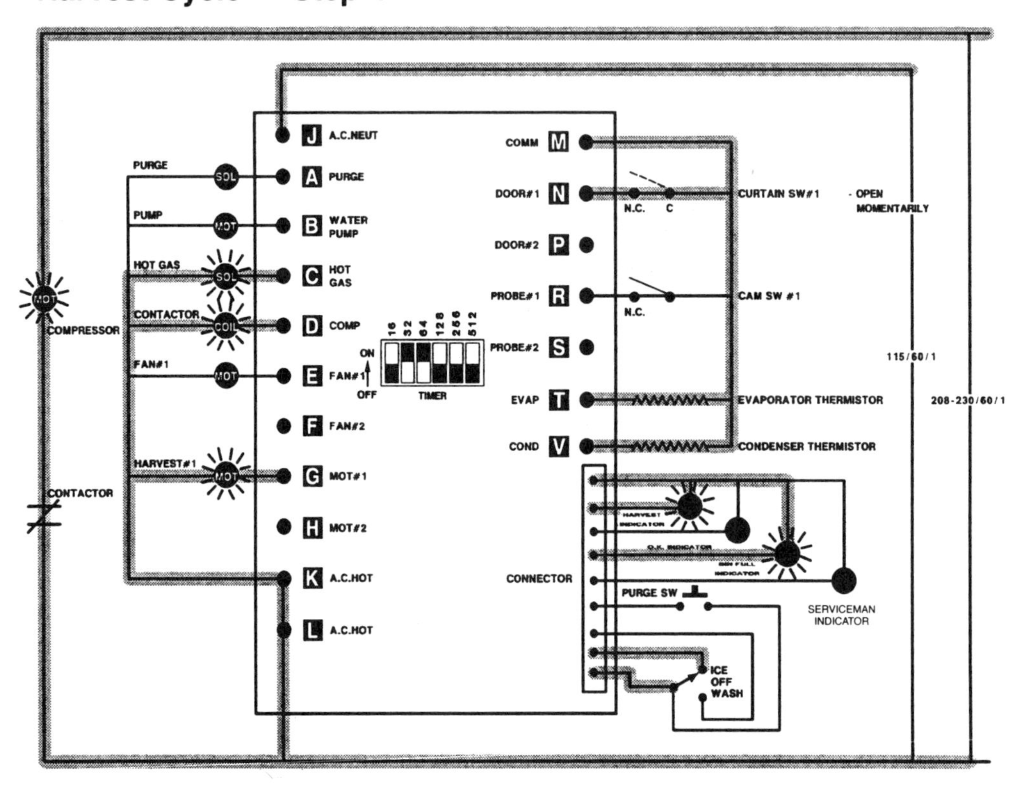

STEP-BY-STEP WIRING DIAGRAMS
C-81/121-C

Freeze Cycle - Step 1

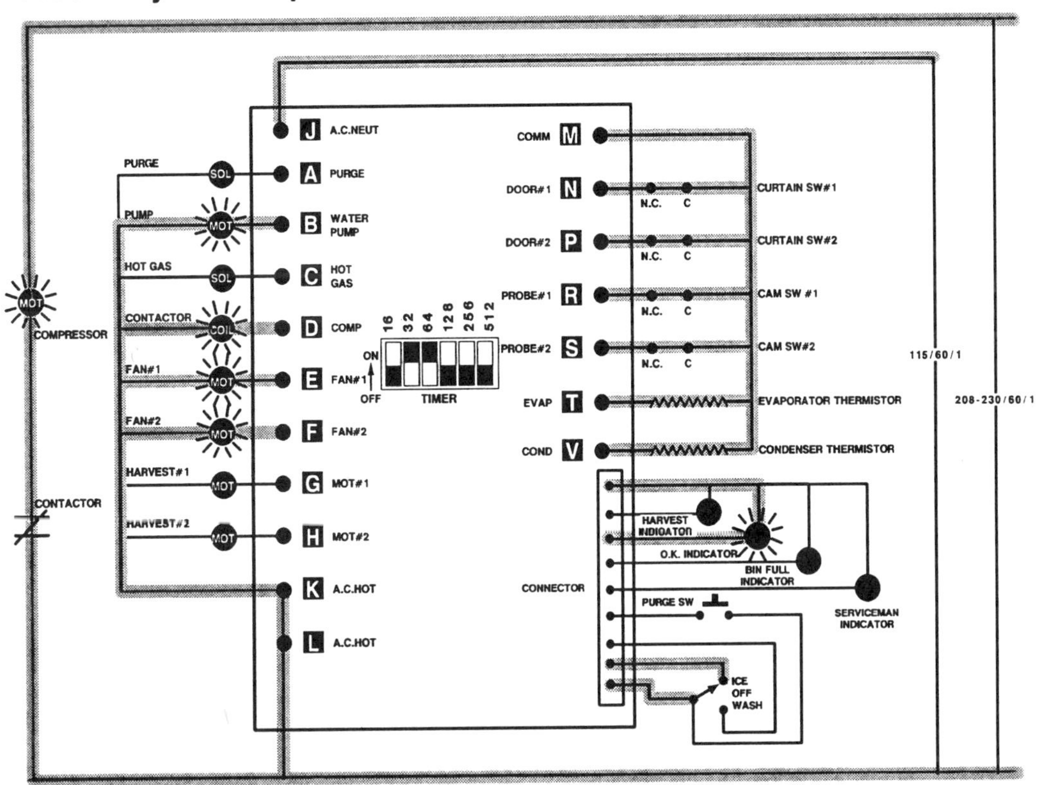

Freeze Cycle — Step 2

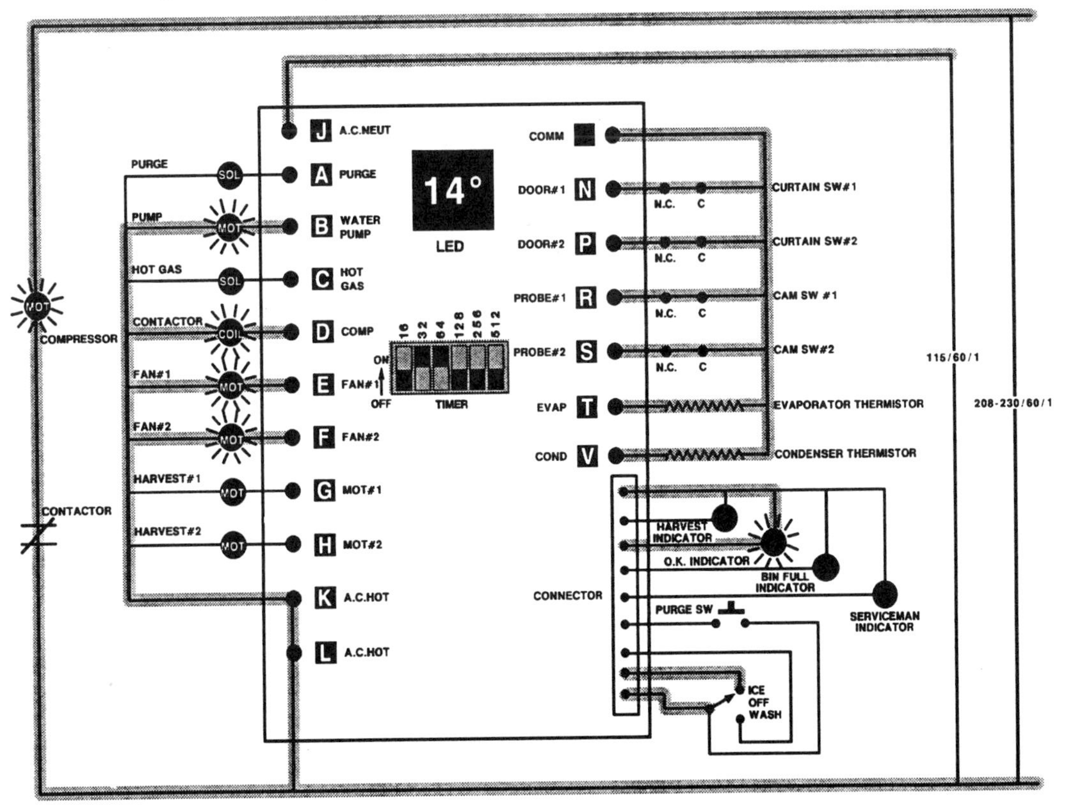

STEP-BY-STEP WIRING DIAGRAMS
C-81/121-C

Freeze Cycle — Step 3

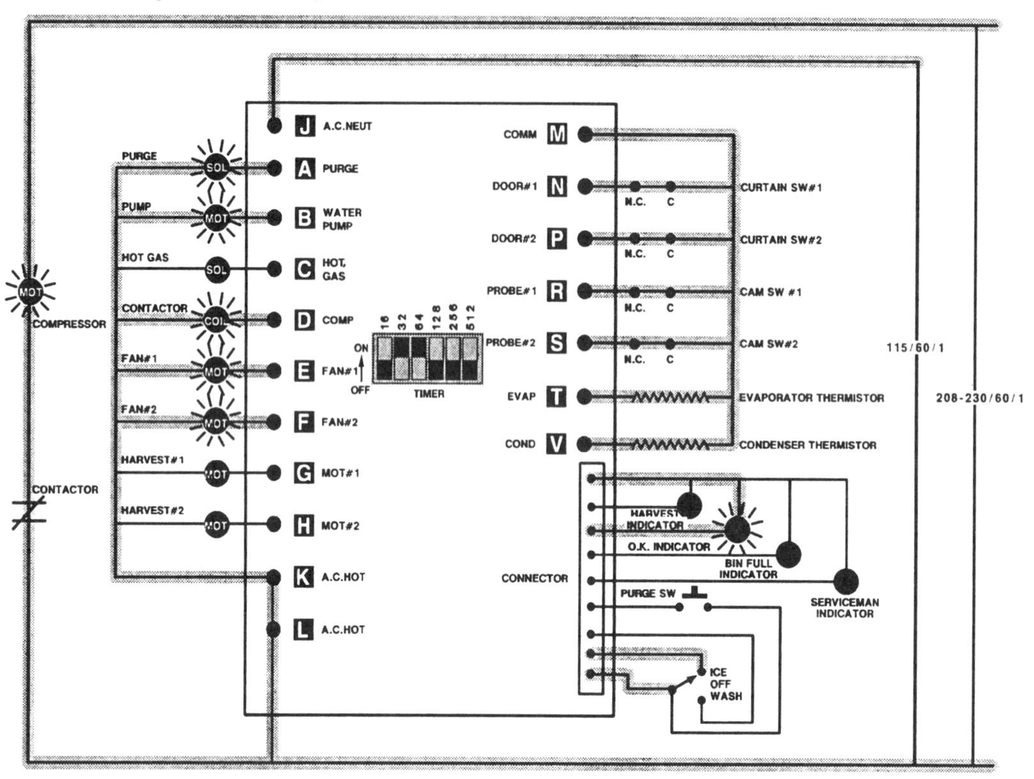

Harvest Cycle — Step 4

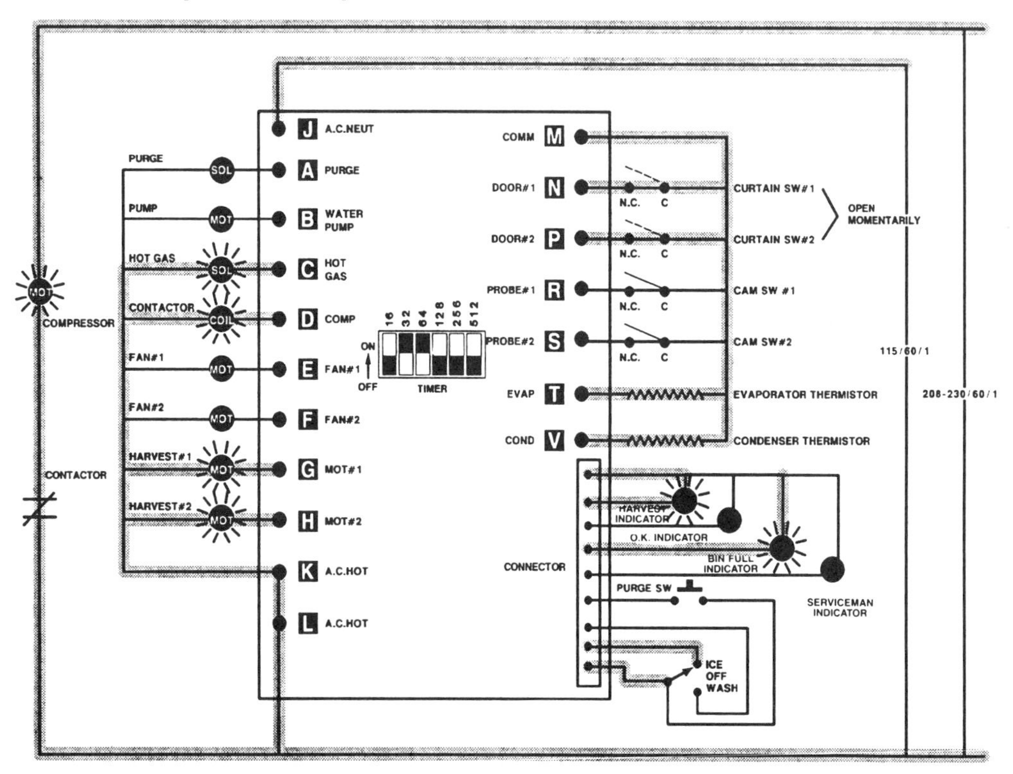

THERMISTOR DIAGNOSIS

The evaporator and condenser thermistors continuously sense temperature and send a DC voltage signal to the electronic controller board. Thermistor temperatures are translated by the controller board and appear on the LED display.

To determine if the thermistors are functioning properly and the electronic controller is receiving the correct DC voltage signal from the thermistors according to the temperature displayed on the LED, the following procedure must be followed:

NOTE: The evaporator and condenser thermistors can be checked at any time while the machine is running. The only piece of equipment required is a multimeter that can measure low range DC voltages.

1. Be certain machine is running.
2. Set multimeter to low DC voltage range (no lower than 5 volts DC).
3. Place the black multimeter lead on the terminal labeled COMM on the 5 volt DC (input) side of the controller board. Refer to Figure 1.
4. To check the evaporator thermistor, place the red multimeter lead on the terminal labeled EVAP on the 5 volt DC side of the controller board. Refer to controller board drawing on page C-8.
5. The multimeter should now display a DC voltage.
6. Observe the voltage on the multimeter and the temperature on the LED display on the controller board. Compare these observed figures with evaporator figures listed in the voltage/temperature thermistor chart on page D-16.

   Example: Checking an evaporator thermistor. 1.70 volts DC is observed on the multimeter and 30° is observed on the LED display on the electronic controller. Referring to the voltage/temperature thermistor chart shows that the evaporator thermistor is functioning properly.

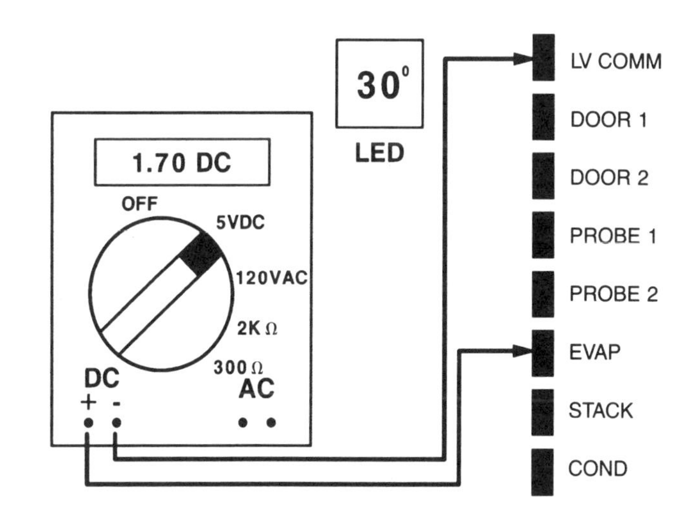

Figure 1 — Evaporator Thermistor Diagnosis

7. To check the condenser thermistor, leave the black multimeter lead on the COMM terminal. Place the red multimeter lead on the terminal labeled COND on the 5 volt DC side of the controller board. Refer to controller board drawing on page C-6.
8. Follow steps 5 and 6 from above. Compare observed figures with condenser figures listed in the voltage/temperature thermistor chart on page D-16.

   Example: Checking a condenser thermistor. 2.30 volts DC is observed on the multimeter and 90° is observed on the LED display on the controller board. Referring to the voltage/temperature thermistor chart shows that the condenser thermistor is functioning properly.

VOLTAGE/TEMPERATURE THERMISTOR CHART

EVAPORATOR THERMISTOR

DC VOLTS	°F TEMP	DC VOLTS	°F TEMP
.00	OPEN	2.60	6.08
.10	170.56	2.70	3.42
.20	133.70	2.80	0.52
.30	113.35	2.90	− 2.65
.40	99.50	3.00	− 5.12
.50	88.90	3.10	− 7.13
.60	80.32	3.20	− 9.37
.70	73.09	3.30	−11.87
.80	66.66	3.40	−14.68
.90	61.14	3.50	−17.87
1.00	56.20	3.60	−20.90
1.10	51.30	3.70	−23.86
1.20	47.38	3.80	−26.31
1.30	43.42	3.90	−29.21
1.40	39.67	4.00	−32.69
1.50	36.32	4.10	−36.94
1.60	32.78	4.20	−43.80
1.70	30.09	4.30	−44.35
1.80	27.54	4.40	−48.70
1.90	24.80	4.50	−54.80
2.00	21.92	4.60	−62.00
2.10	18.80	4.70	−69.50
2.20	15.49	4.80	−77.00
2.30	12.87	4.90	−86.00
2.40	10.78	5.00	SHORT
2.50	8.52		

CONDENSER THERMISTOR

DC VOLTS	°F TEMP	DC VOLTS	°F TEMP
.00	OPEN	2.60	79.87
.10	310.00	2.70	76.42
.20	253.00	2.80	73.31
.30	225.00	2.90	69.91
.40	205.20	3.00	66.60
.50	191.15	3.10	63.43
.60	179.55	3.20	59.92
.70	169.80	3.30	56.65
.80	158.33	3.40	53.23
.90	153.90	3.50	49.51
1.00	147.13	3.60	46.13
1.10	140.98	3.70	42.24
1.20	135.54	3.80	38.51
1.30	130.18	3.90	34.44
1.40	125.45	4.00	30.43
1.50	120.74	4.10	26.63
1.60	116.47	4.20	21.87
1.70	112.08	4.30	15.74
1.80	108.22	4.40	10.50
1.90	104.19	4.50	4.28
2.00	100.64	4.60	− 4.54
2.10	96.94	4.70	−12.65
2.20	93.41	4.80	−25.43
2.30	90.00	4.90	−44.67
2.40	86.37	5.00	SHORT
2.50	83.20		

C-61-C-AIR WIRING DIAGRAM

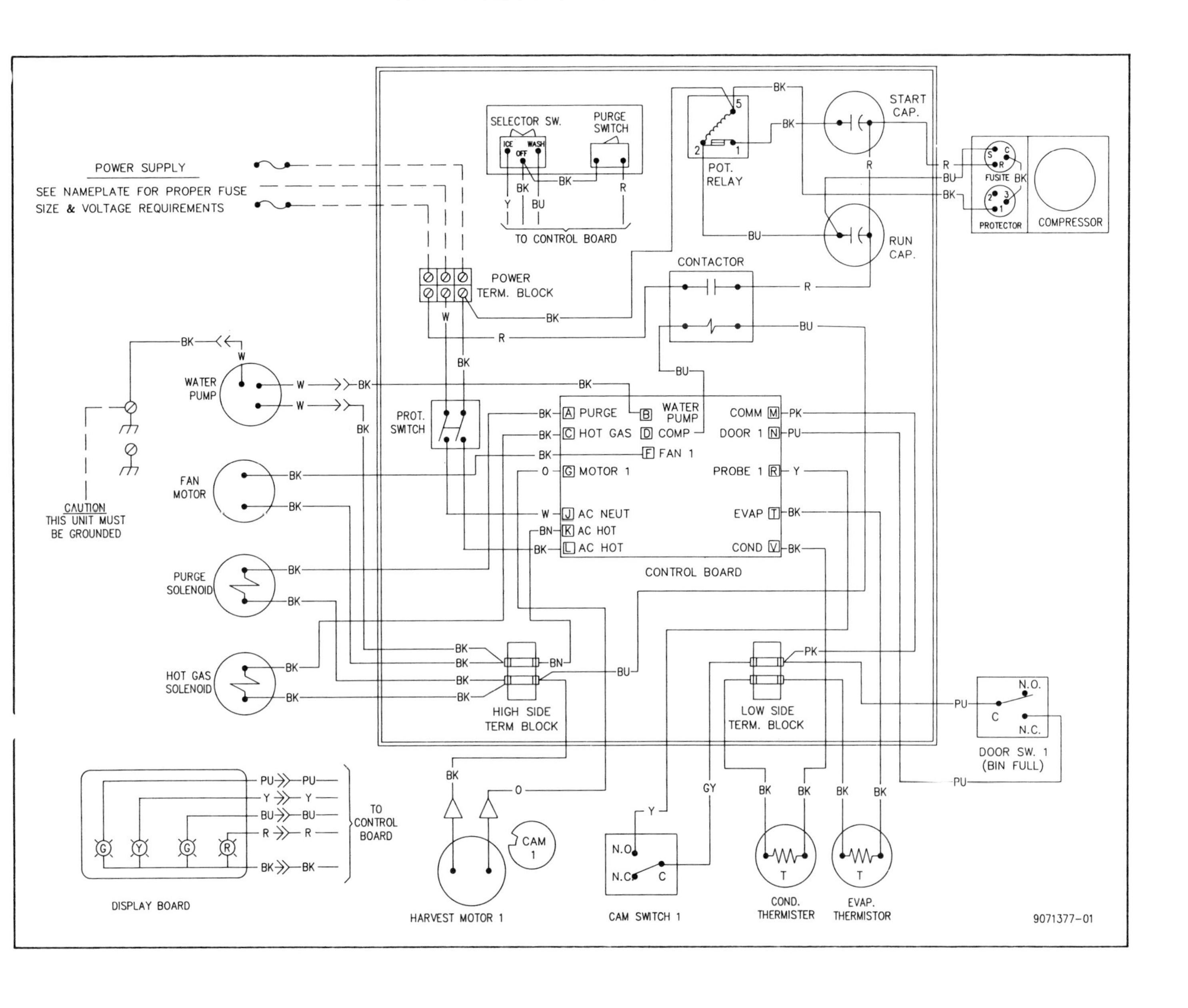

C-61-C-WATER WIRING DIAGRAM

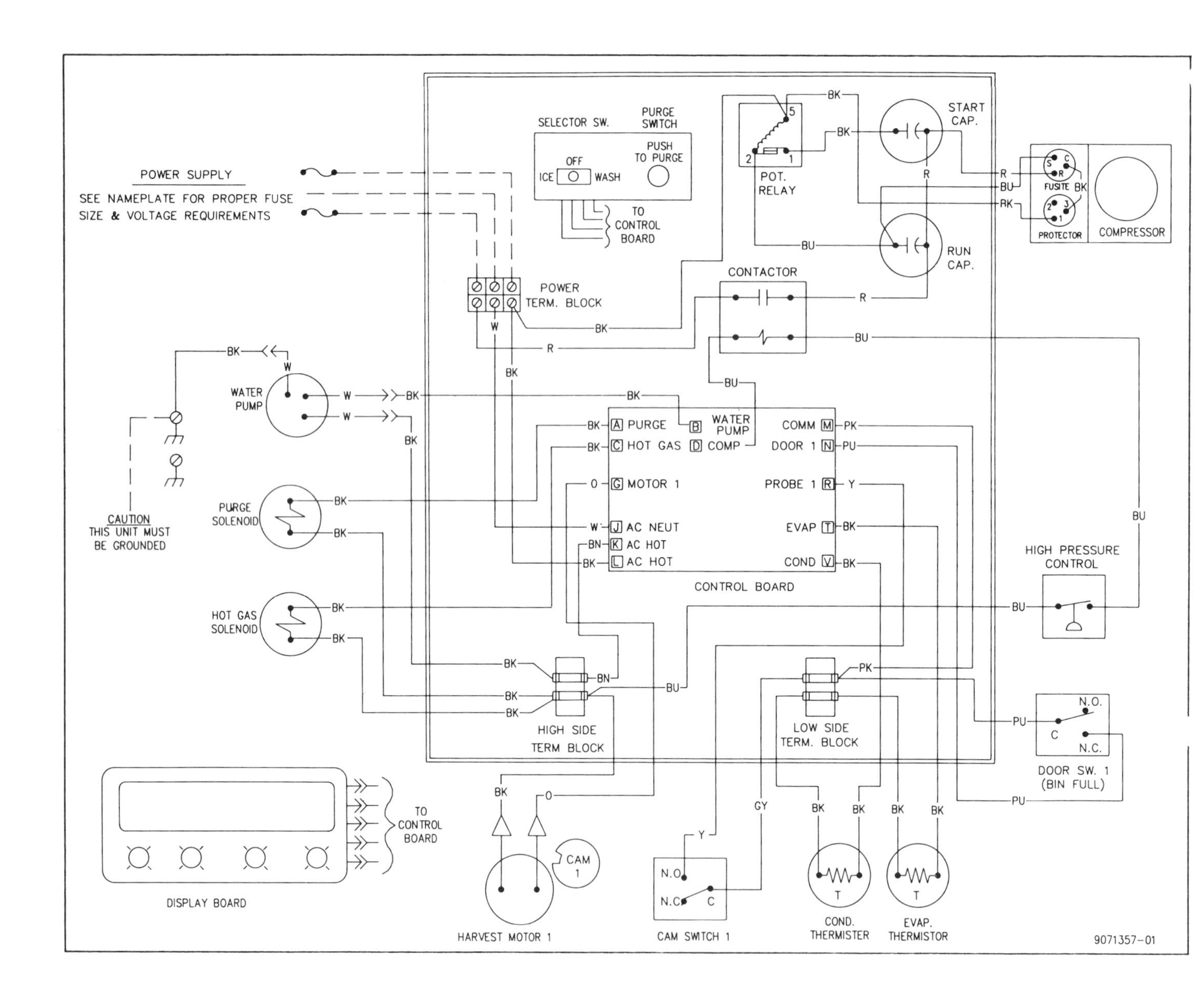

C-81/121-C-AIR WIRING DIAGRAM

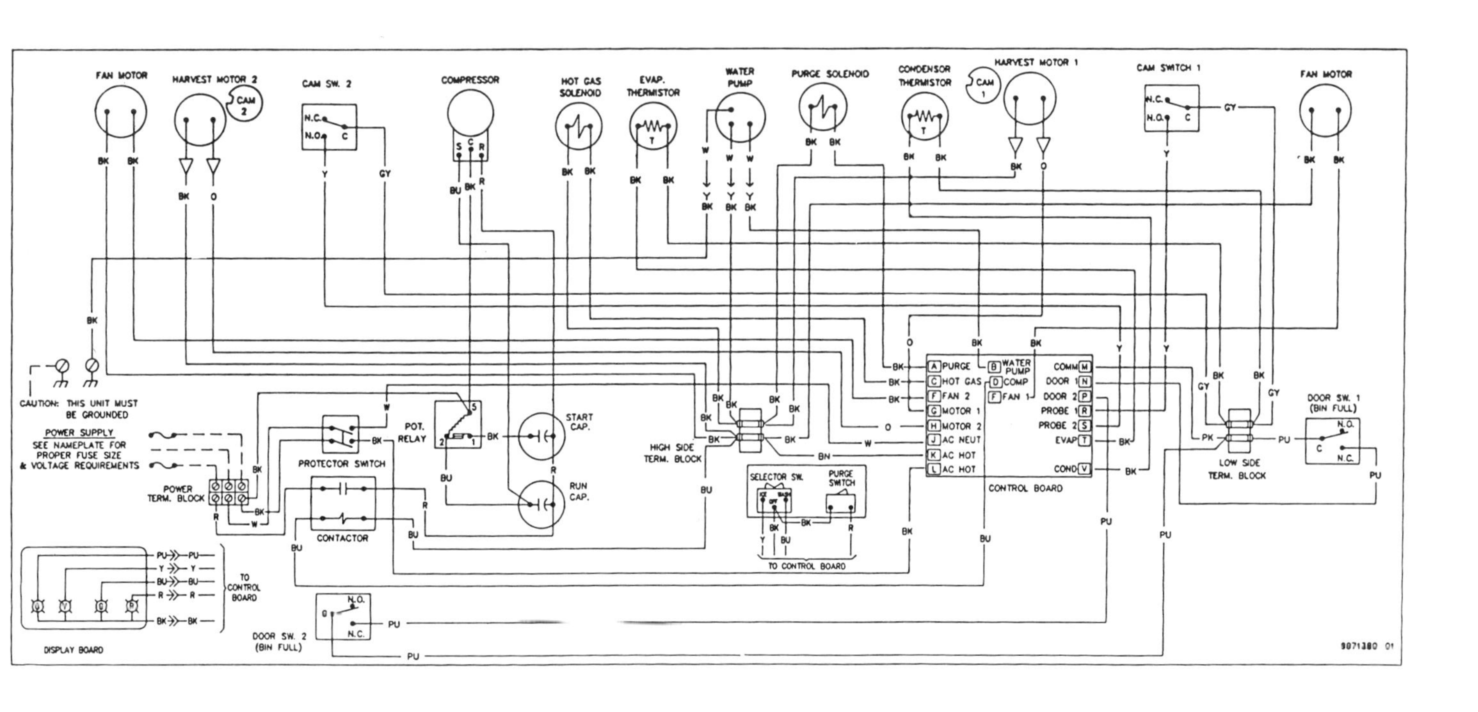

C-81/121-C-WATER WIRING DIAGRAM

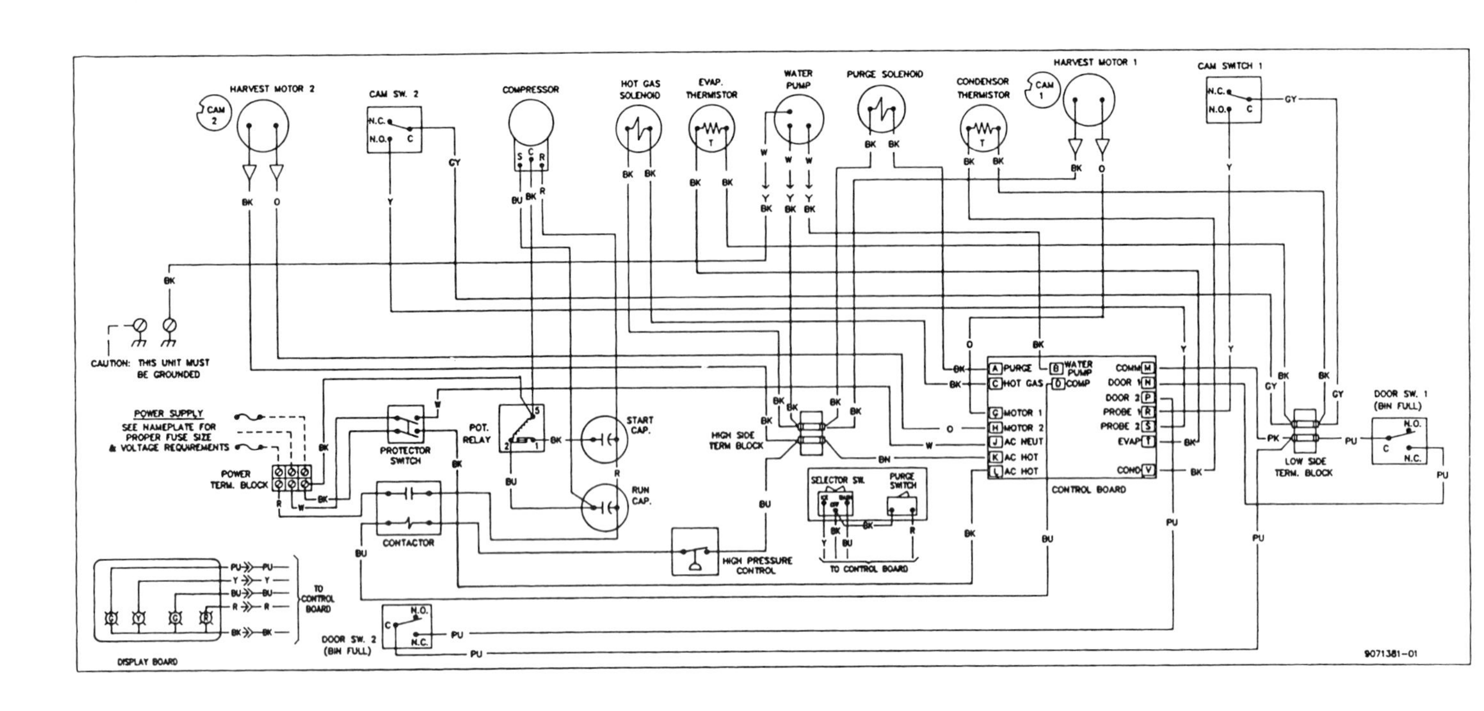

C-84/124-C-AIR & WATER WIRING DIAGRAM

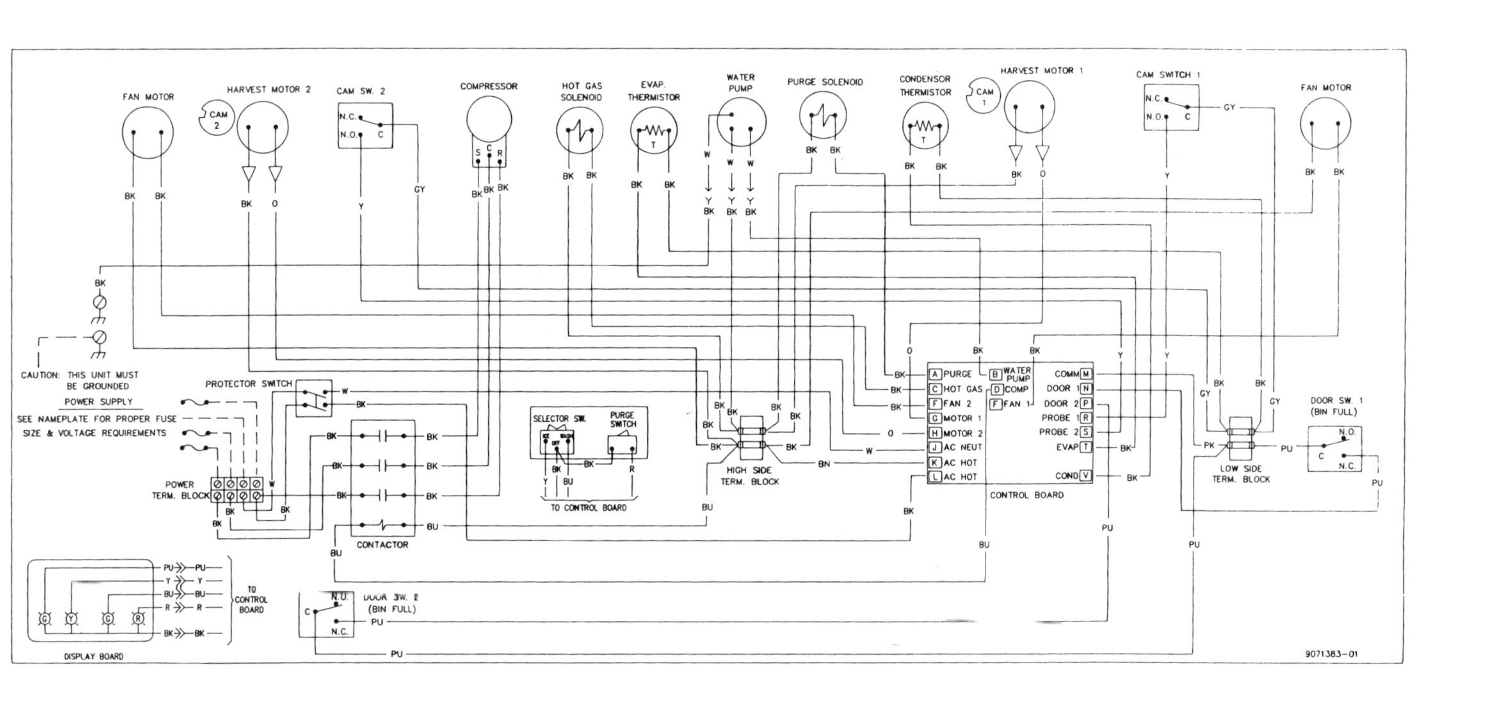

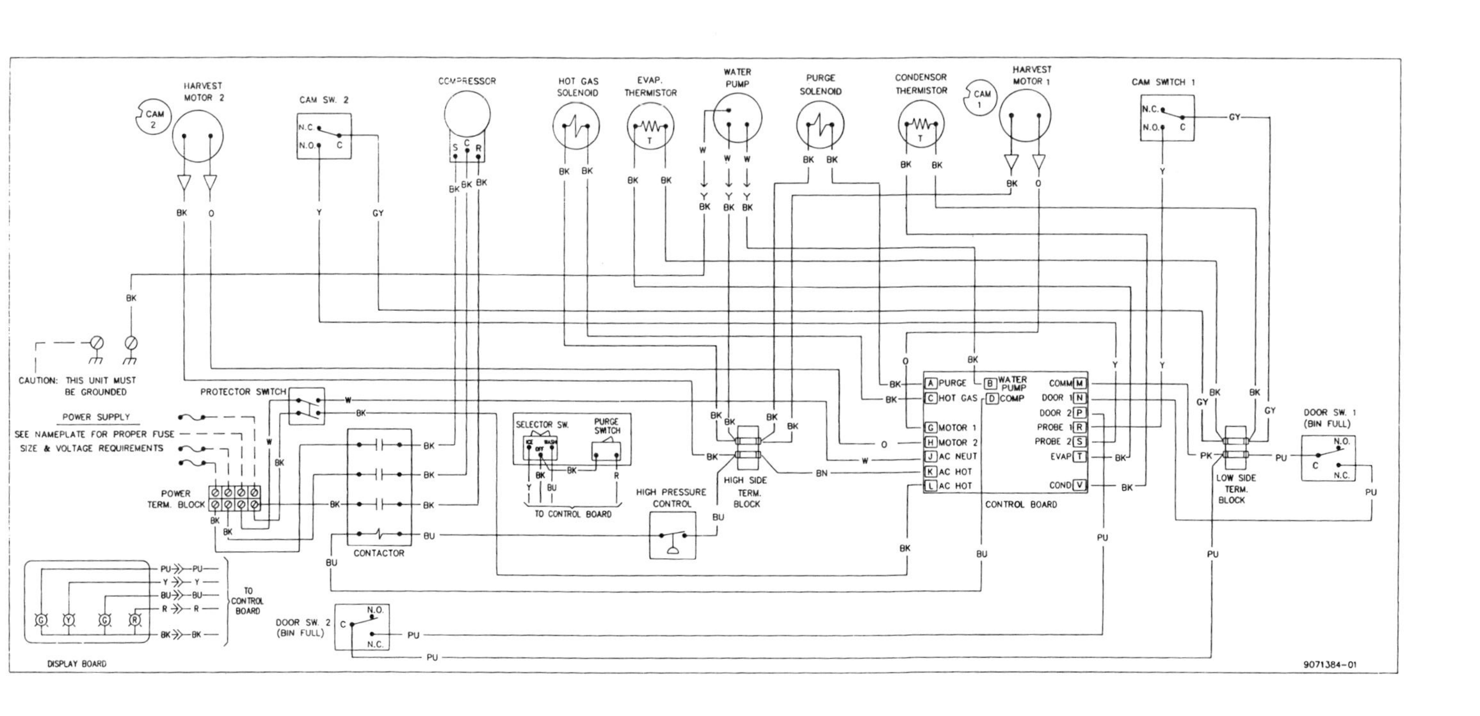

GENERAL INFORMATION
WATER DISTRIBUTION TUBE ASSEMBLY

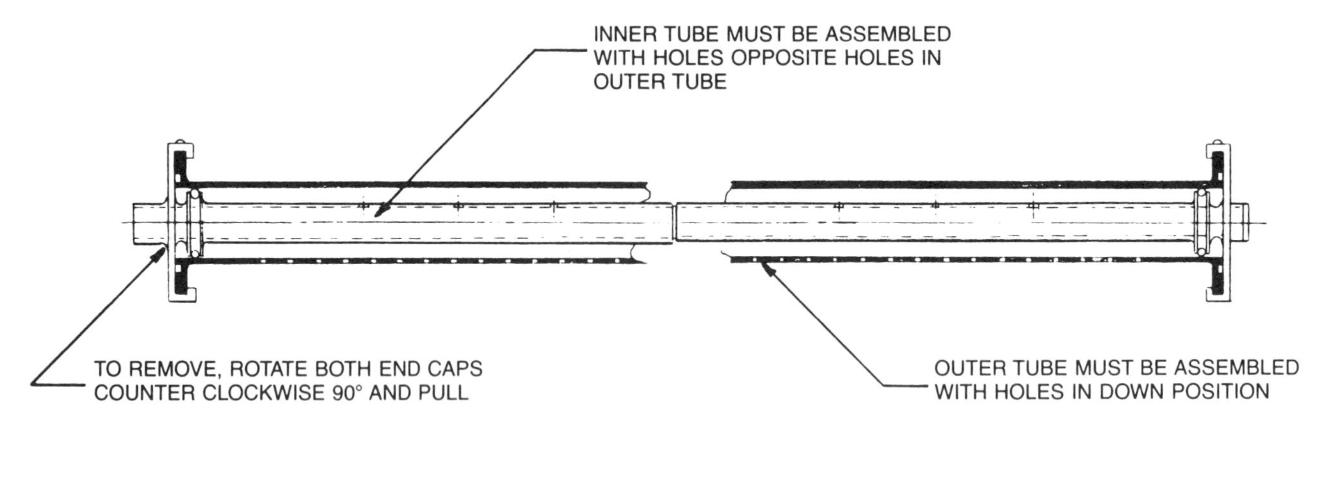

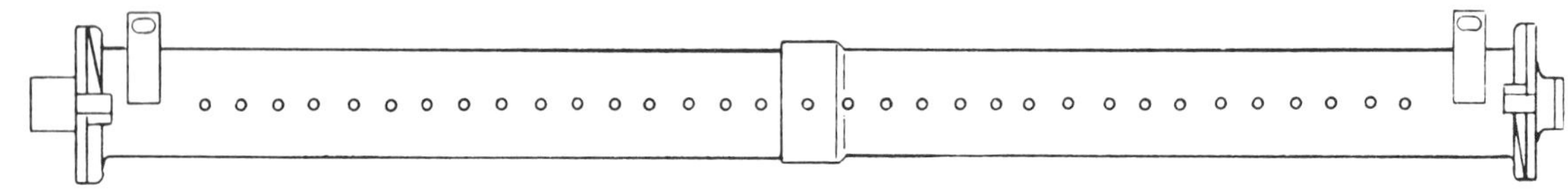

Water System

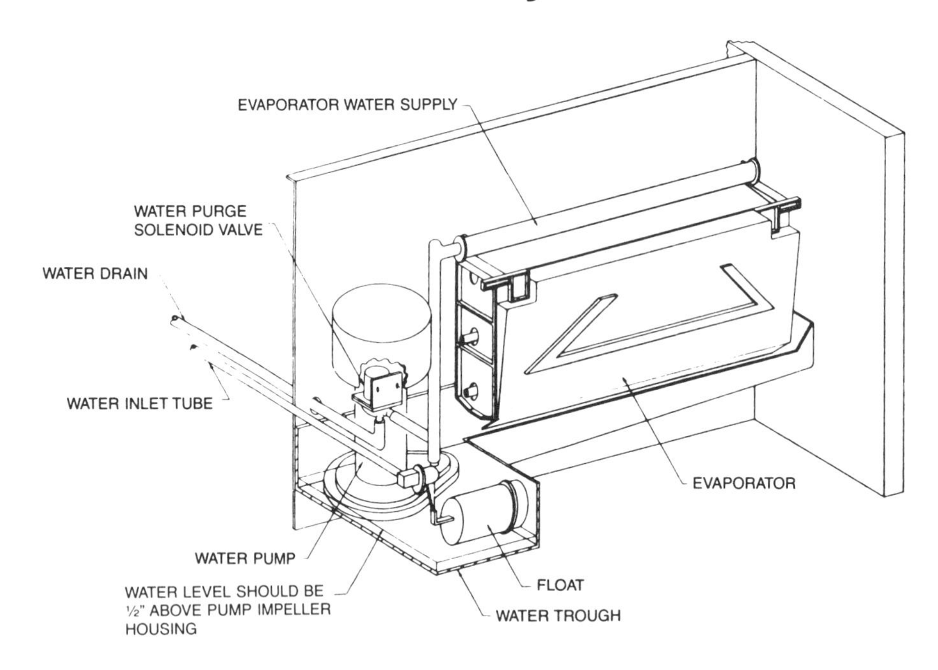

Water is circulated, by a water pump, from the water trough over the evaporator freeze plate(s) to be frozen. Any water that is not frozen returns to the trough to be re-circulated. Near the end of the freeze cycle, a water purge valve opens and the water pump continues to run. This empties all water from the trough out through the purge valve drain. Completely dumping the water in this manner keeps the cubes clear, and lengthens the time between cleanings. This allows the water trough to refill in preparation for the next freeze cycle.

GENERAL INFORMATION
REFRIGERATION SYSTEM FUNDAMENTALS
(Air and Water Cooled)

The below figure shows the Refrigeration System Schematic. During the freezing cycle high temperature and pressure liquid refrigerant is fed from the condenser through a drier and heat exchanger to a thermostatic expansion valve. This valve meters the refrigerant to the coils on the back side of the evaporator. The refrigerant absorbs heat causing a portion of the water flowing over the evaporator to freeze into ice. The refrigerant is maintained at a low pressure by the action of the compressor.

The low temperature and pressure refrigerant leaving the evaporator is directed through the heat exchanger and is returned to the compressor. There it is compressed to a high temperature and pressure gas. It is then directed to a condenser to be converted again to high temperature and pressure liquid.

During the harvest cycle, the hot gas solenoid valve (normally closed during the freezing cycle) opens to direct the high temperature gas leaving the compressor into the evaporator ahead of the expansion valve. This gas rapidly warms up the evaporator to above freezing. The ice slab is thawed loose from the evaporator and then released by the probe assist motor. At the end of the harvest cycle the hot gas solenoid valve closes and a freezing cycle starts.

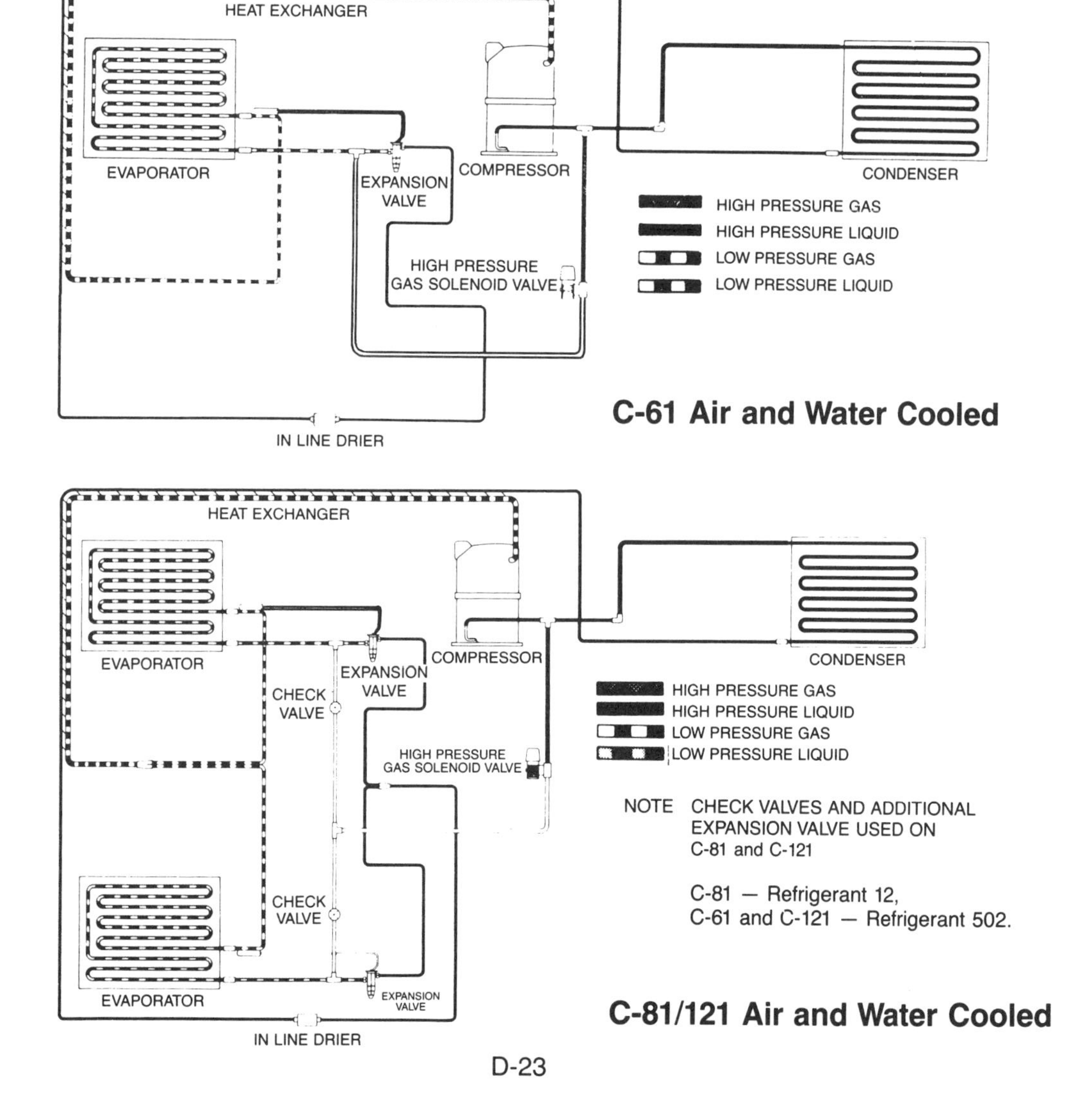

C-61 Air and Water Cooled

C-81/121 Air and Water Cooled

NOTES

REMOTE CONDENSER MODELS
INFORMATION, INSTALLATION AND SERVICE

RC-60-2
RC-80
RC-80-2
RC-120
RC-120-2

ICE-O-MATIC REMOTE CONDENSER SYSTEMS

INTRODUCTION

Ice-O-Matic Remote Condenser Systems are made up of three packages. The remote condenser, the ice making unit, and the pre-charged line set. Pre-charged line sets are available in either 25 or 40 foot lengths.

Normal installation of the ice making unit should be followed. Please see installation instructions included with the ice maker.

In any installation, the pre-charged line sets, consisting of a liquid line (3/8" dia.) and a discharge line (1/2" dia.) are used as a one time, initial charge type of installation. Once the sealed couplings are connected and the internal seal is broken, the lines cannot be disconnected without losing the refrigerant charge. They are, however, reusable and when they are removed and re-connected the complete refrigeration system must be evacuated and re-charged with the proper amount of refrigerant. See page A6 for proper refrigerant charges.

GENERAL DESCRIPTION

The remote condenser should be used in areas where sufficient airflow is not available in the area the ice maker is being installed or the heat being rejected by the condenser coil is un-desirable.

The condenser coil should not be exposed to temperatures below −20°F or above 120°F.

The remote condenser functions as a normal refrigeration system until the temperature at the condenser coil drops below 70° farenheit. At this time the mixing valve will begin to bypass enough hot gas from the discharge line directly into the receiver to keep the liquid line feeding the expansion valve at a steady 125 p.s.i. on R-12 (225 p.s.i. on R-502 units). This bypassing is done in spurts through the mixing valve. The amount of gas being bypassed will depend on the temperature at the condenser coil (e.g. the colder the temperature at the condenser coil, the more gas will bypass, the warmer the line between the valve and the receiver will become). For the complete explanation of this valve please refer to the schematic on page E4 of this manual.

The condenser fan will run throughout the cycle. The bypassing action of the mixing valve eliminates the need for a fan control.

There is an off-on switch provided at the condenser coil. This switch turns off only the condenser fan and should always be in the ON position when the icemaker is running. For proper wiring of the condenser to the ice maker please refer to the wiring diagrams included in this manual beginning on page E-10.

For individual part description of the components in the remote condenser refer to page E3 of this service manual.

GENERAL SERVICE INFORMATION
COMPONENT DESCRIPTION

1. **Mixing Valve** — This valve serves as the head pressure regulating valve. It contains a pre-determined charge of nitrogen in the bellows. Refer to page E4 for diagram and description.

2. **Receiver** — The amount of liquid in the receiver will vary with the temperature at the condenser coil.

3. **Liquid solenoid valve and pump-down low pressure control** — When the bin control opens or the on-off switch is turned to the OFF position, the liquid line solenoid valve closes and the system begins to pump down. When the low pressure reaches 5 p.s.i. on R-12 and 502 units, the dual pump-down low pressure control opens the contactor and shuts the machine off. After the bin control closes or the on-off switch is turned back on, the liquid line solenoid valve opens and the high and low side pressures start to equalize. When the low side pressure reaches 55 p.s.i. on R-12 and 502 units the pump-down low pressure control closes and the machine starts into a freeze cycle. When the machine is off for extended periods of time (e.g. overnight) it will pump itself down approximately once every hour. This is due to the equalizing of the pressures during the off cycle resetting the pump-down low pressure control. It is a normal function of the system. The high pressure control opens at 250 p.s.i. on R-12 units (400 p.s.i. on R-502 units) in the event of extremely high discharge pressure.

4. **Condenser fan motor** — Is a single-speed, permanent-split capacitor motor and should be wired according to the wiring diagram beginning on page E-10 in this manual. The motor should be oiled every six months with 5 ccs of 20 weight SAE non-detergent oil. Apply oil through the neoprene oil hole in the center of the motor. The on-off switch mounted on the outside of the condenser cabinet should be left in an on position for proper operation. Turning this switch off cuts power only to the fan motor.

   Note: If two machines are being run on an (RC-60-2, RC-80-2, or RC-120-2, the fan motor should be wired to a separate power source to run 100% of the time.

GENERAL INFORMATION
MIXING VALVE OPERATION

The mixing valve is used to maintain a constant head pressure. There is a nitrogen charge contained in the bellows of the valve. This charge works against liquid line pressure going to the receiver.

When the temperature at the condenser is above 70° the refrigerant flow through the valve is from the condenser to the receiver.

When the temperature at the condenser is below 70° the pressure in the bellows of the mixing valve overcomes the pressure in the liquid line going to the receiver, the valve opens and allows discharge high pressure gas, as well as liquid from the condenser, to flow through the valve and mix in the receiver to maintain head pressure and liquid line temperature and pressure.

NOTE: If it becomes necessary to evacuate and recharge the machine, the entire charge (see refrigerant charge on page A6 of this manual) must be added before accurate pressure readings can be taken.

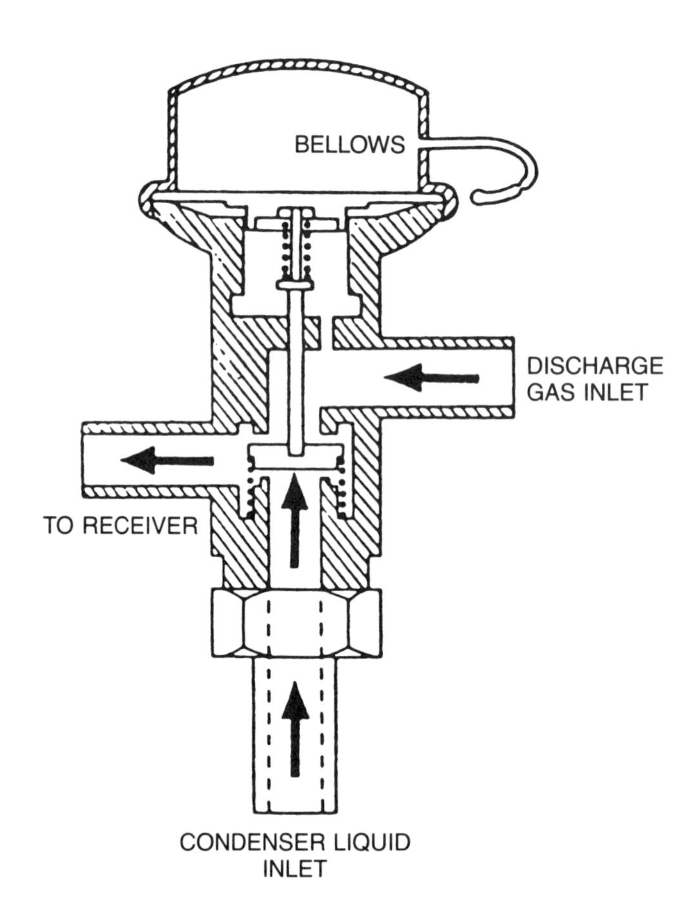

GENERAL INSTALLATION INSTRUCTIONS

GENERAL INSTALLATION INSTRUCTIONS

REMOTE CONDENSER

1. Choose a location that is protected from the extremes of dirt, dust, rain, sun and prevailing winds.
2. Vertical discharge mounting of the condenser is recommended.
3. Unit must be leveled.
4. Installation must meet local building, plumbing and electrical codes.
5. Condenser coil should not be exposed to temperatures above 120°F or below −20°F.

PRE-CHARGED LINE SET

1. Install the 3/8" liquid line and 1/2" discharge line to the proper ports, as labeled, being sure the service port is towards the remote condenser end. See attached diagram on page E6.
2. See attached diagram on pages E6 & E7 for proper installation techniques and tube routings of the line sets.
3. Lubrication (refrigerant oil) is recommended for both coupling halves to insure proper trouble-free assembly of line sets.
4. Thread coupling valves together by hand to insure proper fit. Using a wrench on both halves, to prevent tubes from twisting, tighten until snug. Then an extra 1/4 turn to ensure a leak-free joint.
5. Plan ahead when routing lines so that excess tubing remains inside building.

ELECTRICAL CONNECTIONS

1. The ice maker and the remote condenser both require a solid earth ground that meets national, state and local electrical requirements.
2. See nameplate for current requirements to determine wire size to be used for electrical hook up.
3. Make sure supply voltage is the same as the rated voltage shown on nameplate.
4. See wiring diagram beginning on page E10 for proper connections between remote condenser and ice machine.

RULES FOR RUNNING REMOTE LINES

1. Loops in excess tubing should be run vertically and contained inside building.
2. Condenser should be above ice-maker, lowest part of system should be the ice-maker, lines should not run below the line connections on the ice maker (e.g., lines routed below — then up to connections on ice maker result in oil traps.) See pages E6 & E7.

GENERAL INFORMATION
REMOTE TUBING ROUTING

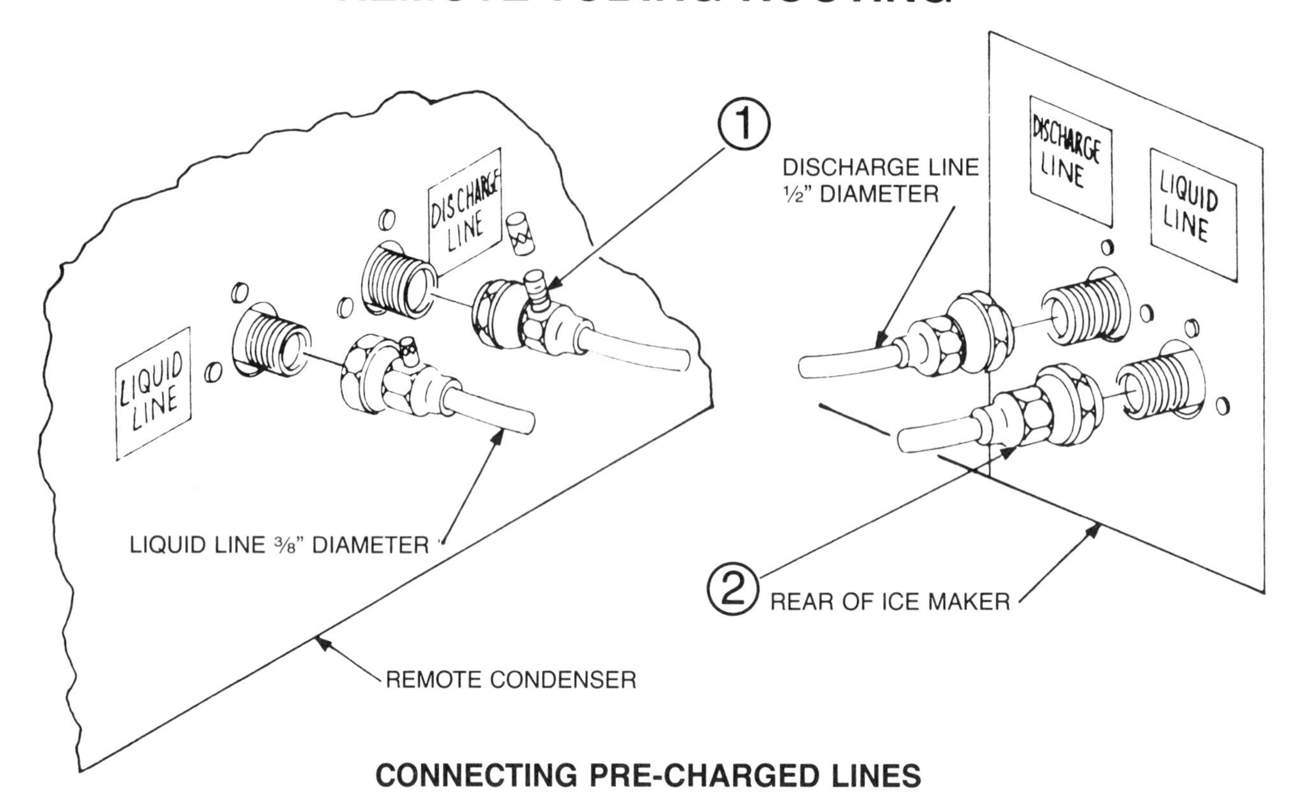

CONNECTING PRE-CHARGED LINES

Both the discharge and liquid lines come equipped with Schraeder fittings on one end of the tubing. This Schraeder should be connected to the remote condenser on both the discharge and liquid lines for access to pressure readings at the condenser as shown in #1. When connecting the quick connect fittings, #2, always lubricate fittings with refrigerant oil. Fittings should be tightened until snug — and then given another quarter turn. ALWAYS LEAK CHECK AROUND FITTINGS AFTER INSTALLATION HAS BEEN MADE.

PROPER TUBING ROUTING

When installing discharge and liquid lines from remote condenser to the ice making head please use the following guidelines:

1. Remote condenser #3 should always be installed above the icemaker #5 as shown.
2. All excess tubing #4 should be routed inside the building and coiled in a vertical spiral as shown in #6 to prevent oil trapping in the lines. Any tubing run outside should be insulated to protect it from surrounding ambient conditions. Tubing should follow straight line routing whenever possible. The lowest spot in the remote tubing run should be the connection at the back of the ice maker head.

FOR DUAL CIRCUITED CONDENSER

The dual pass condenser is two separate condensers contained in one shroud. When routing lines insure that the discharge and liquid lines from each machine go to one condenser. Do not connect the discharge line from one machine and the liquid line from another machine to the same condenser. When running two machines on the same condenser the remote fan motor should be wired to run 100% of the time.

GENERAL INFORMATION
REMOTE TUBING ROUTING (continued)

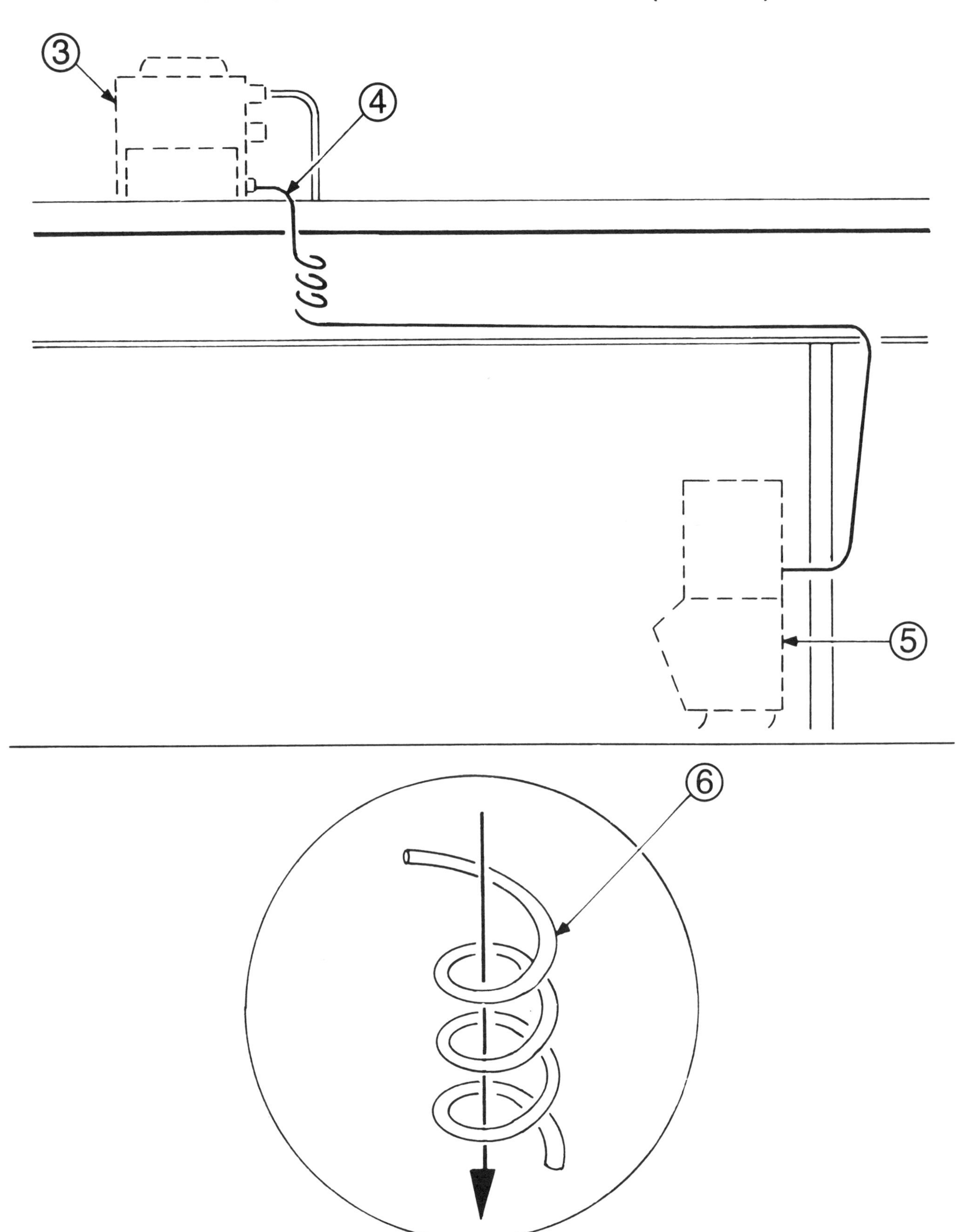

CONDENSER DIMENSIONS
RC-60
RC-80
RC-120

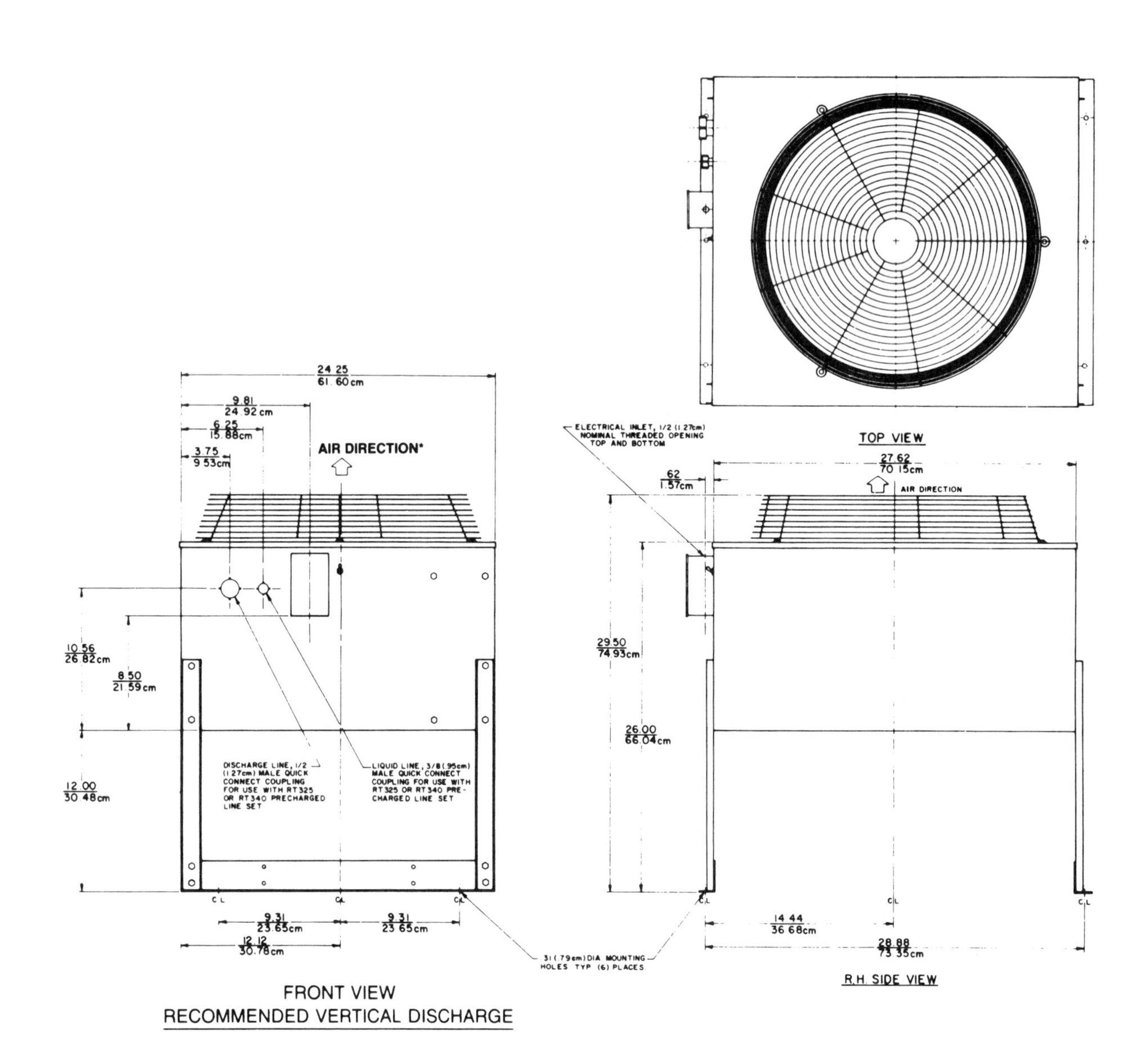

FRONT VIEW
RECOMMENDED VERTICAL DISCHARGE

*NOTE AIR FLOW

CONDENSER DIMENSIONS
RC-60-2
RC-80-2
RC-120-2

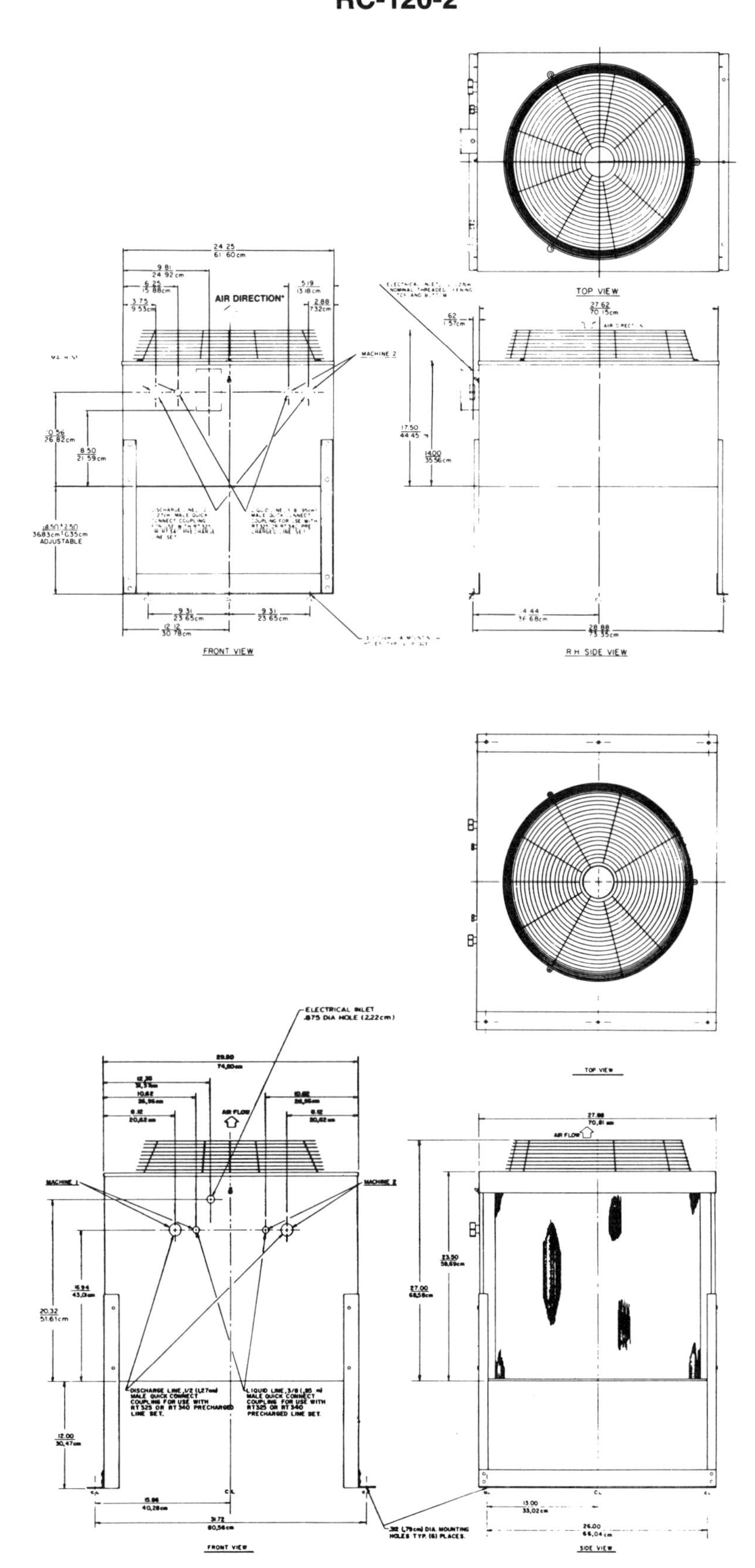

C-61-C-Remote Wiring Diagram

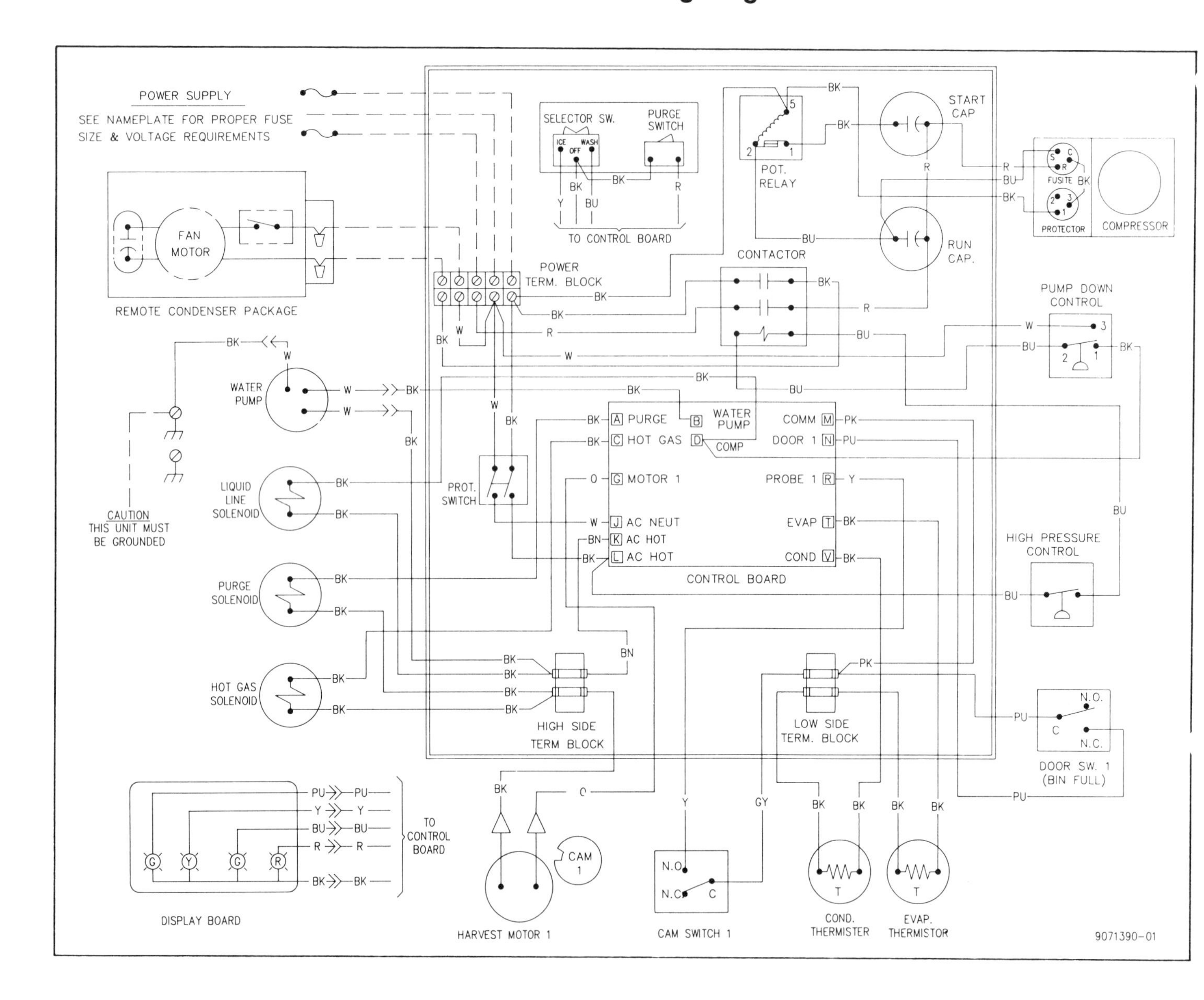

C-81/121-C-Remote & C-84/124-C-Remote Wiring Diagram

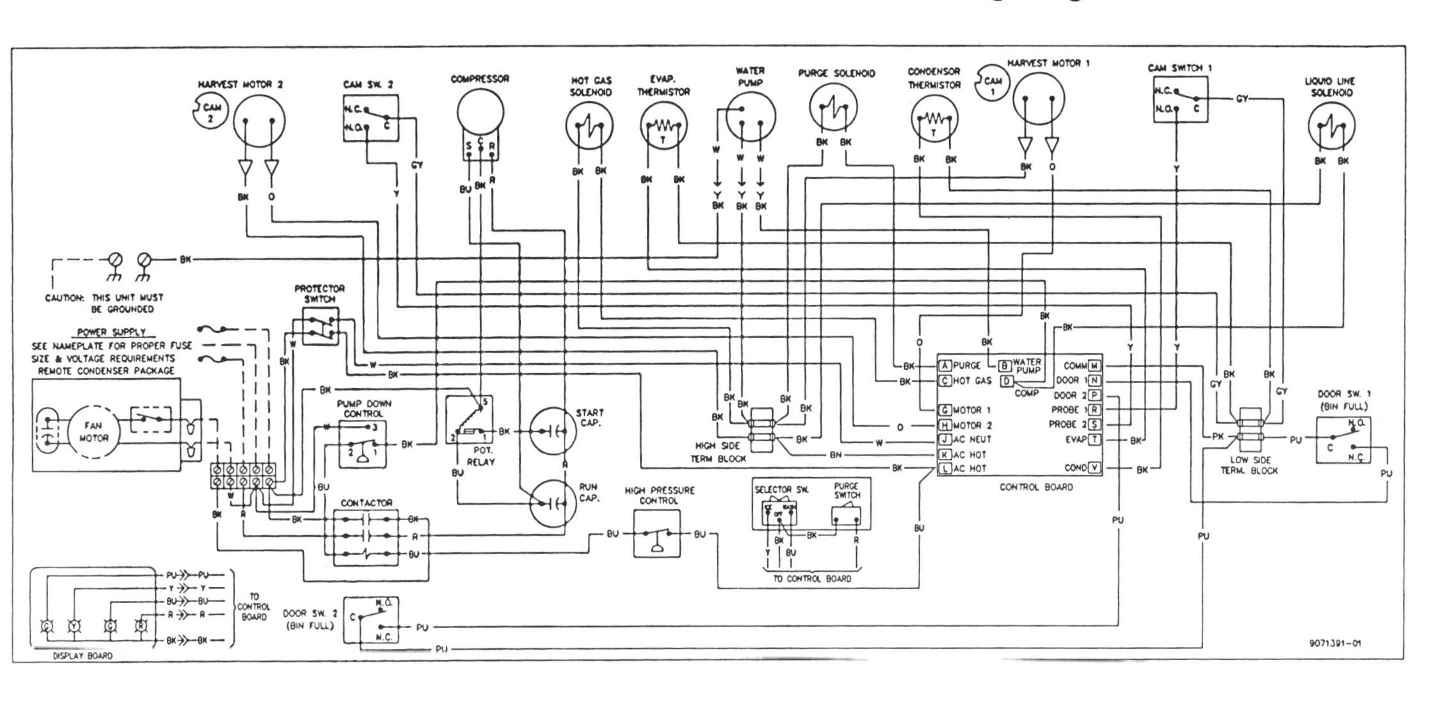

C-81/121-C Remote

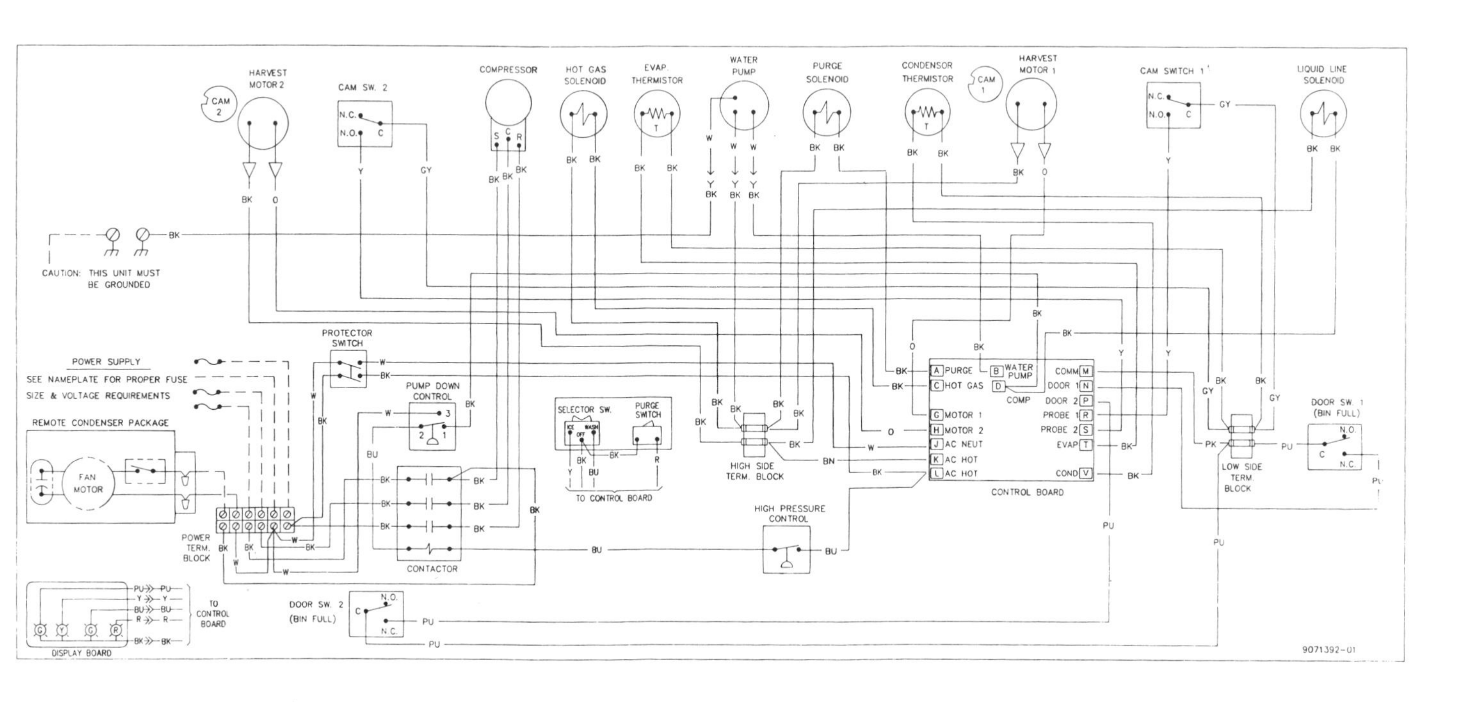

C-84/124-C Remote

GENERAL INFORMATION
REFRIGERATION SYSTEM FUNDAMENTALS
(Remote Air Condenser Cooled)

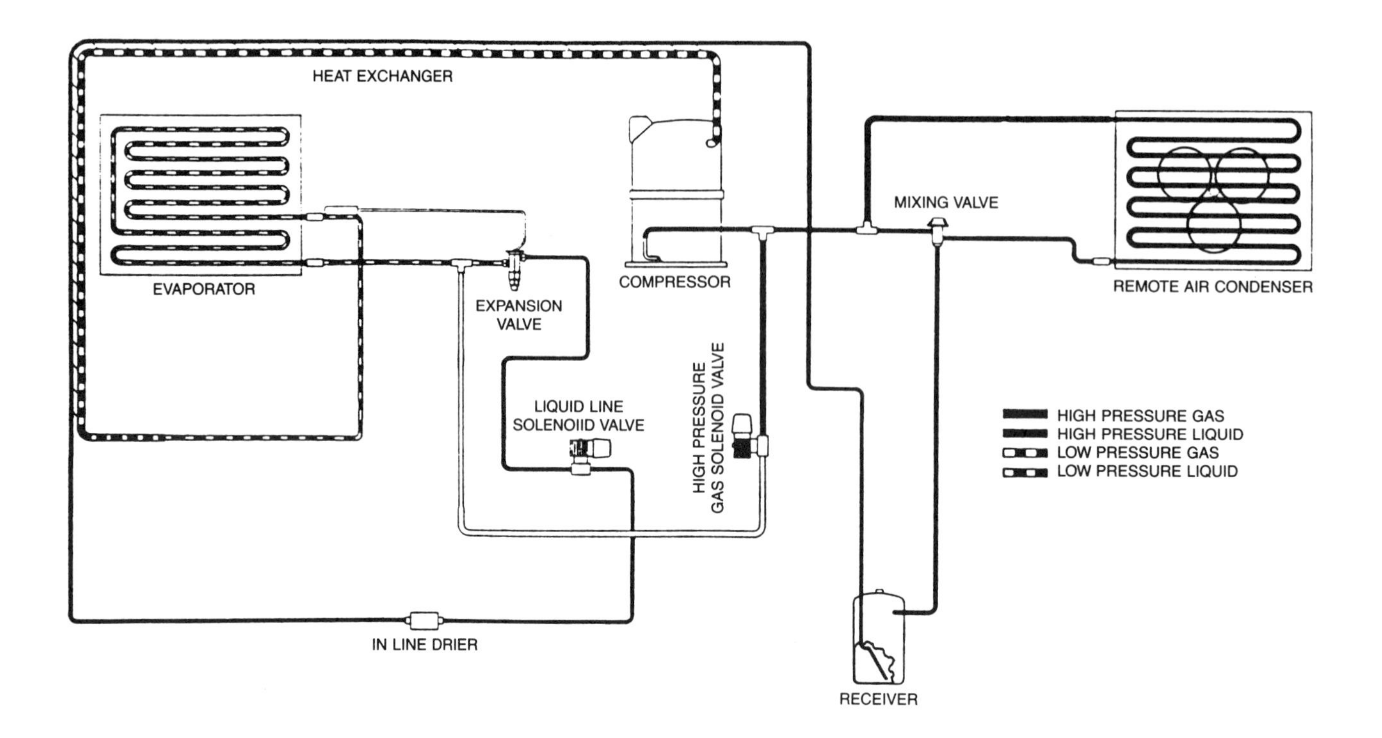

The figure above shows the Remote Refrigeration Schematic. During the freezing cycle high temperature and pressure liquid refrigerant is fed from the condenser through a receiver, drier and heat exchanger to a thermostatic expansion valve. This valve meters the refrigerant to the coils on the back side of the evaporator. The refrigerant absorbs heat, causing a portion of the water flowing over the evaporator to freeze into ice. The refrigerant is maintained at a low pressure by the action of the compressor.

In low ambient conditions (below 70°F air temperature at the condenser) the mixing valve opens to mix discharge gas with liquid returning from the condenser, in the receiver, to maintain discharge and liquid line pressures.

The low temperature and pressure refrigerant leaving the evaporator is directed through the heat exchanger and is returned to the compressor. There it is compressed to a high temperature and pressure gas. It is then directed to a condenser to be converted again to high temperature and pressure liquid.

During the harvest cycle, the hot gas solenoid valve (normally closed during the freeze cycle), opens and directs the high temperature gas leaving the compressor into the evaporator. Once the evaporator has reached approximately 40°F, the harvest motor overcomes the capillary attraction of the ice and the evaporator, and releases the ice from the evaporator. At the end of the harvest cycle, the hot gas solenoid valve closes and another freeze cycle begins.

During the off cycle the liquid line solenoid valve is closed to prevent refrigerant migration to the roof-top condenser.

PARTS SECTION

When Ordering Parts Please Specify Model and Serial Number①

NOTE: All replacement parts ordered or returned must come through an authorized Ice-O-Matic Distributor.

MILE HIGH EQUIPMENT DENVER, COLORADO 80239 U.S.A.
MODEL ______ SERIAL NO. ______
VOLTS ______ Hz. ______ Ph. ______ WIRE ______
COMPRESSOR ...... R.L.A. ______ L.R.A. ______
MOTORS: (EA) H.P./WATTS F.L.A.
AUGER ......
FAN(S) ......
HARVEST(S) ......
PUMP ......
BIN DISPENSER ......

DESIGN PRESSURES		
	PSI	BARS
HI		
LO		

①
SAMPLE
MINIMUM CIRCUIT AMPACITY ......
MAXIMUM BRANCH CIRCUIT FUSE SIZE OR HACR TYPE CIRCUIT BREAKER ......
BEVERAGE FAUCETS (EA) ...... WATTS AT 24 VOLTS
USE WITH ICE-O-MATIC REMOTE CONDENSER MODEL NO. ......
TOTAL SYSTEM REFRIGERANT CHARGE TYPE: ______ Oz.: ______ Grams: ______
MILE HIGH EQUIPMENT COMPANY DENVER, COLORADO 80239 U.S.A.

ORDERING PROCEDURE

STOCKING ORDERS

- All stocking orders should be mailed, telefaxed, or telexed.
- Specify routing requests on all purchase orders
- Routing will be at Ice-O-Matic's discretion for prepaid freight orders.
- Orders will be shipped complete, less back orders, via one carrier. No split shipments allowed under one purchase order number.

EMERGENCY ORDERS

- All emergency orders should be telephoned to Ice-O-Matic.
- Specify routing requests when placing order.

REPLACEMENT PARTS WARRANTY

DURATIONS OF WARRANTY PERIOD

- All service parts carry a 90-day replacement warranty.
- All service compressors, not included in the 5 year compressor warranty, carry a 1 year warranty.
- All 1 year service compressor warranty is determined by compressor serial number

WARRANTY RETURN POLICY

I. All in Warranty parts except compressors, water pumps, gear reducers, cuber expansion valves and auger drive motors are to be in-field scrapped and not returned to Ice-O-Matic.
- All field scrap items should be listed on the Field Scrap Form, FSF #186①. Return this form along with the white (top) copy of R.G. Tag③ to Ice-O-Matic.
- R.G. Tag③ must have all information correctly filled in for proper warranty transactions.
- Designate on R.G. Tag③ and FSF #186② for in warranty replacement or warranty credit. If no box is checked, a replacement part will be sent to you automatically.
- All field scrap items should be held by distributor for 45 days after return of R.G. Tag③.
- Field scrap items may be requested to be returned for factory evaluation within the 45 day limit.
- Do not submit warranty requests for assemblies which have component failure. Component only will be honored.
- See Section III for compressors which may be in-field scrapped.

II. In Warranty water pumps, gear reducers, cuber expansion valves and auger motors must be returned to Ice-O-Matic under the following conditions:
- Parts must be returned freight prepaid.
- R.G. Tag③ must be completely filled out with hard copy and white (top) copy attached to the part.
- Parts must be returned to Ice-O-Matic within 90 days of removal for proper credit.
- Part must be properly packed to avoid damage. Warranty parts which are damaged due to improper packing will be returned to you with no credit issued.

III. All Warranty compressors, except those noted in Section IV, are to be field scrapped and compressor serial plate only returned to Ice-O-Matic.
- Remove compressor serial plate and return to Ice-O-Matic with a completed R.G. Tag③.
- Return hard copy and white (top) copy of R.G. Tag③ with compressor and serial plate.

IV. In warranty compressors which must be returned to Ice-O-Matic, with serial plate intact, are:
- Any Copeland compressor which is less than 18 months old and the compressor serial number ends in a 1 or 7.
- All Tecumseh compressors which are less than 18 months old must be returned to Ice-O-Matic.
- All date code reference is compressor serial plate, not machine serial number.
- All returned compressors must be plugged or sealed before being returned for warranty. A $25 fee will be imposed for any returned compressor whose package is oil saturated.

V. In warranty Copeland compressors may be exchanged even up through a Copeland wholesaler within these guidelines:
- Compressor must be less than 18 months old by serial date.
- Compressor has been installed less than 12 months.
- Exchange must be an even swap. We are not liable for any additional costs incurred.
- Compressors must be returned to Ice-O-Matic if there is a dispute with the Copeland wholesaler.

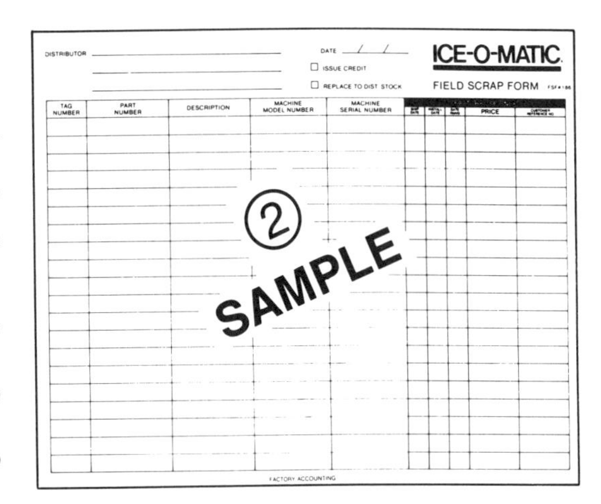
ICE-O-MATIC
FIELD SCRAP FORM
DISTRIBUTOR ______ DATE __/__/__
☐ ISSUE CREDIT
☐ REPLACE TO DIST STOCK
TAG NUMBER | PART NUMBER | DESCRIPTION | MACHINE MODEL NUMBER | MACHINE SERIAL NUMBER | PRICE
②
SAMPLE

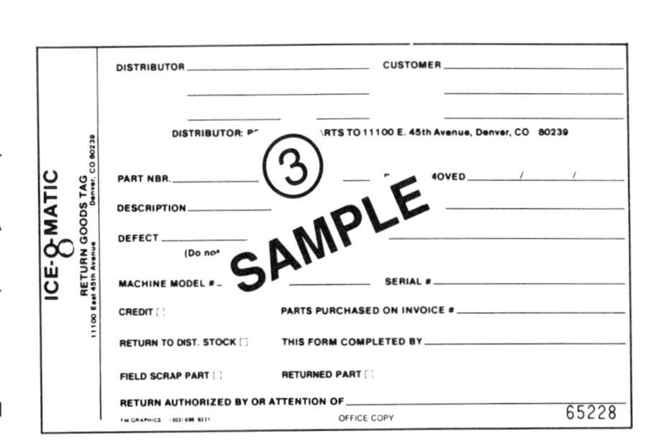
ICE-O-MATIC
RETURN GOODS TAG
DISTRIBUTOR ______ CUSTOMER ______
PART NBR. ______
DESCRIPTION ______
DEFECT ______
③
SAMPLE
MACHINE MODEL # ______ SERIAL # ______
CREDIT ☐ PARTS PURCHASED ON INVOICE # ______
RETURN TO DIST. STOCK ☐ THIS FORM COMPLETED BY ______
FIELD SCRAP PART ☐ RETURNED PART ☐
RETURN AUTHORIZED BY OR ATTENTION OF ______
OFFICE COPY
65228

ICE-O-MATIC
11100 E. 45th Avenue
Denver, Colorado 80239
(303) 371-3737
Telex: 386 453
Telefax: (303) 371-6296

Exterior Panels

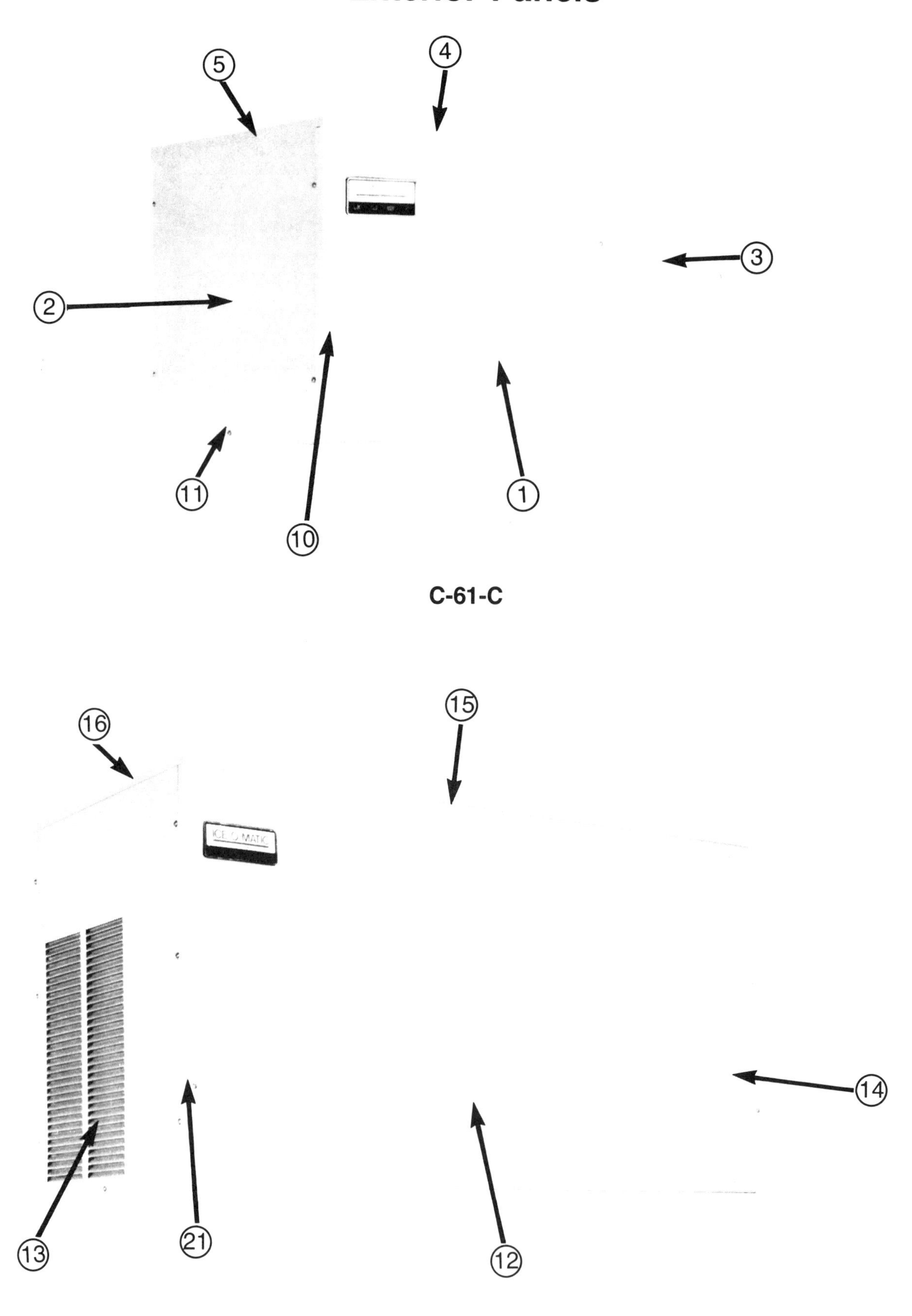

C-61-C

C-81/121-C

PARTS SECTION
EXTERIOR PANELS
C-61/81/121-C

Index #	Description	Part Number	Notes	Model C-61	Model C-81	Model C-121
1	Front Panel (Plastic)	2031321-03	w/Indicator LEDs	X		
2	Left Hand Side Panel	3021378-03	Painted (No Louvers)	X		
3	Right Hand Side Panel	3021378-01	Painted (Louvered)	X		
4	Top Panel	3021375-01	Painted	X		
5	Rear Panel	3021377-01	Galvanized	X		
6	Front Panel (S/S) (Not Shown)	2031321-01	w/Indicator LEDs	X		
7	Left Hand Side Panel (S/S) (Not Shown)	3021378-04	Stainless Steel (No Louvers)	X		
8	Right Hand Side Panel (S/S) (Not Shown)	3021378-02	Stainless Steel (Louvered)	X		
9	Top Panel (S/S) (Not Shown)	3021433-01	Stainless Steel	X		
10	Panel Screws	9031074-02	Front Panel	X		
11	Panel Screws	9031074-01	Side, Back, Top Panel	X	X	X
12	Front Panel (Plastic)	2031321-04	w/Indicator LEDs		X	X
13	Left Hand Side Panel	3021394-03	Painted (Louvered)		X	X
14	Right Hand Side Panel	3021394-01	Painted (Louvered)		X	X
15	Top Panel	3021393-01	Painted		X	X
16	Rear Panel	3021396-01	Water Cooled & Remote Only		X	X
17	Front Panel (S/S) (Not Shown)	2031321-02	w/Indicator LEDs		X	X
18	Left Hand Side Panel (S/S) (Not Shown)	3021394-04	Painted (Louvered)		X	X
19	Right Hand Side Panel (S/S) (Not Shown)	3021394-02	Stainless Steel (Louvered)		X	X
20	Top Panel (S/S) (Not Shown)	3021432-01	Stainless Steel		X	X
21	Panel Screws (Not Shown)	9031074-03	Front Panel		X	X

ELECTRICAL CONTROL BOX
C-61-C

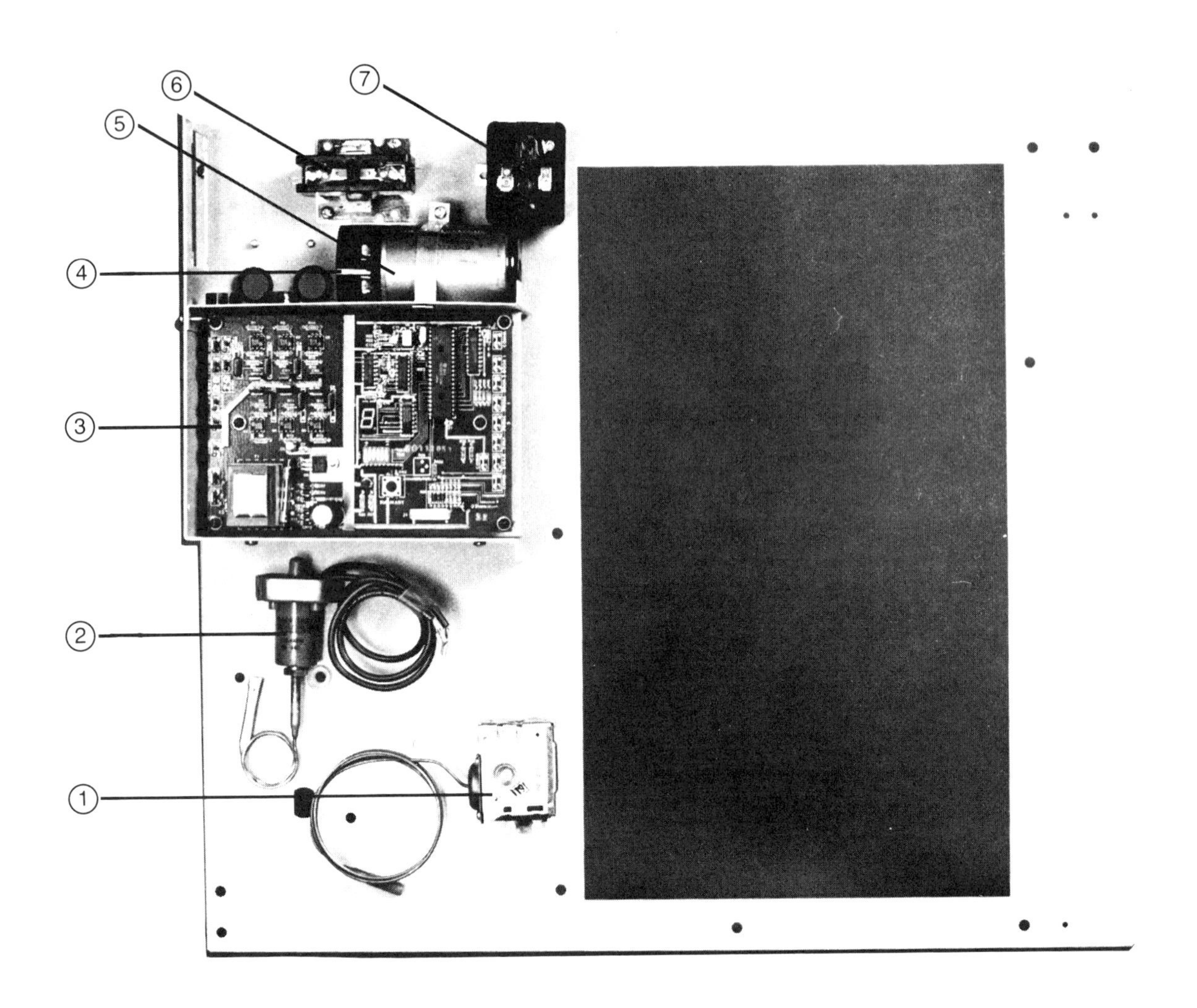

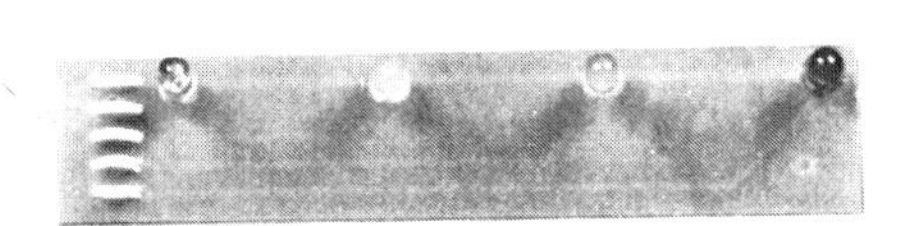

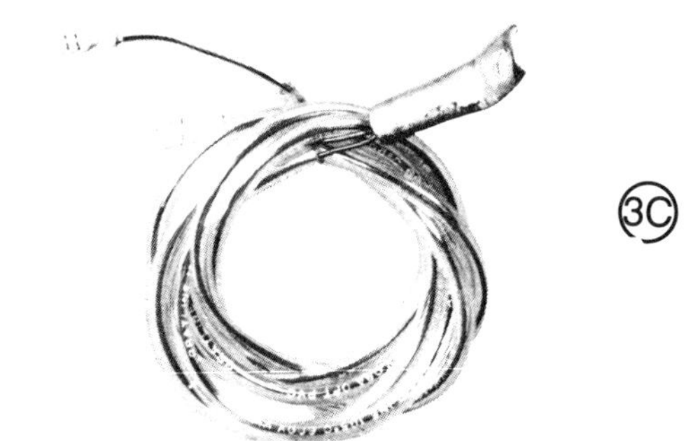

3C

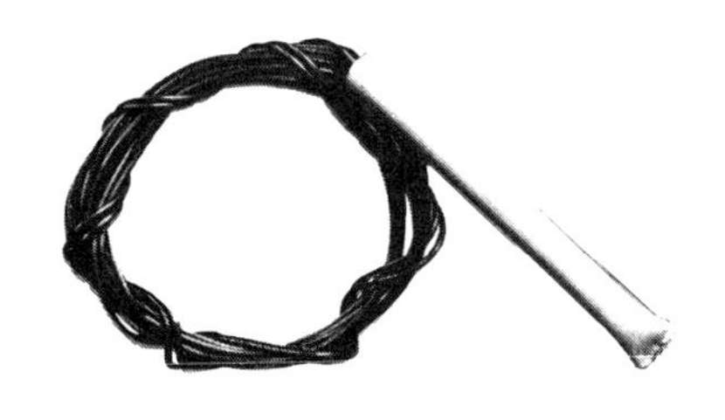

3D

PARTS SECTION
ELECTRICAL CONTROL BOX
C-61-C

Index #	Description	Part Number	Notes	Model C-61	C-81	C-121	C-84	C-124
1	Pumpdown Control	9041040-01	Remotes Only	X				
2	High Pressure Control	9041051-02	Water Cooled and Remote	X				
3	Controller Board	9101128-01		X				
3A	Indicator Light Assy.	9101128-02		X				
3B	Selector/Purge Switch & Connector Cable Assy.	9101128-03	Not Shown	X				
3C	Condensor Thermistor	9101131-01		X				
3D	Evaporator Thermistor	9101131-02		X				
4	Start Capacitor	9181003-08	Air, Water and Remote	X				
5	Run Capacitor	9181009-08	Air, Water and Remote	X				
6	Contactor, Single Pole	9101002-03	Air, Water and Remote	X				
7	Potential Relay	9181010-04	Air, Water and Remote	X				
	Terminal Board (2 Positions)	9101129-01	Not Shown	X				
	Terminal Board (3 Positions)	9101037-01	Air and Water Cooled (Not Shown)	X				
	Terminal Board (5 Positions)	9101037-03	Remote (Not Shown)	X				
	Fan Control	9041003-01	Air Cooled Only (Not Shown)	X				

ELECTRICAL CONTROL BOX
C-81/121-C

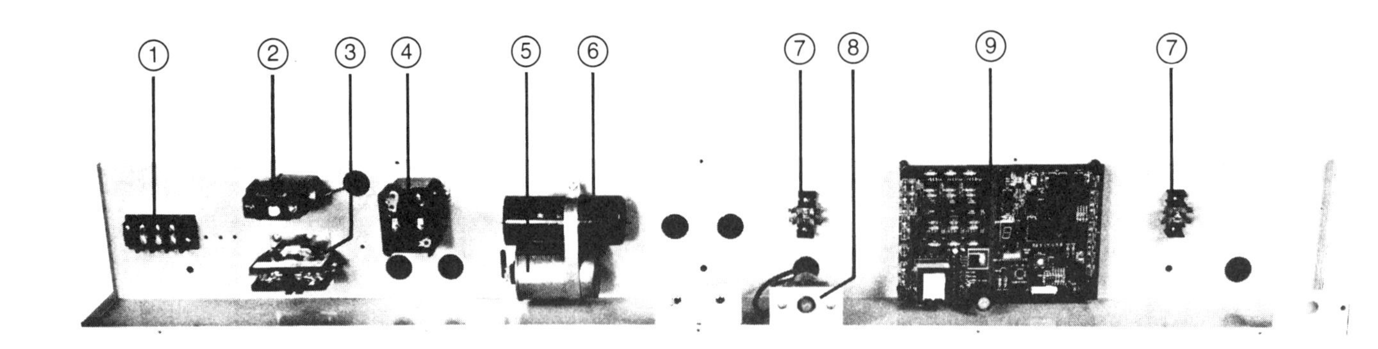

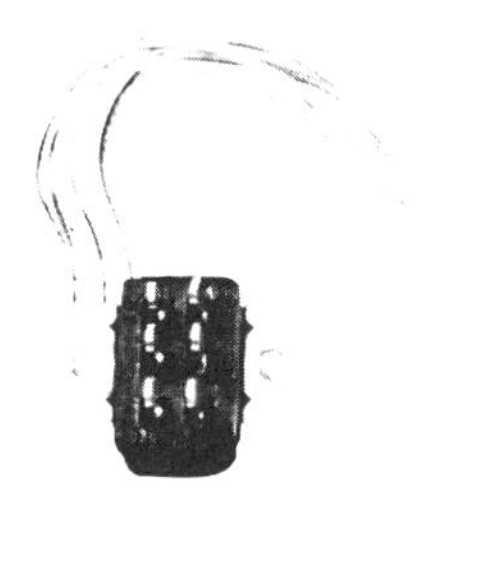

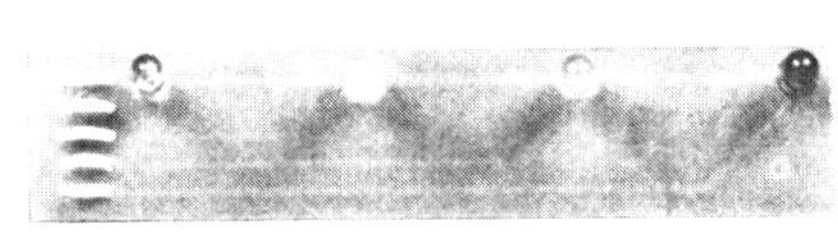

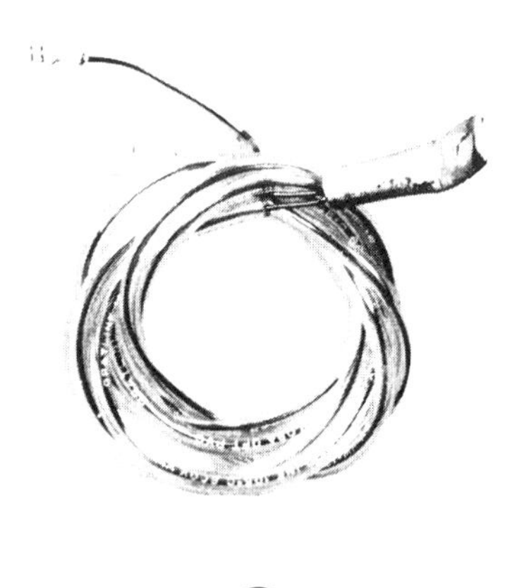

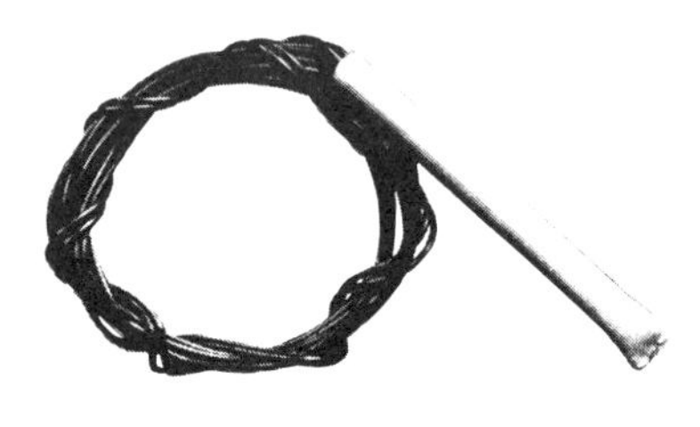

PARTS SECTION
ELECTRICAL CONTROL BOX
C-81/121-C

Index #	Description	Part Number	Notes	Model C-61	Model C-81	Model C-121
1	Terminal Board (3 Positions)	9101037-01	Air and Water Cooled		X	X
2	Fan Control	9041003-01	Air Cooled Only		X	X
3	Contactor, Single Pole	9101002-03	Air, Water and Remote		X	X
4	Potential Relay	9181010-06	Air, Water and Remote		X	X
5	Run Capacitor	9181009-04	Air, Water and Remote		X	X
6	Start Capacitor	9181003-11	Air, Water and Remote		X	X
7	Terminal Board (2 Positions)	9101129-01			X	X
8	High Pressure Control	9041051-01	Water Cooled and Remote		X	
9	Controller Board	9101128-01			X	X
9A	Indicator Light Assy.	9101128-02			X	X
9B	Selector/Purge Switch (Not Shown) & Connector Cable Assy.	9101128-03	Not Shown		X	X
9C	Condenser Thermistor	9101131-01			X	X
9D	Evaporator Thermistor	9101131-02			X	X
	Terminal Board (4 Positions) (Not Shown)	9101037-02	Air and Water Cooled 3 Phase Units			
	Terminal Board (5 Positions) (Not Shown)	9101037-03	Remote		X	X
	Terminal Board (6 Positions) (Not Shown)	9101037-04	Remote 3 Phase Units			
	Contactor, Three Pole (Not Shown)	9101079-01	Air, Water and Remote 3 Phase Units			
	High Pressure Control (Not Shown)	9041051-02	Water Cooled and Remote			X
	Pumpdown Control (Not Shown)	9041040-01	Remote Only		X	X

EVAPORATOR & WATER SYSTEM
C-61-C

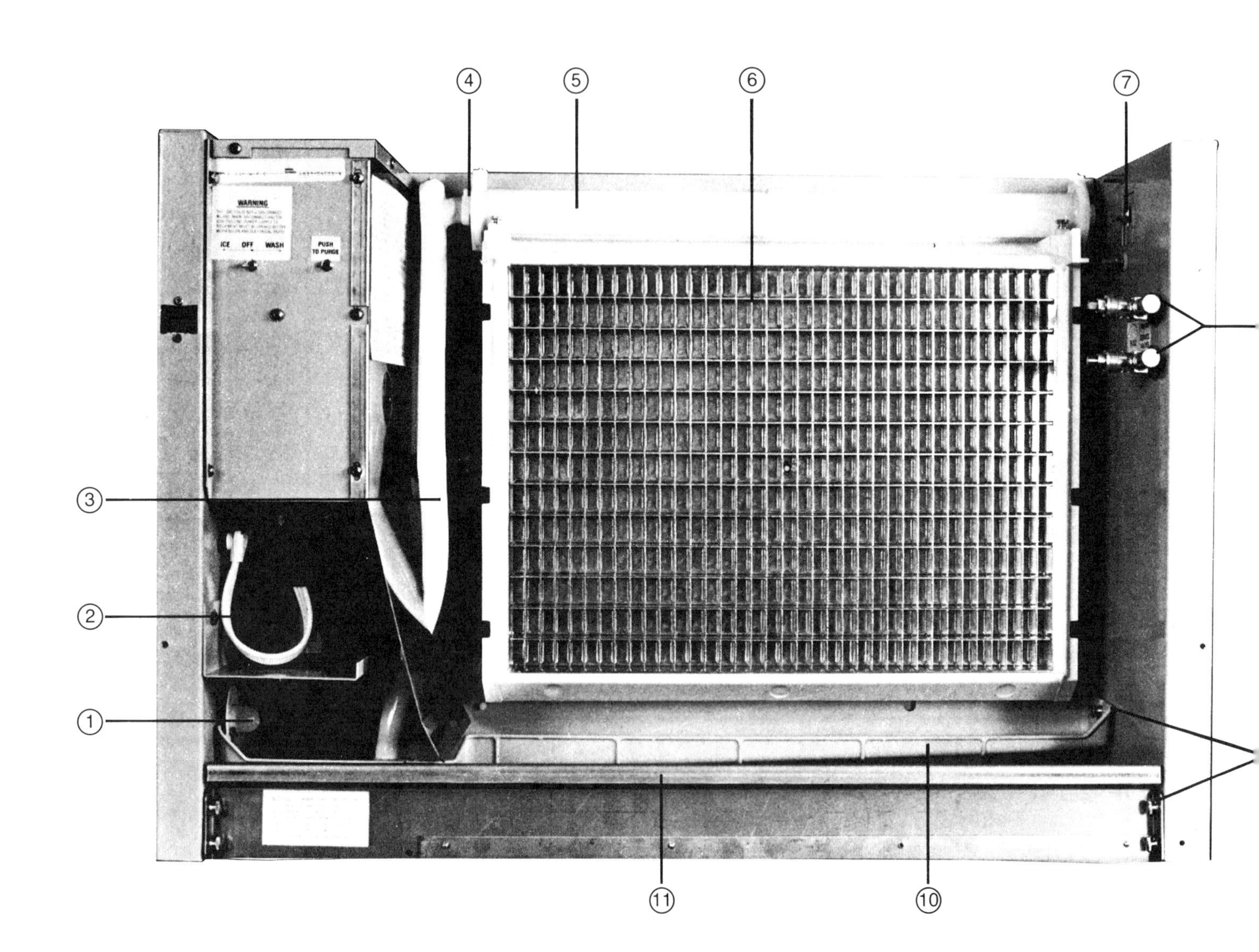

PARTS SECTION
EVAPORATOR & WATER SYSTEM
C-61-C

Index #	Description	Part Number	Notes	Model C-61	C-81	C-121
1	Float Ball and Stem	9131127-01		X		
2	Water Pump	9161076-01	Plug-in Type	X		
3	Water Pump Tube	9051142-01	Pump to Dist Tube	X		
4	Hose Clamp	9021010-04		X		
5	Water Distribution Assy.	2041338-01		X		
6	Evaporator	2051110-05	C-61, Half Cube	X		
7	Bin Control Switch	9101124-01		X		
8	Service Valve	9091064-01		X		
9	Thumbscrews	9031051-01	8-32x3/8	X		
10	Water Return Trough	9051153-01		X		
11	Ice Deflector	2021319-01		X		
	Evaporator (Not Shown)	2051110-04	C-61, Full Cube	X		
	Splash Curtain Assy.	9051154-02	Not Shown	X		

EVAPORATOR & WATER SYSTEM
C-81/121-C

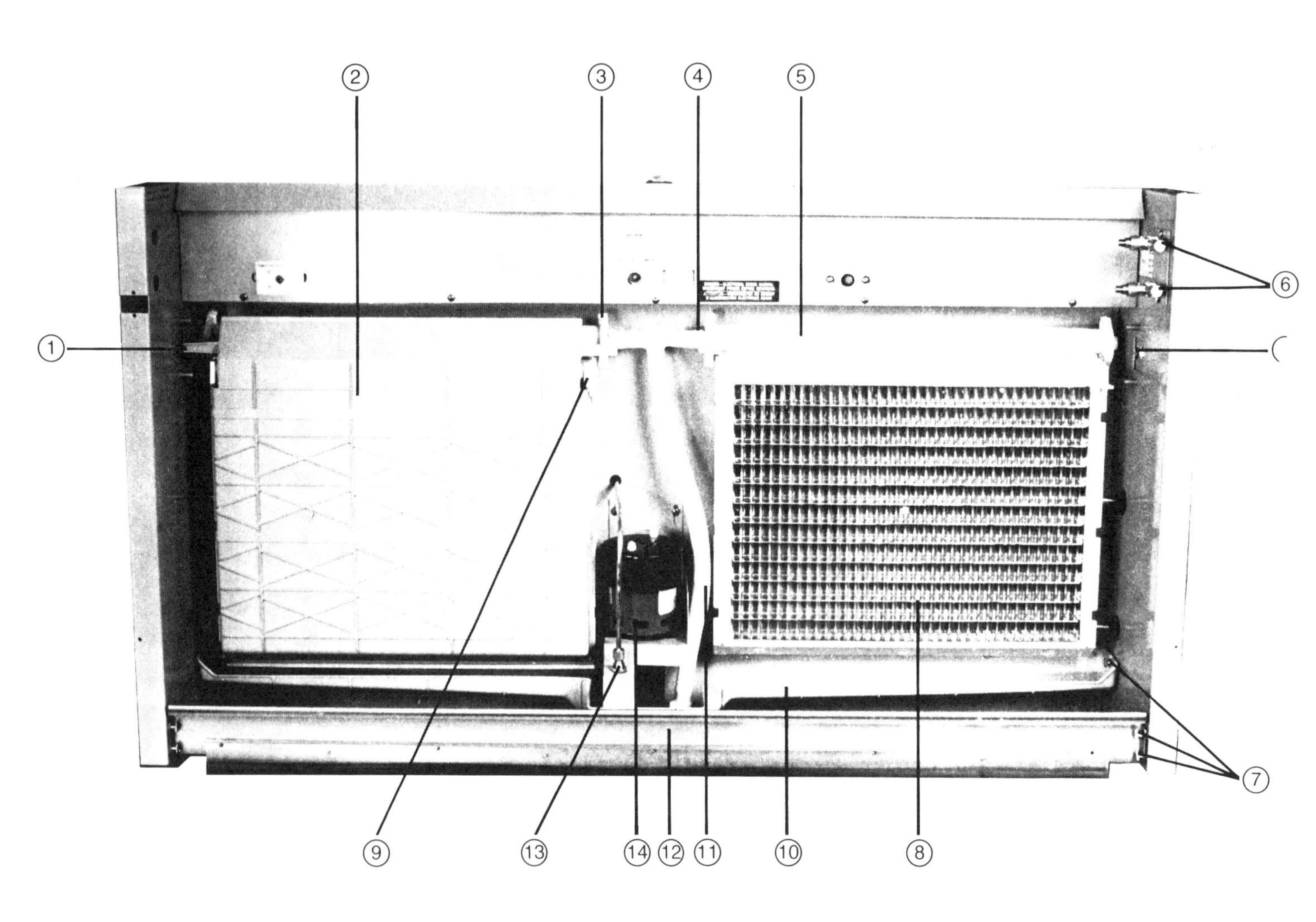

PARTS SECTION
EVAPORATOR & WATER SYSTEM
C-81/121-C

Index #	Description	Part Number	Notes	Model C-61	C-81	C-121
1	Bin Control Switch	9101124-01			X	X
2	Splash Curtain Assy.	9051154-02			X	X
3	Water Distribution Assy. L/H	2041338-02	Left Hand		X	X
4	Hose Clamp	9021010-04			X	X
5	Water Distribution Assy. R/H	2041338-01	Right Hand		X	X
6	Service Valve	9091064-01			X	X
7	Thumbscrews	9031051-01	8-32x3/8		X	X
8	Evaporator R/H	2051111-02	C81/121, Half Cube		X	X
9	Evaporator L/H	2051110-05	C81/121, Half Cube		X	X
10	Water Return Trough	9051153-02			X	X
11	Water Pump Tube	9051148-01	Pump to Dist Tube		X	X
12	Ice Deflector	2021326-01			X	X
13	Float Ball and Stem	9131111-01			X	X
14	Water Pump	9161079-01	Plug-in Type		X	X
	Evaporator R/H (Not Shown)	2051111-01	C81/121, Full Cube		X	X
	Evaporator L/H (Not Shown)	2051110-04	C81/121, Full Cube		X	X

HARVEST PROBE ASSEMBLY
C-61/81/121-C

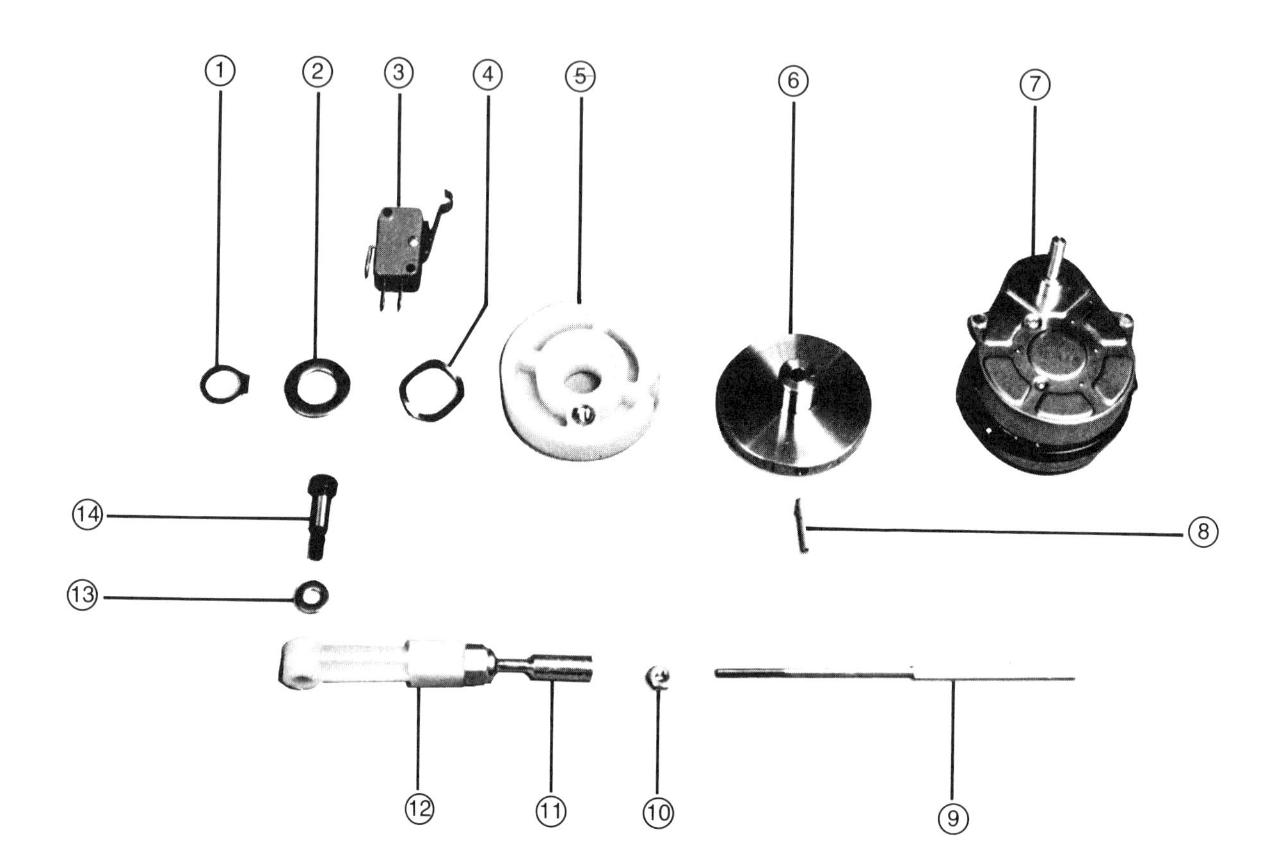

PARTS SECTION
HARVEST PROBE ASSEMBLY
C-61/81/121-C

Index #	Description	Part Number	Notes	Model C-61	C-81	C-121
1	Snap Ring	9021038-02		X	X	X
2	Flat Washer	9031003-03		X	X	X
3	Cam Switch	9101133-01		X	X	X
4	Wave Washer	9031030-01		X	X	X
5	Cam Driven Clutch Half	9051031-02	Plastic	X	X	X
6	Cam Driver Clutch Half	3991001-01	Stainless Steel	X	X	X
7	Probe Motor	9161030-01	115V	X	X	X
8	Screw-Set ¾"	9031033-04		X	X	X
9	Probe Shaft	9021053-02		X	X	X
10	Lock Nut	9031005-09		X	X	X
11	Swivel	9021017-01		X	X	X
12	Connecting Rod	9051057-01		X	X	X
13	Washer	9031004-03	Flat Washer	X	X	X
14	Allen Screw	9031029-01		X	X	X
	Probe Motor Bracket	3011857-01	Not Shown	X	X	X
	Motor Mounting Screw	9031008-28	Not Shown	X	X	X
	Motor Mounting Nut	9031005-06	Not Shown	X	X	X
	Motor Mounting Lockwasher	9031006-02	Not Shown	X	X	X
	Cam Switch Mounting Screw	9031008-16	Not Shown	X	X	X
	Cam Switch Mounting Nut	9031005-08	Not Shown	X	X	X

CONDENSING UNIT
C-61-C

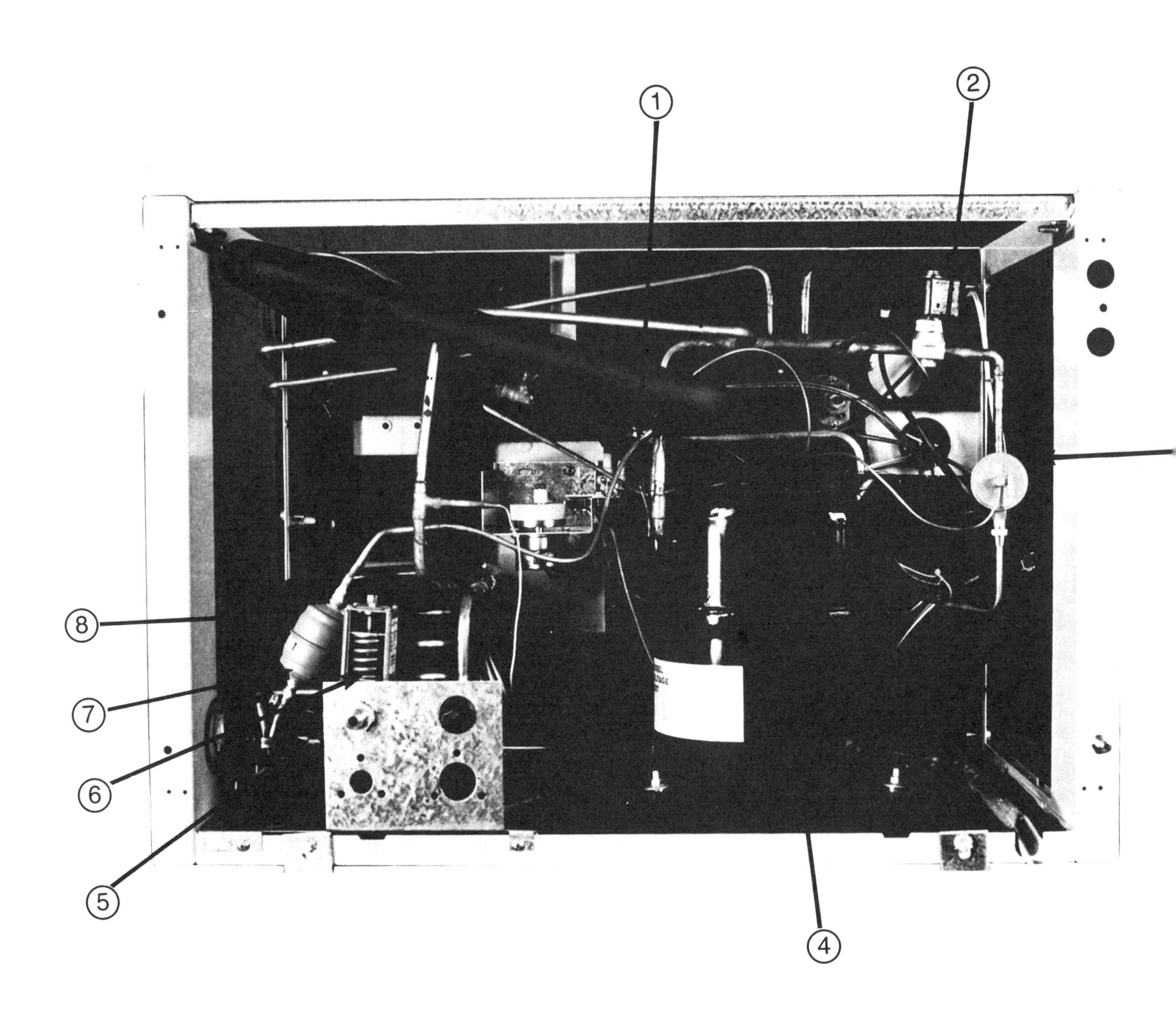

PARTS SECTION
CONDENSING UNIT
C-61-C

Index #	Description	Part Number	Notes	Model C-61	C-81	C-121
1	Evaporator Thermistor	9101131-02		X		
2	Hot Gas Valve Assy.	9151053-01	Assembly	X		
2A	Hot Gas Valve Coil	9151053-91	Coil Only	X		
3	Thermostatic Expansion Valve	9151006-01	R-502	X		
4	Compressor	9181054-01	Copeland ¾ HP R-502	X		
5	Condenser Thermistor	9101131-01		X		
6	Water Regulating Valve	9041010-01	Water Cooled Only	X		
7	Filter/Drier	9151004-02		X		
8	Water Condenser	9141024-01	Water Cooled Only	X		
	Air Condenser (Not Shown)	9141050-01	Air Cooled Only	X		
	Fan Motor (Not Shown)	9161078-01	Air Cooled Only	X		
	Fan Blade (Not Shown)	9131053-03	Air Cooled Only	X		
	Fan Motor Bracket (Not Shown)	9131054-01	Air Cooled Only	X		
	Fan Shroud (Not Shown)	9131122-01	Air Cooled Only	X		
	Water Purge Valve Assy. (Not Shown)	9041053-01	Assembly	X		

CONDENSING UNIT
C-81/121-C

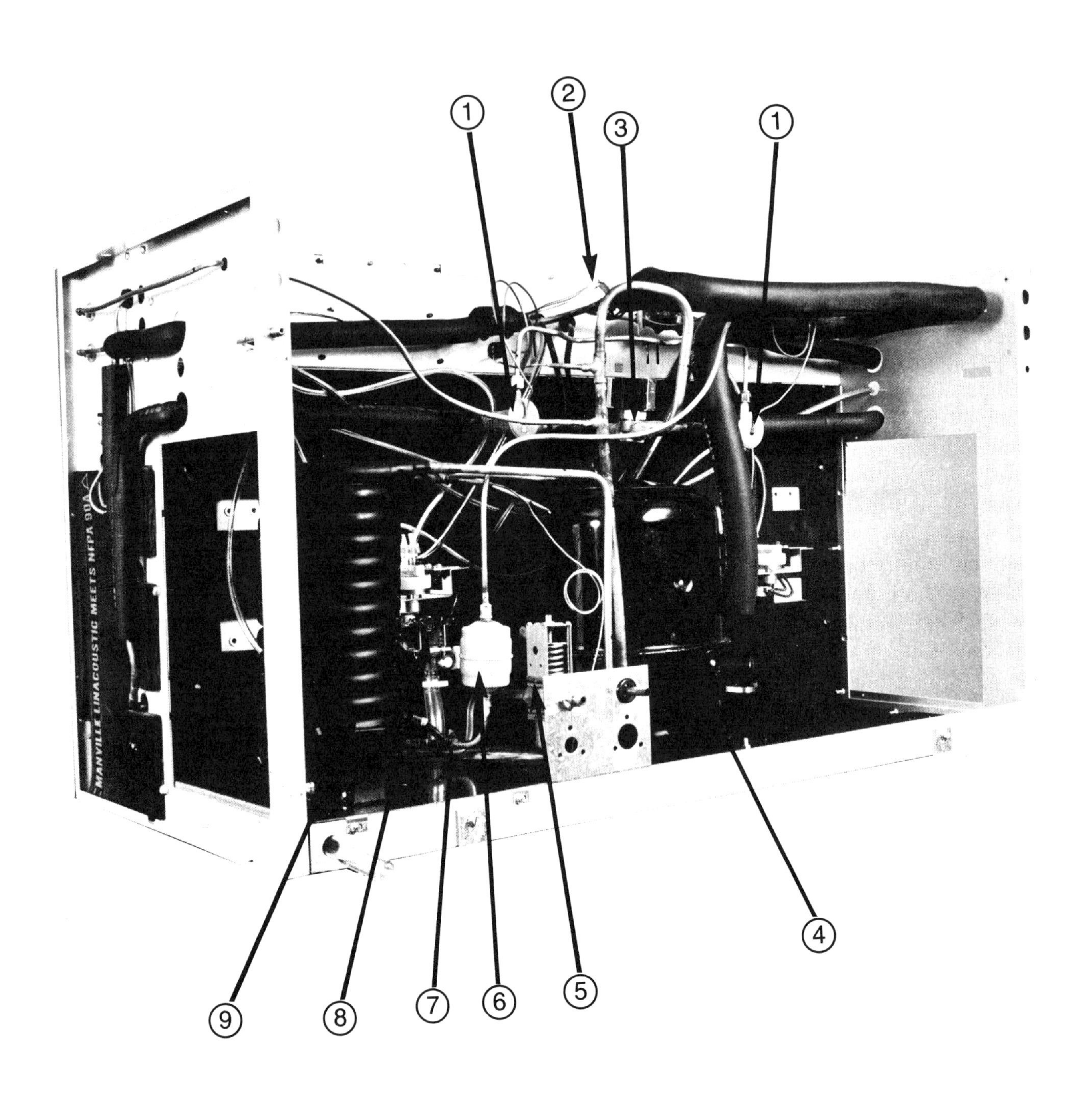

PARTS SECTION
CONDENSING UNIT
C-81/121-C

Index #	Description	Part Number	Notes	Model C-61	Model C-81	Model C-121
1	Thermostatic Expansion Valve	9151006-01	R-502			X
2	Evaporator Thermistor	9101131-02			X	X
3	Hot Gas Valve Assy.	9151012-02	Assembly		X	X
3A	Hot Gas Valve Coil	9151012-91	Coil Only		X	X
3B	Hot Gas Valve Internal Parts Kit	9151012-93			X	X
4	Compressor	9181038-91	Tecumseh 1½ HP R-502			X
5	Water Regulating Valve	9041010-01	Water Cooled Only		X	X
6	Filter/Drier	9151004-04			X	X
7	Condenser Thermistor	9101131-01			X	X
8	Water Purge Valve Assy.	9041014-02	Assembly		X	X
8A	Water Purge Valve Coil	9041053-90	Coil Only		X	X
9	Water Condenser	9141020-01	Water Cooled Only			X
	Water Condenser (Not Shown)	9141024-01	Water Cooled Only		X	
	Air Condenser (Not Shown)	9141052-01	Air Cooled Only		X	X
	Fan Motor (Not Shown)	9161045-04	Air Cooled Only		X	X

PARTS SECTION
CONDENSING UNIT (continued)
C-81/121-C

Index #	Description	Part Number	Notes	Model		
				C-61	C-81	C-121
	Fan Blade (Not Shown)	9131053-01	Air Cooled Only		X	X
	Fan Motor Bracket (Not Shown)	9131054-01	Air Cooled Only		X	X
	Fan Shroud (Not Shown)	9131131-01	Air Cooled Only		X	X
	Thermostatic Expansion Valve (Not Shown)	9151020-01	R-12		X	
	Compressor (Not Shown)	9181038-92	Tecumseh 1½ HP 3 phase R-502			
	Compressor (Not Shown)	9181060-91	Tecumseh 1 HP R-12		X	
	Compressor (Not Shown)	9181060-92	Tecumseh 1 HP 3 phase R-12			

NOTES

REMOTE CONDENSING UNIT

REMOTE CONDENSING UNIT

Index #	Description	Part Number	Notes	Model C-61	Model C-81	Model C-121
1	Remote Condenser Coil only	9141025-01	RC-60-2	X		
1a	Remote Condenser Coil only	9141021-01	RC-80-2, 120-2		X	X
1b	Remote Condenser Coil only	9141001-01	RC-80		X	
2	Remote Fan Switch	9101001-02	On/Off	X	X	X
3	Remote Fan Blade	9131006-01	Set of 4 (RC-60-2, 80, 120)	X	X	X
3	Remote Fan Blade	9131006-02	Set of 4 (RC-80-2, RC-120-2)		X	X
4	Remote Fan Shroud	9131005-01	RC-60-2, 80,120	X	X	X
4a	Remote Fan Shroud	9131126-01	RC-80-2, 120-2		X	X
	Remote Fan Motor (Not Shown)	9161001-01	115 Volt (RC-60-2, 80,120)	X	X	X
	Remote Fan Motor (Not Shown)	9161001-03	115 Volt (RC-80-2, 120-2)		X	X
	Remote Fan Motor Capacitor (Not Shown)	9101036-02	8MFD@236V (RC-60-2,80,120)	X	X	X
	Remote Fan Motor Capacitor (Not Shown)	9101036-04	12.5MFD@440V (RC-80-2, 120-2)		X	X
	Receiver (Not Shown)	9151026-01		X		
	Receiver (Not Shown)	9151060-01			X	X
	Liquid Line Pumpdown Solenoid Assy. (Not Shown)	9151039-01		X	X	X
	Liquid Line Pumpdown Solenoid Coil (Not Shown)	9151039-91		X	X	X
	Liquid Line Internal Parts Kit (Not Shown)	9151039-93		X	X	
	Mixing Valve (Not Shown)	9151027-01	R-12 Units		X	
	Mixing Valve (Not Shown)	9151027-02	R-502 Units	X		
	Pumpdown Control (Not Shown)	9041040-03		X	X	X

SERVICE AND MAINTENANCE RECORD

Date Purchased______________________________

Date Installed______________________________

Model Number______________________________

Serial Number______________________________

Service Policy Number__________________________

Have you filled out and mailed in the warranty card?

Service and Cleaning Record

Date	Maintenance Performed	By Whom

Service & Parts Manual

Cuber
UC-20-B

11100 East 45th Avenue
Denver, CO 80239
Phone (303) 371-3737
FAX (303) 371-6296

988 by Mile High Equipment Company.
:-O-MATIC 8/88

9081059-01

Parts Warranty

Mile High Equipment Company (the "Company") warrants the Ice-O-Matic ice machine to the original user-customer (the "Customer") against defects in material and factory workmanship, if properly installed, cared for and operated under normal conditions with adequate supervision, for a period of twenty-four (24) months from the original date of installation on all parts, and an additional thirty-six (36) months on all compressors, cuber evaporator plates and cuber computer control boards, from the date of original installation, when the Warranty Registration Card is returned to the Company. The Company will replace, F.O.B. the Company plant or F.O.B. the Company authorized distributor, without cost to the Customer, that part of any such machine that becomes defective, except that no replacement will be made for a part or assembly which has been (I) subject to any alteration or accident; (II) used in any way which, in the Company's opinion, adversely affects the performance, or (III) is from a machine on which the serial number has been altered or removed.

Labor Warranty

In addition to the parts warranty, Ice-O-Matic will pay straight time labor (effective on all equipment installed after January 1, 1988) to correct defects in material or factory workmanship for twenty-four (24) months on these key components: compressors, cuber evaporator plates, cuber harvest motors, water pumps, computer control boards, air/water/remote condenser, condenser fan motors, flaker gear reducers, auger motors, and flaker evaporator barrels and on all other parts for twelve (12) months from the original date of installation. This warranty is valid only when installation, service and preventative maintenance are performed by a Company authorized distributor, a Company authorized service agency, or a Company Regional Marketing Manager.

The decision of the Company's Service Department with respect to repair or replacement of a part shall be final. Any defective part to be repaired or replaced must be returned to the Company, transportation charges prepaid, and properly sealed and tagged. The serial and model number of the machine, and the date of original installation of such machine must be included. The Company does not assume any responsibility for any expenses incurred in the field incidental to the repair or replacement of equipment covered by this warranty.

This warranty does not apply to destruction or damage caused by alterations, by unauthorized service, using other than identical replacements, risks of transportation, damage resulting from adverse water conditions, accidents, misuse, abuse, damage by fire, flood or acts of God.

The liability of the Company for breach of this warranty shall, in any case, be limited to the cost of a new part to replace any part which proves to be defective. The Company makes no representations or warranties of any character as to accessories or auxiliary equipment not manufactured by the Company.

THIS WARRANTY AND THE COMPANY'S OBLIGATION HEREUNDER IS IN LIEU OF ALL WARRANTIES, EXPRESSED OR IMPLIED OR ARISING BY CUSTOM OR TRADE USAGE. WITHOUT LIMITING THE GENERALITY OF THE FOREGOING, **THE IMPLIED WARRANTIES OF MERCHANTABILITY AND FITNESS FOR ANY UNIQUE, SPECIAL OR PARTICULAR PURPOSE ARE EXPRESSLY DISCLAIMED,** TOGETHER WITH ALL OTHER REPRESENTATIONS TO THE ORIGINAL USER-CUSTOMER AND ALL OTHER OBLIGATIONS OR LIABILITIES ON THE PART OF THE COMPANY WITH RESPECT TO THE SALE OR USE OF THE EQUIPMENT, INCLUDING WITHOUT LIMITATION LIABILITY FOR DAMAGES WHETHER GENERAL OR SPECIAL, DIRECT OR INDIRECT, INCIDENTAL, CONSEQUENTIAL AND EXEMPLARY OR FOR ANY CLAIM FOR THE LOSS OF PROFITS OR BUSINESS OR DAMAGE TO GOOD WILL. IT IS EXPRESSLY AGREED THAT THE ABOVE WARRANTY SHALL BE THE SOLE AND EXCLUSIVE REMEDY OF THE CUSTOMER.

No person is authorized to give any other warranties or to assume any other liability on the Company's behalf unless done in writing by an officer of the Company.

Mile High Equipment Company 11100 E. 45th AVE, DENVER, CO 80239 303 371-3737 FAX 303 371-6296

Uncrating & Set Up

Step by Step Instructions

1. Inspect exterior of shipping carton and skid for any signs of shipping damage.
2. Using a pry bar or large screwdriver, remove all staples from bottom of carton. Do not pry against carton.
3. Lift carton straight up, entirely off machine. Do not discard carton until satisfactory inspection of machine is complete.
4. Inspect machine cabinet for any signs of shipping damage. If any damage is noted, stop installation and follow Freight Claim Procedures on the following page.
5. Remove the four 3/4-inch skid bolts located underneath skid. Lift machine off skid and place on bin.
6. Make all water and drain connections.
7. Make certain that unit is level side-to-side and front-to-back. Accurate leveling is essential for proper operation.
8. Carefully follow Installation and Start-Up instructions in Sections B and C of this manual.

Freight Claim Procedures

Important

Inspect Promptly

This merchandise has been carefully inspected and packed in accordance with the carrier's packing specifications. Responsibility for safe delivery has been assumed by the carrier. If loss or damage occurs, you as consignee must file a claim with the carrier, and hold container for carrier's inspection.

Visible Loss or Damage

Any external evidence of loss or damage must be fully described and noted on your freight bill or express receipt and signed by the carrier's agent. Claim should then be filed on a form available from the carrier.

Concealed Loss or Damage

If loss or damage does not appear until merchandise has been unpacked, make a written request for inspections by the carrier within 15 days of the delivery date. Then file a claim on a form available from the carrier.

File Claim Without Delay

Do Not Return Damaged Merchandise to ICE-O-MATIC

Foreword

We provide this service manual to aid the service men and users in the installation, operation, and maintenance of your ICE-O-MATIC equipment.

If at any time, you encounter conditions that are not answered in this manual, write to:

Service Department
ICE-O-MATIC
11100 East 45th Avenue
Denver, CO 80239

All Service Communications Must Include:

- Model Number
- Serial Number
- A Detailed Explanation of the Problem or Condition.

We will give your letter our immediate attention and will reply promptly.

Notes

Table of Contents

General Information

UC-20 Cuber

General Information

Dimensions

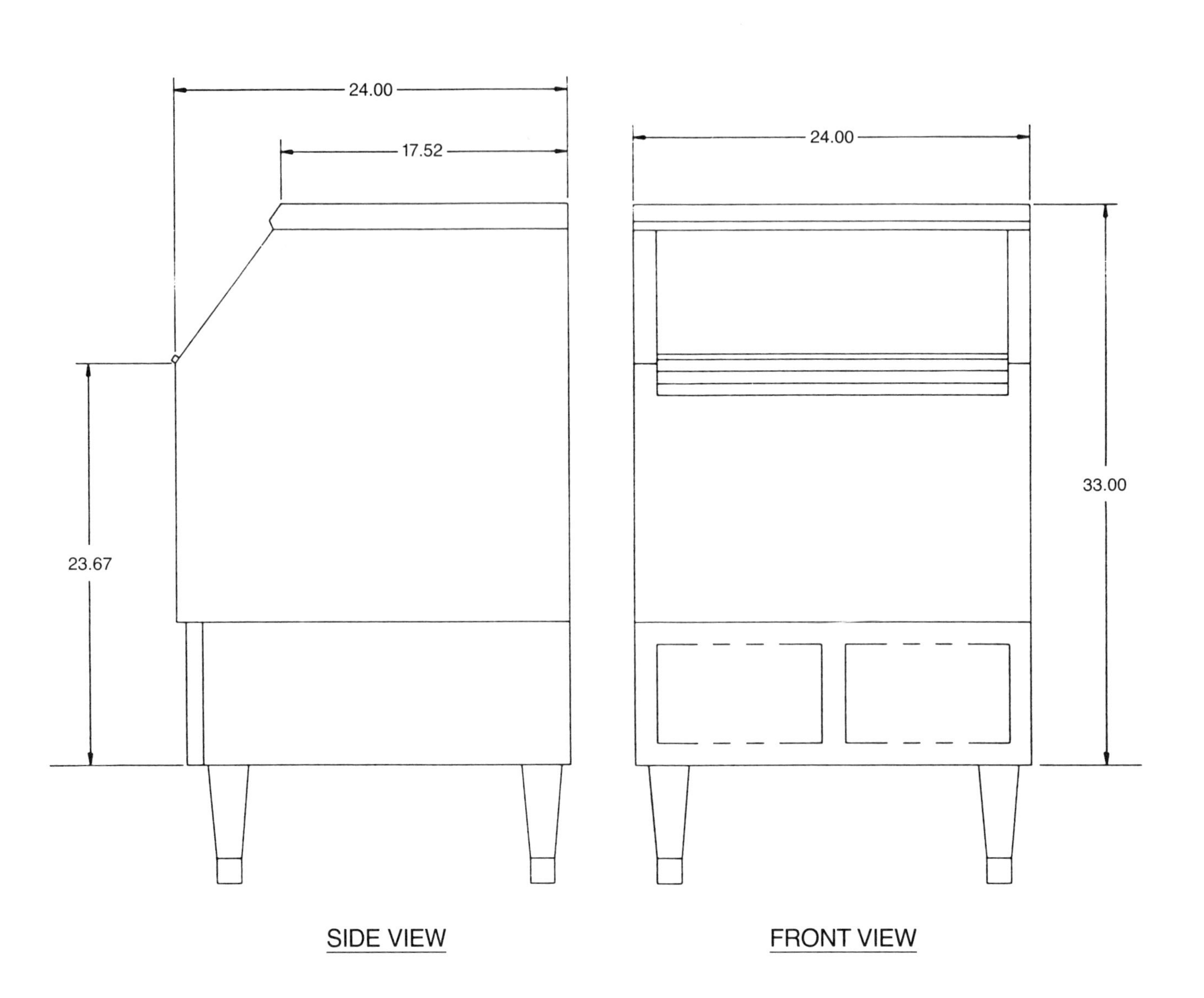

General Information

Electrical & Mechanical Specifications
Ice Production Chart

Electrical & Mechanical Specifications

Model Number	Condensing Unit	Comp. H.P.	Voltage Requirements	No. Of Wires	Operating Amps @ Rated Voltage	Max. Amp Fuse Size	Refrigerant Type	Refrigerant Charge (oz.)	Approx. Shipping Weight (lbs.)
UC-20-A	Air	1/3	115/60/1	3	7.6	15	R-12	14	176
UC-20-W	Water	1/3	115/60/1	3	7.0	15	R-12	13	176

Note:
This machine requires a separate ground.

Ice Production Chart

Condensing Unit	Air Temp. °F	Water Temp. °F 80°	70°	50°
Air	70°	170	181	210
	80°	142	153	189
	90°	116	127	149
Water	70°	166	187	215
	80°	155	172	205
	90°	142	161	202
Model Number		UC-20		

Notes

General Information

How the UC-20 Cuber Works

A general description of how the UC-20 cuber works is given below. The remainder of Section A provides more detail about the cycles, components, and systems.

During the freezing cycle water is circulated over the evaporator freezing plate where the ice cubes are formed. When the low side has pulled down to 13 p.s.i. on the UC-20, the timer initiate control contacts close energizing the time delay module. The timer is preset at the factory to achieve a 3⁄16" bridge thickness. It may be necessary to readjust the timer at start-up for proper thickness (see page C-3).

There are two steps to the freeze cycle. The time it takes for the low side pressure to pull down to 13 p.s.i., and also the time delay setting to build the bridge across the front of ice slab (3⁄16" thick for proper harvest). The time it takes to achieve the thickness of this bridge depends on surrounding air and incoming water temperatures.

Once the time mechanism in the time delay has finished timing, power is supplied to the (#1) relay coil. This change supplies power to the hot gas solenoid valve, water purge valve, and the harvest assist motor. The water purge valve opens, and allows the water pump to purge the water from the water trough removing all impurities and sediment from the trough. This allows the unit to make clear ice cubes, and keep mineral build-up at a minimum in the water system.

The hot gas solenoid opens allowing hot gas to go directly to the evaporator, breaking the bond between the evaporator and the slab of ice for harvesting. Once the evaporator has reached approximately 40°F in temperature, the harvest assist motor overcomes the bonding of the ice to the evaporator, and releases the slab of ice off the evaporator and into the bin.

As the harvest assist motor turns, it activates a cam switch which is riding on the outside edge of the cam. The cam is driven by the harvest assist motor. When the cam switch goes from the normally closed position, to the normally open position, it resets the timer to allow the machine to go back into another freeze cycle after the harvest assist motor has made a complete rotation. If the ice level in the storage bin is high enough to contact bin thermostat capillary tube the bin thermostat opens and the machine will shut off.

General Information

Operating Cycle

Freeze Cycle (Step 1)

The compressor is energized through the contactor. The water pump is energized through the normally closed contacts of the cam switch. The fan motor is energized through the normally closed contacts of relay #1.

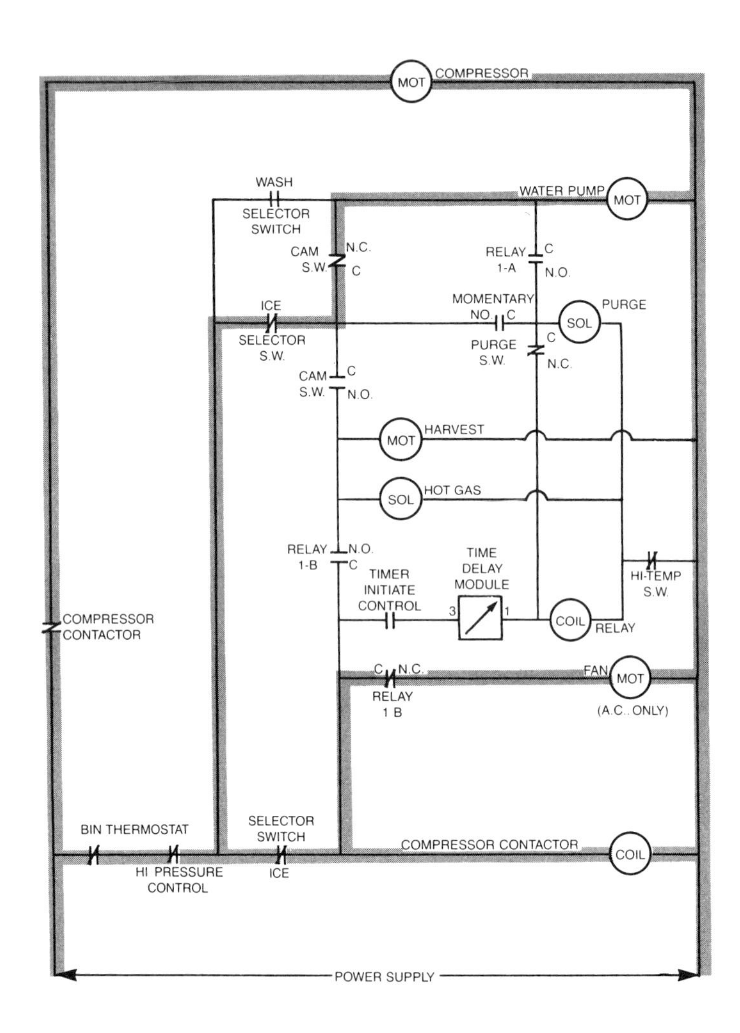

Freeze Cycle (Step 2)

The compressor, water pump and fan motor are energized as described in Cycle 1. The Timer Initiate Control has closed (13 p.s.i.) energizing the Time Delay Module.

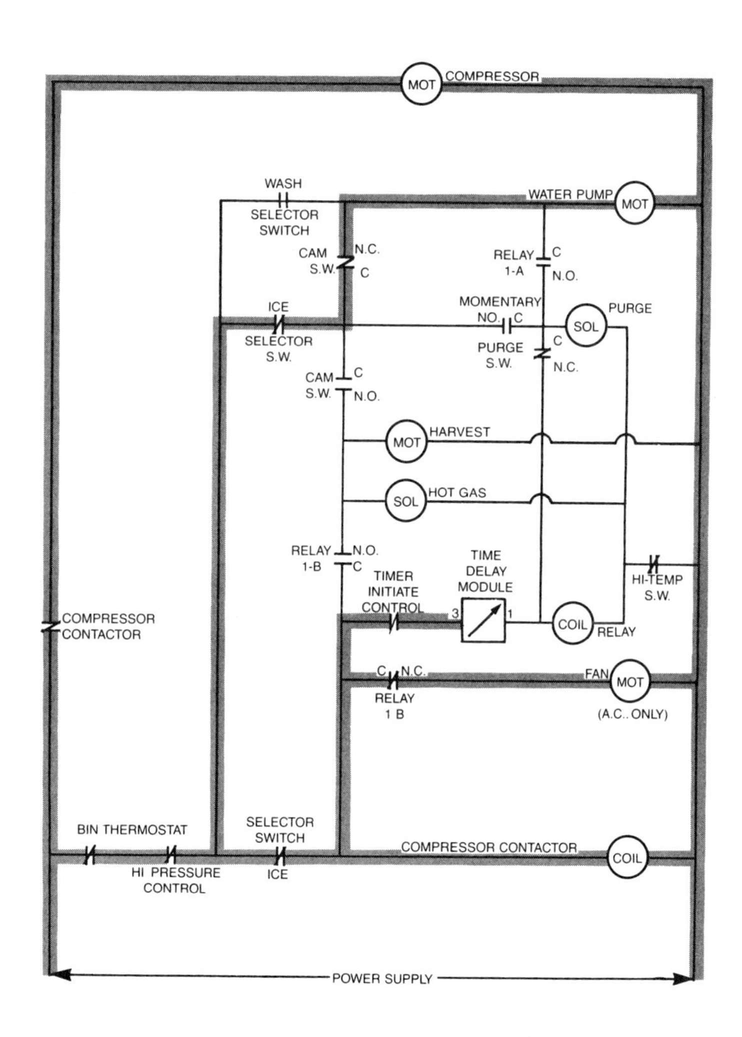

Operating Cycle

Defrost Cycle

When the Time Delay Switch closes, power is supplied to relay #1. Relay #1, through its normally open contacts, energizes the hot gas solenoid, harvest assist motor and water purge valve. The fan motor is de-energized through the normally closed contacts of relay #1.

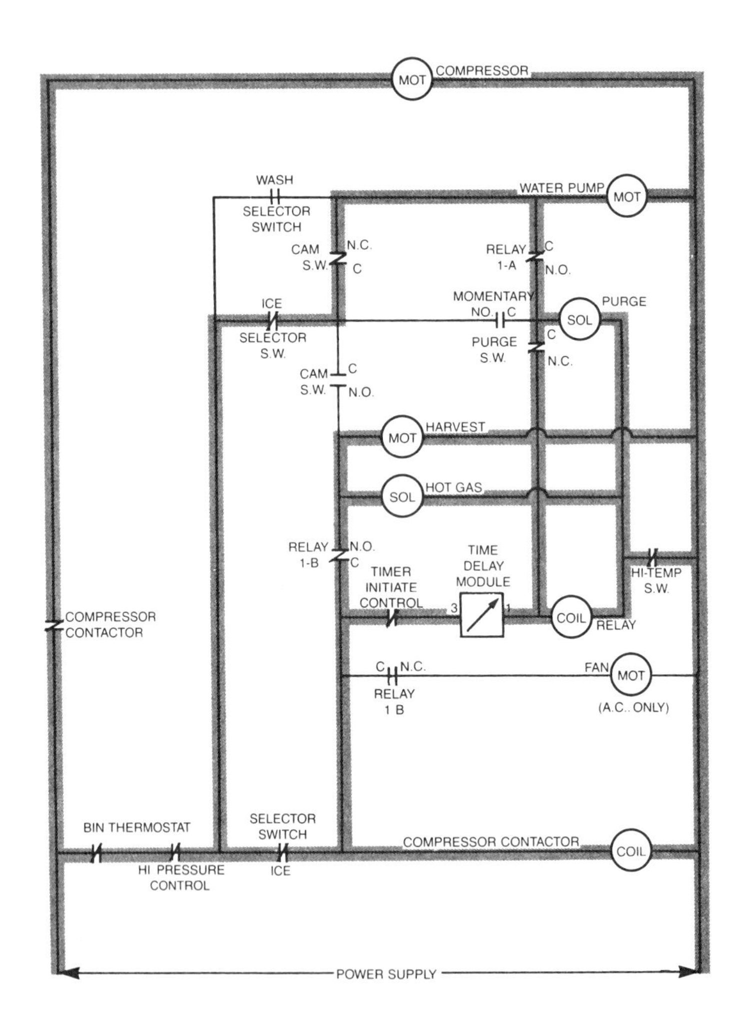

General Information

Operating Cycle

Harvest Cycle

The harvest assist motor has overcome the capillary attraction of the ice to the evaporator plate. The motor has moved and changed the position of the cam switch from normally closed to normally open. This has de-energized #1 relay, the water pump and water purge valve. The hot gas solenoid and harvest assist motor remain energized through the normally open contacts of the cam switch. The timer initiate control has opened due to the rise in suction pressure, de-energizing the time delay module. After the harvest assist motor has returned to its original position and the cam switch contacts have switched back to normally closed, the machine will begin another freeze cycle.

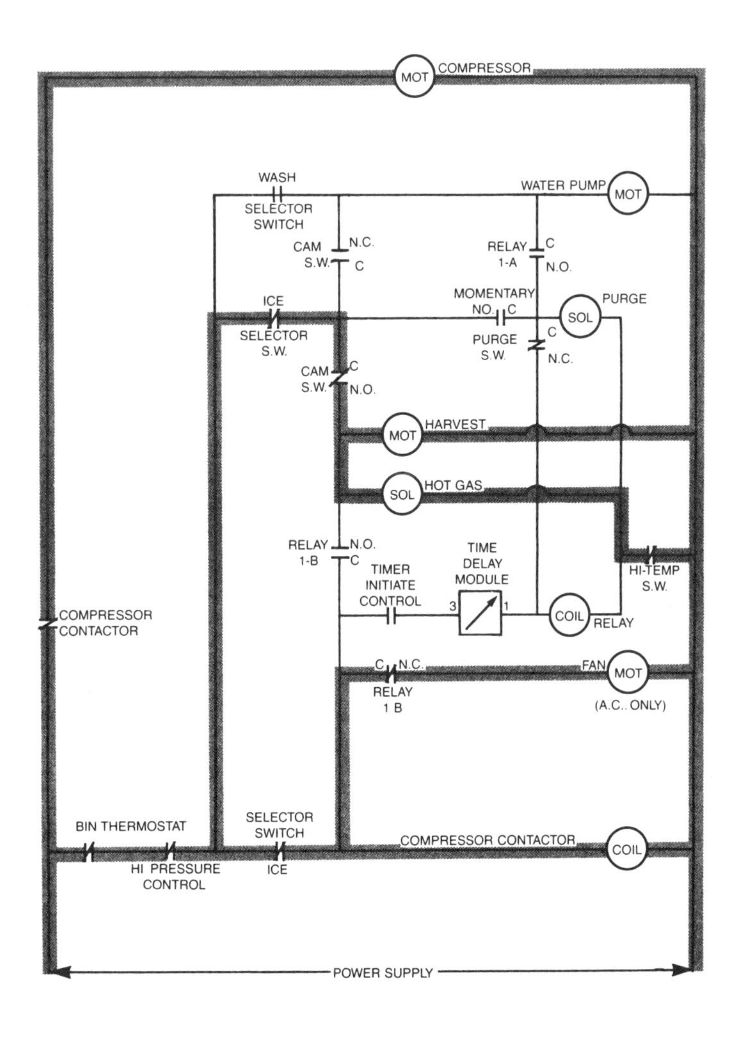

General Information

Component Descriptions

Danger

Electrical shock and/or injury from moving parts inside this machine can cause serious injury. Disconnect electrical supply voltage to machine prior to performing any adjustments or repairs.

1. Cam Switch

Actuated by the driven cam on harvest assist motor. N.C. contacts supply power to the water pump during freeze cycle. N.O. contacts supply power to harvest assist motor and hot gas solenoid valve during harvest cycle. Adjustment is made by bending the switch arm in or out.

2. Bin Thermostat

When the ice level in the storage bin is high enough to contact bin thermostat capillary tube, the bin thermostat opens. The sensitivity of thermostat can be adjusted. See page C-2 for procedure.

3. Timer Initiate Control

Energizes the solid-state timer. This control closes at 13 p.s.i. See page C-2 for adjustment procedure.

4. Timer/Bridge Thickness

The timer is located in the control box. To change settings — By moving the switch next to the number to an "ON" position (note the On-Off marking on the timer) that number is activated. The numbers indicate seconds. To determine total time delay add up the number turned to the ON position. To increase the bridge thickness add time. To decrease bridge thickness subtract time. See page C-3 for diagram.

5. #1 Relay

Located in control box. When energized, it supplies power to hot gas valve, water purge valve and harvest assist motor during defrost cycle.

6. Ice/Off/Wash Switch

A three position switch located on the front of the control box. Power is supplied to the ice-making components when the switch is in the ICE position. With the switch in the WASH position, power is supplied to the pump only, for cleaning purposes. The center position (OFF) shuts the machine off completely.
Note:
This switch is not a power disconnect. Before servicing unit you must turn off ice machine circuit breaker, or unplug ice machine.

7. Water Purge Switch

Located next to the ICE/OFF/WASH switch. By holding in the purge switch, with selector switch in the ICE or WASH position, it opens the water purge solenoid valve and completely empties the water trough. This switch is used after cleaning the ice machine to flush the ice machine cleaning solution out of the water trough to the drain.

8. High Pressure Control

Used on all models. This control turns machine off if air flow across condenser is interrupted or water supply to water cooled condenser is interrupted. The control is pre-set at the factory and is non-adjustable. Cut-out pressure is set at 350 p.s.i. on air cooled models and 250 p.s.i. on water cooled models. The control is located in control box and requires a manual reset.

9. Harvest Assist Motor

A slip clutch type motor. When the clutch pressure can overcome the capillary attraction of the ice to the evaporator plate the harvest assist motor pushes the ice slab out of the evaporator and into the bin. There are no adjustments to be made to the clutch portion of this assembly. The length of the probe shaft is adjustable; it is set at the factory. If it is necessary to adjust the length of this shaft, it can be done by loosening the nut on the shaft, and threading the shaft in, or out of the coupler in which it is inserted. For proper adjustment, plastic probe tip in the freeze cycle position should be 1/32 of an inch back into the bushing it rides on, so it does not become frozen into the ice. Also see probe alignment on page C-4.

10. High Temperature Safety Switch

Located on suction line — Installed on all models. If the suction line temperature exceeds 120°F during defrost, this Bi-Metal switch opens and returns the machine to a freeze cycle by interrupting power to the hot gas solenoid. The switch will close when the suction line temperature drops to 80°F.

11. Water Regulating Valve

An adjustable valve. The head pressure on the water cooled models should be set at 125 p.s.i.

Water Distribution System

Water Distribution Tube

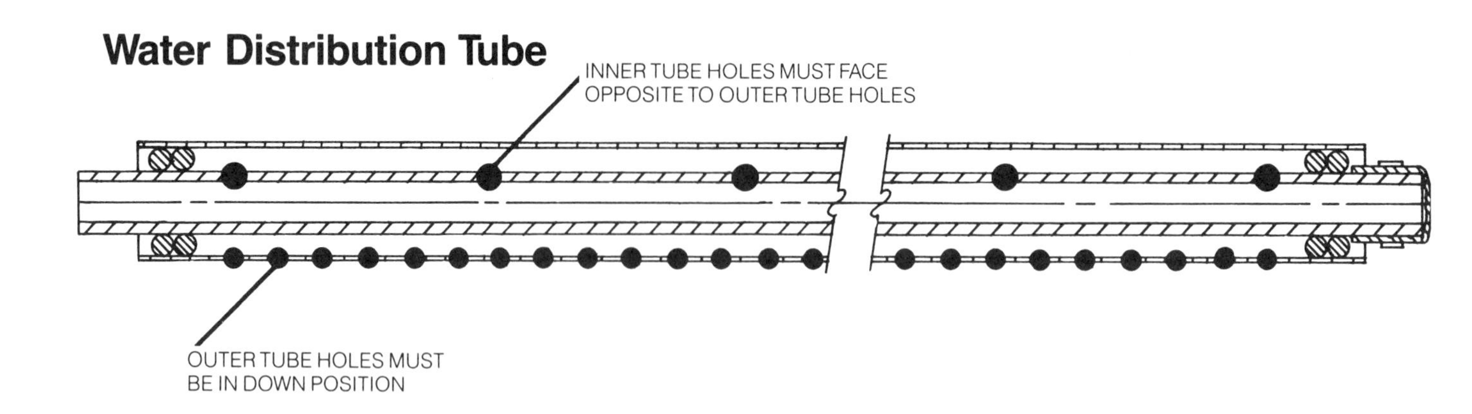

Water System

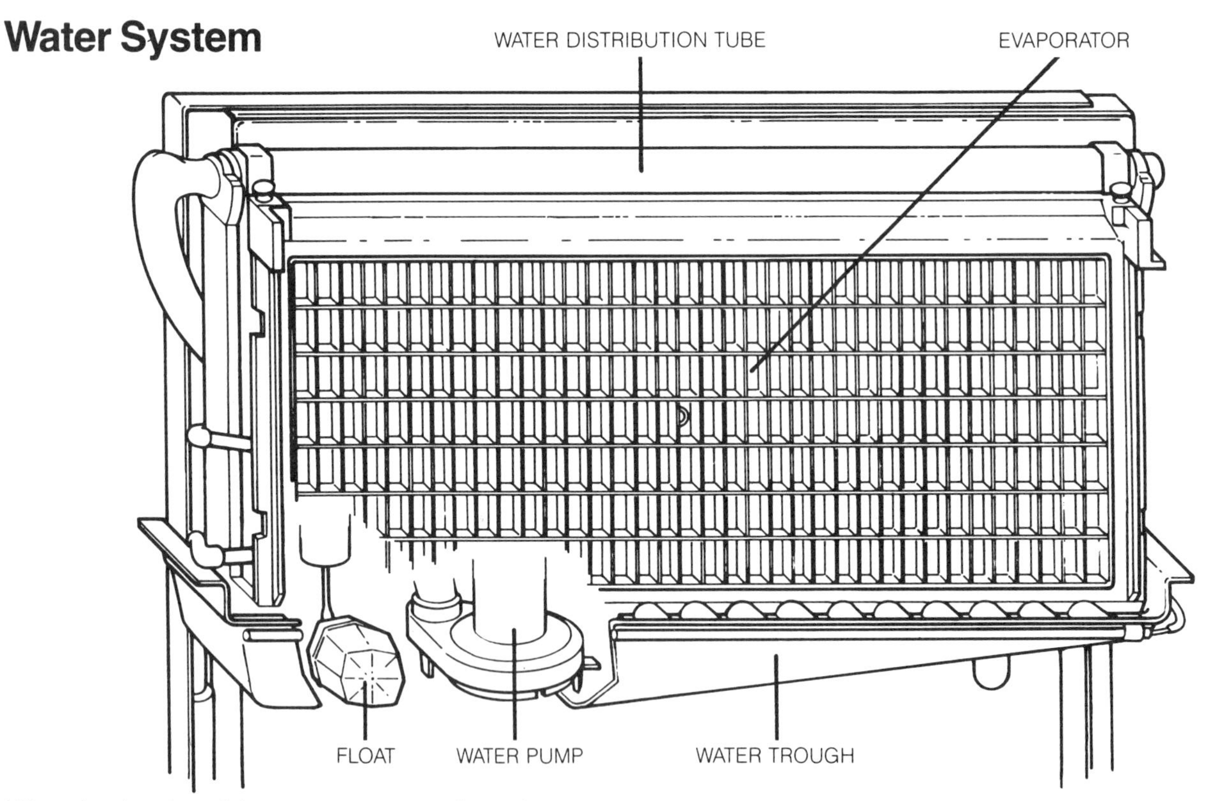

Water is circulated, by a water pump, from the water trough over the evaporator freeze plate to be frozen. Any water that is not frozen returns to the water trough to be re-circulated. At the end of the freeze cycle, when the machine is in hot gas defrost, a water purge valve opens and the water pump continues to run. This expels all water from the water trough through the purge valve drain. Completely dumping the water in this manner keeps the cubes clear, and lengthens the time between cleanings. The purge valve closes, and the water pump stops when the harvest assist motor moves the cam switch to its n.o. contacts. This allows the water trough to re-fill in preparation for the next freeze cycle.

General Information

Refrigeration System

Normal Refrigeration Cycle

The figure below shows the Refrigeration System Schematic. During the freezing cycle, high pressure liquid refrigerant is fed from the condenser through a heat exchanger and drier to a thermostatic expansion valve. This valve meters the refrigerant to the coils on the back side of the evaporator. The refrigerant absorbs heat causing a portion of the water flowing over the evaporator to freeze into ice. The coils are maintained at a low pressure by the action of the compressor.

The low temperature and pressure refrigerant leaving the coils is directed through the heat exchanger and is returned to the compressor. There it is compressed to a high temperature and pressure gas. It is then directed to a condenser to be converted again to high pressure liquid.

During the harvest cycle, a solenoid valve (normally closed during the freezing cycle) is opened to direct the high temperature gas leaving the compressor into the freezing plate coils ahead of the expansion valve. This gas rapidly warms up the plate to above freezing. The ice slab is thawed loose from the plate and then released by the harvest assist motor. At the end of the harvest cycle the solenoid valve is closed and a freezing cycle is again started.

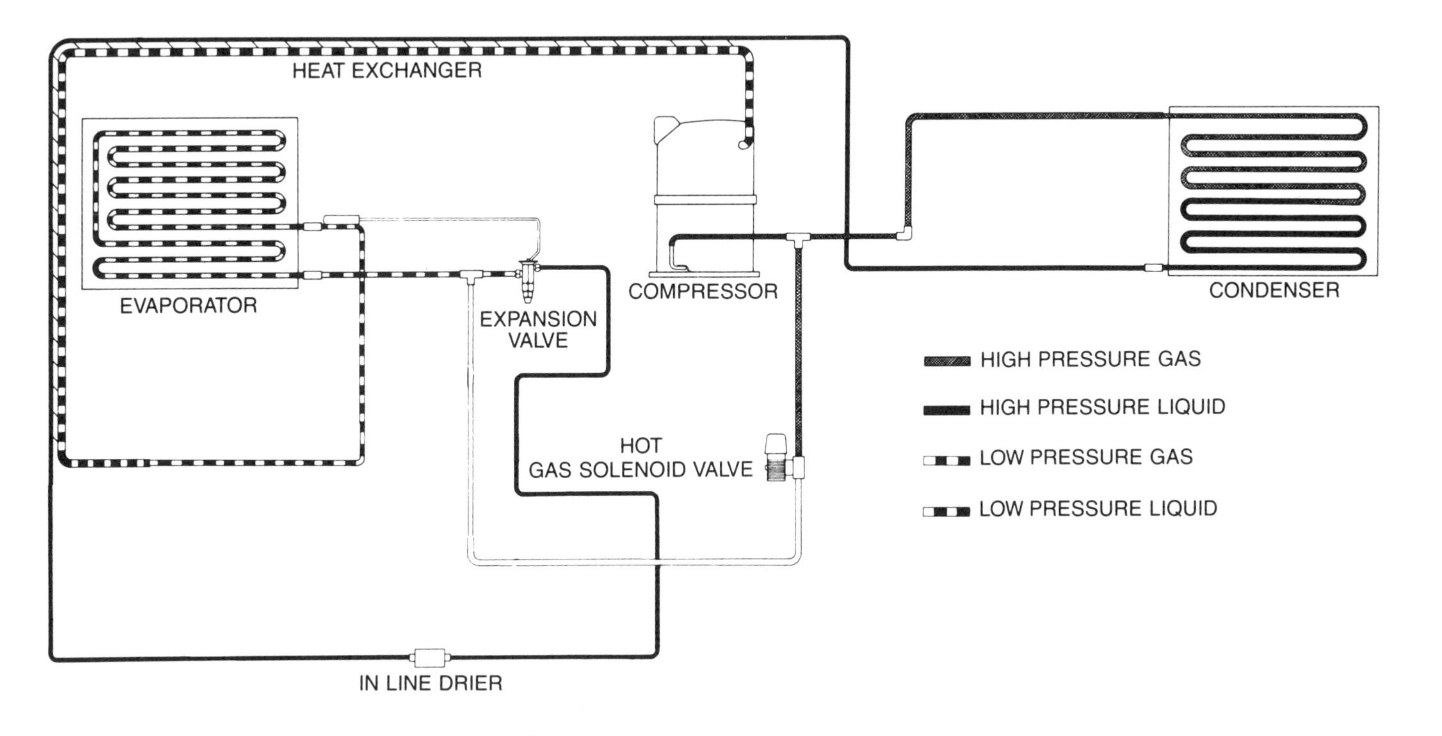

Electrical Instructions & Diagram

Danger

Electrical shock and/or injury from moving parts inside this machine can cause serious injury. Disconnect electrical supply voltage to machine prior to performing any adjustments or repairs.

Important

All installations must comply with local and national electrical codes.

Instructions

1. Check for proper voltage, wire, and fuse size according to specifications listed on serial plate, located on machine cabinet.
2. Supply power to ice machine using a separate electrical circuit. Only the ice machine should be on this circuit.
3. Make electrical connections in the control box, located in back of machine, per wiring diagram below.
4. Secure ground wire to green ground screw provided in the control box.

Electrical Diagram

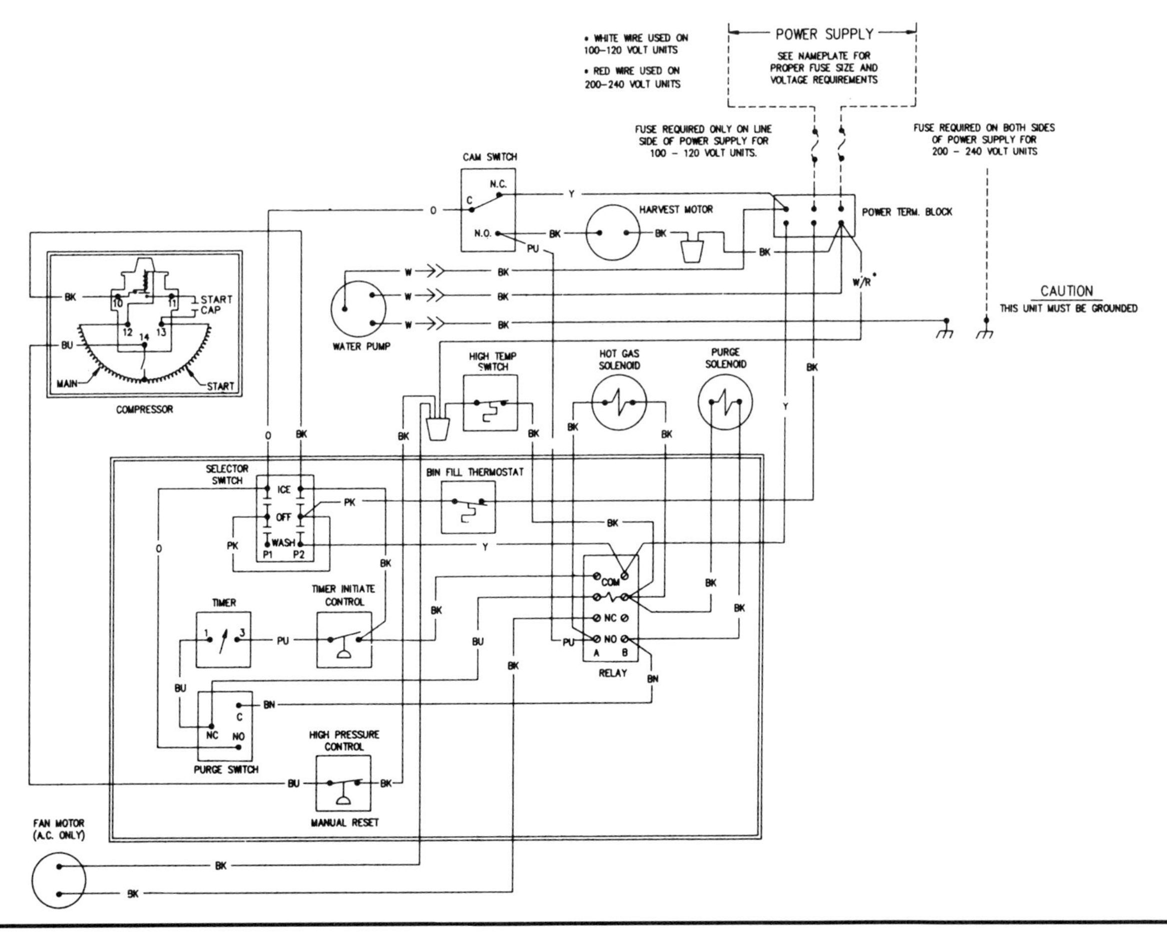

Plumbing Instructions

Danger
Electrical shock and/or injury from moving parts inside this machine can cause serious injury. Disconnect electrical supply voltage to machine prior to performing any adjustments or repairs.

Important
All installations must comply with local and national plumbing codes.

Water Connections
1. See Page B-3 for location and size of water inlet fittings.
2. Check for correct water pressure, water pressure must be between 20 p.s.i.g. and 60 p.s.i.g. for proper operation.
3. Connect ice making water supply line to ¼ male Flare at rear of machine.
4. Connect separate condenser water supply line (water-cooled models only) to ⅜ O.D. tubing at rear of machine.

Drain Connections
1. See Page B-3 for location and size of drain fittings.
2. Insure that an open or trapped drain is available within 10 feet of ice machine.
3. Connect a separate well-pitched purge drain line to the ⅝ I.D. tubing at rear of machine.
4. Connect a separate, well-pitched condenser drain line (water-cooled models only) to the ⅜″ O.D. tubing at rear of machine.
5. Connect a separate, well-pitched bin drain line to the ¾ F.P.T. fitting at rear of machine.

Installation

Plumbing Diagrams

Rear View — Air Cooled/Water Cooled

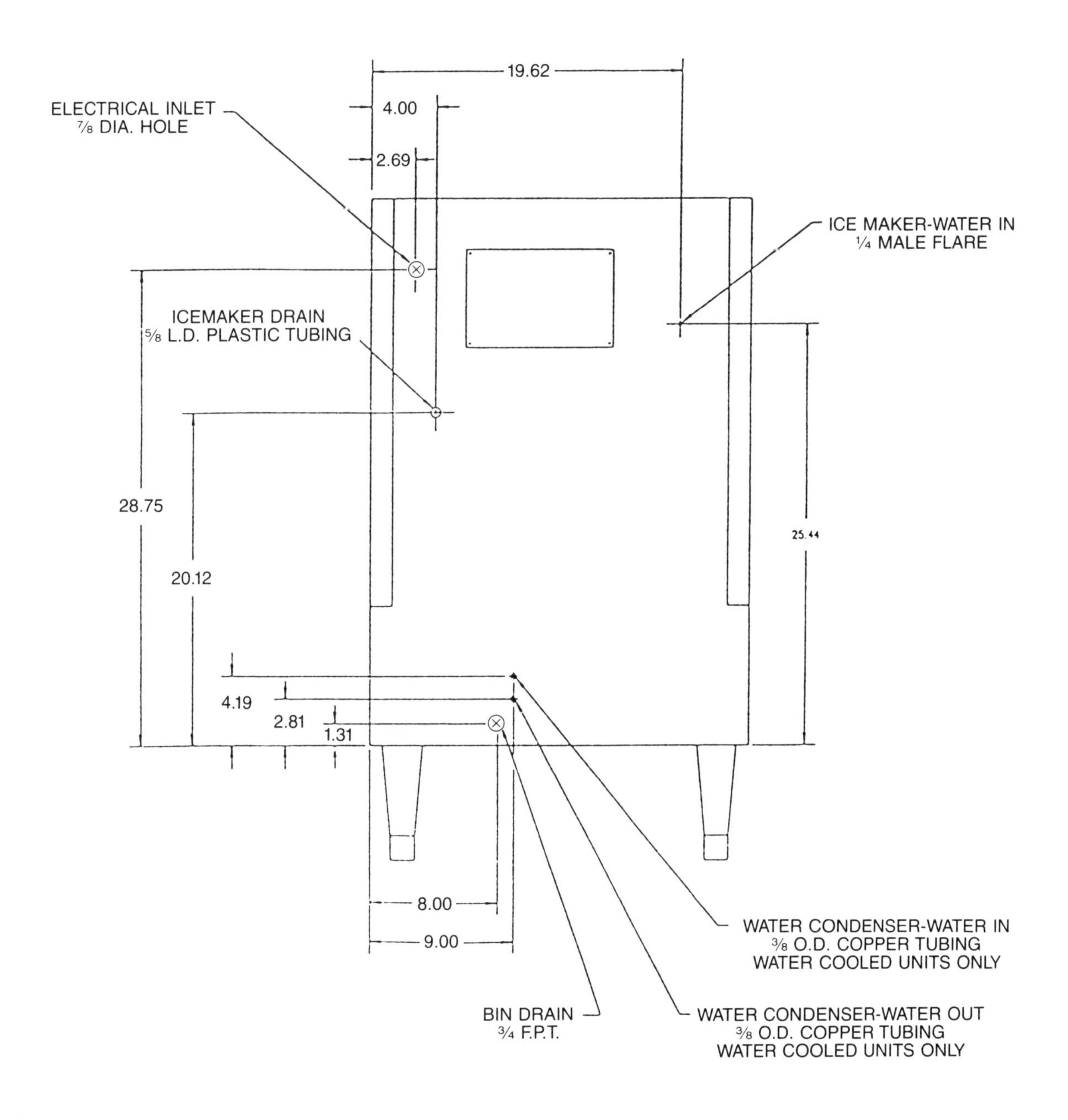

Operation

Start-Up

Please Follow These Instructions Carefully.

1. Check operation of ICE/OFF/WASH selector switch and test purge function (Step C).

   a. ICE Position: If the bin is not full, all ice-making components will turn on and the machine will begin a freeze cycle.

      If the bin is full, and the bin thermostat is OPEN, all components will be turned off. The components will remain off until the bin thermostat CLOSES, then the unit will begin a freeze cycle. Ice making cycles will continue until the bin becomes full or the selector switch is moved to either the OFF or WASH position.

   b. OFF Position: Entire machine and all components will be off when selector switch is in this position.

   c. WASH Position: All machine components will be off except the water pump. At this time the purge function can be tested by depressing and holding down the purge switch. The purge valve will open allowing all water in the water trough to be removed by the water pump.

      Note:
      If suction line temperature is above 120°F, the hi-temp safety control will be open. This will prevent purge valve from operating. The suction line must reach 80°F, to close hi-temp safety control.

2. Move selector switch to ICE position for initial freeze cycle.

3. Observe first slabs of ice for 3⁄16″ bridge thickness. If adjustments are necessary, follow procedures on page C-3.

4. Fill out the Quality Control Card and return to factory.

5. Fill in information on Page F-1. The full model and serial numbers will be required information when ordering parts or requesting factory service assistance.

Operation

Bin Thermostat Adjustment and Timer Initiate Control Adjustment

Danger

Electrical shock and/or injury from moving parts inside this machine can cause serious injury. Disconnect electrical supply voltage to machine prior to performing any adjustments or repairs.

Bin Thermostat

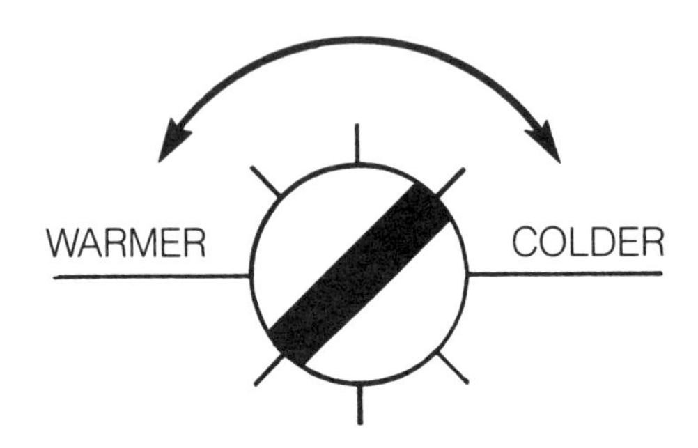

Adjustment

Hold ice to the bin thermostat capillary tube. Within 1 to 2 minutes the machine should shut off. Turning the adjustment screw to warmer position will shorten this time, and turning the adjustment screw to colder position will increase this time.

Timer Initiate Control

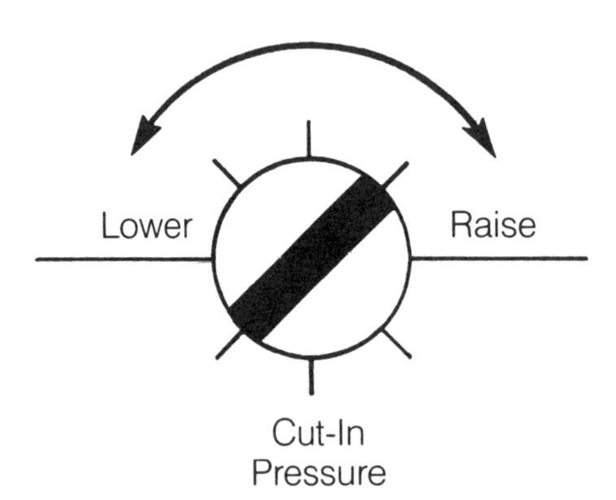

Adjustment

The proper cut-in pressure on UC-20 is 13. p.s.i. To lower cut-in pressure turn adjustment screw counter-clockwise. To raise cut-in pressure turn adjustment screw clockwise.

Operation

Bridge Thickness Adjustment

Danger

Electrical shock and/or injury from moving parts inside this machine can cause serious injury. Disconnect electrical supply voltage to machine prior to performing any adjustments or repairs.

Time Delay Module Adjustment

To change timer settings — By moving the switch next to the number to an "ON" position (note the ON-OFF marking on the timer) that number is activated. The numbers indicate seconds. To determine total time delay add up the numbers turned to the ON position.

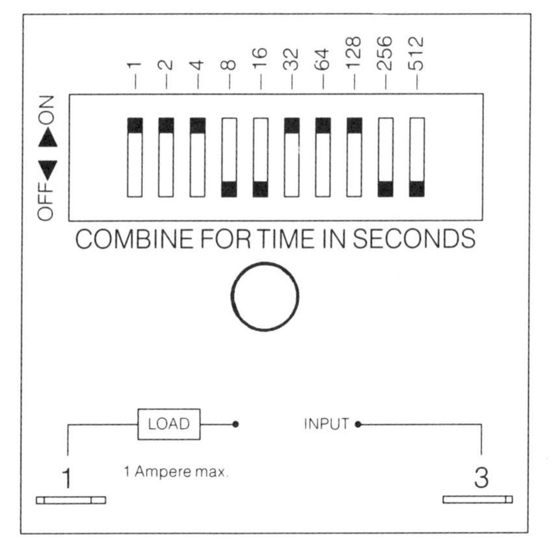

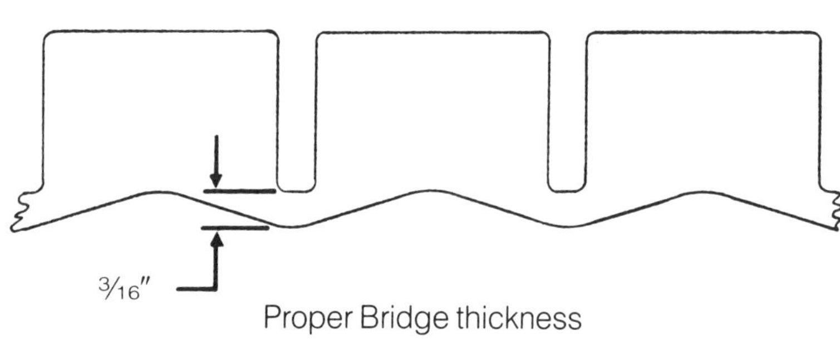

Proper Bridge thickness

Changing Ice Bridge Thickness

The ice bridge thickness is controlled by the time delay module. To increase the bridge thickness, increase the timer setting. To decrease the bridge thickness, decrease the timer setting.

Operation

Probe Alignment

Danger

Electrical shock and/or injury from moving parts inside this machine can cause serious injury. Disconnect electrical supply voltage to machine prior to performing any adjustments or repairs.

It is possible for the evaporator to shift in transit. It is recommended that alignment of the probe be checked at the time of installation. This should be done as follows:

1. Remove allen screw ② from the swivel part of the probe shaft. Rotate probe by hand in direction of arrow following the clutch surface as if it were in normal operation.

If there is any binding during rotation, re-adjust mounting bracket by loosening mounting screws on the assembly bracket. Once shaft is free (there should be no drag at all), re-tighten mounting screws, and re-check.

Note:
When aligning probe, check that probe lock nut (pointed out by arrow marked ①) is tight and that tip of probe does not extend out of the evaporator bushing into the freezing surface of the evaporator.

2. Check probe clutch allen screw in stainless steel portion of clutch (pointed out by arrow marked ③), to insure it's tight against the flat portion of the harvest motor drive shaft.

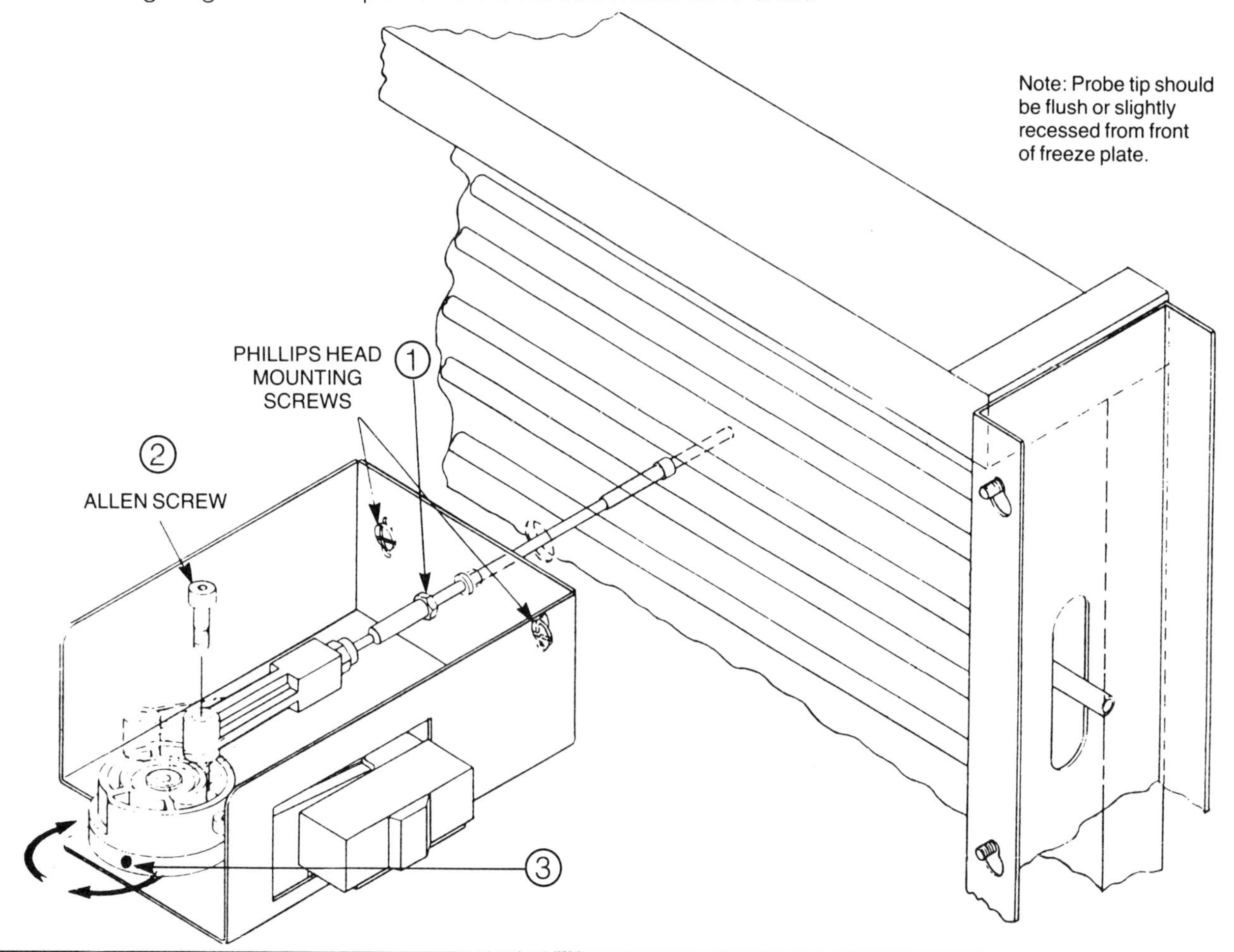

Routine Maintenance

Maintenance Record

Date Purchased ______________________________

Date Installed ______________________________

Model Number ______________________________

Serial Number ______________________________

Service Policy Number ______________________________

Date	Maintenance Performed	By Whom

Routine Maintenance

Quarterly Maintenance Procedure

Danger

Electrical shock and/or injury from moving parts inside this machine can cause serious injury. Disconnect electrical supply voltage to machine prior to performing any adjustments or repairs.

To insure economical, trouble-free operation of your machine, it is recommended that the following maintenance be performed every 3 months.

1. Check the Cam Switch Setting.
2. Check to see that the T.X. Valve Bulb is securely fastened.
3. Check for leaks of any kind; water, refrigerant, oil, etc.
4. Check the Bin Control, it should shut the machine off. See bin thermostat adjustment on page C-2.
5. Check all electrical connections.
6. Clean the condenser to insure unobstructed air flow across the fins.
7. Clean the ice-making section, if necessary, per instructions on page D-3, (Local water conditions may require that cleaning be performed more often than 6 month intervals.).

Semi-Annual Cleaning Procedure

IMPORTANT

Harvest problems may occur if these procedures are not performed every 6 months. It may be necessary to clean machine more often than every 6 months depending on local water conditions.

Cleaning Procedure

Harvest problems may occur if these procedures are not performed every 6 months.

1. Turn main electrical service switch to the ice maker "OFF."
2. Rcmove front panel, and turn the ICE/OFF/WASH switch to the "OFF" position.
3. Remove all ice from the ice bin.
4. Depress float valve and allow fresh water to fill water trough.
5. Add 4 ounces of an approved liquid ice machine cleaner to water trough. **Do not** use nickel safe ice machine cleaner.
6. Turn ICE/OFF/WASH switch to the "WASH" position and re-install panel (safety precaution).
7. Allow the solution to circulate 30 minutes.
8. Repeat Step 1 and see that all visible deposits are removed from the evaporator. If not, purge water per step 10, then repeat Steps 4 through 7.
9. If necessary, remove the water distribution tube assembly. Use a brush to remove any scale buildup inside either the inner or outer tube. Refer to page A-12 for assembly and dis-assembly.
10. Push purge valve switch with ICE/OFF/WASH switch in "WASH" position. This will flush ice machine cleaner out through drain. Allow water trough to fill with fresh water.
11. Repeat Step 1 and wipe out any scale deposits in the water trough.
12. Re-install the water distributor tube and wipe down all areas of the ice machine head and water trough area with a sanitizing solution.
13. Wipe out the ice storage bin with warm, soapy water; rinse and sanitize.
14. Turn ICE/OFF/WASH switch to the "ON" position and replace front panel.
15. Discard the first two batches of cubes.

Winterizing Procedure

Danger

Electrical shock and/or injury from moving parts inside this machine can cause serious injury. Disconnect electrical supply voltage to machine prior to performing any adjustments or repairs.

Important

Whenever the ice machine is taken out of operation during the winter months, the procedure below must be performed. Failure to do so will void warranties.

1. Shut the water off to the machine.
2. Turn the ICE/OFF/WASH Switch on the Control box to the "OFF" position.
3. Let the machine stand for one hour (or as long as necessary) to melt any ice on the evaporator plate.
4. Disconnect the tubing between the pump discharge and the water distributor manifold.
5. Drain water system completely. Do not replace the tubes.
6. On water cooled machines, use compressed air to blow all water out of condenser while holding water regulating valve open using a screwdriver.
7. Wipe out the storage bin.

Service Record

Date	Description	Technician & Company

Service

Bin Removal

Procedure

1. Remove 2 Phillip head screws from top panel.
2. Remove lower front grille.
3. Using a 11/32 open end wrench, remove 2 hex nuts from mounting studs on bin.
4. Lift front of bin upward, and slide bin outward towards front.
5. Remove Phillip head screw holding base plate and remove base plate for access to refrigeration and component sections.

Note:
All machines manufactured prior to April 1, 1988, the bin thermostat was attached to the bin wall.

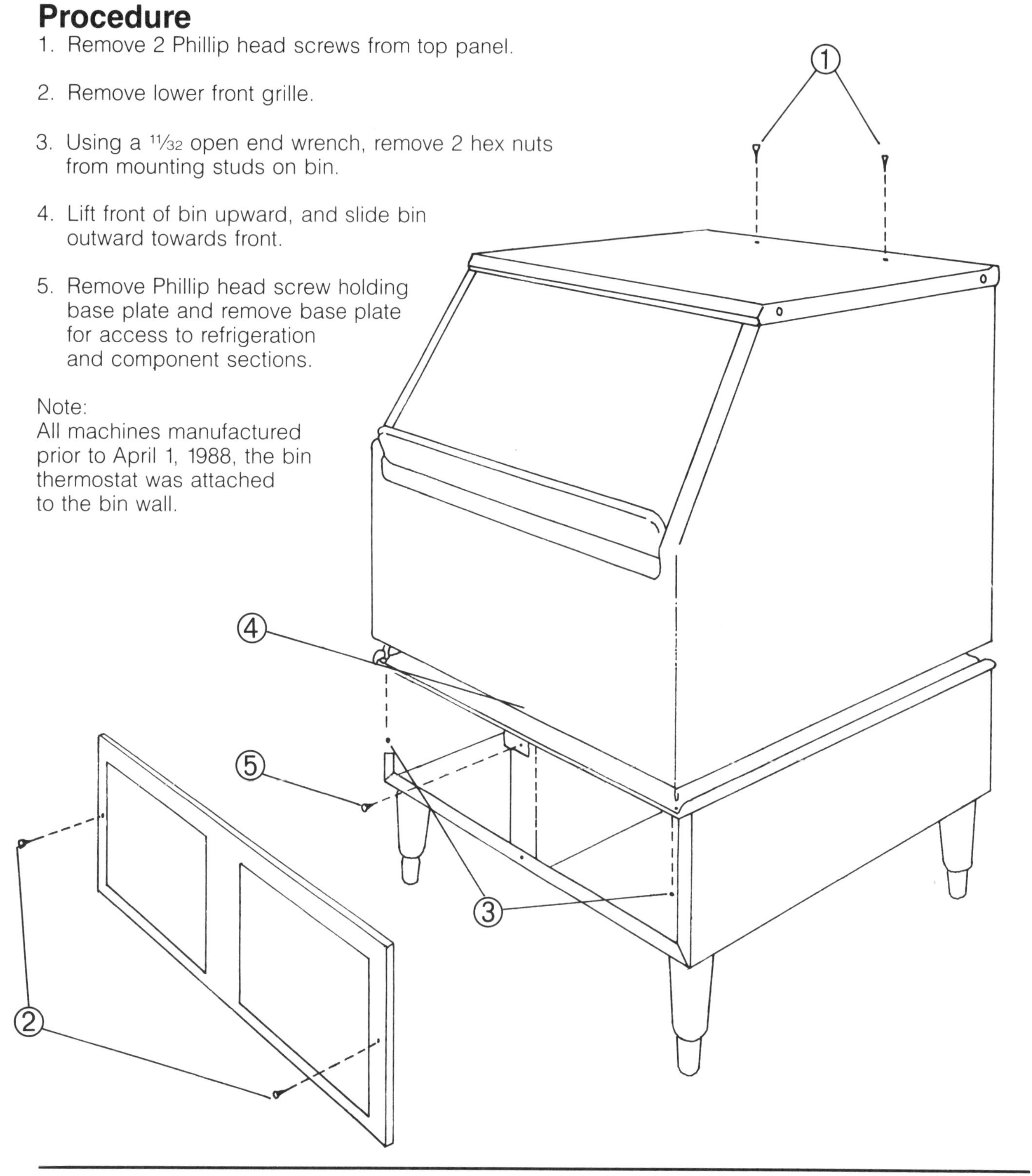

Service

Troubleshooting

Danger

Electrical shock and/or injury from moving parts inside this machine can cause serious injury. Disconnect electrical supply voltage to machine prior to performing any adjustments or repairs.

Troubleshooting an ice machine can be difficult. The technician must find the source of the problem, then make the necessary repairs. The following pages contain some of the problems and solutions that we believe are the most common.

The troubleshooting guide will only give an indication of where a possible problem can be and what repairs are typically needed to correct that problem. In many cases, more or other repair work will be necessary beyond the recommendations in the guide.

Always keep in mind the fact that a problem is not normally caused by only one part, but by the relation of one part with other parts in the system. This guide, obviously, cannot address all possible problems or corrections.

If the guide does not help to correct a problem, the technician should contact the local Ice-O-Matic distributor.

Problems, Causes, Remedies

Problem	Cause	Remedy
1. Unit runs but nc ice production	a. water shut off	a. turn on water and reset high pressure control
	b. hot gas valve stuck open	b. clean or replace valve
	c. unit out of gas	c. repair leak, evacuate and recharge
	d. compressor not pumping	d. replace compressor
	e. switch set to "WASH"	e. set switch to "ICE"
	f. refrigerant drier plugged	f. replace drier
	g. TX valve not feeding	g. clean or replace valve
	h. probe motor not running	h. replace motor
	i. cam switch not working	i. adjust or replace switch
	j. water pump not pumping	j. remove obstruction or replace pump
	k. timer initiate not closing	k. adjust or replace control
	l. timer not properly set	l. adjust or replace timer
	m. high temp safety open	m. replace safety switch
	n. leaking purge valve	n. clean or replace purge valve
2. Low ice production	a. dirty condenser	a. clean condenser
	b. low line voltage	b. ± 10% of rating
	c. bad compressor valves	c. replace compressor
	d. over or under charge of refrigerant	d. evacuate and weigh in correct charge
	e. leak in water system	e. repair leak
	f. leaking purge valve	f. clean or replace purge valve
	g. hot gas valve not seating	g. replace hot gas valve
	h. dirty evaporator	h. clean evaporator
3. Dimples in ice too large	a. timer initiate closing too soon	a. adjust or replace control
	b. timer not properly set	b. adjust or replace timer
	c. poor refrigerant feed	c. low charge, dirty condenser, TXV, restriction, etc.
4. Cloudy ice	a. water system needs cleaning	a. clean water system (per page D-3)
	b. water pump not pumping	b. remove obstruction or replace pump
	c. purge system malfunction	c. clean purge drain, replace purge valve
5. Machine does not shut off when bin is full	a. bin control out of adjustment or defective	a. adjust or replace control
6. Machine is noisy	a. defective compressor	a. replace compressor
	b. loose components	b. check and tighten
	c. fan hitting shroud	c. adjust fan/shroud position

Problems, Causes, Remedies

Problem	Cause	Remedy
7. Probe motor runs constantly	a. cam switch out of adjustment	a. adjust or replace switch
	b. relay #1 defective	b. replace relay
8. Machine does not harvest	a. defective hot gas valve	a. replace coil or valve
	b. defective probe motor	b. replace motor
	c. timer initiate not closing	c. adjust or replace control
	d. defective timer	d. replace timer
	e. high temp safety open	e. replace high temp safety
	f. relay #1 defective	f. replace relay
9. Machine in defrost too long	a. low refrigerant charge	a. evacuate and weigh in correct charge
	b. defective probe motor	b. replace motor
	c. defective probe clutch	c. replace clutch
	d. bent or sticking probe	d. check, adjust, or replace probe
10. Compressor cuts out on overload	a. low line voltage	a. ± 10% of rating
	b. high head pressure	b. clean condenser, adjust water valve to 125 psig head pressure on water cooled units
	c. defective compressor or starting component	c. replace defective compressor/ component
11. Motor compressor runs but condenser fan does not	a. loose electrical connections	a. make connections secure
	b. fan blade cannot turn due to obstruction	b. remove obstructions
	c. defective fan motor	c. replace motor
12. Machine not operating at all	a. power supply off	a. turn on breaker/disconnect
	b. ICE/OFF/WASH switch in "OFF" position or defective	b. turn to "on" or replace if defective
	c. bin control out of adjustment	c. adjust or replace control
	d. loose electrical connections	d. make connections secure
	e. out of refrigerant	e. repair leak, evacuate, weigh in correct charge
	f. water supply shut off (water cooled)	f. turn on water supply and reset high pressure control

Problems, Causes, Remedies

Problem	Cause	Remedy
13. High head pressure	a. dirty condenser	a. clean condenser
	b. system is overcharged	b. evacuate and weigh in proper charge
	c. air in refrigeration system	c. evacuate and weigh in proper charge
	d. defective fan motor	d. replace fan motor
	e. water regulator out of adjustment (water cooled units only)	e. adjust to 125 psig head pressure or replace valve
14. Low suction pressure	a. restricted flow through drier	a. replace drier
	b. system is low on refrigerant	b. evacuate and weigh in proper charge
	c. moisture in refrigeration system	c. replace drier, evacuate, weigh in proper charge
	d. water distribution restricted	d. clear obstruction
	e. defective T.X.Valve	e. replace T.X.Valve
15. Ice slabs break when harvesting	a. bridging across front of ice slab too thin. Must be a minimum of 3/16" thick for proper harvest	a. adjust bridging per page C-3
	b. probe shaft is loose and has worked out into evaporator	b. adjust probe per page C-4
16. Water dripping into ice storage area from freezing plage	a. cam switch out of adjustment	a. adjust switch per page A-10, #1
	b. probe assembly out of adjustment	b. adjust probe per page C-4
	c. machine low on charge making ice only on lower half of plate	c. repair leak, evacuate and weigh in proper charge
	d. water tube turned back allowing water flow down back side of evaporator	d. turn tube with outer holes in down position
	e. water trough cracked or incorrectly mounted	e. replace water trough or mount correctly

Model & Serial Number Form

Important

Please refer to serial plate on back of machine to obtain model and serial numbers.

Serial Plate

MILE HIGH EQUIPMENT
DENVER, COLORADO
80239 U.S.A.

UL · NSF

MODEL .... ____________________

SERIAL NO. ... ____________________

VOLTAGE ..... ____________________

HZ.__________ PH.________ WIRE________

COMPRESSOR - R.L.A.________ L.R.A.________

MOTORS (EA.)	HP/WATTS	F.L.A.
AUGER		
FAN		
HARVEST		
PUMP		
DISPENSE		

MINIMUM CIRCUIT AMPACITY ..... ________
MAXIMUM BRANCH CIRCUIT FUSE
OR HACR CIRCUIT BREAKER ..... ________
USE WITH ICE-O-MATIC REMOTE CONDENSER

____

TOTAL REFRIGERANT CHARGE:
TYPE__________ OZ________ GRAM________

DESIGN PRESSURES		
	P.S.I.	BARS
HIGH		
LOW		

CSA

LR23798

Parts

Model Number Format

Cuber Modeling Format

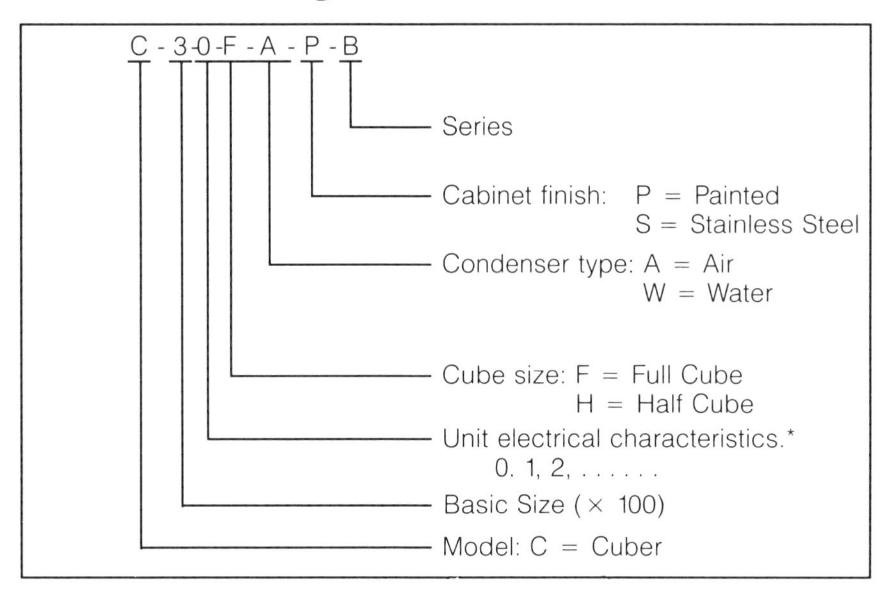

Bin Modeling Format

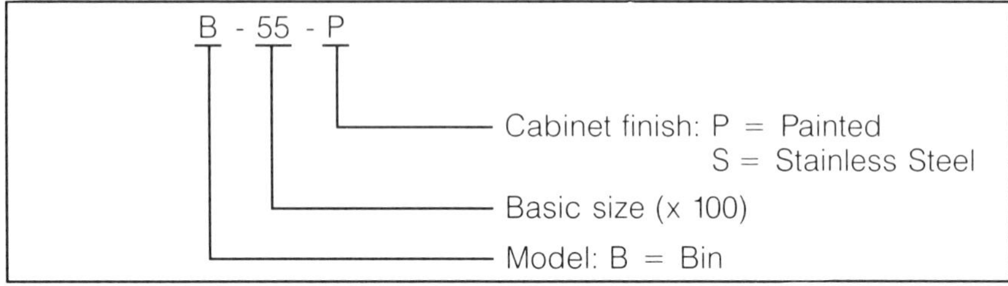

*Electrical Characteristics

Code Number	*Volts	**Cycles	Phase	No. of Service Wires Required
0	115	60	1	3
1	208-230/115	60	1	4
2	208	60	1	3
3	230	60	1	3
4	230/115	60	3	5
5	230	50	1	3
6	208-230	60	1	4
7	208-230	60	3	4
8	208-230	50-60	1	3

*Two voltage values separated by a hyphen (-) means that the unit is capable of operating over this voltage range. Two voltage values separated by a slant line (/) means that both voltages must be supplied at the installation. This is done by a neutral wire.

**Two cycle values separated by a hyphen (-) means that the unit is capable of operating at either cycle value.

Parts

Serial Number Format

Serial Number Example:

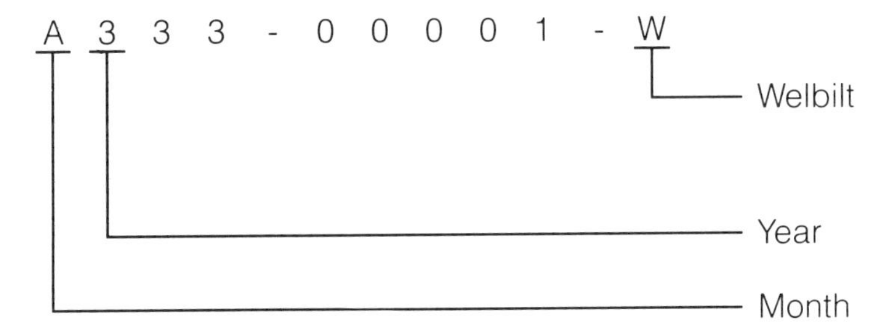

Month	Year
A - January	3 - 1983
B - February	4 - 1984
C - March	5 - 1985
D - April	6 - 1986
E - May	7 - 1987
F - June	8 - 1988
G - July	9 - 1989
H - August	0 - 1990
I - September	
J - October	
K - November	
L - December	

Parts

Evaporator & Water System

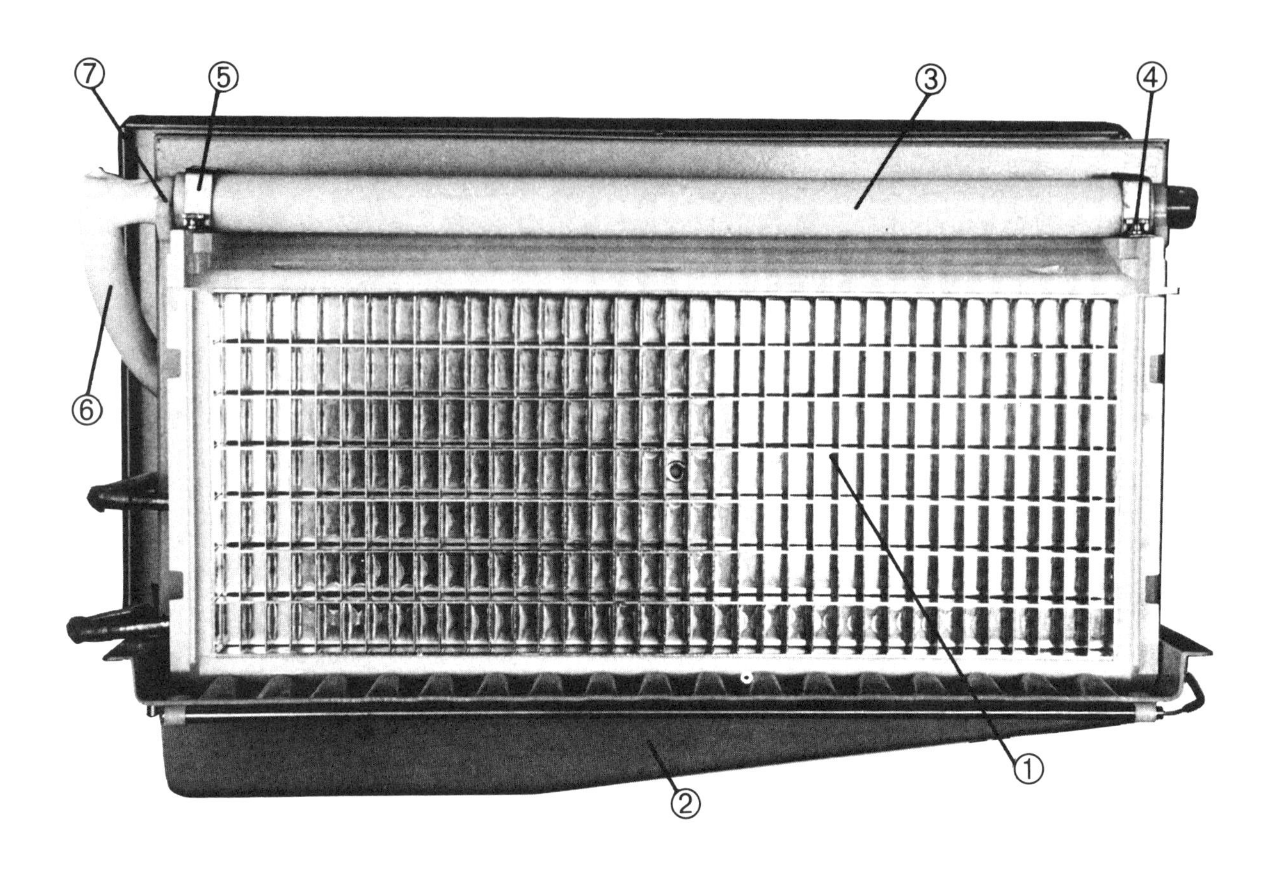

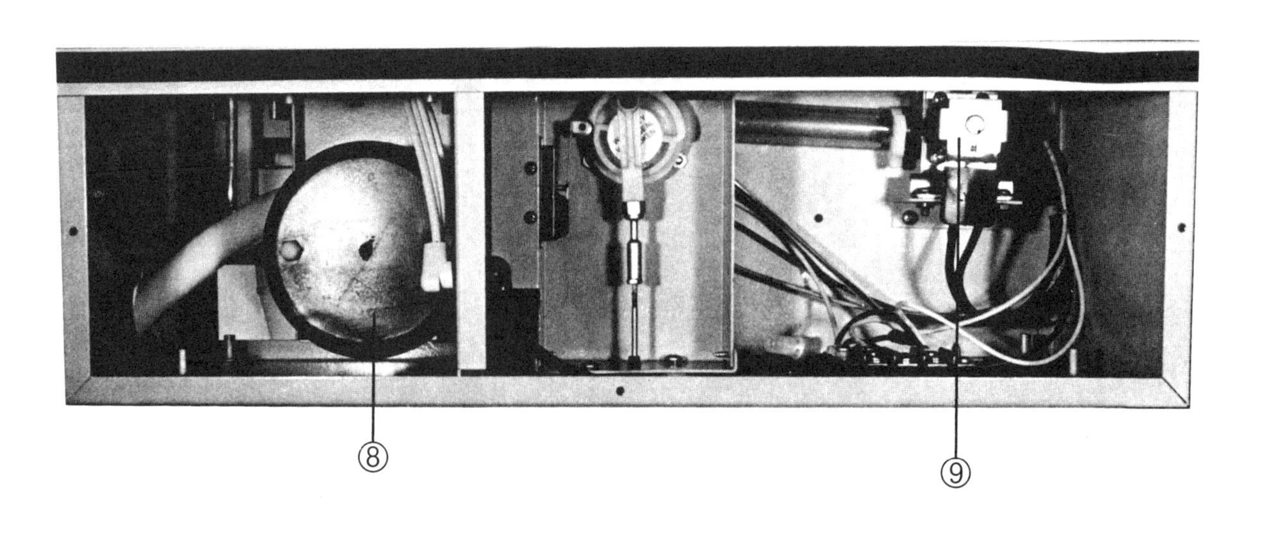

Parts

Evaporator & Water Section

Index #	Description	Part Number	Notes
1	Evaporator	2051095-02	Half Cube
1A	Evaporator	2051095-01	Full Cube
2	Water Trough	9051149-01	
3	Water Distribution Tube	2041013-01	
4	Thumb Screws	9031051-01	
5	Water Tube Bracket	3011874-01	
6	Water Pump Tube	9051142-02	
7	Clamp	9021010-03	
8	Water Pump	9161076-01	
9	Purge Valve Assy.	9041053-01	Body & Coil
9A	Purge Valve Coil	9041053-91	Coil Only
10	Float Ball & Stem	9131111-01	Not Shown

Parts

Condensing Section

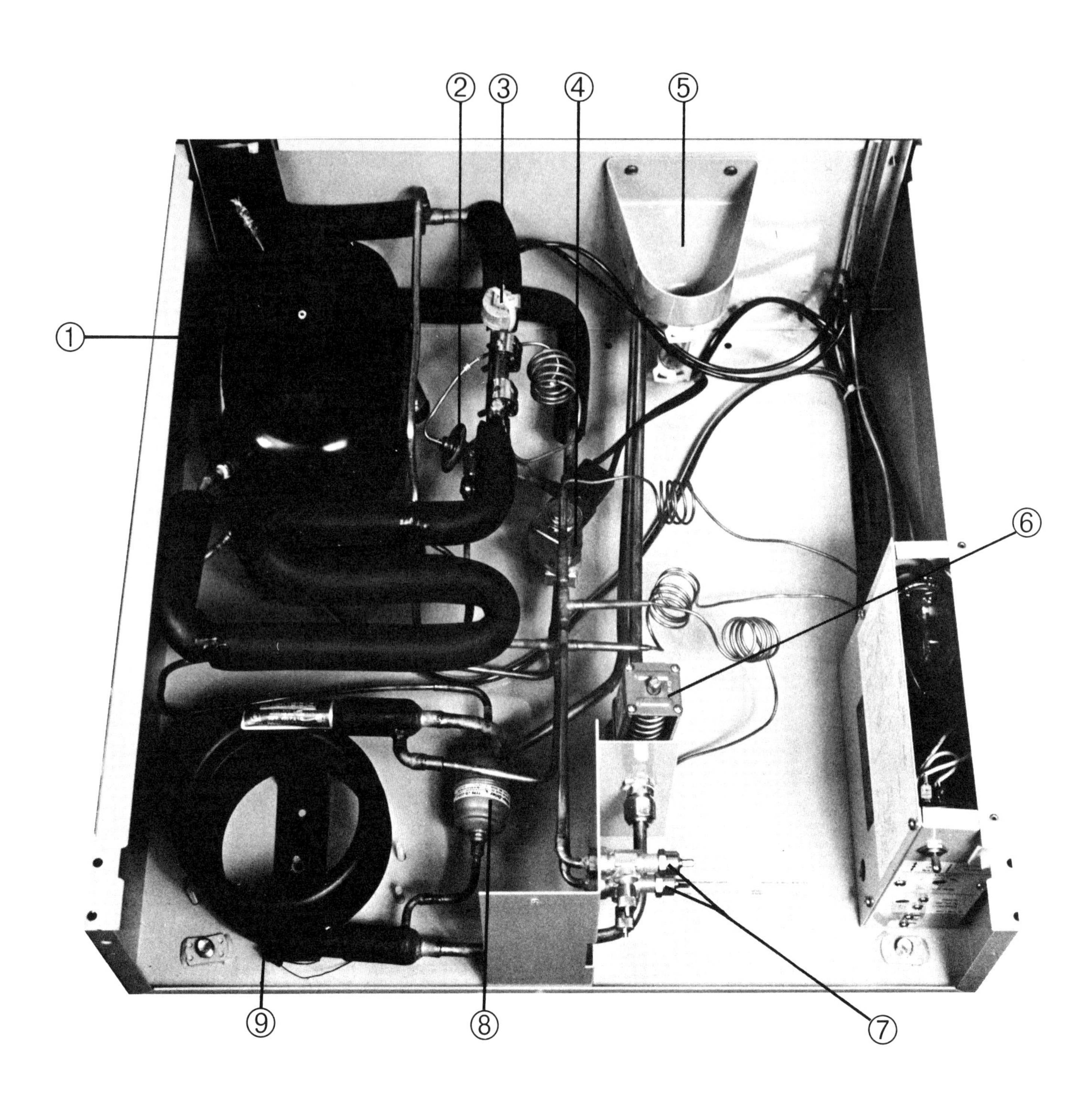

Parts

Condensing Section

Index #	Description	Part Number	Notes
1	Compressor	9181061-01	Dan Foss ⅓ h.p.
1A	Start Relay	9181004-08	
1B	Start Capacitor	9181061-60	
2	Expansion Valve	9151020-01	
3	High Temperature Safety	9041058-01	
4	Hot Gas Valve Assy.	9151053-01	Body & Coil
4A	Hot Gas Coil	9151053-91	Coil Only
5	Drain Tray	9051152-01	
6	Water Regulating Valve	9041010-01	Water-cooled Only
7	Service Valve	9091064-01	
8	Drier	9151004-02	
9	Water-cooled Condenser	9141022-01	Water-cooled Only
9A	Air-cooled Condenser	9141051-01	Not Shown
10	Fan Motor	9161028-01	Air-cooled Only
11	Fan Blade	9131046-01	Air-cooled Only
12	Fan Bracket	9131054-02	Air-cooled Only
13	Fan Shroud	9131128-01	Air-cooled Only

Parts

Control Box Section

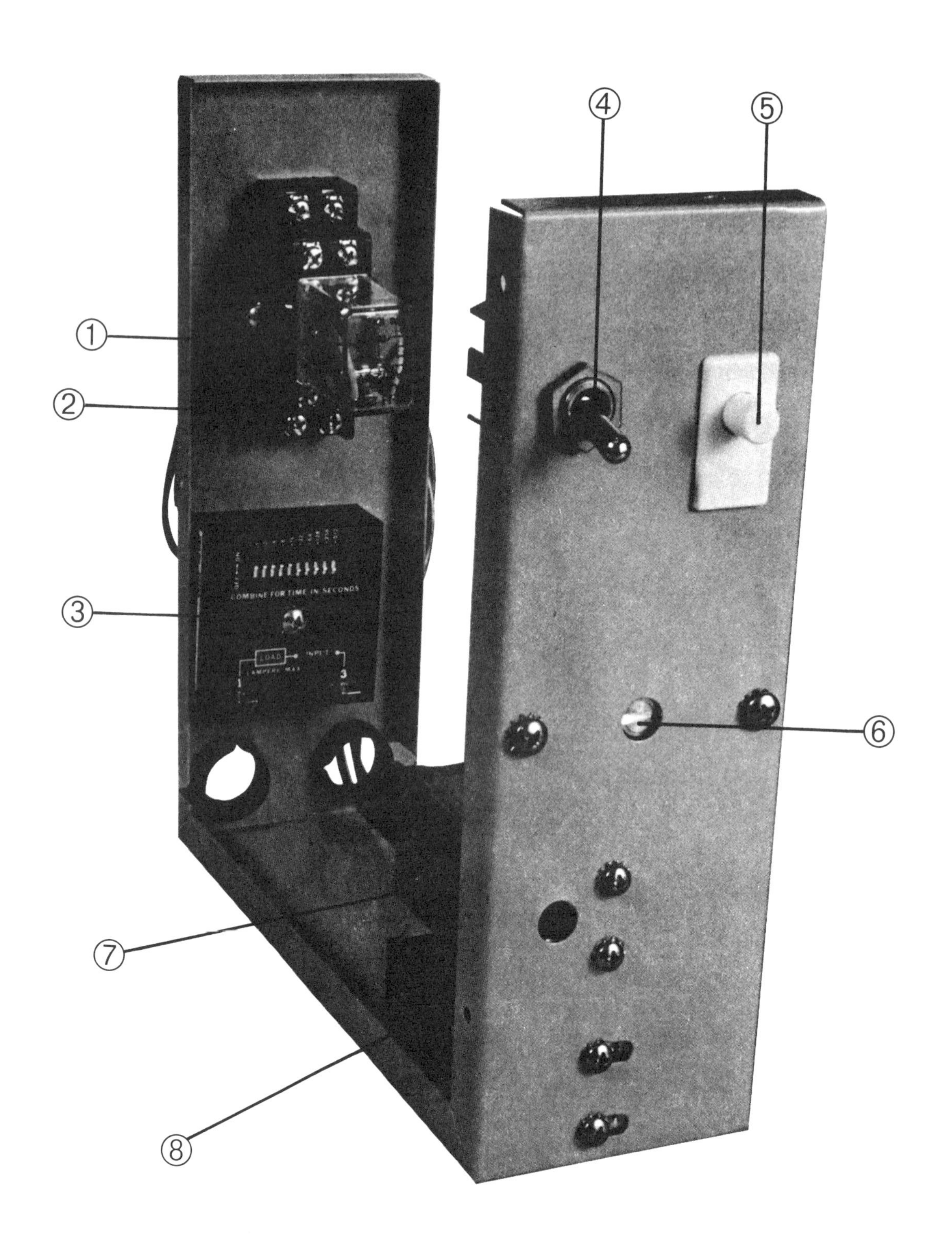

Parts

Control Box Section

Index #	Description	Part Number	Notes
1	Relay	9101084-01	
2	Relay Socket	9101083-01	
3	Time Delay Module	9101148-01	
4	ON/OFF/WASH Switch	9101086-01	
5	Purge Switch	9101125-01	
6	Bin Thermostat	9041004-03	
7	Timer Initiate	9041054-01	After Dec. 1986
7A	Slab Thickness Control	1051024-01	Prior to Dec. 1986
8	High Pressure Control	9041051-01	Water-cooled
8A	High Pressure Control	9041051-02	Air-cooled
9	Terminal Board	9101037-01	

Parts

Harvest Assembly

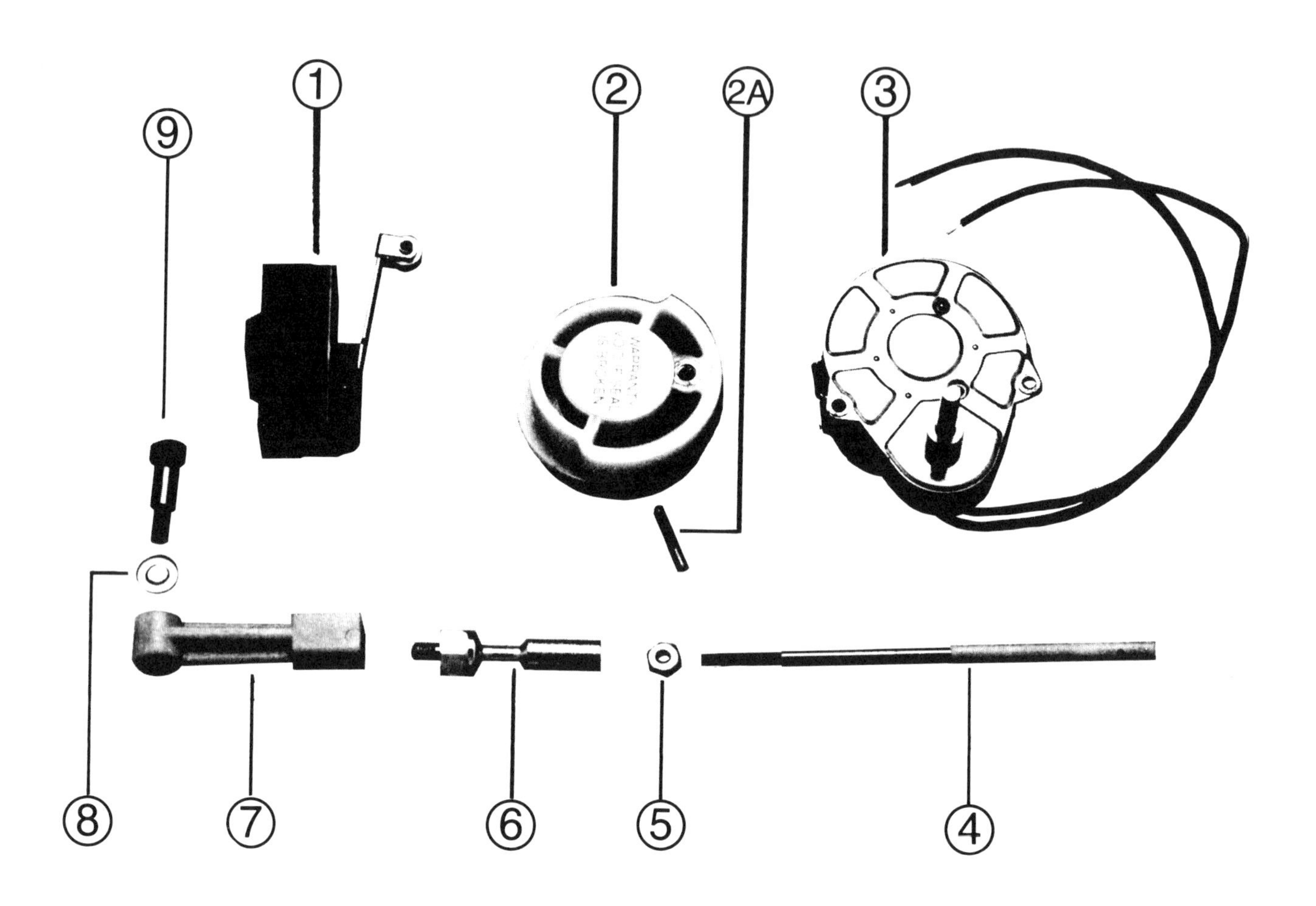

Parts

Harvest Assembly

Index #	Description	Part Number	Notes
1	Cam Switch	9101023-01	
2	Clutch Assembly	2081035-03	
2A	Set Screw	9031033-04	
3	Probe Motor	9161030-01	
4	Probe Shaft	9021053-02	
5	Lock Nut	9031005-09	
6	Swivel	9021017-01	
7	Connecting Rod	9051057-01	
8	Washer	9031003-03	
9	Allen Screw	9031029-01	

Parts

Panel

Parts

Panels

Index #	Description	Part Number	Notes
1	Top Panel	3021397-01	
2	Bin Door	2991016-01	
3	Bin Assy.	2101130-01	
4	Front Grille Panel	3021400-01	
5	Front Panel Screws	9031074-02	4 Required
6	Top Panel Screws	9031074-01	8 Required
7	Access Panel	3011868-01	On Rear Panel
8	Base Plate Cover	3011872-01	Not Shown

Kold-Draft

KOLD-DRAFT

AUTOMATIC ICE CUBER

SERVICE MANUAL

Products of the
KOLD-DRAFT DIVISION

Uniflow Manufacturing Co.
ERIE, PA. 16514
ESTABLISHED 1920
Area Code 814—Phone 453-6761
TELEX 91-4434

GBB-01949
10/87
*Trademark Reg. U.S. Pat. Off.
©1987, Uniflow, Erie, Pa.

ELECTRONIC CUBER MODEL NUMBER BREAKDOWN

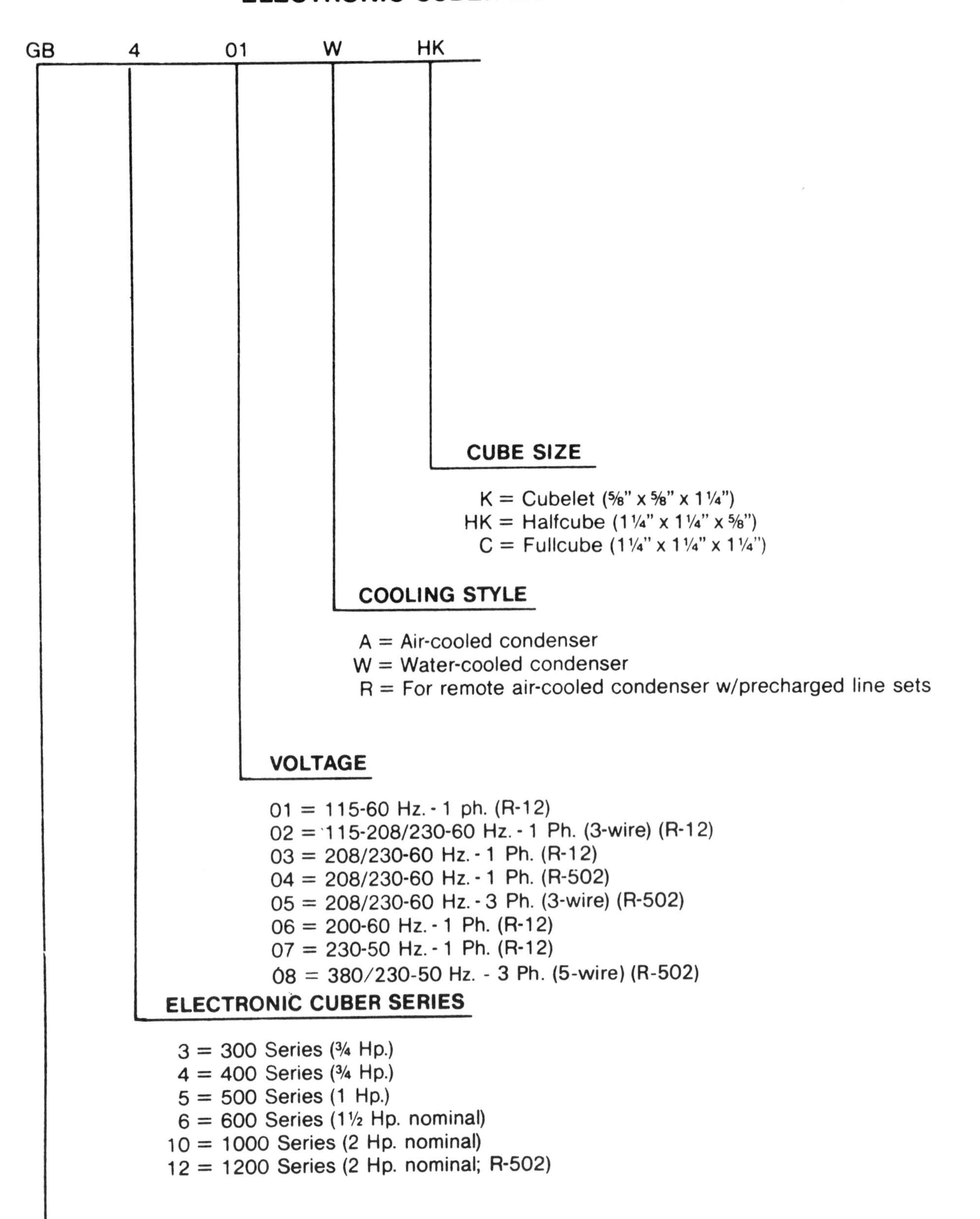

GT Slimline unit, mounts on GTN bins (28 1/2" wide)
GB Horizontal unit, mounts on GBN bins (42" wide)
IS Self-Contained Cuber Dispenser Unit

Check current literature for availability or models.

The name you can count on.
Kold-Draft Division
Uniflow Manufacturing Company
1525 East Lake Road, Erie, Pennsylvania 16514
Tel 814/453-6761
Telex 91/4434

TABLE OF CONTENTS

ELECTRONIC

ELECTRO-MECHANICAL CUBER SECTION

TABLE OF CONTENTS — CONTINUED

CUBER INSTALLATION SPECS
(60 Cycle Electronic Cubers)

MODEL	HP	MINIMUM AMPACITY	MINIMUM INCOMING WIRE SIZE	*FUSE OR HACR TYPE CIRCUIT BREAKER SIZE RECOMMENDED	MAXIMUM*
GT301(A)	3/4	22.7	10	25	35
(W)	3/4	17	12	20	25
GT401(A)	3/4	25	10	25	40
(W)	3/4	18	12	20	25
(R)	3/4	20.9	10	25	35
GT402(A)	3/4	16	12	20	20
(W)	3/4	15	14	15	20
(R)	3/4	16.9	12	20	25
GT503(W)	1	13.3	14	15	20
(R)	1	12.2	14	15	20
GT603(W)	11,000 BTU	15	14	15	20
(R)	11,000 BTU	18.9	12	20	30
GB401(A)	3/4	25	10	25	35
(W)	3/4	18	12	20	25
(R)	3/4	20.9	10	25	35
GB402(A)	3/4	16	12	20	20
(W)	3/4	15	14	15	20
(R)	3/4	16.9	12	20	25
GB503(A)	1	14.4	14	15	20
(W)	1	13.3	14	15	20
(R)	1	12.2	14	15	20
GB603(A)	11,000 BTU	20	12	20	30
(W)	11,000 BTU	15	14	15	20
(R)	11,000 BTU	18.9	12	20	30
GB1003(W)	14,000 BTU	17	12	20	25
(R)	14,000 BTU	19.3	12	20	30
GB1204(W)	27,600 BTU	28.2	10	30	45
(R)	27,600 BTU	28.2	10	30	45
GB1205(W)	22,500 BTU	17.8	12	20	30
(R)	22,500 BTU	17.9	12	20	30
IS401(A)	3/4	30	10	30	40
(W)	3/4	23	10	25	30
IS503(W)	1	15	14	15	20

(60 Cycle Thermostatic Cubers)

MODEL	HP	MINIMUM AMPACITY	MINIMUM INCOMING WIRE SIZE	RECOMMENDED	MAXIMUM*
GS6(A)	1/5	15	14	15	15
(W)	1/5	15	14	15	15
GY3(A)	1/3	15	14	15	15
(W)	1/3	15	14	15	15
GT7(A)	3/4	22.7	10	25	35
(W)	3/4	17.0	12	20	25

NOTE: MAXIMUM BRANCH CIRCUIT FUSE OR HACR TYPE CIRCUIT BREAKER SIZE is dependent on the size of the conductors supplying the Ice Maker. They must be no less than the minimum ampacity rating and no more than "Maximum" rating on the Nameplate.

SUPPLEMENT FUSES (INSTALLED ON THE ICE MAKER) do NOT provide primary protection and may be sized as required for continuous operation without nuissance blowing, up to the indicated Nameplate "Maximum" rating to compensate for ambient conditions.

SPECIAL NOTE - Wire sizes on this chart are good up till 25 feet.
Anything 25 feet to 100 feet increase wire 1 size.
Any runs 100 feet or more increase the wire 2 sizes.

CHART FOR ADAPTING LOCAL POWER SUPPLY FOR CUBERS

MODIFICATIONS NECESSARY

RATED VOLTAGE **BOLD VOLTAGES** ARE PREFERRED	CUBER MODELS: GB1, GB503, GT1, GT503, IS1, IS503, *GB7D, GB402	CUBER MODELS: GB7, GB401, GT7, GT301, GT8, GT401, IS7, IS401, GY3, GS6	CUBER MODELS: GB4, GB103, GB2, GB603, GB1204	CUBER MODELS: GB1205
60 CYCLE SINGLE PHASE	60 CYCLE, SINGLE PHASE	60 CYCLE SINGLE PHASE	60 CYCLE SINGLE PHASE	60 CYCLE THREE PHASE
240 & Higher	Transform to 230 Volt-2 KVA	Transform to 115 Volt - 1.5 KVA	Transform to 230 Volt - 3 KVA	
220 to 230	None	Transform to 110-115 Volt - No. 5-1004 Transformer	None	
208	Reconnect Internal Transformer by moving incoming brown wire from 230 volt to 208 volt terminal. Leave wire to compressor on 230 volt terminal.	See 208/120 Volt, 4 wire, Phase below or transform to 115 Volt - 2 KVA	Move Brown Wire to 208 Volt Terminal	
120-208 or 230 3-Wire	Connect to 2 Hot Wires Only. *For GB7D & GB402 Only, connect 2 hot wires & white neutral wire.	Connect Black Cuber wire to either hot wire, & connect White Cuber wire to White neutral wire.	Connect to 2 hot wires only.	
110 to 120 (115 nominal)	Transform to 230 Volt-2 KVA No. 5-1044 Transformer	None	Transfer to 230 Volt - 3 KVA	
208 or 230 V. 3 phase	Connect Cuber wires to 3 hot wires of either 3 phase power system pictured at right. For 208 V. only, reconnect internal transformer by moving both brown wires from 230 V. to 208 V. term.			

FOR 60 CYCLE APPLICATIONS & VOLTAGES NOT LISTED ABOVE — CONSULT THE FACTORY.

THREE PHASE

Diagrams below are the two most common power systems.

208/120 VOLT
WYE SYSTEM

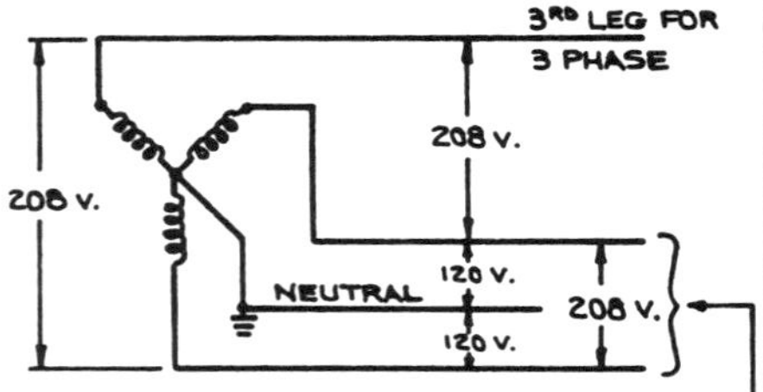

3RD LEG FOR 3 PHASE
208 V.
208 V.
NEUTRAL
120 V.
120 V.
208 V.

230 VOLT
DELTA SYSTEM

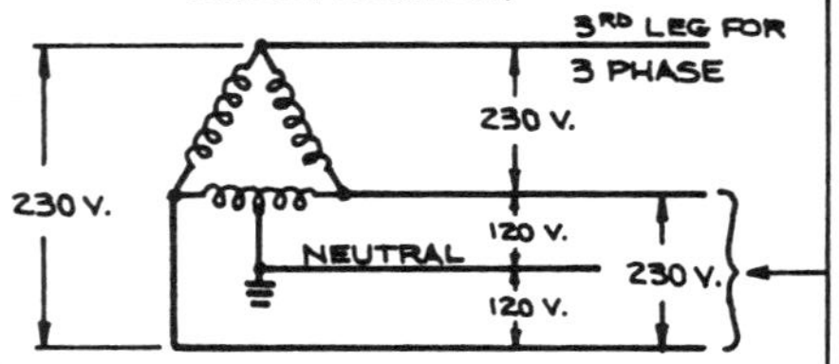

3RD LEG FOR 3 PHASE
230 V.
230 V.
NEUTRAL
120 V.
120 V.
230 V.

Either of these single phase services are commonly supplied to residential and commercial buildings.

208/120 VOLT, 4 WIRE-3 PHASE

Use any two legs of 208 Volt, 60 cycle, for GB1, GB2, GB4, IS1, MD1 with modifications as per chart. Use one leg and neutral for 120 Volt, 60 cycle, single phase to GY3, GY7, GT7, GB7, GT8, GS6, MD5, IS5 & IS7. Use two legs and neutral for 3-wire GB7D & GB402 only.

CHART OF WATER LEVELS, PRESSURES, CYCLES AND CHARGE

(Higher Than Average Temperatures Increase Pressures and Cycle Times)

Model Number Electronic Control	GB1200			GB1000			GB600, GT600			GB500, GT500, IS500			GB400, GT400, IS400			GT300		
	C	HK	K	C	HK	K	C	HK	K	C	HK	K	C	HK	K	C	HK	K
Full Water Level Distance Below Top of Tank	2⅝″	2¾″	3½″	2⅝″	2¾″	3½″	2⅝″	2¾″	3½″	2⅝″	2¾″	3½″	2½″	2¾″	3½″	2″		3″
Suction Pressure After Defrost	55-60 psig.		55-60 psig.	20 psig.		20 psig.	20 psig.		20 psig.	15, 20 psig.		15, 20 psig.	20, 25 psig.		20, 25 psig.	11, 13 psig.		11, 13 psig.
Suction Pressure Before Defrost	10-15 psig.		10-15 psig.	0-2 psig.		0-2 psig.	0-2 psig.		0-2 psig.	0	0	0	3 psig.		3 psig.	0		0
Defrost Press	70 psig. max. with CPR valve		100-110 psig.	55 psig. max. with CPR valve		40-60 psig.	40-60 psig.		40-60 psig.	40-60 psig.		40-60 psig.	40-60 psig.		40-60 psig.	40 psig.		40 psig.
Cycle Time Approximate	19 min.	17 min.	11 min.	27 min.	22 min.	15 min.		17 min.		26 min.	20 min.	13 min.	28 min.	25 min.	14 min.	28 min.	21 min.	15 min.
Refrigerant Charge Remotes: See note below.	3 lb.			3 lb.			3 lb.			3 lb.		3 lb.	3 lb.		3 lb.	13 oz. GT301A 10 oz. GT301W		13 oz. GT301A 10 oz. GT301W
Type	R 502			R-12			R-12			R-12		R-12	R-12		R-12	R-12		R-12
Approximate Lb. Ice Per Batch	15 lbs.	14 lbs.	8 lbs.	15 lb.	14 lb.	8 lb.	7½ lb.	7 lb.	4 lb.	7½ lb.	7 lb.	4 lb.	7½ lb.	7 lb.	4 lb.	3¾ lb.	3½ lb.	2 lb.
Compressor Size	GB1205 22500 BTU TXV Controlled GB1204 27600 BTU TXV Controlled			14000 BTU TXV Controlled			11000 BTU TXV Controlled			9800 BTU TXV Controlled			6800 BTU TXV Controlled			6800 BTU Cap. Tube Controlled		

NOTE: Remote condenser application cubers require a total minimum charge of 10 1/2 lbs. GB1200 Series cubers use R-502, all others use R-12.

NOTE: Maximum fuse size check electrical rating plate on left rear of cuber.

CONSTRUCTION OF THE CUBER: KOLD-DRAFT ELECTRONIC CUBERS

SKINS - The skins consist of the top, the left end, back panel, right end and front inspection panels.

CONDENSING UNITS - Varies with each model. For compressor size and charge, refer to Table of Chart on Water Level, Pressures and Cycles.

EVAPORATOR - The GB Series full cube (C) evaporator is made up of 108 cells, 1 1/4" each way. The cubelet (K) evaporator is made up of 216 cells, 1 1/4" x 1 1/4" x 5/8" deep. The half cube (HK) evaporator is made up of 216 cells, 5/8" x 1 1/4" x 1 1/4" deep. The material is copper and the entire assembly is tinned, thus preventing corrosion and making its use acceptable to any sanitary board. A seal is not required between the evaporator and water plate. Normally there is about 1/32" clearance between them. Refer to Index for Water Plate-Evaporator Alignment for adjustments.
NOTE: GT300 and GT7 series are half the number of cells.

REFRIGERANT CONTROL

1. Thermostatic expansion valves are used on all GB and IS models and GT400, GT500, GT600 models. For replacement of the valve consult the Parts Price List for type of expansion valve. It has been regulated properly on tests before being shipped to give minimum superheat and maximum flooding of evaporator and should not require adjustment.

   Sometimes after shipping or storage, the expansion valve sticks and allows more refrigerant to pass than necessary, increasing the low side pressure and temperature, thus excessive frostback and a long cycle. If this condition does not correct itself during the second cycle, it will be necessary to adjust the superheat on the expansion valve, closing the valve clockwise 1/8 to 1/4 turn to increase superheat, reducing the suction pressure and preventing frostback. If valve hunts (varies suction pressure up and down) more than 2 to 3 lbs. when suction pressure is below 13 lbs., it indicates that the valve should be opened more.

2. The GT7 and GT300 Series are capillary tube systems and do not have an expansion valve.

WATER PLATE

The water plate is made of approved plastics and used to distribute the water through jet holes into the freezing cells and also to return water through two drain holes in each cell in the water plate. On the front of the water plate is a control stream. That should be set at the base of the dam once the fill cycle is complete. Backing the screw out increases flow. At the end of a cycle when cubes are virtually fully frozen there is an increasing water pressure in the system causing the stream to rise and go over the dam dumping dreg water.

CIRCULATION TANK

The circulation tank is secured to the bottom of the water plate. It is just a reservoir to hold enough water to make one batch of ice.

CIRCULATION STRAINER

The circulation strainer has a large screen inserted in the tank outlet to prevent dirt or particles of precipitated mineral from clogging the jet holes or the control stream. It also protects the pump impeller. If the screen becomes clogged with precipitated minerals, it is advisable to clean the whole circulation system with the ice machine cleaner.

WATER PUMP

The water pump is of the centrifugal type, direct with sealed bearings that require no lubrication. The inlet tube is at the bottom of the water circulation tank; and the outlet at the top of the pump is connected to the header of the water plate. It is made entirely of molded plastic.

WATER LEVEL PROBE ASSEMBLY (GBR-03170)

This probe is connected to the main water circulation tank by means of flexible tubing. The height of water in the probe assembly indicates the height of water in the circulation tank. Thermistor probes determine water valve on-off levels. Refer to page 4 for correct setting per model for upper probe. The bottom probe is positioned so that 15-30 seconds after the water flows over the dam on control stream box on the water plate the water level is below the glass tip and harvest is initiated.

PUMP AND DEFROST SWITCH

A spring loaded switch is mounted in the control box. It controls the defrost circuit and water pump and is operated by an adjustable screw in the lift plate attached to the water plate. As the water plate closes after defrost, the adjustable screw in the lift plate pushed up the pump switch to cut one connection to the defrost circuit and start the circulation water pump.

WATER INLET VALVE

The water inlet valve is mounted in the front left of the freezing compartment and controls the rate of flow of water into the water tank when the cuber is filling. It is of a constant flow type requiring minimum 15 PSI. An external "Y" type strainer is used with valves. For pressures over 100 PSI or if there is water hammer, use a pressure regulator, part #55R-01108.

ACTUATOR MOTOR & CAM ASSEMBLY

The assembly is located just to the right and front of the evaporator. The actuator motor drives the cam shaft directly and is reversible. A cam on each end of the shaft forces the water plate down, separating the water plate from the ice in the evaporator. The two springs on the cams pull the water plate up at the end of the defrost cycle, and hold the water plate against the bottom of the cams during the freeze cycle. To prevent the water plate from opening after current has stopped flowing to the actuator motor, a drift stop spring with a plastic end (GBR-00965) presses against the actuator motor shaft, on the front of the motor. If the drift stop is not aligned with the motor shaft it can be removed easily and bent into shape.

EVAPORATOR PROBE (GBR-03176)

Resistance values of the evaporator probe change due to temperature changes on the evaporator. These changes are transmitted to the printed circuit card and allows one probe to perform three functions - cold water control, actuator control and thermal - over temperature function.

1. COLD WATER CONTROL opens the hot gas valve during the fill cycle when the evaporator cools to 45° F. (Red L.E.D. comes on bright on P.C.Card). The hot gas will shut-off when the water fill is complete or when the evaporator warms to 50° F (Red L.E.D. off or dimly lit).

2. ACTUATOR CONTROL resets cold when the evaporator reached 26° F (Orange L.E.D. comes on bright on P.C. Card), and will send power through the actuator toggle switch to actuator motor at harvest. It also sends power to the actuator motor when the evaporator warms to 40° F after harvest to return the water plate (Orange L.E.D. off on P.C. Card).

3. THERMAL-OVER TEMPERATURE FUNCTION shuts the cuber down when the evaporator temperature reaches 140° F, should the hot gas valve stick open. It will re-energize the cuber when the evaporator cools to 120° F.

BIN PROBE (GBR-03177)

The Bin Probe is mounted to a flexible probe holder on the ice chute or through a bracket in the Bin. Its resistance values change with temperature fluctuations at probe tip due to proximity of ice. These changes are transmitted to Bin Control on the P.C. Card and turn the Cuber on or off.

TO SET BIN CONTROL

1. Do not attempt to set Bin Control while unit is filling with water. The P.C. Card has a circuit to fill the water tank before shutting off on "Full Bin Condition."

2. Once the water fill is complete, hold ice to the tip of the probe. The Cuber should shut off in 15-30 seconds. If longer, adjust the Bin potentiometer on the P.C. Card slowly counter-clockwise until the unit shuts off.

3. After the Cuber stops, remove the cubes from the probe tip. The Cuber should start within 1 minute if ambient air is above 50° F.

The bin probe is standard length but when stacking units check the chart below to determine whether a stacked cuber will require a bin probe extension (GBR-03240) in addition to the bin probe (GBR-03177) to reach into the bin.

GBR-03177 Probe & GBR-03240 Extension	GBR-03177 Probe & GBR-03240 Extension
GBR-03177 Probe & GBR-03240 Extension	GBR-03177 Probe & GBR-03240 Extension
GBR-03177 Probe & GBR-03240 Extension	GBR-03177 Probe & GBR-03240 Extension
GBR-03177 Probe	GBR-03177 Probe
GBR-03177 Probe	T-100 Crusher
Any Kold-Draft Bin	Any Kold-Draft Bin

KOLD-DRAFT PRINTED CIRCUIT CARD

GBR-13135-02 type P.C.Cards have been used since August, 1980. See Engineering Bulletin #10-80 for pre-August, 1980 details, but note that all of the older P.C.Cards should have been updated with retrofit kits offered by the factory after August, 1980.

In October, 1982, and through the time of this printing, an updated version identified as GBR-03135-02-E has been employed. The "E" cards can easily be recognized by the presence of 3 wires (black, brown & white) extending from its transformer winding as illustrated below. To replace a GBR-03135-02 P.C. Card with a GBR-03135-02-E, a GBR-03222 wiring adapter kit is required.

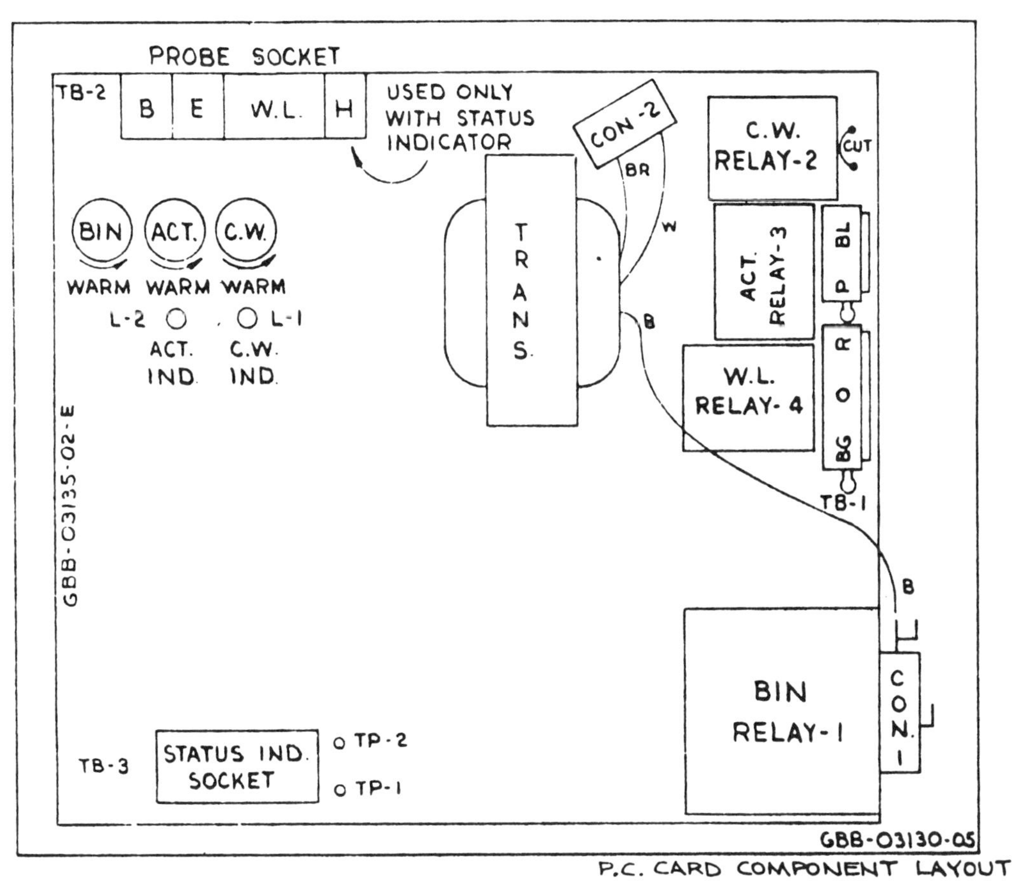

P.C. CARD COMPONENT LAYOUT

B	E	W.L.	H

TB-2 UPPER LEFT PORTION P.C. CARD

1. B - Pins 1 & 2, location of bin probe which turns the cuber on and off according to ice demand.

2. E - Pins 3 & 4, location to connect evaporator probe which relays evaporator temperature to P.C. Card, allowing card to set to the cold or warm side, the cold water control, actuator control, and thermal over temperature protector.

3. W.L. - Pins 5, 6, 7 & 8 location to connect water level probes, which turn the water inlet valve on and off (W). It initiates harvest when water falls below the lower probe (L) when actuator control set cold.

4. H - Pins 9 & 10, location of harvest switch when optional status indicator is used. Senses harvest of ice.

ADJUSTABLE POTENTIOMETERS AND L.E.D.'S - Located Beneath TB-2.

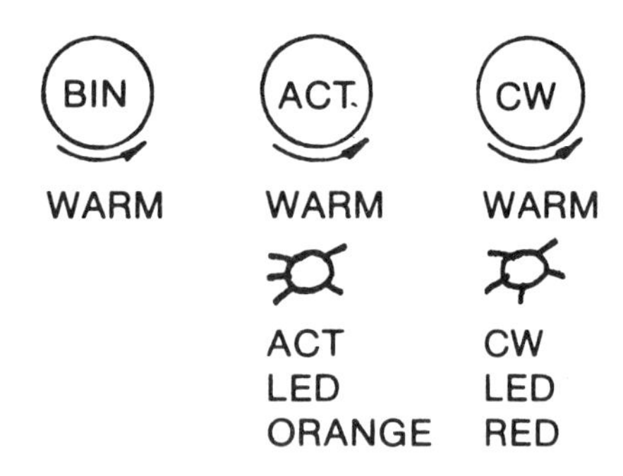

A. Bin Potentiometer - Adjustment to turn unit off on contact of Bin Probe with ice in 15-30 seconds, provided the water fill is satisfied. (Refer to Page 7)

B. Acutator Potentiometer - Adjust to raise water plate 15-30 seconds after ice harvest.

C. Cold Water Potentiometer (C.W.) - Adjustment to open hot gas valve should evaporator cool to 45° F. during water fill cycle.

D. Cold Water LED - (RED) illuminates when evaporator colder than 45° F. Off when C.W. control set warm at 50° F. or above.

E. Actuator LED - (ORANGE) illuminates when evaporator temperature colder than 26° F. Off when actuator control set warm at 40° or above.

STATUS INDICATOR PLUG & TEST PINS (TP1 & TP2) - Location - Lower Left Corner of P.C. Card.

The STATUS INDICATOR OPTION is a light display that will visually show two functions and two fault conditions on the Front Panel. For description of Status Indicator refer to Page 10.

BIN RELAY - Location Lower Right Corner of the P.C. Card.

On P.C. Cards made since January 1984, this relay is replaceable, should it ever fail. This style Card is easily distinguished by the 2 wire leads coming through a hole in the P.C. Card and connecting to the back-side of the relay.

OUTPUT SOCKETS & RELAYS 2, 3, 4 - Location Upper Right of P.C. Card.

The connections for output supply from the P.C. Card to the small motors and valves when sequenced by the relays.

TRANSFORMER - Location Upper Center of P.C. Card.

One printed Circuit Card is used on all Electronic Cubers. The transformer has 2 leads (brown and white) attached to a plug which mates to a plug from front wire channel. On 115 volt Cubers, the plugs mate white to white (brown open on transformer plug). On 208/230 and 115-230 (3 wire) the plugs mate brown to brown (white open on transformer plug).

SPEEDY ELECTRONIC CUBER FUNCTIONAL TEST - Page 25.

STATUS INDICATOR OPTION - FOR GB ELECTRONIC SERIES ONLY

The status indicator option consisting of a small printed circuit card and harvest sensor switch. The user will tell at a glance whether or not his cuber is operating properly by the light display on the front panel.

The status indicator P.C. Card is secured to the front panel and the power ribbon is connected to terminal block 3 in the lower left hand corner of the main P.C. Card.

GREEN LIGHT

This light blinks when the cuber is making ice.

YELLOW LIGHT

This light is on when the cuber is off on a full bin condition.

GREEN & YELLOW LIGHTS

These lights are on when the cuber has shut itself off due to an overheated evaporator (140° F). The cuber will return itself to service when the evaporator cools down (120° F).

RED & YELLOW LIGHTS

These lights are on when the cuber has shutdown due to a "time-out" condition. This occurs when the ice has not triggered the ice harvest switch, thus a "Service Required" condition. A cuber left in the wash position or any failure to harvest will preceed the time out condition. A cuber can be put back into operation by tripping the ice harvest switch, removing power temporarily or by unplugging the status indicator. The time in seconds is equal to the time in minutes that the time-out timer is set to. Adjustment is made from the back of the inspection panel through the small hole in the plastic status indicator P.C. Card cover.

TIMEOUT PONTENTIOMETER

Mounted on the Statis Indicator P.C. Card, is adjustable from 15 minutes to 1 1/2 hours. This timeout period can be accurately measured by timing 17 pulses of the Green light when the Cuber is operating. The time in seconds is equal to the time in minutes that the timeout timer is set to. Adjustment is made from the back of the inspection panel through the small hole in the plastic indicator P.C. Card cover.

The time setting should exceed the normal cycle by 50% to allow for longer cycle on startup. Doubling this time period for cubelet machine will eliminate possible "No Harvest Trips", due to the small size of the cubes not tripping the switch the first time. Doubling the time gives the switch a second try to sense harvest.

ICE HARVEST SWITCH

Mounted on the support brace at the end of the water plate, this switch resets a lockout time on the status indicator P.C. Card. Falling ice during harvest operates the switch.

To set the ice harvest switch:

With water plate in full "UP" position and the adjustment screw pointing straight down:

1. Turn adjusting screw clockwise until the yellow light stays ON.
2. Turn counter-clockwise until light goes out. Then continue turning 1/8 turn.

Allow the cuber to make one batch of cubes, and carefully observe the yellow status indicator light at the moment the ice hits the water plate. The light should blink briefly several times, and then go out. If the light does not blink, turn the adjustment screw in to increase sensitivity. If the yellow light remains on (and the cuber is not off on bin control), turn the adjustment out to decrease sensitivity (CCW).

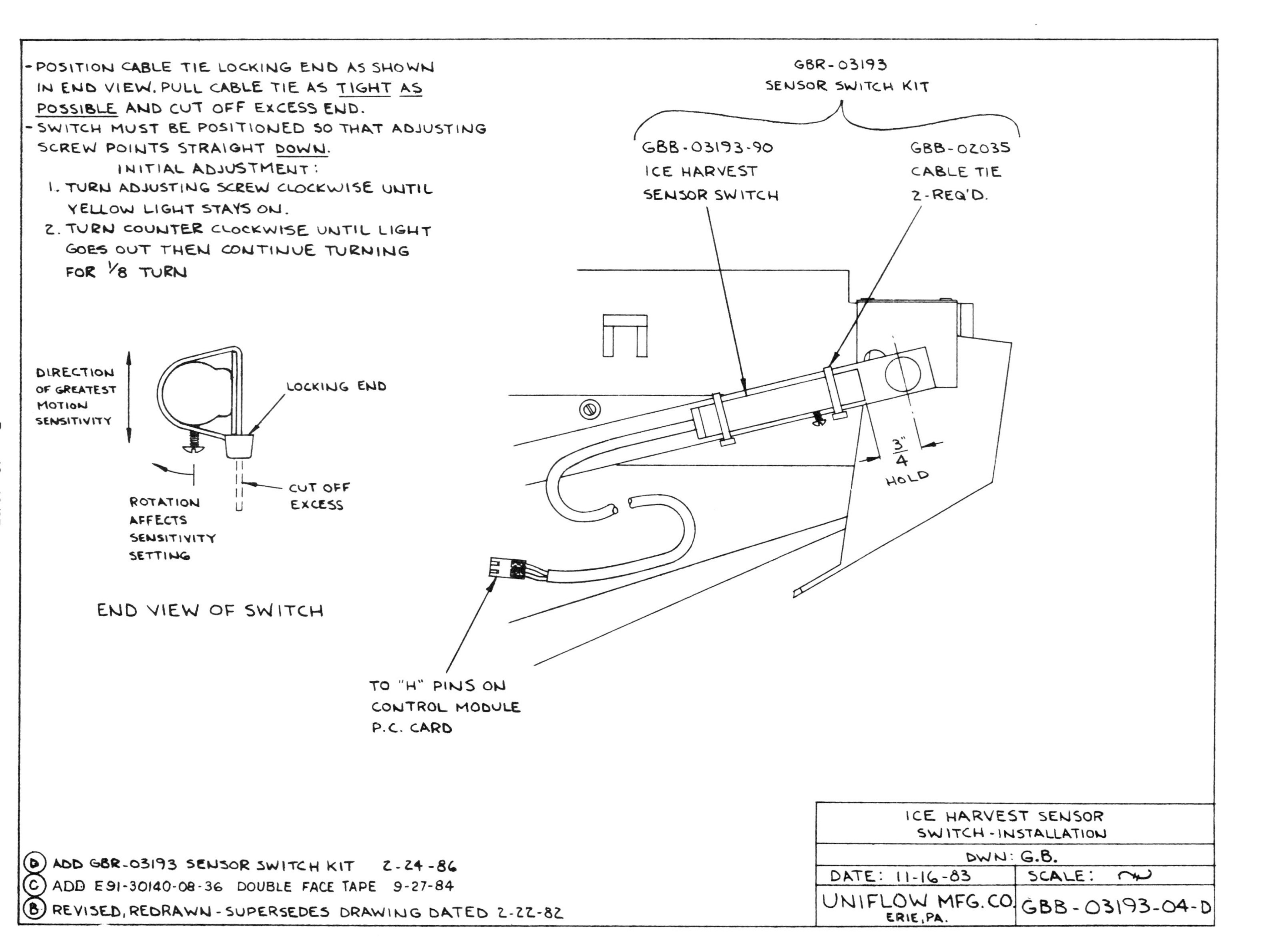

- POSITION CABLE TIE LOCKING END AS SHOWN IN END VIEW. PULL CABLE TIE AS TIGHT AS POSSIBLE AND CUT OFF EXCESS END.
- SWITCH MUST BE POSITIONED SO THAT ADJUSTING SCREW POINTS STRAIGHT DOWN.
INITIAL ADJUSTMENT:
1. TURN ADJUSTING SCREW CLOCKWISE UNTIL YELLOW LIGHT STAYS ON.
2. TURN COUNTER CLOCKWISE UNTIL LIGHT GOES OUT THEN CONTINUE TURNING FOR 1/8 TURN
GBR-03193
SENSOR SWITCH KIT
GBB-03193-90
ICE HARVEST
SENSOR SWITCH
GBB-02035
CABLE TIE
2-REQ'D.
DIRECTION OF GREATEST MOTION SENSITIVITY
LOCKING END
CUT OFF EXCESS
ROTATION AFFECTS SENSITIVITY SETTING
END VIEW OF SWITCH
3"/4 HOLD
TO "H" PINS ON CONTROL MODULE P.C. CARD
D ADD GBR-03193 SENSOR SWITCH KIT 2-24-86
C ADD E91-30140-08-36 DOUBLE FACE TAPE 9-27-84
B REVISED, REDRAWN - SUPERSEDES DRAWING DATED 2-22-82
ICE HARVEST SENSOR
SWITCH-INSTALLATION
DWN: G.B.
DATE: 11-16-83
SCALE:
UNIFLOW MFG. CO.
ERIE, PA.
GBB-03193-04-D

SEQUENCE OF OPERATION

With the Electronic Cuber mounted to a Bin, the water turned on, the electricity connected and the wash switch in the "ON" position, we can follow the Cuber through one complete cycle.

The Bin, actuator and cold water controls are set to the warm side (L.E.D.'S OFF - dimly lit on P.C. Card). The water plate is in the "UP" position. The pump and defrost switch is held up by the lift bolt on the water plate energizing the pump. The water inlet valve is open. The evaporator is cooling. The water level in the main water tank is rising and the corresponding water level is indicated in the water level probe assembly. Water is being pumped from the bottom of the circulation tank into the header at the left end of the water plate, through the lateral tubes under each row of cells, through the squirt holes in the center of each cell. The water stream hits the top of each cell then cascades down the 4 sides of each cell and returns to the main water tank through drain holes on each side of the squirt hole. When enough water to make one batch of ice is in the main water tank and the water level probe assembly, water will touch the upper thermistor probe and the water inlet valve shuts off. The control stream at the front left corner of the water plate should be flowing at the base of the dam. The screw at the inlet of the control stream may be adjusted to obtain proper setting.

When the evaporator temperature reaches 45° F. the evaporator thermistor probe signals the Printed Circuit Card to set the cold water control cold - the Red L.E.D. comes on bright.

The evaporator continues to cool, at 32° F. minute layers of ice form on the top and the 4 sides of each cell. As this process continues, the water level will decrease in the main water tank and the water level probe assembly.

NOTE: When the circulating water reaches a temperature of 32° F., it MAY be supercooled and it MAY partially crystalize in the water tank. If this occurs, the flow of water in the control stream nozzle will stop or fluctuate considerably and most of the circulation will stop for about 30 seconds. This is STRICTLY A NORMAL OPERATION AND THE CONTROL STREAM SHOULD NOT BE ADJUSTED AT THIS TIME.

When the evaporator temperature reaches 26° F., the evaporator thermistor probe signals the Printed Circuit Card to set the actuator control cold - the Orange L.E.D. comes on bright.

The water continues to freeze until each cell is almost completely full of ice. As the ice comes closer to the jet stream in the center of each cell, the head pressure in the water plate increases causing the control stream to rise and flow over the dam in the control stream box. This water will be dissipated to the drain pan, lowering the level in the water level probe assembly. Within 15-30 seconds the water will leave the bottom thermistor probe and expose it to air.

The harvest cycle is initiated - every circuit in the system is simultaneously energized. The hot gas valve opens sending hot gas to warm the evaporator. The water inlet valve opens and water rinses the water plate. The actuator circuit is energized through the actuator control, through the actuator toggle switch to the actuator motor causing the actuator motor shaft to turn counter-clockwise and seperate the water plate from the evaporator. When the water plate opens approximately one inch the pump and defrost swich drops, turning the water pump off. The pump switch now completes a second circuit to the defrost valve to keep it open during the harvest cycle. The actuator motor shaft continues to turn a 1/2 revolution until the trip lever on the actuator motor snaps the actuator toggle to the right and breaks the actuator circuit stopping the actuator motor in the stop down position.

As the evaporator warms, the cubes on the left edge of the evaporator slide out of the evaporator and rest on the water plate. There is a small fin connecting the bottom of the cubes together so that they will come down in unison and clear the water plate and slide into the Bin where they will break apart.

After the ice is out of the evaporator, the evaporator and the evaporator thermistor probe warm up rapidly. When the evaporator senses a temperature between 40°-45° F., its resistance change signals the Printed Circuit Card and the actuator control sets warm (Orange L.E.D. goes out - dimly lit), and completes the circuit from the defrost circuit to the reversing side of the actuator toggle switch and actuator motor. (NOTE: The actuator control does not start the defrost cycle; it only ends it after the ice falls out). The motor revolves clockwise raising the water plate. When the water plate is almost closed the lift bolt on the water plate pushes up the pump and defrost switch lever starting the pump and breaking one circuit to the hot gas valve. The water plate continues to the full "UP" position when the actuator toggle arm will snap the actuator toggle to the left causing the actuator motor to stop. When the evaporator temperature reaches 50° F., the cold water control will switch to the warm side shutting off the hot gas valve (Red L.E.D. goes out - dimly lit). When the water level is high enough to touch the upper thermistor probe it breaks the circuit to the water valve.

NOTE: Should some ice cubes be left on the water plate keeping it partially open when the water plate comes up, the actuator motor will continue to operate and the springs will stretch to allow the cams to take their vertical position and snap the actuator toggle switch. Since the lift plate cannot push up the defrost switch, the circuit is complete through this switch to the stop up side of the actuator toggle switch. When the actuator toggle arm snaps the actuator toggle switch to the stop up, the actuator motor will immediately reverse itself and open the plate. The captive ice then falls off. When the plate is in the lowest position, the actuator toggle arm will again reverse the actuator toggle switch and the actuator motor and cause the plate to close. This will continue until the water plate is clear and the lift plate can push up the defrost switch, breaking the circuit through this switch to the stop up side of the actuator toggle switch, so that the motor will stop with cams up when the actuator toggle arm pushes the actuator toggle to the stop up position.

The Cuber has now completed its full cycle and started another freezing cycle. This will be regularly repeated until the Bin is full and the Bin control shuts off the Cuber automatically. When some ice is removed from the Bin, the Cuber will start up and refill the Bin.

A detailed description, including wiring circuits can be seen, starting on Page 14, which shows and explains, in detail, the Sequence of Operation. Comparison is made to the thermostatic Cubers.

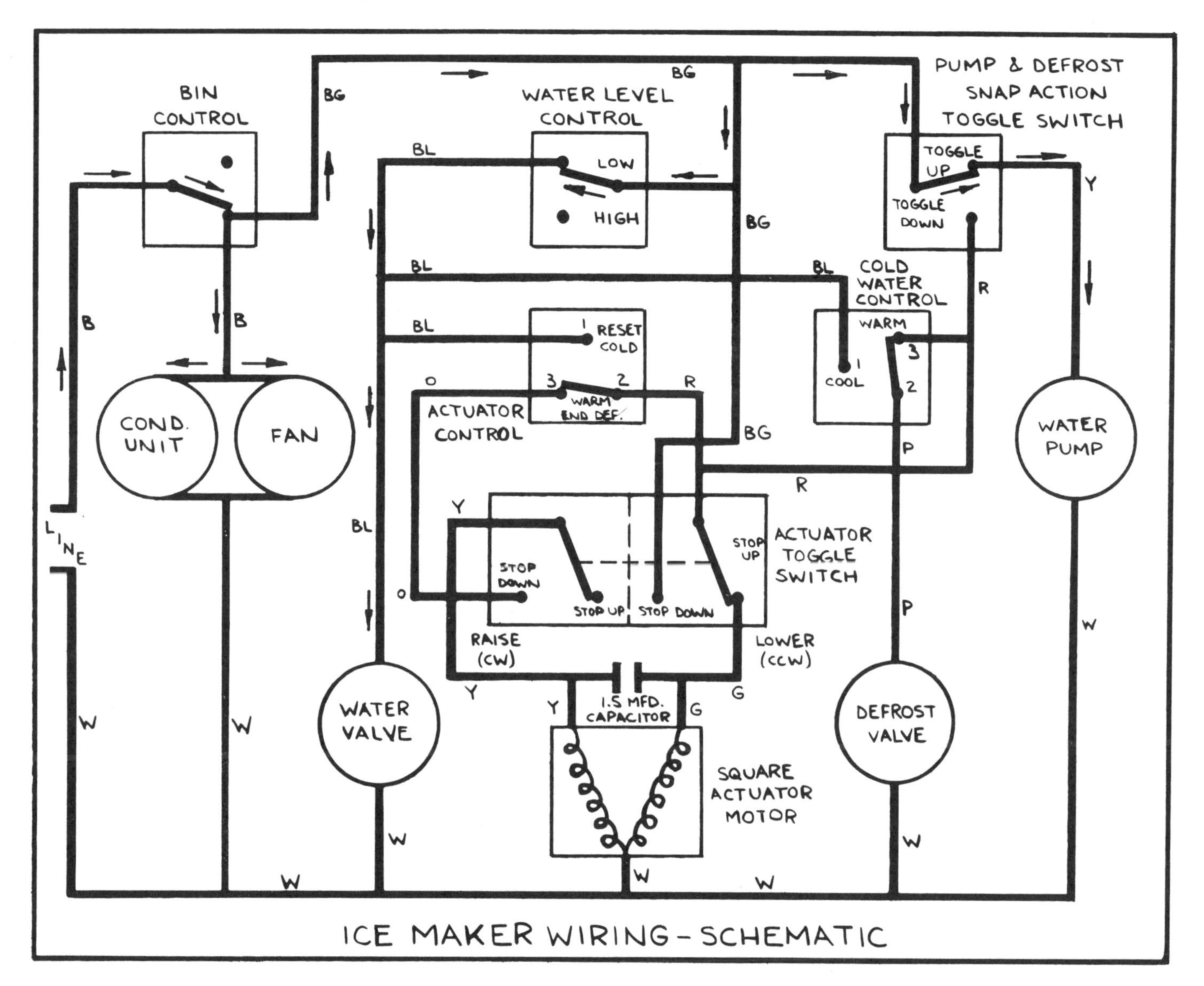

SEQUENCE OF ELECTRICAL CIRCUITS

A color schematic wiring diagram with moveable switches is provided at the end of the electronic section and is an excellent tool to follow the electrical sequence.

The following schematic wiring diagrams show the flow of electric current at the several steps in the operation of the machine. The arrows beside wires show current flow on all wiring diagrams. Below is a description of the operation specifically related to the circuits that are used at each step. These circuits apply to all machines built 1964 and after and those in the field which have been revised to the 1964 wiring.

NOTE 1. — Paragraph headings "EC" refer to thermostatically controlled cubers.
NOTE 2. — Paragraph headings "EEC" refer to electronically controlled cubers.

EC 1 — Water fill (circulating water above 45 degrees) — Current to condensing unit, water circulating pump, and water inlet valve. Water filling circulation tank and control tank. Water pump circulating water through water header, distribution laterals and jet holes to individual evaporator cells. Evaporator is cooling water. (approximately 2 minutes)

EEC 1 — Water fill (circulating water above 45 degrees) — Current to condensing unit, water circulating pump, and water inlet valve. Water filling circulation tank and water level probe assembly. Water pump circulating water through water header, distribution laterals and jet holes to individual evaporator cells. Evaporator is cooling water. (approximately 2 minutes)

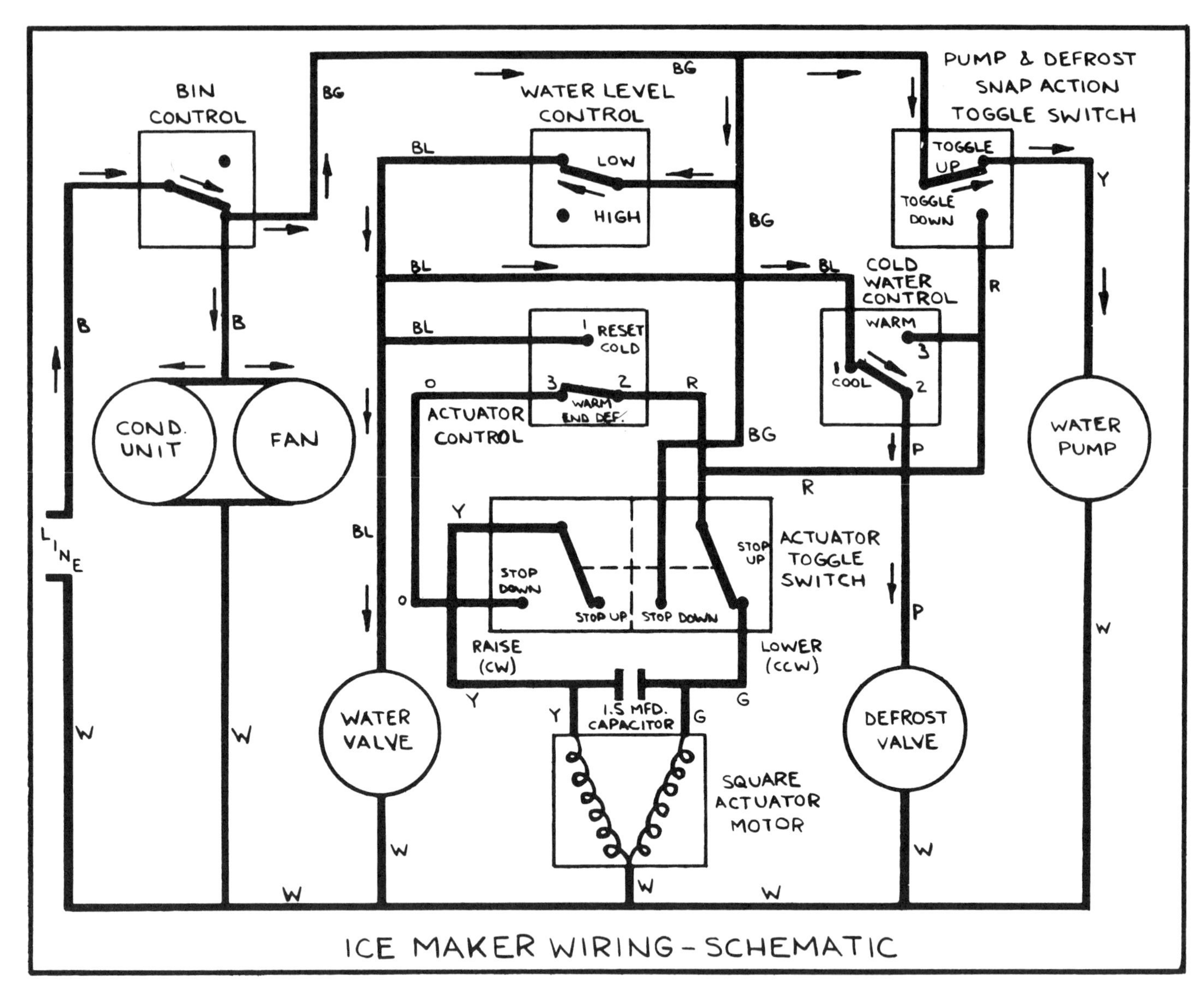

ICE MAKER WIRING - SCHEMATIC

EC 2 — Water fill (cuber with cold water thermostat — circulating water below 40 degrees) — current to condensing unit, water pump, water valve and through blue circuit through cold water thermostat, pink circuit to defrost valve. When the incoming water is cold and the compressor can cool the water below 40 degrees during the "Fill Cycle" the cold water thermostat will switch to the cool side connecting the pink and blue circuits giving power to the defrost valve allowing hot gas from the compressor to go through the evaporator warming up the circulating water. GS models have no cold water thermostat.

EC 2A — Water fill (cuber with cold water thermostat — circulating water warms up to 50 degrees). If the water warms up to 50 degrees before the water fill is complete and the weight control switch drops, the cold water thermostat will switch to the warm side shutting the defrost valve and the compressor will start to cool the water again.

EEC 2 — Water fill control — (circulating water below 40 to 45 degrees) current to condensing unit, water pump, water valve and through blue circuit through cold water control, pink circuit, to defrost valve. When the incoming water is cold and the compressor can cool the water below 40 degrees during the "Fill Cycle", the cold water control will switch to the cool side connecting the pink and blue circuits giving power to the defrost valve allowing hot gas from the compressor to go through the evaporator warming up the circulting water.

EEC 2A — Water fill control — (circulating water warms up to 50 to 55 degrees). If the water warms up to 50 degrees before the water fill is complete and water touches the top thermistor water level probe, the cold water control will switch to the warm side shutting the defrost valve and the compressor will start to cool the water again. A red L.E.D. is mounted on the control card below the cold water control potentiometer as a service aid. It is on bright when the cold water control is in the cold position.

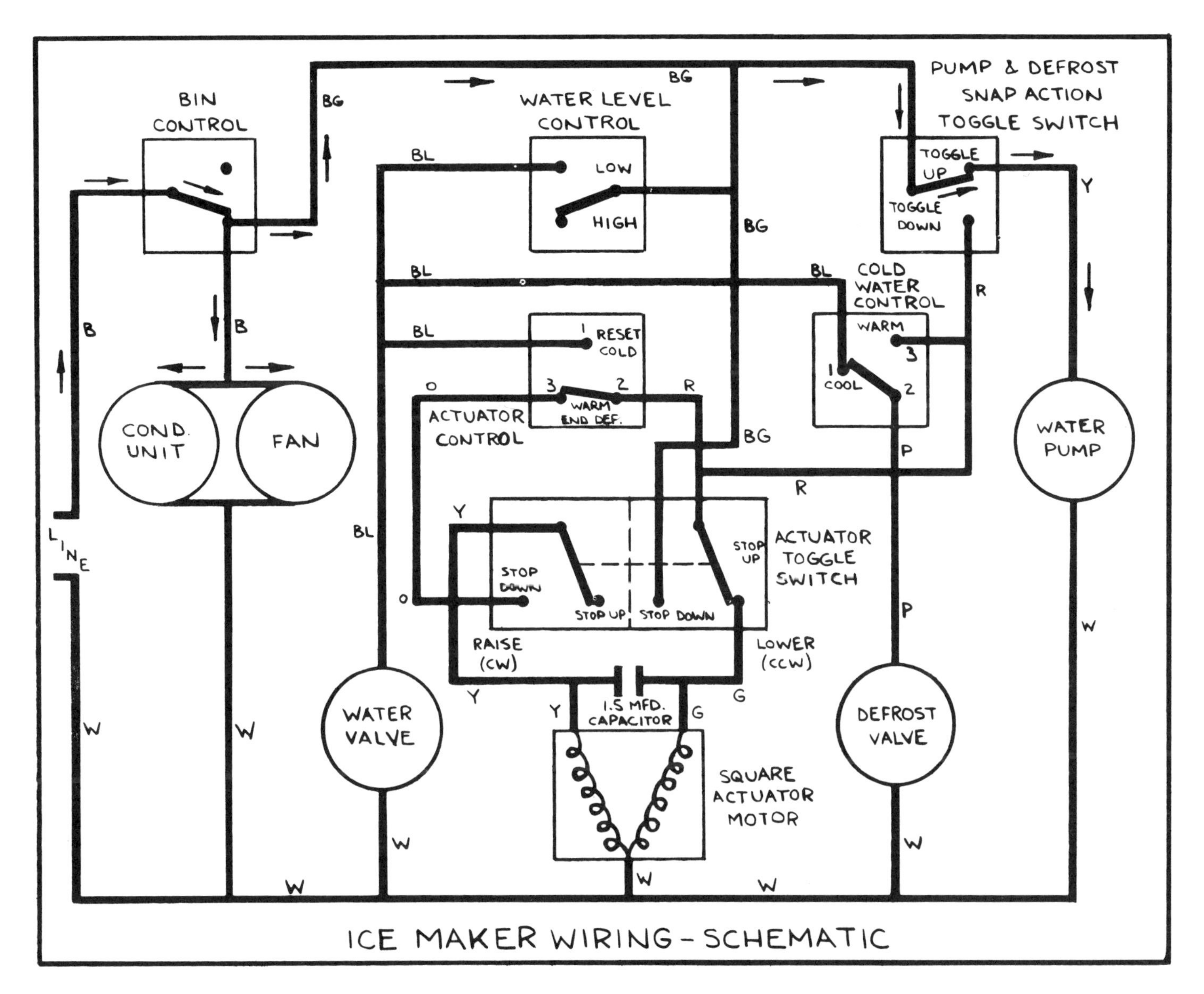

ICE MAKER WIRING - SCHEMATIC

EC3 — Tanks Full, Weight Switch Drops — Freeze Cycle Evaporator Above 20 Degrees — Sufficient water is in the circulation tank and control tank so that weight of water in control tank pulls weight control switch down shutting water inlet valve. (If the cold water thermostat is still on the cool side and the circulating water is still being warmed as the weight control switch drops, this switch cuts off power to the blue circuit and through the cold water thermostat to the defrost valve shutting the defrost valve). Water level should settle in the control tank at a preset distance below the top of the main circulation tank. (See chart of Water levels). Current to condensing unit and water circulating pump. Water freezing in cells. Springs hold water plate edges against bottom of cams. Cams and hinge leaves hold water plate approximately 1/32" from evaporator to maintain adequate fin between the cubes.

EEC 3 — Tank and water level probe assembly full-freeze cycle evaporator above 26 degrees — sufficient water is in the circulation tank and water level probe assembly so that water touches the top thermistor probe and the water inlet valve shuts off.

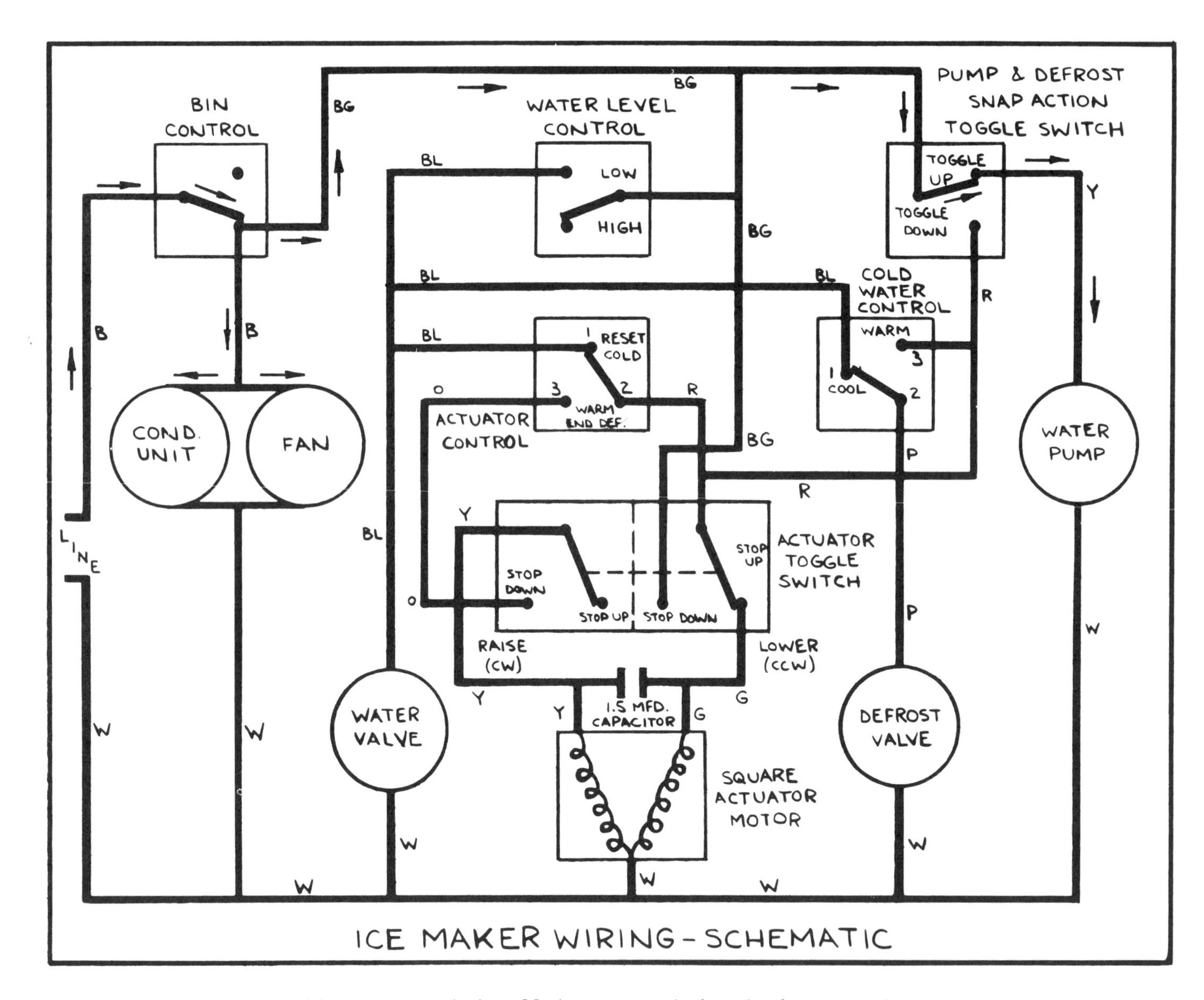

EC 4 — Freeze cycle with evaporator below 20 degrees — during the freeze cycle, the evaporator and the actuator thermostat get cold enough (approximately 20 degrees) to reset the actuator control ready for the next defrost cycle. The particular time when this happens is unimportant since the water control stays off and prevents defrost until the cubes are full and the water is used up, and there is no change in the current flow at this step. During this period the water is being used up in the formation of ice and the water level drops slowly in the circulation tank and the water level probe tube.

EEC 4 — Freeze cycle with evaporator below 26 degrees — during the freeze cycle, the evaporator and the evaporator thermistor probe get cold enough (approximately 26 degrees) to reset the actuator control ready for the next defrost cycle. The particular time when this happens is unimportant since the water control stays off and prevents defrost until the cubes are full and the water is used up, and there is no change in the current flow at this step. During this period the water is being used up in the formation of ice and the water level drops slowly in the circulation tank and the level probe tube. An orange L.E.D. is mounted on the printed circuit card below the actuator control potentiometer as a service aid. It is on (bright) when the actuator control is in the cold position.

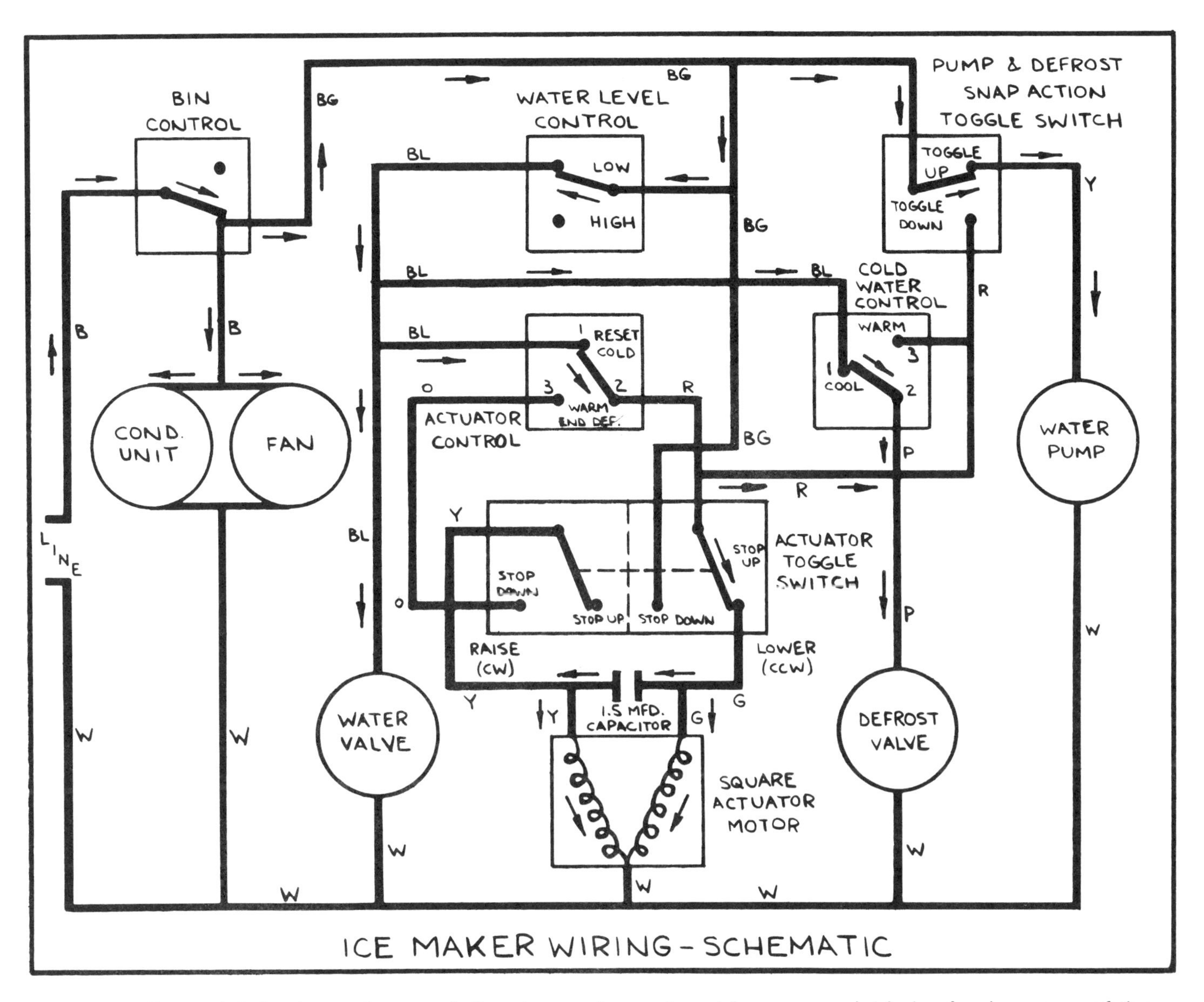

ICE MAKER WIRING - SCHEMATIC

EC 5 — Start of defrost — cubes are full and some have almost frozen over jet holes forcing some of the remaining water faster through the control stream over the control stream dam, lowering water level rapidly in the control tank until it is light enough for the weight control switch to snap up. Current flows to condensing unit, water pump, through weight control switch and the blue circuit to water valve, also from blue through actuator thermostat, first connection is complete to the Red defrost circuit. Also from Red, current flows through the actuator toggle switch giving 115 volts on the gray actuator motor circuit; current through the capacitor which changes phase and boosts the voltage to 200 volts on the yellow actuator motor circuit giving counter-clockwise rotation. Current from the Blue circuit flows through the water thermostat to the Pink circuit and defrost valve. Cams start rotating counter-clockwise pressing on the teflon brackets on the water plate to release it from the ice (approximately 15 seconds). The open defrost valve allows hot refrigerant gas from the compressor to go through the evaporator coil to start releasing the cubes.

EEC 5 — Start of defrost — cubes are full and some have frozen over jet holes forcing some of the remaining water faster through the control stream over the control stream dam lowering water level rapidly in the water level probe assembly until the bottom thermistor probe is exposed to air and trips the water level control on in 5-20 seconds. Current flows to condensing unit, water pump, through water level control and Blue circuit to water valve, also from Blue through actuator control; first connection is complete the Red defrost circuit. Also from Red, current flows through the actuator toggle switch giving 115 volts on the Gray actuator motor circuit current through the capacitor which changes phase and boosts the voltage to 200 volts on the Yellow actuator motor circuit giving counter-clockwise rotation. Current from the Blue circuit flows through the cold water control to the Pink circuit and the defrost valve. Cams start rotating counter-clockwise pressing on the teflon brackets on the water plate to release it from the ice. (approximately 15 seconds). The open defrost valve allows hot refrigerant gas from the compressor to go through the evaporator coil to start releasing the cubes.

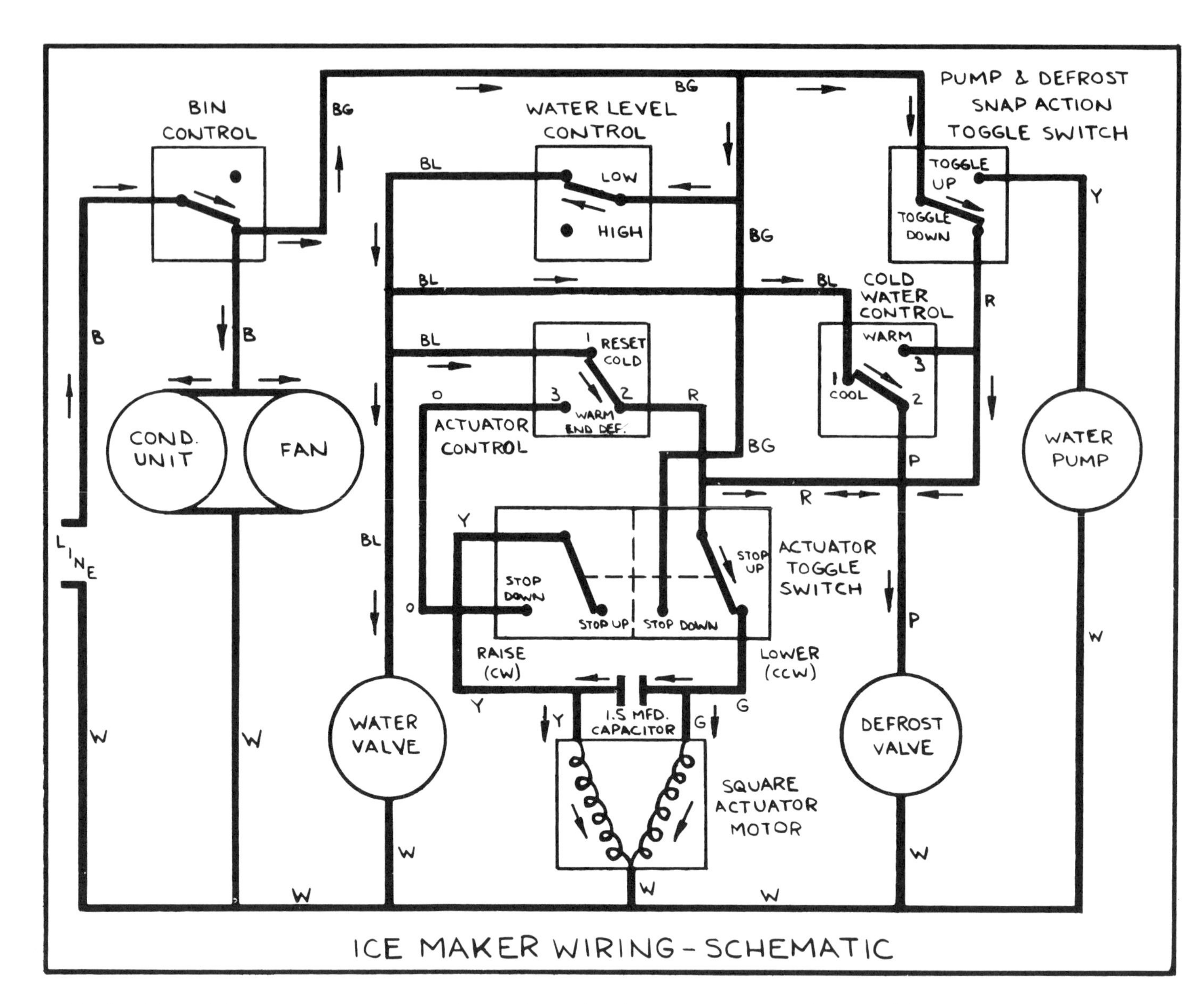

EC 6 — Water Plate Lowering — Cams have released water plate from ice. Then the lift plate on the water plate allows pump and defrost switch to drop, stopping pump and completing second connection to the Red defrost circuit. Acutator motor and cams continue to rotate counter-clockwise lowering the water plate. The open water valve allows water to begin rinsing off the water plate. Current to condensing unit actuator motor, water valve, and defrost solenoid. (Approximately 20 seconds).

EEC 6 — Same as EC 6.

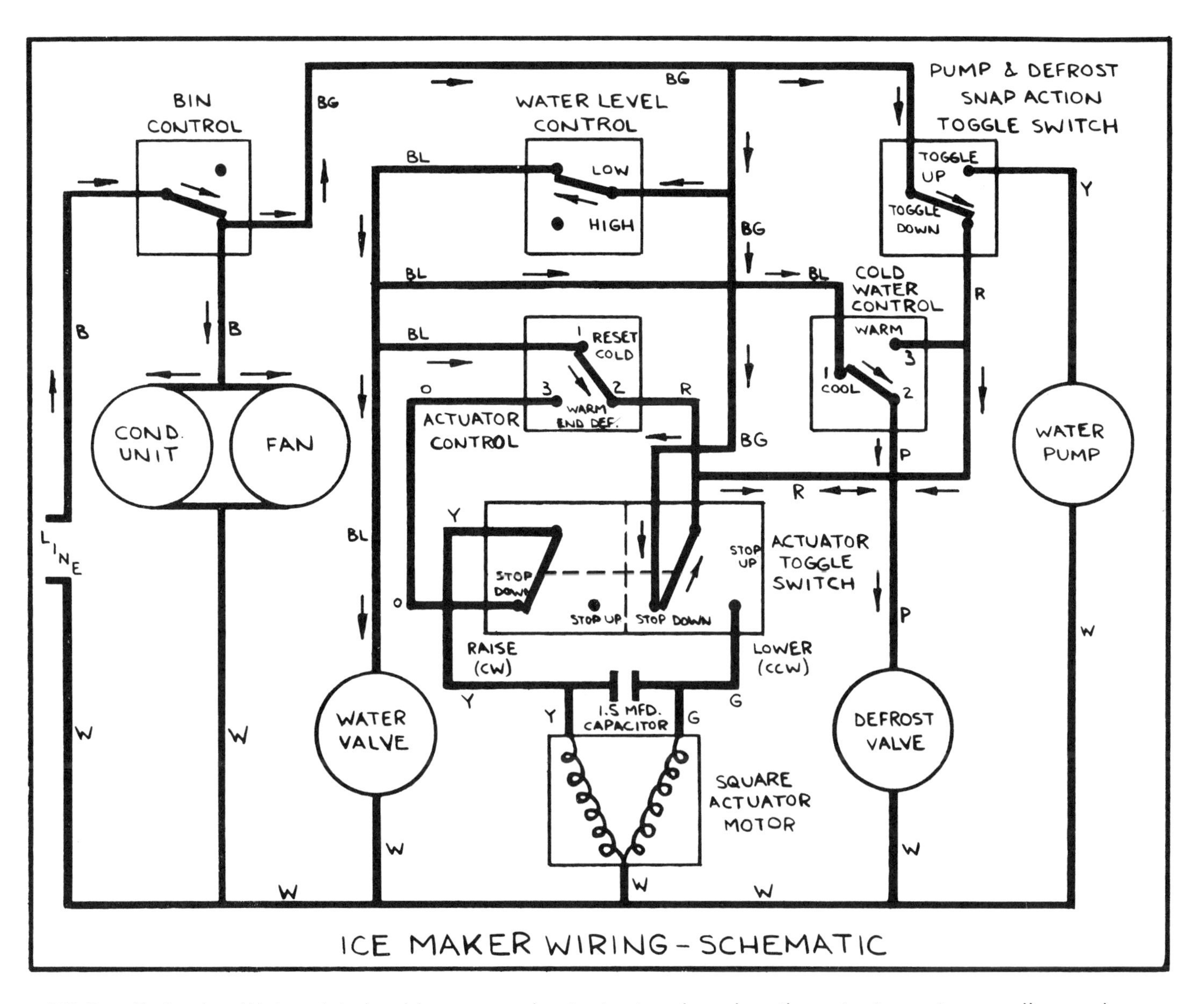

ICE MAKER WIRING - SCHEMATIC

EC 7 — Defrost — Water plate is wide open and actuator toggle rod on the actuator motor coupling pushes actuator toggle switch to the left, stopping the motor with the water plate in "down" position and completing the third connection to the Red defrost circuit. Current to condensing unit, water valve and defrost solenoid. Excess water concentrated with minerals drains from water tank. Fresh water washes water plate and tank. Hot refrigerant gas continues flowing through evaporator releasing ice slowly. Edge of the evaporator and actuator thermostat bulb remain cool (32 degrees to 35 degrees) as long as ice remains in the evaporator. (Approximately 2 to 4 minutes depending on ambient and hot gas temperature). Ice releases on left side first and rests on water plate, then the ice falls out substantially all at one time as fin between cubes tends to hold them together until they drop into the bin. After ice falls out, side of evaporator warms up rapidly to 45 degrees. (Approximately 30 seconds). Actuator toggle pushed left also completes orange-yellow circuit but with no power in the orange circuit the actuator motor gets no power during the defrost period.

EEC 7 — Defrost — Water plate is wide open and actuator toggle lever on the actuator motor rear shaft pushes actuator toggle switch to the right stopping the motor with the water plate in "down" position and completing the third connection to the Red defrost circuit. Current to condensing unit, water valve and defrost solenoid. Excess water concentrated with minerals drains from water tank. Fresh water washes water plate and tank. Hot refrigerant gas continues flowing through evaporator releasing ice slowly. Edge of the evaporator and evaporator probe remain cool (32 degrees to 35 degrees) as long as ice remains in the evaporator. (Approximately 2 to 4 minutes depending on ambient and hot gas temperature). Ice releases on left side first and rests on water plate, then the ice falls out substantially all at one time as fin between cubes tends to hold them together until they drop into the bin. After ice falls out, side of evaporator warms up rapidly to 45 degrees. (15-30 seconds) Actuator toggle pushed right also completes orange-yellow circuit but with no power in the orange circuit the actuator motor gets no power during the defrost period.

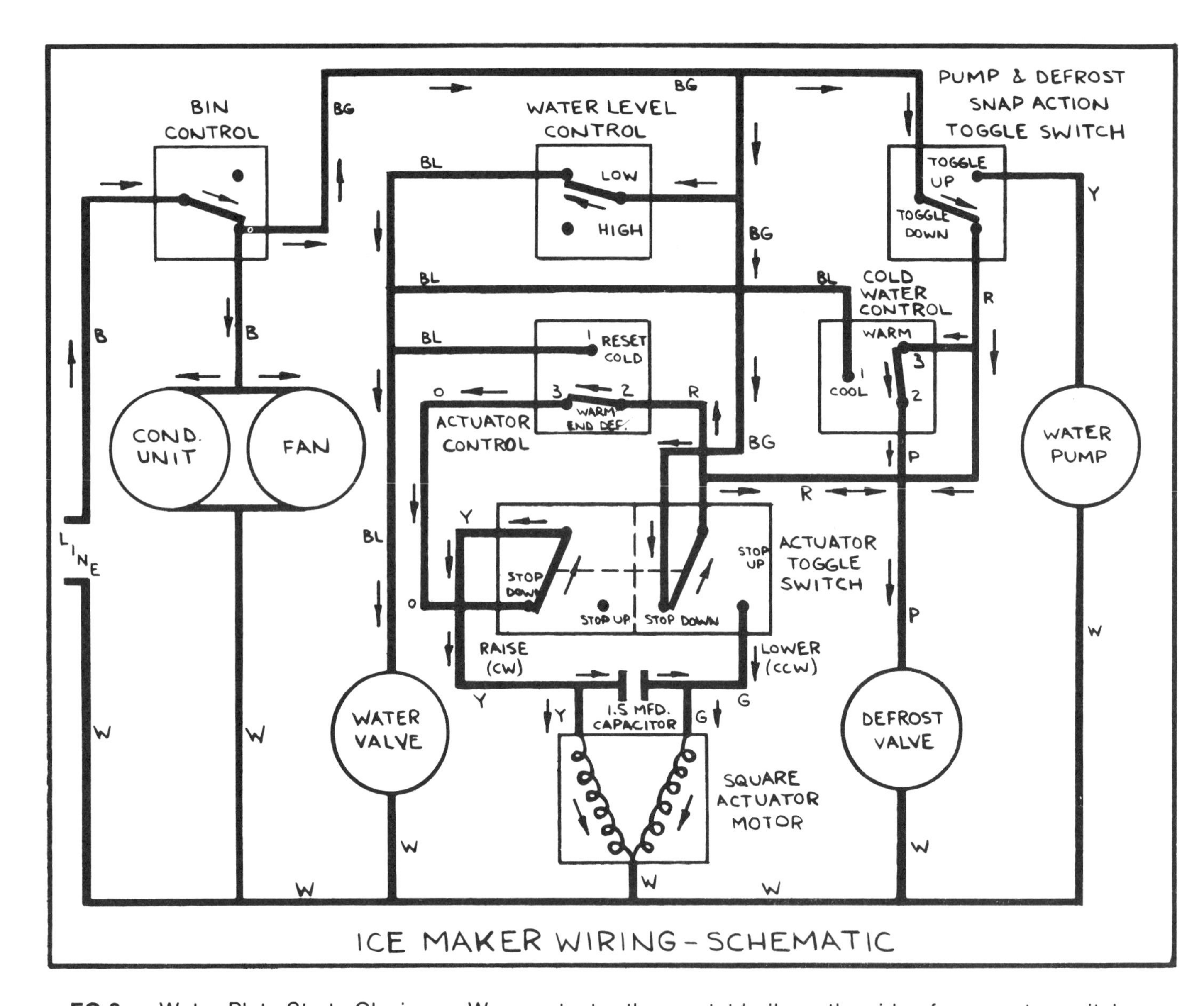

EC 8 — Water Plate Starts Closing — Warm actuator thermostat bulb on the side of evaporator switches actuator thermostat from cold position disconnecting the first circuit to Red; the current from green to the Red circuit continues through the actuator toggle and pump toggle switches. The actuator thermostat switching to the warm position completes the red-orange circuit and through the actuator toggle gives 115 volts to the yellow winding on the actuator motor. The capacitor change the phase and boosts the voltage to 200 volts to the gray winding of the motor to give clockwise rotation. Cams rotating clockwise pull up the springs and water plate. Current to condensing unit, defrost water valve and actuator motor. (Approximately 20 seconds). The cold water thermostat bulb on the edge of the evaporator will warm up at about this time or during closing and switch the cold water thermostat to the warm side connecting the red to pink circuit; since red is energized the defrost valve stays open.

EEC 8 — Water Plate Starts Closing — Warm evaporator probe on the side of evaporator switches actuator control from cold position disconnecting the first circuit to red; the current from green to the red circuit continues through the actuator toggle and pump toggle switches. The actuator control switching to the warm position completes the red-orange circuit and through the actuator toggle gives 115 volts to the yellow winding on the actuator motor. The capacitor changes the phase and boosts the voltage to 200 volts to the gray winding of the motor to give clockwise rotation. Cams rotating clockwise pull up the springs and water plate. Current to condensing unit, defrost water valve and actuator motor. (Approximately 20 seconds). The cold water control will warm up at about this time or during closing and switch to the "warm" side turning off the red "L.E.D." Models with GBB-03135-02-E cards will turn off the hot gas valve at this time. Regardless of water plate position. Refer to model wiring diagrams for circuit details.

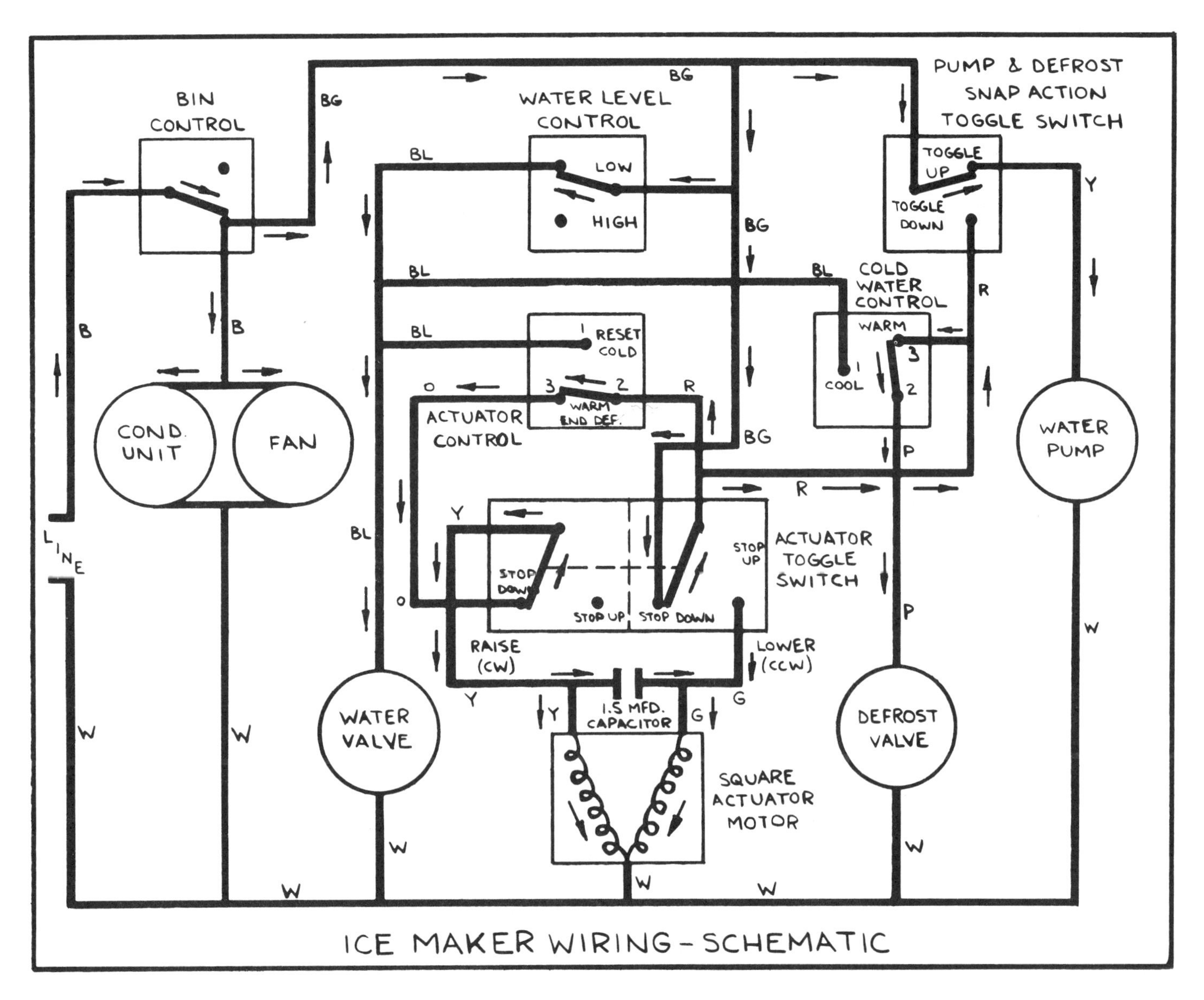

EC 9 & EEC 9 — Water Plate Almost Closed — Lift plate pushes up the pump and defrost toggle switch, starting pump and disconnecting second circuit to Red, but current from green to Red continues through actuator toggle switch. Current to all operating parts of the cuber. Water starting to fill tank. Hot gas from the defrost valve keeps the evaporator warm to melt any small piece of ice that may be left on the water plate as it closes. Cams continue rotating to upright position. (Approximately 5 seconds).

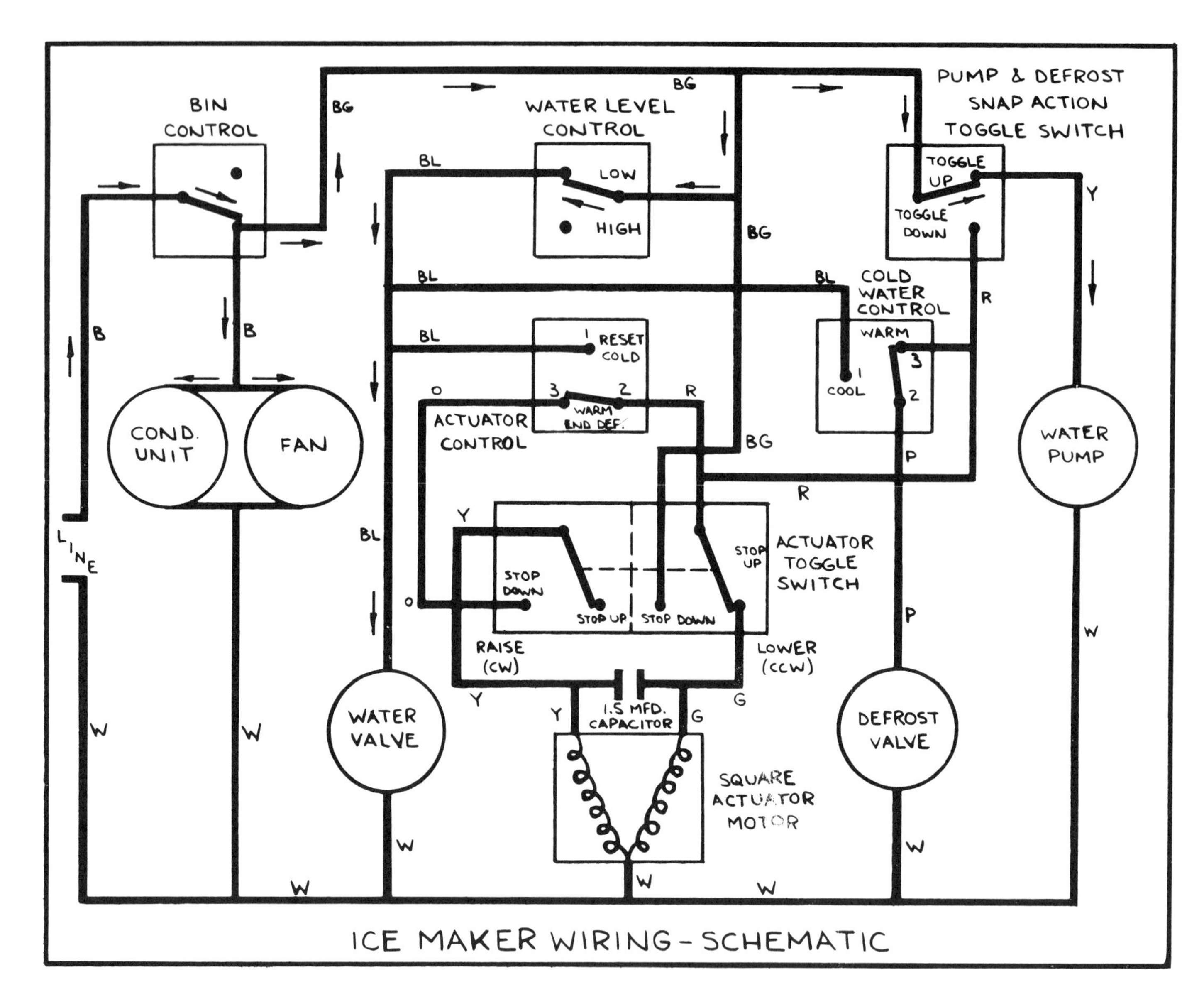

EC 10 — Water Plate Closed — End Defrost — Front cam in upright position pushes actuator toggle switch to the right disconnecting the third and final circuit to Red, stopping the actuator motor with the water plate up and closing the defrost valve (cycle completed - same circuit as EC 1 — Water Fill). Current to condensing unit, water circulating pump, and water inlet valve, cycle starts.

EC 11 — (Circuit not shown) — Bin Full — Ice against bin thermostat tube opens bin thermostat shutting off all parts of the machine. When ice is removed, bin thermostat closes; machine will start up and operate regardless of what part of the cycle it was in when it was shut off.

EEC 10 — Water Plate Closed — End Defrost — Front cam in upright position pushes actuator toggle switch to the left disconnecting the third and final circuit to Red, stopping the actuator motor with the water plate up and closing the defrost valve (cycle completed — same circuit as EC 1 — Water Fill). Current to condensing unit, water circulating pump and water inlet valve, new cycle starts.

EEC 11 — (Circuit not shown) — Bin Full — Ice against bin probe after water fill cycle is completed, opens bin control relay shutting off all parts of machine. When ice is removed, bin control relay closes; machine will start up and operate regardless of what part of the cycle it was in when it was shut off.

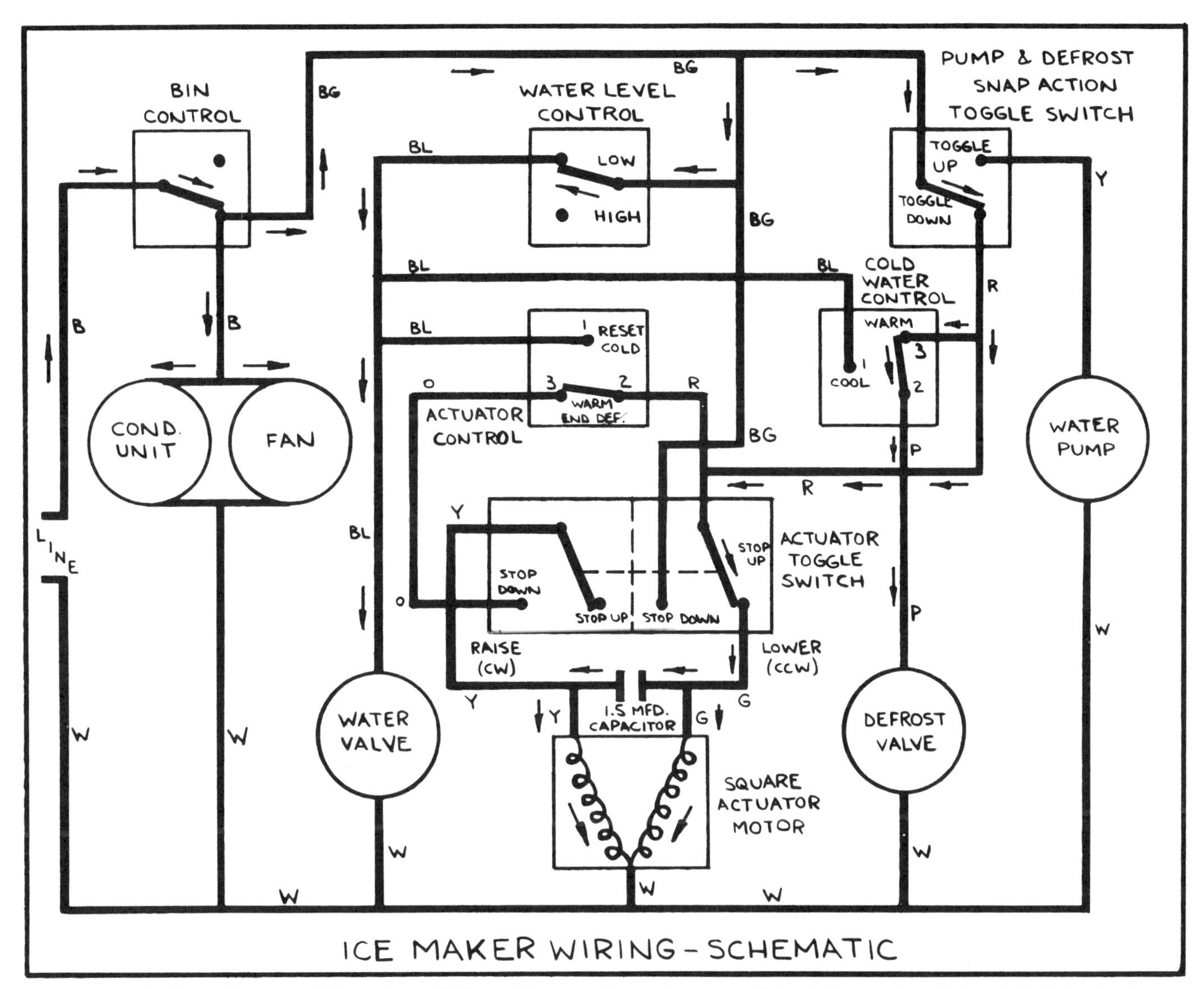

EEC & EC 12 — Abnormal Opening of Water Plate — If a cube ever sticks to the water plate while it is closing (Step EC 8) and by stretching the springs the cube prevents the water plate from closing enough to push pump and defrost toggle switch up as it normally does in Step EC 9, the circuit through the pump and defrost toggle switch will remain complete to the defrost "Red" circuit to the lowering (counter-clockwise) side of the actuator toggle switch so that, when the front cam pushes the actuator toggle switch to the right, the actuator motor will immediately reverse and, with cams rotating counter-clockwise, the water plate will re-open. Current to condensing unit, water inlet valve and through pump and defrost toggle switch to actuator motor. Any other obstruction between the water plate and the evaporator can cause the same effect as a cube on the water plate. If the collar on the pump and defrost switch lift rod is set incorrectly so the switch will not go all the way up, abnormal opening will occur. Further, if the water level is set much too high and/or spring is unhooked, allowing the water plate to sag during the water fill so that the defrost switch goes down, the circuit will be completed to the actuator motor and the plate will open. Likewise abnormal opening can be created during the water fill or beginning of the freeze cycle if the water plate is pulled down by hand, stretching the springs until the pump and defrost switch goes down, completing the circuit to the actuator motor. This is done to rinse the machine after using ice machine cleaner. Opening by hand is also used to observe the jet streams by allowing the cams to go down to horizontal then pushing up on the pump toggle rod. Further opening by hand allows a quick partial check of the weight control switch as the water drains from the control tank and the switch snaps up with a small amount of water left in the tank. (During this check the control tank tube must be a full 11 inches long to prevent binding between the control tank and the control stream box on the water plate).

EEC & EC 13 — (See Diagram EC 8) — Water Plate Closing After Abnormal Opening — Since the evaporator and actuator control are warm during abnormal opening, a circuit is complete through the actuator control to the orange circuit of the actuator toggle switch. As soon as the water plate is wide open and the actuator toggle rod on the motor coupling pushes the actuator toggle switch to the left, the motor will immediately reverse and, with cams turning clockwise, close the water plate.

If the obstruction remains, Steps EC 12 and EC 13 will repeat. If the obstruction is removed, such as the cube falling out, Step EC 13 will be followed by Steps EC 9 and EC 10 and normal operation will be resumed.

SPEEDY ELECTRONIC CUBER FUNCTIONAL TESTS

Without the use of special tools or testers, these tests can quickly determine any major faults with either the control module card or probes. An ordinary pocket screwdriver can be ground down to fit slots on potentiometers which have been designed so they cannot be adjusted with a standard screwdriver.

BEFORE STARTING TEST, BE SURE THERE IS POWER TO CUBER AND CHECK FUSES IN CONTROL MODULE BOX.

Kold-Draft Electronic Cuber

Speedy Test

(A Screwdriver May Be Used To Short Pins)

Upper Left Corner P.C. Card

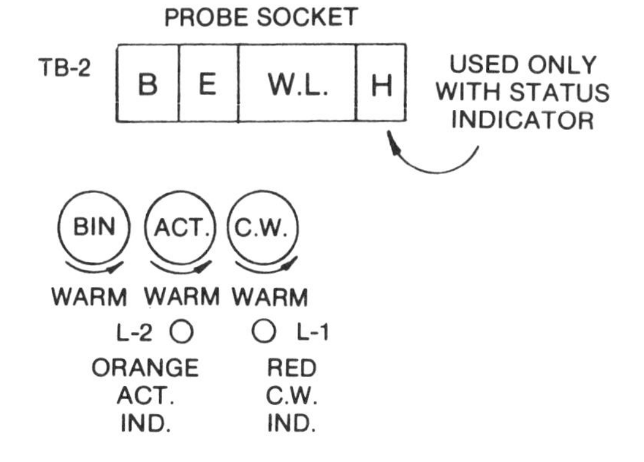

A. "Open" terminals produce a "Cold" signal to the P.C. Card.
B. "Short" terminals produce a "Hot" signal to the P.C. Card.
C. Turn wash switch to "wash" to prevent compressor short cycling during tests. To perform speedy test induce the following conditions on TB-2.
D. A screwdriver may be used to short pins.
NOTE: When lights are on they are bright, when off, a faint glow.

	B	E	W	L	H	
1	Open	Open	Open	Open	Open	Cuber Stops, Orange & Red L.E.D. ON
2	Short	Open	Open	Open	Open	Cuber Runs, Orange & Red L.E.D. ON
3	Short	Short	Open	Open	Open	Orange & red L.E.D.s OFF. Cuber Stops.
4	Short	Open	Open	Short	Open	Harvest Begins (Allow plate to open fully), Both L.E.D.'s ON
5	Open	Short	Open	Open	Open	Press Bin Relay Plunger, Water Plate Closes, Both L.E.D.'s OFF
	— STATUS INDICATOR TESTS — IF OPTION IS ON CUBER					
6	Short	Open	Open	Open	Short	Yellow L.E.D. on Status Indicator ON
7	Open	Open	Open	Open	Open	Yellow L.E.D. on Status Indicator ON
8	Short	Open	Open	Open	Open	Green L.E.D. on Status Indicator BLINKING
9	Short	Short	Open	Open	Open	Yellow & Green L.E.D. on Status Indicator ON

E. To test Status Indicator "time out", short "B" pins (or insert good, warm probe), and short the test pins (TP1 and TP2) at the Status indicator socket on the P.C. Card. Green L.E.D. will blink 1 to 4 times, then cuber will shut off. Status Indicator Red and Yellow L.E.D.'s will turn On.

Above Tests confirm a good P.C. Card.

TO TEST PROBES:

A. To test probes, connect each in turn to the "E" pins with all other pins left open. Turn cold water pot mid range (12:00 o'clock). A warm probe will turn the L.E.D. off. Submerging the probe in ice water will turn the Red L.E.D. on. Reaction time 5-20 seconds. The most accurate method is to test resistance of probe in ice water. 5650 Ohms ± 2% evaporator probe, ± 4% bin probe, ± 20% water level probe.
B. Be sure to return cold water pot to original position after probe tests.

TEST PROCEDURE WITH CUBER ANALYZER

The following test procedure will thoroughly check all operations of the control module P.C. Card by simulating correctly operating bin, evaporator and water level probe signals. If a particular probe is suspect of causing trouble, the proper operation of that probe may be simulated by using the Cuber analyzer. If a particular part of the Cuber does not respond to one or more of these tests, either the part itself or the P.C. Card may be defective. The suspect part should always be checked before a replacement P.C. Card is installed. If a new probe is required, it should be installed and the Cuber watched through a minimum of one cycle to assure proper operation.

1. Turn wash switch to wash position.
2. Remove all probe plugs from the upper left hand plug. Once removed, install the analyzer plug. There is an interlocking lip which will mate with another lip on the P.C. Card plug. At this time the thermistor probes must be connected to the back of the analyzer plug. The correct order is, from left to right, Bin, Evaporator, Water level, and Harvest switch (if a status indicator is being used).
3. Now that the analyzer is correctly hooked up, you may send the correct logic signal to the associated circuit by moving a switch or group of switches.
4. On the analyzer plug a small slide switch is incorporated to allow testing of all P.C. Cards. This switch interchanges the water level probe connections to the analyzer. GBR-03135 position used for all cards with GBR-03135-02-DC, GBR-03135-02-FC, GBR-03135-02-P or GBR 03130-P inked on P.C. Card or transformer. The GBR-03130 position used for GBR-03130 cards. (Prior 1980)

BIN THERMOSTAT SWITCH — When switched to the "Full"side, the Cuber will shut down. If a Status indicator is attached, the yellow L.E.D. will come on. If switch is pushed to the "Full" side during the fill cycle, the Cuber may not shut down until the water reservoir is full.

WATER LEVEL SWITCH — When taking on water you can move the switch to the high position. This will shut off the water valve. To continue filling, move the switch to the low position and then release it. If you move the switch to the low position, the water valve will open, however, if the red L.E.D. is on, the water inlet valve and defrost valve will open and if the orange and red L.E.D.'s are both on, the water valve, defrost valve and actuator motor will all energize.

ACTUATOR THERMOSTAT SWITCH — Moving this switch "Cold" will simulate the actuator and Cold Water Thermistors Cold. This will be evident by the orange and red L.E.D.'s being lighted. Once the switch is released, the L.E.D.'s will go out, unless the evaporator is cold (below 40°), in which case red or both L.E.D.'s will stay on; pushing the switch up (warm) will cause the L.E.D.'s to go out. If you would like to initiate harvest, hold the actuator switch cold and move the water level switch low, which will energize the water inlet valve, defrost valve and actuator motor. Moving this switch to warm wil raise the water plate to the full up position after harvest.

COLD WATER THERMOSTAT SWITCH — Moving this switch to the cold position will turn the red L.E.D. on and close the circuit for the defrost valve. If you energize this circuit during the period when water is coming in, the defrost valve will open. You can also perform this function by moving the cold water switch to cold and moving the water level switch to low. If the evaporator probe is good and evaporator near 32° orange L.E.D. may also come on.

SAFETY THERMOSTAT SWITCH — Move this to the warm position and it will shut down the cuber on overheated evaporator. If a status indicator is attached, the yellow and green L.E.D.'s will come on.

TIMEOUT — Timeout of the cuber may be simulated by connecting the two alligator clip leads to the two test pins located adjacent to the status indicator terminal block. On P.C. Cards manufactured after July, 1980, these test points will be labeled TP1 and TP2. The status indicator P.C. Card must be connected to simulate this feature. (On old circuit boards without test points connect alligator clips to left and right pins of status indicator terminal block). With the clips connected, move the switch to "Timeout". within a few blinks of the green L.E.D., the cuber will time out and the red and yellow L.E.D.'s on the status indicator will turn on. Releasing the switch should restart the Cuber, turn off the red and yellow L.E.D.'s and start the green L.E.D. blinking. Moving the same switch to the trip position should turn the green L.E.D. off and the yellow L.E.D on (on new cards). For more information, refer to Trouble, Cause and Remedy.

TROUBLE, CAUSE AND REMEDY ELECTRONIC

Note:
1. Common to electronic & thermostatically controlled cubers (regular type)
2. Electronic only **(bold type)**
3. Thermostatically controlled cuber only *(italicized type)*

TROUBLE	CAUSE/SYMPTOM	REMEDY
1. Cuber will not start.	1. Line fuse blown.	1. Check circuit for short or ground. Replace fuse.
	2. Bin full of ice.	2. Use some ice.
	3. Open circuit in cord or feed wires.	3. Repair or replace.
	4. No money in meter if meter is used.	4. Feed meter.
	5. Room too cold (below 45 degrees).	5. Warm room. Consult factory for cold room adaptation.
	6. Overheated evaporator.	6. If evaporator is hot, allow to cool. Check defrost circuit.
	7. Blown fuse on P.C. Card.	**7. Replace fuse.**
	8. Bin probe disconnected or loose. Set too warm counter-clockwise.	**8. Install bin probe properly. Set slightly clockwise.**
	9. Defective bin probe.	**9. Jumper bin probe pins (B) on P.C. Card, if cuber starts, replace probe.**
	10. Defective P.C. Card or Bin Relay.	**10. Check with cuber analyzer. Replace P.C. Card or Bin Relay.**
	11. Status Indicator (if used) timed out.	**11. See Status Indicator Operation page to reset.**
	12. Shorted Evaporator Probe.	**12. Replace.**
	13. Bin potentiometer locked out in full counter-clockwise position.	**13. Turn clockwise (colder).**
	14. Bin relay coil defective.	**14. A good coil has 100 ohms resistance. Replace if defective.**
2. Condensing Fan operates but not the compressor.	1. Compressor stuck.	1. Jar with Mallet.
	2. Inoperative capacitors or relay.	2. Replace capacitors or relay.
	3. Overload switch defective.	3. Replace overload switch or compressor with internal overload.
	4. Open wash switch.	4. Switch to on or replace.
	5. Open high or low pressure cut-out.	5. Check charge and condenser.
	6. Defective compressor	6. Replace compressor.
3. Compressor operating but fan off.	1. Circuit not complete.	1. Check circuit.
	2. Fan motor burned out.	2. Replace motor.
4. Condenser fan operating, but condensing unit operating intermittently during freezing cycle, wait till end of defrost to see if unit returns to normal operation.	1. Dirty condenser coil.	1. Clean coil.
	2. Low voltage.	2. Correct to proper voltage - not less than 5% below that stated on nameplate. Install automatic brownout voltage booster number 5-1320 for 115 volt cuber.
	3. Excessive refrigerant.	3. Bleed off some refrigerant.
	4. Fuse blown one leg of 3-wire -electronic system.	**4. Replace fuse.**

TROUBLE	CAUSE/SYMPTOM	REMEDY
5. Compressor cuts out.	1. Defective run capacitor.	1. Run capacitor should draw 1 to 3 amps. GB2 & GB4, 4 amps.
	2. Open high or low pressure cutout.	2. Check refrigeration system pressure.
6. Water plate closes and opens constantly. Water plate closes all the way when cams are up but defrost valve stays open and pump does not run.	1. Maladjusted Pump & Defrost Switch.	**1a. Adjust lift bolt on water plate to push switch lever up, closing hot gas valve and starting pump when water plate is up.**
		1b. Adjust lift rod collar to push pump switch up closing hot valve & starting pump when water plate is up.
	2. Water plate does not close all the way.	2. Remove obstruction. Adjust hinge for clearance between evaporator, and water plate. Make sure teflon brackets on water plate are tight against cams. Check springs.
7. Water plate opens before water probe assembly tube is full.	1. Spring missing or springs weak allowing water plate to lower slightly (as water fills tank) until pump switch drops and plate opens under power.	1. Replace springs.
	2. Drift stop not adjusted. Cams drift counter-clockwise until water plate lowers slightly and pump switch drops.	2. Drift stop on front of actuator motor. Remove drift stop and bend spring for more tension on motor shaft.
	3. Slow fill cold incoming water. Orange and Red L.E.D.'s on P.C. Card go on.	**3. Adjust cold water control on P.C. Card or replace P.C. Card. Improve water supply. Clean strainer.**
	4. Slow Fill-Cold in incoming water.	*4. Adjust or replace cold water thermostat.*
8. Water plate will not completely close.	1. Obstruction between evaporator and water plate.	1. Remove obstruction. Check clearance between water plate and evaporator.
	2. Lift bolt for pump toggle on water plate too high, holding plate away from cams.	**2. Adjust lift bolt so water plate comes up against cams and lift bolt holds pump and defrost toggle switch up without binding and holding water plate down.**
	3. Collar on lift rod too low, holding water plate away from cams.	*3. Adjust collar so water plate comes up against cams & lift rod holds pump & defrost switch up without binding & holding water plate down.*

TROUBLE	CAUSE/SYMPTOM	REMEDY
9. Water plate closes before cubes dropped.	**1. Actuator pot on P.C. Card adjusted too cold.**	**1. Adjust to warmer position (counter-clockwise). Water plate should remain in down position 10 to 30 seconds after ice drops.**
	2. Faulty evaporator probe.	**2. Replace probe.**
	3. Faulty P.C. Card.	**3. Check with cuber analyzer. Replace P.C. Card.**
	4. Actuator thermostat adjusted too cold.	*4. Adjust to warmer position (counter-clockwise). Water plate should remain in down position 10 to 30 seconds after ice drops.*
10. Water plate stays wide open after defrost and all ice is out of evaporator.	**1. Orange L.E.D. stays lit on P.C. Card.**	**1. Adjust actuator control slightly clockwise. Check and replace the evaporator probe or P.C. Card if adjustment has no effect.**
	2. Orange L.E.D. is off, but no voltage to the yellow actuator motor lead.	**2. Check wiring. Replace actuator toggle switch GB-897.**
	3. No voltage to No. 3 (orange lead) of actuator thermostat.	*3. Actuator Thermostat adjusted too warm or bulb has lost charge.*
	4. Voltage to No. 3 (orange lead) of Actuator Thermostat, but no voltage to yellow actuator motor lead.	*4. Wiring loose or defective actuator toggle switch.*
11. Water plate open - evaporator will not defrost.	1. Refrigerant charge low.	1. Check for leaks and recharge.
	2. Inadequate hot gas volume.	2. Check for tube obstruction or cold condenser.
	3. Defective hot gas valve.	3. Replace valve or coil.
	4. Red L.E.D. on P.C. Card is on.	**4. Check voltage at defrost valve coil, if not 115 V. change P.C. Card.**
	5. Red L.E.D. is off.	**5. Check evaporator probe (5600 ohms at 32° F.) If probe is functional, replace P.C. Card.**
	6. Cold water thermostat not making contact.	*6. Tap cold water thermostat or short across pink & red leads. If defrost valve opens, change thermostat.*
12. Water pump does not operate.	1. Fuse blown in transformer box, or incontrol module box.	1. Replace fuse.
	2. Pump bearings defective.	2. Replace pump motor.
	3. Pump windings burned out or off on thermal overload.	3. Allow to cool, or replace motor, check for 115 V, plus or minus 10%.
	4. Circuit incomplete between water pump & pump - defrost switch.	4. Check circuit and switch.

TROUBLE	CAUSE/SYMPTOM	REMEDY
13. Water pump motor running but not pumping water.	1. Impeller loose. 2. Strainer in tank outlet to pump clogged. 3. Impeller broken.	1. Replace impeller. 2. Clean or replace screen. 3. Replace impeller, replace screen in tank outlet.
14. Most cubes not fully formed.	1. Not enough pressure from water pump. 2. Clogged strainer in tank outlet to pump. 3. Leak in water circulation system. 4. Water plate not aligned.	1. Check bearings. Check voltage. Replace pump. 2. Clean strainer. 3. Fix leak or replace water plate. 4. Check alignment with evaporator.
15. A few cloudy cubes, others okay.	1. Some holes in water plate clogged.	1. Unplug with 1/16" drill. Flush laterals by removing plugs.
16. Holes in left hand cubes (evaporator inlet).	1. Expansion valve too far open.	1. Close 1/8 turn at a time (clockwise).
17. Holes in right hand cubes (evaporator outlet).	1. Shortage of refrigerant. 2. Expansion valve too far closed.	1. Check for leak and recharge. 2. Open 1/8 turn at a time (counter-clockwise).
18. Holes in all cubes sometimes and solid cubes most of the time.	1. Power shut off while water is filling tank or temporary power shut off near end of freeze cycle. **2. Bin control shuts the cuber during water fill.** *3. Bin thermostat shuts the Cuber off during fill.*	1. Correct power source if possible. **2. Interlock between water fill control and bin control not operating. Replace P.C. Card.** *3. This condition will happen occasionally on any machine but the frequency can be reduced by turning bin thermostat warmer counter-clockwise until machine turns off very shortly after ice is against bulb. After resetting, place cubes on bulb to shut off, remove and machine should restart within five minutes.*
19. Holes in cubes all of the time. Control stream does not go over the dam at end of freeze cycle.	1. Water level too low. **2. Lower water level probe too high.**	1. Measure from the top of the circulation tank down to the water level in the water level control tube, sight carefully across water in the control tube. See "Chart of Water Levels, etc." **2. Adjust low water level probe to remain immersed in water in control tube at least 10 seconds after control stream starts "going over the dam".**

TROUBLE	CAUSE/SYMPTOM	REMEDY
	3. Leak in water system.	3. Water dripping steadily off the circulation tank indicates a leak which should be located & repaired. Make sure all lateral plugs are in place.
20. Holes in cubes all of the time. Control stream does go over the dam.	1. Control stream too high allowing water to splash over the dam during freeze cycle. (It should only go over the dam after cubes are fully formed.	1. Lower control stream, turn adjusting screw clockwise.
21. Cuber will not harvest, water plate will not come down.	1. Control stream obstructed.	1. Loosen adjusting screw to flush out foreign matter.
	2. Actuator motor problem.	2. Check motor and circuit.
	3. Warm air infiltration from compressor compartment or room.	3. A. Secure all skin panels. B. Skin gaskets must seal. C. All panels must seal to prevent air from compressor compartment getting into ice making compartment. Check especially, top cover over partition.
	4. Orange L.E.D. does not come on.	**4. Check or replace evaporator probe (5600 ohms at 32° F).**
	5. Orange L.E.D. will not come on but probe is okay.	**5. Replace P.C. Card.**
	6. Inoperative lower probe.	**6. Check water level probe connections or replace probe assembly. Replace P.C. Card.**
	7. Control tank won't snap up.	*7. Adjust switch differential.*
	8. Actuator thermostat out of adjustment.	*8. Adjust slightly counter-clockwise.*
	9. Defective actuator toggle switch.	*9. Replace switch.*
	10. Power supply has failed.	*10. Check power.*
22. Cuber stops when bin is not full.	1. Bin control adjusted too warm.	1. Readjust bin control slightly clockwise.
	2. Bin probe connector loose or or dirty.	**2. Clean connector and install properly.**
	3. Defective bin probe.	**3. Replace if considerably more than 5600 ohms at 32° F.**
	4. Intermittent Evaporator Probe.	**4. Replace.**
	5. Shorted Evaporator Probe.	**5. Replace.**
	6. Power supply has failed.	*6. Check power.*
	7. Bin bulb too low.	*7. Attach bin bulb to plastic chute.*
	8. Bin thermostat touching some cold section of the machine.	*8. Trace whole thermostat capillary tube.*

TROUBLE	CAUSE/SYMPTOM	REMEDY
23. Cubes do not harvest in a slab but some cubes hang up in the evaporator and become distorted after others fall out.	1. Fin too thin.	1. Adjust water plate hinges to 1/32" fin thickness.
	2. Deformed evaporator cells.	2. Straighten cells with smooth jaw pliers.
24. Slab does not break up into individual cubes.	1. Fin too thick.	1. Adjust hinges up or evaporator down. Leave 1/32" space between water plate and evaporator.
25. Unusually long cycles.	1. Voltage below required potential at the cuber.	1. Check power source for full voltage. Run at least No. 12 wire directly to cuber to prevent line loss.
	2. Dirty condenser.	2. Clean.
	3. Hot air leaks between condensing unit compartment and freezing compartment.	3. Check for leaks and close with permagum, or presstite tape. All skin parts must be tight.
	4. Expansion valve too far open.	4. Close valve 1/8 turn at a time so that there will be no frost back to compressor and pressures are according to "Chart of Water Levels, etc."
	5. Expansion valve too far shut and large holes in right hand rows of evaporator.	5. Open expansion valve 1/8 turn, but recheck to see that there is no frost-back to compressor at end of freeze cycle. (See No. 2 and No. 3 above).
	6. Water level too high after water fill.	6. Adjust water level according to "Chart of Water Levels, etc." — See Page 4.
	7. Refrigerant low.	7. Check for leak and add Refrigerant. See Page 4.
	8. Compressor defective.	8. Replace compressor.
	9. Control stream too low.	9. Adjust control stream up, but not so high that it goes over dam at beginning of cycle.
	10. Fan not operating.	10. Check fan wires, replace motor if necessary.
26. Some cubes do not form in right hand corners of evaporator.	1. Jet holes in ends of laterals frozen shut and will not thaw because of very low incoming water temperature.	1. Thaw out by shutting off unit, and adjust cold water control warmer CCW. Adjust expansion valve 1/8 turn closed.
27. Ice freezes to water plate causing shear pins to break.	1. Water fill level too high.	1. Adjust fill level per page 4 (electronic cubers). page 74 (electro-mechanical cubers).
	2. Incorrect clearance between water plate and evaporator.	2. Refer to page 60.
	3. Misadjusted control stream.	3. Adjust per page 83(electronic cubers), page 54 (electro-mechanical cubers).
	4. Control stream will not go over dam.	4. Cracked lateral tube on bottom of water plate. Reglue or replace water plate.

TROUBLE	CAUSE/SYMPTOM	REMEDY
28. Water plates out of synchronization on GB4, GB1003, or GB1205 Cubers.		1. See GB1000 section.
29. Water valve stays closed.	**1. Water level probe connector loose or dirty.**	**1. Clean connector and install properly, low probe lead to right (close). Use NC123 or any electrical contact cleaner.**
	2. Defective water probe assembly.	**2. Check with probes standing in 32° F. ice-water mixture. Resistance should be 5600 ohms 20%.**
	3. Circuits okay, 115 volts to water valve. Coil open. Flow control jammed cockeyed.	**3. Replace coil. Clear valve passages or replace valve.**
	4. Defective P.C. Card.	**4. Replace P.C. Card.**
30. Water valve stays open after upper probe covered, will not shut off.	**1. Water level probe connector loose or dirty.**	**1. Clean connector and install properly, low probe lead to right (close). Use NC123 or any electrical contact cleaner.**
	2. Defective water probe assembly.	**2. Check with probes standing in 32° F. ice mixture. Resistance should be 5600 ohms =20%.**
	3. Defective P.C. Card.	**3. Replace P.C. Card.**
	4. Water pressure below 15 P.S.I.	
	5. Defective water valve.	
31. Water valve stays open more than 5 seconds after upper probe covered, then shuts off.	**1. Upper probe covered with scale.**	**1. Clean cuber with ice machine cleaner. If necessary, remove and clean probe carefully.**
32 Status indicator shuts cuber off on "service required" but Cuber operates normally most of the time and timer resets when ice harvest switch triggered manually (cubers built after June, 1980 - yellow light flashes when ice drops).	**1. Cubelet ice does not trip harvest switch.**	**1. Trip wire on switch too high. Bend trip wire until it is only 3/8" higher than water plate surface.**
	2. Maladjusted impact switch.	**2. Turn adjusting screw in to increase sensitivity.**
	3. Time out period too short.	**3. Lengthen time out period by adjusting status indicator pot slightly clockwise.**
33. Status indicator shut down sometimes occurs after being off on full bin with status indicators built before March, 1980.	**1. Circuit voltage borderline, timer does not reset upon start up after bin shut-down.**	**1. Replace status indicator P.C. Card (new style revised).**

GB1000 - GB4

The GB1000 and GB4 Series Ice Cubers differ from othe Kold-Draft Cubers in that two ice making sections are refrigerated by one 14,000 B.T.U. Condesing Unit. GB1000 is electronic version of GB4.

The Upper Ice Making Section, called the "Master", contains the controls necessary to operate both sections simultaneously, while the lower or "Slave" section contains only those switches required to operate the actuator motor and water pump, and to provide synchronization between the two water plates.

The GB4 uses an Electrical Synchronization System to ensure that both water plates are fully closed before the freeze cycle begins, and to prevent repeated false harvesting caused by "out of time" actuator motors.

There are two key parts in the GB4 synchronization system: A modified rotor in the top actuator motor to slow the motor by 5 seconds, and a resistor between the upper and lower actuator toggle switches to stall the lower motor in the closed position at the end of harvest.

Synchronization is achieved by the lower actuator motor running slightly faster than the upper motor, stalling upon direction reversal in the full upright (12 o'clock) position, thus allowing the upper motor to catch up and synchronize. Upon upper motor direction reversal, both motors continue to the end of the travel limit in the other direction and repeat the same synchronization procedure. With ice in the evaporators, both plates go down and stop together in the full down position automatically synchronized. When dry cycling the cuber without ice, the lower actuator motor trips the lower actuator toggle switch and raises ahead of the upper actuator motor. Synchronization occurs when the lower actuator motor stalls after raising to the 12 o'clock position and waits for the upper actuator motor to catch up.

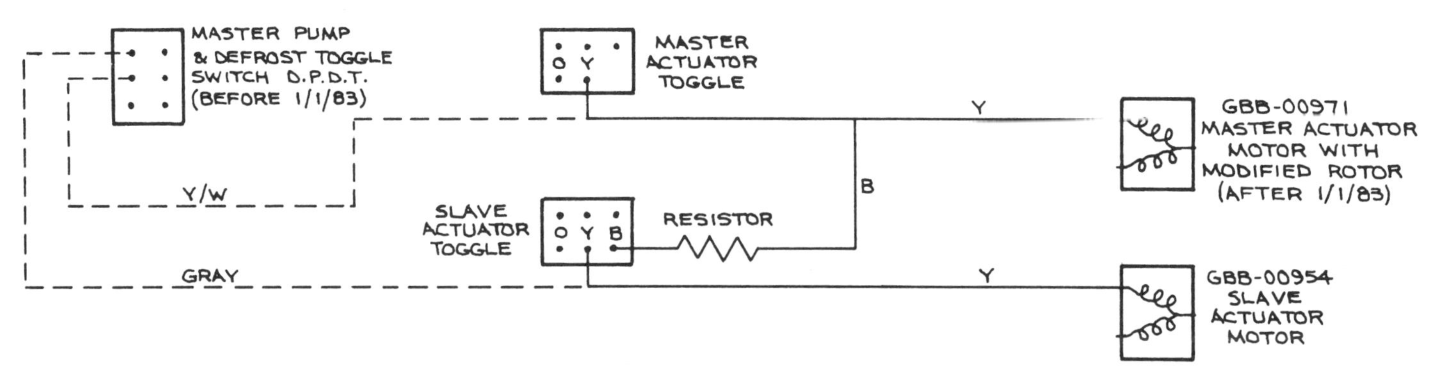

The Drawing above shows the synchronization circuit in GB4 Cubers. The dotted lines indicate an interim sub-circuit which was used from 9/81 until 1/83. To update any GB4, cut the Yellow/White wire in the Top Channel, tape the ends, and install a modified rotor (Kit No. GB-3110-71) in the top actuator motor. Cubers manufactured before 9/81 do not have the Yellow/White wire in the Top Channel (unless they have been field modified) and only the modified rotor is required for updating.

Request Engineering Bulletin 16-82 for complete information concerning the update Kit.

The GB4 cuber uses two water pumps (Master & Slave) and are connected to the transformer independently to decrease the load on the 115 V. tap.

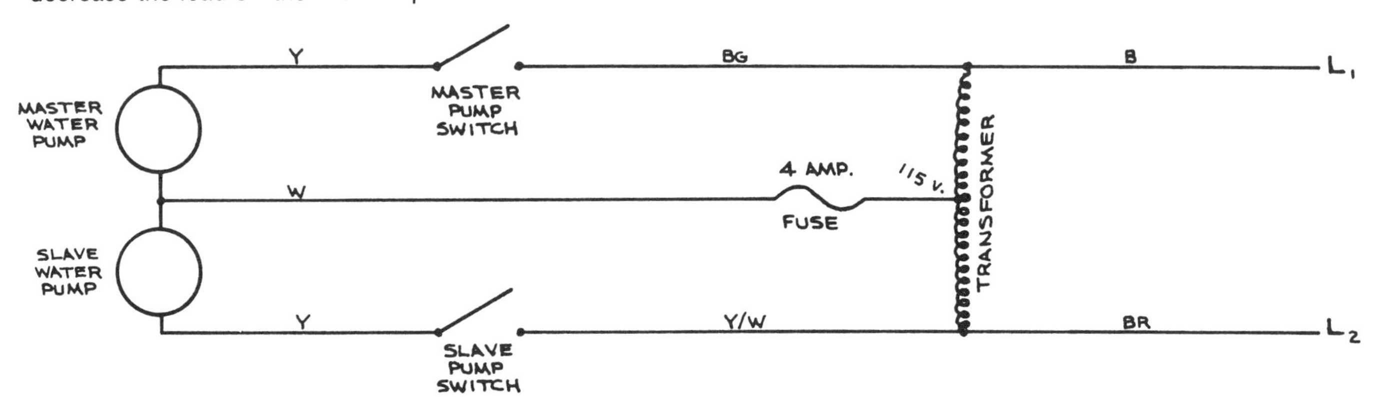

As can be seen from the above Diagram, the 4-amp fuse protects transformer as it does in all GB Series cubers, but the pumps will continue to operate if the fuse opens as the pumps are actually connected in Series across the line taps of the transformer.

To allow the water pumps to operate independently, a D.P.D.T. Pump and Defrost Toggle Switch is used in the Slave channel. The left contacts power the pump when the water plate is closed, and the right contacts permit caught-cube false harvesting through the red circuit.

The transformer box contains high and low pressure cut-outs, as do all 1 HP and larger Kold-Draft cubers, and also a Hi and Low cut-out relay which is unique to the GB4.

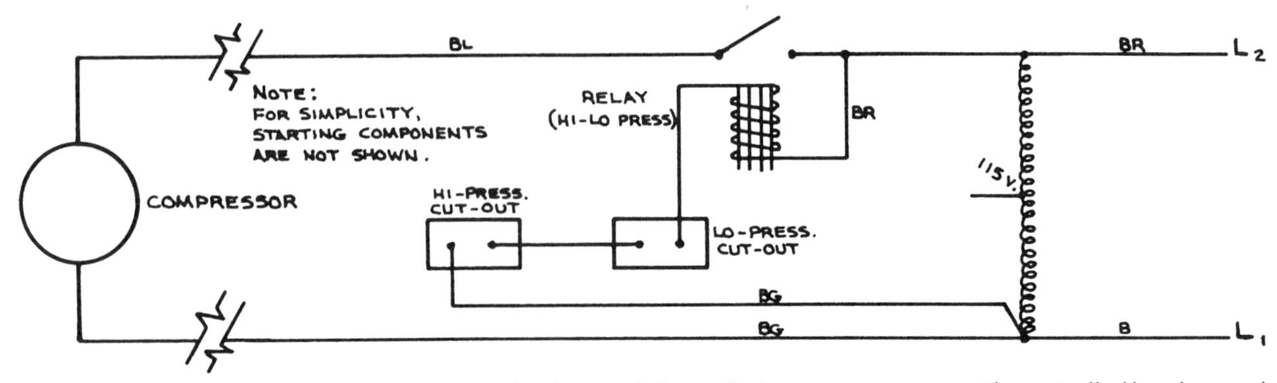

The above Diagram illustrates the cut-out circuitry, and shows that compressor current is controlled by a heavy-duty relay rather than the cut-out contacts which are not heavy enough to reliably handle the amperage required of the GB4 compressor. The High and Low pressure cut-outs supply voltage only to the coil of the control relay, which will open if either of the cut-outs open due to excessively high or low refrigerant pressure.

The GB4 contains two evaporators and therefore two expansion valves, but only one Hot Gas Defrost Valve.

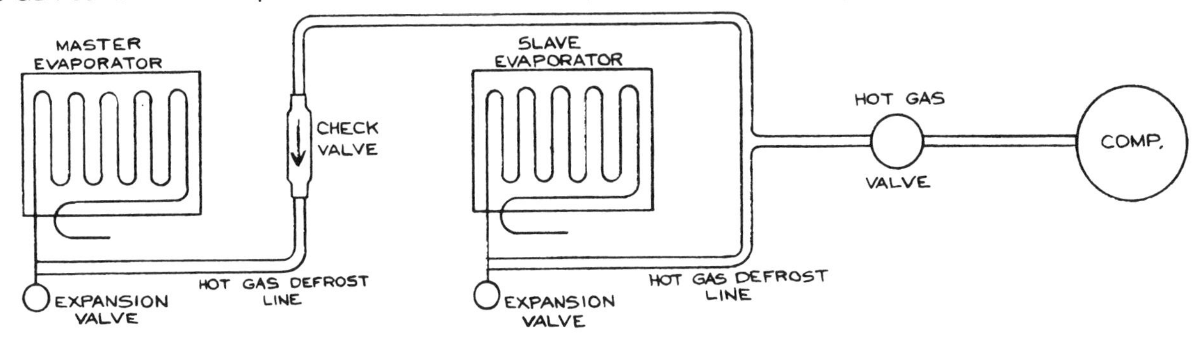

As can be seen by the simplifed refrigeration Diagram above, the Hot Gas Defrost tubing from each evaporator is joined at the Hot Gas Valve, and provides a refrigeration circuit (during the freeze cycle) between the evaporators. As the evaporator inlet pressures are never exactly equal, refrigerant can flow from one evaporator inlet to the other through the Hot Gas tubing, causing frost on the tubing. To the service man not familiar with the GB4, this can be a startling phenomenon (hot gas lines aren't supposed to be cold), but is perfectly normal and has little or no effect on the operation of the cuber. To minimize frost on the Hot Gas tubing (which turns to water during the defrost cycle) a ball check valve was added 8/82 to the Hot Gas line feeding the upper evaporator. During harvest, the check valve opens fully and allows unrestricted Hot Gas flow to both evaporators. During the freeze cycle the top evaporator expansion valve, which is set at a slightly lower superheat than the lower expansion valve, will cause the check valve to close and block refrigerant flow through the Hot Gas tubing.

Theoretically, both TXV's can be set at the same superheat, as the gravity-operated ball check valve requires about 1/4 P.S.I. to unseat. Hot Gas line frost will not occur as long as the inlet pressure in the top evaporator is equal to or greater than the inlet pressure in the lower evaporator.

GB1204, GB1205

The GB-1200 Series Cubers are derived from the (GB1003) Series Cubers, and therefore share the same synchronization circuits, water pump control wiring, and refrigeration circuit.

The compressor used inthe GB-1205 is 208/230 3 phase, the GB-1204 is 208/230 1 phase, and therefore uses a line contactor compressor operation. The drawing below illustrates the contactor circuit, and also shows the transformer connection to provide single phase 115 volt power for the water pumps and solenoid valves.

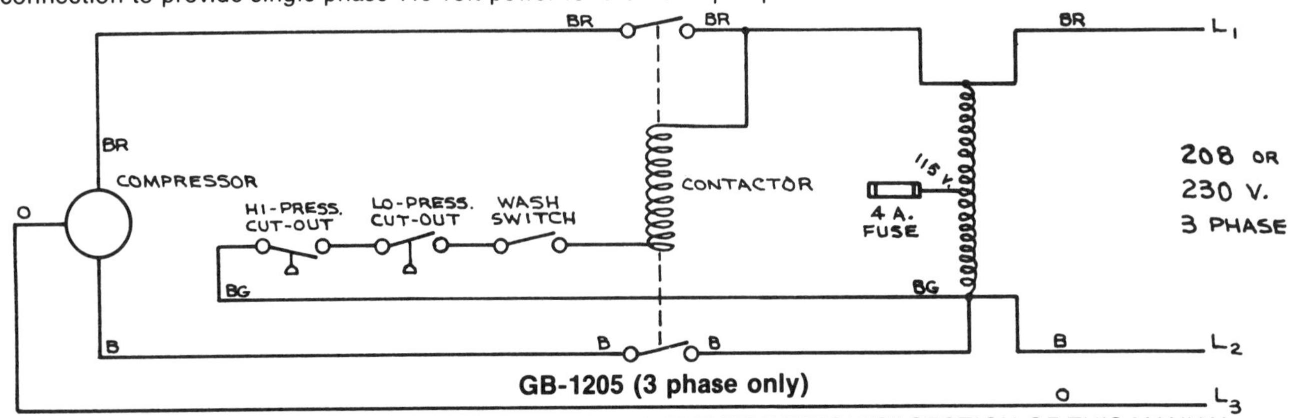

GB-1205 (3 phase only)

A COMPLETE WIRING DIAGRAM MAY BE FOUND IN THE "ELECTRONIC" SECTION OF THIS MANUAL.

Although the refrigeration circuit is identical to the GB4 (GB-1003), the compressor in the GB-1200 series uses Refrigerant 502 along with R-502 expansion valves.

SERVICE INFORMATION

ACTUATOR MOTOR — DESCRIPTION & TEST PROCEDURE

The motor has a two-coil field. One is fed with 115 volts. A small capacitor changes the phase and boosts the voltage to the other coil to give the motor direction of rotation.

It is located: 1. Thermostatically controlled cubers in wiring channel.
2. Electronic cubers: A. Behind P.C. Card before Sept. 1984
B. In CRC Network after Sept. 1984

When the actuator toggle switch is thrown, the connection to the capacitor is changed so that the second coil now gets 115 volts and the first coil gets the out of phase boosted voltage from the capacitor to change the direction of rotation.

The small motor bearings are self-aligning and, when properly aligned, the motor will produce 30 pounds force at the end of the cam arm and will be difficult to stall by pressure offingers on the cam arm. Being a self-aligning bearing, it can get jarred out of alignment in shipment so that the power produced by the motor is reduced and the motor can be stalled easily with a couple of fingers, or it may not run at all. The bearings can be easily re-aligned by tapping the gear case near the motor while the power is on (pull down on the water plate). After the motor starts running, tap it again. A tap from a small hammer or crescent wrench will provide sufficient shock so that the bearings will re-align and the motor come up to full power. Once re-aligned, the proper alignment will be maintained.

Precautionary procedure on every installation would be to pull on the water plate so that the actuator motor will run, then tap the gear case near the motor to get perfect alignment. Check with fingers to be sure cam cannot be stalled easily.

1. Use a 250 volt voltmeter — Make sure all wire connections to capacitor and actuator toggle switch are good.
2. Do not accidentally ground capacitor leads or meter leads while making this test as this will blow a fuse on a 115 volt or 115-230 volt, three wire cuber or burn out the transformer on a 230 volt two wire cuber. Take readings from on lead to the other of the actuator motor capacitor (yellow &gray). If water plate is closed, block water plate down about 1" and make sure control tank is up and evaporator above 40° to provide incoming power to both orange and red leads on the actuator toggle switch. Switch the toggle switch left and right slowly and read meter.
3. Note: Motor coils can be checked out of the circuit with an ohm meter. Yellow to white should read approximately 450 ohms. Gray to white should read apporximately 450 ohms.
4. Note: The capacitor is not acessible to connect meter probes in early electronic models. Attach GB-02865 tap connectors to the gray and yellow actuator motor wires for test points. Meter probes can be inserted into the holes in the connectors to obtain the voltage ratings described below. Do not remove the connectors after testing.

Voltage Reading	Capacitor	Motor	Remedy
180 v. to 240 v. in both switch positions	Good	Good	With power on, check for binding bearing, tap gear case near motor for proper alignment of self-aligning bearing.
115 v. in both positions	Open	Good	Change capacitor.
115 v. in one position, 0 v. in other position	Open	1 motor winding open	Change both motor coil and capacitor.
200 v. in one position, 0 v. in other position	Good	1 motor winding open	Change motor coil.
0 v. reading in both positions. To be sure you have incoming power (about 115 v.) leave one probe on a capacitor lead and move the other probe of the meter to the white motor lead. If you have incoming power:	Cap. shorted	or both windings open	Change motor coil — recheck; if no results change capacitor.

On thermostatically controlled cubers, capacitors have been improved with additional moisture protection. However, it is recommended that the BX connections at the right and left ends of the wiring channel be plugged with Permagum to prevent the entrance of warm, moist air.

TESTING C/R-C NETWORK (GB-3244) ON ELECTRONIC CUBERS ONLY.

CAUTION: Short all terminals together before testing. This prevents possible damage to meter.

1. Set ohm meter on RX1000 (lk) scale and check zero set.

2. W to G - Connect test leads. Reverse test leads and the meter should deflect. A good component will cause the meter to drop to approximately 200k ohms and then climb back to infinity.

3. G to Y - Connect test leads and then reverse. The meter should drop to about 150k ohms and then climb back to infinity.

4. If the meter goes to the approximate ohm readings listed above and stays there, the capacitor is shorted and should be replaced.

5. If the meter doesn't deflect at all, the capacitor is open and should be replaced.

BIN PROBE POSTIONING - ELECTRONIC CUBERS.

1. GB Series Electronic Cubers when ice chute is used.

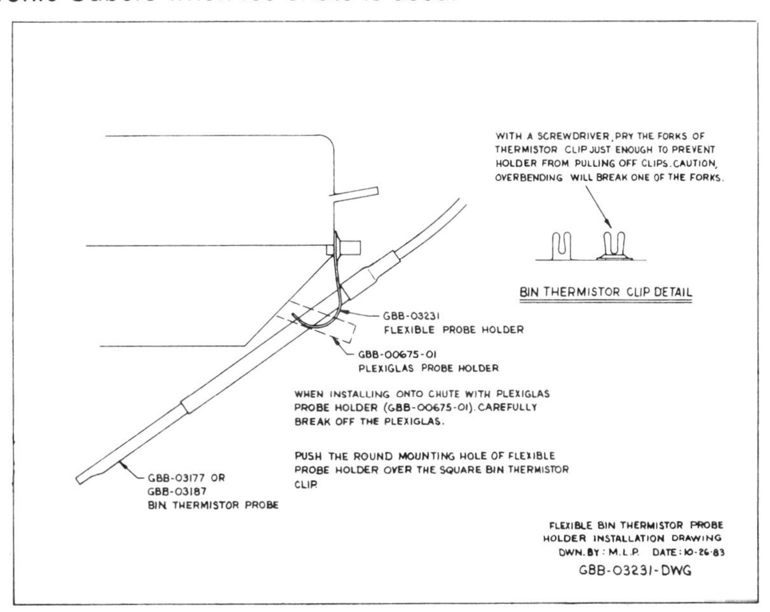

2. GB Series Cubers that are placed on bins with a **sleeve and the front and rear deflector** used instead of an ice chute. The flexible probe holder is secured to one of 3 holes on the front of the drain pan, with a plastic bushing, and the probe thru the other two holes pointing into the ice drop zone.

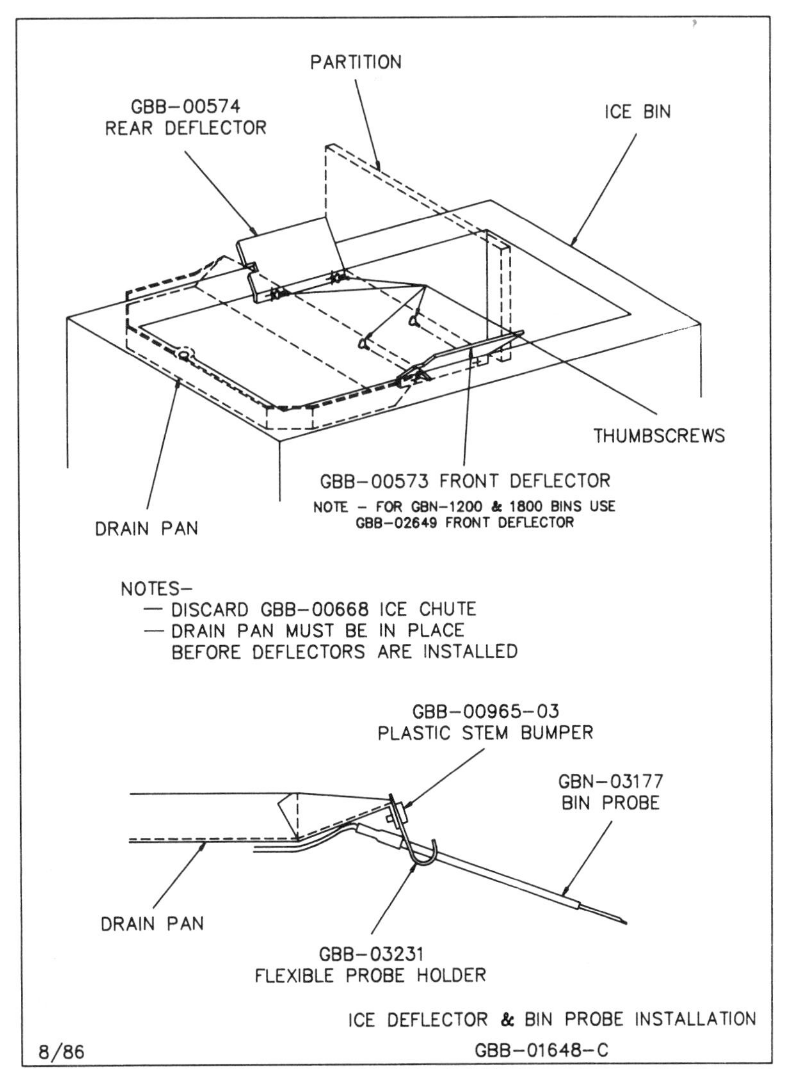

3. IS Series Cubers when the bin probe is positioned as shown below.

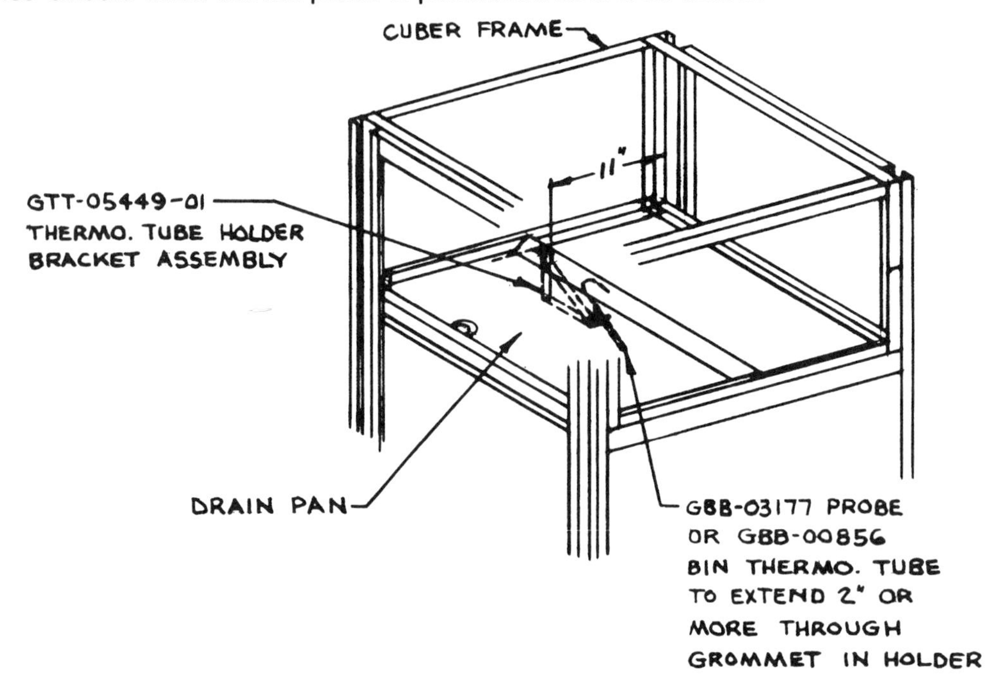

4. GT Series Cubers on GT model bins when the probe is positioned as shown below.

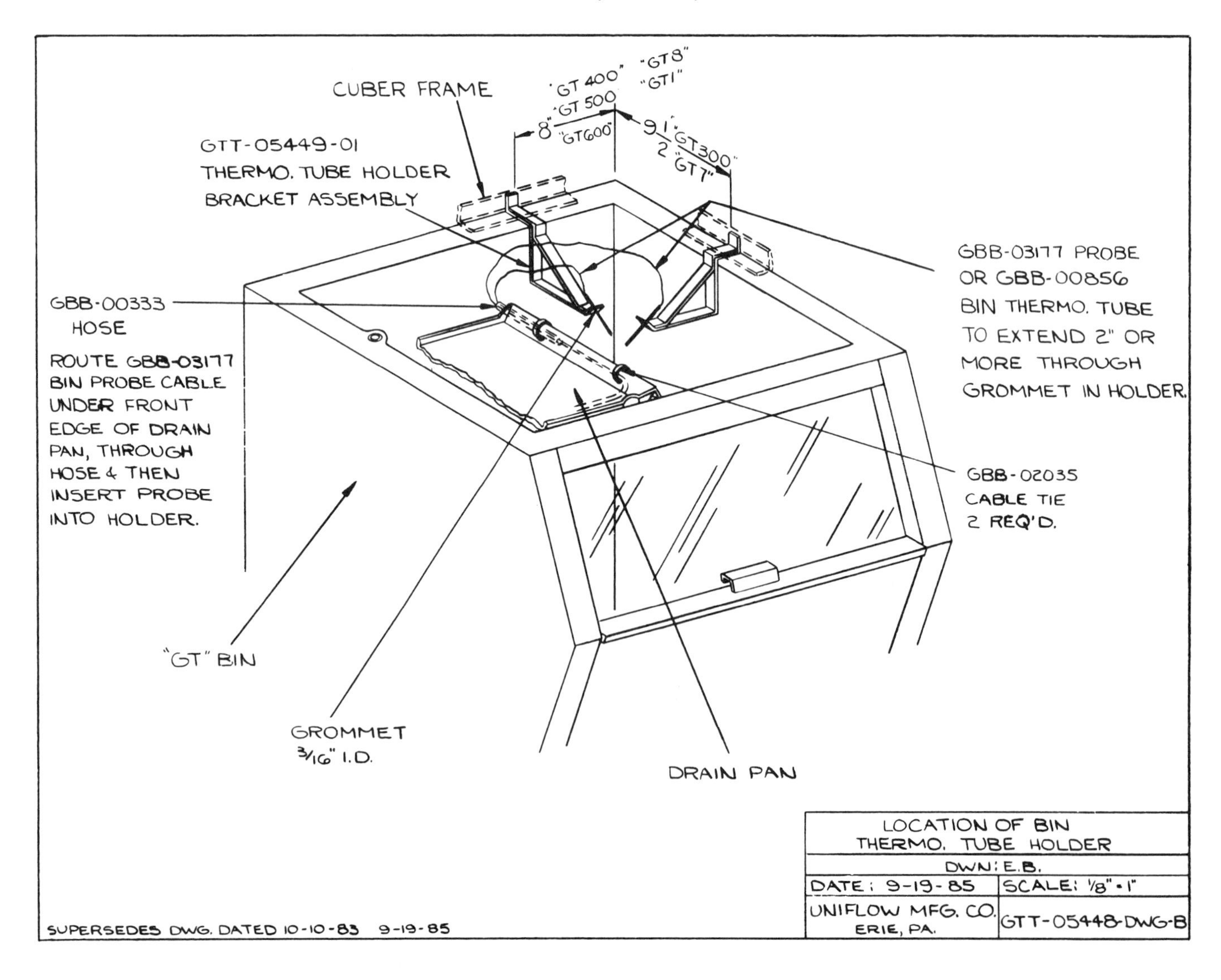

REMOTE CONDENSERS

The following pages may be used as a guide to installing, troubleshooting, and specifying Kold-Draft Remote Condenser installations.

When installed as a U.L. listed, factory designed system consisting of a Kold-Draft Cuber, Kold-Draft line sets, and a Kold-Draft supplied condenser, few operational difficulties will be encountered.

If system components are modified or substituted for components not specified by Kold-Draft, proper operation can be compromised to the point of system failure.

If the line set(s) and /or condenser is supplied by others, we must assume that the specifying engineer has properly applied the non-Kold-Draft accessories to our Ice Makers.

Kold-Draft reserves the right to disallow any warrantee claims which result from the use of non-Kold-Draft condensers and/or line sets.

REMOTE CONDENSERS

REMOTE CONDENSER APPLICATION CHART

PART NUMBER	CONFIGURATION	BTU @1°F TD EACH CIRCUIT	CFM	REFRIGERANT VOLUME EACH CIRCUIT	NORMAL AMBIENT APPLICATION (Under 110°F)	HIGH AMBIENT APPLICATION (Over 110°F)
GB-2222	Single Circuit 1½ Ton	787	2300	3.9#	GB400, 500, 600, 1000 GT400, 500, 600 FT800, FB1500	GB400, 500 600 GT400, 500, 600 FT800, FB1500
GB-2223	Two Circuit 2 Ton	505	2050	2.5#	GB400, 500 GT400, 500 FT800, FB1500	GB400 GT400 FT800
GB2224	Three Circuit 3 Ton	485	2150	3.45#	GB400, 500 GT400, 500 FT800, FB1500	GB400 GT400 FT800
GB-2418	Single Circuit 2 Ton	1010	2050	5#	GB1200	GB1000
GB-2413	Two Circuit 3 Ton	728	2150	5.2#	GB600, 1000	GB500 GT500 FB1500
GB-2425	Single Circuit 3 Ton	1455	2150	10.4#*		GB1200
GB-2426	Two Circuit 5 Ton	1215	5050	10.7#*	GB1200	GB1000

*Due to the high internal volume of these condensers, only double-evaporator Cubers with 22# receivers can be used with GB-2425 and GB-2426 condensers.

Each condenser is vapor pre-charged, and is equipped with Aero-Quip re-usable fittings on each circuit. Each ice machine contains the proper liquid refrigerant charge required to provide proper head pressure control. No additional charge is required. Each ice machine also contains a head pressure control valve, preset to 120 psi (R-12) or 180 psi (R-502). No additional control is required. Remote condenser ice makers, condensers, and line sets are available in precharged configuration only.

NOTE: Serious operational problems can occur of the above chart is not followed, or if non-Kold-Draft condensers and line sets are substituted for factory equipment. If you must substitute, be sure to contact the factory before committing yourself to an installation which can cause loss of sleep.

Work Sheets and Guidelines are available upon request to assist you in designing a non-Kold-Draft System.

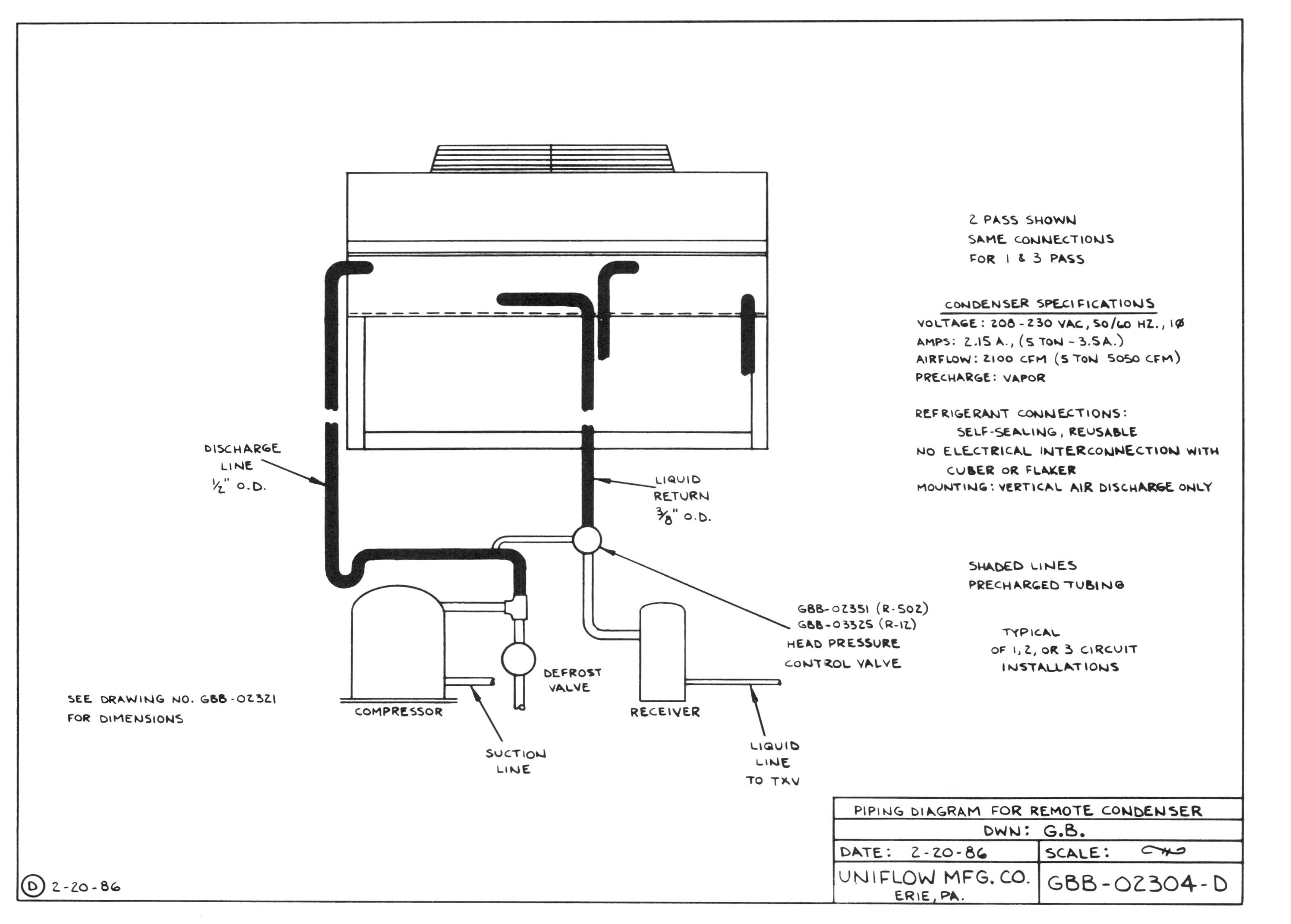
2 PASS SHOWN
SAME CONNECTIONS
FOR 1 & 3 PASS
CONDENSER SPECIFICATIONS
VOLTAGE: 208-230 VAC, 50/60 HZ., 1Ø
AMPS: 2.15 A., (5 TON - 3.5 A.)
AIRFLOW: 2100 CFM (5 TON 5050 CFM)
PRECHARGE: VAPOR
REFRIGERANT CONNECTIONS:
SELF-SEALING, REUSABLE
NO ELECTRICAL INTERCONNECTION WITH
CUBER OR FLAKER
MOUNTING: VERTICAL AIR DISCHARGE ONLY
SHADED LINES
PRECHARGED TUBING
TYPICAL
OF 1, 2, OR 3 CIRCUIT
INSTALLATIONS
DISCHARGE
LINE
1/2" O.D.
LIQUID
RETURN
3/8" O.D.
GBB-02351 (R-502)
GBB-03325 (R-12)
HEAD PRESSURE
CONTROL VALVE
DEFROST
VALVE
COMPRESSOR
RECEIVER
SUCTION
LINE
LIQUID
LINE
TO TXV
SEE DRAWING NO. GBB-02321
FOR DIMENSIONS
PIPING DIAGRAM FOR REMOTE CONDENSER
DWN: G.B.
DATE: 2-20-86
SCALE:
UNIFLOW MFG. CO.
ERIE, PA.
GBB-02304-D
D 2-20-86

DESIGN CRITERIA FOR NON-KOLD-DRAFT REMOTE CONDENSER INSTALLATIONS

The Kold-Draft Remote Condenser-Cuber package is intended to be purchased as a system consisting of an "R" Model Cuber, a properly sized "GB" Series Pre-Charged Condenser, and Pre-Charged Refrigerant Lines supplied by the factory with the necessary self-sealing connectors required to couple the Cuber to the Condenser. The complete package provides a balanced system which has been designed and tested to operate properly under all ambient temperatures. At ambients above 120° F. the High-Pressure Cut-Out may interrupt operation for awhile as it is set to prevent damage to the system under extreme conditions. If extra high ambients are expected to be routinely encountered, extreme ambient condensers are available for most Models.

We strongly recommend the use of Kold-Draft Condensers and Line Sets, but also realize that some installations require the use of a Remote Condenser supplied by others. If your customer insists on using a Non-Kold-Draft Condenser, a qualified refrigeration engineer should be consulted to examine the proposed installation to minimize operational problems.

The points to be addressed include, but are not limited to, the following:

1) The Condenser should be properly matched to the refrigeration load. The Peak Total Heat of Rejection of 400 Series Cubers is 9,150 BTUH, 500 Series is 11,360 BTUH, 600 Series is 17,000 BTUH, 1000 Series is 19,800 BTUH, and the 1200 Series rejects 24,000 BTUH (with CPR). The THR of 800 Series Flakers is 7,100 BTUH, and the 1500 Series Flaker rejects 12,500 BTUH. (These figures are based on a condensing temperature of 130° F.) The THR of the Condenser should match the Ice Maker at 20° F. T.D., but no higher than 30° F. T.D.

2) Check the internal volume of the Condenser. Total volume per circuit should not exceed the equivalent of 5 lbs. of refrigerant. Generally speaking, a Condenser which uses 3/8" O.D. internal tubing and is properly sized for the THR of the Cuber will not exceed the capacity of the receiver. Larkin FCA and FCB Model Condensers, and Bohn URD Series Condensers use 3/8" tubing, and are known to operate properly on Kold-Draft equipment. GB-1000-R and Gb-1200-R Series Cubers use an oversized receiver tank which allows the use of condensers with an equivalent volume of 10 lbs. refrigerant per circuit.

3) Line Sets must be properly sized and routed. Diameters should be 3/8" O.D. and 1/2" O.D. for the liquid and discharge lines respectively. Line length should be as short as possible. Line lengths exceeding 100' may require an excessive refrigerant charge for cool weather operation. Try to avoid cold areas when routing the discharge line. A "P" trap should be installed in the discharge line close to the compressor to minimize oil trapping.

4) Always try to locate the Condenser above the receiver. If absolutely necessary, Series Circuited Condensers can be installed down to 10' below the receiver. If the Condenser is constructed with a manifold using two or more tubes in parallel, the Condenser must be located a minimum of 3' higher than the receiver plus one additional foot for each 10' of run.

5) System Charge must be calculated. To determine the proper Charge, add together the Condenser volume at the lowest expected ambient, the weight of Refrigerant in the liquid line, the weight of refrigerant in any part of the discharge line which will be cooler than 70° F. during shut-off, and the weight of refrigerant required to operate the ice machine. Single evaporator Models require about 2# of refrigerant, double evaporator Models use 3½#; 100 ft. of 3/8" liquid line contains 4.4#, 100 ft. of 1/2" discharge line can hold up to 8.7# liquid refrigerant. If a 5/8" discharge line is recommended by your consultant, the possible liquid volume of the line equals 14.5# per 100'.

   Condenser liquid requirements can be supplied by the manufacturer of the Condenser - be sure that you specify that you need the total capacity per circuit at the lowest ambient expected.

The receiver tank flooded volume for single evaporator Models (and the double evaporator Flaker) is 15#, and 22# for the dual evaporator Cubers. If the total System Charge calculations approach or exceed receiver capacity plus liquid line capacity and 10% of Condenser capacity, a larger or additional receiver will be required.

A Remote Condenser System charge work sheet is included in this section to help you determine proper refrigerant charge.

6) Each Cuber or Flaker contains a head pressure control valve. The valve is designed to maintain a minimum high side pressure during cold weather by flooding the Condenser to modulate effective cooling surface area. No additional controls are required or recommended.

7) It is preferable to allow the Cuber to cycle "OFF" on bin control rather than shutting off with a manual switch. Cycling on bin level will prevent extended "OFF" periods, and maintain Compressor temperature to reduce liquid accumulation in the discharge line and Compressor. A crankcase heater should be added if the ice machine is to be shut down for extended periods, or if the Compressor is located in a refrigerated room.

Adherence to the above rules should result in a properly operating Remote Condenser installation. If it is absolutely necessary to violate any of these points, please consult the Engineering Department at Kold-Draft before quoting on the installation.

Kold-Draft cannot, of course, be held responsible for any operational problems encountered when using Condensing Systems not supplied by the factory, nor can we guarantee U.L. compliance on anything but a complete Kold-Draft Pre-Charged Remote Condenser System.

If you find it necessary to contact us concerning Remote Condenser operational problems, we will need the following information to properly analyze your installation:

DISTRIBUTOR/DEALER ____________________

LOCATION OF INSTALLATION: NAME ____________________

ADDRESS ____________________

CITY ____________ STATE ____________

MACHINE:

MODEL ____________

SERIAL NUMBER ____________

TYPE OF HEAD PRESSURE VALVE ____________

SIZE & VOLUME OF RECEIVER: SIZE ____________ VOLUME ____________

CHARGE OF GAS IN SYSTEM: LBS. ____________ TYPE: F12 ____________ F502 ____________

CPR VALVE: YES ____________ NO ____________ SETTING ____________

CONDENSER:

TYPE OR MODEL ____________

BTU RATING ____________ AT TD ____________

AMBIENT TEMPERATURE: MIN. ____________ MAX. ____________

LIQUID & DISCHARGE LINE SIZE & LENGTH: L ____________ D ____________ LENGTH ____________

LENGTH OF DISCHARGE LINE EXPOSED TO TEMPERATURES BELOW 70° ____________

FAN CYCLING DEVICE? YES ____________ NO ____________

TYPE OF DEVICE: TEMP. ____________ PRESSURE ____________

SETTINGS OF DEVICE: HIGH ____________ LOW ____________

FLOODED VOLUME OF CONDENSER ____________ LBS.

LOCATION OF CONDENSER IN RELATION TO MACHINE: ABOVE ____________ BELOW ____________

HEAT RECLAIM UNIT? YES ____________ NO ____________

VOLUME OF HEAT RECLAIM ____________ LBS.

DESCRIBE PROBLEM:

REMOTE CONDENSER SYSTEM CHARGE WORK SHEET

SYSTEM CHARGE = Cuber Charge + Condenser Charge (Volume) + Liquid Line Charge + Discharge Line Charge (Vapor and Liquid)

SYSTEM CAPACITY = Receiver Tank Volume + Liquid Line Volume + 10% of Condenser Volume

System Capacity should be 125% of System Charge, or greater.

SYSTEM CHARGE

CUBER VOLUME ____________#

CONDENSER VOLUME ____________#

LIQUID LINE ____________#

DISCHARGE LINE, VAPOR ____________#

DISCHARGE LINE, LIQUID ____________#*

TOTAL CHARGE ____________#

EACH CIRCUIT

__________" O.D., __________' long

__________" O.D., __________' long

SYSTEM CAPACITY

RECEIVER CAPACITY ____________# (15.5# Single Evap., 22# Dual Evap.)

LIQUID LINE CAPACITY ____________#

10% OF CONDENSER VOLUME ____________#

TOTAL CAPACITY ____________#

WEIGHT OF REFRIGERANT IN COPPER LINES

#/100 FEET, R-12

O.D.	Vapor	Liquid
3/8	.15	4.4
1/2	.28	8.1
5/8	.45	12.9
7/8	.93	26.9

(Use .96 Multiplier for R-502)

KOLD-DRAFT REMOTE CONDENSER

VOLUMES

1½	TON	-	3.88#
2	TON	-	5.02#
3	TON	-	10.36#
5	TON	-	21.4#

DIVIDE ABOVE WEIGHTS BY NUMBER OF CIRCUITS IN CONDENSER.

*Note: The discharge line capacity can be assumed to be pounds of VAPOR only if the line remains at or above room temperature. Any part of the line which can fall below 70° when the ice maker is "OFF" can increase LIQUID charge requirements and must be added to the system charge.

DETERMINING REFRIGERANT CHARGE:

A) The head pressure control valve operates by flooding the condenser with varying amounts of refrigerant to change effective condensing area. Bypass port maintains receiver pressure at 120 PSIG R-12 or 180 PSIG R-502.

B) Find the volume of the condenser, totally flooded, in pounds of Refrigerant.

C) Calculate the liquid weight of the 3/8" return line, and the liquid weight in that part of the discharge which can fall below 70°F when the ice maker is "OFF".

D) Assume the refrigerant requirement for the Cuber to be 3#, and add to the condenser and line requirements. The result is total system charge.

E) Liquid storage ability is the sum of receiver capacity, liquid line volume, and 10% of condenser volume. If the total system charge is greater than liquid storage ability, a larger or additional receiver will be required.

THINGS TO AVOID

A) Try to keep line set lengths under 50'. Runs as long as 100' have been used successfully with selected condensers and increased refrigerant charge. Line sizes must be 3/8" liquid, 1/2" discharge regardless of run length.

B) Try to locate the condenser above the Cuber. Low Condensers can collect oil and refrigerant during the OFF cycle, causing problems when the Cuber re-starts.

C) Try to keep the compressor warmer than the Condenser. In most installations, the Cuber runs often enough that residual motor heat minimizes liquid migration to the crankcase. If the Cuber is in a cool location, and is OFF for extended periods, a crankcase heater should be added.

D) Try to use a Condenser that contains 3/8" diameter internal tubing. Condensers manufactured with 1/2" tubing have been used successfully, but usually require an overcharge to accommodate the increase in volume. Avoid Condensers which use parallel 1/2" tube connections, and any Condenser which contains 5/8" or larger tubing.

E) Avoid the use of a fan cycling switch. If the customer insists on replacing the head pressure control valve with a fan switch, expect operational problems in ambients below 32°. If the customer insists on using a fan switch in addition to the head pressure control valve, set the switch to cut in at 20 PSIG higher than the valve setting, with a differential of 30 PSIG. Remember that a separate fan switch will be required for each circuit. Fan cycling switches can be used above 32° to allow use of oversized condensers which have a liquid volume too large to allow use of a head pressure control valve. If fan cycling is used in lieu of increasing receiver capacity, the head pressure control valve must be removed.

F) Avoid placing the Condenser in the exhaust air stream of other roof-top equipment. Stay away from kitchen exhaust fans to prevent grease accumulation on the fins. Use a curb which extends above the deepest expected pond in the Condenser area of the roof. A 3" pond can cut air flow in half if a curb is not used.

G) Liquid return lines can sweat indoors in cool weather. Avoid running the lines through heated areas above a finished ceiling. If alternate routing is not possible, use "Armaflex" to insulate the liquid line.

H) Avoid the use of any heat reclaim device which taps into the Cubers refrigeration system.

RETROFITS:

A) When converting a water cooled or self-contained air cooled ice maker to remote condenser application, the receiver capacity must be increased and a head pressure control valve must be installed in the ice maker. R-12 models use the GB-3325 (120#) head pressure control valve, and R-502 models require a GB-2351 valve. Single evaporator cubers and all flakers require a 15# receiver (GB-1329); double evaporator cubers require a 22# receiver (GB-3346).

B) If you are installing a new cuber or flaker on an older condenser, check for the presence of a head pressure control valve on the condenser. As each ice maker contains its own head pressure control, any valves which are located on the condenser must be removed.

C) When adding an additional ice maker to an existing multi-circuit condenser, be sure that any fan cycling device which may have been added to the condenser won't interfere with the operation of the new cuber or flaker. Fan cycling is not required with Kold-Draft equipment, but if the customer insists on retaining the device, be sure to use parallel connected relays when more than one ice maker is connected to the same condenser.

AIR-COOLED CONDENSERS

1. Maintain adequate ventilation around the Cuber. It is also important to prevent re-circulation of the hot condenser air from the side to the back. This would occur particularly if the condenser end were placed in a corner of a room. If this must be the location then an external air block should be built to prevent the hot outlet air from returning to the inlet of the condenser.

2. Cleaning Air-Cooled Condenser - The air-cooled condenser should be cleaned weekly with a stiff brush and a vacuum cleaner to remove dust and dirt for efficient operation of unit. To determine that the condenser is clean, a light which is held at one side of the condenser will be clearly visible from the other side.

WATER-COOLED CONDENSERS

1. Are factory set for 115 pounds head pressure on all but GB-1200 Series (R-502) which is set at 230-235 pounds head pressure. To reduce water consumption, head pressure can be increased but ice capacity will be slightly reduced.

2. If head pressure is excessive, and water usage is higher than normal, condenser must be cleaned.

3. Cleaning Water-Cooled Condenser - Halstead Mitchell condensers have end plates which can be removed for mechanical cleaning if necessary.

   Water-Cooled condensers without end plates can be cleaned by flushing with a condenser cleaning solution. Acid re-circulating pumps and solution are available at refrigeration supply houses.

COOLING TOWER APPLICATION

	AVERAGE COOLING TOWER LOAD			MAXIMUM COOLING TOWER LOAD		
MODEL #	B.T.U. PER HOUR	G.P.M.	PRESSURE DROP P.S.I.	B.T.U. PER HOUR	G.P.M.	PRESSURE DROP P.S.I.
42" Frame						
GB401 w	4372	.8	6.5	9655	1.42	17.75
GB402 w	4372	.8	6.5	9655	1.42	17.75
GB-503 w	5543	1.0	.5	11359	1.72	1.5
GB-603 w	6558	1.2	.8	15619	2.4	2.5
GB-1003 w	11713	1.89	.8	25559	3.05	5.79
GB-1204 w	14056	1.97	2.42	39475	4.91	15.22
GB-1205 w	14056	1.97	2.7	31949	4.91	15.22
28¼" Frame						
GT-7 w	2342	.7	6.0	9655	1.42	17.75
GT-301 w	3096	.7	6.0	9655	1.42	17.75
GT-401 w	4372	.8	6.5	9655	1.42	17.75
GT-402 w	4372	.8	6.5	9655	1.42	17.75
GT-503 w	5543	1.0	.5	11359	1.72	1.5
GT-603 w	6558	1.2	.8	15619	2.4	2.5

Above table: Incoming water 85°F (from cooling tower)
Water out 100°F (R12; 117 PSIG+ : R502; 215 PSIG+)

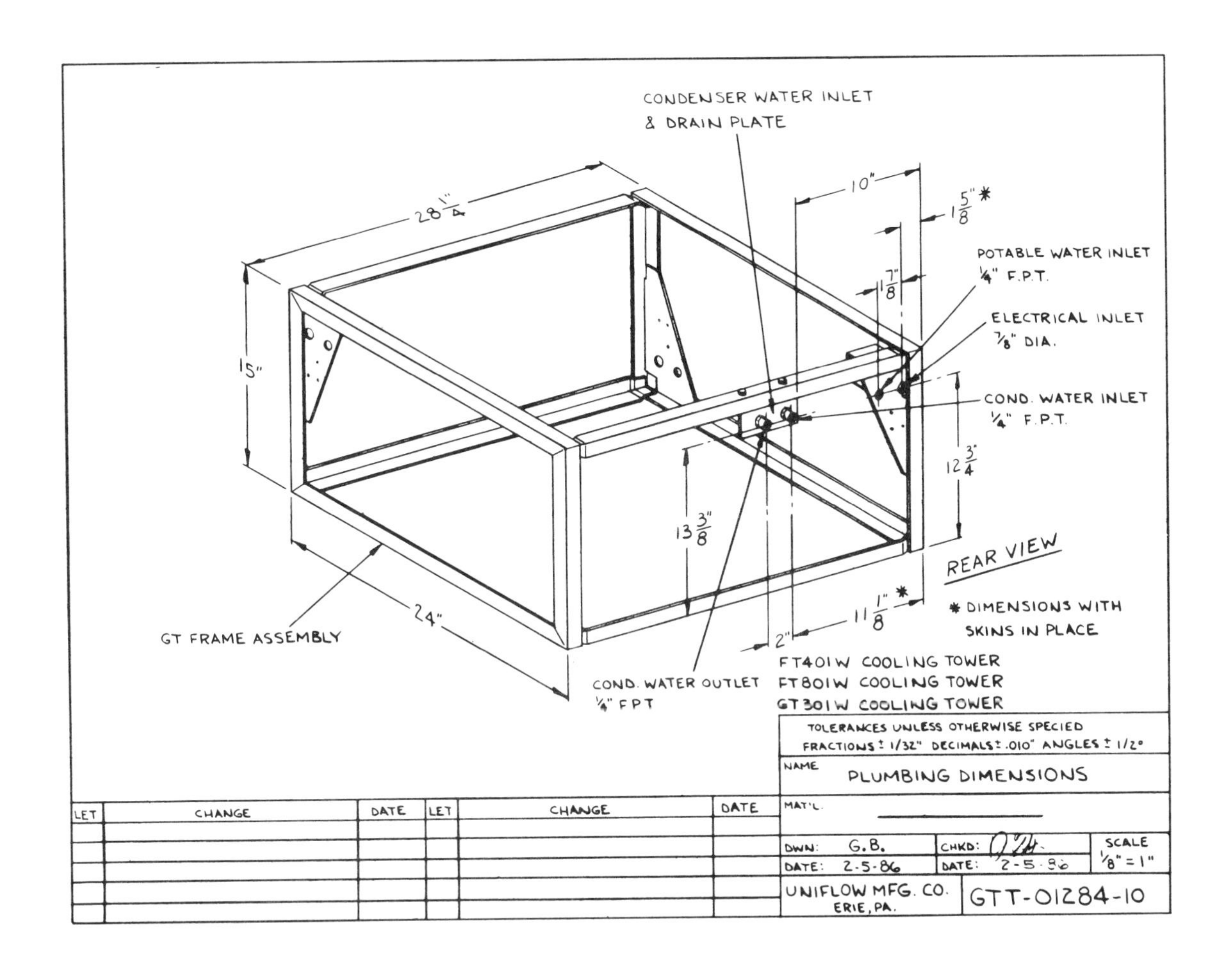
CONDENSER WATER INLET
& DRAIN PLATE
28 1/4"
10"
1 5/8" *
POTABLE WATER INLET
1/4" F.P.T.
1 7/8"
ELECTRICAL INLET
7/8" DIA.
COND. WATER INLET
1/4" F.P.T.
15"
12 3/4
13 3/8"
REAR VIEW
11 1/8" *
* DIMENSIONS WITH
SKINS IN PLACE
24"
2"
GT FRAME ASSEMBLY
COND. WATER OUTLET
1/4" FPT
FT401W COOLING TOWER
FT801W COOLING TOWER
GT301W COOLING TOWER
TOLERANCES UNLESS OTHERWISE SPECIED
FRACTIONS ± 1/32" DECIMALS ± .010" ANGLES ± 1/2°
NAME PLUMBING DIMENSIONS
MAT'L.
LET CHANGE DATE LET CHANGE DATE
DWN: G.B.
CHKD:
SCALE 1/8" = 1"
DATE: 2-5-86
DATE: 2-5-86
UNIFLOW MFG. CO.
ERIE, PA.
GTT-01284-10

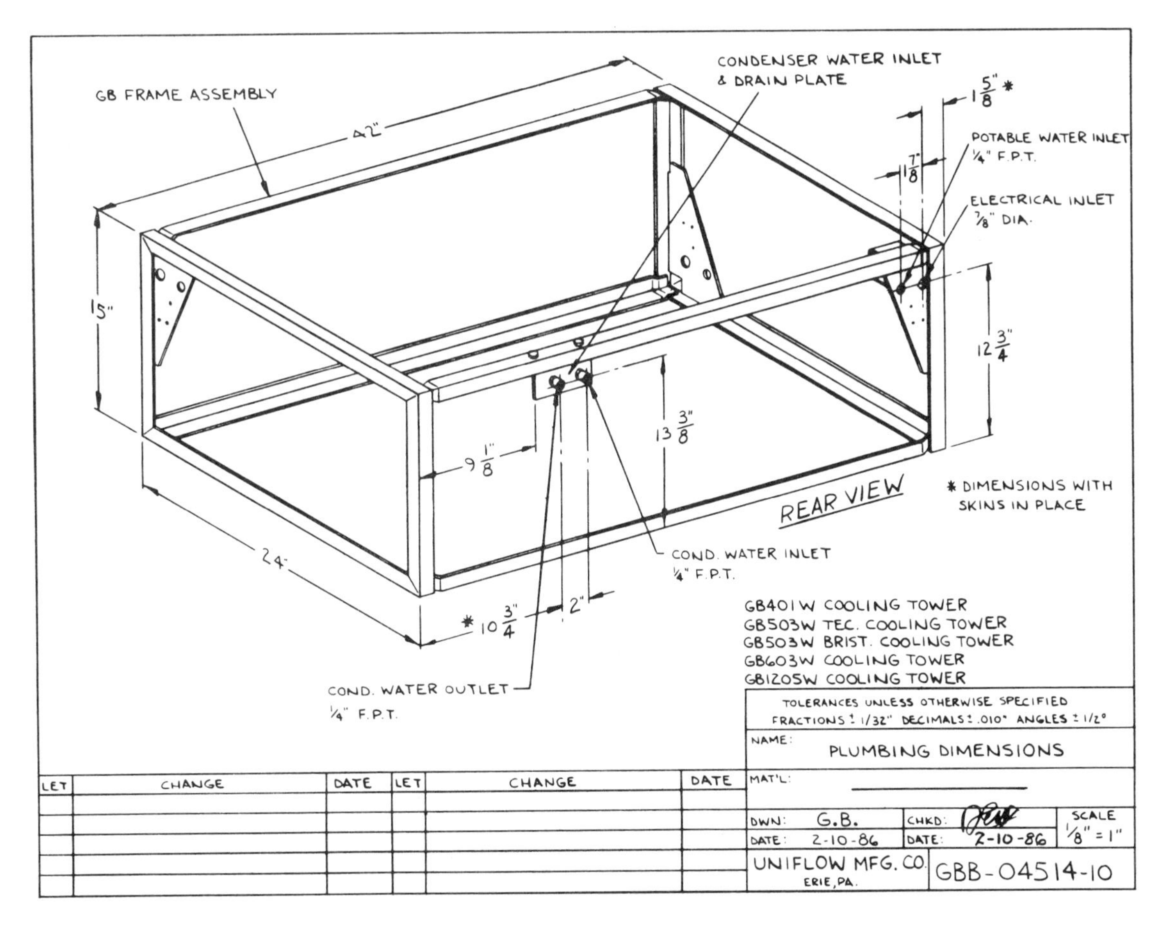
GB FRAME ASSEMBLY
42"
CONDENSER WATER INLET
& DRAIN PLATE
1 5/8" *
POTABLE WATER INLET
1/4" F.P.T.
1 7/8"
ELECTRICAL INLET
7/8" DIA.
15"
12 3/4"
13 3/8"
9 1/8"
REAR VIEW
* DIMENSIONS WITH
SKINS IN PLACE
24"
COND. WATER INLET
1/4" F.P.T.
* 10 3/4"
2"
GB401W COOLING TOWER
GB503W TEC. COOLING TOWER
GB503W BRIST. COOLING TOWER
GB603W COOLING TOWER
GB1205W COOLING TOWER
COND. WATER OUTLET
1/4" F.P.T.
TOLERANCES UNLESS OTHERWISE SPECIFIED
FRACTIONS ± 1/32" DECIMALS ± .010" ANGLES ± 1/2°
NAME: PLUMBING DIMENSIONS
MAT'L:
LET CHANGE DATE LET CHANGE DATE
DWN: G.B.
CHKD:
SCALE 1/8" = 1"
DATE: 2-10-86
DATE: 2-10-86
UNIFLOW MFG. CO.
ERIE, PA.
GBB-04514-10

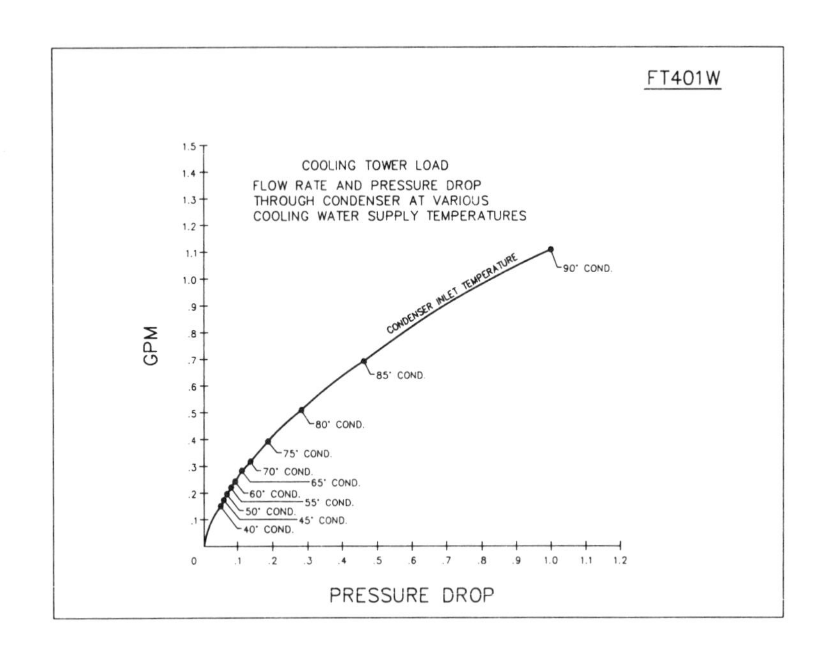

FT401W
COOLING TOWER LOAD
FLOW RATE AND PRESSURE DROP
THROUGH CONDENSER AT VARIOUS
COOLING WATER SUPPLY TEMPERATURES
CONDENSER INLET TEMPERATURE
90° COND.
85° COND.
80° COND.
75° COND.
70° COND.
65° COND.
60° COND.
55° COND.
50° COND.
45° COND.
40° COND.
GPM
PRESSURE DROP

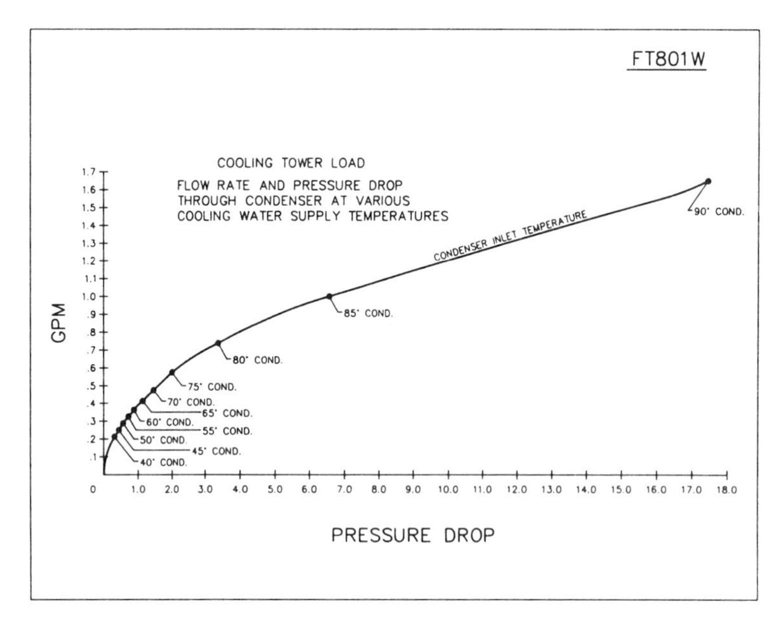

FT801W
COOLING TOWER LOAD
FLOW RATE AND PRESSURE DROP
THROUGH CONDENSER AT VARIOUS
COOLING WATER SUPPLY TEMPERATURES
CONDENSER INLET TEMPERATURE
90° COND.
85° COND.
80° COND.
75° COND.
70° COND.
65° COND.
60° COND.
55° COND.
50° COND.
45° COND.
40° COND.
GPM
PRESSURE DROP

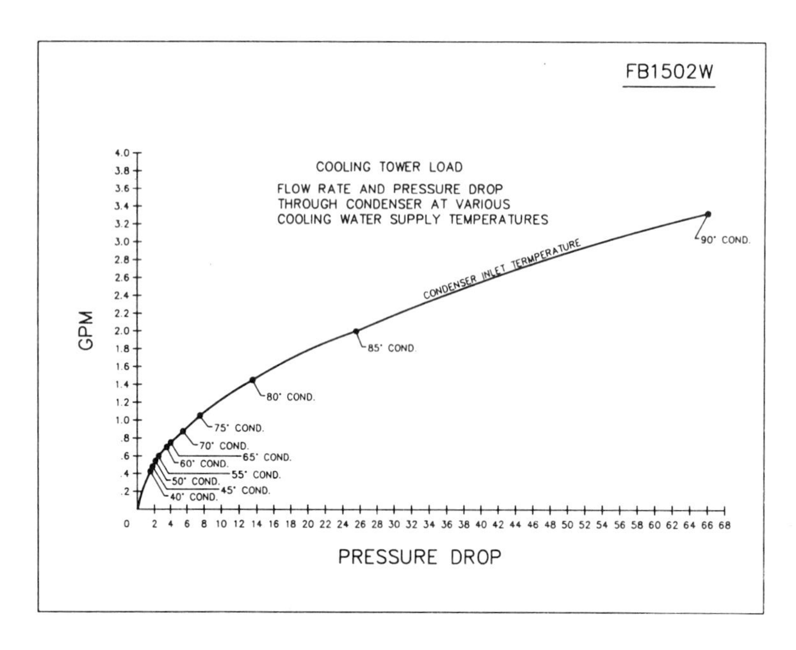

FB1502W
COOLING TOWER LOAD
FLOW RATE AND PRESSURE DROP
THROUGH CONDENSER AT VARIOUS
COOLING WATER SUPPLY TEMPERATURES
CONDENSER INLET TEMPERATURE
90° COND.
85° COND.
80° COND.
75° COND.
70° COND.
65° COND.
60° COND.
55° COND.
50° COND.
45° COND.
40° COND.
GPM
PRESSURE DROP

FUSES

1. Each Ice Maker must be on its own circuit. Individually fused or on HAC-R Type circuit breaker. The maximum allowable voltage fluctuation should not exceed 10 per cent of the nameplate rating even under starting conditions. Low voltage will cause erratic operation, low capacity, and serious damage to the overload switch and motor windings.

2. Normal conditions will permit the use of a nominal fuse size that is rated at or slightly larger than the "minimum wire size" listed for each model. Higher than average ambient conditions will dictate the use of a larger fuse size which should never exceed the "maximum fuse size" listed for each model.

3. Most cubers have supplemental fusing. These fuses are located in the electrical junction box at the left rear side of the Cuber and should be installed at time of original installation.

NOTE: For fuse sizes, refer to page 3.

INTERNAL OVERLOADS

Each motor used on this ice maker has overload protection. None should ever burn out due to an overload. However, voltage fluctuating more than 5% for dual voltage application may cause burn out trouble. To allow a safety factor, make sure voltage is at least 95% of rated voltage while the cuber is operating. Also, voltage lower than rated will reduce ice production.

CONTROL STREAM — After the circulating water is cold, note the control stream and compare with figures below. Adjust screw "A" so that control stream strikes as shown in the illustration. (NOTE: After water is cold, it may sub-cool and form crystals in circulation tank for a minute or two, partially stopping circulation and control stream temporarily. Do not adjust control stream at this time.) If control stream adjusting screw is changed at end of freeze cycle to make stream go over the dam, recheck it during the beginning of the next freeze cycle to be sure it does not go over the dam until all cubes are ready for harvest.

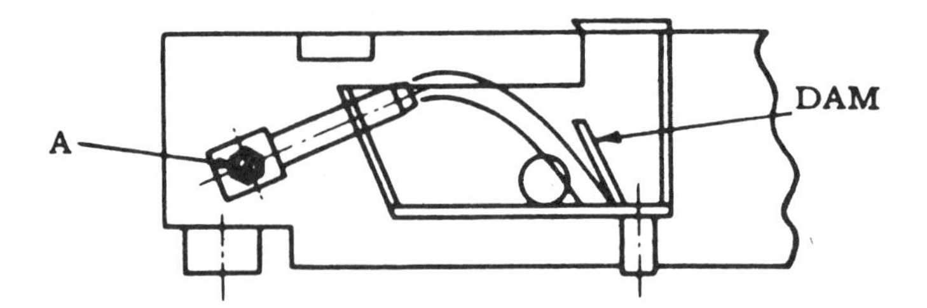

CONTROL STREAM AT THE END OF FREEZE CYCLE — When cubes are fully frozen, they freeze over some of the jet holes, increasing the water pressure in the circulation system. This makes the control stream rise and go over the dam, dumping water on the drain pan. Within 15-30 seconds the lower probe should be exposed to air, a resistance change relayed to the printed circuit card and the harvest cycle initiated.

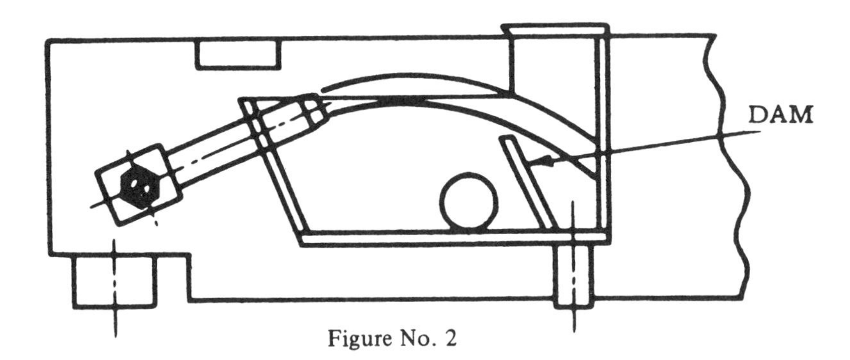

Figure No. 2

CUBE QUALITY

1. If cubes are full with only small dimples (1/8" - 3/16" deep), the exact minimum amount of water was taken in.

2. If control stream does not go over the dam and control tube initiates defrost, check cubes. If many have large holes, there was insufficient water taken in at the beginning of the cycle or there is a leak in the system. Eliminate the leak and/or adjust the high water level probe up to increase water on next fill cycle. NOTE: Leak may be internal (cracked lateral on water plate).

3. If control stream goes over the dam more than 1 minute before water clears the low probe, check cubes. If many cubes have large holes, control stream was set too high. Reset slightly lower during beginning of next freeze cycle. If ice is frozen to waterplate, adjust the lower water level up slightly so harvest initiates within 15-30 seconds from control stream rising over the dam.

4. A Holes at the right or left two rows of the evaporator on expansion valve models - Holes at the left - close the valve 1/8 turn at a time. Holes at the right and front - open the expansion valve 1/8 turn at a time.

   B The capillary tube GT models. Holes at the left indicate overcharge. Holes at the right indicate undercharge.

PUMP INTAKE HOSE - COLLAPSING

1. The surgical tubing is too long or it has been twisted, the tubing will collapse. The two conditions can be corrected very easily by checking the length so that the tubing is approximately 3 3/8″ long, or by twisting the tubing to make certain it is straight.

2. Dirty water tank outlet screen. In this instance, the tubing will collapse and remain collapsed until the screen is cleaned. This could cause cloudy cubes since not enough water is being circulated.

3. Occasionally at the beginning of a cycle, if the water is super cooled and crystalization occurs in the water, the crystals would plug up the screen and cause a collapse of the surgical tubing. This condition would last only a matter of less than a minute until the crystals thaw and the machine would begin functioning properly. This condition is not serious and would not damage any of the ice cuber parts. The condition cures itself and it is of short duration.

4. The hose may also collapse due to aging. The hose should be replaced and a spring (GBR-00217-01) installed.

Water Level Probe Assembly GBR-01370 is connected to the main water circulation tank by means of flexible tubing. The height of water in the probe assembly indicates the height of water in the circulation tank. Thermistor probes determine water valve on-off levels. Refer to page 4 for correct water setting for upper probe. The bottom probe is positioned so that 15-30 seconds after the water flows over the dam harvest is initiated.

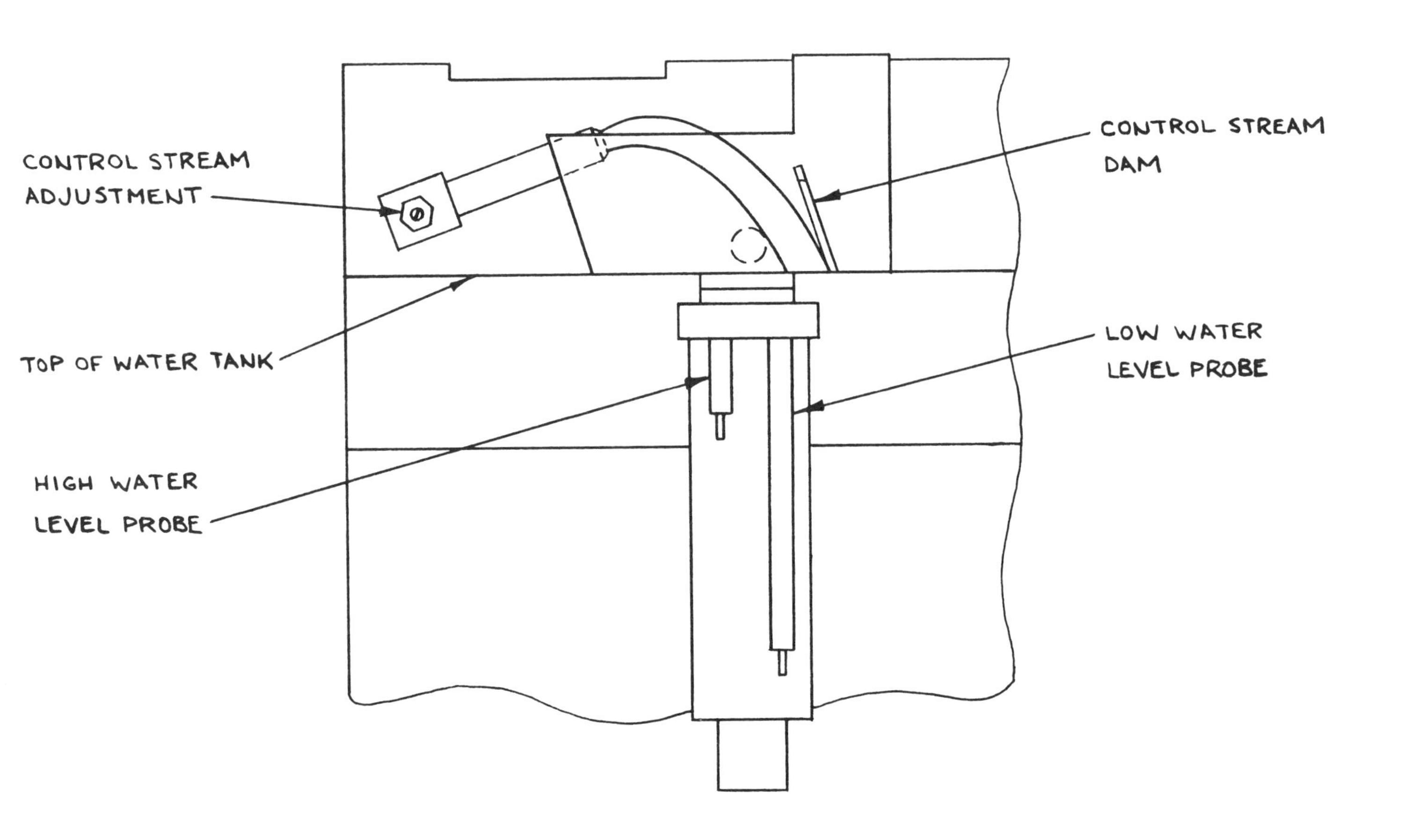

WATER PLATES

WATER PLATE PROBLEMS

1. Improper setting of the fin thickness. Under tolerance fin thickness may cause icing of injection and return holes on the water plate surface. Refer to alignment page 60.

2. If the water injection holes on the water plate become blocked, some of the freezing cells will be void of ice at time of harvest. This condition may be cured by using a 1/16" drill bit to clean the injection holes.

NOTE: Freezing cells void of ice may collapse due to expansion of ice in surrounding cells.

3. If the jet stream holes are cleaned and no water flows through them to form ice in the cells, the laterals have become clogged. To clean the laterals, remove the plastic plugs at the end of the water plate. Run a small brush through the laterals and flush out the foreign material. When replacing the plug use new ones as the old ones have aged and may not seal properly. To replace the plug, push it in with an Allen wrench or similar blunt end tool. This stretches the plug so it will be tight when the tool is removed.

4. A crack in the lateral strips on the underside of the water plate may prevent the control system from rising and going over the dam. This could extend the freeze cycle and freeze water into the water plate. If the laterals are cracked, the water plate should be replaced.

5. Water Plate Silicone Treatment - In certain areas of the country where water is unusually pure with pratically no dissolved solids, the water plates must have fairly frequent applications of silicone in order to prevent excessive ice adherence. Application of Silicone - Treatment is recommended every three months, or whenever Cuber is serviced, where water conditions cause ice to stick to the water plate.

   1. Defrost the unit by lifting the control tank or pulling down on the water plate. With wash switch to "Wash", allow the unit to refill to warm up the plate and melt off any accumulated ice. Open the plate to dump out the tank.

   2. Turn off the power with water plate down. Wipe water drops off evaporator and with several rags, wipe the water plate as dry as possible.

   3. Apply Kold-Draft water plate spray(5-1052) to the water plate being sure to avoid the last half-inch on the right side. If the sillicone coating is on this edge, the water will run off into the bin during defrost.

   4. Turn on the power and wash switch to "ON".

NOTE: Use Kold-Draft silicone water plate spray. Others may contain substances detrimental to water plate finish.

STEPS TO CHANGE WATER PLATE

REMOVAL OF WATER PLATE AND TANK:

1. Turn off the water and allow the plate to close; then turn off power.

2. Remove the control stream drain hose.

3. Electronic Cubers:

   a. Remove water level probe assembly by sliding to the right beyond control stream box and lift. Disconnect tube assembly from main tank.

   Electro-Mechanical Cuber:

   b. Remove the water control tank from the wire in hanger. Disconnect tube assembly from main tank.

4. Remove the pump mounting screw holding the water plate brace and inlet and outlet hoses from pump.

5. a. Cubers with plastic pump and rear hinge brackets, pry brackets away from the plate.

   b. Cubers with metal brackets (prior 1977) remove the four lower screws and lock washers holding the water plate to the pump bracket and rear hinge. Remove the nuts and lock washers holding the top of the water plate to the pump bracket and rear hinge (5/16" socket).

6. Run the cam down to the 9 o'clock position by hand or with power. Unhook the main springs from the water plate.

7. Electronic cubers:

   a. Remove the screws that mount the control module box. Pull module forward to disconnect front cam. Rotate module up for clearance (if a status indicator exists,unplug harvest switch from control module).

   Electro-Mechanical cubers:

   b. Remove the cotter pin out of the pump toggle switch and remove the lift rod and collar assembly.

8. Slide the water plate and tank to the right without turning and slide it forward out of the machine.

WATER PLATE REPLACEMENT ON CIRCULATION TANK:

1. Remove the spring bosses, water plate brace, water deflector and teflon brackets and the four screws holding tank to water plate.

    a. Electronic cubers: Remove plastic bolt from shoulder on water plate and place in new plate.

2. Attach the tank to new water plate with four screws, put on teflon brackets, water deflector, water plate brace and spring bosses. Bosses should be tight but not with excessive force that would strip out the plastic threads.

    a. Electro-Mechanical Cubers: Install the hinge screws and plates.

TO REINSTALL THE WATER PLATE AND TANK ASSEMBLY:

1. With the open end of the water plate to the right, slide it back into the cuber and to the left of its normal position.

    a. Elector-Mechanical: Install the lift rod and collar assembly and cotter pin with the small end of the crook to the right of the pump toggle switch.

2. Hoop-up the main springs to the water plate, rear spring first.

3. Plastic Hinge Brackets:

    a. Hook on and snap in place.
    GB Metal Hinge Brackets: Holes have to be drilled and tapped per drawing SKK-04576

    b. Run the top screws through the hinges, put on the lock washers and nuts (not tight). Run the bottom screws and lock washers through the hinges into the plate (then tighten the nuts and screws). It may be easier tc tighten the rear screws if the cams are run up to the 12 o'clock position.

4. Secure the pump mounting screw holding the water plate brace. Install the inlet and outlet hoses to the pump. When installing the tank to pump hose,it is usually easier to put it on the tank first with one finger inside the hose. Then slip it onto the pump connection, taking care to avoid any twist or kink. The hoses will slide on easier if wet.

5. Secure water control assembly.

    a. Electronic Cubers: Remount water level probe assembly, reposition control stream drain tube. Position and secure control module box to mounting bracket. Check adjustment of lift bolt. Reconnect ice harvest switch if present.

    b. Electro-mechanical: Reconnect control stream drain tube and water control tank.

6. Water plate must be aligned with the evaporator. Refer to page 60.

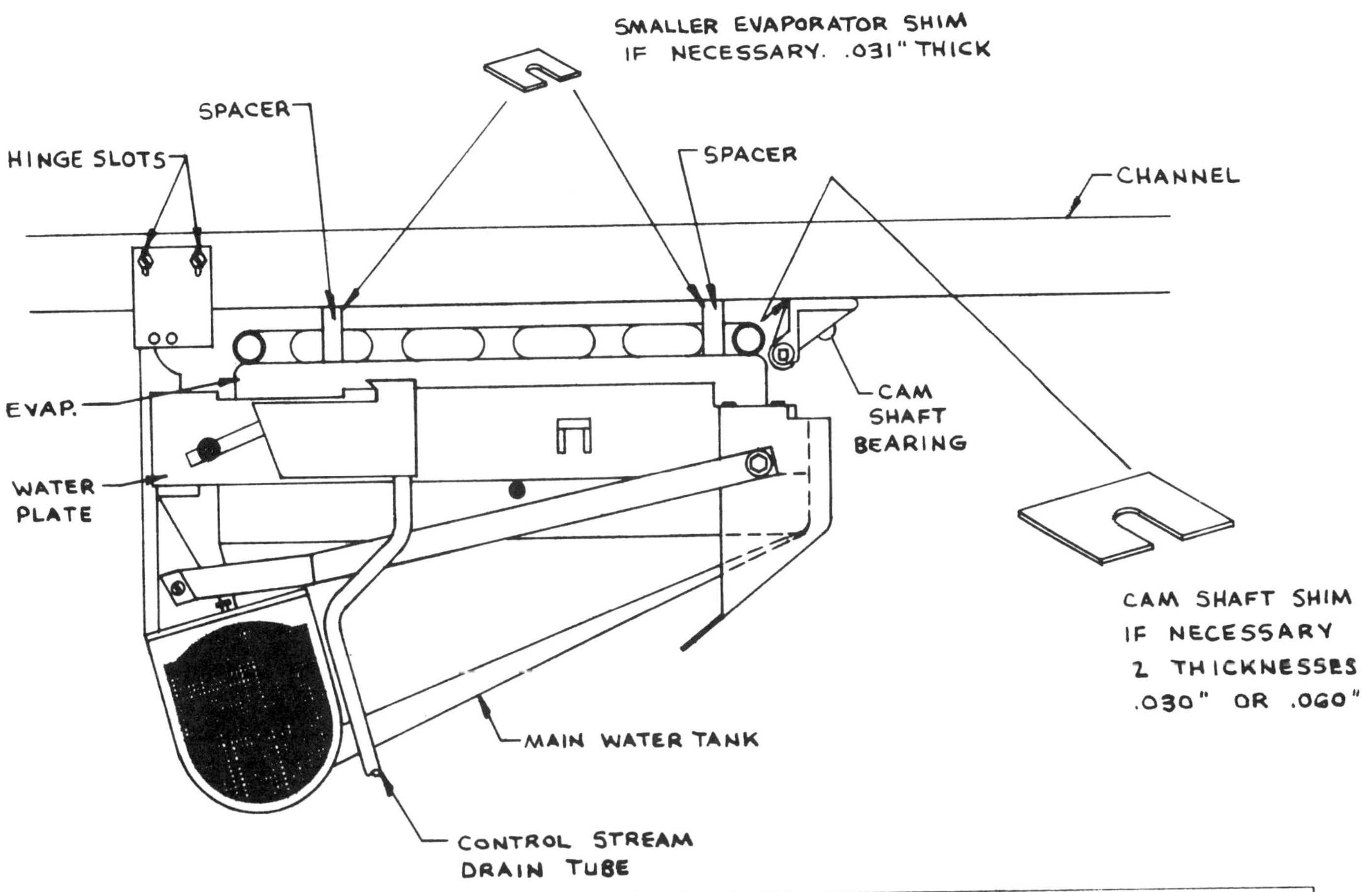
SMALLER EVAPORATOR SHIM
IF NECESSARY. .031" THICK
SPACER
HINGE SLOTS
SPACER
CHANNEL
EVAP.
CAM
SHAFT
BEARING
WATER
PLATE
CAM SHAFT SHIM
IF NECESSARY
2 THICKNESSES
.030" OR .060"
MAIN WATER TANK
CONTROL STREAM
DRAIN TUBE

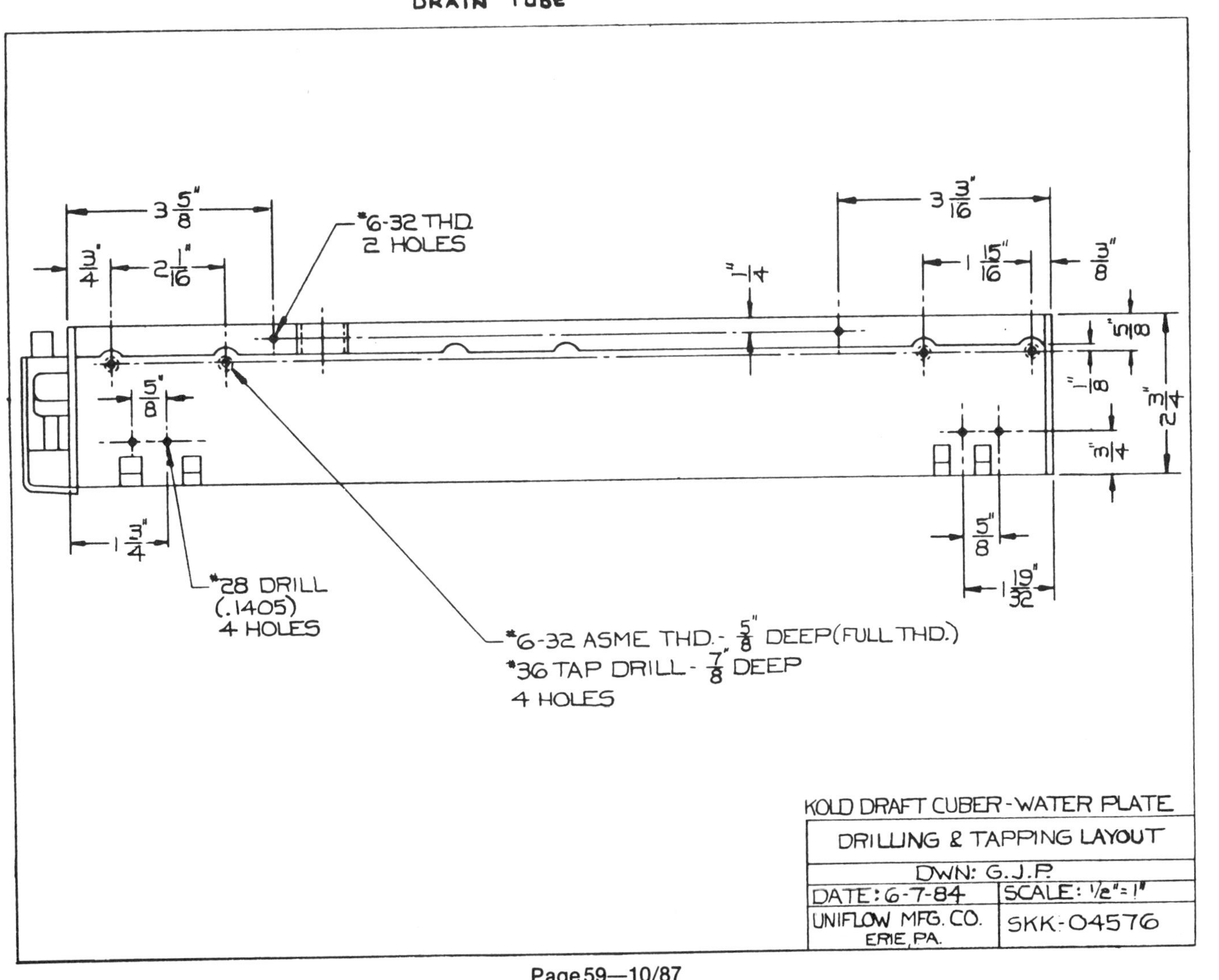
3 5/8"
*6-32 THD.
2 HOLES
3 3/16"
3/4"
2 1/16"
1/4"
1 15/16"
3/8"
5/8"
5/8"
1/8"
2 3/4"
3/4"
1 3/4"
5/8"
1 19/32"
*28 DRILL
(.1405)
4 HOLES
*6-32 ASME THD.- 5/8" DEEP(FULL THD.)
*36 TAP DRILL- 7/8" DEEP
4 HOLES
KOLD DRAFT CUBER-WATER PLATE
DRILLING & TAPPING LAYOUT
DWN: G.J.P.
DATE: 6-7-84
SCALE: 1/2"=1"
UNIFLOW MFG. CO.
ERIE, PA.
SKK-04576

ALIGNMENT OF WATER PLATE AND EVAPORATOR

(A) MAINTAIN 1 1/2" CLEARANCE ON LEFT SIDE BETWEEN WATER PLATE AND EVAPORATOR.
(B) MAINTAIN 3/8" CLEARANCE ON FRONT BETWEEN WATER PLATE AND EVAPORATOR.
(C) MAINTAIN A DIME CLEARANCE BETWEEN THE EVAPORATOR AND WATER PLATE SHIM IF NECESSARY.

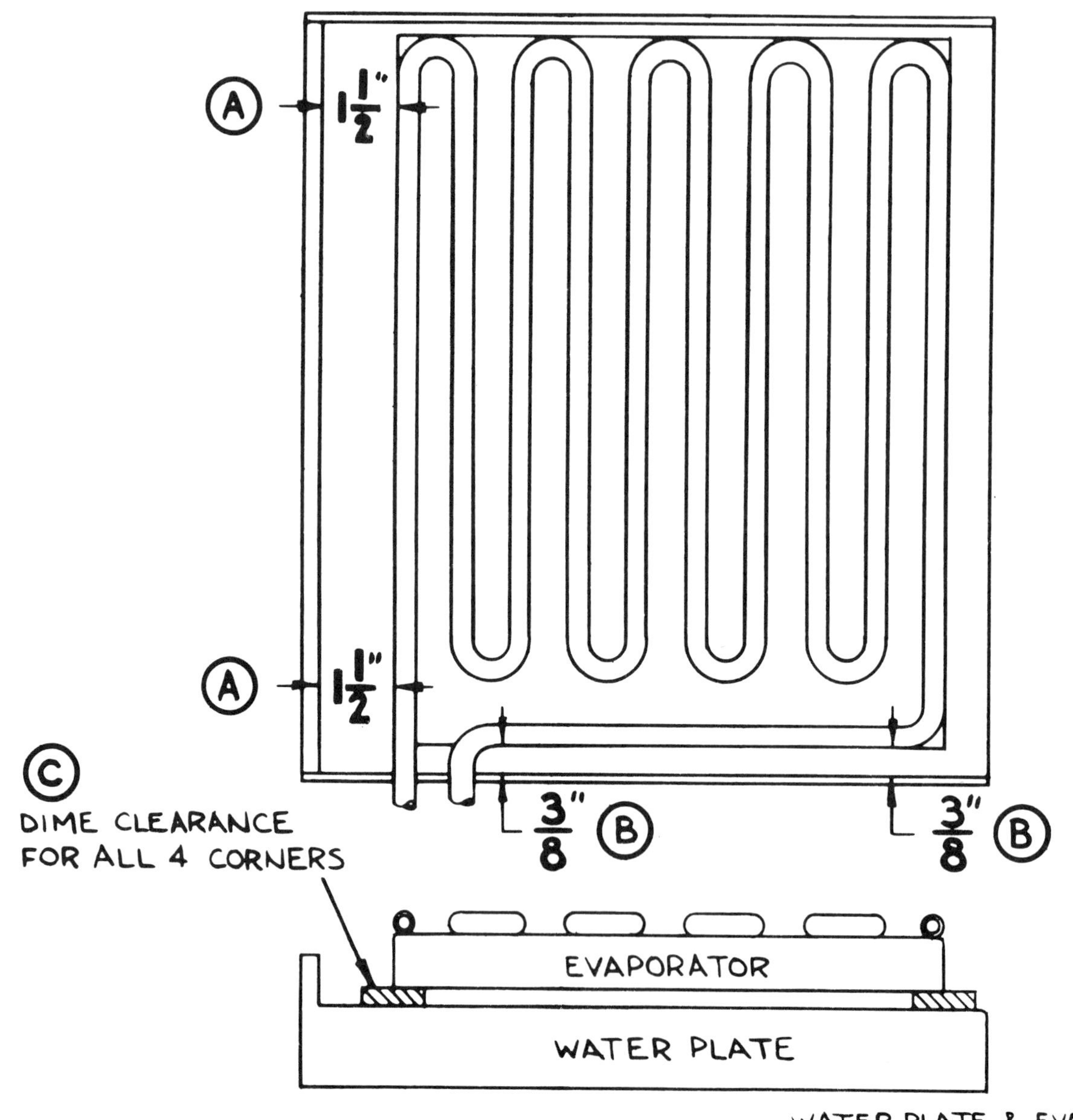

WATER PLATE & EVAP. ALIGNMENT

G.B. 7-25-85 GBB-03380

CLEANING INSTRUCTIONS FOR KOLD-DRAFT ICE CUBERS

Use rubber gloves and eye protection, and an apron is recommended.

1. Mix one bag of Kold-Draft Ice Machine Cleaner (55R-1000) in two quarts of warm (approx. 130° F.) water fro GB or GT400, -500, & -600 Series (two batches required for 1000 and 1200 Models) or 1/2 bag of cleaner in one quart water for GS, GY, or GT7/GT300 Series Cubers. Use the 55R-01002 washing bottle assembly for mixing and pouring solutions.

2. If the Cuber is operating wait until the Harvest Cycle occurs, then trip the Wash Switch to "WASH" as soon as the water plate begins to close.

3. Empty all ice from the Storage Bin and shut off other machines on the same Bin.

4. After the water fill is completed remove the cap from the water level control and pour about 1/2 of the mixed cleaning solution into the control tank or tube.

   CAUTION: Do not remove the water level probe assembly completely or lower the control tube far enough to overflow in Electronic Cubers.

   Replace the cap, then pour the remaining solution into the control stream box.

5. Allow the cleaner to circulate approximately 15 minutes, then pull the right end of the water plate down far enough to stop the waterpump and hold open until the pump will not re-start when the plate is released. The water plate will continue to open and dump the cleaning solution.

6. The water plate will close immediately and the water tank will refill. Repeat this dumping and refilling three times to rinse out all of the cleaner.

7. After thorough rinsing of the Cuber, leave the wash switch in the "WASH" position, then mix a sanitizing solution of 2 oz. 5-1/4% sodium hypochlorite (household strength laundry bleach) and one quart of clean water.

8. Pour about 1/2 of this solution into the water level control tank or tube as above, then pour the remaining solution into the control stream box.

9. Allow the sanitizing solution to circulate at least 5 minutes, then dump and rinse 2 times as described above.

10. While the cleaning and sanitizing solutions are circulating, clean, rinse, and sanitize all accessible parts of the Cuber with clean cloths. Use a cleaning solution of 8 tablespoons (1/2 cup) baking soda per gallon of warm water, and a sanitizing solution of no less than 1 teaspoonful (5 ml.) 5-1/4% sodium hypochlorite per quart of clean water. Clean and sanitize the storage bin last.

11. Trip the Wash Switch to "ON", check to be sure that the Cuber is operating properly, then re-assemble all enclosures.

NOTE: DO NOT USE AMMONIA SOLUTIONS IN CLEANING ANY PART OF THE ICE MAKER.

WARNING

Be sure to completely rinse all ice machine cleaner from cuber before adding sanitizing solution (bleach). Mixing of ice machine cleaner with bleach can cause a chemical reaction which will release poisonous fumes.

ICE BIN CLEANING INSTRUCTIONS

The bin should be cleaned periodically. If bin drain has any horizontal run, remove ice from left side of bin and flush with two quarts of hot water monthly. (Long drain lines should be flushed weekly.) The inner door of the GBN-2 bin may be removed by first removing the triangular guards on each side and then by holding in the two sides of the inner door, so they will slide past the stops until the door is horizontal. The hinge pins on the bottom of the inner door can be lifted and pushed back out of their slots at the bottom front of the bin, and the inner door then slid forward and free the outer door.

Clean exterior of bin frequently.

To clean the interior:

1. Empty the storage area and disconnect the electrical power supply to the ice maker (s).
2. Remove the ice maker inspection panel, top, left and right end panels, and drain pan. Sliding bin doors may be removed by lifting them up, then pulling out from the bottom.
3. Wash interior with a solution of 2 tablespoons of baking soda per quart of warm water.
4. Rinse with clean water.
5. Flush the drain with at least one quart of HOT water. Do not use drain cleaners in the ice storage area.
6. Sanitize the interior by wiping with a solution of 1 teaspoon 5-1/4% Sodium Hypochlorite (household laundry bleach) per quart of clean water. Pour unused sanitizing solution down the storage area drain.
7. When cleaning the ice maker, follow the ice maker cleaning instructions and clean the bin last.
8. Replace all enclosure panels before re-connecting the electrical supply.

DO NOT USE STRONG DETERGENTS OR AMMONIA SOLUTIONS ON ANY PART OF THE MACHINE OR BIN.

NEVER USE APPLIANCE POLISHERS OR OTHER FINISH PRESERVATIVES OR CLEANERS IN ICE STORAGE AREAS.

WINTER CONDITIONING

Ice machines that are idle in the winter months require preparation to prevent damage from freezing. The following procedure should insure the safety of the machine so that it can be started easily the following year:

1. Shut off the water supply to the Ice Machine.
2. Detach the water supply line to the Ice Machine.
3. If the machine is a combination water and air or a straight water cooled machine, it should be running while air is introduced through the water inlet connection to blow water out of the condenser coils.

   NOTE: If the condensing unit is not warm enough to open the condenser water valve on combination air-water cooled machine, then the air cooled condenser must be blocked to raise the temperature, or machine should be run long enough to open condenser water valve. EXTREME CARE MUST BE USED TO ENSURE ALL WATER IS BLOWN OUT.

4. The machine should then be run into defrost to drain the water tank and the air blown into the water inlet to blow out the water solenoid.

PROBLEM WATER AND THE KOLD-DRAFT CUBERS

The basic raw material we work with in cubers is water, the supply varying from municipal to private throughout the world, and the quality of the water will vary according to the supply.

The Kold-Draft Cuber is specifically designed and manufactured to work well and provide good cubes with water on which other ice machinery cannot function. Occasionally, some water supplies are encountered which will require special treatment, even with the Kold-Draft machine.

There is no hard and fast rule when special treatment will be needed, but through experience we have found certain water carried elements which may require treatment. For the most part, these elements include hardness, algae, dirt, and total dissolved solids.

1. High hardness may cause sludge build-up which slows the circulation system, causing cloudy cubes and necessitates frequent cleaning. A a rule of thumb, when machine cleaning is necessary more than once every three months, special treatment is suggested. The table which follows is supplied to help you determine when this special treatment will be necessary. (Special treatment may include a scale inhibitor feeder 555-1003 or a water conditioner).

Degree of Hardness in Grains per gallon	Pre-Treatment Suggested
0 to 9	None
10 to 14	Treatment unlikely
15 to 19	K.D. Scale Inhibitor may be required (555-1005)
20 to 24	K.D. Scale Inhibitor (555-1005)
25 to 29	K.D. Scale Inhibitor with occasional cleaning, or water softener may be necessary (555-1005)
30 and up	Water Softener is recommended.

If a softener is used, consider the possibility of softening all the water in the establishment or at least the hot water for dishwater and coffee maker, as well as the water for the cuber. If a softener is used with a water cooled cuber, a separate hard water line should be run to the water cooled condenser.

2. HIGH TOTAL DISSOLVED SOLIDS (TDS)
Water with over 1500 ppm Total Dissolved Solids in a 3/4 H.P. Cuber (or over 1000 ppm in a larger Cuber) may produce cubes with some cloudiness even in a clean machine. If clouds are sufficient to be intolerable, another source of water is the only solution.

3. SEDIMENT TREATMENT
Should the water supply be heavily laden with suspended matter to such an extent that the water inlet screen in water solenoid clogs frequently causing slow water fill at the beginning of the cycle, it may be necessary to put a special filter on the inlet water line. Sediment filtration finer than 20 micron is not recommended unless a peculiar local problem, such as giardia cyst, is present. Consult local water conditioning firm for equipment adequately suited to flow rate and life requirements.

4. ALGAE TREATMENT
In the event algae, some forms of bacteria, or other growing matter are found in the water supply, chlorination equipment will be necessary.

We urge you to consult our Water Service Department (supply a copy of Bulletin No. 2162 and a sample of the water) for advice and suggested treatment when water or scaling problems are encountered. On request, we will provide special bottles and containers suitable for shipping samples of water for test, free water analysis and pre-treatment suggestions (when a regular container is not available, be certain to provide us with at least 1/2 pint of water). We urge that you take full advantage of this service and discuss with the owners the benefits of conditioned water for the entire supply.

WATER QUESTIONNAIRE – KOLD-DRAFT ICE MAKER

(Please provide a sample of the water to be used, (at least 1/2 pint) or water in use with the information that follows). Send to UNIFLOW MFG. CO., DEPT. W.S., P.O. BOX 10335 ERIE, PENNA. 16514.

User's Name ______________________________

Address ______________________________

Dealer Name ______________________________

Address ______________________________

1. Type Ice Machine ______________________________
   Model Number ______________________________
   SERIAL NUMBER (If Known) ______________________________
   Number of Machines ______________________________
2. Source of the Water: Private Well________ City________ Pond________ River________ Cistern________
3. Is the water Chlorinated? Yes________ No________
   If yes, what type chlorination equipment? ______________________________
4. Is the water pre-treated? Yes________ No________
   If yes, what type of equipment: Filter________ Feeder________ Softener________
   (Specify type, size and make of equipment) ______________________________
   ____
5. Is Uniflow machine installed? Yes________ No________
   (Specify type and Model) ______________________________
   What problems are being encountered? ______________________________
   ____
   ____
6. Are there other makes of machines installed at the present time? Yes________ No________
   If yes, specify type, make and model ______________________________
   ____
7. Are other make machines having problems? Yes________ No________
   If yes, specify ______________________________
   ____
   How treated?______________________________
   ____
8. Are problems apparent with other water using appliances: Yes________ No________
   (Please describe) ______________________________
   ____
   ____
   ____
   ____
9. Total water usage for establishment? Monthly____________ Gallons.
   What is flow rate at peak usage ____________GPM.

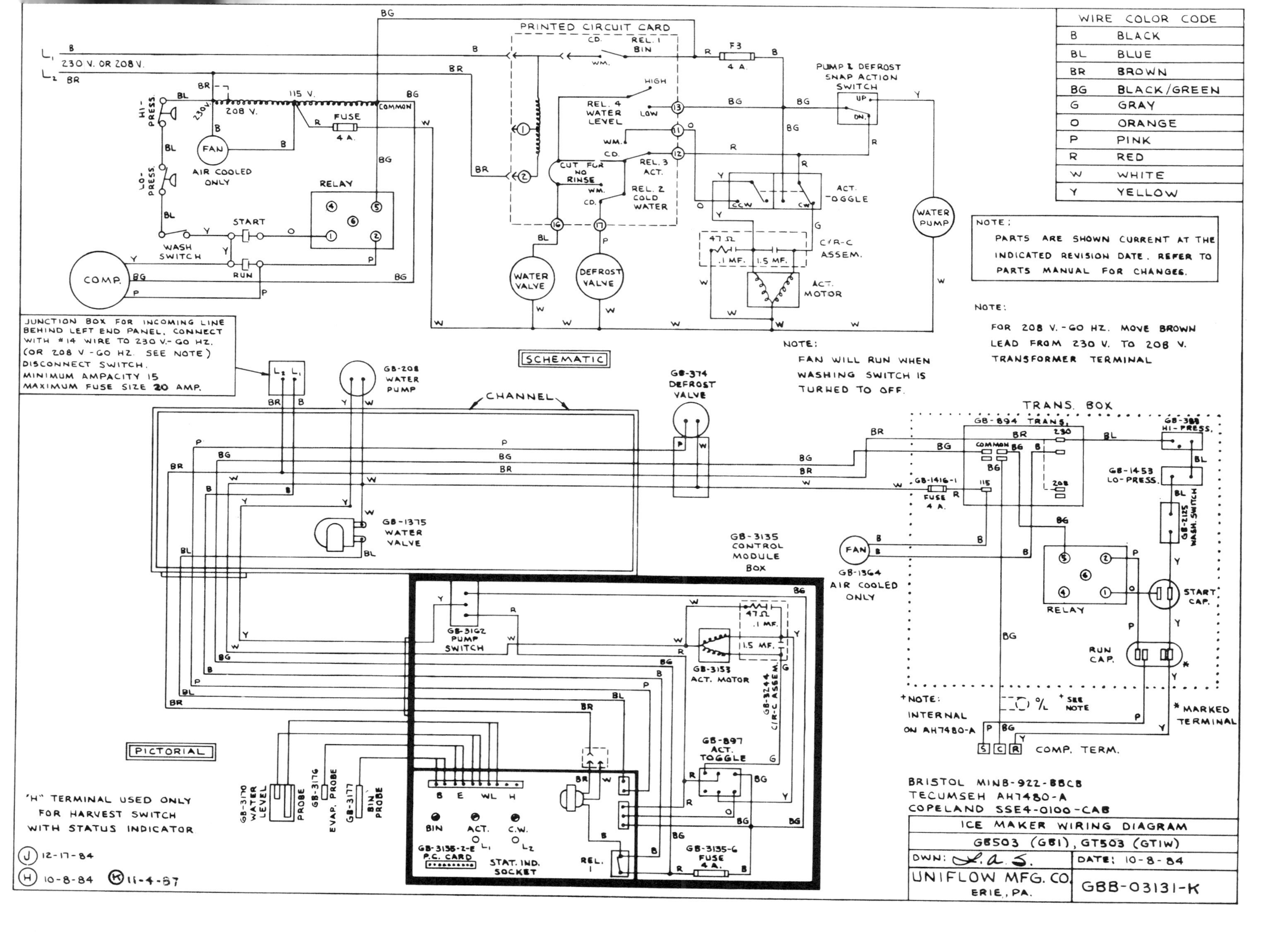
WIRE COLOR CODE
B BLACK
BL BLUE
BR BROWN
BG BLACK/GREEN
G GRAY
O ORANGE
P PINK
R RED
W WHITE
Y YELLOW
NOTE: PARTS ARE SHOWN CURRENT AT THE INDICATED REVISION DATE. REFER TO PARTS MANUAL FOR CHANGES.
NOTE: FOR 208 V.-60 HZ. MOVE BROWN LEAD FROM 230 V. TO 208 V. TRANSFORMER TERMINAL
NOTE: FAN WILL RUN WHEN WASHING SWITCH IS TURNED TO OFF.
JUNCTION BOX FOR INCOMING LINE BEHIND LEFT END PANEL. CONNECT WITH #14 WIRE TO 230 V.-60 HZ. (OR 208 V-60 HZ. SEE NOTE) DISCONNECT SWITCH. MINIMUM AMPACITY 15 MAXIMUM FUSE SIZE 20 AMP.
L1 230 V. OR 208 V.
L2 BR
PRINTED CIRCUIT CARD
REL. 1 BIN
REL. 4 WATER LEVEL
REL. 3 ACT.
REL. 2 COLD WATER
CUT FOR NO RINSE
HIGH
LOW
F3 4 A.
PUMP & DEFROST SNAP ACTION SWITCH
ACT. TOGGLE
C/R-C ASSEM.
ACT. MOTOR
WATER PUMP
WATER VALVE
DEFROST VALVE
HI-PRESS.
LO-PRESS.
FAN AIR COOLED ONLY
FUSE 4 A.
RELAY
START
RUN
WASH SWITCH
COMP.
115 V.
208 V.
230 V.
COMMON
SCHEMATIC
CHANNEL
GB-208 WATER PUMP
GB-374 DEFROST VALVE
GB-1375 WATER VALVE
TRANS. BOX
GB-894 TRANS.
GB-388 HI-PRESS.
GB-1453 LO-PRESS.
GB-2125 WASH. SWITCH
GB-1416-1 FUSE 4 A.
GB-1364 AIR COOLED ONLY
START CAP.
RUN CAP.
+NOTE: INTERNAL ON AH7480-A
COMP. TERM.
*MARKED TERMINAL
GB-3135 CONTROL MODULE BOX
GB-3162 PUMP SWITCH
GB-3153 ACT. MOTOR
GB-3244 C/R-C ASSEM.
GB-897 ACT. TOGGLE
GB-3135-6 FUSE 4 A.
GB-3135-2-E P.C. CARD
STAT. IND. SOCKET
REL. 1
BIN ACT. C.W.
GB-3170 WATER LEVEL PROBE
GB-3176 EVAP. PROBE
GB-3177 BIN PROBE
PICTORIAL
"H" TERMINAL USED ONLY FOR HARVEST SWITCH WITH STATUS INDICATOR
J 12-17-84
H 10-8-84
K 11-4-87
BRISTOL MINB-922-BBCB
TECUMSEH AH7480-A
COPELAND SSE4-0100-CAB
ICE MAKER WIRING DIAGRAM
GB503 (GB1), GT503 (GT1W)
DATE: 10-8-84
UNIFLOW MFG. CO. ERIE, PA.
GBB-03131-K

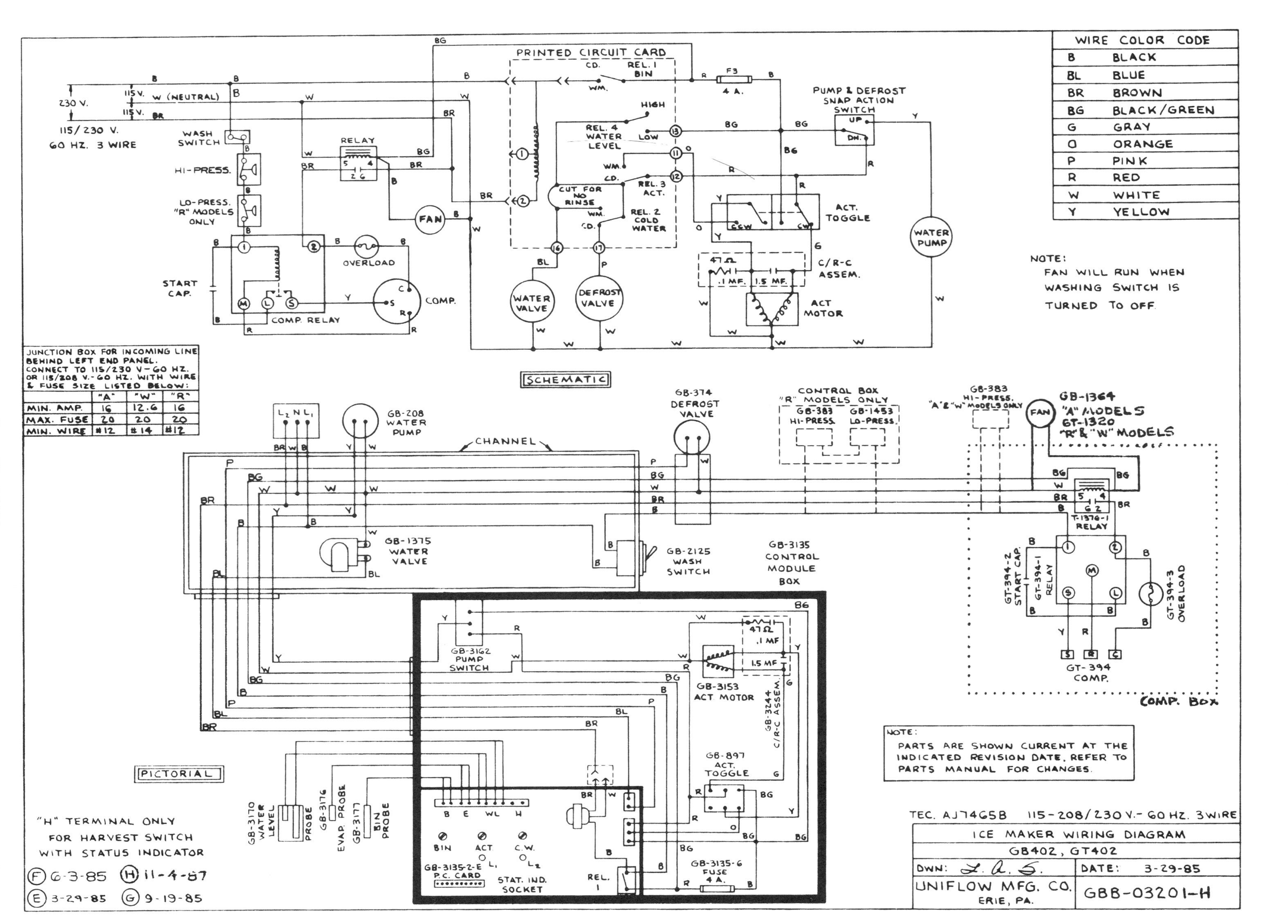
WIRE COLOR CODE
B BLACK
BL BLUE
BR BROWN
BG BLACK/GREEN
G GRAY
O ORANGE
P PINK
R RED
W WHITE
Y YELLOW
PRINTED CIRCUIT CARD
REL. 1 BIN
REL. 4 WATER LEVEL
REL. 3 ACT.
REL. 2 COLD WATER
CUT FOR NO RINSE
HIGH
LOW
230 V.
115 V. W (NEUTRAL)
115 V.
115/230 V.
60 HZ. 3 WIRE
WASH SWITCH
HI-PRESS.
LO-PRESS. "R" MODELS ONLY
START CAP.
COMP. RELAY
OVERLOAD
COMP.
RELAY
FAN
WATER VALVE
DEFROST VALVE
F3
4 A.
PUMP & DEFROST SNAP ACTION SWITCH
ACT. TOGGLE
C/R-C ASSEM.
ACT MOTOR
WATER PUMP
NOTE: FAN WILL RUN WHEN WASHING SWITCH IS TURNED TO OFF.
JUNCTION BOX FOR INCOMING LINE BEHIND LEFT END PANEL. CONNECT TO 115/230 V - 60 HZ. OR 115/208 V. - 60 HZ. WITH WIRE & FUSE SIZE LISTED BELOW:
"A" "W" "R"
MIN. AMP. 16 12.6 16
MAX. FUSE 20 20 20
MIN. WIRE #12 #14 #12
SCHEMATIC
GB-208 WATER PUMP
CHANNEL
GB-374 DEFROST VALVE
CONTROL BOX "R" MODELS ONLY
GB-383 HI-PRESS.
GB-1453 LO-PRESS.
GB-383 HI-PRESS. "A" & "W" MODELS ONLY
GB-1364 "A" MODELS
GT-1320 "R" & "W" MODELS
T-1376-1 RELAY
GT-394-2 START CAP.
GT-394-1 RELAY
GT-394-3 OVERLOAD
GT-394 COMP.
COMP. BOX
GB-1375 WATER VALVE
GB-2125 WASH SWITCH
GB-3135 CONTROL MODULE BOX
GB-3162 PUMP SWITCH
GB-3153 ACT MOTOR
GB-3244 C/R-C ASSEM.
GB-897 ACT. TOGGLE
GB-3135-6 FUSE 4 A.
GB-3135-2-E P.C. CARD
STAT. IND. SOCKET
REL. 1
BIN ACT C.W.
GB-3170 WATER LEVEL PROBE
GB-3176 EVAP. PROBE
GB-3177 BIN PROBE
PICTORIAL
NOTE: PARTS ARE SHOWN CURRENT AT THE INDICATED REVISION DATE. REFER TO PARTS MANUAL FOR CHANGES.
"H" TERMINAL ONLY FOR HARVEST SWITCH WITH STATUS INDICATOR
F 6-3-85 H 11-4-87
E 3-29-85 G 9-19-85
TEC. AJ7465B 115-208/230 V.- 60 HZ. 3 WIRE
ICE MAKER WIRING DIAGRAM
GB402, GT402
DWN: L. A. S.
DATE: 3-29-85
UNIFLOW MFG. CO.
ERIE, PA.
GBB-03201-H

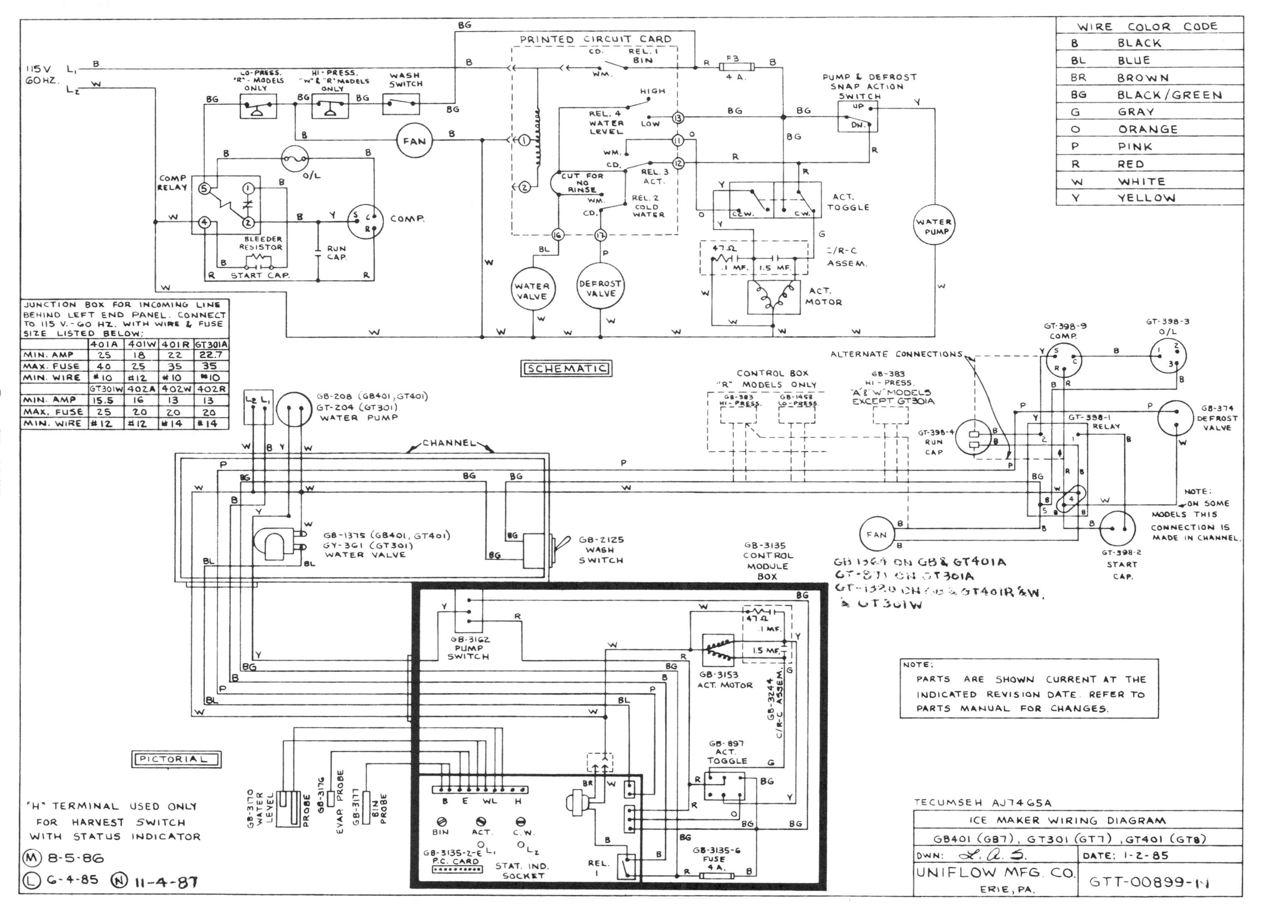

JUNCTION BOX FOR INCOMING LINE BEHIND LEFT END PANEL. CONNECT TO 115 V.-60 HZ. WITH WIRE & FUSE SIZE LISTED BELOW:

	401A	401W	401R	GT301A
MIN. AMP	25	18	22	22.7
MAX. FUSE	40	25	35	35
MIN. WIRE	#10	#12	#10	#10
	GT301W	402A	402W	402R
MIN. AMP	15.5	16	13	13
MAX. FUSE	25	20	20	20
MIN. WIRE	#12	#12	#14	#14

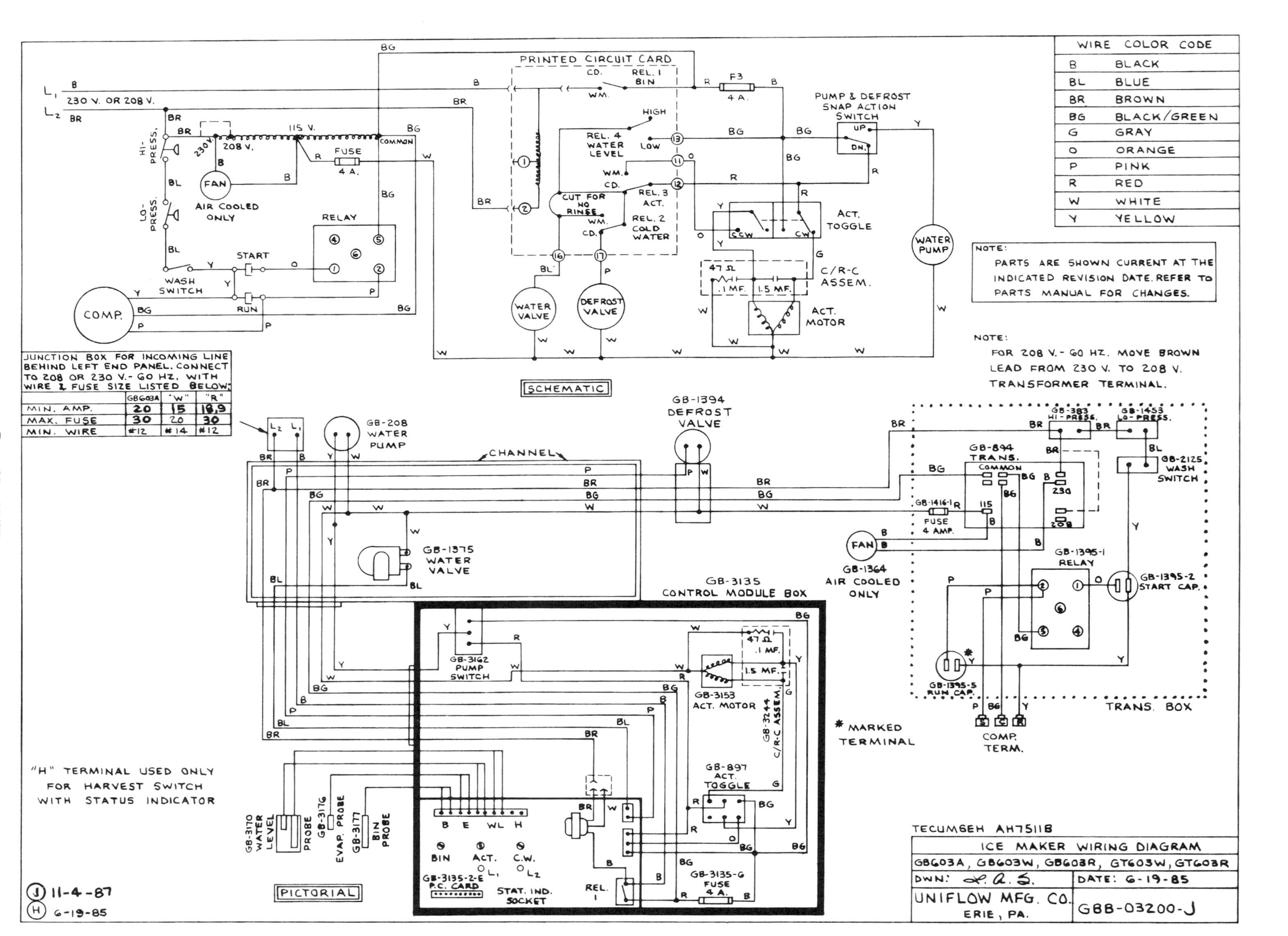

	GB603A	"W"	"R"
MIN. AMP.	20	15	18.9
MAX. FUSE	30	20	30
MIN. WIRE	#12	#14	#12

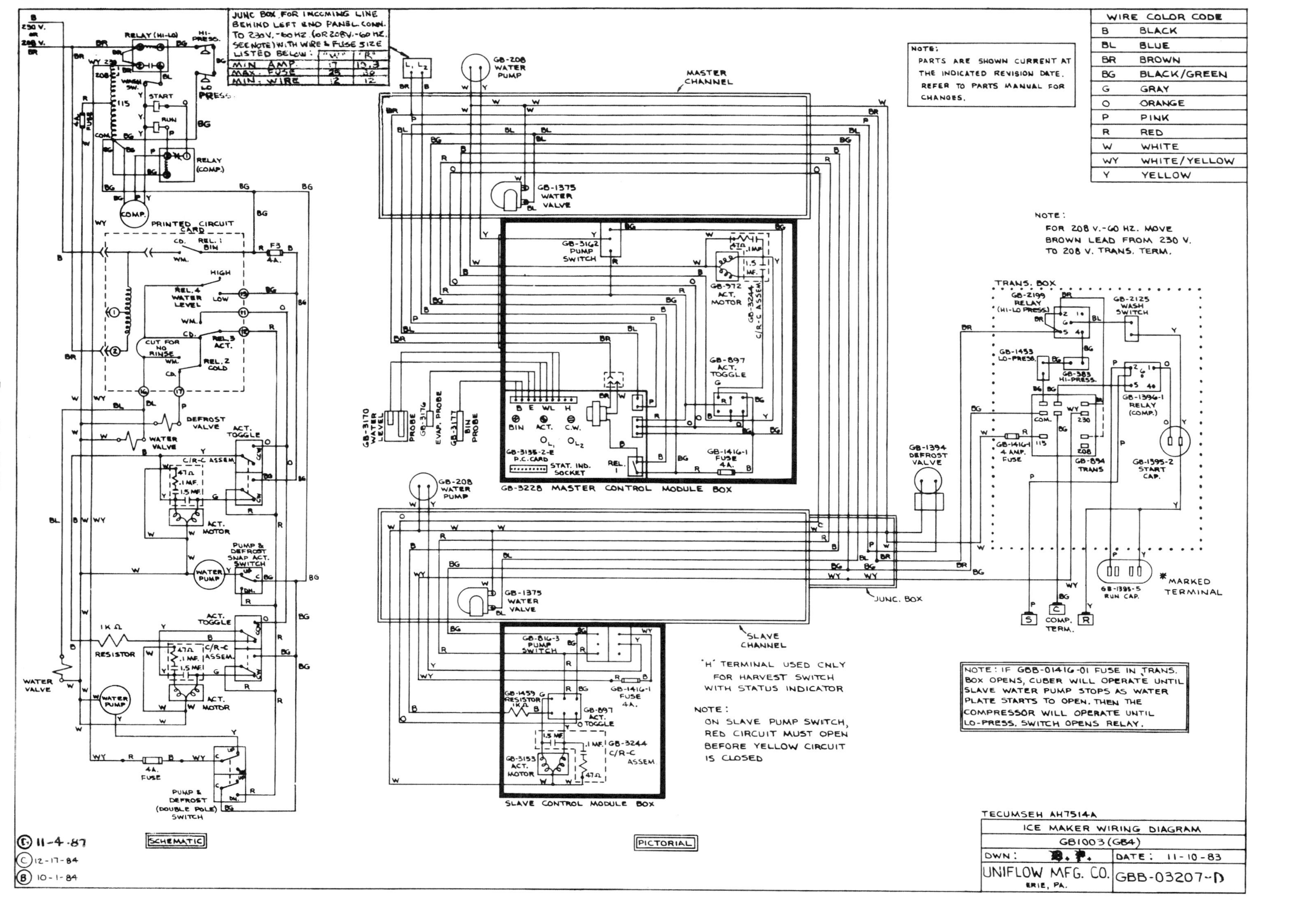
JUNC BOX FOR INCOMING LINE BEHIND LEFT END PANEL CONN. TO 230V.-60 HZ. (OR 208V.-60 HZ. SEE NOTE) WITH WIRE & FUSE SIZE LISTED BELOW:
	"W"	"R"
MIN. AMP.	17	12.3
MAX. FUSE	25	30
MIN. WIRE	12	12
NOTE: PARTS ARE SHOWN CURRENT AT THE INDICATED REVISION DATE. REFER TO PARTS MANUAL FOR CHANGES.
WIRE COLOR CODE
B BLACK
BL BLUE
BR BROWN
BG BLACK/GREEN
G GRAY
O ORANGE
P PINK
R RED
W WHITE
WY WHITE/YELLOW
Y YELLOW
NOTE: FOR 208 V.-60 HZ. MOVE BROWN LEAD FROM 230 V. TO 208 V. TRANS. TERM.
MASTER CHANNEL
GB-208 WATER PUMP
GB-1375 WATER VALVE
GB-3162 PUMP SWITCH
GB-972 ACT. MOTOR
GB-3244 C/R-C ASSEM
GB-897 ACT. TOGGLE
GB-1416-1 FUSE 4A.
GB-3135-2-E P.C. CARD
STAT. IND. SOCKET
GB-3228 MASTER CONTROL MODULE BOX
GB-3170 WATER LEVEL PROBE
GB-3176 EVAP. PROBE
GB-3177 BIN PROBE
TRANS. BOX
GB-2199 RELAY (HI-LO PRESS.)
GB-2125 WASH SWITCH
GB-1453 LO-PRESS.
GB-383 HI-PRESS.
GB-1396-1 RELAY (COMP.)
GB-1416-1 4 AMP. FUSE
GB-894 TRANS
GB-1395-2 START CAP.
GB-1394 DEFROST VALVE
GB-1395-5 RUN CAP.
* MARKED TERMINAL
COMP. TERM.
JUNC. BOX
SLAVE CHANNEL
GB-816-3 PUMP SWITCH
GB-1459 RESISTOR 1KΩ
GB-3153 ACT. MOTOR
SLAVE CONTROL MODULE BOX
'H' TERMINAL USED ONLY FOR HARVEST SWITCH WITH STATUS INDICATOR
NOTE: ON SLAVE PUMP SWITCH, RED CIRCUIT MUST OPEN BEFORE YELLOW CIRCUIT IS CLOSED
NOTE: IF GBB-01416-01 FUSE IN TRANS. BOX OPENS, CUBER WILL OPERATE UNTIL SLAVE WATER PUMP STOPS AS WATER PLATE STARTS TO OPEN. THEN THE COMPRESSOR WILL OPERATE UNTIL LO-PRESS. SWITCH OPENS RELAY.
RELAY (HI-LO)
HI-PRESS.
LO PRESS.
WASH SW.
START
RUN
RELAY (COMP.)
COMP.
PRINTED CIRCUIT CARD
REL. 1 BIN
REL. 4 WATER LEVEL
CUT FOR NO RINSE
REL. 3 ACT.
REL. 2 COLD
DEFROST VALVE
WATER VALVE
ACT. TOGGLE
C/R-C ASSEM.
ACT. MOTOR
PUMP & DEFROST SNAP ACT. SWITCH
WATER PUMP
RESISTOR
PUMP & DEFROST (DOUBLE POLE) SWITCH
SCHEMATIC
PICTORIAL
(D) 11-4-87
(C) 12-17-84
(B) 10-1-84
TECUMSEH AH7514A
ICE MAKER WIRING DIAGRAM
GB1003 (GB4)
DATE: 11-10-83
UNIFLOW MFG. CO. ERIE, PA.
GBB-03207-D

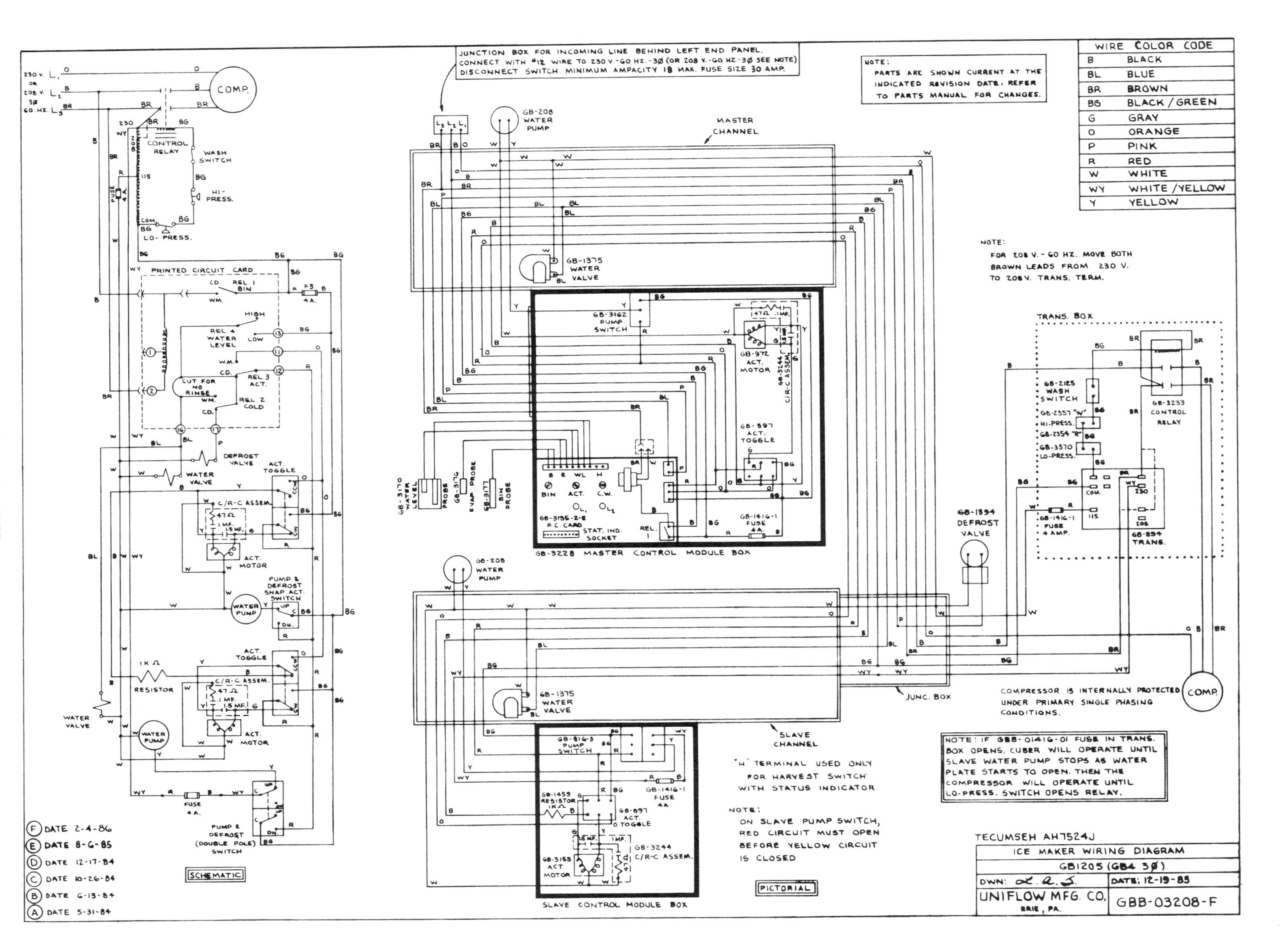

WIRE COLOR CODE
B BLACK
BL BLUE
BR BROWN
BG BLACK/GREEN
G GRAY
O ORANGE
P PINK
R RED
W WHITE
WY WHITE/YELLOW
Y YELLOW
JUNCTION BOX FOR INCOMING LINE BEHIND LEFT END PANEL. CONNECT WITH #12 WIRE TO 230 V.-60 HZ.-3Ø (OR 208 V.-60 HZ.-3Ø SEE NOTE) DISCONNECT SWITCH MINIMUM AMPACITY 18 MAX. FUSE SIZE 30 AMP.
NOTE: PARTS ARE SHOWN CURRENT AT THE INDICATED REVISION DATE, REFER TO PARTS MANUAL FOR CHANGES.
230 V. OR 208 V. 3Ø 60 HZ. L1 L2 L3
COMP.
CONTROL RELAY
WASH SWITCH
HI-PRESS.
LO-PRESS.
FUSE 4A.
PRINTED CIRCUIT CARD
REL. 1 BIN
HIGH
LOW
REL 4 WATER LEVEL
REL. 3 ACT.
REL. 2 COLD
CUT FOR NO RINSE
DEFROST VALVE
WATER VALVE
ACT. TOGGLE
C/R-C ASSEM.
ACT. MOTOR
PUMP & DEFROST SNAP ACT. SWITCH
WATER PUMP
1K Ω RESISTOR
PUMP & DEFROST (DOUBLE POLE) SWITCH
SCHEMATIC
GB-208 WATER PUMP
MASTER CHANNEL
GB-1375 WATER VALVE
GB-3162 PUMP SWITCH
GB-972 ACT. MOTOR
GB-3244 C/R-C ASSEM.
GB-897 ACT. TOGGLE
GB-1416-1 FUSE 4A.
GB-3170 WATER LEVEL PROBE
GB-3176 EVAP PROBE
GB-3177 BIN PROBE
BIN ACT. C.W.
GB-3155-2-E P.C. CARD
STAT. IND. SOCKET
REL. 1
GB-3228 MASTER CONTROL MODULE BOX
GB-1394 DEFROST VALVE
NOTE: FOR 208 V.-60 HZ. MOVE BOTH BROWN LEADS FROM 230 V. TO 208 V. TRANS. TERM.
TRANS. BOX
GB-2125 WASH SWITCH
GB-2357 "W" HI-PRESS.
GB-2354 "R"
GB-3370 LO-PRESS.
GB-3233 CONTROL RELAY
GB-1416-1 FUSE 4 AMP.
GB-894 TRANS.
JUNC. BOX
COMPRESSOR IS INTERNALLY PROTECTED UNDER PRIMARY SINGLE PHASING CONDITIONS.
SLAVE CHANNEL
GB-816-3 PUMP SWITCH
GB-1459 RESISTOR 1KΩ
GB-897 ACT. TOGGLE
GB-1416-1 FUSE 4A.
GB-3244 C/R-C ASSEM.
GB-3155 ACT. MOTOR
SLAVE CONTROL MODULE BOX
"H" TERMINAL USED ONLY FOR HARVEST SWITCH WITH STATUS INDICATOR
NOTE: ON SLAVE PUMP SWITCH, RED CIRCUIT MUST OPEN BEFORE YELLOW CIRCUIT IS CLOSED
PICTORIAL
NOTE: IF GBB-01416-01 FUSE IN TRANS. BOX OPENS, CUBER WILL OPERATE UNTIL SLAVE WATER PUMP STOPS AS WATER PLATE STARTS TO OPEN. THEN THE COMPRESSOR WILL OPERATE UNTIL LO-PRESS. SWITCH OPENS RELAY.
F DATE 2-4-86
E DATE 8-6-85
D DATE 12-17-84
C DATE 10-26-84
B DATE 6-13-84
A DATE 5-31-84
TECUMSEH AH7524J
ICE MAKER WIRING DIAGRAM
GB1205 (GB4 3Ø)
DATE: 12-19-83
UNIFLOW MFG. CO. ERIE, PA.
GBB-03208-F

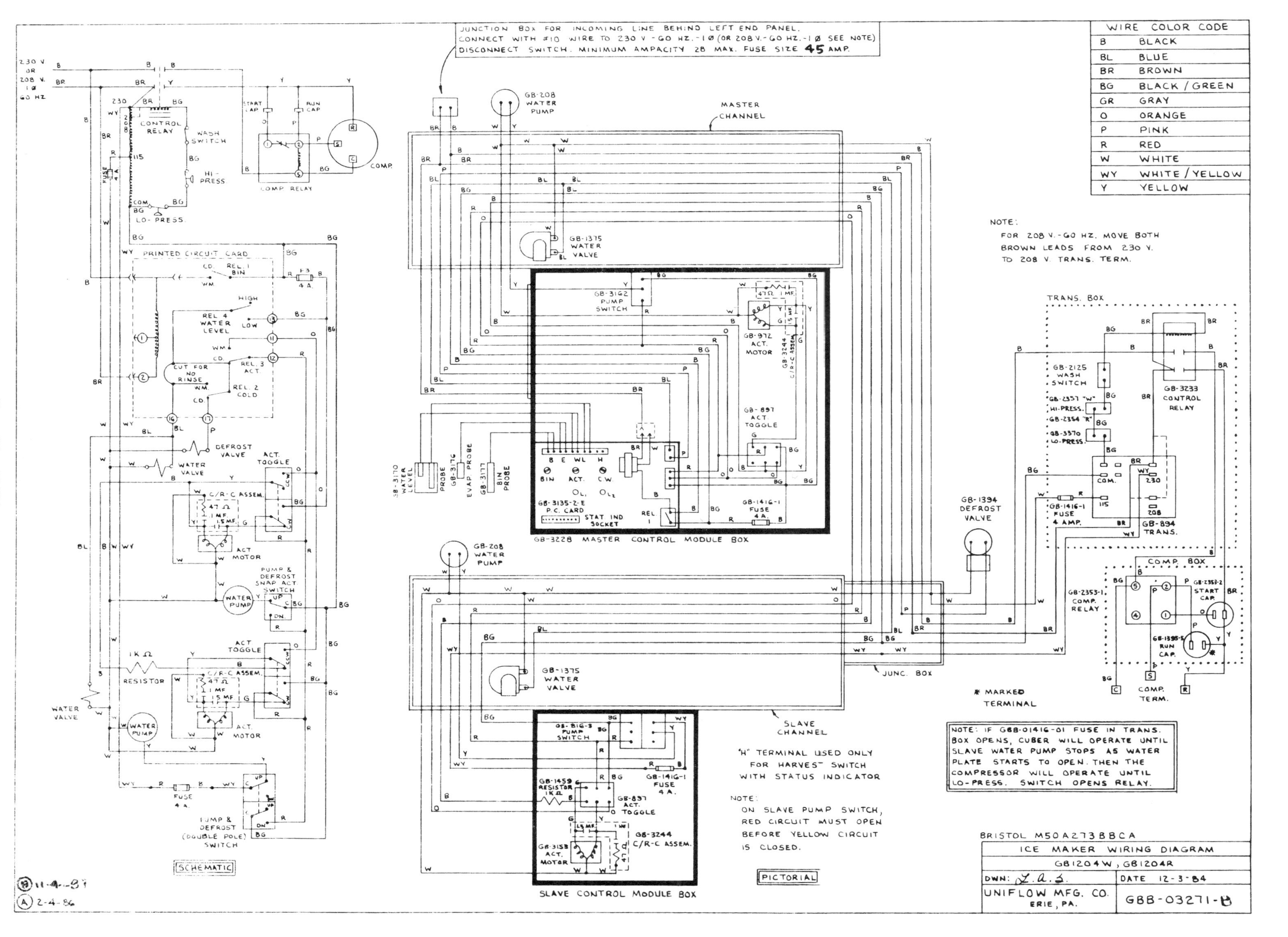
JUNCTION BOX FOR INCOMING LINE BEHIND LEFT END PANEL.
CONNECT WITH #10 WIRE TO 230 V - 60 HZ. -1Ø (OR 208 V.- 60 HZ. -1Ø SEE NOTE)
DISCONNECT SWITCH. MINIMUM AMPACITY 28 MAX. FUSE SIZE 45 AMP.
WIRE COLOR CODE
B BLACK
BL BLUE
BR BROWN
BG BLACK / GREEN
GR GRAY
O ORANGE
P PINK
R RED
W WHITE
WY WHITE / YELLOW
Y YELLOW
NOTE: FOR 208 V.-60 HZ. MOVE BOTH BROWN LEADS FROM 230 V. TO 208 V. TRANS. TERM.
230 V OR 208 V. 1Ø 60 HZ
CONTROL RELAY
WASH SWITCH
HI-PRESS.
LO-PRESS.
START CAP
RUN CAP
COMP. RELAY
COMP.
PRINTED CIRCUIT CARD
REL. 1 BIN
REL. 4 WATER LEVEL
REL. 3 ACT.
REL. 2 COLD
CUT FOR NO RINSE
DEFROST VALVE
WATER VALVE
ACT. TOGGLE
C/R-C ASSEM.
ACT. MOTOR
PUMP & DEFROST SNAP ACT SWITCH
WATER PUMP
RESISTOR
PUMP & DEFROST (DOUBLE POLE) SWITCH
SCHEMATIC
MASTER CHANNEL
GB-208 WATER PUMP
GB-1375 WATER VALVE
GB-3162 PUMP SWITCH
GB-972 ACT. MOTOR
GB-3244 C/R-C ASSEM.
GB-897 ACT. TOGGLE
GB-3135-2-E P.C. CARD
STAT IND SOCKET
GB-1416-1 FUSE 4 A.
GB-3170 WATER LEVEL PROBE
GB-3176 EVAP PROBE
GB-3177 BIN PROBE
GB-3228 MASTER CONTROL MODULE BOX
TRANS. BOX
GB-2125 WASH SWITCH
GB-2357 "W" HI-PRESS.
GB-2354 "R"
GB-3370 LO-PRESS.
GB-3233 CONTROL RELAY
GB-1416-1 FUSE 4 AMP.
GB-894 TRANS.
GB-1394 DEFROST VALVE
COMP. BOX
GB-2353-1 COMP. RELAY
GB-2353-2 START CAP.
GB-1395-5 RUN CAP.
COMP. TERM.
* MARKED TERMINAL
JUNC. BOX
SLAVE CHANNEL
GB-816-3 PUMP SWITCH
GB-1459 RESISTOR 1K Ω
GB-897 ACT. TOGGLE
GB-3158 ACT. MOTOR
GB-3244 C/R-C ASSEM.
SLAVE CONTROL MODULE BOX
"H" TERMINAL USED ONLY FOR HARVEST SWITCH WITH STATUS INDICATOR
NOTE: ON SLAVE PUMP SWITCH, RED CIRCUIT MUST OPEN BEFORE YELLOW CIRCUIT IS CLOSED.
PICTORIAL
NOTE: IF GBB-01416-01 FUSE IN TRANS. BOX OPENS, CUBER WILL OPERATE UNTIL SLAVE WATER PUMP STOPS AS WATER PLATE STARTS TO OPEN. THEN THE COMPRESSOR WILL OPERATE UNTIL LO-PRESS. SWITCH OPENS RELAY.
BRISTOL M50A273BBCA
ICE MAKER WIRING DIAGRAM
GB1204W, GB1204R
DATE 12-3-84
UNIFLOW MFG. CO. ERIE, PA.
GBB-03271-B

ELECTRONIC ICE CUBER COMPRESSOR CHART
60 HERTZ ONLY

H.P.	MODEL	PART #	NAME	SUPPLIER	SUPPLIER #	ALT.	ELECTRICAL RATING
¾	GT301, GT7-E	GT398	Compressor	Tecumseh	AJ7465A	—	115 Volt - 60 Hertz
	GT401, GT8-E	GT398-1	Relay	G.E.	3ARR3A2J6	820ARRB92	—
	GB401, GB7-E	GT398-3	Overload	Tecumseh	8300CSTA96	CST-17AEN-181	—
	IS401	GT398-2	Start Cap.	G.E.	3628730	85PS110009	270-324 Mfd. 110 Volt
	IS7-E	GT398-4	Run Cap.	G.E.	97F5050	85PR370E31	15 Mfd. 370 Volt
¾	GB402	GT394	Compressor	Tecumseh	AJ7465B	—	230/208 Volt-60 Hertz
	GB7D-E	GT394-1	Relay	Tecumseh	82481-1	3ARR2KCR1725	—
	GT402	GT394-2	Start Cap.	Tecumseh	85PS330A20	35F0888BC9	88-108 Mfd. 330 Volt
		GT394-3	Overload	Tecumseh	83512	MST20A1N-112	—
1	GB503, GB1-E	GB1347	Compressor	Tecumseh	AH7480A	—	230/208 Volt-60 Hertz
	GT503, GT1-E	GB1347-1	Relay	G.E.	3ARR3K13B5	82765	—
	IS500	GB1347-3	Start Cap.	Mepco	35F1622BC2	85577-1	135-155 Mfd. 330 Volt
	IS1-E	GT398-4	Run Cap.	G.E.	97F5050	85561-2	15 Mfd. 370 Volt
1	GB503	GB2350	Compressor	Bristol	MINB-922-BBCB	—	230/208 Volt-60 Hertz
	GT503	GB1385-1	Relay	G.E.	3ARR3CT20V5	04-0001-19	—
	IS500	GB-1347-3	Start Cap.	Tecumseh	84477-1	35F1622BC2	135-155 Mfd. 330 Volt
	IS1-E	GB-1385-3	Run Cap.	Jard	SPB3720-A14	014-0002-02	20 Mfd. 370 Volt
11000 BTU	GB603, GB2-E	GB1395	Compressor	Tecumseh	AH7511B	—	230 Volt - 60 Hertz
	GT603	GB1395-1	Relay	G.E.	3ARR3K13P5	82457	—
		GB-1395-2	Start Cap.	G.E.	35F1558BC2	85668	88-106 Mfd. 250 Volt
		GB-1395-5	Run Cap.	Jard	SPB4435-A14	85PR370857	35 Mfd. 440 Volts
14000 BTU	GB1003	GB1396	Compressor	Tecumseh	AH7514A	—	230/208 Volt-60 Hertz
	GB4-E	GB1396-1	Relay	G.E.	3ARR3A13P2	B2477	—
		GB1395-2	Start Cap.	G.E.	35F1558BC2	85668	88-106 Mfd. 250 Volt
		GB1395-5	Run Cap.	Jard	SPB4435-A14	85PR370B57	35 Mfd. 440 Volts
22500 BTU	GB1205	GB2358	Compressor	Tecumseh	AH7524J	—	230/208 Volt-60 Hertz-3 Phase
		GB3233	Control Relay	G.E.	3ARR8AAR-DPST	—	240 Volts
27600 BTU	GB1204	GB02353	Compressor	Bristol	M50A273BBCA	—	230/208 Volt-60 Hertz-1 Phase
		2353-01	Start Relay	Bristol	650409	3ARR3W4A3	230 Volt - 60 Hertz
		2353-02	Start Cap.	Bristol	650573	—	189-227 Mfd. 330 Volt
		1395-05	Run Cap.	Jard	SPB4435-A14	85PR370B57	35 Mfd. 440 Volt

KEY TO ICE MAKER MODEL NUMBER

GB 1 A N 4 K E

ELECTRONIC

CUBE SIZE

C = Full-Cube (1¼" x 1¼" x 1¼")
K = Cubelet (⅝" x ⅝" x 1¼")
HK = Half-Cube (1¼" x 1¼" x ⅝")
KK = Half-Cube (1¼" x 1¼" x ⅝")

ICE MAKING PRODUCTION PER 24 HOURS (NOMINAL)

64 = 75 lbs. w/40 lb. storage bin (self-contained)
1 = 110 lbs. w/65 lb. storage bin (self-contained)
2 = 200 lbs.
3 = 300 lbs.
4 = 400 lbs.
5 = 500 lbs.
6 = 600 lbs.
7 = 700 lbs.
8 = 800 lbs.
9 = 900 lbs.
10 = 1000 lbs.
11 = 1100 lbs.
12 = 1200 lbs.
16 = 1600 lbs.
20 = 2000 lbs.

CODE LETTER

N-NSF = (National Sanitation Foundation) Listed
D = 115/230 Volt, 60 Hz, 3-Wire, NSF Listed
F = 230 Volt, 50 Hz, NSF Listed

COOLING STYLE

A = Air-cooled condenser
W = Water-cooled condenser
R = Remote air-cooled condenser w/Self-sealing condenser fittings, Precharged
X = Remote air-cooled condenser

COMPRESSOR HORSEPOWER RATING

0 = No condensing unit (vendors only)
1 = 1 HP
2 = 11,000 BTU
3 = ⅓ HP
4 = 14,000 BTU
5 = ½ HP
6 = ⅕ HP
7 = ¾ HP
8 = ¾ HP

MACHINE TYPE

GB = Horizontal unit, mounts on bin (42" wide)
GT = Slim-line unit, mounts on bin (28½" wide)
GY = Self-contained compact unit (30" wide)
GS = Self-contained compact unit (24" wide)
IS = Self-contained dispenser (31-¾" wide)

CHART OF WATER LEVELS, PRESSURES, CYCLES AND CHARGE

(Higher Than Average Temperatures Increase Pressures and Cycle Times)

Model Number Thermostatic Control	GB4				GB2	GB1, GT1		ISI MD1		GB5 MD5		IS5		IS7 GB7		GT8		GT7				GY3				GS6
	C	HK	K	KK	HK	C	HK	K	KK	C	HK	K	KK	C	HK	K	KK	C	HK	K	KK	C	HK	K	KK	HK
Full Water Level Distance Below Top of Tank	2⅝"	2¾"	3½"		2¾"	2⅝"	2¾"	3½"		2⅝"	2¾"	2½"		2½"	2¾"	3½"		2"		3"		2"		3"		1⅝"
Tank Insert or Hanger Extension Loop	Insert		Loop		Insert	Insert		Loop		Insert		Loop		Insert		Loop		Insert		Loop		Insert		Loop		
Approximate Trip-Up Level Above Tank Bottom	½" For Thermostatic Cubers																									
Suction Pressure After Defrost	20 psig.		20 psig.		20 psig.	15, 20 psig.		15, 20 psig.		30 psig.	30 psig.	30 psig.	30 psig.	20, 25 psig.		20, 25 psig.		11, 13 psig.		11, 13 psig.		27 psig.		27 psig.		25 psig.
Suction Pressure Before Defrost	0-2 psig.		0-2 psig.		0-2 psig.	0	0	0		7-3 psig.	9-5 psig.	7-3 psig.	9-5 psig.	3 psig.		3 psig.		0		0		6, 3 psig.		7, 4 psig.		8-4 psig.
Defrost Press	40-60 psig.		40-60 psig.		40-60 psig.	40-60 psig.		40-60 psig.		40-60 psig.	40-60 psig.	40-60 psig.	40-60 psig.	40-60 psig.		40-60 psig.		40 psig.		40 psig.		40-60 psig.		40-60 psig.		40-60 psig.
Cycle Time Approximate	32 min.	28 min.	19 min.	23 min.	21 min.	28 min.	26 min.	15 min.	18 min.	44 min.	40 min.	27 min.	32 min.	40 min.	33 min	20 min.	23 min.	30 min.	26 min.	15 min.	18 min.	61 min.	58 min.	32 min.	38 min.	48 min.
Refrigerant Charge Remote: See note below Type	3 lb. R-12		3 lb. R-12		3 lb. R-12	3 lb. R-12		3 lb. R-12		3 lb. R-12	3 lb. R-12	3 lb. R-12		3 lb. R-12		3 lb. R-12		13 oz. GT7A 10 oz. GT7W R-12		13 oz. GT7A 10 oz. GT7W R-12		15½ oz. GY3A 10 oz. GY3W R-12		15½ oz. GY3A 10 oz. GY3W R-12		10 oz. GS6A 10 oz. GS6W R-12
Approximate Lb. Ice Per Batch	15 lb.	14 lb.	8 lb.	9½ lb.	7 lb.	7½ lb.	7 lb.	4 lb.	4¾ lb.	7½ lb.	7 lb.	4 lb.	4¾ lb.	7½ lb.	7 lb.	4 lb.	4¾ lb.	3¾ lb.	3½ lb.	2 lb.	2⅜ lb.	3¼ lb.	3½ lb.	2 lb.	2⅜ lb.	2 1/3 lb.
Compressor Size	14,000 BTU				11,000 BTU	1 Hp.				1/2 Hp.				3/4 Hp.				3/4 Hp.				1/3 Hp.				1/5 Hp.

NOTE: REMOTE CONDENSER APPLICATION CUBERS REQUIRE MINIMUM CHARGE OF 10 1/2 LBS.

CONSTRUCTION OF THERMOSTAT CONTROLLED CUBERS

The construction of thermostat controlled cubers is very similar to the electric cubers but differs in a few controls which are described below:

REFRIGERANT CONTROL

1. Thermostatic expansion valve was used on GB, IS, GT, and GT8 models. For replacement of the valve consult the parts price list for the correct expansion valve.
2. GT, GY, GS models are capillary tube systems and do not have an expansion valve.

CONTROL TANK AND SWITCH

The control tank is mounted on a weight switch and is connected to the main water circulation tank by means of flexible tubing. The height of water in the control tank indicates the height of water in the circulation tank and controls the switch action.

ACTUATOR MOTOR AND CAM ASSEMBLY

The assembly is located just to the right and front of the evaporator. The actuator motor drives the cam shaft directly and is reversible. The cam on each end of the shaft forces the water plate down, separating the water plate from the ice in the evaporator. The two springs on the cams pull the water plate up at the end of the defrost cycle, and hold the water plate against the bottom of the cams during the freeze cycle. To prevent the water plate from opening after current has stopped flowing to the actuator motor, a drift stop spring with a plastic end (No. GB-965) presses against the actuator motor shaft, on the front of the motor. Prior to mid 1972, drift stop No. GB-946 was used. If the drift stop spring is not aligned with the motor shaft it can be removed easily and bent into shape.

ACTUATOR THERMOSTAT

The actuator motor is controlled by a thermostatic switch in the front channel to the left of the actuator motor, with its control bulb in a tube on the evaporator, and the toggle switch mounted in the channel and operated by the front cam.

This control must be set to the cold side when the weight tank trips up after running out of water for the actuator motor to receive power to rotate the plate to the full open position.

ACTUATOR THERMOSTAT ADJUSTMENT

After cubes fall out, the cams should start rotating clockwise to close the water plate within 10 seconds to 30 seconds. If water plate stays down too long, adjust actuator thermostat slightly clockwise until cams start to turn. If water plate starts to close before ice falls out, turn actuator thermostat all the way counter-clockwise to stop cams, then after ice falls out, turn slowly clockwise until cams start to close water plate again. If the actuator thermostat is adjusted, it is advisable to check it again after another cycle, if possible.

COLD WATER THERMOSTAT

This is mounted in the evaporator support channel. The control bulb is inserted into a tube on the evaporator, with the actuator thermostat control bulb. Its function is to prevent the water plate from dropping prematurely due to cold inlet water which could cause resetting of the actuator thermostat before the weight control tank drops. When incoming water is below 50 degrees and circulating water drops to 40 degrees the action of the cold water thermostat opens the defrost valve, allowing hot gas to warm up the evaporator, thus warming the water, and preventing resetting of the actuator thermostat until the "Water Fill" cycle is completed. If the circulating water warms to 50 degrees the cold water thermostat cuts off current to the defrost valve; when the "Water Fill" cycle is completed, the weight control tank drops, the current to the cold water thermostat is cut off and the defrost valve closes.

The Cold Water Control when set to the cold side at harvest gives us our first path for power to energize the defrost valve at harvest.

BIN THERMOSTAT

This is mounted in the channel to the left of the control tank switch. Its function is to turn off the entire cuber when the bin fills up with ice. The capillary tube controlling this switch is mounted in the bin near ice drop zone.

Bin Thermostat Setting - Hold two or more ice cubes against bin thermostat bulb holder if used. The whole cuber should shut off in a minute or two. If it does not turn the thermostat slowly counter-clockwise.

After cuber stop, take the cube away from the bulb. The cuber should start in a minute or two unless ambient temperature is quite low.

CAUTION: Do not make the above check or shut off cuber when it is filling with water and the control tank is nearly full to the proper level. If the cuber is shut off then, the circulating pump will stop, allowing the water in the distribution tubes to drain back to the circulation tank and control tank and give a false shut-off on the control switch. The next freeze will then have insufficient water to fill the cubes, although subsequent cycle will be normal.

BIN PROBE CONTROLS

Uniflow stocks two bin controls with capillary tube lengths of 66″ and 120″.

1st Machine GB-856	66″ bin control	w/crusher GB-856	66″ bin control
2nd Machine GB-856	66″ bin control	w/crusher GB-813	120″ bin control
3rd Machine GB-813	120″ bin control	w/crusher GB-813	120″ bin control
4th Machine GB-813	120″ bin control		
5th Machine GB-813	120″ bin control		
6th Machine Gb-813	120″ bin control		

MULTIPLEX UNITS	MULTIPLEX UNITS
Special Bin Control GB-813 120″ Capillary	GB-813 120″ Capillary
Special Bin Control GB-813 120″ Capillary	Special Bin Control GB-813 120″ Capillary
Special Bin Control GB-813 120″ Capillary	Standard Bin Control GB-856 66″ Capillary
Standard Bin Control GB-813 120″ Capillary	T-100 Crusher
Standard Bin Control GB-856 66″ Capillary	Any Kold-Draft Bin
Standard Bin Control GB-856 66″ Capillary	
Any Kold-Draft Bin	

SEQUENCE OF OPERATION

After the cuber is installed, the water turned on, and the electricity connected, the water flows through the water valve, the water solenoid valve being open, and the water plate in the "up" position. The defrost and pump toggle switch is held up by the lift rod on the water plate causing the pump to operate and the defrost valve to be closed. Water will gradually come into the control tank and, as the water level rises in the control tank, the tank starts downward. When a sufficient amount of water is in the circulation tank and the control tank, the weight of the water snaps the switch downward, breaking the circuit to the water valve, thus closing the valve. The water is being pumped from the bottom of the circulation tank into the water header in the left of the water plate, through the laterals and through the small openings in the laterals, in a jet stream up in the center of each evaporator cell. As the evaporator cools, the water is cooled and in about 5 or 10 minutes the water will be sufficiently cold, so that it will form minute layers of ice on the inside of the evaporator cells. As this process continues, the water level will decrease in the circulation tank and control tank.

When the circulating water reaches a temperature of 32 degrees F., it MAY be subcooled and it MAY partially crystallize in the water tank. When and if this occurs, the flow of water in the control nozzle will stop or flunctuate considerably and most of the circulation will stop for about 30 seconds. This is STRICTLY A NORMAL OPERATION AND THE CONTROL STREAM SHOULD NOT BE ADJUSTED AT THAT TIME.

The control stream on the front left corner of the water plate, should operate so that it does not go over the dam and under normal freezing conditions. Should it be too low or too high, the screw at the inlet of the control stream may be adjusted until the stream is set for proper operation. (See Figure No. 1, Page 54). The water continues to freeze until each cell is almost completely full of ice. As the ice comes closer to the jet streams, in the center of each cell, the head pressure in the water circulation system increases to a point where the control stream will rise and flow over the dam (see Figure No. 2, Page 54) in the control stream box. This water will be dissipated to the drain pan and cause the entire level of the water to decrease, therefore, the **weight of the water in the control tank will decrease until it can be overcome by the spring in the control tank switch.**

When this occurs, the control tank switch snaps up completing the circuit through the actuator thermostat and through the actuator toggle switch to the actuator motor, causing the actuator shaft to turn counter-clockwise (looking from the front) and separate the water plate from the evaporator. (Also the water inlet and hot gas circuits are complete). The cams will be seen rotating down on the left of the actuator motor. The actuator shaft continues to turn for 1/2 revolution at which time the trip lever on the front cam snaps the toggle switch to the left and breaks the actuator motor circuit, stopping the actuator.

As the water plate was lowered to its harvesting position, the defrost and pump toggle switch dropped, causing the pump to stop. This toggle switch also completes a second circuit to the defrost valve to keep it open until the water plate closes all the way at the end of defrost. The machine is now in the defrost cycle and the hot gas from the compressor is passing through the evaporator coil, warming up the evaporator so that the cubes will drop down on the plate as a sheet. There is a small fin connecting the bottom of the cubes together so that they will come down in unison and clear the water plate without any difficulty and slide into the bin where they will break apart.

As long as ice is in the evaporator, it remains cool (32 degrees to 35 degrees), keeping the actuator thermostat bulb cool and keeping the cuber in defrost.

After the ice is out of the evaporator, the evaporator and the actuator thermostat bulb warm up rapidly. (40 degrees to 45 degrees) and the actuator thermostat snaps over, completing the circuit from the weight control tank switch to the "reversing" side of the actuator toggle switch and the actuator motor. (Note: The actuator thermostat does not start the defrost cycle; it only ends it after the ice falls out). The motor revolves clockwise (cams will be seen rotating up on the left of the actuator motor) until the water plate is in its full "up" position when the front cam will snap the actuator toggle over to the right, causing the actuator to stop. When the water plate rises to the "up" position, the lift plate and lift rod push up the defrost and pump switch, starting the pump to begin another freezing cycle. When the water level is high enough to snap down the control tank switch, it breaks the circuit to the water valve.

Should some ice cubes be left on the water plate keeping it partially open when the water plate comes up, the actuator motor will continue to operate and the springs will stretch to allow the cams to take their vertical position and snap the actuator toggle switch. Since the lift plate cannot push up the lift rod and the defrost switch, the circuit is complete through this switch to the right side of the actuator toggle switch. When the cam snaps the actuator toggle switch to the right, the actuator motor will immediately reverse itself and open the plate. The captive ice then generally falls off. When the plate is in the lowest position, the trip lever on the front cam will again reverse the actuator toggle switch and the actuator motor and cause the plate to close. This will continue until the water plate is clear and the lift plate can push up the lift rod and defrost switch, breaking the circuit through this switch to the right side of the actuator toggle switch, so that the motor will stop with cams up when the cam pushes the actuator toggle to the right.

The cuber has now completed its full cycle and started on another freezing cycle. This will be regularly repeated until the bin is full and the bin thermostat shuts off the cuber automatically. When some ice is removed from the bin, the cuber will start up and refill the bin.

For complete electrical sequence refer to electrical circuits in electronic section on page 14.

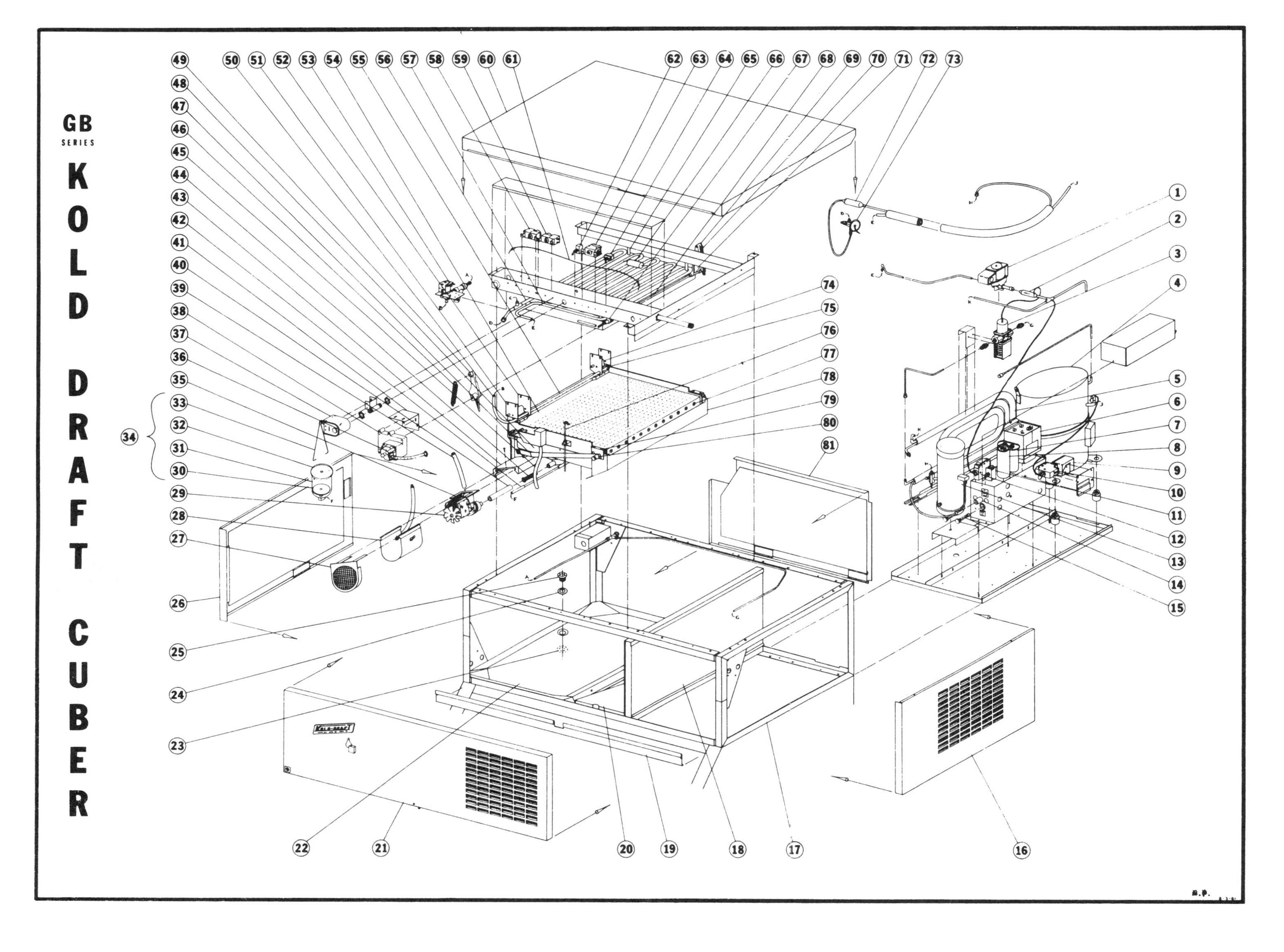
GB SERIES
KOLD DRAFT CUBER

No.	Part No.	Description
1.	GBB-00374-B	Defrost Solenoid Valve-Complete (Alco)
	GBB-00374-C	Defrost Solenoid Valve-Complete (Jackes-Evans)
2.	GBB-00375	Strainer-Defrost Solenoid Valve Inlet
3.	GAA-00701	Water Regulator Valve-Complete (Penn)
	GAA-00701-B	Water Regulator Valve-Complete (Singer)
4.	GBB-01347	Compressor Dome-Tecumseh AH7480-A
	GBB-02350	Compressor Dome-Bristol M1NB-922-BBCB
5.	GBB-00300-B	Water Cooled Condenser
6.	GBB-00894	Transformer
7.	GBB-00377-04	Run Capacitor for Tecumseh 15 MFD. 370 V.
	GBB-01385-03	Run Capacitor for Bristol 20 MFD. 370 V.
8.	GBB-01347-03	Start Capacitor for Tecumseh & Bristol
9.	GBB-01347-01	Relay for Tecumseh
	GBB-01385-01	Relay for Copeland
10.	GBB-01453	Low Pressure Cut-Out Switch
11.	GBB-02125	Washing Switch
12.	GBB-00383	High Pressure Cut-Out Switch
13.	GBB-00379	Receiver Tank (4″ Dia.)
14.	GBB-01416	Fuse Holder w/4 Amp. Fuse
15.	GBB-01416-01	Fuse Only 4 Amp.
16.	GBB-00655	Skin, Ice Maker Right Side-Painted
	GBB-00655-50	Skin, Ice Maker Right Side-Stainless
17.	GBB-00557	Frame, Ice Maker
18.	GBB-00509	Partition Assembly
19.	GBB-00613	Bottom Angle Cover Between Cuber & Bin-Painted
	GBB-00613-50	Bottom Angle Cover Between Cuber & Bin-Stainless
20.	GBB-00668	Ice Chute
21.	GBB-00621	Inspection Plate-Painted
	GBB-00621-50	Inspection Plate-Stainless
22.	GBB-00535	Evaporator Drain Pan
23.	GBB-01065	Drain Washer, Thick
24.	GBB-01061	Drain Washer, Thin
25.	GBB-01548	Drain Yoke-Cuber on Bin
26.	GBB-00652	Skin, Ice Maker Left Side-Painted
	GBB-00652-50	Skin, Ice Maker Left Side-Stainless
27.	GBB-00282-03	Screen Assembly
28.	GBB-00279	Pump Drip Shield
29.	GYY-00208	Fan Blade w/Hub for Water Pump
30.	GBB-00308	Control Tank Insert
31.	GAA-00338	Control Tank
32.	GAA-00305	Control Tank Cover
33.	GBB-01369	Hanger Wire
34.	GBB-01307	Control Tank Assembly (Tank, Cover, Insert, Wire & Loop)
35.	GBB-00208	Water Pump and Motor with Fan
36.	GBB-00965-A	Drift Stop
37.	GBB-00385-03	Penn Weight Control Kit
38.	GBB-00954	Square Actuator Motor and Front Cam Assembly
39.	GBB-00211	Surgical Hose-Pump Outlet (¾″ x 6″ Long)
40.	GBB-00212	Surgical Hose-Pump Inlet (1″ x 3½″ Long)
41.	GBB-00910	Actuator Motor Bracket
42.	GBB-00318	Control Tank Tube Assembly
43.	GBB-00282-04	Water Pump Bracket (Upper and Lower Half Only)
44.	GBB-00245	Screen-Tank Outlet
45.	GBB-00333	Vinyl Tube-Overflow (⅜″ x 8″ Long)
46.	GBB-00908	Spring to Raise Water Plate-Front
47.	GBB-00285	Water Pump Brace
48.	GAA-00203	Water Tank-Main
49.	GBB-00956	Front Cam Assembly
50.	GBB-00403-02	Surgical Hose-Singer Valve (⅜″ x 7″ Long)
51.	GBB-00520	Hinge Plate (Front of Front Channel-Rear of Rear Channel)
52.	GBB-00539	Hinge Plate (Rear of Front Channel-Front of Rear Channel)
53.	GBB-01375-02	Complete Hays Replacement Assembly
54.	GBB-00200	Water Plate (Large Cube & KK Half Cube)
	GBB-00270	Water Plate(Cube-Let & HK Half Cube)
55.	GBB-00403	Water Distribution Tube Assembly
56.	GBB-00148	Kit-Full Cube Evaporator
	GBB-00153	Kit-Cube-Let Evaporator
	GBB-00161	Kit-HK Half Cube Evaporator
57.	GBB-00113	Evaporator Spacer-Front (Large Cube & HK Half Cube)
	GBB-00126	Evaporator Spacer-Front (Cube-Let & KK Half Cube), Rear (Large Cube & HK Half Cube)
58.	GBB-00856	Bin Thermostat (66″ Cap. Tube)
	GBB-00813	Bin Thermostat (120″ Cap. Tube)
59.	GBB-00837	Cold Water Thermostat
60.	GBB-00622	Top, Ice Maker-Painted
	GBB-00622-50	Top, Ice Maker-Stainless
61.	GBB-01411	Channel Heater
62.	GBB-01498	Pump Toggle Switch (Snap-Action, Spring Loaded)
63.	GBB-00814	Actuator Thermostat
64.	GBB-00897	Actuator Toggle Assembly
65.	GBB-00127	Evaporator Spacer-Rear (Cube-Let & KK Half Cube)
66.	GBB-00826	Capacitor 1.2 MFD. for GBB-00954 Actuator Motor
67.	GBB-00937	Bearing Bracket-Cam Shaft (Front)
68.	GBB-00942	Cam Shaft Assembly
69.	GBB-00949	Rear Cam Assembly
70.	GBB-00909	Spring to Raise Water Plate-Rear (Blue)
71.	GBB-00933	Bearing Bracket-Cam Shaft (Rear)
72.	K64-00494	Filter-Drier
73.	GBB-00301-E	Expansion Valve-Sporlan
	GBB-00301-F	Expansion Valve-Singer
74.	GBB-00280	Rear Hinge
75.	GBB-00403-04	Plug-Distribution Tube
76.	GAA-00805	Lift Rod and Collar Assembly
77.	GBB-00223	Water Plate Plug
78.	GBB-00202	Water Deflector
79.	GBB-00235	Teflon Bracket
80.	GBB-00951	Water Plate Boss-Front & Rear
81.	GBB-00631	Skin, Ice Maker Rear-Galvanized

SERVICE INFORMATION

CAP TUBE LOCATION

Cold water and actuator thermostat cap tube location on the evaporator.

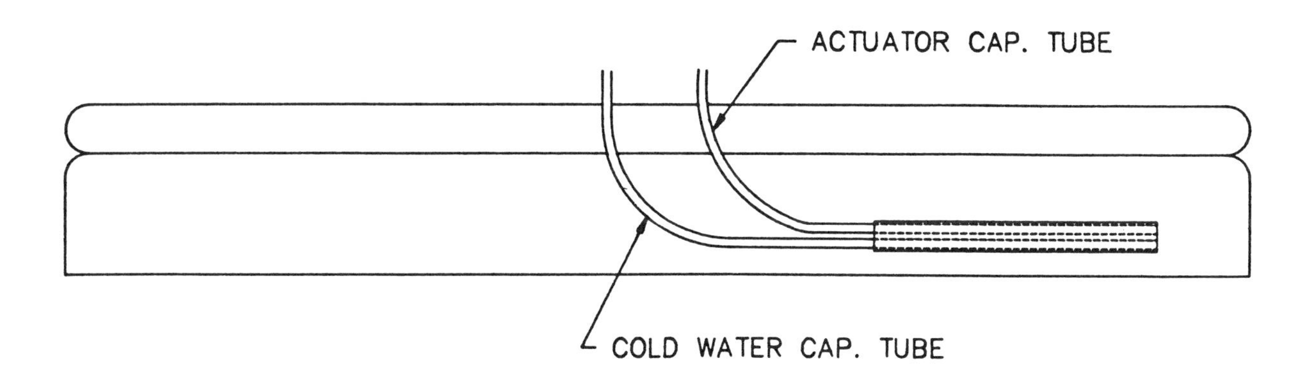

BIN PROBE LOCATION

GB - Bin thermostat capillary tube should reach 4" into bin pointing toward the ice drop zone.

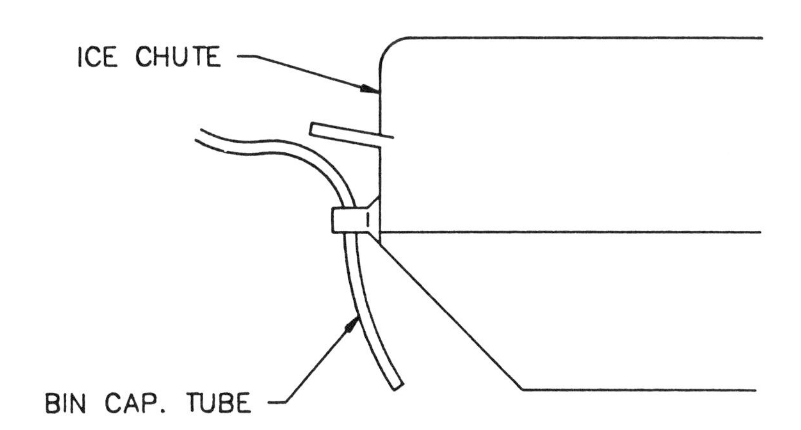

WHEN BIN CAP. TUBE IS PUSHED INTO CLIP.

NORMALLY

CLIP DETAIL

GT - Bin thermostat cap tube location.

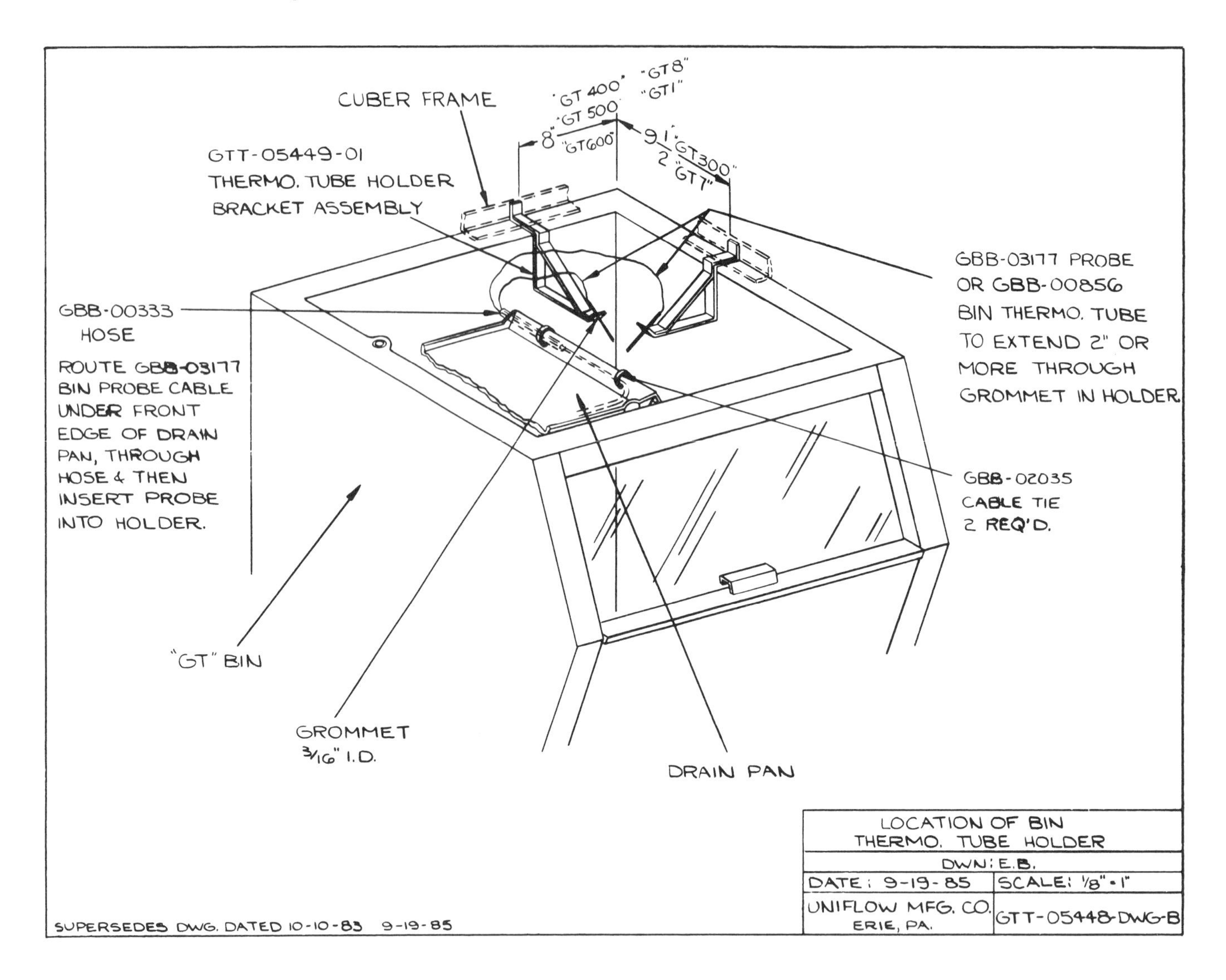

GY cubers - The bin thermostat cap tube is positioned in the tube mounted to bin wall.

GS cubers - The bin thermostat cap tube is positioned 4" into right rear corner of liner.

IS models - The bin thermostat cap tube is run through a thermo tube holder (like GT above) and the holder is positioned 11" from right rear edge of bin.

CONTROL STREAM - After the circulating water is cold, note the control stream and compare the figures below. Adjust screw "A" so that control stream strikes as shown in the illustrations. (NOTE: After water is cold, it may sub-cool and form crystals in circulation tank for a minute or two, partially stopping circulation and control stream temporarily. Do not adjust control stream at this time). If control stream adjusting screw is changed at end of freeze cycle to make stream go over the dam, recheck it during the beginning of the next freeze cycle to be sure it does not go over the cam until all cubers are ready for harvest.

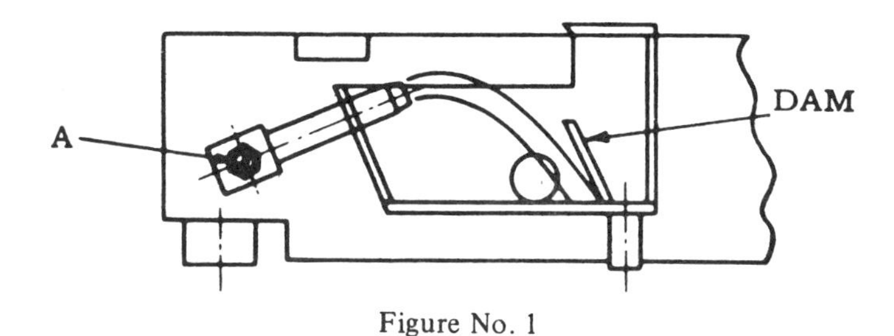

Figure No. 1

CONTROL STREAM AT THE END OF FREEZE CYCLE - When cubes are fully frozen, they freeze over some of the jet holes, increasing the water pressure in the circulation system. This makes the control stream rise and go over the dam, dumping water down the drain and lowering the water level rapidly so that the control tank can snap up and start defrost without delay. Note control stream at the end of the cycle.

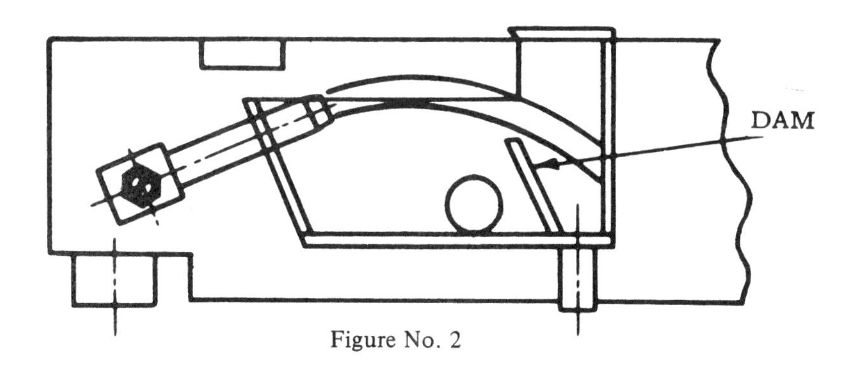

Figure No. 2

CUBE QUALITY

1. If cubes are full with only small dimples (1/8" - 3/16" deep), the exact minimum amount of water was taken in.

2. If control stream does not go over the dam and control tank initiates defrost, check cubes. If many have large holes, there is insufficient water taken in at the beginning of the cycle or there is a leak in the system. Refer to "chart of Water Levels, etc." and adjust weight control switch setting per page 74. If leak present, eliminate leak.

3. If the control stream goes over the dam more than 1 minute before the weight tank snaps up and harvest initiates, check cubes. If many cubes have large holes, the control stream was set too high. Reset per figure no. 1. If ice is frozen to water plate, adjust the differential screw on weight switch (refer page 74) for earlier trip up.

4. A. Holes at the right or left two rows of the evaporator on expansion valve models: Holes at the left - close the valve 1/8 turn at a time. Holes at the right and front - open the expansion valve 1/8 turn at a time.

   B. The capillary tube models GY, GT, GS: Holes at the left indicate over charge. Holes at the right hand side indicate under charge.

WEIGHT CONTROL SWITCH SETTING AND ADJUSTMENT

WATER LEVEL - After water fill when control tank has gone down shutting off water inlet valve, the water level will equalize between the circulation tank and small clear control tank. Measure from top of white circulation tank down to water level. Sight water level carefully across water in control tank. Level should be according to "Chart of Water Levels, etc." The external screw at bottom of weight control switch adjusts water level (counter-clockwise increases, clockwise decreases water level during subsequent fill cycles). Turn only 1/8 turn at a time.

The following is an easy way to set the differential and trip-up point (after the trip-down level is properly set with the range screw). Turn off inlet water and put unit through defrost (pull down water or, if near end of freeze, raise control tank). Make sure most of the water drains from circulation tank (if water plate closes before most of the water is out; open the water plate again and hold down control tank until main tank is empty). Allow water plate to close. Put a one ounce check weight on or in the control tank ($1.50 in silver coins no pennies or nickels, is close enough to one ounce). While turning differential screw slowly clockwise (about 1/4 to 3/4 of a turn), push down and release control tank until it stays down because of the check weight. Then slowly turn differential screw counter-clockwise until control tank snaps up. Remove check weight and turn on water. Control tank will now trip up normally at end of cycle when water level is about halfway between insert disc and bottom of the tank.

WATER LEVEL AT END OF FREEZE CYCLE - The water level after cubes are frozen will normally be near or below the top of the control tank insert as seen in the insert tube, for C & HK machines. (Refer to the "Chart of Water Levels, etc.) If it is above this point when the control stream arises and if the control stream goes over the dam more than a minute, it indicates water level may be lowered if desired to shorten cycle. (Turn weight switch screw 1/8 turn clockwise).

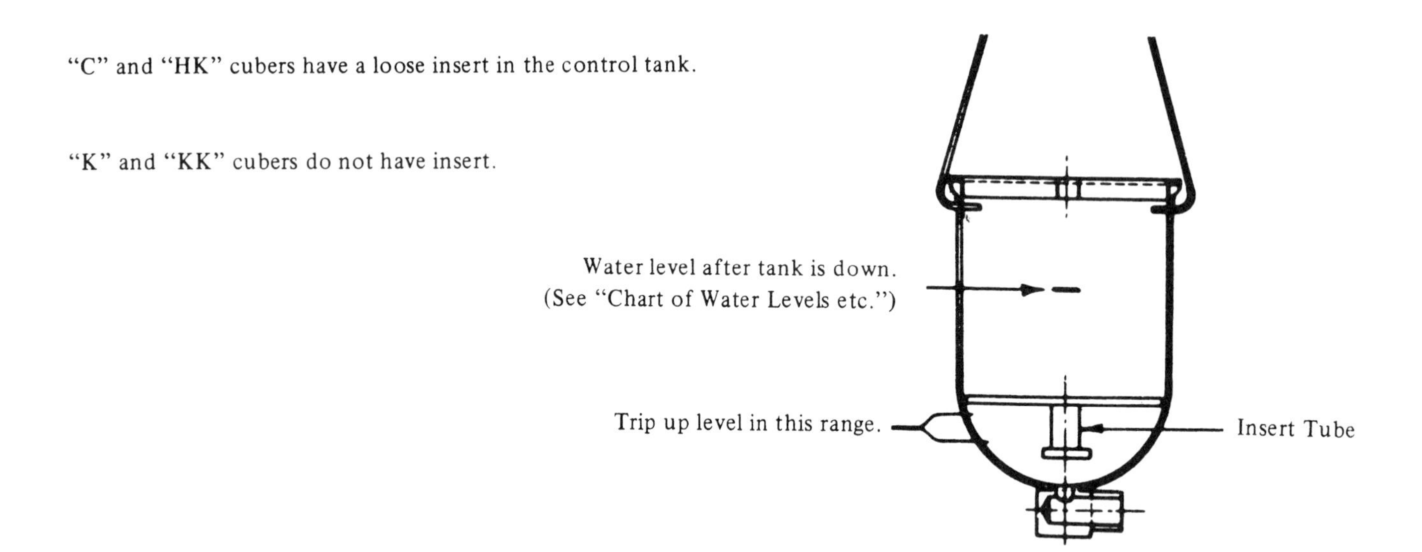
"C" and "HK" cubers have a loose insert in the control tank.
"K" and "KK" cubers do not have insert.
Water level after tank is down.
(See "Chart of Water Levels etc.")
Trip up level in this range.
Insert Tube

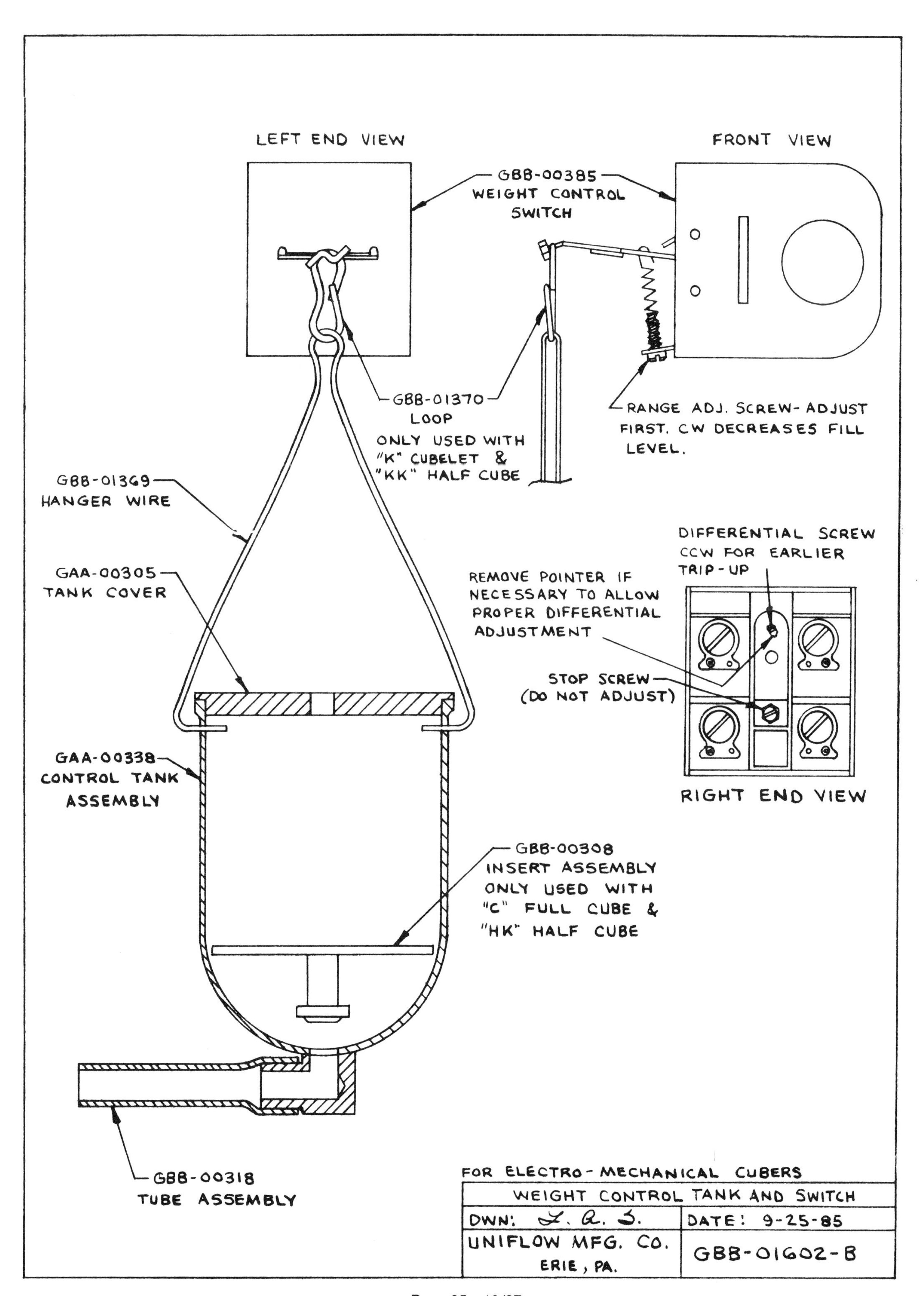
LEFT END VIEW
FRONT VIEW
GBB-00385
WEIGHT CONTROL
SWITCH
GBB-01370
LOOP
ONLY USED WITH
"K" CUBELET &
"KK" HALF CUBE
RANGE ADJ. SCREW- ADJUST
FIRST. CW DECREASES FILL
LEVEL.
GBB-01369
HANGER WIRE
GAA-00305
TANK COVER
DIFFERENTIAL SCREW
CCW FOR EARLIER
TRIP-UP
REMOVE POINTER IF
NECESSARY TO ALLOW
PROPER DIFFERENTIAL
ADJUSTMENT
STOP SCREW
(DO NOT ADJUST)
GAA-00338
CONTROL TANK
ASSEMBLY
RIGHT END VIEW
GBB-00308
INSERT ASSEMBLY
ONLY USED WITH
"C" FULL CUBE &
"HK" HALF CUBE
GBB-00318
TUBE ASSEMBLY
FOR ELECTRO-MECHANICAL CUBERS
WEIGHT CONTROL TANK AND SWITCH
DWN: L. Q. S.
DATE: 9-25-85
UNIFLOW MFG. CO.
ERIE, PA.
GBB-01602-B

Sequence of Electrical Circuits - Refer to Pages 15-25.

Trouble, Cause and Remedy - Refer to Pages 28-34.

Actuator Motor Testing - Refer to Page 37.

Condensers - Refer to Pages 41-52.

Fuses - Refer to Page 53.

Internal Overload on Motors - Refer to Page 53.

Pump Intake Hose Collapsing - Refer to Page 54.

Water Plates - Refer to Pages 57-60.

Problem Water - Refer to Pages 63-64.

Ice Machine Cleaning - Refer to Page 61-62.

Bin Cleaning - Refer to Page 62.

Winter Conditioning - Refer to Page 62.

GS6 CUBERS

GS6 air and water-cooled "Mini-Cubers" operate exactly as all other Kold-Draft cubers except they are not equipped with cold-water thermostats and are only available with "HK" evaporators. Refrigerant control is by capillary tube and the R-12 charge is critical.

Models equipped with power cords must be plugged into a receptacle with no other load. Air-cooled models must have left side and rear skins in place to operate.

The triangular weight-control tank was designed so that the same switch could be used as in all other models.

NOTE: DO NOT ATTEMPT TO DECREASE THE INCOMING WATER LEVEL WITHOUT TESTING THROUGH A NORMAL FREEZE CYCLE. TOO LOW A LEVEL MAY PREVENT THE WEIGHT-CONTROL SWITCH FROM PULLING UP THE TANK TO START HARVEST.

THE CONTROL STREAM SHOULD GO OVER THE DAM FOR SEVERAL MINUTES PRIOR TO START OF HARVEST, BUT BE SURE THERE IS AT LEAST 1/2" OF WATER IN THE CONTROL TANK WHEN THE SWITCH SNAPS UP.

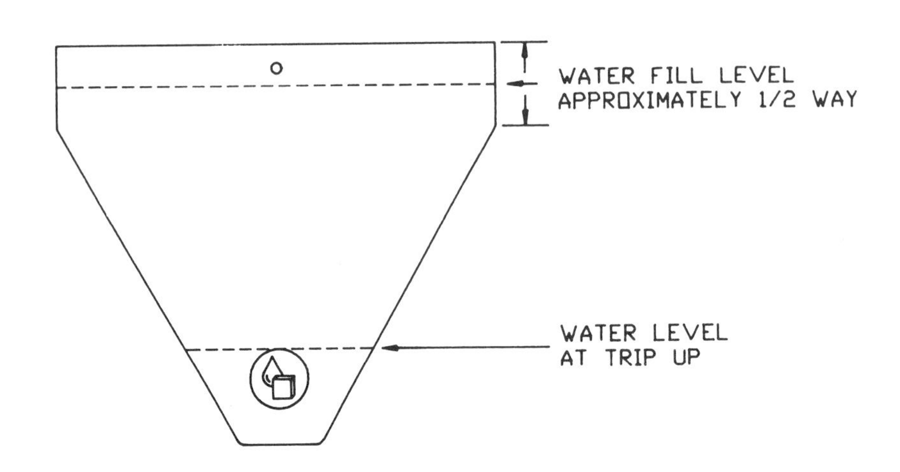

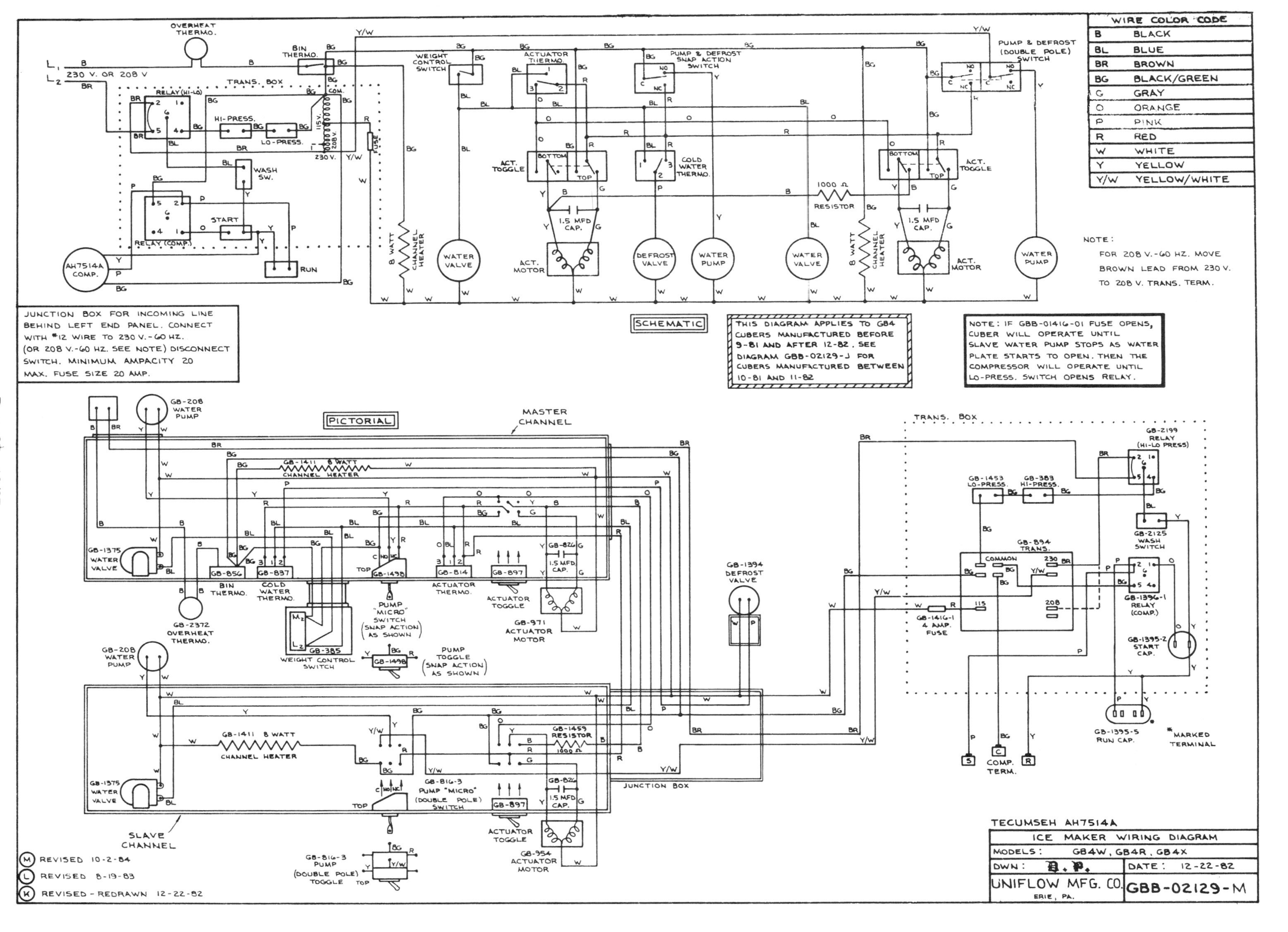
WIRE COLOR CODE
B BLACK
BL BLUE
BR BROWN
BG BLACK/GREEN
G GRAY
O ORANGE
P PINK
R RED
W WHITE
Y YELLOW
Y/W YELLOW/WHITE
NOTE: FOR 208 V.-60 HZ. MOVE BROWN LEAD FROM 230 V. TO 208 V. TRANS. TERM.
SCHEMATIC
THIS DIAGRAM APPLIES TO GB4 CUBERS MANUFACTURED BEFORE 9-81 AND AFTER 12-82. SEE DIAGRAM GBB-02129-J FOR CUBERS MANUFACTURED BETWEEN 10-81 AND 11-82
NOTE: IF GBB-01416-01 FUSE OPENS, CUBER WILL OPERATE UNTIL SLAVE WATER PUMP STOPS AS WATER PLATE STARTS TO OPEN. THEN THE COMPRESSOR WILL OPERATE UNTIL LO-PRESS. SWITCH OPENS RELAY.
JUNCTION BOX FOR INCOMING LINE BEHIND LEFT END PANEL. CONNECT WITH #12 WIRE TO 230 V.-60 HZ. (OR 208 V.-60 HZ. SEE NOTE) DISCONNECT SWITCH. MINIMUM AMPACITY 20 MAX. FUSE SIZE 20 AMP.
PICTORIAL
MASTER CHANNEL
SLAVE CHANNEL
TRANS. BOX
JUNCTION BOX
OVERHEAT THERMO.
BIN THERMO.
WEIGHT CONTROL SWITCH
ACTUATOR THERMO.
PUMP & DEFROST SNAP ACTION SWITCH
PUMP & DEFROST (DOUBLE POLE) SWITCH
ACT. TOGGLE
COLD WATER THERMO.
1000 Ω RESISTOR
1.5 MFD CAP.
ACT. MOTOR
WATER VALVE
DEFROST VALVE
WATER PUMP
8 WATT CHANNEL HEATER
RELAY (HI-LO)
HI-PRESS.
LO-PRESS.
WASH SW.
START
RUN
RELAY (COMP.)
AH7514A COMP.
L1
L2
230 V. OR 208 V
GB-208 WATER PUMP
GB-1375 WATER VALVE
GB-2372 OVERHEAT THERMO.
GB-856 BIN THERMO.
GB-837 COLD WATER THERMO.
GB-385 WEIGHT CONTROL SWITCH
GB-1498 PUMP "MICRO" SWITCH (SNAP ACTION AS SHOWN)
PUMP TOGGLE (SNAP ACTION AS SHOWN)
GB-814 ACTUATOR THERMO.
GB-897 ACTUATOR TOGGLE
GB-826 1.5 MFD CAP.
GB-971 ACTUATOR MOTOR
GB-1411 8 WATT CHANNEL HEATER
GB-1459 RESISTOR 1000 Ω
GB-816-3 PUMP "MICRO" (DOUBLE POLE) SWITCH
GB-816-3 PUMP (DOUBLE POLE) TOGGLE
GB-954 ACTUATOR MOTOR
GB-1394 DEFROST VALVE
GB-2199 RELAY (HI-LO PRESS)
GB-1453 LO-PRESS.
GB-383 HI-PRESS.
GB-2125 WASH SWITCH
GB-894 TRANS.
GB-1396-1 RELAY (COMP.)
GB-1416-1 4 AMP. FUSE
GB-1395-2 START CAP.
GB-1395-5 RUN CAP.
* MARKED TERMINAL
COMP. TERM.
TECUMSEH AH7514A
ICE MAKER WIRING DIAGRAM
MODELS: GB4W, GB4R, GB4X
DWN: B. P.
DATE: 12-22-82
UNIFLOW MFG. CO. ERIE, PA.
GBB-02129-M
(M) REVISED 10-2-84
(L) REVISED 8-19-83
(K) REVISED - REDRAWN 12-22-82

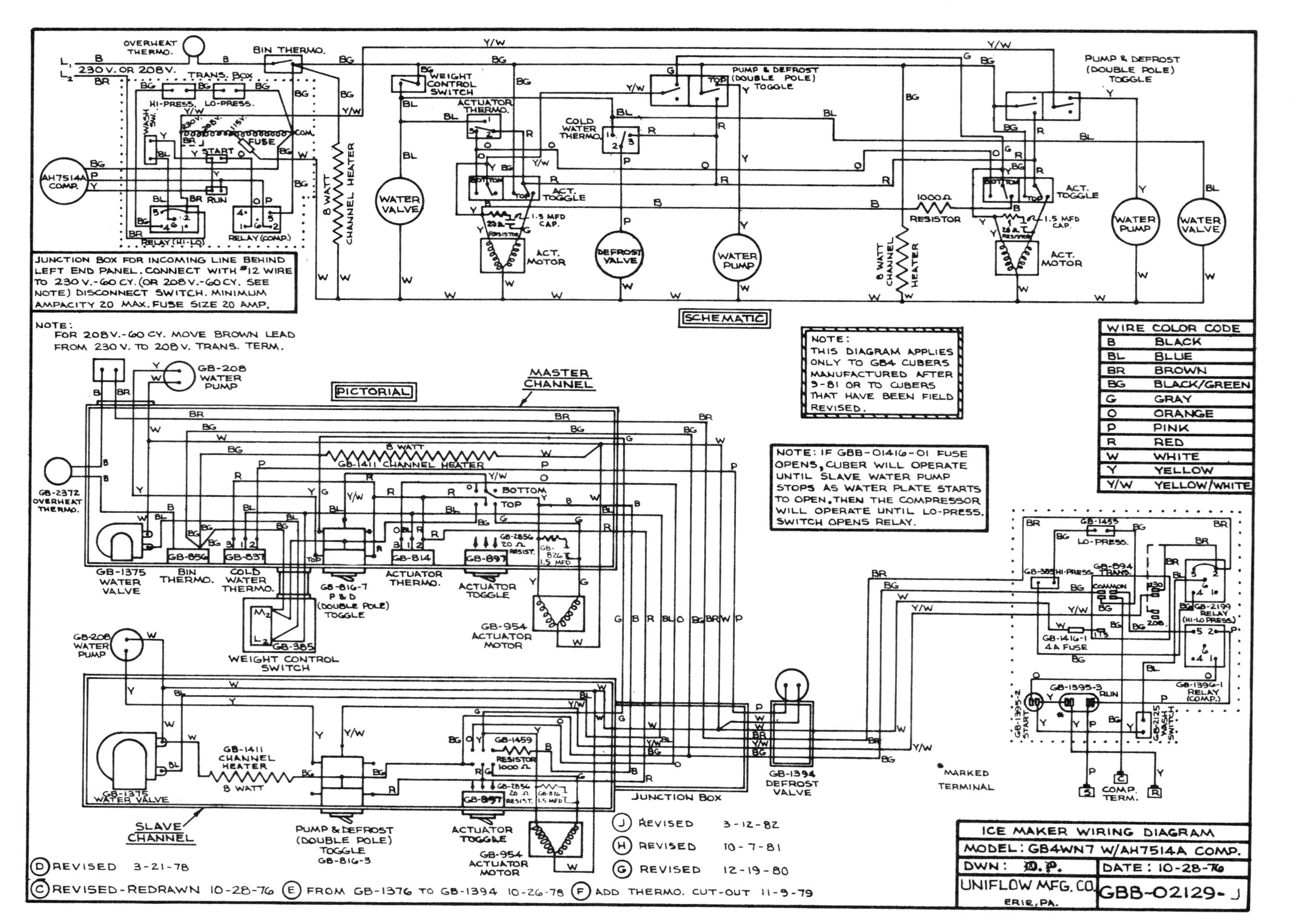
SCHEMATIC
PICTORIAL
OVERHEAT THERMO.
BIN THERMO.
L1 230V. OR 208V.
L2
TRANS. BOX
HI-PRESS.
LO-PRESS.
WASH SW.
FUSE
START
RUN
RELAY (HI-LO)
RELAY (COMP.)
AH7514A COMP.
8 WATT CHANNEL HEATER
WEIGHT CONTROL SWITCH
ACTUATOR THERMO.
WATER VALVE
ACT. TOGGLE
1.5 MFD CAP.
ACT. MOTOR
COLD WATER THERMO.
DEFROST VALVE
PUMP & DEFROST (DOUBLE POLE) TOGGLE
WATER PUMP
1000 Ω RESISTOR
JUNCTION BOX FOR INCOMING LINE BEHIND LEFT END PANEL. CONNECT WITH #12 WIRE TO 230V.-60CY. (OR 208V.-60CY. SEE NOTE) DISCONNECT SWITCH. MINIMUM AMPACITY 20 MAX. FUSE SIZE 20 AMP.
NOTE: FOR 208V.-60CY. MOVE BROWN LEAD FROM 230V. TO 208V. TRANS. TERM.
NOTE: THIS DIAGRAM APPLIES ONLY TO GB4 CUBERS MANUFACTURED AFTER 9-81 OR TO CUBERS THAT HAVE BEEN FIELD REVISED.
NOTE: IF GBB-01416-01 FUSE OPENS, CUBER WILL OPERATE UNTIL SLAVE WATER PUMP STOPS AS WATER PLATE STARTS TO OPEN, THEN THE COMPRESSOR WILL OPERATE UNTIL LO-PRESS. SWITCH OPENS RELAY.
WIRE COLOR CODE
B BLACK
BL BLUE
BR BROWN
BG BLACK/GREEN
G GRAY
O ORANGE
P PINK
R RED
W WHITE
Y YELLOW
Y/W YELLOW/WHITE
MASTER CHANNEL
GB-208 WATER PUMP
GB-2372 OVERHEAT THERMO.
GB-1411 CHANNEL HEATER
GB-1375 WATER VALVE
GB-856 BIN THERMO.
GB-837 COLD WATER THERMO.
GB-816-7 P & D (DOUBLE POLE) TOGGLE
GB-814 ACTUATOR THERMO.
GB-897 ACTUATOR TOGGLE
GB-2856 20 Ω RESIST.
GB-826 1.5 MFD
GB-954 ACTUATOR MOTOR
GB-385 WEIGHT CONTROL SWITCH
SLAVE CHANNEL
GB-1411 CHANNEL HEATER 8 WATT
GB-1459 RESISTOR 1000 Ω
PUMP & DEFROST (DOUBLE POLE) TOGGLE GB-816-3
JUNCTION BOX
GB-1394 DEFROST VALVE
GB-1455 LO-PRESS.
GB-385 HI-PRESS.
GB-894 TRANS.
COMMON
GB-2199 RELAY (HI-LO PRESS.)
GB-1416-1 4A FUSE
GB-1395-3 RUN
GB-1395-2 START
GB-1396-1 RELAY (COMP.)
GB-2125 WASH SWITCH
COMP. TERM.
*MARKED TERMINAL
(J) REVISED 3-12-82
(H) REVISED 10-7-81
(G) REVISED 12-19-80
(D) REVISED 3-21-78
(C) REVISED-REDRAWN 10-28-76
(E) FROM GB-1376 TO GB-1394 10-26-78
(F) ADD THERMO. CUT-OUT 11-9-79
ICE MAKER WIRING DIAGRAM
MODEL: GB4WN7 W/AH7514A COMP.
DWN: D.P.
DATE: 10-28-76
UNIFLOW MFG. CO. ERIE, PA.
GBB-02129-J

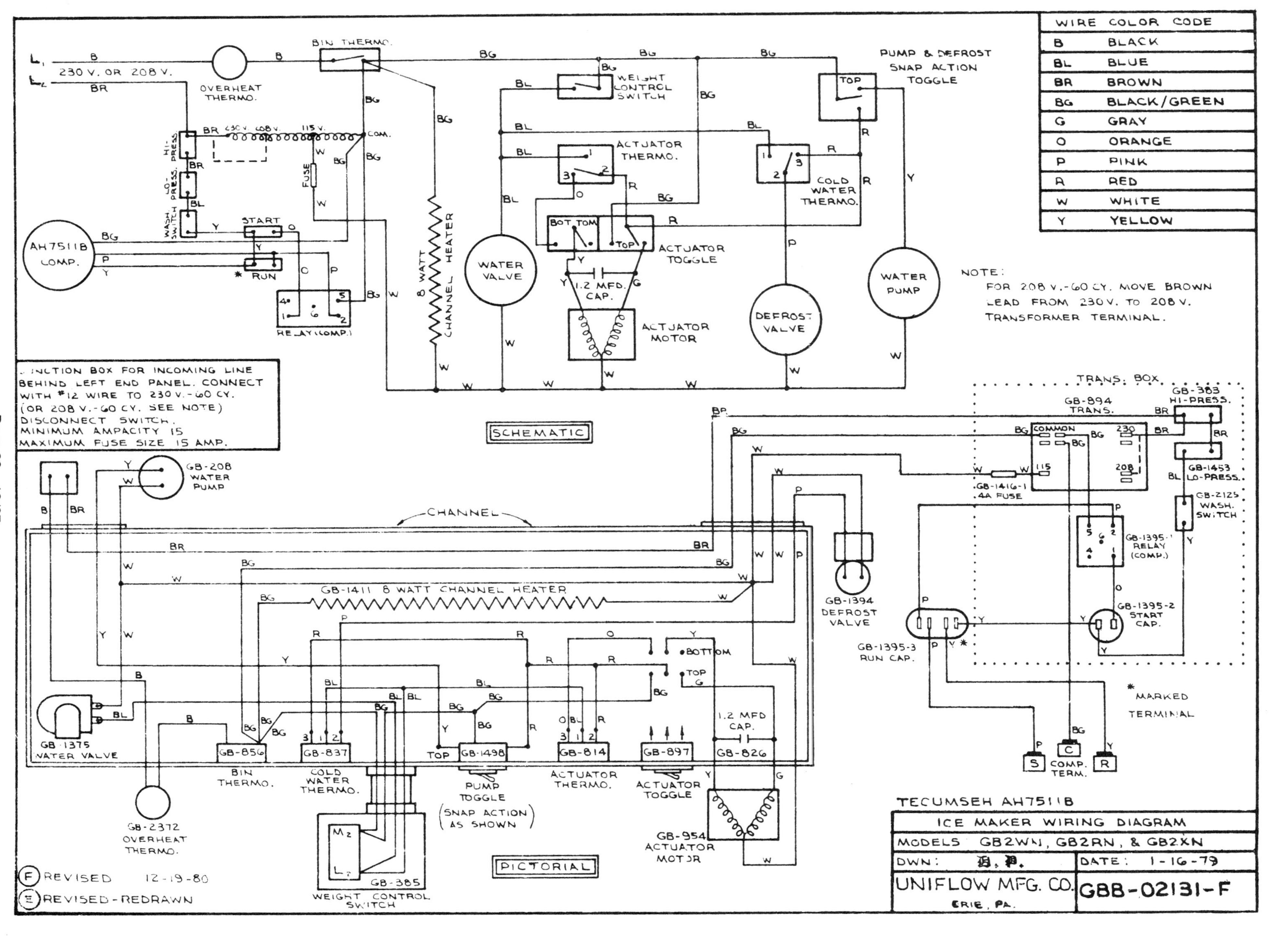

WIRE COLOR CODE
B BLACK
BL BLUE
BR BROWN
BG BLACK/GREEN
G GRAY
O ORANGE
P PINK
R RED
W WHITE
Y YELLOW
NOTE: FOR 208 V.-60 CY. MOVE BROWN LEAD FROM 230 V. TO 208 V. TRANSFORMER TERMINAL.
L1 230 V. OR 208 V.
OVERHEAT THERMO.
BIN THERMO.
HI-PRESS.
LO-PRESS.
WASH SWITCH
AH7511B COMP.
START
RUN
FUSE
RELAY (COMP.)
8 WATT CHANNEL HEATER
WEIGHT CONTROL SWITCH
ACTUATOR THERMO.
ACTUATOR TOGGLE
1.2 MFD. CAP.
ACTUATOR MOTOR
WATER VALVE
COLD WATER THERMO.
DEFROST VALVE
PUMP & DEFROST SNAP ACTION TOGGLE
WATER PUMP
JUNCTION BOX FOR INCOMING LINE BEHIND LEFT END PANEL. CONNECT WITH #12 WIRE TO 230 V.-60 CY. (OR 208 V.-60 CY. SEE NOTE) DISCONNECT SWITCH. MINIMUM AMPACITY 15 MAXIMUM FUSE SIZE 15 AMP.
SCHEMATIC
TRANS. BOX
GB-894 TRANS.
GB-383 HI-PRESS.
GB-1453 LO-PRESS.
GB-2125 WASH. SWITCH
GB-1416-1 4A FUSE
GB-1395-1 RELAY (COMP.)
GB-1395-2 START CAP.
GB-1395-3 RUN CAP.
* MARKED TERMINAL
COMP. TERM.
GB-208 WATER PUMP
CHANNEL
GB-1411 8 WATT CHANNEL HEATER
GB-1394 DEFROST VALVE
GB 1375 WATER VALVE
GB-856 BIN THERMO.
GB-837 COLD WATER THERMO.
GB-1498 PUMP TOGGLE (SNAP ACTION AS SHOWN)
GB-814 ACTUATOR THERMO.
GB-897 ACTUATOR TOGGLE
1.2 MFD CAP.
GB-826
GB-954 ACTUATOR MOTOR
GB-2372 OVERHEAT THERMO.
GB-385 WEIGHT CONTROL SWITCH
PICTORIAL
F REVISED 12-19-80
E REVISED-REDRAWN
TECUMSEH AH7511B
ICE MAKER WIRING DIAGRAM
MODELS GB2WN, GB2RN, & GB2XN
DWN:
DATE: 1-16-79
UNIFLOW MFG. CO. ERIE, PA.
GBB-02131-F

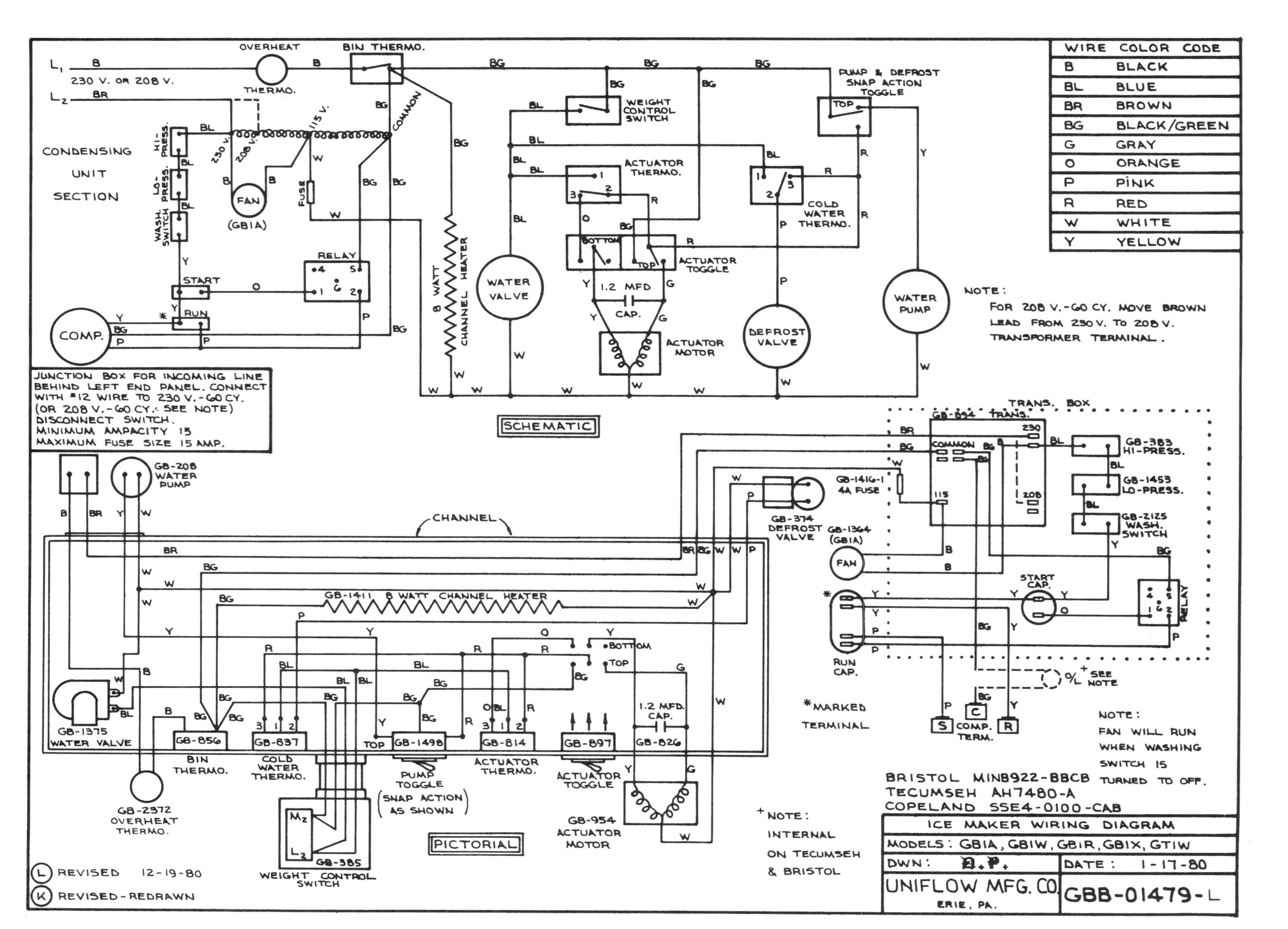
WIRE COLOR CODE
B BLACK
BL BLUE
BR BROWN
BG BLACK/GREEN
G GRAY
O ORANGE
P PINK
R RED
W WHITE
Y YELLOW
L1 B
230 V. OR 208 V.
L2 BR
OVERHEAT THERMO.
BIN THERMO.
COMMON
CONDENSING UNIT SECTION
HI-PRESS.
LO-PRESS.
WASH. SWITCH
FAN (GB1A)
FUSE
115 V.
230 V.
208 V.
RELAY
START
RUN
COMP.
8 WATT CHANNEL HEATER
WEIGHT CONTROL SWITCH
ACTUATOR THERMO.
ACTUATOR TOGGLE
BOTTOM
TOP
1.2 MFD CAP.
ACTUATOR MOTOR
WATER VALVE
COLD WATER THERMO.
DEFROST VALVE
PUMP & DEFROST SNAP ACTION TOGGLE
WATER PUMP
NOTE: FOR 208 V.-60 CY. MOVE BROWN LEAD FROM 230 V. TO 208 V. TRANSFORMER TERMINAL.
JUNCTION BOX FOR INCOMING LINE BEHIND LEFT END PANEL. CONNECT WITH #12 WIRE TO 230 V.-60 CY. (OR 208 V.-60 CY.-SEE NOTE) DISCONNECT SWITCH. MINIMUM AMPACITY 15 MAXIMUM FUSE SIZE 15 AMP.
SCHEMATIC
TRANS. BOX
GB-894 TRANS.
GB-383 HI-PRESS.
GB-1453 LO-PRESS.
GB-2125 WASH. SWITCH
GB-1416-1 4A FUSE
GB-374 DEFROST VALVE
GB-1364 (GB1A)
FAN
START CAP.
RUN CAP.
SEE %L NOTE
COMP. TERM.
*MARKED TERMINAL
GB-208 WATER PUMP
CHANNEL
GB-1411 8 WATT CHANNEL HEATER
GB-1375 WATER VALVE
GB-856 BIN THERMO.
GB-837 COLD WATER THERMO.
GB-1498 PUMP TOGGLE (SNAP ACTION AS SHOWN)
GB-814 ACTUATOR THERMO.
GB-897 ACTUATOR TOGGLE
GB-826 1.2 MFD. CAP.
GB-954 ACTUATOR MOTOR
GB-2372 OVERHEAT THERMO.
GB-385 WEIGHT CONTROL SWITCH
PICTORIAL
+NOTE: INTERNAL ON TECUMSEH & BRISTOL
NOTE: FAN WILL RUN WHEN WASHING SWITCH IS TURNED TO OFF.
BRISTOL MINB922-BBCB
TECUMSEH AH7480-A
COPELAND SSE4-0100-CAB
ICE MAKER WIRING DIAGRAM
MODELS: GB1A, GB1W, GB1R, GB1X, GT1W
DWN: D.P.
DATE: 1-17-80
UNIFLOW MFG. CO. ERIE, PA.
GBB-01479-L
(L) REVISED 12-19-80
(K) REVISED-REDRAWN

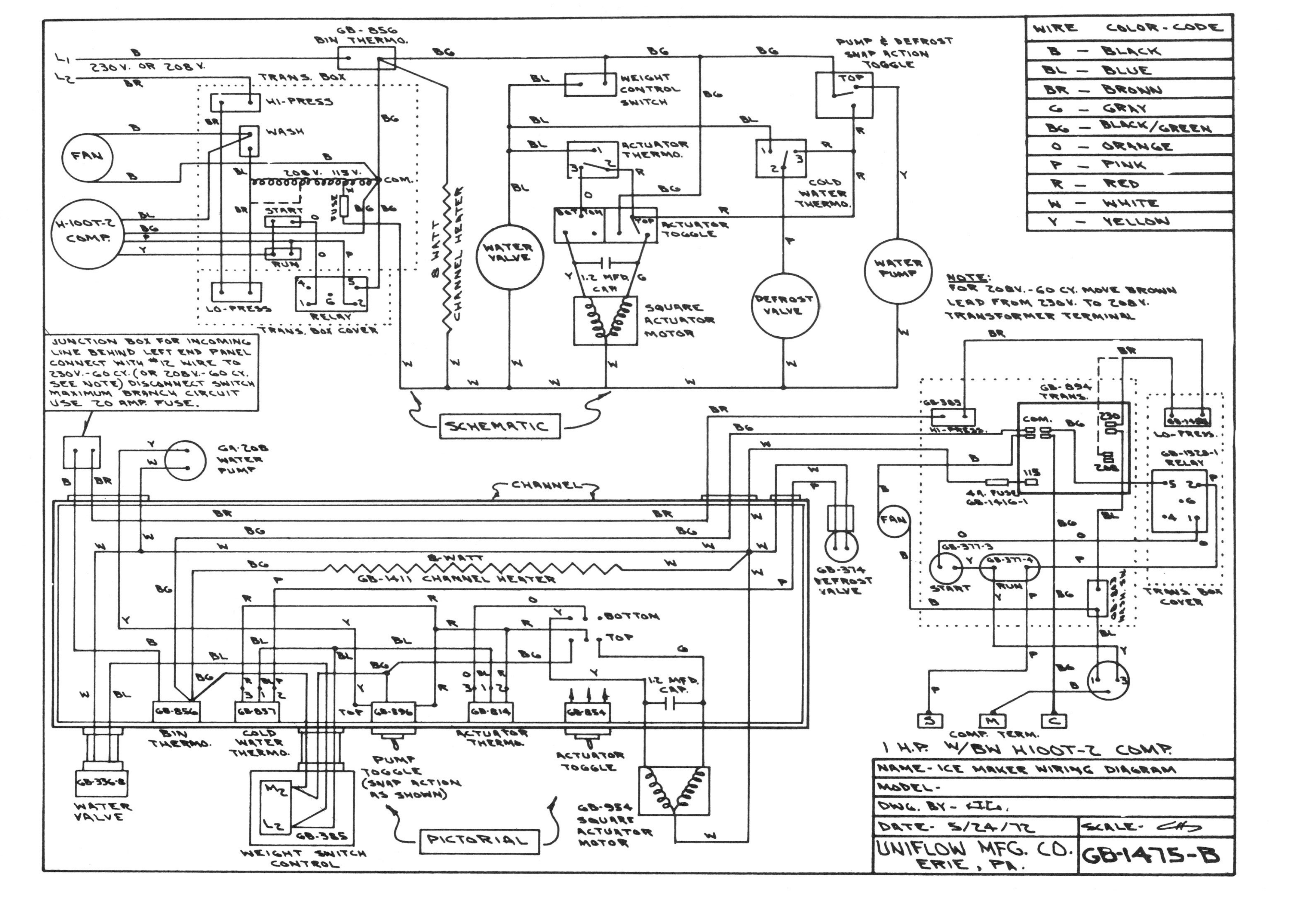

WIRE COLOR-CODE
B – BLACK
BL – BLUE
BR – BROWN
G – GRAY
BG – BLACK/GREEN
O – ORANGE
P – PINK
R – RED
W – WHITE
Y – YELLOW
L1
L2
230V. OR 208V.
GB-856 BIN THERMO.
TRANS. BOX
HI-PRESS
WASH
FAN
H-100T-2 COMP.
208V. 115V.
COM.
FUSE
START
RUN
LO-PRESS
RELAY
TRANS. BOX COVER
8 WATT CHANNEL HEATER
WEIGHT CONTROL SWITCH
ACTUATOR THERMO.
ACTUATOR TOGGLE
BOTTOM
TOP
WATER VALVE
1.2 MFD CAP.
SQUARE ACTUATOR MOTOR
PUMP & DEFROST SNAP ACTION TOGGLE
COLD WATER THERMO.
DEFROST VALVE
WATER PUMP
NOTE: FOR 208V.-60 CY. MOVE BROWN LEAD FROM 230V. TO 208V. TRANSFORMER TERMINAL
JUNCTION BOX FOR INCOMING LINE BEHIND LEFT END PANEL CONNECT WITH #12 WIRE TO 230V.-60 CY. (OR 208V.-60 CY. SEE NOTE) DISCONNECT SWITCH MAXIMUM BRANCH CIRCUIT USE 20 AMP. FUSE.
SCHEMATIC
GA-208 WATER PUMP
CHANNEL
8-WATT GB-1411 CHANNEL HEATER
GB-383 HI-PRESS.
GB-894 TRANS.
LO-PRESS.
GB-1528-1 RELAY
4A. FUSE GB-1416-1
GB-374 DEFROST VALVE
FAN
GB-377-3 START
GB-377-4 RUN
TRANS. BOX COVER
COMP. TERM.
GB-336-8 WATER VALVE
GB-856 BIN THERMO.
GB-837 COLD WATER THERMO.
GB-385 WEIGHT SWITCH CONTROL
GB-896 PUMP TOGGLE (SNAP ACTION AS SHOWN)
GB-814 ACTUATOR THERMO.
GB-854 ACTUATOR TOGGLE
1.2 MFD. CAP.
GB-954 SQUARE ACTUATOR MOTOR
PICTORIAL
1 H.P. W/BW H100T-2 COMP.
NAME- ICE MAKER WIRING DIAGRAM
MODEL-
DWG. BY-
DATE- 5/24/72
SCALE-
UNIFLOW MFG. CO. ERIE, PA.
GB-1475-B

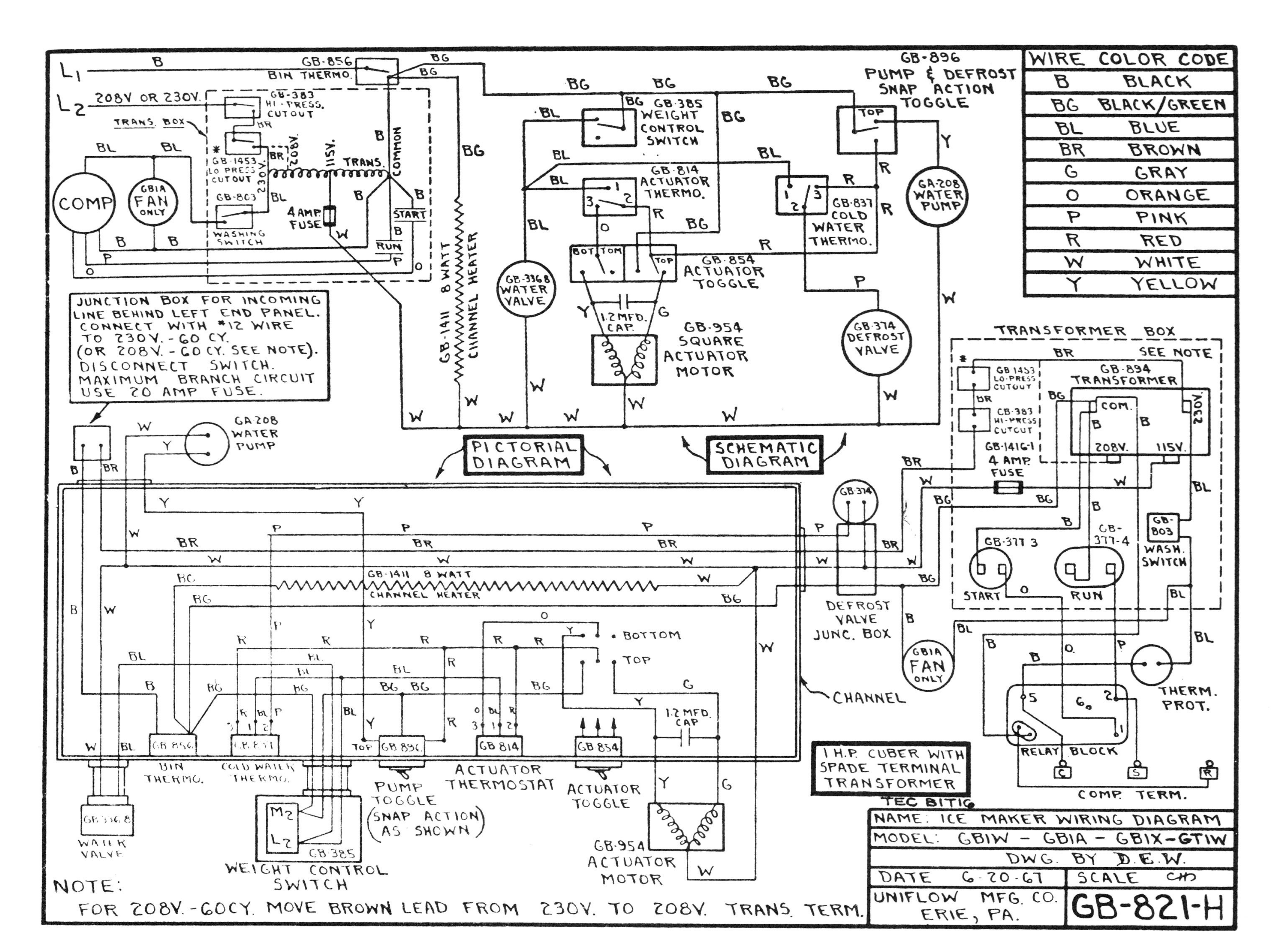
WIRE COLOR CODE
B BLACK
BG BLACK/GREEN
BL BLUE
BR BROWN
G GRAY
O ORANGE
P PINK
R RED
W WHITE
Y YELLOW
GB-856 BIN THERMO.
208V OR 230V.
TRANS. BOX
GB-383 HI-PRESS. CUTOUT
GB-1453 LO PRESS CUTOUT
GB-803 WASHING SWITCH
COMP
GBIA FAN ONLY
TRANS.
COMMON
START
RUN
4 AMP. FUSE
GB-1411 8 WATT CHANNEL HEATER
GB-385 WEIGHT CONTROL SWITCH
GB-814 ACTUATOR THERMO.
GB-3368 WATER VALVE
GB-854 ACTUATOR TOGGLE
1.2 MFD. CAP.
GB-954 SQUARE ACTUATOR MOTOR
GB-896 PUMP & DEFROST SNAP ACTION TOGGLE
GA-208 WATER PUMP
GB-837 COLD WATER THERMO.
GB-374 DEFROST VALVE
JUNCTION BOX FOR INCOMING LINE BEHIND LEFT END PANEL. CONNECT WITH #12 WIRE TO 230V.-60 CY. (OR 208V.-60 CY. SEE NOTE). DISCONNECT SWITCH. MAXIMUM BRANCH CIRCUIT USE 20 AMP FUSE.
PICTORIAL DIAGRAM
SCHEMATIC DIAGRAM
TRANSFORMER BOX
SEE NOTE
GB-894 TRANSFORMER
GB-1416-1 4 AMP. FUSE
GB-377 3
GB-377-4
WASH. SWITCH
DEFROST VALVE JUNC. BOX
CHANNEL
RELAY BLOCK
THERM. PROT.
COMP. TERM.
BIN THERMO.
COLD WATER THERMO.
PUMP TOGGLE (SNAP ACTION AS SHOWN)
ACTUATOR THERMOSTAT
ACTUATOR TOGGLE
GB-954 ACTUATOR MOTOR
WEIGHT CONTROL SWITCH
WATER VALVE
1 H.P. CUBER WITH SPADE TERMINAL TRANSFORMER
NAME: ICE MAKER WIRING DIAGRAM
MODEL: GBIW - GBIA - GBIX-GTIW
DWG. BY D.E.W.
DATE 6-20-67
SCALE
UNIFLOW MFG. CO. ERIE, PA.
GB-821-H
NOTE:
FOR 208V.-60CY. MOVE BROWN LEAD FROM 230V. TO 208V. TRANS. TERM.

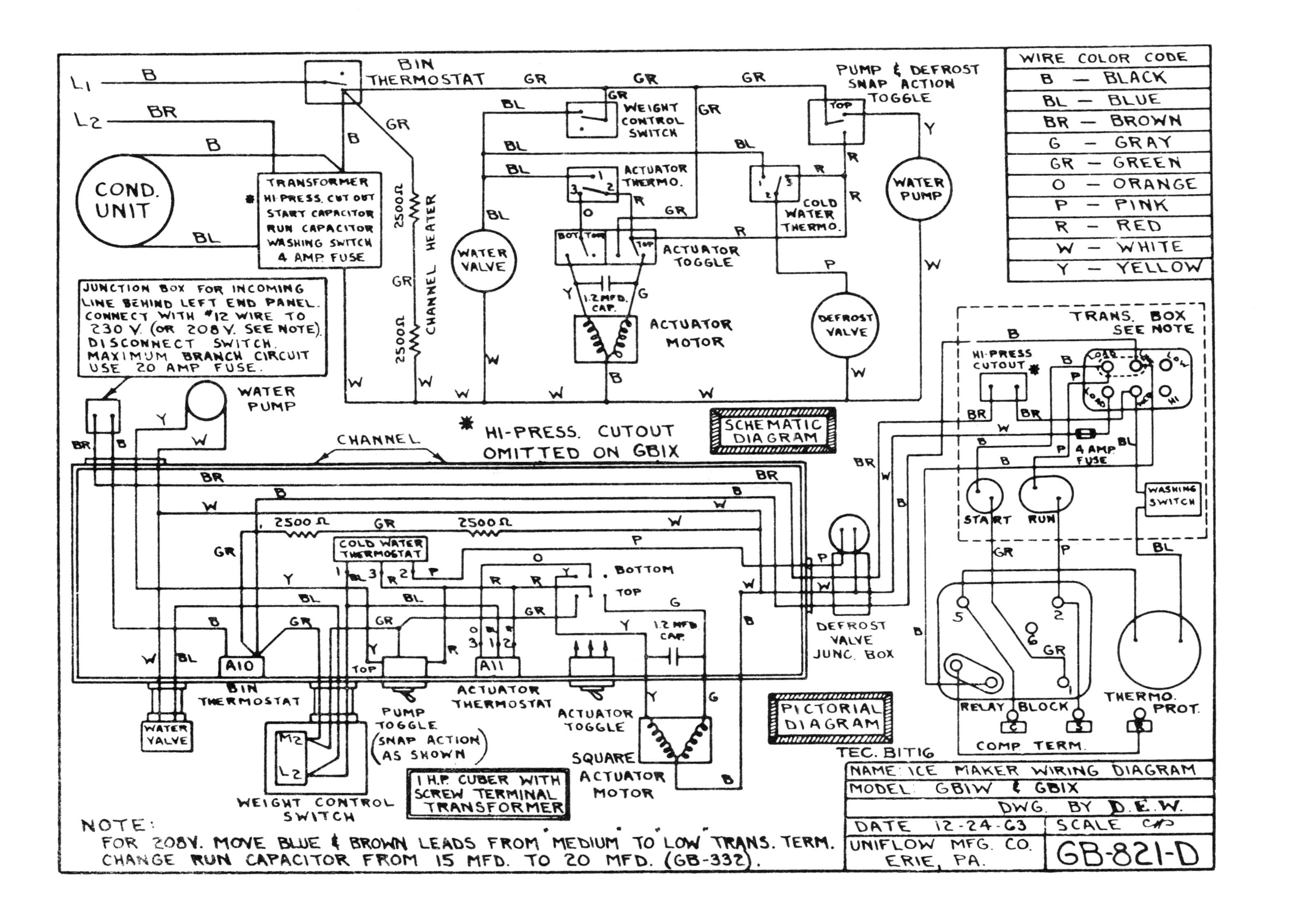
WIRE COLOR CODE
B — BLACK
BL — BLUE
BR — BROWN
G — GRAY
GR — GREEN
O — ORANGE
P — PINK
R — RED
W — WHITE
Y — YELLOW
L1
L2
COND. UNIT
BIN THERMOSTAT
TRANSFORMER
* HI-PRESS. CUT OUT
START CAPACITOR
RUN CAPACITOR
WASHING SWITCH
4 AMP. FUSE
WEIGHT CONTROL SWITCH
ACTUATOR THERMO.
COLD WATER THERMO.
PUMP & DEFROST SNAP ACTION TOGGLE
WATER PUMP
WATER VALVE
CHANNEL HEATER
2500Ω
ACTUATOR TOGGLE
1.2 MFD. CAP.
ACTUATOR MOTOR
DEFROST VALVE
JUNCTION BOX FOR INCOMING LINE BEHIND LEFT END PANEL. CONNECT WITH #12 WIRE TO 230 V. (OR 208 V. SEE NOTE). DISCONNECT SWITCH. MAXIMUM BRANCH CIRCUIT USE 20 AMP FUSE.
* HI-PRESS. CUTOUT OMITTED ON GBIX
SCHEMATIC DIAGRAM
CHANNEL
TRANS. BOX SEE NOTE
HI-PRESS CUTOUT
LOAD
LOW
MED
HI
4 AMP FUSE
WASHING SWITCH
START
RUN
COLD WATER THERMOSTAT
BOTTOM
TOP
DEFROST VALVE JUNC. BOX
A10
BIN THERMOSTAT
A11
ACTUATOR THERMOSTAT
PUMP TOGGLE (SNAP ACTION AS SHOWN)
ACTUATOR TOGGLE
SQUARE ACTUATOR MOTOR
WEIGHT CONTROL SWITCH
M2
L2
1 H.P. CUBER WITH SCREW TERMINAL TRANSFORMER
RELAY BLOCK
COMP. TERM.
THERMO. PROT.
PICTORIAL DIAGRAM
TEC. BIT16
NAME: ICE MAKER WIRING DIAGRAM
MODEL: GBIW & GBIX
DWG. BY D.E.W.
DATE 12-24-63
SCALE
UNIFLOW MFG. CO. ERIE, PA.
GB-821-D
NOTE:
FOR 208V. MOVE BLUE & BROWN LEADS FROM "MEDIUM" TO "LOW" TRANS. TERM.
CHANGE RUN CAPACITOR FROM 15 MFD. TO 20 MFD. (GB-332).

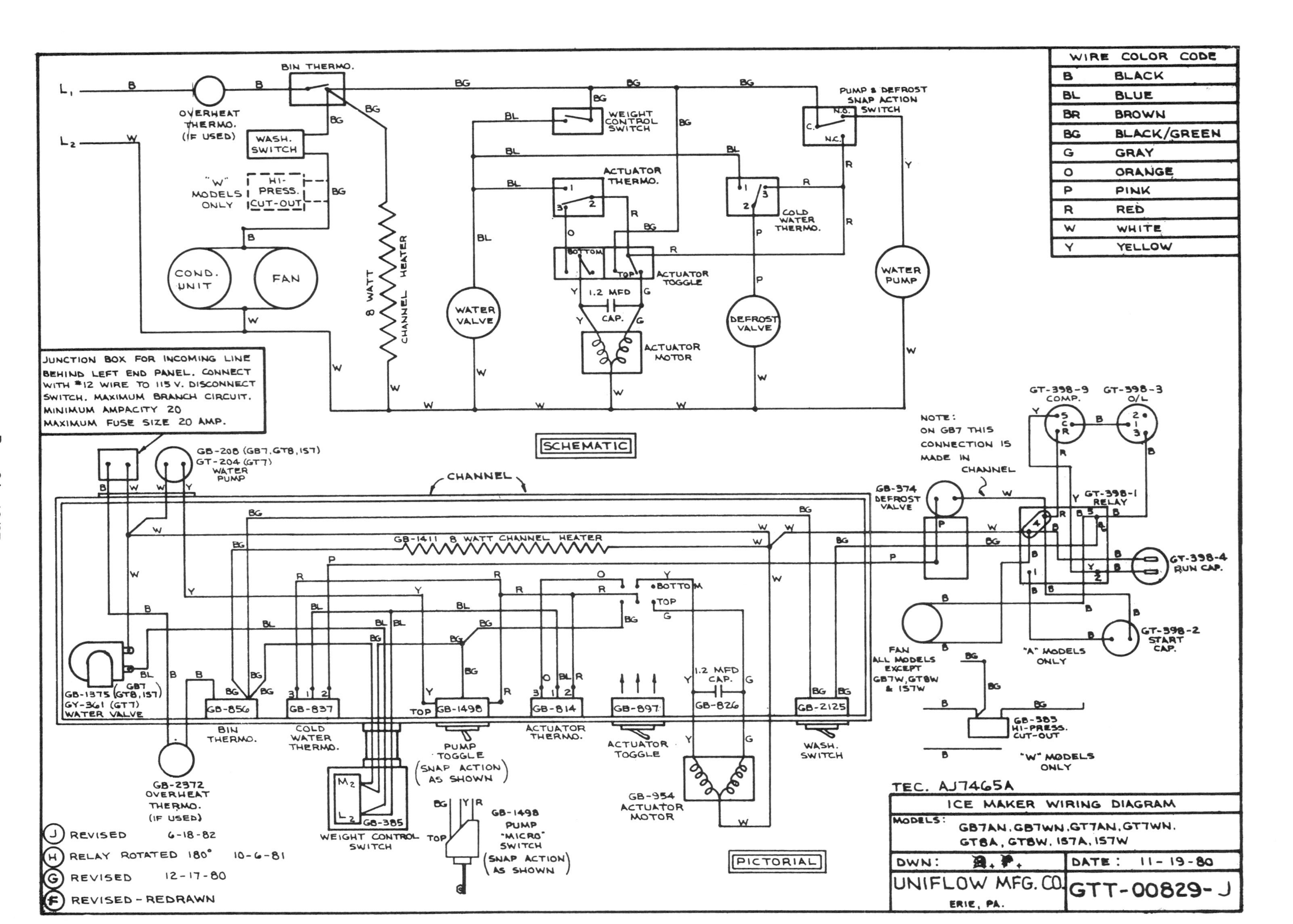

WIRE COLOR CODE
B BLACK
BL BLUE
BR BROWN
BG BLACK/GREEN
G GRAY
O ORANGE
P PINK
R RED
W WHITE
Y YELLOW
L1
L2
OVERHEAT THERMO. (IF USED)
BIN THERMO.
WASH. SWITCH
"W" MODELS ONLY
HI-PRESS. CUT-OUT
COND. UNIT
FAN
8 WATT CHANNEL HEATER
WEIGHT CONTROL SWITCH
ACTUATOR THERMO.
ACTUATOR TOGGLE
BOTTOM
TOP
1.2 MFD CAP.
ACTUATOR MOTOR
WATER VALVE
COLD WATER THERMO.
DEFROST VALVE
PUMP & DEFROST SNAP ACTION SWITCH
N.O.
N.C.
WATER PUMP
JUNCTION BOX FOR INCOMING LINE BEHIND LEFT END PANEL. CONNECT WITH #12 WIRE TO 115 V. DISCONNECT SWITCH. MAXIMUM BRANCH CIRCUIT. MINIMUM AMPACITY 20 MAXIMUM FUSE SIZE 20 AMP.
SCHEMATIC
GB-208 (GB7, GT8, 157) GT-204 (GT7) WATER PUMP
CHANNEL
GB-1411 8 WATT CHANNEL HEATER
NOTE: ON GB7 THIS CONNECTION IS MADE IN CHANNEL
GB-374 DEFROST VALVE
GT-398-9 COMP.
GT-398-3 O/L
GT-398-1 RELAY
GT-398-4 RUN CAP.
GT-398-2 START CAP.
"A" MODELS ONLY
FAN ALL MODELS EXCEPT GB7W, GT8W & 157W
GB-385 HI-PRESS. CUT-OUT
"W" MODELS ONLY
GB-1375 (GB7 GT8, 157) GY-361 (GT7) WATER VALVE
GB-856 BIN THERMO.
GB-837 COLD WATER THERMO.
GB-1498 PUMP TOGGLE (SNAP ACTION AS SHOWN)
GB-814 ACTUATOR THERMO.
GB-897 ACTUATOR TOGGLE
GB-826
GB-2125 WASH. SWITCH
GB-954 ACTUATOR MOTOR
GB-2372 OVERHEAT THERMO. (IF USED)
GB-385 WEIGHT CONTROL SWITCH
GB-1498 PUMP "MICRO" SWITCH (SNAP ACTION AS SHOWN)
PICTORIAL
J REVISED 6-18-82
H RELAY ROTATED 180° 10-6-81
G REVISED 12-17-80
F REVISED - REDRAWN
TEC. AJ7465A
ICE MAKER WIRING DIAGRAM
MODELS: GB7AN, GB7WN, GT7AN, GT7WN, GT8A, GT8W, 157A, 157W
DWN:
DATE: 11-19-80
UNIFLOW MFG. CO. ERIE, PA.
GTT-00829-J

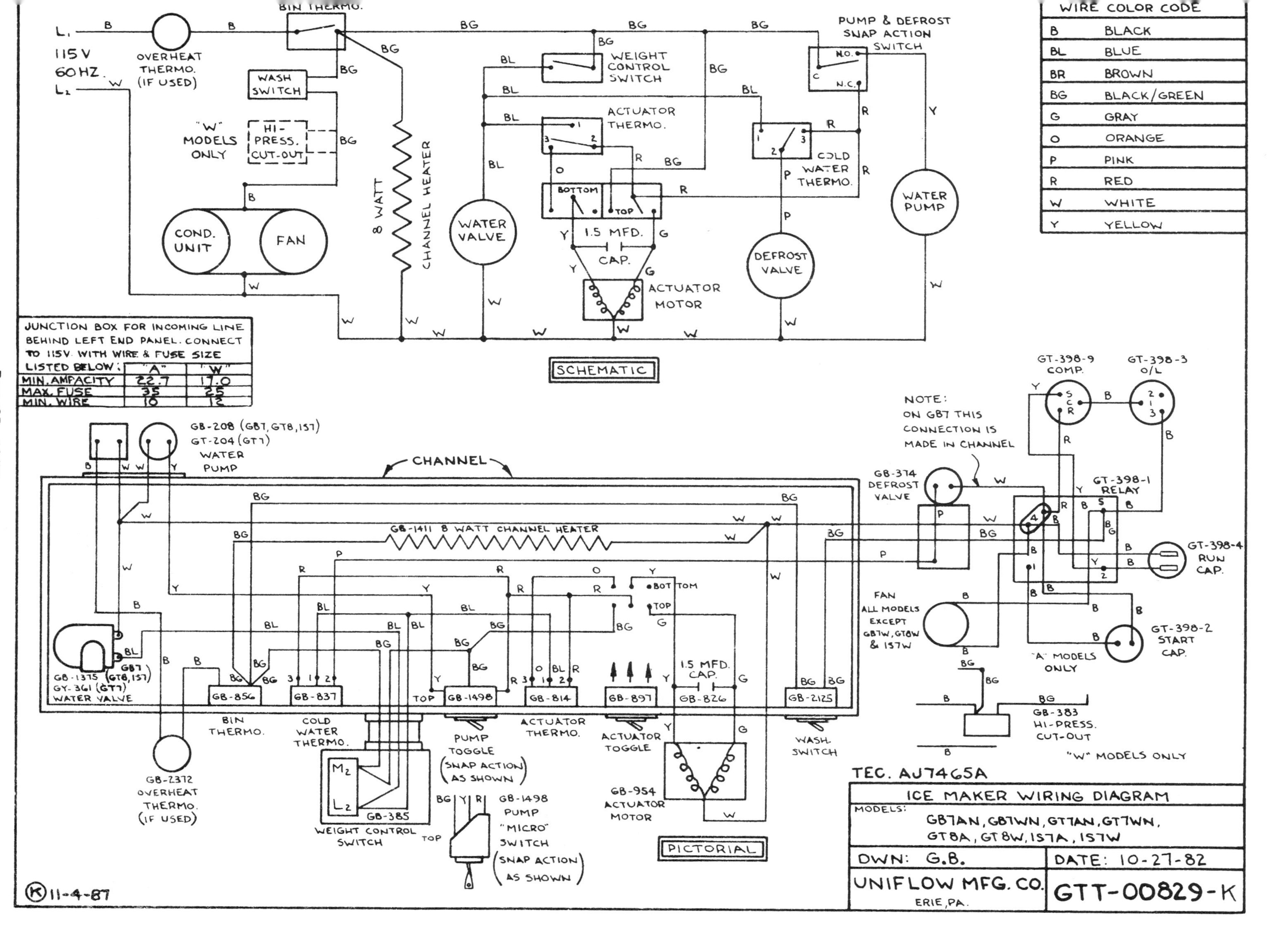
WIRE COLOR CODE
B BLACK
BL BLUE
BR BROWN
BG BLACK/GREEN
G GRAY
O ORANGE
P PINK
R RED
W WHITE
Y YELLOW
L1 115V 60HZ. L2
OVERHEAT THERMO. (IF USED)
BIN THERMO.
WASH SWITCH
"W" MODELS ONLY
HI-PRESS. CUT-OUT
COND. UNIT
FAN
8 WATT CHANNEL HEATER
WEIGHT CONTROL SWITCH
ACTUATOR THERMO.
WATER VALVE
BOTTOM
TOP
1.5 MFD. CAP.
ACTUATOR MOTOR
PUMP & DEFROST SNAP ACTION SWITCH
N.O.
N.C.
COLD WATER THERMO.
DEFROST VALVE
WATER PUMP
SCHEMATIC
JUNCTION BOX FOR INCOMING LINE BEHIND LEFT END PANEL. CONNECT TO 115V. WITH WIRE & FUSE SIZE LISTED BELOW:
"A" "W"
MIN. AMPACITY 22.7 17.0
MAX. FUSE 35 25
MIN. WIRE 10 12
GB-208 (GB7, GT8, IS7)
GT-204 (GT7)
WATER PUMP
CHANNEL
GB-1411 8 WATT CHANNEL HEATER
GB-1375 (GB7)
(GT8, IS7)
GY-361 (GT7)
WATER VALVE
GB-856
BIN THERMO.
GB-837
COLD WATER THERMO.
GB-1498
PUMP TOGGLE (SNAP ACTION AS SHOWN)
GB-814
ACTUATOR THERMO.
GB-897
ACTUATOR TOGGLE
GB-826
1.5 MFD. CAP.
GB-2125
WASH SWITCH
GB-2372
OVERHEAT THERMO. (IF USED)
GB-385
WEIGHT CONTROL SWITCH
GB-1498
PUMP "MICRO" SWITCH (SNAP ACTION AS SHOWN)
GB-954
ACTUATOR MOTOR
PICTORIAL
NOTE: ON GB7 THIS CONNECTION IS MADE IN CHANNEL
GB-374
DEFROST VALVE
GT-398-9
COMP.
GT-398-3
O/L
GT-398-1
RELAY
GT-398-4
RUN CAP.
GT-398-2
START CAP.
"A" MODELS ONLY
FAN
ALL MODELS EXCEPT GB7W, GT8W & IS7W
GB-383
HI-PRESS. CUT-OUT
"W" MODELS ONLY
TEC. AJ7465A
ICE MAKER WIRING DIAGRAM
MODELS: GB7AN, GB7WN, GT7AN, GT7WN, GT8A, GT8W, IS7A, IS7W
DWN: G.B.
DATE: 10-27-82
UNIFLOW MFG. CO.
ERIE, PA.
GTT-00829-K
Ⓚ 11-4-87

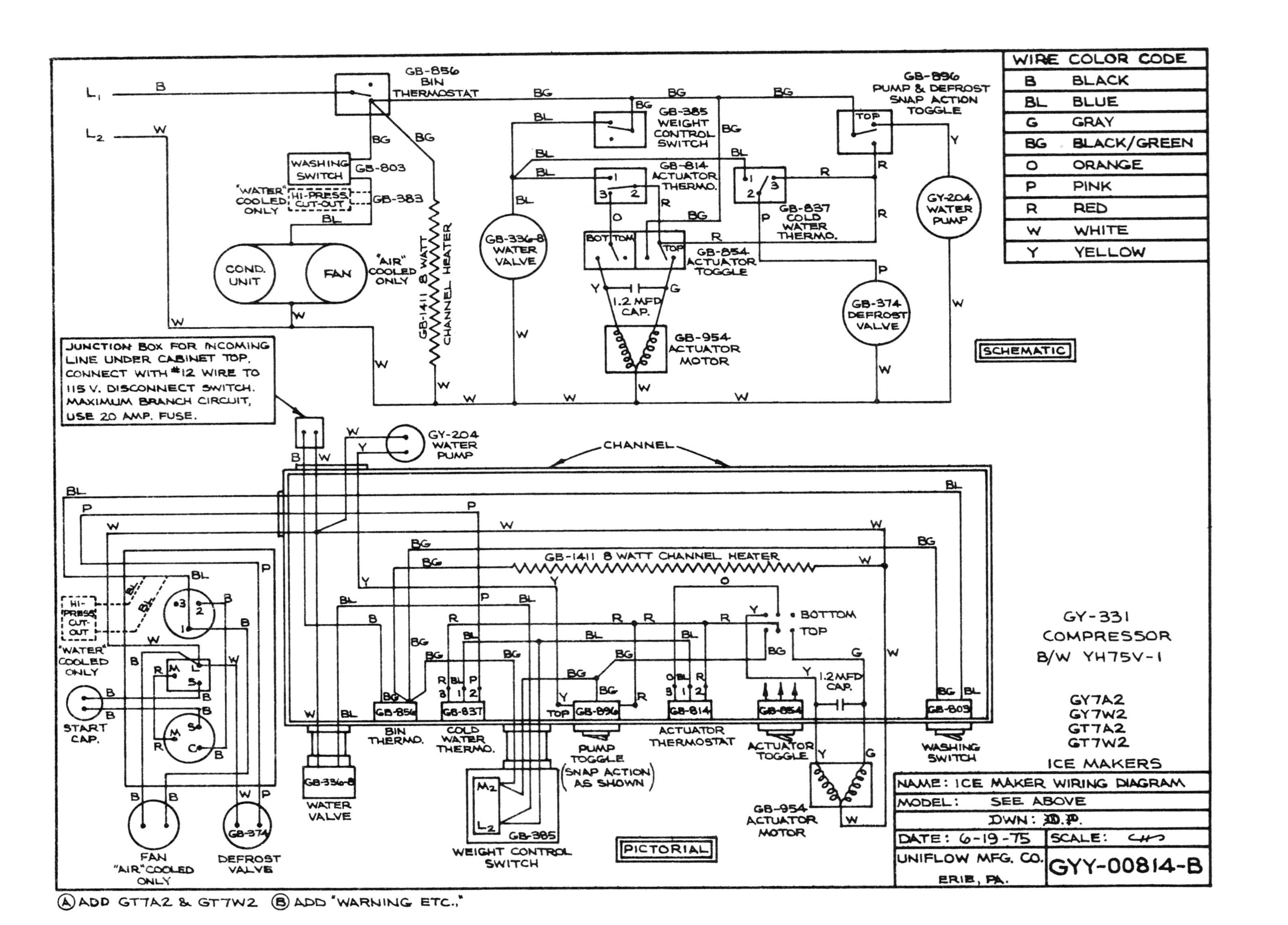
WIRE COLOR CODE
B BLACK
BL BLUE
G GRAY
BG BLACK/GREEN
O ORANGE
P PINK
R RED
W WHITE
Y YELLOW
GB-856 BIN THERMOSTAT
WASHING SWITCH GB-803
"WATER" COOLED ONLY HI-PRESS. CUT-OUT GB-383
COND. UNIT
FAN "AIR" COOLED ONLY
GB-1411 8 WATT CHANNEL HEATER
GB-336-8 WATER VALVE
GB-385 WEIGHT CONTROL SWITCH
GB-814 ACTUATOR THERMO.
GB-854 ACTUATOR TOGGLE
1.2 MFD CAP.
GB-954 ACTUATOR MOTOR
GB-837 COLD WATER THERMO.
GB-896 PUMP & DEFROST SNAP ACTION TOGGLE
GY-204 WATER PUMP
GB-374 DEFROST VALVE
SCHEMATIC
JUNCTION BOX FOR INCOMING LINE UNDER CABINET TOP. CONNECT WITH #12 WIRE TO 115 V. DISCONNECT SWITCH. MAXIMUM BRANCH CIRCUIT, USE 20 AMP. FUSE.
GY-204 WATER PUMP
CHANNEL
GB-1411 8 WATT CHANNEL HEATER
BOTTOM
TOP
HI-PRESS CUT-OUT
"WATER" COOLED ONLY
START CAP.
FAN "AIR" COOLED ONLY
DEFROST VALVE
WATER VALVE
BIN THERMO.
COLD WATER THERMO.
PUMP TOGGLE (SNAP ACTION AS SHOWN)
ACTUATOR THERMOSTAT
ACTUATOR TOGGLE
WASHING SWITCH
GB-954 ACTUATOR MOTOR
WEIGHT CONTROL SWITCH
PICTORIAL
GY-331 COMPRESSOR B/W YH75V-1
GY7A2
GY7W2
GT7A2
GT7W2
ICE MAKERS
NAME: ICE MAKER WIRING DIAGRAM
MODEL: SEE ABOVE
DWN: D.P.
DATE: 6-19-75
UNIFLOW MFG. CO. ERIE, PA.
GYY-00814-B
(A) ADD GT7A2 & GT7W2 (B) ADD "WARNING ETC.,"

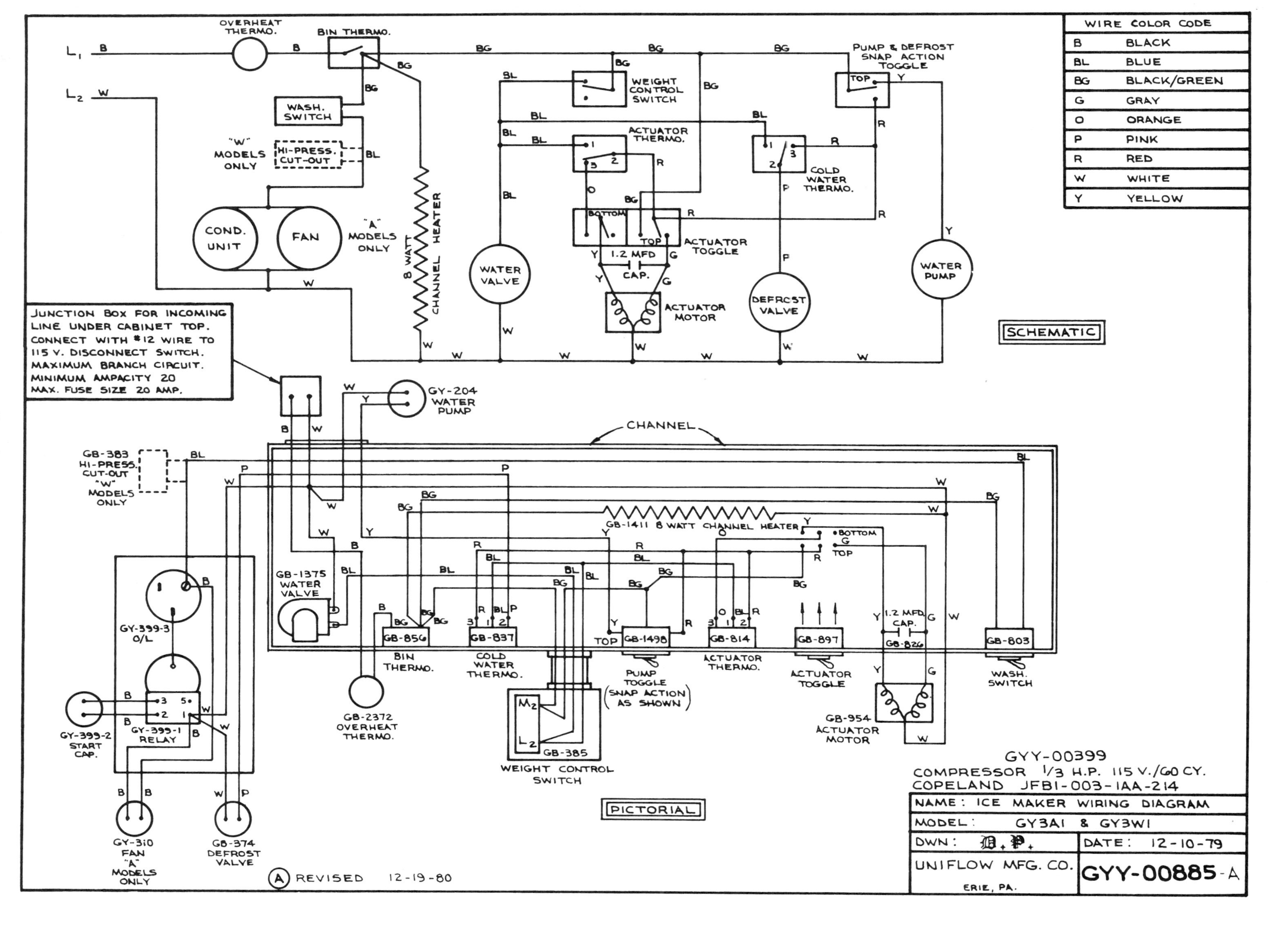
WIRE COLOR CODE
B BLACK
BL BLUE
BG BLACK/GREEN
G GRAY
O ORANGE
P PINK
R RED
W WHITE
Y YELLOW
SCHEMATIC
OVERHEAT THERMO.
BIN THERMO.
WASH. SWITCH
"W" MODELS ONLY
HI-PRESS. CUT-OUT
COND. UNIT
FAN
"A" MODELS ONLY
8 WATT CHANNEL HEATER
WEIGHT CONTROL SWITCH
ACTUATOR THERMO.
ACTUATOR TOGGLE
1.2 MFD CAP.
ACTUATOR MOTOR
WATER VALVE
COLD WATER THERMO.
DEFROST VALVE
PUMP & DEFROST SNAP ACTION TOGGLE
WATER PUMP
JUNCTION BOX FOR INCOMING LINE UNDER CABINET TOP. CONNECT WITH #12 WIRE TO 115 V. DISCONNECT SWITCH. MAXIMUM BRANCH CIRCUIT. MINIMUM AMPACITY 20 MAX. FUSE SIZE 20 AMP.
GY-204 WATER PUMP
CHANNEL
GB-383 HI-PRESS. CUT-OUT "W" MODELS ONLY
GB-1411 8 WATT CHANNEL HEATER
GB-1375 WATER VALVE
GB-856 BIN THERMO.
GB-837 COLD WATER THERMO.
GB-1498 PUMP TOGGLE (SNAP ACTION AS SHOWN)
GB-814 ACTUATOR THERMO.
GB-897 ACTUATOR TOGGLE
1.2 MFD CAP. GB-826
GB-803 WASH. SWITCH
GY-399-3 O/L
GY-399-1 RELAY
GY-399-2 START CAP.
GB-2372 OVERHEAT THERMO.
GB-385 WEIGHT CONTROL SWITCH
GB-954 ACTUATOR MOTOR
GY-310 FAN "A" MODELS ONLY
GB-374 DEFROST VALVE
PICTORIAL
(A) REVISED 12-19-80
GYY-00399
COMPRESSOR 1/3 H.P. 115 V./60 CY.
COPELAND JFB1-003-1AA-214
NAME: ICE MAKER WIRING DIAGRAM
MODEL: GY3A1 & GY3W1
DATE: 12-10-79
UNIFLOW MFG. CO.
ERIE, PA.
GYY-00885-A

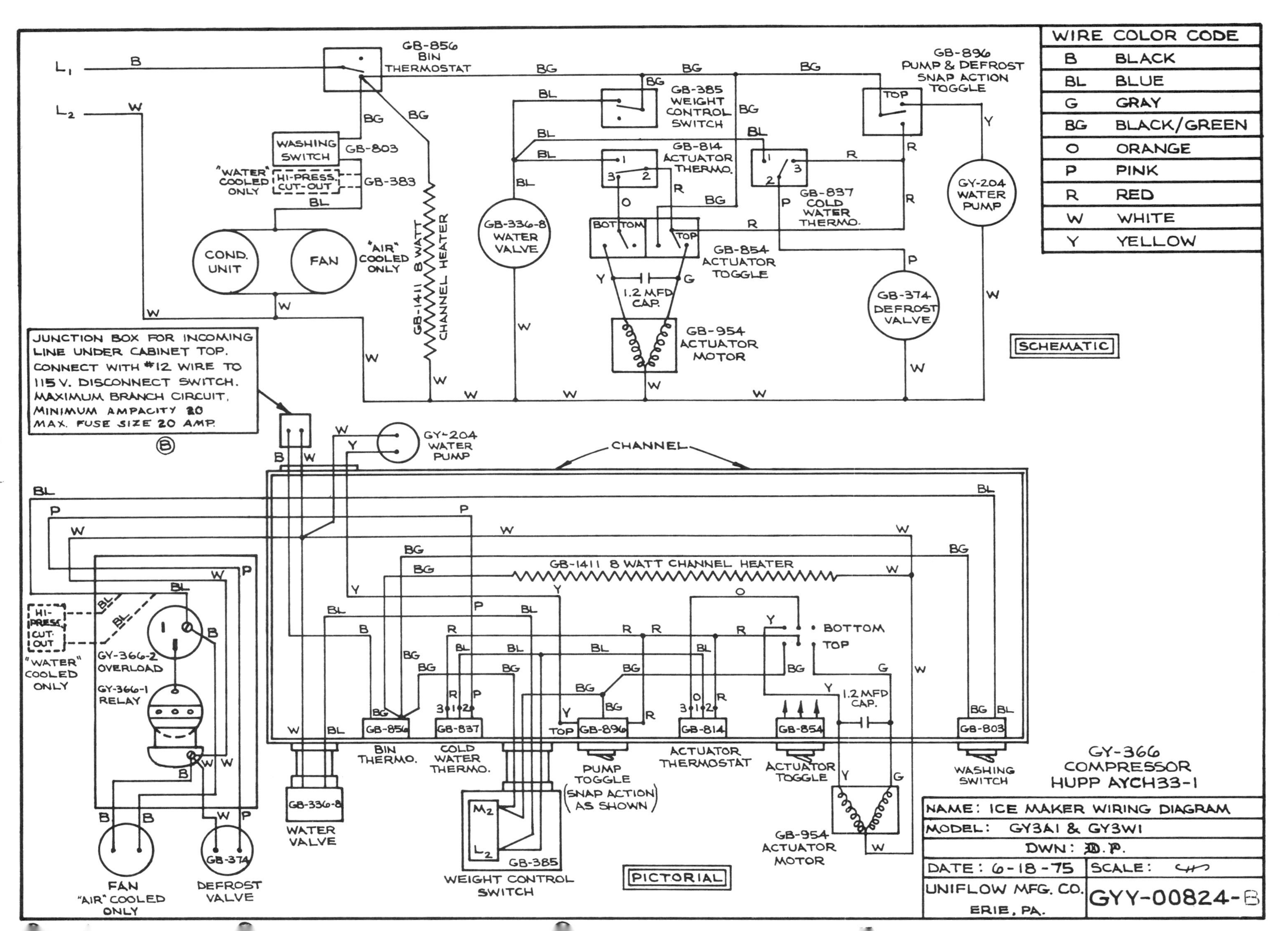
WIRE COLOR CODE
B BLACK
BL BLUE
G GRAY
BG BLACK/GREEN
O ORANGE
P PINK
R RED
W WHITE
Y YELLOW
SCHEMATIC
GB-856 BIN THERMOSTAT
GB-385 WEIGHT CONTROL SWITCH
GB-896 PUMP & DEFROST SNAP ACTION TOGGLE
WASHING SWITCH
GB-803
"WATER" COOLED ONLY
HI-PRESS. CUT-OUT
GB-383
GB-814 ACTUATOR THERMO.
GB-837 COLD WATER THERMO.
GY-204 WATER PUMP
COND. UNIT
FAN
"AIR" COOLED ONLY
GB-1411 8 WATT CHANNEL HEATER
GB-336-8 WATER VALVE
GB-854 ACTUATOR TOGGLE
1.2 MFD CAP.
GB-374 DEFROST VALVE
GB-954 ACTUATOR MOTOR
JUNCTION BOX FOR INCOMING LINE UNDER CABINET TOP. CONNECT WITH #12 WIRE TO 115 V. DISCONNECT SWITCH. MAXIMUM BRANCH CIRCUIT, MINIMUM AMPACITY 20 MAX. FUSE SIZE 20 AMP.
CHANNEL
GY-204 WATER PUMP
GB-1411 8 WATT CHANNEL HEATER
BOTTOM
TOP
GY-366-2 OVERLOAD
GY-366-1 RELAY
HI-PRESS. CUT-OUT
"WATER" COOLED ONLY
BIN THERMO.
COLD WATER THERMO.
PUMP TOGGLE (SNAP ACTION AS SHOWN)
ACTUATOR THERMOSTAT
ACTUATOR TOGGLE
WASHING SWITCH
GB-336-8 WATER VALVE
GB-385 WEIGHT CONTROL SWITCH
PICTORIAL
GB-954 ACTUATOR MOTOR
FAN "AIR" COOLED ONLY
GB-374 DEFROST VALVE
GY-366 COMPRESSOR HUPP AYCH33-1
NAME: ICE MAKER WIRING DIAGRAM
MODEL: GY3A1 & GY3W1
DWN: D.P.
DATE: 6-18-75
SCALE:
UNIFLOW MFG. CO. ERIE, PA.
GYY-00824-B

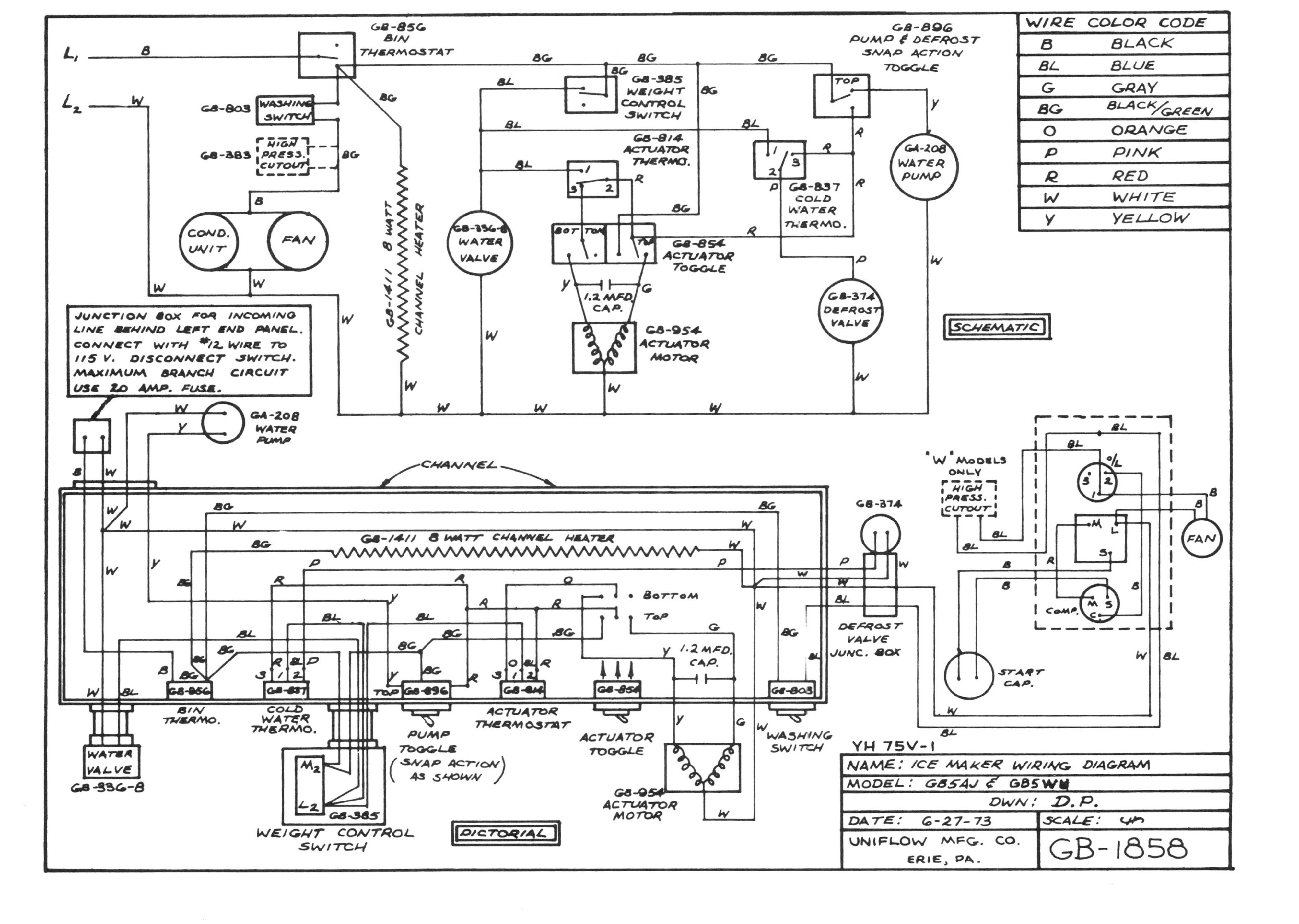
WIRE COLOR CODE
B BLACK
BL BLUE
G GRAY
BG BLACK/GREEN
O ORANGE
P PINK
R RED
W WHITE
Y YELLOW
GB-856 BIN THERMOSTAT
GB-803 WASHING SWITCH
GB-383 HIGH PRESS. CUTOUT
COND. UNIT
FAN
GB-1411 8 WATT CHANNEL HEATER
GB-336-B WATER VALVE
GB-385 WEIGHT CONTROL SWITCH
GB-814 ACTUATOR THERMO.
GB-854 ACTUATOR TOGGLE
1.2 MFD. CAP.
GB-954 ACTUATOR MOTOR
GB-896 PUMP & DEFROST SNAP ACTION TOGGLE
GB-837 COLD WATER THERMO.
GA-208 WATER PUMP
GB-374 DEFROST VALVE
SCHEMATIC
JUNCTION BOX FOR INCOMING LINE BEHIND LEFT END PANEL. CONNECT WITH #12 WIRE TO 115 V. DISCONNECT SWITCH. MAXIMUM BRANCH CIRCUIT USE 20 AMP. FUSE.
CHANNEL
'W' MODELS ONLY
HIGH PRESS. CUTOUT
COMP.
START CAP.
DEFROST VALVE JUNC. BOX
BOTTOM
TOP
BIN THERMO.
COLD WATER THERMO.
PUMP TOGGLE (SNAP ACTION AS SHOWN)
ACTUATOR THERMOSTAT
ACTUATOR TOGGLE
WASHING SWITCH
WATER VALVE
GB-336-B
WEIGHT CONTROL SWITCH
GB-385
PICTORIAL
YH 75V-1
NAME: ICE MAKER WIRING DIAGRAM
MODEL: GB5AJ & GB5WJ
DWN: D.P.
DATE: 6-27-73
SCALE: —
UNIFLOW MFG. CO. ERIE, PA.
GB-1858

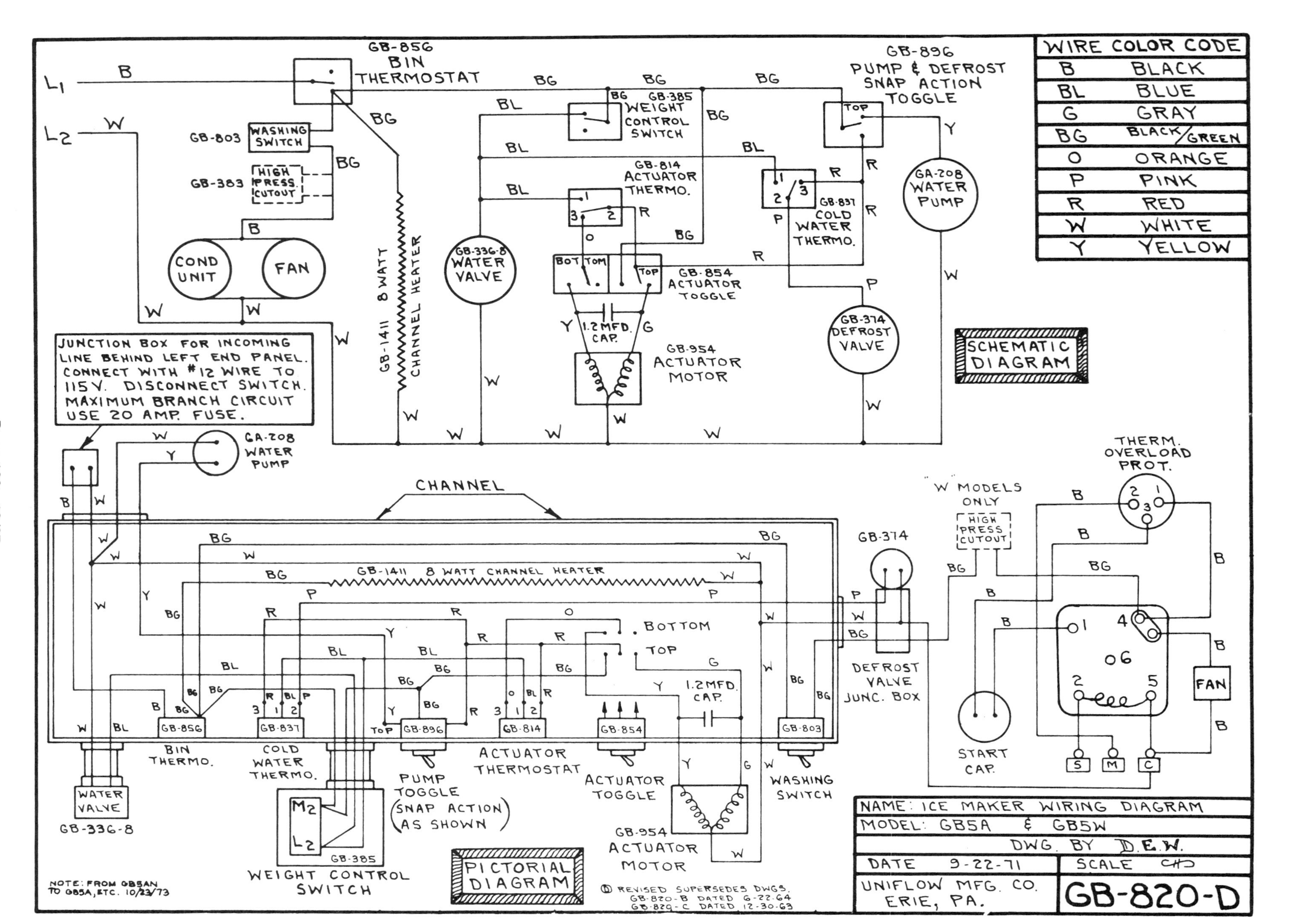
WIRE COLOR CODE
B BLACK
BL BLUE
G GRAY
BG BLACK/GREEN
O ORANGE
P PINK
R RED
W WHITE
Y YELLOW
SCHEMATIC DIAGRAM
PICTORIAL DIAGRAM
NAME: ICE MAKER WIRING DIAGRAM
MODEL: GB5A & GB5W
DWG. BY D.E.W.
DATE 9-22-71
SCALE CHD
UNIFLOW MFG. CO.
ERIE, PA.
GB-820-D
JUNCTION BOX FOR INCOMING LINE BEHIND LEFT END PANEL. CONNECT WITH #12 WIRE TO 115V. DISCONNECT SWITCH. MAXIMUM BRANCH CIRCUIT USE 20 AMP. FUSE.
GB-856 BIN THERMOSTAT
GB-803 WASHING SWITCH
GB-383 HIGH PRESS. CUTOUT
COND UNIT
FAN
GB-1411 8 WATT CHANNEL HEATER
GB-336-8 WATER VALVE
GB-385 WEIGHT CONTROL SWITCH
GB-814 ACTUATOR THERMO.
GB-854 ACTUATOR TOGGLE
1.2 MFD. CAP.
GB-954 ACTUATOR MOTOR
GB-837 COLD WATER THERMO.
GB-896 PUMP & DEFROST SNAP ACTION TOGGLE
GA-208 WATER PUMP
GB-374 DEFROST VALVE
CHANNEL
"W" MODELS ONLY
THERM. OVERLOAD PROT.
START CAP.
DEFROST VALVE JUNC. BOX
BOTTOM
TOP
BIN THERMO.
COLD WATER THERMO.
PUMP TOGGLE (SNAP ACTION AS SHOWN)
ACTUATOR THERMOSTAT
ACTUATOR TOGGLE
WASHING SWITCH
GB-336-8 WATER VALVE
WEIGHT CONTROL SWITCH
NOTE: FROM GB5AN TO GB5A, ETC. 10/23/73
REVISED SUPERSEDES DWGS. GB-820-B DATED 6-22-64 GB-820-C DATED 12-30-63

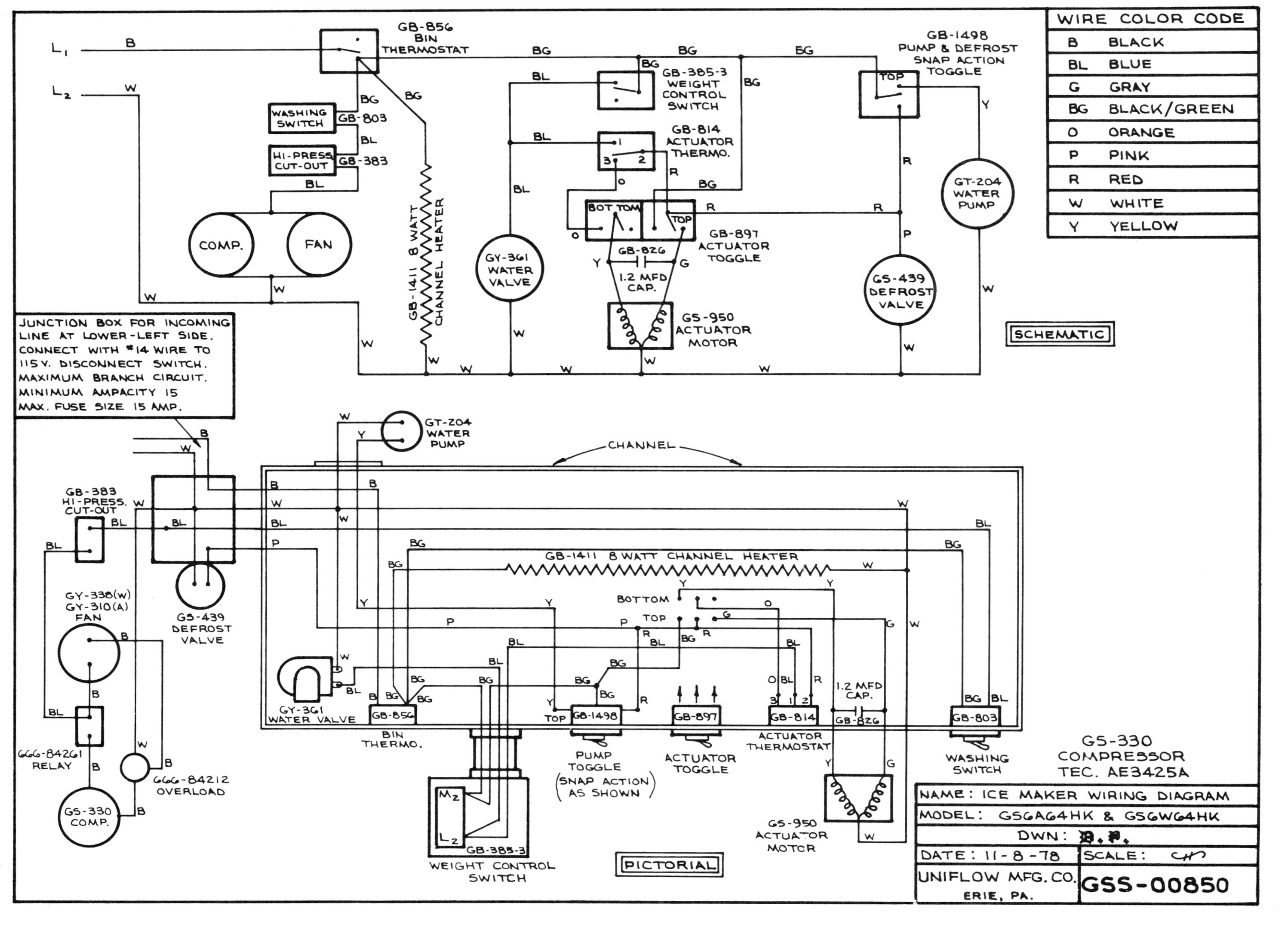
WIRE COLOR CODE
B BLACK
BL BLUE
G GRAY
BG BLACK/GREEN
O ORANGE
P PINK
R RED
W WHITE
Y YELLOW
SCHEMATIC
GB-856 BIN THERMOSTAT
GB-803 WASHING SWITCH
GB-383 HI-PRESS CUT-OUT
COMP.
FAN
GB-1411 8 WATT CHANNEL HEATER
GY-361 WATER VALVE
GB-385-3 WEIGHT CONTROL SWITCH
GB-814 ACTUATOR THERMO.
GB-897 ACTUATOR TOGGLE
GB-826 1.2 MFD CAP.
GS-950 ACTUATOR MOTOR
GB-1498 PUMP & DEFROST SNAP ACTION TOGGLE
GT-204 WATER PUMP
GS-439 DEFROST VALVE
JUNCTION BOX FOR INCOMING LINE AT LOWER-LEFT SIDE. CONNECT WITH #14 WIRE TO 115V. DISCONNECT SWITCH. MAXIMUM BRANCH CIRCUIT. MINIMUM AMPACITY 15 MAX. FUSE SIZE 15 AMP.
CHANNEL
GB-383 HI-PRESS. CUT-OUT
GY-338(W) GY-310(A) FAN
666-84261 RELAY
666-84212 OVERLOAD
GS-330 COMP.
GS-439 DEFROST VALVE
GT-204 WATER PUMP
GB-1411 8 WATT CHANNEL HEATER
BOTTOM
TOP
GY-361 WATER VALVE
GB-856 BIN THERMO.
GB-1498 PUMP TOGGLE (SNAP ACTION AS SHOWN)
GB-897 ACTUATOR TOGGLE
GB-814 ACTUATOR THERMOSTAT
1.2 MFD CAP. GB-826
GB-803 WASHING SWITCH
GS-950 ACTUATOR MOTOR
GB-385-3 WEIGHT CONTROL SWITCH
PICTORIAL
GS-330 COMPRESSOR TEC. AE3425A
NAME: ICE MAKER WIRING DIAGRAM
MODEL: GS6A64HK & GS6W64HK
DATE: 11-8-78
UNIFLOW MFG. CO. ERIE, PA.
GSS-00850

THERMOSTATICALLY CONTROLLED ICE CUBER COMPRESSOR CHART

Bold H.P. Indicates Models in Current Production.

60 HERTZ ONLY

H.P.	MODEL #	PART #	NAME	SUPPLIER	SUPPLIER'S #	MFR'S. #	ELECTRICAL RATING
1/5		GS-330	Compressor	Tecumseh	AE3425A/AE170AT	—	115 Volt - 60 Hertz
	GS6A	666-84261	Relay	Tecumseh	82626	—	—
	GS6W	666-84212	Overload	Tecumseh	—	—	—
1/3		GY-366	Compressor	Hupp	AYCH33-1-2537	—	115 Volt - 60 Hertz
	GY3A1	GY-366-1	Relay	Hupp	1456-3033	3CR-203-186	—
	GY3W1	GY-366-2	Overload	Hupp	1456-3043	MRA-1728	—
1/3		GY-399	Compressor	Copeland	JFB1-003-1AA-214	—	115 Volt - 60 Hertz
	GY3A1	GY-399-1	Relay	Copeland	040-0090-01	3ARR12PB203	—
	GY3W1	GY-399-2	Start Cap.	Copeland	014-0032-00	—	233-280 Mfd. 110 Volt
		GY-399-3	Overload	Copeland	071-0369-22	MRA-7991-160	—
1/2		GB-368-4	Compressor	B/W	H45-1-1804	—	115 Volt - 60 Hertz
	GB5AN	GB-368-1	Relay	B/W	1456-637	3ARR3B17M5	—
	GB5WN	GB-368-2	Start Cap.	B/W	1509-51	35F125BC2	324-388 Mfd. 110 Volt
		GB-368-3	Overload	B/W	1456-626	CRB3917	—
1/2		GB-1358	Compressor	B/W	H50-1-1807	—	115 Volt - 60 Hertz
	GB5AN	GB-368-1	Relay	B/W	1456-637	3ARR3B17M5	—
	GB5WN	GB-368-2	Start Cap.	B/W	1509-51	35F125BC2	324-388 Mfd. 110 Volt
		GB-368-3	Overload	B/W	1456-626	CRB3917	—
3/4	GY7A	GY-331	Compressor	B/W	YH75V-1-2348	—	115 Volt - 60 Hertz
	GY7W	GY-331-1	Relay	B/W	1456-966	6409-25-933	—
	GT7A	GY-331-2	Overload	B/W	1456-813	MST00AFN-136	—
	GT7W	GY-331-3	Start Cap.	B/W	1509-102	35F163BC2	189 Mfd. 110 Volt
	GB5ANJ						
3/4		GT-373	Compressor	Copeland	RSL2-0075-CAA-204	—	115 Volt - 60 Hertz
	GT7AN	GT-373-1	Relay	Copeland	040-0001-25	3ARR3CT15M1	
	GT7WN	GT-373-2	Start Cap.	Copeland	014-0008-79	35F1277BC8	270-324 Mfd. 165 Volt
	GB7AN	GT-373-3	Run Cap.	Copeland	014-0001-00	97F5066	10 Mfd. 370 Volt
	GB7WN	GT-373-4	Overload	Copeland	071-0127-06	CRA-1719-138	—
3/4	GT7A	GT398	Compressor	Tecumseh	AJ7465A	—	115 Volt - 60 Hertz
	GT7W	GT398-1	Relay	Tecumseh	820ARR3892	3ARR3A2J6	—
	GB7A	GT398-3	Overload	Tecumseh	8300CSTA96	CST-17AEN-181	—
	GB7W	GT398-2	Start Cap.	Tecumseh	85PS110C09	—	270-324 Mfd. 110 Volt
		GT398-4	Run Cap.	Tecumseh	85561-2	97F5050	14 Mfd. 370 Volt
1		GB-377	Compressor	Tecumseh	B1T16 1212-68-4	—	230 Volt - 60 Hertz
	GB1A	GB-377-2	Relay	Tecumseh	82135-1	3ARR3A13D3	—
	GB1W	GB-377-3	Start Cap.	Tecumseh	85551-3	35F296BC2	141-170 Mfd. 220 Volt
	GB1X	GT-398-4	Run Cap.	Tecumseh	85561-2	97F5050	15 Mfd. 370 Volt
		GB-377-5	Overload	Tecumseh	83223	CRA 3971	—
1	GT1W	GB-1345	Compressor	B/W	H100T-2-2155	—	230 Volt - 60 Hertz
	GB1A	GB-1345-1	Overload	B/W	1456-618	CEB 2998	—
	GB1W	GB-1328-1FF	Relay	B/W	1456-673	3ARR3B3C3	—
	GB1X	GB-377-3	Start Cap.	Tecumseh	85551-3	35F296BC2	141-170 Mfd. 220 Volt
		GT-398-4	Run Cap.	Tecumseh	85561-2	97F5050	15 Mfd. 370 Volt
1	GT1W	GB-1347	Compressor	Tecumseh	AH7480A/AH160FT	—	230 Volt - 60 Hertz
	GB1A	GB-1347-1	Relay	Tecumseh	82765	3ARR3K13B5	—
	GB1W	GB-1347-3	Start Cap.	Tecumseh	85577-1	35F1622BC2	135-155 Mfd. 330 Volt
	GB1X	GT-398-4	Run Cap.	Tecumseh	85561-2	97F5050	15 Mfd. 370 Volt
1		GB-1385	Compressor	Copeland	SSE4-0100-CAB-214	—	230 Volt - 60 Hertz
	GT1W	GB-1385-1	Relay	Copeland	040-0001-19	3ARR3CT20V5	—
	GB1A	GB-1385-2	Overload	Copeland	071-0127-28	CRA-1783-138	—
	GB1W	GB-1385-3	Run Cap.	Copeland	014-0002-02	97F5052	20 Mfd. 370 Volt
	GB1X	GB-1357-3FF	Start Cap.	Copeland	014-0008-51	35F1316BC8	145-175 Mfd. 220 Volt
1	GT1W	GB-2350	Compressor	Bristol	M1NB-922-BBCB	—	230 Volt - 60 Hertz
	GB1A	GB-1385-1	Relay	Copeland	040-0001-19	3ARR3CT20V5	—
	GB1W	GB-1347-3	Start Cap.	Tecumseh	85577-1	35F1622BC2	135-155 Mfd. 330 Volt
	GB1X	GB-1385-3	Run Cap.	Copeland	014-0002-02	97F5052	20 Mfd. 370 Volt
11,000 B.T.U.	GB2WN5HK	GB-1395	Compressor	Tecumseh	AH7511B/AH311GT	—	230 Volt - 60 Hertz
		GB-1395-1	Relay	Tecumseh	82457	3ARR3K13P5	—
		GB-1395-2	Start Cap.	Tecumseh	85668	35F1558BC2	88-106 Mfd. 250 Volt
		GB-1395-5	Run Cap.	Tecumseh	85PR370B57	97F5165	35 Mfd. 370 Volt
14,000 B.T.U.	GB4WN7	GB-1396	Compressor	Tecumseh	AH7514A/AH333ET	—	230 Volt - 60 Hertz
		GB-1396-1	Relay	Tecumseh	82477	3ARR3A13P2	—
		GB-1395-2	Start Cap.	Tecumseh	85668	35F1558BC2	88-106 Mfd. 250 Volt
		GB-1395-5	Run Cap.	Tecumseh	85PR370B57	97F5165	35 Mfd. 370 Volt

For obsolete compressor cross reference refer to Parts Price Section, Page 4

ICE CUBER COMPRESSOR CHART
50 CYCLE ONLY

*Indicates Models in Current Production.

H.P.	MODEL #	PART #	NAME	SUPPLIER	SUPPLIER'S #	MFR'S. #	ELECTRICAL RATING
1/5	GS6A *	GS-331FF	Compressor	Tecumseh	AE3425A/AE170JT	—	240/220 Volt - 50 Cycle
	GS6W	666-84263	Relay	Tecumseh	82622	—	—
		666-84264	Overload	Tecumseh	—	—	—
1/3		GY-323FF	Compressor	B/W	YCH33V-8 BM 2219	—	230 Volt - 50 Cycle
	GY3A1FF	GY-323-1FF	Overload	B/W	BW 1456-880	MRP39HK-85	—
	GY3W1FF	GY-323-2FF	Relay	B/W	BW 1456-891	6409-28-156	—
		GY-323-3FF	Start Cap.	B/W	BW 1509-148	35F1154BC2	47 Mfd. 220 Volt
1/3	*	GB-1324-1FF-B	Compressor	Tecumseh	AJ4461A/AJ311JT-204-B4	—	230 Volt - 50 Cycle
	GY3A1FF	GB-1324-6FF	Overload	Tecumseh	83929	MRT36ALX-112	—
	GY3W1FF	GB-1324-3FF	Relay	Tecumseh	82487-1/82003CRB39	3ARR2-CP176S	—
		GB-1324-4FF	Start Cap.	Tecumseh	85PS250B38	35F1520-BC9	64-77 Mfd. 250 Volt
1/2		GB-1324-1FF	Compressor	Tecumseh	CAJ2612/AJH10-141-4J	—	230 Volt - 50 Cycle
	GB5ANFF	GB-1324-3FF	Relay	Tecumseh	82487-1/82003CRB39	3ARR2-CP176S	—
	GB5WNFF	GB-1324-6FF	Overload	Tecumseh	83929	MRT36ALX-112	—
		GB-1324-4FF	Start Cap.	Tecumseh	85PS250B38	35F1520-BC9	64-77 Mfd. 250 Volt
1/2		GB-1324-1FF-A	Compressor	Tecumseh	AJ4461A/AJ311JT-141-C1	—	230 Volt - 50 Cycle
	GB5ANFF	GB-1324-3FF	Relay	Tecumseh	82487-1/82003CRB39	3ARR2-CP176S	—
	GBWNFF	GB-1324-6FF	Overload	Tecumseh	83929	MRT36ALX-112	—
		GB-1324-4FF	Start Cap.	Tecumseh	85PS250C32	35F1520-BC9	64-77 Mfd. 250 Volt
3/4		GY-341FF	Compressor	B/W	BW YH75V-8-2393	—	230 Volt - 50 Cycle
	GY7AFF	GY-341-1FF	Overload	B/W	1456-956	MST28ALN-122	—
	GY7WFF	GY-341-2FF	Relay	B/W	1456-957	6409-11-188	—
		GY-341-3FF	Start Cap.	B/W	1509-143	35F1121BC	72-88 Mfd. 220 Volt
3/4	GB7ANFF *	GT-372FF	Compressor	Tecumseh	AJ7465B/AJ201JT-206-B4	—	230 Volt - 50 Cycle
	GB7WNFF	GT-372-1FF	Relay	Tecumseh	82772-1	3ARR2 KCP171S	Spade Terminals
						3ARR2 KCR171S	Screw Terminals
	GT7ANFF	GT-372-2FF	Overload	Tecumseh	83984	MST16A1K-112	—
	GT7WNFF	GT-372-3FF	Start Cap.	Tecumseh	855S330A59	35F887-BC9	108-130 Mfd. 330 Volt
1		GB-1328FF	Compressor	B/W	BW H100t-8-2330	—	230 Volt - 50 Cycle
	GB1ANFF	GB-1328-1FF	Relay	B/W	1456-673	3ARR3-B3C3	—
		GB-1328-2FF	Overload	B/W	1456-962	CEB1780	—
	GB1WNFF	GB-377-3	Start Cap.	Tecumseh	85551-3	35F296BC2	141-170 Mfd. 220 Volt
		GT-398-4	Run Cap.	Tecumseh	85561-2	97F4525	15 Mfd. 370 Volt
1		GB-1357FF	Compressor	Copeland	SSE4-0100-CAG 207BM	—	230 Volt - 50 Cycle
		GB-1357-1FF	Relay	Copeland	040-0001-10	3ARR3CT6AB5	—
	GB1ANFF	GB-1357-2FF	Overload	Copeland	071-0127-13	CRA-1735-138	—
	GB1WNFF	GB-1357-3FF	Start Cap.	Copeland	014-0008-51	35F1316BC8	145-175 Mfd. 220 Volt
		GB-1357-4FF	Run Cap.	Copeland	014-0001-04	45F258FB	15 Mfd. 370 Volt
		GT-398-4	(Alternate)	Tecumseh	85561-2	97F4525	15 Mfd. 370 Volt
1	*	GB-2352FF	Compressor	Copeland	SSE4-0100-CAZ-204	—	230 Volt - 50 Cycle
	GB1ANFF	GB-2352-1FF	Relay	Copeland	040-0001-27	3ARR3CT10A55	—
	GB1WNFF	GB-2352-2FF	Overload	Copeland	071-0127-10	CST-31ALN-138	—
		GB-1324-4FF	Start Cap.	Tecumseh	85P5250B38	35F1520-BC9	64-77 Mfd. 250 Volt
		GB-1385-3	Run Cap.	Copeland	014-0002-02	97F4527	20 Mfd. 370 Volt

ICE STATIONS

The Ice Station is a self-contained Cuber Dispenser.

1. The Ice Making Compartment is secured to the top of a 200# Stainless Steel Storage Bin. It is connected to the condensing unit which is located on the base of the Ice Station.

2. The ice falls into the insulated Storage Bin and is dispensed into the sink on demand.

3. Every time ice is dispensed the dispensing motor turns both the paddles to dispense the cubes and rotates the agitator to insure the ice is dispensed as cubes.

4. The ice is dispensed into the sink area through an ice safety chute into the ice bucket placed on the grate.

5. The water inlet fitting, condenser and gravity drain connections are located on the back of the Ice Station at the base. There are plugs for bottom connections on each Ice Station.

6. There are 3 gravity drains that tee together and drain to the back bottom base. The drains are for:

   A) Drain pan for the dreg water that is discarded at the end of each cycle.

   B) Bin drain for any meltage that occurs in the Bin.

   C) Sink drain as a means to dispell water from the meltage of cubes that land in the sink should someone overfill the bucket or dispense ice without a bucket.

BASIC MODELS

Although there are many Models, they fall under three (3) Basic Models.

1. CONTINUOUS DISPENSE MODEL - Ice will dispense as long as the rocker switch is pushed.

2. PORTION CONTROL - Continuous Dispense Model. The end user will determine which method of ice dispensing they desire. Pushing the portion control receiver switch, the unit will dispense a pre-determined amount of ice. Pushing the continuous dispense rocker switch it will dispense ice as long as the switch is held.

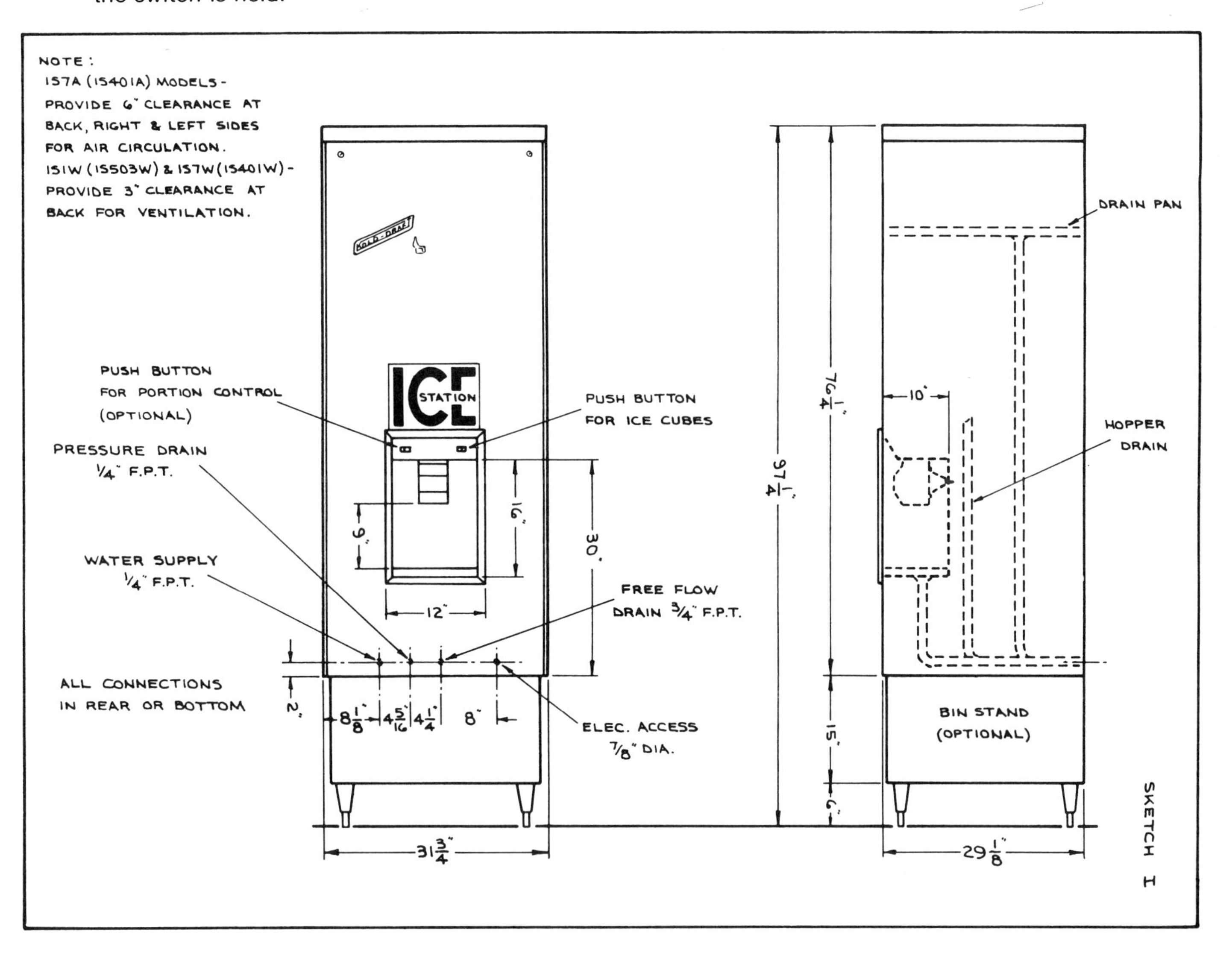

3. DELUXE ICE STATION which may be coin operated, key operated or a push button Model. These Models have standard features not available on othe Models. They are:

   1. Bin Status Indicator which tells at a glance whether the Bin is full or empty. Green light indicates ice is ready to be vended. A Red light which indicates an empty Bin and ice will not be vended.

   2. Photo Electric Eye to end the problem of wasted ice as ice will not be dispensed until electric eye "sees" the container in place.

   3. Lift up Safety Door where ice will not be vended until the door is completely down.

   4. Adjustable portion control from 8 ounces to 5 lbs.

   5. Five digit vend counter to keep exact record of how many times the machine is used.

   6. A safety lock on the hinged door.

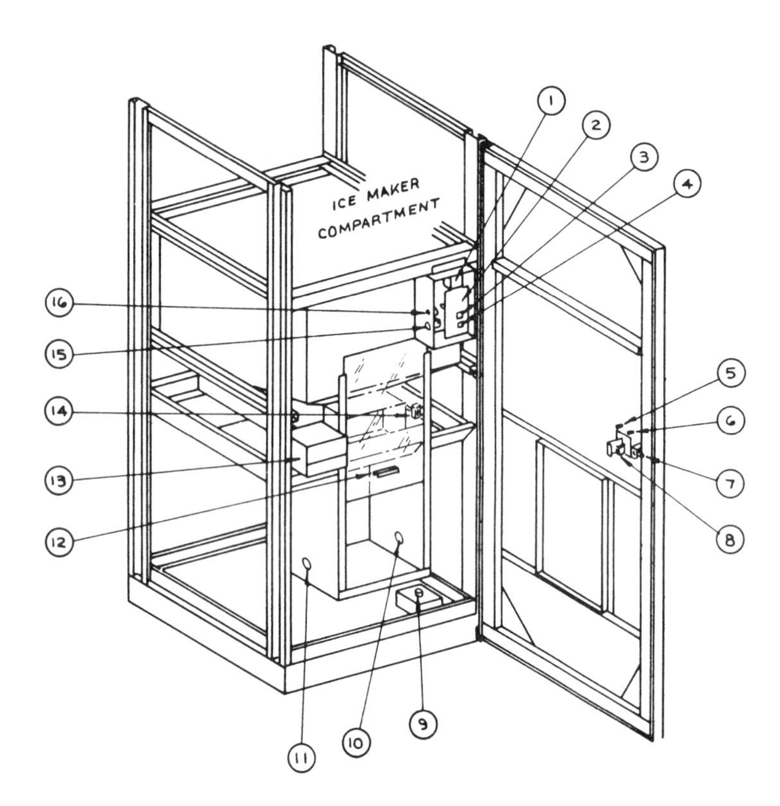

1. Coin Counter (555-01098)
2. Dispenser Control Printed Circuit Card (MDD-00319)
3. Dispenser Delay Timer Adjustment
4. Photo Eye Sensitivity Adjustment
5. Bin Empty Light (Red) (MDD-00313)
6. Coin Accept. Light (Green) (MDD-00312)
7. Front Panel Latch (MDD-00320-08)
8. Coin Acceptor Mechanism (MDD-00352)

8A. Push Button Mechanism (MDD-00468)

8B. Key-Op Mechanism (Ours) (TTT-02813)

8C. Key-Op Mechanism (Theirs)

9. Dispenser Line Fuse 6-1/2 Amp. (MDD-00132)
10. Photo Cell (MDD-00311)
11. Photo Cell Light (MDD-00315)
12. Lift-Up Door (MDD-00323)
13. Coin Box (MDD-00326)
14. Safety Door Switch (MDD-00111)
15. Bin Thermostat Adjustment (MDD-00325)
16. Dispenser Timer Adjustment (MDD-00317)

INSTRUCTIONS PERTINENT TO COIN-OP OPTION ONLY

The Coin-Op Option is a separately fused, independently operated circuit with no interacting circuitry connected to the cuber section of the Dispenser.

The Dispenser has its own bin thermostat with an adjustment accessible from the front of the control card box to the right of the dispensing mechanism.

The coin acceptor is in the middle left front section of the Dispenser Door and has two lights, a coin reject relay and a coin accepting microswitch. A coin box is mounted under the acceptor on the Dispenser frame. The Red light lights when the bin is low on ice causing the bin thermostat to shut off, indicating that the Dispenser will not dispense ice and will reject coins or tokens. A full bin will trip the bin thermostat and turn the Red light off.

Two safety controls are functioning in the Dispenser which will not allow the acceptor to accept coins or tokens and dispense ice until the safety controls are satisfied. The safety controls are a photo eye which senses the presence of a bucket in the Dispenser and a door switch which senses the front door being completely closed. When these conditions are met, the Green light on the acceptor will light if the Red light is off, and indicates that the acceptor will accept a coin or token.

An adjustment of the photo eye sensitivity is available on the printed circuit control card which is factory set, but may require field adjustment is the Green light comes on when no bucket is in the Dispenser. To adjust the photo eye sensitivity, the Red light must be off, the Yellow light in the Dispenser on, and no bucket present. Remove the control box cover and turn the lower thumbwheel potentiometer on the printed circuit card fully clockwise, looking from the Right side, and notice that the Green light comes on. Start to turn the thumbwheel counter-clockwise until the Green light goes out and continue to turn the wheel 20° or the width of one notch of the wheel past that point when the Green light goes out. The thumbwheel to the top of the box is factory set and should not be reset in the field.

The ice portion dispensed is adjustable from one to eight pounds. A timer on the left side of the control card box is adjustable over a 270° rotation which allows from 2 to 25 seconds of dispensing which is sufficient to meet the required weight portion. This adjustment is made in the field and requires several trial dispensing operations to meet the desired portion.

CAUTION: The MDD-00329 is a dual operating voltage printed circuit card. When exchanging cards, refer to MDD-00309 for location of jumpers for correct incoming line voltage of 115 or 230.

"Quarter Substitute Tokens" (QST) are available from Van Brook of Lexington, Inc.
P.O. Box 5044, Lexington, Ky.
(606) 255-5990 40505

April 6, 1978
Revised - June 2, 1978
Oct. 18, 1978

SERVICE INFORMATION

1. Front Panel Access

    1. Continuous dispense or portion control and continuous dispense Models - Loosen 2 locking screws and lift up on panel to clean lip on door bottom.

    2. Deluxe Models simply unlock door and door will pivot on hinge.

2. Ice Maker produces ice in the same manner asany 3/4 H.P. or 1 H.P. Kold-Draft Cuber. For service to ice maker section refer to Electrical Circuits, Service Information or Trouble, Cause and Remedy Sections for the Electronic or Electro-Mechanical Cubers.

3. Access to Condensing Unit

    1. Remove unit from the wall and remove rear panels.

    2. If access to the rear is impractical, the sink may be removed for front access. There are only 4 mounting nuts and washers to remove.

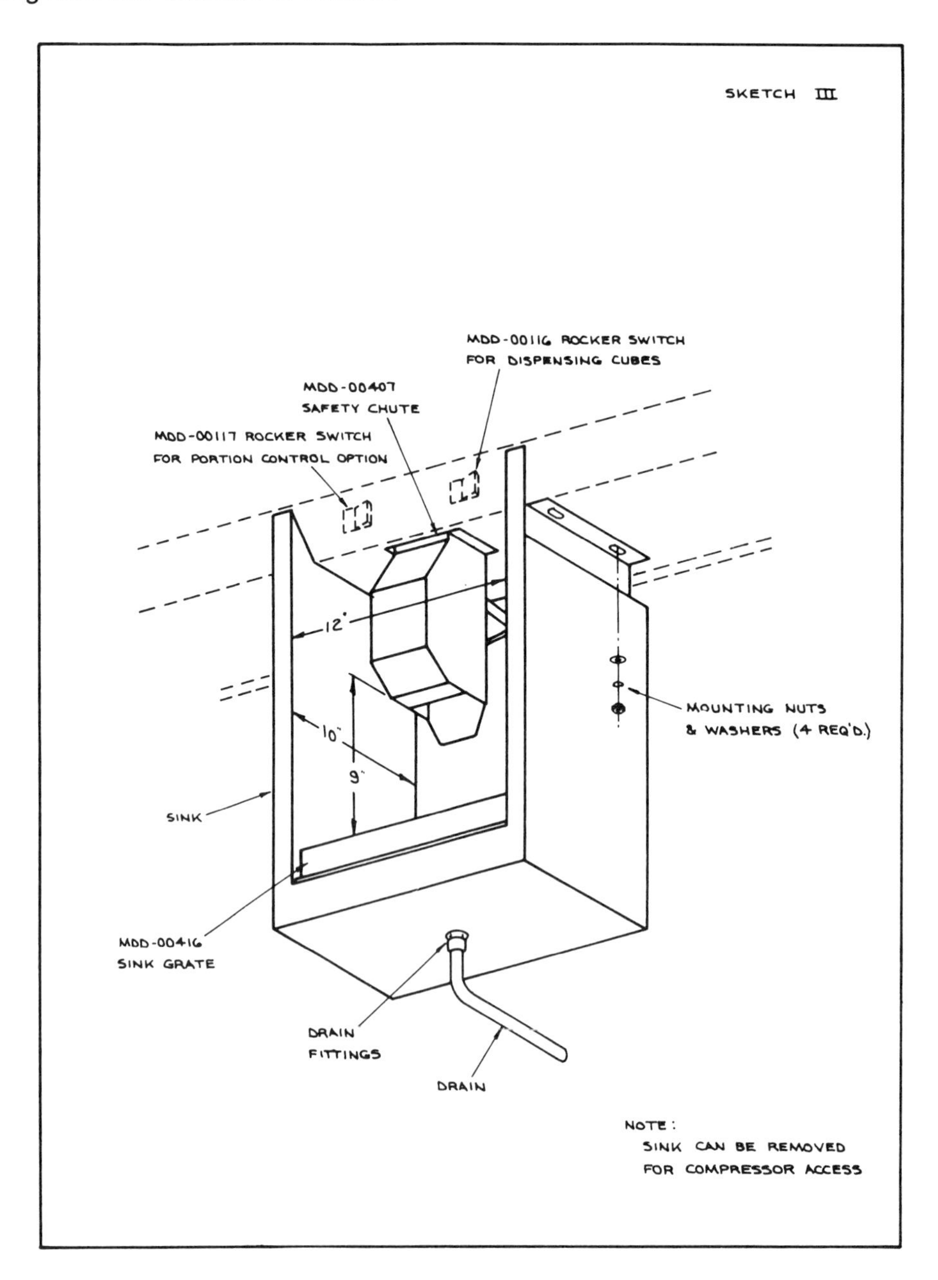

4. Bridging

Although the Kold-Draft Ice Station (formerly the "Medi-Cuber") contains a built-in agitator to prevent ice bridging, ice jams can still occur in the throat area after a prolonged period of inactivity.

Twice-daily use of the Ice Station should provide enough agitation to keep the ice in the bin in a "fluid" state, however there are a few adjustments which can be made and points that should be checked to minimize ice bridging.

1) Check the drain pan and drain yoke for leaks - water splashing on the ice in the bin can freeze freshly harvested cubes together.

2) A prolonged freeze cycle can "sub-cool" the cubes. We all know that water freezes at 32°, but ice can be any temperature below 32°. If the harvested ice is considerably below 32°, the wet surfaces which develop during harvest will re-freeze in the bin, "welding" the cubes into clumps.

   Keep the freeze cycle as short as possible. Increase the dimple size by raising the control stream and/or the water level at the start of defrost. Decrease the fill level if the control stream goes over the dam more than 20 seconds. Keep the low side pressure above 4 PSIG at the end of the freeze cycle. Adjust the TXV to provide an even dimple pattern over the entire ice slab. (Check your Service Manual for specific adjustment procedures.)

3) In some cases, Ice Station throats have been blocked by shards of ice which are actually fins which have broken off the cubes. Keep the fin thickness as small as possible. Shards formed from thin fins are generally melted before they get to the throat.

4) Check the ice stop. If the ice stop has dropped, it can block ice flow to the paddles. If a small amount of "overcast" is not objectionable, the ice stop can be removed.

5) Check the agitator. Be sure it revolves when the dispense button is pressed. Check for broken welds between the band and the rods, and between the rods and the shaft.

6) This has only happened once that I've heard of, and it's probably not worth mentioning, but it is possible to reverse the direction of the Dispenser motor. One serviceman, when installing a new gear motor, inadvertently reversed the rotation of the motor which resulted in crushing the ice as the paddles attempted to force ice through the small space under the paddle wheels. Some of the crushed ice came through, but most of it backed up into the throat blocking flow before any serious damage occurred to the drive train.

One final comment: If you are servicing an Ice Station which has jammed, it is very important to completely empty the bin and wash out all ice shards before the unit is put back into service.

ELECTRO-MECHANICAL ICE MAKING SECTION ONLY

DIAGRAMS ILLUSTRATING OPTIONS MAY BE FOUND ON THE FOLLOWING PAGES.

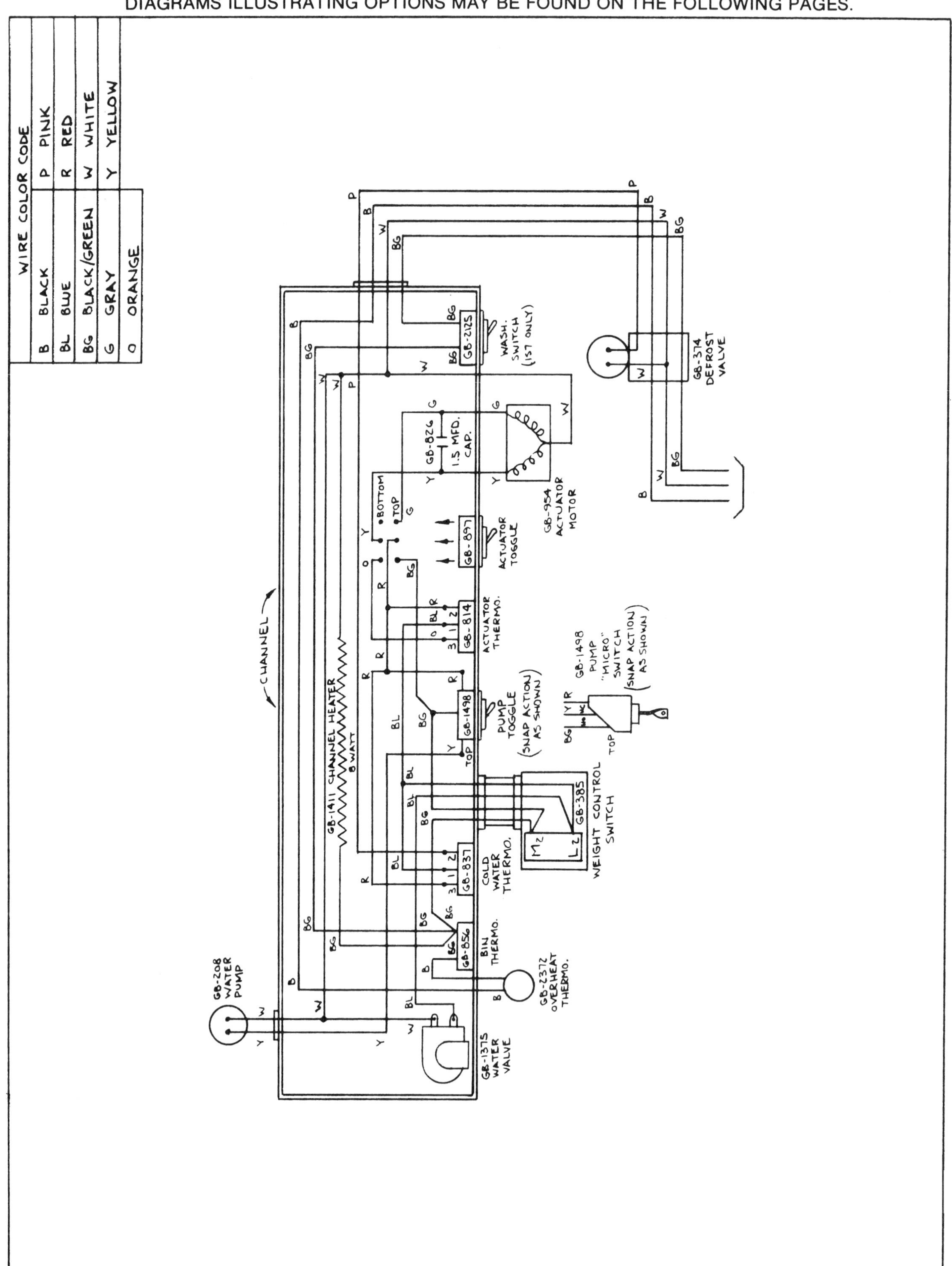

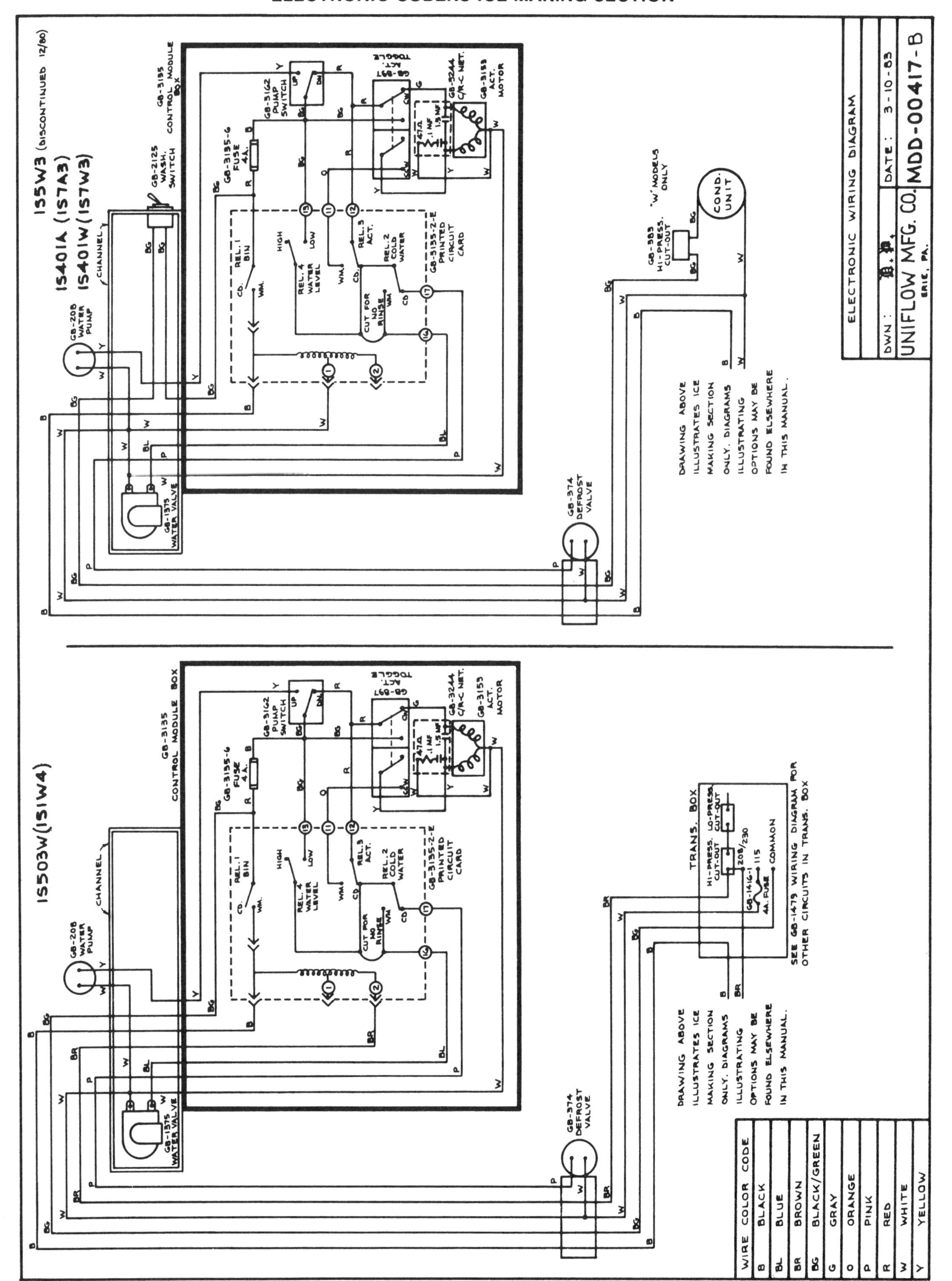
IS5W3 (DISCONTINUED 12/80)
IS401A (IS7A3)
IS401W (IS7W3)
IS503W (IS1W4)
GB-3135 CONTROL MODULE BOX
GB-3162 PUMP SWITCH
GB-3135-6 FUSE 4A.
GB-2125 WASH. SWITCH
GB-897 ACT. TOGGLE
GB-3244 C/R-C NET.
GB-3153 ACT. MOTOR
GB-3135-2-E PRINTED CIRCUIT CARD
REL. 1 BIN
REL. 2 COLD WATER
REL. 3 ACT.
REL. 4 WATER LEVEL
CUT FOR NO RINSE
HIGH
LOW
CHANNEL
GB-208 WATER PUMP
GB-1375 WATER VALVE
GB-374 DEFROST VALVE
GB-383 HI-PRESS. CUT-OUT
'W' MODELS ONLY
COND. UNIT
TRANS. BOX
HI-PRESS. CUT-OUT
LO-PRESS. CUT-OUT
208/230
115
COMMON
GB-1416-1 4A. FUSE
SEE GB-1479 WIRING DIAGRAM FOR OTHER CIRCUITS IN TRANS. BOX
DRAWING ABOVE ILLUSTRATES ICE MAKING SECTION ONLY. DIAGRAMS ILLUSTRATING OPTIONS MAY BE FOUND ELSEWHERE IN THIS MANUAL.
ELECTRONIC WIRING DIAGRAM
DWN: D.P.
DATE: 3-10-83
UNIFLOW MFG. CO. ERIE, PA.
MDD-00417-B
WIRE COLOR CODE
B BLACK
BL BLUE
BR BROWN
BG BLACK/GREEN
G GRAY
O ORANGE
P PINK
R RED
W WHITE
Y YELLOW

WIRE COLOR CODE	
B	BLACK
BR	BROWN
BG	BLACK/GREEN
W	WHITE

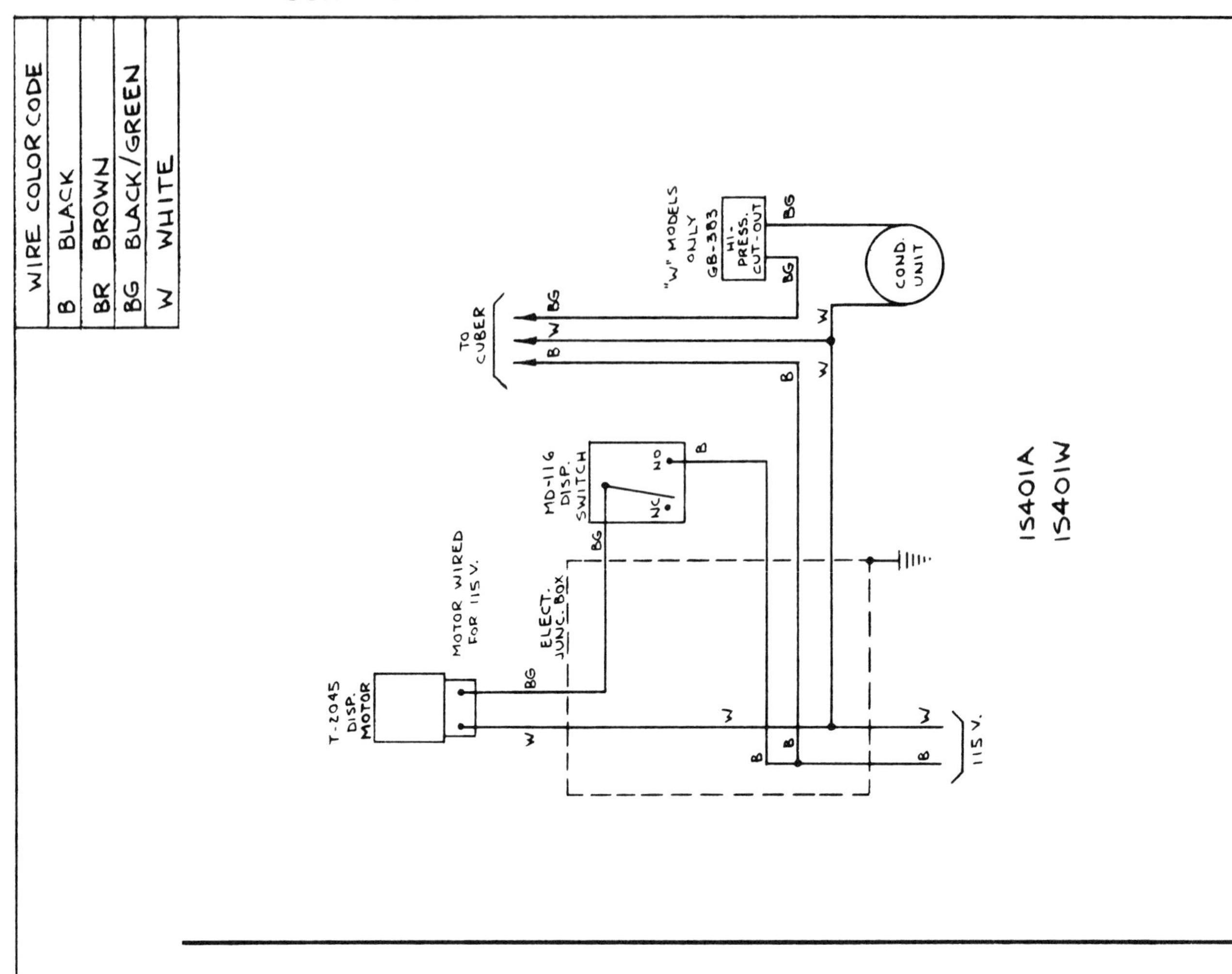

TO CUBER
TRANS. BOX
COM.
115
208/230
GB-1416-1
FUSE
HI-PRESS. CUT-OUT
LO-PRESS. CUT-OUT
SEE GB-1479 WIRING DIAGRAM FOR OTHER CIRCUITS IN TRANS. BOX
MD-116 DISP. SWITCH
MOTOR WIRED FOR 230 V.
ELECT. JUNC. BOX
T-2045 DISP. MOTOR
230 V. - 60 HZ. OR 208 V. - 60 HZ.
1S503W

NOTE: ON 208 V. MOVE BROWN WIRE FROM 230 TO 208 TERMINAL ON TRANSFORMER.

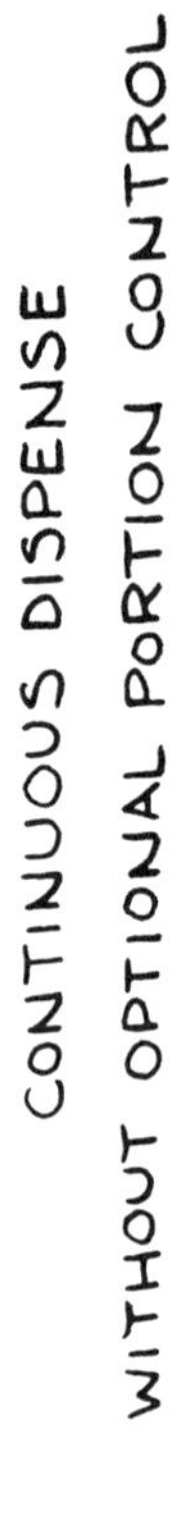

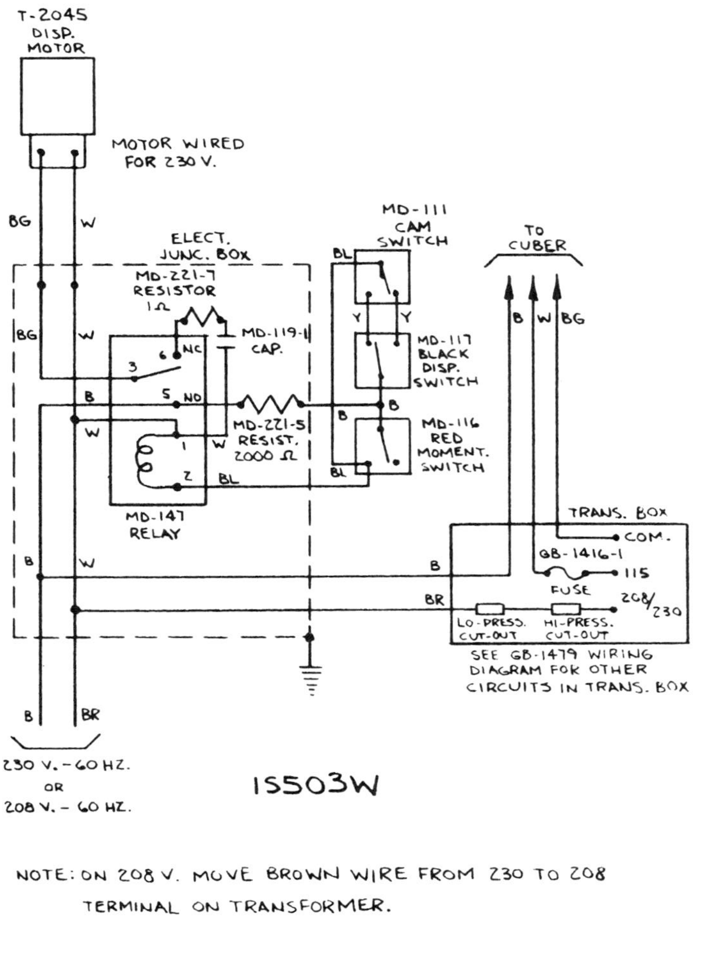

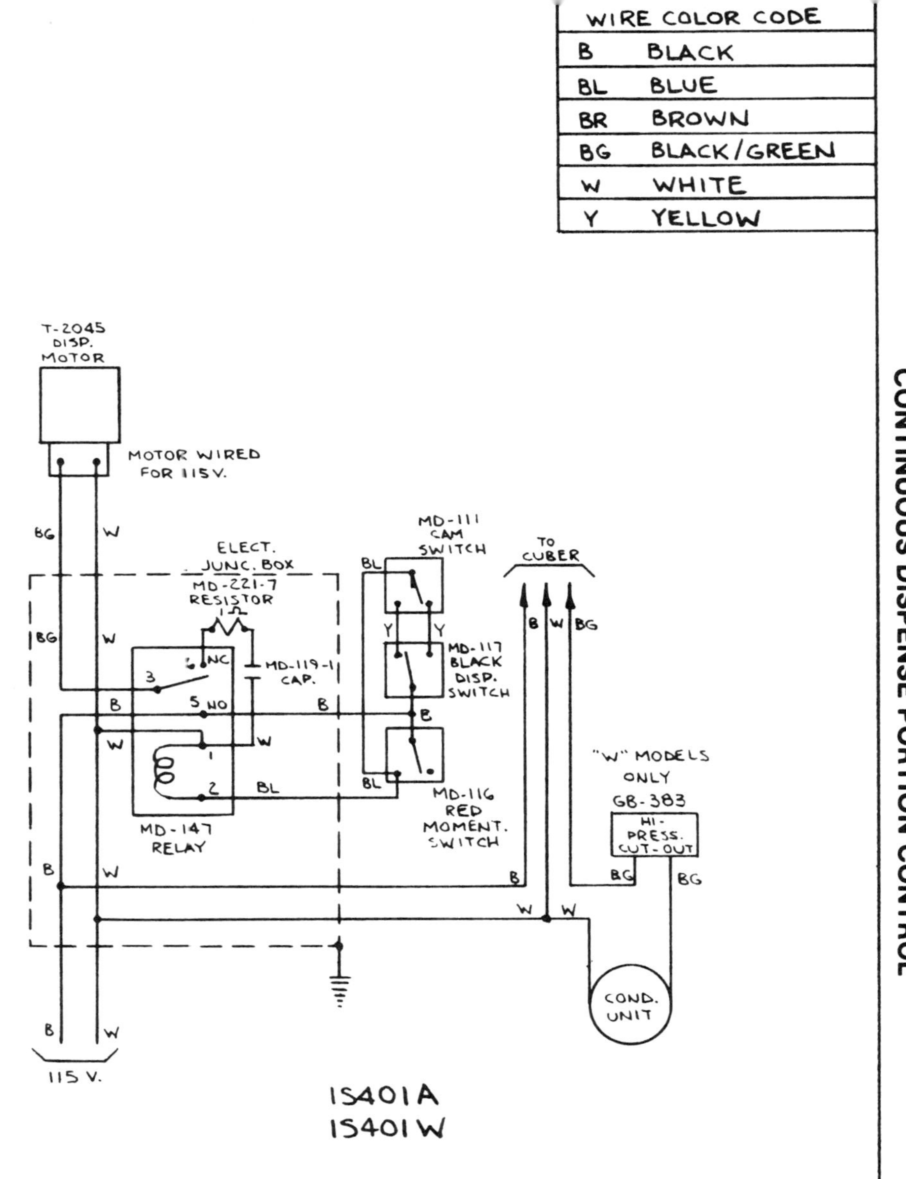

CONTINUOUS DISPENSE PORTION CONTROL

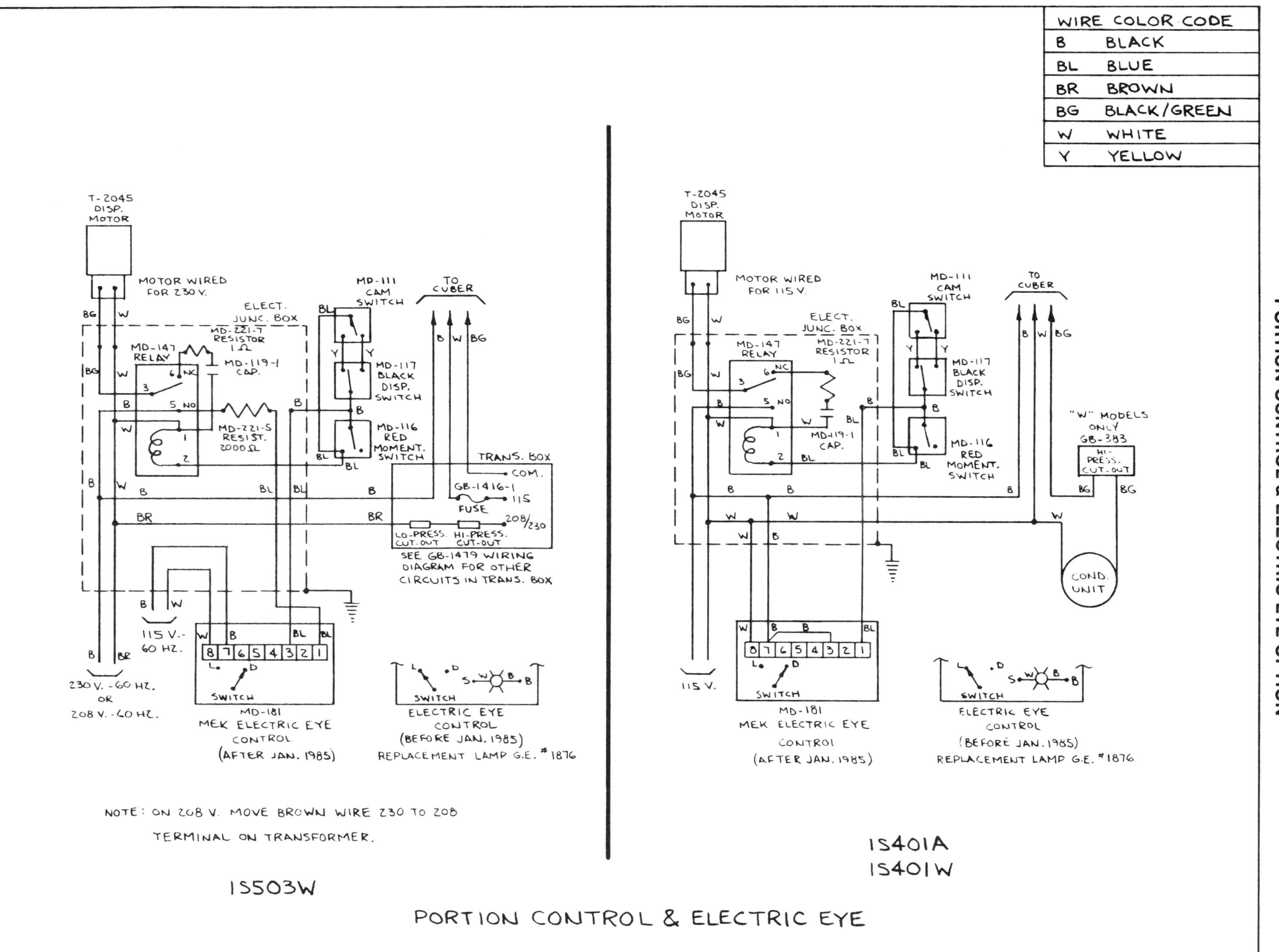

WIRE COLOR CODE
B BLACK
BL BLUE
BR BROWN
BG BLACK/GREEN
W WHITE
Y YELLOW
T-2045 DISP. MOTOR
MOTOR WIRED FOR 230 V.
ELECT. JUNC. BOX
MD-221-7 RESISTOR 1 Ω
MD-147 RELAY
MD-119-1 CAP.
MD-221-5 RESIST. 2000 Ω
MD-111 CAM SWITCH
MD-117 BLACK DISP. SWITCH
MD-116 RED MOMENT. SWITCH
TO CUBER
TRANS. BOX
COM.
GB-1416-1 FUSE
115
208/230
LO-PRESS. CUT-OUT
HI-PRESS. CUT-OUT
SEE GB-1479 WIRING DIAGRAM FOR OTHER CIRCUITS IN TRANS. BOX
115 V.- 60 HZ.
230 V.- 60 HZ. OR 208 V.- 60 HZ.
SWITCH
MD-181 MEK ELECTRIC EYE CONTROL (AFTER JAN. 1985)
ELECTRIC EYE CONTROL (BEFORE JAN. 1985) REPLACEMENT LAMP G.E. #1876
NOTE: ON 208 V. MOVE BROWN WIRE 230 TO 208 TERMINAL ON TRANSFORMER.
IS503W
MOTOR WIRED FOR 115 V.
"W" MODELS ONLY GB-383
HI-PRESS. CUT-OUT
COND. UNIT
115 V.
IS401A
IS401W
PORTION CONTROL & ELECTRIC EYE

ELECTRIC EYE WIRING DIAGRAM

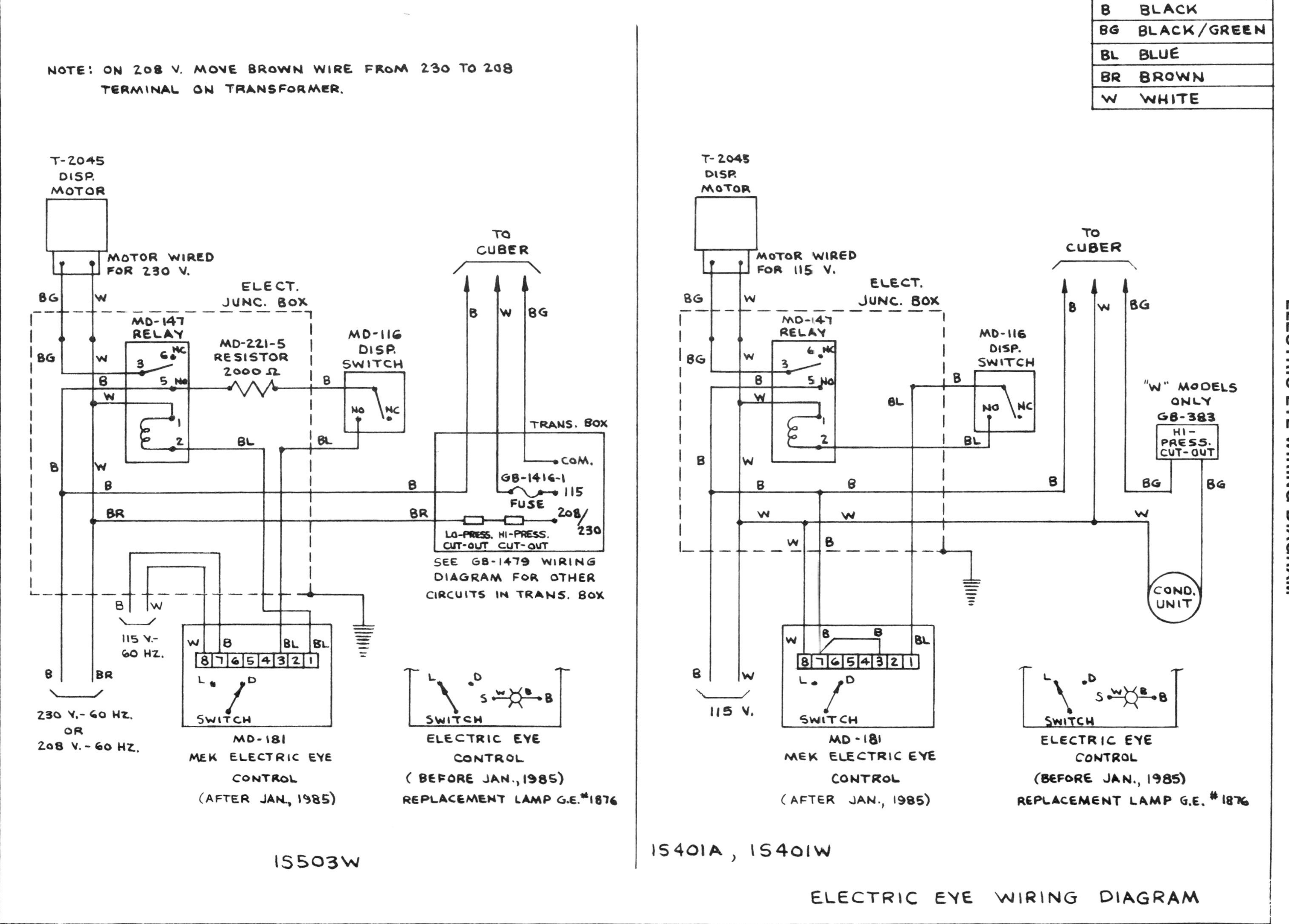

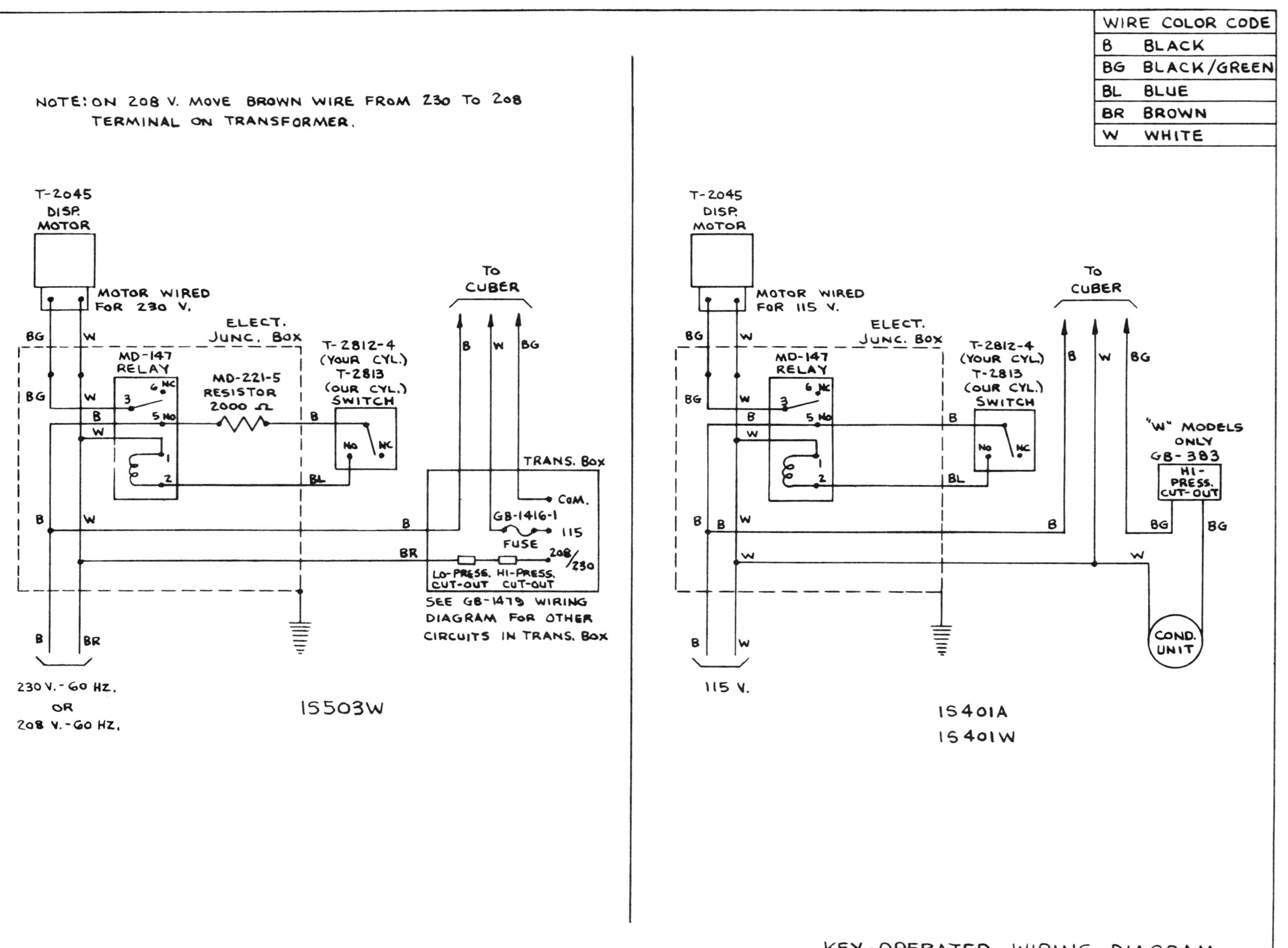

NOTE: ON 208 V. MOVE BROWN WIRE FROM 230 TO 208 TERMINAL ON TRANSFORMER.
WIRE COLOR CODE
B BLACK
BG BLACK/GREEN
BL BLUE
BR BROWN
W WHITE
T-2045 DISP. MOTOR
MOTOR WIRED FOR 230 V.
ELECT. JUNC. BOX
MD-147 RELAY
MD-221-5 RESISTOR 2000 Ω
T-2812-4 (YOUR CYL.) T-2813 (OUR CYL.) SWITCH
TO CUBER
TRANS. BOX
COM.
GB-1416-1
FUSE
115
208/230
LO-PRESS. CUT-OUT
HI-PRESS. CUT-OUT
SEE GB-1479 WIRING DIAGRAM FOR OTHER CIRCUITS IN TRANS. BOX
230 V.-60 HZ. OR 208 V.-60 HZ.
IS503W
T-2045 DISP. MOTOR
MOTOR WIRED FOR 115 V.
ELECT. JUNC. BOX
MD-147 RELAY
T-2812-4 (YOUR CYL.) T-2813 (OUR CYL.) SWITCH
TO CUBER
"W" MODELS ONLY GB-383
HI-PRESS. CUT-OUT
COND. UNIT
115 V.
IS401A
IS401W
KEY-OPERATED WIRING DIAGRAM

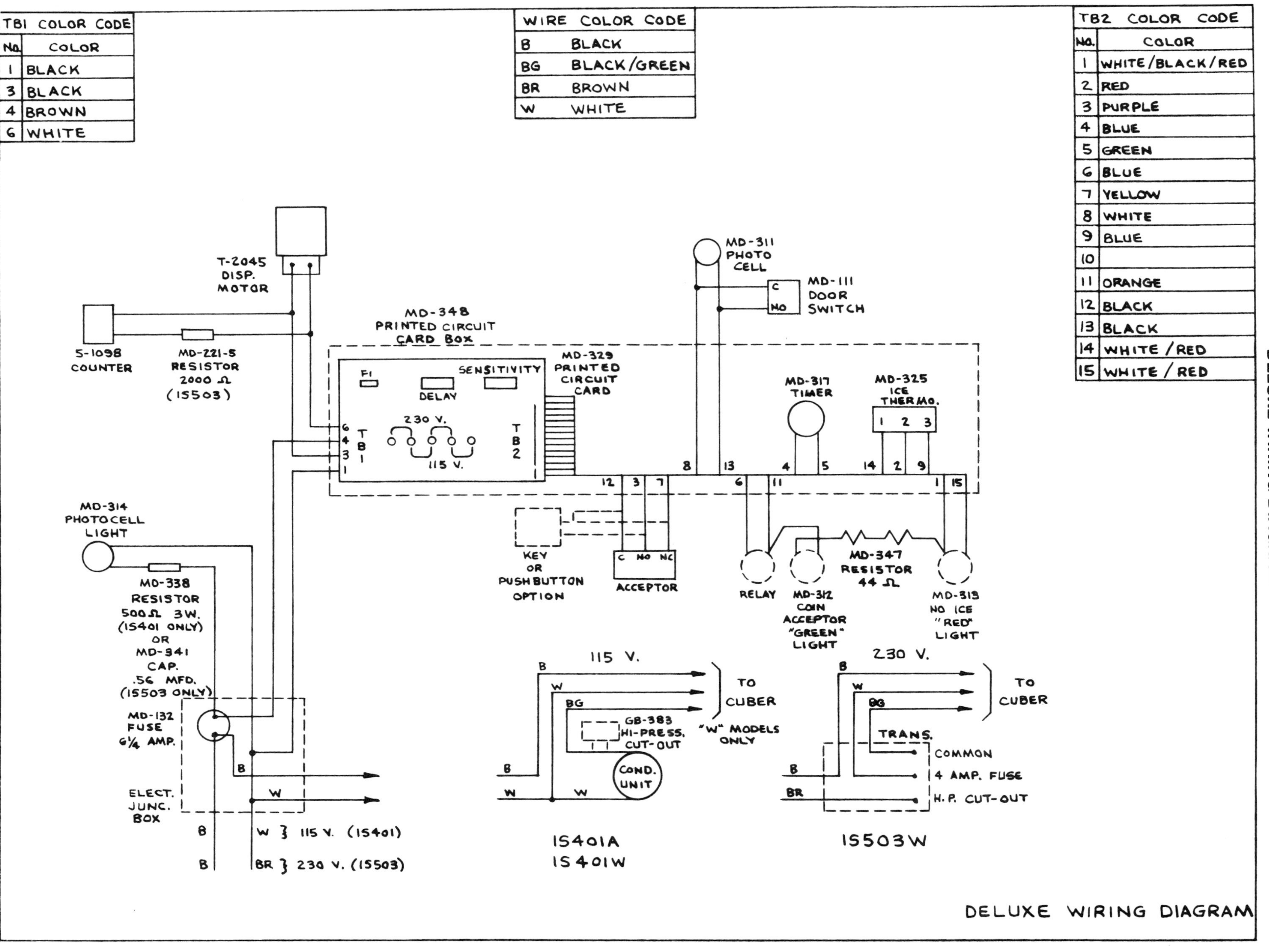

TB1 COLOR CODE	
NO.	COLOR
1	BLACK
3	BLACK
4	BROWN
6	WHITE

WIRE COLOR CODE	
B	BLACK
BG	BLACK/GREEN
BR	BROWN
W	WHITE

TB2 COLOR CODE	
NO.	COLOR
1	WHITE/BLACK/RED
2	RED
3	PURPLE
4	BLUE
5	GREEN
6	BLUE
7	YELLOW
8	WHITE
9	BLUE
10	
11	ORANGE
12	BLACK
13	BLACK
14	WHITE / RED
15	WHITE / RED

ICE DISPENSER CLEANING INSTRUCTIONS

Clean exterior frequently and empty bottom drain pan when provided.

To clean interior and parts in contact with ice:

1. Empty the storage area, then disconnect the electrical power supply.
2. Remove the top cover or remove the ice maker inspection panel, top, left and right end panels, and drain pan.
3. TK AND TDA MODELS ONLY: Remove the auger.
   A. TK - Unclamp the upper bearing and pull the auger up at least 1" on its axis to clear the lower driving mechanism, then out through the top or through the left side of the ice maker with the water plate up. DO NOT TWIST THE AUGER WHILE IT IS ENGAGED ON THE DRIVE MECHANISM.
   B. TDA — Remove the dispensing mechanism hood, then the wing nuts, and pull the mechanism out of the hopper. FIRST BE SURE THAT THE WIRING IS FREE OR HAS ADEQUATE SLACK.
4. Wash interior and parts with a solution of 2 tablespoons baking soda per quart of warm water.
5. Rinse with clean water.
6. Flush the drain with at least one quart of HOT water. Do not use drain cleaners in the ice storage area.
7. Sanitize with a solution of 1 teaspoon 5 1/4% Sodium Hypochlorite (household laundry bleach) per quart of clean water. Pour unused sanitizing solution down the storage area drain.
8. Replace the auger. Be sure that the TK auger is fully engaged onto the lower drive mechanism.
9. If an ice cuber is installed, lower the water plate to gain access for cleaning the dispensing spout and upper bearing clamp area in TK models.
10. When cleaning the ice maker, follow the ice maker cleaning instructions and clean the dispenser last.
11. Replace all enclosure parts before re-connecting the electrical supply.

CAUTION: When removing the auger, be sure the bin is empty and **do not** attempt to turn the auger manually in either direction.

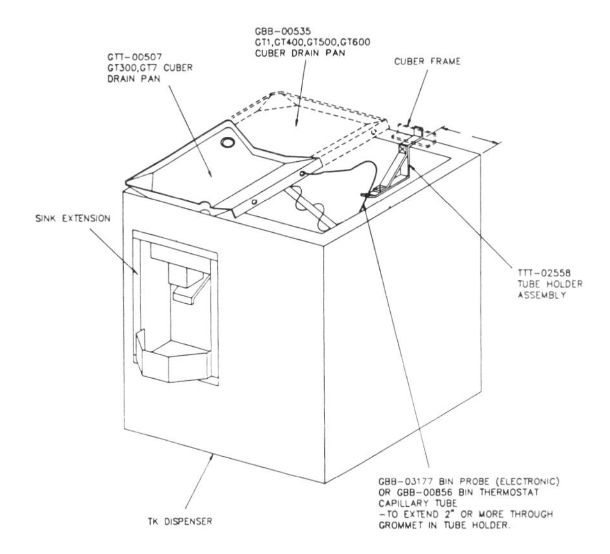

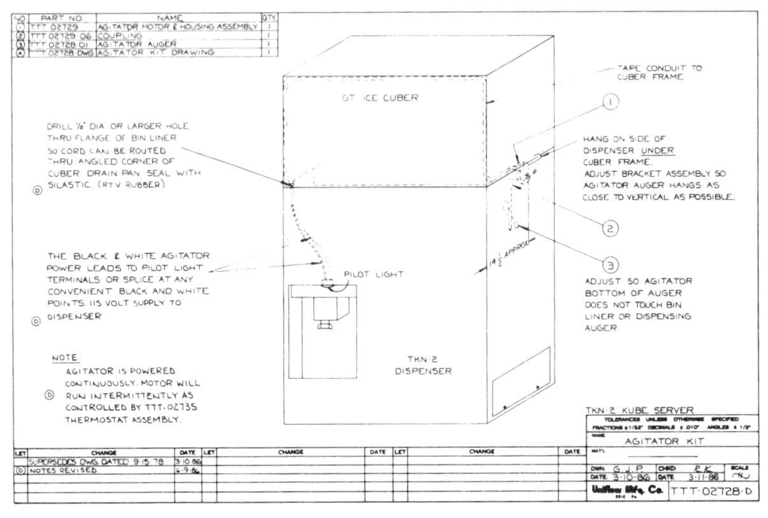

GENERAL INFORMATION & SPECIFICATIONS

Storage capacity:	120 lbs. cube ice
Insulation:	Polyurethane foam and fiberglass
Electrical:	120 volt, 60 Hz. single phase, 200 watts
Gearmotor:	Thermally protected P.S.C.
Portion control (optional):	Timed cycle, adjustable
Drains, All Models:	Two 3/4" female pipe on bottom. (Storage area drain is isolated). Provision for Kold-Draft GT Series ice cuber drain at top.
Mobile Model:	Plastic pan slides in tracks on bottom to catch all drain water.
Water Supply (optional):	1/4" Female pipe on bottom.
Wiring:	Junction box inside access panel on bottom of right side, 7/8" knockout on rear.
Installation:	Install the Kube-Server level and allow space at sides for access to switches, motor and wiring. If an ice cuber is installed on top, set the bin thermostat at or below the "full" line and clear of the dispensing auger.

To dispense ice cubes:

1. Standard model: Move push bar under spout straight back so container is under spout. The auger will run when the bar is all the way back and continue running until the bar is released.
2. Portion control: The portion control unit is equipped with a selector switch which bypasses the timer circuit to the gear-motor when in the "on" position to give operation exactly as described under standard model.

   When the selector is "off", operation is the same as described under standard model except that the auger will stop after a preset time on the portion control. The push bar must be released and pushed again to reactivate the auger once the timer circuit stops it. To increase the portion size (auger runs longer) move dial on portion control to a higher number.

INSTRUCTIONS FOR ADJUSTING THE TKN-2 DISPENSING SWITCH/SPOUT CLOSURE

LARGE, STRAIGHT-SIDED SPOUT WITH SPRING-LOADED CLOSURE (TTT-02554)

Bend Push Bar forward to position smaller containers in center of spout when closure is fully opened.

TAPERED GLASS FILLING SPOUT WITH HEAVY BACK-PLATE PUSH-BAR (TTT-02577)

If switch action is not free, remove spout and push-bar. Increase the opening space on push-bar to allow more clearance for back of spout — leave push-bar hooked onto back of spout and force the push-bar back and forth beyond its present travel until the push-bar moves freely about 1/2£ back and forth.

Note: For dispensing large cubes and/or when very tall ice containers (carafes) are used, to avoid ice jamming the spout use the TTT-02554 spout and mount the push-button operator in the hole which is covered by the "Kube Server" label. If overfilling of container is a problem, an additional TTT-02716-03 ice disc may be installed on the Auger to slow the dispensing rate.

TROUBLE SHOOTING GUIDE

TROUBLE	CAUSE/SYMPTOM	REMEDY
1. Drive Motor will not run A. Motor hot	A. 1) Low Voltage 2) Defective Capacitor 3) Auger or drive linkage bound 4) Defective Motor 5) Overfilled Ice bin 6) Ice fused together in bin	A. 1) Requires minimum of 105 volts 2) Replace 3) Clean, lube & adjust as required 4) Replace or take to factory service center 5) Set Bin thermostat 6) Lower bin thermostat or add agitator kit
B. Motor not hot	B. 1) Portion control set too low 2) Open circuit 3) Defective Portion Control 4) Dispensing switch not closing	B. 1) Increase setting 2) Check blue to white for line voltage or blue to black for continuity. 3) Replace 4) Adjust push-bar or replace switch, be sure switch opens when push bar is released.
2. Drive Motor will not run on portion control but runs on continuous bypass.	1) Loose connection 2) Portion control set too low 3) Defective portion control	1) Check all terminals on portion control. 2) Increase setting 3) Replace
3. Inconsistent Portions	1) Portion will vary as level of ice in bin changes. 2) Large chunks of ice or irregular shapes 3) Ice fused together in bin	1) Some variation is normal 2) Keep cuber bin thermostat at or below "Full" line, do not use bin for prolonged storage, set fin thickness. 3) See 2 above-keep silicone water plate coating away from edge of plate to prevent raw water sliding into bin during harvest cycle.
4. Pilot light out	1) Power off, fuse blown 2) Lamp burned out	1) Check circuit 2) Replace
5. Auger disengages from top locking clamp	1) Bearing adjusted too tight 2) Ice jamming against clamp bar	1) Loosen 2) Be sure rubber disc is in place at top of auger - remove any large fused masses from bin. (Refer to 3)
6. Ice Jams in spout	1) User overfilling container 2) Large chunks or irregular shapes	1) Add portion control or additional rubber disc at top of Auger to reduce dispensing rate. (Refer to 3)
7. Water "leaking" from mobile model	1) Drain pan full	1) Empty frequently
8. Water glass filler flow rate too fast or too slow	1) Line pressure	1a) Use regulator or increase line size 1b) Slight adjustment can be made by turning nozzle on filler.

WIRING DIAGRAMS

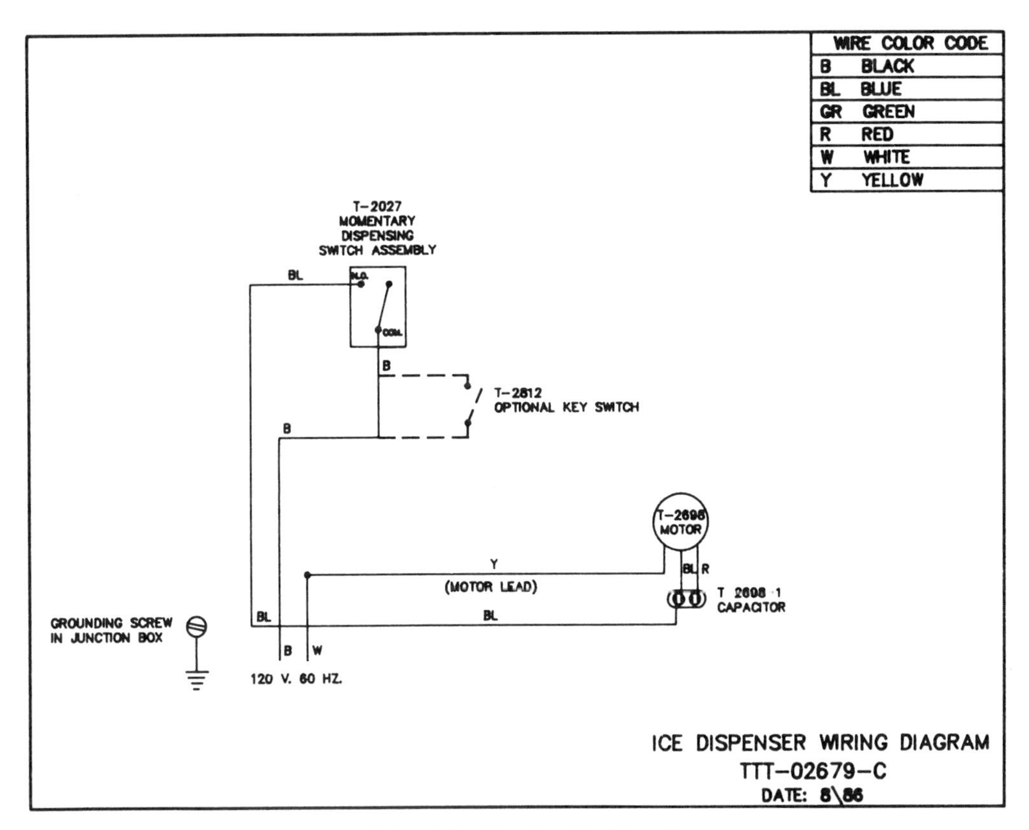

STANDARD TKN-2

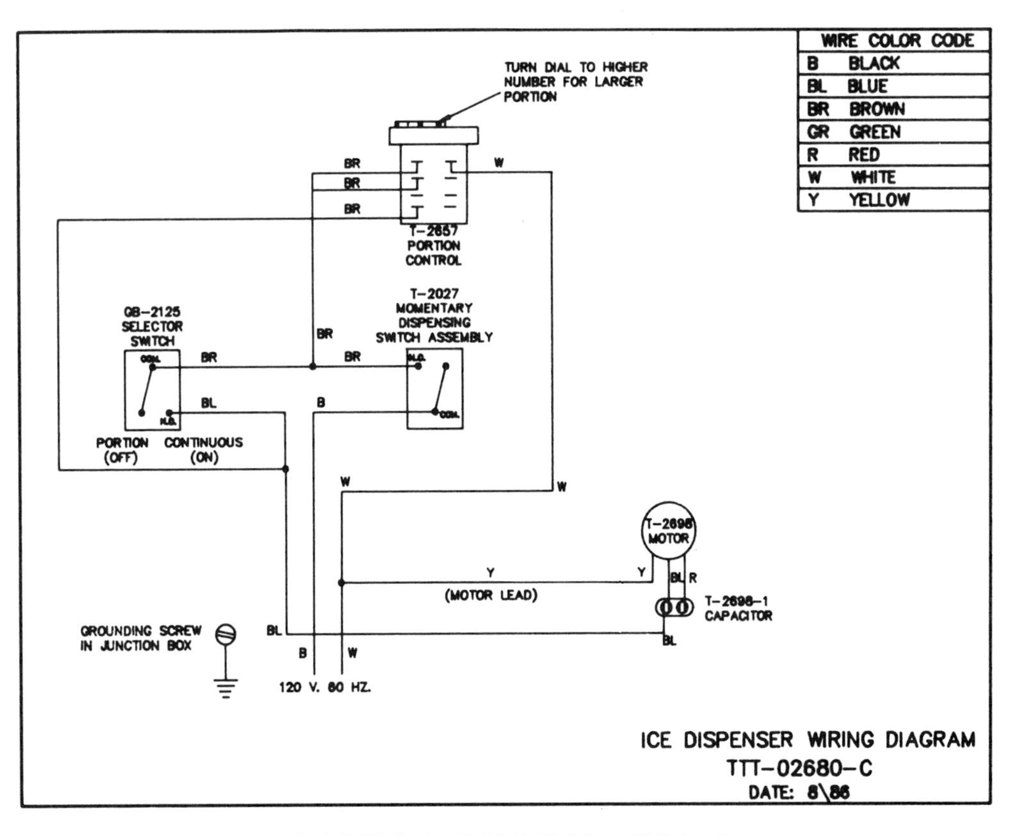

PORTION CONTROL TKN-2

T-100 CRUSHER

MAINTENANCE EVERY 6 MONTHS THE FOLLOWING SHOULD BE DONE

1. Moter should be oiled

2. Timing belt checked moderate tension

3. Drum shaft bearings greased

T-100 CRUSHER SIZE BAR 10/1/84

The size bar is held firmly in place by two bolts which pass through the housing, and thread directly into the cast-aluminum size bar.

To allow adjustment for ice granule size, several holes are punched in the housing, and ice size can be changed from the factory setting of "fine" to "medium" or "coarse". If the change in ice size is necessary, remove the two bolts, slide the ice size bar to the desired position, and reinstall the bolts in the closest hole to the tapped hole in the bar.

Although adjustment is most easily accomplished before the crusher is installed under a cuber, removal of the ice chute and drain bay from the cuber will provide full access to the rear adjustment bolt after cuber installation.

BIN PROBE KITS FOR CRUSHERS WITH ELECTRONIC CUBERS FOR CONTROL OF BIN ICE LEVELS

1. ALL CRUSHED ICE in an undivided bin, for T-100-B and TGT-100 crushers use kit TTR-01847. Refer to page 128 for installation.

2. CRUSHED AND CUBED ICE in a divided bin for T-100-B crushers use kit TTr-01841. Refer to page 126-127 for specifics.

BIN THERMOSTAT CAP TUBES - ELECTRO-MECHANICAL CUBERS

When installing bin thermostat cap tubes from the cubers, run the tubes through the crusher grommet (extra grommets may be removed from cuber drain pans) to keep them away from the belt. Then run about 22 inches of the tubes through the lower grommet into the bin.Slip the 16 inch plastic tube over the thermostat tubes all the way to the grommet. Bend the tails of the tubes enough to hold the plastic in place and slide the tubes through the grommet in the bin divider. This last step may be easier if the cube deflector panels on the right side are removed, particularly if the crusher is on a GB4 Bin this is necessary. The bin thermostats should be adjusted internally one turn warmer and externally turned all the way counter-clockwise to the warmest position. Then if, on an occasional thermostat, it is necessary to adjust it colder to prevent shut off due to spray ice, it can be turned part way clockwise externally.

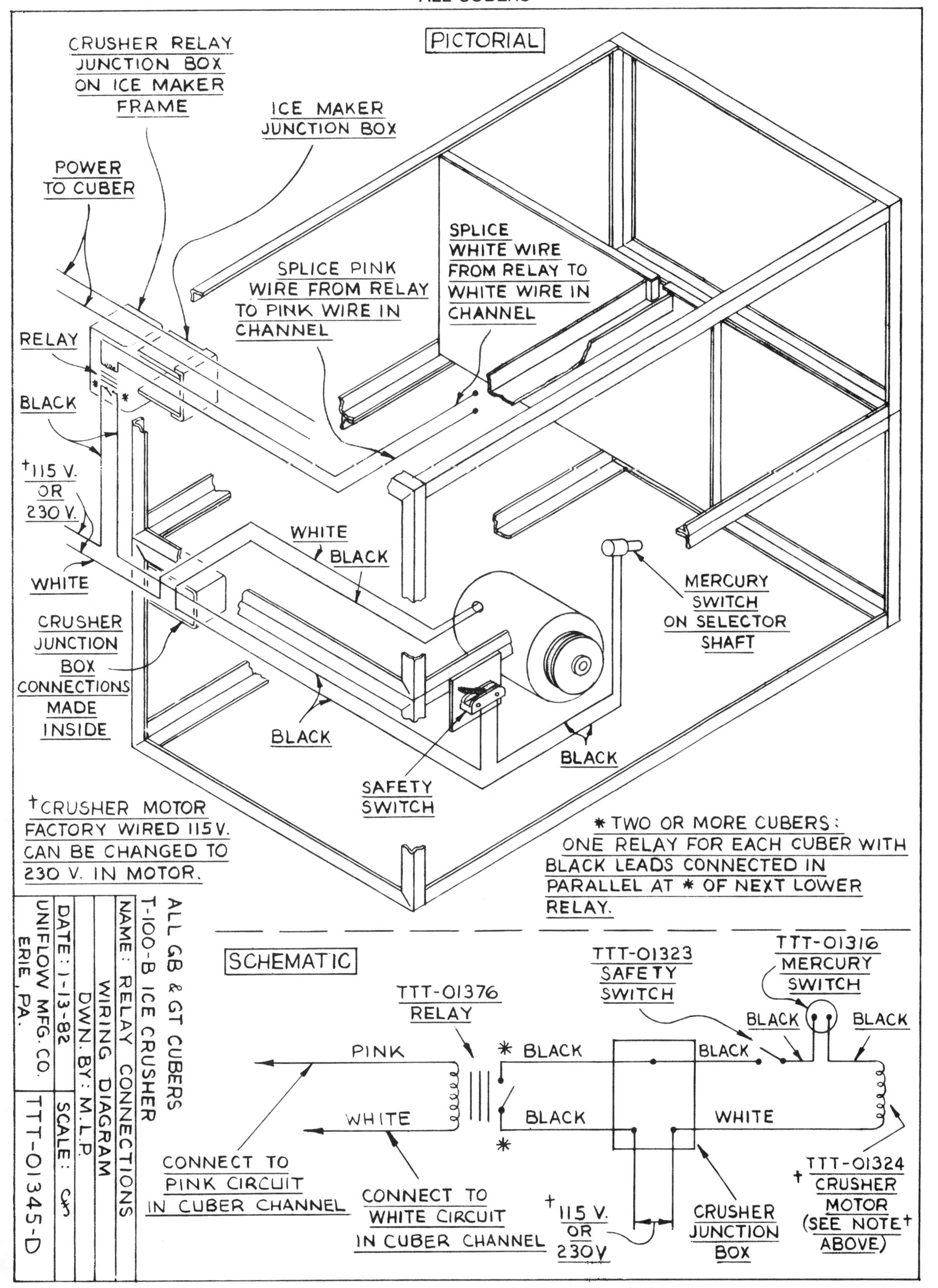
PICTORIAL
CRUSHER RELAY JUNCTION BOX ON ICE MAKER FRAME
ICE MAKER JUNCTION BOX
POWER TO CUBER
SPLICE PINK WIRE FROM RELAY TO PINK WIRE IN CHANNEL
SPLICE WHITE WIRE FROM RELAY TO WHITE WIRE IN CHANNEL
RELAY
BLACK
†115 V. OR 230 V.
WHITE
BLACK
WHITE
MERCURY SWITCH ON SELECTOR SHAFT
CRUSHER JUNCTION BOX CONNECTIONS MADE INSIDE
BLACK
BLACK
SAFETY SWITCH
†CRUSHER MOTOR FACTORY WIRED 115V. CAN BE CHANGED TO 230 V. IN MOTOR.
* TWO OR MORE CUBERS: ONE RELAY FOR EACH CUBER WITH BLACK LEADS CONNECTED IN PARALLEL AT * OF NEXT LOWER RELAY.
SCHEMATIC
TTT-01376 RELAY
TTT-01323 SAFETY SWITCH
TTT-01316 MERCURY SWITCH
PINK
* BLACK
BLACK
BLACK
BLACK
WHITE
BLACK
WHITE
CONNECT TO PINK CIRCUIT IN CUBER CHANNEL
CONNECT TO WHITE CIRCUIT IN CUBER CHANNEL
†115 V. OR 230V
CRUSHER JUNCTION BOX
† TTT-01324 CRUSHER MOTOR (SEE NOTE† ABOVE)
ALL GB & GT CUBERS
T-100-B ICE CRUSHER
NAME: RELAY CONNECTIONS WIRING DIAGRAM
DATE: 1-13-82
DWN. BY: M.L.P.
SCALE:
UNIFLOW MFG. CO. ERIE, PA.
TTT-01345-D

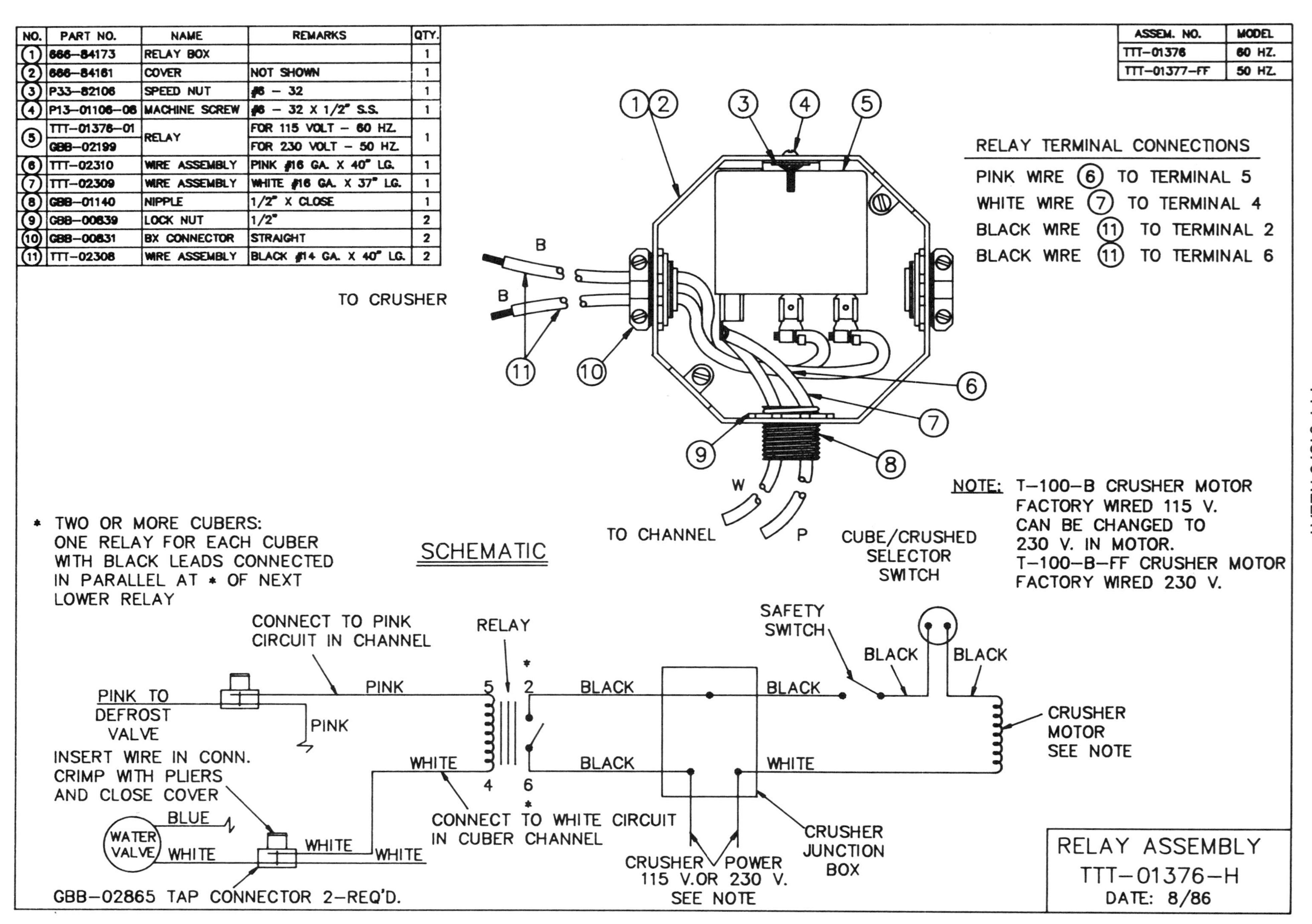

NO.	PART NO.	NAME	REMARKS	QTY.
1	666-84173	RELAY BOX		1
2	666-84161	COVER	NOT SHOWN	1
3	P33-82106	SPEED NUT	#8 – 32	1
4	P13-01106-06	MACHINE SCREW	#8 – 32 X 1/2" S.S.	1
5	TTT-01376-01	RELAY	FOR 115 VOLT – 60 HZ.	1
	GBB-02199		FOR 230 VOLT – 50 HZ.	
6	TTT-02310	WIRE ASSEMBLY	PINK #16 GA. X 40" LG.	1
7	TTT-02309	WIRE ASSEMBLY	WHITE #16 GA. X 37" LG.	1
8	GBB-01140	NIPPLE	1/2" X CLOSE	1
9	GBB-00639	LOCK NUT	1/2"	2
10	GBB-00631	BX CONNECTOR	STRAIGHT	2
11	TTT-02308	WIRE ASSEMBLY	BLACK #14 GA. X 40" LG.	2

ASSEM. NO.	MODEL
TTT-01376	60 HZ.
TTT-01377-FF	50 HZ.

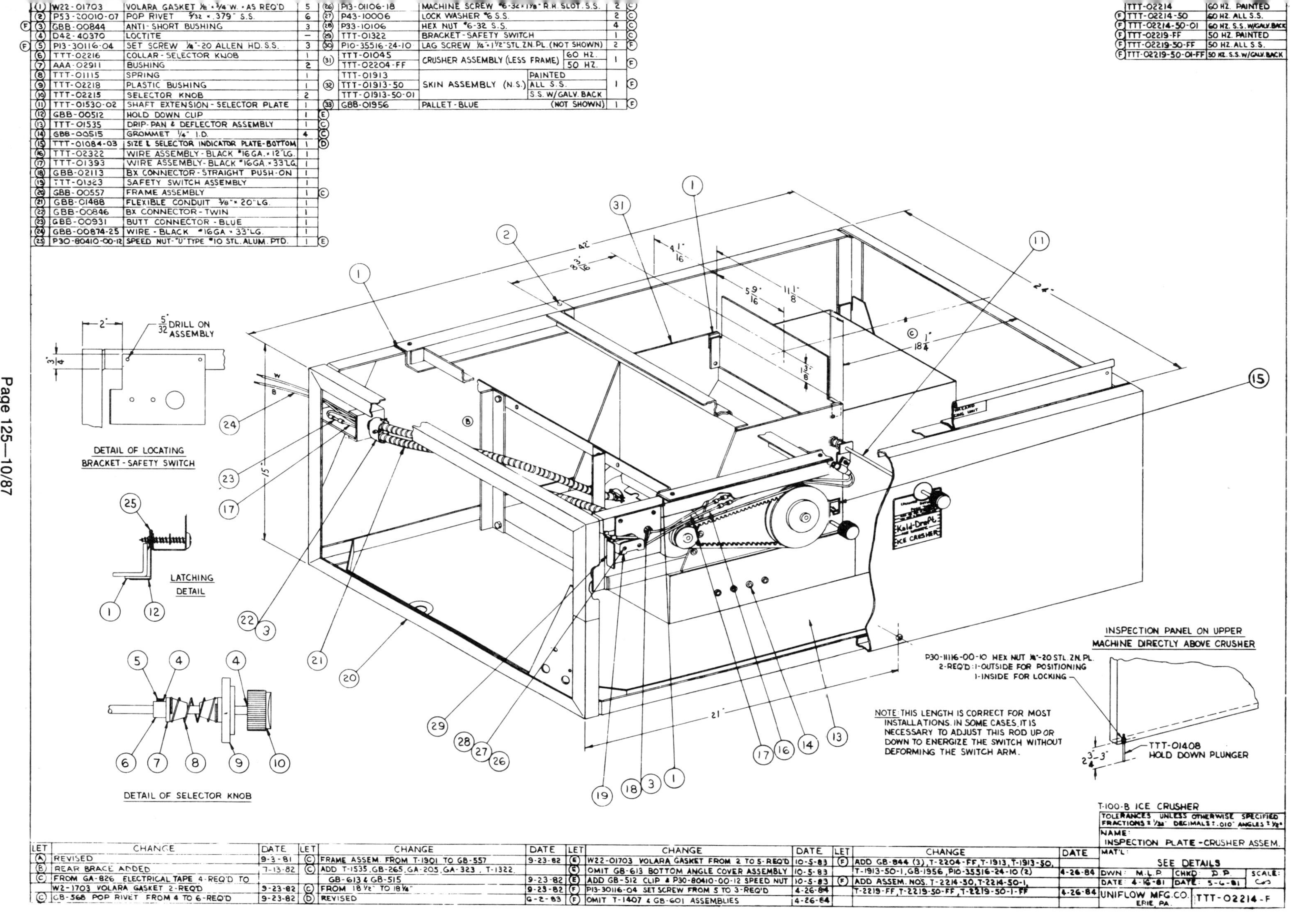

(1) W22-01703 VOLARA GASKET 1/8 × 3/4 W. × AS REQ'D 5
(2) P53-20010-07 POP RIVET 5/32 × .379" S.S. 6
(3) GBB-00844 ANTI-SHORT BUSHING 3
(4) D42-40370 LOCTITE —
(5) P13-30116-04 SET SCREW 1/4"-20 ALLEN HD. S.S. 3
(6) TTT-02216 COLLAR-SELECTOR KNOB 1
(7) AAA-02911 BUSHING 2
(8) TTT-01115 SPRING 1
(9) TTT-02218 PLASTIC BUSHING 1
(10) TTT-02215 SELECTOR KNOB 2
(11) TTT-01530-02 SHAFT EXTENSION-SELECTOR PLATE 1
(12) GBB-00512 HOLD DOWN CLIP 1
(13) TTT-01535 DRIP-PAN & DEFLECTOR ASSEMBLY 1
(14) GBB-00515 GROMMET 1/4" I.D. 4
(15) TTT-01084-03 SIZE & SELECTOR INDICATOR PLATE-BOTTOM 1
(16) TTT-02322 WIRE ASSEMBLY-BLACK #16GA. × 12" LG. 1
(17) TTT-01393 WIRE ASSEMBLY-BLACK #16GA. × 33" LG. 1
(18) GBB-02113 BX CONNECTOR-STRAIGHT PUSH-ON 1
(19) TTT-01323 SAFETY SWITCH ASSEMBLY 1
(20) GBB-00557 FRAME ASSEMBLY 1
(21) GBB-01488 FLEXIBLE CONDUIT 3/8" × 20" LG. 1
(22) GBB-00846 BX CONNECTOR-TWIN 1
(23) GBB-00931 BUTT CONNECTOR-BLUE 1
(24) GBB-00874-25 WIRE-BLACK #16GA × 33" LG. 1
(25) P30-80410-00-12 SPEED NUT-"U" TYPE #10 STL. ALUM. PTD. 1
(26) P13-01106-18 MACHINE SCREW #6-32 × 1 1/8" R.H. SLOT. S.S. 2
(27) P43-10006 LOCK WASHER #6 S.S. 2
(28) P33-10106 HEX NUT #6-32 S.S. 4
(29) TTT-01322 BRACKET-SAFETY SWITCH 1
(30) P10-35516-24-10 LAG SCREW 1/4 × 1 1/2" STL. ZN. PL. (NOT SHOWN) 2
(31) TTT-01045 / TTT-02204-FF CRUSHER ASSEMBLY (LESS FRAME) 60 HZ. / 50 HZ. 1
(32) TTT-01913 / TTT-01913-50 / TTT-01913-50-01 SKIN ASSEMBLY (N.S.) PAINTED / ALL S.S. / S.S. W/GALV. BACK 1
(33) GBB-01956 PALLET-BLUE (NOT SHOWN) 1
TTT-02214 60 HZ. PAINTED
TTT-02214-50 60 HZ. ALL S.S.
TTT-02214-50-01 60 HZ. S.S. W/GALV. BACK
TTT-02219-FF 50 HZ. PAINTED
TTT-02219-50-FF 50 HZ. ALL S.S.
TTT-02219-50-01-FF 50 HZ. S.S. W/GALV. BACK
5/32 DRILL ON ASSEMBLY
DETAIL OF LOCATING BRACKET-SAFETY SWITCH
LATCHING DETAIL
DETAIL OF SELECTOR KNOB
Kold-Draft ICE CRUSHER
INSPECTION PANEL ON UPPER MACHINE DIRECTLY ABOVE CRUSHER
P30-11116-00-10 HEX NUT 1/4"-20 STL. ZN. PL. 2-REQ'D: 1-OUTSIDE FOR POSITIONING 1-INSIDE FOR LOCKING
TTT-01408 HOLD DOWN PLUNGER
NOTE: THIS LENGTH IS CORRECT FOR MOST INSTALLATIONS. IN SOME CASES, IT IS NECESSARY TO ADJUST THIS ROD UP OR DOWN TO ENERGIZE THE SWITCH WITHOUT DEFORMING THE SWITCH ARM.
T-100-B ICE CRUSHER
TOLERANCES UNLESS OTHERWISE SPECIFIED FRACTIONS ± 1/32" DECIMALS ± .010" ANGLES ± 1/2°
NAME: INSPECTION PLATE-CRUSHER ASSEM.
MAT'L: SEE DETAILS
DWN: M.L.P. CHKD: D.P. SCALE:
DATE 4-16-81 DATE 5-6-81
UNIFLOW MFG. CO. ERIE, PA. TTT-02214-F
LET CHANGE DATE
(A) REVISED 9-3-81
(B) REAR BRACE ADDED 7-13-82
(C) FROM GA-826 ELECTRICAL TAPE 4-REQ'D TO W2-1703 VOLARA GASKET 2-REQ'D 9-23-82
(C) GB-568 POP RIVET FROM 4 TO 6-REQ'D 9-23-82
(C) FRAME ASSEM. FROM T-1901 TO GB-557 9-23-82
(C) ADD T-1535, GB-265, GA-205, GA-323, T-1322, GB-613 & GB-515 9-23-82
(C) FROM 18 1/2" TO 18 1/4" 9-23-82
(D) REVISED 6-2-83
(E) W22-01703 VOLARA GASKET FROM 2 TO 5-REQ'D 10-5-83
(E) OMIT GB-613 BOTTOM ANGLE COVER ASSEMBLY 10-5-83
(E) ADD GB-512 CLIP & P30-80410-00-12 SPEED NUT 10-5-83
(F) P13-30116-04 SET SCREW FROM 5 TO 3-REQ'D 4-26-84
(F) OMIT T-1407 & GB-601 ASSEMBLIES 4-26-84
(F) ADD GB-844 (3), T-2204-FF, T-1913, T-1913-50, T-1913-50-1, GB-1956, P10-35516-24-10 (2) 4-26-84
(F) ADD ASSEM. NOS. T-2214-50, T-2214-50-1, T-2219-FF, T-2219-50-FF, T-2219-50-1-FF 4-26-84

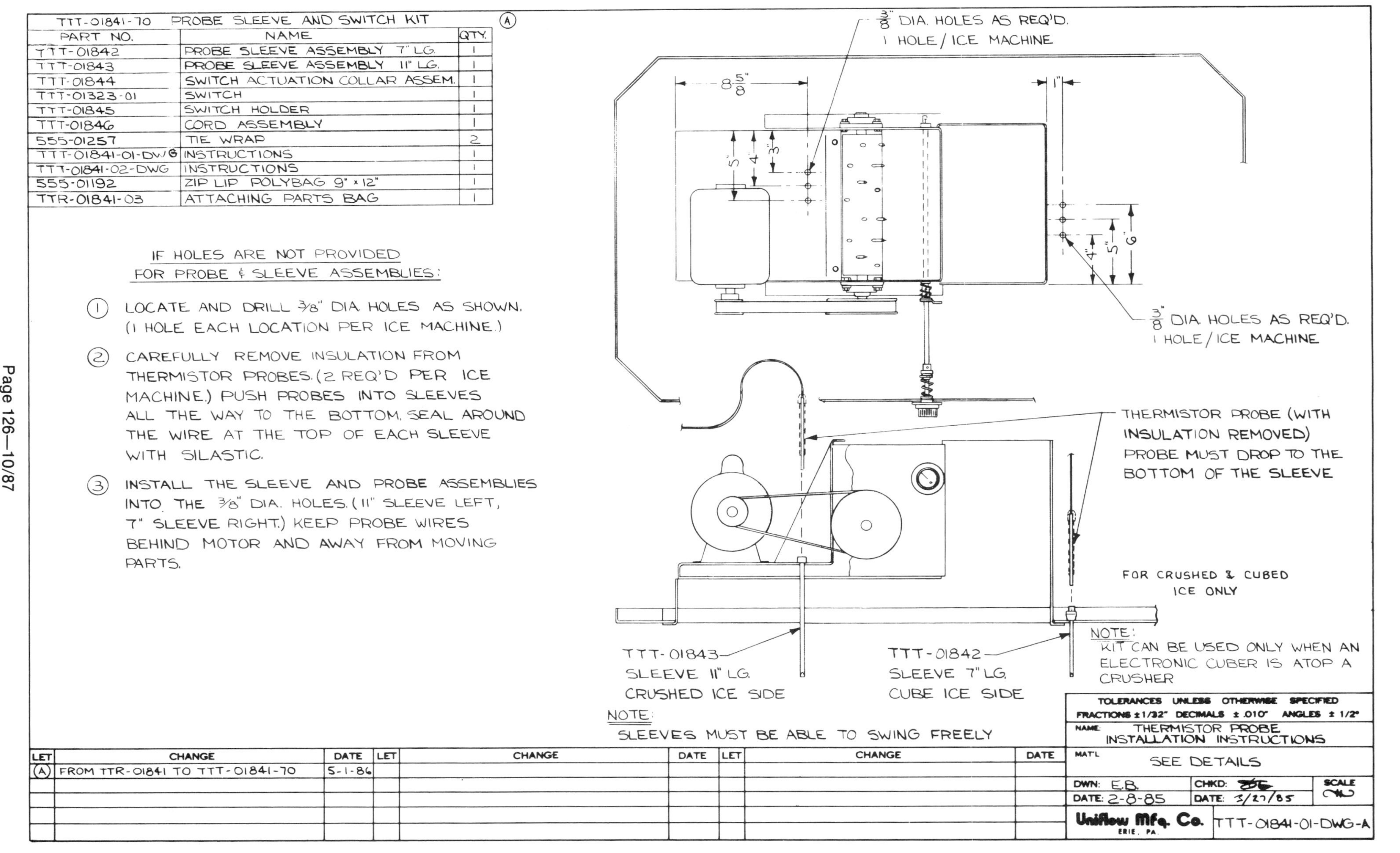

TTT-01841-70 PROBE SLEEVE AND SWITCH KIT (A)

PART NO.	NAME	QTY.
TTT-01842	PROBE SLEEVE ASSEMBLY 7" LG.	1
TTT-01843	PROBE SLEEVE ASSEMBLY 11" LG.	1
TTT-01844	SWITCH ACTUATION COLLAR ASSEM.	1
TTT-01323-01	SWITCH	1
TTT-01845	SWITCH HOLDER	1
TTT-01846	CORD ASSEMBLY	1
555-01257	TIE WRAP	2
TTT-01841-01-DWG	INSTRUCTIONS	1
TTT-01841-02-DWG	INSTRUCTIONS	1
555-01192	ZIP LIP POLYBAG 9" x 12"	1
TTR-01841-03	ATTACHING PARTS BAG	1

IF HOLES ARE NOT PROVIDED
FOR PROBE & SLEEVE ASSEMBLIES:

1. LOCATE AND DRILL 3/8" DIA. HOLES AS SHOWN. (1 HOLE EACH LOCATION PER ICE MACHINE.)
2. CAREFULLY REMOVE INSULATION FROM THERMISTOR PROBES. (2 REQ'D PER ICE MACHINE.) PUSH PROBES INTO SLEEVES ALL THE WAY TO THE BOTTOM. SEAL AROUND THE WIRE AT THE TOP OF EACH SLEEVE WITH SILASTIC.
3. INSTALL THE SLEEVE AND PROBE ASSEMBLIES INTO THE 3/8" DIA. HOLES. (11" SLEEVE LEFT, 7" SLEEVE RIGHT.) KEEP PROBE WIRES BEHIND MOTOR AND AWAY FROM MOVING PARTS.

LET	CHANGE	DATE	LET	CHANGE	DATE	LET	CHANGE	DATE
(A)	FROM TTR-01841 TO TTT-01841-70	5-1-86						

TOLERANCES UNLESS OTHERWISE SPECIFIED
FRACTIONS ±1/32" DECIMALS ±.010" ANGLES ±1/2°
NAME: THERMISTOR PROBE INSTALLATION INSTRUCTIONS
MAT'L: SEE DETAILS
DWN: E.B. DATE: 2-8-85
CHKD: [illegible] DATE: 3/27/85
SCALE: [illegible]
Uniflow Mfg. Co. ERIE, PA.
TTT-01841-01-DWG-A

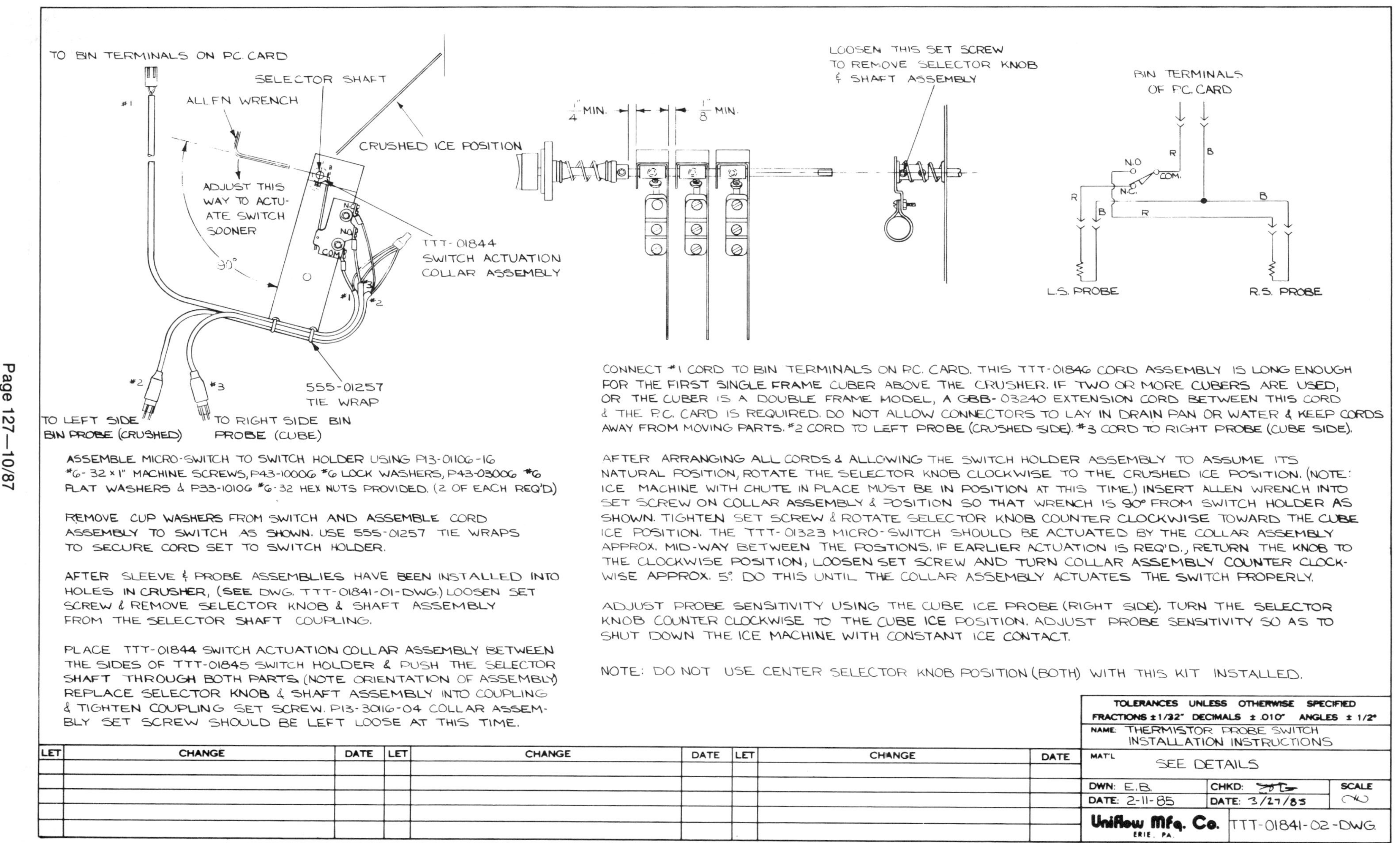

ASSEMBLE MICRO-SWITCH TO SWITCH HOLDER USING P13-01106-16 #6-32 x 1" MACHINE SCREWS, P43-10006 #6 LOCK WASHERS, P43-03006 #6 FLAT WASHERS & P33-10106 #6-32 HEX NUTS PROVIDED. (2 OF EACH REQ'D)

REMOVE CUP WASHERS FROM SWITCH AND ASSEMBLE CORD ASSEMBLY TO SWITCH AS SHOWN. USE 555-01257 TIE WRAPS TO SECURE CORD SET TO SWITCH HOLDER.

AFTER SLEEVE & PROBE ASSEMBLIES HAVE BEEN INSTALLED INTO HOLES IN CRUSHER, (SEE DWG. TTT-01841-01-DWG.) LOOSEN SET SCREW & REMOVE SELECTOR KNOB & SHAFT ASSEMBLY FROM THE SELECTOR SHAFT COUPLING.

PLACE TTT-01844 SWITCH ACTUATION COLLAR ASSEMBLY BETWEEN THE SIDES OF TTT-01845 SWITCH HOLDER & PUSH THE SELECTOR SHAFT THROUGH BOTH PARTS. (NOTE ORIENTATION OF ASSEMBLY) REPLACE SELECTOR KNOB & SHAFT ASSEMBLY INTO COUPLING & TIGHTEN COUPLING SET SCREW. P13-30116-04 COLLAR ASSEMBLY SET SCREW SHOULD BE LEFT LOOSE AT THIS TIME.

CONNECT #1 CORD TO BIN TERMINALS ON P.C. CARD. THIS TTT-01846 CORD ASSEMBLY IS LONG ENOUGH FOR THE FIRST SINGLE FRAME CUBER ABOVE THE CRUSHER. IF TWO OR MORE CUBERS ARE USED, OR THE CUBER IS A DOUBLE FRAME MODEL, A GBB-03240 EXTENSION CORD BETWEEN THIS CORD & THE P.C. CARD IS REQUIRED. DO NOT ALLOW CONNECTORS TO LAY IN DRAIN PAN OR WATER & KEEP CORDS AWAY FROM MOVING PARTS. #2 CORD TO LEFT PROBE (CRUSHED SIDE). #3 CORD TO RIGHT PROBE (CUBE SIDE).

AFTER ARRANGING ALL CORDS & ALLOWING THE SWITCH HOLDER ASSEMBLY TO ASSUME ITS NATURAL POSITION, ROTATE THE SELECTOR KNOB CLOCKWISE TO THE CRUSHED ICE POSITION. (NOTE: ICE MACHINE WITH CHUTE IN PLACE MUST BE IN POSITION AT THIS TIME.) INSERT ALLEN WRENCH INTO SET SCREW ON COLLAR ASSEMBLY & POSITION SO THAT WRENCH IS 90° FROM SWITCH HOLDER AS SHOWN. TIGHTEN SET SCREW & ROTATE SELECTOR KNOB COUNTER CLOCKWISE TOWARD THE CUBE ICE POSITION. THE TTT-01323 MICRO-SWITCH SHOULD BE ACTUATED BY THE COLLAR ASSEMBLY APPROX. MID-WAY BETWEEN THE POSITIONS. IF EARLIER ACTUATION IS REQ'D., RETURN THE KNOB TO THE CLOCKWISE POSITION, LOOSEN SET SCREW AND TURN COLLAR ASSEMBLY COUNTER CLOCKWISE APPROX. 5°. DO THIS UNTIL THE COLLAR ASSEMBLY ACTUATES THE SWITCH PROPERLY.

ADJUST PROBE SENSITIVITY USING THE CUBE ICE PROBE (RIGHT SIDE). TURN THE SELECTOR KNOB COUNTER CLOCKWISE TO THE CUBE ICE POSITION. ADJUST PROBE SENSITIVITY SO AS TO SHUT DOWN THE ICE MACHINE WITH CONSTANT ICE CONTACT.

NOTE: DO NOT USE CENTER SELECTOR KNOB POSITION (BOTH) WITH THIS KIT INSTALLED.

LET	CHANGE	DATE	LET	CHANGE	DATE	LET	CHANGE	DATE

TOLERANCES UNLESS OTHERWISE SPECIFIED
FRACTIONS ±1/32" DECIMALS ±.010" ANGLES ±1/2°

NAME: THERMISTOR PROBE SWITCH INSTALLATION INSTRUCTIONS

MAT'L: SEE DETAILS

DWN: E.B.	CHKD: [signature]	SCALE
DATE: 2-11-85	DATE: 3/27/85	[illegible]

Uniflow Mfg. Co. ERIE, PA. — TTT-01841-02-DWG.

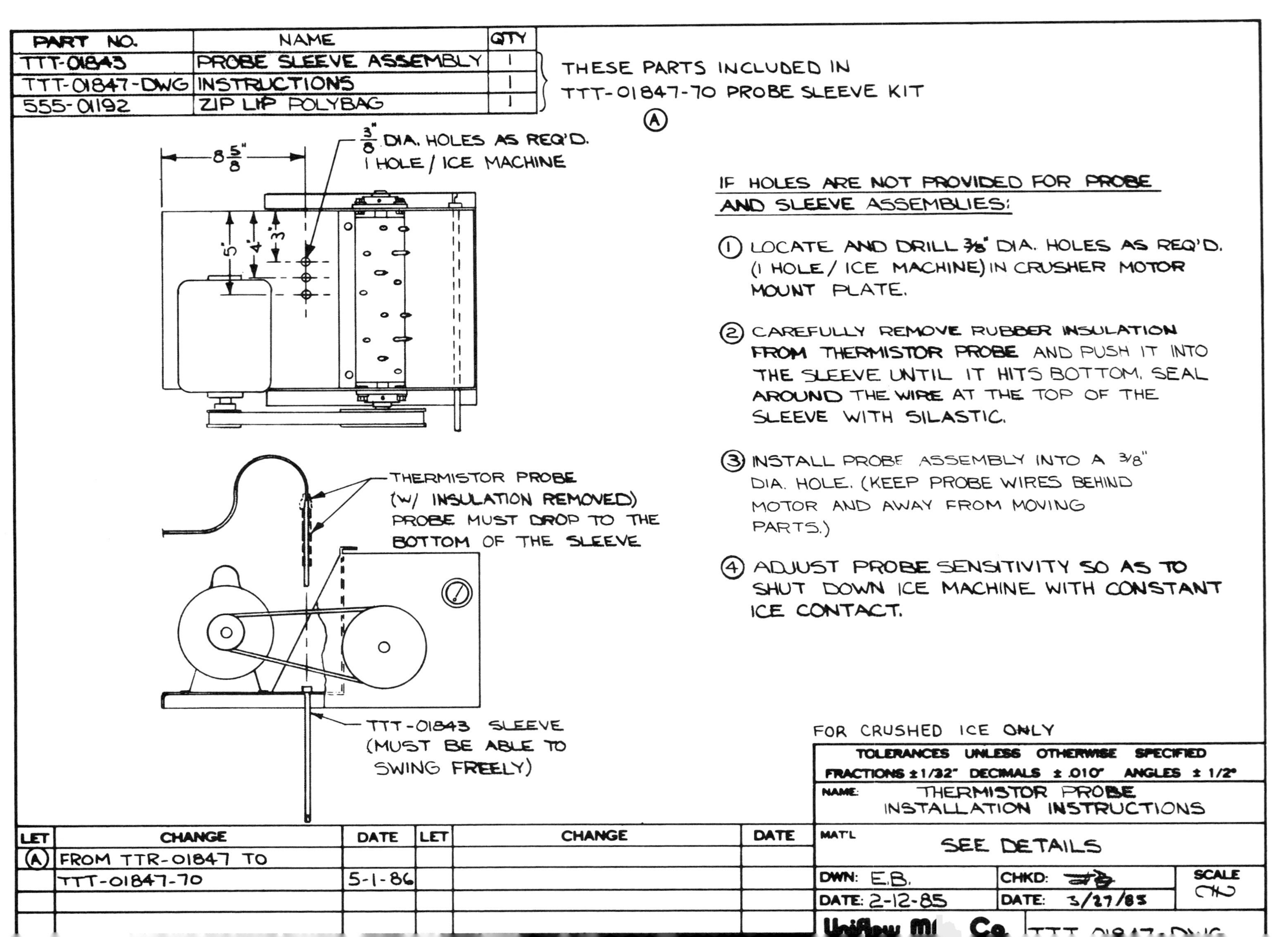

PART NO. | NAME | QTY
TTT-01843 | PROBE SLEEVE ASSEMBLY | 1
TTT-01847-DWG | INSTRUCTIONS | 1
555-01192 | ZIP LIP POLYBAG | 1
THESE PARTS INCLUDED IN
TTT-01847-70 PROBE SLEEVE KIT
(A)
3/8" DIA. HOLES AS REQ'D.
1 HOLE / ICE MACHINE
8 5/8"
5"
4"
3"
IF HOLES ARE NOT PROVIDED FOR PROBE AND SLEEVE ASSEMBLIES:
(1) LOCATE AND DRILL 3/8" DIA. HOLES AS REQ'D. (1 HOLE / ICE MACHINE) IN CRUSHER MOTOR MOUNT PLATE.
(2) CAREFULLY REMOVE RUBBER INSULATION FROM THERMISTOR PROBE AND PUSH IT INTO THE SLEEVE UNTIL IT HITS BOTTOM, SEAL AROUND THE WIRE AT THE TOP OF THE SLEEVE WITH SILASTIC.
(3) INSTALL PROBE ASSEMBLY INTO A 3/8" DIA. HOLE. (KEEP PROBE WIRES BEHIND MOTOR AND AWAY FROM MOVING PARTS.)
(4) ADJUST PROBE SENSITIVITY SO AS TO SHUT DOWN ICE MACHINE WITH CONSTANT ICE CONTACT.
THERMISTOR PROBE
(W/ INSULATION REMOVED)
PROBE MUST DROP TO THE BOTTOM OF THE SLEEVE
TTT-01843 SLEEVE
(MUST BE ABLE TO SWING FREELY)
FOR CRUSHED ICE ONLY
TOLERANCES UNLESS OTHERWISE SPECIFIED
FRACTIONS ±1/32" DECIMALS ±.010" ANGLES ±1/2°
NAME: THERMISTOR PROBE INSTALLATION INSTRUCTIONS
MAT'L SEE DETAILS
DWN: E.B.
CHKD:
DATE: 2-12-85
DATE: 3/27/85
SCALE
LET | CHANGE | DATE | LET | CHANGE | DATE
(A) | FROM TTR-01847 TO TTT-01847-70 | 5-1-86

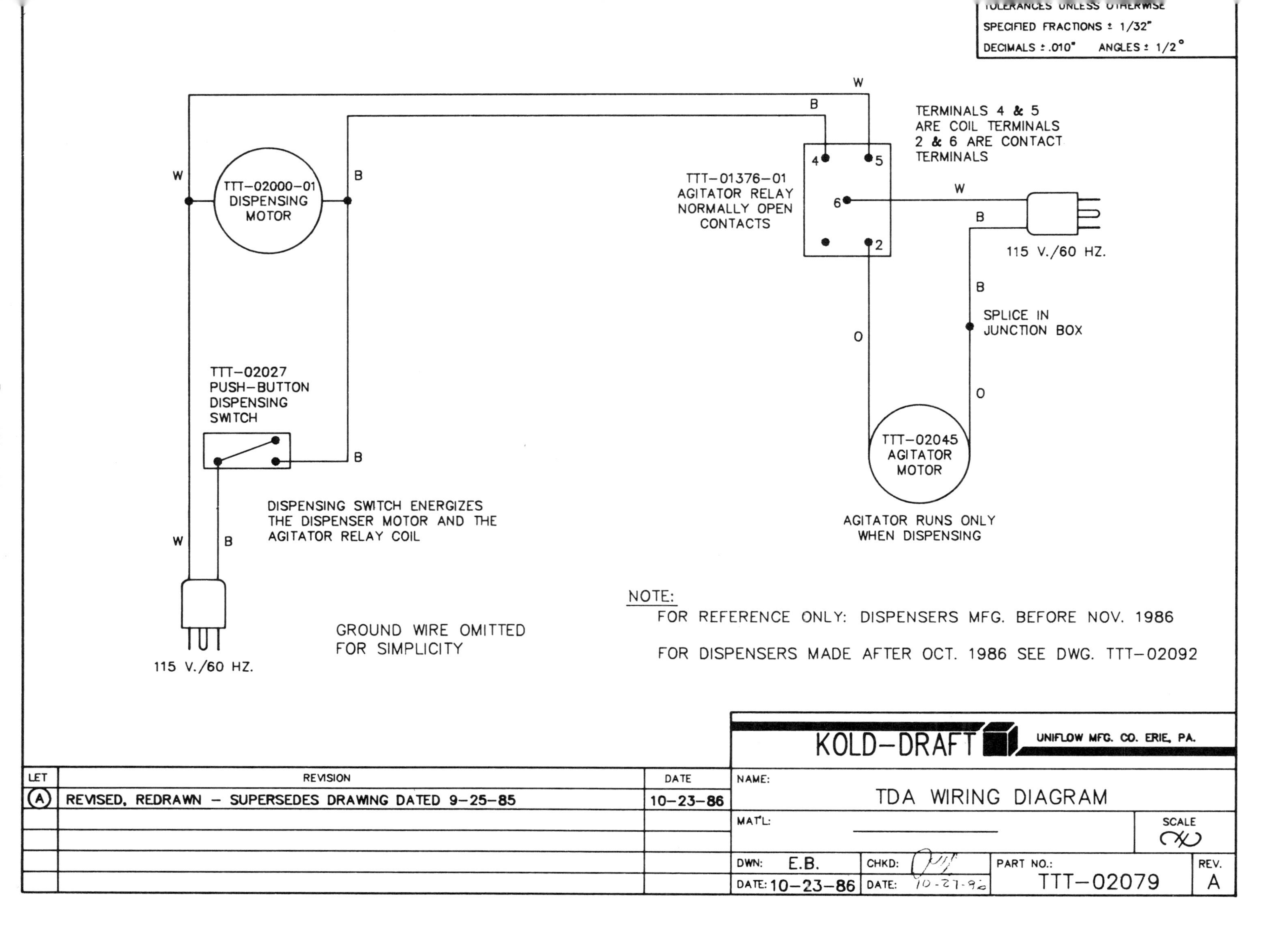
TOLERANCES UNLESS OTHERWISE
SPECIFIED FRACTIONS ± 1/32"
DECIMALS ± .010" ANGLES ± 1/2°
TERMINALS 4 & 5
ARE COIL TERMINALS
2 & 6 ARE CONTACT
TERMINALS
TTT-02000-01
DISPENSING
MOTOR
TTT-01376-01
AGITATOR RELAY
NORMALLY OPEN
CONTACTS
115 V./60 HZ.
SPLICE IN
JUNCTION BOX
TTT-02027
PUSH-BUTTON
DISPENSING
SWITCH
TTT-02045
AGITATOR
MOTOR
DISPENSING SWITCH ENERGIZES
THE DISPENSER MOTOR AND THE
AGITATOR RELAY COIL
AGITATOR RUNS ONLY
WHEN DISPENSING
115 V./60 HZ.
GROUND WIRE OMITTED
FOR SIMPLICITY
NOTE:
FOR REFERENCE ONLY: DISPENSERS MFG. BEFORE NOV. 1986
FOR DISPENSERS MADE AFTER OCT. 1986 SEE DWG. TTT-02092
KOLD-DRAFT
UNIFLOW MFG. CO. ERIE, PA.
LET
REVISION
DATE
A
REVISED, REDRAWN – SUPERSEDES DRAWING DATED 9-25-85
10-23-86
NAME:
TDA WIRING DIAGRAM
MAT'L:
SCALE
DWN: E.B.
CHKD:
PART NO.:
REV.
DATE: 10-23-86
DATE:
TTT-02079
A

TDA DISPENSER

The TDA is a high volume, high speed dispenser with either pushrod or foot pedal control.

Service Information

1. Large volume output the large pulley (4") is onthe drive motor and small pulley (2") on the gear reducer.
2. Small volume output the small pulley (2") is on the drive motor and the large pulley(4") on the gear reducer. The tapered spout is used for small volume as pitcher fulling.
3. Drains - Do not tee side and center drains. Run them separate to prevent the water from the evaporator drain pan from backing into the drip sink and/or the bottom of the hopper.
4. An optional motorized agitator is available to eliminate ice bridging caused by infrequent use. This option can be field installed if necessary, ask for kit No. TTR-02071-01.

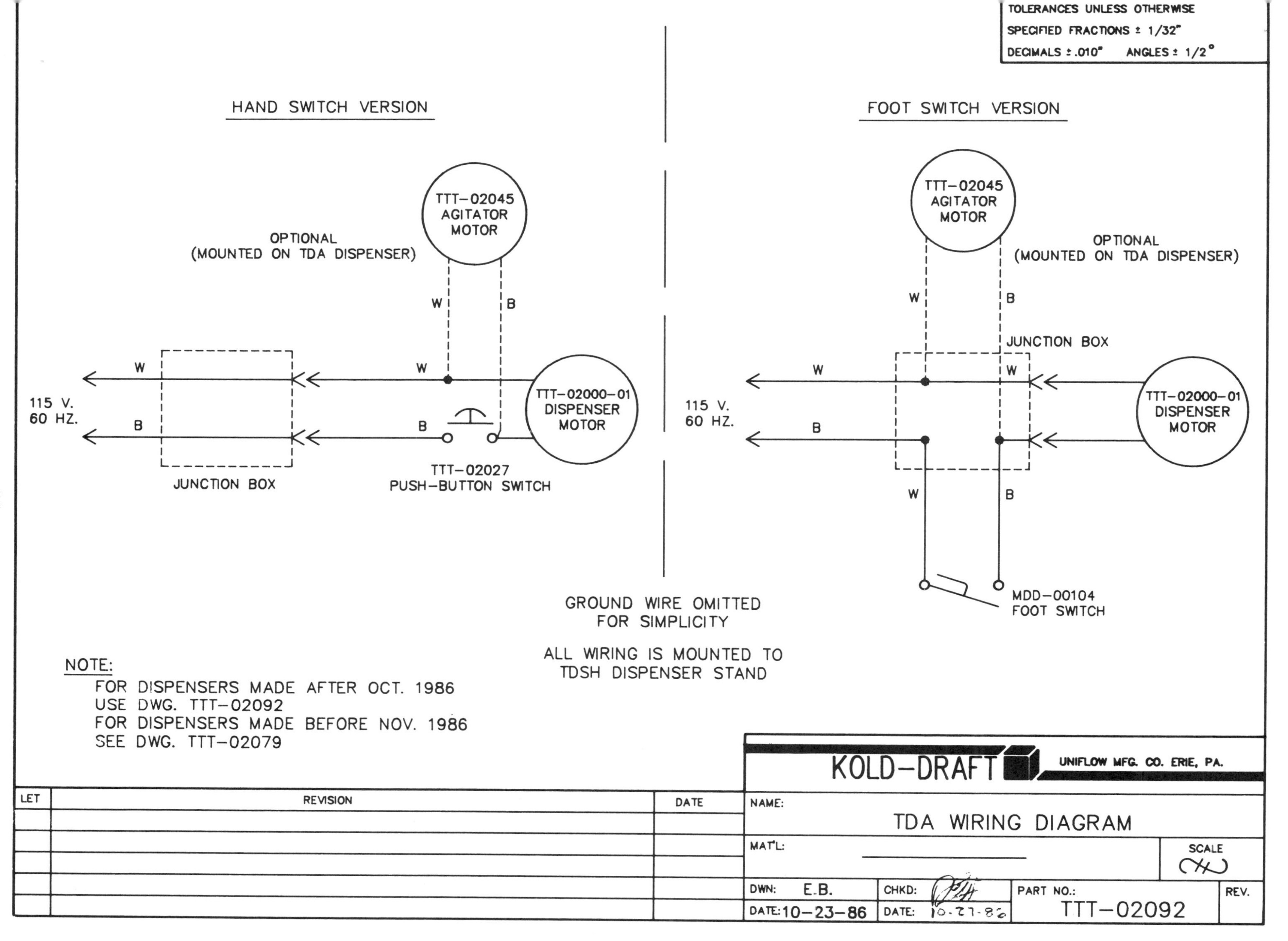
TOLERANCES UNLESS OTHERWISE
SPECIFIED FRACTIONS ± 1/32"
DECIMALS ± .010" ANGLES ± 1/2°
HAND SWITCH VERSION
TTT-02045
AGITATOR
MOTOR
OPTIONAL
(MOUNTED ON TDA DISPENSER)
W
B
115 V.
60 HZ.
JUNCTION BOX
TTT-02000-01
DISPENSER
MOTOR
TTT-02027
PUSH-BUTTON SWITCH
FOOT SWITCH VERSION
JUNCTION BOX
MDD-00104
FOOT SWITCH
GROUND WIRE OMITTED
FOR SIMPLICITY
ALL WIRING IS MOUNTED TO
TDSH DISPENSER STAND
NOTE:
FOR DISPENSERS MADE AFTER OCT. 1986
USE DWG. TTT-02092
FOR DISPENSERS MADE BEFORE NOV. 1986
SEE DWG. TTT-02079
KOLD-DRAFT
UNIFLOW MFG. CO. ERIE, PA.
LET
REVISION
DATE
NAME:
TDA WIRING DIAGRAM
MAT'L:
SCALE
DWN: E.B.
DATE: 10-23-86
CHKD:
DATE:
PART NO.:
TTT-02092
REV.

GS SERIES INSTRUCTIONS

1) CHECK FOR FREIGHT DAMAGE. If damage is present, contact freight carrier now and file claim.

2) Remove bolts that secure Cuber to the pallet.

3) Slide the Cuber off the pallet.

4) Remove leg pack from bin and screw in legs.

5) Position Cuber: Insure 4" air space on left, right and backsides.

6) Level Cuber by adjusting legs.

7) ELECTRICAL INSTRUCTIONS

   A) Each Cuber must be on its own circuit. Individually fused or HACR type circuit breaker. Maximum fuse size 15 AMP. All wiring must meet National & Local wiring requirements.

   B) GS Cuber is factory wired for 115 volt, 60 or 50 hz., single phase and is equipped with a ground type power cord. Voltage must not vary more than plus or minus 10%.

8) PLUMBING INSTRUCTIONS

   A) A 1/2" water supply line is connected to 1/4" IPS female fitting at lower left side. Water supply must have a capacity to maintain at least 15 PSI under maximum service load. (Maximum supply pressure 100 lb.) Before final connection is made, flush water line.

   B) Drains: Two - 3/4" IPS female drains located on lower left side. Use at least 5/8" I.D. drain hose or 3/4" O.D. tubing to carry off water and prevent water backing up into bin or drain pan.

9) Remove the top cover and upper front panel.

   Remove top cover by lifting straight up. Grip bottom of upper front panel and lift up to remove.

   NOTE: If you ever need access to sides, pull up until clips are disengaged. For access to lower front, remove (2) screws, then pull bottom straight down, then out to clear the bin opening bezel strip.

10) Remove wire and tag, securing the water plate in the UP position for shipping purposes.

11) Remove tape securing water control tank and hanger wire. Position wire and tank on Penn Weight Switch. Insure that the weight tank hangs straight down and the tubing from the main water tank is straight and not kinked.

12) Factory settings are made but shipping may have changed these settings - (Check).

    A) That the drain pan is properly positioned.

    B) The bin thermostat cap tube is secured in the bin.

    C) The hinges are tight positioning the water plate.

    D) The fin thickness is correct (with the water plate in the full-upright position - should be able to slide a nickel between the evaporator and the water plate).

13) Turn wash switch to "WASH" position. It is located on right side of front wiring channel.

14) Turn on water and power. Observe for water leaks as unit fills. Eliminate leaks, if any exist.

15) Turn the wash switch to "ON" position and observe the Cuber as it completes 2 cycles to insure it is operating properly.

16) Check the bin thermostat by holding ice up to the capillary tube. Cuber should shut off in 30-60 seconds.

17) Attach the wiring diagram and ice machine cleaning instructions to inside of top cover.

18) Remove any protective plastic from panels and replace upper front and top panels. CUBER MUST NOT BE OPERATED WITH PANELS REMOVED.

19) Complete and mail in WARRANTY CARD.

20) For service information, contact your local Kold-Draft Distributor or Dealer or Kold-Draft's Complete Ice Machine Service Manual.

NOTE: A WASHABLE AIR FILTER IS INSTALLED AT THE CONDENSER AIR INLET LOWER FRONT - LEFT SIDE. FOR ACCESS - REMOVE LOWER FRONT PANEL BY REMOVING THE TWO SCREWS AND PULLING OUT ON THE BOTTOM OF THE PANEL. CHECK FILTER AT LEAST ONCE EACH MONTH - WASH IF NECESSARY, SHAKE OUT EXCESS WATER AND REPLACE.

CAUTION: When operating and servicing a Kold-Draft GY or GS Model Ice Cuber care must be taken to prevent possible personal injury or damage to the unit.

WARNING: Before operating the Cuber, check to verify:

1. Each Cuber is on its own circuit, protected with fuses or HACR type circuit breaker.
2. The water pump screen must be in place and secured with its screw.
3. The front electrical channel must be secured in place with hold-down clip and screw.
4. Compressors must have covers secured in place to prevent access to compressor wiring and/or start components.
5. The hot gas defrost valve cover must be secured in place.
6. On thermostatically controlled Cubers the water weight control switch cover must be secured with screw.
7. Cuber must not be operated with the exterior panels removed without a qualified service mechanic present.

WARNING: When servicing Kold-Draft Cubers safe service practices must be followed.

1. DANGER: Do not braze refrigerant lines until refrigerant is removed from system.
2. WARNING: Beware of potential electrical shock hazard when testing for voltage at electrical connections within the system.
3. CAUTION: Exhibit safe practices when working around moving parts such as the pump fan blade, the front and rear cams, front and rear water hinges, the actuator motor and/or condenser fan blades.

GY INSTALLATION INSTRUCTIONS

1) CHECK FOR FREIGHT DAMAGE. If damage is present, contact freight carrier now and file claim.

2) Remove bolts that secure the Cuber to the pallet.

3) Slide the Cuber off the pallet.

4) Remove the leg pack from bin and screw in legs into bin bottom.

5) Position Cuber: Insure 4" air space on left and right side of Cuber, 2 1/2" minimum on back for water, electrical, and air space.

6) Level Cuber by adjusting legs.

7) ELECTRICAL INSTRUCTIONS:

   A) Each Cuber must be on its own circuit. Individually fused or HACR type circuit breaker. Maximum fuse size is 15 AMP. There is a 15 AMP in line fuse installed in the junction box at upper left rear of Cuber (except on 50 cycle Models).

   B) GY Cubers are factory wired for 115 volt, 60 hz., single phase, or 50 cycle. Use No. 12 wire for installation of GY Cuber.

   C) All connections must meet National and Local Electronic Code Requirements.

8) PLUMBING INSTRUCTIONS:

   A) A 1/2" water supply line is connected to 1/4" IPS female fitting at upper left rear of Cuber. Water supply must have a capacity to maintain at least 15 PSI under maximum service load. (Maximum supply pressure 100#) Before final connection is made, flush water line.

   B) Drains - are two 3/4" IPS female drains located on lower left side. Use at least 5/8" I.D. drain hose or 3/4" O.D. tubing to carry off water and prevent water backing up into bin or drain pan.

9) To gain access to ice making section, lift top straight up. To remove upper front panel, remove (4) screws - 2 located on each side.

   NOTE: For service access to condensing unit, remove screw holding in front louvered access panel. Once panel is removed, the condensing unit after removing 1 screw will virtually slide out. For access to the left side, remove (2) screws holding access panel.

10) Remove wire and tag securing the water plate in the UP position for shipping purposes.

11) Remove tape securing the weight control tank and hanger wire. Position wire and tank on the Penn Weight switch. Insure that the weight tank hangs straight down and the tubing from the tank to the main water tank is straight and not kinked.

12) Uncoil the bin thermostat cap tube and slide it through the tube holder mounted on the upper left side of the bin, with 3" of cap tube extending out the tube holder bending down and toward the center of bin.

13) Remove ice scoop from the bin.

14) Attach the wiring diagram and ice machine cleaning instructions to under side of top cover.

15) Turn wash switch to the "WASH" position. It is located on the right side of the front channel.

16) Turn on water and power. Check for water leaks as unit fills. Eliminate leaks if present.

17) Turn the wash switch to "ON" position and observe the Cuber as it completes 2 cycles to insure it is operating properly.

18) Check the bin thermostat by holding ice up to the capillary tube Cuber should shut off in 30-60 seconds.

19) Remove any protective plastic from panels. Replace the front and top panel. CUBER MUST NOT BE OPERATED WITH PANELS REMOVED.

20) Complete and mail WARRANTY CARD.

21) For service information contact your local Kold-Draft Distributor or Dealer or Kold-Draft's Complete Ice Machine Service Manual.

CAUTION: When operating and servicing a Kold-Draft GY or GS Model Ice Cuber care must be taken to prevent possible personal injury or damage to the unit.

WARNING: Before operating the Cuber, check to verify:

1. Each Cuber is on its own circuit, protected with fuses or HACR type circuit breaker.
2. The water pump screen must be in place and secured with its screw.
3. The front electrical channel must be secured in place with hold-down clip and screw.
4. Compressors must have covers secured in place to prevent access to compressor wiring and/or start components.
5. The hot gas defrost valve cover must be secured in place.
6. On thermostatically controlled Cubers the water weight control switch cover must be secured with screw.
7. Cuber must not be operated with the exterior panels removed without a qualified service mechanic present.

WARNING: When servicing Kold-Draft Cubers safe service practices must be followed.

1. DANGER: Do not braze refrigerant lines until refrigerant is removed from system.
2. WARNING: Beware of potential electrical shock hazard when testing for voltage at electrical connections within the system.
3. CAUTION: Exhibit safe practices when working around moving parts such as the pump fan blade, the front and rear cams, front and rear water hinges, the actuator motor and/or condenser fan blades.

INSTALLATION INSTRUCTIONS
FOR
GB & GT SERIES CUBERS

1) CHECK FOR FREIGHT DAMAGE. If damage is present, contact freight carrier immediately and file claim.

2) Position Bin per Figure A.

3) Level Bin by adjusting leg bottoms up or down. If bin is sealed to the floor (no legs) and it is necessary to shim the bin level, install a cove molding around the bottom of the bin. Seal the bin or molding to the floor with RTV Sealant.

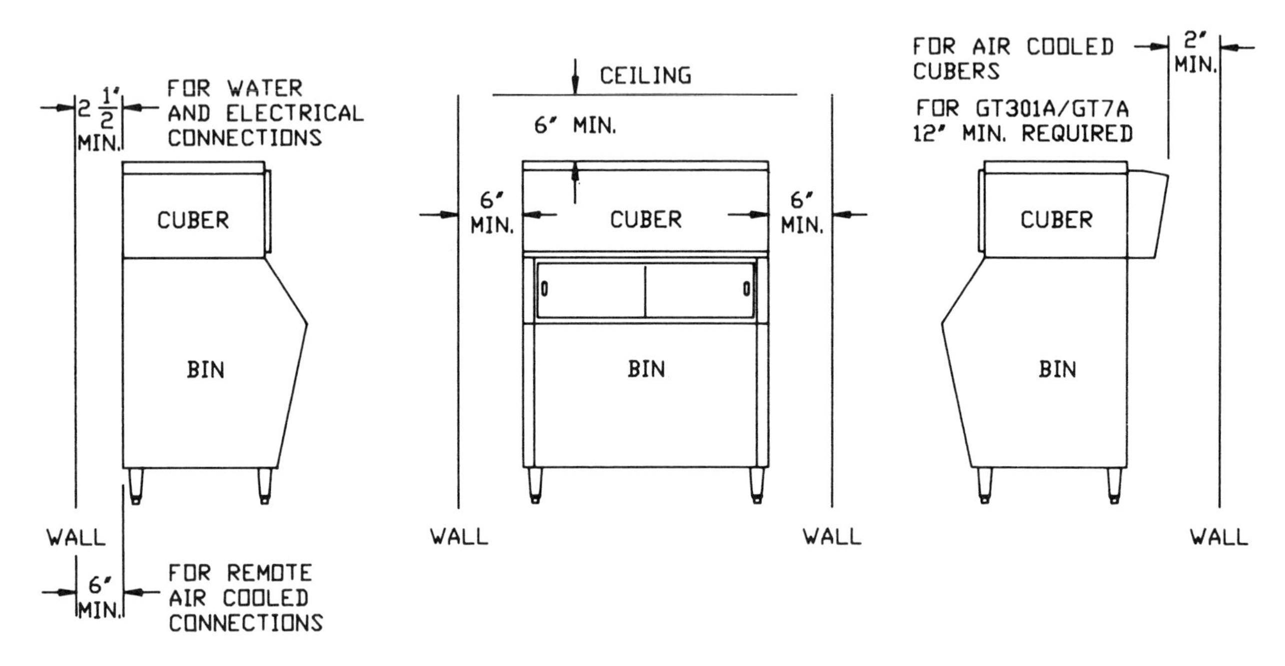

FIGURE A

4) If panels are on the Cuber, remove them now. To remove:

   1. Front Panel - loosen screw, (do not remove screw), lift up on panel.
   2. Top Panel Lift - bring forward until the rear clip clears the frame.
   3. Side Panels - pull forward and lift to clear clips.
   4. Rear Panel - Lift straight up.
   5. Remove drain pan and ice chute.

5) Place pressure sensitive gasketing on top outside edge of Bin except the front. Front gasket is already installed on bottom of front rail.

6) Remove shipping bolts in condensing unit pan that protrude through bottom frame to prevent damage to sealing tape on bin top as unit is positioned. Condensing unit pan will be held in place by cuber mounting bolts.

7) Lift and position Cuber on Bin.

8) Center front angle cover between the frame and Bin.

9) Locate and drill four 5/32" holes in the top of the Bin (except GBN-950 Bin) then secure Cuber to Bin using 1/4" lag screws provided per Figure B. Drilling is not required on GBN-950 Bin. See Instructions packaged with Bin.

FIGURE B

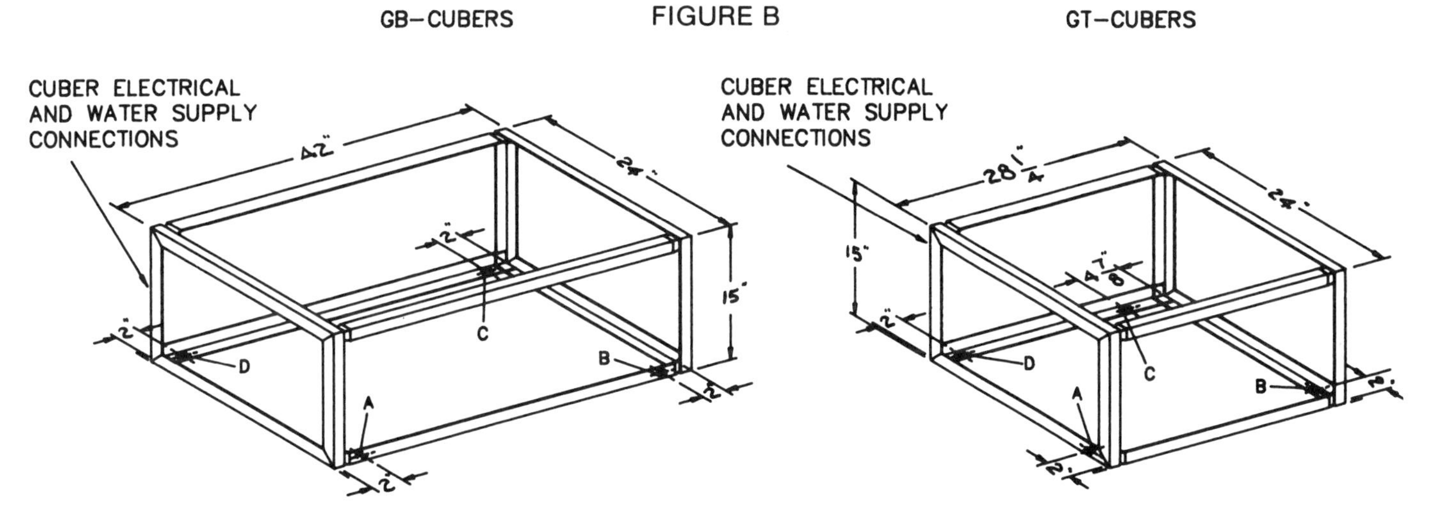

NOTE:
1. GB500, GB600, GB1000, GB1200 Series – Holes A and B require 1 1/4" lag bolt and washer. Holes C and D require 2" lag bolt and washer. Air cooled GB500 or GB600 models omit lag bolt at hole C.
2. GB400 Series holes A and D require 1 1/4" lag bolt and washer. Holes B and C require 2" lag bolt and washer.

NOTE:
1. 1 1/4" lag bolt at holes A and B
2. 2" lag bolt at holes C and D

* DO NOT USE LONG LAG BOLT AT HOLES THAT REQUIRE SMALL LAG BOLT AS THEY WILL PROTRUDE THROUGH TRACK AND INTERFERE WITH OPENING AND CLOSING BIN DOORS.

10) ELECTRICAL INSTRUCTIONS

A) Each cuber must be on its own circuit, protected with fuses or HACR type circuit breaker. Refer to electrical rating plate on rear of Cuber for fuse size.

B) There is a supplemental fuse Kit (except 3 phase models) packed with the cuber which should be installed in the junction box at left side of the Cuber.

C) Connect Cuber leads per Instructions on Tag attached to leads in junction box on left side of Cuber.

D) On GB500, GB600, GB1000, GB1200, GT500 and GT600 Series Cubers, you must connect the Brown lead(s) to the transformer at either the 208 or 230 volt terminal to match supply voltage. Transformer box is located next to the compressor.

11) Position drain pan and screw drain yoke into drain fitting at top of bin, thin gasket on pan, thick gasket beneath drain pan, Figure C.

FIGURE C

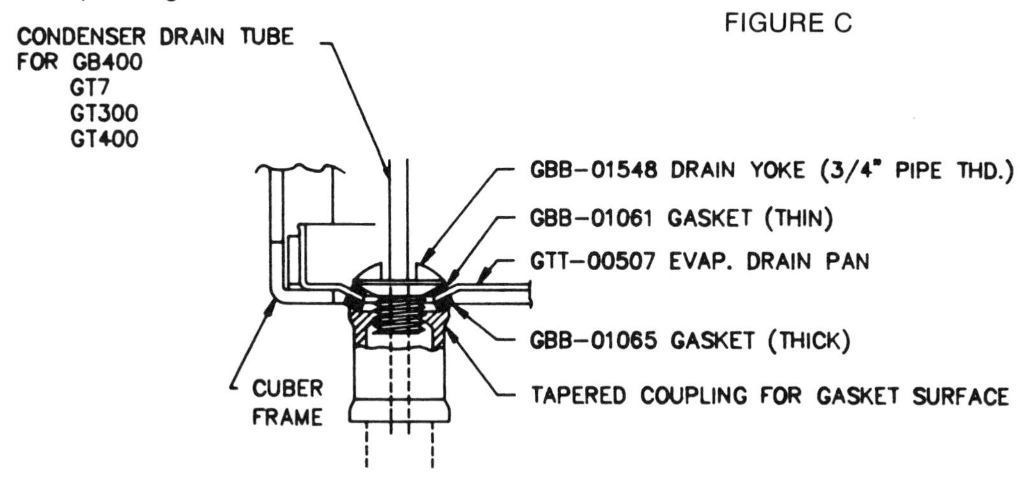

12) PLUMBING INSTRUCTIONS

A) Check local code for backflow preventer requirement. This item is not provided with the cuber and may be obtained locally if required. A 1/2" water supply line is connected to 1/4" IPS female fitting in rear of Cuber in upper left hand corner. The water supply must have a capacity to maintain 15 PSI under maximum service load, (maximum supply pressure 100 lb.), before final connection is made, flush water line.

B) Condenser Drains — (Water-Cooled)

1. GB400, GT300, GT400, GT7 Series Cubers, the condenser drains are run into the drain as shown in Figure C.

2. On all other GB or GT Series water-cooled Cubers, the condenser drain is 1/2" SAE male flare. Use 1/2" O.D. copper tubing to route the condenser drain water out the back of the Cuber to a suitable drain.

C) Bin drains are (2) 3/4" IPS female drains. Use at least 5/8" I.D. drain hose or 3/4" O.D. tubing to carry off water and prevent back-up in the Bin or drain pan. NOTE — BIN DRAIN CONNECTION MUST COMPLY WITH ALL SANITARY REQUIREMENTS FOR A GOOD STORAGE AREA. CONSULT LOCAL CODE.

13) Remove strap and tag securing water plate in upright position.

14) BIN PROBE PLACEMENT

A) GB Series — Remove bin probe and flexible probe holder from water control tank. Run bin probe under the front of the drain pan and secure flexible probe holder to clip on ice chute. Pry forks open to prevent the holder from pulling off the clips, Figure D. Position ice chute in Cuber.

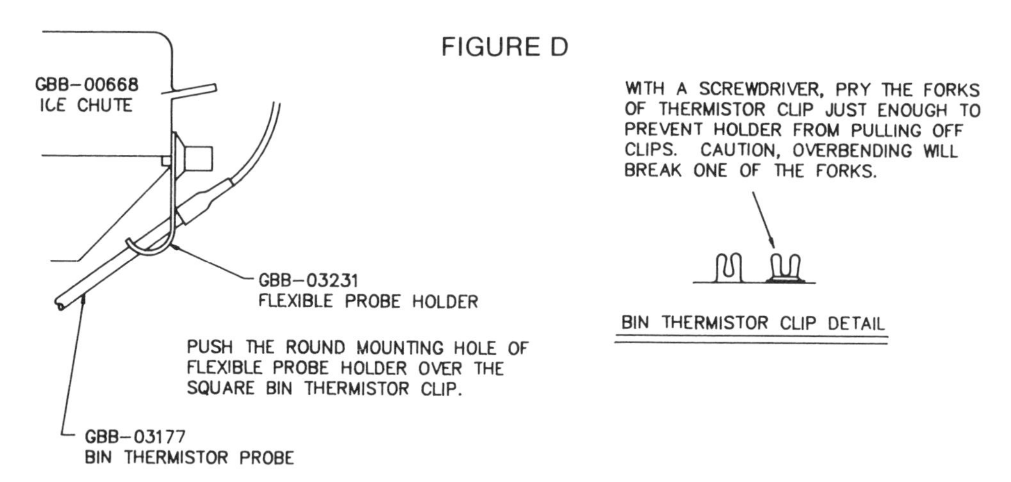

B) On Bins with front and rear deflectors provided, refer to Figure E.

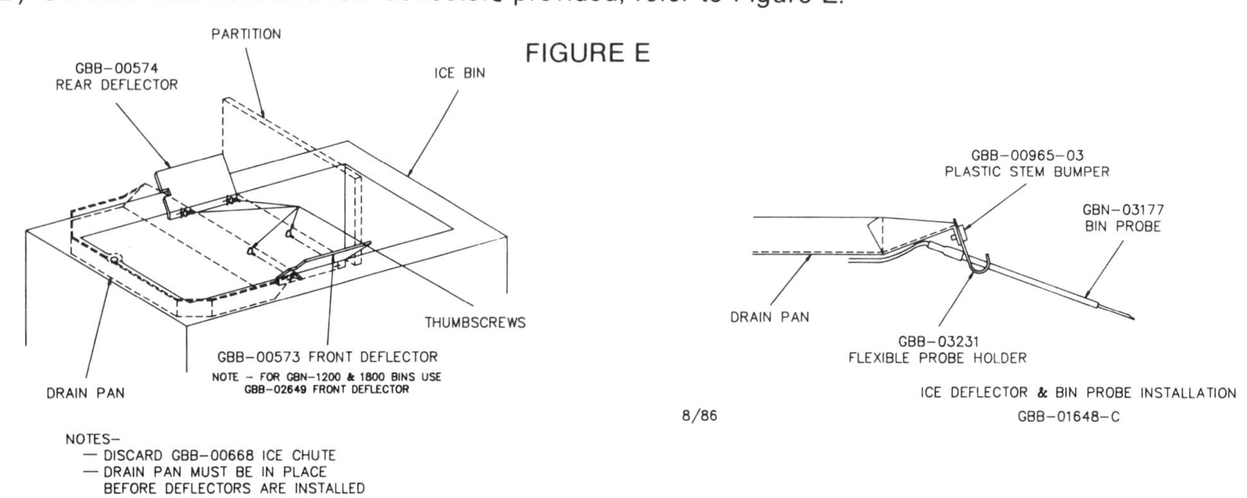

C) GT Series — Route Bin probe or Bin thermostat capillary tube and secure as shown in Figure F. Position ice chute.

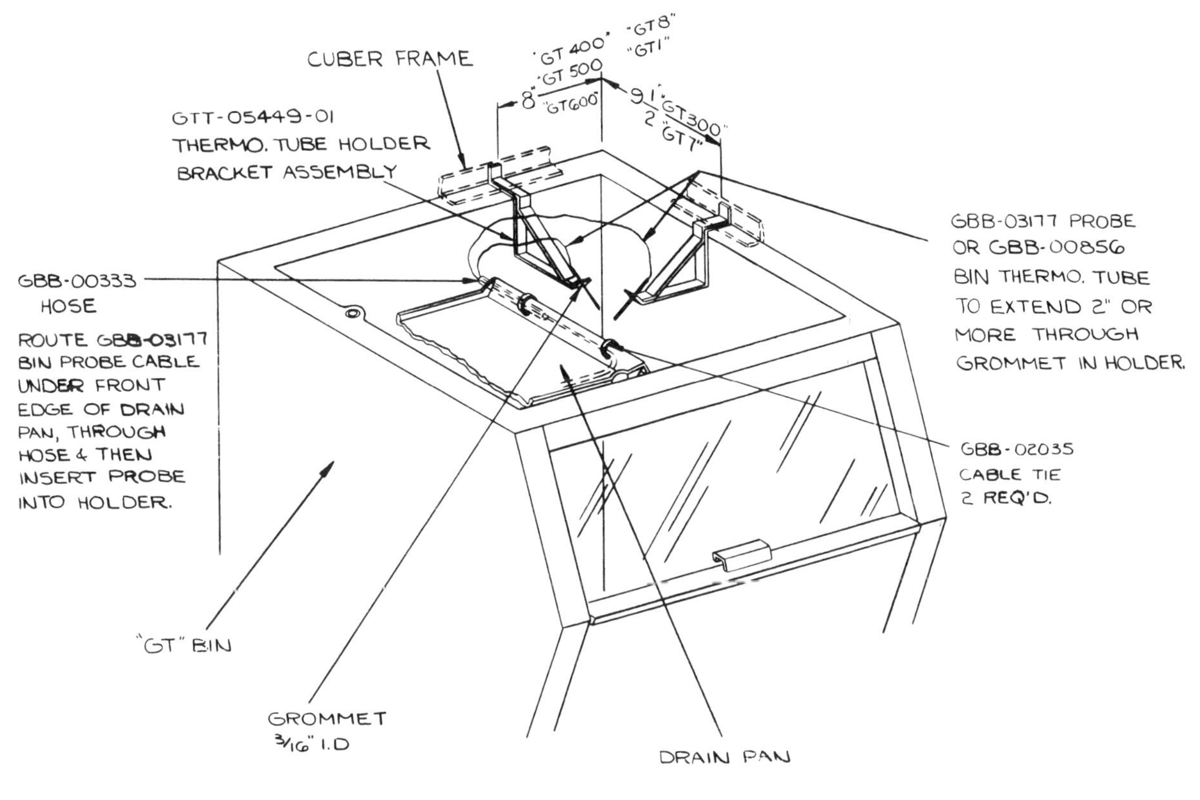

FIGURE F

15) On GB1000 or GB1200 Series install Cube gate over camshaft rod between front and rear support channels on bottom ice making section. This will direct ice into drop zone and prevent ice from falling onto lower drain pan.

16) Turn the wash switch to "WASH" position.

  A) On GB400, GT300, GT400, GT7 Series Cubers, the wash switch is located on the right side of front wiring channel.

  B) On all other GB or GT Series Cubers, the wash switch is located on the transformer box next to the compressor.

17) Turn on water and power and check for leaks as Cuber fills.

18) Turn the wash switch to "ON" position. Observe the Cuber through 2 cycles to insure it is operating properly.

19) Check bin probe function by holding ice up to probe tip. Cuber will shut off within 30 seconds (test after water fill is completed as Cuber will not shut off during fill cycle).

20) Install wiring diagram and ice machine cleaning instructions on inside of front panel.

21) Install panels. CUBER MUST HAVE PANELS IN PLACE DURING NORMAL OPERATION.

22) Complete and mail in WARRANTY CARD.

23) For service information, contact your local Kold-Draft Distributor or Dealer or Kold-Draft's Complete Ice Machine Service Manual.

CAUTION: When operating and servicing a Kold-Draft GB or GT Series Ice Cuber care must be taken to prevent possible personal injury or damage to the unit.

WARNING: Before operating the Cuber, check to verify:

1. Each Cuber is on its own circuit, protected with fuses or HACR type circuit breakers.
2. All Cubers (except 3 phase Models) must have a supplemental fuse kit installed in junction box at the left side of the Cuber (Refer to Engineering Bulletin #22-84).
3. The water pump screen must be in place and secured with its screw.
4. The front electrical channel must be secured in place with hold-down clip and screw.
5. Electronic Cubers (except GT300 Series) must have the control module box cover in place and secured with screw. GT300 series Cubers - it is built into the front panel and must be secured in place when the machine is not being serviced.
6. Cubers with transformers must have cover in place and secured with a screw.
7. Compressors must have covers secured in place to prevent access to compressor wiring and/or start components.
8. The hot gas defrost valve cover must be secured in place.
9. On thermostatically controlled Cubers the water weight control switch cover must be secured with screw.
10. Cuber must not be operated with the exterior panels removed without a qualified service mechanic present.

WARNING: When servicing Kold-Draft Cubers safe service practices must be followed.

1. DANGER: Do not braze refrigerant lines until refrigerant is removed from system.
2. WARNING: Beware of potential electrical shock hazard when testing for voltage at electrical connections within the system.
3. CAUTION: Exhibit safe practices when working around moving parts such as the pump fan blade, the front and rear cams, front and rear water plate hinges, the actuator motor and/or condenser fan blades.

GB AND GT MULTIPLEXING
INSTALLATION INSTRUCTIONS

1) CHECK FOR FREIGHT DAMAGE. If damage is present, contact freight carrier now and file claim.

2) Turn water and power OFF to existing Cuber.

3) Remove the front and top panels from the existing Cuber.

4) Check minimum clearance per Figure A.

5) Place pressure sensitive gasket on the top inside edge of the existing Cuber except the front. Front gasket is already installed on bottom of the front double rail.

FIGURE A

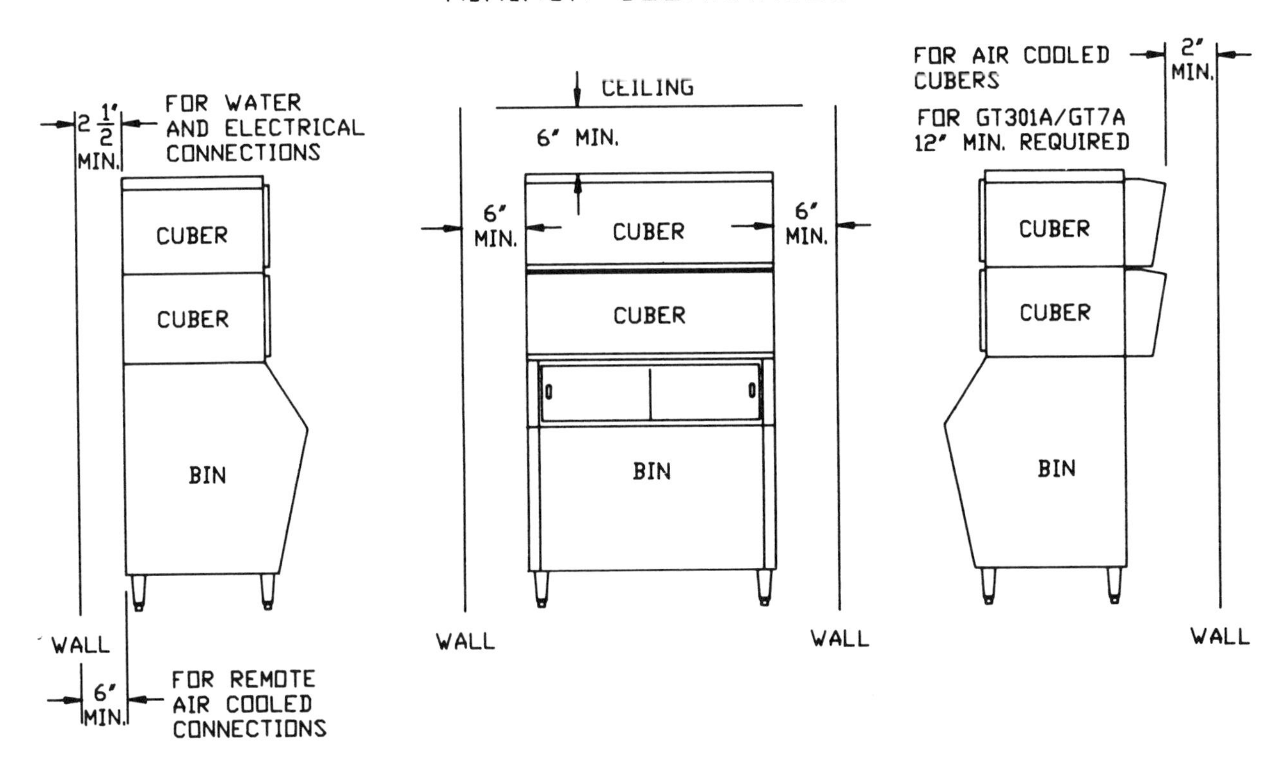

6) If panels are on the Cuber to be stacked, remove them now. To remove:

   A) Front Panel - Loosen screw (do not remove screw), lift up on panel.

   B) Top Panel - Bring forward until the rear clip clears the frame.

   C) Side Panels - Pull forward and lift to clear clips.

   D) Rear Panel - Lift straight up.

   E) Remove drain pan and ice chute.

7) Lift Cuber and position on top of existing Cuber.
   NOTE: The Cuber must be lifted and not slid into place to prevent damage to cams and gasketing on lower Cuber.

8) Install front double rail. Insure that it is centered between Cuber frames.

9) Locate holes and bolt frames together. Refer to Figure B.

FIGURE B

GB SERIES CUBERS

GT SERIES CUBERS

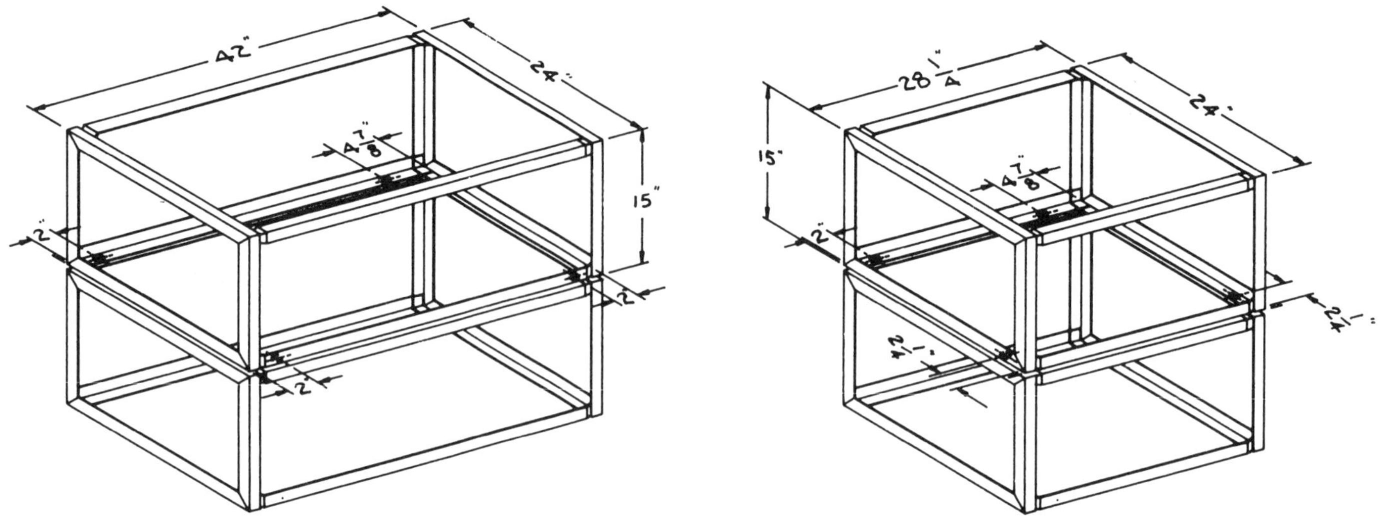

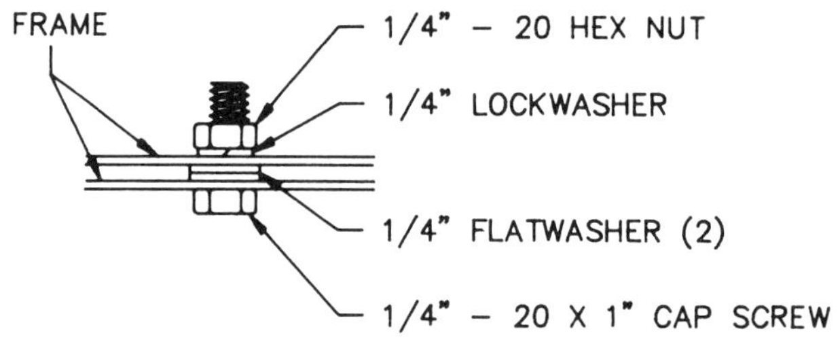

FRAME

1/4" – 20 HEX NUT

1/4" LOCKWASHER

1/4" FLATWASHER (2)

1/4" – 20 X 1" CAP SCREW

NOTE:

1. Stacking a GB400 Series cuber use (2) 1/4–20x2 1/2" bolts on the right side. (They are not provided in kit)
2. When stacking a GB500 or GB600 air–cooled model omit bolt at right rear as condenser is in way.
3. When stacking electronic and electromechanical cubers it is advisable for service accessibility to put electronic cuber on bottom.

NOTE:

1. Never stack any GT cuber on a GT400, GT500, GT600, GT1 or GT8 cuber as ice cannot fall into bin.
2. Multiplexing a GT300 or GT7 requires a GTR–00645 ice chute and (2) 1/4–20x2 1/2" bolts are not supplied in kit to secure cubers in back.
3. Multiplexing a GT400, GT500, GT600, GT1 or GT7 cuber requires a GTR–00600 ice chute.

10) PLUMBING INSTRUCTIONS

A) Remove water inlet fitting from bottom Cuber. Install tee.

B) Install 90° Elbow in top Cuber's water inlet fitting. Before completely tightening, position copper tube connecting tee to 90° Elbow to interconnect the water supply to both Cubers.

C) Connect 1/2" copper water supply tube to bottom of tee. Water supply must be able to maintain a 15# PSI minimum under maximum service load (maximum supply pressure - 100# PSI).

D) Condenser drains - water-cooled units only.

1. GB400, GT300, GT400, GT7 Series - the condenser drain is run into bin drain as in Figure C.

2. GB500, GB600, GB1000, GB1200, GT500, GT600 Series - the condenser drains are run out the back to a floor drain.

E) Bin Drains - (2) 3/4" IPS female drains. Use at least 5/8" I.D. drain hose or 3/4" O.D. tubing to carry off water and prevent back-up in the Bin or drain pan. Drains are located on the left side but there is bottom access to drains if desired.

11) Position drain assembly in the upper unit through drain pan as in Figure C.

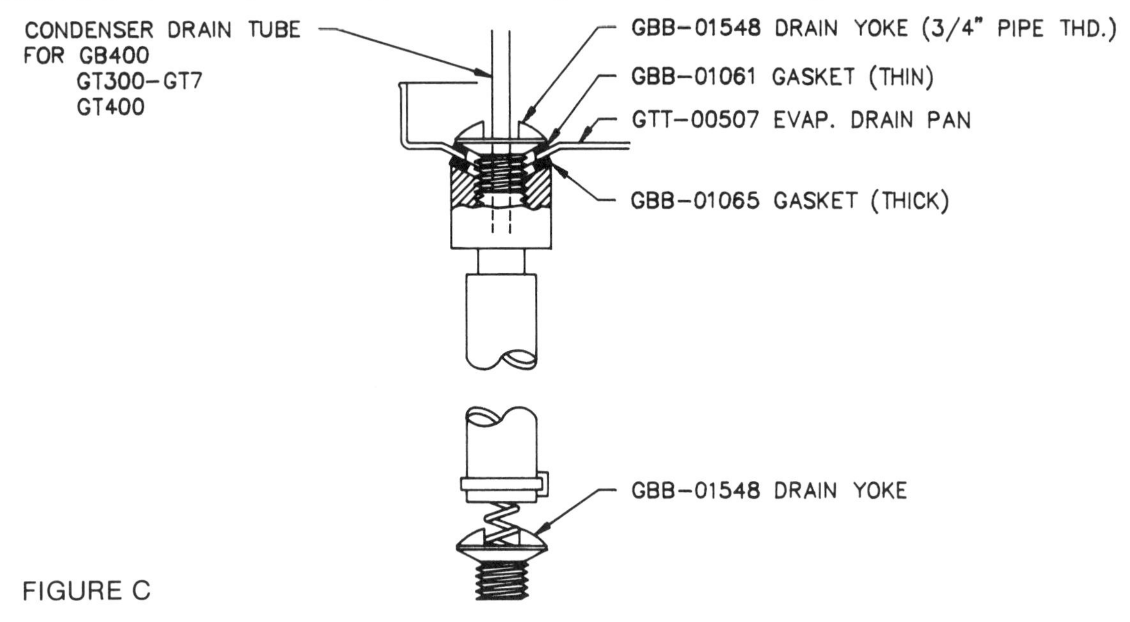

FIGURE C

12) ELECTRICAL INSTRUCTIONS

A) Each Cuber must be on its own circuit, protected with fuses or on HACR Type circuit breakers. Refer to electrical rating plate on rear of Cuber.

B) There is a supplemental fuse Kit (except 3 phase models) packed with the Cuber which should be installed in the junction box at the left side of the Cuber.

C) Connect Cuber leads per Instructions on Tag attached to leads in junction box on left side of Cuber.

D) On GB500, GB600, GB1000, GB1200, GT500, GT600 Series, you must connect the Brown lead(s) on the transformer to either 208 or 230 volt terminal as per power supply readings at the Cuber. Transformer box is located next to compressor.

13) Remove wire and tag securing the water plate in upright position.

14) BIN PROBE PLACEMENT

A) GB Series - Remove the bin probe and flexible probe holder from water level tube. Run the upper bin probe straight down along the control box, then between the drain pan and ice chute on both Cubers. Secure the flexible bin probe holder to clip on ice chute as shown in Figure D.

WITH A SCREWDRIVER, PRY THE FORKS OF THERMISTOR CLIP JUST ENOUGH TO PREVENT HOLDER FROM PULLING OFF CLIPS. CAUTION, OVERBENDING WILL BREAK ONE OF THE FORKS.

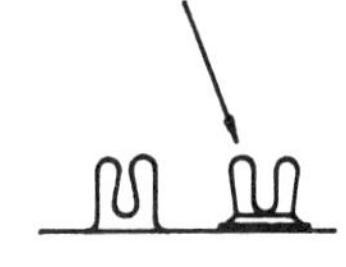

BIN THERMISTOR CLIP DETAIL

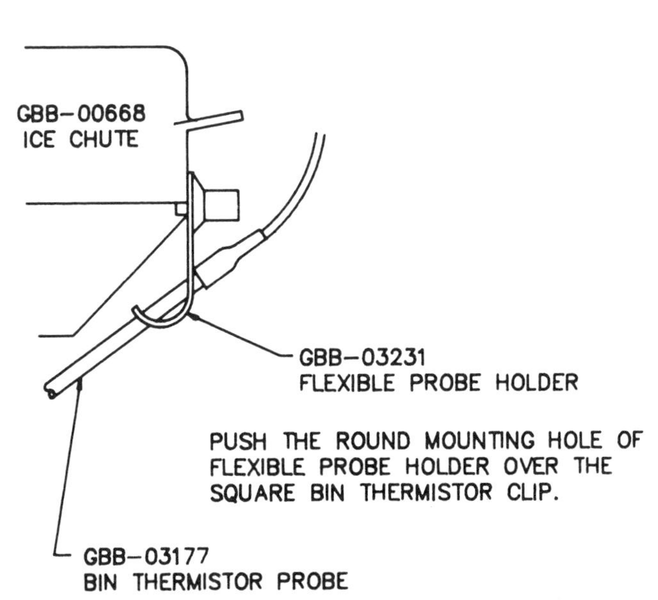

PUSH THE ROUND MOUNTING HOLE OF FLEXIBLE PROBE HOLDER OVER THE SQUARE BIN THERMISTOR CLIP.

FIGURE D

B) On Bins with front and rear deflectors, refer to Figure E.

FIGURE E

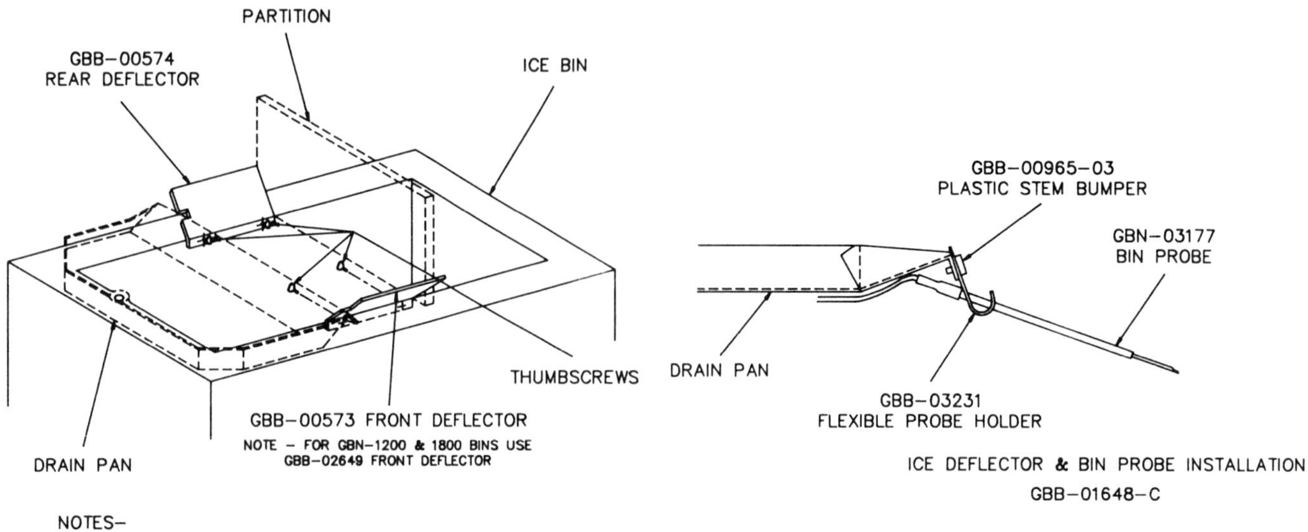

NOTES–
— DISCARD GBB–00668 ICE CHUTE
— DRAIN PAN MUST BE IN PLACE
BEFORE DEFLECTORS ARE INSTALLED

C) GT Series Cubers - Uncoil bin probe or bin thermostat capillary tube, route upper bin control out of the way of moving components. Run it under lower drain pan and position per Figure F.

NOTE: 1. GTR-00645 Ice Chute used in upper unit when Multiplexing GT300 or GT7 Cubers.

2. GTR-00600 Ice Chute used in GT400, GT500, GT600 Cubers on GT300 or GT7 Cubers.

FIGURE F

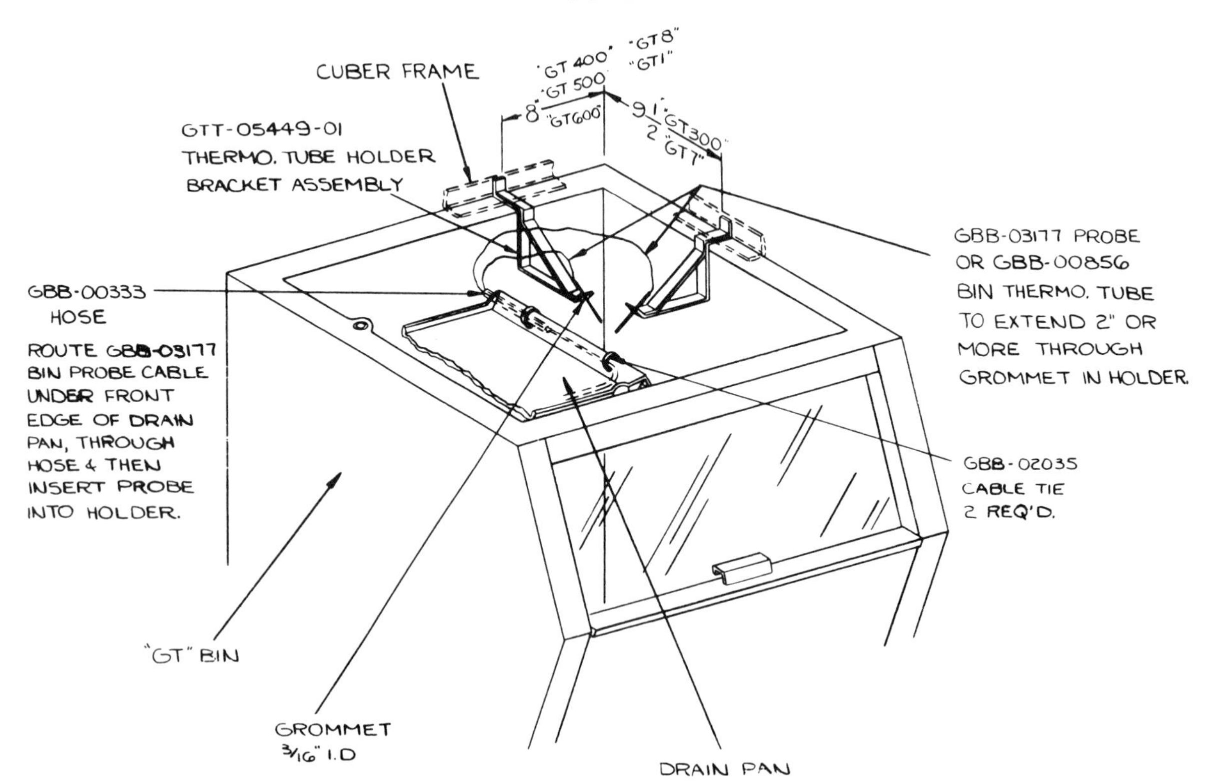

15) Position Cuber Gate(s) over cam shaft rod between front and rear support channels on all ice making sections except the top one. This will direct ice into drop zone and prevent ice from falling onto lower drain pans.

16) Turn the wash switch to "WASH" position.

A) GB400, GT300, GT400, GT7 Series is located at the right side of the front wiring channel.

B) All other GB and GT Cubers - it is located on the transformer box next to the compressor.

17) Turn on the water and power. As the Cuber is filling, check for water leaks. Eliminate if present.

18) Turn wash switch to "ON" position and observe the Cuber through two (2) complete cycles to insure it is operating properly.

19) Check Bin Probe function by holding ice up to probe tip. Cuber will shut off within 30 seconds (on electronic-test after water-fill is completed as Cuber will not shut off during fill cycle).

20) Install wiring diagram and ice machine cleaning instructions on inside of front panel.

21) Install panels. CUBER MUST NOT BE OPERATED WITHOUT PANELS IN PLACE.

22) Complete and mail in WARRANTY CARD.

23) For service information, contact your local Kold-Draft Distributor or Dealer or Kold-Draft's Complete Ice Machine Service Manual.

CAUTION: When operating and servicing a Kold-Draft GB or GT Series Ice Cuber care must be taken to prevent possible personal injury or damage to the unit.

WARNING: Before operating the Cuber, check to verify:

1. Each Cuber is on its own circuit, protected with fuses or HACR type circuit breakers.
2. All Cubers (except 3 phase Models) must have a supplemental fuse kit installed in junction box at the left side of the Cuber (Refer to Engineering Bulletin #22-84).
3. The water pump screen must be in place and secured with its screw.
4. The front electrical channel must be secured in place with hold-down clip and screw.
5. Electronic Cubers (except GT300 Series) must have the control module box cover in place and secured with screw. GT300 Series Cubers - it is built into the front panel and must be secured in place when the machine is not being serviced.
6. Cubers with transformers must have cover in place and secured with a screw.
7. Compressors must have covers secured in place to prevent access to compressor wiring and/or start components.
8. The hot gas defrost valve cover must be secured in place.
9. On thermostatically controlled Cubers the water weight control switch cover must be secured with screw.
10. Cuber must not be operated with the exterior panels removed without a qualified service mechanic present.

WARNING: When servicing Kold-Draft Cubers safe service practices must be followed.

1. DANGER: Do not braze refrigerant lines until refrigerant is removed from system.
2. WARNING: Beware of potential electrical shock hazard when testing for voltage at electrical connections within the system.
3. CAUTION: Exhibit safe practices when working around moving parts such as the pump fan blade, the front and rear cams, front and rear water plate hinges, the actuator motor and/or condenser fan blades.

INSTALLATION INSTRUCTIONS
FOR MOUNTING
KOLD-DRAFT SLEEVES ON KOLD-DRAFT BINS

GBN-210 OR GBN-290 BIN SLEEVE MOUNTING ON GBN-240 (GBN-2) OR GBN-550 (GBN-5) BINS.

1. Mount the sleeve on the bin and insert rods through at the vertical openings in the sleeve.
2. Tap each rod lightly.
3. Remove rods and sleeves.
4. Drill 1/2" holes at the indentation made by tapping the rods.
5. Insert rods and washers through sleeve holes and screw on the expanding nuts.
6. Place a continuous bead of silastic on the inner and outer edges of the bin.
7. Place sleeve on top of bin, aligning expanding nuts with the holes and drop in place.
8. Insert drain pipe and start thread, using thread sealant.
9. Tighten rods and drain pipe, and seal the inner and outer seams with silastic. Make special effort to seal around drain fittings on bin.

GBN-210 OR GBN-290 BIN SLEEVE MOUNTING TO GBN-450 (GBN-4) OR GBN-760 (GBN-7) BINS OR ANY SLEEVE WHICH HAS BEEN INSTALLED IN THE FIELD USING TIE-RODS.

1. Remove the present tie-rods.
2. Place continuous beads of silastic on the inner and outer edges of the existing top surface.
3. Place new sleeve so that the vertical holes line up with the vertical holes from which rods have been removed.
4. Assemble new rod and connector in such a manner that the rod will thread into the expanded nut from which the old rod was removed.
5. Insert drain pipe and start thread, using thread sealant.
6. Tighten four rods and drain pipe, and seal the inner and outer seams with silastic.

GTN-130 BIN SLEEVE MOUNTING ON GTN-350 (GTN-3) OR GTN-200 (GTN-17) BIN.

1. Mount the sleeve on the bin and insert rods through at the vertical openings in the sleeve.
2. Tap each rod lightly.
3. Remove rods and sleeves.
4. Drill 1/2" holes at the indentations made by tapping the rods.
5. Insert rods and washers through sleeve holes and screw on the expanding nuts.
6. Place a continuous bead of silastic on the inner and outer edges of bin.
7. Place sleeve on top of bin, aligning expanding nuts with the holes and drop in place.
8. Insert drain pipe and start thread, using thread sealant.
9. Tighten rods and drain pipe, and seal the inner and outer seams with silastic. Make special effort to seal around drain fittings on bin.

SOME COMBINATIONS ARE LISTED BELOW TO SHOW THE PROPER METHOD OF ASSEMBLING PARTS. SEE GB1696.

1. GBN-210 sleeve on a bin requires 1 rod and 1 nut per hole.
2. A GBN-290 sleeve on a bin requires 1 rod, 1 coupling and one 5" bolt per hole.
3. A GBN-290 sleeve on a GBN-210 (GBN-1) sleeve on a bin requires 2 rods, 1 coupling and 1 nut per hole.
4. Two GBN-210 sleeves on a bin requires 2 rods, 1 coupling, and 1 nut per hole.
5. Two GBN-290 sleeves on a bin requires 2 rods, 2 couplings, and one 5" bolt per hole.

NOTE: When installing a sleeve, use ice deflectors front and back and discard the chute.

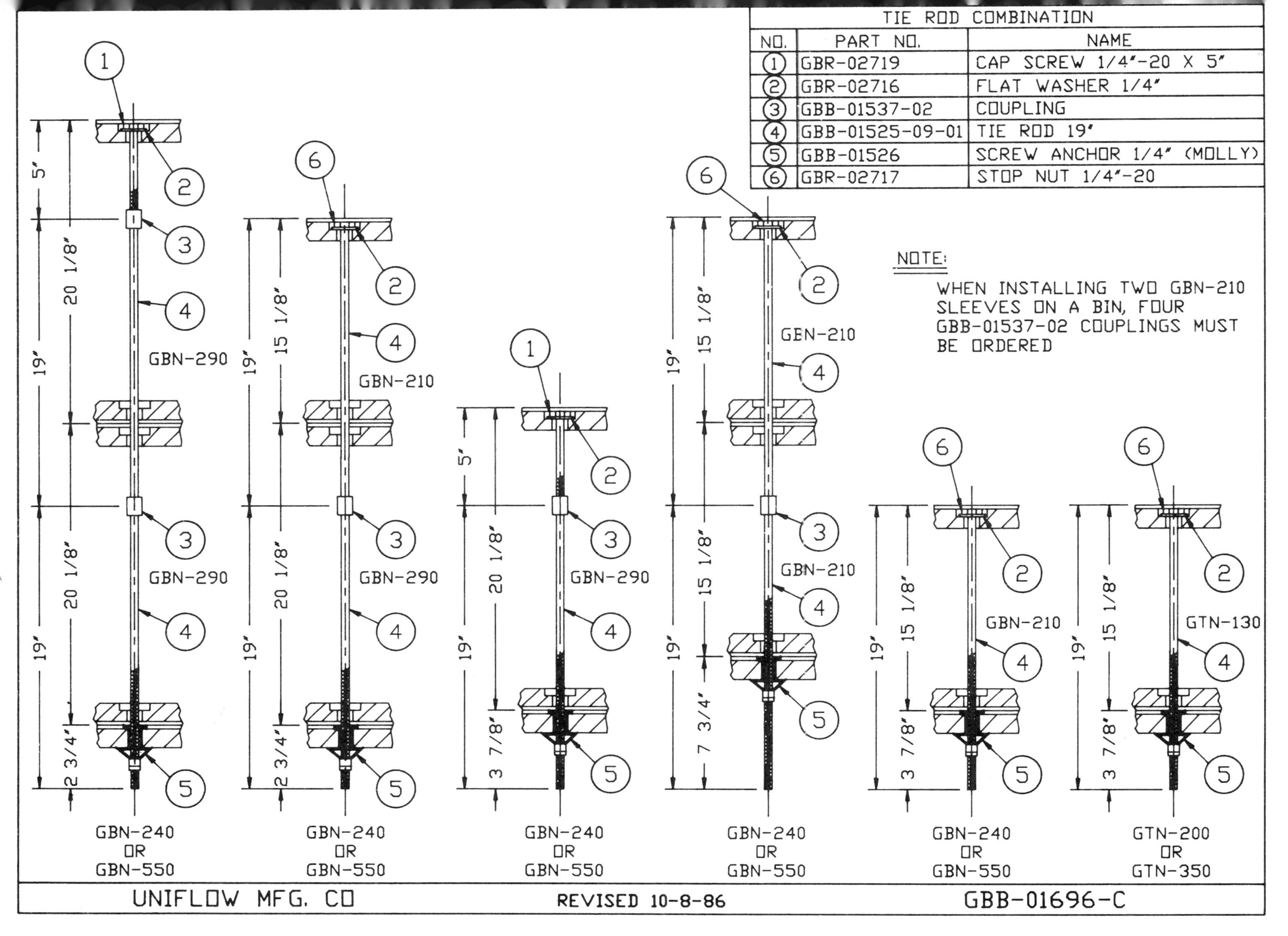

NO.	PART NO.	NAME
1	GBR-02719	CAP SCREW 1/4"-20 X 5"
2	GBR-02716	FLAT WASHER 1/4"
3	GBB-01537-02	COUPLING
4	GBB-01525-09-01	TIE ROD 19'
5	GBB-01526	SCREW ANCHOR 1/4" (MOLLY)
6	GBR-02717	STOP NUT 1/4"-20

BIN DIVIDER INSTALLATION

INSTRUCTIONS FOR PLACEMENT OF BIN DIVIDER IN GBN-240 BIN (GBN-2)

Parts Included in Kit:

1 - T-1619 Divider
3 - -T-1625 Well Nut
4 - GB-1522 Thumb Screw
1 - T-1620 Divider Angle Clip
3 - GB-515 Rubber Grommet
1 - GA-104 Hex Nut

Insert divider into the bin 18 1/2" from right side, so that it rests on the bottom and is lined up with the back. Mark the divider mounting hole locations on the liner for drilling 5/16" holes, two in back, two in front. Fasten the T-1620 divider angle clip to the bottom front of the divider using a thumb screw and hex nut.

Place the well nuts into the holes, align the divider holes with the well nuts and insert the thumb screws. After tightening the divider in place, solder the divider angle clip to the bottom of the bin liner.

Put grommets in holes provided in top of divider for bin thermostat tube.

INSTRUCTIONS FOR PLACEMENT OF BIN DIVIDER IN GBN-450 BIN (GBN-4)

Parts included in Kit:

1 - T-1617 Divider
1 - T-1619 Divider
7 - T-1625 Well Nut
8 - GB-1522 Thumb Screw
1 - T-1620 Divider Angle Clip
3 - GB-515 Rubber Grommet
1 - GA-104 Hex Nut

Follow instructions above for placement of T-1619 divider. Place T-1617 divider in the sleeve so that the bottom of it straddles the top of the T-1619 divider. Drill two holes in front and two holes in back, 5/16" in diameter and insert the well nuts. Line up the holes in the divider with well nuts in sleeve and insert the thumb screws. Tighten thumb screws in place.

Put grommets in holes provided in top of divider for bin thermostat tube.

INSTRUCTIONS FOR PLACEMENT OF BIN DIVIDER IN GBN-550 BIN (GBN-5)

Parts included in Kit:

1 - T-1622 Divider
6 - T-1625 Well Nut
6 - GB-1522 Thumb Screw
3 - GB-515 Rubber Grommet

Insert divider into bin 18 1/2" from right side so that the top of divider is 1/8" below the top surface of bin. Mark the divider mounting hole locations on the liner for drilling 5/16" holes, three in back, and two in front. Place the well nuts into the holes, align the divider holes with the well nuts and insert and tighten the thumb screws.

Put grommets in holes provided in top of divider for bin thermostat tube.

INSTRUCTIONS FOR PLACEMENT OF BIN DIVIDER IN GBN-740 BIN (GBN-6)

Parts included in Kit:

1 - T-1617 Divider
1 - T-1619 Divider
1 - T-1623-1 Divider
11 - T-1625 Well Nut
12 - GB-1522 Thumb Screw
1 - T-1620 Divider Angle Clip
1 - GA-104 Hex Nut
3 - GB-515 Rubber Grommet

Follow instruction for placement of T-1619 divider in GBN-2 bin, next, insert the T-1623-1 divider in the center sleeve with door using 4 well nuts and thumb screws. The bottom of the T-1623-1 divider is to straddle the top of the T-1619 divider. Drill two 5/16" holes in front, and two in back and insert the well nuts. Tighten the thumb screws in place.

Place the T-1617 divider in the top sleeve so the bottom straddles the top of the T-1623-1 divider. Drill two 5/16" holes in front and two in back. Insert the well nuts, align the holes in the divider with well nuts and insert and tighten the thumb screws.

Put grommets in hole provided in top of divider for bin thermostat tube.

INSTRUCTIONS FOR PLACEMENT OF BIN DIVIDER IN GBN-760 BIN (GBN-7)

Parts included in Kit:

1 - T-1622 Divider
1 - T-1617 Divider
10 - T-1625 Well Nut
10 - T-1625 Well Nut
10 - GB-1522 Thumb Screw
3 - GB-515 Rubber Grommet
3 - GB-515 Rubber Grommet

Follow instructions above for placement of T-1622 divider. Place T-1617 divider in the sleeve so the bottom straddles the top of the T-1622 divider. Drill two holes in front and two holes in back 5/16" in diameter and insert the well nuts. Line up the holes in the divider with the well nuts in sleeve and insert the thumb screws. Tighten thumb screws in place.

INSTRUCTIONS FOR PLACEMENT OF BIN DIVIDER IN GBN-310 BIN (GBN-27)

Parts included in Kit:

1 - T-1641 Divider
5 - T-1625 Well Nut
5 - GB-1522 Thumb Screw
3 - GB-515 Rubber Grommet

Insert divider into bin 18 1/2" from right side so that the top of divider is 1/8" below the top surface of bin. Mark the divider mounting hole locations on the liner for drilling 5/16" holes, two in back, and two in front. Place the well nuts into the holes, align the divider holes with the well nuts and insert and tighten the thumb screws.

Put grommets in holes provided in top of divider for bin thermostat tube.

INSTALLATION INSTRUCTIONS
FOR GBB-02222, GBB-02223, GBB-02224, GBB-02413, GBB-02418, GBB-02425 & GBB-02426
REMOTE PRECHARGED CONDENSERS

The following pages may be used as a guide to installing Kold-Draft Remote Condenser installations.

When installed as a U.L. listed factory designed system consisting of a Kold-Draft Cuber, Kold-Draft line sets, and a Kold-Draft supplied condenser, few operational difficulties will be encountered.

If system components are modified or substituted for components not specified by Kold-Draft, proper operation can be compromised to the point of system failure.

If the line set(s) and/or condenser is supplied by others, we must assume that the specifying engineer has properly applied the non-Kold-Draft accessories to our Ice Makers.

Kold-Draft reserves the right to disallow any warrantee claims which result from the use of non-Kold-Draft condensers and/or line sets.

1) Uncrate the condenser, and install the mounting legs. Be sure to install the leg stabilizer between the legs.

2) Fasten the condenser to the roof or wall using whatever method that will satisfy the building codes in your area. Condenser must not be lower than 5′ below the receiver.

3) The line sets are packed separately, with the quantity and length marked on the carton. Be sure that the lines are correct for your installation.

4) A single pass condenser installation, which uses one line set, will require a 1 3/4″ diameter hole to pass the lines through a ceiling or wall. The lines for a 2-pass condenser require a 2″ hole, and the 3-pass requires a 2 1/4″ diameter opening for the 3 line sets involved.

5) Each line set consists of a 3/8″ liquid line, and a 1/2″ discharge line. Connect the 3/8″ line to the lower (liquid) fitting on the condenser, and to "REF.IN" on the ice maker. The 1/2″ line connects to the upper (inlet) fitting on the condenser and the "REF. OUT" on the ice maker.

6) Each fitting on the line sets, condenser, and ice maker is self-sealing, and should be tightened 1/4″ turn more than hand tight. Always use a backup wrench to prevent tubing twist when tightening these fittings!

7) The condenser fan motor operates on 208/230, 50/60 cycle, single phase. The 5 ton (GB-2426) condenser uses a 1/3 H.P. Fan Motor, rated at 3.5 amps. All others use a 1/4 H.P. Motor, rated at 2.15 amps. Wire size should be 14 ga. or larger. The power line should be enclosed in a properly grounded conduit wherever exposed to the weather. The fan motor is designed to operate continuously, and therefore requires no interconnection with the ice maker(s). If necessary, the condenser fan motor can be cycled with the ice maker(s) through the use of relays.

8) The refrigerant lines should be routed inside the building wherever possible to prevent vandalism, and to minimize the "condenser effect" that exposed lines can produce in very cold weather.

   In cool weather operation, the liquid return line (3/8″) may sweat inside the building. If this is objectionable, "Armaflex" sponge insulation can be used to minimize condensation.

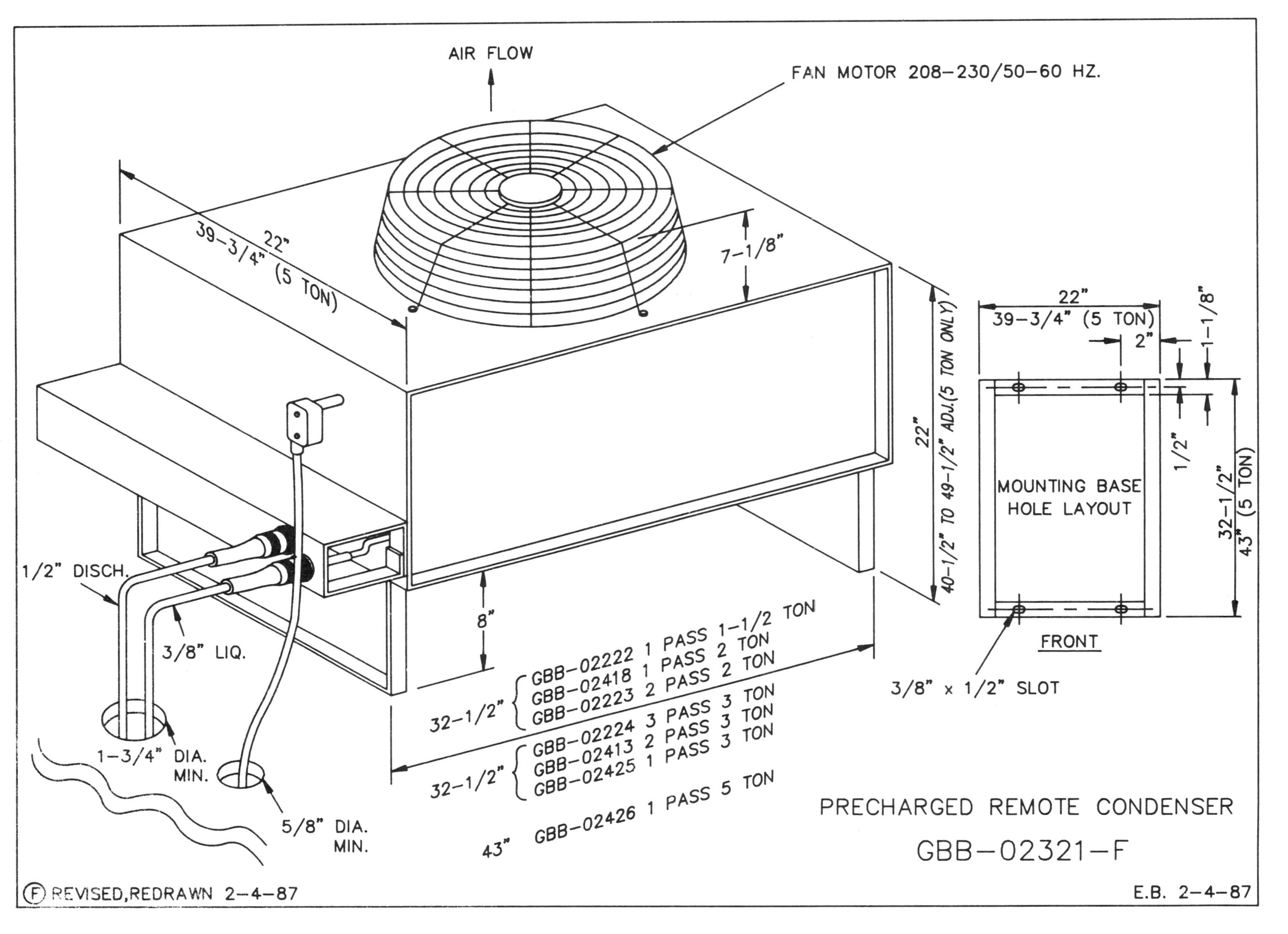
AIR FLOW
FAN MOTOR 208-230/50-60 HZ.
22"
39-3/4" (5 TON)
7-1/8"
22"
40-1/2" TO 49-1/2" ADJ.(5 TON ONLY)
8"
1/2" DISCH.
3/8" LIQ.
1-3/4" DIA. MIN.
5/8" DIA. MIN.
32-1/2" GBB-02222 1 PASS 1-1/2 TON
GBB-02418 1 PASS 2 TON
GBB-02223 2 PASS 2 TON
32-1/2" GBB-02224 3 PASS 3 TON
GBB-02413 2 PASS 3 TON
GBB-02425 1 PASS 3 TON
43" GBB-02426 1 PASS 5 TON
22"
39-3/4" (5 TON)
2"
1-1/8"
1/2"
32-1/2"
43" (5 TON)
MOUNTING BASE HOLE LAYOUT
FRONT
3/8" x 1/2" SLOT
PRECHARGED REMOTE CONDENSER
GBB-02321-F
(F) REVISED,REDRAWN 2-4-87
E.B. 2-4-87

ICE STATION INSTALLATION INSTRUCTIONS

1) CHECK FOR FREIGHT DAMAGE. If damage is present, contact carrier now and file claim.

2) Open door or remove front panel and remove box of legs that are located on right middle of unit.

   A) WITHOUT OPTIONAL STAND - Screw legs into bottom of unit.

   B) WITH OPTIONAL STAND - Screw legs into stand bottom. Set Ice Station on stand and secure it with 4 bolts that secured Ice Station to pallet.

3) Position Ice Station allowing 4" minimum clearance at the rear of Ice Station. If built in, allow for air space at rear, top, and bottom. Allow 6" at sides and top for air-cooled Models.

4) Level unit by screwing legs up or down.

5) ELECTRICAL INSTRUCTIONS:

   A) Each Cuber must be on its own circuit. Individually fused or HACR type circuit breaker. Refer to electrical rating plate located inside on right frame.

   B) Electrical line brought through plug on bottom right rear of unit. Connect per directions on tag attached to leads (2 wires #10 plus ground) for IS400 Series, #12 wire for IS500 Series.

   C) For IS503W Models, you must connect the Brown lead in transformer to either 208 or 230 volt terminal as per installation power supply reading.

6) PLUMBING INSTRUCTIONS:

   A) Connections are made in either rear or bottom of unit.

   B) Connect 3/4" FPT Free Flow Drain (water-cooled Models 1/4" FPT pressure drain connections also).

   C) A 1/2" water supply line is connected to water inlet fitting - 1/4" FPT. Insure that supply maintains a 15 PSI minimum under maximum service load. (Maximum supply pressure 100 PSI)

7) Remove wire and tag securing water plate in upright position and tape securing ice deflector for shipment.

8) Check to insure the ON/WASH switch is in "WASH" position.

   A) IS400 Series - the wash switch is on the right side of front channel.

   B) IS500 Series - the wash switch is located in transformer box at the bottom left of the Cuber.

9) Turn on water and power. Check for water leaks as unit fills. Eliminate leaks if present.

10) Turn the wash switch to "ON" position. Observe the unit to insure it is operating properly as it completes the first 2 cycles.

11) Check the bin probe function by holding ice up to probe tip. Cuber will shut off within 30 seconds. (Test after water fill is completed as the Cuber will not shut off during fill cycle.)

12) Check dispensing function of unit.

13) Remove any protective plastic from panels. Reinstall the front panel. CUBER MUST NOT BE OPERATED WITH PANELS REMOVED.

14) Complete and mail WARRANTY CARD.

15) For service information, contact your local Kold-Draft Distributor or Dealer or Kold-Draft's Complete Ice Machine Service Manual.

CAUTION: When operating and servicing a Kold-Draft Ice Station care must be taken to prevent possible personal injury or damage to unit.

WARNING: Before operating the Ice Station, check to verify:

1. Each Ice Station is on its own circuit, protected with fuses or HACR type circuit breaker.
2. Dispenser motor electrical access cover must be secured in place.
3. Plexiglass access panel to dispensing paddles must be in place and secured with two nuts.
4. The water pump screen must be in place and secured with its screw.
5. The front electrical channel must be secured in place with hold-down clip and screw.
6. Electronic Ice Stations must have the control module box cover in place and secured with screw.
7. Models with electronic controlled dispensing unit must have Plexiglass cover installed and secured with screw.
8. Ice Stations with transformer box, relay box, and/or fuse box must have cover(s) in place and each secured with a screw.
9. Compressors must have covers secured in place to prevent access to compressor wiring and/or start components.
10. The hot gas defrost valve cover must be secured in place.
11. On thermostatically controlled Cubers the water weight control switch cover must be secured with screw.
12. Ice safety chute must be secured in place.
13. Cuber must not be operated with the exterior panels removed without a qualified service mechanic present.

WARNING: When servicing Kold-Draft Cubers safe service practices must be followed.

1. DANGER: Do not braze refrigerant lines until refrigerant is removed from system.
2. WARNING: Beware of potential electrical shock hazard when testing for voltage at electrical connections or replacing components without turning the power off.
3. CAUTION: Exhibit safe practices when working around moving parts such as the pump fan blade, the front and rear cams, front and rear water plate hinges, the actuator motor, condenser fan blades, dispensing paddles, drive chains and/or agitator.

INSTALLATION INSTRUCTIONS FOR GT SERIES CUBER ON TKN-2

1) CHECK FOR FREIGHT DAMAGE. If damage is present, contact freight carrier now and file claim.

2) Install legs on base of dispenser using the 5/8-11 tapped holes provided. The legs are adjustable for leveling.

3) If stand is used, install legs on base of stand.

4) Place TKN-2 on stand and secure per Figure A.

FIGURE A

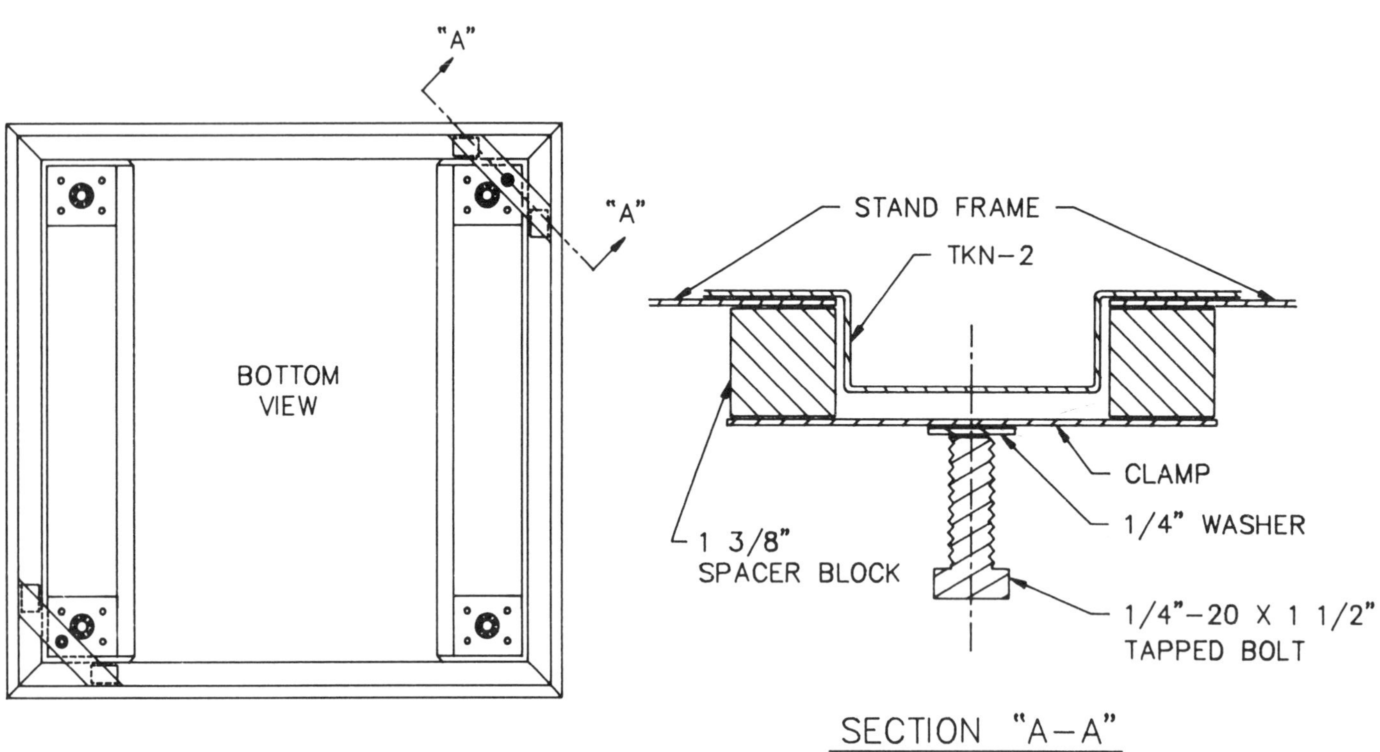

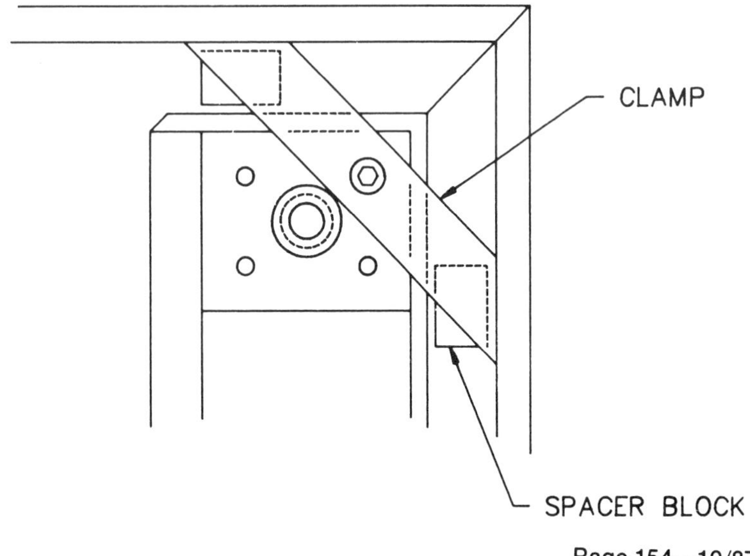

MINIMUM CLEARANCES

FIGURE B

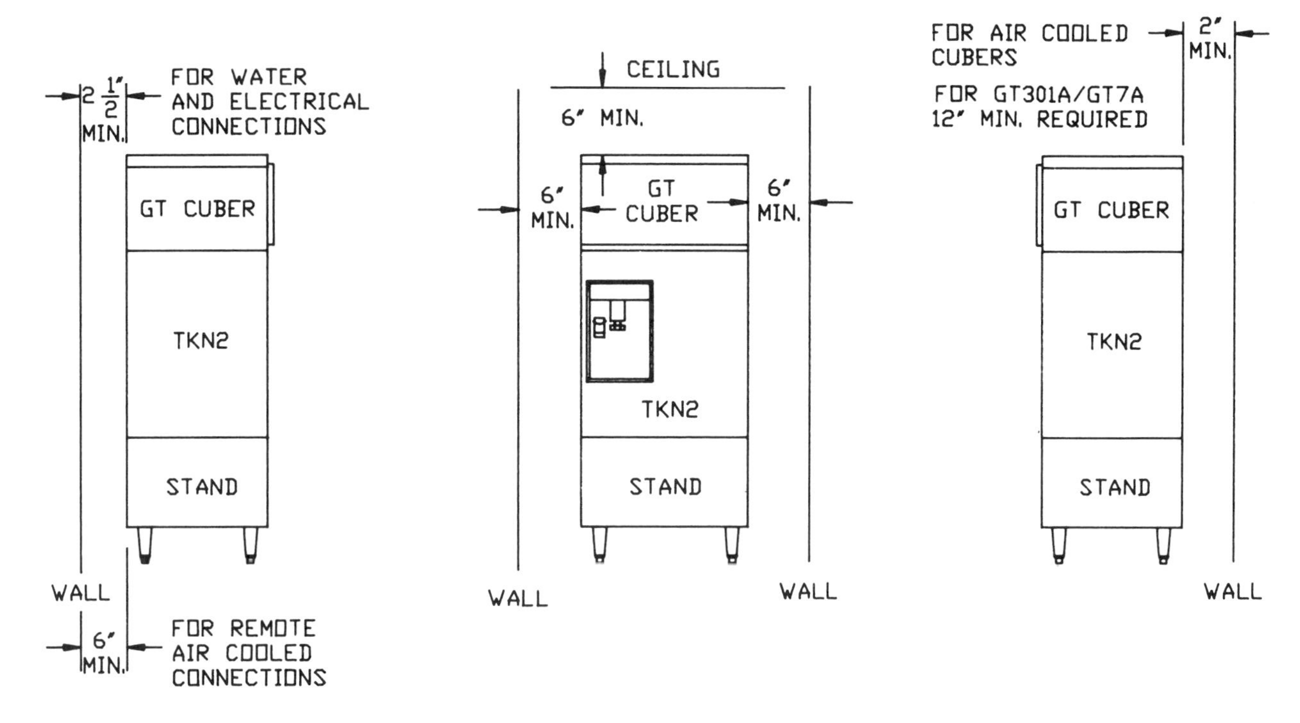

5) Position stand and TKN-2 with minimum clearance, Figure B.

6) Level by adjusting leg bottoms up or down.

7) If panels are on the Cuber, remove them now. To remove:

   1) Front Panel - loosen screw, (do not remove screw), lift up on panel.

   2) Top Panel Lift - bring forward until clip is beyond rear frame.

   3) Side Panels - pull forward and lift to clear clips.

   4) Rear Panel - lift straight up.

   5) Remove drain pan and ice chute.

8) Lift and position Cuber on top of dispenser.

9) Locate holes in Cuber frame over the notches on top of the left and right dispenser sides.

10) Locate holes in Cuber frame over notches on right and left Bin flange. Mark Center.

11) Remove Cuber and drill spotted holes using 7/16" diameter drill.

12) Insert hollow wall anchors into holes and tighten screws gently until the anchors are set. Then remove screws.

13) Place pressure sensitive gasket on top outside edge of Bin, except the front. Front gasket is already installed on bottom of front rail.

14) Lift and position Cuber on dispenser.

15) Position bottom rail between Cuber frame and dispenser.

16) Tighten screws in anchors ONLY UNTIL CUBER IS SECURE.

17) ELECTRICAL INSTRUCTIONS

A) TKN-2

1. Electrical supply connection is made in the "Handy Box" located inside the lower access cover. A female pipe thread coupling on the rear of the dispenser will accept standard 3/8" or 1/2" threaded male conduit fittings. Replace the box cover after connections are made. See nameplate for ratings.

B) GT Series Cuber

1. Each Cuber must be on its own circuit, individually fused or on a HACR type circuit breaker. Refer to electrical rating plate on rear of Cuber for fuse size.

2. There is a supplemental fuse Kit on the drain pan that should be installed in the junction box at the left side of the Cuber.

3. Connect Cuber leads per Instructions on Tag attached to leads in conduit box on left side of Cuber.

4. On GT500 or GT600 Series Cubers, you must connect the Brown lead(s) to the transformer at either the 208 or 230 volt terminal to match supply voltage. Transformer box is located next to the compressor.

18) PLUMBING INSTRUCTIONS

A) TKN-2

1. Bin drains are (2) 3/4" IPS female drains located on dispenser bottom. Use at least 5/8" I.D. drain hose or 3/4" O.D. tubing to carry off water and prevent back-up in the Bin or drain pan.
NOTE: BIN DRAIN CONNECTION MUST COMPLY WITH ALL SANITARY REQUIREMENTS FOR A FOOD STORAGE AREA. CONSULT LOCAL CODE.

2. Optional Water Faucet - connect supply line to 1/4" FPT on dispenser bottom.

B) GT Series Cuber

1. A 1/2" water supply line is connected to 1/4" IPS female fitting in rear of Cuber in upper left hand corner. The water supply must have a capacity to maintain 15 PSI under maximum service load, (maximum supply pressure 100 lb.), before final connection is made, flush water line.

2. Condenser Drains - (Water-Cooled)

A. GB400, GT300, GT400, GT7 Series Cubers, the Condenser drains are run into the Bin drain as shown in Figure C.

B. On all other GT Series water-cooled Cubers, the condenser drain is 1/2" SAE male flare. Use 1/2" O.D. copper tubing to route the condenser drain water out the back of the Cuber to a floor drain.

19) Remove strap and tag securing water plate in upright position.

20) BIN PROBE REPLACEMENT per Figure D.

FIGURE C

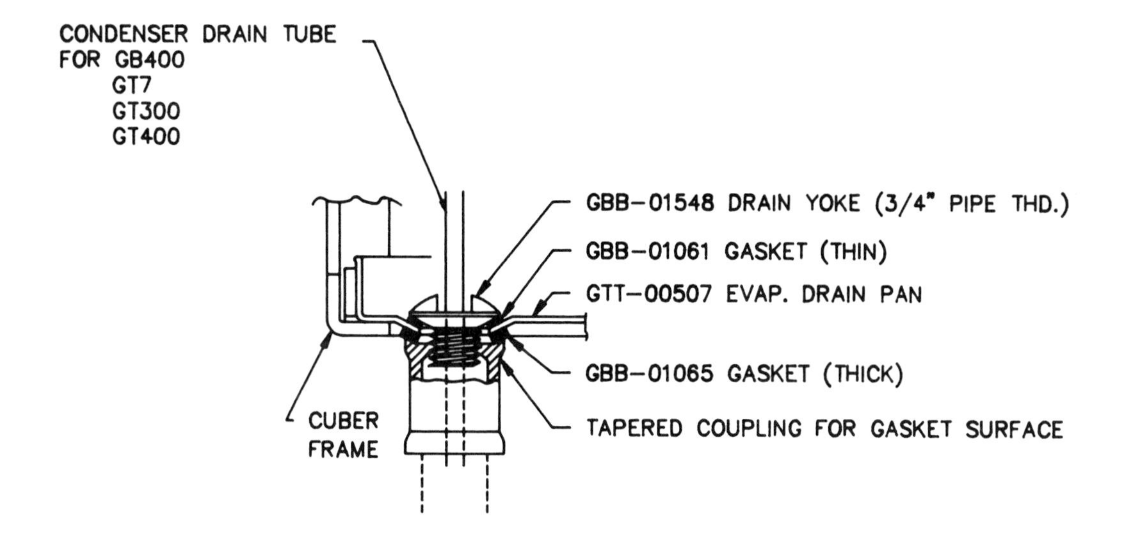

FIGURE D

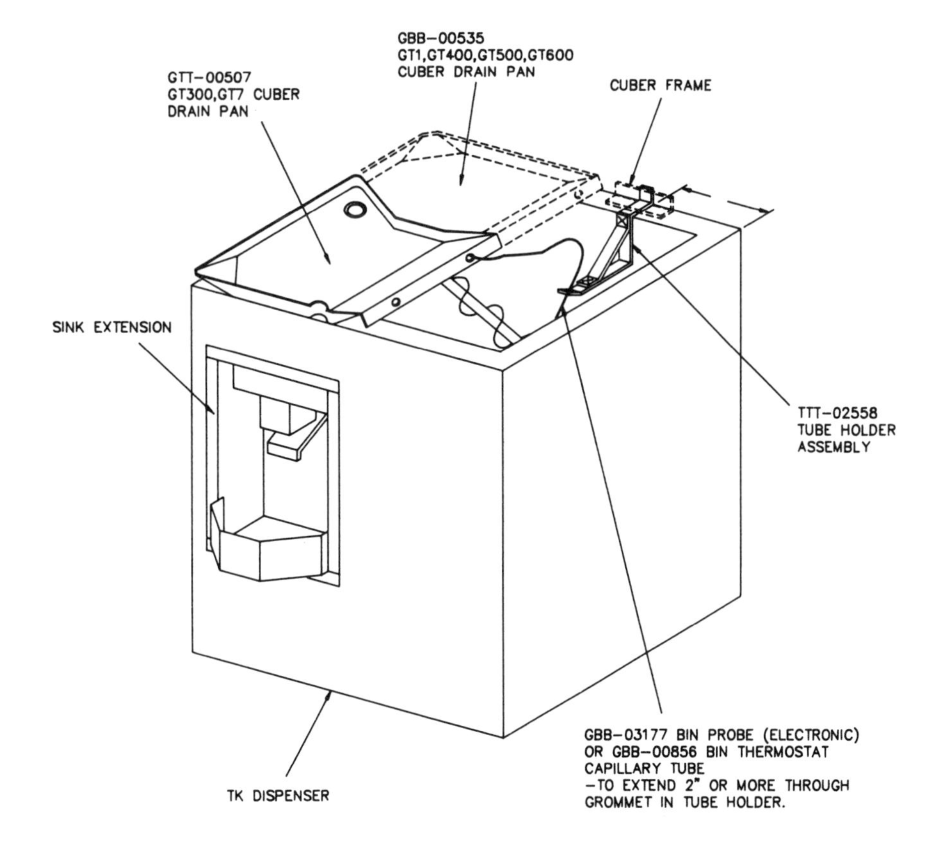

21) Install ice chute in Cuber.

22) Install the sink extension in sink by sliding it over the front edge.

23) Turn the wash switch to "WASH" position.

A) GT300, GT400, GT7 Series is located at the right side of the front wiring channel.

B) All other GT Cubers - it is located on the transformer box next to the compressor.

24) Turn on the water and power. As the Cuber is filling, check for water leaks. Eliminate if present.

25) Turn wash switch to "ON" position and observe the Cuber through two (2) complete cycles to insure it is operating properly.

26) Check Bin Probe function by holding ice up to probe tip. Cuber will shut off within 30 seconds (on electronic-test after water-fill is completed as Cuber will not shut off during fill cycle).

27) Install wiring diagram and ice machine cleaning instructions on inside of front panel.

28) Install panels. CUBER MUST NOT BE OPERATED WITHOUT PANELS IN PLACE.

29) Complete and mail in WARRANTY CARD.

30) For service information, contact your local Kold-Draft Distributor or Dealer or Kold-Draft's Complete Ice Machine Service Manual.

CAUTION: When operating and servicing a Kold-Draft GT Ice Cuber and/or TKN2 Dispenser, care must be taken to prevent possible personal injury or damage to the unit.

WARNING: Before operating the GT Ice Cuber/TKN2 Dispenser Combination, check to verify:

1. Each Cuber as well as TKN-2 is on its own circuit, protected with fuses or HACR type circuit breakers.
2. All Cubers (except 3 phase Models) must have a supplemental fuse kit installed in junction box at the left side of the Cuber (Refer to Engineering Bulletin #22-84).
3. The water pump screen must be in place and secured with its screw.
4. The front electrical channel must be secured in place with hold-down clip and screw.
5. Electronic Cubers (except GT300 Series) must have the control module box cover in place and secured with screw. GT300 Series Cubers - it is built into the front panel and must be secured in place when the machine is not being serviced.
6. Cubers with transformers must have cover in place and secured with a screw.
7. Compressors must have covers secured in place to prevent access to compressor wiring and/or start components.
8. The hot gas defrost valve cover must be secured in place.
9. On thermostatically controlled Cubers the water weight control switch cover must be secured with screw.
10. TKN-2 left and right access panels must be locked in place.
11. TKN-2 ice spout must be secured in place.
12. Cuber must not be operated with the exterior panels removed without a qualified service mechanic present.

WARNING: When servicing Kold-Draft GT Cubers and/or TKN2 Dispensers safe service practices must be followed.

1. DANGER: Do not braze refrigerant lines until refrigerant is removed from system.
2. WARNING: Beware of potential electrical shock hazard when testing for voltage at electrical connections within the system.
3. CAUTION: Exhibit safe practices when working around moving parts such as the Cuber pump fan blade the front and rear cams, the actuator motor and/or condenser fan blades, as well as the TKN2 gear motor, auger, and if present agitator.

INSTALLATION INSTRUCTIONS FOR T-100 OR TGT-100 CRUSHER

1) CHECK FOR FREIGHT DAMAGE. If present, contact freight carrier IMMEDIATELY and file claim.

2) If retrofit to an existing installation, disconnect power and water to existing Cuber and remove Cuber from Bin.

3) Position Bin maintaining minimum clearances. Be sure to allow a 4" minimum clearance behind the Bin for mounting the relay box assembly. Refer to Figure A.

4) Level Bin by adjusting leg bottoms up or down. If Bin is sealed to the floor (no legs and it is necessary to shim the Bin level, install a cove molding around the bottom of the Bin. Seal the Bin or molding to the floor with RTV sealant.

FIGURE A

MINIMUM CLEARANCES

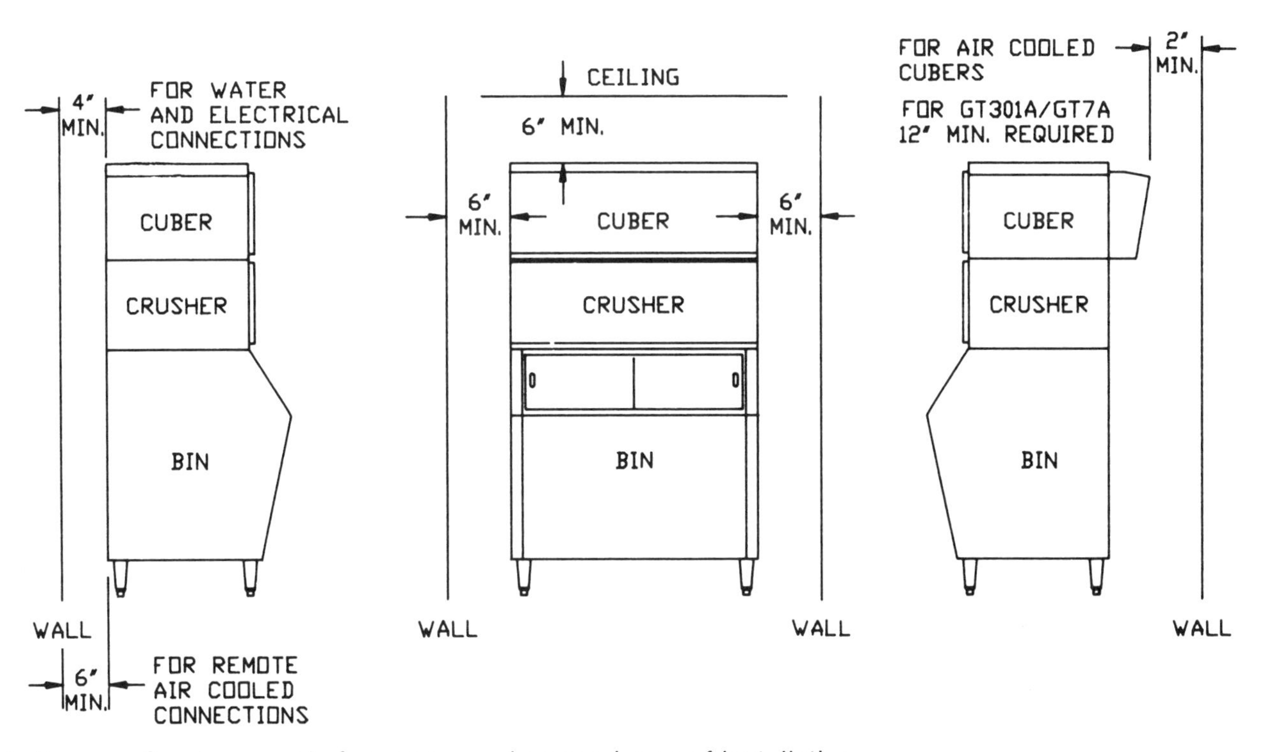

5) Remove Crusher panels for access and convenience of installation.

   A. FRONT PANEL - Loosen screw (do not remove screw), lift up on panel and move forward taking care to prevent damage to selector knob assembly.

   B) SIDE PANELS - Pull forward and lift to clear clips.

   C) REAR PANEL - Lift straight up.

   D) Remove top and bottom angle covers.

6) A) New Installation - Place pressure sensitive gasketing on top outside edge of Bin except the front. Front gasket is already installed on bottom of angle cover.

   B) Retrofit to existing installation gasket should already be in place.

7) Lift the Crusher in place on the Bin.

8) Center the angle cover to the Crusher frame.

9) Secure Drain Pan - Screw drain yoke into drain fitting on the top edge of the Bin, thin gasket on pan, thick gasket beneath drain pan, Figure B.

FIGURE B

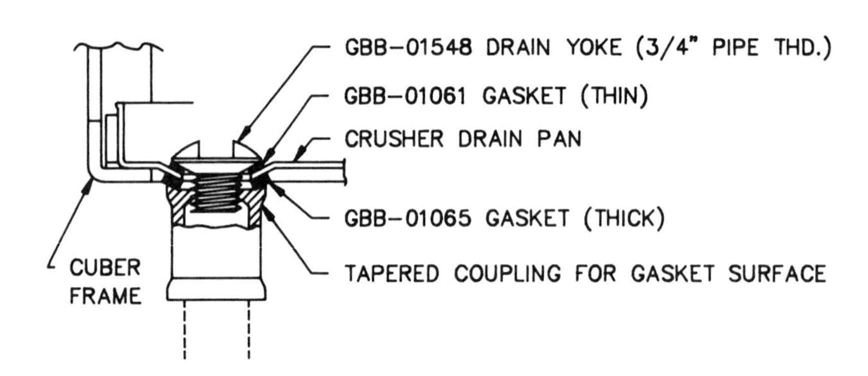

10) Locate and drill four 5/32" holes in the top of the Bin, then secure Crusher to Bin using 1/4" lag bolts (drilling is not required on GBN-950 Bin - see Instructions packaged with Bin), Refer to Figure C.

FIGURE C

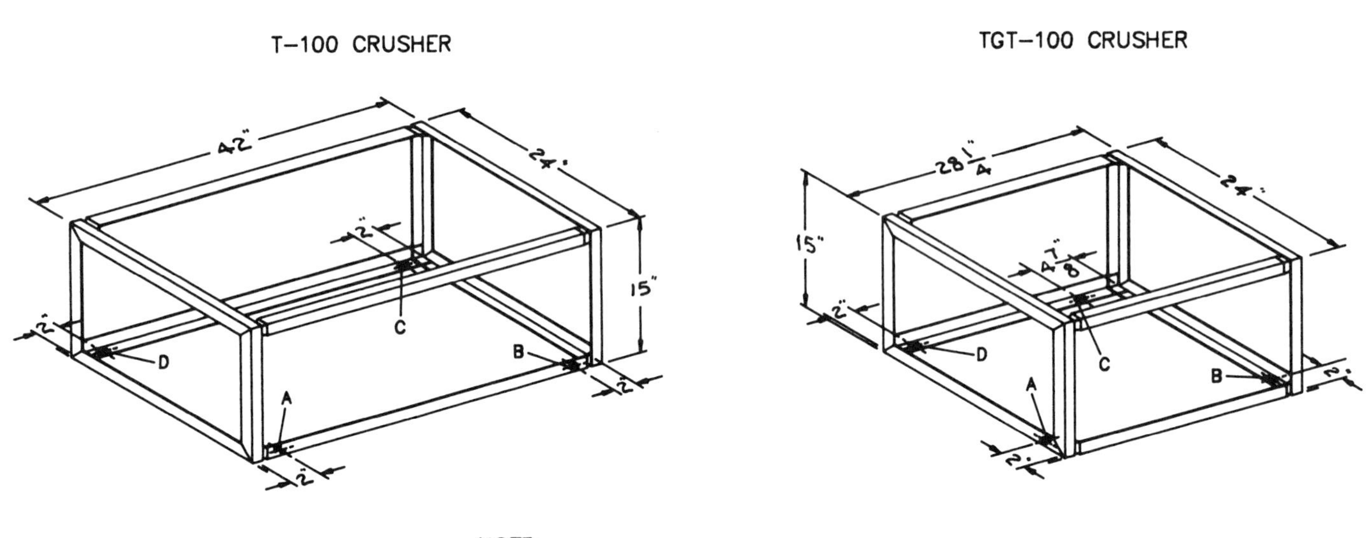

NOTE:
1. 1 1/4" lag bolt at holes A and B
2. 2" lag bolt at holes C and D

11) Place pressure sensitive gasket on top inside edge of Crusher except the front. Front gasket is already installed on bottom of double angle cover.

12) Lift Cuber and postion on top of the Crusher.

13) Install double angle cover. Insure that it is centered in the Cuber and Crusher frames.

14) Bolt units together - Refer to Figure D.

FIGURE D

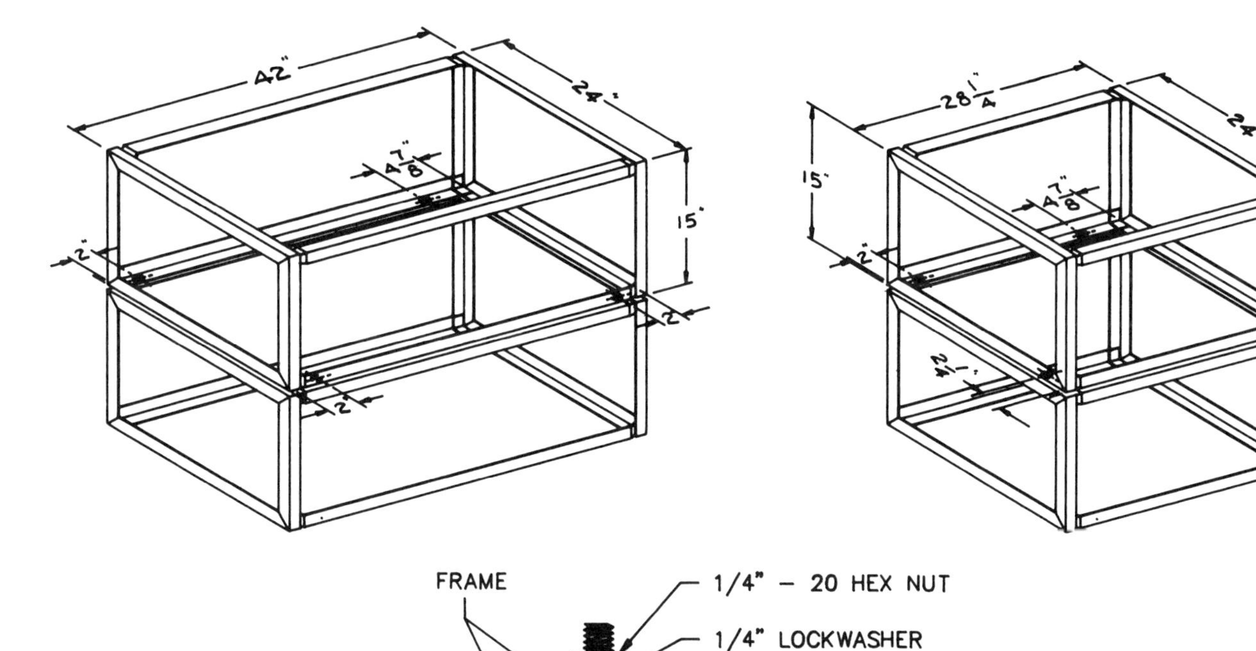

NOTE:

1. Stacking a GB400 Series cuber use (2) 1/4-20x2 1/2" bolts on the right side. (They are not provided in kit)
2. When stacking a GB500 or GB600 air-cooled model omit bolt at right rear as condenser is in way.

NOTE:

1. Mounting a GT300 or GT7 requires TTT-01521 deflector and (2) 1/4-20x2 1/2" bolts not supplied in kit.
2. GT400, GT500, GT600, GT1 or GT8 require TTT-01528 deflector.

15) Electrical

A) Refer to Figure E for Crusher Cuber inter-connections.

FIGURE E

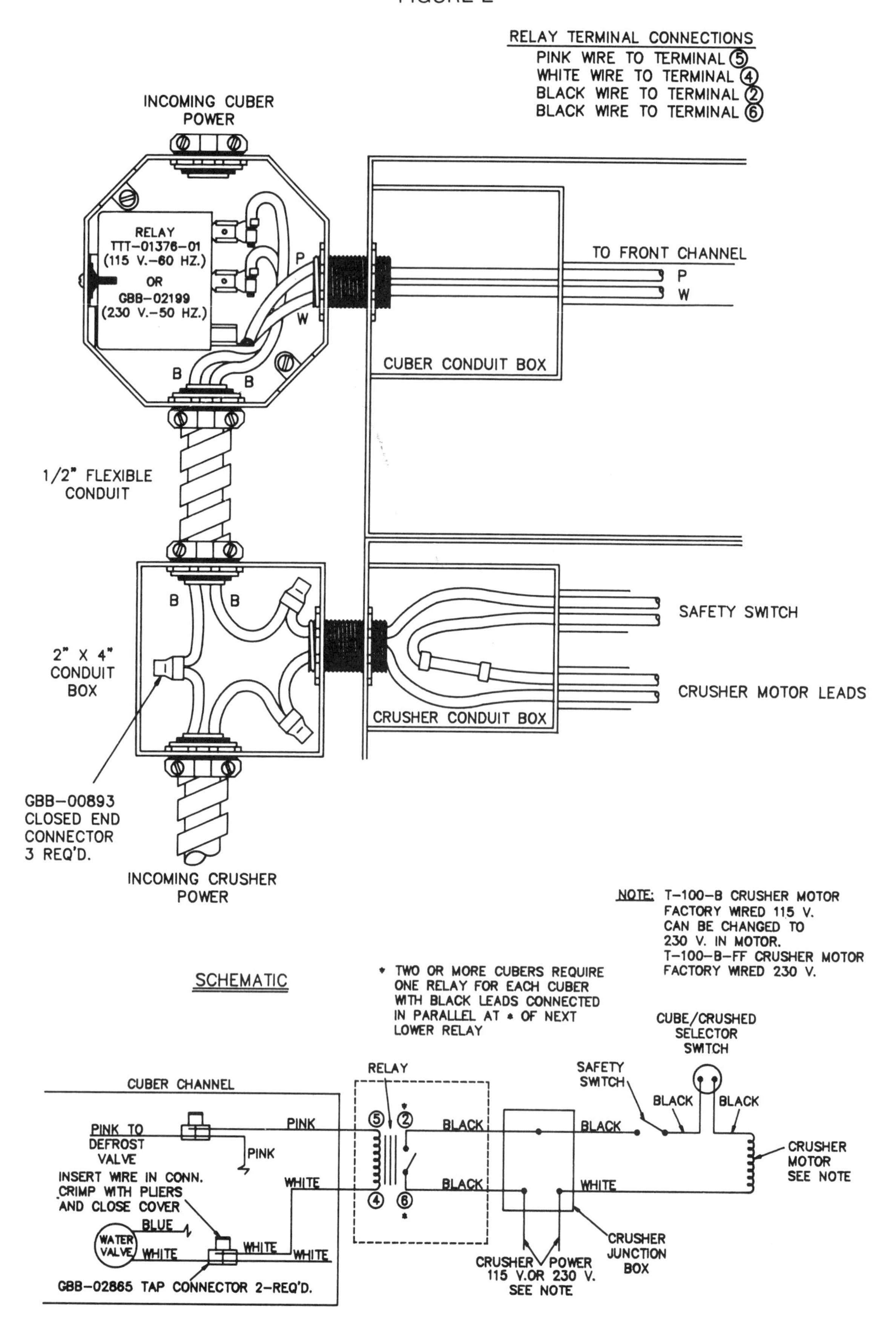

B) Each Cuber must be on its own circuit, protected with fuse(s) or HACR type circuit breakers. Refer to electrical rating plate on rear of Cuber for fuse size.

C) There is a supplemental fuse Kit (except 3 phase models) packed with the Cuber which should be installed in the junction box at left side of the Cuber.

D) Connect Cuber leads per Instructions on Tag attached to leads in junction box on left side of Cuber.

E) On GB500, GB600, GB1000, GB1200, GT500, or GT600 Series Cubers, you must connect the Brown lead(s) to the transformer at either the 208 or 230 volt terminal to match supply voltage. Transformer box is located next to the compressor.

16) Plumbing Instructions

A) Position drain assembly in the Cuber through drain pan as in Figure F.

B) Check local code for backflor preventer requirement. This item is not provided with the Cuber and may be obtained locally if required. A 1/2" water supply line is connected to 1/4" IPS female fitting in rear of Cuber in upper left hand corner. The water supply must have a capacity to maintain 15 PSI under maximum service load, (maximum supply pressure 100 lb.), before final connection is made, flush water line.

FIGURE F

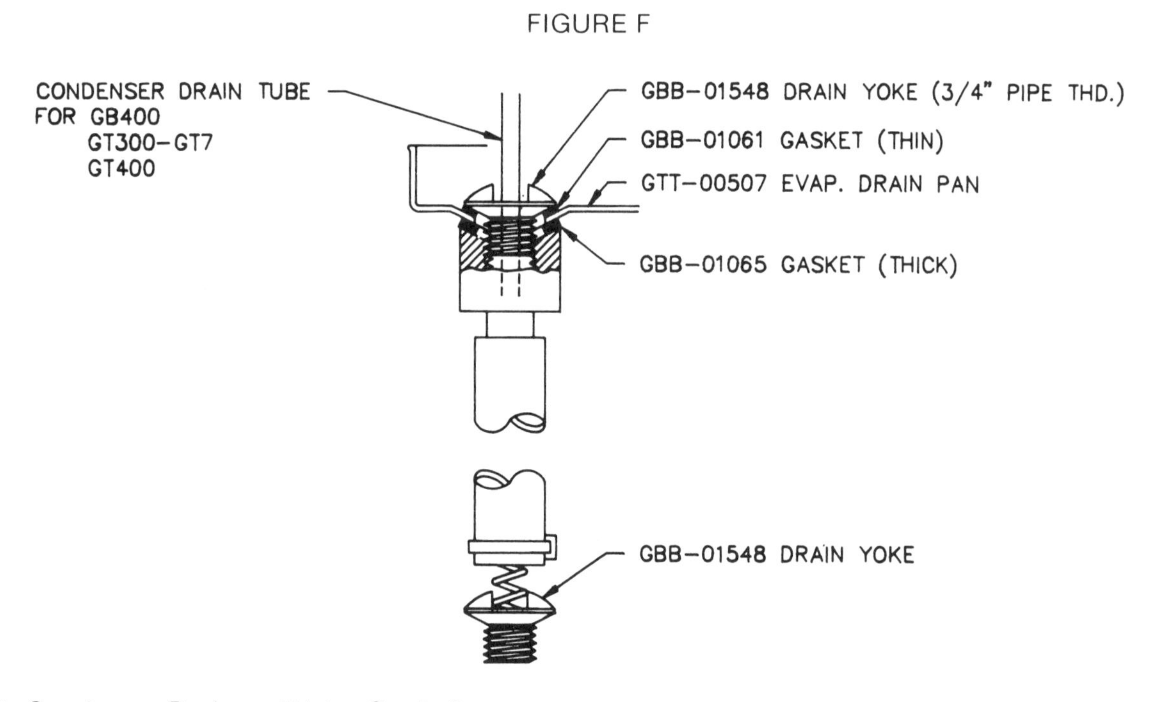

C) Condenser Drains - (Water-Cooled)

1. GB400, GT300, GT400, GT7 Series Cubers, the condenser drains are run into the drain as shown in Figure F.

2. On all other GB or GT Series water-cooled Cubers, the condenser drain is 1/2" SAE male flare. Use 1/2" O.D. copper tubing to route the condenser drain water out the back of the Cuber to a suitable drain.

D) Bin drains are (2) 3/4" IPS female drains. Use at least 5/8" I.D. drain hose or 3/4" O.D. tubing to carry off water and prevent back-up in the Bin or drain pan. NOTE: BIN DRAIN CONNECTION MUST COMPLY WITH ALL SANITARY REQUIREMENTS FOR A FOOD STORAGE AREA. CONSULT LOCAL CODE.

17) Remove strap and tag securing Cuber water plate in upright position.

18) Bin Probe Positioning

A) TGT-100 Crusher (Crushed Ice Only) with GT Electronic Cuber - use Probe Kit, TTT-01847. Refer to Figure G.

FIGURE G

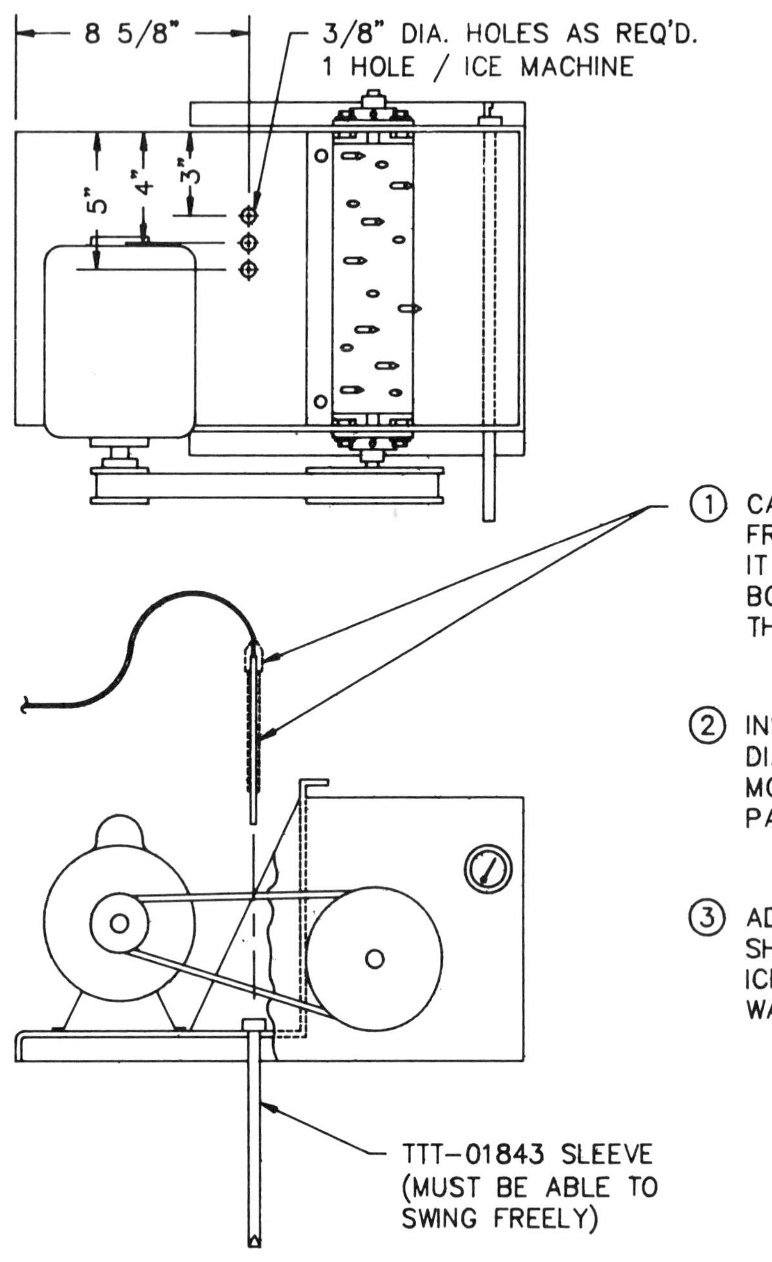

With themostat controlled GT Cuber, refer to Figure H.

FIGURE H

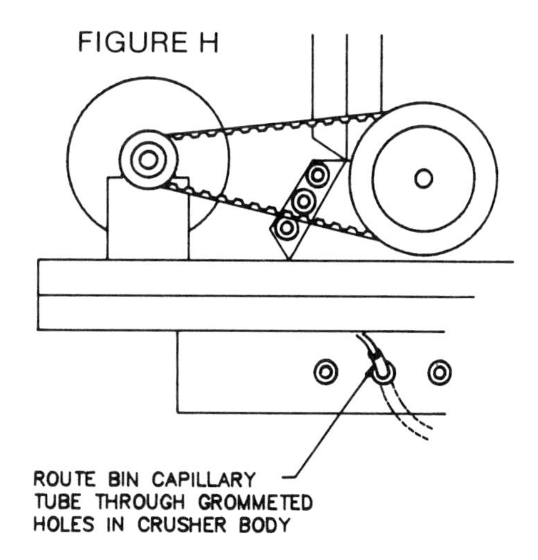

B) T-100 Crusher

1. Electronic Cuber - (Crushed and Cubed Ice) Refer to Figures I & J to install Probe Kit, TTT-01841.

FIGURE I

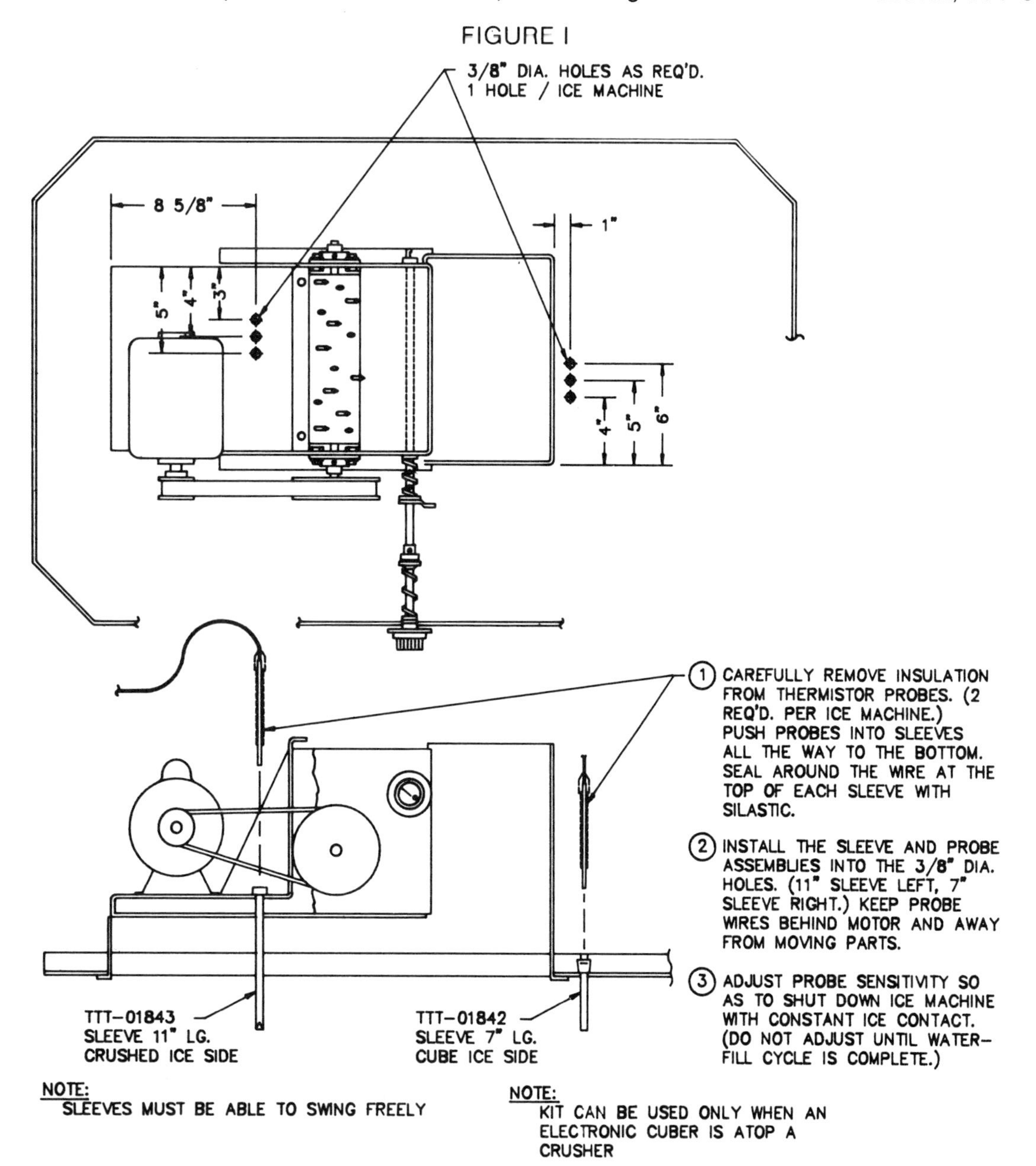

FIGURE J

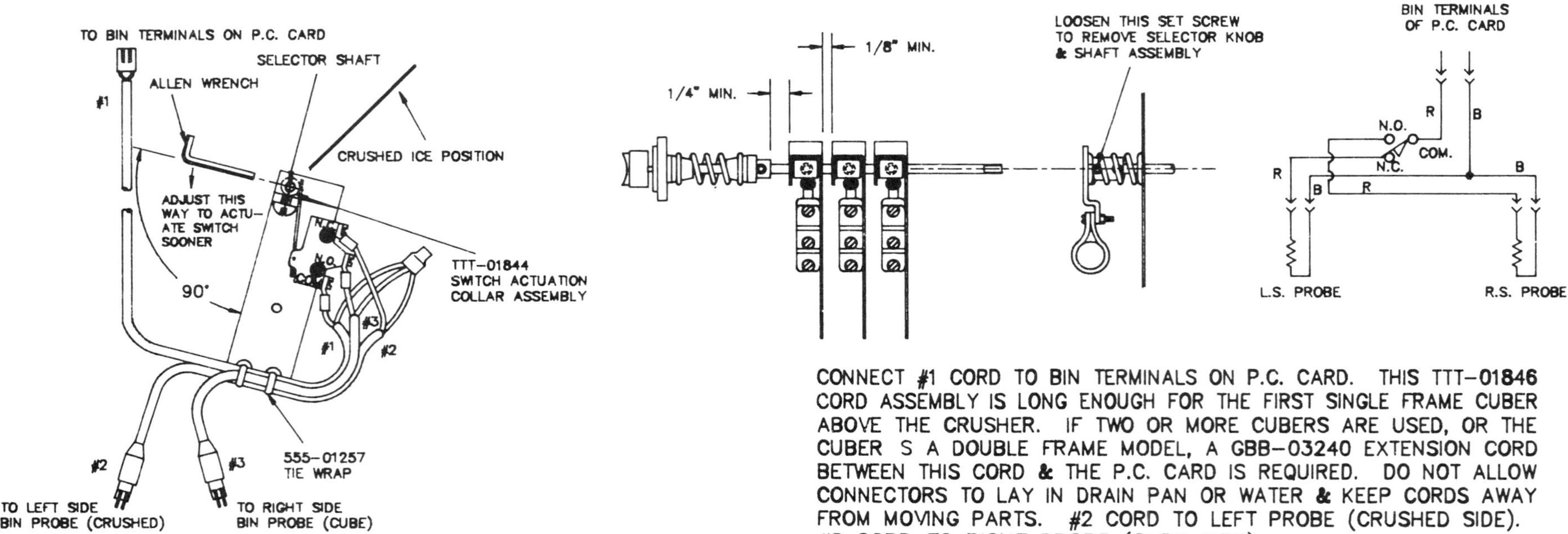

ASSEMBLE MICRO-SWITCH TO SWITCH HOLDER USING P13-01106-16 #6-32 X 1" MACHINE SCREWS, P43-10006 #6 LOCK WASHERS, P43-03006 #6 FLAT WASHERS & P33-10106 #6-32 HEX NUTS PROVIDED. (2 OF EACH REQ'D.)

REMOVE CUP WASHERS FROM SWITCH AND ASSEMBLE CORD ASSEMBLY TO SWITCH AS SHOWN. USE 555-01257 TIE WRAPS TO SECURE CORD SET TO SWITCH HOLDER.

AFTER SLEEVE & PROBE ASSEMBLIES HAVE BEEN INSTALLED INTO HOLES IN CRUSHER, (SEE DWG. TTT-01841-01-DWG) LOOSEN SET SCREW & REMOVE SELECTOR KNOB & SHAFT ASSEMBLY FROM THE SELECTOR SHAFT COUPLING.

PLACE TTT-01844 SWITCH ACTUATION COLLAR ASSEMBLY BETWEEN THE SIDES OF TTT-01845 SWITCH HOLDER & PUSH THE SELECTOR SHAFT THROUGH BOTH PARTS. (NOTE ORIENTATION OF ASSEMBLY) REPLACE SELECTOR KNOB & SHAFT ASSEMBLY INTO COUPLING & TIGHTEN COUPLING SET SCREW. P13-30116-04 COLLAR ASSEMBLY SET SCREW SHOULD BE LEFT LOOSE AT THIS TIME.

CONNECT #1 CORD TO BIN TERMINALS ON P.C. CARD. THIS TTT-01846 CORD ASSEMBLY IS LONG ENOUGH FOR THE FIRST SINGLE FRAME CUBER ABOVE THE CRUSHER. IF TWO OR MORE CUBERS ARE USED, OR THE CUBER S A DOUBLE FRAME MODEL, A GBB-03240 EXTENSION CORD BETWEEN THIS CORD & THE P.C. CARD IS REQUIRED. DO NOT ALLOW CONNECTORS TO LAY IN DRAIN PAN OR WATER & KEEP CORDS AWAY FROM MOVING PARTS. #2 CORD TO LEFT PROBE (CRUSHED SIDE). #3 CORD TO RIGHT PROBE (CUBE SIDE).

AFTER ARRANGING ALL CORDS & ALLOWING THE SWITCH HOLDER ASSEMBLY TO ASSUME ITS NATURAL POSITION, ROTATE THE SELECTOR KNOB CLOCKWISE TO THE CRUSHED ICE POSITION. (NOTE: ICE MACHINE WITH CHUTE IN PLACE MUST BE IN POSITION AT THIS TIME.) INSERT ALLEN WRENCH INTO SET SCREW ON COLLAR ASSEMBLY & POSITION SO THAT WRENCH IS 90°FROM SWITCH HOLDER AS SHOWN. TIGHTEN SET SCREW & ROTATE SELECTOR KNOB COUNTER CLOCKWISE TOWARD THE CUBE ICE POSITION. THE TTT-01323 MICRO-SWITCH SHOULD BE ACTUATED BY THE COLLAR ASSEMBLY APPROX. MID-WAY BETWEEN THE POSITIONS. IF EARLIER ACTUATION IS REQ'D., RETURN THE KNOB TO THE CLOCKWISE POSITION, LOOSEN SET SCREW AND TURN COLLAR ASSEMBLY COUNTER CLOCKWISE APPROX. 5° DO THIS UNTIL THE COLLAR ASSEMBLY ACTUATES THE SWITCH PROPERLY.

ADJUST PROBE SENSITIVITY USING THE CUBE ICE PROBE (RIGHT SIDE). TURN THE SELECTOR KNOB COUNTER CLOCKWISE TO THE CUBE ICE POSITION. ADJUST PROBE SENSITIVITY SO AS TO SHUT DOWN THE ICE MACHINE WITH CONSTANT ICE CONTACT.

NOTE: DO NOT USE CENTER SELECTOR KNOB POSITION (BOTH) WITH THIS KIT INSTALLED.

19) Safety Switch Hold-Down Plunger - is installed on front panel on Cuber directly above Crusher, per Figure K.

FIGURE K

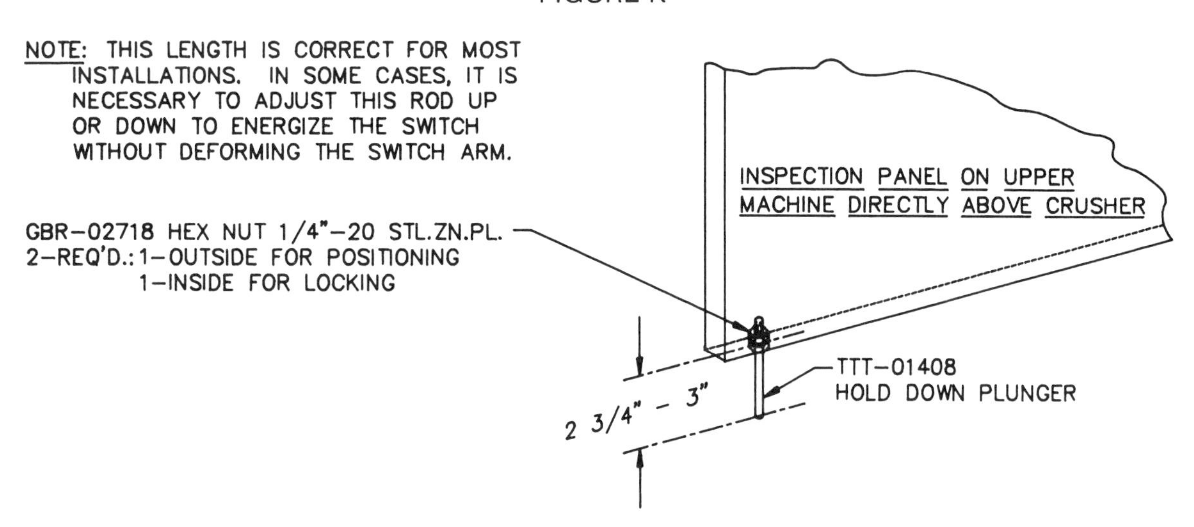

20) Install appropriate Ice Chute or Deflector. On T-100 Crusher, push in and turn Ice Selector Switch to position Selector Plate in Ice Chute Drop Zone.

21) Turn the Wash Switch to "WASH" position.

    A) Gb400, GT300, GT400, GT7 Series is located at the right side of the front wiring channel.

    B) All other GB and GT Cubers - It is located on the transformer box next to the compressor.

22) Turn on the water and power. As the Cuber is filling, check for water leaks. Eliminate if present.

23) Turn wash switch to "ON" position and observe the Cuber through two (2) complete cycles to insure Crusher and Cuber are operating properly.

24) Check Bin Probe function by holding ice up to probe tip. Cuber will shut off within 30 seconds (on electronic-test after water-fill is completed as Cuber will not shut off during fill cycle).

25) Install wiring diagram and ice machine cleaning instructions on inside of front panel.

26) Install panels on Crusher then Cuber. Be careful not to damage or bind the safety switch in the upper left-hand corner of Crusher when installing Cuber front panel.

27) Install panels. CUBER MUST NOT BE OPERATED WITHOUT PANELS IN PLACE.

28) Complete and mail in WARRANTY CARD.

29) For service information, contact your local Kold-Draft Distributor or Dealer or Kold-Draft's Complete Ice Machine Service Manual.

NOTE: 1. Every 6 months the Crusher motor should be oiled, the timing belt checked, and the drum shaft bearings greased.

2. To allow adjustment for ice granule size, several holes are punched in the Housing, and ice size can be changed from the factory setting of "fine" to "medium" or "coarse". If a change in ice size is necessary, remove the two bolts, slide the Ice Size Bar to the desired position, and re-install the bolts in the closest hole to the tapped hole in the Bar.

Although adjustment is most easily accomplished before the Crusher is installed under a Cuber, removal of the Ice Chute and Drain Pan from the Cuber will provide full access to the rear adjustment bolt after Cuber installation.

CAUTION: When operating and servicing a Kold-Draft Ice Cuber with the appropriate T-100 or TGT-100 Crusher, care must be taken to prevent possible personal injury or damage to the units.

WARNING: Before operating the Crusher-Cuber combination, check to verify:

1. Each Cuber as well as each Crusher is on its own circuit, protected with fuses or HACR type circuit breaker.
2. All Cubers (except 3 phase Models) must have a supplemental fuse kit installed in junction box at the left side of the Cuber (Refer to Engineering Bulletin #22-84).
3. The water pump screen must be in place and secured with its screw.
4. The front electrical channel must be secured in place with hold-down clip and screw.
5. Electronic Cubers (except GT300 Series) must have the control module box cover in place and secured with screw. GT300 Series Cubers - it is built into the front panel and must be secured in place when the machine is not being serviced.
6. Cubers with transformers must have cover in place and secured with a screw.
7. Compressors must have covers secured in place to prevent access to compressor wiring and/or start components.
8. The hot gas defrost valve cover must be secured in place.
9. On thermostatically controlled Cubers the water weight control switch cover must be secured with screw.
10) Crusher relay box and junction box covers are secured in place.
11. Crusher motor electrical access cover is secured in place.
12. On T-100 Crushers - the ice selector plate is positioned in the ice chute drop zone.
13. Safety switch hold-down plunger is installed in front panel on Cuber directly above Crusher.
14. Cuber and Crusher must not be operated with the exterior panels removed without a qualified service mechanic present.

WARNING: When servicing Kold-Draft Crusher-Cuber Combinations safe service practices must be followed.

1. DANGER: Do not braze refrigerant lines until refrigerant is removed from system.
2. WARNING: Beware of potential electrical shock hazard when testing for voltage at electrical connections within the system.
3. CAUTION: Exhibit safe practices when working around moving parts such as the pump fan blade, the front and rear cams, front and rear water plate hinges, the actuator motor condenser fan blades, crusher drum, and/or timing belt.

INSTALLATION INSTRUCTIONS FOR TDA-15, TDA-43D AND MIP-43D ICE CUBE DISPENSERS

REFER TO FIGURE A, B, OR C TO FIND EQUIPMENT PURCHASED.

1. If optional adjustable legs are used, screw legs onto stand bottom. Legs have 1 1/8″ level adjustment.
2. Secure Dispensing Unit to stand per appropriate Figure A, B, or C.

FIGURE A

DISPENSING UNIT MOUNTING ON TDSH

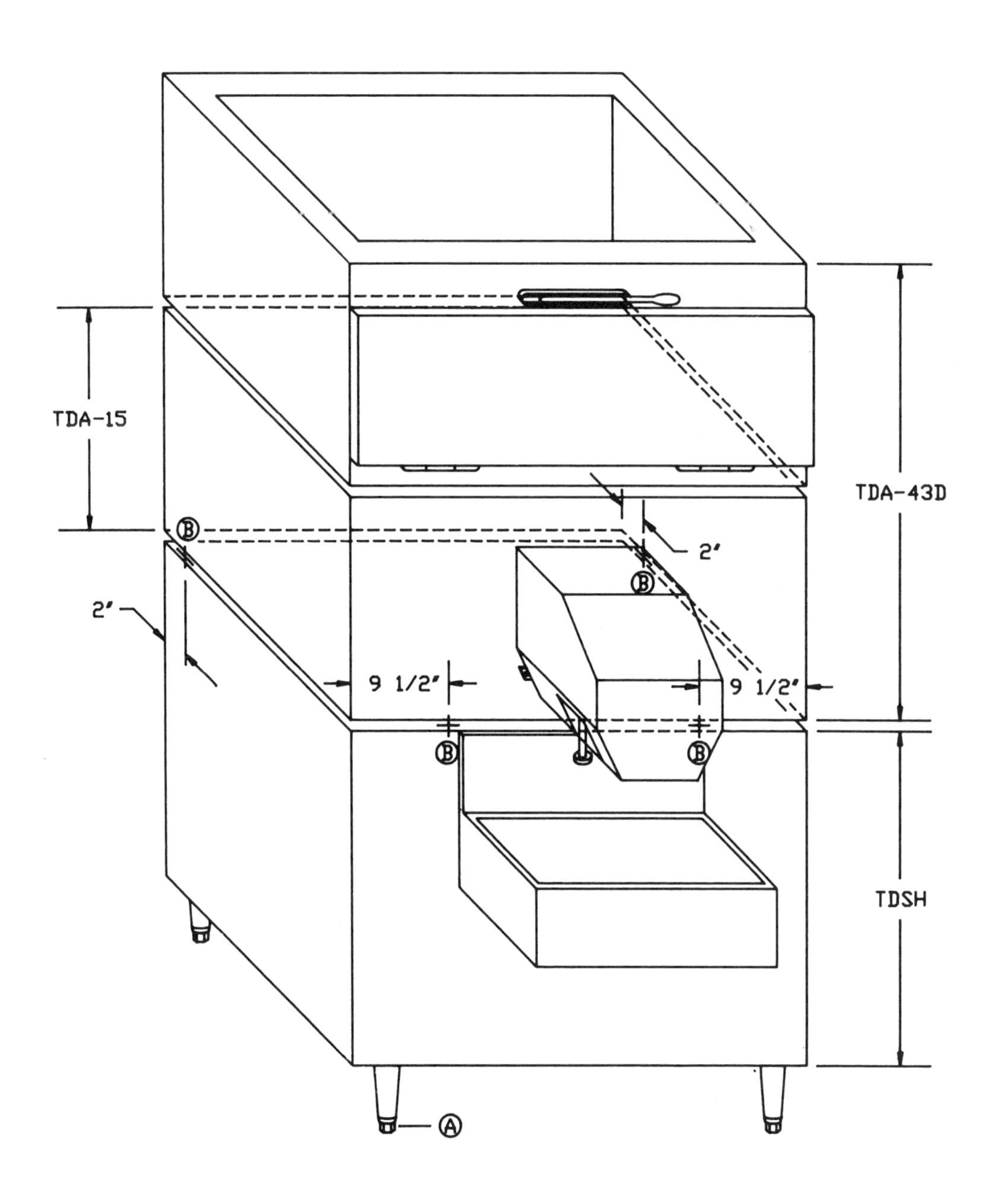

Ⓐ 1 1/8″ LEVEL ADJUSTMENT ON LEGS

Ⓑ 1/4″ X 1 1/2″ LAG SCREW

FIGURE B

DISPENSING UNIT MOUNTING ON TDSD AND 15″ STAND

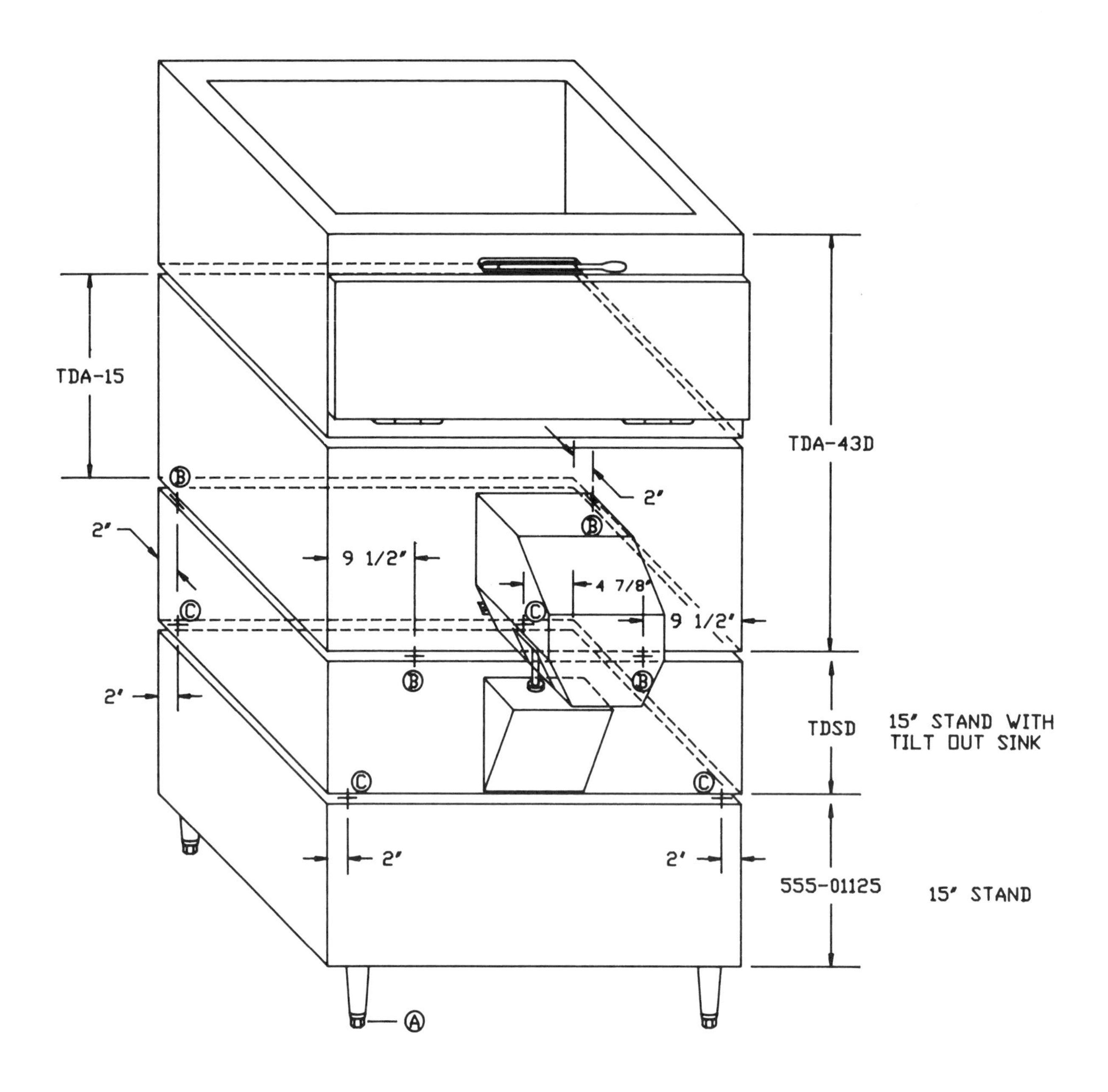

Ⓐ 1 1/8″ LEVEL ADJUSTMENT OF LEGS

Ⓑ 1/4″ X 1 1/2″ LAG SCREW

Ⓒ

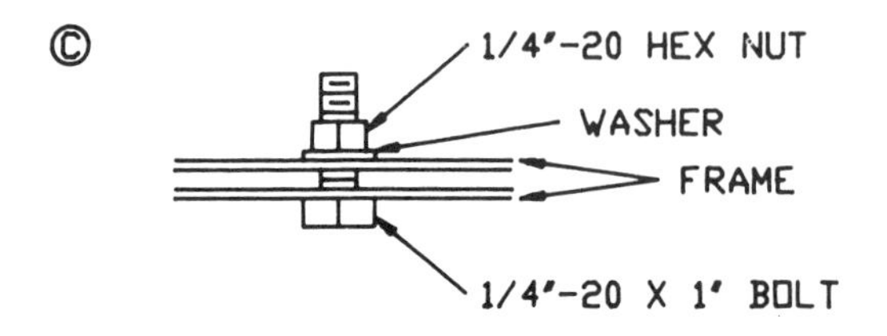

FIGURE C

DISPENSING UNIT MOUNTING ON MINI-ICE PLANT

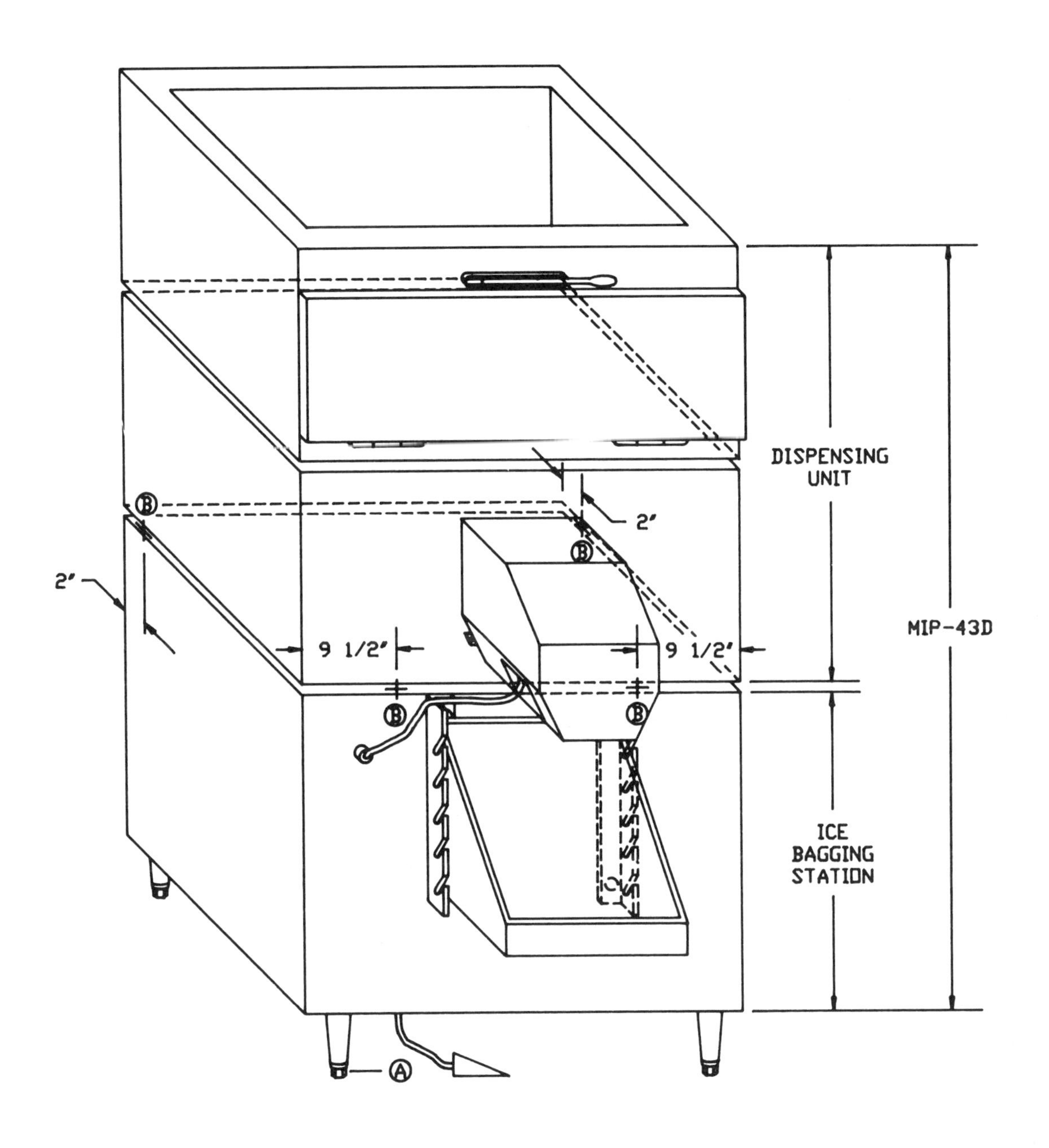

Ⓐ 1 1/8" LEVEL ADJUSTMENT ON LEGS

Ⓑ 1/4" X 1 1/2" LAG SCREW

Ⓒ

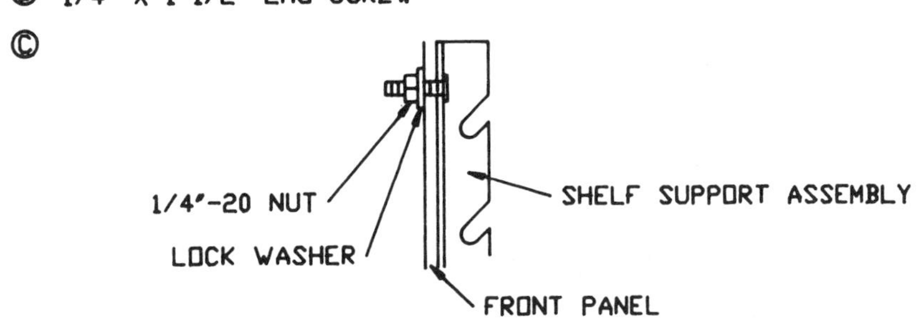

Ⓓ INSERT SINK ASSEMBLY IN SLOTS PER DESIRED HEIGHT

FIGURE D

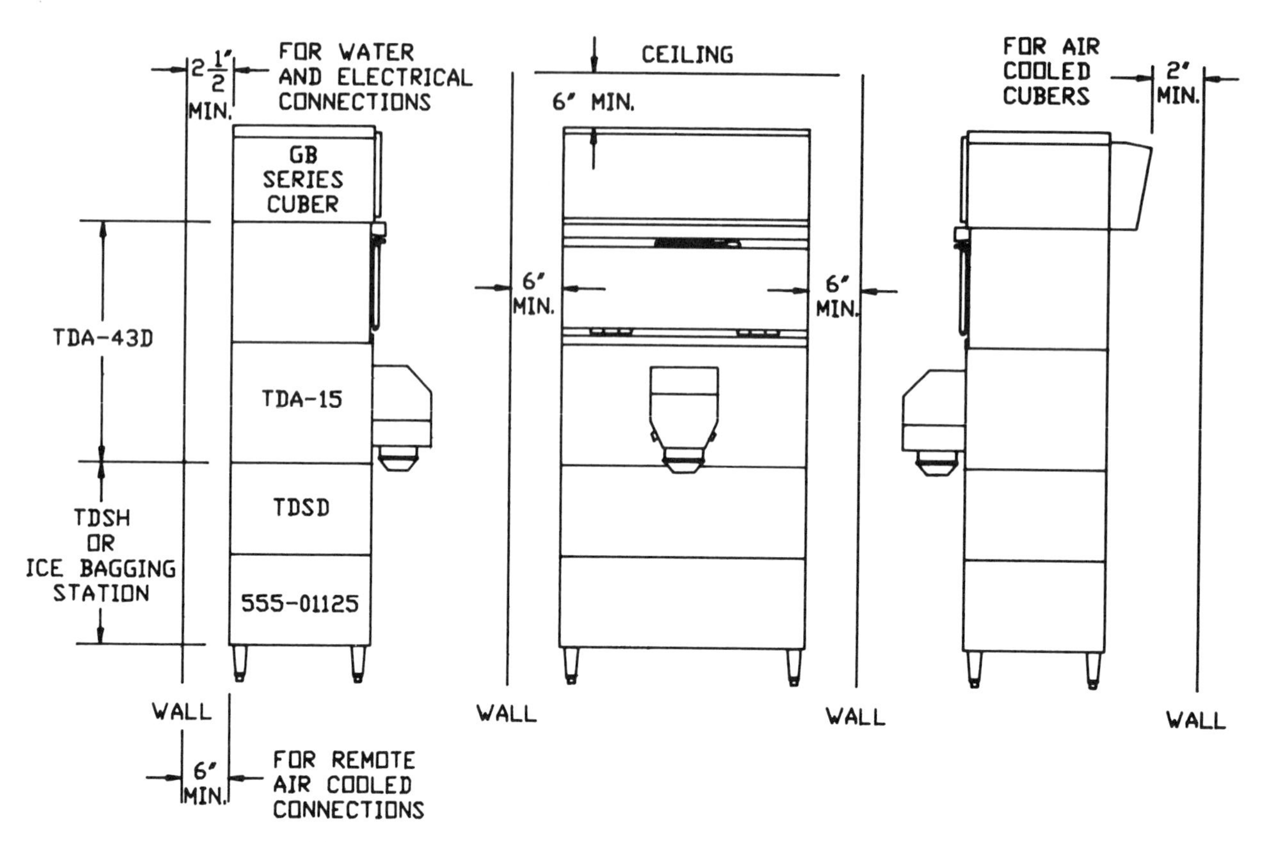

3. Position Dispenser and Stand. Be sure to maintain minimum clearance on Left and Right sides, Rear and Top per Figure D.
4. Level unit by adjusting Legs or with shims if Legs are not purchased.

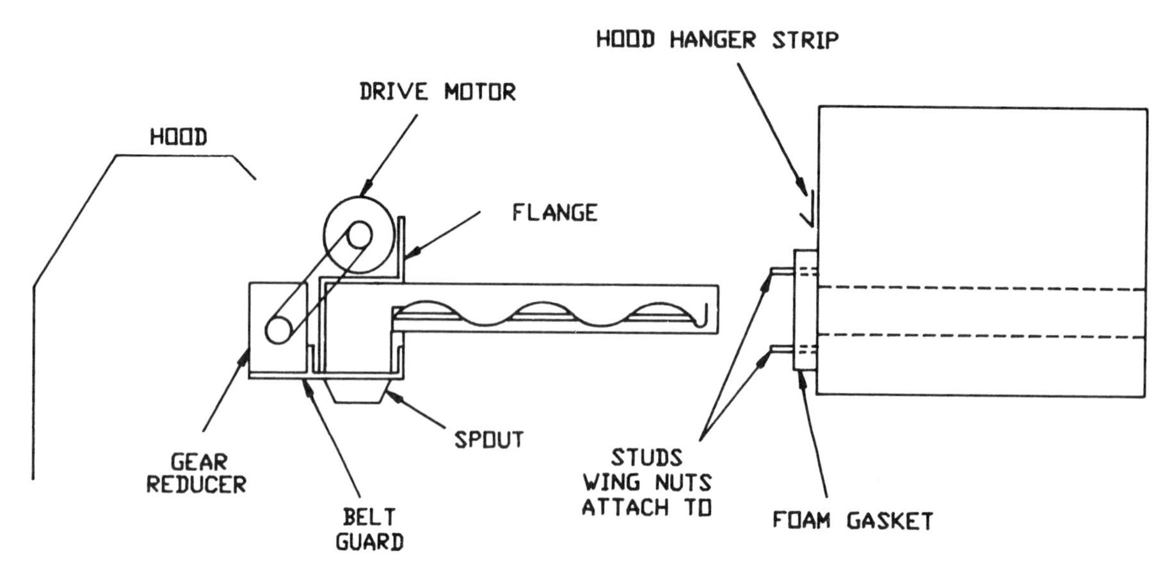

5. Unpack the Dispenser mechanism. Remove tape and rotate spout and snap it in place. Wipe out any dirt from the tube and auger if present.
CAUTION: DO NOT OPERATE WITH SPOUT REMOVED.

6. Remove 4 wing nuts from around Dispenser hole. Leave the foam gasket in place.

7. Slide the Dispenser mechanism into the hole, lining up the belt guard on the top right stud and tighten the flange nuts with 4 wing nuts.

   NOTE: A. All units are factory assembled with the large pulley on the drive motor and the small pulley on the gear reducer for large volume output. (1 pound of ice per second.)

   B. To reduce ice volume dispensed (1 pound per 4 seconds) move the small pulley to the drive motor and the large pulley to the gear reducer.

8. Refer to Drawing TTT-02092 for proper electrical wiring.

9. Install the hood over the Dispenser mechanism using hood hanger strip, then screw the hood to the dispensing unit.

10. A. On Push-Rod activated Models: Through the opening on the left side of the hood attach the Push-Rod onto the Shaft.

    B. On Mini-Ice Plants: Put the plug through the hood, making sure that the gray gasket is in place around the cut-out.

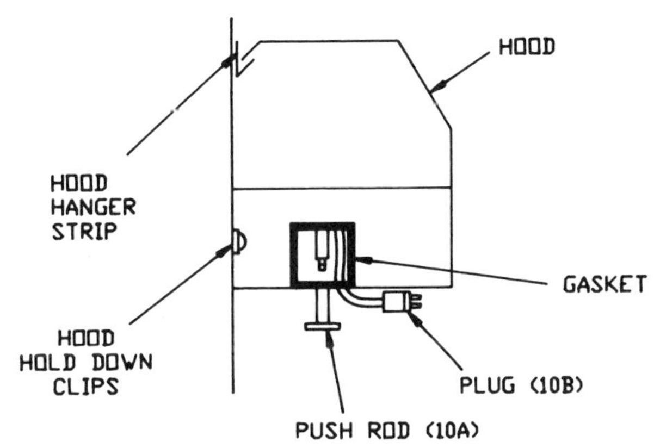

DRAINS - Refer to Figure E.

11. Connect sink and hopper drains and run drain line to a suitable drain. (Bagging Station does not have sink drains.)

12. Connect 3/4" IP Ice Cuber Drain to Drain Pipe on bottom left of TDA and run to a suitable Drain.

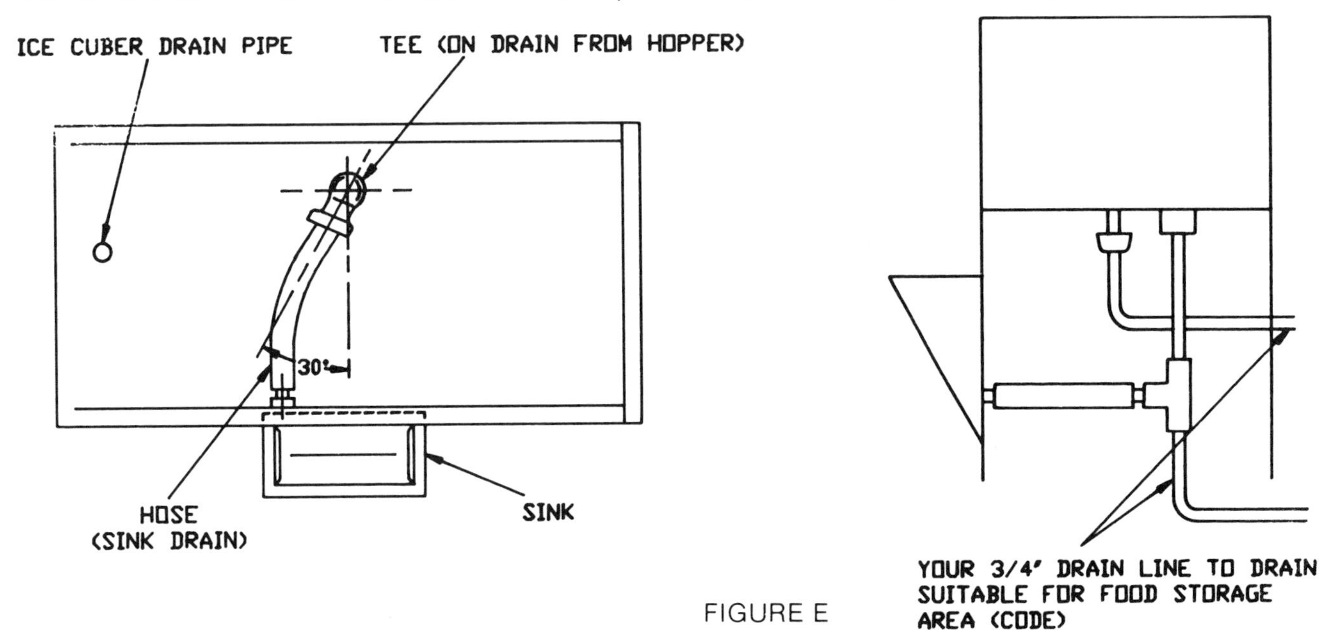

FIGURE E

13. Mount Cuber to TDA-15, TDA-43D or MIP-43D per Cuber Installation Instructions.

14. Complete and mail in Warranty Card.

15. For Service contact your local Kold-Draft Distributor or Dealer.

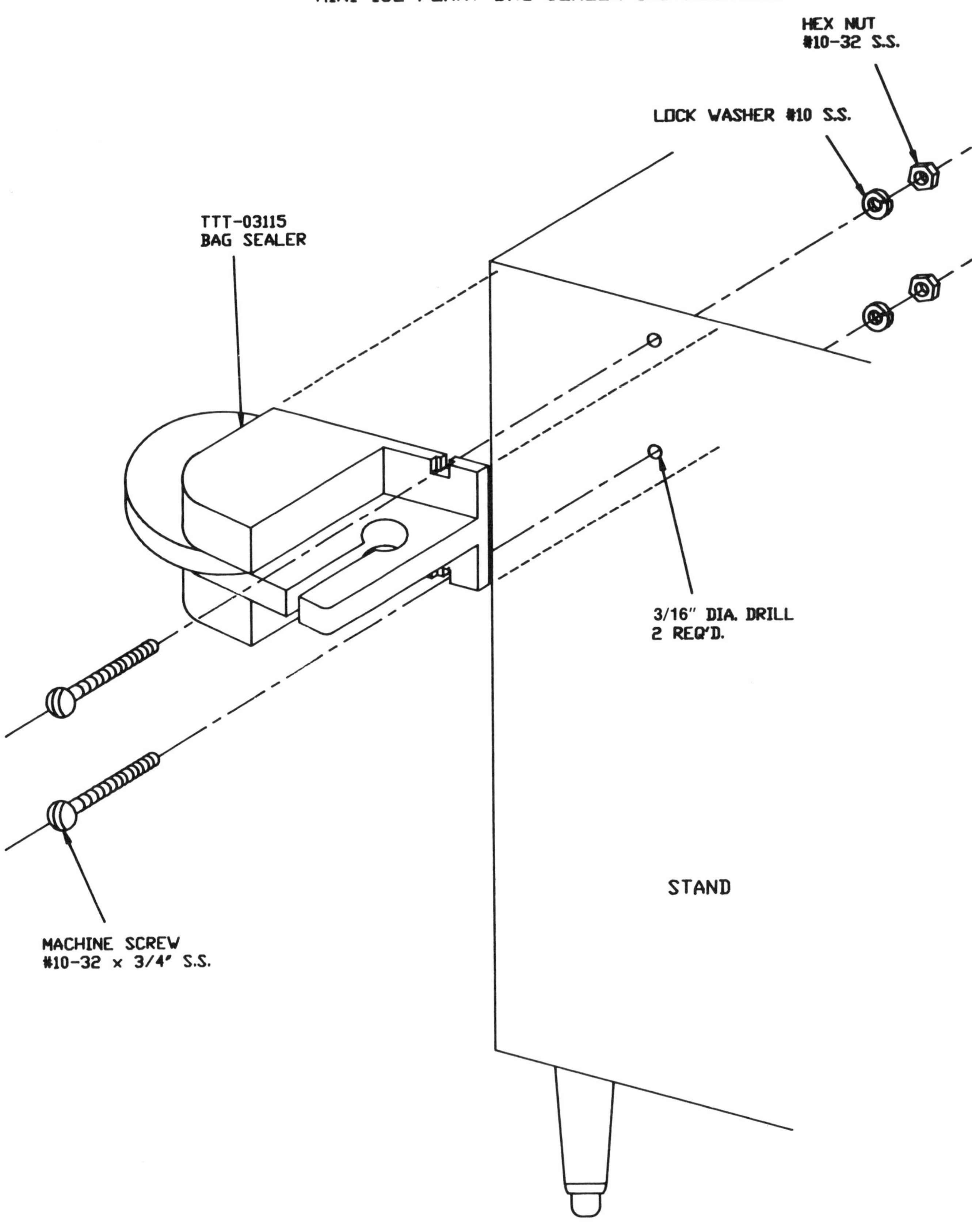

A. Position the bag sealer in the top left corner of the stand.
B. Mark the 2 positions where the holes are to be drilled.
C. Drill the 2 marks with a 3/16" dia. drill.
C. Mount the bag sealer.

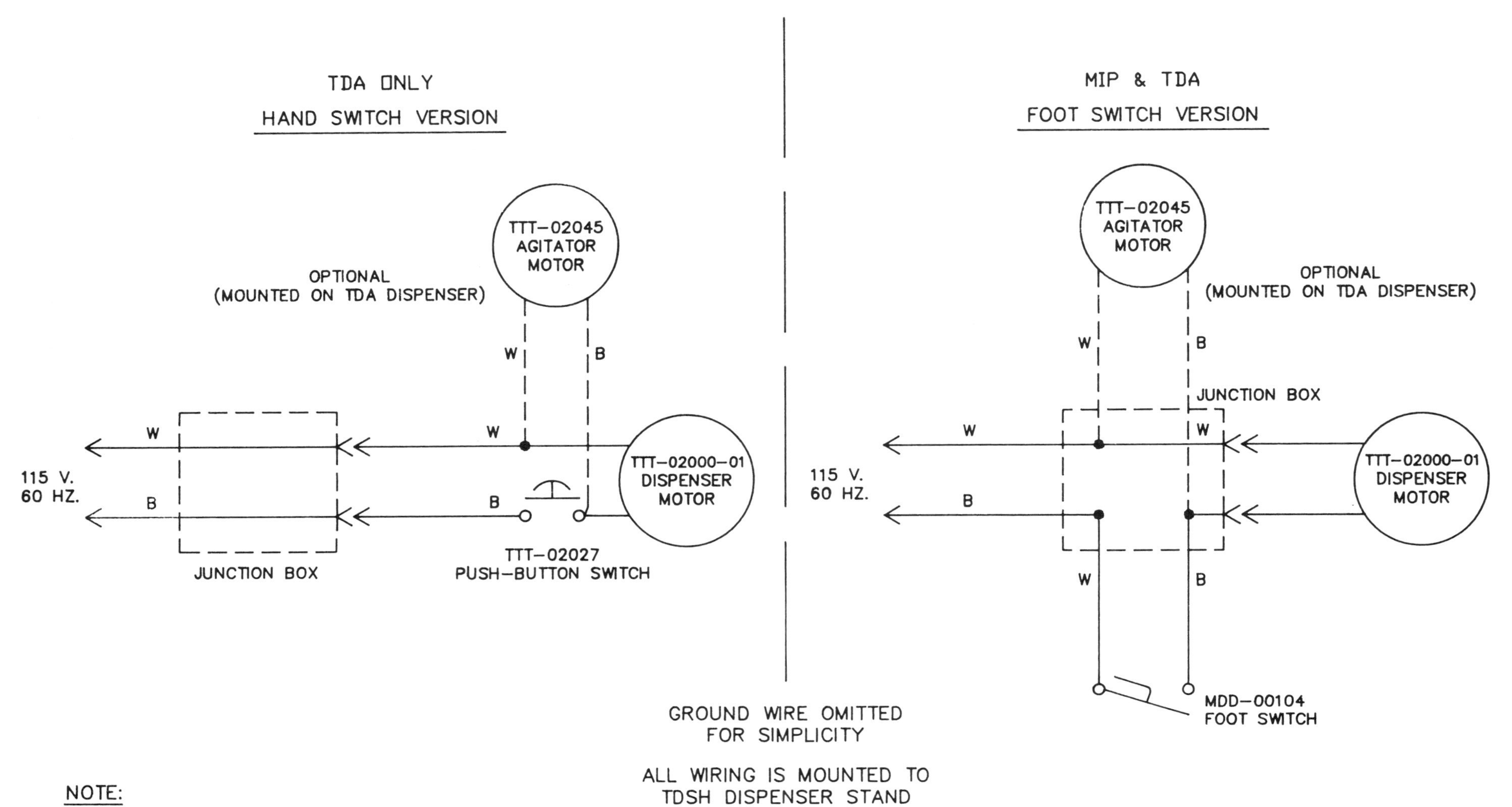

NOTE:
FOR DISPENSERS MADE AFTER OCT. 1986
USE DWG. TTT-02092
FOR DISPENSERS MADE BEFORE NOV. 1986
SEE DWG. TTT-02079

MIP & TDA WIRING DIAGRAM

TTT-02092

CAUTION: When operating and servicing a Kold-Draft GB Cuber and TDA Dispenser combination, care must be taken to prevent possible personal injury or damage to the units.

WARNING: Before operating the GB Cuber-TDA Dispenser combination, check to verify:

1. Each Cuber as well as each Dispenser is on its own circuit, protected by fuses or HACR type circuit breakers.
2. All cubers (except 3 phase Models) must have a supplemental fuse kit installed in junction box at the left side of the Cuber (Refer to Engineering Bulletin #22-84).
3. The water pump screen must be in place and secured with its screw.
4. The front electrical channel must be secured in place with hold-down clip and screw.
5. Electronic Cubers (except GT300 Series) must have the control module box cover in place and secured with screw. GT300 Series Cubers - it is built into the front panel and must be secured in place when the machine is not being serviced.
6. Cubers with transformers must have cover in place and secured with a screw.
7. Compressors must have covers secured in place to prevent access to compressor wiring and/or start components.
8. The hot gas defrost valve cover must be secured in place.
9. On thermostatically controlled Cubers the water weight control switch cover must be secured with screw.
10. On the TDA Dispenser, the hood as well as the ice spout are in place.
11. Dispenser drive motor connection junction box cover secured in place.
12. Cuber must not be operated with the exterior panels removed without a qualified service mechanic present.

WARNING: When servicing Kold-Draft GB Cubers and TDA Dispensers, safe service practices must be followed.

1. DANGER: Do not braze refrigerant lines until refrigerant is removed from system.
2. WARNING: Beware of potential electrical shock hazard when testing for voltage at electrical connections within the system.
3. CAUTION: Exhibit safe practices when working around moving parts such as the Cuber's pump fan blade, the front and rear cams, front and rear water plate hinges, the actuator motor, condenser fan blades, as well as the TDA's motor and auger and, if present, agitator.

FT & FB SERIES FLAKERS
SERVICE MANUAL

Model Shown: FT401 on GTN-200 Bin.

Products of the
KOLD-DRAFT DIVISION

Uniflow Manufacturing Co.
ERIE, PA. 16514
ESTABLISHED 1920
Area Code 814 — Phone 453-6761
TELEX 91-4434
FAX 814/455-6336
800/548-9392
CANADA 800/548-8929

555-01385
5/89
*Trademark Reg. U.S. Pat. Off.
© 1987, Uniflow, Erie, Pa.

TABLE OF CONTENTS

FLAKER MODEL NUMBER BREAKDOWN

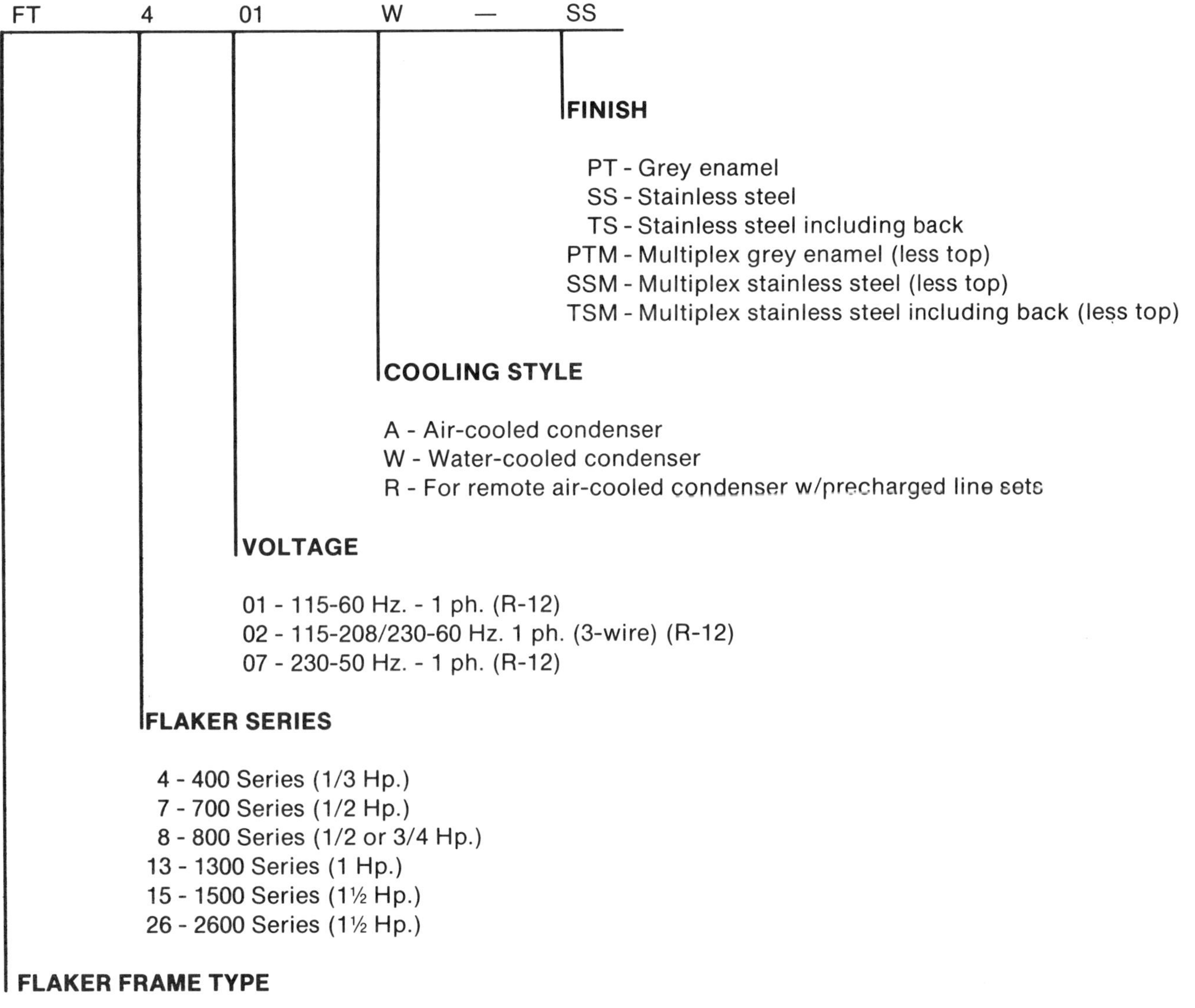

FLAKER FRAME TYPE

FT (28 1/2" wide)
FB (42" wide)

FLAKER GENERAL SPECIFICATIONS

Model	H.P.	Voltage /Phase	Minimum Ampacity	Minimum Incoming Wire Size	Fuse or HACR Type Circuit Breaker Size Recommended	Maximum	Suction Pressure	Charge R-12	Water Usage Gal./Day at 70°F Ambient
FT401	(A) 1/3	115/1	10	14	15	15	14-17	17 oz.	60
	(W) 1/3	115/1	10	14	15	15	14-17	14 oz.	375
FT701	(A) 1/2	115/1	15	12	15	15	12-15		N/A
	(W) 1/2	115/1	15	12	15	15	12-15		N/A
	(R) 1/2	115/1	15	12	15	15	12-15		N/A
Prior 3/84									
FT801	(A) 1/2	115/1	15	12	15	15	9-11	17 oz.	85
	(W) 1/2	115/1	15	12	15	15	9-11	14 oz.	748
	(R) 1/2	115/1	15	12	15	15	9-11		85
Since 3/84 - without EPR Valve									
FT801	(A) 3/4	115/1	N/A	12	20	20	7-9	17 oz.	85
	(W) 3/4	115/1	N/A	12	20	20	7-9	14 oz.	748
	(R) 3/4	115/1	N/A	12	20	2	7-9	10 lb. Min.	85
	- with EPR Valve								
FT801	(A) 1/2	115/1	N/A	12	20	20	9-12	17 oz.	85
	(W) 1/2	115/1	N/A	12	20	20	9-12	14 oz.	748
	(R) 1/2	115/1	N/A	12	20	20	9-12	10 lb. Min.	85
FB1302	(A) 1	115/208-230/1	13.6	14	20	20	10-15	3 lb.	N/A
	(W) 1	115/208-230/1	11.0	14	15	15	10-15	3 lb.	N/A
	(R) 1	115/208-230/1	14.9	12	20	25	10-15	10 lb. Min.	N/A
	- without EPR Valve								
FB1502	(A) 1½	115/208-230/1	N/A	12	20	20	6-7	3½ lb.	190
	(W) 1½	115/208-230/1	N/A	12	20	20	6-7	2½ lb.	1075
	(R) 1½	115/208-230/1	N/A	12	20	20	6-7	10 lb. Min.	190
	- with EPR Valve								
FB1502	(A) 1½	115/208-230/1	N/A	12	20	20	9-12	3½ lb.	190
	(W) 1½	115/208-230/1	N/A	12	20	20	9-12	2½ lb.	1075
	(R) 1½	115/208-230/1	N/A	12	20	20	9-12	10 lb. Min.	190
FB2602	(A) 1½	115/208-230/1	24.9	10	30	35	12-15		N/A
	(W) 1½	115/208-230/1	16.5	12	20	25	12-15		N/A
	(R) 1½	115/208-230/1	16.5	12	20	25	12-15		N/A

NOTE: MAXIMUM BRANCH CIRCUIT FUSE OR HACR TYPE CIRCUIT BREAKER SIZE is dependent on the size of the conductors supplying the Flake Ice Machine. They must be no less than the minimum ampacity rating and no more than the maximum rating on the name plate.

SUPPLEMENTAL FUSES DO NOT provide primary protection and may be sized as required for continuous operation without nuisance blowing, up to the indicated NAMEPLATE maximum rating to compensate for ambient conditions.

SPECIAL NOTE - Wire sizes on this Chart are good up to 80 feet. Anything 80 feet to 150 feet increase wire 1 size. Any runs 150 feet to 250 feet increase the wire 2 sizes.

Sight glass can be used to determine a refrigerant undercharge after 15 minutes of operation.

CONSTRUCTION OF KOLD-DRAFT FLAKERS

Exterior Panels:

Consist of removable top, left end, right end, back panel(s), and front inspection panels.

Condensing Units:
Varies with each Model.

Control System:
Totally contained within a front-accessible control box(s). The Control Box contains:

1. 115 Volt receptacle(s) on bottom for gear motor plug(s).
2. Gear thermostat controls the gear motor energization.
3. Bin thermostat - that turns compressor off and on.
4. Defrost thermostat - that energizes the hot gas defrost valve should the bottom evaporator temperature drop below 32° F. to prevent an overfreeze condition.
5. Three (3) indicator lights that allow visual checking of components energized.
   A) Green - Gear motor energized.
   B) Red - Defrost Valve energized.
   C) Yellow - Unit off on Full Bin condition.

Freezing Unit:

1. Evaporator - Refrigerant coil wrapped cylinder. (Expansion valve controlled.)
2. Bottom Bearing - that locates bottom of auger and segregates water from the bearing and freezing surfaces.
3. Stainless Steel Auger - which peels ice from evaporator wall in vertical direction.
4. Ice Deflector - that changes ice direction from vertical to horizontal as it forces it through the output spout into the ice delivery tube.
5. Deflector Bearing - Replaceable Bearing in deflector which positions upper portion of auger.
6. Gear Motor - (1/10 H.P. 15 RPM 115 Volt) that drives the auger.

Water Level Control:

Adjustable float type water inlet control.

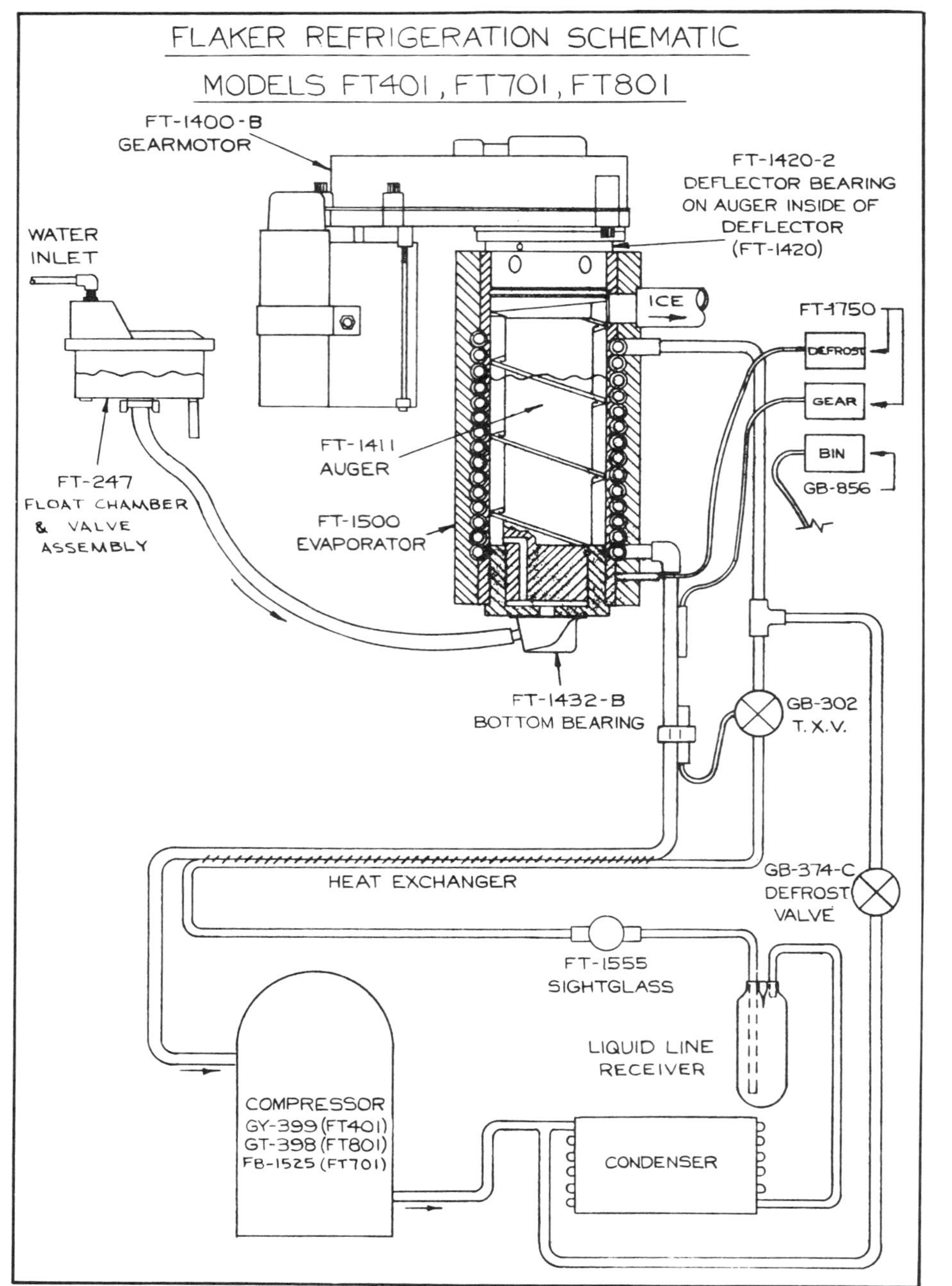

FLAKER REFRIGERATION SCHEMATIC
MODELS FT401, FT701, FT801
FT-1400-B
GEARMOTOR
FT-1420-2
DEFLECTOR BEARING
ON AUGER INSIDE OF
DEFLECTOR
(FT-1420)
WATER
INLET
ICE
FT-1750
DEFROST
GEAR
BIN
GB-856
FT-1411
AUGER
FT-247
FLOAT CHAMBER
& VALVE
ASSEMBLY
FT-1500
EVAPORATOR
FT-1432-B
BOTTOM BEARING
GB-302
T.X.V.
HEAT EXCHANGER
GB-374-C
DEFROST
VALVE
FT-1555
SIGHTGLASS
LIQUID LINE
RECEIVER
COMPRESSOR
GY-399 (FT401)
GT-398 (FT801)
FB-1525 (FT701)
CONDENSER

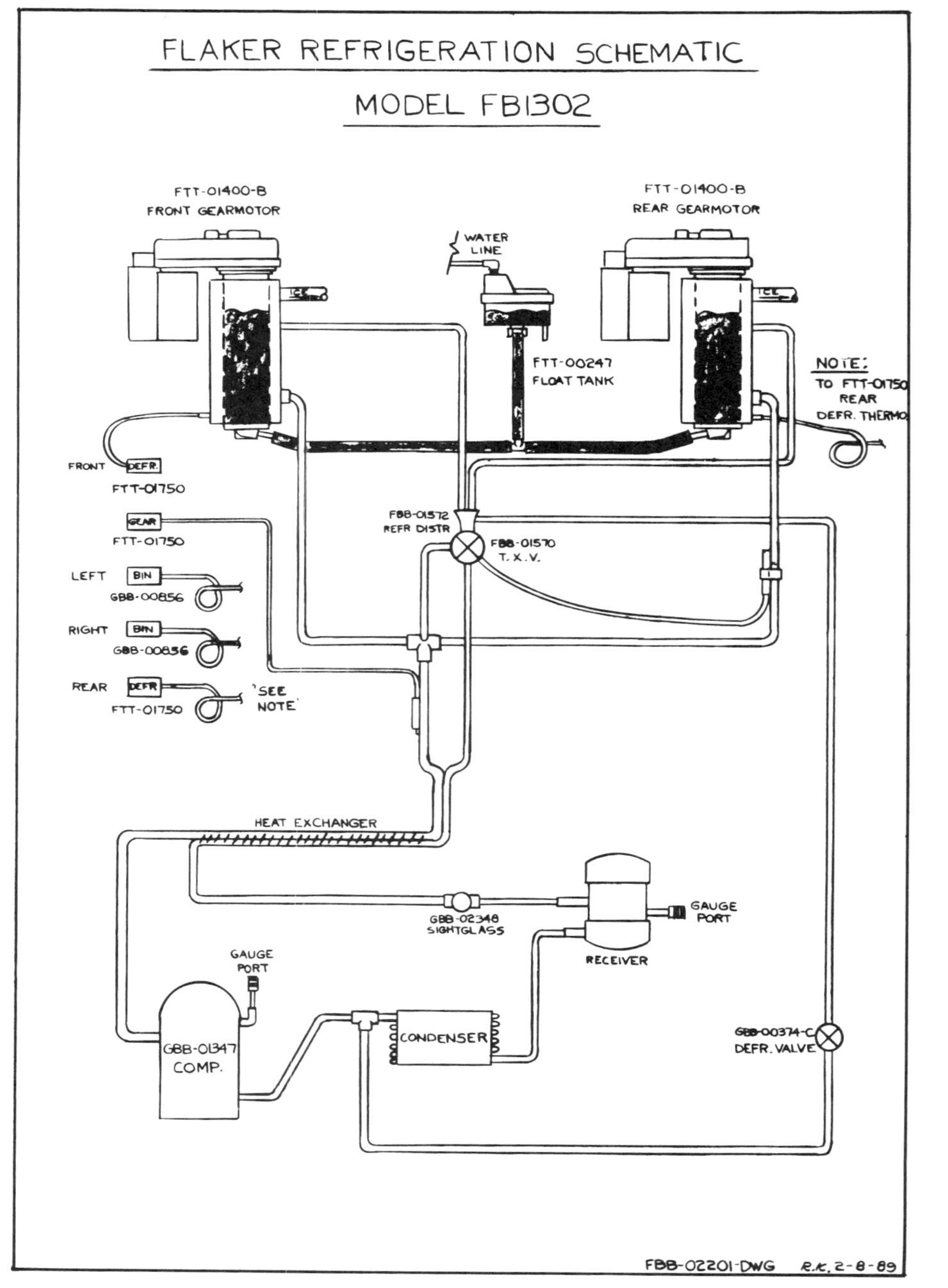

FLAKER REFRIGERATION SCHEMATIC
MODEL FB1302
FTT-01400-B
FRONT GEARMOTOR
FTT-01400-B
REAR GEARMOTOR
WATER
LINE
FTT-00247
FLOAT TANK
NOTE:
TO FTT-01750
REAR
DEFR. THERMO
FRONT
DEFR.
FTT-01750
GEAR
FTT-01750
FBB-01572
REFR DISTR
FBB-01570
T.X.V.
LEFT
BIN
GBB-00856
RIGHT
BIN
GBB-00856
REAR
DEFR
FTT-01750
'SEE
NOTE'
HEAT EXCHANGER
GBB-02348
SIGHTGLASS
GAUGE
PORT
RECEIVER
GAUGE
PORT
GBB-01347
COMP.
CONDENSER
GBB-00374-C
DEFR. VALVE
FBB-02201-DWG
R.K. 2-8-89

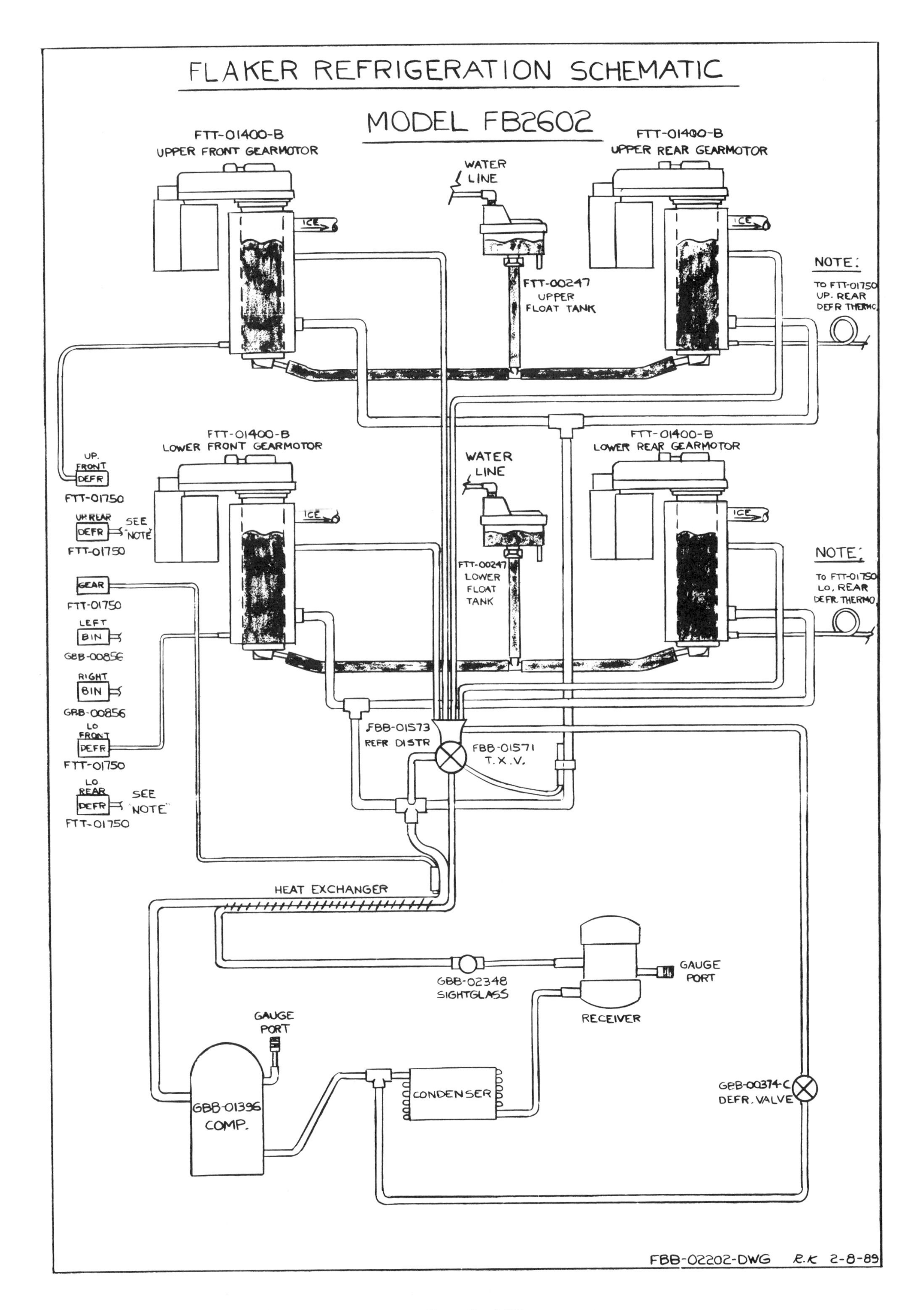
FLAKER REFRIGERATION SCHEMATIC
MODEL FB2602
FTT-01400-B
UPPER FRONT GEARMOTOR
FTT-01400-B
UPPER REAR GEARMOTOR
WATER LINE
ICE
ICE
NOTE:
TO FTT-01750 UP. REAR DEFR THERMO.
FTT-00247 UPPER FLOAT TANK
FTT-01400-B
LOWER FRONT GEARMOTOR
FTT-01400-B
LOWER REAR GEARMOTOR
WATER LINE
UP. FRONT DEFR
FTT-01750
UP. REAR DEFR
SEE "NOTE"
FTT-01750
ICE
ICE
NOTE:
TO FTT-01750 LO. REAR DEFR. THERMO.
GEAR
FTT-01750
FTT-00247 LOWER FLOAT TANK
LEFT BIN
GBB-00856
RIGHT BIN
GBB-00856
LO FRONT DEFR
FTT-01750
FBB-01573 REFR DISTR
FBB-01571 T.X.V.
LO REAR DEFR
SEE "NOTE"
FTT-01750
HEAT EXCHANGER
GBB-02348 SIGHTGLASS
GAUGE PORT
RECEIVER
GAUGE PORT
GBB-01396 COMP.
CONDENSER
GBB-00374-C DEFR. VALVE
FBB-02202-DWG R.K 2-8-89

SEQUENCE OF OPERATION

1. At start-up all thermostats are warm, electric power flows from fuse thru Bin Thermostat directly to the compressor (thru High and Low Pressure Cut-Outs as applicable) and thru the warm side of the Gear Thermostat to the Gearmotor (and in parallel to the Green light).

   1a. Original pull down of Flaker may be slow, 5 to 10 minutes before ice is produced. During the last half of pull down the suction line at the compressor will be frosted, and the sight glass will show bubbles during pull down. After ice starts to be produced, frost will leave the compressor suction line and within 5 minutes the sight glass will show clear or only a small bubble in the top, which will disappear slowly.

2. As ice begins to form the suction line from the evaporator (location of Gear Thermostat cap tube) gets cold enough (below 32°) to trip Gear Thermostat to the cold side bypassing the Bin Thermostat (resetting for later Ice Unloading period) so power flows directly thru the Gear Thermostat to the Gearmotor (and Green light); no change is apparent.

   2a. [On rare occasions with some particle-free waters, the water will subcool substantially below freezing, then suddenly crystallize and stall the Gearmotor (designed to withstand a stall). The evaporator will continue to get colder until the bottom (location of the Defrost Thermostat cap tube) gets below freezing, tripping the Defrost Thermostat cold, sending power to the Defrost Valve (and Red light) which sends hot refrigerant thru the evaporator, releasing the auger and gearmotor until the bottom of the evaporator reaches about 45° tripping the Defrost Thermostat open and normal refrigeration restarts. Since subcooling with crystallization occurs only 2 or 3 times out of 100 it is unlikely to happen and very unlikely to repeat.]

3. During ice formation ice is drawn to the top of the evaporator, lowering the water level, opening the float valve. When water-flow is seen, it indicates ice formation even before ice appears in the bin coming from the ice tube. Water flows thru the bottom of the auger and from a hole near the bottom flight into the ice chamber, cools to 32°, then freezes in thin layers on the wall of the evaporator cylinder where the auger flights remove it and spiral it upward. The ice thickens until it is a fairly dense mass as it rotates under the helical path of the deflector until it is forced into the discharge spout by the deflector. The ice then flows by gravity thru the ice tube to give it velocity which spreads it in the bin.

4. When ice piles over about half the length of the 1/4" stainless tube which holds the Bin Thermostat cap tube, the thermostat opens shutting off power to the compressor. (Turning on the Yellow light which is connected in parallel with the thermostat). The Green light stays on as the Gear Thermostat is still cold supplying power from the fuse to the Gearmotor (and Green light) to Unload Ice from the evaporator until the freezing surface is above 32°. The Gearmotor must remain on a couple minutes to complete unloading and usually stays on 5 to 15 minutes until the suction line out of the evaporator warms to about 45°. Then the Gear Thermostat trips warm connecting the Gearmotor to the output side of the bin thermostat (resetting for the next start-up) shutting off the Gearmotor and Green light.

5. When the bin is full, only the Yellow light is illuminated indicating power is on and fuse okay. Only a trickle of power is used thru Yellow neon-type light during flaker-off periods. Flaker may start-up while a small amount of ice is still on the 1/4" stainless steel tube. This is okay as the deflector can push ice thru the ice tube creating some spread around the top of the bin until the stainless steel tube is fairly well covered again.

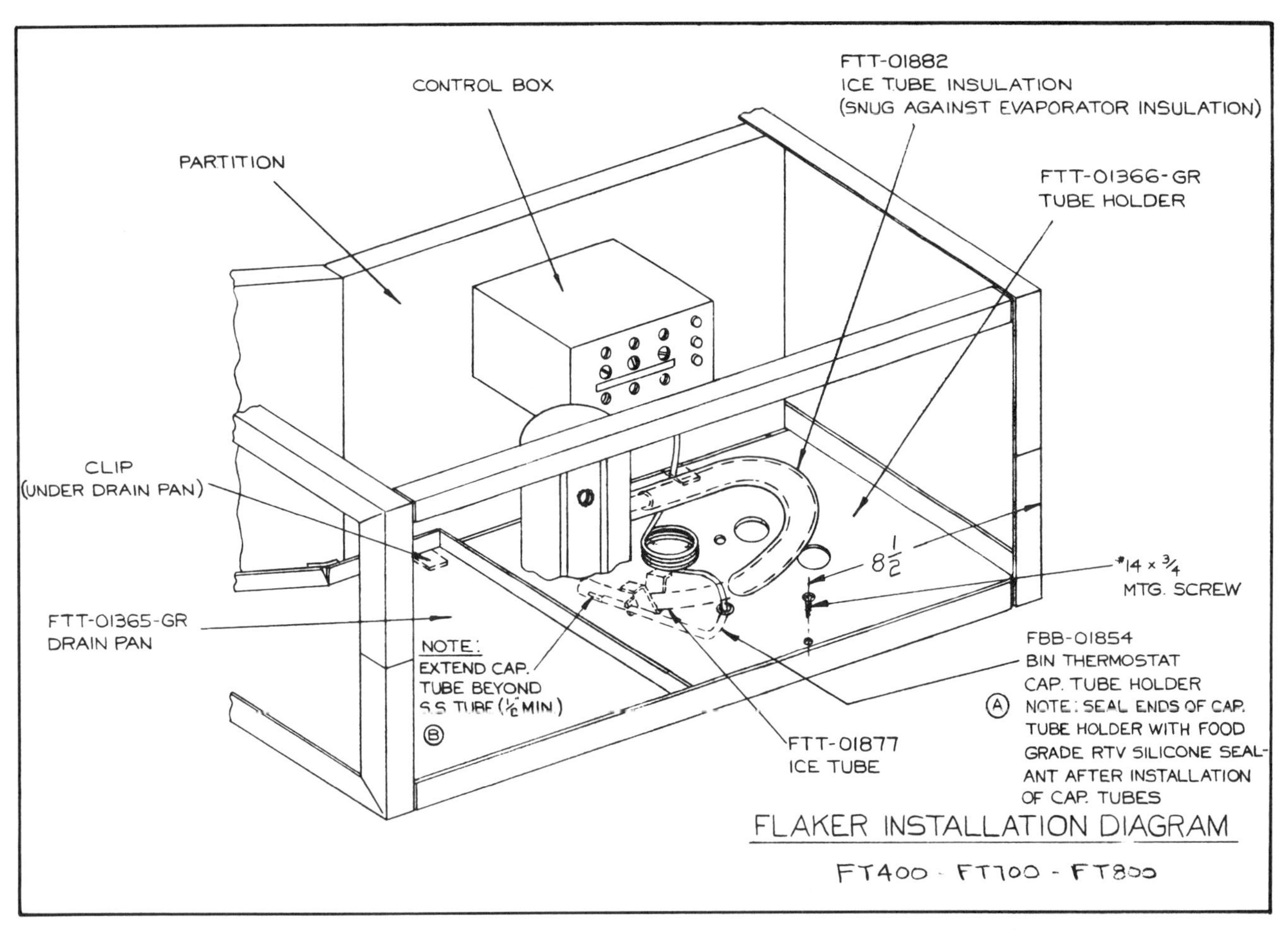
CONTROL BOX
FTT-01882
ICE TUBE INSULATION
(SNUG AGAINST EVAPORATOR INSULATION)
PARTITION
FTT-01366-GR
TUBE HOLDER
CLIP
(UNDER DRAIN PAN)
$8\frac{1}{2}$
#14 x 3/4
MTG. SCREW
FTT-01365-GR
DRAIN PAN
NOTE:
EXTEND CAP.
TUBE BEYOND
S.S. TUBE (1/2 MIN)
B
FBB-01854
BIN THERMOSTAT
CAP. TUBE HOLDER
A NOTE: SEAL ENDS OF CAP.
TUBE HOLDER WITH FOOD
GRADE RTV SILICONE SEAL-
ANT AFTER INSTALLATION
OF CAP. TUBES
FTT-01877
ICE TUBE
FLAKER INSTALLATION DIAGRAM
FT400 - FT700 - FT800

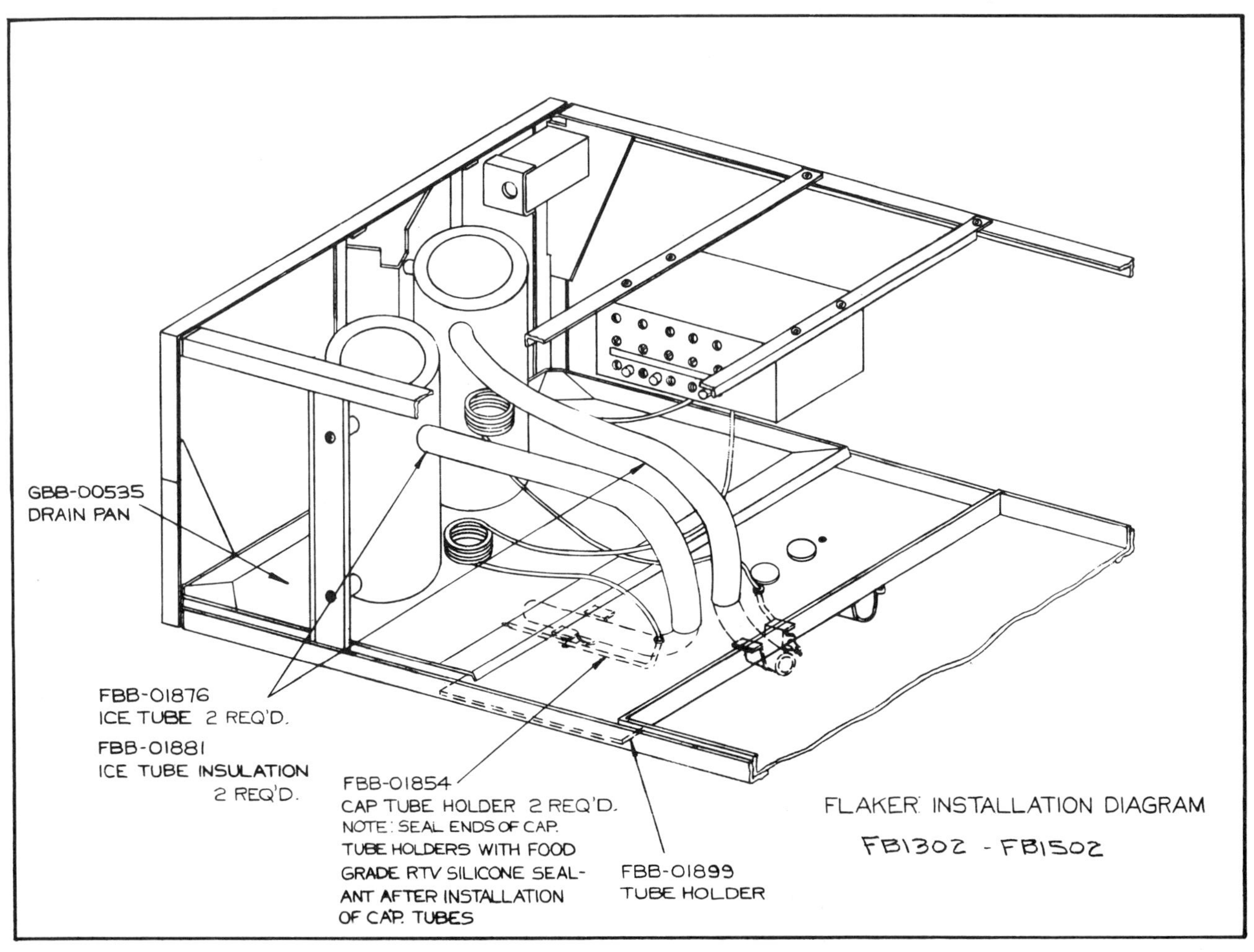
GBB-00535
DRAIN PAN
FBB-01876
ICE TUBE 2 REQ'D.
FBB-01881
ICE TUBE INSULATION
2 REQ'D.
FBB-01854
CAP TUBE HOLDER 2 REQ'D.
NOTE: SEAL ENDS OF CAP.
TUBE HOLDERS WITH FOOD
GRADE RTV SILICONE SEAL-
ANT AFTER INSTALLATION
OF CAP. TUBES
FBB-01899
TUBE HOLDER
FLAKER INSTALLATION DIAGRAM
FB1302 - FB1502

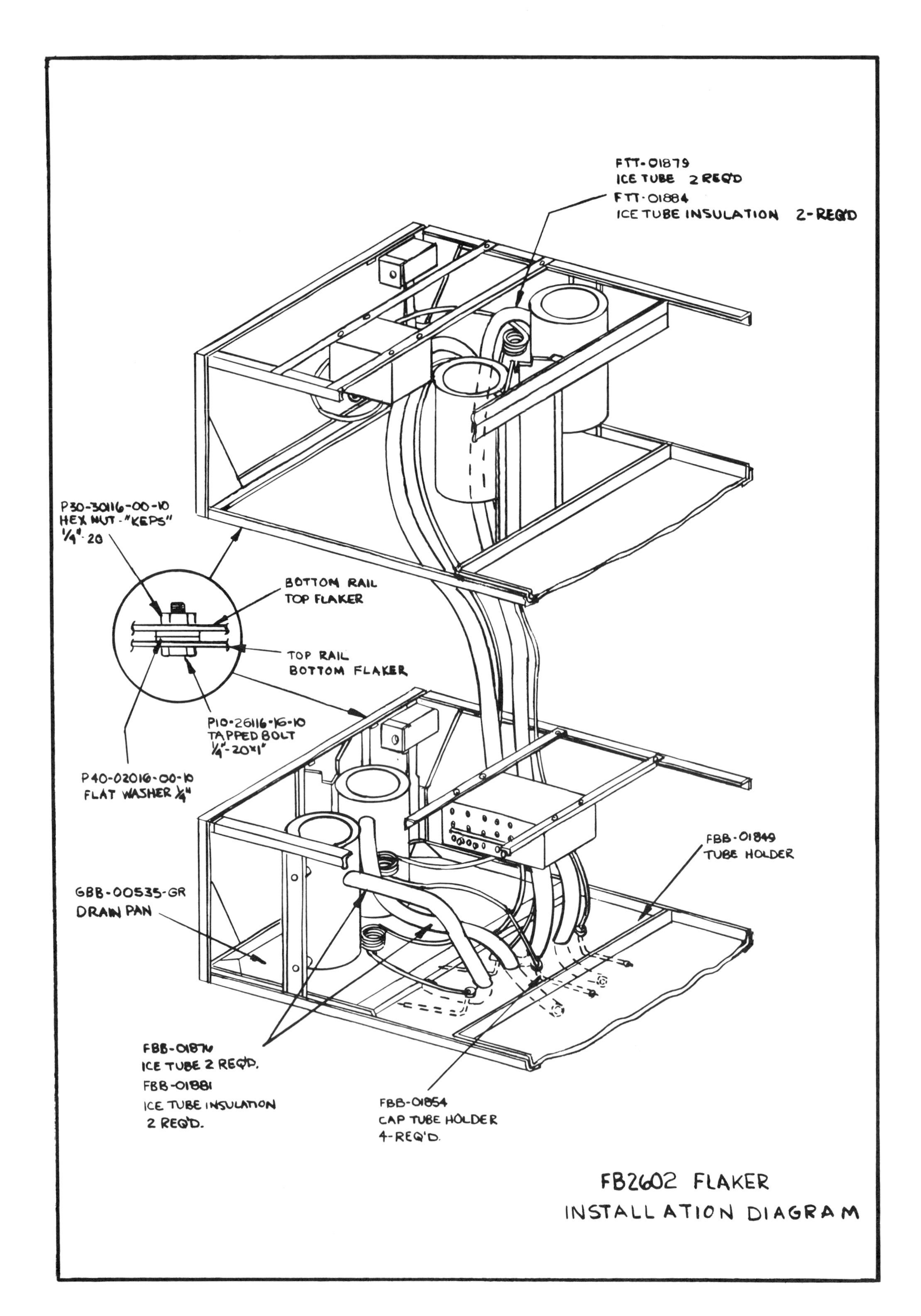

FTT-01879
ICE TUBE 2 REQ'D
FTT-01884
ICE TUBE INSULATION 2-REQ'D
P30-30116-00-10
HEX NUT-"KEPS"
1/4"-20
BOTTOM RAIL
TOP FLAKER
TOP RAIL
BOTTOM FLAKER
P10-26116-16-10
TAPPED BOLT
1/4"-20x1"
P40-02016-00-10
FLAT WASHER 1/4"
FBB-01849
TUBE HOLDER
GBB-00535-GR
DRAIN PAN
FBB-01876
ICE TUBE 2 REQ'D.
FBB-01881
ICE TUBE INSULATION
2 REQ'D.
FBB-01854
CAP TUBE HOLDER
4-REQ'D.
FB2602 FLAKER
INSTALLATION DIAGRAM

EXPANSION VALVES

1. Types

   A. FT400, FT700, FT800, FB1500 - Internally equalized TXV (1/2 ton) - GBR-00300
   B. FB1300 - Externally equalized TXV (1 ton) - FBR-01570
   C. FB2600 - Externally equalized TXV (1½ ton) - FBR-01571

2. Adjustment

   A. Do not attempt to adjust TXV until the Flaker has been operating for at least 20 minutes to stabilize system pressures and temperatures. During the first 5 minutes of operation, some frost may occur at the compressor - this is a normal operating characteristic of the Flaker and does **not** indicate a mis-adjusted TXV.

   B. The TXV is adjust to maintain a cool but not frosted suction line near the compressor (about 50° to 60° F.). This will correspond to the low side suction pressure as indicated on suction pressure and charge chart in manual.

   NOTE: 1) Most FT800 & FT1500 Model Flakers were equipped with an E.P.R. valve. Evaporator pressure is factory set at 9 PSIG and should not be changed without consulting the factory.

   2) FB1500 expansion valves were set at the factory by way of ice production. The difference in ice production (lb/day) between evaporators should not exceed 50 lbs.

   REFER TO GENERAL SPECIFICATIONS

WATER LEVEL CONTROL

1. Float type inlet water control. The float tank is attached to the frame with screws in slotted holes.

2. The float tank though adjustable is recommended to be set to the highest position. This will result in the most water in the evaporator and greatest ice making capacity.

   NOTE: In February 1986 the slotted holes in the water level mounting bracket were enlongated from 1" to 1⅝" to eliminate any necessity of bending the float arm which in some cases though increasing the water level started a siphoning effect.

3. It is not recommended that the float tank be adjusted downward. This will decrease the water level and produce drier harder ice, decrease ice making capacity and can result in an increased number of defrost cycles and/or stalling of the gear motor.

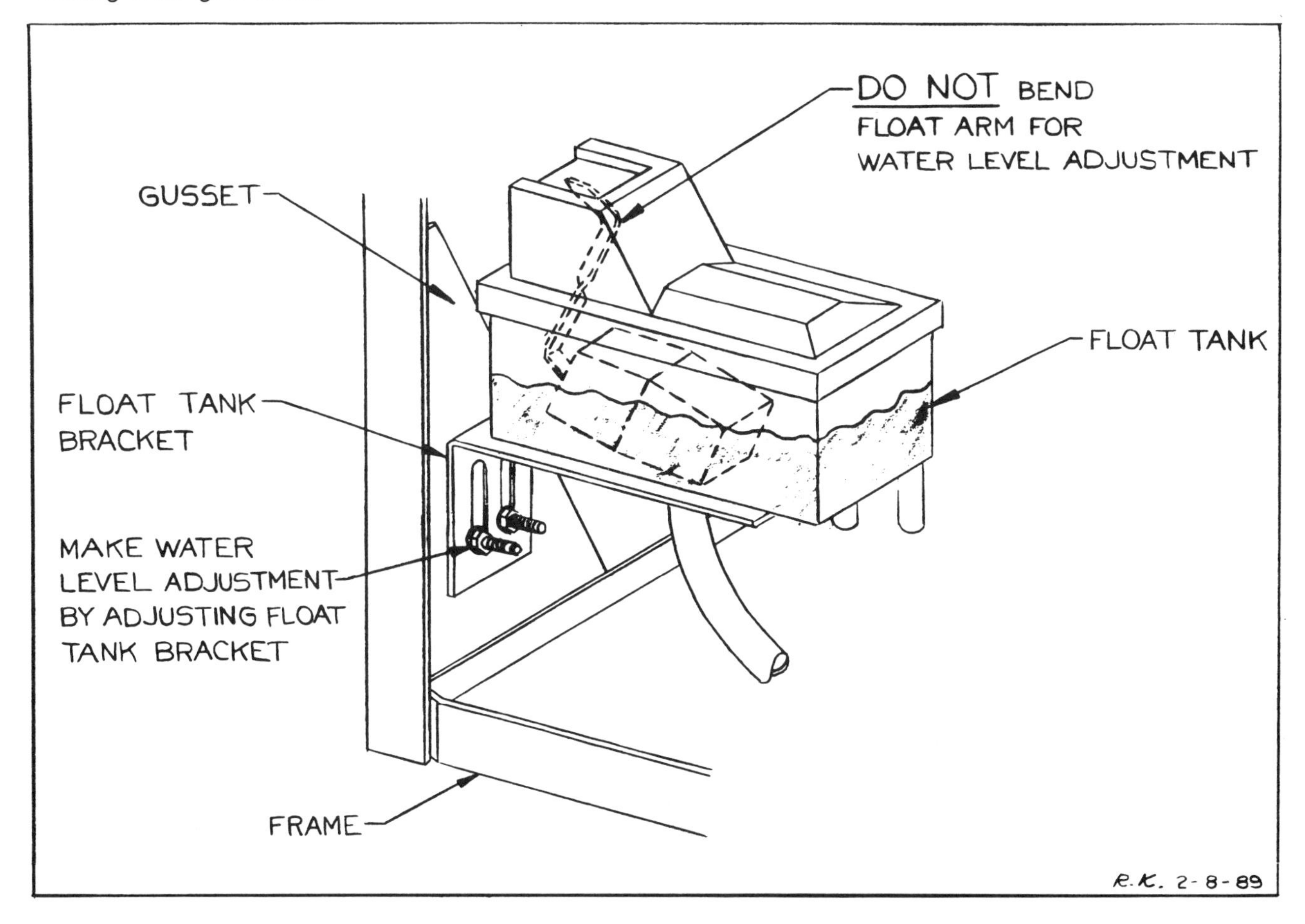

DEFLECTOR

The Deflector is positioned in the evaporator and should be set in the wide open position and locked in place. This will allow a minimum amount of pressure build up in the evaporator and allow the ice out of the evaporator as shown below.

DEFLECTOR ALIGNMENT LOOKING THROUGH THE EVAPORATOR OUTPUT SPOUT

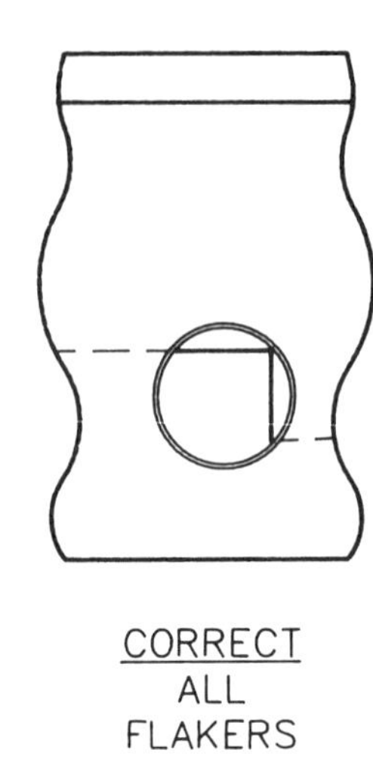

CORRECT
ALL
FLAKERS

NOTE: 1. Since May 1988 Deflectors are locked in the FULL OPEN position as shown above with the upper evaporator mounting screw and 2 set screws to prevent restriction of the output spout.

2. Prior to May 1988 the Deflectors were positioned and secured in place with 2 set screws located to the right and left of the evaporator mounting bracket upper screw.

ICE DISCHARGE TUBE INSTALLATION

The plastic discharge tube should be replaced with a reinforced tube (see Parts Manual for appropriate part numbers). Be sure that there are no kinks or collapsed areas in the tube, and that the interior of the tube is clean and smooth. **Also,** check the ice discharge tube below the tube holder plate to be sure that there is no "sag" in the ice tube which is lower than the open end. This condition can set up a water siphon from the evaporator to the bin which will continuously melt the stored ice.

Be sure that the bin thermostat cap-tube is routed with a small loop under the evaporator spout to insure shut-down of the Flaker if the ice discharge tube comes off of the spout. If the tube comes off of the spout due to loose fit, not kinks or ice backup from an over-filled bin, installation of a plastic ratcheting-type clamp, **finger tight only,** is acceptable. Packs of 4 clamps are available as Part No. FTR-01918. The bin thermostat holder and tube must be positioned far enough past the end of the ice tube so that the ice can contact it to stop the Flaker before ice piles up against the end of the ice tube and causes ice backup in the tube.

REFER TO INSTALLATION DIAGRAM.

AIR-COOLED CONDENSERS

1. It is imperative to maintain adequate ventilation around the Flaker - 12" minimum behind air-cooled Flakers along with 6" on top, right and left sides.

   NOTE: It is necessary to prevent re-circulation of the hot condenser air from the sides to the back and may require external air blocks to prevent the hot outlet condenser air from returning to the inlet of the condenser. (Particularly in stacked units or units placed in corners.)

2. Cleaning Air-Cooled Condenser. The air-cooled condenser should be cleaned weekly with a stiff brush and a vacuum cleaner to remove dust and dirt for efficient operation of unit. To determine that the condenser is clean, a light which is held at one side of the condenser will be clearly visible from the other side.

WATER-COOLED CONDENSERS

1. Are factory set at 115 to 120 pounds head pressure.

   NOTE: To reduce water consumption, head pressure can be increased but ice capacity will be reduced.

2. Excessive head pressure and higher than normal water usage indicates the condenser needs to be cleaned. Water-cooled condensers can be cleaned by flushing with a condenser cleaning solution. Acid recirculating pumps and solution are available at refrigeration supply houses.

CONDENSER WATER REGULATOR

Adjust the regulator to maintain a constant operating head pressure of 115-120 PSIG, or 95° to 100° condenser outlet water temperature.

FUSES

1. Each flaker must be on its own circuit. Either individually fused or on HAC-R type circuit breaker. The maximum allowable voltage fluctuation should not exceed 10 per cent of the nameplate rating even under starting conditions. Low voltage will cause erratic operation, low capacity, and serious damage to the overload switch and motor windings.
2. All flakers have supplemental fusing. These fuses are located in the electrical junction box at the left rear side of the flaker.

INTERNAL OVERLOADS

Each motor used on these flakers have overload protection. None should ever burn out due to an overload. However, voltage fluctuating more than 10% less than normal may cause burn-out trouble. To allow a safety factor, make sure voltage is at least 95% of rated voltage while the Flaker is operating.

BIN THERMOSTATS:

1. Routing
   A. The stainless steel bin thermostat capillary tube holder is locked in place in the metal tube holder.

   B. The bin thermostat capillary tube is inserted in the stainless steel bin thermostat capillary tube holder until 1/2" to 1" of the capillary tube extends beyond the stainless steel tube in the bin. Both ends of the stainless steel tube holder should be sealed with food grade RTV silicone sealant after installation of the capillary tube.

   C. Any excess bin thermostat capillary tube is coiled under the corresponding evaporator output spout and ice delivery tube connection. This is done to shut off the condensing unit should either the output spout or ice delivery tube come off due to ice backing up in a kinked ice delivery tube. This will prevent excessive ice build-up in the ice making section.

   NOTE: REFER TO INSTALLATION DIAGRAMS

2. Setting
   A. The bin thermostat is normally set in the warmest position counter-clockwise (CCW). This will shut off the condensing unit with 60 seconds of when ice covers the exposed 1/2" to 1" of bin thermostat capillary tube and stainless steel bin thermostat capillary tube holder.

   B. After the ice is removed from the bin thermostat capillary tube and it warms to 40° F., the condensing unit will re-energize with 60 seconds starting the next ice making cycle.

   C. If the Flaker is a single unit mounted on a larger bin that extends beyond the Flaker on the right side, it may be desirable to set the bin thermostat up to a 1/4 of a turn colder than the warmest position. Ice can then pile up higher, to the outlet of the ice tube, and more of the bin will be filled before shutdown. If this is done put a quantity of ice on the exposed coil of the thermostat cap tube and the Flaker should shut off within 60 seconds, if it does not adjust the bin thermostat slightly warmer (CCW).

GEAR THERMOSTAT

1. Routing of Capillary Tube - Positioned in tube mounted on the suction line coming off the evaporator coil. Excess is coiled under control box.
2. Setting - Normally set somewhere between the mid-position and the coldest position (CW), but must be adjusted so that the gear motor runs until the suction line warms to 40° F (at least 2 minutes) after the compressor is OFF on a full bin condition.

   NOTE:

   A. All Flakers manufactured since December 1985 - the gear motor is de-energized during the hot gas defrost cycle.

   B. Flakers manufactured before December 1985 were factory wired to keep the gear motor energized during hot gas defrost cycle (Green and Red indicator lights "ON" during defrost). This occasionally resulted in the sub-cooling of water and a flash freeze when the refrigeration cycle resumed after the defrost cycle.

   If such flash freezing occurs, a wiring change to place the gear thermostat in series with the defrost thermostat (Reference appropriate FT401, FT801, FB1502 wiring diagrams and/or Engineering Bulletin #8-85) for the following Flakers may still require the wiring change:

   FT400 before Serial #053044
   FT800 before Serial #053020
   FB1500 before Serial #052992

DEFROST THERMOSTAT

1. Routing of Capillary Tube - Push the capillary tube all the way into the tube that wraps around the bottom of the evaporator. Position the excess capillary coil under the control box.
2. Normal Setting & Checking Procedure

   A. FT400, FT700 & FT800 Models

   1. Checking or setting the defrost thermostat requires the use of a low side gauge on all **except** FT400 Models. To prevent refrigerant loss do **not** connect high side gauge.
   2. This control can be checked while the Flaker is producing ice normally by removing the water supply hose from the bottom of the evaporator and slipping the end of the hose over the float tank overflow tube.
   3. Check the time, after most of the water drains out of the evaporator. The defrost thermostat should energize the defrost valve between 1 and 4 minutes (on FT700 and FT800 Models, the suction pressure should not go below 1 PSIG).
   4. If the Flaker energizes the defrost valve too soon adjust the thermostat slightly colder (CW). Except on FT700 or FT800 Models, if the suction pressure goes to 1 PSIG.
   5. If the thermostat is adjusted colder recheck after re-attaching water supply and making ice for a few minutes.
   6. If the time goes beyond 4 minutes (or the FT700 or FT800 pressure goes to 1 PSIG) without defrost going on, adjust the thermostat warmer (CCW) until it does go on in the proper range. After adjusting, re-attach the water supply and watch the unit as it again begins to make ice to be sure that the defrost valve does not come on before ice is produced.

2. Normal Setting & Checking Procedure

   B. FB1300, FB1500, FB2600 Models:

   1. Checking or setting the defrost thermostat(s) requires the use of a low side pressure gauge. To prevent refrigeration loss, do **not** connect high side gauge.
   2. Remove defrost thermostat capillary tube(s) from the bottom of the tube holder(s) located on the bottom of each evaporator whose defrost thermostat is **not** being tested. (This will bypass these defrost thermostats.)
   3. With the Flaker running, remove the water supply hose from the bottom of the evaporator whose defrost thermostat is being checked or tested and slip the hose over the float tank overflow tube.
   4. Check the time after most of the water drains out of this evaporator. This defrost thermostat should energize the hot gas defrost valve between 1 and 4 minutes and the suction pressure should not drop below 1 PSIG.
   5. If the Flaker turns on defrost too soon adjust the thermostat slightly colder (CW): except if the unit suction pressure drops to 1 PSIG when it kicks into defrost. If this is the case, do not adjust colder. If the thermostat is adjusted colder recheck after re-attaching water supply and making ice for a few minutes.

6. If the time goes beyond 4 minutes (or if the suction pressure drops below 1 PSIG) without the defrost thermostat going on, adjust the defrost thermostat warmer (CCW) until it does kick on in the proper range. After adjusting, re-attach the water supply and watch the unit as it again begins to make ice to be sure that the defrost valve does not kick on before ice is produced.
7. If you need to check another defrost thermostat, repeat the above steps after installing the next defrost thermostat to be checked and re-connect the supply hose and remove the defrost thermostat capillary tube from the evaporator just checked or adjusted.
8. When you are finished checking or adjusting, insure that all defrost thermostat capillary tubes are in place and all water supply hoses are connected.

ABNORMAL DEFROST CYCLING

If the gearmotor has power and defrost is going on and off every couple of minutes without producing ice, it is possible that:

1. The defrost thermostat is set much too warm (Gearmotor is turning).
2. The water entry plastic fitting in the bottom of the lower bearing is turned in one turn too far.
3. The water path has been frozen up and the defrost thermostat has been set too cold to defrost it (Gearmotor is running or hot and stalled).

TO CHECK FOR THESE THREE POSSIBILITIES: FOR FT MODELS

1. Remove the water inlet hose from the bottom bearing fitting and watch for water drainage from the fitting. If a substantial amount of water drains out, the defrost thermostat is set too warm. Reconnect the hose and set the defrost thermostat colder (cw) and see the earlier part of the Service Manual for defrost thermostat setting.
2. If only a few drops of water come from the bottom bearing, allow the flaker to go through one more defrost period (Red light on). If water does not flow then the water path is frozen and not defrosting. Turn the defrost thermostat to the warmest position (ccw) and allow the flaker to go through another defrost period. If water still does not flow into the evaporator it will be necessary to reset the internal adjustment on the defrost thermostat with ice (see #3).

   NOTE: Early bottom bearing had an elbow screwed into the bottom of the bearing. If this is screwed in so it protrudes into the bottom bearing water flow can be restricted. Remedy: Unscrew the elbow (1) turn and re-attach the water supply hose and check defrost setting.
3. Pull the defrost thermostat cap tube out of the 1/4" copper tube in the bottom of the evaporator until a couple of inches can be covered with ice. Set external thermostat adjustment to the coldest position (cw), turn off power, open control box, pry plastic thermostat cover toward the rear while lifting the back edge, then with ice on the cap tube turn **internal** adjustment warmer (cw) until the switch will just stay in defrost position with the contact away from the front (as you turn the screw you tend to push the switch control down which tends to trip the contact points back, continue to do this until the contact points stay back when you remove your screwdriver). Be sure ice has remained on the thermostat cap tube during this adjustment. Replace plastic cover on the thermostat and the control box front, turn on the power and with ice still on the cap tube turn external adjustments slowly warmer (ccw) until defrost goes on. If you have seen water flow during this adjustment the water path has defrosted. To double check you can remove hose from the elbow and see if water flows from the evaporator and then reconnect the hose to see that water flows back in. After defrost shuts off and later after ice is being made the defrost thermostat setting can be rechecked according to the instructions in the earlier part of the Service Manual.

If the gearmotor is not running after the evaporator is thawed out refer back to #5 under Trouble, Cause and Remedy. If the gearmotor is running it can be detected by rotating the motor ccw and noticing either a light or strong torque.

TO CHECK FOR THESE THREE POSSIBILITIES: FOR FB MODELS

1. Remove the water inlet hoses from the bottom bearing fittings on both evaporators and watch for water drainage from both evaporators. If a substantial amount of water drains from both evaporators, one or both of the defrost thermostats are set too warm. See the earlier part of the Service Manual for defrost thermostat settings.
2. If only a few drops of water drain from one or both of the elbow fittings, let the machine go through one more defrost cycle. After the defrost remove hoses from the elbows. If there still isn't sufficient drainage from one or both of the evaporators, the defrost thermostat that controls the frozen evaporator(s) must be adjusted to the warmest position (ccw) then allow the flaker to go through another defrost period. If water still does not drain from the evaporator after removal of the hose, it will be necessary to reset the internal adjustment on the defrost thermostat with ice (see #3).

3. If both of the evaporators are frozen, both defrost thermostat cap tubes must be pulled from the 1/4" copper tube around the bottom of each evaporator until a couple of inches can be covered with ice. If only one of the evaporators is frozen, you only have to pull the cap tube out a couple of inches on the evaporator that is frozen.

   Next you must turn power off, open control box and find the thermostat which controls the evaporator you are adjusting and set the external thermostat adjustment to the coldest (cw position). Pry the white plastic thermostat cover toward the rear while lifting the back edge, then with ice on the cap tube turn **internal** adjustment warmer (cw) until the switch will just stay in defrost position with the contact away from the front (as you turn the screw you tend to push the switch control down which tends to trip the contact points back, continue to do this until the contact points stay back when you remove your screwdriver). Be sure ice has remained on the thermostat cap tube during this adjustment. Replace plastic cover on the thermostat and the control box front, turn on the power and with ice still on the cap tube turn external adjustments slowly warmer (ccw) until defrost goes on. If you have seen water flow during this adjustment the water path has defrosted. To double check you can remove hose from the bearing and see if water flows from the evaporator and then reconnect the hose to see that water flows back in. After defrost shuts off and later after ice is being made the defrost thermostat setting can be rechecked according to the instructions in the earlier part of the Service Manual.

If the gearmotor is not running after the evaporator is thawed out refer back to #5 under Trouble, Cause and Remedy. If the gearmotor is running it can be detected by rotating the motor ccw and noticing either a light or strong torque.

BOTTOM BEARING

Whenever the Kold-Draft Flaker gearmotor and auger is removed from the evaporator, also remove and inspect the lower auger bearing. Remove the bottom auger O-rings and slide the bearing onto the auger to check for tightness. Replace the bearing if it is loose or shows signs of wear. When re-installing the lower bearing be sure it is clean, dry, and properly lubricated. O-ring replacement is recommended, and a new Kit, Part No. FTR-01917 which contains all of the necessary O-rings.

Effective 9/1/87 the lower bearing lubrication system in production Flakers and replacement parts will be changed to eliminate the felt wick. The new bearings will employ helical grooves for lubricant retention and distribution, and will **require a special food grade lubricant** for replacement or re-lubrication.

Additionally the O-Ring material has been changed. **The new O-Rings are not compatible with FTR-01888 (Food Grade Grease) or FTR-01890 (Mineral Oil)** lubricants which have been used for several years. This is critical only when re-lubricating (not replacing) factory-installed rings. All except FTR-01420-03 factory-supplied **replacement** rings will be the types which are compatible with old or new lubricants to avoid necessity to replace wick-lubricated lower bearing if bottom auger O-Rings are replaced. FTR-01888 food grade grease (petro gel) will no longer be available, but replacement wicks (FTR-01432-03) and mineral oil will remain available. Replacement rings and some parts will be provided with a supply of special lubricant.

Flakers which are factory-equipped with the new O-Rings can be easily identified by the Black-colored lower auger rings rather than Opaque Yellow, or by the helical groove inside the lower bearing. **Do not use food grade grease (FTR-01888) to lubricate the grooved lower bearngs regardless of the O-Ring type.**

The special lubricant will be available separately as follows:

GBR-02747, 3 Gram "Lube-Packs" (4)
GBR-02748, 5.3 Oz. Tubes

Complete Instructions are provided with either type.

One 3 Gram Pack is sufficient for lubrication of the grooved lower bearing and installation of O-Rings. The 5.3 Oz. Tube will be required for deflector bearing installation (less than 1 oz. will be used for each assembly).

Recommended procedure is to use the new lubricant in all cases except wick-lubricated lower bearings. Follow the instructions carefully and **especially avoid an excessive amount of lubricant in the lower bearing.**

REMOVAL OF GEARMOTOR AND OTHER PARTS

The Kold Draft Flaker can be easily disassembled for service requiring a minimum of tools. It is recommended that the serviceman obtain a 1/8" "Ball-End" Allen wrench (Part #55R-01375) for the deflector set screws.

TO REMOVE THE GEARMOTOR — AUGER ASSEMBLY:

1. Place ice on bin thermostat coil until ice is unloaded from evaporator, then turn off power.
2. Unplug gearmotor.
3. Remove water hose from evaporator, and place on float tank overflow tube to stop water flow.
4. Loosen top evaporator mounting screw **one turn.**
5. Loosen deflector set screws 5 turns.
6. Rotate Gearhead 90° clockwise while lifting up. Rotate as necessary until deflector is out of evaporator. (If necessary to release a tight deflector, insert a large screwdriver between the insulation of the deflector and the evaporator, so the tip is over the edge of the evaporator and the shaft of the screwdriver is under the gearmotor screw and pry up on the gearmotor).

TO REMOVE BOTTOM BEARING:

1. (If applicable) Remove water inlet elbow.
2. Retighten top evaporator mounting screw.
3. Loosen bottom evaporator mounting screw 5 turns.
4. Insert a 12" long wooden dowel, hammer handle, screwdriver handle, etc. into the evaporator barrel and drive downward to push bottom bearing clear from evaporator.

TO DISASSEMBLE AUGER:

1. Lay Gearmotor Auger Assembly on flat surface. Be careful not to damage Auger flight or bottom Auger bearing surface.
2. Insert 1/4" rod or Phillips head screwdriver into 1/4" diameter hole immediately above top "O" ring on bottom of Auger.
3. Strike rod sharply with your hand or mallet to rotate Auger **clockwise.** NOTE: The Auger mounting threads are **left hand.**
4. Unscrew Auger (clockwise) and remove from Gearmotor.
5. Lift deflector assembly from Auger and set aside. The top stainless bearing should now be visible. NOTE: Do not remove stainless bearing if there is no sign of damage.

NOTE: FOR STACKED FLAKERS WITHOUT TOP ACCESS

When Flakers are stacked, the two front holes on the left and right ends are used to bolt the units together.

The following procedure will enable a serviceman to pivot the top Flaker frame in order to remove the auger from the Flaker beneath. It will be necessary to slide the whole assembly out from the wall so that it clears the right wall by 7" and the rear wall by 7½" in order to remove the gear head without lifting off the upper machine.

1. Remove skins from frames as necessary.
2. Remove 1/4-20 bolts from left end, and front, then loosen bolt on right end.
3. Disconnect water supply lines from upper Flakers.
4. Disconnect drain fittings from upper drain pan.
5. Remove any parts which will interfere with pivoting of the upper Flaker frame.
6. Pivot Flaker frame to the right - push left end back until the frame clears the gear head to be removed.

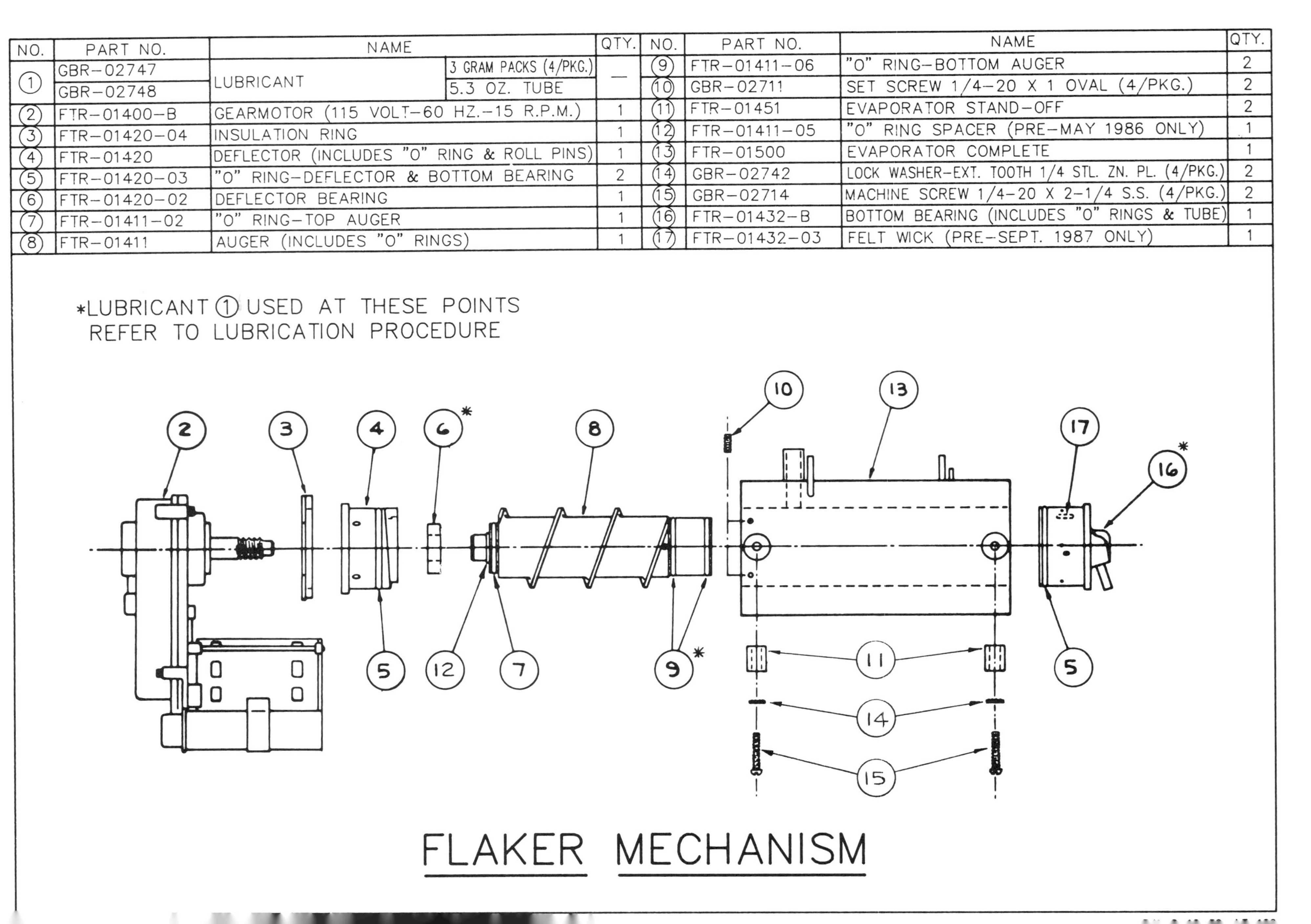

NO.	PART NO.	NAME		QTY.	NO.	PART NO.	NAME	QTY.
①	GBR-02747	LUBRICANT	3 GRAM PACKS (4/PKG.)	—	⑨	FTR-01411-06	"O" RING-BOTTOM AUGER	2
	GBR-02748		5.3 OZ. TUBE		⑩	GBR-02711	SET SCREW 1/4-20 X 1 OVAL (4/PKG.)	2
②	FTR-01400-B	GEARMOTOR (115 VOLT-60 HZ.-15 R.P.M.)		1	⑪	FTR-01451	EVAPORATOR STAND-OFF	2
③	FTR-01420-04	INSULATION RING		1	⑫	FTR-01411-05	"O" RING SPACER (PRE-MAY 1986 ONLY)	1
④	FTR-01420	DEFLECTOR (INCLUDES "O" RING & ROLL PINS)		1	⑬	FTR-01500	EVAPORATOR COMPLETE	1
⑤	FTR-01420-03	"O" RING-DEFLECTOR & BOTTOM BEARING		2	⑭	GBR-02742	LOCK WASHER-EXT. TOOTH 1/4 STL. ZN. PL. (4/PKG.)	2
⑥	FTR-01420-02	DEFLECTOR BEARING		1	⑮	GBR-02714	MACHINE SCREW 1/4-20 X 2-1/4 S.S. (4/PKG.)	2
⑦	FTR-01411-02	"O" RING-TOP AUGER		1	⑯	FTR-01432-B	BOTTOM BEARING (INCLUDES "O" RINGS & TUBE)	1
⑧	FTR-01411	AUGER (INCLUDES "O" RINGS)		1	⑰	FTR-01432-03	FELT WICK (PRE-SEPT. 1987 ONLY)	1

FLAKER MECHANISM

FLAKER MECHANISM ASSEMBLY INSTRUCTIONS:
CAUTION: CAREFULLY READ AND FOLLOW ALL LUBRICATION INSTRUCTIONS:

1. Thoroughly clean and dry all parts.
2. If there is any visible wear or damage to any of the "O" Rings, they must be replaced. Use lubricant when replacing "O" Rings. (See Lubrication Details in the following Instructions).
3. If top bearing was removed, lay a 1/8" bead of GBR-02748 around the top of Auger core, then slip the bearing onto the Auger shaft.
4. Lay a 1/8" bead of GBR-02748 on the top of the bearing, and drop the deflector over the bearing. Remove excess lubricant from the threaded shaft hole in the Auger which can cause a hydraulic lock when the Auger is threaded onto the gearhead shaft.
5. Screw Auger partially onto gearmotor shaft (left-hand thread). Center the deflector roll pins to the left and right of the stop boss on the gear case. Continue to tighten the Auger until snug. If the deflector can be slid up or down more than 1/32" remove the Auger and clean excess lubricant from shaft hole. When properly assembled the clearance between the top of the auger and the deflector must be between 0.030" and 0.040". If clearance is less than 0.030", remove the auger and install one shim (FTR-01411-12).

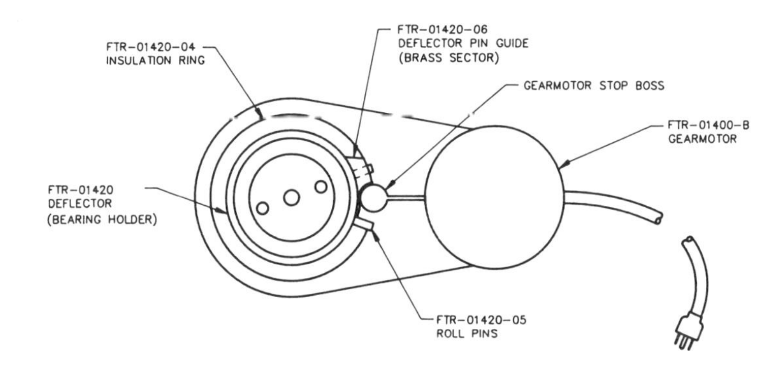

BOTTOM VIEW

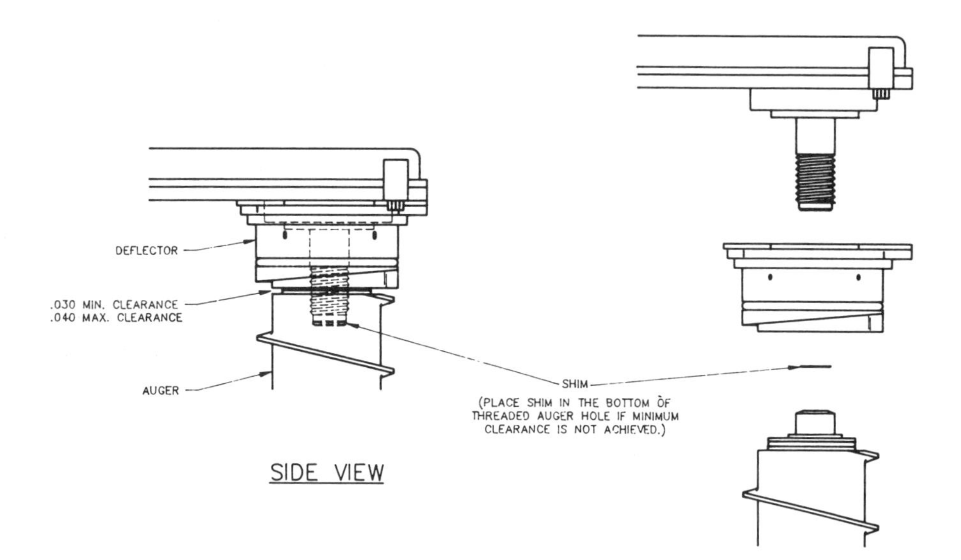

SIDE VIEW

6. Carefully clean evaporator barrel with a damp rag. If necessary, a solution of ice machine cleaner may be used to remove deposits from bore.
7. Clean bottom bearing, and examine inside surface for any evidence of metal to metal contact which indicates failure of the plastic liner. If the plastic bearing liner has a vertical slot with a lubricating wick **and the bottom Auger O-Rings are not Black colored material,** use FTR-01890 (mineral oil) to re-lubricate the bearing. If the bearing liner has a helical groove and no wick, re-lubricate the bearing with GBR-02747 or GBR-02748 (silicone compound). Fill the groove and wipe the I.D. so that only a **thin film** of lubricant remains on the bearing surface. **CAUTION:** An excessive amount of lubricant could work its way into the water supply hole in the Auger and restrict the water flow. The GBR-02747 pack contains a sufficient amount for lower bearing and O-Ring lubrication. Replace bottom bearing if any severe score marks are evident.

8. Lubricate the bottom bearing O-Ring and slide into the bottom of the evaporator barrel being careful to align the 3/8" hole with the bottom evaporator mounting screw. Tighten bottom screw. Note: Overtightening the bottom evaporator mounting screw can distort the bottom bearing causing binding. Also, some flakers are equipped with a shim between the bottom bearing and the evaporator to adjust end clearance. Do not remove the shim.
9. Reinstall the water inlet elbow if applicable **NO MORE THAN 3 TURNS** into the bottom bearing. Overtightening the elbow will restrict water flow to the auger.
10. Clean the auger bottom bearing surface. If the plastic bearing liner has a vertical slot with a lubricating wick **and the bottom Auger O-Rings are not Black colored material,** coat the auger bottom bearing surface generously with FTR-01890 oil. If the bearing liner has a helical groove and no wick, be sure the bearing is lubricated as specified in #7 above and **do not** coat the auger bottom bearing surface with lubricant. A small amount of GBR-02747 or GBR-02748 should be used to lubricate the O-Rings only for installation on the auger. Loosen the top mounting screw in evaporator mounting bracket one turn. Clean the outside surface of the deflector, lubricate the O-Ring, and reinstall the Gearmotor-Auger assembly into the evaporator.
11. Remove ice tube, and rotate gearmotor assembly until vertical line at the end of the deflector helix is in center of the ice discharge spout opening. Note: The ice tube may adhere so tightly to the spout that attempting to pull it off would cause excessive strain on the spout. If the ice tube is shoved over some of the black evaporator insulation coating, it is possible to loosen the end of the ice tube from the black coating with a small screwdriver. A better method is to cut the ice tube around the spout 1" from the end of the spout. The ice tube can then be removed by twisting, leaving the tightly adhering piece on the spout.
12. Tighten right set screw clockwise until snug, then counterclockwise one full turn. Tighten left set screw to lock the deflector in place. Do not use excessive force. Check the deflector opening (can be seen easily with a mirror and flashlight).
13. Tighten top evaporator mounting screw.
14. Reinstall water inlet hose and the ice tube.
15. Plug in Gearmotor.

GEARMOTOR & AUGER ASSEMBLY, REMOVAL & INSPECTION

1. Unplug gearmotor.
2. Loosen top evaporator mounting screw 2 or 3 times.
3. Loosen set screws approx. 5 turns.
4. Check for rotation of gearmotor approx. 45° each way to see if all screws are loose. If not, back out screws while checking for rotation.
5. While rotating gearmotor lift upwards and out of evaporator. If gearmotor will not lift out, place a 2 x 3 piece of wood under gearcase and across frame and pry upwards. Gearmotor may also be pried upwards between evaporator and gear-housing but extreme care must be taken that pry bar does not go between gearmotor and deflector. If this happens, then gearmotor will never come out.
6. Place gearmotor and auger assembly on bench. Place on a wooden or padded surface being careful of auger flights.
7. Remove auger using spanner wrench, 2 ¼ bolts and bar or large screwdriver or a ½" rod in auger side hole unscrewing clockwise.
8. Set auger aside and remove deflector, bearing, and insulation ring.
9. Check condition of auger for major nicks, broken flights, or heavy scoring.
10. Check condition of deflector for nicks, gouges, loss of plating. Check for roll pins.
11. Check condition of bearing.
12. Plug in motor and check for operation.
13. Remove 2 screws on motor end plate and remove plate and motor housing. Check condition of armature and bearings. Clean armature and replace bearings if necessary.
14. Remove Allen screws on gearcase and remove cover.
15. Check for grease packing gearcase, check condition of gears, and bearings on output shaft. Replace gears and/or bearings if needed.

16. Pack gearcase with grease and re-assemble.
17. Re-assemble motor.
18. Place insulation ring on gearmotor.
19. Place deflector on, align roll pins over boss on gearmotor.
20. Place bearing over shaft and in deflector.
21. Place a small bead of grease on bearing.
22. Replace "O" ring on top of auger and screw auger onto shaft being careful not to get any grease into threaded hole. Screw auger counterclockwise until it is tight.
23. Check gap between top of auger and deflector - it must be between .030" and .040". If it is not at least .030" then the auger will have to be removed and a shim added.
24. Tighten auger with spanner wrench, 2 ¼" bolts, and bar or ¼" rod. One sharp blow with hand is enough.
25. Replace "O" rings on auger and deflector and put a very light coat of mineral oil or grease on "O" rings (Wick bearing - use mineral oil only).
26. Check condition of bottom bearing, replace if badly scored.
27. To remove bottom bearing, tighten top evaporator mtg. screw and back bottom mtg. screw all the way out. Place a wooden hammer handle in the bottom bearing and strike the hammer with another to drive the bearing out the bottom of the evap.
28. If bottom bearing is to be re-used then replace "O" ring.
29. To replace bottom bearing, coat "O" ring lightly with mineral oil or grease and mark top edge of bearing above mounting hole. Work bearing into evaporator by rocking slightly and turning until bearing is fully seated. Be careful not to cock bearing. Align mark with mounting screws and tighten bottom mounting screw making sure it engages hole in bottom bearing (Wick bearing - use mineral oil only).
30. Replace gearmotor and auger assembly first making sure top mtg. screw and set screws are backed out. Drop assembly into evaporator turning gear head back and forth until assembly is fully seated. Align deflector opening wide open and tighten set screws - first the right one and then the left until they are just snug. Then, tighten them fully. Tighten top mounting screw. Note on the new deflectors a longer mtg. screw is used and it engages a hole in the deflector.
31. Re-check deflector opening and plug in gearmotor.

QUICK FLAKER OPERATION CHECK

1. Observe machine for any broken or missing parts and excessive dirt buildup. Repair and clean as necessary, especially check and clean air condenser.
2. Grasp gearmotor and check for free rotation 10°-12°. If gearmotor does not rotate or rotates too far, pull gearmotor.
3. Turn on machine. Check condition of water and water level in float tank.
4. Check sight glass for refrigerant and control box for Green light.
5. Grasp gearmotor and check for rotation 10°-12° counterclockwise. There should be a slight resistance and motor should rotate clockwise when released.
6. Observe suction line at compressor. There will usually be frost and the sight glass will show bubbles. Flaker may defrost once or twice.
7. After 10-15 minutes frost should leave compressor and sight glass should clear up and ice will start being produced.
8. Check rotation of gearmotor - it should rotate 10°-12° counterclockwise with moderate pressure and snap back when released. If not, gearmotor and auger will have to be pulled and most probably a shim added to auger.
9. Observe quality and quantity of ice.
10. Check position of bin cap tubes, ice discharge tubes, defrost and gear cap tubes.

FLAKER GEARMOTOR REPLACEMENT PARTS

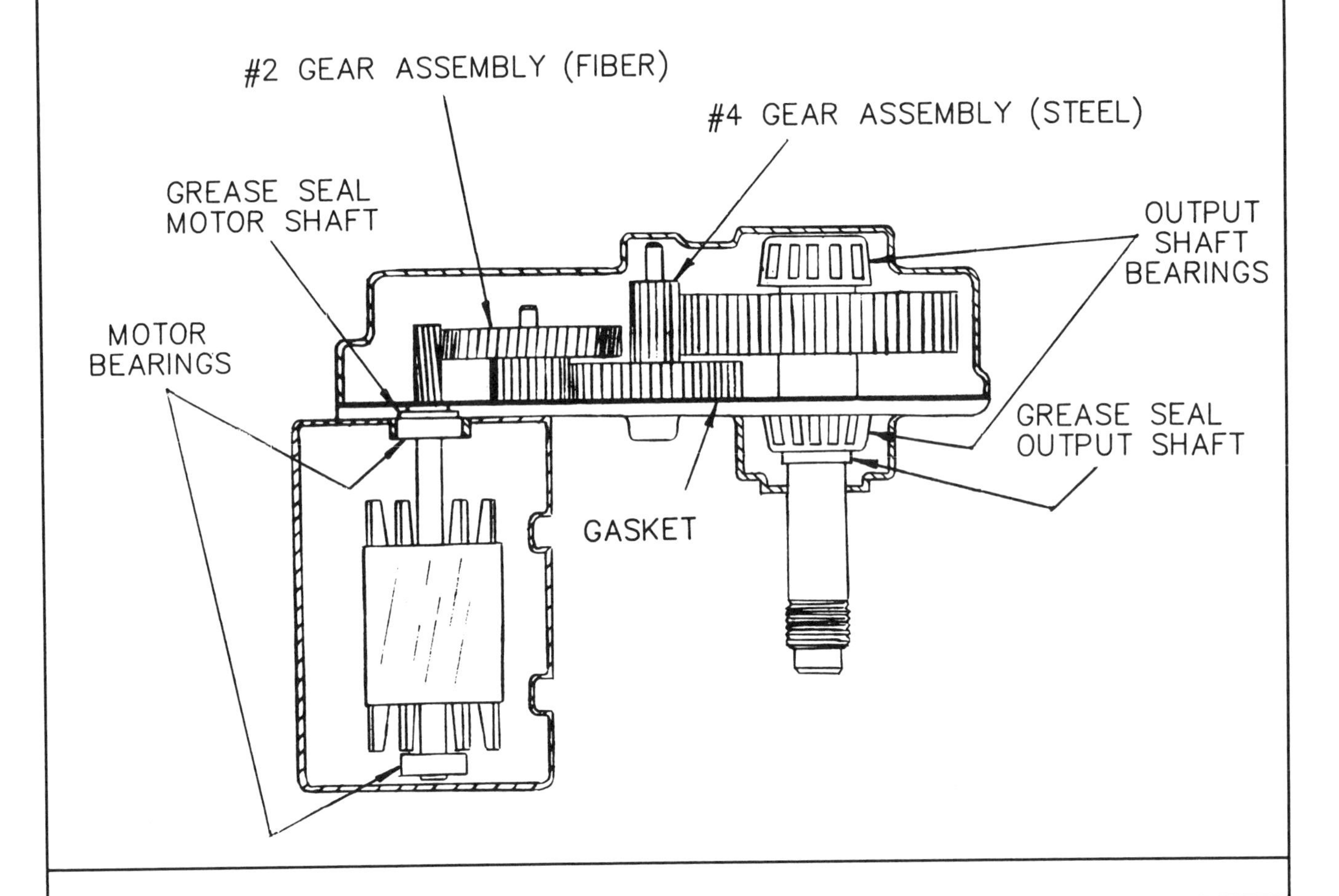

REPLACEMENT PARTS

- FTR-01400-11 MOTOR BEARING AND SEAL KIT
  INCLUDES:
  MOTOR BEARINGS (2)
  GREASE SEAL - MOTOR SHAFT
- FTR-01400-10 GEAR KIT (REQUIRED TO UPDATE 23 R.P.M. GEARMOTOR TO 15 R.P.M.)
  INCLUDES:
  #2 FIBER GEAR ASSEMBLY
  #4 STEEL GEAR ASSEMBLY
- MMR-00221-04-23 #2 FIBER GEAR ASSEMBLY (FOR 15 R.P.M. GEARMOTOR ONLY)
- FTR-01400-12 OUTPUT SHAFT BEARING AND SEAL KIT
  INCLUDES:
  OUTPUT SHAFT BEARING ASSEMBLY (2)
  GREASE SEAL - OUTPUT SHAFT
- FTR-01400-15 GASKET - GEARMOTOR COVER

DWG. BY TEK 8-31-88 CD-305

CLEANING INSTRUCTIONS FOR KOLD-DRAFT ICE FLAKERS

Use rubber gloves and eye protection, and an apron is recommended.

1. Mix 1/2 bag (2 ounces) ice machine cleaner (Kold-Draft Part No. 55R-01000) in 1/2 gallon of warm water. Mix thoroughly until crystals dissolve. Use the 55R-01002 washing bottle assembly for mixing and pouring solutions.
2. Remove Flaker front panel, and float tank cover. Wash or soak the float tank cover in a small container with ice machine cleaner solution while completing the steps below.
3. Remove ice discharge tube(s) from the holder(s) in the bin, and place in a separate container to prevent contamination of ice in bin with cleaner.
4. Use a screwdriver or wire to lift float to stop fresh water supply.
5. Pour cleaning solution into the float tank, **with the Flaker operating,** at whatever rate is necessary to keep the float tank full to the overflow level.
6. After the cleaner is collected in the container as slush ice, remove the float block and allow the evaporator(s) to fill with fresh water. Hold the float down to overfill the system and overflow the float tank.
7. Remove the water inlet hose(s) from the bottom of the evaporator(s) to drain residual cleaner from the system. Repeat filling and draining three (3) times to flush all traces of the cleaner from the system. Replace hose(s).
8. Prepare a sanitizing solution (100 to 200 ppm free Cl_2) as follows: For FT Models, mix one teaspoon (5 ml.) of 5¼% Sodium Hypochlorite (household strength laundry bleach) to 1 qt. water. For FB Models, mix two teaspoonfuls (10 ml.) bleach to two qts. water. Soak the float tank cover in a small amount of bleach solution while compeleting the next step.
9. Allow Flaker to make ice while feeding solution into float tank with fresh water shut off as above. Keep the float tank full to overflow level and continue catching ice in a separate container. When bleach solution is used up, remove the float block and operate the Flaker for about 10 minutes with fresh water while catching ice in the separate container.
10. Replace float tank cover, reinstall ice tube(s) into tube holder(s) with 1/2" extending beyond holder flange(s), and replace Flaker front panel.

NOTE: DO NOT USE AMMONIA SOLUTIONS IN CLEANING ANY PART OF THE ICE FLAKERS.

WARNING
BE SURE TO COMPLETELY RINSE ALL ICE MACHINE CLEANER FROM FLAKER BEFORE ADDING SANITIZING SOLUTION (BLEACH). MIXING OF ICE MACHINE CLEANER WITH BLEACH CAN CAUSE A CHEMICAL REACTION WHICH WILL RELEASE POISONOUS FUMES.

TROUBLE, CAUSE AND REMEDY

TROUBLE	CAUSE	REMEDY
1. Flaker will not operate. No lights on in control box.	1. Line fuse blown.	1. Check circuits for short or ground. Replace fuse.
	2. Loose connection in control box or in power supply line.	2. Check for power supply at controls in control box. Check connections to bin thermostat.
1a. Flaker will not operate. Yellow light on, but no ice on bin thermostat.	1a. Bin control set too warm in a cold room between 45° and 55°.	1a. Set bin thermostat colder (cw) but recheck with ice to be sure it will shut off.
	2a. Room below 45°.	2a. Add heat to the room.
	3a. Bin control has lost charge.	3a. Replace bin control.
2. Condensing fan and gearmotor operate (green light on) but not the compressor.	1. Inoperative capacitors or relay.	1. Replace capacitors or relay.
	2. Overload switch defective.	2. Replace overload switch.
	3. Loose connections or defective compressor.	3. Check for power at compressor C-R terminals, C-S terminals. With power off, remove "C" connection, check ohms between C and R, also C and S.
2a. Water-cooled flaker: gearmotor operates (green light on) but not the condensing fan or compressor.	1a. High pressure cut-out open; inadequate water supply.	1a. Check water supply and condenser water valve.
	2a. Water supply okay, high pressure cut-out won't close with condenser cool.	2a. Replace defective high pressure cut-out.
3. Compressor operating but fan off.	1. Circuit not complete.	1. Check circuit.
	2. Fan motor burned out.	2. Replace motor.
4. Condenser fan operating but compressor unit operating intermittently.	1. Dirty Condenser Coil.	1. Clean coil.
	2. High or low voltage.	2. Correct to proper voltage within 10% of nameplate.
	3. Excessive refrigerant.	3. Bleed off some refrigerant, check sight glass.

TROUBLE	CAUSE	REMEDY
5. Intermittent defrost, red light cycling on and off. Water level normal in float tank.	1. Water line elbow in bottom bearing in too far.	1. Back out elbow 1 turn.
	2. Defrost thermostat mis-adjusted, very cold supply water.	2. See service manual for defrost thermostat setting. Also see Abnormal Defrost Thermostat Setting in the manual
	3. TXV too far open.	3. Close TXV 1/4 turn see service manual for proper setting.
	4. Deflector partially closed.	4. See service manual for correct deflector setting.
	5. Gearmotor not running.	5. Check power to gearmotor receptacle in bottom of control box.
	6. Gearmotor stalled with power on, bottom evaporator mounting screw too tight.	6. Back off evaporator mounting screw 1 turn to see if gearmotor will operate after it cools down sufficiently for overload to cut back in. Also see #7.
	7. Gearmotor stalled.	7. Remove gearmotor and auger assembly and check for operation on workbench (take care not to damage auger flights). Remove bottom bearing, check for condition and check for snug but not a tight fit on bottom of auger. Check clearance between top of auger and bottom of deflector (see Flaker Mechanism Assem. Instr. #5).
6. Wet ice.	1. High water level.	1. Lower water level.
	2. Undercharge, bubbles going thru sight glass.	2. Check for leaks, add R-12.
	3. Mis-adjusted TXV.	3. Adjust TXV (see "Expansion Valve Adjustment").
7. Ice too hard.	1. Low water level.	1. Raise water level.
	2. Deflector incorrectly adjusted.	2. SEE DEFLECTOR ADJUSTMENTS PROCEDURE.
	3. TXV closed much too far.	3. See service manual for proper suction pressure and suction line temperature.
	4. Moisture in system and TXV partially frozen shut.	4. Dehydrate and recharge system.
8. No ice with gearmotor, compressor and condenser fan operating. (Red light not on no power to defrost valve).	1. Very low refrigerant.	1. Repair leak, see service manual for proper charge.
	2. Stuck defrost valve, defrost line warm, suction pressure above 20 pounds.	2. Repair or replace defrost valve.

TROUBLE	CAUSE	REMEDY
9. Flaker does not turn off.	1. Mis-adjusted bin thermostat.	1. Adjust bin thermostat (ccw), check with ice on coil.
	2. Bin thermostat will not open when set warmest with ice on thermostat cap tube.	2. Replace bin thermostat.
	3. Mis-located bin thermostat cap tube.	3. Check location of 1/4" stainless cap tube holder parallel and below ice path coming from ice tube. Check thermostat cap tube located in the 1/4" stainless tube.
10. Flaker cycles off and on.	1. Ice falling on bin thermostat capillary tube.	1. Relocate Tube.
	2. Ice tube not on outlet spout.	2. Remount ice tube and check for restriction.
11. Low Production.	1. High head pressure.	1a. Clean condenser. Improve water supply to water-cooled condenser. Improve ventilation.
		1b. Replace head pressure control valve on remote condenser models.
	2. Inadequate water supply.	2. Check and Clean filters. Raise water level.
	3. TXV mis-adjusted.	3. Adjust TXV.
	4. High ambient.	4. Decrease ambient to 90°F. Max.
	5. Deflector closed.	5. SEE DEFLECTOR ADJUSTMENTS PROCEDURE.
	6. Low head pressure.	6a. Adjust water regulating valve.
		6b. Add R-12 to remote condenser models.
		6c. Refer to Remote Condenser section of Service Manual
12. Flaker spouts coming off.	1. I.D. of discharge tubing to small.	1. Change to braided nylon tubing.
	2. Roll pins breaking on deflector.	2. Readjust defrost thermostat.
	3. Hose kinking.	3. Change to braided nylon tubing reroute presently used tubing.

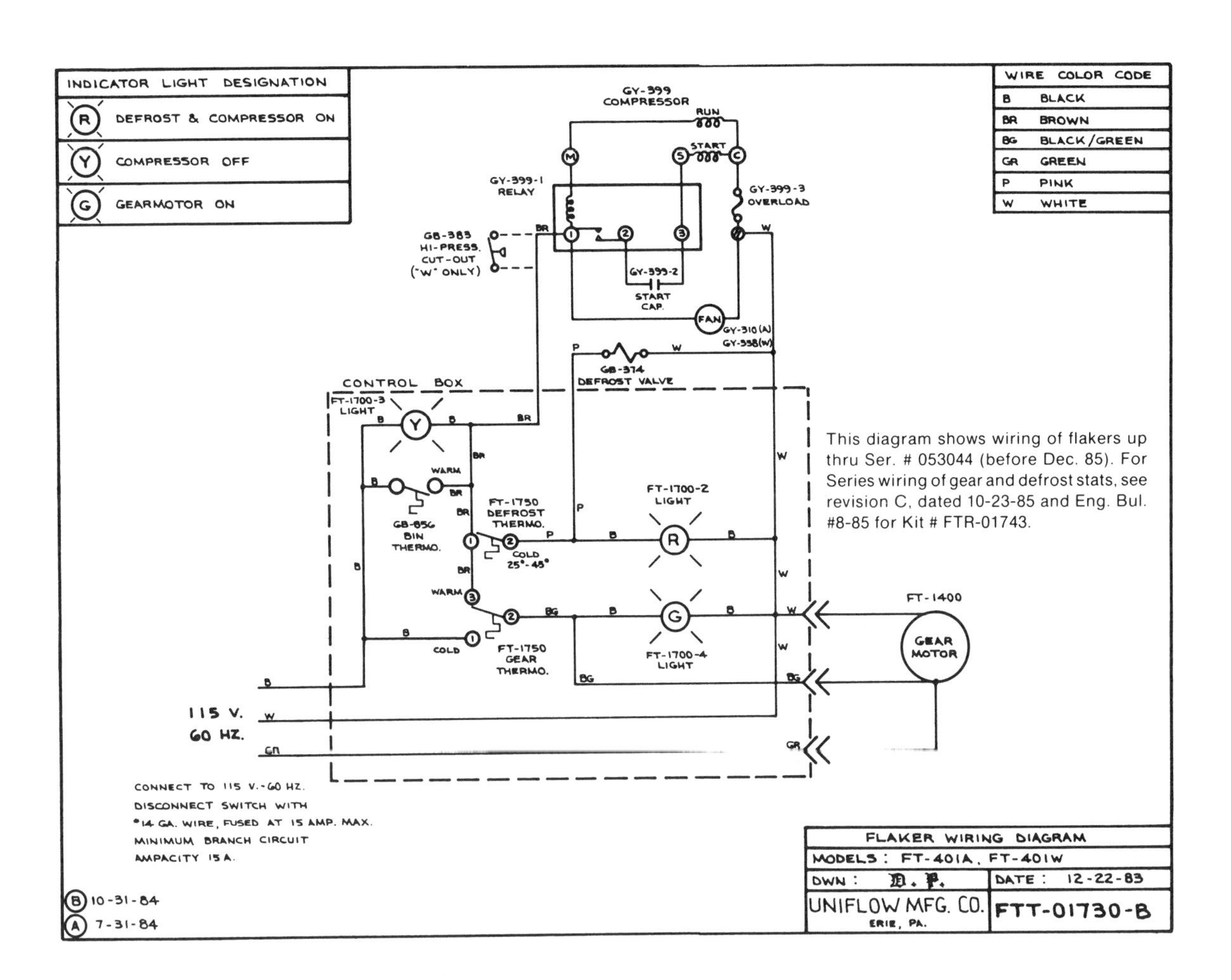

INDICATOR LIGHT DESIGNATION
R DEFROST & COMPRESSOR ON
Y COMPRESSOR OFF
G GEARMOTOR ON
WIRE COLOR CODE
B BLACK
BR BROWN
BG BLACK/GREEN
GR GREEN
P PINK
W WHITE
GY-399 COMPRESSOR
RUN
START
GY-399-1 RELAY
GY-399-3 OVERLOAD
GB-383 HI-PRESS. CUT-OUT ("W" ONLY)
GY-399-2 START CAP.
FAN
GY-310(A) GY-338(W)
GB-374 DEFROST VALVE
CONTROL BOX
FT-1700-3 LIGHT
WARM
GB-856 BIN THERMO.
FT-1750 DEFROST THERMO.
COLD 25°-45°
FT-1700-2 LIGHT
FT-1750 GEAR THERMO.
FT-1700-4 LIGHT
FT-1400
GEAR MOTOR
115 V. 60 HZ.
This diagram shows wiring of flakers up thru Ser. # 053044 (before Dec. 85). For Series wiring of gear and defrost stats, see revision C, dated 10-23-85 and Eng. Bul. #8-85 for Kit # FTR-01743.
CONNECT TO 115 V.-60 HZ. DISCONNECT SWITCH WITH #14 GA. WIRE, FUSED AT 15 AMP. MAX. MINIMUM BRANCH CIRCUIT AMPACITY 15 A.
B 10-31-84
A 7-31-84
FLAKER WIRING DIAGRAM
MODELS: FT-401A, FT-401W
DWN: D. P.
DATE: 12-22-83
UNIFLOW MFG. CO. ERIE, PA.
FTT-01730-B

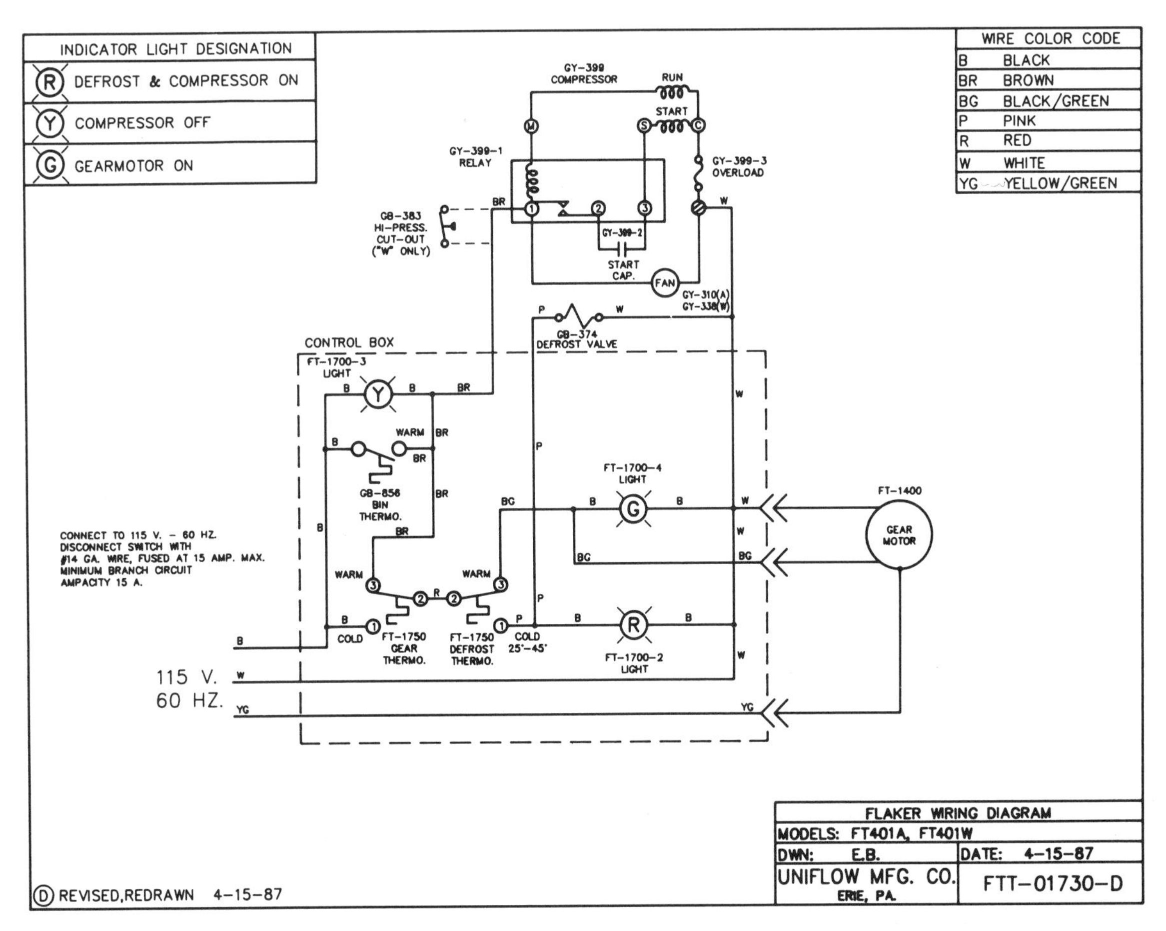

INDICATOR LIGHT DESIGNATION
R DEFROST & COMPRESSOR ON
Y COMPRESSOR OFF
G GEARMOTOR ON
WIRE COLOR CODE
B BLACK
BR BROWN
BG BLACK/GREEN
P PINK
R RED
W WHITE
YG YELLOW/GREEN
GY-399 COMPRESSOR
RUN
START
GY-399-1 RELAY
GY-399-3 OVERLOAD
GB-383 HI-PRESS. CUT-OUT ("W" ONLY)
GY-399-2 START CAP.
FAN
GY-310(A) GY-338(W)
GB-374 DEFROST VALVE
CONTROL BOX
FT-1700-3 LIGHT
WARM
GB-856 BIN THERMO.
FT-1700-4 LIGHT
FT-1400
GEAR MOTOR
CONNECT TO 115 V. – 60 HZ. DISCONNECT SWITCH WITH #14 GA. WIRE, FUSED AT 15 AMP. MAX. MINIMUM BRANCH CIRCUIT AMPACITY 15 A.
COLD
FT-1750 GEAR THERMO.
FT-1750 DEFROST THERMO.
25°-45°
FT-1700-2 LIGHT
115 V. 60 HZ.
FLAKER WIRING DIAGRAM
MODELS: FT401A, FT401W
DWN: E.B.
DATE: 4-15-87
UNIFLOW MFG. CO. ERIE, PA.
FTT-01730-D
D REVISED, REDRAWN 4-15-87

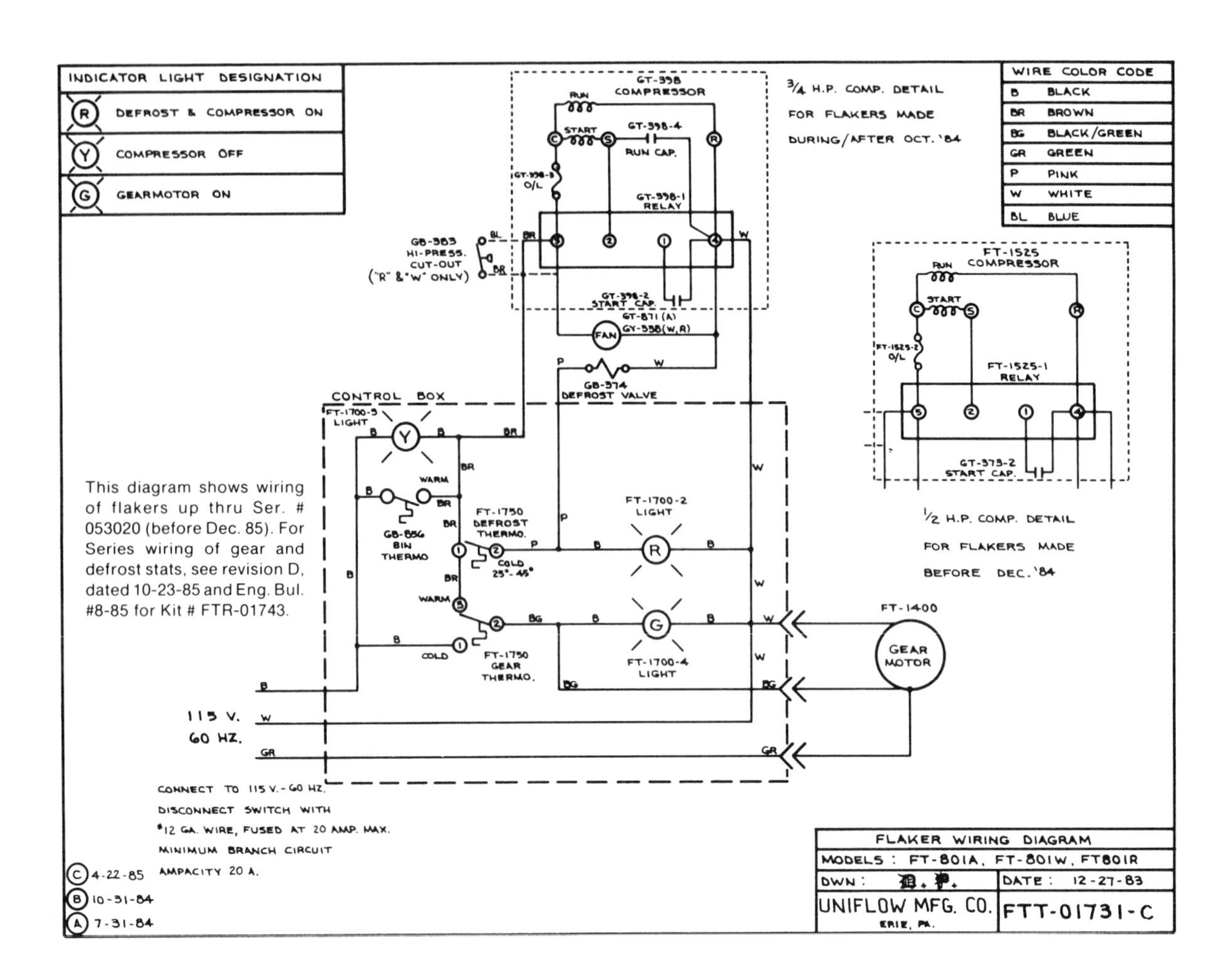

INDICATOR LIGHT DESIGNATION
R DEFROST & COMPRESSOR ON
Y COMPRESSOR OFF
G GEARMOTOR ON
GT-398 COMPRESSOR
RUN
START
GT-398-4
RUN CAP.
GT-398-3 O/L
GT-398-1 RELAY
GB-383 HI-PRESS. CUT-OUT ("R" & "W" ONLY)
GT-398-2 START CAP.
GT-871 (A) GY-338 (W, R)
FAN
GB-374 DEFROST VALVE
3/4 H.P. COMP. DETAIL FOR FLAKERS MADE DURING/AFTER OCT. '84
WIRE COLOR CODE
B BLACK
BR BROWN
BG BLACK/GREEN
GR GREEN
P PINK
W WHITE
BL BLUE
FT-1525 COMPRESSOR
FT-1525-2 O/L
FT-1525-1 RELAY
GT-373-2 START CAP.
1/2 H.P. COMP. DETAIL FOR FLAKERS MADE BEFORE DEC. '84
CONTROL BOX
FT-1700-3 LIGHT
WARM
GB-856 BIN THERMO
FT-1750 DEFROST THERMO.
COLD 25°-45°
FT-1700-2 LIGHT
FT-1750 GEAR THERMO.
FT-1700-4 LIGHT
FT-1400
GEAR MOTOR
This diagram shows wiring of flakers up thru Ser. # 053020 (before Dec. 85). For Series wiring of gear and defrost stats, see revision D, dated 10-23-85 and Eng. Bul. #8-85 for Kit # FTR-01743.
115 V. 60 HZ.
CONNECT TO 115 V.-60 HZ. DISCONNECT SWITCH WITH #12 GA. WIRE, FUSED AT 20 AMP. MAX. MINIMUM BRANCH CIRCUIT AMPACITY 20 A.
C 4-22-85
B 10-31-84
A 7-31-84
FLAKER WIRING DIAGRAM
MODELS: FT-801A, FT-801W, FT801R
DWN: B.P.
DATE: 12-27-83
UNIFLOW MFG. CO. ERIE, PA.
FTT-01731-C

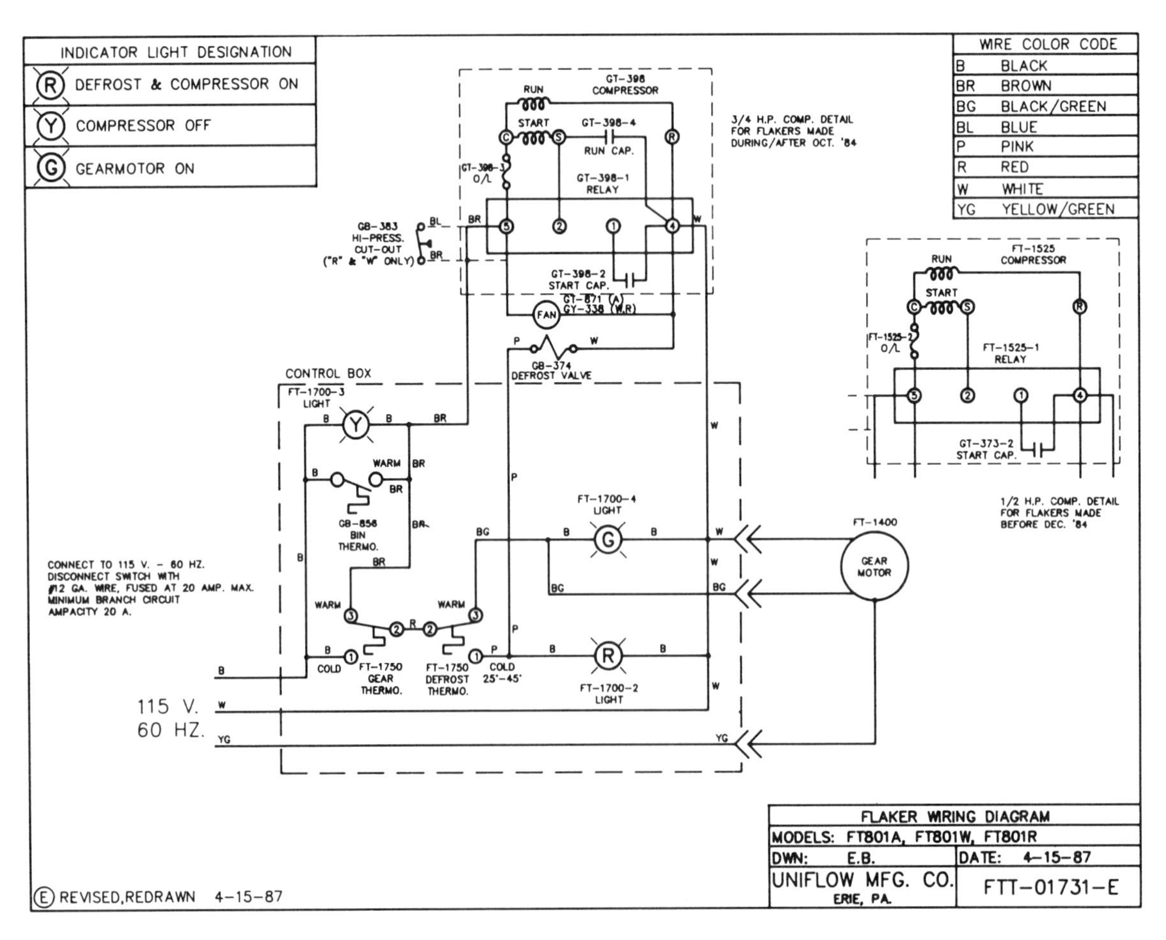

INDICATOR LIGHT DESIGNATION
R DEFROST & COMPRESSOR ON
Y COMPRESSOR OFF
G GEARMOTOR ON
GT-398 COMPRESSOR
RUN
START
GT-398-4
RUN CAP.
GT-398-3 O/L
GT-398-1 RELAY
GB-383 HI-PRESS. CUT-OUT ("R" & "W" ONLY)
GT-398-2 START CAP.
GT-871 (A) GY-338 (W,R)
FAN
GB-374 DEFROST VALVE
3/4 H.P. COMP. DETAIL FOR FLAKERS MADE DURING/AFTER OCT. '84
WIRE COLOR CODE
B BLACK
BR BROWN
BG BLACK/GREEN
BL BLUE
P PINK
R RED
W WHITE
YG YELLOW/GREEN
FT-1525 COMPRESSOR
FT-1525-2 O/L
FT-1525-1 RELAY
GT-373-2 START CAP.
1/2 H.P. COMP. DETAIL FOR FLAKERS MADE BEFORE DEC. '84
CONTROL BOX
FT-1700-3 LIGHT
WARM
GB-856 BIN THERMO.
FT-1700-4 LIGHT
FT-1400
GEAR MOTOR
CONNECT TO 115 V. - 60 HZ. DISCONNECT SWITCH WITH #12 GA. WIRE, FUSED AT 20 AMP. MAX. MINIMUM BRANCH CIRCUIT AMPACITY 20 A.
COLD
FT-1750 GEAR THERMO.
FT-1750 DEFROST THERMO.
COLD 25°-45°
FT-1700-2 LIGHT
115 V. 60 HZ.
FLAKER WIRING DIAGRAM
MODELS: FT801A, FT801W, FT801R
DWN: E.B.
DATE: 4-15-87
UNIFLOW MFG. CO. ERIE, PA.
FTT-01731-E
E REVISED, REDRAWN 4-15-87

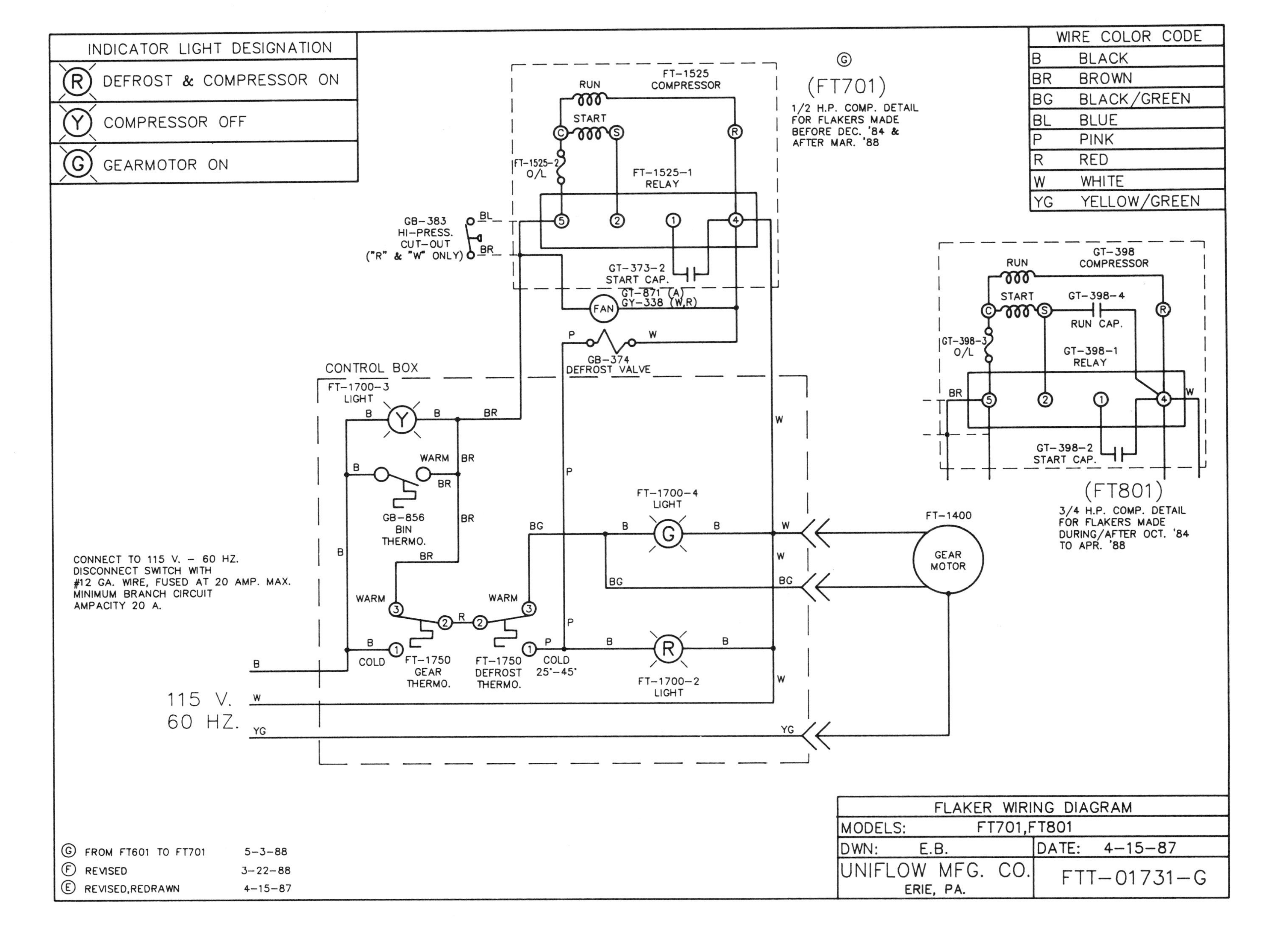

INDICATOR LIGHT DESIGNATION
R DEFROST & COMPRESSOR ON
Y COMPRESSOR OFF
G GEARMOTOR ON
WIRE COLOR CODE
B BLACK
BR BROWN
BG BLACK/GREEN
BL BLUE
P PINK
R RED
W WHITE
YG YELLOW/GREEN
FT-1525 COMPRESSOR
RUN
START
FT-1525-2 O/L
FT-1525-1 RELAY
GB-383 HI-PRESS. CUT-OUT ("R" & "W" ONLY)
GT-373-2 START CAP.
GT-871 (A) GY-338 (W,R)
FAN
GB-374 DEFROST VALVE
(FT701)
1/2 H.P. COMP. DETAIL FOR FLAKERS MADE BEFORE DEC. '84 & AFTER MAR. '88
GT-398 COMPRESSOR
GT-398-4 RUN CAP.
GT-398-3 O/L
GT-398-1 RELAY
GT-398-2 START CAP.
(FT801)
3/4 H.P. COMP. DETAIL FOR FLAKERS MADE DURING/AFTER OCT. '84 TO APR. '88
CONTROL BOX
FT-1700-3 LIGHT
WARM
GB-856 BIN THERMO.
FT-1700-4 LIGHT
FT-1400
GEAR MOTOR
WARM
COLD
FT-1750 GEAR THERMO.
FT-1750 DEFROST THERMO.
COLD 25°-45°
FT-1700-2 LIGHT
CONNECT TO 115 V. - 60 HZ. DISCONNECT SWITCH WITH #12 GA. WIRE, FUSED AT 20 AMP. MAX. MINIMUM BRANCH CIRCUIT AMPACITY 20 A.
115 V. 60 HZ.
G FROM FT601 TO FT701 5-3-88
F REVISED 3-22-88
E REVISED,REDRAWN 4-15-87
FLAKER WIRING DIAGRAM
MODELS: FT701,FT801
DWN: E.B.
DATE: 4-15-87
UNIFLOW MFG. CO. ERIE, PA.
FTT-01731-G

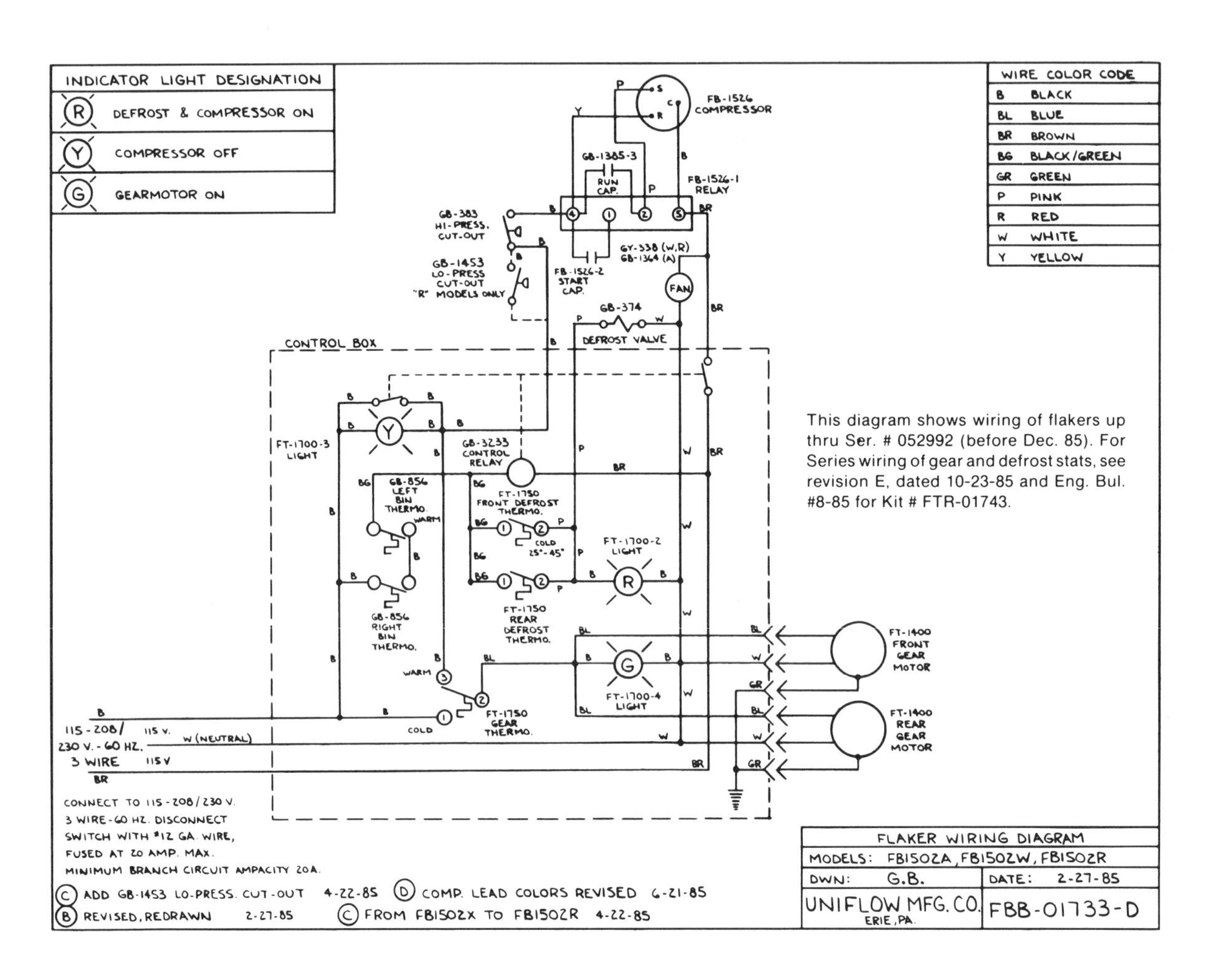

INDICATOR LIGHT DESIGNATION
R DEFROST & COMPRESSOR ON
Y COMPRESSOR OFF
G GEARMOTOR ON
WIRE COLOR CODE
B BLACK
BL BLUE
BR BROWN
BG BLACK/GREEN
GR GREEN
P PINK
R RED
W WHITE
Y YELLOW
FB-1526 COMPRESSOR
GB-1385-3 RUN CAP.
FB-1526-1 RELAY
GB-383 HI-PRESS. CUT-OUT
GB-1453 LO-PRESS CUT-OUT "R" MODELS ONLY
FB-1526-2 START CAP.
GY-338 (W,R) GB-1364 (A)
FAN
GB-374 DEFROST VALVE
CONTROL BOX
FT-1700-3 LIGHT
GB-3233 CONTROL RELAY
GB-856 LEFT BIN THERMO
WARM
FT-1750 FRONT DEFROST THERMO.
COLD 25°-45°
FT-1700-2 LIGHT
GB-856 RIGHT BIN THERMO.
FT-1750 REAR DEFROST THERMO.
FT-1700-4 LIGHT
FT-1750 GEAR THERMO.
COLD
FT-1400 FRONT GEAR MOTOR
FT-1400 REAR GEAR MOTOR
115-208/230 V.-60 HZ. 3 WIRE
115 V.
W (NEUTRAL)
115 V
This diagram shows wiring of flakers up thru Ser. # 052992 (before Dec. 85). For Series wiring of gear and defrost stats, see revision E, dated 10-23-85 and Eng. Bul. #8-85 for Kit # FTR-01743.
CONNECT TO 115-208/230 V. 3 WIRE-60 HZ. DISCONNECT SWITCH WITH #12 GA. WIRE, FUSED AT 20 AMP. MAX. MINIMUM BRANCH CIRCUIT AMPACITY 20A.
C ADD GB-1453 LO-PRESS. CUT-OUT 4-22-85
D COMP. LEAD COLORS REVISED 6-21-85
B REVISED, REDRAWN 2-27-85
C FROM FB1502X TO FB1502R 4-22-85
FLAKER WIRING DIAGRAM
MODELS: FB1502A, FB1502W, FB1502R
DWN: G.B.
DATE: 2-27-85
UNIFLOW MFG. CO. ERIE, PA.
FBB-01733-D

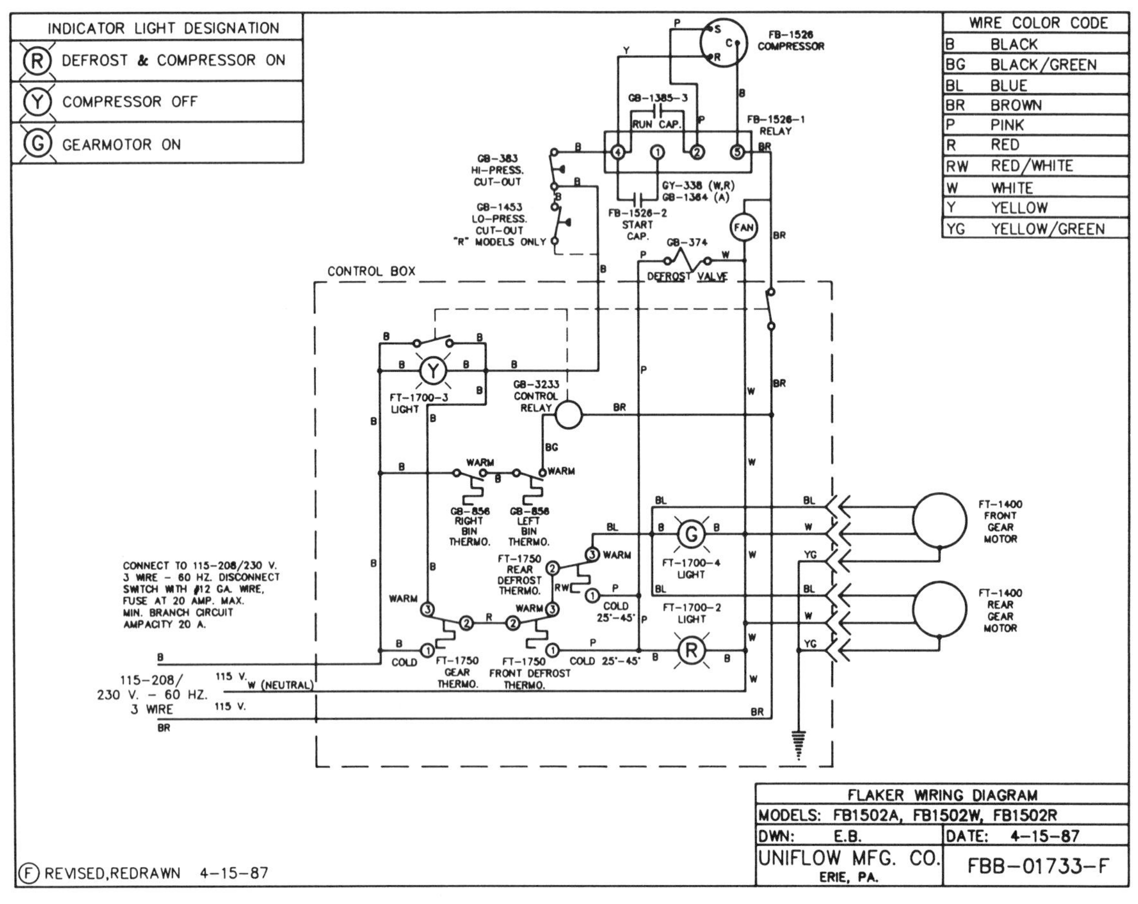

INDICATOR LIGHT DESIGNATION
R DEFROST & COMPRESSOR ON
Y COMPRESSOR OFF
G GEARMOTOR ON
WIRE COLOR CODE
B BLACK
BG BLACK/GREEN
BL BLUE
BR BROWN
P PINK
R RED
RW RED/WHITE
W WHITE
Y YELLOW
YG YELLOW/GREEN
FB-1526 COMPRESSOR
GB-1385-3 RUN CAP.
FB-1526-1 RELAY
GB-383 HI-PRESS. CUT-OUT
GB-1453 LO-PRESS. CUT-OUT "R" MODELS ONLY
GY-338 (W,R) GB-1364 (A)
FB-1526-2 START CAP.
FAN
GB-374 DEFROST VALVE
CONTROL BOX
FT-1700-3 LIGHT
GB-3233 CONTROL RELAY
WARM
GB-856 RIGHT BIN THERMO.
GB-856 LEFT BIN THERMO.
FT-1750 REAR DEFROST THERMO.
COLD 25°-45°
FT-1700-4 LIGHT
FT-1700-2 LIGHT
FT-1750 GEAR THERMO.
FT-1750 FRONT DEFROST THERMO.
COLD
FT-1400 FRONT GEAR MOTOR
FT-1400 REAR GEAR MOTOR
CONNECT TO 115-208/230 V. 3 WIRE - 60 HZ. DISCONNECT SWITCH WITH #12 GA. WIRE. FUSE AT 20 AMP. MAX. MIN. BRANCH CIRCUIT AMPACITY 20 A.
115-208/230 V. - 60 HZ. 3 WIRE
115 V.
W (NEUTRAL)
115 V.
FLAKER WIRING DIAGRAM
MODELS: FB1502A, FB1502W, FB1502R
DWN: E.B.
DATE: 4-15-87
UNIFLOW MFG. CO. ERIE, PA.
FBB-01733-F
F REVISED, REDRAWN 4-15-87

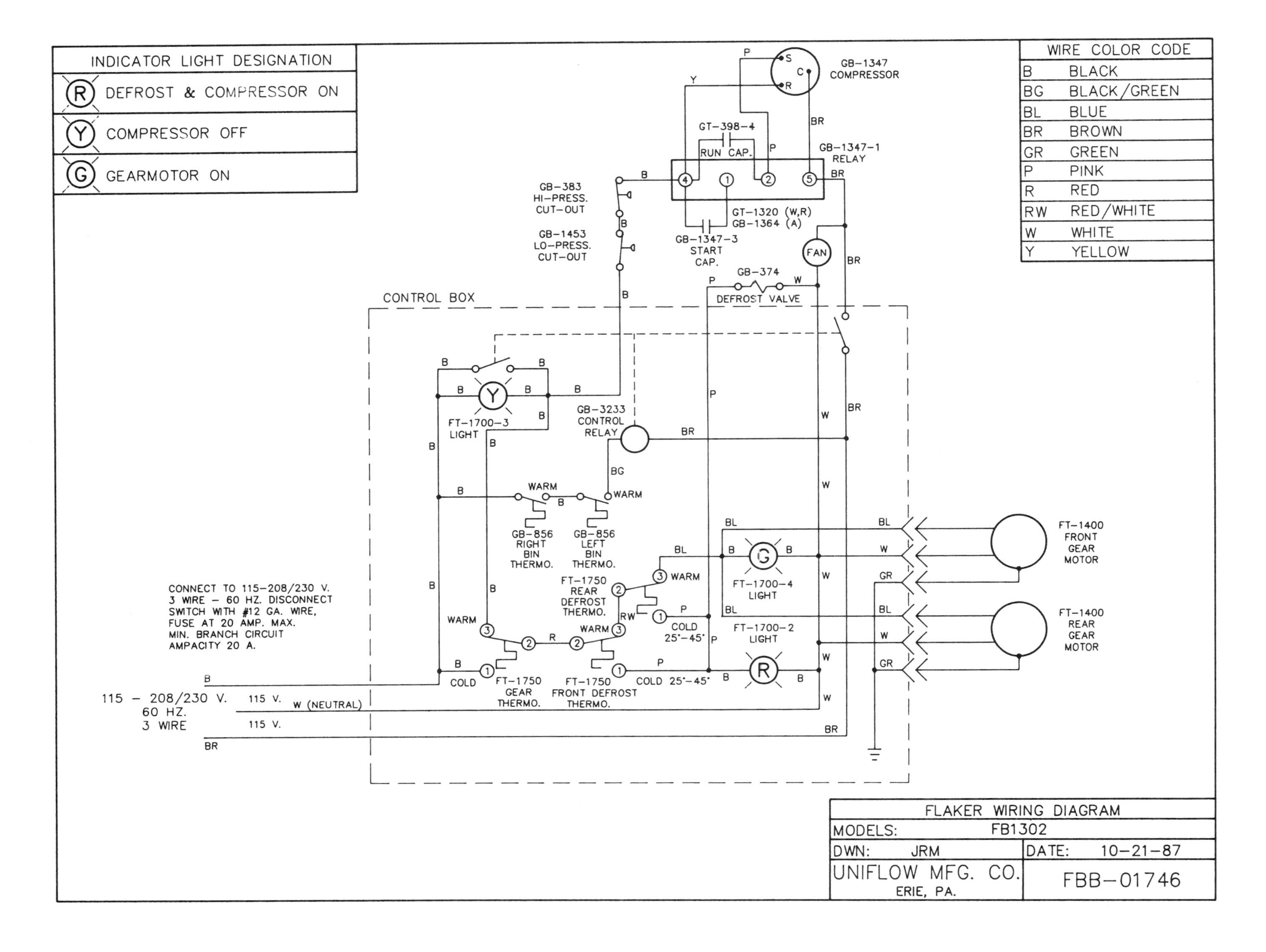
INDICATOR LIGHT DESIGNATION
R DEFROST & COMPRESSOR ON
Y COMPRESSOR OFF
G GEARMOTOR ON
WIRE COLOR CODE
B BLACK
BG BLACK/GREEN
BL BLUE
BR BROWN
GR GREEN
P PINK
R RED
RW RED/WHITE
W WHITE
Y YELLOW
GB-1347 COMPRESSOR
GT-398-4 RUN CAP.
GB-1347-1 RELAY
GT-1320 (W,R)
GB-1364 (A)
GB-1347-3 START CAP.
GB-383 HI-PRESS. CUT-OUT
GB-1453 LO-PRESS. CUT-OUT
FAN
GB-374 DEFROST VALVE
CONTROL BOX
FT-1700-3 LIGHT
GB-3233 CONTROL RELAY
GB-856 RIGHT BIN THERMO.
GB-856 LEFT BIN THERMO.
FT-1750 REAR DEFROST THERMO.
FT-1700-4 LIGHT
FT-1700-2 LIGHT
FT-1750 GEAR THERMO.
FT-1750 FRONT DEFROST THERMO.
COLD 25°-45°
WARM
COLD
FT-1400 FRONT GEAR MOTOR
FT-1400 REAR GEAR MOTOR
CONNECT TO 115-208/230 V. 3 WIRE - 60 HZ. DISCONNECT SWITCH WITH #12 GA. WIRE, FUSE AT 20 AMP. MAX. MIN. BRANCH CIRCUIT AMPACITY 20 A.
115 - 208/230 V. 60 HZ. 3 WIRE
115 V.
W (NEUTRAL)
115 V.
FLAKER WIRING DIAGRAM
MODELS: FB1302
DWN: JRM
DATE: 10-21-87
UNIFLOW MFG. CO. ERIE, PA.
FBB-01746

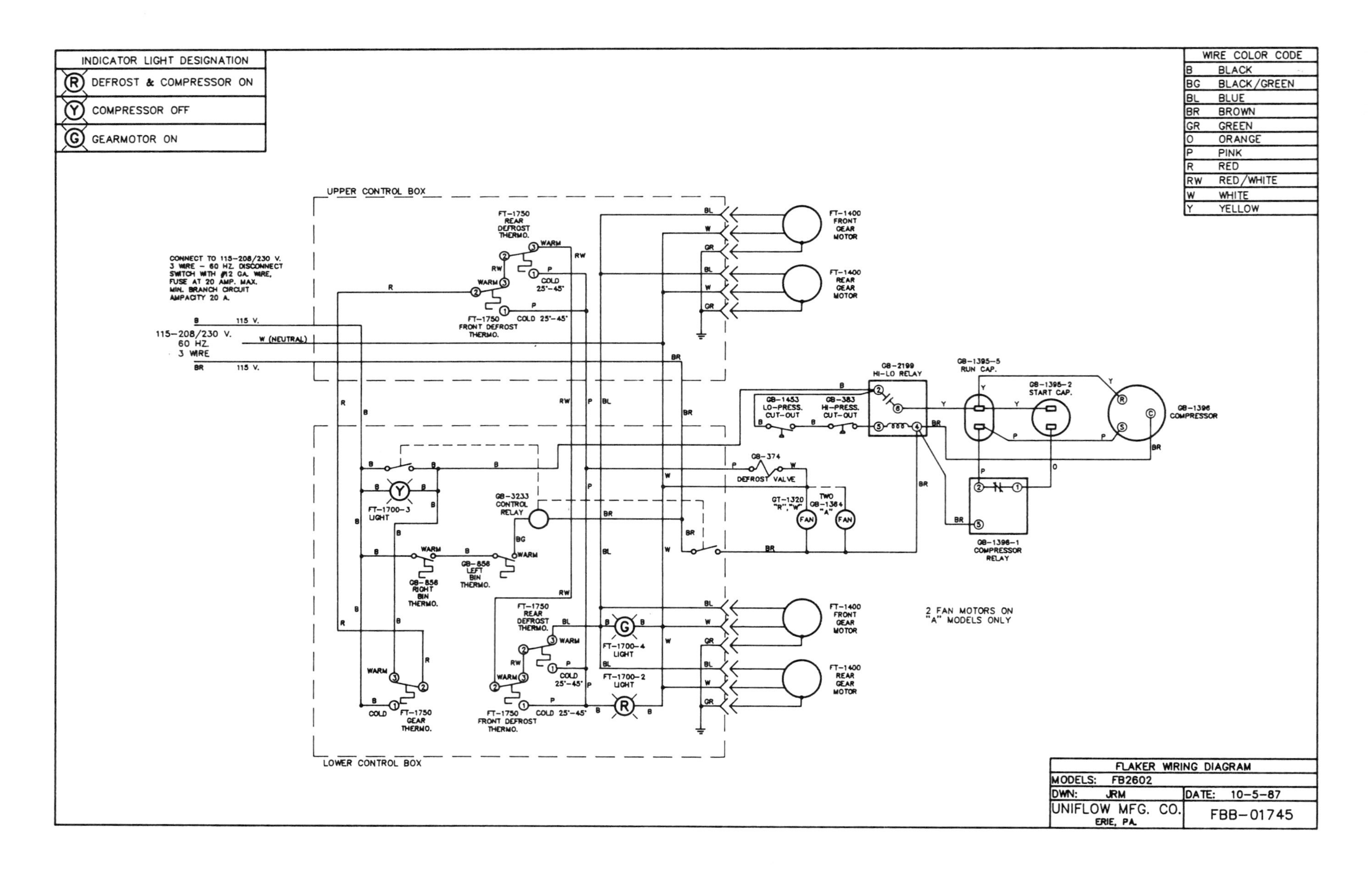
INDICATOR LIGHT DESIGNATION
R DEFROST & COMPRESSOR ON
Y COMPRESSOR OFF
G GEARMOTOR ON
WIRE COLOR CODE
B BLACK
BG BLACK/GREEN
BL BLUE
BR BROWN
GR GREEN
O ORANGE
P PINK
R RED
RW RED/WHITE
W WHITE
Y YELLOW
UPPER CONTROL BOX
LOWER CONTROL BOX
CONNECT TO 115-208/230 V. 3 WIRE - 60 HZ. DISCONNECT SWITCH WITH #12 GA. WIRE, FUSE AT 20 AMP. MAX. MIN. BRANCH CIRCUIT AMPACITY 20 A.
115-208/230 V. 60 HZ. 3 WIRE
115 V.
W (NEUTRAL)
FT-1750 REAR DEFROST THERMO.
FT-1750 FRONT DEFROST THERMO.
COLD 25°-45°
WARM
FT-1400 FRONT GEAR MOTOR
FT-1400 REAR GEAR MOTOR
GB-2199 HI-LO RELAY
GB-1453 LO-PRESS. CUT-OUT
GB-383 HI-PRESS. CUT-OUT
GB-1395-5 RUN CAP.
GB-1395-2 START CAP.
GB-1396 COMPRESSOR
GB-1396-1 COMPRESSOR RELAY
GB-374 DEFROST VALVE
GT-1320 "R", "W"
TWO GB-1364 "A"
FAN
GB-3233 CONTROL RELAY
FT-1700-3 LIGHT
GB-856 RIGHT BIN THERMO.
GB-856 LEFT BIN THERMO.
FT-1700-4 LIGHT
FT-1700-2 LIGHT
FT-1750 GEAR THERMO.
2 FAN MOTORS ON "A" MODELS ONLY
FLAKER WIRING DIAGRAM
MODELS: FB2602
DWN: JRM
DATE: 10-5-87
UNIFLOW MFG. CO. ERIE, PA.
FBB-01745

Manitowoc

Service Technician's Handbook

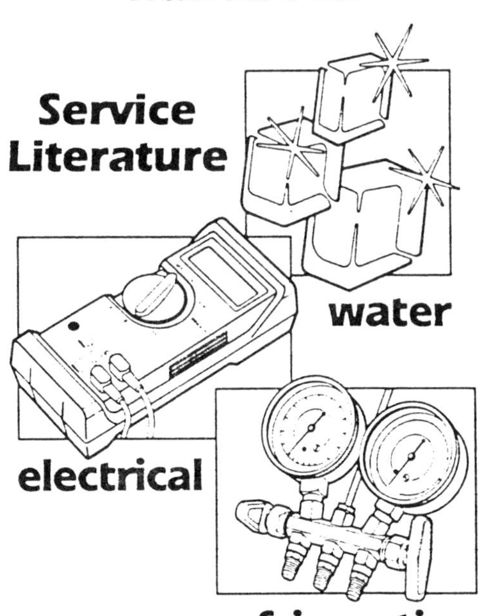

MANITOWOC EQUIPMENT WORKS
Division of The Manitowoc Company, Inc.
2110 South 26th Street, P.O. Box 1720
Manitowoc, WI 54221-1720 U.S.A.
Phone: 414-682-0161

P/N 83-5164-3
© Manitowoc 1988

TABLE OF CONTENTS
50/60 Cycle Units

(Continued on next page)

Table of Contents (continued)

ICE MACHINE CLEANING/SANITIZING PROCEDURES

NOTE: Manitowoc Ice Machine Cleaner and Manitowoc Ice Machine Sanitizer are the **only** solution recommended for use with Manitowoc products. The chemicals used are compatible with Manitowoc cleaning and sanitizing procedures and with the materials used in the construction of Manitowoc products.

ICE MACHINE CLEANING

If the frequency of cleaning exceeds once every 6 months, the incoming water must be treated to reduce cleaning to two times a year. A properly operating flush system minimizes cleaning frequency. (See Dump System Review.) The following chart lists the proper amount of cleaner to be used for Series cuber cleaning:

Amount of cleaner	Series
2 oz.	100, 200, 400, 600
3 oz.	800, 1100, 1200
4 oz.	1700

CLEANING INSTRUCTIONS

Clean machine when evaporator is free of ice.

1. Place toggle switch in "off" position.
2. Shut off water supply.
3. Remove all ice from bin.
4. Remove water trough, water curtain(s), water distribution tube(s), bridge thickness control probe, and water pump.
5. Mix 3 oz. ice machine cleaner per gallon of warm water in plastic container and place components in solution. Soak the components until they are free of deposits. (A stronger solution of up to 16 oz. per gallon of water may be used for excessively dirty components.)
6. Use above solution to clean storage bin and other areas where deposits have collected, then rinse cleaned areas with fresh water.
7. Use cleaning solution to remove deposits from top, bottom and side extrusions and other components.
8. Replace cleaned components, turn on water, and place toggle switch in "water pump" position.
9. Add proper amount of cleaning solution (see chart) to water trough and allow solution to circulate a maximum of 10 minutes. (Do not run excessive cleaner over evaporator for prolonged periods.) Insure weep holes are open.
10. Drain water trough by removing drain plug and allowing fresh water to flush water trough for a minimum of 30 seconds. After flushing, reinstall drain plug.
11. Sanitize ice machine. (Follow "Sanitizing Instructions" below.)
12. Thoroughly rinse bin with clean water after all components are cleaned.
13. Place toggle switch in "ice" position.
14. Discard first batch of ice.

SANITIZING INSTRUCTIONS

Sanitize after cleaning or when evaporator is free of ice

1. Place toggle switch in "water pump" position.
2. Add 1 oz. Manitowoc Ice Machine Sanitizer to water trough; allow to circulate a minimum of 1 minute.
3. Drain water trough by removing drain plug and allow fresh water to flush trough for a minimum of 30 seconds. Reinstall drain plug after flushing is complete.
4. Wash all ice contact surfaces, bins, and other surfaces requiring sanitizing with a solution of 1 oz. sanitizer and up to 4 gallons water. Rinse thoroughly with fresh water after sanitizing is complete.
5. Place toggle switch in "ice" position.
6. Discard first batch of ice.

MANITOWOC ICE MACHINE CLEANER — 94-0546-3

MANITOWOC ICE MACHINE SANITIZER — 94-0565-3
For slime and algae after cleaning.

WATER LEVEL ADJUSTMENTS

1. Adjust water levels when pump is off.
2. Inlet water pressure should be between 40-60 psig.
3. The water level should be even with the top of the offset in the water trough. A tolerance of 6⅛" above or below this point is acceptable.
4. The float arm may be adjusted if proper water level cannot be obtained.

WATER LEVEL ADJUSTMENTS
G0800 SERIES

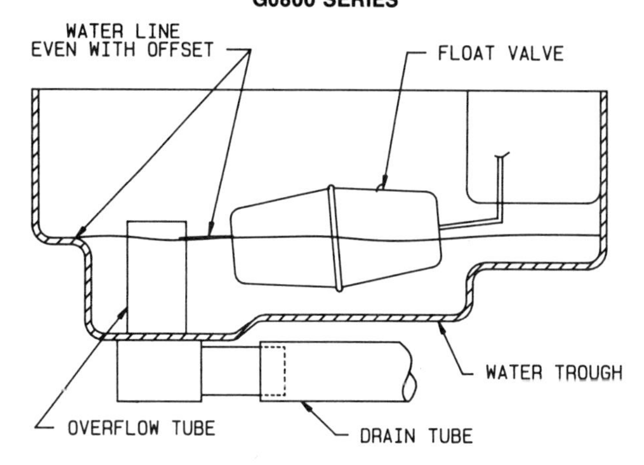

G1200 AND G1700 SERIES

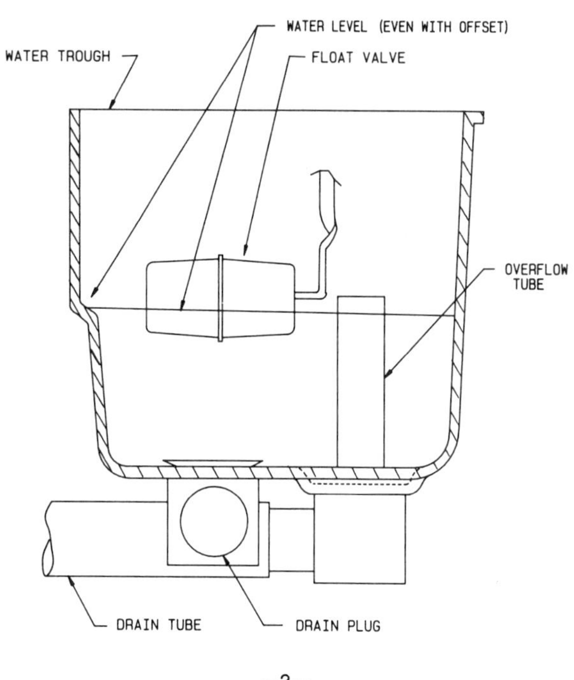

WATER LEVEL ADJUSTMENTS
E0200, H0200, E0400, E0600, AND G0600 SERIES

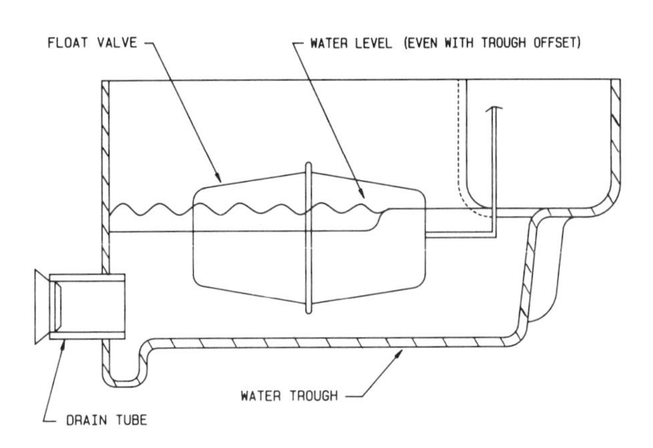

WATER FLOW CLAMP ADJUSTMENTS

The water flow clamps between the pump and distributor are to be set to maintain the following dimensions. For ice machines manufactured before use of the Manitowoc pump using Beckett or Hartel pumps, use the regular cube adjustments. For those machines with Manitowoc style water pumps, use:

	Dice & Half Dice	Regular
E & H-200	not used	5/16 - 3/8
E0400	not used	5/16 - 3/8
G0600	not used	5/16 - 3/8
G0800	not used	5/16 - 3/8[1]
G01200	not used	5/16 - 3/8[1]
G01700[2]	5/16 ± 1/32[1]	5/16 ± 1/32[1]

[1]Because of variations in wall thickness of water pump discharge tube, the water flow clamp setting on these machines may be varied in order to obtain adequate ice sensor probe operation.

[2]The G1700 uses only Beckett or Hartell style pumps. The Manitowoc style pump is not to be used on G1700 and E1100 series cubers.

BRIDGE THICKNESS ADJUSTMENT

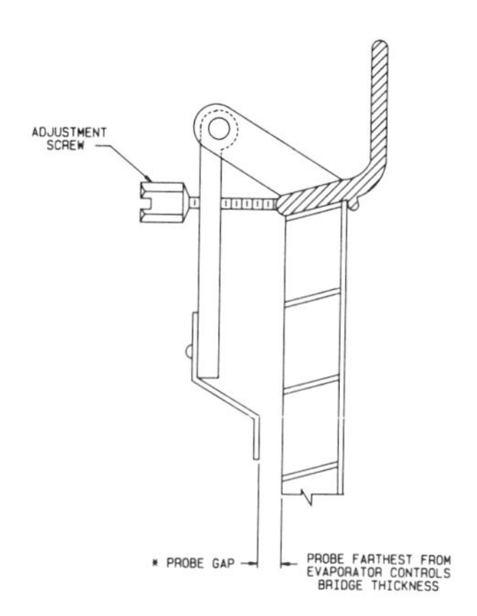

Ice bridge thickness should be approximately ⅛" thick. The probe gap will normally be approximately 1/16" greater than the actual ice bridge thickness. Turn adjustment screw clockwise to increase bridge thickness. (1/3 turn on the screw will change bridge thickness 1/16").

Check wires and alignment of sensor. It must move freely and return to the correct "gap" position following each harvest operation.

BIN SWITCH ADJUSTMENT
Single Evaporator Cubers

1. Initially set switch activating pin "X" dimension at 1/4".

2. Pull water curtain away from evaporator. While slowly returning curtain to its free position, the switch must activate (hear a click) when bottom edge of curtian has passed the sump and up to 3/16" inside.

NOTE One turn on switch activating pin corresponds to 1/4" movement at bottom of water curtain.

3. Tighten locking nut on switch activating pin.

4. Check water curtain for free movement and final position.

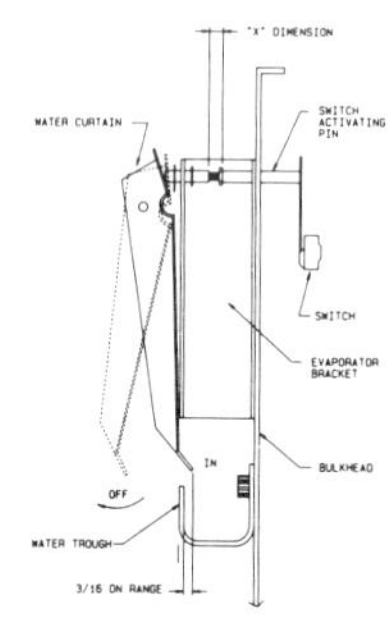

G1200 and G1700 SERIES

1. Check water curtain position. Pull curtain away from evaporator. While slowly returning curtain to its free position, the switch must activate (hear a click) when bottom edge of curtain has passed the sump and up to 3/16" inside.

2a. Early Production
If adjustment must be made, bend switch arm until correct curtain position is set.

2b. Late Production
If adjustment must be made, loosen the appropriate locking screw and move the lever until the correct curtain position is set. Retighten locking screw.

3. Check curtain for free movement and final position.

4. Adjust rear curtain in identical manner.

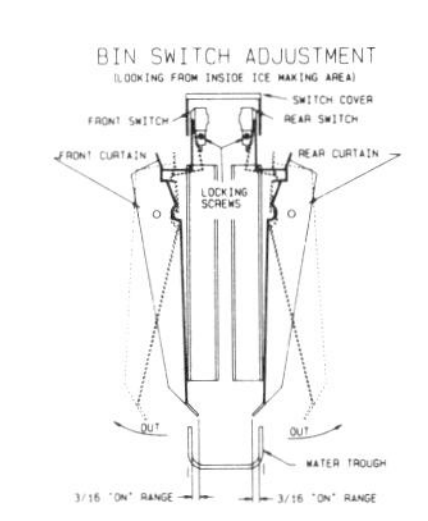

F1100 BIN SWITCH CHECKING PROCEDURE

The ice maker should only operate when bottoms of water curtains are inside water trough ensuring all recirculating water will be retained. Check by operating ice machine in the freezing mode. Pull curtains away from evaporators stopping the ice machine. Slowly return curtains to their normal positions. The ice machine should start only after curtains have passed over water trough.

ADJUSTMENT PROCEDURE

1. Disconnect all electrical supplies to ice machine (on water-cooled units, the high pressure reset control must be closed).

2. Connect battery operated test light (or ohmmeter) to terminal #12 on board with #3 relay and to terminal #35 of small terminal board.

3. Loosen thumbscrew on rear switch bracket, move bracket inward until light comes on. Note graduation mark at thumbscrew. Move bracket inward 1½ to 2 graduations and tighten thumbscrew. (Use threaded hole which best accommodates bracket position.)

4. Adjust front switch in identical manner, except move test light lead from terminal #12 to terminal #3 of sensor transformer board.

ICE PRODUCTION CHECK

Capacity Formulas*

1. Freeze time + harvest time = total cycle time
2. 1440 ÷ total cycle time = cycles/day.
3. Weight of ice slab × cycles/day = pounds of ice/24 hours.

*All times are in minutes.

Capacity Formula Example

Model HY0204A with 70° Air and 50° Water Temperatures

1. 15 min. + 1.7 min. = 16.7 min. total cycle time.
2. 1440 ÷ 16.7 = 86 cycles/day.
3. 2.5 lbs. × 86 = 215 lbs./24 hours.

Compare your known facts to the Ice Production Charts.

NOTE: These results are based on the presumption the ice machine is allowed to operate 24 hours. If in doubt of run time, use a voltage recorder to verify the ice machine is allowed to function 24 hours.

NORMAL SLAB WEIGHTS

E or H0200 — 2 lb.-6 oz./3 lb.
E0400 — 4 lb.-2 oz./4 lb.-12 oz.
E0600 — 6 lb./7 lb.
E1100 — 13 lb./14 lb. (both slabs)
G0600 — 4 lb.-2 oz./4 lb.-12 oz.
G0800 — 6 lb.-4 oz./7 lb.-4 oz.
G1200 — 8 lb.-2 oz./9 lb.-4 oz. (both slabs)
G1700 — 12 lb.-4 oz./14 lb.-8 oz. (both slabs)

CONTROL SPECIFICATIONS

	G-Series Controls	E-Series Controls
1. High Pressure Cutout (all units)	440 psi-MR	300 psi-MR (remotes)
2. Low Pressure Pump Down Contol (remotes)	15 psi CO 40 psi CI	2 psi CO 27 psi CI
3. Harvest Pressure Limiting Control (remotes)	110 psi CO 85 psi CI	85 psi CO 40 psi
4. Head Master Control Valve (remotes)	185 psi ±10 psi	125 psi ±5 psi
5. Fan Cycling Control (air-cooled)	175 psi CO 225 psi CI	100 psi CO 140 psi CI
6. Harvest Pressure Regulator (remotes)	76 psi - 84 psi* (G1200) 71 psi - 79 psi* (G1700)	(E1100 uses a restrictor tube in place of valve)

*Non-adjustable

WATER REGULATING VALVE SETTINGS

E0200 — 125 ±5
H0200 — 125 ±5
E0400 — 125 ±5
E0600 — 125 ±5
E1100 — 125 ±5
G0600 — 220 ±5
G0800 — 240 ±5
G1200 — 230 ±10
G1700 — 220 ±10

COMPRESSOR OHM VALUES AND LOCKED-ROTOR AMPS

SINGLE PHASE COMPRESSORS

Series Cuber	Compressor		OHMS C-S	OHMS C-R	OHMS R-S	Locked Rotor-amps
A100	Copeland	JFC1-0025-IAA	5.0-6.3	1.7-2.2	6.7-8.5	25.5
H200	Tecumseh	AK9434A	4.1-5.2	0.6-0.8	4.7-6.0	48.
E400	Copeland	RRS4-0075-IAA	2.1-2.6	0.4-0.5	2.5-3.1	66.3
G600	Tecumseh	AB5519G	4 - 8	1 - 2	6 - 9	53
G600	Bristol	M52B123BBCB	4.4-5.6	1.1-1.4	5.5-7.0	54
G800	Tecumseh	AB5524G	3-6	1	4-7	64
G800	Bristol	M52B143BBCB	3.4-4.3	0.8-1.1	4.2-5.4	65
G1200	Bristol	M53A223BBCA	1.6-2.1	0.5-0.7	2.1-2.8	97
G1700	Bristol	M50A273BBCA	3.0-3.8	0.5-0.7	3.5-4.5	96

3 PHASE COMPRESSORS

Series Cuber	Compressor		OHMS L1-L2; L2-L3; L1-L3	Locked Rotor-amps
G800	Tecumseh	AB5524H	1.8	50
G800	Bristol	M52B143DBDB	1.8 - 2.3	48
G1200	Bristol	M53A223DBLA	1.1 - 1.4	70
G1700	Bristol	M50A273DBLA	1.1 - 1.4	70

REFRIGERANT CHARGE

Important: Refer to machine serial tag to verify system charge.

Series			
A-100 Series			
Air-Cooled		10 oz.	R-12
Water-Cooled		8 oz.	R-12
A-200 Series			
Air-Cooled		21 oz.	R-12
Water-Cooled		10 oz.	R-12
A-400 Series			
Air-Cooled		28 oz.	R-12
Water-Cooled		15 oz.	R-12
A-600 Series			
Air-Cooled		42 oz.	R-12
Water-Cooled		41 oz.	R-12
A-1100 Series			
Air-Cooled		65 oz.	R-12
Water-Cooled		45 oz.	R-12
Remote Air		16 lb.	R-12
A-2200 Series			
Air-Cooled		65 oz.	R-12
Water-Cooled		45 oz.	R-12
Remote Air		16 lb.	R-12
C-200 Series			
Air-Cooled		22 oz.	R-12
Water-Cooled		9 oz.	R-12
C-400 Series			
Air-Cooled		22 oz.	R-12
Water-Cooled		15 oz.	R-12
C-600 Series			
Air-Cooled		42 oz.	R-12
Water-Cooled		30 oz.	R-12
C-1100 Series			
Air-Cooled		65 oz.	R-12
Water-Cooled	55 oz.[2]	45 oz.[3]	R-12
Remote Air		16 lb.	R-12
E-200 Series			
Air-Cooled	24 oz.[4]	22 oz.[5]	R-12
Water-Cooled	16 oz.[4]	9 oz.[5]	R-12
E-400 Series			
Air-Cooled		22 oz.	R-12
Water-Cooled		15 oz.	R-12
Remote Air[1]		16 lb.	R-12
E-600 Series			
Air-Cooled		32 oz.	R-12
Water-Cooled		32 oz.	R-12
Remote Air		16 lb.	R-12
E-1100 Series			
Air-Cooled	55 oz.[2]	48 oz.[3]	R-12
Water-Cooled	50 oz.[2]	38 oz.[3]	R-12
Remote Air[1]		16 lb.	R-12

G-600 Series			
Air-Cooled	34 oz.[2]	34 oz.[3]	R-502
Water-Cooled	30 oz.[2]	28 oz.[3]	R-502
Remote Air[1]		16 lb.	R-502
G-800 Series			
Air-Cooled	46 oz.[2]	42 oz.[3]	R-502
Water-Cooled	38 oz.[2]	38 oz.[3]	R-502
Remote Air[1]		16 lb.	R-502
G-1200 Series			
Air-Cooled		56 oz.	R-502
Water-Cooled		38 oz.	R-502
Remote Air[1]		18 lb.	R-502
G-1700 Series			
Water-Cooled	62 oz.[2]	50 oz.[3]	R-502
Remote Air[1]		22 lb.	R-502
H-200 Series			
Air-Cooled		24 oz.	R-12
Water-Cooled		18 oz.	R-12

1 Remote air charge is total charge including lines and Manitowoc Condenser.
2 With accumulator.
3 Without accumulator.
4 1/2HP Compressor
5 1/3HP Compressor

MANITOWOC O.E.M. FILTER DRIERS

The filter driers used on Manitowoc ice machines are special O.E.M. filter driers manufactured to our specifications.

We found that although there are many good driers on the market which handle different problems common in various refrigeration systems, ice machines have unique filter drier needs that over-the-counter filter driers did not address.

The biggest difference between our filter driers and off-the-shelf type driers is in filtration. Manitowoc's have dirt retaining filtration with fiberglass filters on both the inlet and outlet ends. This is very important because ice machines have a back-flushing action which takes place every time the unit goes through a harvest cycle. The Manitowoc filter driers have very high moisture removal capability and good acid removal capacity.

In the event of a compressor burnout, we recommend the use of a good quality suction line filter to address the higher acid condition. This would be installed in conjunction with the Manitowoc drier. We do not recommend molded core driers for ice machines because they lack dirt retaining filtration.

The size of the filter drier is important. On small machines, 400 Series and smaller, the refrigerant charge is critical and use of an oversized drier will cause the unit to be undercharged. On larger machines, standardizing on one size larger drier will not create problems.

Listed below are the recommended field replacement driers:

Machine Size	Drier Size	End Connection Size	MEW Part No.
100	"032"	¼"	12-3005-1
200	"032"	¼"	12-3005-1
400	"032"	¼"	12-3005-1
600	"082"	¼"	89-3010-3
800	"082"	¼"	89-3010-3
1200	"082"	¼"	89-3010-3
1700	"163"	⅜"	89-3016-3

Driers are covered as a warranty part and are to be replaced any time the system is opened for repairs.

REMOTE ICE MAKERS REFRIGERANT LINE SETS (Rev. 9/86)

PRECHARGED LINE SETS: Manitowoc recommends the use of precharged refrigeration line sets for general installation practices. This type of line set occasionally comes under criticism due to the mechanical joint made by the quick connect fitting compared to a welded joint. The concern of a higher percentage of leaks with quick connects is unwarranted when the fitting is properly lubricated and tightened according to manufacturer's specifications. The benefits of such fittings are prevention of air, moisture and other contaminants being introduced into the refrigeration system. Precharged line sets also eliminate the cost of recharging the system.

LENGTHENING OR REDUCING LINE SETS: Occasionally, due to certain job sites, it may be necessary to increase or decrease the length of the precharged line sets. Do this prior to connecting lines to the ice machine and condenser. The quick connect fittings on the line sets are equipped with Schrader valves...use these to discharge the line set. Lengthen or shorten the lines while applying good refrigeration practices and insulate new tubing. Do not change the tube sizes. Evacuate the lines and place a vapor charge of approximately 5 oz. in each line. Leak check all joints and then proceed with connection to the ice maker and condenser.

FIELD CONSTRUCTED LINE SETS: Refrigerant lines constructed in the field are made with refrigeration grade copper tubing only — hard or soft tubing may be used. Tubing sizes of ½" discharge and ⅜" liquid are used on the

400, 600, 800, 1100 and 2200 series units; the 1200 and 1700 series use ⅝" discharge and ⅜" liquid lines. The tubing must be insulated. The use of quick connect fittings is recommended and stub kits are available through the Manitowoc Parts Department. The line set, when completed, is then evacuated and a vapor holding charge of approximately 5 oz. of the proper refrigerant is added to each line. Leak check all joints and then proceed with connections to the ice maker and condenser.

REMOVAL OF QUICK COUPLERS: If it is necessary to remove the quick couplers, remove all refrigerant from the ice maker, including that in the receiver, before attempting to remove the fittings from the ice machine and condenser. After completion of the line installations, leak check all joints and then evacuate and recharge the complete system per Manitowoc Service procedures for that particular ice maker.

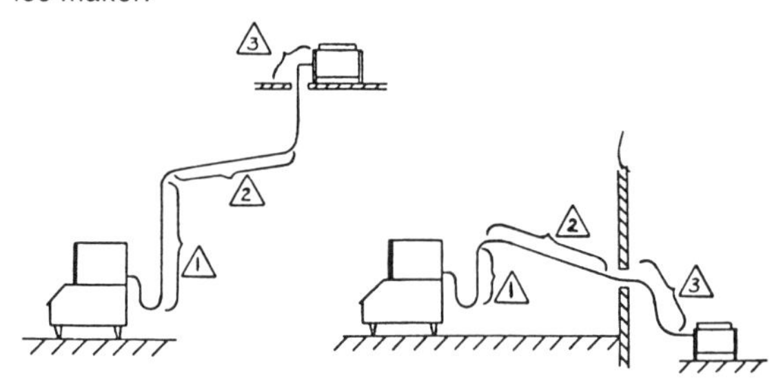

ROUTING OF LINE SETS: The following practices should be followed when routing refrigerant lines to insure the proper performance and service accessibility to the ice machine.

1. Make service loop in line sets as shown, permitting easy access to ice machine for cleaning and service. Hard draw copper should not be used at this location. If ice maker is moved frequently, use flexible line sections. Flexible line sections are available through Manitowoc's Parts Department.
2. This section of lines should always slope upward toward the condenser or downward when condenser is below ice maker. Never form a trap in the discharge line, refrigerant should always be free to drain toward the ice maker or condenser. The trap formed by the service loop is part of the ice maker's design.
3. Refrigerant lines located outdoors should be kept as short as possible. Also form tubes to prevent discharge line traps.

Select the proper length precharged line set. Excessive tubing must be cut out of the system. Do not coil up excess tubing!

CAUTION: *Never form traps!*

REMOTE CONDENSERS
MAXIMUM LOCATION DISTANCE (2/85)

Remote condenser installations consist of vertical and horizontal distances to the condenser that, when combined, must fit within approved guidelines. The guidelines, drawings and calculation method below will help the installer verify a proper remote condenser installation.

MAXIMUM LINE LENGTH: This should never exceed **100 calculated feet**. The pressure drop due to refrigerant flowing through the lines and condenser will exceed the design limits of the head pressure control.

MAXIMUM HEIGHT: This should never exceed 35 feet. The vapor pressure in the condenser and the weight of liquid refrigerant which accumulates in the discharge line during the off cycle will exceed the vapor pressure generated by the Thermoblock or Vaporblock device installed in the compressor discharge line.

MAXIMUM DROP: The condenser should never be positioned more than 15 feet below the ice machine. The liquid refrigerant pushed from the condenser up to the ice machine will contribute to an overall pressure drop.

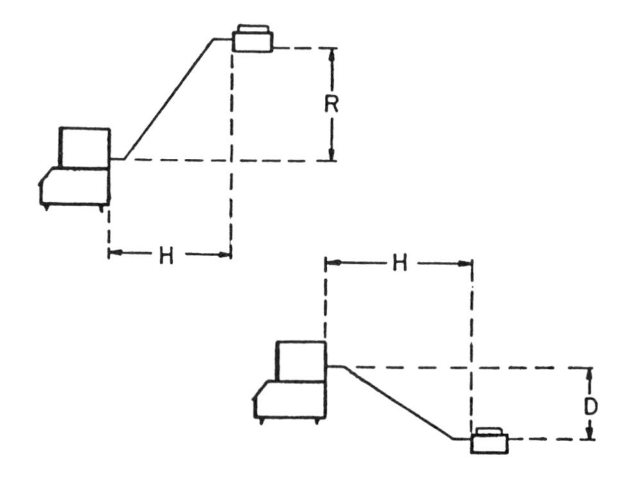

R (rise), H (horizontal) and D (drop) are measured in feet.
Multiply R by 1.7, add H, do not exceed 100 feet.
Multiply D by 6.6, add H, do not exceed 100 feet.
R maximum is 35 feet.
D maximum is 15 feet.
H maximum is 100 feet.

ICE MACHINE HEAT OF REJECTION (9/86)

Ice machines, like other refrigeration equipment, reject heat through the condenser. It is helpful to know the amount of heat rejected to accurately size air conditioning equipment when self-contained air-cooled ice machines are installed in air conditioned environments. This heat rejection information is also necessary to evalutate the benefits of using water-cooled or remote condensers to reduce air conditioning loads. The amount of heat added to an air conditioned environment by an ice maker using a water-cooled or remote condenser is negligible. Knowing the amount of heat rejected is also important when sizing a cooling tower for a water-cooled condenser unit. The heat of rejection varies during the ice making production cycle. The figures shown under air conditioning are the average for this cycle. Peak figures are to be used for sizing cooling towers.

Model Series	Heat Rejected (BTU's/HR) Air Conditioning	Peak
100	2,500	3,100
200	4,300	5,500
400	6,800	9,400
600	10,400	14,400
800	13,200	20,700
1200	20,000	31,700
1700*	25,000	40,500

*Self-contained air-cooled is not available.

SUBJECT: EVACUATION AND RECHARGING RECOMMENDATIONS FOR REMOTE AND SELF-CONTAINED SYSTEMS

To ensure proper equipment operation in any ambient condition, the refrigerant charge in Manitowoc self-contained and remote systems must be nameplate charge.

When repairs require opening the sealed system, four primary functions must be accomplished before the system can be returned to normal operation:

1. **Total** refrigerant purge.
2. Filter drier replacement.
3. **Complete** evacuation.
4. Measured charge

EVACUATION AND CHARGING OF SELF-CONTAINED SYSTEMS

EVACUATION

(Refer to Illustration A-Self-Contained Evacuation.)
Evacuation of self-contained equipment requires connections at 2 points as follows:

1. Suction side of compressor through suction service valve.
2. Discharge side of compressor through discharge service valve.

Procedures for Self-Contained Evacuation:
(NOTE: Manitowoc suggests that an acid oil test is performed if burnout or severe contamination is suspected. Manitowoc also recommends triple evacuation as well as the installation of an acid core suction line drier in burnouts or severe contamination. The suction line drier must be removed within 24-72 hours after system is back in operation.)

1. Purge all refrigerant from system.
2. Position service valves fully open. (Backseated.)
3. Install a good quality two-stage vacuum pump to manifold gauges. A single stage pump will slow the evacuation process.
4. Pull system down to 250 microns. When 250 microns have been achieved, allow pump to run for ½ hour.
5. Close valves on manifold gauges, turn off and disconnect vacuum pump prior to charging. (NOTE: A standing vacuum test may be performed at this time as a preliminary means of leak checking; however, Manitowoc recommends the use of a halide or electronic leak detector after the system has been charged.)

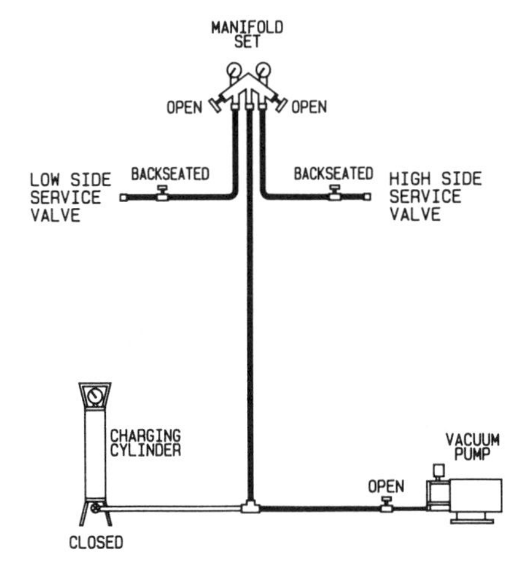

ILLUSTRATION A
SELF-CONTAINED EVACUATION HOOK-UP

SELF-CONTAINED CHARGING

The charge is critical on all Manitowoc Series cubers; therefore, use weight or charging cylinder to determine proper charge.

Charging Procedures:
Refer to Illustration B-Self-Contained Charging Hook-Up.

1. "Dump" measured charge into discharge service valve. (Insure charging hose has been purged of air.)
2. Allow system charge to "settle" for 2 or 3 minutes.
3. Start system, trim charge through suction service valve using vapor **only.**
4. Insure all vapor in charging hoses is drawn into machine before disconnecting manifold surges. The procedures with cuber in freeze mode is as follows:
   a. Frontseat high side service valve, open high side of manifold gauges.
   b. Backseat low side service valve, open low side of manifold gauges.
   c. When refrigerant is drawn into machine, close high side of manifold gauges, close low side of manifold gauges. Frontseat low side service valve.
   d. Remove hoses from machine.

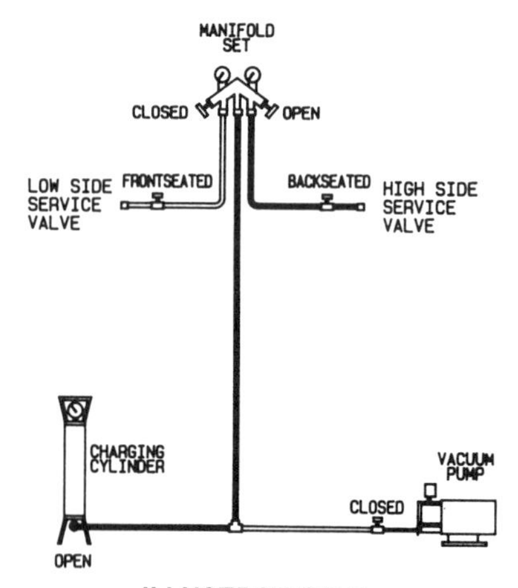

ILLUSTRATION B
SELF-CONTAINED CHARGING HOOK-UP

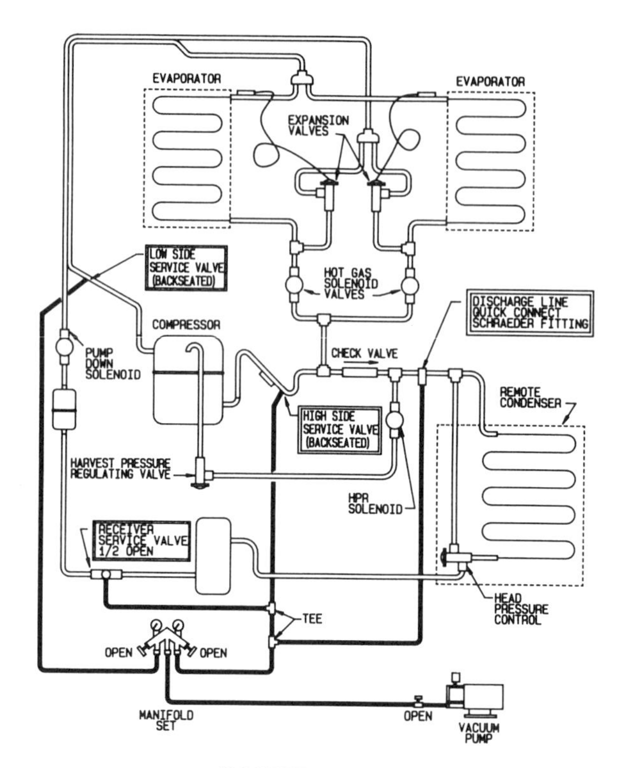

ILLUSTRATION C
4-POINT EVACUATION HOOK-UP

EVACUATION AND CHARGING OF REMOTE SYSTEMS

EVACUATION

Refer to Illustration C-4-Point Evacuation Hook-Up

Evacuation of remote systems requires connections at **4-points** for complete system evacuation as follows:

1. Suction side of compressor through suction service valve.
2. Discharge side of compressor through discharge service valve.

3. Receiver outlet service valve (Evacuates area between head pressure control valve in condenser and pump-down solenoid.)
4. Access valve on quick connect fitting (discharge line) on outside of compressor/evaporator compartment. This connection is necessary to evacuate the condenser. Without this connection, the magnetic check valve would close upon the drop in pressure produced by evacuation prohibiting complete condenser evacuation.

(NOTE: Manitowoc recommends using an access valve core removal and installation tool on the discharge line quick connect fitting. The tool permits removal of the access valve core for faster evacuation and charging without removing the manifold gauge hose.)

Procedures for Remote System Evacuation:
Refer to the "Self-Contained System" procedures for evacuation with the following exceptions:

1. In Step No. 2, position receiver outlet service valve ½ open, along with fully open suction and discharge valves. If access valve core removal and installation tool is used, remove discharge line quick-connect valve core at this time.

   (NOTE: It is not necessary to energize the liquid line solenoid for evacuation, as it is a pilot-operated valve. Pilot-operated valves require actual system pressures to open the main valve. Since there is no positive pressure during evacuation, the valve remains closed.)
2. In Step No. 4, allow vacuum pump to run for **one full hour** after 250 microns are achieved.

CHARGING

Refer to Illustration D-Remote Charging Hook-Up.

1. Frontseal low side service valve, backseat high side service valve, position receiver outlet service valve ½ open.
2. Close low side of manifold gauges open high side of manifold gauges.
3. "Dump" liquid refrigerant into receiver until pressures equalize. (A heated charging cylinder is recommended to increase refrigerant transfer.)
4. If receiver does not take entire charge, close off high side of manifold gauges and move charging hose from bottom to top of charging cylinder.

Backseat low side service valve, start machine, and add remaining charge through low side **in vapor form** until machine is fully charged.

5. Insure all vapor in charging hoses is drawn into the machine before disconnecting manifold gauges as described in Step 4 of "Self-Contained Charging".

(NOTE: Insure receiver outlet service valve is backseated after charging is complete and before system is operated. If access valve core removal and installation tool is used on the discharge line quick-connect fitting, reinstall Schraeder valve core before disconnecting access tool and hose.)

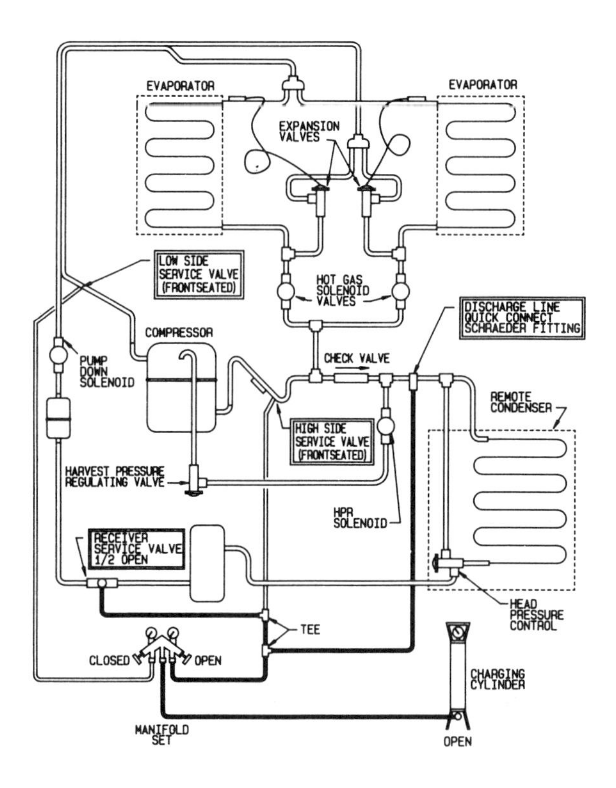

ILLUSTRATION D
REMOTE CHARGING HOOK-UP

DUMP VALVE & SUMP TROUGH OVERFLOW TUBE OPERATION

The dump valve opens during the harvest mode on the E or H0200, E0400, and G0600 Series ice machines and opens for the first 20 seconds of the freeze cycle on the G0800, G1200, and G1700 Series ice machines.

Dump Valve System Operation on the E or H0200, E0400, and G0600 Series Ice Machines

On these machines, the dump valve is closed during the complete freeze cycle. This will allow water to flow up to the distribution tube which supplies water over the evaporator.

When the machine cycles into the harvest mode, the dump valve coil is energized and the dump valve opens. It remains energized and open throughout the complete harvest mode.

While in the harvest mode, the water pump will continue to run. The water pump will now pump the sump water through the "tee" in the pump discharge line, through the open dump valve, and down the drain line.

When the harvest mode is terminated, the dump valve is de-energized. As the new freeze mode begins, the water pump will delay through the use of an adjustable delay timer. During this delay, the evaporator will per-chill delay settings R-12 —30 seconds; R-502 —20 seconds.

DUMP VALVE SYSTEM OPERATION ON THE G0800, G1200, and G1700 SERIES ICE MACHINES

At the start of each freeze mode, the dump valve is energized for approximately 20 seconds through use of an adjustable 20 second delay timer. At this time, sump water is pumped through the "tee" in the pump discharge line, then through the open dump valve and down the drain. **The evaporator is pre-chilled while the sump water is dumped.**

The dump valve will de-energize and close upon completion of the adjustable 20 second delay time. At this time, the sump water will be forced to flow up to the distribution tube.

When ice forms to the pre-set thickness, the unit will cycle into harvest mode. **The water pump will stay *de-energized* throughout the complete harvest mode.**

SUMP TROUGH OVERFLOW TUBE ON G0800, G1200, & G1700 SERIES ICE MACHINES

These units now have an overflow tube located in the sump trough. The overflow tube prevents water in the sump trough from overflowing when the water pump is de-energized. On earlier models, the overflow tube is removable for each sump trough flushing. On later models, the overflow tube is permanently installed. An external drain with a plug has been added for flushing

IMPORTANT: *A leaking dump valve during the freeze mode will cause long freeze times and uneven ice fill. There should be no water running through the dump valve during freeze mode. Check this* ***before*** **diagnosing the refrigeration system.**

SENSOR CONTROL ASSEMBLY

The E & G Series Models use a sensor assembly to control the ice bridge thickness and initiate the harvest mode. This assembly includes the following:

Bridge Thickness Control Probe — 76-1996-3, all models

The bridge thickness control probe hangs over the front of the evaporator, and is used to adjust the ice bridge thickness. This is done when water comes into constant contact with the two probes for 6-10 seconds.

Sensor Module — 25-1041-3, all models

The sensor module plugs into the transformer board, and are both orange and blue. They are interchangeable as they will operate in either 120V or 240V transformer boards.

Internally the sensor module works quite simply by providing us with four basic functions:

1. A mini relay that provides the switching for the electrical components which are energized or de-energized during the harvest mode.
2. The electronics that sense when water is in contact with the probes.
3. A 6-10 second non-adjustable timer to ensure the water is completing a circuit over the bridge thickness control.
4. A 4-5 minute safety timer to ensure the machine cannot stay in harvest for any prolonged period of time.

Transformer Board — 25-1039-3, 120 VAC units
25-1040-3, 208/230 VAC units

We use two transformer boards; one which is 120 volts and the other is 240 volts. Be sure you are using the proper board matching the primary voltage to the machine's line voltage. The transformer board is used to transform the line voltage down to the lower voltage (12 VAC) which the sensor module works on.

ELECTRICAL SEQUENCE OF OPERATION
All E and G Series Single Evaporator Cubers
Before Dump Valve Water Flush System

1. Freeze Mode

   Placing the toggle switch in the ice making position energizes the compressor, transformer primary, and the fan motor. After a short delay, the water pump will energize as it is wired to a delay timer. On remote units, the liquid line solenoid will also energize and cuber will start after suction pressure rises.

   As ice begins to form, water will eventually come into contact with the bridge thickness control probes. After the sensor module waits 6-10 seconds (to assure circuit is constant) the sensor module relay will energize. This will switch the relay's contact positions.

2. Harvest Mode

   At this point, the normally closed contact #5 will open de-energizing the condenser fan motor and water pump. On E0400 and E0600 remotes, the condenser fan will continue to run. Normally open contact #3 closes energizing the hot gas valve. During the harvest mode, the water will siphon out the sump trough and go down the drain. This is needed to remove minerals from the sump trough.

   The hot gas flowing through the energized hot gas valve raises the evaporator temperature releasing the ice from the evaporator. The ice drops causing the water curtain to momentarily open the bin switch. This momentarily interrupts the power to the primary side of the transformer causing the sensor module to reset back into the freeze mode by de-energizing the relay coil inside the sensor module.

3. Bin Full of Ice

   a. All E Series Self Contained Cubers
   If the bin is full the ice falling off the evaporator will hold the water curtain open which in turn opens the bin switch de-energizing the cuber. The cuber will not start again until ice is removed from the bin allowing the curtain to fall back in toward the evaporator closing the bin switch.

   b. G Series Self Contained Units
   When the bin is full the bin switch remains open, energizing the compressor off-delay timer. After 7 seconds the timer opens its normally closed switching position (SCR) thus de-energizing the contactor, compressor, water pump and condenser fan motor.

   The off-delay timer remains energized while the cuber is off. Once the curtain closes due to ice being removed from the bin, the timer is de-energized causing the contactor to initiate a new ice making cycle.

   c. E and G Series Remote Cubers
   When the bin is full, the bin switch remains open, de-energizing the liquid solenoid and sensor module. With the liquid solenoid closed, the cuber will pump down the low side. The low pressure cut-out control contacts open, de-energizing the contactor (G series only), compressor, condenser fan motor, water pump and transformer primary.

ELECTRICAL SEQUENCE OF OPERATION
E0200, H0200, E0400, and G0600 Series Cubers
After Dump Valve Water Flush System

E0600 series out of production at time of dump valve system introduction.

1. Freeze Mode
   Place the main toggle switch in the ice position; this energizes the compressor, water pump (on delay of a timer), condenser fan motor, and the primary side of the transformer. On remote units, the liquid line solenoid will also energize and cuber will start after suction pressure rises. As ice begins to form, water will eventually come into contact with the bridge thickness control probes. After the sensor module waits its 6-10 seconds (to assure circuit is constant), the sensor module relay will energize. This will switch the relay's contact positions.

2. Harvest Mode
   At this point, the normally closed contact #5 will open de-energizing the condenser fan motor, normally open contact #6 will close energizing the hot gas solenoid and the dump valve solenoid and normally open contact #3 will close keeping the water pump running.

   During the harvest mode, the water pump pumps water out through the energized dump valve to remove minerals from the sump through. The hot gas flowing through the energized hot gas valve raises the evaporator temperature releasing the ice from the evaporator. The ice drops causing the water curtain to momentarily open the bin switch. This momentarily interrupts the power to the primary side of the transformer causing the sensor module to reset back into the freeze mode by de-energizing the relay coil inside the sensor module.

3. Bin Full of Ice
   a. E0200, H0200, E0400 Self Contained Cubers
      If the bin is full the ice falling off the evaporator will hold the water curtain open which in turn opens the bin switch de-energizing the cuber. The cuber will not start again until ice is removed from the bin allowing the curtain to fall back in toward the evaporator closing the bin switch.
   b. G0600 Self Contained
      When the bin is full the bin switch remains open, energizing the compressor off-delay timer. After 7 seconds the timer opens its normally closed switching position (SCR) thus de-energizing the contactor, compressor, water pump, and condenser fan motor.

      The off-delay timer remains energized while the cuber is off. Once the curtain closes due to ice being removed from the bin, the timer is de-energized causing the contactor to initiate a new ice making cycle.
   c. E0400 and G0600 Remotes
      When the bin is full, the bin switch remains open, de-energizing the liquid solenoid and sensor module. With the liquid solenoid closed, the cuber will pump down the low side. The low pressure cut-out control contact opens, de-energizing the contactor (G series only), compressor, condenser fan motor, water pump, and transformer primary.

ELECTRICAL SEQUENCE OF OPERATION
G1200 and G1700 Air and Water

1. Freeze Mode
   Place the toggle switch in the ice making position. This energizes the contactor coil, fan motor, water pump, transformer primary, and the dump valve. After 20 seconds the dump valve is de-energized.

   As ice forms to the preset thickness, the water comes in constant contact with the ice bridge thickness probe. Within 6 seconds the sensor module relay will be energized, opening normally closed contact #5, de-energizing water pump and fan motor. Normally open contact #3 closes to keep the contactor coil energized as the bin switches open later in the harvest cycle. Normally open contact #6 closes energizing both hot gas solenoids (front and rear) and relay "C" including LED light (light emitting diode) on relay board. The cuber is now in the harvest mode.

2. Harvest Mode
   During the harvest mode, the hot gas raises the evaporator temperature, causing release of the ice from the evaporators. When the ice drops, it causes the water curtains, front and rear, to momentarily open their corresponding bin switches.

   For this explanation, let's say the front evaporator drops its ice before the rear. This will momentarily open the front bin switch energizing relay "A". As the bin switch opens interlock through #3 sensor module contacts will hold the contactor energized to keep the compressor running without momentary interruption. The N.C. contact of Relay "A" will open de-energizing the front hot gas valve. One of the N.O. contacts will close acting as a holding circuit for relay "A" to keep it energized. The other N.O. contact will close to allow the compressor off-delay timer to start timing after the ice falls off the back evaporator. At this point in time, the ice machine is still in harvest with hot gas only running through the back evaporator.

   As the ice falls off the rear evaporator, the rear bin switch will energize relay "B". Relay "B's" N.C. contacts will open de-energizing the rear hot gas solenoid. One of relay "B's" N.O. contacts will close acting as a holding circuit keeping relay "B" energized as the bin switch returns to its original energized position. The other N.O. contact closes allowing the compressor-off delay timer to be activated. After approximately 7 seconds the S.C.R. switch will open causing power interruption to the transformer primary to reset the sensor module contacts back into their normal position. The cuber is now back into the freeze mode.
3. Full Bin
   As a full bin results, the water curtains front or rear will be held open and so will the corresponding bin switch. The holding contact (#3) of the sensor module opens, de-activates the contactor, which in turn, shuts off the ice cuber. Upon ice removal and return of water curtain, the bin switches close and the cuber is re-activated immediately.

ELECTRICAL SEQUENCE OF OPERATION
G1200 and G1700 Remotes

To fully understand this sequence of operation, first read sequence of operation for the G1200 and G1700 air and water.

1. Freeze Mode

   Place main toggle switch in ice position. This energizes the liquid solenoid and sensor module. As the low pressure equalizes, the low pressure cut-out control contacts close, energizing the compressor, condenser fan motor(s), water pump, and dump valve. After 20 seconds the dump valve is de-energized.

   As ice forms to the preset thickness, the water comes in constant contact with the dual sensor probe. Within 6 seconds the sensor module relay will be energized, opening contact #5, de-energizing water pump and fan motor(s). Contacts #3 and #6 close, thus energizing both hot gas solenoids (front and rear), the H.P.R. solenoid, and relay "C" including L.E.D. light on relay board. The cuber is now in the harvest mode.

2. Harvest Mode

   During harvest mode, the hot gas raises evaporator temperature, thus causing release of the ice from the evaporators. When the ice drops, it causes the water curtains, front and rear, to momentarily open the bin switches.

   The momentary interruption of the bin switches energizes relay "A" (front) and relay "B" (rear) and the corresponding L.E.D. lights. As the hot gas solenoids are de-energized in sequence with the bin switches. The delay timer (7 seconds) is activated, causing the reset of the sensor module, de-energizing the relays and the H.P.R. solenoid, re-activating the water pump, dump valve and fan motor(s).

3. Full Bin

   As a full bin results, the water curtains front or rear will be held open and so will the corresponding bin switch. The holding contact (#3) of the sensor module opens and de-activates the liquid solenoid.

   With the liquid solenoid closed, the cuber will pump down the low side to approximately 15 P.S.I.G. The low pressure cut-out control contact open, de-energizing the contactor, condenser, fan motors(s), and water pump.

A SYSTEMATIC APPROACH TO DIAGNOSTIC PROCEDURES

Ice machine problem symptoms may often be misleading and at times, may simulate a refrigeration system problem, while the actual cause may be a commonly overlooked basic area such as load conditions and water side problems.

Ice machine problems can have many causes. By following a systematic approach to problem analysis starting from the simple, frequently overlooked basics and progressing systematically to the refrigeration system, you will be able to quickly arrive at a correct diagnosis.

This process of elimination will save both time and grief and may prevent the needless replacement of electrical and/or refrigeration system parts resulting from a wrong initial diagnosis.

TROUBLESHOOTING SENSOR CIRCUIT

Possible Problem:

Unit freezes slab of ice but will not harvest.

Note: CT-3W Tester will only work with 600K modules and jumper resistor removed from #20 and #21 if being used)

Step #1 Clean bridge thickness control probe to remove dirt, scale, etc. Check sensor probe for free movement and adjustment. (See page 5 for adjustment of probe).

Step #2 Check voltage between terminal #1 and terminal #2 on transformer board. Must read line voltage (120 or 240 volt, as indicated on serial # tag). Does voltmeter read ±10% of line voltage?

Yes, proper voltage: Proceed to Step #3.

No voltage: Proceed to "Possible Problem: No voltage at transformer board".

Step #3 Clip the black and green leads of the sensor circuit tester CT-3W on to the ice sensor probes. Did the cuber go into harvest? ***Note:*** Before clipping tester on to terminals, disconnect jumper resistor across #20 and #21 if being used.

Yes, cuber goes into harvest in 6-10 seconds: Sensor circuit operating properly. Proceed to Step #7.

No, cuber does not harvest: Proceed to Step #4.

Note: If tester is unavailable, place a 510 k-ohm resistor across the sensor probe. The resistance of the water across the sensor probes must be below 600 k-ohm for the sensor module to energize. Testing with a 510 k-ohm resistor or CT-3W tester will simulate very closely, the resistance of the water. A faulty module may sense and energize at a low resistance level, but not at its maximum range level of 600 k-ohm. Therefore, using a screwdriver is not a good test, as its resistance level is very low. This low of a resistance level may not indicate whether the sensor module is responding within its maximum range of 600 kohm. Example: a bad Sensor module could work with a shorting of a screwdriver across the probe and not with flowing water shorting the probe).

Step #4 Disconnect wires on terminals #20 and #21 on transformer board. Clip the green and black leads of tester or 500 k-ohm resistor to terminals #20 and #21 on the transformer board. Does cuber go into harvest cycle?

Yes, cuber goes into harvest in 6-10 seconds: Your transformer board and sensor module are ok. Proceed to Step #6.

No, cuber does not go into harvest: Proceed to Step #5.

Step #5 With toggle switch in ice-making position, remove the sensor module (orange or blue) from the transformer board. The module socket is numbered 1 through 8. Check these socket voltages with voltmeter. Sockets #6 to #7 = 20 to 30 volts AC, #5 to #6 = 9 to 15 volts AC, #5 to #7 = 9 to 15 volts AC. Do voltages read within the specified range?

NOTE: *When unplugging module, the water pump and condenser fan motor will not run. You may need to disconnect the compressor from power supply as the high side pressure could become excessive.*

Yes, voltages are within range: Replace the module (orange or blue).

No, voltages not within ranges: Replace transformer board if primary voltage is correct.

Step #6 The circuit is not being triggered by ice sensor probes. Check for open leads, bad solder connection on the sensor probe, or poor cabinet ground. Replace probe if needed.

Step #7 The circuit is not being triggered by water, check following 3 items:
—Probe adjustment (See STEP #1)
—Check for DC voltage generated across ice sensor. Place one probe of DC voltmeter in the water flow from distributor tube, other probe to chassis ground.

If DC voltage is **above 6-8** volts, you may have an exciter circuit by laying the DC voltage over the AC on the probe, the module will not react.

Correction: Placing a stainless jumper wire from the thickness control mounting bracket to the chassis will cancel this DC voltage and allow cuber to operate properly.

If DC voltage is not generated, the water to your ice machine may be too pure and not offer enough resistance across the probes for proper operation. To check, drop a small amount of salt into sump area. This will lower the resistance level of the water. If cuber goes into harvest after dropping salt into sump, it is still possible the sensitivity of the module may be set too low for your water. Order part number 76-2266-3 from your local distributor. This is a simple resistor to put across terminals #20 and #21 on the transformer board to increase the sensitivity of the module. Then when the water comes in contact with the probe, the sensor module will energize, putting the cuber into a harvest cycle.

Possible Problem:
Cuber prematurely goes into harvest with no ice formation on evaporator.

Step #1 Check following items (if not found to be a problem, proceed to Step #2):

—Sensor probe too close to evaporator and/or scale build-up on sensor probe. Clean and adjust as needed. (See probe adjustment page 5).

—Water running too fast over evaporator splashing on ice thickness control. (See water flow clamp adjustments page 4).

—Slush build-up hitting sensor probes. Check evaporator prechill for proper operation and settings. R-12 units approximately 30 seconds; R-502 units approximately 20 seconds.

Step #2 Disconnect wires on terminals #20 and #21 on the transformer board. Did the cuber stay in the freeze cycle?

Yes, cuber stays in freeze cycle: The transformer board and sensor module (orange or blue) are ok. Proceed to Step #4.

No, cuber still goes into harvest: Proceed to Step #3.

Step #3 Check for moisture on the backof transformer board. Correction: Dry board and spray with any fast-drying lacquer. If this procedure does not work, perform transformer board check as outlined in Step #5, page 28.

Step #4 If sensor wires connect to grommet(s) in bulkhead, check for moisture within grommet(s). Correction: Dry and seal with food-grade RTV. If this is not found to be the cause and Step #1 in this process was performed correctly, replace the white sensor probe.

Possible Problem:
No voltage at transformer board terminal #1 and #2. Cuber freezes although it will not harvest. E1100, G1200, and G1700 Series remotes. (Models with 3-relay board).

CAUTION: *Turn machine off at circuit breaker before hooking voltmeter leads up for each step.*

Step #1 Clip one lead of voltmeter to terminal #1 on transformer board and other to terminal #19 on 3-relay board; (bottom board). Leave all wires connected to **their original terminals**. Do you read line voltage?

Yes, have line voltage: Check for proper installation of jumper wire between terminal #2 on transformer board & terminal #19 on 3-relay board; (bottom board).

No, do not have line voltage: Proceed to Step #2.

Step #2 Clip one lead of voltmeter to terminal #1 on transformer board, the other to terminal #1 on the compressor off-delay timer. Leave all wires connected to their original terminals. Do you read line voltage?

Yes, have line voltage; Burnt or loose connections on jumper wire from #1 on delay timer to #19 on 3 relay board.

No, do not have line voltage: Proceed to Step #3.

Step #3 Clip one lead of voltmeter to terminal #1 on transformer board, the other lead to terminal #2 on the compressor off-delay timer. Do you read line voltage?

Yes, have line voltage: Change compressor off delay timer. It has an opened S.C.R. switch.

No, do not have line voltage:
1. Self-contained unit — proceed to check main contactor for proper operation.
2. Remote unit — proceed to Step #4.

Step #4 Clip one lead of voltmeter to terminal #1 on the transformer board and the other lead to terminal #40. Do you read line voltage?

Yes, have line voltage; Harvest pressure limiter is open. See page 7 for settings.

No, do not have line voltage: Proceed to check main contactor for proper operation.

DIAGNOSING REFRIGERATION PROBLEMS
and Problems With Similar Symtoms

Manitowoc provides step-by-step diagnostic procedures used before condemning refrigeration components. The following Six-Step Procedure is to be used when diagnosing refrigeration problems. Each step is designed to assist in identifying the cause as well as any problem in the shortest period of time. After each step you evaluate your findings. If no problem is found, proceed to the following step. As each additional step is completed, evaluate data from previous steps together with the last step completed. Continue to do so until all abnormalities are corrected.

If steps are not performed in order, it may result in a misdiagnosis. An example: the water step is before the refrigeration step. Water related problems can simulate a refrigeration problem, such as a leaking water flush dump valve has some of the same characteristics of a malfunctioning thermal expansion valve.

To review:
1. After each step, evaluate data together with the last steps completed.
2. Do not condemn refrigeration components until all steps are completed and evaluated.

PROPER DIAGNOSIS

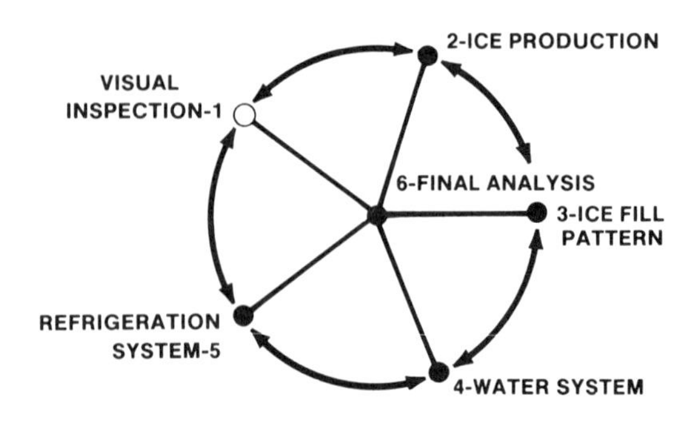

Observe the following items before proceeding to Diagnostic Procedures!
1. Don't make adjustments or turn the machine off until you have identified the malfunction. The problem may not repeat itself and you may not have another opportunity to check it out in the failed mode.
2. Follow the systematic approach throughout the diagnosis and write down information as you collect it. This will keep you organized.

STEP 1 VISUAL INSPECTION

This information could be the determining factor in your final diagnosis.

Possible Problem	Actual Finding	Corrective Measure
1. Machine not properly installed		reinstall properly (see installation manual)
2. Location, air temperatures, air flow restrictions, etc.		reinstall properly (see installation manual)
3. Air space clearances at back and sides of machine		must have minimum of 5" around back and sides of machine
4. Machine not level side to side, back to front		level machine
5. Air-cooled condenser dirty		clean condenser

Steps 2, 3, and 4 can be done in conjunction with each other. Be careful not to interfere with ice production check.

STEP 2 ICE PRODUCTION

Determine and compare operating times and load conditions (ambient and water temperatures) to 24 hour capacity ratings for the specific ice machine model.

I. OPERATING CONDITIONS

1. Condenser inlet air temperatue: ______________
2. Water inlet temperature (taken at float valve): ____
3. Ambient air temperatue at machine: ___________
4. From Ice Production Charts, the published 24 hour ice production at the above conditions: _________

II. ICE PRODUCTION CAPACITY CHECK

1. Freeze Time, Minutes: ______ + Harvest Time, Minutes ______ = Total cycle time ______________
2. 1440 ÷ total cycle time = cycles/day ___________.
3. Weight of ice slab(s)______ × cycles/day = ice production in 24 hours. ______________lbs./24 hours.

Ice machines make less ice as load conditions (air & water tempratures) increase. If the machine is producing to specifications, no further diagnosis is required. If production is low, continue through the checklist.

STEP 3 ICE FILL PATTERN ON EVAPORATOR

Compare the ice fill pattern to the normal uniform bridge thickness of ⅛". The bridge is the inter-connecting waffle between cubes.

1. Single evaporator ice normal.
   ______ Yes ______ No
2. Dual evaporator; ice fill normal on both front and rear evaporators. ______ Yes ______ No
3. If no, what is fill pattern? ________________

   ____

Ice fill patterns will be used in your final diagnostic procedures.

Examples of ice fill pattern characteristics: thick on bottom, thin on top; thin on bottom, thick on top; ice fill spotty such as a corner not filling, etc.

STEP 4 THE WATER SYSTEM

Water related problems can simulate a refrigeration problem. An example would be a leaking water flush dump valve leaking simulating the characteristics of a malfunctioning thermal expansion valve.

Possible Problem	Actual Finding	Corrective Measure
1. Water area (evap) dirty		see "Ice Cleaning Procedures", p. 1
2. Water inlet pressure, should 20-80 psi		install regulator valve
3. Incoming water supply temperature should be 35° - 90°		too hot — check hot water line check valves in other store equipment.
4. Water filter (if used) plugged		replace filter
5. Dump valve malfunctioning		see "Dump Valve Operation", p. 20
6. Dump valve/ siphon leaking water down drain during freeze		Clean dump valve/replace as needed
7. Vent tube not installed on water outlet drain		see "Installation Manual"
8. Sump trough hoses leaking water		install properly/ replace
9. Water float valve stuck open or out of adjustment		see "Adjustments", p. 2
10. Water freezing behind evaporator		check water flow
11. Water freezing between white plastic extrusions & evaporator		seal with food-grade RTV
12. Water flow uneven across evaporator		clean ice-machine/ check water flow rate
13. Tubing separating from back of evaporator		replace evaporator

STEP 5 THE REFRIGERATION SYSTEM

Determine and compare operating pressures to normal pressure charts for the specific ice machine model. If pressures are normal proceed to Step 6.
CAUTION — Gauges must be put on and off using proper procedures to assure no refrigerant is lost.

OPERATING PRESSURES	Discharge	Suction
1. Beginning of freeze mode:		
2. End of freeze mode:		
3. Beginning of harvest mode:		
4. End of harvest mode:		

Always check in order listed.

Possible Problem	Actual Findings	Corrective Measure
A. HIGH DISCHARGE PRESSURE		
1. Load conditions high (air/water temps)		move cuber to location within guidelines
2. Dirty condenser		clean
3. Obstruction in high side lines a) kinked		repair, see "Evacuation/Charging Procedures", p. 14
b) plugged drier		
c) remote couplings not pierced		
d) remote receiver service valve not opened		backseat receiver service valve
4. Water regulating valve (water-cooled condenser)		
a) too small supply water line		replace with proper size line
b) defective regulating valve		replace
c) out of adjustment		see "Proper Adjustment", p. 7
5. Fan motor/fan cycling switch defective		replace

STEP 5 Continued

Possible Problem	Actual Findings	Corrective Measure
6. Headmaster defective (remotes)		see "Headmaster Diagnostics" p. 42
7. Improper refrigerant charge		Continue through entire procedures before concluding a machine is improperly charged
8. Non-condensables in system		see Evacuation/Charging Procedures" p. 14
B. LOW DISCHARGE PRESSURE		
1. Load conditions high (air/water temps)		relocate cuber to location within guidelines
2. Water regulating valve (water-cooled condensers)		
a) defective		replace
b) out of adjustment		see "Proper Adjustment", p. 7
c) leaking water through during harvest		readjust/replace if necessary
3. Headmaster valve (remotes)		
a) non-approved valve		see "Proper Headmaster Valves",
b) defective headmaster		see "Proper Diagnostics", p. 42
4. Low refrigerant charge		continue through entire procedures before concluding a machine is improperly charged
C. LOW SUCTION PRESSURE		
1. Low load conditions		relocate cuber to location with guidelines
2. No water flow over evaporator		see "Electrical Operation", p. 22

STEP 5 Continued

Possible Problem	Actual Findings	Corrective Measure
3.Tubing separating from backside of evaporator		replace evaporator
4. Plugged drier/restriction in liquid line		see "Evacuation/Charging Procedures" p. 14
5. TX valve starving		see "TXV Diagnostics", p. 40
6. Improper refrigerant charge		see "Proper Refrigerant Charges", p. 14
D. HIGH SUCTION PRESSURE		
1. High discharge pressure affective low side		see "High Discharge Pressure",
2. H.P.R. Solenoid leaking (remotes)		replace
3. Hot gas valve leaking (Stuck open)		see "Hot Gas Valve Diagnostics", p. 39
4. TX valve flooding		see "TXV Diagnostics", p. 40
5. Inefficient compressor? (Do not perform pumpdown test)		see "Compressor Diagnostics", p. 43

STEP 6 FINAL DIAGNOSIS

Diagnosing refrigeration components is not always black and white. Before condemning a refrigeration component, you must evaluate all data and symptoms collected.

YOUR FINDINGS

1. Ice production: ______________________
2. Ice fill pattern: ______________________
3. Water system: ______________________
4. Operating pressures: ______________________

COMPARE CHARACTERISTICS OF ITEMS 1-4 TO SYMPTONS OF REFRIGERATION COMPONENTS A THROUGH D.

A. Hot gas valve check page: ______________
B. TXV check page : ______________
____
C. Headmaster check page : ______________
D. Compressor check page : ______________
____

Based on information gathered, your final diagnosis (the problem if any) is: ______________________

____

Reason for your answer: ______________________

____

____

DIAGNOSING HOT GAS VALVE(S)

Before diagnosing a hot gas valve, you must first eliminate non-refrigerant related area's that simulate hot gas valve failures. Follow Step 6 Procedure.

Possible Problem:

1. Non-approved valve.
   A hot gas valve requires a specific orifice size to allow the proper amount of hot gas flow into the evaporator during the harvest mode. Remember to replace defective hot gas valves with "original" Manitowoc replacement parts only. Check your Parts Manual for proper valve application.
2. Not opening electrically.
   A. Ohm coil: if open, replace
   B. See Electrical Sequence of Operation
3. Stuck in harvest — voltage at coil?
   YES — see Electrical Sequence of Operation
   NO — foreign material in valve — clean
   — valve seat faulty — replace with O.E.M. Seat Kit only
   — valve sticking — replace valve
4. Leaking by during freeze cycle.
   A. SYMPTOMS OF A LEAKING HOT GAS VALVE:
      1. **Capacity** reduction will be minimal, small leak; no ice production; stuck wide open
      2. **Ice fill** on evaporator(s) will be normal. This includes both evaporators on a dual evaporator system with one hot gas valve leaking.
      3. **Suction pressure** at the end of the freeze will be slightly higher. This 1-4 psi increase can be difficult to detect.

   B. A good hot gas valve **inlet** line will be too hot to touch during the harvest cycle and be cool enough to touch after approximately 5 minutes into the freeze cycle. On a leaking hot gas valve, the inlet will be close to the discharge line temperature; too hot to touch during both the freeze and harvest cycle.

   C. **Check procedures**
      1. Check all external sources first.
      2. Feel hot gas valve inlet **after 5 minutes** into freeze cycle.
      3. If the Inlet is the same temperature as the discharge line (too hot to touch) replace valve.
      4. If cooler then discharge line — valve is OK.

DIAGNOSING TXV'S

Before diagnosing the TXV, you must first eliminate non-refigerant related areas that simulate TXV failures. Follow Step 6 procedure.

Possible problem:

1. Non-approved valve.
   Manitowoc O.E.M. expansion valves have special bulb gas charges, port sizes, and stroke lengths to control the refrigerant flow to the evaporator under all conditions. Replace defective expansion valves with original Manitowoc replacement parts only. Manitowoc O.E.M. valves must never be adjusted.
2. TXV not installed properly.
   - A. Bulb location — 10:00 or 2:00 position.
   - B. Bulb clamp location — two stainless steel clamps mounted flush with each end of bulb.
   - C. Bulb clamp tightness — properly tightened clamps can slightly deform bulb.
   - D. Bulb insulation — insulation must be at least ⅜" thick.
   - E. Valve overheated during installation — when soldering valve, wrap with wet rag.

   NOTE: Service Bulletin SB-29-88 indicates proper TXV valve usuage and installation procedures.
3. TXV starving
   - A. SYMPTONS OF A STARVING VALVE:
     1. Lower than normal suction pressures.
     2. Ice fill pattern thick on bottom — thin on top.
     3. Will have capacity reduction.
   - B. External sources that cause a TXV to starve:
     1. Tubing separating from back of evaporator —TXV ok — replace evaporator.
     2. Non-approved valve — replace with O.E.M. valve.
     3. TXV improperly installed — reinstall properly.
     4. TXV bulb lost charge — replace TXV.
     5. Foreign material in valve — clean/replace.
     6. Moisture in system — TXV ok — replace drier, evacuate and recharge.
   - C. Diagnostic tip:
     A properly operating expansion valve will react if the bulb is taken off the suction line and heated by hand, then cooled down in an ice bath. The suction pressure should rise when the bulb is warmed and drop when chilled.
4. TXV Flooding
   - A. SYMPTONS OF FLOODING VALVE(S):
     1. **Suction pressure** will be high.
     2. **Ice fill:**
        Single evaporator cubers — the ice fill pattern will be non-uniform. Typically the ice will be thin at the bottom of the evaporator and thick on the top
        Dual evaporator cubers — it is extremely rare for two TXV's to fail at the same time. One evaporator (good TXV) will have normal ice fill, the other (flooding valve) will have non-uniform ice fill as described in single evaporator cubers.
     3. **Capacity** — there is a large capacity reduction. Freeze times will be extremely long.
     4. At higher ambient conditions, the cuber may not make sufficient ice to initiate a harvest cycle.
   - B. Diagnostic tip
     The steps below are to be used along with all information gathered, such as load conditions. Basing your final diagnosis on these checks only can result in a misdiagnosis.

     SINGLE EVAPORATOR CUBERS
     1. Measure the inlet and outlet temperatures of evaporator copper lines at least 5 minutes into the freeze cycle. Take temperatures as close to the evaporator as possible.
     2. A properly operating valve should maintain the inlet and outlet temperatures within approximately 5° of each other.
     3. If the temperatures are not within approximately 5°, the valve must be replaced.

     DUAL EVAPORATOR CUBERS
     1. Measure temperature at least 5 minutes into freeze cycle. Take temperatures as close to TXV outlets as possible.
     2. If temperatures between valves are within 5° of each other, both TXV's are ok.
     3. If temperatures between valves vary more than 5°, suspect one bad TXV.
        - a. The TXV with the highest temperature is the defect valve.
        - b. Only change the one faulty valve.

HEADMASTER CONTROL VALVE

Manitowoc remote systems require special headmaster control valves manufactured to our specifications. Remember to replace defective headmaster control valves with original Manitowoc replacement parts only.

The R-12 headmaster control valve has a non-adjustable setting of 125 psi ± 5 psi; while the R-502 valve is set for 180 psi ± 10 psi. Normal flow pattern through the valve is from the condenser to the receiver inlet. This flow pattern is normal at ambient temperatures of 70°F and above. At temperatures below 70°F, the head pressure control dome pressure which holds a nitrogen charge, will force the valve portage to change, closing the condenser port and opening the bypass port from the compressor discharge line. In this modulating mode, the valve is maintaining minimum head pressure by building up liquid in the condenser and bypassing discharge gas directly to the receiver.

DIAGNOSING HEADMASTER VALVE

To diagnose a headmaster, the system must be properly charged. A low-charged system can cause the headmaster to stick in the by-pass mode.

Low head pressure

1. Add refrigerant in 2 lb. increments (do not exceed 6 lbs.). If the cuber was low on charge, the headmaster function and discharge pressure will return to normal after adding charge.
2. If you added refrigerant, as a diagnostic procedure you must change drier, evacuate and recharge, to assure proper operation in all ambient conditions. (See SB-18-84.)

	Possible Problem	Solution:
1.	Non-approved valve	Install O.E.M. headmaster
2.	Stuck in by-pass	If stuck in by-pass, the liquid line returning to the receiver will be hot to touch
3.	Not by-passing	at approximately 70° and below, the headmaster must by-pass and maintain its proper setting. if discharge pressure is below setting, the valve is not by-passing.
4.	Aero-quip fitting not pierced	On new installations, make sure the diaphragms are completely folded back.
5.	Pressure drop in line sets	The pressure drop from the discharge line (leaving cuber) to the liquid line (leaving condenser) m ıst not exceed 20 psi. Check for obstructions, kinked lines, and aero-quip fitting not pierced.

MANITOWOC REPLACEMENT HEADMASTER CONTROL VALVES

83-6806-3	All R-502
83-6804-9	R-12 with sweat fittings
76-2236-3	R-12 with aero-quip couplings — for mounting outside of condenser

NOTE: When replacing a defective headmaster valve in a 1100 or 2200 Series cuber and the valve is inside the compressor compartment, install the replacement valve outside at the remote condenser.

DIAGNOSING COMPRESSOR

1. "Inefficient" compressor (suction valves).
   An inefficient compressor can be hard to detect. The best method in determining the compressor condition is to systematically check other components and rule them out one-by-one.

   Components or problems that are not directly related to the compressor can simulate the characteristics below. Checking all other sources first will save time and minimize the replacement of the compressor resulting from a misleading diagnosis. Follow step 6 procedure.

   SYMPTOMS OF AN INEFFICIENT COMPRESSOR:
   NOTE: A low capacity compressor will not "pump" as much refrigerant as it was designed for.
   1. **Capacity** reduction will be noticeable at lower ambient conditions and become more pronouned as ambient temperatures increase. Freeze times will be longer then normal.
   2. **Suction pressures** at the end of freeze cycle will be abnormally high and become more pronounced as ambient temperature increases.
   3. **Ice fill** on the evaporator(s) will usually be normal, although in extreme high ambient cases, there may be little to no ice formation. On dual evaporator machines, the ice fill will be normal on both evaporators.
   4. There may be intermittent flooding by the TXV(s).
   5. An inefficient compressor may "pull down" and hold, therefore this type of test must not be used as a determining factor for changing out compressors.

2. Discharge valves
   NOTE: Discharge valve related problems are extremely rare in this type of application.
   1. The compressor shell will become hot and compressor may cycle on overload.
   2. Suction pressure will be extremely high.
   3. Discharge pressure will be lower than normal.
   4. **Check procedure** for discharge valves:
      a. Ensure compressor is running.
      b. Turn cuber off.
      c. Immediately feel suction line — it will turn hot if the discharge valve is leaking or broken.

3. Electrical failures
   Compressor will not start or will trip repeatedly on overload.
   A. CHECK RESISTANCE (OHM) VALUES
   Compressors winding can have very low OHM values. (See Table of Contents, Compressor OHM Values). The use of a properly calibrated meter is recommended.
   The resistance test is done after the compressor is cool. The compressor dome should be cool enough

to touch (approximately 120°F) to assure overload is closed and resistance readings will be accurate.

1. Single phase compressors
   a. Disconnect power to cuber; wires from compressor terminals.
   b. With wires removed, the resistance values must be within guidelines for the compressor. (See Table of Contents, Compressor OHM Values). The resistance value from C to S and C to R added together, should equal value from S to R.
   c. An open overload will give a resistance reading from S to R and a "O" reading from C to S and C to R. Allow the compressor to cool, then recheck readings.
2. Three phase compressors
   a. Disconnect power to cuber; remove wires from compressor terminals.
   b. With wires remvoed, the resistance values must be within guidelines for the compressor (see Table of Contents, Compressor OHM Values). L1 to L2; L2 to L3; and L1 to L3, should all be equal to each other.
   c. An open overload will give a resistance reading of "O" from L1 to L2; L2 to L3; and L1 to L3. Allow compressor to cool, then recheck reading.

B. CHECK MOTOR WINDINGS TO GROUND

Check continuity between all three terminals and the compressor shell or copper refrigeration line (be sure to scrape metal surface clean to get good contact). If continuity is present, the compressor windings are gounded and the compressor should be replaced.

C. DETERMINE IF THE COMPRESSOR IS "SEIZED"

Check amp draw while compressor is trying to start.

1. Compressor drawing locked rotor, the two likely causes would be a defective starting component or a mechanically seized compressor. To determine which you have:
   a. Install high and low side gauges.
   b. Try to start compressor (watch pressures closely).
   c. If pressures do not move, compressor is seized up. Replace compressor.
   d. If pressures move, the compressor is turning slowly and is not seized. Check capacitors and start relay.
2. Compressor drawing high amps

   The continuous amperage draw on start-up should not near the maximum fuse size as indicated on the serial tag.

   Check the following:

   Low voltage — The voltage at the time the compressor is trying to start must be within ±10% of the nameplate voltage (exception — E200 Series 1/3 HP compressor and 1100 Series using Tecumseh compressor are ±5%).

D. DIAGNOSING CAPACITORS AND RELAYS

1. Capacitors

   If the compressor attempts to start, or hums and trips the overload protector, you must check the starting components before replacing a compressor.
   a. Capacitors can show visual evidence of failure, such as a bulged terminal end or a ruptured membrane. **Do not assume a capacitor is good** if no visual signs are evident.
   b. A good test is to install a known good substitute capacitor.
   c. Use of a capacitor tester is recommended when checking a suspect capacitor. Remember to clip the bleed resistor off the capacitor terminals before testing.

2. Relays

   Potential type:

   Potential relay contacts are closed during the initial starting cycle, and open as the compressor comes up to speed.

   Check Procedure:
   1. Disconnect power supply.
   2. Remove wires from relay.
   3. Use a high voltage OHM meter to check the relay coil — open — replace; continuity — ok.
   4. Use an OHM meter to check across the contacts. Potential relay contacts are normally closed.

   Current type:

   Current relay contacts are normally open.

   Check Procedure:
   1. Disconnect power supply
   2. If relay is on the compressor, pull off.
   3. Keeping relay upright, check continuity with OHM meter. Closed — replace.

      NOTE: Turning relay upside down will give a closed reading.
   4. Check continuity through relay coil, replace if no continuity.

A0100 SERIES AIR-COOLED

These are approximate characteristics that may vary depending on operating conditions.

CYCLE TIMES
TOTAL CYCLE TIME
Freeze time plus harvest time
Water Temp.

Ambient Temp.	70°	80°	90°	Harvest Time
50°	16-21.5	18-24	20-27	1.5-2.5
70°	17-23	19-25	21-30	
90°	18-24	20-27	23-31	

Based on average ice slab weight of 1lb 2oz to 1lb 8oz. Times in minutes.

24 HOUR ICE PRODUCTION

Air Temp. °F	WATER TEMP. °F		
	70°	80°	90°
50°	100	90	80
70°	95	85	75
90°	90	80	70

OPERATING PRESSURES

	FREEZE CYCLE		HARVEST CYCLE	
Ambient Temp. °F	Head Pressure PSI	Suction Pressure PSI	Head Pressure PSI	Suction Pressure PSI
70°	101-135	18-7	70-80	40-44
90°	140-180	22-8	70-80	40-44
105°	165-215	24-9	70-80	42-44

A0100 SERIES WATER-COOLED

These are approximate characteristics that may vary depending on operating conditions.

CYCLE TIMES
TOTAL CYCLE TIME
Freeze time plus harvest time
Water Temp.

Ambient Temp.	70°	80°	90°	Harvest Time
50°	16-21.5	17-23	18-24	1.5-2.5
70°	17-23	18-24	19-24	
90°	18-24	19-25	20-27	

Based on average ice slab weight of 1lb 2oz to 1lb 8oz. Times in minutes.

24 HOUR ICE PRODUCTION

Air Temp. °F	WATER TEMP. °F		
	70°	80°	90°
50°	100	95	90
70°	95	90	85
90°	90	85	80

OPERATING PRESSURES

	FREEZE CYCLE		HARVEST CYCLE	
Ambient Temp. °F	Head Pressure PSI	Suction Pressure PSI	Head Pressure PSI	Suction Pressure PSI
70°	120-125	20-4	70-80	35-44
90°	120-125	21-4	70-80	35-44
105°	121-126	22-5	70-80	35-44

A0100 SERIES AIR AND WATER
FAN ON AIR-COOLED ONLY
SHOWN AT BEGINNING OF FREEZE CYCLE
115 VAC 60 CYCLE OR 230 VAC 50 CYCLE

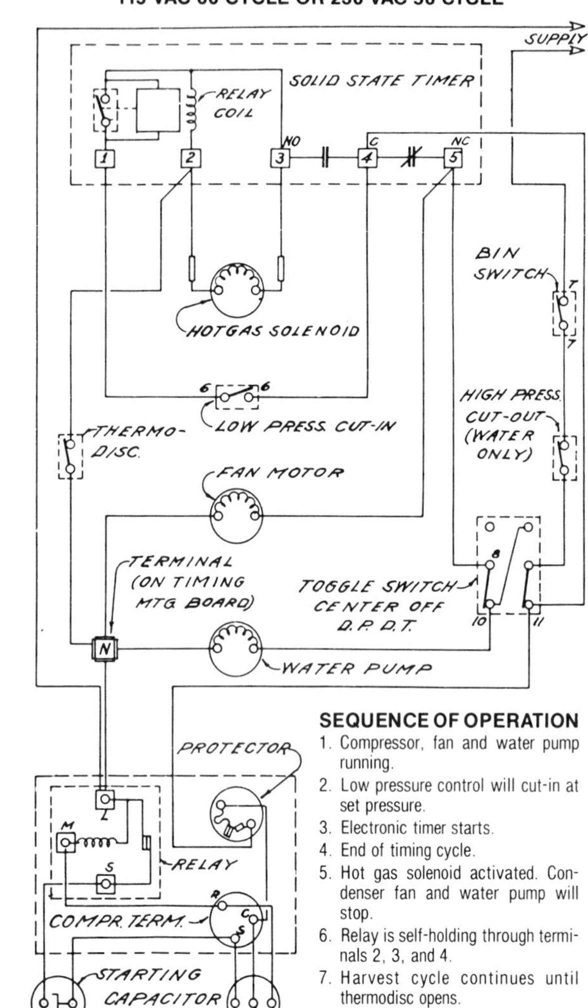

SEQUENCE OF OPERATION

1. Compressor, fan and water pump running.
2. Low pressure control will cut-in at set pressure.
3. Electronic timer starts.
4. End of timing cycle.
5. Hot gas solenoid activated. Condenser fan and water pump will stop.
6. Relay is self-holding through terminals 2, 3, and 4.
7. Harvest cycle continues until thermodisc opens.
8. Thermodisc opening de-energizes timer relay and hot gas solenoid.
9. New cycle; terminals 3 and 4 open, 4 and 5 close.
10. Fan and water pump start.

E0200 SERIES AIR-COOLED
(1/3 hp Compressor)

These are approximate characteristics that may vary depending on operating conditions.

CYCLE TIMES
TOTAL CYCLE TIME
Freeze time plus harvest time
Water Temp.

Ambient Temp.	50°	70°	90°	Harvest Time
70°	16-20	16-21	17.5-22	1.5-2.5
80°	17.5-22	18.5-23	20-25	
90°	19-24	20-25	22-28	
100°	22.5-29	23.5-30	26-33	

Based on average ice slab weight of 2lb 6oz to 3lb 0oz.
Times in minutes

24 HOUR ICE PRODUCTION

Air Temp. °F	WATER TEMP. °F		
	50°	70°	90°
70°	216	208	194
80°	192	183	171
90°	177	169	154
100°	151	145	131

OPERATING PRESSURES

	FREEZE CYCLE		HARVEST CYCLE	
Ambient Temp. °F	Head Pressure PSI	Suction Pressure PSI	Head Pressure PSI	Suction Pressure PSI
50°	100-135	22-5.5	65-80	35-65
70°	100-135	22-5.5	70-85	35-65
80°	110-135	22-5.5	80-95	35-65
90°	120-140	22-5.5	85-100	35-65
100°	135-155	22-5.5	90-115	35-65

E0200 SERIES WATER-COOLED
(1/3 hp Compressor)

These are approximate characteristics that may vary depending on operating conditions.

CYCLE TIMES
TOTAL CYCLE TIME
Freeze time plus harvest time
Water Temp.

Ambient Temp.	50°	70°	90°	Harvest Time
70°	18-23	20-25	22-28	1.5-2.5
80°	18.5-23.5	20.5-25.5	23-29	
90°	19-24	21-26	23.5-29.5	
100°	19-24	21-26	23.5-29.5	

Based on average ice slab weight of 2lb 6oz to 3lb 0oz.
Times in minutes

24 HOUR ICE PRODUCTION

Air Temp. °F	WATER TEMP. °F		
	50°	70°	90°
70°	190	172	155
80°	183	166	150
90°	178	161	145
100°	176	160	143

Condenser Water Consumption	WATER TEMP. °F		
	50°	70°	90°
Gal/24 Hrs	153	264	934
At 125 psi head pressure			

OPERATING PRESSURES

	FREEZE CYCLE		HARVEST CYCLE	
Ambient Temp. °F	Head Pressure PSI	Suction Pressure PSI	Head Pressure PSI	Suction Pressure PSI
50°	120-130	22-5.5	70-90	35-64
70°	120-130	22-5.5	70-90	35-64
80°	120-130	22-5.5	70-90	35-64
90°	120-130	22-5.5	70-90	35-64
100°	120-130	22-5.5	70-90	35-64

E0200 & H0200 SERIES AIR-COOLED
(½ hp compressor)

These are approximate characteristics that may vary depending on operating conditions.

CYCLE TIMES
TOTAL CYCLE TIME
Freeze time plus harvest time
Water Temp.

Ambient Temp.	50°	70°	90°	Harvest Time
70°	15-18.5	16.5-21	19-24	1.5-2.5
80°	16-20	18-22.5	20-25	
90°	17.5-22	20-25	23-29	
100°	21-27	24-31	27-34.5	

Based on average ice slab weight of 2lb 6oz to 3lb 0oz.
Times in minutes.

24 HOUR ICE PRODUCTION

Air Temp. °F	WATER TEMP. °F		
	50°	70°	90°
70°	230	205	180
80°	215	190	170
90°	195	170	150
100°	160	140	125

OPERATING PRESSURES

	FREEZE CYCLE		HARVEST CYCLE	
Ambient Temp. °F	Head Pressure PSI	Suction Pressure PSI	Head Pressure PSI	Suction Pressure PSI
50°	100-124	25-12	65-85	40-50
70°	100-125	25-12	65-85	40-50
80°	120-150	28-14	75-95	45-55
90°	140-170	30-14	85-105	55-65
100°	155-200	38-14	105-125	70-80
110°	170-230	45-15	120-140	80-90

E0200 & H0200 SERIES WATER-COOLED
(½ hp Compressor)

These are approximate characteristics that may vary depending on operating conditions.

CYCLE TIMES
TOTAL CYCLE TIME
Freeze time plus harvest time
Water Temp.

Ambient Temp.	50°	70°	90°	Harvest Time
70°	15-18.5	17.5-22	21-26	1.5-2.5
80°	15-19	18-22.5	22-28	
90°	15.5-19.5	18.5-23	23-29	
100°	16-20	19.5-24.5	23.5-30	

Based on average ice slab weight of 2lb 6oz to 3lb 0oz.
Times in minutes

24 HOUR ICE PRODUCTION

Air Temp. °F	WATER TEMP. °F		
	50°	70°	90°
70°	230	195	160
80°	225	190	155
90°	220	185	150
100°	215	175	145

Condenser Water Consumption	WATER TEMP. °F		
	50°	70°	90°
Gal/24 Hrs	153	264	934
At 125 psi head pressure			

OPERATING PRESSURES

Ambient Temp. °F	FREEZE CYCLE		HARVEST CYCLE	
	Head Pressure PSI	Suction Pressure PSI	Head Pressure PSI	Suction Pressure PSI
50°	120-130	22-10	75-95	40-50
70°	120-130	22-12	80-100	45-55
80°	120-130	22-12	80-100	45-55
90°	120-130	22-12	80-100	45-55
100°	120-130	22-12	80-100	45-55
110°	120-130	24-14	80-105	50-60

E0200 AND H0200 AIR AND WATER
1 PHASE — 50/60 CYCLE

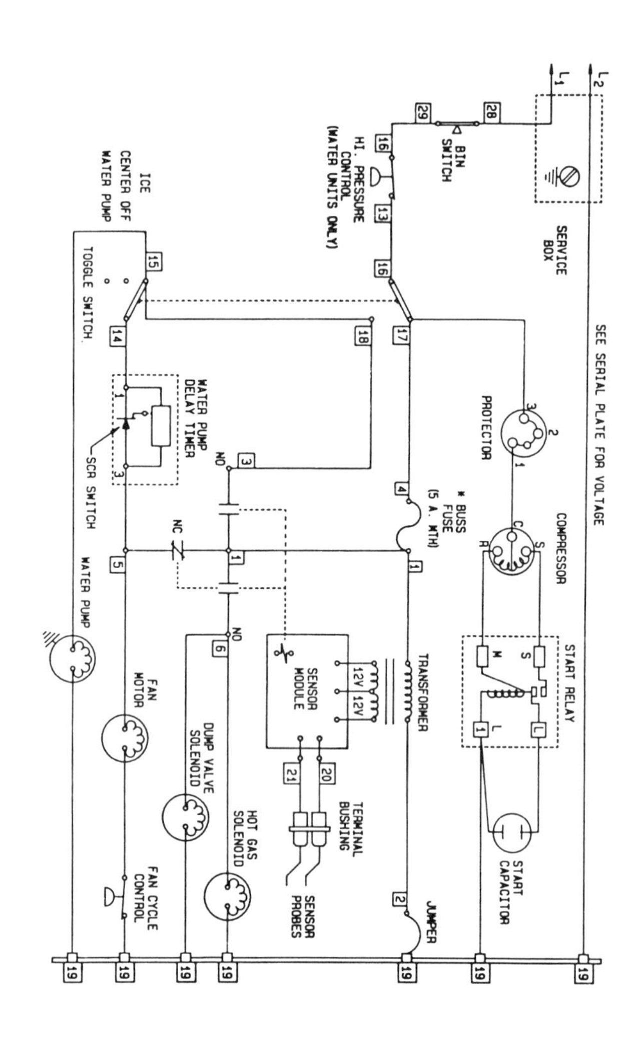

E0400 SERIES AIR-COOLED

These are approximate characteristics that may vary depending on operating conditions.

CYCLE TIMES
TOTAL CYCLE TIME
Freeze time plus harvest time
Water Temp.

Ambient Temp.	50°	70°	90°	Harvest Time
70°	14.5-16.5	16.5-19	18-20.5	1.5-2.5
80°	15.5-18	18-20.5	20-23	
90°	17-19.5	20-23	22-25	
100°	20-23	23-26	27-31	

Based on average ice slab weight of 4lb 2oz to 4lb 12oz.
Times in minutes

24 HOUR ICE PRODUCTION

Air Temp. °F	WATER TEMP. °F		
	50°	70°	90°
70°	410	360	330
80°	380	330	300
90°	350	300	270
100°	300	260	230

OPERATING PRESSURES

Ambient Temp. °F	FREEZE CYCLE		HARVEST CYCLE	
	Head Pressure PSI	Suction Pressure PSI	Head Pressure PSI	Suction Pressure PSI
50°	100-135	22-5.5	65-80	35-65
70°	100-135	22-5.5	70-85	35-65
80°	110-135	22-5.5	80-95	35-65
90°	120-140	22-5.5	85-100	35-65
100°	135-155	22-5.5	90-115	35-65

E0400 SERIES WATER COOLED

These are approximate characteristics that may vary depending on operating conditions.

CYCLE TIMES
TOTAL CYCLE TIME
Freeze time plus harvest time
Water Temp.

Ambient Temp.	50°	70°	90°	Harvest Time
70°	16-18.5	17.5-20	18.5-21	1.5-2.5
80°	16-18.5	17.5-20	19-22	
90°	16-18.5	17.5-20	19-22	
100°	16-18.5	18-20.5	20-23	

Based on average ice slab weight of 4lb 2oz to 4lb 12oz.
Times in minutes

24 HOUR ICE PRODUCTION

Air Temp. °F	WATER TEMP. °F		
	50°	70°	90°
70°	370	340	320
80°	370	340	310
90°	370	340	310
100°	370	330	300

Condenser Water Consumption	WATER TEMP. °F		
	50°	70°	90°
Gal/24 Hrs	255	410	1120
At 125 psi head pressure			

OPERATING PRESSURES

Ambient Temp. °F	FREEZE CYCLE		HARVEST CYCLE	
	Head Pressure PSI	Suction Pressure PSI	Head Pressure PSI	Suction Pressure PSI
50°	125-135	22-5.5	70-90	34-64
70°	125-135	22-2.5	70-90	34-64
80°	125-135	22-2.5	70-90	34-64
90°	125-135	22-2.5	70-90	34-64
100°	125-135	22-2.5	70-90	34-64

E0400 SERIES REMOTE
(With AC0495A Condenser)

These are approximate characteristics that may vary depending on operating conditions.

CYCLE TIMES
TOTAL CYCLE TIME
Freeze time plus harvest time

Water Temp.

Ambient Temp.	50°	70°	90°	Harvest Time
-20° to 70°	15-17	16.5-19	19-22	1.5-2.75
90°	15-17.5	17-19.5	20-23	
100°	15.5-18	17.5-20	21-24	
110°	18.5-21	20.5-23.5	23-26	

Based on average ice slab weight of 4lb 2oz to 4lb 12oz.
Times in minutes

24 HOUR ICE PRODUCTION

Air Temp. °F	WATER TEMP. °F		
	50°	70°	90°
-20° to 70°	400	360	310
90°	390	350	300
100°	380	340	280
110°	320	290	260
Based on 70° air at ice maker			

OPERATING PRESSURES
With accumulator - up to S/N 880962503

Ambient Temp. °F	FREEZE CYCLE		HARVEST CYCLE	
	Head Pressure PSI	Suction Pressure PSI	Head Pressure PSI	Suction Pressure PSI
-20° to 50°	110-130	22-12	65-85	40-50
70°	120-140	24-14	70-90	45-55
80°	120-140	24-14	70-90	45-55
90°	130-150	24-14	80-100	50-60
100°	140-160	25-15	90-110	50-60
110°	150-170	25-15	100-120	60-70
120°	170-190	25-16	110-130	70-80

OPERATING PRESSURES
With Harvest Cycle refrigerant limiter after S/N 880962503

Ambient Temp. °F	FREEZE CYCLE		HARVEST CYCLE	
	Head Pressure PSI	Suction Pressure PSI	Head Pressure PSI	Suction Pressure PSI
-20° to 50°	110-130	22-12	65-85	35-48
70°	120-140	24-14	70-90	38-48
80°	120-140	24-14	70-90	40-50
90°	130-150	24-14	80-100	40-50
100°	140-160	25-15	90-110	42-52
110°	150-170	25-15	90-115	42-52
120°	170-190	25-16	100-130	44-54

E0400 AIR AND WATER
1 PHASE — 60 CYCLE

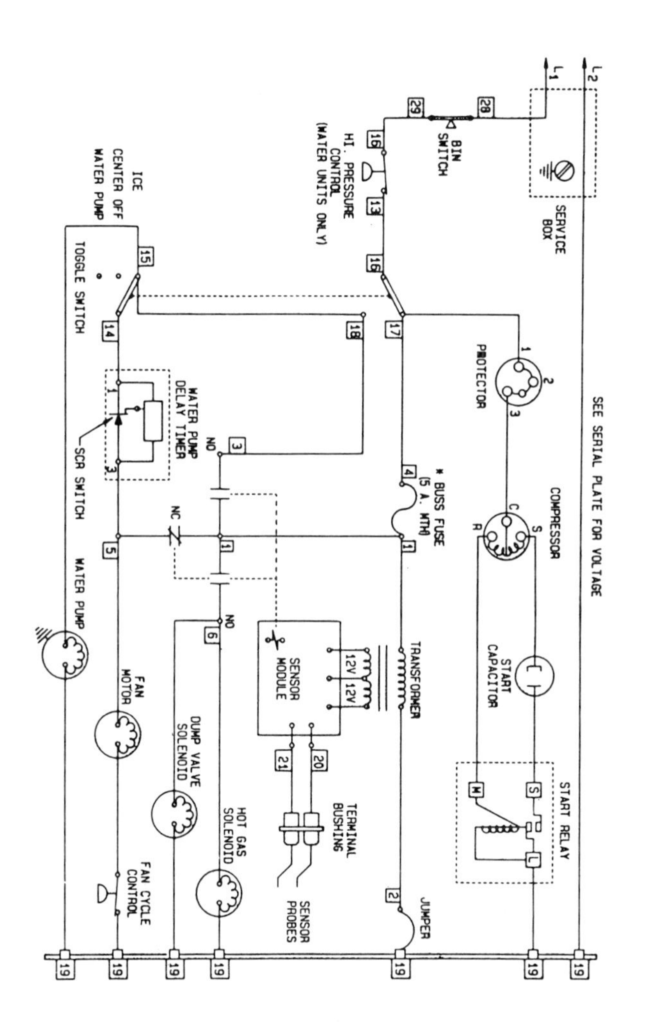

E0400 REMOTE
1 PHASE — 60 CYCLE

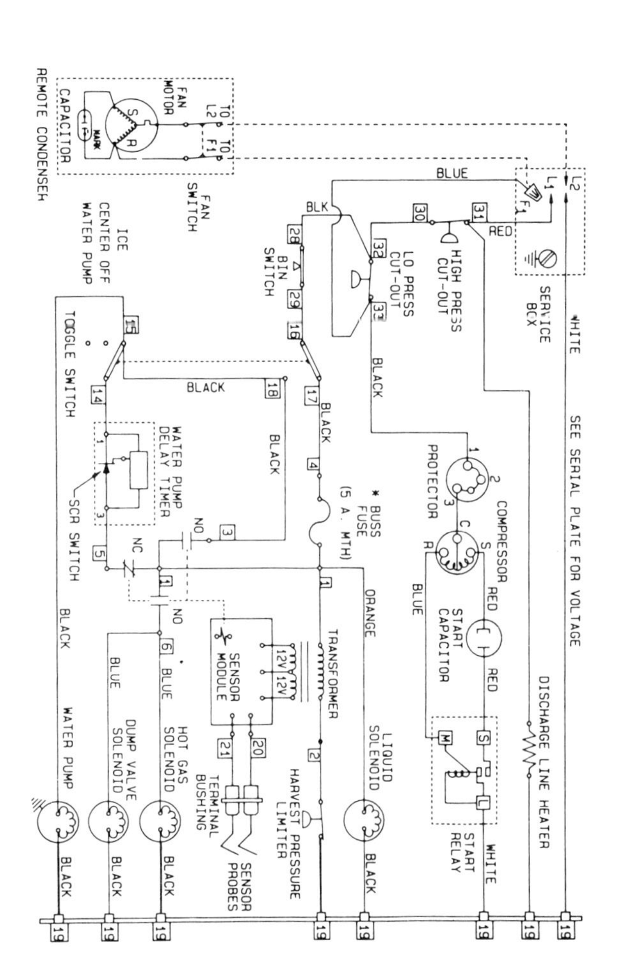

E0400 AIR AND WATER
1 PHASE — 50 CYCLE

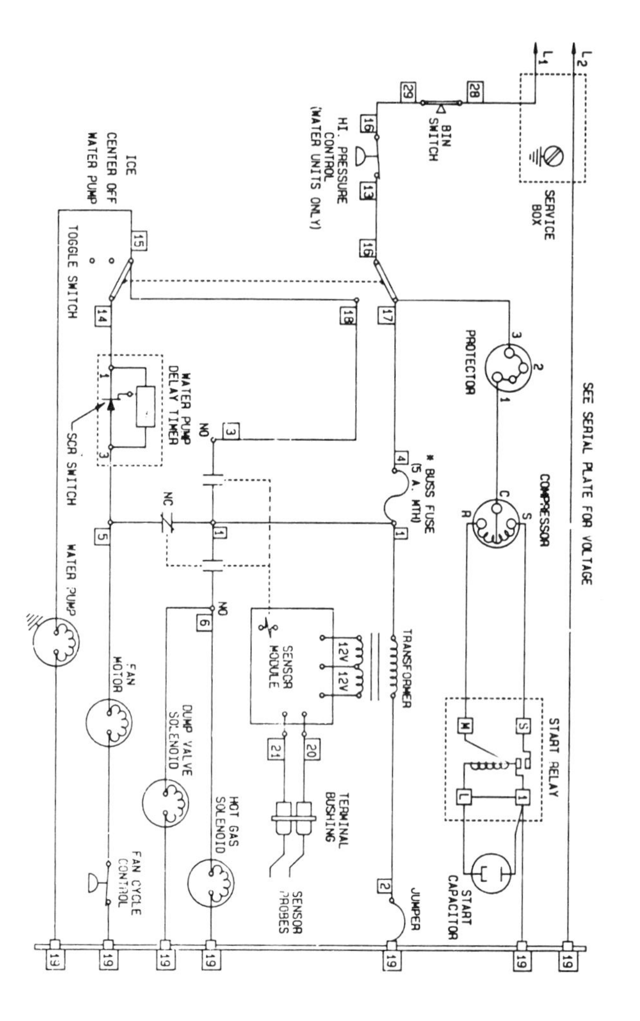

E0600 SERIES AIR-COOLED

These are approximate characteristics that may vary depending on operating conditions.

CYCLE TIMES
TOTAL CYCLE TIME
Freeze time plus harvest time
Water Temp.

Ambient Temp.	70°	80°	90°	Harvest Time
70°	14.75-17	16.5-19	17-20	1.5-3.0
80°	15.25-17.5	17-20	18-20.75	
90°	15.75-18.25	18-20.75	20-23	
100°	17.5-20.5	19.5-22.75	22-25.5	

Based on average ice slab weight of 6lb 4oz to 7lb 4oz. Times in minutes.

24 HOUR ICE PRODUCTION

Air Temp. °F	WATER TEMP. °F		
	50°	70°	90°
70°	610	540	520
80°	590	520	500
90°	570	500	450
100°	510	460	410

OPERATING PRESSURES

Ambient Temp. °F	FREEZE CYCLE		HARVEST CYCLE	
	Head Pressure PSI	Suction Pressure PSI	Head Pressure PSI	Suction Pressure PSI
50°	110-135	18-6	75-90	34-44
70°	110-135	18-6	75-90	35-45
80°	125-150	20-8	80-95	40-50
90°	140-180	21-8	90-105	50-60
100°	170-210	22-9	110-120	55-65

E0600 SERIES WATER-COOLED

These are approximate characteristics that may vary depending on operating conditions.

CYCLE TIMES
TOTAL CYCLE TIME
Freeze time plus harvest time
Water Temp.

Ambient Temp.	50°	70°	90°	Harvest Time
70°	15-17.5	17-19.25	19.5-22.75	1.5-3.0
80°	15.25-17.75	17-20	20-23	
90°	15.5-18	17.5-20.5	20.5-23.75	
100°	17-19.75	19-22	21-24.5	

Based on average ice slab weight of 6lb 4oz to 7lb 4oz.
Times in minutes

24 HOUR ICE PRODUCTION

Air Temp. °F	WATER TEMP. °F		
	50°	70°	90°
70°	600	530	460
80°	590	520	450
90°	580	510	440
100°	530	470	410

Condenser Water Consumption	WATER TEMP. °F		
	50°	70°	90°
Gal/24 Hrs	400	470	1650
At 125 psi head pressure			

OPERATING PRESSURES

	FREEZE CYCLE		HARVEST CYCLE	
Ambient Temp. °F	Head Pressure PSI	Suction Pressure PSI	Head Pressure PSI	Suction Pressure PSI
50°	125-130	18-5	80-90	38-45
70°	125-130	19-6	80-90	38-45
80°	125-130	20-7	80-90	38-45
90°	125-130	21-8	80-90	38-45
100°	125-130	22-8	80-90	38-45

E0600 SERIES REMOTE (with AC0695A condenser)

These are approximate characteristics that may vary depending on operating conditions.

CYCLE TIMES
TOTAL CYCLE TIME
Freeze time plus harvest time
Water Temp.

Ambient Temp.	50°	70°	90°	Harvest Time
-20° to 70°	15-17.5	16.5-19	18.5-23	1.5-3.0
90°	16-18.5	17.5-20.5	18.5-23	
100°	17-20	17.5-20.5	19.5-22.75	
110°	18.5-23	19.5-22.75	21.5-25	

Based on average ice slab weight of 6lb 4oz to 7lb 4oz.
Times in minutes

24 HOUR ICE PRODUCTION

Air Temp. °F	WATER TEMP. °F		
	50°	70°	90°
-20° to 70°	600	540	490
90°	560	510	490
100°	520	510	460
110°	490	460	420
Based on 70° air at ice marker.			

OPERATING PRESSURES

	FREEZE CYCLE		HARVEST CYCLE	
Ambient Temp. °F	Head Pressure PSI	Suction Pressure PSI	Head Pressure PSI	Suction Pressure PSI
-20° to 50°	100-130	28-6	70-90	40-50
70°	100-130	18-6	70-90	40-50
80°	100-140	20-8	70-95	40-50
90°	110-145	21-8	75-100	50-60
100°	110-145	22-9	90-110	55-65

E0600 AIR AND WATER
1 PHASE — 50/60 CYCLE

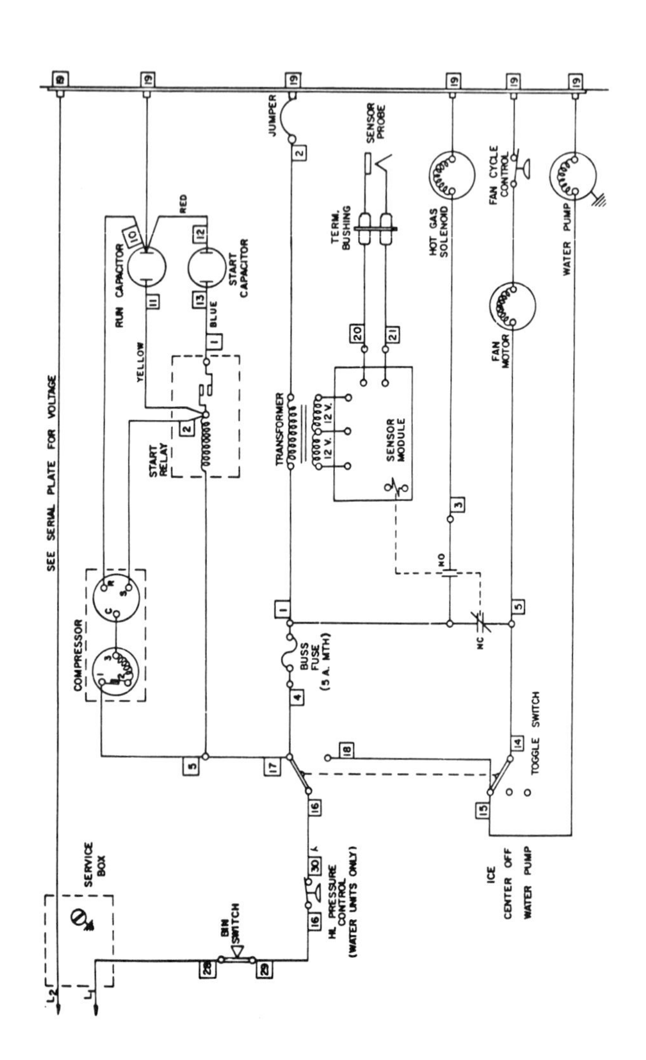

E0600 REMOTE
1 PHASE — 50/60 CYCLE

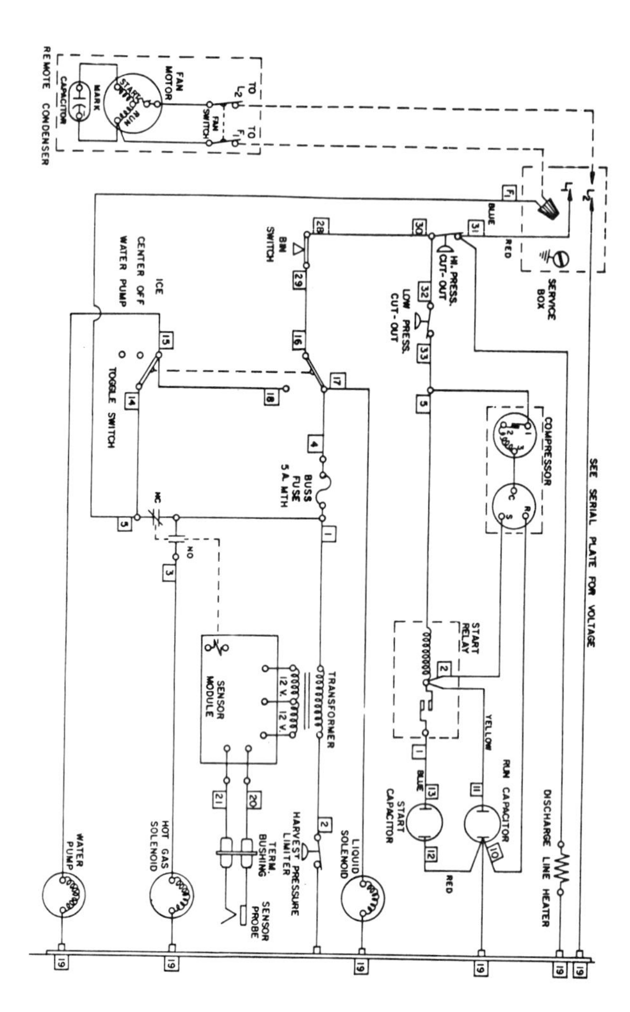

E1100 SERIES AIR-COOLED

These are approximate characteristics that may vary depending on operating conditions.

CYCLE TIMES
TOTAL CYCLE TIME
Freeze time plus harvest time
Water Temp.

Ambient Temp.	50°	70°	90°	Harvest Time
70°	15-18	17-20	19-23	1.5-2.5
80°	16-19	18-21.5	20.5-24	
90°	17.5-20.75	19.5-23	22-26	
100°	18.75-22	21-25	23.75-28	

Based on average ice slab weight of 12lb 4oz to 14lb 8oz. Times in minutes.

24 HOUR ICE PRODUCTION

Air Temp. °F	WATER TEMP. °F		
	50°	70°	90°
70°	1160	1030	910
80°	1090	970	860
90°	1010	900	800
100°	940	840	740

OPERATING PRESSURES

Ambient Temp. °F	FREEZE CYCLE		HARVEST CYCLE	
	Head Pressure PSI	Suction Pressure PSI	Head Pressure PSI	Suction Pressure PSI
50°	105-140	25-8	70-85	35-65
70°	105-140	25-8	80-95	35-65
80°	120-150	25-8	90-100	35-65
90°	140-160	25-8	100-120	35-65
100°	160-190	25-8	105-120	35-65
110°	170-200	25-8	110-125	35-65

E1100 SERIES WATER-COOLED

These are approximate characteristics that may vary depending on operating conditions.

CYCLE TIMES
TOTAL CYCLE TIME
Freeze time plus harvest time
Water Temp.

Ambient Temp.	50°	70°	90°	Harvest Time
70°	16-19	17.5-20.75	19-23	1.5-2.5
80°	16-19	17.75-21	19.75-23.5	
90°	16.5-19.5	18-21.5	20-24	
100°	17-20	18.5-22	21-25	

Based on average ice slab weight of 12lb 4oz to 14lb 8oz. Times in minutes

24 HOUR ICE PRODUCTION

Air Temp. °F	WATER TEMP. °F		
	50°	70°	90°
70°	1100	1010	910
80°	1090	990	890
90°	1070	970	870
100°	1050	950	850

Condenser Water Consumption	WATER TEMP. °F		
	50°	70°	90°
Gal/24 Hrs	761	1400	3640 (est.)
At 135 psi head pressure			

OPERATING PRESSURES

Ambient Temp. °F	FREEZE CYCLE		HARVEST CYCLE	
	Head Pressure PSI	Suction Pressure PSI	Head Pressure PSI	Suction Pressure PSI
50°	125-140	25-8	60-80	30-40
70°	125-140	25-8	60-80	30-40
80°	125-140	25-8	80-100	35-70
90°	125-140	25-8	90-110	35-70
100°	125-140	25-8	100-125	35-70
110°	125-140	25-8	110-130	35-70

E1100 SERIES REMOTE
(with AC1195A condenser)

These are approximate characteristics that may vary depending on operating conditions.

CYCLE TIMES
TOTAL CYCLE TIME
Freeze time plus harvest time
Water Temp.

Ambient Temp.	50°	70°	90°	Harvest Time
-20° to 70°	15.5-18	18-21	20-23.75	1.5-3.5
90°	16-19	18.5-21.75	21-25	
100°	17.5-20.75	18.5-21.75	21-25	
110°	19-23	21.5-25.5	23-27.5	

Based on average ice slab weight of 12lb 4oz to 14lb 8oz. Times in minutes

24 HOUR ICE PRODUCTION

Air Temp. °F	WATER TEMP. °F		
	50°	70°	90°
-20° to 70°	1140	980	880
90°	1100	960	840
100°	1010	960	840
110°	910	820	760
Based on 90° air at ice maker			

OPERATING PRESSURES

Ambient Temp. °F	FREEZE CYCLE		HARVEST CYCLE	
	Head Pressure PSI	Suction Pressure PSI	Head Pressure PSI	Suction Pressure PSI
-20° to 50°	120-130	25-8	60-80	30-40
70°	120-130	25-8	60-80	30-40
80°	130-150	25-8	80-100	30-70
90°	140-170	25-8	90-110	30-70
100°	160-190	25-8	100-125	30-70
110°	190-230	25-8	110-130	30-70

E1100 AIR AND WATER
1 PHASE — 50/60 CYCLE

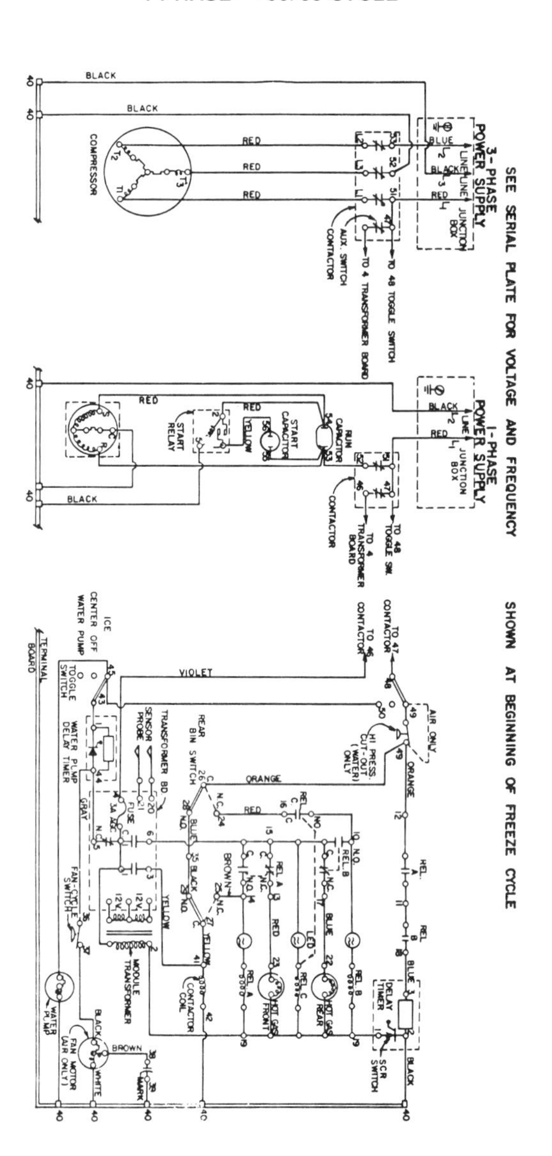

E1100 REMOTE
1 or 3 PHASE — 50/60 CYCLE

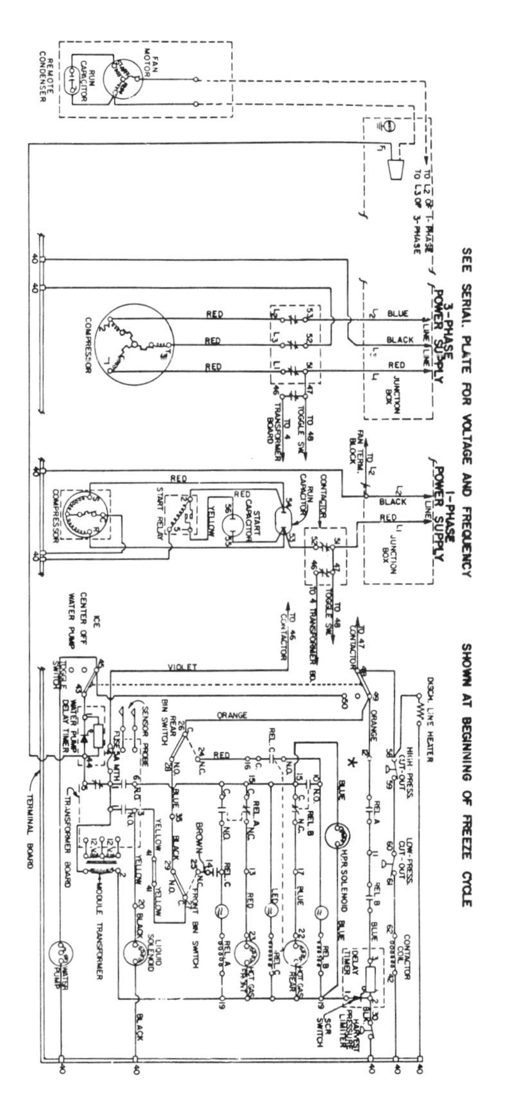

* Not Shown H.P.R. delay timer

G0600 SERIES AIR-COOLED

These are approximate characteristics that may vary depending on operating conditions.

CYCLE TIMES
TOTAL CYCLE TIME
Freeze time plus harvest time
Water Temp.

Ambient Temp.	50°	70°	90°	Harvest Time
70°	10-11.5	11.5-13	12.25-14.25	1.25-2.25
80°	10.5-12	12-13.75	13-15	
90°	12-13.75	13-15	14.75-17	
100°	13.75-16	15.5-18	16.5-19	

Based on average ice slab weight of 4lb 2oz to 4lb 12oz. Times in minutes

24 HOUR ICE PRODUCTION

Air Temp. °F	WATER TEMP. °F		
	50°	70°	90°
70°	600	520	480
80°	560	500	450
90°	500	460	400
100°	430	380	360

OPERATING PRESSURES

Ambient Temp. °F	FREEZE CYCLE		HARVEST CYCLE	
	Head Pressure PSI	Suction Pressure PSI	Head Pressure PSI	Suction Pressure PSI
50°	175-220	36-20	120-150	55-80
70°	180-225	40-22	140-170	65-80
80°	200-250	42-24	160-180	70-85
90°	240-280	44-26	170-200	85-100
100°	260-300	46-26	200-220	100-115
110°	300-350	48-28	225-250	120-130

G0600 SERIES WATER-COOLED

These are approximate characteristics that may vary depending on operating conditions.

CYCLE TIMES
TOTAL CYCLE TIME
Freeze time plus harvest time
Water Temp.

Ambient Temp.	50°	70°	90°	Harvest Time
70°	10.25-11.75	12-13.75	14-15.5	1.25-2.5
80°	10.5-12	12-14	13.75-16	
90°	10.5-12	12.25-14.25	14-16.25	
100°	10.75-12.5	12.25-14.25	14.75-17	

Based on average ice slab weight of 4lb 2oz to 4lb 12oz.
Times in minutes

24 HOUR ICE PRODUCTION

Air Temp. °F	WATER TEMP. °F		
	50°	70°	90°
70°	580	500	440
80°	570	490	430
90°	560	480	420
100°	550	480	400

Condenser Water Consumption	WATER TEMP. °F		
	50°	70°	90°
Gal/24 Hrs	425	700	2575
At 220 psi head pressure			

OPERATING PRESSURES

Ambient Temp. °F	FREEZE CYCLE		HARVEST CYCLE	
	Head Pressure PSI	Suction Pressure PSI	Head Pressure PSI	Suction Pressure PSI
50°	215-225	38-24	130-160	70-85
70°	215-225	40-26	140-160	70-85
80°	215-225	42-26	150-170	70-85
90°	215-225	42-26	150-170	70-85
100°	215-225	44-26	155-175	75-90
110°	215-225	44-26	160-180	75-90

G0600 SERIES REMOTE
(With AC0895A Condenser)

These are approximate characteristics that may vary depending on operating conditions.

CYCLE TIMES
TOTAL CYCLE TIME
Freeze time plus harvest time
Water Temp.

Ambient Temp.	50°	70°	90°	Harvest Time
-20° to 70°	9.5-11.25	11-13	12.25-14.25	1.25-2.25
90°	10.5-12	12.25-14.25	13.75-16	
100°	12-13.75	13.75-16	15.25-17.5	
110°	14-15.5	15.5-18	17-19.5	

Based on average ice slab weight of 4lb 2oz to 4lb 12oz.
Times in minutes

24 HOUR ICE PRODUCTION

Air Temp. °F	WATER TEMP. °F		
	50°	70°	90°
-20° to 70°	620	530	480
90°	560	480	430
100°	500	430	390
110°	440	380	350
Based on 70° air at ice maker			

OPERATING PRESSURES
With accumulator - up to S/N 880961220

	FREEZE CYCLE		HARVEST CYCLE	
Ambient Temp. °F	Head Pressure PSI	Suction Pressure PSI	Head Pressure PSI	Suction Pressure PSI
-20° to 50°	170-190	38-24	150-170	70-90
70°	170-190	38-24	150-170	70-90
80°	180-220	38-24	160-180	80-100
90°	200-240	40-24	170-190	80-100
100°	230-280	42-26	190-220	90-110
110°	260-320	42-26	220-250	110-130
120°	280-340	44-28	240-270	120-140

OPERATING PRESSURES
With H.P.R. Harvest System after S/N 880961220

	FREEZE CYCLE		HARVEST CYCLE	
Ambient Temp. °F	Head Pressure PSI	Suction Pressure PSI	Head Pressure PSI	Suction Pressure PSI
-20° to 50°	170-190	38-32	150-170	75-85
70°	180-210	38-22	150-170	80-90
90°	210-250	40-24	160-180	85-95
100°	240-280	40-24	160-180	85-95
110°	270-320	42-26	170-190	90-100
120°	300-360	46-28	180-200	95-105

G0600 AIR AND WATER
1 PHASE — 50/60 CYCLE

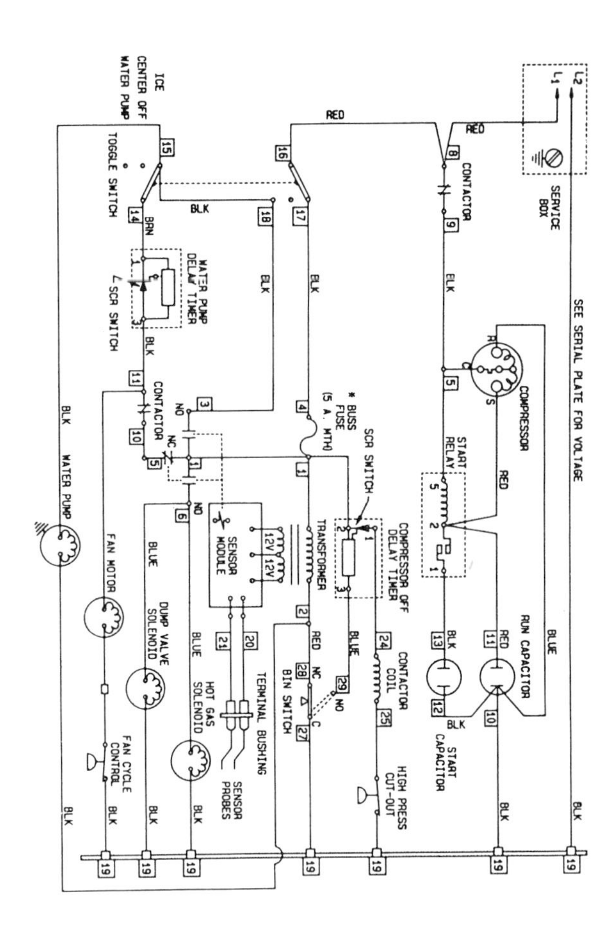

G0600 REMOTE
1 PHASE
accumutator system

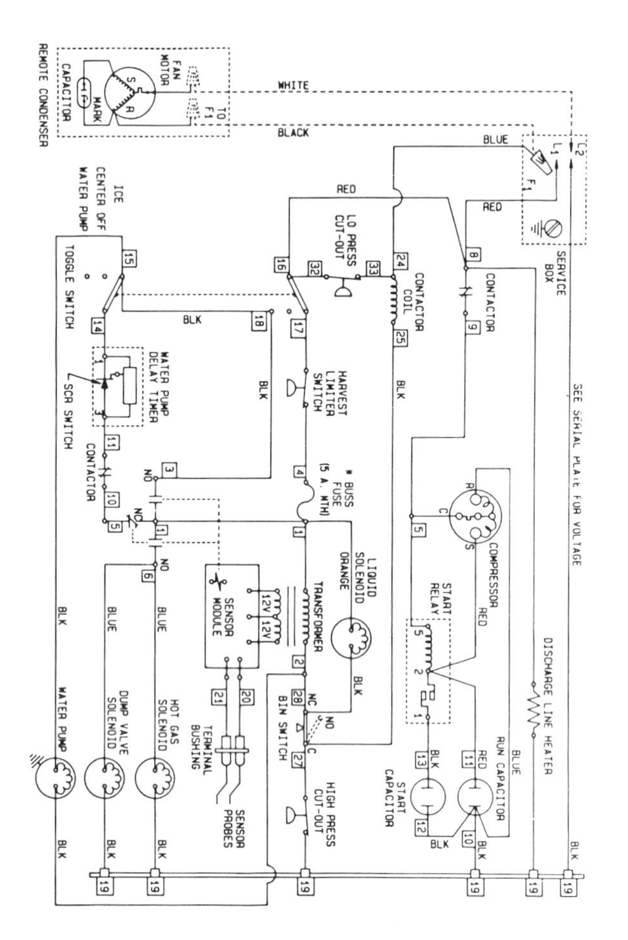

G0800 SERIES AIR-COOLED

These are approximate characteristics that may vary depending on operating conditions.

CYCLE TIMES
TOTAL CYCLE TIME
Freeze time plus harvest time
Water Temp.

Ambient Temp.	50°	70°	90°	Harvest Time
70°	11.25-13	12.5-14.5	14-16.25	1.25-2.25
80°	13.25-15.25	13.5-15.5	15-17.5	
90°	13.5-15.5	15-17.5	17-19.75	
100°	15.5-18	17-20	18.25-21.25	

Based on average ice slab weight of 6lb 4oz to 7lb 4oz.
Times in minutes

24 HOUR ICE PRODUCTION

Air Temp. °F	WATER TEMP. °F		
	50°	70°	90°
70°	800	720	640
80°	740	670	600
90°	670	600	530
100°	580	520	490

OPERATING PRESSURES

Ambient Temp. °F	FREEZE CYCLE		HARVEST CYCLE	
	Head Pressure PSI	Suction Pressure PSI	Head Pressure PSI	Suction Pressure PSI
50°	175-220	38-24	125-150	65-80
70°	175-225	40-24	140-170	70-85
80°	220-270	42-24	160-180	75-90
90°	250-300	44-26	185-215	80-105
100°	275-325	46-28	210-230	105-120
110°	310-360	48-28	225-250	120-130

G0800 SERIES WATER-COOLED

These are approximate characteristics that may vary depending on operating conditions.

CYCLE TIMES
TOTAL CYCLE TIME
Freeze time plus harvest time
Water Temp.

Ambient Temp.	50°	70°	90°	Harvest Time
70°	11.25-13	13.25-15.25	15-17.5	1.5-2.5
80°	12-14	13.5-15.5	15-17.5	
90°	12.25-14.25	13.5-15.75	15-17.5	
100°	13-15	14.25-16.5	15.75-18.25	

Based on average ice slab weight of 6lb 4oz to 7lb 4oz.
Times in minutes

24 HOUR ICE PRODUCTION

Air Temp. °F	WATER TEMP. °F		
	50°	70°	90°
70°	800	680	600
80°	750	670	600
90°	730	660	600
100°	700	630	570

Condenser Water Consumption	WATER TEMP. °F		
	50°	70°	90°
Gal/24 Hrs	500	800	1700
At 240 psi head pressure			

OPERATING PRESSURES

Ambient Temp. °F	FREEZE CYCLE		HARVEST CYCLE	
	Head Pressure PSI	Suction Pressure PSI	Head Pressure PSI	Suction Pressure PSI
50°	235-245	38-24	130-170	65-80
70°	235-245	40-24	130-170	65-80
80°	235-245	40-24	140-180	65-80
90°	235-245	42-24	150-190	65-80
100°	235-245	44-24	160-200	65-80
110°	235-245	46-26	170-200	70-90

G0800 SERIES REMOTE

These are approximate characteristics that may vary depending on operating conditions.

CYCLE TIMES
TOTAL CYCLE TIME
Freeze time plus harvest time
Water Temp.

Ambient Temp.	50°	70°	90°	Harvest Time
-20° to 70°	11-12.75	12.5-14.5	13.25-15.25	1.25-2.25
90°	11.5-13.25	13-15	14.25-16.5	
100°	12.25-14.25	13.75-16	15.5-18	
110°	15-17.5	16.25-19	18.25-21.25	

Based on average ice slab weight of 6lb 4oz to 7lb 4oz.
Times in minutes

24 HOUR ICE PRODUCTION

Air Temp. °F	WATER TEMP. °F		
	50°	70°	90°
-20° to 70°	810	720	680
90°	780	700	630
100°	730	650	580
110°	600	550	490
Based on 70° air at ice maker			

OPERATING PRESSURES
With accumulator - up to S/N 880960000

	FREEZE CYCLE		HARVEST CYCLE	
Ambient Temp. °F	Head Pressure PSI	Suction Pressure PSI	Head Pressure PSI	Suction Pressure PSI
-20° to 50°	170-190	38-24	150-200	75-95
70°	180-210	40-26	175-200	75-95
80°	220-250	40-26	175-200	75-95
90°	230-260	42-26	180-210	80-100
100°	250-300	42-26	200-225	90-110
110°	275-325	44-28	225-250	115-135
120°	300-350	46-28	250-275	130-150

OPERATING PRESSURES
With H.P.R. Harvest System after S/N 880960000

	FREEZE CYCLE		HARVEST CYCLE	
Ambient Temp. °F	Head Pressure PSI	Suction Pressure PSI	Head Pressure PSI	Suction Pressure PSI
-20° to 50°	170-190	38-24	140-160	75-85
70°	180-210	40-26	150-170	75-85
90°	220-260	42-26	160-180	80-90
100°	240-290	44-28	160-180	80-90
110°	270-330	44-28	160-180	85-95
120°	300-360	46-30	160-180	85-95

G0800 AIR AND WATER
1 PHASE — 50/60 CYCLE

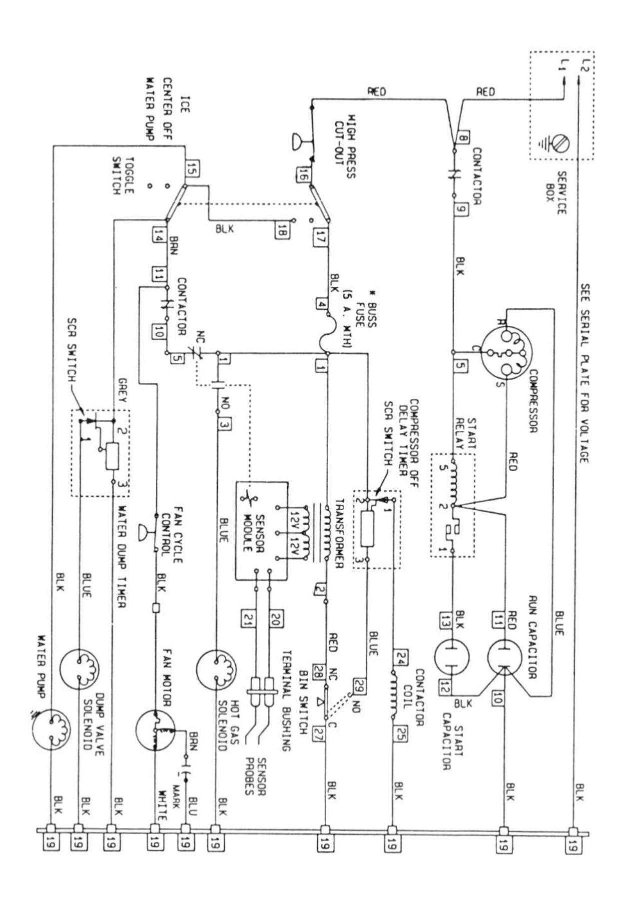

G0800 AIR AND WATER
3 PHASE — 50/60 CYCLE

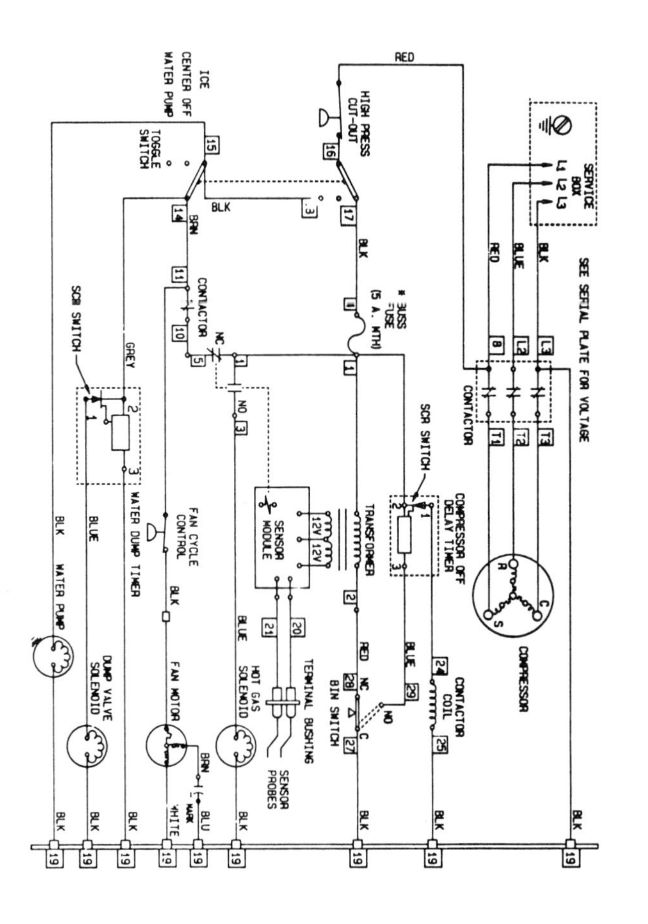

G0800 REMOTE
1 PHASE — 50/60 CYCLE

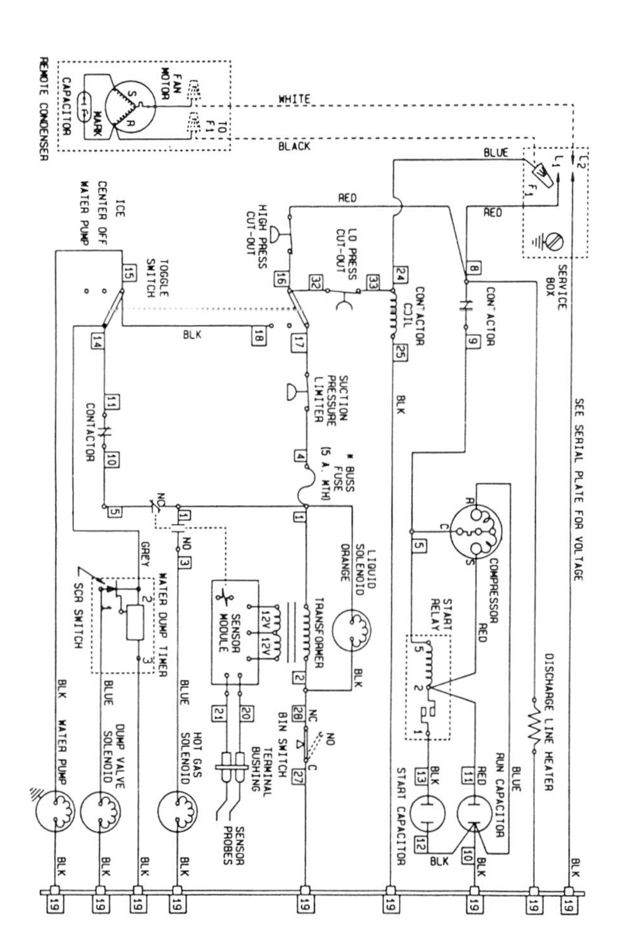

G0800 REMOTE
3 PHASE — 50/60 CYCLE

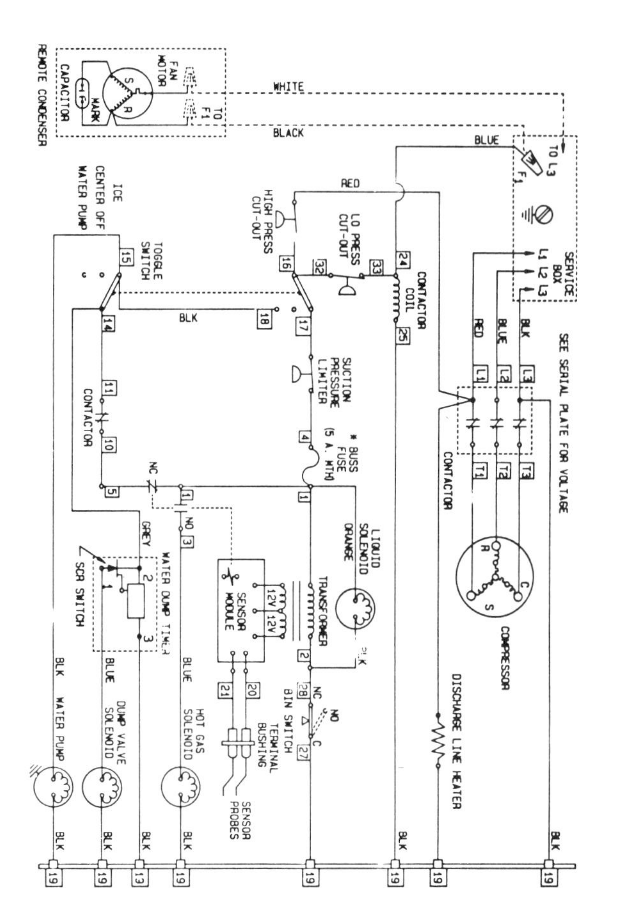

G1200 SERIES AIR-COOLED

These are approximate characteristics that may vary depending on operating conditions.

CYCLE TIMES
TOTAL CYCLE TIME
Freeze time plus harvest time
Water Temp.

Ambient Temp.	50°	70°	90°	Harvest Time
70°	9.25-10.5	10-11.5	10.75-12.25	1.00-2.5
80°	9.75-11	10.5-12	11.75-13.25	
90°	10-11.5	11.25-12.75	12.5-14.25	
100°	11.25-12.75	12.5-14.25	14-15.75	

Based on average ice slab weight of 8lb 2oz to 9lb 4oz. Times in minutes.

24 HOUR ICE PRODUCTION

Air Temp. °F	WATER TEMP. °F 50°	70°	90°
70°	1260	1170	1080
80°	1200	1100	1000
90°	1150	1040	930
100°	1040	930	840

OPERATING PRESSURES

	FREEZE CYCLE		HARVEST CYCLE	
Ambient Temp. °F	Head Pressure PSI	Suction Pressure PSI	Head Pressure PSI	Suction Pressure PSI
50°	175-225	38-24	125-150	65-75
70°	175-225	40-26	125-150	66-76
80°	200-250	42-27	140-165	72-82
90°	220-270	44-28	150-175	80-90
100°	275-325	46-30	175-200	100-110
110°	300-350	48-30	185-210	105-115

G1200 SERIES WATER-COOLED

These are approximate characteristics that may vary depending on operating conditions.

CYCLE TIMES
TOTAL CYCLE TIME
Freeze time plus harvest time
Water Temp.

Ambient Temp.	50°	70°	90°	Harvest Time
70°	9-10.25	9.75-11	11.5-13	1.0-2.5
80°	9-10.5	10.25-11.75	11.75-13.25	
90°	9.25-10.5	10.25-11.75	.75-13.5	
100°	9.25-10.5	10.5-12	12-13.5	

Based on average ice slab weight of 8lb 2oz to 9lb 4oz.
Times in minutes

24 HOUR ICE PRODUCTION

Air Temp. °F	WATER TEMP. °F		
	50°	70°	90°
70°	1290	1500	1010
80°	1280	1140	1000
90°	1270	1130	990
100°	1260	1120	980

Condenser Water Consumption	WATER TEMP. °F		
	50°	70°	90°
Gal/24 Hrs	985	1950	5280
At 230 psi head pressure			

OPERATING PRESSURES

Ambient Temp. °F	FREEZE CYCLE		HARVEST CYCLE	
	Head Pressure PSI	Suction Pressure PSI	Head Pressure PSI	Suction Pressure PSI
50°	225-235	38-24	130-160	72-76
70°	225-235	38-24	135-165	74-78
80°	225-235	38-24	135-165	76-80
90°	225-235	40-24	140-170	78-82
100°	225-235	44-24	145-175	78-82
110°	230-240	45-24	145-175	80-84

G1200 SERIES REMOTE (with AC1295A condenser)

These are approximate characteristics that may vary depending on operating conditions.

CYCLE TIMES
TOTAL CYCLE TIME
Freeze time plus harvest time
Water Temp.

Ambient Temp.	50°	70°	90°	Harvest Time
-20° to 70°	9.75-11	10.5-12	11.75-13.25	1.-2.5
90°	10.25-11.75	11.25-13	12.5-14.25	
100°	11-12.5	12-13.75	13.5-15.5	
110°	11.75-13.5	13-15	14.75-17	

Based on average ice slab weight of 8lb 2oz to 9lb 4oz.
Times in minutes

24 HOUR ICE PRODUCTION

Air Temp. °F	WATER TEMP. °F		
	50°	70°	90°
-20° to 70°	1200	1100	1000
90°	1130	1030	930
100°	1060	960	860
110°	990	890	790
Based on 70° air at ice maker			

OPERATING PRESSURES

Ambient Temp. °F	FREEZE CYCLE		HARVEST CYCLE	
	Head Pressure PSI	Suction Pressure PSI	Head Pressure PSI	Suction Pressure PSI
-20° to 50°	170-200	36-22	110-140	76-84
70°	175-200	36-22	110-140	76-84
80°	200-225	38-22	120-150	76-84
90°	225-275	40-24	130-170	76-84
100°	250-300	41-24	140-180	76-84
110°	280-330	44-26	150-190	76-84
120°	325-375	48-28	150-190	76-84

G1700 SERIES WATER-COOLED

These are approximate characteristics that may vary depending on operating conditions.

CYCLE TIMES
TOTAL CYCLE TIME
Freeze time plus harvest time
Water Temp.

Ambient Temp.	50°	70°	90°	Harvest Time
70°	10.25-12.25	11.5-13.5	12.5-15	1.5-2.5
80°	10.25-12.25	11.5-13.75	12.75-15	
90°	10.5-12.5	11.5-13.75	12.75-15.25	
100°	10.5-12.5	11.75-14	13-15.5	

Based on average ice slab weight of 12lb 4oz to 14lb 8oz.
Times in minutes

24 HOUR ICE PRODUCTION

Air Temp. °F	WATER TEMP. °F 50°	70°	90°
70°	1710	1530	1390
80°	1700	1520	1380
90°	1690	1510	1370
100°	1680	1500	1360

Condenser Water Consumption	WATER TEMP. °F 50°	70°	90°
Gal/24 Hrs	1500	2600	7500
At 220 psi head pressure			

OPERATING PRESSURES

Ambient Temp. °F	FREEZE CYCLE Head Pressure PSI	FREEZE CYCLE Suction Pressure PSI	HARVEST CYCLE Head Pressure PSI	HARVEST CYCLE Suction Pressure PSI
50°	210-230	40-26	110-150	65-80
70°	210-230	40-26	110-150	65-80
80°	210-230	42-28	120-160	70-85
90°	210-230	42-28	120-160	70-85
100°	210-230	42-28	120-160	75-90
110°	210-230	42-28	120-160	75-90

G1700 SERIES REMOTE (with AC1795A condenser)

These are approximate characteristics that may vary depending on operating conditions.

CYCLE TIMES
TOTAL CYCLE TIME
Freeze time plus harvest time
Water Temp.

Ambient Temp.	50°	70°	90°	Harvest Time
-20° to 70°	10-11.75	11-13	12-14.25	1.5-2.5
90°	10-12	11.5-13.5	12.5-15	
100°	10.5-12.5	12-14	13.25-17	
110°	11.25-13.5	12.5-15	14.25-17	

Based on average ice slab weight of 12lb 4oz to 14lb 8oz.
Times in minutes

24 HOUR ICE PRODUCTION

Air Temp. °F	WATER TEMP. °F 50°	70°	90°
-20° to 70°	1780	1600	1460
90°	1750	1530	1390
100°	1660	1480	1320
110°	1550	1390	1230
Based on 90° air at ice maker			

OPERATING PRESSURES

Ambient Temp. °F	FREEZE CYCLE Head Pressure PSI	FREEZE CYCLE Suction Pressure PSI	HARVEST CYCLE Head Pressure PSI	HARVEST CYCLE Suction Pressure PSI
-20° to 50°	170-190	40-25	90-140	71-79
70°	175-200	40-25	90-140	71-79
80°	200-225	40-25	100-150	71-79
90°	220-250	42-26	110-160	71-79
100°	250-230	44-28	120-170	71-79
110°	290-230	46-30	120-180	71-79
120°	330-360	48-32	130-190	71-79

G1200 AND G1700* AIR AND WATER
1 or 3 PHASE — 60 CYCLE
Before Dump Valve System
**G1700 Comes in Water-Cooled and Remote Only*

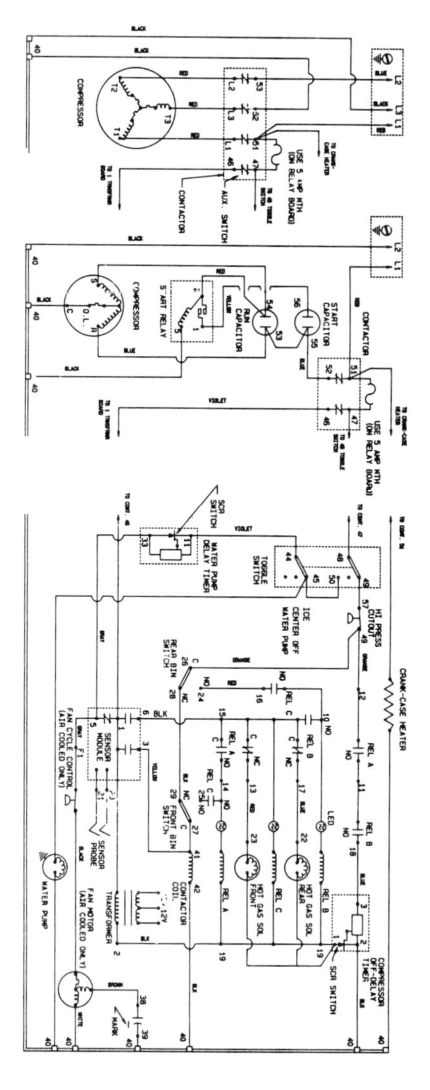

G1200 AND G1700 REMOTE
1 or 3 PHASE — 60 CYCLE
Before Dump Valve System

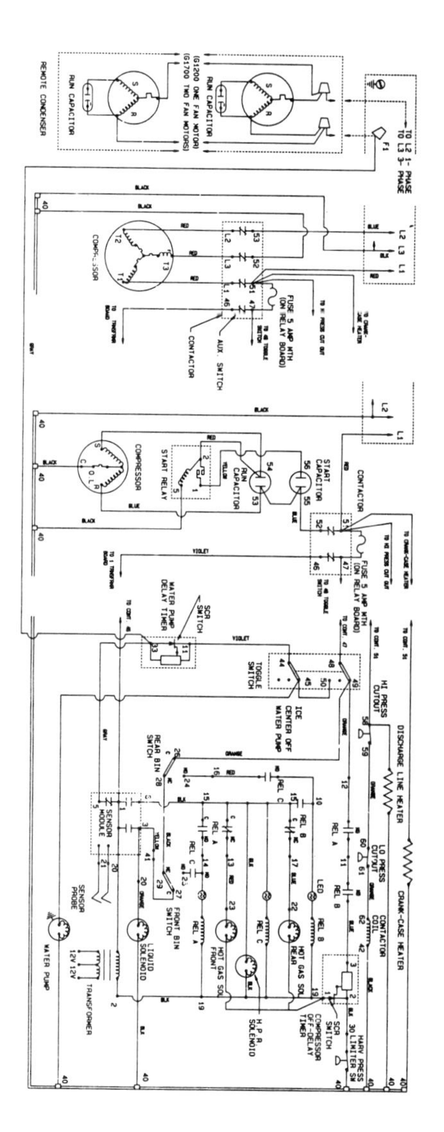

G1200 AND G1700* AIR AND WATER
1 or 3 PHASE — 60 CYCLE
After Dump Valve System

**G1700 Comes in Water-Cooled and Remote Only*

L2 L3 L1
53 52 51 47
L2 L3 L1 46
USE 5 AMP MTH (ON RELAY BOARD)
TO CRANK-CASE HEATER
TO 48 TOGGLE SWITCH
AUX. SWITCH
CONTACTOR
T1 T2 T3
COMPRESSOR
TO 1 TRANSFORMER BOARD
L2 L1
CONTACTOR
51 52 47 46
USE 5 AMP MTH (ON RELAY BOARD)
START CAPACITOR
56 55
54 53
RUN CAPACITOR
START RELAY
C^MPRESSOR
S R O.L. C
VIOLET
TO CONT. 51
CRANK-CASE HEATER
TO CONT. 47
HI PRESS CUTOUT
57 49 12
REL A 11 REL B 18
NO NO
COMPRESSOR OFF-DELAY TIMER
SCR SWITCH
48 50 45 44
TOGGLE SWITCH
ICE CENTER OFF WATER PUMP
REL C
LED
REL B
HOT GAS SOL REAR
REL C
HOT GAS SOL FRONT
REL A
REL C
CONTACTOR COIL
REAR BIN SWITCH
FRONT BIN SWITCH
TRANSFORMER
SENSOR MODULE
SENSOR PROBE
FAN MOTOR (AIR COOLED ONLY)
FAN CYCLE CONTROL (AIR COOLED ONLY)
HARV
WATER PUMP
SCR SWITCH
DUMP VALVE SOLENOID
TO CONT. 46

G1200 AND G1700 REMOTE
1 or 3 PHASE — 60 CYCLE
After Dump Valve System

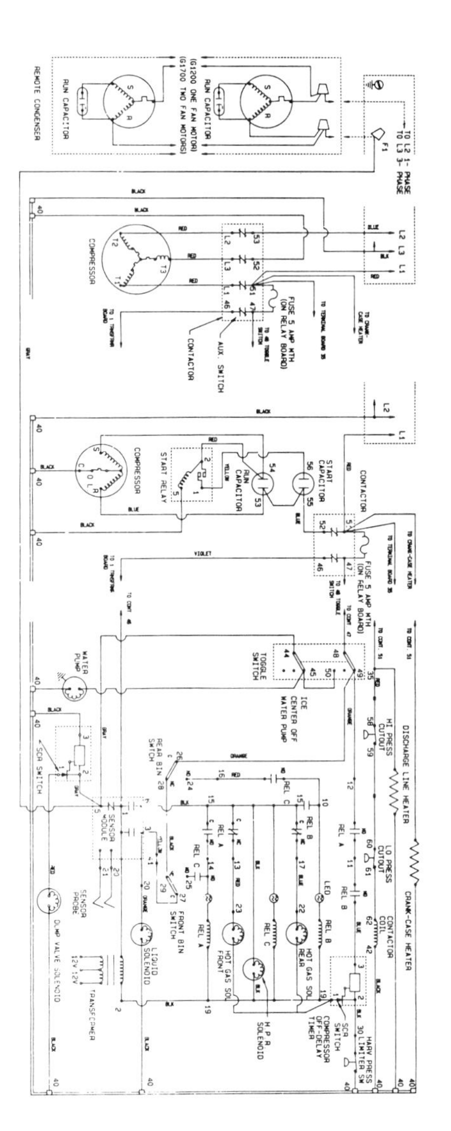

G1200 AND G1700* AIR AND WATER
3 PHASE — 50 CYCLE
4-Wire Hook Up 380/220 VAC

**G1700 Comes in Water-Cooled and Remote Only*

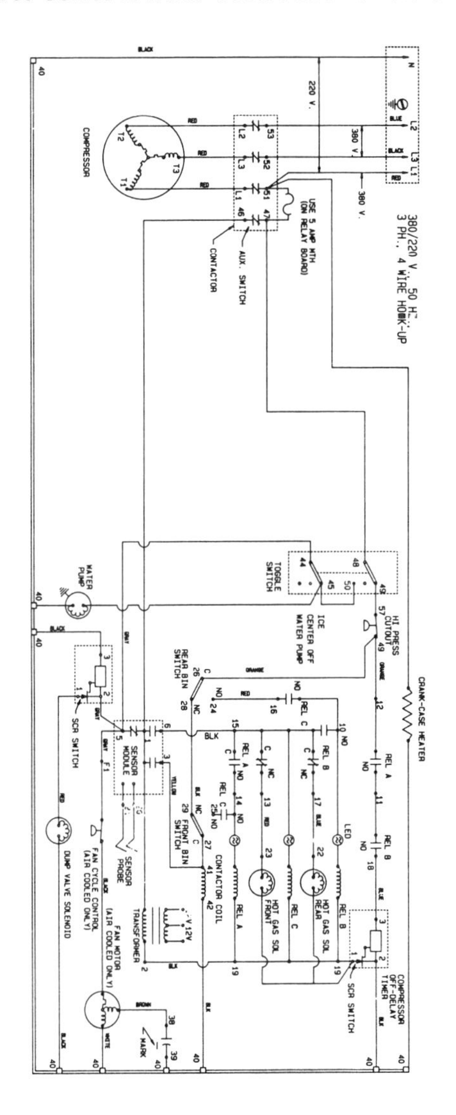

G1200 AND G1700 REMOTE
3 PHASE — 50 CYCLE
4-Wire Hook Up 380/220 VAC

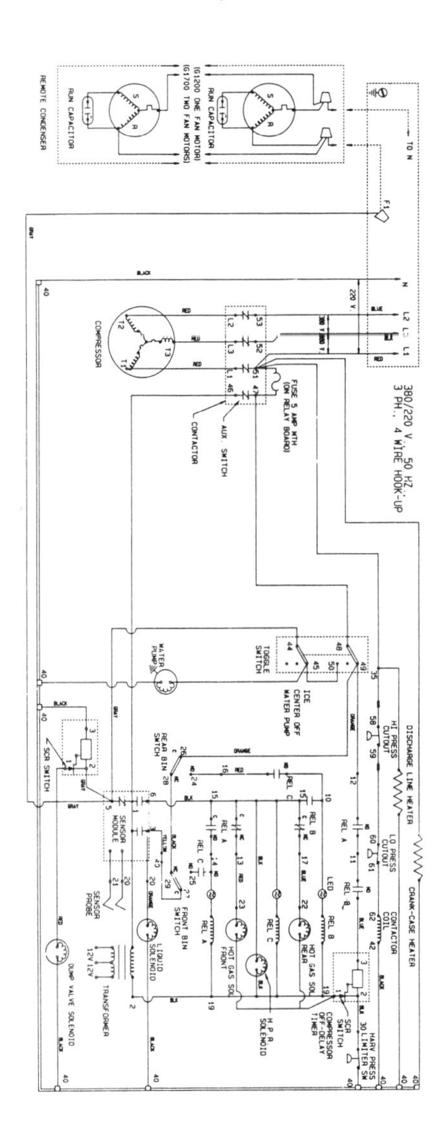

TUBING SCHEMATIC
E0200, E0400, AND E0600 AIR AND WATER

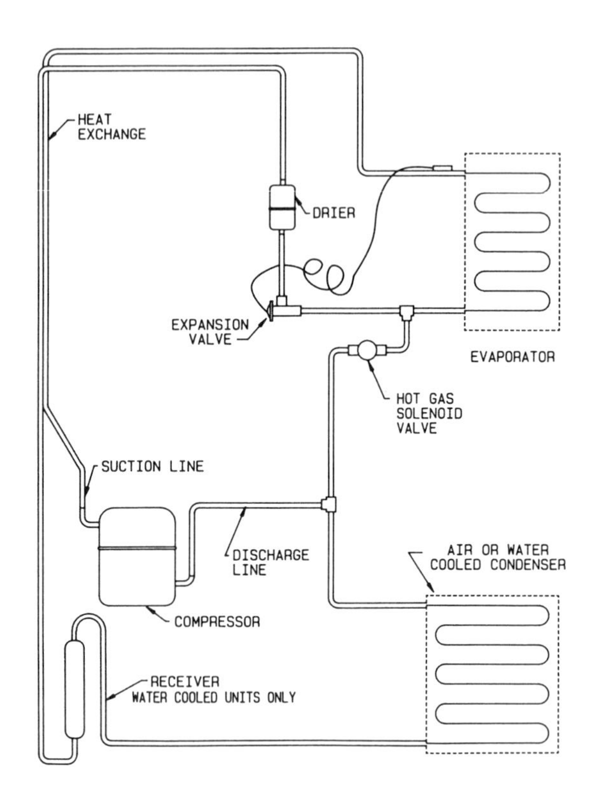

TUBING SCHEMATIC
E0400 AND E0600 REMOTE
(With Accumulator)

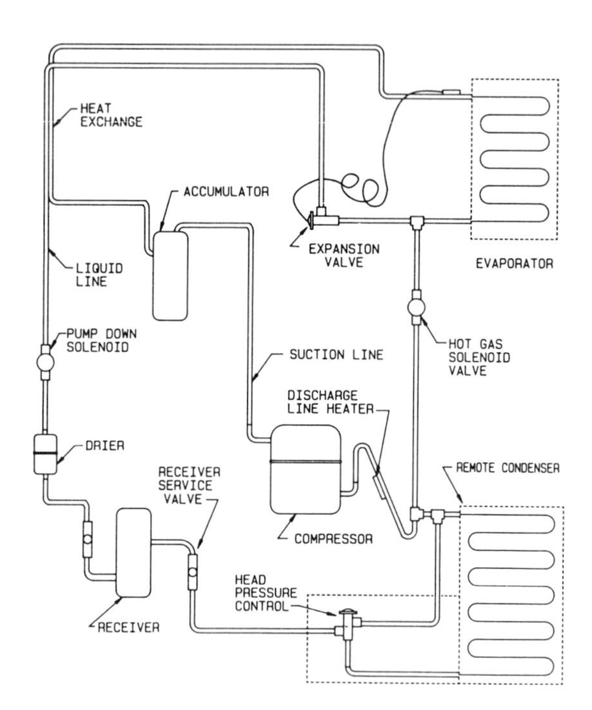

* THE HEAD PRESSURE CONTROL IS LOCATED IN THE REMOTE CONDENSER UNIT FOR THE E0400

TUBING SCHEMATIC
E1100 AIR AND WATER

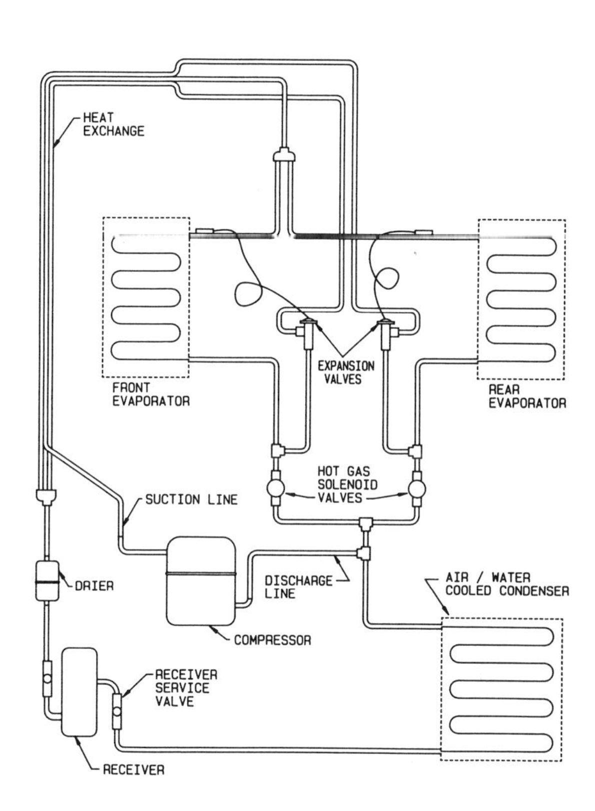

TUBING SCHEMATIC E1100 REMOTE

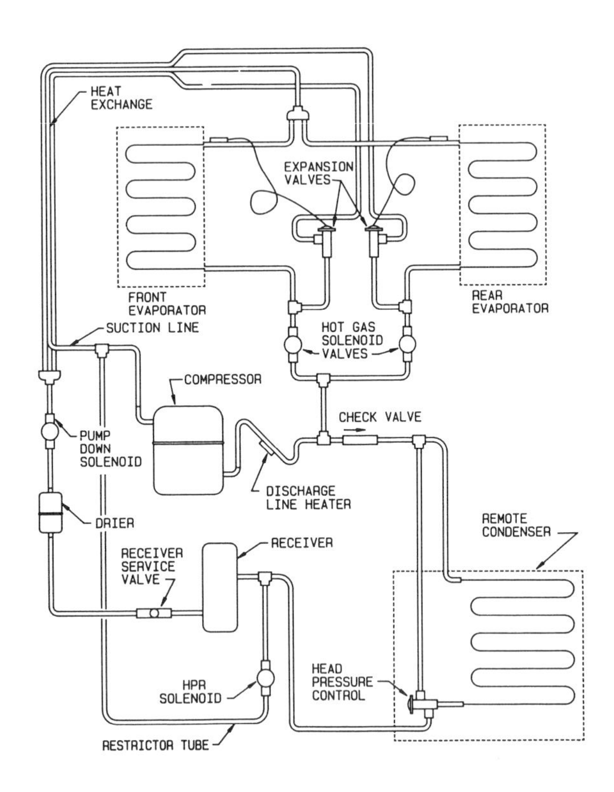

TUBING SCHEMATIC G0600 AND G0800 AIR AND WATER

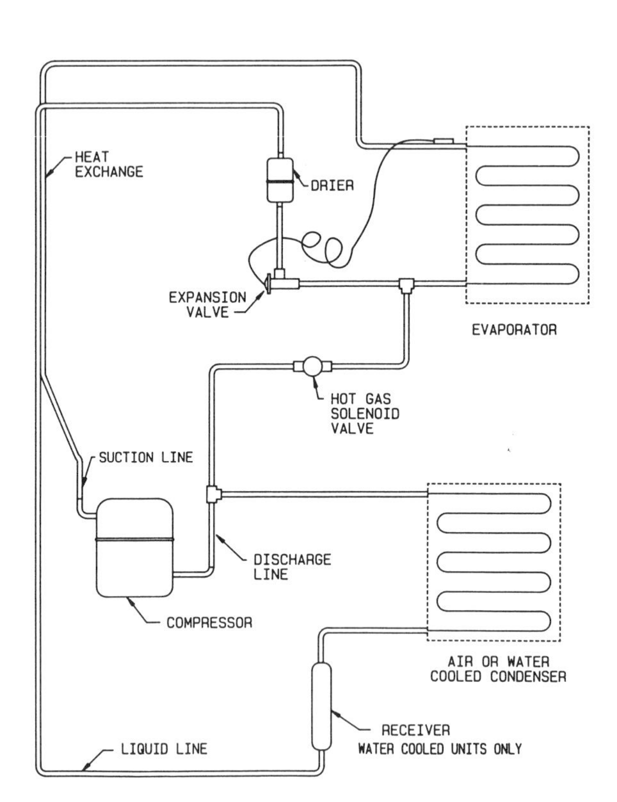

TUBING SCHEMATIC
G0600 AND G0800 REMOTE
(With Accumlator)

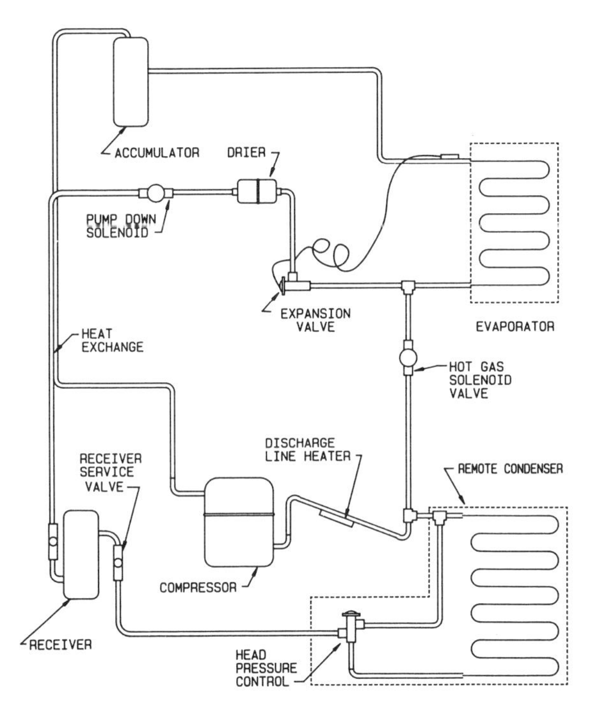

* THE HEAD PRESSURE CONTROL IS LOCATED IN THE REMOTE CONDENSER UNIT.

TUBING SCHEMATIC
G0600 AND G0800 REMOTE
(With HPR System)

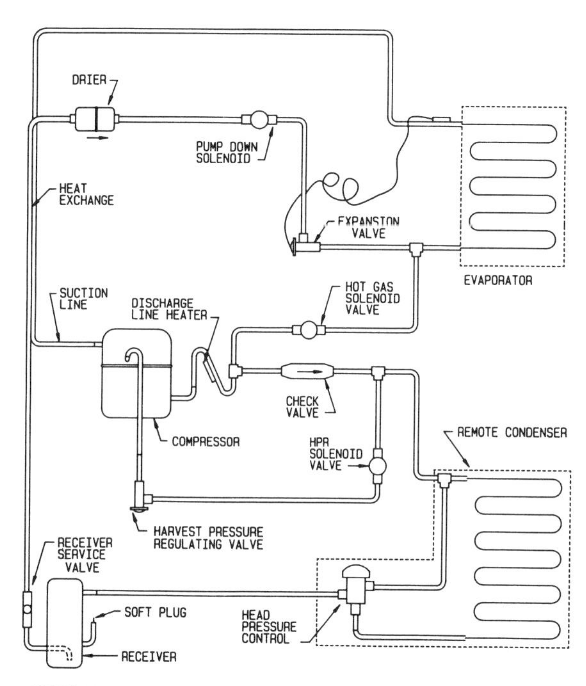

* THE HEAD PRESSURE CONTROL IS LOCATED IN THE REMOTE CONDENSER UNIT.

TUBING SCHEMATIC G1200 AIR AND WATER

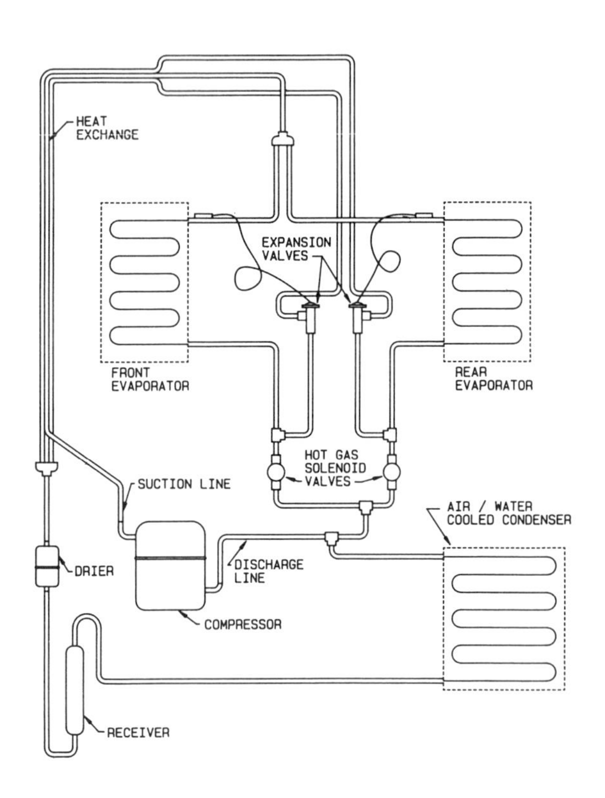

TUBING SCHEMATIC G1200 REMOTE

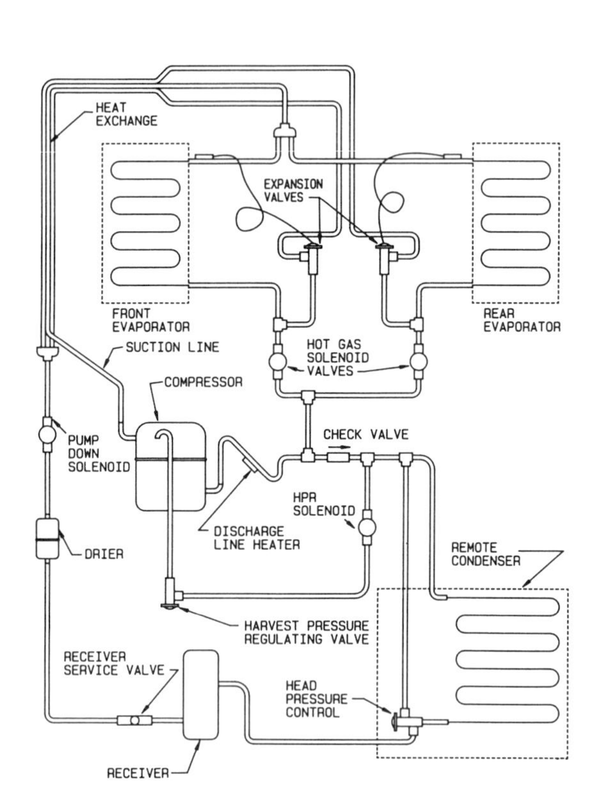

TUBING SCHEMATIC
G1700 WATER
Accumulator out of late models

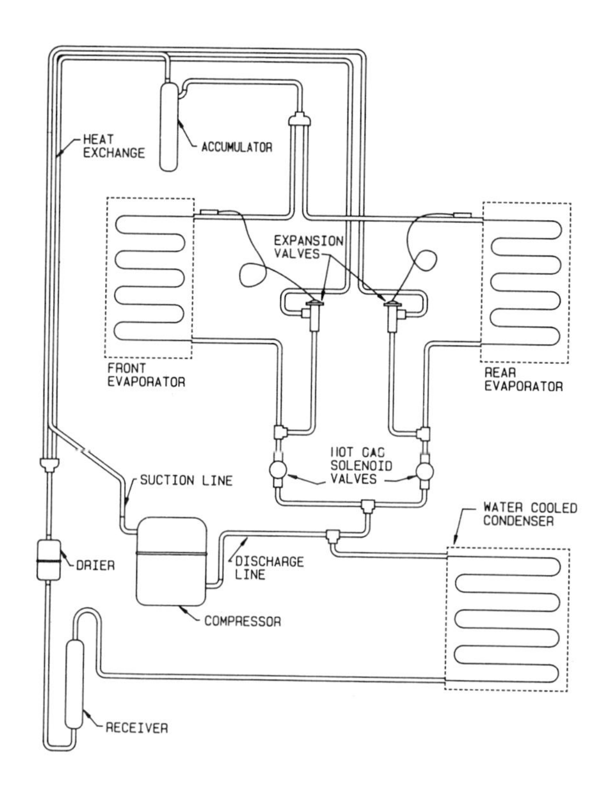

TUBING SCHEMATIC
G1700 REMOTE

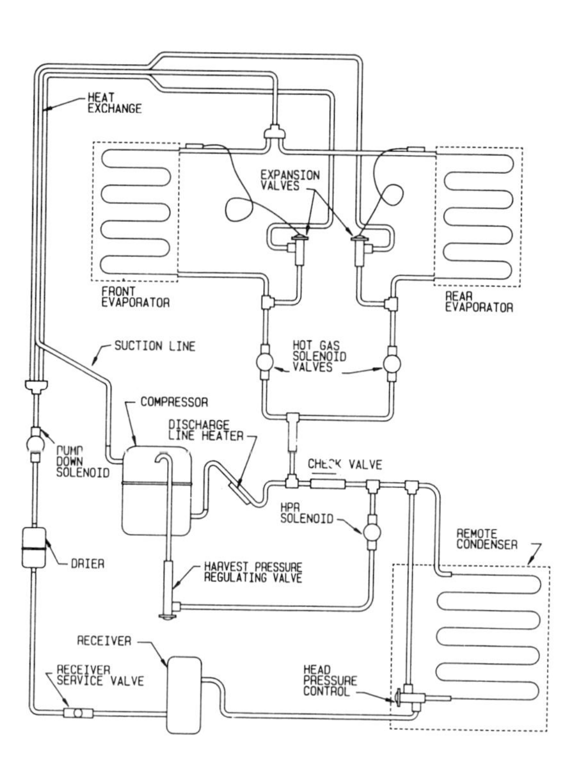

TIMER-PRESSURE SEQUENCE OF OPERATION A-C SERIES

1. The main ON-OFF toggle switch for the cuber is located on the control box. The toggle switch has three positions; the center "OFF" position, the "WATER PUMP" position (only the water pump runs), and the "ICE" position.

2. Freeze Mode
   Placing the toggle switch in the Ice position will start the compressor, the condenser fan motor (air-cooled cubers), and the water pump through terminals #4 and #5 on the solid state timer. The cuber is now in the freeze cycle. As water flows over the evaporator, ice will begin to form on the cube plate.

   As the ice forms the suction pressure will continually decrease. At a preset point the reverse acting low pressure cut-in control will close its contacts. Aproximately 8-12 psig.

   Closing of the low pressure cut-in control contacts energizes the time delay circuit of the solid state timer through terminal #1. The remaining freeze time is dependent upon the timer setting. At the end of the timing sequence the timer relay will energize. This will open the relay contacts between terminals #4 and #5 shutting off the water pump and condenser fan motor and closing its contacts between terminals #3 and #4 energizing the hot gas solenoid valve. The cuber is now in the harvest cycle.

3. Harvest Mode
   Shortly after the hot gas solenoid valve opens, the suction pressure will rise. This opens the low pressure control contacts. The solid state timer relay remains energized through the timer relay interlock circuit.

   As the harvest cycle progresses the hot gas will warm the evaporator allowing the ice to fall out of the evaporator and through the ice chute, opening the damper door. Opening of the damper trips the bin switch, momentarily de-energizing the entire cuber. The timer relay will return to its normal position, de-energizing the hot gas solenoid and energizing the water pump and condenser fan motor (air-cooled cubers). The cuber is now in a new freeze cycle.

   The ice cuber will continue to cycle until the ice storage bin is full. The cuber automatically discontinues ice production as the ice fills the chute, holding the damper door and the bin switch open.

   If, for some reason, the bin switch would fail when the ice falls through the ice chute, the cuber would remain

in the harvest cycle until the hot gas raised the suction temperature to 70° ± 5°. At this time the safety thermodisc, located on the suction line, would open its contacts, de-energizing the timer relay and hot gas solenoid valve, placing the cuber back into the freeze cycle. As the evaporator begins to cool, the thermodisc will reclose its contacts at 40° ± 5°.

A high pressure cut-out control is used on the water-cooled models only. This control shuts the entire cuber off should the condensing pressure become too high.

TIMER — PRESSURE CONTROLS

Low Pressure Cut-In Control

This is a low pressure reverse acting control that closes its contacts on a drop in pressure and opens its contacts on a rise in pressure. The low pressure control contacts close at approx. 8-12 psig. The differential is fixed at 12 psig so the contacts will open at approximately 22 psig.

The low pressure control is used to initiate the time delay circuit of the solid state timer. If the timer is set at its maximum or minimum setting and the proper bridge thickness cannot be obtained, the low pressure control cut-in point must be recalibrated.

Check Out Procedure:

1. Install a service gauge on the suction service valve and voltmeter probes on terminals #1 and the common binding post "19". Disconnect the water pump.
2. Turn the cuber on. As it runs with the water pump disconnected, the suction pressure will fall. When the suction pressure reaches 12 psig the contacts should close giving you a 208/230 volt reading.
3. To change the cut-in point, turn the phillips head screw clockwise to increase the cut-in point and counterclockwise to decrease the cut-in point.
4. Allow the timer to time out and place the cuber in harvest. The suction pressure will rise, opening the low pressure control contacts at approximately 22 psig.

Solid State Timer:

The primary function of the timer is to control the ice bridge thickness and to initiate the harvest cycle by energizing the hot gas solenoid valve and de-energizing the water pump and condenser fan motor. The timer is activated by the low pressure cut-in control and is de-energized by the opening of the bin switch or thermodisc.

Solid State Timer *(Continued)*

The Model C0600 Ice Cubers are manufactured with several different makes of solid state timers. All these timers are interchangeable. The wiring is the same. Be sure the numbered wires are placed on the terminals with corresponding numbers. When replacing the timer check for proper voltage setting. This setting is determined by the location of the jumper wire.

For all dual voltage timers, 24-0623-3, place a jumper wire across terminals #2 and A for 230 VAC or across terminals #2 and B for 115 VAC units. The jumper wire must be in place for the timer to operate.

TIMER ADJUSTMENT TO INCREASE OR DECREASE BRIDGING BETWEEN CUBES

To adjust the solid state timer:

1. Remove control box timer.
2. Locate the timer and the dial.
3. To increase bridge thickness, rotate dial clockwise.
4. To decrease bridge thickness, rotate dial counterclockwise.

   Adjustments should not be greater than 10° to 15° at one time. After each adjustment allow the cuber to produce two ice harvests and observe bridge thickness. Repeat this procedure as required.

Check Out Procedure for Solid State Timer:

1. Check wiring of timer to wiring diagram.
2. Set toggle switch to OFF position. Place a jumper wire across the low pressure control.
3. Mark approximate position of timing dial and turn fully counterclockwise to get a short time delay period (30 to 60 seconds).
4. Place the toggle switch into ice making position. The timer, fan motor, water pump, and compressor should be energized. With the low pressure control jumpered, check for a line voltage reading at terminals #1 and #19 or N with a voltmeter. ***NOTE:*** *After the hose gas valve opens, the suction pressure will rise opening the low pressure contacts.*
5. At the end of a predetermined timing period, the timer relay contacts should transfer, turning the fan motor and water pump OFF, and energizing the hot gas solenoid. Using a voltmeter there should be a line voltage reading at terminals #3 and #2 confirming proper operation.
6. If the relay contacts fail to transfer after 2-3 minutes with the timer set at minimum setting, check to see if the thermodisc and bin switch are in normally closed position. If the bin switch and thermodisc are closed, replace timer.

7. With the cuber in the harvest sequence, trip the bin switch or place the toggle switch in the "OFF" position and back into "ICE" position. Either of these actions should deactivate the timer relay placing the cuber back in freeze. If proper sequence of all operations is accomplished, the timer is in working order. If not, replace timer.
8. Place toggle switch in "OFF" position. Remove jumper wire from low pressure control. Reset timer dial to previous position. Test cuber for proper operation.

   NOTE: *If needed, the solid state timers can be manually tripped to place the cuber in harvest.*

   Timers with H and H415 terminals can be placed in harvest by shorting across H and H415 terminals. This overrides the timing sequence. Timers without H and H415 terminals, place in harvest by shorting across terminals #3 and #4 on the timer. This overrides the timer completely.

SAFETY THERMODISC

The thermodisc is a safety control located on the suction line. The control is temperature sensitive and opens at 70° ± 5° and closes at 40° ± 5°. The thermodisc prevents the cuber from overheating. If the bin damper door switch should fail when the ice harvests, the cuber hot gas valve would remain open. The suction temperature would rise and the thermodisc would open at 70°. This will place the cuber back into the freezing cycle by de-energizing the timer and hot gas solenoid.

Check Out Procedure:

1. The thermodisc is a normally open control at 70° and above, and recloses at 40°. Closing is accomplished by operating the cuber for approximately 5 minutes. The thermodisc will open only if the cuber sits idle in room temperature for a period of time, or the cuber bin switch fails in harvest.
2. Disconnect the thermodisc leads at terminals #2 and N of the timer, after the cuber has been operating. Check continuity with ohm meter. If there is no reading, check for contact between the suction line and the thermodisc. If contact is adequate, replace thermodisc.
3. To check the cut-out temperature of the thermodisc, remove the damper door from the ice chute. Start the cuber and wait for the evaporator to harvest. With the door removed, the unit will remain in its harvest cycle.

THERMODISC Check Out Procedure *(continued)*
This will increase the suction line temperature sharply until the thermodisc reaches its calibrated opening point (70° ± 5°). From the time the ice has harvested to the point the thermodisc opens, under normal ambient temperatures, it should take no longer than 2-3 minutes.

TROUBLESHOOTING TIMER-PRESSURE OPERATION

Problem:
Unit continues in freezing cycle with the water pump running and the hot gas solenoid valve de-energized. The machine does not electrically go into the harvest cycle.

Troubleshooting Sequence:
To determine the cause the machine must be checked when the evaporator is frozen up. When this condition is found, do not switch the unit off or defrost the evaporator until the service sequence below has been followed. Be careful not to bump or disturb any control which would switch the unit into the harvest cycle before the service sequence is completed.

1. Check for line voltage between terminals #1 and #2 on the timer. If line voltage is there, check for proper installation of the jumper wire and turn the timer to the minimum setting. If the unit does not switch into the harvest mode at this time, replace the timer.
2. If you do not get line voltage between terminals #1 and #2 check for voltage between terminals #2 and #4. Without a line voltage reading, the thermodisc is electrically open. To verify this place the jumper between terminals #19 and #2. If the unit now switches into the harvest mode, replace the thermodisc.
3. If you had line voltage between terminals #2 and #4, the thermodisc is electrically closed and is not the cause of the problem. Check for voltage between #19 and #1. If you do not have line voltage, the low pressure control is open. At this time, it is necessary to install a low side gauge on the unit and determine the suction pressure.
4. If the suction pressure is below the cut-in point of the control, the problem is with the control itself. The control will require readjustment or replacement.
5. If the suction pressure is above the cut-in point of the low pressure control, the refrigeration system operation must be analyzed. Check the following:
   a. Excessive head pressure
   b. Improper charge
   c. Expansion valve flooding
   d. Hot gas solenoid valve leaking
   e. Inefficient compresser

VAPOR PRESSURE, PSI

Temp.	FREON* 11	FREON* 12	FREON* 13	FREON* 22	FREON* 113	FREON* 114	FREON* 500	FREON* 502
-50	28.9	15.4	57.0	6.1		27.2		0.0
-45	28.7	13.3		2.7		26.7		2.0
-40	28.4	11.0	72.7	0.5		26.1	7.9	4.3
-35	28.1	8.4		2.5		25.5	4.8	6.7
-30	27.8	5.5	90.9	4.8	29.3	24.7	1.4	9.4
-25	27.4	2.3		7.3	29.2	23.9	1.1	12.3
-20	27.0	0.6	111.7	10.1	29.1	22.9	3.1	15.5
-15	26.5	2.4		13.1	28.9	21.8	5.4	19.0
-10	26.0	4.5	135.4	16.4	28.7	20.6	7.8	22.8
-5	25.4	6.7		20.0	28.5	19.3	10.4	26.9
0	24.7	9.2	162.2	23.9	28.2	17.8	13.3	31.2
5	24.0	11.8		28.1	27.9	16.1	16.4	36.0
10	23.1	14.6	192.2	32.7	27.6	14.3	19.8	41.1
15	22.1	17.7		37.7	27.2	12.3	23.4	46.6
20	21.1	21.0	225.8	43.0	26.8	10.1	27.3	52.4
25	19.9	24.6		48.7	26.3	7.6	31.6	58.7
30	18.6	28.5	263.3	54.8	25.8	5.0	36.1	65.4
35	17.2	32.6		61.4	25.2	2.1	41.0	72.6
40	15.6	37.0	305.0	68.5	24.5	0.5	46.2	80.2
45	13.9	41.7		76.0	23.8	2.2	51.8	87.7
50	12.0	46.7	351.2	84.0	22.9	4.0	57.8	96.9
55	10.0	52.0		92.5	22.1	6.0	64.1	109.7
60	7.7	57.7	402.4	101.6	21.0	8.1	71.0	115.6
65	5.3	63.8		111.2	19.9	10.4	78.1	125.8
70	2.6	70.2	458.8	121.4	18.7	12.9	85.8	136.6
75	0.1	77.0		132.2	17.3	15.5	93.9	147.9
80	1.6	84.2	521.0	143.6	15.9	18.3	102.5	159.9
85	3.2	91.8		155.6	14.3	21.4	111.5	172.5
90	5.0	99.8		168.4	12.5	24.6	121.2	185.8
95	6.8	108.3		181.8	10.6	28.0	131.3	199.7
100	8.9	117.2		195.9	8.6	31.7	141.9	214.4
105	11.1	126.6		210.7	6.4	35.6	153.1	229.7
110	13.4	136.4		226.3	4.0	39.7	164.9	245.8
115	15.9	146.8		242.7	1.4	44.1	177.4	266.1
120	18.5	157.7		259.9	0.7	48.7	190.3	280.3
125	21.3	169.1		277.9	2.2	53.7	204.0	298.7
130	24.3	181.0		296.8	3.7	58.8	218.2	318.0
135	27.4	193.5		316.5	5.4	64.3	233.2	338.1
140	30.8	206.6		337.2	7.2	70.1	248.8	359.2
145	34.4	220.3		358.8	9.2	76.3		381.1
150	38.2	234.6		381.5	11.2	82.6		404.0

Shaded Sections—Inches Hg. Below 1 ATM
*FREON is Du Pont's registered trademark for its fluorcarbon refrigerants.

Remcor

REMCOR

SPIRAL ICEMAKER—DISPENSER
S.I.D.

OWNER'S MANUAL

ALL S.I.D. WITH 250LB. STORAGE

MODELS:
SID851A/250S
SID851W/250S
SID851A/250S—BC
SID851W/250S—BC

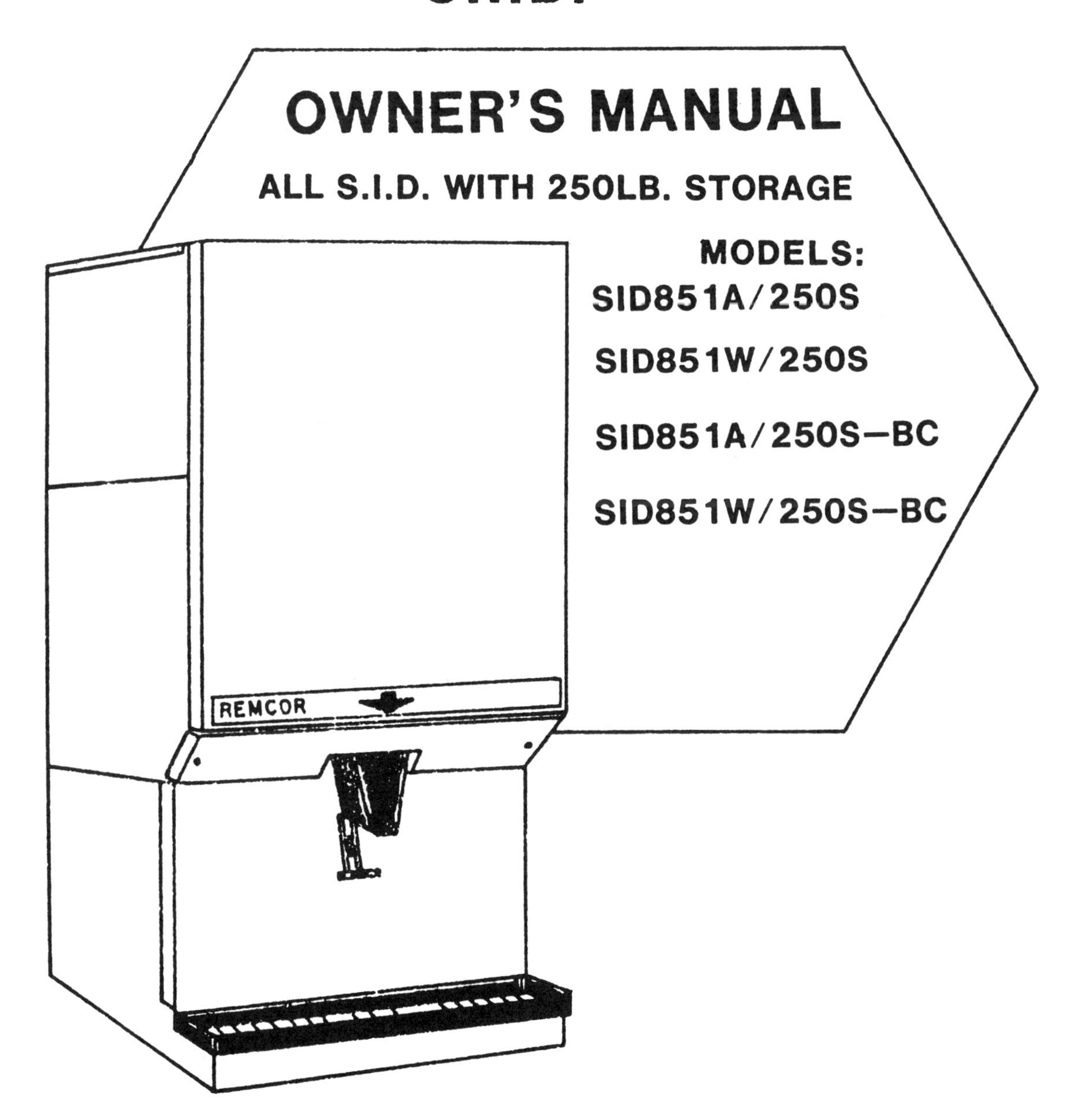

P/N 90914 REV. B

REMCOR PRODUCTS CO.
500 REGENCY DRIVE
GLENDALE HEIGHTS, IL 60139-2268
(312)980-6900

IMPORTANT

INSTALLATION NOTICE

An Everpure Model 9320-42, Systems IV Model B1000, or equal, icemaker quality water treatment unit MUST BE INSTALLED in the water supply line to the icemaker. Failure to do so may result in poor quality ice, low production output, and may cause premature failure of the icemaker evaporator and void the extended evaporator warranty.

This icemaker is provided with a stainless steel evaporator, designed to last the life of the product. But some of the chemicals in treated and untreated water, specifically chlorine and sulfur (sulfide), have the ability to attack stainless steel and cause premature failure. An initial investment in proper water treatment will pay for itself in increased production, quality and long life of the product.

CONTENTS

DESCRIPTION

The Remcor S.I.D. (Spiral Icemaker-Dispenser) is a unique, self-contained, countertop style unit which automatically makes hard, clear cube-quality ice and stores it in a sealed hopper for sanitary dispensing. The ice is made by a new, patented process on a spiral-shaped, stainless steel evaporator and produces true cube quality ice on the outside of the tubes. There are no augers, no compressing of flaked-ice, no bearings and no high gear motor loads in the ice making process. The unit has been designed to be simple, yet effective, to provide many years of trouble free operation.

SPECIFICATIONS

COMPRESSOR: HP: 1

REFRIGERANT: R-502/2.25 lbs (air-cooled), 2.75 lbs (water-cooled)

VOLTAGE: 208-230/115, 1 phase, 60 hz.

AMPS: 18

CIRCUIT AMPACITY: 20

FUSE SIZE: 20A Time-Delay

ICE STORAGE CAPACITY: 225 lbs.

ICE MAKING CAPACITY: up to 964 lbs/24 hrs.

Lbs/24hr Ice Production

Air Temp	Water Temp					
	40°	50°	60°	70°	80°	90°
60°	964	901	861	806	755	712
70°	916	858	820	777	728	687
80°	817	756	731	700	656	619
90°	725	680	650	630	590	556

SHIPPING WEIGHT: 550 lbs.

UNPACKING

1. With the unit upright, carefully remove the shipping crate. Inspect for shipping damage and report any such damage to the shipper immediately.

2. Open the hinged service door and remove shipping tape from the ice drop cover, storage hopper cover, and agitator in storage hopper.

3. Remove shipping tape from air inlet filter on top panel and from sink grill.

INSTALLATION

1. LOCATION

Locate the icemaker-dispenser indoors in a well ventilated area. Avoid exposure to direct sunlight and/or heat caused by radiation. Ambient room temperature must be in the range of 60 to 90°F. Do not install unit in an enclosed area where heat build-up could be a problem. For proper airflow for the refrigeration system, allow 6 in. clearance at the back of the unit and 12. in above the top panel.

Consult Fig. 1 for utility connection locations.

Consult Fig. 2 for dimensions for mounting unit to the counter with the hardware provided. Note that the unit must be level for proper operation.

The unit must be sealed to the counter. The template drawing (Fig. 2) indicates the openings which must be cut in the counter. Locate the desired position for the unit, then mark the outline dimensions and cut-out locations using the template drawing. Cut openings in counter.

Apply a continuous bead of NSF approved silastic sealant (Dow 732 or equal) approximately ¼" inside of the unit outline dimensions, and around all openings. Then position the unit on the counter within the outline dimensions. All excess sealant must be wiped away immediately.

2. PLUMBING

Connect the icemaker to a cold, potable water source, suitable for drinking. Do not install unit on a water softener line. It is recommended that a hand shut-off valve and strainer be used on the incoming supply line. A 3/8 OD compression tube fitting is provided at the back of the unit for the water supply hook-up (see Fig 1). For proper operation the incoming water supply pressure must be in the range of 30-120 PSIG. Install a pressure regulating valve if above this range!

IMPORTANT

To insure proper icemaker operation and also to reduce the frequency of water-related service problems, a water filter should be installed. Remcor recommends the use of one of the following basic systems:

1. Everpure Inc.
   660 N. Blackhawk Drive
   Westmont, IL 60559
   (312)654-4000

   InsurIce Twin System #9320-42

2. Systems IV
   16632 Burke Lane
   Huntington Beach, CA 92647
   (714)842-4221

   Basic Water System #B1000

For specific recommendations on these filter systems for your local water conditions, consult with a distributor in your area or contact the filter manufacturer.

Connect two 3/4" IPS (or equal) drain lines to the 3/4" threaded drain connections at the lower rear of the unit. These lines _must pitch downward_ to an open drain, and must contain _no traps_, or improper drainage will result.

NOTE: In areas where consistently warm water temperatures are encountered, the use of a Remcor precooler in the water line is recommended to maximize the ice production of this unit. Contact Remcor for more information on this product.

3. <u>ELECTRICAL</u>

A 4 X 2 junction box is located at the rear of the unit for the supply hook-up. Connect the icemaker to its own individual circuit per the national electric code and local code (see specification section for ampacity and fuse size).

IMPORTANT: The wire size must be adequate for the ampacity rating and the supply voltage must be within a range of ± 10% for proper icemaker operation.

Note that unit requires a <u>3-wire</u> system plus earth ground for proper operation.

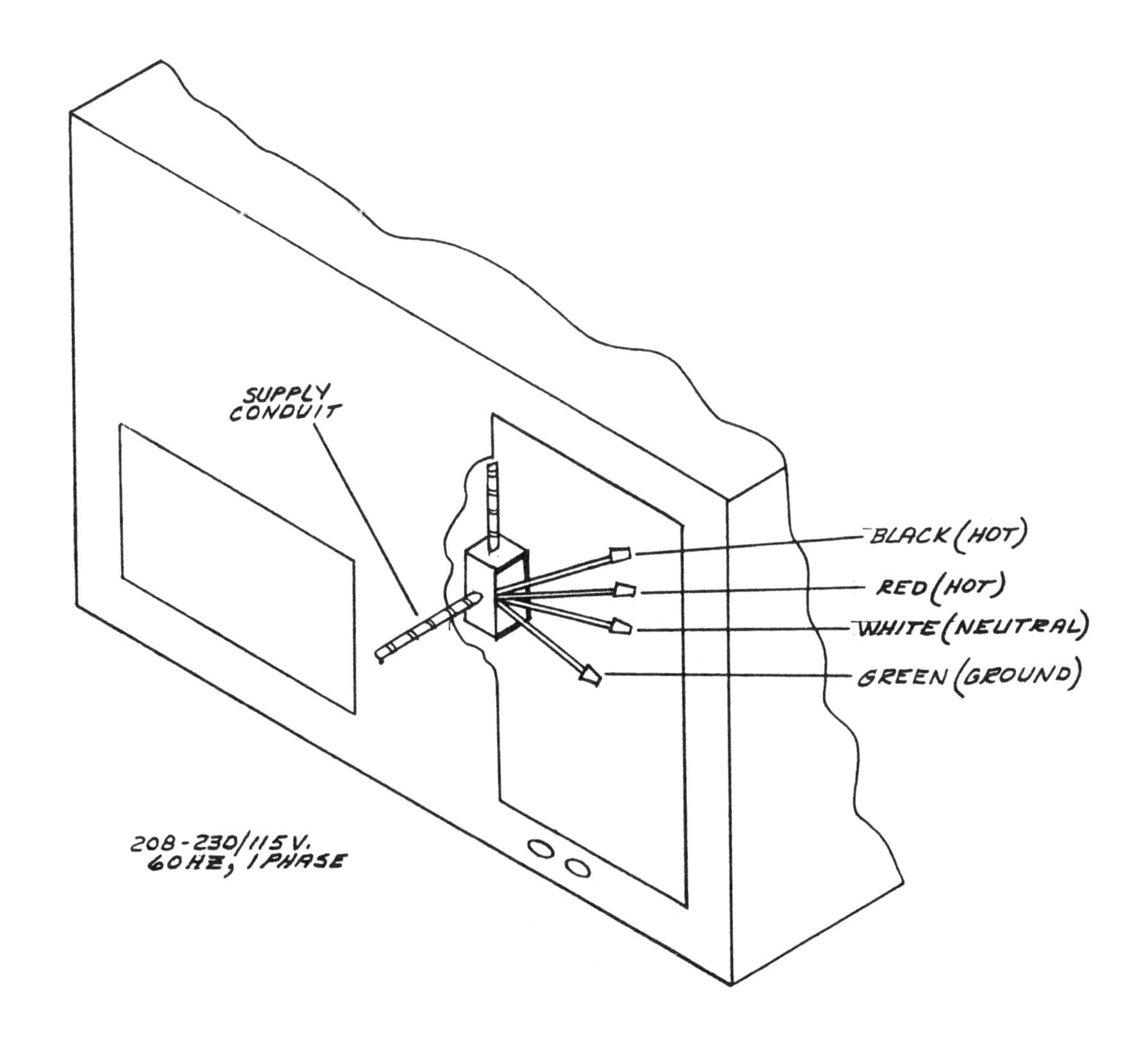

FIG. 1
INSTALLATION DIMENSIONS

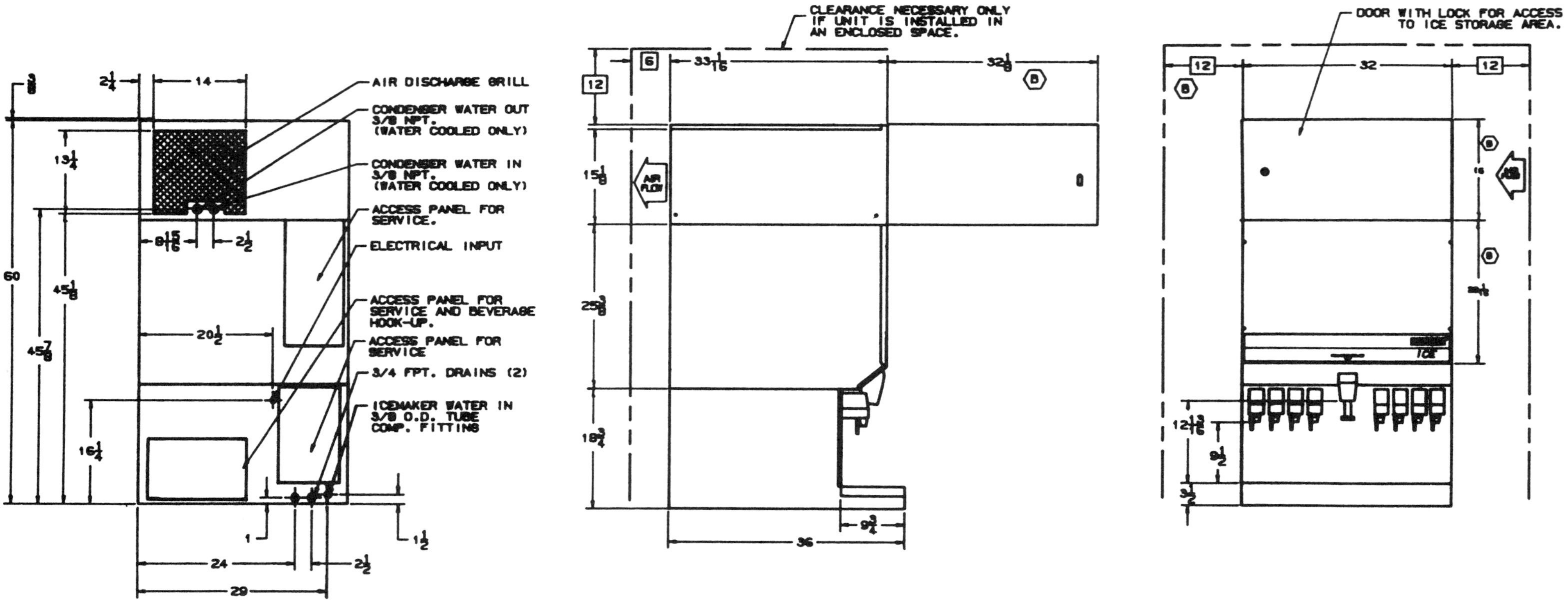

AIR DISCHARGE GRILL

CONDENSER WATER OUT
3/8 NPT.
(WATER COOLED ONLY)

CONDENSER WATER IN
3/8 NPT.
(WATER COOLED ONLY)

ACCESS PANEL FOR SERVICE.

ELECTRICAL INPUT

ACCESS PANEL FOR SERVICE AND BEVERAGE HOOK-UP.

ACCESS PANEL FOR SERVICE

3/4 FPT. DRAINS (2)

ICEMAKER WATER IN
3/8 O.D. TUBE
COMP. FITTING

REAR

CLEARANCE NECESSARY ONLY IF UNIT IS INSTALLED IN AN ENCLOSED SPACE.

LEFT SIDE

DOOR WITH LOCK FOR ACCESS TO ICE STORAGE AREA.

FRONT

FIG. 2

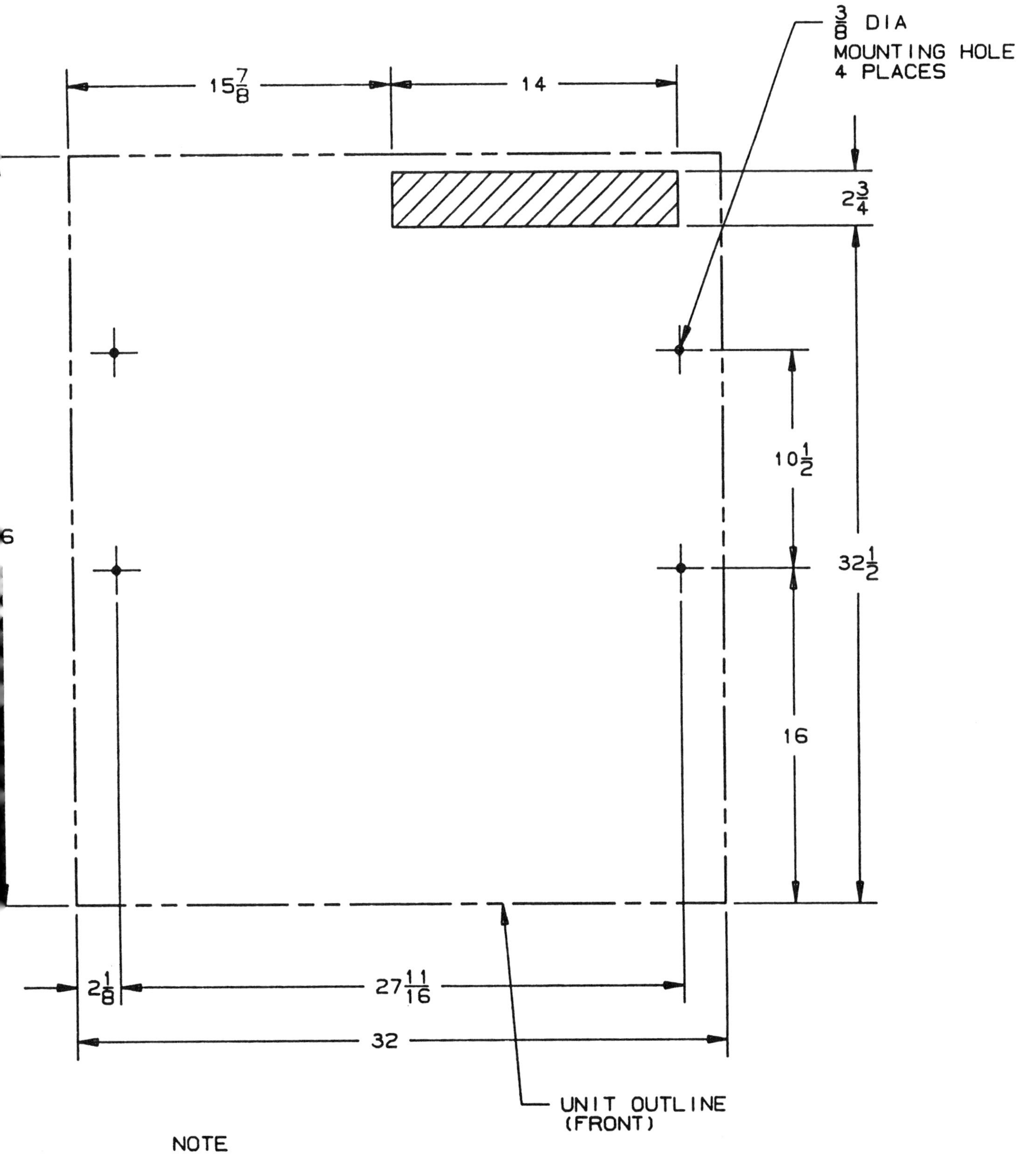

NOTE

SHADED AREA INDICATES OPENING IN BOTTOM FOR BEVERAGE TUBING.

BEVERAGE SYSTEM

"B" models contain beverage faucets only and must be supplied with cold product from any remote coldplate or refrigerated soda factory. "BC" units have a built-in coldplate in addition to the beverage faucets and are designed to be supplied direct from syrup tanks and carbonator, with no additional cooling required.

INSTALLATION

1. Locate the required openings in the counter top for the beverage lines as shown in Fig 2.

2. For "B" models, carefully pull the beverage tubes through the bottom opening in the unit and through the clearance opening in the counter.

3. For "BC" models, tube fittings are provided at the rear of the unit on the coldplate for syrup and water line hook-up.

4. Connect the beverage system product lines as indicated in Figures 3 (-B units) and 4 (-BC units). This work should be done by a qualified serviceman. Note that the hoses are marked with nos. (1-6) for syrup connections and "CW" for carbonated water connection.

FIG. 3

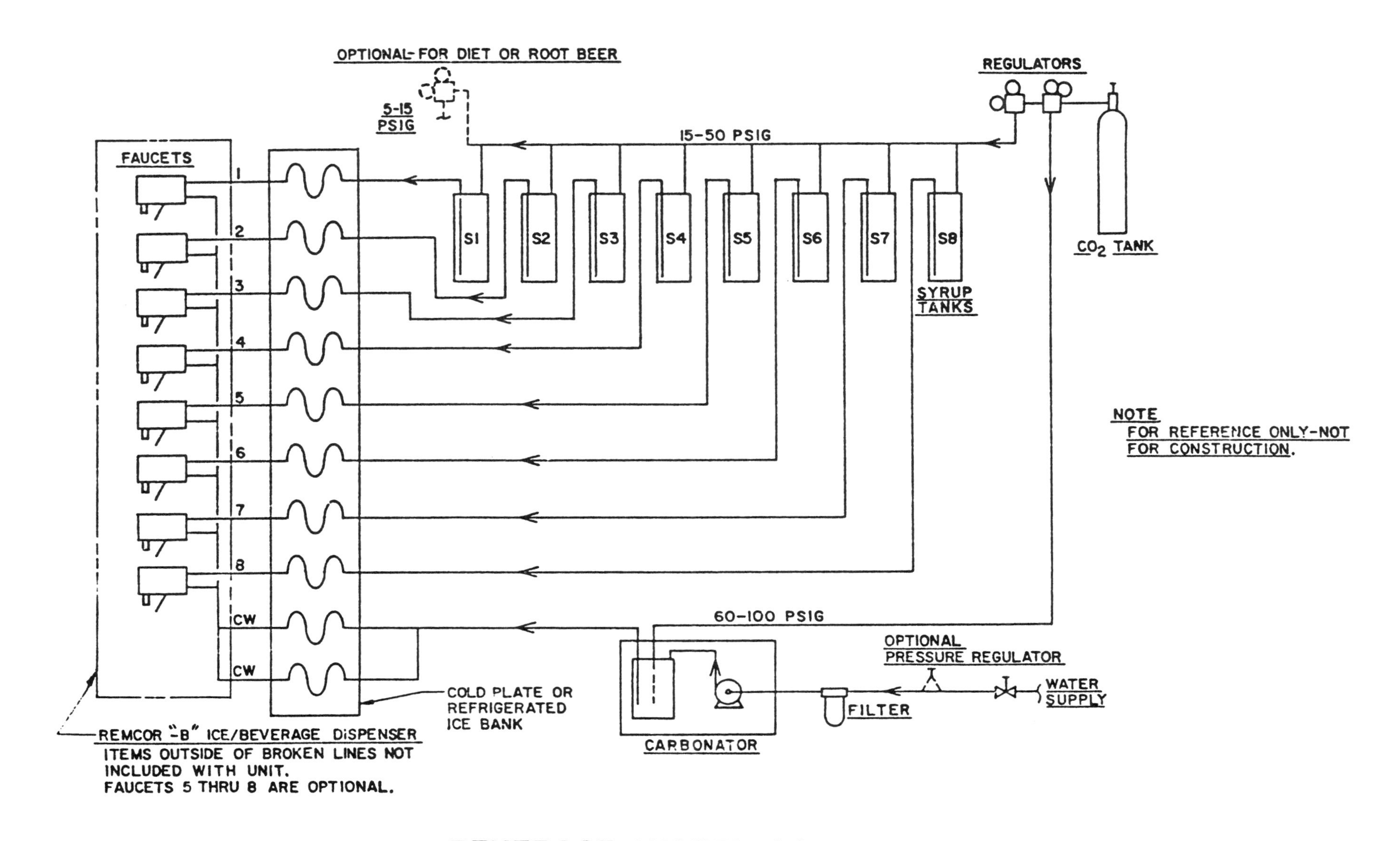

BEVERAGE SYSTEM SCHEMATIC
"—B" MODELS

FIG. 4

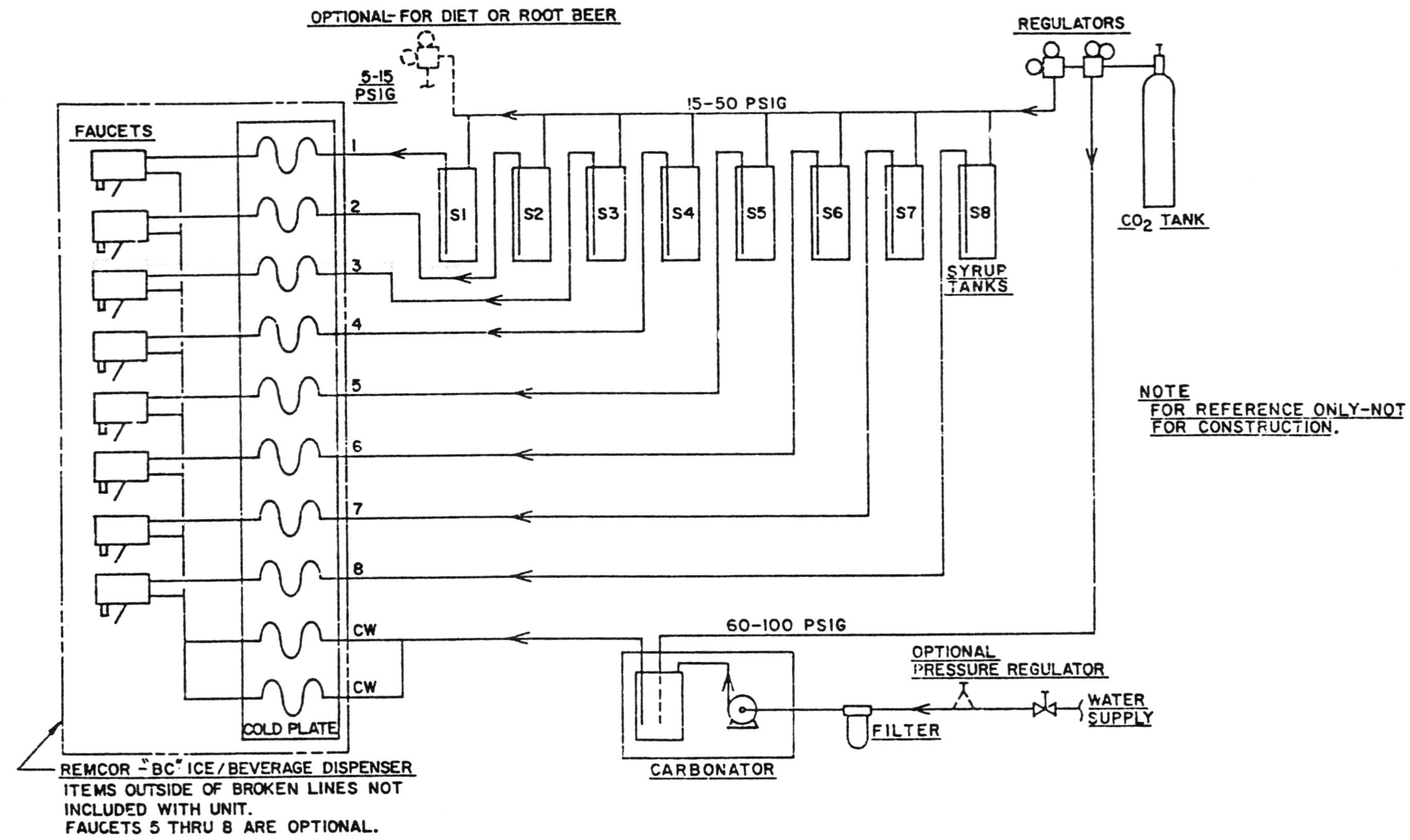

BEVERAGE SYSTEM SCHEMATIC
"–BC" MODELS

START—UP

1. Open the hinged service door. Remove ice drop cover and storage hopper cover.

2. Turn on water to icemaker.

3. Depress the flush switch to verify that water dump valve operates and that water drain lines are open and not plugged.

WARNING: To prevent possible injury, do not stick fingers or hand into icemaker nozzle or hopper with power applied to unit.

4. Put the (stop/run) switch in the run position. Observe that the icemaker goes through proper icemaking and harvest cycle. If unit malfunctions, consult trouble shooting section.

   Note: Due to meltage loss because of a warm storage hopper, it will take longer to fill the hopper the first time than when the icemaker has been operating continuously.

5. Depress the vend switch lever. Check that both the gate solenoid and atitator motor are energized simultaneously to lift the gate slide and rotate the agitator in the storage hopper, respectively. If either component malfunctions, consult the trouble shooting section. Replace the ice drop and hopper covers.

6. For beverage units, start up the beverage system and adjust the faucets to the proper brix. Contact your local syrup distributor for complete information on the beverage system. For units with build-in coldplate, it will take approximately 1 hour from initial machine start-up for coldplate operation to be at full capacity.

7. The bin and harvest thermostats are calibrated at an atmospheric pressure equivalent at 500 feet above sea level. For locations at higher elevations, it may be necessary to readjust these controls. Consult the maintenance/ adjustment procedures section.

OPERATION

A temperature-sensing control bulb located in the storage hopper starts and stops the icemaking process in response to ice level in the hopper. With this ice-level control "calling" for ice (hopper ice level is low), ice begins to form on the stainless-steel tubing coils in the evaporator section of the icemaker. Ice continues to "grow" on the evaporator coils until it contacts the ice thickness probe (low voltage conductivity sensor). At this point, the conductivity probe triggers the harvest timer motor. The harvest timer contains five cam operated switches which function as detailed in the following table:

HARVEST CYCLE

TIME	CAM SWITCH	ACTION
0-86 sec	#1	Timer motor energized
1-13 sec ±1 sec	#4	Water dump valve open
1-37 sec ±2 sec	#2	Hot gas solenoid valve open Air pump off Condenser fan motor off
37-90 sec	#2	Air pump on Condenser fan motor on Hot gas solenoid valve closed
36-69 sec	#3	Harvest motor on
43-47 sec	#5	Hopper agitator motor operates

When ice contacts the ice-level control bulb in the storage hopper, the control will shut down the refrigeration system. If this signal occurs during the harvest cycle, the harvest cycle will be completed before shut-down occurs.

To dispense ice, push the lever located on the lower front panel. Ice will flow from the ice chute until the lever is released.

For units with a built in coldplate, ice will automatically fill the coldplate cabinet. Allow 1 hour for the coldplate to reach its maximum capacity. Start up the beverage system and adjust the faucets to the proper brix. Pushing the lever on any faucet will provide beverage of the appropriate flavor.

FIG. 5

WIRING SCHEMATIC
SID 850/250

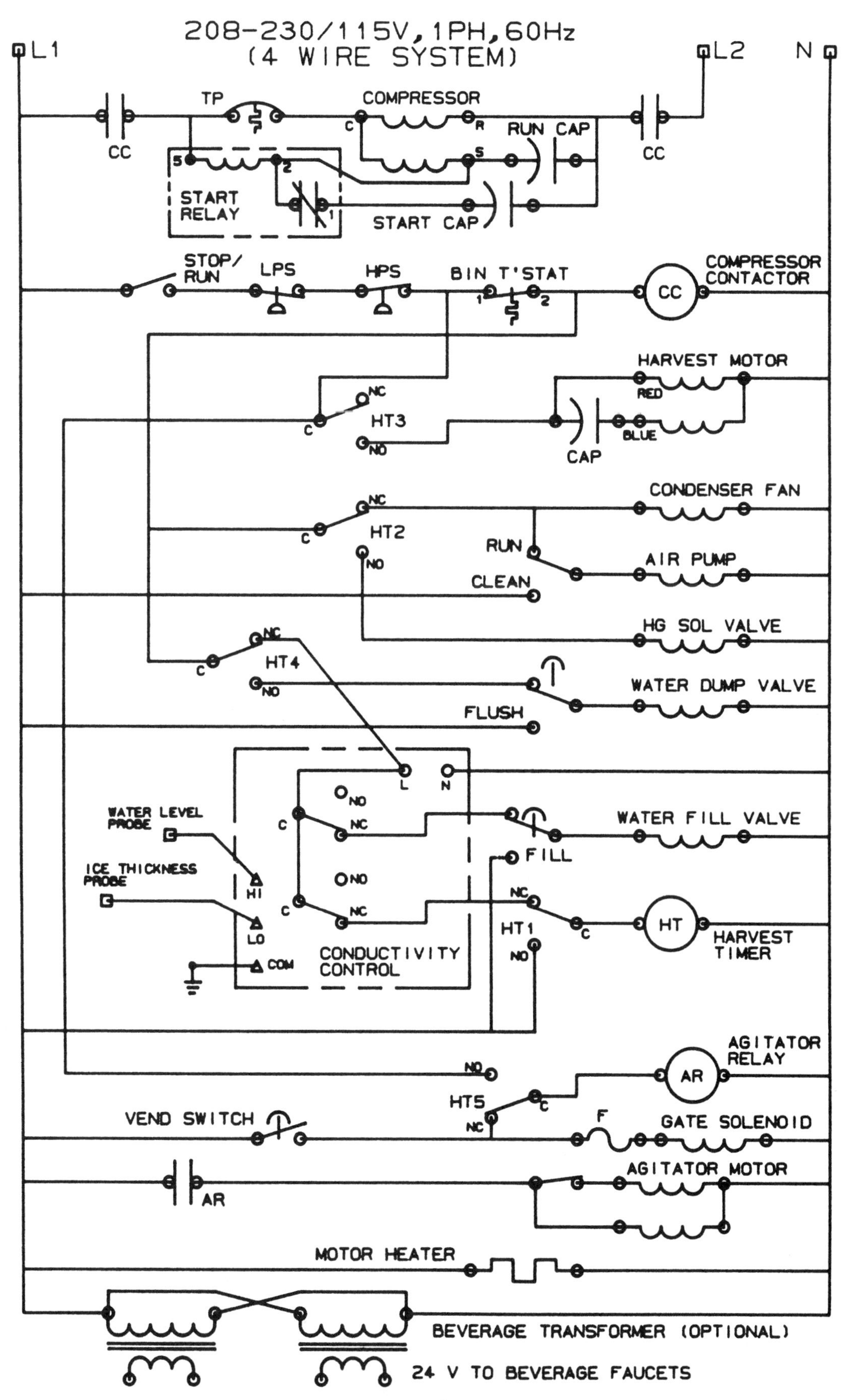

FIG. 5A

WIRING DIAGRAM
SID 850/250

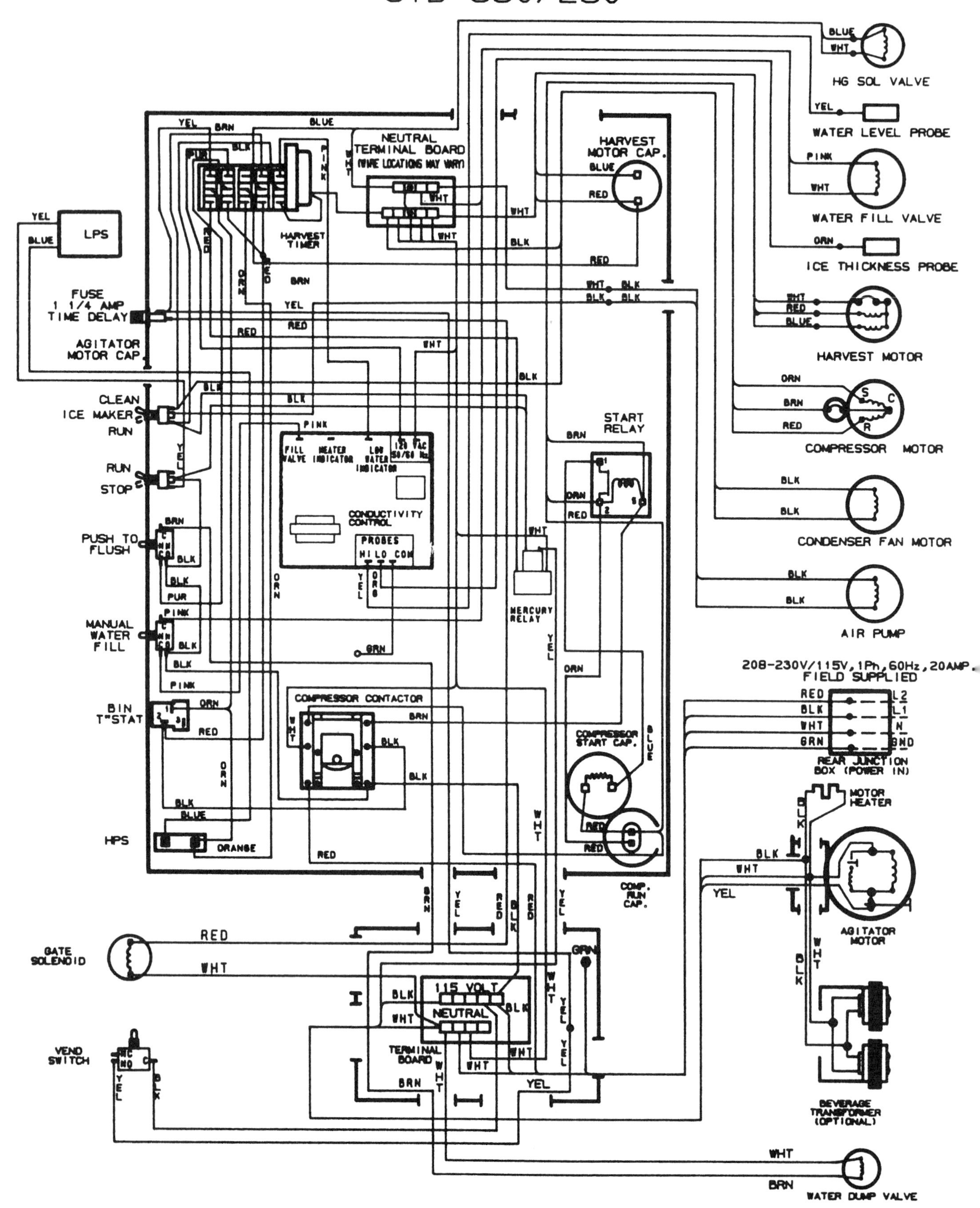

FIG. 6

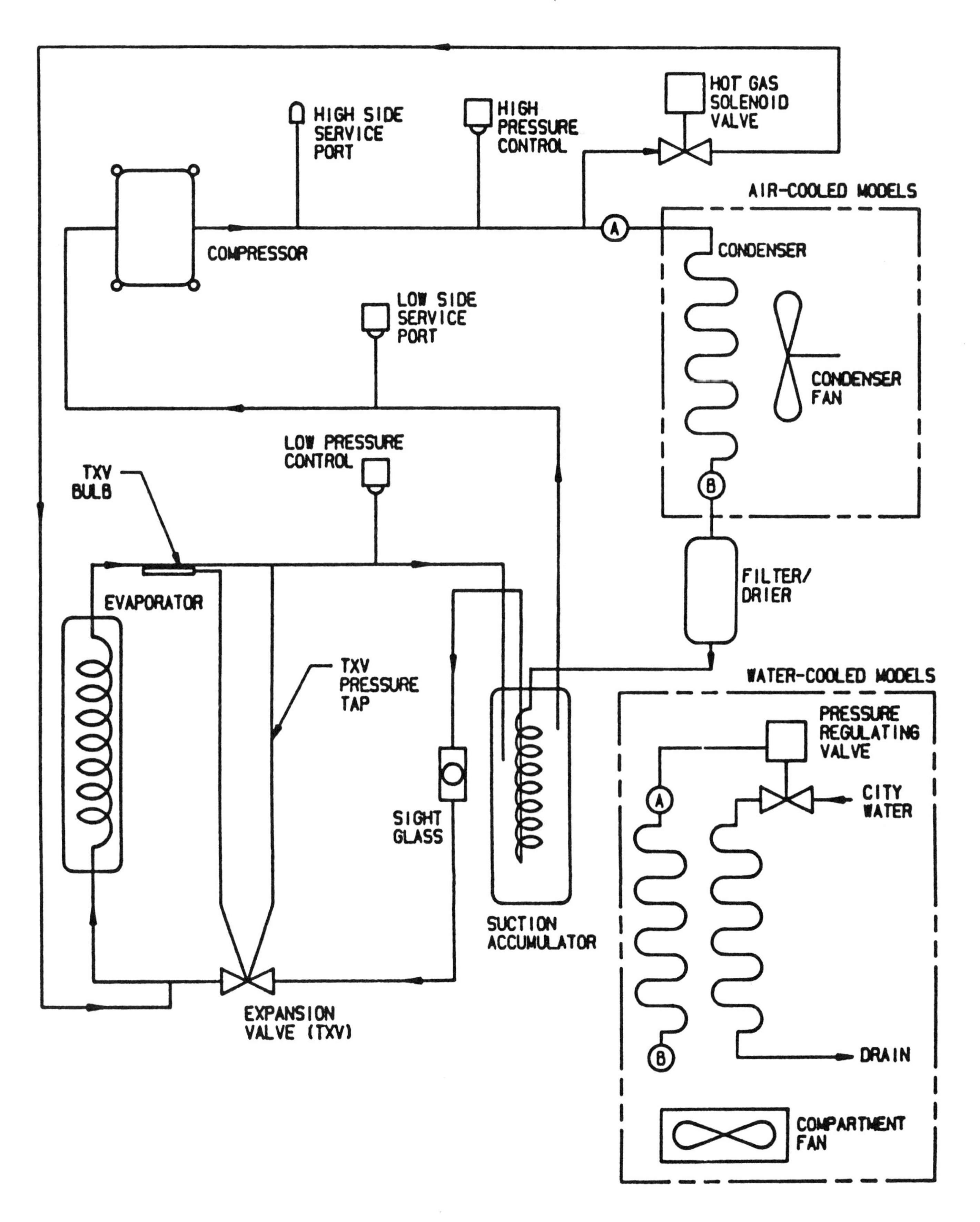

REFRIGERATION SCHEMATIC
HIGH SIDE SERVICE PORT
HIGH PRESSURE CONTROL
HOT GAS SOLENOID VALVE
AIR-COOLED MODELS
COMPRESSOR
CONDENSER
LOW SIDE SERVICE PORT
CONDENSER FAN
LOW PRESSURE CONTROL
TXV BULB
FILTER/ DRIER
EVAPORATOR
TXV PRESSURE TAP
WATER-COOLED MODELS
PRESSURE REGULATING VALVE
CITY WATER
SIGHT GLASS
SUCTION ACCUMULATOR
EXPANSION VALVE (TXV)
DRAIN
COMPARTMENT FAN
A
B

MAINTENANCE

It is recommended that the air inlet filter be cleaned every 3 months or sooner depending on the operating environment for proper refrigeration system performance. On air-cooled units, also check that the condenser is free of dirt/foreign material that could cause air flow blockage. Consult the maintenance/adjustment procedures for cleaning these items.

Cleaning of the icemaker is recommended on a regular basis not only for sanitary reasons but also to maintain the performance of the unit. Buildup of lime and scale can hinder icemaking production rates and interfere with proper dispensing of the ice. See the cleaning section for the recommended procedure.

Periodically, check the vending area sink for proper water drainage. Remove any foreign material from the sink to prevent drain blockage.

CLEANING INSTRUCTIONS

IMPORTANT: The ice maker should be cleaned at a minimum of 3 month intervals or more frequently depending on local water conditions. The storage hopper interior should be cleaned at least once a month.

WARNING: Do not use metal scrapers, sharp objects or abrasives on the surface of the storage hopper, as damage may result. Do not use solvents or other cleaning agents as they may attack the plastic surface. Use only the recommended chemicals and solutions for both the icemaker and hopper.

ICEMAKER SECTION

1. Open the hinged service door.
2. Put the (stop/run) switch in the "stop" position at the end of the harvest cycle.

WARNING: Electrical power is on to the unit during icemaker-section cleaning. Do not reach into hopper. Do not contact exposed electrical wiring and components.

3. Remove the ice drop cover from evaporator and the storage hopper cover.
4. Seal the evaporator outlet with the plastic plug provided with the unit and replace the ice drop cover.
5. Remove cleaning fill plug and add 4 oz. of Virginia Ice Machine Cleaner to the evaporator and replace cleaning fill plug.

CAUTION: Virginia Ice Machine Cleaner is a mild acid. Normal care should be taken - Keep out of eyes and cuts. Read warnings on package before using. Do not operate unit in the cleaning mode without the ice drop cover in place. There may be some overflow of cleaning solution through the evaporator vent tube during the cleaning cycle.

6. Push manual water fill switch and fill evaporator with water (approx. 5 sec).
7. Put the (clean/run) switch in the "clean" position. Allow unit to run in the cleaning mode for at least 30 minutes.
8. Put the (clean/run) switch in the "run" position.
9. Depress the flush switch pushbutton and drain evaporator for about 1½ minutes. Release pushbutton. Push manual water fill switch and allow evaporator to refill with water. Repeat step 9 three times to thoroughly remove cleaning solution from evaporator.
10. Depress the flush switch pushbutton for 1½ minutes to drain the evaporator.
11. Remove the evaporator plug.
12. Put the (stop/run) switch in the "run" position and allow unit to run through at least 3 complete icemaking and harvest cycles, and until ice is free of "sweet" taste.

WARNING: If unit fails to harvest ice, put the (stop/run) switch in the "stop" position. Depress the flush switch pushbutton for 1½ minutes to drain the evaporator. Flush the evaporator with hot water to thoroughly melt all the ice in the evaporator. Repeat step 9 to remove all traces of the cleaning solution from the evaporator.

13. Dispense all ice out of storage hopper and discard.

DISPENSER SECTION

14. Turn off main electrical power supply to machine.
15. Remove agitator assembly from storage hopper and wash and rinse it thoroughly.
16. Wash down all inside surfaces of the ice storage area including the top cover and ice drop cover with a mild detergent solution, and rinse thoroughly to remove all traces of detergent.
17. Replace agitator.
18. With brush provided, clean the inside of the ice chute with a mild detergent solution and rinse thoroughly to remove all traces of detergent.
19. Sanitize the inside of the hopper, agitator, the ice chute, and the hopper and ice drop covers with a solution of 1 oz. of household bleach in 1 gallon of water.
20. Replace the hopper cover and ice drop cover. Replace cabinet panels and turn on the electrical power supply. The icemaker is ready for normal operation.

FOR UNITS WITH BEVERAGE SYSTEM

COLDPLATE

1. Carefully remove the lower front panel.
2. Remove coldplate cover by loosening thumbscrew on the ice drop chute and lowering chute from plastic drop tube. Then remove cover by lifting slightly in front and slide forward.
3. Wash down the inside of the coldplate, tray, and cover with mild detergent solution and rinse. A small long handled brush will be found helpful in reaching the corners.
4. Replace the cover taking care that it is securely positioned in coldplate tray.
5. Replace ice drop chute.
6. Replace the lower front panel, carefully feeding the tubing and wires into the cabinet. Be sure not to pinch any tubing or wires between the panel and cabinet.

BEVERAGE SYSTEM

1. Remove faucet spouts, wash in mild detergent, rinse, and replace.

2. Disconnect electrical power to the carbonator. Shut off the water supply and close the CO_2 regulator to the carbonator.

3. Disconnect the syrup tanks from the system.

4. Energize the beverage faucets to purge the remaining soda water in the system.

5. Use a clean 5 gallon tank for each of the following:

    1. Cleaning Tank - Fill with hot (120-140°F) potable water.

    2. Sanitizing Tank - Fill with a chlorine sanitizing solution in the strength of 1 ounce of household bleach (sodium hypochlorite) to 1 gallon of cold (ambient) potable water (410 PPM).

6. Repeat the following procedure on each of the unit's syrup product lines:

    A. Connect the cleaning tank to the syrup line to be sanitized and to the CO_2 system.

    B. Energize the beverage faucet until the liquid dispensed is free of any syrup.

    C. Disconnect the cleaning tank and hook-up the sanitizing tank to the syrup line and CO_2 system.

    D. Energize the beverage faucet until the chlorine sanitizing solution is dispensed through the faucet. Flush at least 2 cups of liquid to insure that the sanitizing solution has filled the entire length of the syrup line.

    E. Disconnect the sanitizing tank. Hook-up the product tank to the syrup line and to the CO_2 system.

    F. Energize the faucet to flush the sanitizing solution from the syrup line and faucet. Continue draw on faucet until only syrup is dispensed.

7. Repeat step #2 in reverse order to turn on the carbonator. Dispense at least 1 cup of beverage from each faucet. Check taste. Continue to flush if needed to obtain satisfactory tasting drink.

TROUBLE SHOOTING GUIDE

The following pages contain trouble shooting charts designed to aid an experienced serviceman in diagnosing any operating problems which may be experienced. It is assumed that normal service techniques and skills are familiar to the person doing the trouble shooting. In order to gain maximum benefit from these charts, please note:

1. Start at the beginning of the chart and supply the appropriate answer to each question.

2. Do not skip any section, unless instructed to do so. You might miss the solution to your problem.

3. Evaluate the possible problem causes in the sequence in which they are presented. In general, they begin with the mose likely or easiest to check, and proceed to the less likely or more complicated.

4. If, after checking all indicated causes, the problem is not resolved, it is recommended that you retry a second time, carefully evaluating the symptoms and modifying your answers as necessary.

5. If you are unable to resolve a problem after several attempts, contact Remcor customer service for assistance.

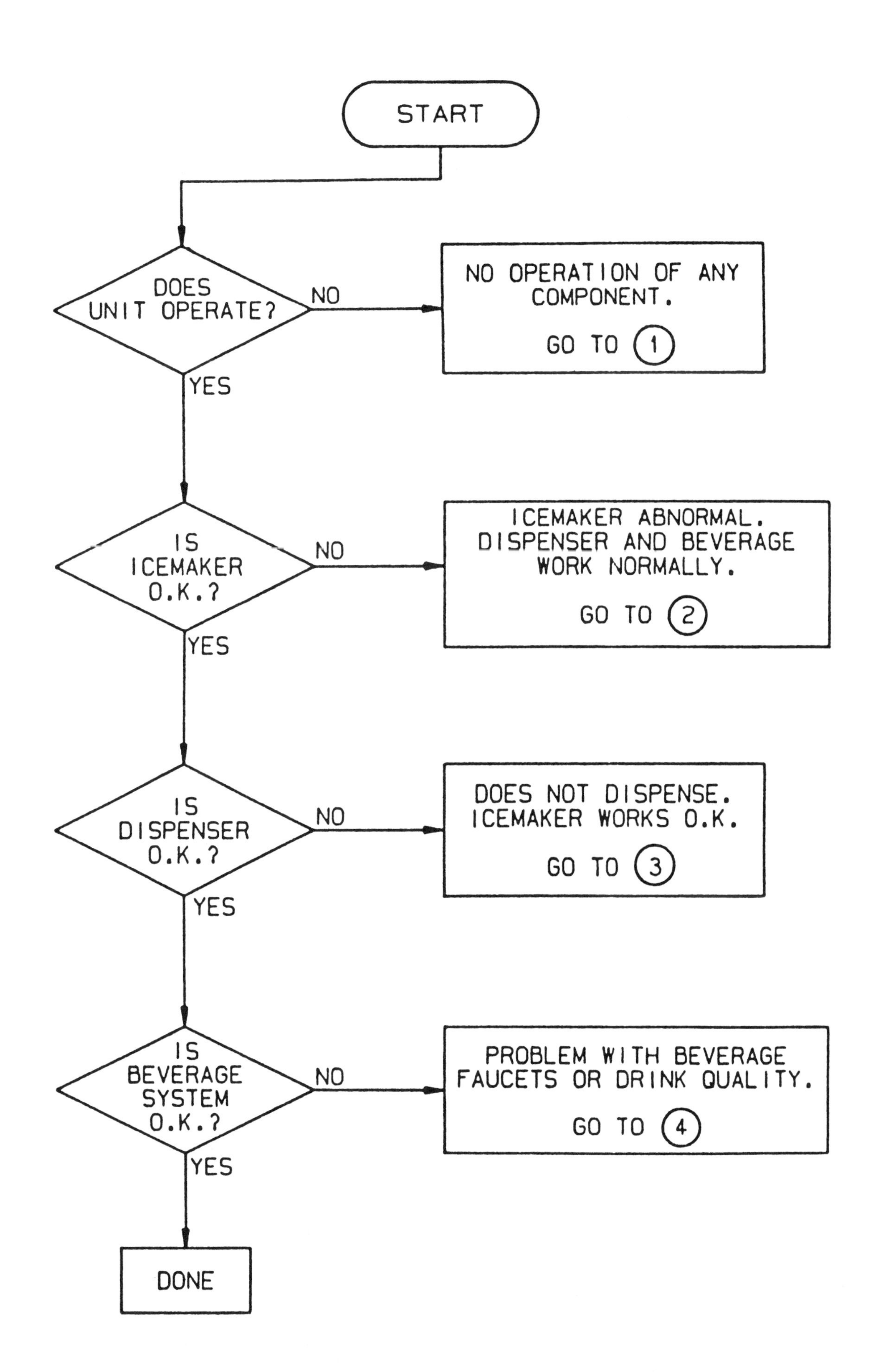
START
DOES UNIT OPERATE?
NO
NO OPERATION OF ANY COMPONENT.
GO TO 1
YES
IS ICEMAKER O.K.?
NO
ICEMAKER ABNORMAL. DISPENSER AND BEVERAGE WORK NORMALLY.
GO TO 2
YES
IS DISPENSER O.K.?
NO
DOES NOT DISPENSE. ICEMAKER WORKS O.K.
GO TO 3
YES
IS BEVERAGE SYSTEM O.K.?
NO
PROBLEM WITH BEVERAGE FAUCETS OR DRINK QUALITY.
GO TO 4
YES
DONE

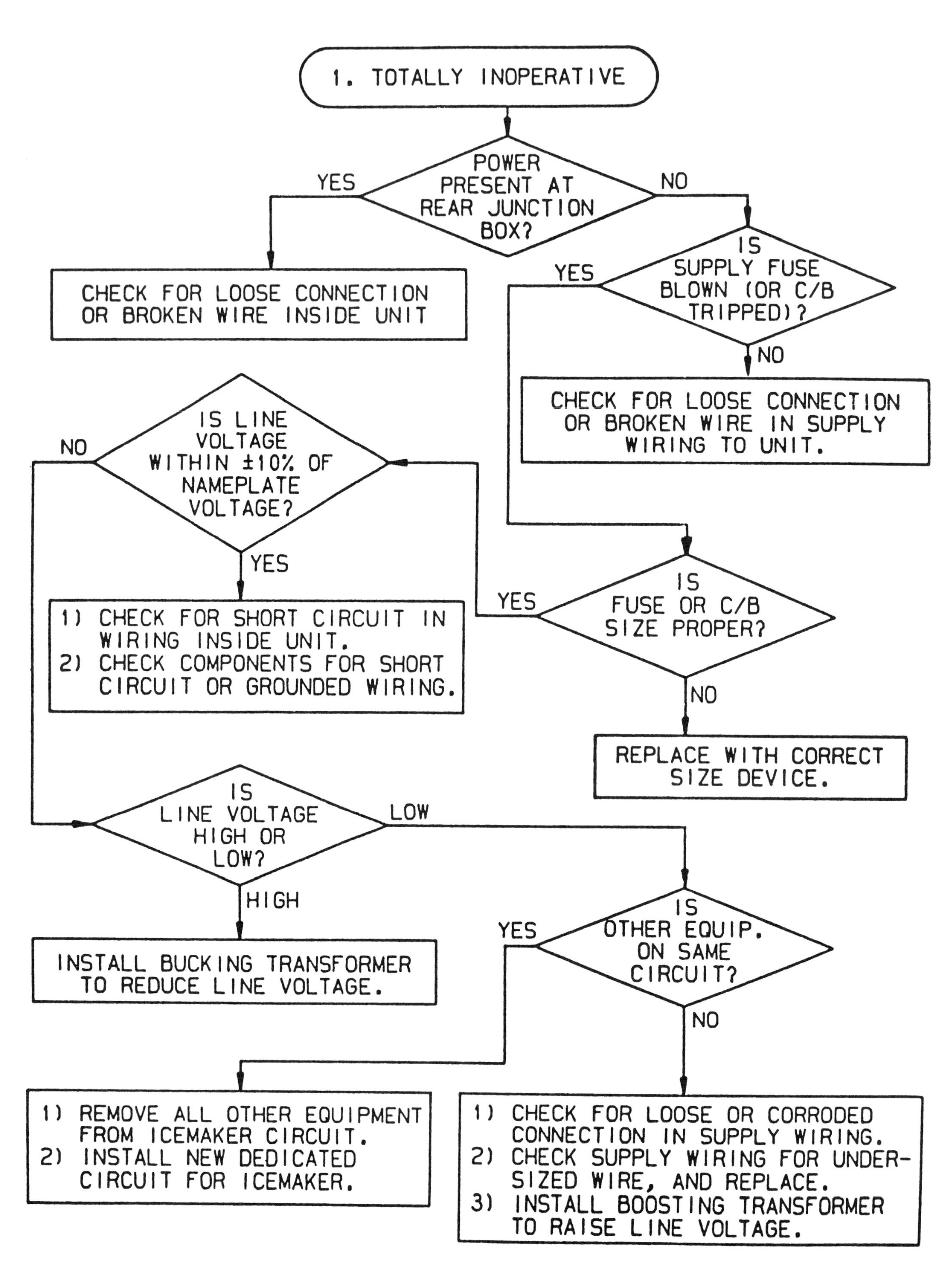

1. TOTALLY INOPERATIVE
POWER PRESENT AT REAR JUNCTION BOX?
YES
NO
CHECK FOR LOOSE CONNECTION OR BROKEN WIRE INSIDE UNIT
IS SUPPLY FUSE BLOWN (OR C/B TRIPPED)?
YES
NO
CHECK FOR LOOSE CONNECTION OR BROKEN WIRE IN SUPPLY WIRING TO UNIT.
IS LINE VOLTAGE WITHIN ±10% OF NAMEPLATE VOLTAGE?
NO
YES
1) CHECK FOR SHORT CIRCUIT IN WIRING INSIDE UNIT.
2) CHECK COMPONENTS FOR SHORT CIRCUIT OR GROUNDED WIRING.
IS FUSE OR C/B SIZE PROPER?
YES
NO
REPLACE WITH CORRECT SIZE DEVICE.
IS LINE VOLTAGE HIGH OR LOW?
LOW
HIGH
INSTALL BUCKING TRANSFORMER TO REDUCE LINE VOLTAGE.
IS OTHER EQUIP. ON SAME CIRCUIT?
YES
NO
1) REMOVE ALL OTHER EQUIPMENT FROM ICEMAKER CIRCUIT.
2) INSTALL NEW DEDICATED CIRCUIT FOR ICEMAKER.
1) CHECK FOR LOOSE OR CORRODED CONNECTION IN SUPPLY WIRING.
2) CHECK SUPPLY WIRING FOR UNDER-SIZED WIRE, AND REPLACE.
3) INSTALL BOOSTING TRANSFORMER TO RAISE LINE VOLTAGE.

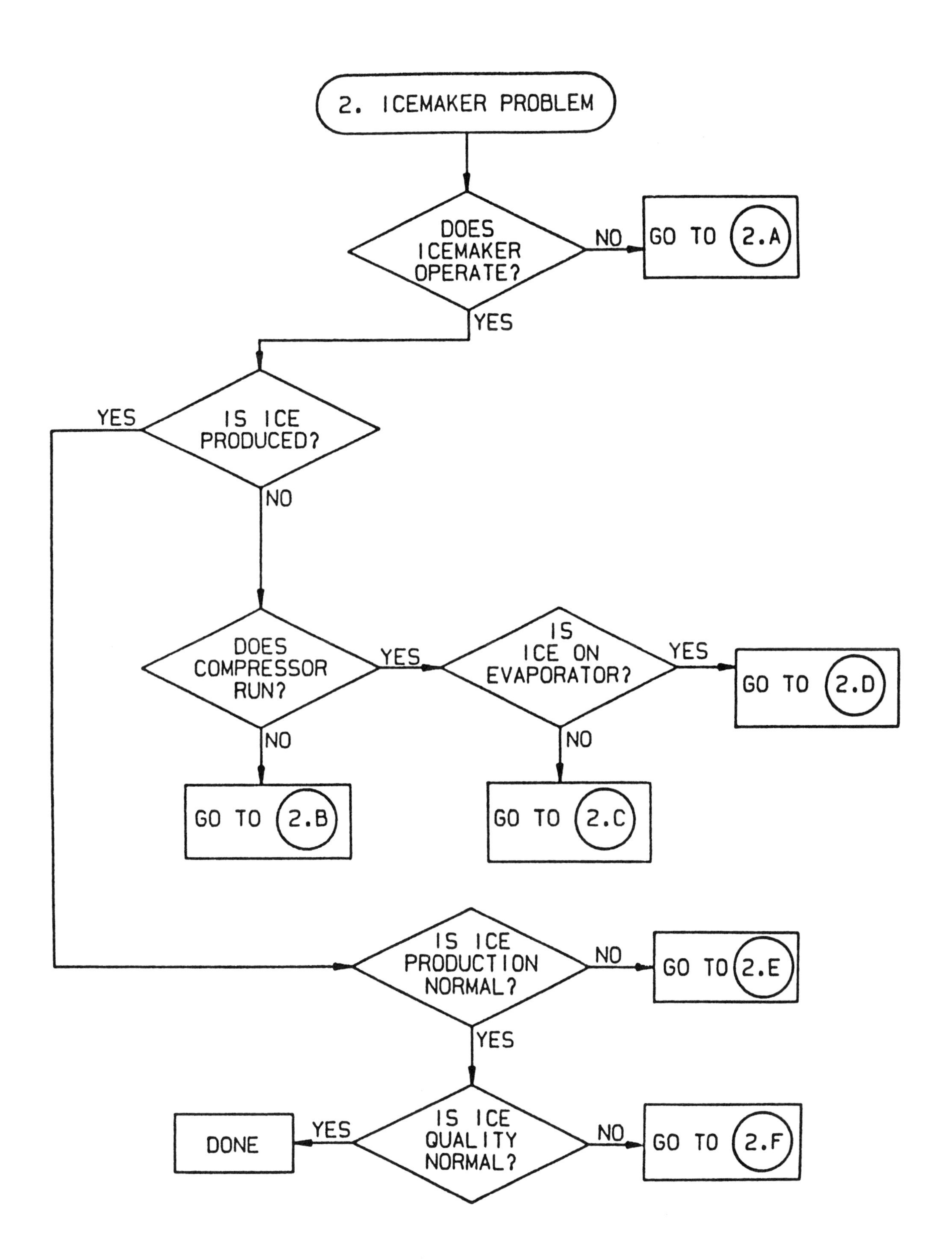
2. ICEMAKER PROBLEM
DOES ICEMAKER OPERATE?
NO
GO TO 2.A
YES
IS ICE PRODUCED?
YES
NO
DOES COMPRESSOR RUN?
YES
IS ICE ON EVAPORATOR?
YES
GO TO 2.D
NO
GO TO 2.B
NO
GO TO 2.C
IS ICE PRODUCTION NORMAL?
NO
GO TO 2.E
YES
DONE
YES
IS ICE QUALITY NORMAL?
NO
GO TO 2.F

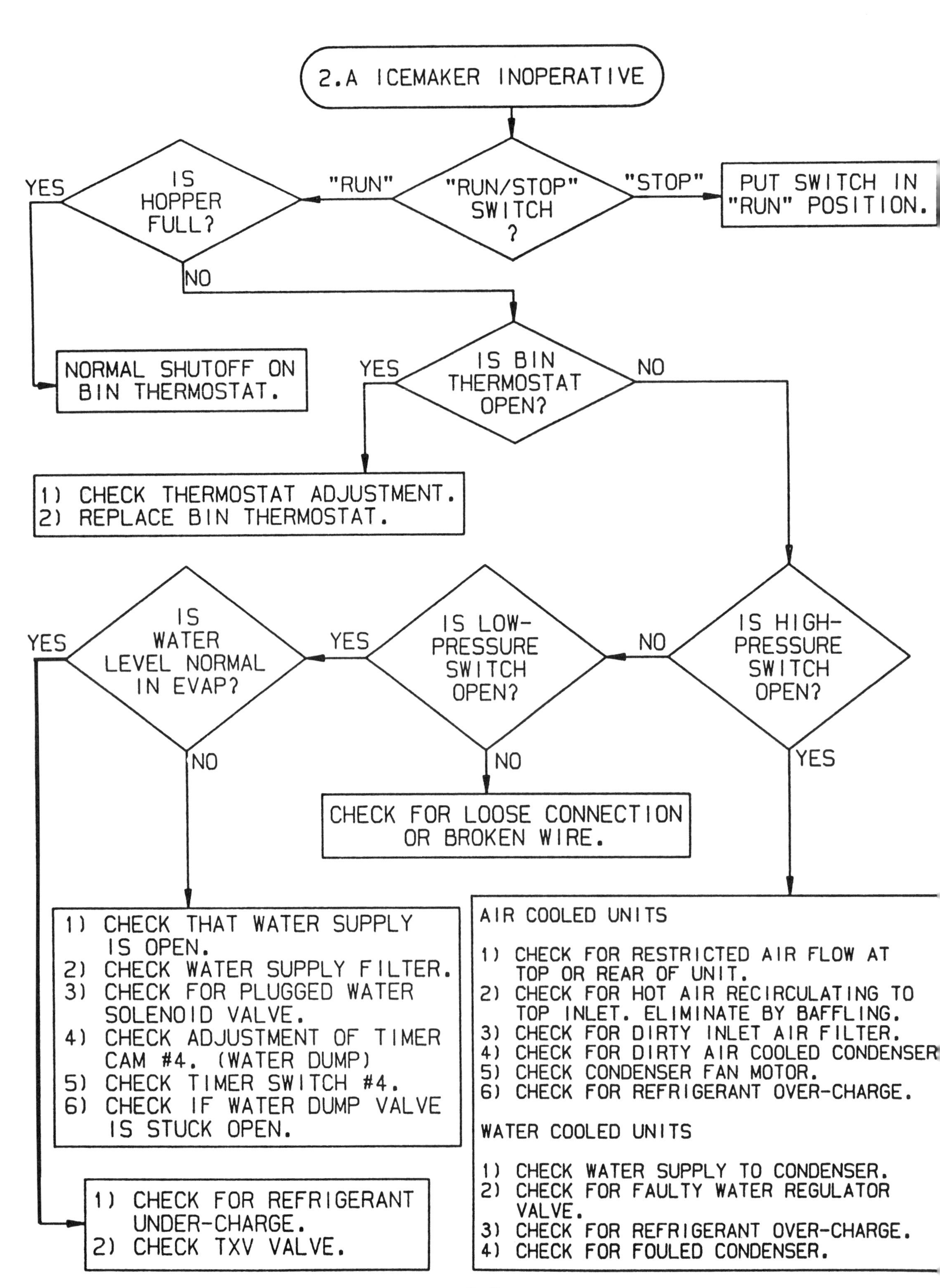
2.A ICEMAKER INOPERATIVE
"RUN/STOP" SWITCH ?
"STOP"
PUT SWITCH IN "RUN" POSITION.
"RUN"
IS HOPPER FULL?
YES
NORMAL SHUTOFF ON BIN THERMOSTAT.
NO
IS BIN THERMOSTAT OPEN?
YES
1) CHECK THERMOSTAT ADJUSTMENT.
2) REPLACE BIN THERMOSTAT.
NO
IS HIGH-PRESSURE SWITCH OPEN?
YES
AIR COOLED UNITS
1) CHECK FOR RESTRICTED AIR FLOW AT TOP OR REAR OF UNIT.
2) CHECK FOR HOT AIR RECIRCULATING TO TOP INLET. ELIMINATE BY BAFFLING.
3) CHECK FOR DIRTY INLET AIR FILTER.
4) CHECK FOR DIRTY AIR COOLED CONDENSER
5) CHECK CONDENSER FAN MOTOR.
6) CHECK FOR REFRIGERANT OVER-CHARGE.
WATER COOLED UNITS
1) CHECK WATER SUPPLY TO CONDENSER.
2) CHECK FOR FAULTY WATER REGULATOR VALVE.
3) CHECK FOR REFRIGERANT OVER-CHARGE.
4) CHECK FOR FOULED CONDENSER.
NO
IS LOW-PRESSURE SWITCH OPEN?
NO
CHECK FOR LOOSE CONNECTION OR BROKEN WIRE.
YES
IS WATER LEVEL NORMAL IN EVAP?
NO
1) CHECK THAT WATER SUPPLY IS OPEN.
2) CHECK WATER SUPPLY FILTER.
3) CHECK FOR PLUGGED WATER SOLENOID VALVE.
4) CHECK ADJUSTMENT OF TIMER CAM #4. (WATER DUMP)
5) CHECK TIMER SWITCH #4.
6) CHECK IF WATER DUMP VALVE IS STUCK OPEN.
YES
1) CHECK FOR REFRIGERANT UNDER-CHARGE.
2) CHECK TXV VALVE.

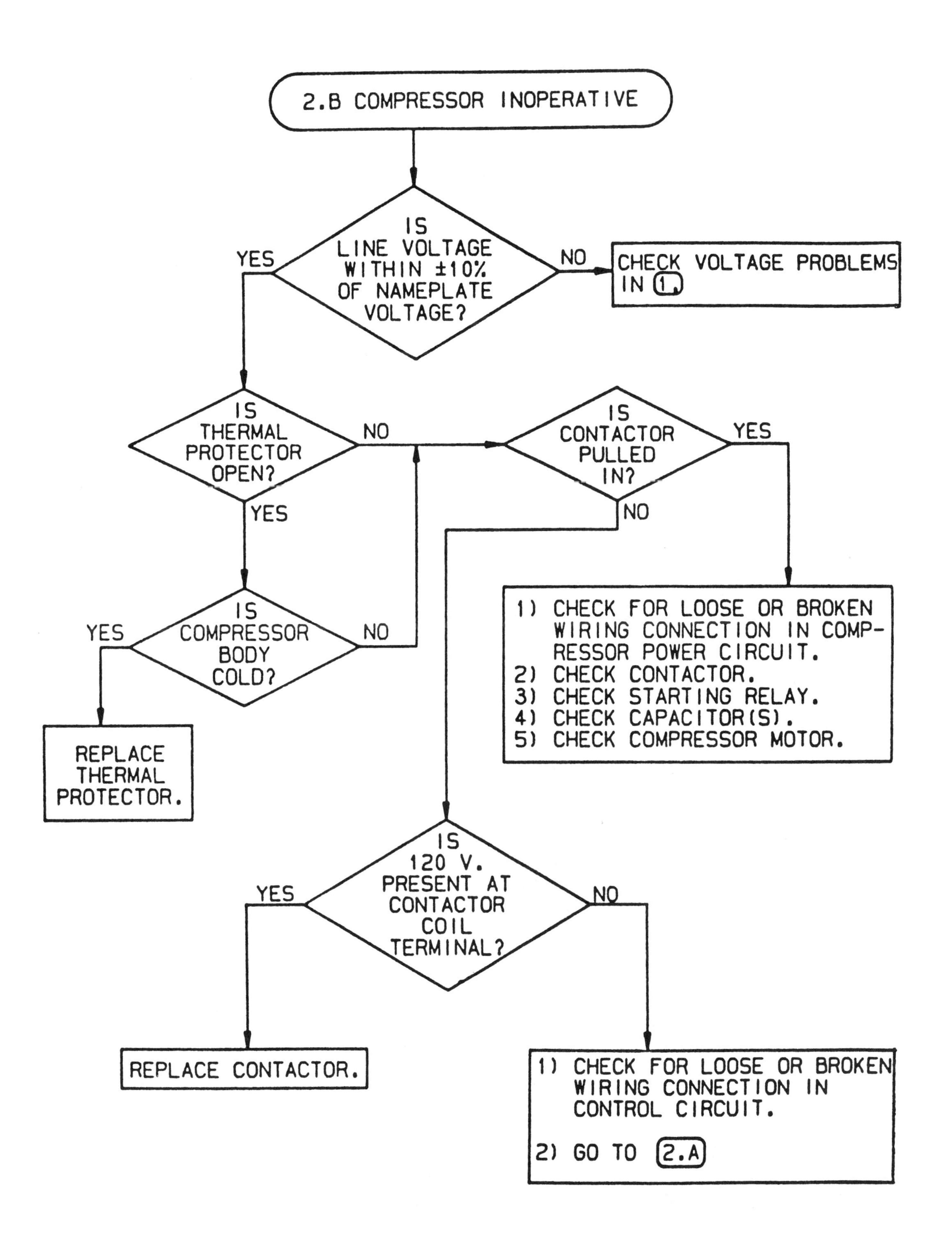
2.B COMPRESSOR INOPERATIVE
IS LINE VOLTAGE WITHIN ±10% OF NAMEPLATE VOLTAGE?
YES
NO
CHECK VOLTAGE PROBLEMS IN 1.
IS THERMAL PROTECTOR OPEN?
NO
YES
IS CONTACTOR PULLED IN?
YES
NO
IS COMPRESSOR BODY COLD?
YES
NO
REPLACE THERMAL PROTECTOR.
1) CHECK FOR LOOSE OR BROKEN WIRING CONNECTION IN COMPRESSOR POWER CIRCUIT.
2) CHECK CONTACTOR.
3) CHECK STARTING RELAY.
4) CHECK CAPACITOR(S).
5) CHECK COMPRESSOR MOTOR.
IS 120 V. PRESENT AT CONTACTOR COIL TERMINAL?
YES
NO
REPLACE CONTACTOR.
1) CHECK FOR LOOSE OR BROKEN WIRING CONNECTION IN CONTROL CIRCUIT.
2) GO TO 2.A

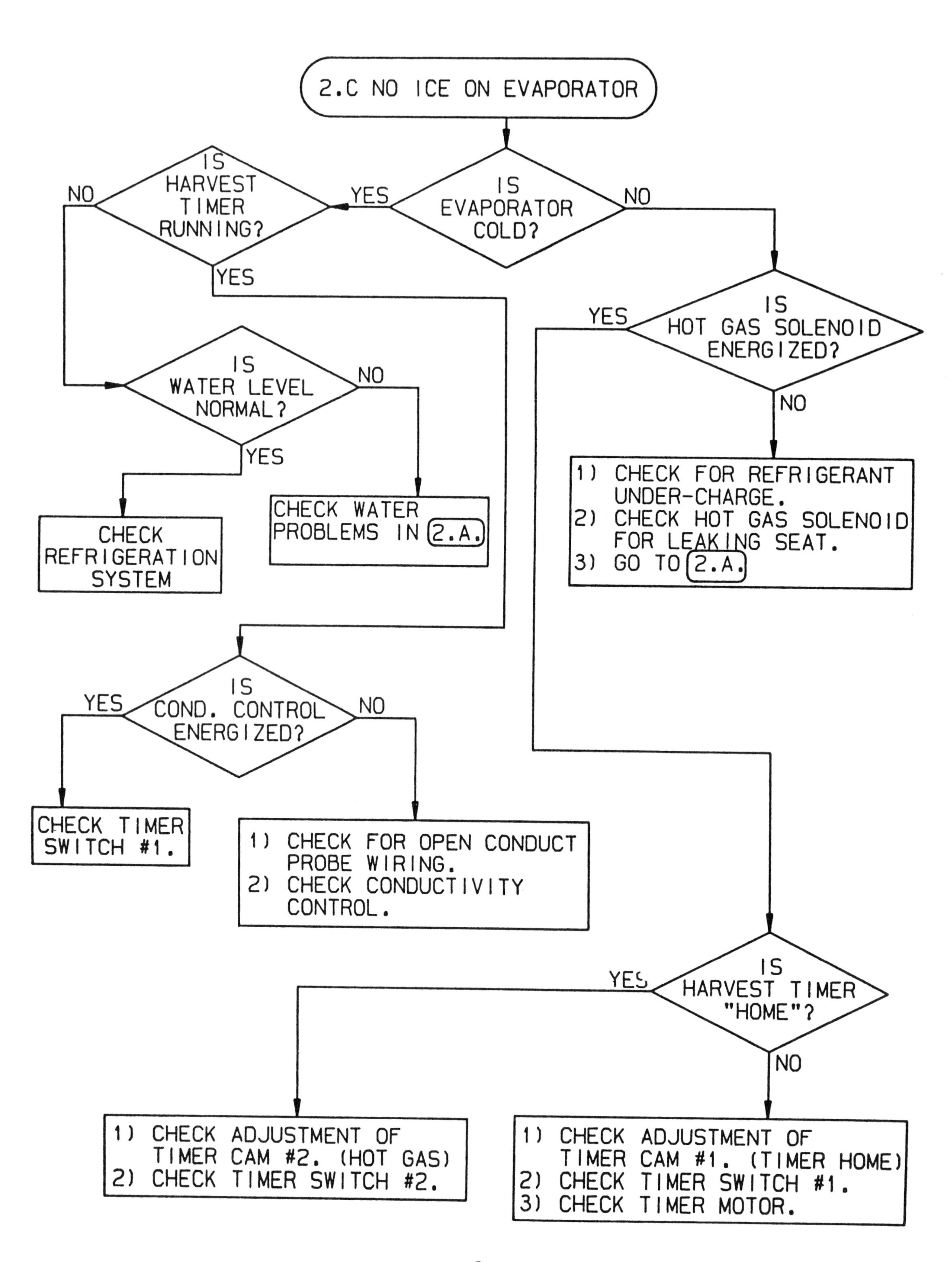
2.C NO ICE ON EVAPORATOR
IS EVAPORATOR COLD?
YES
NO
IS HARVEST TIMER RUNNING?
NO
YES
IS HOT GAS SOLENOID ENERGIZED?
YES
NO
IS WATER LEVEL NORMAL?
NO
YES
1) CHECK FOR REFRIGERANT UNDER-CHARGE.
2) CHECK HOT GAS SOLENOID FOR LEAKING SEAT.
3) GO TO 2.A.
CHECK REFRIGERATION SYSTEM
CHECK WATER PROBLEMS IN 2.A.
IS COND. CONTROL ENERGIZED?
YES
NO
CHECK TIMER SWITCH #1.
1) CHECK FOR OPEN CONDUCT PROBE WIRING.
2) CHECK CONDUCTIVITY CONTROL.
IS HARVEST TIMER "HOME"?
YES
NO
1) CHECK ADJUSTMENT OF TIMER CAM #2. (HOT GAS)
2) CHECK TIMER SWITCH #2.
1) CHECK ADJUSTMENT OF TIMER CAM #1. (TIMER HOME)
2) CHECK TIMER SWITCH #1.
3) CHECK TIMER MOTOR.

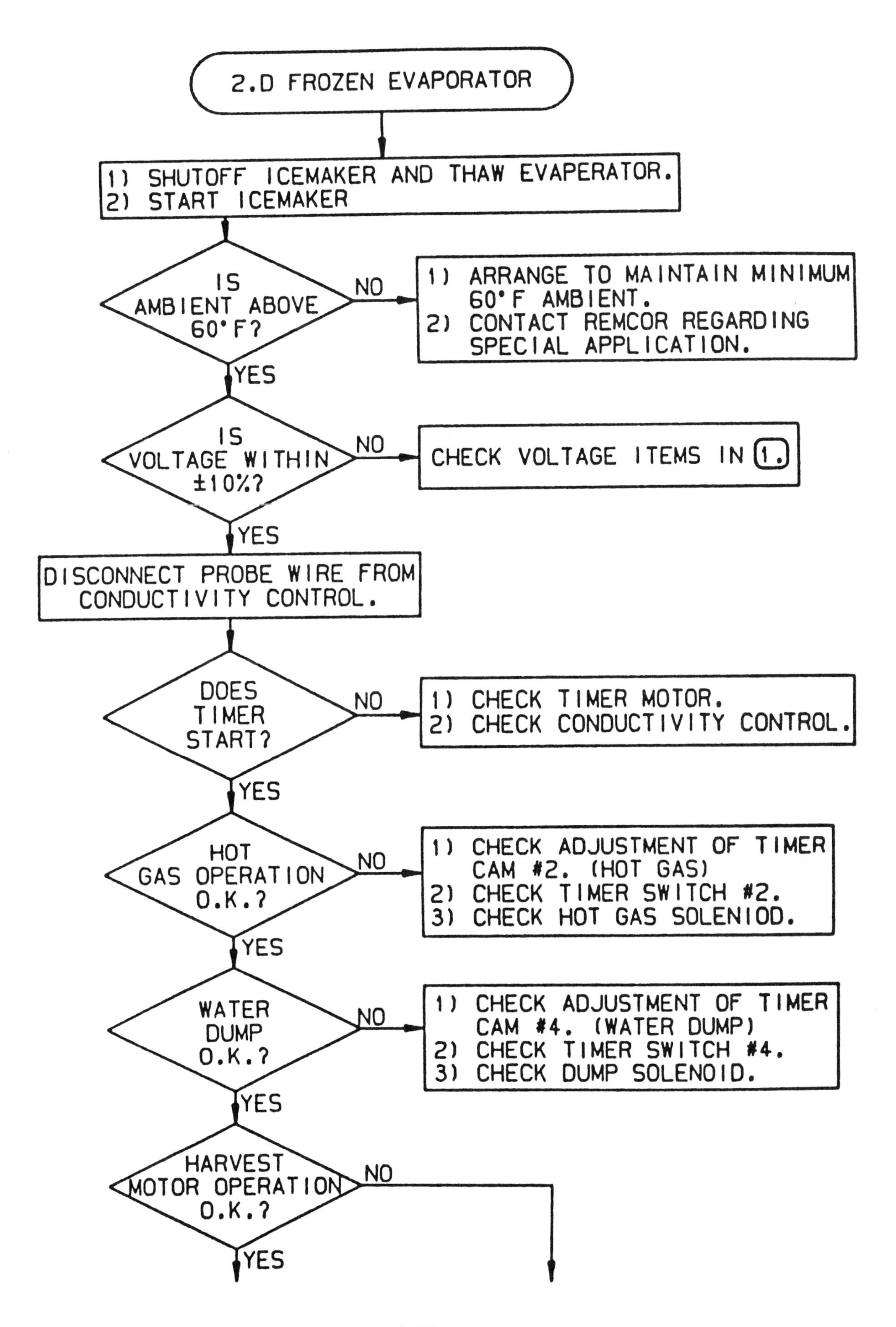

2.D FROZEN EVAPORATOR
1) SHUTOFF ICEMAKER AND THAW EVAPERATOR.
2) START ICEMAKER
IS AMBIENT ABOVE 60°F?
NO
1) ARRANGE TO MAINTAIN MINIMUM 60°F AMBIENT.
2) CONTACT REMCOR REGARDING SPECIAL APPLICATION.
YES
IS VOLTAGE WITHIN ±10%?
NO
CHECK VOLTAGE ITEMS IN 1.
YES
DISCONNECT PROBE WIRE FROM CONDUCTIVITY CONTROL.
DOES TIMER START?
NO
1) CHECK TIMER MOTOR.
2) CHECK CONDUCTIVITY CONTROL.
YES
HOT GAS OPERATION O.K.?
NO
1) CHECK ADJUSTMENT OF TIMER CAM #2. (HOT GAS)
2) CHECK TIMER SWITCH #2.
3) CHECK HOT GAS SOLENIOD.
YES
WATER DUMP O.K.?
NO
1) CHECK ADJUSTMENT OF TIMER CAM #4. (WATER DUMP)
2) CHECK TIMER SWITCH #4.
3) CHECK DUMP SOLENOID.
YES
HARVEST MOTOR OPERATION O.K.?
NO
YES

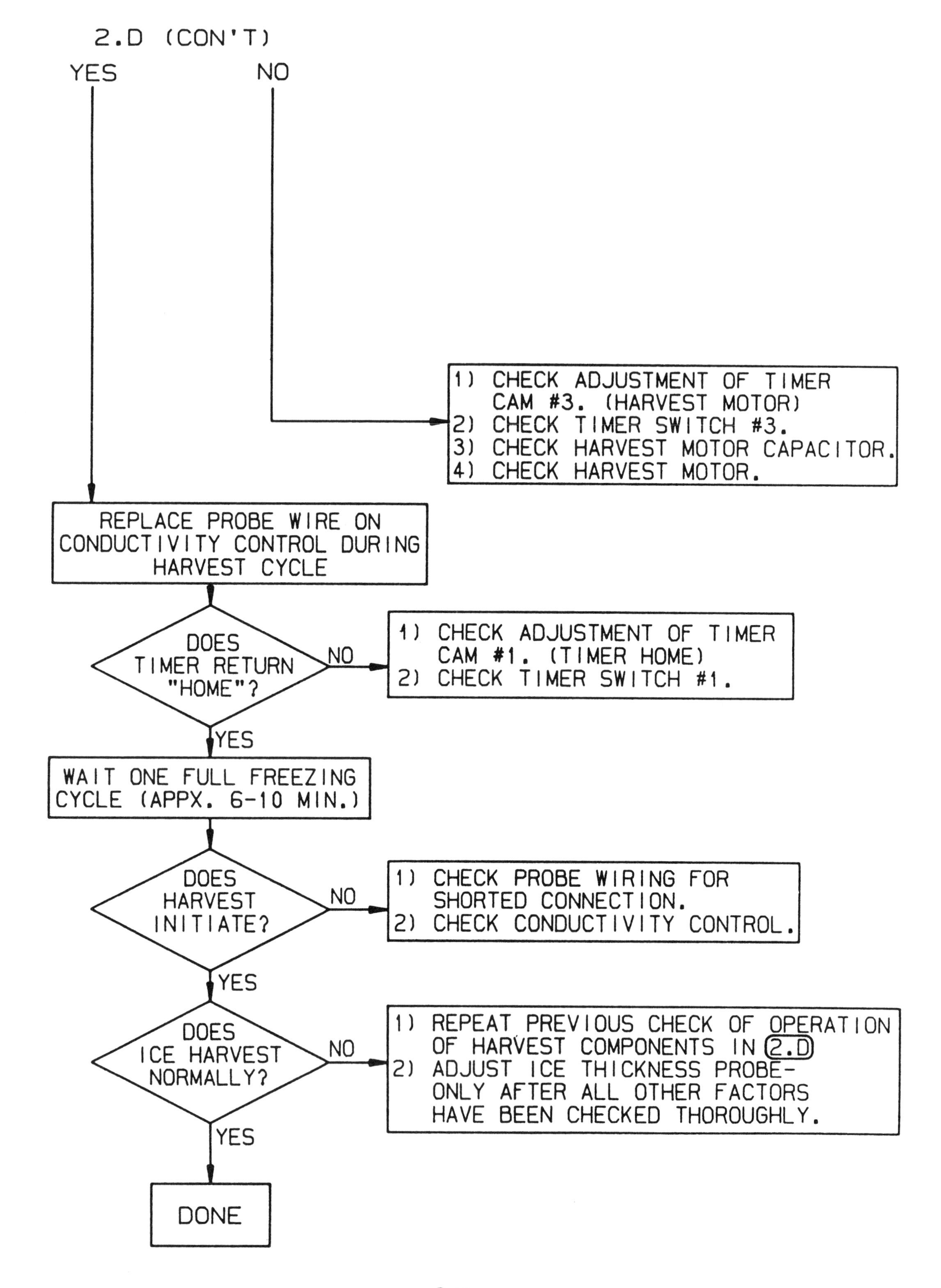

2.D (CON'T)
YES
NO
1) CHECK ADJUSTMENT OF TIMER CAM #3. (HARVEST MOTOR)
2) CHECK TIMER SWITCH #3.
3) CHECK HARVEST MOTOR CAPACITOR.
4) CHECK HARVEST MOTOR.
REPLACE PROBE WIRE ON CONDUCTIVITY CONTROL DURING HARVEST CYCLE
DOES TIMER RETURN "HOME"?
NO
1) CHECK ADJUSTMENT OF TIMER CAM #1. (TIMER HOME)
2) CHECK TIMER SWITCH #1.
YES
WAIT ONE FULL FREEZING CYCLE (APPX. 6-10 MIN.)
DOES HARVEST INITIATE?
NO
1) CHECK PROBE WIRING FOR SHORTED CONNECTION.
2) CHECK CONDUCTIVITY CONTROL.
YES
DOES ICE HARVEST NORMALLY?
NO
1) REPEAT PREVIOUS CHECK OF OPERATION OF HARVEST COMPONENTS IN 2.D
2) ADJUST ICE THICKNESS PROBE- ONLY AFTER ALL OTHER FACTORS HAVE BEEN CHECKED THOROUGHLY.
YES
DONE

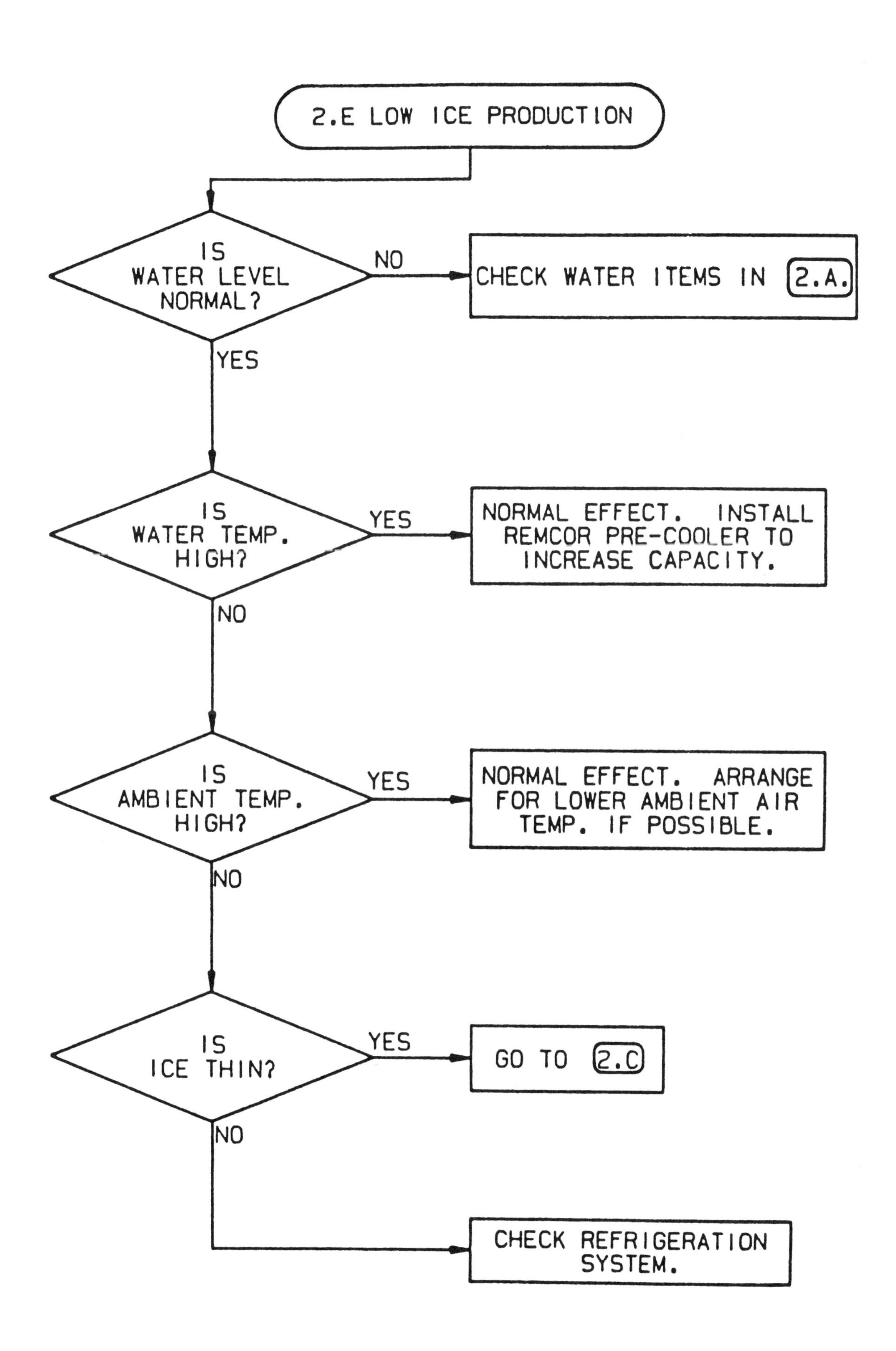
2.E LOW ICE PRODUCTION
IS
WATER LEVEL
NORMAL?
NO
CHECK WATER ITEMS IN 2.A.
YES
IS
WATER TEMP.
HIGH?
YES
NORMAL EFFECT. INSTALL
REMCOR PRE-COOLER TO
INCREASE CAPACITY.
NO
IS
AMBIENT TEMP.
HIGH?
YES
NORMAL EFFECT. ARRANGE
FOR LOWER AMBIENT AIR
TEMP. IF POSSIBLE.
NO
IS
ICE THIN?
YES
GO TO 2.C
NO
CHECK REFRIGERATION
SYSTEM.

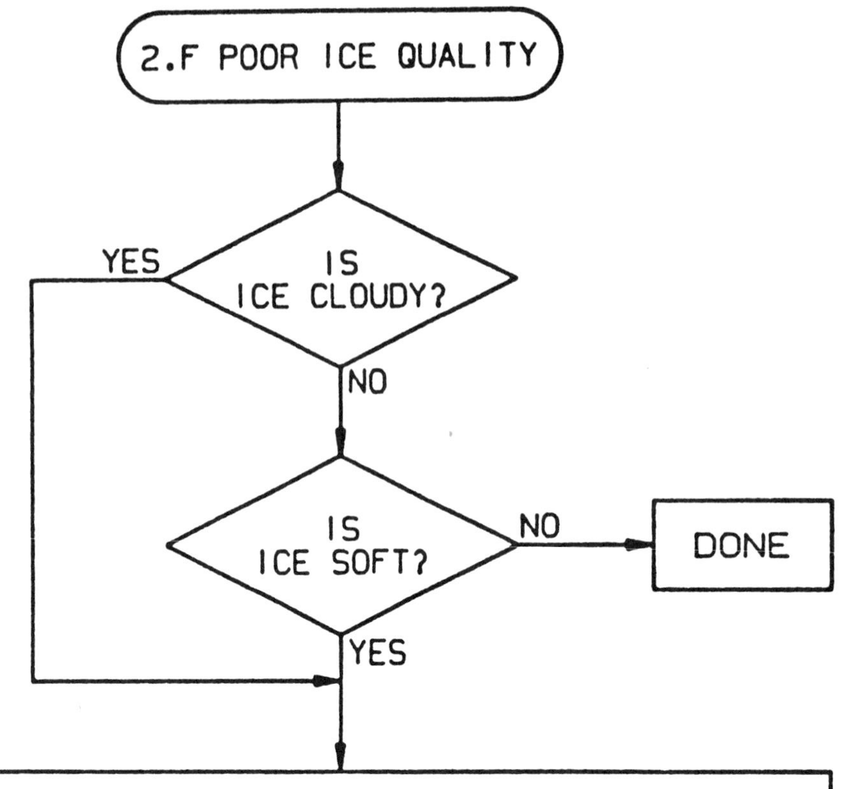

1) CHECK AIR PUMP.
2) CHECK ADJUSTMENT OF TIMER CAM #4. (WATER DUMP)
3) CHECK TIMER SWITCH #4.
4) CHECK WATER DUMP VALVE.
5) CHECK WATER FILTER.
6) INCREASE WATER DUMP BY ADJUSTING TIMER CAM #4.
7) INSTALL ADDITIONAL WATER TREATMENT DEVICES FOR SPECIFIC "PROBLEM" WATER.

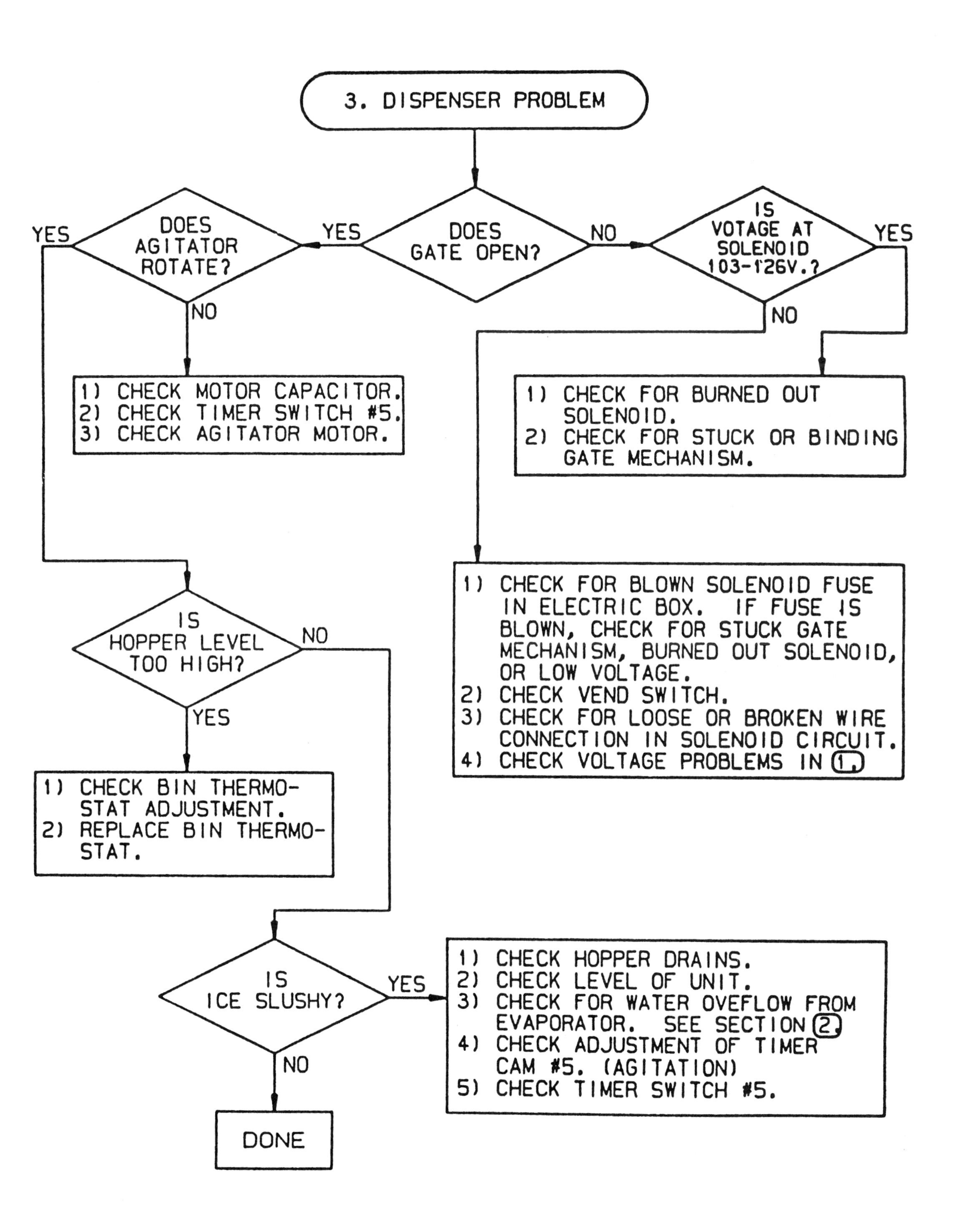
3. DISPENSER PROBLEM
DOES GATE OPEN?
YES
NO
DOES AGITATOR ROTATE?
YES
NO
IS VOTAGE AT SOLENOID 103-126V.?
YES
NO
1) CHECK MOTOR CAPACITOR.
2) CHECK TIMER SWITCH #5.
3) CHECK AGITATOR MOTOR.
1) CHECK FOR BURNED OUT SOLENOID.
2) CHECK FOR STUCK OR BINDING GATE MECHANISM.
1) CHECK FOR BLOWN SOLENOID FUSE IN ELECTRIC BOX. IF FUSE IS BLOWN, CHECK FOR STUCK GATE MECHANISM, BURNED OUT SOLENOID, OR LOW VOLTAGE.
2) CHECK VEND SWITCH.
3) CHECK FOR LOOSE OR BROKEN WIRE CONNECTION IN SOLENOID CIRCUIT.
4) CHECK VOLTAGE PROBLEMS IN 1.
IS HOPPER LEVEL TOO HIGH?
NO
YES
1) CHECK BIN THERMOSTAT ADJUSTMENT.
2) REPLACE BIN THERMOSTAT.
IS ICE SLUSHY?
YES
NO
1) CHECK HOPPER DRAINS.
2) CHECK LEVEL OF UNIT.
3) CHECK FOR WATER OVEFLOW FROM EVAPORATOR. SEE SECTION 2.
4) CHECK ADJUSTMENT OF TIMER CAM #5. (AGITATION)
5) CHECK TIMER SWITCH #5.
DONE

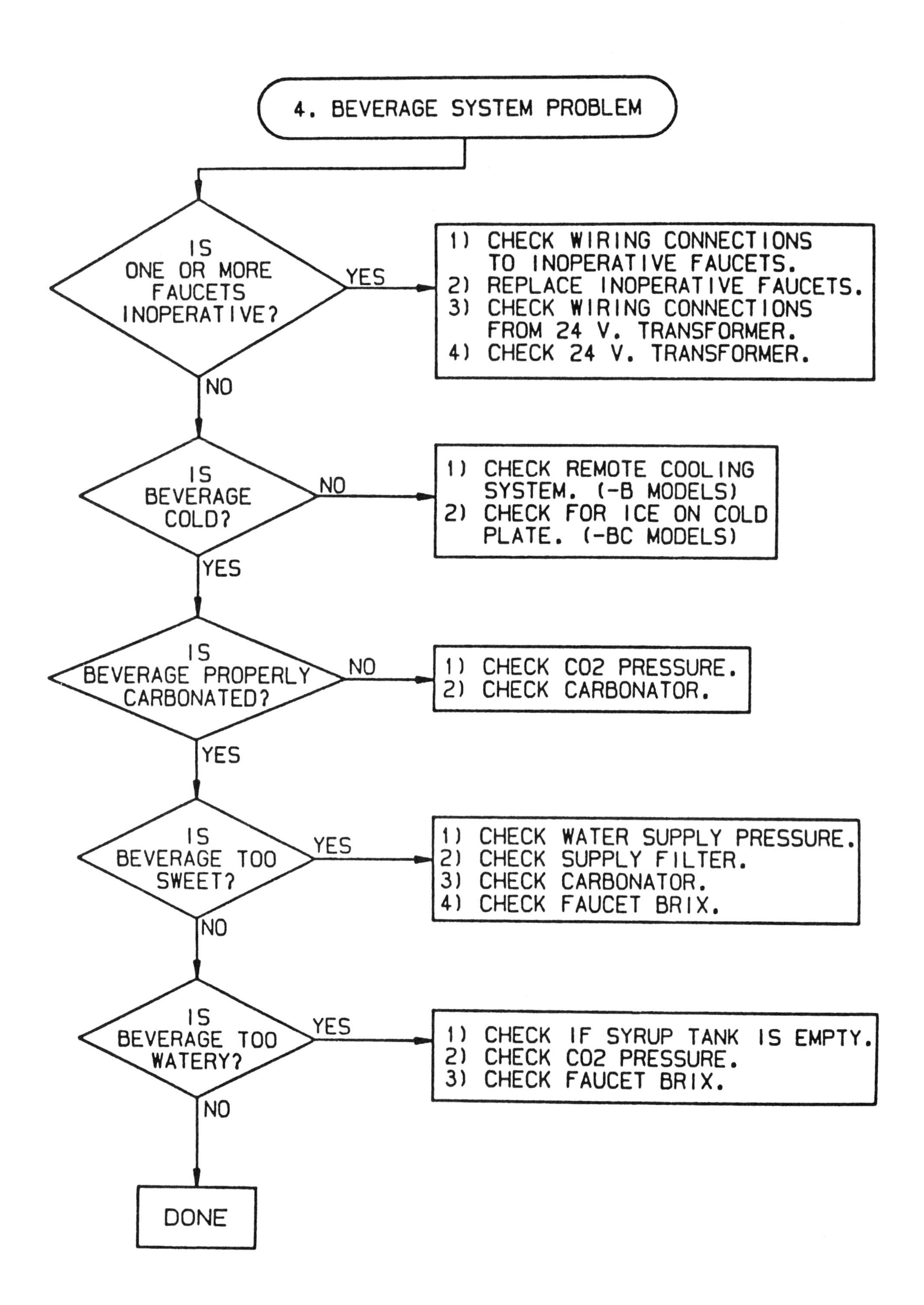
4. BEVERAGE SYSTEM PROBLEM
IS ONE OR MORE FAUCETS INOPERATIVE?
YES
1) CHECK WIRING CONNECTIONS TO INOPERATIVE FAUCETS.
2) REPLACE INOPERATIVE FAUCETS.
3) CHECK WIRING CONNECTIONS FROM 24 V. TRANSFORMER.
4) CHECK 24 V. TRANSFORMER.
NO
IS BEVERAGE COLD?
NO
1) CHECK REMOTE COOLING SYSTEM. (-B MODELS)
2) CHECK FOR ICE ON COLD PLATE. (-BC MODELS)
YES
IS BEVERAGE PROPERLY CARBONATED?
NO
1) CHECK CO2 PRESSURE.
2) CHECK CARBONATOR.
YES
IS BEVERAGE TOO SWEET?
YES
1) CHECK WATER SUPPLY PRESSURE.
2) CHECK SUPPLY FILTER.
3) CHECK CARBONATOR.
4) CHECK FAUCET BRIX.
NO
IS BEVERAGE TOO WATERY?
YES
1) CHECK IF SYRUP TANK IS EMPTY.
2) CHECK CO2 PRESSURE.
3) CHECK FAUCET BRIX.
NO
DONE

THERMOSTAT ALTITUDE ADJUSTMENTS

BIN T'STAT

IMPORTANT: Adjust the bin t'stat setting only if storage hopper overfill is a problem.

1. Open the hinged service door.

2. The adjustment screw is located below the flush switch on the left side of the electrical box.

3. For altitudes up to 6000 ft., turn the adjustment screw COUNTER-CLOCKWISE as follows:

ELEVATION (FT)	CCW TURN
2000	1/13
4000	1/6
6000	1/4

4. For altitudes above 6000 ft., consult the factory.

CLEARING EVAPORATOR FREEZE—UP

WARNING:	To prevent possible injury, do not stick fingers or hand into ice-maker nozzle or hopper with power applied to unit.

1. Open the hinged service door.
2. Put the (stop/run) switch in the "stop" position.
3. Close the water supply valve to the icemaker.
4. Remove the ice drop and hopper covers.
5. Depress the flush switch pushbutton and drain the evaporator.
6. Pour hot water into the evaporator ice exit opening. It will be necessary to use either a funnel or a container with a spout. Fill the evaporator completely.
7. Drain the evaporator. Repeat steps 5 and 6 as required to insure that all the ice in the evaporator is melted.
8. Open the water supply valve. Depress the fill switch and refill evaporator with water.
9. Replace the ice drop and hopper covers.
10. Consult trouble shooting guide to determine cause of freeze-up before putting unit back in service.

ICE THICKNESS ADJUSTMENT

WARNING:	Do not adjust ice thickness probe unless all other problem causes have been evaluated.

1. Open the hinged service door and remove the ice drop and hopper covers.

2. Collect and weigh the ice produced during the harvest cycle. The amount of ice harvested should be approximately 5 pounds. Use the following procedure to adjust the probe to obtain this weight. (A clockwise adjustment will reduce the harvest weight, while counter-clockwise turns will increase the amount)

CAUTION:	Do not turn the screw on the end of the probe. Rotate the plastic probe body only using a 3/8" open end wrench. Make adjustments in 1/8 turn increments.

   A. Put the (stop/run) switch in the "stop" position. (If unit is in the icemaking cycle, stop the unit at the end of the harvest cycle.)

   B. Access to the probe is obtained by removing the rear service panel. (For units without beverage faucets, the probe can be adjusted from the front by removing the lower front panel if rear access is blocked.)

   C. Adjust the probe.

   D. Put the (stop/run) switch in the "run" position.

   E. Collect and weigh the ice harvested. Repeat steps A-E as necessary to obtain the required amount of ice.

3. If making an initial adjustment (for example, if the probe has been removed and replaced for any reason), turn probe clockwise until it just touches the evaporator coil (a slight backpressure will be felt). Turn probe counter-clockwise 2 1/2 turns. Follow the procedure in step 2 to obtain the required ice harvest weight.

CLEANING/REPLACING THE FILTER

1. Remove the filter from the top cabinet panel by sliding it forward toward the front of the unit.

2. Wash the filter in a solution of warm water and a mild detergent. Do not use caustic detergents as they may attack the aluminum filter elements.

3. Dry filter thoroughly.

4. For maximum effectiveness, reactivate the filter with an air filter coating (see the parts section under miscellaneous components).

CLEANING THE CONDENSER (AIR—COOLED UNITS)

1. Disconnect power to the unit.

2. Remove the upper right side panel.

3. Remove all dirt/foreign matter build-up from the condenser fins (fan side). Be carefull not to damage the fins. It is recommended that a power vacuum cleaner with a "crevice" tool attachment be used.

HARVEST TIMER ADJUSTMENT

WARNING: Disconnect electrical power to unit before servicing timer in electric box.

1. Disconnect power to icemaker.

2. Open the hinged service door and remove the electrical control box cover.

3. Put the (stop/run) switch in the "stop" position.

4. Using Fig. 7 as a guide, set the timer cam tabs as follows, starting with cam wheel #1 (all cam tab positions are in relation to #1 left cam tab).

   NOTE: Timer cam wheels can be manually rotated only in the normal direction of rotation-downward as viewed from the front of the unit.

   A. "Manually" adjust the cam tabs by using each "click", as the cam tab is rotated, as equivalent to .75 second.

   B. Set up cam wheel #1 with the left and right cam tabs back-to-back as shown in Fig 7A.

   C. Adjust the cam tabs on wheels 2 through 5 in sequence as shown in the chart. Rotate the cam wheels manually downward to set each wheel.

   D. After the cam tabs are manually set, reconnect power to the icemaker.

   E. Rotate the cam wheels slightly to activate the timer motor (#1 tell-tale down).

   F. Using a stopwatch, time the cam switch telltales. Adjust the cam tabs as necessary for the required cycle times.

FIG. 7

HARVEST TIMER

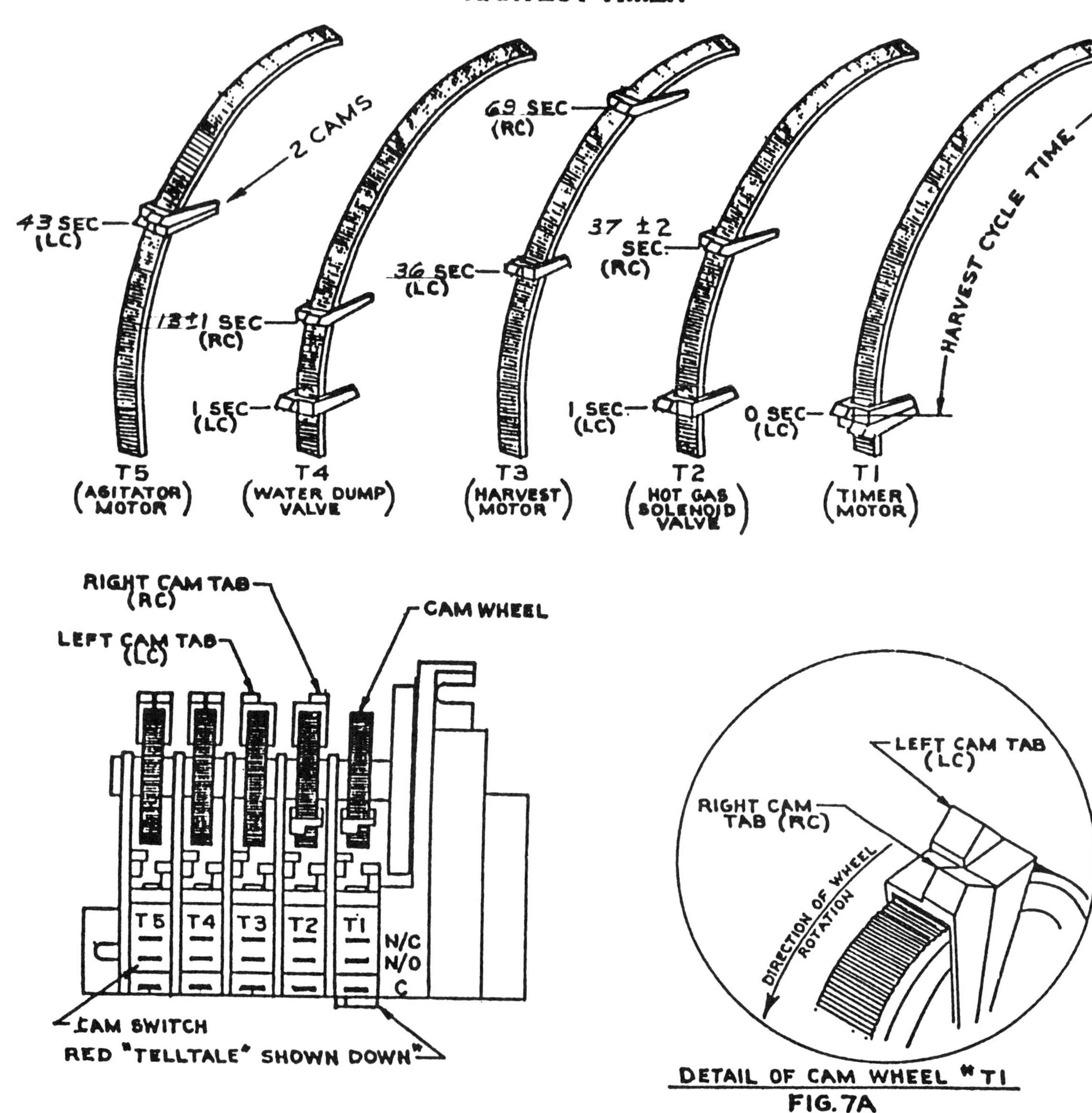

DETAIL OF CAM WHEEL "T1

FIG. 7A

MANUAL FILLING

In the event that the icemaker is not functioning, the hopper may be manually filled with ice.

1. Open the hinged service door.
2. Put the (stop/run) switch in the "stop" position.

WARNING: Electrical power is on to the agitator motor gate solenoid. Avoid contact with these components.

3. Remove the ice drop and storage hopper covers.
4. Fill hopper with ice and replace covers. Unit is now ready for dispensing.

CAUTION

1. Do not use crushed or flaked ice.
2. Use of bagged ice, which has frozen into large chunks can void warranty. The agitator is not designed to be an ice crusher - use of large chunks of ice which "jam-up" inside the hopper will cause failure of the agitator motor and damage to the hopper. If bagged ice is used it must be carefully and completely broken into small, cube-size pieces before filling into the storage hopper. Do not allow foreign material to enter the ice storage hopper.

PARTS LIST

Description	PART NO. Air Cooled -250	Air Cooled -250BC	Water Cooled -250	Water Cooled -250BC
DISPENSER COMPONENTS				
Gate Slide	21491	21491	21491	21491
Depressor Retainer	22644	22644	22644	22644
Agitator	23692	23692	23692	23692
Vend Switch	30895	30895	30895	30895
Switch Boot	31007	31007	31007	31007
Switch Insert	31163	31163	31163	31163
Agitator Motor W/Gaskets	31197-1	31197-1	31197-1	31197-1
Agitator Motor Shaft Seal	50891	50891	50891	50891
Agitator Motor Plate Insulation	50967	50967	50967	50967
Sink	51024	51032	51024	51032
Sink Grill	70496	70496	70496	70496
Ice Chute	50751	50751	50751	50751
Gate Gasket	50770	50770	50770	50770
Gate Solenoid Assy	31470	31470	31470	31470
Gate Rebuilding Kit	70438	70438	70438	70438
Agitator Motor Heater	-----	30794	-----	30794
Agitator Motor Gasket	50481	50481	50481	50481
ELECTRICAL CONTROLS				
Contactor	30379	30379	30379	30379
Toggle Switch	30385	30385	30385	30385
Flush Switch	30895	30895	30895	30895
Bin Thermostat	31001	31001	31001	31001
Fuse, 1¼ Amp	31406	31406	31406	31406
Timer, Harvest	31840	31840	31840	31840
Conductivity Control	31743	31743	31743	31743
Capacitor, Harvest Motor	31673	31673	31673	31673
Harvest Motor Thermostat	31714	31714	31714	31714
Compressor Start Relay	31671	31671	31671	31671
Capacitor, Comp. Start	31741	31741	31741	31741
Relay, Agitator Motor	31375	31375	31375	31375

Description	PART NO. Air Cooled -250	-250BC	Water Cooled -250	-250BC
ELECTRICAL CONTROLS CONT'D				
Capacitor, Comp. Run	31673	31673	31673	31673
Compressor Overload	31742	31742	31742	31742
Hi Press Control	60501	60501	60501	60501
Low Press Control	60369	60369	60369	60369
Transformer, Beverage	31091	31091	31091	31091
REFRIGERATION COMPONENTS				
Compressor	60678	60678	60678	60678
Compressor Mtg. Kit	31607	31607	31607	31607
Air Pump	31568	31568	31568	31568
Hose Adapter 3/8 NPT-3/8 BARB	51189	51189	51189	51189
90° Hose Adapter 3/8 NPT-3/8 BARB	51190	51190	51190	51190
Condenser Fan Motor	31738	31738	-----	-----
Condenser Fan Blade	31739	31739	-----	-----
Water Valve-Inlet	40672	40672	40672	40672
Condenser Air Cooled	60679	60679	-----	-----
Condenser Shroud	51388	51388	-----	-----
Tinnerman Clip for Shroud	70404	70404	-----	-----
Filter (Drier)	60623	60623	60623	60623
Hot Gase Solenoid W/Coil	60620	60620	60620	60620
Hot Gas Solenoid Coil 115V	31717	31717	31717	31717
TXV R-502	60635	60635	60635	60635
Condenser Water Cooled	-----	-----	60309	60309
Water Regulating Valve	-----	-----	40122	40122
Compressor Fan	-----	-----	30763	30763
Compressor Fan Cord	-----	-----	30764	30764
Water Drain Valve	40653	40653	40653	40653
Tubing, Water Drain 1/2 ID	50351	50351	50351	50351
Tubing, Air Pump 3/8 ID	50096	50096	50096	50096
EVAPORATOR COMPONENTS				
Evaporator Assy Comp. W/Motor	60703	60703	60703	60703
Evaporator Housing Foamed W/Gaskets	60704	60704	60704	60704
Evaporator Coil Assy W/Gaskets	60705	60705	60705	60705
Harvest Bar Assy W/Gaskets	51369-1	51369-1	51369-1	51369-1

Description	PART NO. Air Cooled -250	-250BC	Water Cooled -250	-250BC
EVAPORATOR COMPONENTS CONT'D				
Gasket Kit	51356	51356	51356	51356
Ice Thickness Probe	51179	51179	51179	51179
Harvest Motor W/Gaskets	31560-1	31560-1	31560-1	31560-1
Hose Adaptor 1/4 NPT-3/8 BARB	51191	51191	51191	51191
Hose Adaptor 1/4 NPT-1/2 BARB	51192	51192	51192	51192
10-32 X 1/4 Flat Hd. Screw	70536	70536	70536	70536
MISC. COMPONENTS				
Filter	70551	70551	70551	70551
Filter Coating 16 oz.	51355	51355	51355	51355
Label "Press for Ice"	90906	90906	90906	90906
Wiring Diagram	90915	90915	90915	90915
Manual	90914	90914	90914	90914
Cleaning Label	90916	90916	90916	90916

REMCOR PRODUCTS CO.

Rev. 5/30/89

WARRANTY POLICY

Spiral Ice® Icemaker/Dispenser

REMCOR Products Company warrants to the original purchaser of each new REMCOR SPIRAL ICE® ICEMAKER-DISPENSER, for a period of 21 months from date of installation of 24 months from date of shipment, whichever occurs first, that all parts shall be free from defects in material and workmanship under normal use and service. The S.S. icemaker evaporator is specifically warranted for a period of two (2) years, the compressor for a period of five (5) years, and labor cost to repair factory defective parts or workmanship is covered for a period of thirty (30) days from date of installation. Labor warranty does not cover normal installation or start-up.

Under this warranty, a defective part (or parts) is to be returned to REMCOR Products Company, 500 Regency Drive, Glendale Heights, Illinois 60139-2268 (Phone 312/980-6900) and shall be limited exclusively to repairing or replacing F.O.B. Factory, such part or parts which it concludes upon examination to be defective under the terms of the warranty. Return of any part disassembled will void warranty on any part. The decision of our Service Department regarding the warranty of parts will be final.

The warranties defined herein shall not apply to any damage of defects created or arising from accident, misapplication, abuse, misuse, neglect, alteration, acts of vandalism, flood, fire, acts of God or any other occurrences beyond the control of REMCOR.® Warranty validity also requires that all instructions have been followed and adhered to as provided in the Owner's Manual included with each unit.

REMCOR's warranty responsibility ceases if shipment is not received by you in good order and in accordance with quantity shown on Invoice or Packing Sheet; if you accept shipment from the Transportation Company in damaged condition without having a proper notation made by the Station Agent you do so at your own risk. If cartons are in apparent good order, but upon opening, contents are found to be damaged; call agent or adjuster to view same and have him mark the freight bill relative to such concealed damage. The procedure must be executed within fifteen days after delivery.

Prior to returning any material, (part or unit), a Return of Material Authorization (RMA) must be obtained. To obtain this authorization provide the reason for return, model number, serial number and part number. Either call or write REMCOR's Parts Service Department and an Authorization number and tag will then be issued - this tag must accompany the material returned. No representative, dealer, distributor or any person is authorized to make any other decisions regarding the warranty liability in accordance with REMCOR's Warranty. Any material that does not have a pre-issued Return Material Authorization number when received will be refused by REMCOR,® and returned to the sender (freight collect).

REMCOR Products Company will not honor, or assume any responsibility for any expenses (including labor), incurred in the field for the repair of equipment covered in our Warranty unless authorization has been granted from REMCOR's Service Department prior to work being performed. The request for repair Warranty work will only apply to units shipped from REMCOR within a thirty day time period of request. It will further be at the discretion of REMCOR® to decide if said labor will be reimbursed in full, or partial payment. REMCOR Products Company will also control the right to decline payment - all decisions will be based on circumstances prevailing. Charges for a repeat service call on the same unit to rectify the same problem previously corrected shall not be honored. REMCOR Products shall retain the right to select, or recommend another company to complete the necessary repair work.

CAUTION

The inherent nature of ice may cause spillage onto counter or floor areas. The Owner or Operator is cautioned to Maintain these areas in a clean, ice-free condition.

REMCOR®

SPIRAL ICEMAKER-DISPENSER

S.I.D.

OWNER'S MANUAL

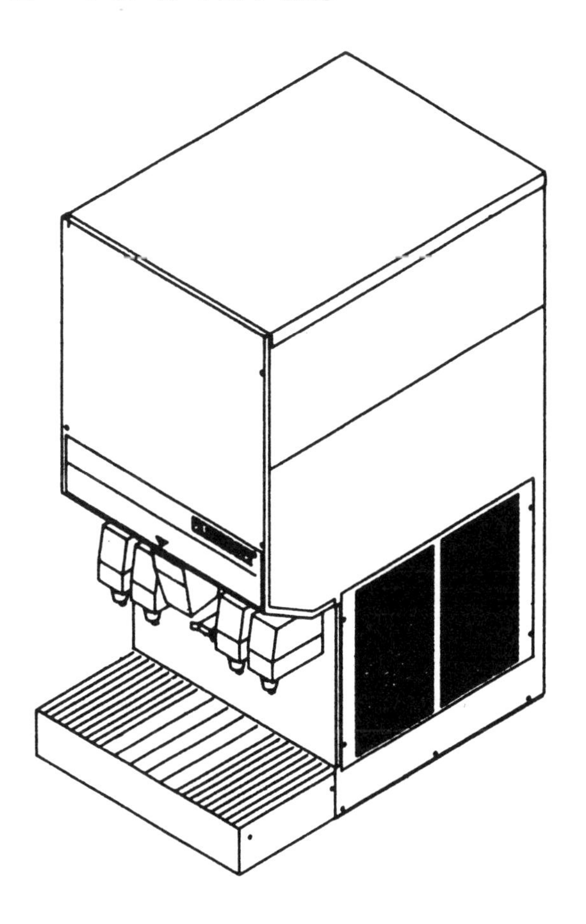

MODEL:
SID350A/35S
SID350A/35S-B

P/N 91002

REMCOR PRODUCTS CO.
500 REGENCY DR.
GLENDALE HEIGHTS, IL.
60139-2268

IMPORTANT

INSTALLATION NOTICE

An Everpure Model 9320-42, Systems IV Model DB900, or equal, icemaker quality water treatment unit MUST BE INSTALLED in the water supply line to the icemaker. Failure to do so may result in poor quality ice, low production output, and may cause premature failure of the icemaker evaporator and void the extended evaporator warranty.

This icemaker is provided with a stainless steel evaporator, designed to last the life of the product. But some of the chemicals in treated and untreated water, specifically chlorine and sulfur (sulfide), have the ability to attack stainless steel and cause premature failure. An initial investment in proper water treatment will pay for itself in increased production, quality and long life of the product.

DESCRIPTION

The Remcor S.I.D. (Sprial Icemaker-Dispenser) is a unique, self-contained, countertop style unit which automatically makes hard, clear cube-quality ice and stores it in a sealed hopper for sanitary dispensing. The ice is made by a new, patented process on a spiral-shaped stainless steel evaporator and produces true cube quality ice on the outside of the tubes. There are no augers, no compressing of flaked-ice, no bearings and no high gear motor loads in the ice making process. The unit has been designed to be simple, yet effective, to provide many years of trouble free operation.

SPECIFICATIONS

COMPRESSOR:	HP: 1/2	
REFRIGERANT:	R-12/14oz. (Air Cooled)	
VOLTAGE:	115/1/60	208-240/1/50
AMPS:	15 AMPS	12 AMPS
CIRCUIT AMPACITY:	18.3 AMPS	16.4 AMPS
FUSE SIZE:	20 AMPS	20 AMPS
ICE STORAGE CAPACITY:	35 lbs.	
ICE MAKING CAPACITY:	up to 350 lbs/24 hours	

Air Temp	Water Temp				
	40°	50°	60°	70°	80°
60°	376	351	336	314	294
70°	339	317	303	286	269
80°	304	285	272	256	240
90°	269	252	241	226	212

SHIPPING WEIGHT: 180 lbs.

UNPACKING

1. With the unit upright, carefully remove the shipping crate. Inspect for shipping damage and report any such damage to the shipper immediately.

2. Unscrew and remove front panel for electrical and start up access.

3. Remove shipping tape from storage hopper cover and agitator in storage hopper.

INSTALLATION

1. LOCATION

Locate the icemaker-dispenser indoors in a well ventilated area. Avoid exposure to direct sunlight and/or heat caused by radiation. Ambient room temperature must be in the range of 60 to 90°F. Do not install unit in an enclosed area where heat build-up could be a problem. Note: Air flow direction and spacing required on Figure #1.

Consult Fig. 1 for utility connection locations.

Consult Fig. 2 for dimensions for mounting unit to the counter with the hardware provided. Note that the unit must be level for proper operation.

The unit must be sealed to the counter. The template drawing (Fig.2) indicates the openings which must be cut in the counter. Locate the desired position for the unit, then mark the outline dimensions and cut-out locations using the template drawing. Cut openings in counter.

Apply a continuous bead of NSF listed silastic sealant (Dow 732 or equal) approximately 1/4" inside of the unit outline dimensions, and around all openings. Then position the unit on the counter within the outline dimensions. All excess sealant must be wiped away immediately.

2. PLUMBING

Connect the icemaker to a cold, potable water source, suitable for drinking. This water source must comply with the basic plumbing code of the Building Officials and Code Administrators International Inc. (BOCA) and the Food Service Sanitation Manual of the Food and Drug Administration. Do not install unit on a water softener line. It is recommended that a hand shut-off valve and strainer be used on the incoming supply line. (See Fig. 1 for size and location.) For proper operation of the incoming water supply pressure must be in the range of 30-90 PSIG. Install a pressure regulating valve if above this range!

IMPORTANT

To insure proper icemaker operation and also to reduce the frequency of water-related service problems, a water filter should be installed. Remcor recommends the use of one of the following basic systems:

1. Everpure Inc.
   660 N. Blackhawk Drive
   Westmont, IL 60559
   (312)654-4000

   InsurIce Twin System #9320-42

2. Systems IV
   16632 Burke Lane
   Huntington Beach, CA 92647

   Basic Water System #B1000

For specific recommendations on these filter systems for your local water conditions, consult with a distributor in your area or contact the filter manufacturer.

Connect separate drain lines to all drain connections. See Fig. 1 for size and location. These lines must pitch downward to an open drain, and must contain no traps, or improper drainage will result. Separate not joined drain lines are required to prevent back fill.

Note: In areas where consistently warm water temperatures are encountered, the use of a Remcor precooler in the water line is recommended to maximize the ice production of this unit. Contact Remcor for more information on this product.

3. ELECTRICAL

This unit is supplied with a 2 wire-with ground-6 foot long conductor and suitable plug for a standard 115 volt 20 amp circuit receptacle. This cord exits the unit thru the base and should be routed per the national electrical code.

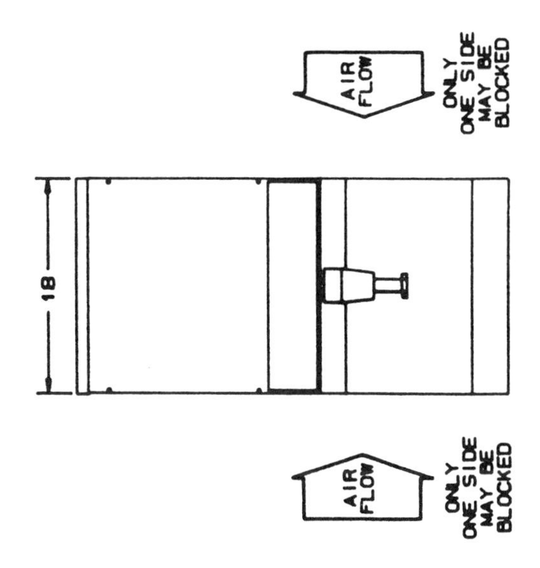

AIR FLOW
ONLY ONE SIDE MAY BE BLOCKED
18
AIR FLOW
ONLY ONE SIDE MAY BE BLOCKED

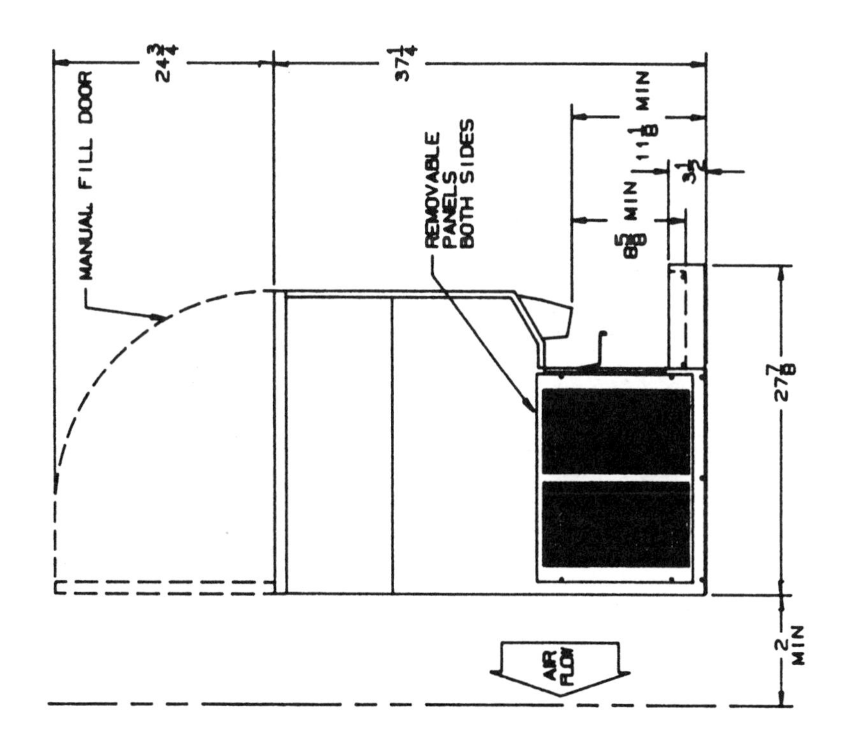

MANUAL FILL DOOR
24 3/4
37 1/4
REMOVABLE PANELS BOTH SIDES
11 1/8 MIN
3 1/2
8 5/8 MIN
27 7/8
AIR FLOW
2 MIN

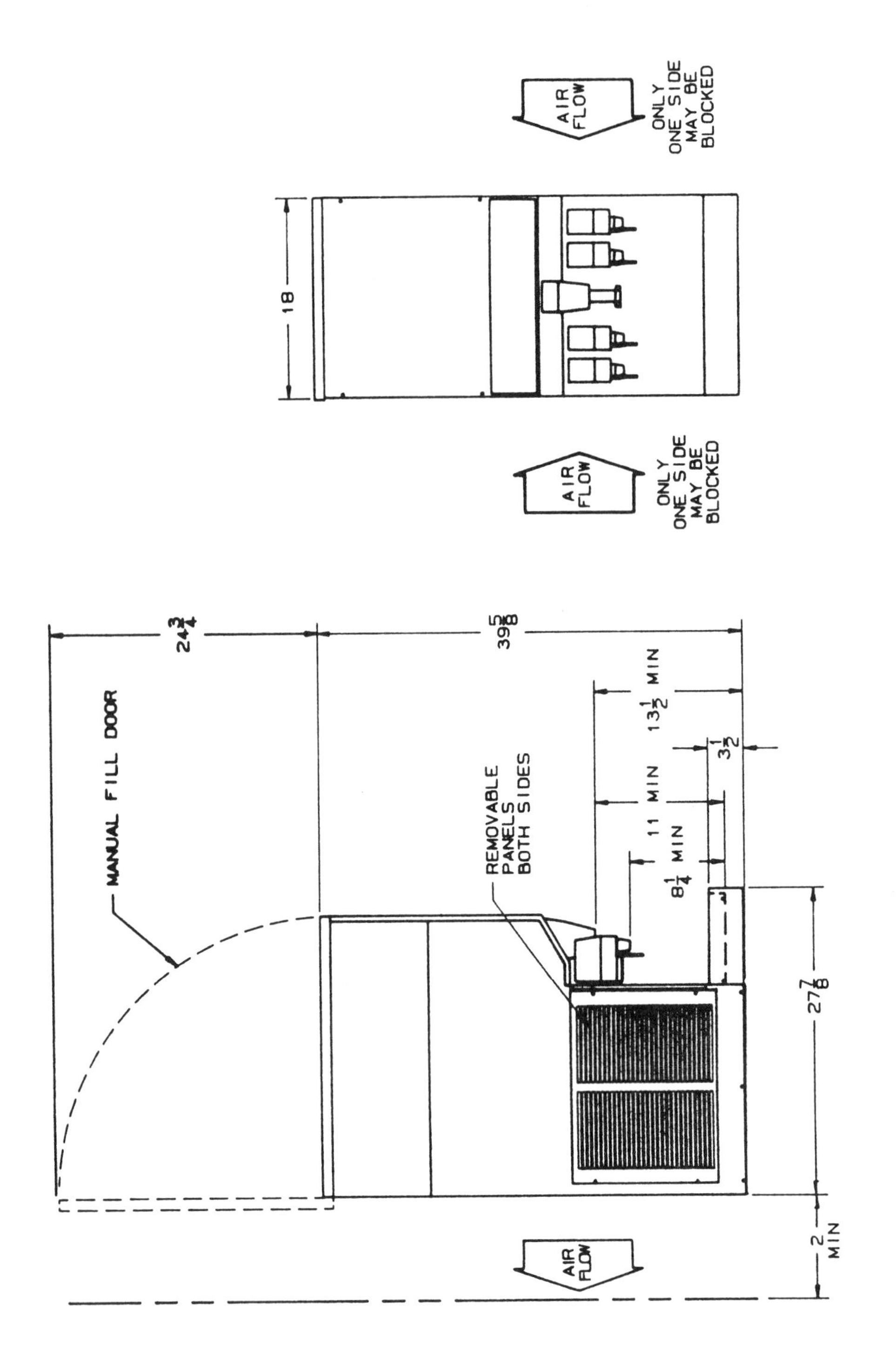
MANUAL FILL DOOR
REMOVABLE PANELS BOTH SIDES
AIR FLOW
ONLY ONE SIDE MAY BE BLOCKED
AIR FLOW
ONLY ONE SIDE MAY BE BLOCKED
AIR FLOW
18
24 3/4
39 5/8
13 1/2 MIN
3 1/2
11 MIN
8 1/4 MIN
27 7/8
2 MIN

SID350/35

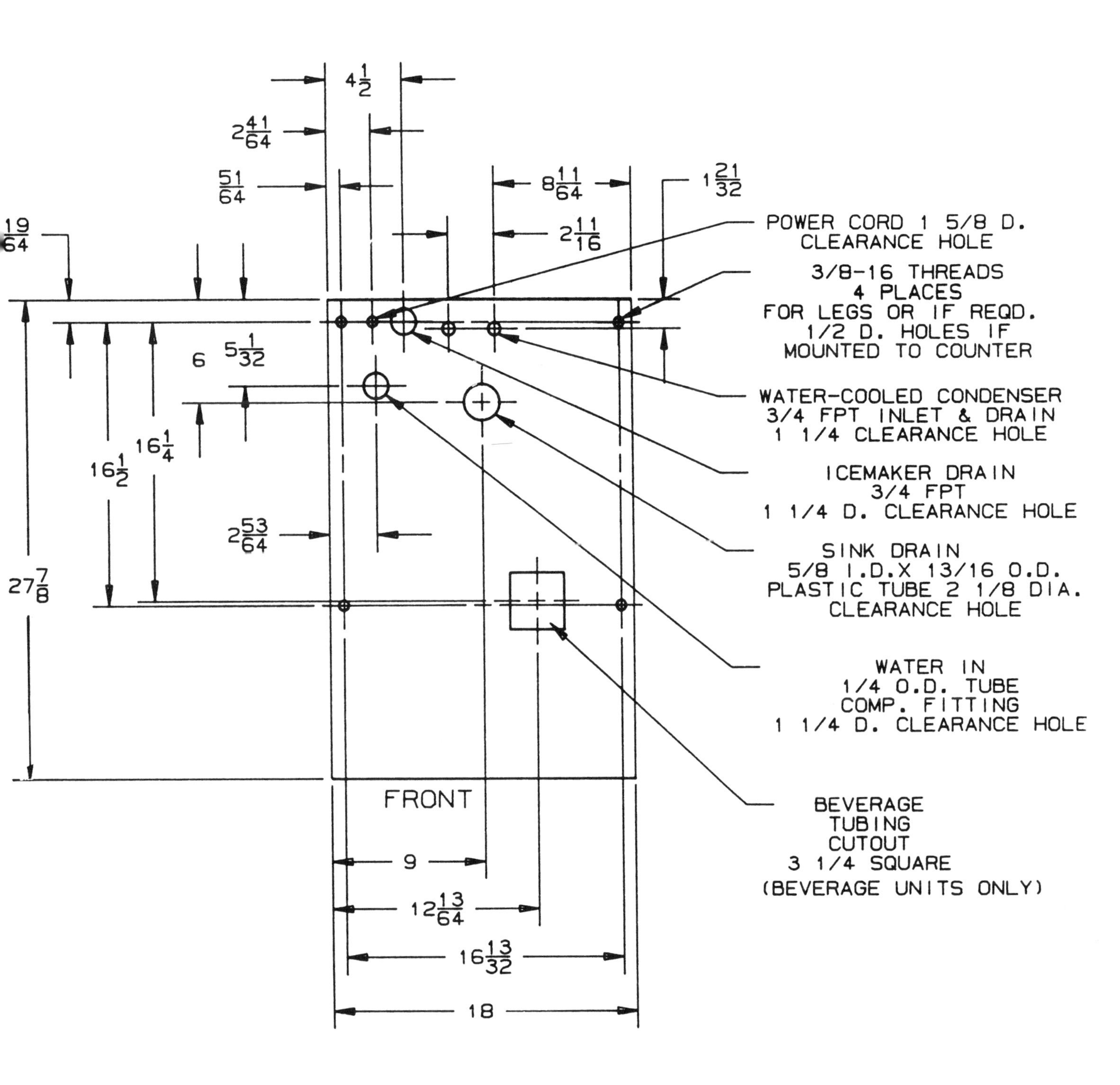

MOUNTING TEMPLATE

MODELS: SID 350A/35
SID 350W/35

BEVERAGE SYSTEM

"B" models contain beverage faucets only and must be supplied with cold product from any remote coldplate or refrigerated soda factory.

INSTALLATION

1. Locate the required openings in the counter top for the beverage lines as shown in Fig. 2.

2. Carefully pull the beverage tubes through the bottom opening in the unit and through the clearance opening in the counter.

3. Connect the beverage system product lines as indicated in Figure 3. This work should be done by a qualified serviceman. Note that the hoses are marked with numbers 1-4 for syrup connections and "CW" for carbonated water connection.

BEVERAGE SYSTEM SCHEMATIC

"-B" MODELS

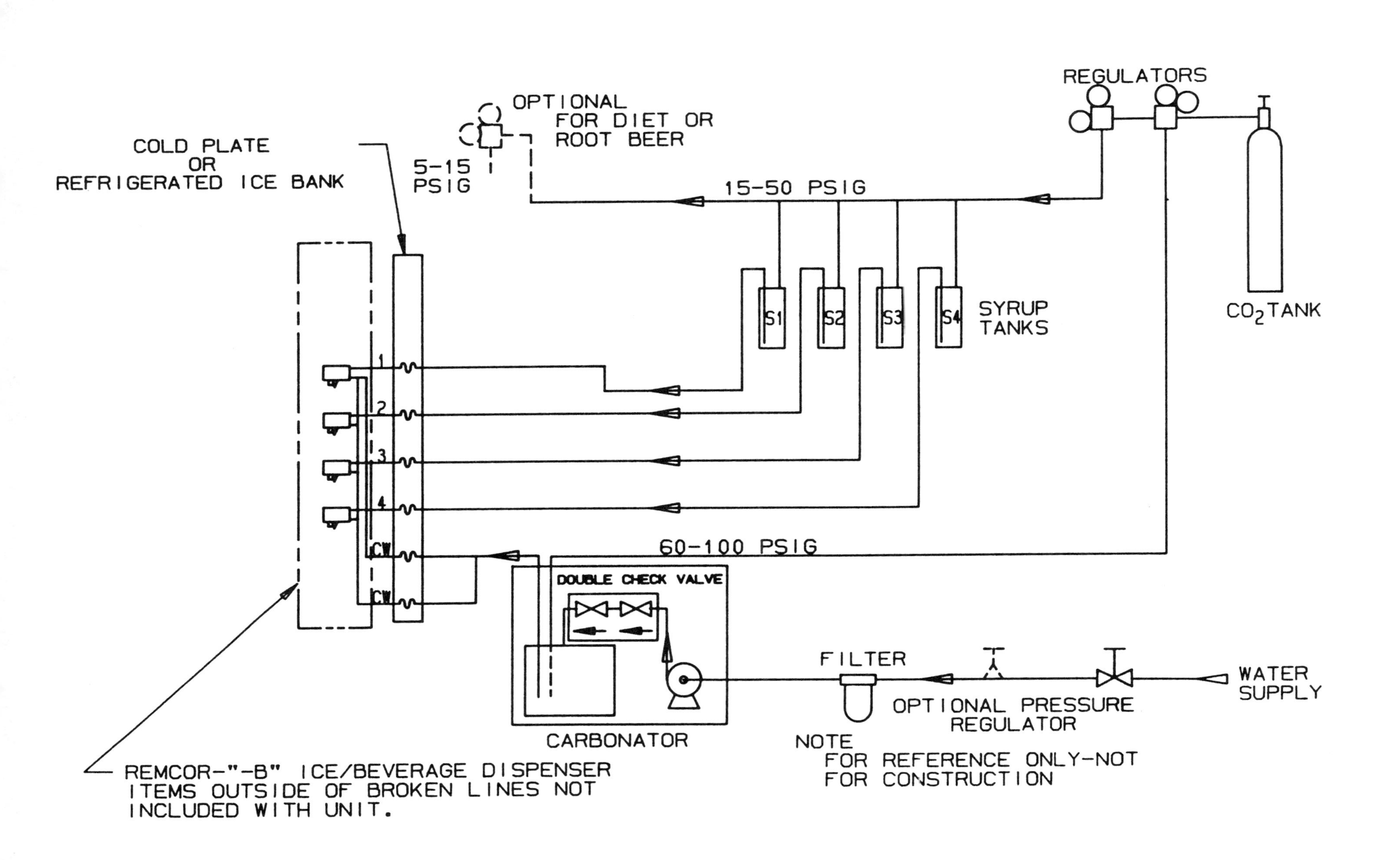

START UP

1. Open the front service door.

2. Turn on water to icemaker. Depress fill switch to fill evaporator with water before starting unit.

3. Put the stop/run switch in the run position. Observe that the icemaker goes through proper icemaking and harvest cycle. If unit malfunctions, consult trouble shooting section.

   Note: Due to meltage loss because of a warm storage hopper, it will take longer to fill the hopper the first time than when the icemaker has been operating continuously.

4. Depress the vend switch lever. Check that both the gate solenoid and agitator motor are energized simultaneously to lift the gate slide and rotate the agitator in the storage hopper, respectively. If either component malfunctions, consult the trouble shooting section.

5. For beverage units, start up the beverage system and adjust the faucets to the proper brix. Contact your local syrup distributor for complete information on the beverage system. For units with built-in coldplate, it will take approximately 1 hour from initial machine start-up for coldplate operation to be at full capacity.

6. The bin thermostat is calibrated at an atmospheric pressure equivalent at 500 feet above sea level. For locations at higher elevations, it may be necessary to re-adjust these controls. Consult the maintenance/adjustment procedures section.

ICE PRODUCTION

1. Open the hopper fill door on top of unit and upper front service door.

2. Collect and weigh the ice produced during the harvest cycle. The amount of ice harvested should be 1.9 to 2.1 pounds.

   A. Put the stop/run switch in the "stop" position. (If unit is in the icemaking cycle, stop the unit at the end of the harvest cycle.)

3. Make adjustments by accessing the RCT timer in the control box and adding or subtracting time to the total cycle time. More time = more ice.

4. Return unit to normal ice making mode and repeat steps 2 and 3 until proper amount of ice is delivered per cycle.

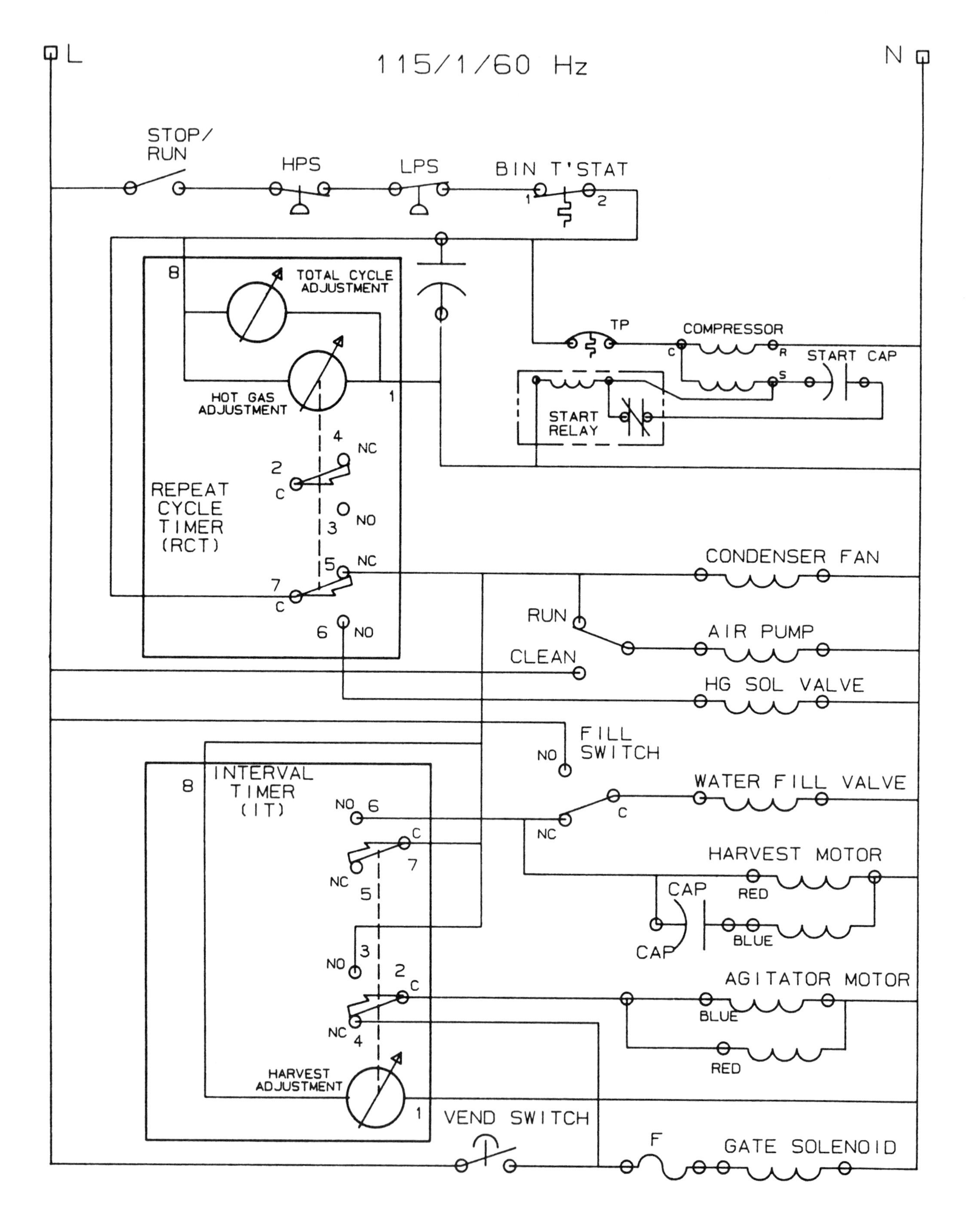
L
115/1/60 Hz
N
STOP/
RUN
HPS
LPS
BIN T'STAT
1
2
8
TOTAL CYCLE
ADJUSTMENT
TP
COMPRESSOR
C
R
START CAP
S
HOT GAS
ADJUSTMENT
1
START
RELAY
4
NC
2
C
REPEAT
CYCLE
TIMER
(RCT)
NO
3
5
NC
7
C
CONDENSER FAN
RUN
6
NO
AIR PUMP
CLEAN
HG SOL VALVE
FILL
SWITCH
NO
INTERVAL
TIMER
(IT)
8
NO 6
WATER FILL VALVE
C
NC
C
7
NC
5
HARVEST MOTOR
CAP
RED
CAP
BLUE
NO
3
2
C
AGITATOR MOTOR
BLUE
NC
4
RED
HARVEST
ADJUSTMENT
1
VEND SWITCH
F
GATE SOLENOID

WIRING DIAGRAM

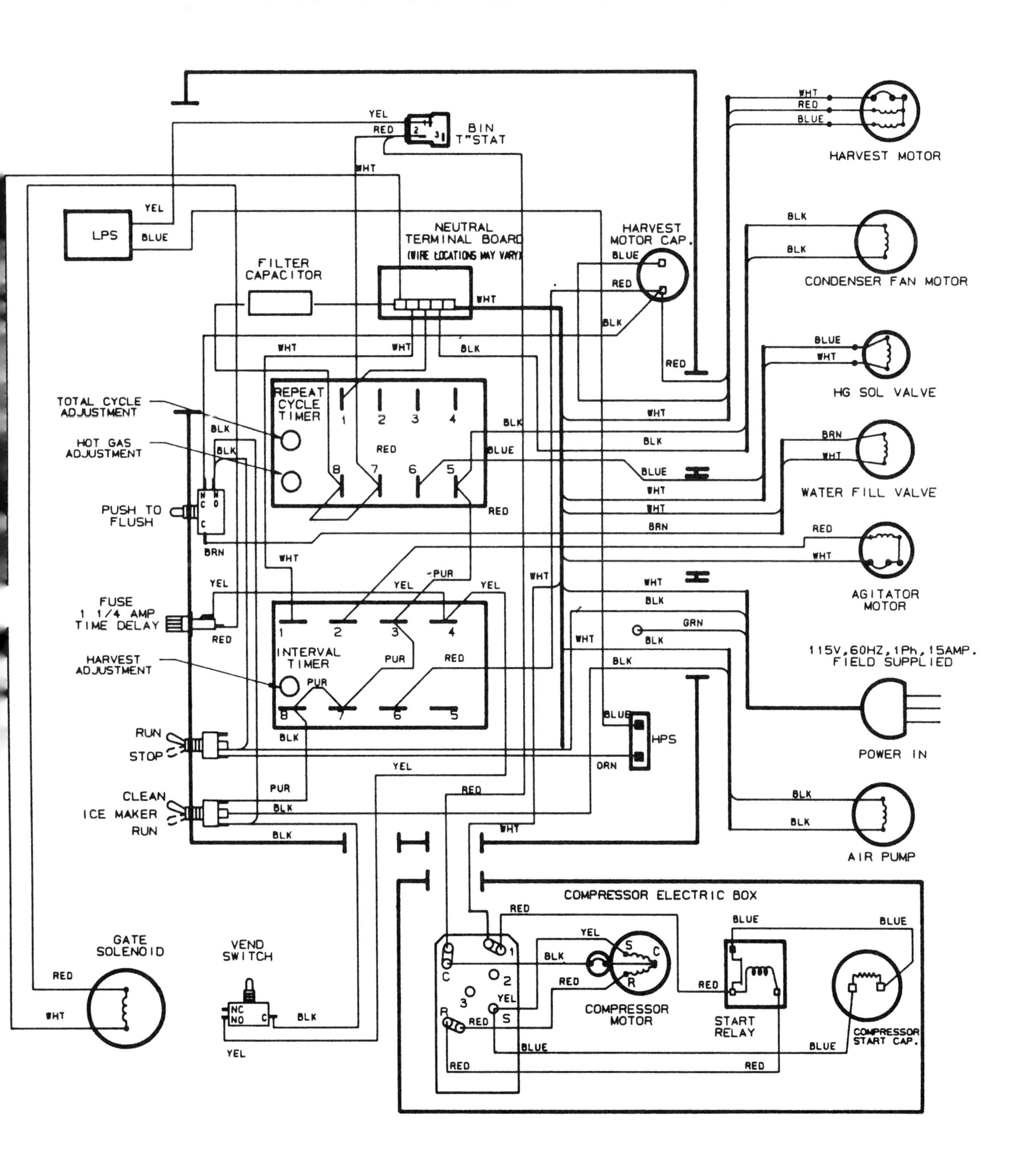

MAINTENANCE

It is recommended that the air cooled condenser be cleaned every 3 months or sooner depending on the operating environment for proper refrigeration system performance. Check that this condenser is free of dirt/foreign material that could cause air flow blockage.

Cleaning of the icemaker is recommended on a regular basis not only for sanitary reasons but also to maintain the performance of the unit. Build-up of lime and scale can hinder icemaking production rates and interfere with proper dispensing of the ice. See the cleaning section for the recommended procedure.

Periodically, check the vending area sink for proper water drainage. Remove any foreign material from the sink to prevent drain blockage.

CLEANING INSTRUCTIONS

IMPORTANT:	The icemaker should be cleaned at a minimum of 3 month intervals or more frequently depending on local water conditions. The storage hopper interior should be cleaned at least once a month.

WARNING:	Do not use metal scrapers, sharp objects or abrasives on the surface of the storage hopper, as damage may result. Do not use solvents or other cleaning agents as they may attack the plastic surface. Use only the recommended chemicals and solutions for both the icemaker and hopper.

ICEMAKER SECTION

1. Open the hinged service door on the upper front of unit.

2. Put the stop/run switch in the "stop" position at the end of the harvest cycle. An alternate method would be to stop the unit during the icemaking cycle and allow ice in the evaporator to melt by waiting for at least 1 hour before beginning the cleaning procedure. The flush switch can be drpressed to bring in warmer water to help the melting process.

WARNING:	Electrical power is on to unit during icemaker-section cleaning. To avoid possible injury, do not reach into hopper or into icemaker nozzle. Do not contact exposed electrical wiring and components.

3. Close the water supply valve to the icemaker.

4. Remove the ice drop cover from evaporator and the storage hopper cover.

5. Tighten clamp around ice maker drain line located at front of unit under hopper.

6. Add 4 oz of Virginia Ice Machine Cleaner through the evaporator outlet.

CAUTION:	Virginia Ice Machine Cleaner is a mild acid. Normal care should be taken - keep out of eyes and cuts. Read warnings on package before using. Do not operate unit in the cleaning mode without the ice drop cover in place. There may be some overflow of cleaning solution through the evaporator vent tube during the cleaning cycle.

7. Seal the evaporator outlet with plastic plug provided with the unit.

8. Push fill switch and fill evaporator (when fluid flows out of evaporator is full) let go of fill switch.

9. Put the clean/run switch in the "clean" position. Allow unit to run in the cleaning mode for at least 30 minutes.

10. Put the clean/run switch in the "run" position.

11. Push fill switch and hold for 6 minutes to flush solution out of evaporator.

12. Open drain line clamp.

13. Push fill switch and hold for 2 minutes to flush solution out of drain line.

14. Put the stop/run switch in the "run" position and allow unit to run through at least 3 complete icemaking and harvest cycles, and until ice is free of "sweet" taste.

WARNING:	If unit fails to harvest put the stop/run switch in the "stop" position and flush the evaporator with hot water to melt remaining ice and repeat steps 5,11,12, and 13 to clean out remaining solution.

15. Dispense all ice out of storage hopper and discard.

<u>DISPENSER SECTION</u>

16. Turn off main electrical power supply to machine.

17. Remove agitator assembly from storage hopper and wash and rinse it thoroughly.

18. Wash down all inside surfaces of the ice storage area including the top cover and ice drop cover with a <u>mild</u> detergent solution, and <u>rinse</u> thoroughly to remove all traces of detergent.

19. Replace agitator.

20. With brush provided, clean the inside of the ice chute with a <u>mild</u> detergent solution and rinse thoroughly to remove all traces of detergent.

21. Sanitize the inside of the hopper, agitator, the ice chute, and the hopper and ice drop covers with a solution of 1 oz. of household bleach in 1 gallon of water.

22. Replace the hopper cover and ice drop cover. Turn on the electrical power supply. The icemaker is ready for normal operation.

BEVERAGE SYSTEM

1. Remove faucet spouts, wash in mild detergent, rinse, and replace.

2. Disconnect electrical power to the carbonator. Shut off the water supply and close the CO_2 regulator to the carbonator.

3. Disconnect the syrup tanks from the system.

4. Energize the beverage faucets to purge the remaining soda water in the system.

5. Use a clean 5 gallon tank for each of the following:

    1. Cleaning Tank - Fill with hot (120-140°F) potable water.

    2. Sanitizing Tank - Fill with a chlorine sanitizing solution in the strength of 1 ounce of household bleach (sodium hypochlorite) to 1 gallon of cold (ambient) potable water (410 PPM).

6. Repeat the following procedure on each of the units syrup product lines:

    a. Connect the cleaning tank to the syrup line to be sanitized and to the CO_2 system.

    b. Energize the beverage faucet until the liquid dispensed is free of any syrup.

    c. Disconnect the cleaning tank and hook-up the sanitizing tank to the syrup line and CO_2 system.

    d. Energize the beverage faucet until the chlorine sanitizing solution is dispensed through the faucet. Flush at least 2 cups of liquid to insure that the sanitizing solution has filled the entire length of the syrup line. Allow sanitizing solution to remain in product line for 15 minutes before proceeding.

    e. Disconnect the sanitizing tank. Hook-up the product tank to the syrup line and to the CO_2 system.

    f. Energize the faucet to flush the sanitizing solution from the syrup line and faucet. Continue draw on faucet until only syrup is dispensed.

7. Repeat step #2 in reverse order to turn on the carbonator. Dispense at least 1 cup of beverage from each faucet. Check taste. Continue to flush if needed to obtain satisfactory tasting drink.

TROUBLESHOOTING GUIDE

The following pages contain troubleshooting charts designed to aid an experienced serviceman in diagnosing any operating problems which may be experienced. It is assumed that normal service techniques and skills are familiar to the person doing the troubleshooting. In order to gain maximum benefit from these charts, please note:

1. Start at the beginning of the chart and supply the appropriate answer to each question.

2. Do not skip any section, unless instructed to do so. You might miss the solution to your problem.

3. Evaluate the possible problem causes in the sequence in which they are presented. In general, they begin with the most likely or easiest to check, and proceed to the less likely or more complicated.

4. If, after checking all indicated causes, the problem is not resolved, it is recommended that you retry a second time, carefully evaluating the symptoms and modifying your answers as necessary.

5. If you are unable to resolve a problem after several attempts, contact Remcor customer service for assistance.

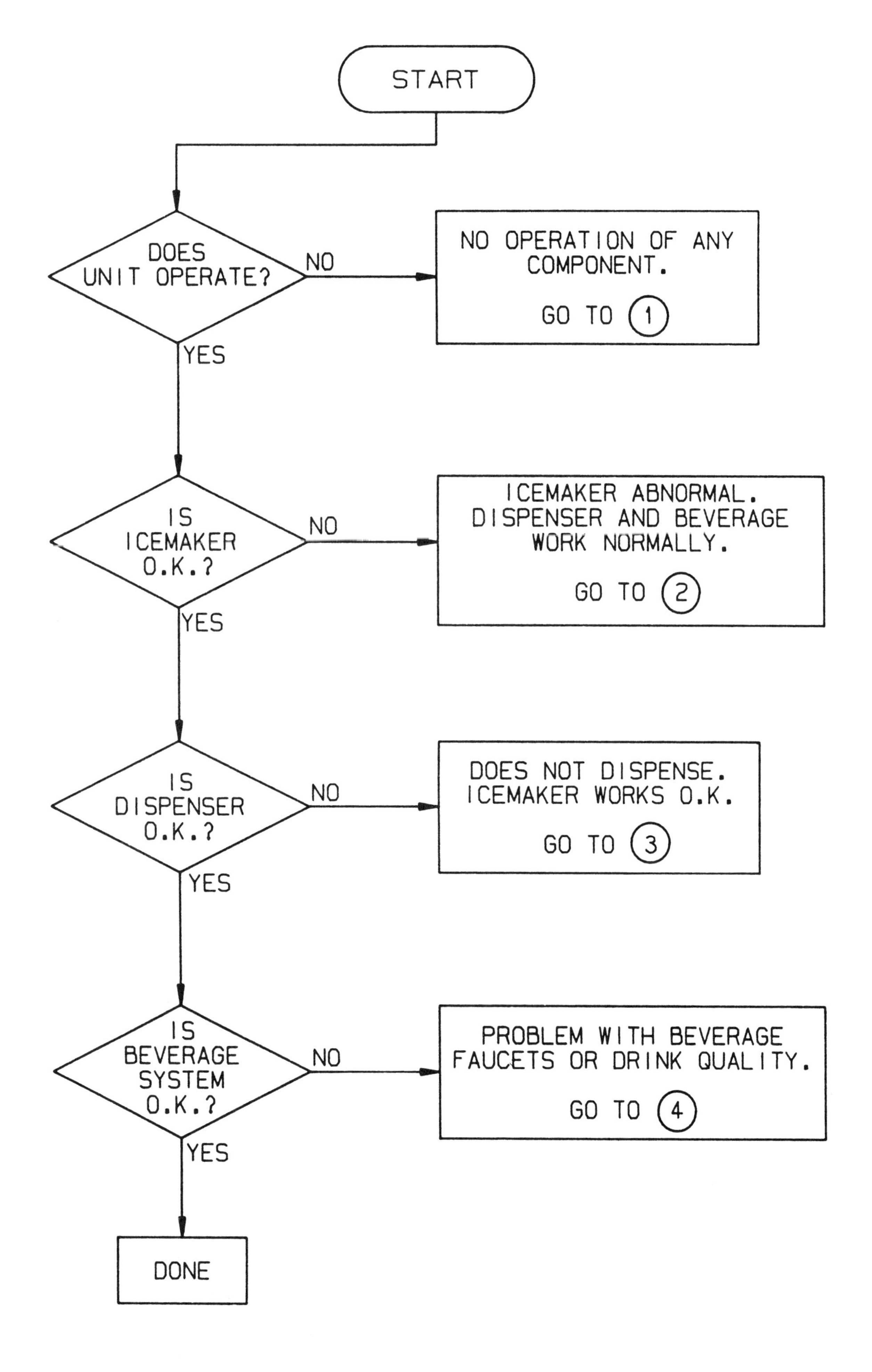
START
DOES
UNIT OPERATE?
NO
NO OPERATION OF ANY
COMPONENT.
GO TO 1
YES
IS
ICEMAKER
O.K.?
NO
ICEMAKER ABNORMAL.
DISPENSER AND BEVERAGE
WORK NORMALLY.
GO TO 2
YES
IS
DISPENSER
O.K.?
NO
DOES NOT DISPENSE.
ICEMAKER WORKS O.K.
GO TO 3
YES
IS
BEVERAGE
SYSTEM
O.K.?
NO
PROBLEM WITH BEVERAGE
FAUCETS OR DRINK QUALITY.
GO TO 4
YES
DONE

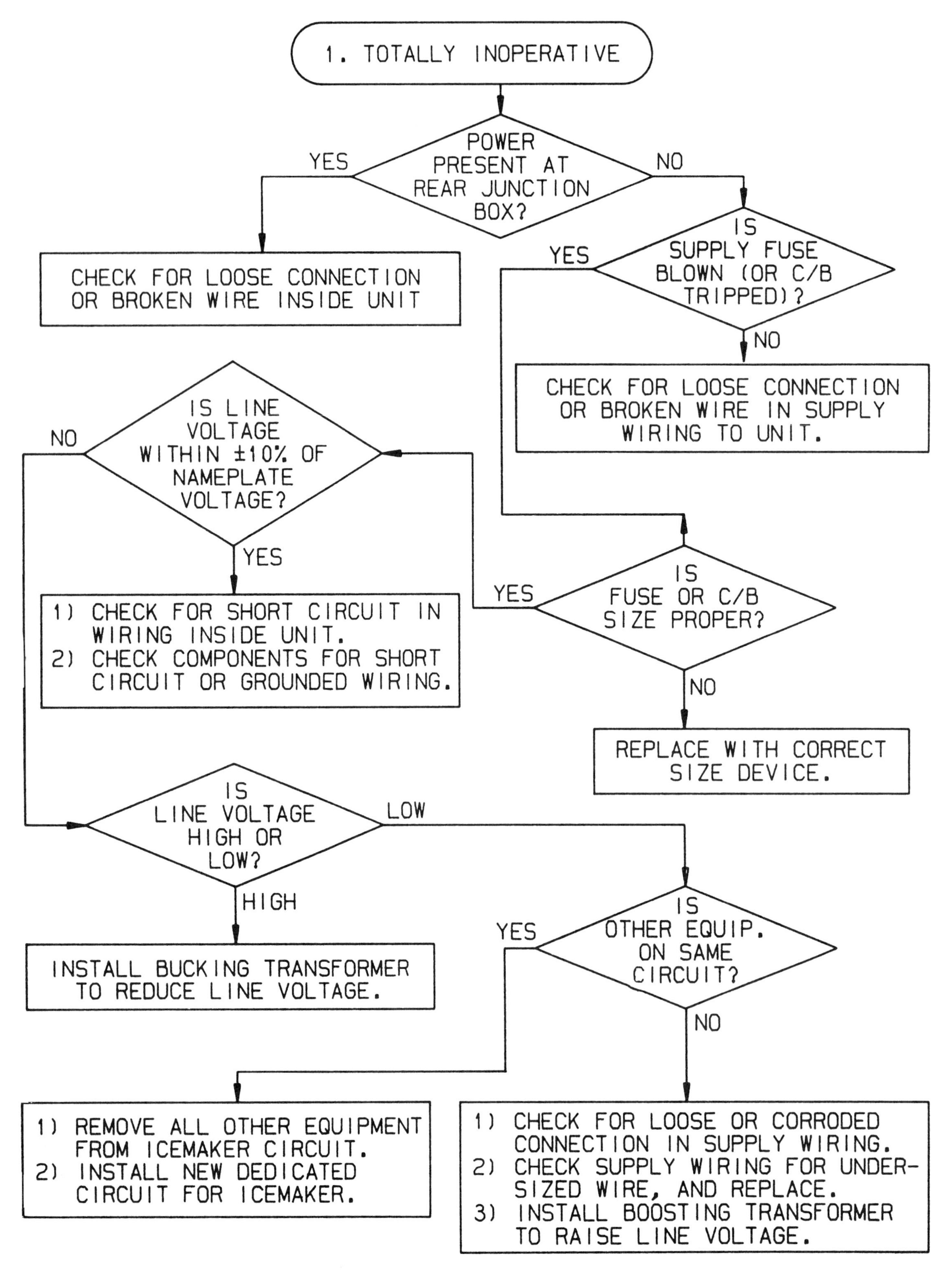

1. TOTALLY INOPERATIVE
POWER PRESENT AT REAR JUNCTION BOX?
YES
NO
CHECK FOR LOOSE CONNECTION OR BROKEN WIRE INSIDE UNIT
IS SUPPLY FUSE BLOWN (OR C/B TRIPPED)?
YES
NO
CHECK FOR LOOSE CONNECTION OR BROKEN WIRE IN SUPPLY WIRING TO UNIT.
IS LINE VOLTAGE WITHIN ±10% OF NAMEPLATE VOLTAGE?
NO
YES
1) CHECK FOR SHORT CIRCUIT IN WIRING INSIDE UNIT.
2) CHECK COMPONENTS FOR SHORT CIRCUIT OR GROUNDED WIRING.
IS FUSE OR C/B SIZE PROPER?
YES
NO
REPLACE WITH CORRECT SIZE DEVICE.
IS LINE VOLTAGE HIGH OR LOW?
LOW
HIGH
INSTALL BUCKING TRANSFORMER TO REDUCE LINE VOLTAGE.
IS OTHER EQUIP. ON SAME CIRCUIT?
YES
NO
1) REMOVE ALL OTHER EQUIPMENT FROM ICEMAKER CIRCUIT.
2) INSTALL NEW DEDICATED CIRCUIT FOR ICEMAKER.
1) CHECK FOR LOOSE OR CORRODED CONNECTION IN SUPPLY WIRING.
2) CHECK SUPPLY WIRING FOR UNDER-SIZED WIRE, AND REPLACE.
3) INSTALL BOOSTING TRANSFORMER TO RAISE LINE VOLTAGE.

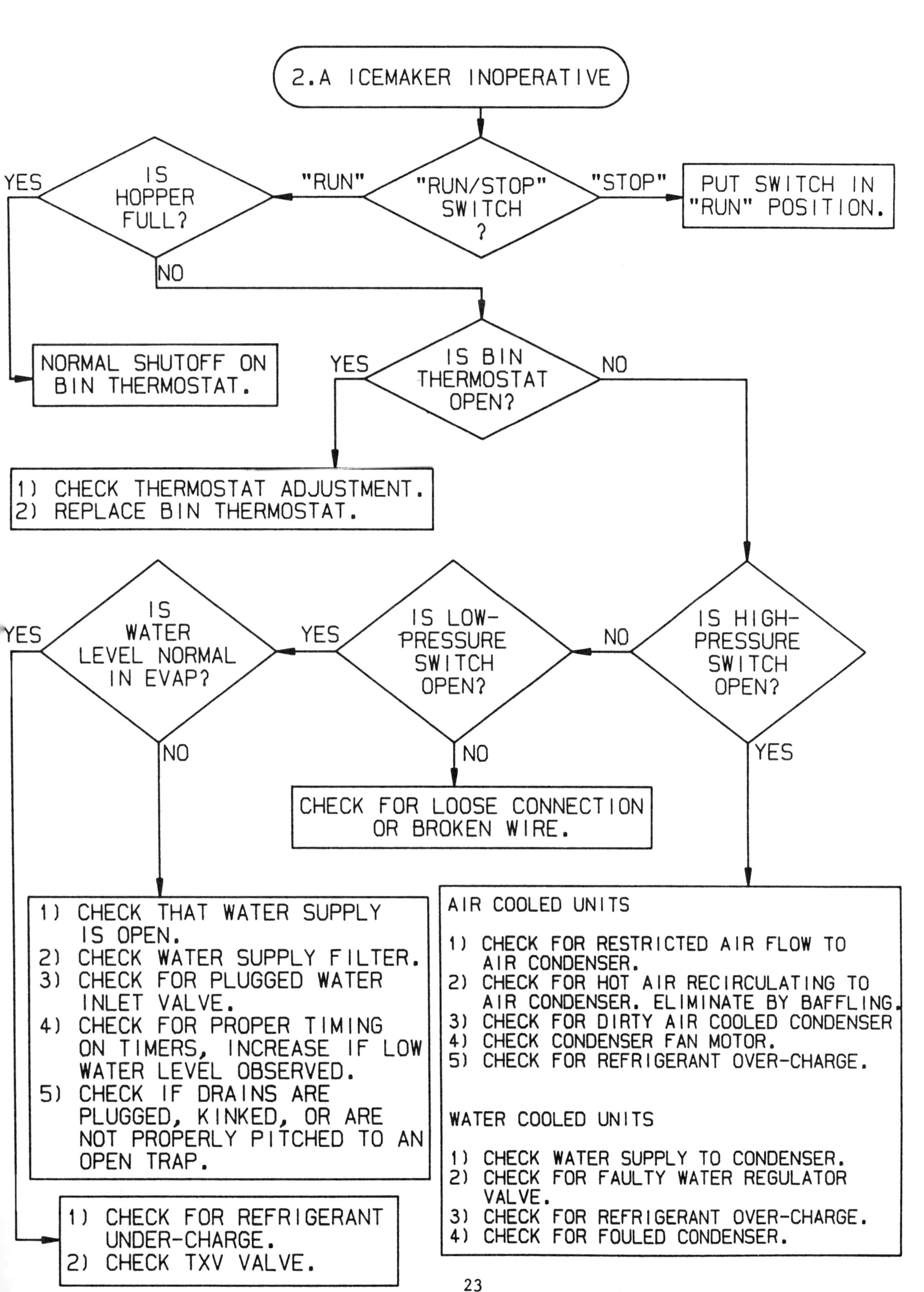
2.A ICEMAKER INOPERATIVE
"RUN/STOP" SWITCH ?
"STOP"
PUT SWITCH IN "RUN" POSITION.
"RUN"
IS HOPPER FULL?
YES
NO
NORMAL SHUTOFF ON BIN THERMOSTAT.
IS BIN THERMOSTAT OPEN?
YES
NO
1) CHECK THERMOSTAT ADJUSTMENT.
2) REPLACE BIN THERMOSTAT.
IS HIGH-PRESSURE SWITCH OPEN?
NO
YES
IS LOW-PRESSURE SWITCH OPEN?
YES
NO
CHECK FOR LOOSE CONNECTION OR BROKEN WIRE.
IS WATER LEVEL NORMAL IN EVAP?
YES
NO
1) CHECK THAT WATER SUPPLY IS OPEN.
2) CHECK WATER SUPPLY FILTER.
3) CHECK FOR PLUGGED WATER INLET VALVE.
4) CHECK FOR PROPER TIMING ON TIMERS, INCREASE IF LOW WATER LEVEL OBSERVED.
5) CHECK IF DRAINS ARE PLUGGED, KINKED, OR ARE NOT PROPERLY PITCHED TO AN OPEN TRAP.
1) CHECK FOR REFRIGERANT UNDER-CHARGE.
2) CHECK TXV VALVE.
AIR COOLED UNITS
1) CHECK FOR RESTRICTED AIR FLOW TO AIR CONDENSER.
2) CHECK FOR HOT AIR RECIRCULATING TO AIR CONDENSER. ELIMINATE BY BAFFLING.
3) CHECK FOR DIRTY AIR COOLED CONDENSER
4) CHECK CONDENSER FAN MOTOR.
5) CHECK FOR REFRIGERANT OVER-CHARGE.
WATER COOLED UNITS
1) CHECK WATER SUPPLY TO CONDENSER.
2) CHECK FOR FAULTY WATER REGULATOR VALVE.
3) CHECK FOR REFRIGERANT OVER-CHARGE.
4) CHECK FOR FOULED CONDENSER.

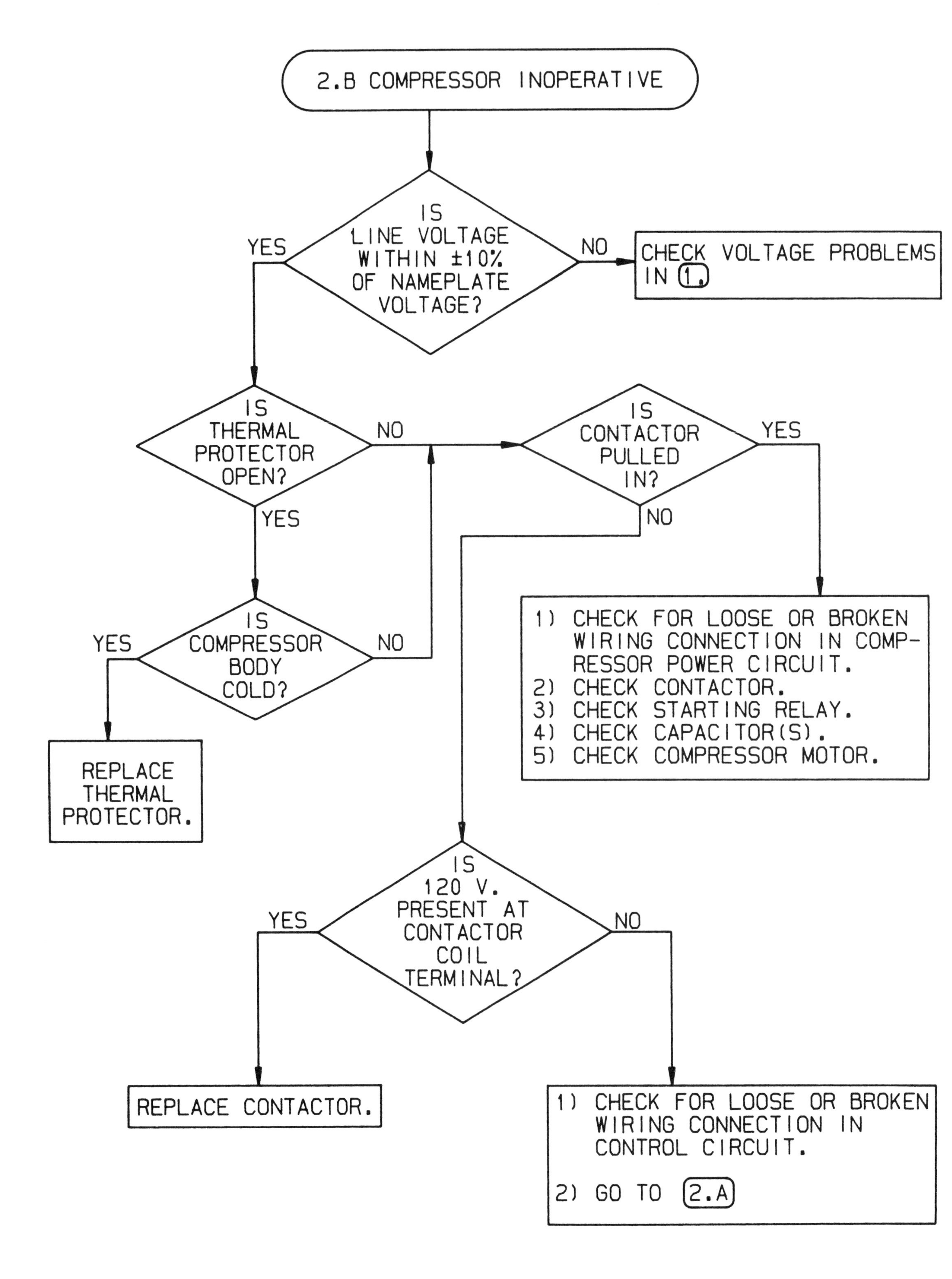
2.B COMPRESSOR INOPERATIVE
IS LINE VOLTAGE WITHIN ±10% OF NAMEPLATE VOLTAGE?
YES
NO
CHECK VOLTAGE PROBLEMS IN 1.
IS THERMAL PROTECTOR OPEN?
NO
YES
IS CONTACTOR PULLED IN?
YES
NO
IS COMPRESSOR BODY COLD?
YES
NO
1) CHECK FOR LOOSE OR BROKEN WIRING CONNECTION IN COMPRESSOR POWER CIRCUIT.
2) CHECK CONTACTOR.
3) CHECK STARTING RELAY.
4) CHECK CAPACITOR(S).
5) CHECK COMPRESSOR MOTOR.
REPLACE THERMAL PROTECTOR.
IS 120 V. PRESENT AT CONTACTOR COIL TERMINAL?
YES
NO
REPLACE CONTACTOR.
1) CHECK FOR LOOSE OR BROKEN WIRING CONNECTION IN CONTROL CIRCUIT.
2) GO TO 2.A

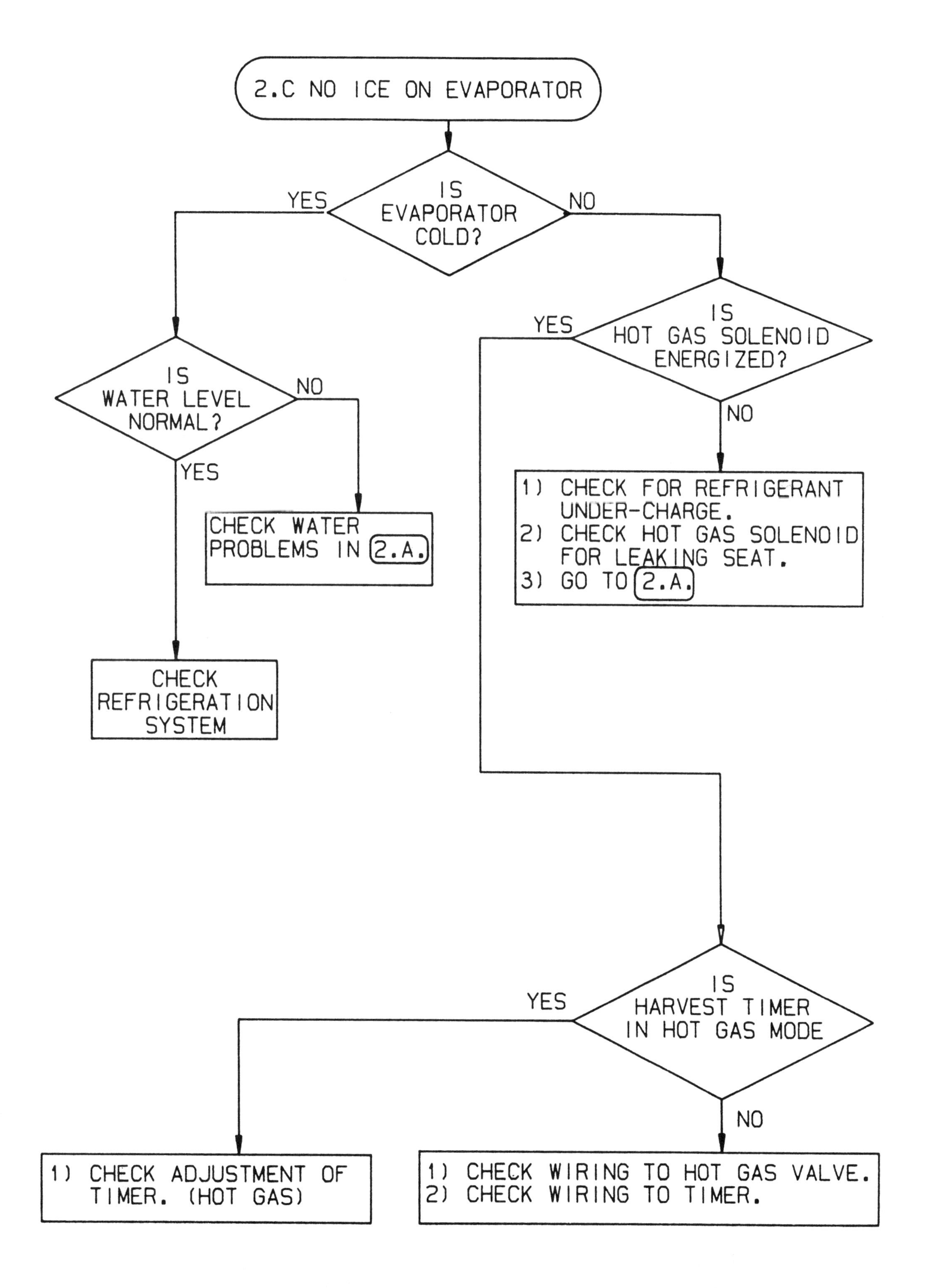

2.C NO ICE ON EVAPORATOR
IS EVAPORATOR COLD?
YES
NO
IS WATER LEVEL NORMAL?
NO
YES
CHECK WATER PROBLEMS IN 2.A.
CHECK REFRIGERATION SYSTEM
IS HOT GAS SOLENOID ENERGIZED?
YES
NO
1) CHECK FOR REFRIGERANT UNDER-CHARGE.
2) CHECK HOT GAS SOLENOID FOR LEAKING SEAT.
3) GO TO 2.A.
IS HARVEST TIMER IN HOT GAS MODE
YES
NO
1) CHECK ADJUSTMENT OF TIMER. (HOT GAS)
1) CHECK WIRING TO HOT GAS VALVE.
2) CHECK WIRING TO TIMER.

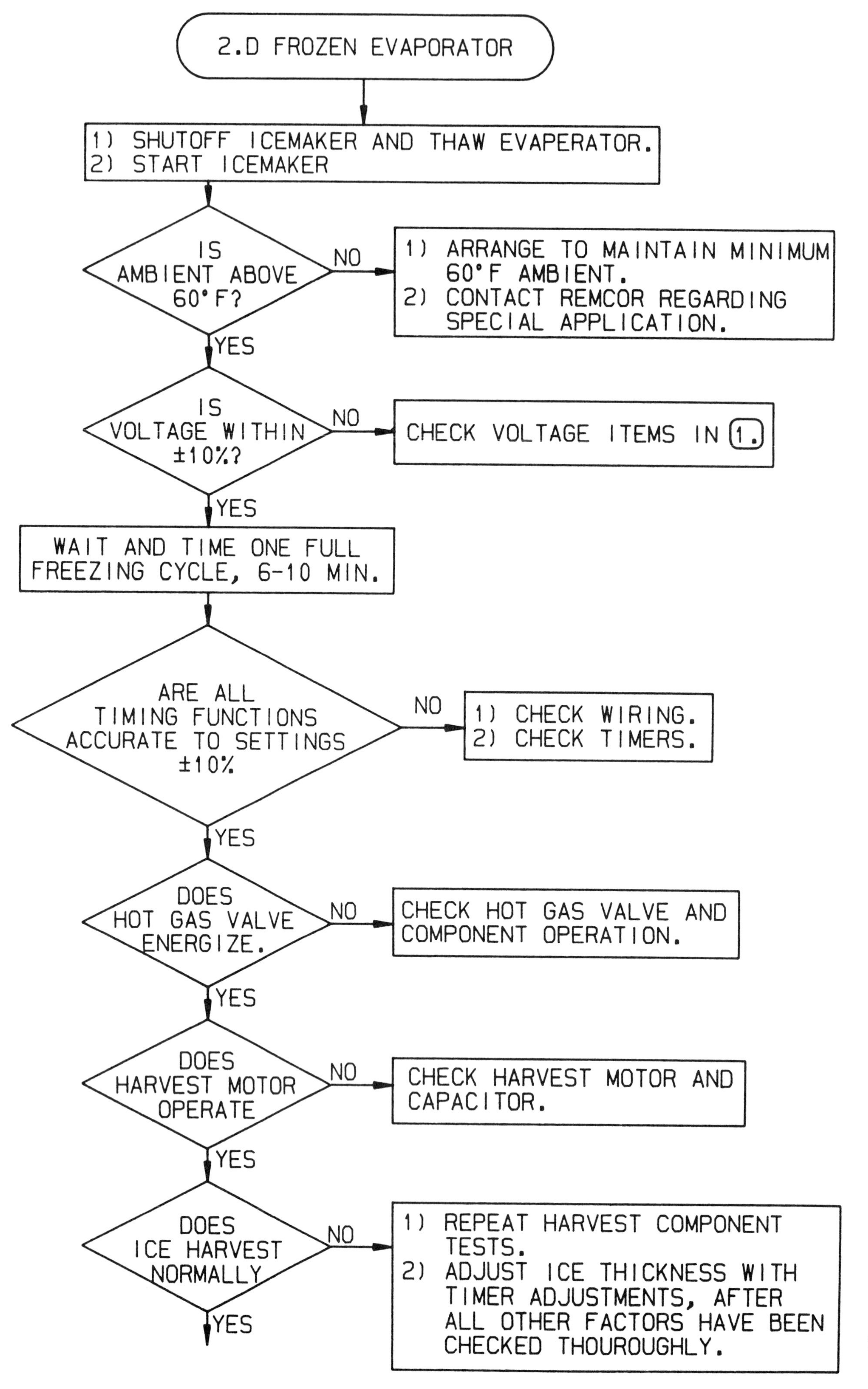
2.D FROZEN EVAPORATOR
1) SHUTOFF ICEMAKER AND THAW EVAPERATOR.
2) START ICEMAKER
IS AMBIENT ABOVE 60°F?
NO
1) ARRANGE TO MAINTAIN MINIMUM 60°F AMBIENT.
2) CONTACT REMCOR REGARDING SPECIAL APPLICATION.
YES
IS VOLTAGE WITHIN ±10%?
NO
CHECK VOLTAGE ITEMS IN 1.
YES
WAIT AND TIME ONE FULL FREEZING CYCLE, 6-10 MIN.
ARE ALL TIMING FUNCTIONS ACCURATE TO SETTINGS ±10%
NO
1) CHECK WIRING.
2) CHECK TIMERS.
YES
DOES HOT GAS VALVE ENERGIZE.
NO
CHECK HOT GAS VALVE AND COMPONENT OPERATION.
YES
DOES HARVEST MOTOR OPERATE
NO
CHECK HARVEST MOTOR AND CAPACITOR.
YES
DOES ICE HARVEST NORMALLY
NO
1) REPEAT HARVEST COMPONENT TESTS.
2) ADJUST ICE THICKNESS WITH TIMER ADJUSTMENTS, AFTER ALL OTHER FACTORS HAVE BEEN CHECKED THOUROUGHLY.
YES

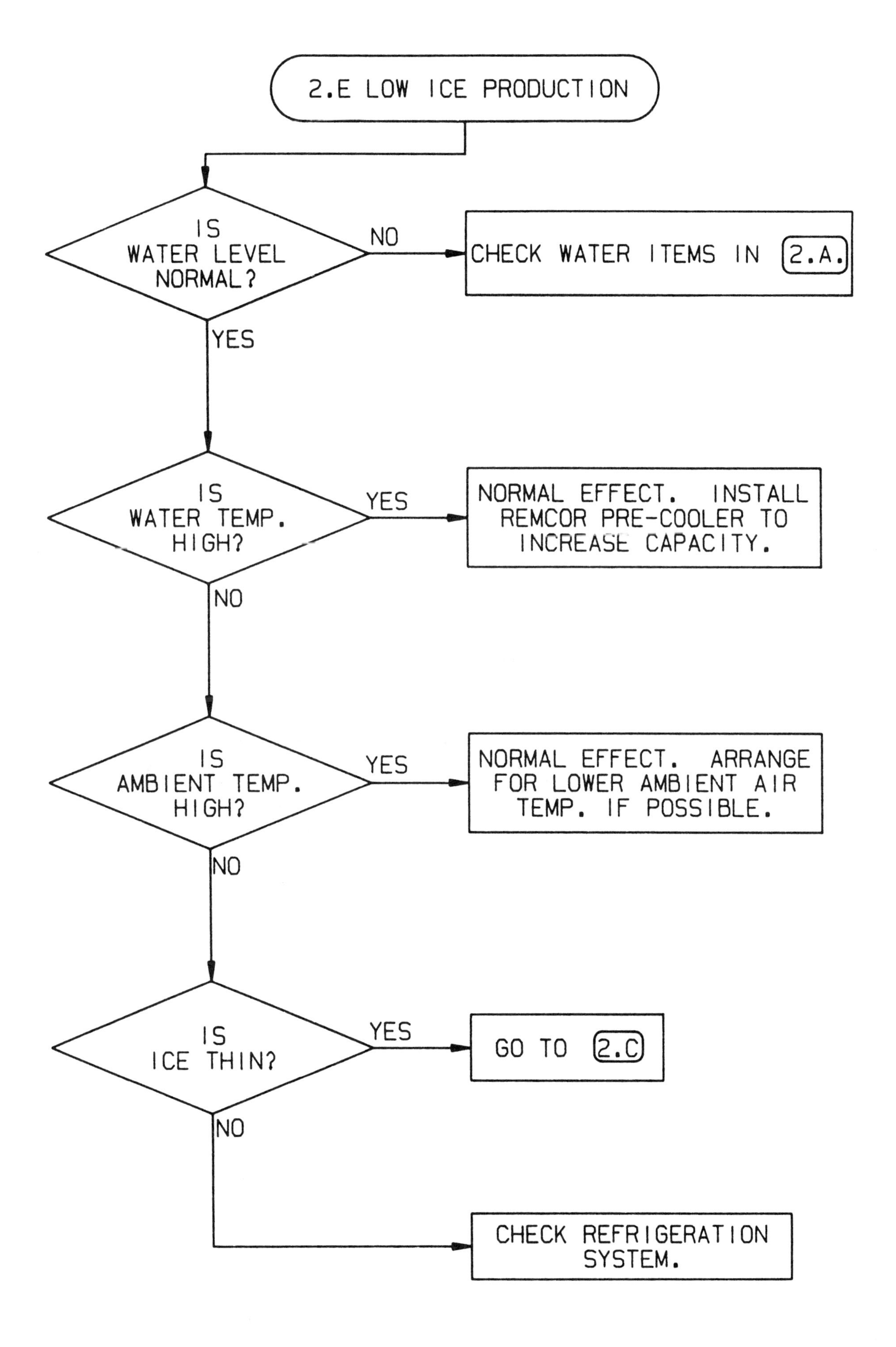
2.E LOW ICE PRODUCTION
IS
WATER LEVEL
NORMAL?
NO
CHECK WATER ITEMS IN 2.A.
YES
IS
WATER TEMP.
HIGH?
YES
NORMAL EFFECT. INSTALL
REMCOR PRE-COOLER TO
INCREASE CAPACITY.
NO
IS
AMBIENT TEMP.
HIGH?
YES
NORMAL EFFECT. ARRANGE
FOR LOWER AMBIENT AIR
TEMP. IF POSSIBLE.
NO
IS
ICE THIN?
YES
GO TO 2.C
NO
CHECK REFRIGERATION
SYSTEM.

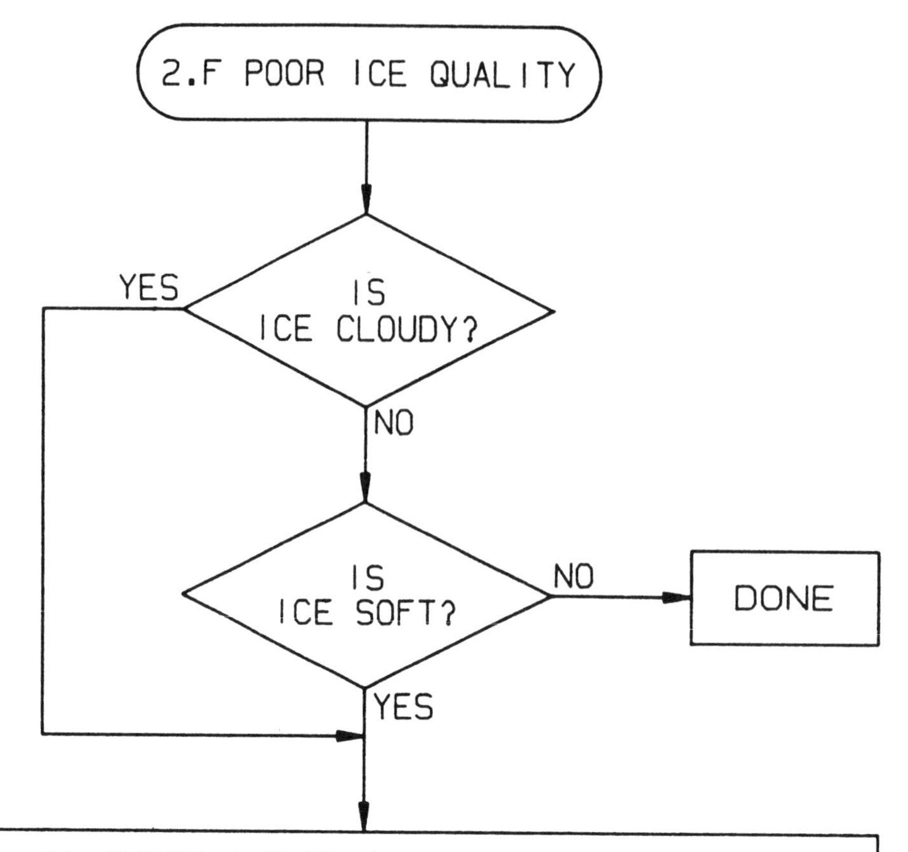

1) CHECK AIR PUMP.
2) CHECK ADJUSTMENT OF TIMER. (WATER DUMP)
3) CHECK WATER FILTER.
4) INCREASE WATER DUMP BY ADJUSTING TIMER.
5) INSTALL ADDITIONAL WATER TREATMENT DEVICES FOR SPECIFIC "PROBLEM" WATER.

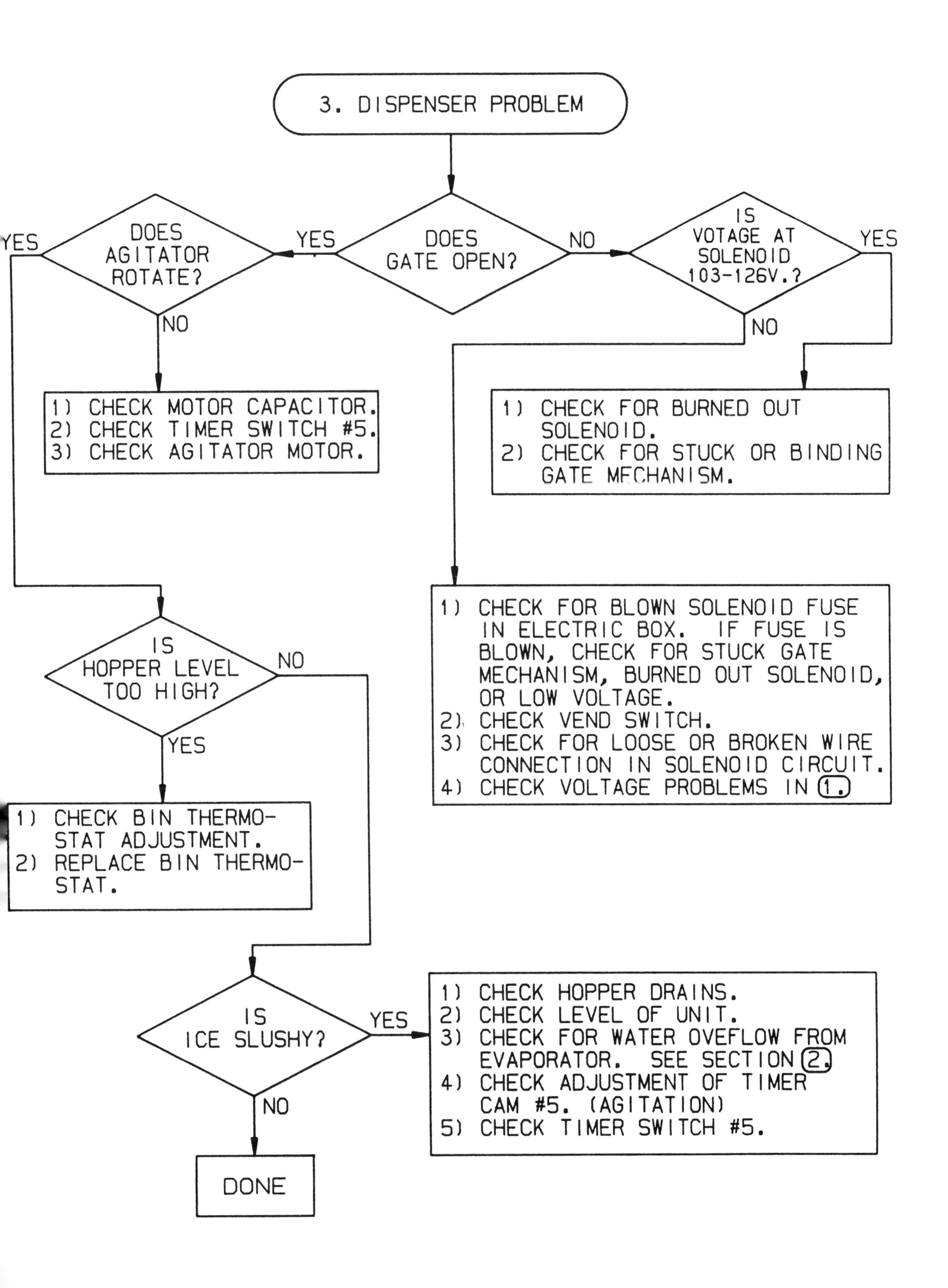
3. DISPENSER PROBLEM
DOES GATE OPEN?
YES
NO
DOES AGITATOR ROTATE?
YES
NO
1) CHECK MOTOR CAPACITOR.
2) CHECK TIMER SWITCH #5.
3) CHECK AGITATOR MOTOR.
IS VOTAGE AT SOLENOID 103-126V.?
YES
NO
1) CHECK FOR BURNED OUT SOLENOID.
2) CHECK FOR STUCK OR BINDING GATE MECHANISM.
1) CHECK FOR BLOWN SOLENOID FUSE IN ELECTRIC BOX. IF FUSE IS BLOWN, CHECK FOR STUCK GATE MECHANISM, BURNED OUT SOLENOID, OR LOW VOLTAGE.
2) CHECK VEND SWITCH.
3) CHECK FOR LOOSE OR BROKEN WIRE CONNECTION IN SOLENOID CIRCUIT.
4) CHECK VOLTAGE PROBLEMS IN 1.
IS HOPPER LEVEL TOO HIGH?
NO
YES
1) CHECK BIN THERMO-STAT ADJUSTMENT.
2) REPLACE BIN THERMO-STAT.
IS ICE SLUSHY?
YES
NO
1) CHECK HOPPER DRAINS.
2) CHECK LEVEL OF UNIT.
3) CHECK FOR WATER OVEFLOW FROM EVAPORATOR. SEE SECTION 2.
4) CHECK ADJUSTMENT OF TIMER CAM #5. (AGITATION)
5) CHECK TIMER SWITCH #5.
DONE

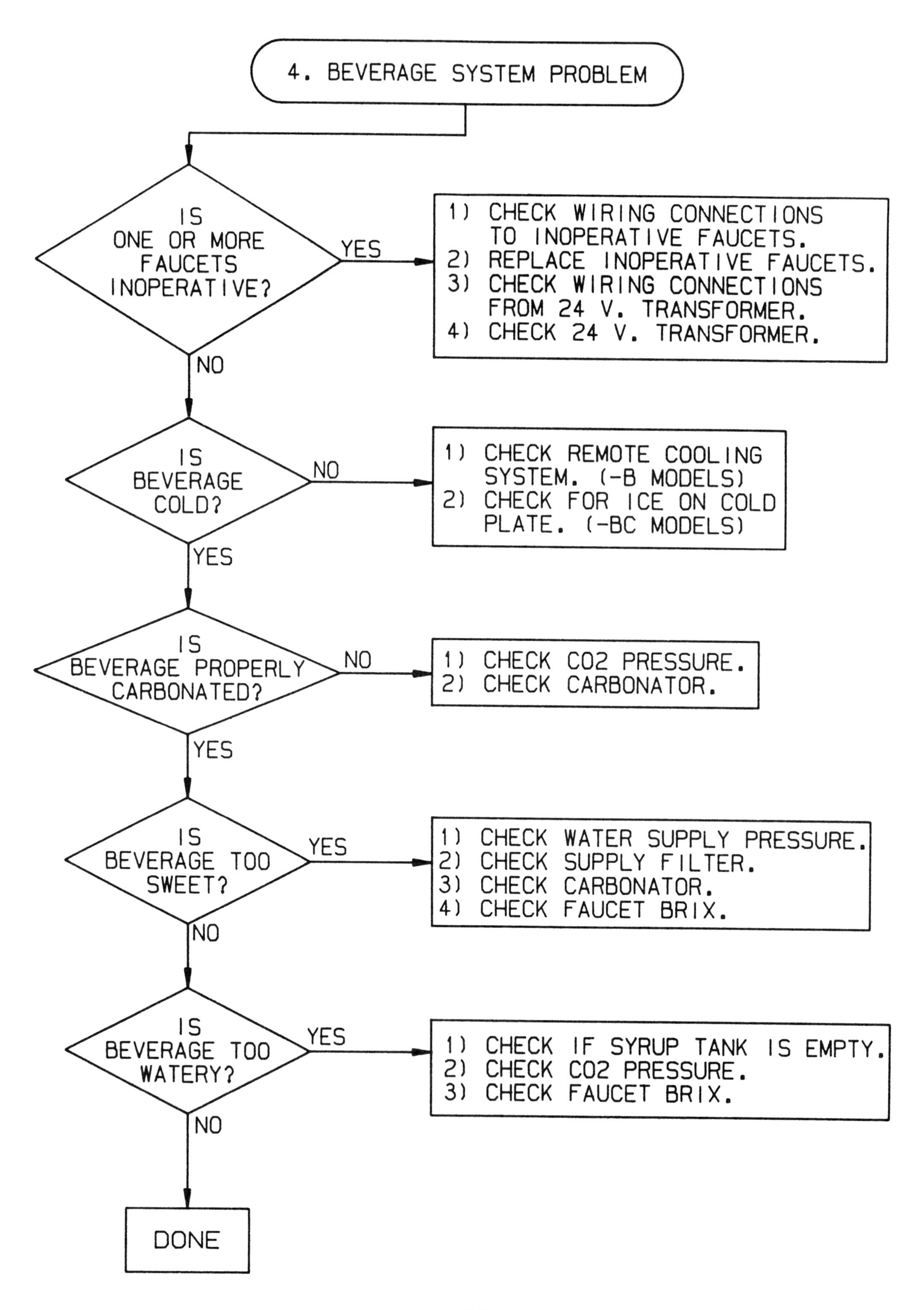
4. BEVERAGE SYSTEM PROBLEM
IS ONE OR MORE FAUCETS INOPERATIVE?
YES
1) CHECK WIRING CONNECTIONS TO INOPERATIVE FAUCETS.
2) REPLACE INOPERATIVE FAUCETS.
3) CHECK WIRING CONNECTIONS FROM 24 V. TRANSFORMER.
4) CHECK 24 V. TRANSFORMER.
NO
IS BEVERAGE COLD?
NO
1) CHECK REMOTE COOLING SYSTEM. (-B MODELS)
2) CHECK FOR ICE ON COLD PLATE. (-BC MODELS)
YES
IS BEVERAGE PROPERLY CARBONATED?
NO
1) CHECK CO2 PRESSURE.
2) CHECK CARBONATOR.
YES
IS BEVERAGE TOO SWEET?
YES
1) CHECK WATER SUPPLY PRESSURE.
2) CHECK SUPPLY FILTER.
3) CHECK CARBONATOR.
4) CHECK FAUCET BRIX.
NO
IS BEVERAGE TOO WATERY?
YES
1) CHECK IF SYRUP TANK IS EMPTY.
2) CHECK CO2 PRESSURE.
3) CHECK FAUCET BRIX.
NO
DONE

MAINTENANCE AND ADJUSTMENTS

THERMOSTAT ALTITUDE ADJUSTMENT

BIN T'STAT

IMPORTANT: Adjust the bin t'stat setting only if storage hopper overfill is a problem.

1. Open the upper front panel. on the upper left side panel.

2. The adjustment screw is located near the top edge of the electrical box.

3. For altitudes up to 6000 ft., turn the adjustment screw COUNTER-CLOCKWISE as follows:

ELEVATION (FT)	CCW TURN
2000	1/13
4000	1/6
6000	1/4

4. For altitudes above 6000 ft., consult the factory.

CLEARING EVAPORATOR FREEZUP

WARNING:	To prevent possible injury, do not stick fingers or hand into icemaker nozzle or hopper with power applied to unit.

1. Open the hinged hopper fill door on top of unit.

2. Put the stop/run switch in the "stop" position.

3. Close the water supply valve to the icemaker.

4. Drain evaporator by disconnecting the water inlet tube at the bottom of the evaporator, when empty reconnect water inlet.

5. Pour hot water into the evaporator ice exit opening. It will be necessary to use either a funnel or a container with a spout. Fill the evaporator completely.

6. Drain the evaporator. Repeat steps 5 and 6 as required to insure that all the ice in the evaporator is melted.

7. Open the water supply valve and refill evaporator.

8. Consult troubleshooting guide to determine cause of freeze-up before putting unit back in service.

REFRIGERATION SCHEMATIC

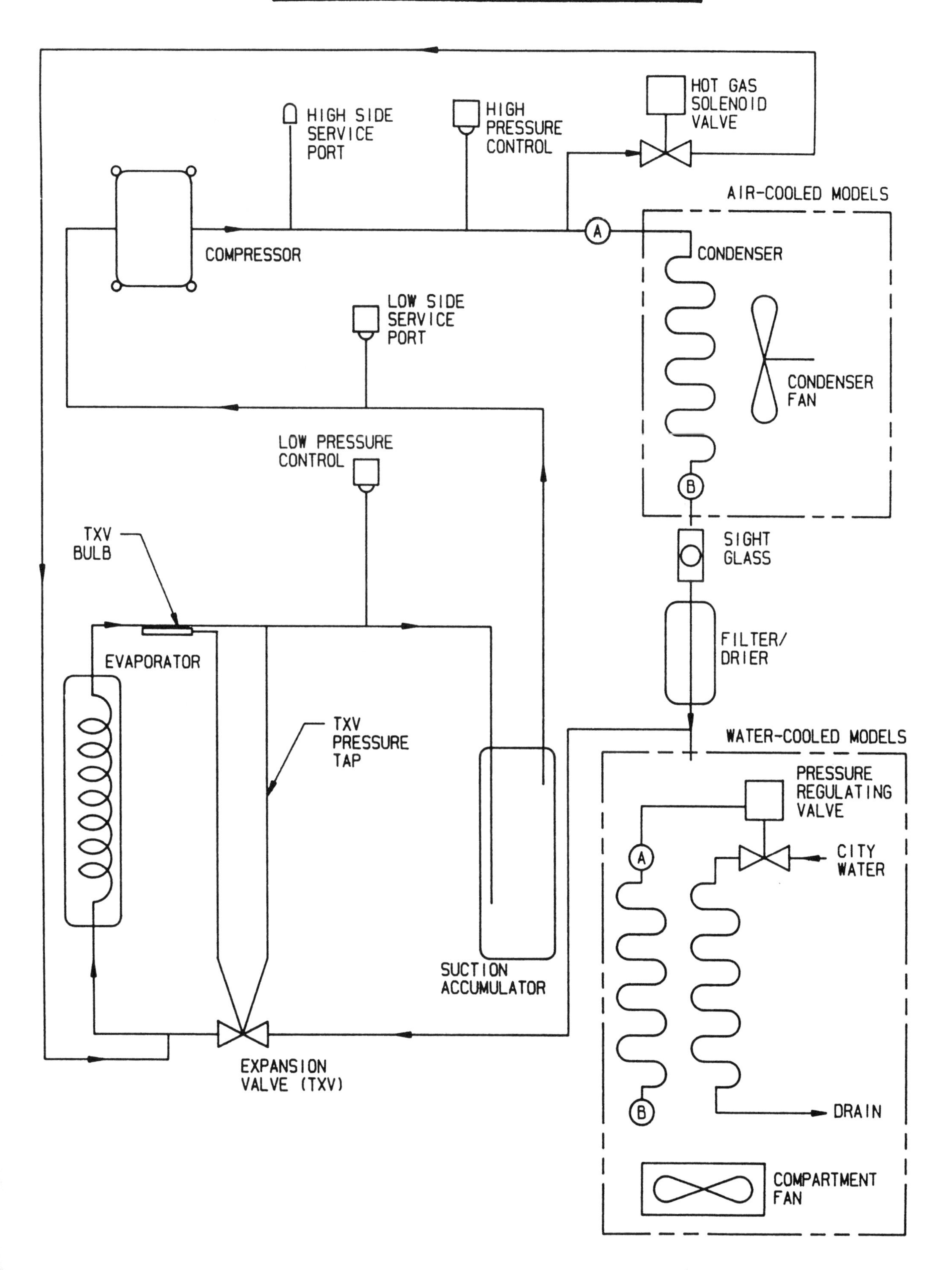

REMCOR PRODUCTS CO.

Rev. 5/30/89

WARRANTY POLICY

Spiral Ice® Icemaker/Dispenser

REMCOR Products Company warrants to the original purchaser of each new REMCOR SPIRAL ICE® ICEMAKER-DISPENSER, for a period of 21 months from date of installation of 24 months from date of shipment, whichever occurs first, that all parts shall be free from defects in material and workmanship under normal use and service. The S.S. icemaker evaporator is specifically warranted for a period of two (2) years, the compressor for a period of five (5) years, and labor cost to repair factory defective parts or workmanship is covered for a period of thirty (30) days from date of installation. Labor warranty does not cover normal installation or start-up.

Under this warranty, a defective part (or parts) is to be returned to REMCOR Products Company, 500 Regency Drive, Glendale Heights, Illinois 60139-2268 (Phone 312/980-6900) and shall be limited exclusively to repairing or replacing F.O.B. Factory, such part or parts which it concludes upon examination to be defective under the terms of the warranty. Return of any part disassembled will void warranty on any part. The decision of our Service Department regarding the warranty of parts will be final.

The warranties defined herein shall not apply to any damage of defects created or arising from accident, misapplication, abuse, misuse, neglect, alteration, acts of vandalism, flood, fire, acts of God or any other occurrences beyond the control of REMCOR.® Warranty validity also requires that all instructions have been followed and adhered to as provided in the Owner's Manual included with each unit.

REMCOR's warranty responsibility ceases if shipment is not received by you in good order and in accordance with quantity shown on Invoice or Packing Sheet; if you accept shipment from the Transportation Company in damaged condition without having a proper notation made by the Station Agent you do so at your own risk. If cartons are in apparent good order, but upon opening, contents are found to be damaged; call agent or adjuster to view same and have him mark the freight bill relative to such concealed damage.. The procedure must be executed within fifteen days after delivery.

Prior to returning any material, (part or unit), a Return of Material Authorization (RMA) must be obtained. To obtain this authorization provide the reason for return, model number, serial number and part number. Either call or write REMCOR's Parts Service Department and an Authorization number and tag will then be issued - this tag must accompany the material returned. No representative, dealer, distributor or any person is authorized to make any other decisions regarding the warranty liability in accordance with REMCOR's Warranty. Any material that does not have a pre-issued Return Material Authorization number when received will be refused by REMCOR,® and returned to the sender (freight collect).

REMCOR Products Company will not honor, or assume any responsibility for any expenses (including labor), incurred in the field for the repair of equipment covered in our Warranty unless authorization has been granted from REMCOR's Service Department prior to work being performed. The request for repair Warranty work will only apply to units shipped from REMCOR® within a thirty day time period of request. It will further be at the discretion of REMCOR® to decide if said labor will be reimbursed in full, or partial payment. REMCOR Products Company will also control the right to decline payment - all decisions will be based on circumstances prevailing. Charges for a repeat service call on the same unit to rectify the same problem previously corrected shall not be honored. REMCOR Products shall retain the right to select, or recommend another company to complete the necessary repair work.

CAUTION

The inherent nature of ice may cause spillage onto counter or floor areas. The Owner or Operator is cautioned to Maintain these areas in a clean, ice-free condition.

REMCOR®

Jan., 1988 | **Sentinel** | Issue # 1

Tom Barnes
National Service Manager

Angela Mitchell
Parts Order Expeditor

Terry Martin
Field Service Supervisor

REMCOR PRODUCTS COMPANY
Parts and Service Technical Publication
312-671-7140

SERVICE BULLETIN # 1

January, 1988

S.I.D. 650/80

Since there have been changes in our S.I.D. 650/80 Harvest Bar Assembly, please insert this notice with your REMCOR PRODUCTS COMPANY Parts and Service Literature.

S.I.D. 650 UP TO SERIAL NUMBER #1300

Original Harvest Bar #51360

This Harvest Bar can be recognized by its —

1. White color;
2. Internal bushing instead of Harvest Bar coupling #10095;
3. Length of 27-13/16"; and
4. The spout that the ice is dispensed out of water jacket is round, signifying a 19 turn evaporator coil.

This Harvest Bar should be substituted with Harvest Bar #51109 with the external coupling #10095. This Harvest Bar can be recognized by its —

1. Tan natural color;
2. Length of 27-5/16" — Coupling #10095 accounts for the 1/2" difference.

S.I.D. 650/80 FROM SERIAL #1300 TO PRESENT PRODUCTION

Original Harvest Bar 51182 can be recognized by its —

1. Tan natural color;
2. External coupling #10095 was supplied with unit;
3. Length of 27-9/16";
4. The spout that the ice is dispensed out of water jacket is square signifying a 20 turn evaporator coil.

This Harvest Bar should be substituted with Harvest Bar. #51423. This Harvest Bar can be recognized by its —

1. Tan natural color;
2. Uses original coupling #10095; and
3. Length of 27-i/16".

S.I.D. 350 Harvest Bar #60741 - 18-13/16"

S.I.D. 850 Harvest Bar #51369 - 36-5/16"

Please take note that failure to match correct Harvest Bar to evaporator assembly can result in a freeze-up occurring.

****

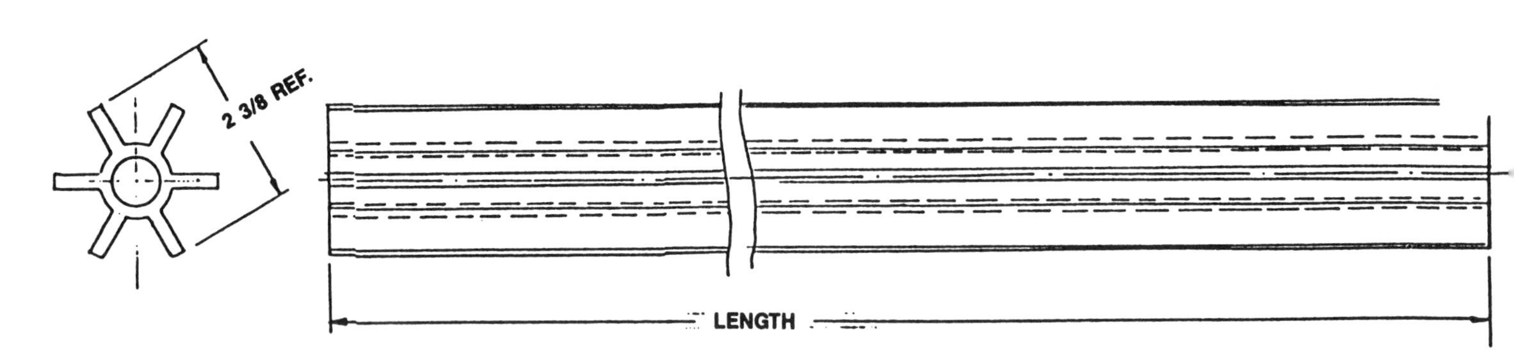

NOTES:

1. REMOVE ALL BURRS.
2. ENDS MUST BE CUT SQUARE.

PART NO.	LENGTH	NEXT. ASSY.	USED ON	REV.	ADDED
51515	18 13/16 ± 1/16	60741	350		9/10/87
51423	27 9/16 ± 1/16	60699	650	D	9/ 8/86
51109	27 5/16 ± 1/16	60624	MACHINED EVAP. 650	D	9/19/85
51369	36 9/16 ± 1/16	60681	850	E	

ITEM NO.	QTY.	PART NO.	DESCRIPTION	
			PRODUCTS CO.	
TITLE		HARVEST BAR		
MATERIAL		50098		
SCALE	NEXT ASSY.	USED ON	REV.	DRAWING NUMBER
	SEE CHART	SEE CHART	D	51369

EVAPORATOR ASSEMBLY

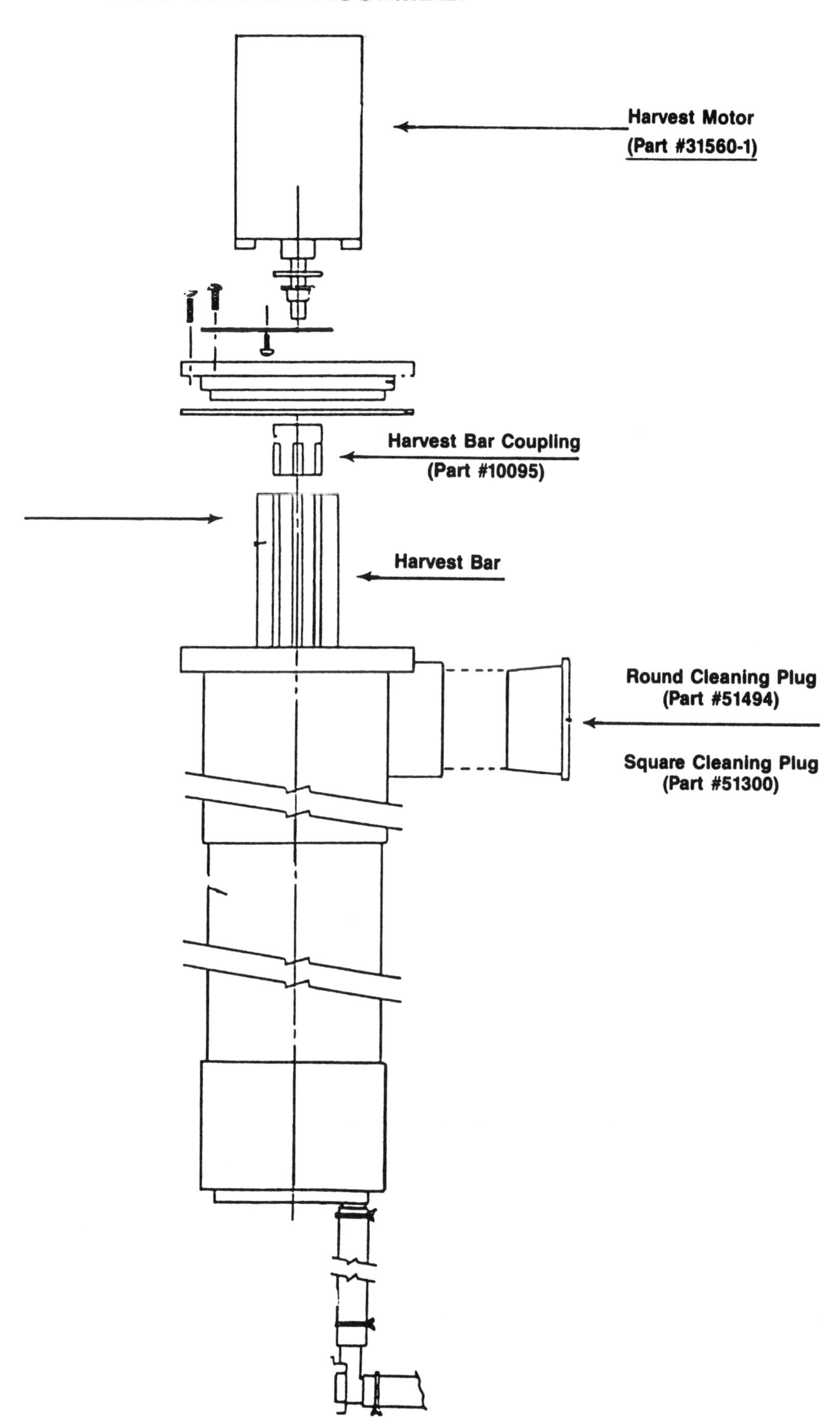

IF AT ANY TIME THE REPLACEMENT OF THE SPIRAL ICEMAKER DISPENSER EVAPORATOR ASSEMBLY IS NECESSARY, PLEASE REFER TO THE FOLLOWING INSTRUCTIONS

S.I.D. 650/80 850/250 MODEL ICEMAKER DISPENSER

WARNING: Disconnect electrical power to unit before proceeding.

1. Remove front, top upper left side, rear and lower front cabinet panels for access to the evaporator assembly and associated "plumbing".
2. Dump refrigerant charge from system.
3. Drain all water from evaporator.
4. Remove harvest motor and top cap from top of the evaporator. Remove harvest bar and coupling from evaporator.
5. Remove rear lower service panel, remove ice thickness probe from evaporator by using a 3/8 wrench and disconnecting probe lead wire.
6. Remove flare fittings from the evaporator coil.
7. Remove jam nut from evaporator tail piece.
8. Remove evaporator coil from evaporator housing.

INSTALLATION OF EVAPORATOR COIL

1. Slide new evaporator coil into evaporator housing.
2. Install jam nut to evaporator tail piece.
3. Tighten flare nuts to evaporator coil.
4. Install harvest bar and coupling in evaporator housing.
5. Install top cap and harvest motor to the top of the evaporator housing.

6. Install ice thickness probe by turning probe clockwise until it just touches the evaporator coil (a slight back pressure will be felt). Turn probe counter-clockwise 2½ turns. Connect probe leads to wire harness.

7. Evacuate and recharge system to name plate specifications leak check system. Reinstate refrigerant lines.

8. Replace cabinet and service panels.

9. Connect electrical power to unit and run icemaker. Check the following items:

a. Check that suction/discharge pressures are in normal operating range;

b. Check for locked sight glass (no bubbles).

ICE THICKNESS ADJUSTMENT

1. Collect and weigh the ice produced during the harvest cycle. The amount of ice harvested should be 3 pounds for S.I.D. 650 and 5 pounds for S.I.D. 850. Use the following procedure to adjust the probe to obtain this weight. (A clockwise adjustment will reduce the harvest weight, while counter-clockwise turns will increase the amount).

CAUTION: Do not turn the screw on the end of the probe. Rotate the plastic probe body only using a 3/8″ open end wrench. Make adjustments in 1/8 turn increments.

a. Put the (stop/run) switch in the "stop" position. (If unit is in the icemaking cycle, stop the unit at end of the harvest cycle.)

b. Adjust the probe.

c. Put the (stop/run) switch in the "run" position.

d. Collect and weigh the ice harvested. Repeat steps a - d as necessary to obtain the required amount of ice.

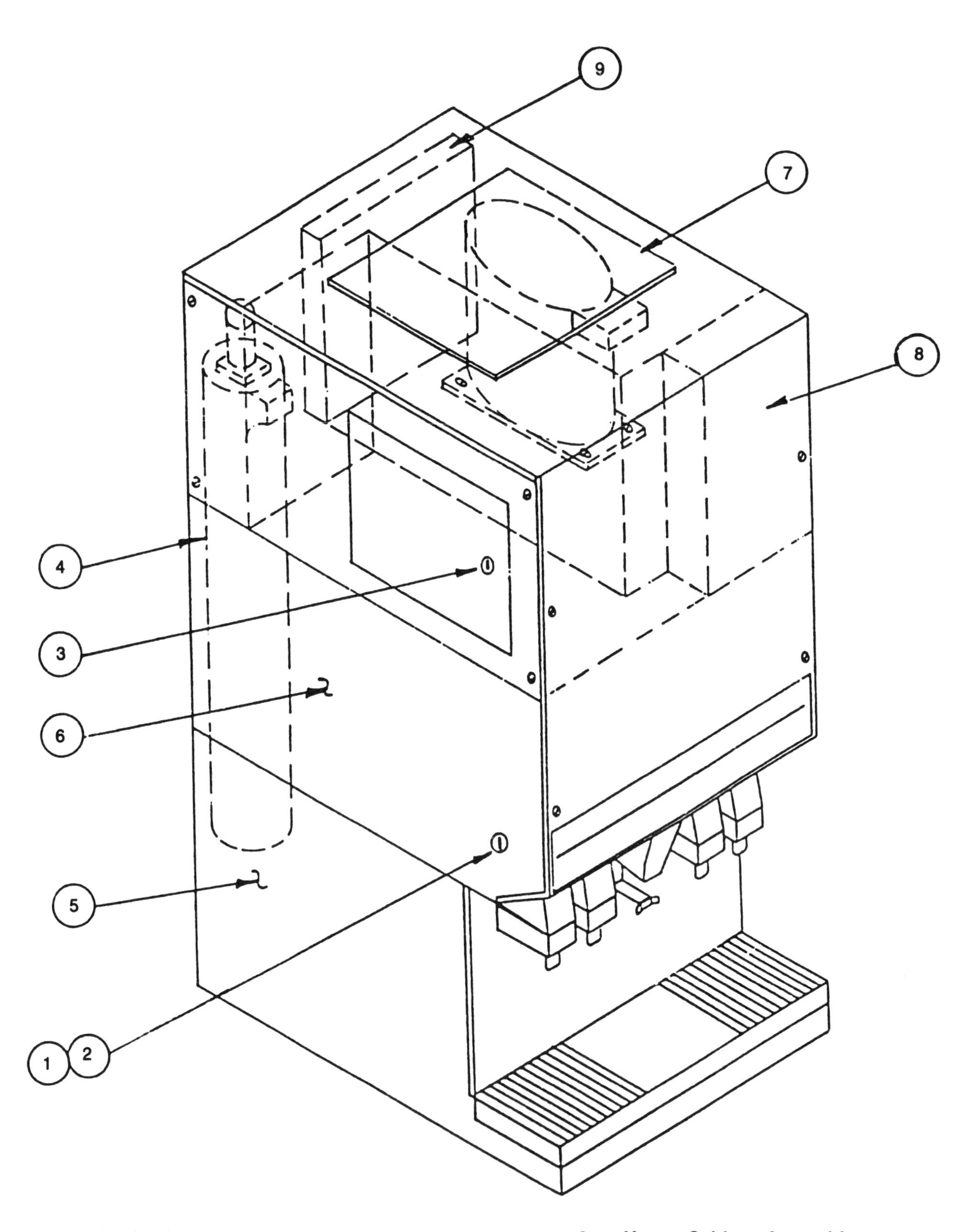

1. Key Lock
2. Contact Block
3. Cabinet Lock
4. Evaporator Assembly
5. Lower Cabinet Assembly
6. Upper Cabinet Assembly
7. Air Filter
8. Electrical Control Box Assembly
9. Condensing Unit Assembly

REFRIGERATION SCHEMATIC

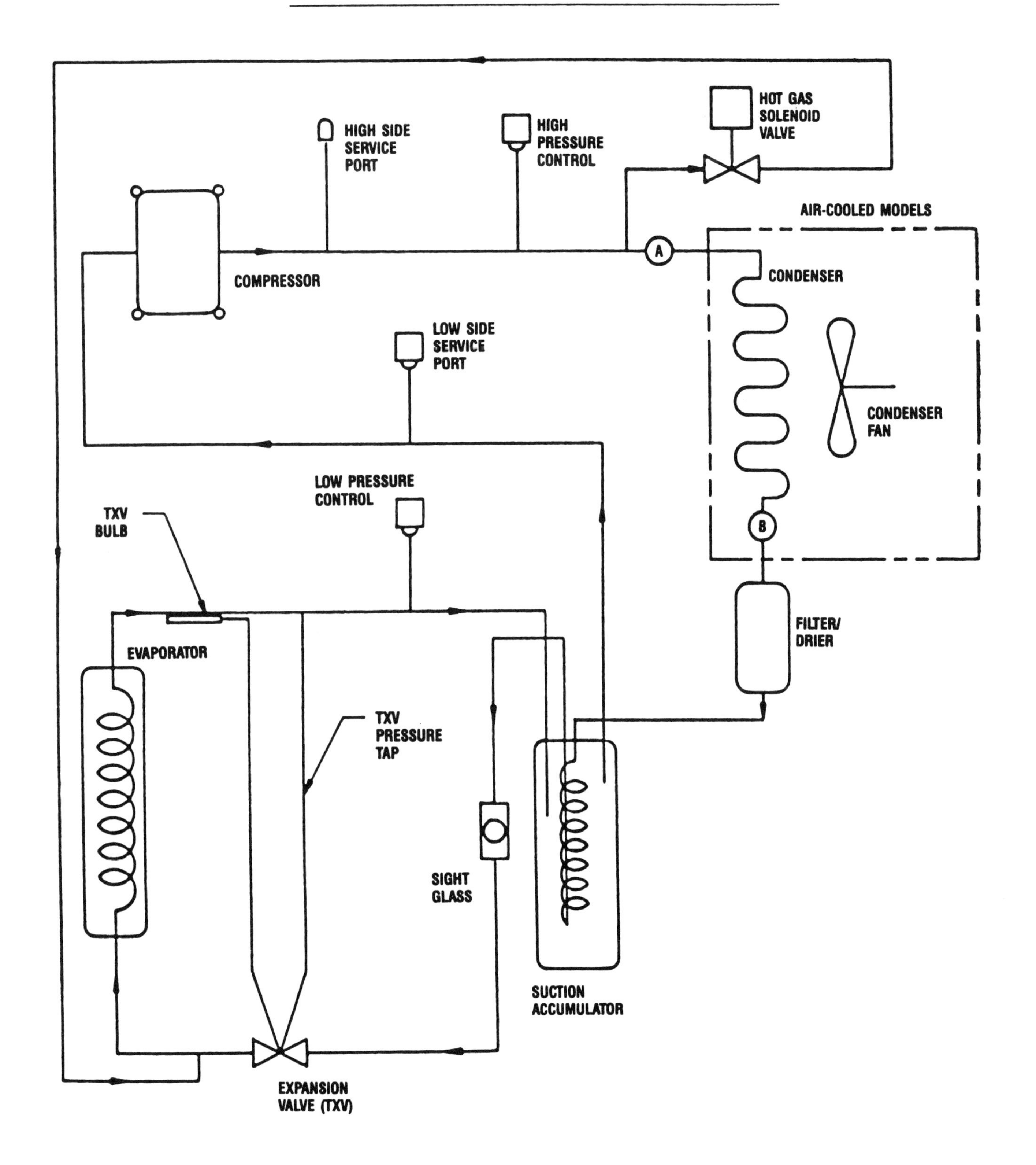

Ross-Temp

ROSS-TEMP

Service Manual

RCU-404-UF

MODULAR CUBED ICE MAKERS

4/89

INTRODUCTION

We have strived to produce a quality product. The design has been kept simple, thus insuring trouble-free operation.

This manual has been prepared to assist servicemen and users with information concerning installation, construction and maintenance of the ice making equipment. The problems of the serviceman and user have been given special attention in the development and engineering of our icemakers.

If you encounter a problem which is not covered in this manual, please feel free to write or call. We will be happy to assist you in any way we can.

When writing, please state the model and serial number of the machine.

Address all correspondence to:

ROSS-TEMP SERVICE DEPARTMENT
2421 15th STREET S. W.
MASON CITY, IOWA 50401

PHONE: (515) 424-6150

TABLE OF CONTENTS

Specifications

A WATER INLET
B DRAIN
C ELECTRICAL CONNECTION
D CONDENSER (WATER OUT) (W/C ONLY)
E CONDENSER (WATER IN) W/C ONLY)
F CONDENSATE DRAIN

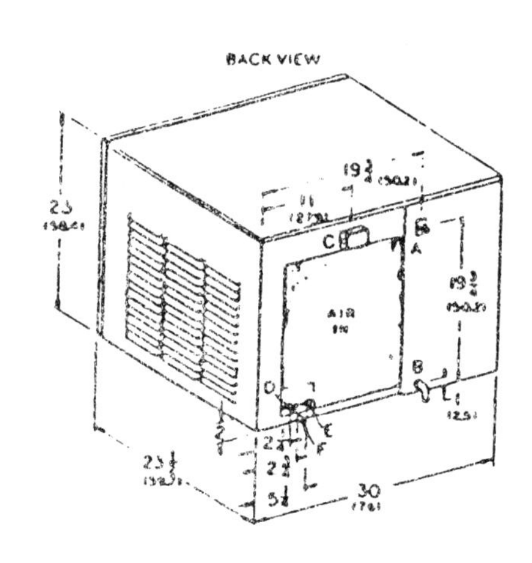

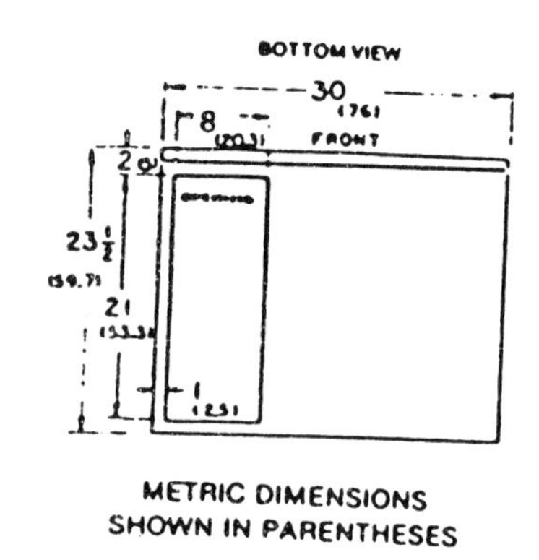

METRIC DIMENSIONS SHOWN IN PARENTHESES

COMPRESSOR ELECTRICAL RATING . . . . 3/4 H.P.
COMPRESSOR MODEL . . . . Tecumseh AJ-7465A
CONDENSER . . . . Air or water cooled
REFRIGERANT CHARGE . . . . Air cooled - 23 oz. R-12
Water cooled - 27 oz. R-12
REFRIGERATION CONTROL . . . . TXV
TXV SUPERHEAT SETTING . . . . 6°
INLET WATER SUPPLY . . . . 3/8" SAE male flare
VOLTAGE. 115V 60 hz. 1 ph.
TOTAL AMP DRAW 10.0

ICE PRODUCTION CAPACITY (approximate):

Model Number (Condenser)	Ambient Temp °F	Incoming Water Temp °F 50°	70°	80°
RCV-404-UF (Air Cooled)	70°	460	400	380
	80°	430	390	360
	90°	400	370	330
RCV-404-UF-W (Water Cooled)	70°	500	450	400
	80°	495	445	395
	90°	490	440	390

INSTALLATION INSTRUCTIONS

A. UNPACKING

1. Uncrate machine and/or bin by removing the staples from around the bottom of cardboard crate and lift off.

2. Remove bolts fastening the crate skid to the bottom of the unit. If auxiliary legs have been purchased for the bin, they should be installed at this time.

B. LEVELING

If legs are used, adjust the leveling legs of the storage bin until the unit is level and all four (4) legs are in solid contact with the floor. Leveling is very important to obtain proper draining and to maintain the proper level in the water pan of the ice cuber.

NOTE: If the bin is to be installed flush to the floor, the machine must be sealed to the floor with an approved mastic such as Sears #3803-0 Caulk, Dow R.T.V. 101, 102 or G. E. 731, 732. This is an N.S.F. requirement and is the responsibility of the installer.

C. UNIT LOCATION

1. Allow at least a minimum of six (6) inches at the rear and side of the ice machine for proper air circulation.

2. This unit has been designed to be installed in an indoor location which is clean and which can be adequately ventilated; the air and water temperatures should never exceed 100 degrees nor fall below 50 degrees. (Temperatures above 100 degrees will cut the ice making capacity below an economical level; temperatures below 50 degrees will cause a malfunction of thermostatic sensors.

3. The unit should be located where air circulation is not restricted. The unit should not be located near a kitchen grill. Air which contains grease vapors will deposit grease on the condenser. The condenser should always be kept clean.

D. UNIT SET-UP

1. Take off front panel of machine and remove hardware bag or service manual envelope with the water strainer enclosed.

2. Mount the ice maker to the top of the ice storage bin or adapter in the proper position over the ice drop opening. The ice maker must then be sealed both on the outside and the inside bottom edges with an approved N.S.F. mastic such as Dow Silastic #732, 734 or General Electric RTV #101, 102. This is an N.S.F. requirement and the responsibility of the installer.

3. Remove shipping tape from evaporator curtains.

E. REMOTE CONDENSERS

Not applicable

F. MAKE ELECTRICAL POWER SUPPLY CONNECTION

Requirements: 115V /60hz. 1 ph. or 220V 50hz. 1 ph. when used

REFER TO SERIAL PLATE FOR MINIMUM CIRCUIT AMPACITY AND MAXIMUM TIME DELAY FUSE SIZE.

ALL WIRING MUST CONFORM TO NATIONAL AND LOCAL ELECTRICAL CODES.

G. MAKE PLUMBING CONNECTIONS

Water supply - (Install per local codes)

The water inlet connection to the unit is a 3/8" flare male connections located at the rear of the ice machine.

NOTE: If the water pressure exceeds 50 pounds, a water pressure regulator should be installed in the water inlet line between the water shut-off valve and the strainer.

Install a reducer fitting on the shut-off valve to accomodate the water strainer, which is supplied with each ice machine and MUST be used. Install the water strainer with the arrow in the proper direction of flow and with the clean out plug down. This is very important for cleaning. Connect either 3/8" or 1/2" copper tubing between the water inlet fitting of the ice machine and the water strainer.

For water cooled units, two water inlet connections are provided; one for the ice making (evaporator) section which is located on the back of the machine and is a 3/8" flared connection. The other is for the water cooled condenser. The reason for the separate water inlet connections is that some installations use a water tower for cooling the water used in the water-cooled condenser and some installations use treated water (filtered) for the ice making inlet water connection. Be sure to install water line (incoming) to the 3/8" male flare connection on the back of the unit that supplies water to the water regulating valve inside.
The setting of the water regulating valve from the factory should be 120 pounds for R-12 units and 250 pounds for R-502 units. NOTE: Always flush out water lines before starting unit. Adjustments, if necessary, should be done at installation.

H. DRAINS

Provide a suitable trapped open drain as close as possible to the area where the ice maker is going to be installed. This may be an existing floor or a 1-1/4" trapped open drain. Two separate drain lines are required for air cooled units, one for the storage bin and one for the Dump Valve drain hose.

An additional separate drain line will be required for water cooled units from the outlet of the condenser coil to the drain. Run all gravity drain lines with a good fall to the open drain.

ALL PLUMBING MUST BE INSTALLED IN ACCORDANCE WITH LOCAL CODES.

NOTE: IN SOME CASES IT MAY BE NECESSARY TO INSULATE THE WATER SUPPLY LINE AND DRAIN LINE. CONDENSATE DRIPPING TO THE FLOOR CAN CAUSE SERIOUS STAINING OF CARPETS OR HARDWOODS.

DRAIN CONNECTION INSTALLATION INSTRUCTIONS

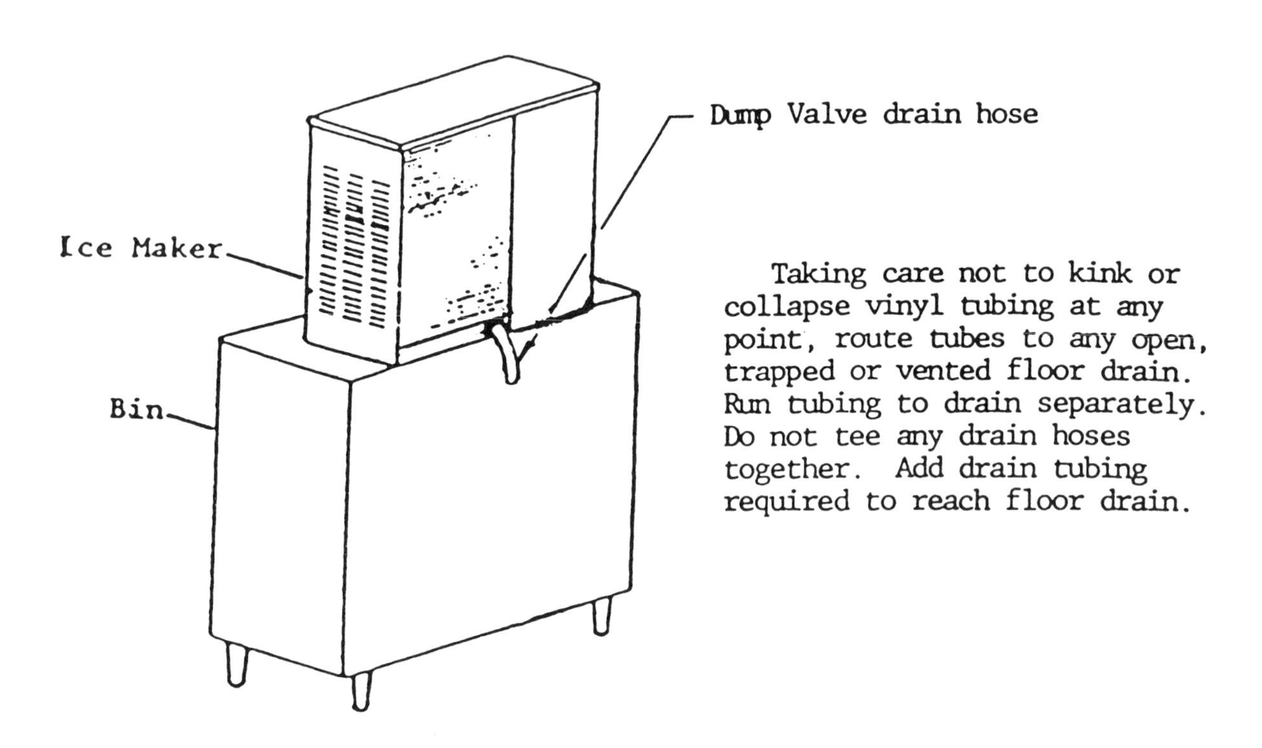

Taking care not to kink or collapse vinyl tubing at any point, route tubes to any open, trapped or vented floor drain. Run tubing to drain separately. Do not tee any drain hoses together. Add drain tubing required to reach floor drain.

I. ADJUSTMENT OF WATER LEVEL IN RESERVOIR

With the water supply turned ON and the power supply OFF'adjust float to maintain water level¼" below the support web insidereservoir

(See Illustration Below)

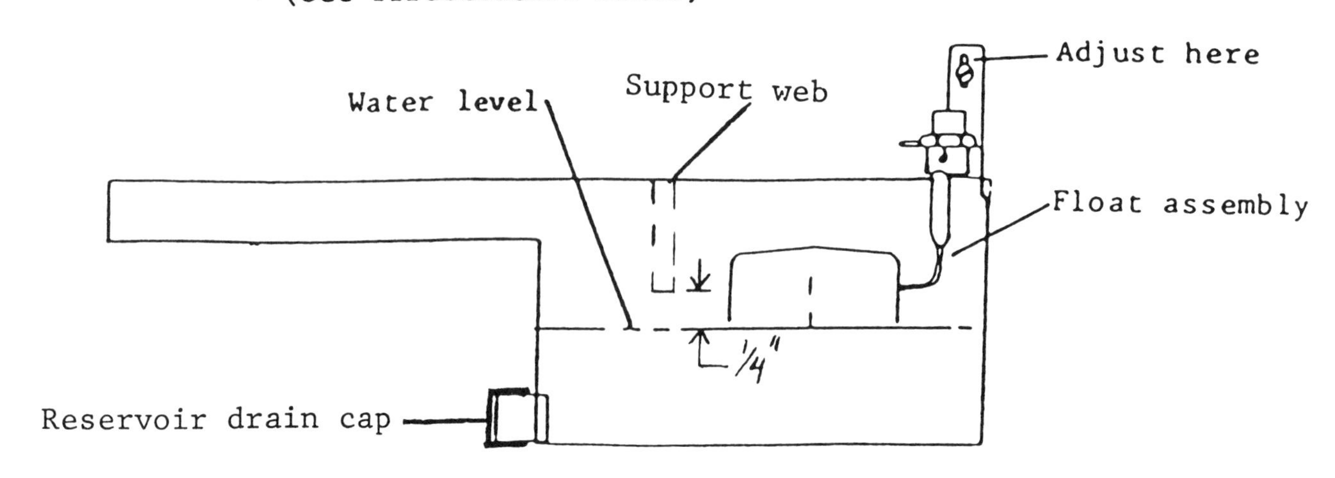

WARNING: Ice Maker will not operate properly when water supply temperature is below 50°F or above 100°F. Water supply pressure must not exceed 50 PSI.

J. STARTING THE UNIT

After the ice cuber has been unpacked and leveled and all plumbing and electrical connections have been made, start the unit and check for proper operation.

A cuber has three separate circuits:

A. The water circuit
B. The refrigerant circuit
C. The electrical circuit

1. Start checking the water circuit by making sure that there are no thread or flare joint leaks, either outside the unit or in the compressor section. Next check the water flow over the evaporator and make sure that all holes in the water distributor are open, and that there is no undue splash or loss of water into the ice bin.

   Also check to see if the float valve is functioning properly and the correct water level is being maintained. Re-adjust if necessary.

2. Check the refrigerant circuit by making sure that the condenser fan is running. (This will be evident by air noise.) Is the compressor running? (Feel the casing for vibration.) Is the evaporator getting cold?

3. Check bin-harvest switch operation. (See proceedure in manual)

5 - #857 screw 8/32 x 3/8
2 - #39449 tie strap

1 - #39450 ice slide diverter
1 - #39454 retainer shield assembly
1 - #39379 interface cable

PLEASE SEE ILLUSTRATIONS FOR REFERENCE FOR THE FOLLOWING INSTALL PROCEDURE

1. After the bottom unit of the stack has been positioned on the bin, the sealing gasket supplied with second unit must be cemented to the base. Nearly any adhesive can be used, however it should not be water soluble. This is an N.S.F. requirement and the responsibility of the installer.

2. Remove the top and front panels from both units.

3. Mount the unit on top of the bottom one of the stack. (It is not recommended to stack more than one unit.)

4. Remove evaporator curtain cover.

5. N/A

6. Put the ice slide diverter through the top unit evaporator section and hook over the front edge of the top unit water reservoir.

7. Secure ice slide diverter to the back wall of the lower unit with a provided screw.

8. Insert the retainer shield from the front of the lower unit evaporator section and align the holes to the diverter and support bracket and secure with the screws provided.

9. Remove electrical box covers from both unit.

10. Remove lowest grommet from the left side of each electrical box .

11. Insert the interface cable through each hole and insert into the split grommets. Note,the cable will fall across the front of both machines with the front covering the cable.

12. Reinstall grommets into electrical box.

13. Plug the interface cable end into the open middle left socket on each solid state board being careful to make sure of a good connection.

14. Replace electrical box covers.

15. Plug the weld nut hole found to the right of the compressor with RTV sealant.

16. Remove the screws from both sides of both units and insert the tie straps. Reinstall screws.

17. When reinstalling the front panels, take care in the positioning of the interface cable.

STACKING KIT ILLUSTRATION

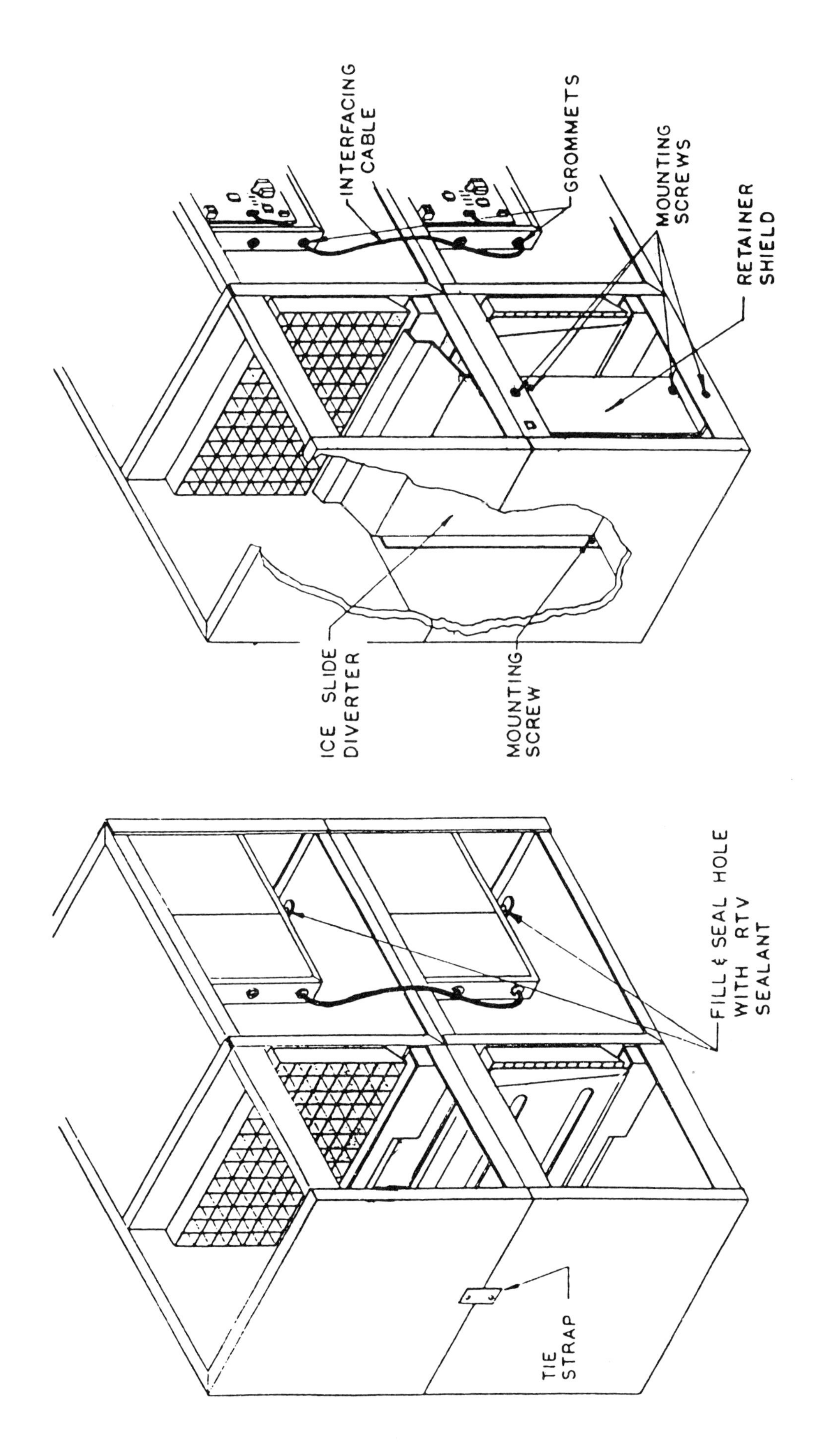

ELECTRICAL CIRCUIT
SEQUENCE OF OPERATION

An L.E.D. digit display mounted on the solid state control board will show a status number 0,1,2,4, &6 and a decimal point to indicate what is happening in the operation of the unit.

The electrical sequence of operation you will see on the digit display for a normal ice making cycle will be as follows:

The status number 0 will be shown telling you the unit is making ice. The solid state control DELAYS the start of the water pump until the evaporator temperatures reaches 20° F. Approximately six minutes after the start up in the freeze cycle a decimal point will appear to the lower right of the "0" to tell you that the evaporator sensor has been switched on. After the evaporator temperature has pulled down low enough for the correct amount of ice to be on the evaporator, the decimal point will begin to flash and stay flashing for approximately 20 seconds. If evaporator stays below the set point, the harvest cycle will start. A number "1" on the digit display will indicate that the machine is in its harvest cycle with the hot gas valve open. The water pump continues to operate and the water dump solenoid valve is now open. The water pump shuts off approximately 15 seconds later after the water reservoir is pumped out.

PLEASE NOTE: During the freeze cycle in low ambient condition the condenser fan motor will be cycled on and off through the condenser sensor and solid state control board. The fan cycling pressures in relation to the temperatures sensed will be approximately 110# for cut out and 150# for cut in of the fan motor.

ADJUSTMENT FOR ICE BRIDGE THICKNESS

An ice bridge connecting all cubes is necessary for a proper harvest or discharge of cubes from the evaporator.

To increase ice "bridge" thickness carefully turn adjustment screw counter clockwise no more than one turn at a time. Wait and check thickness before re-adjusting.

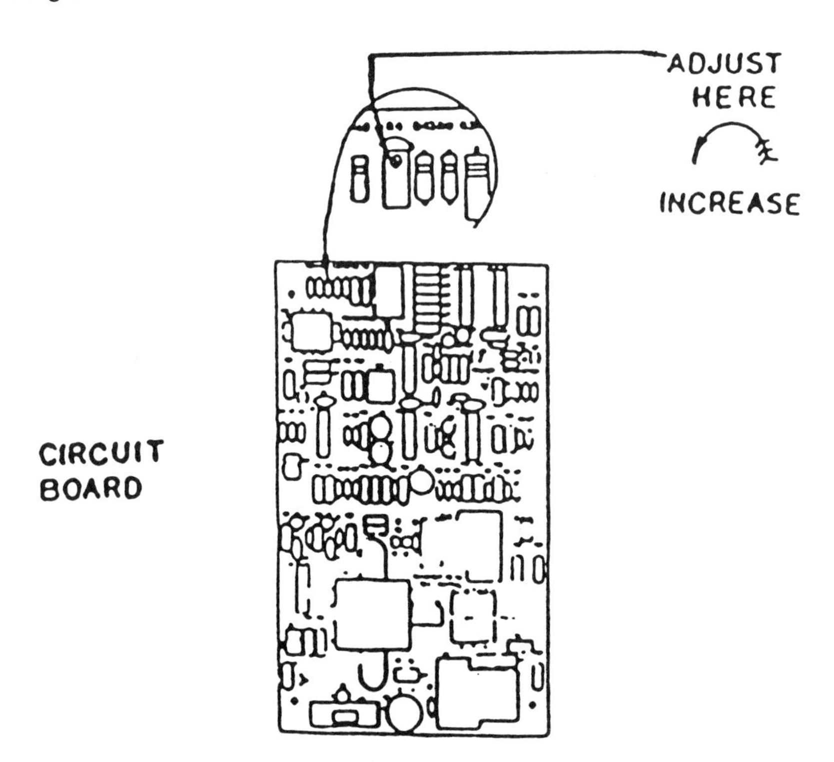

ADJUSTMENT AND CHECK-OUT FOR HARVEST-BIN SWITCHES

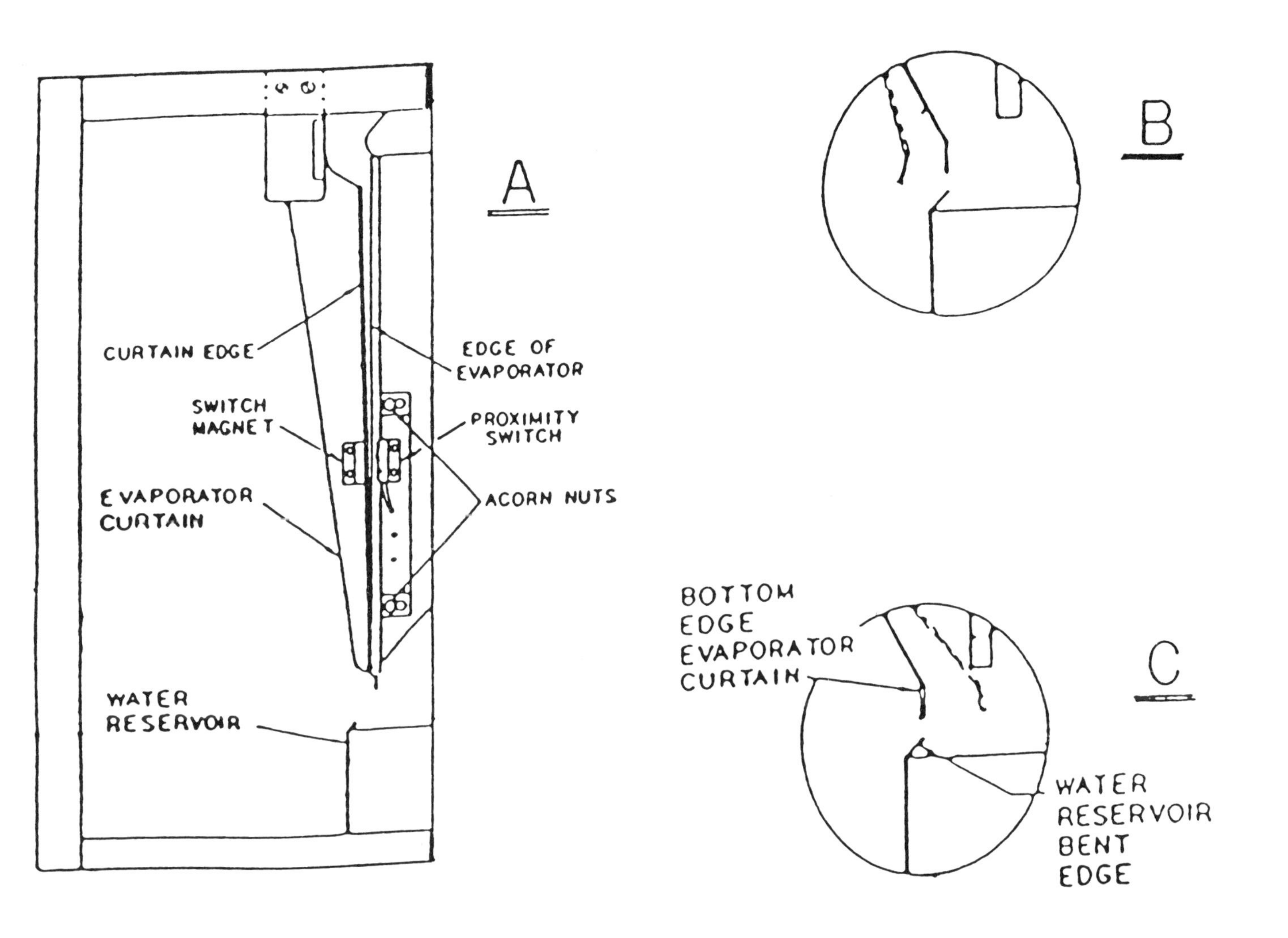

CHECKOUT PROCEUDRE

Turn on the ice machine and move the evaporator curtain(s)away from the evaporator(s). The ice machine should then shutoff in approximitly 8 seconds.(See detail A&B)

Slowly let the evaporator curtain(s) move back toward the evaporator(s) until the bottom edge of the curtain(s) is at least at the bent edge of the water reservoir or closer to the evaporator. With the curtain(s) at that position, the machineshould start.(See detail C)

ADJUSTMENT PROCEDURE

If adjustment is necessary,loosen acorn nutsand move proximity switch closer to the curtain(s)and make sure the curtain is properly mounted. (See detail A)

Re-check per above proceedure.

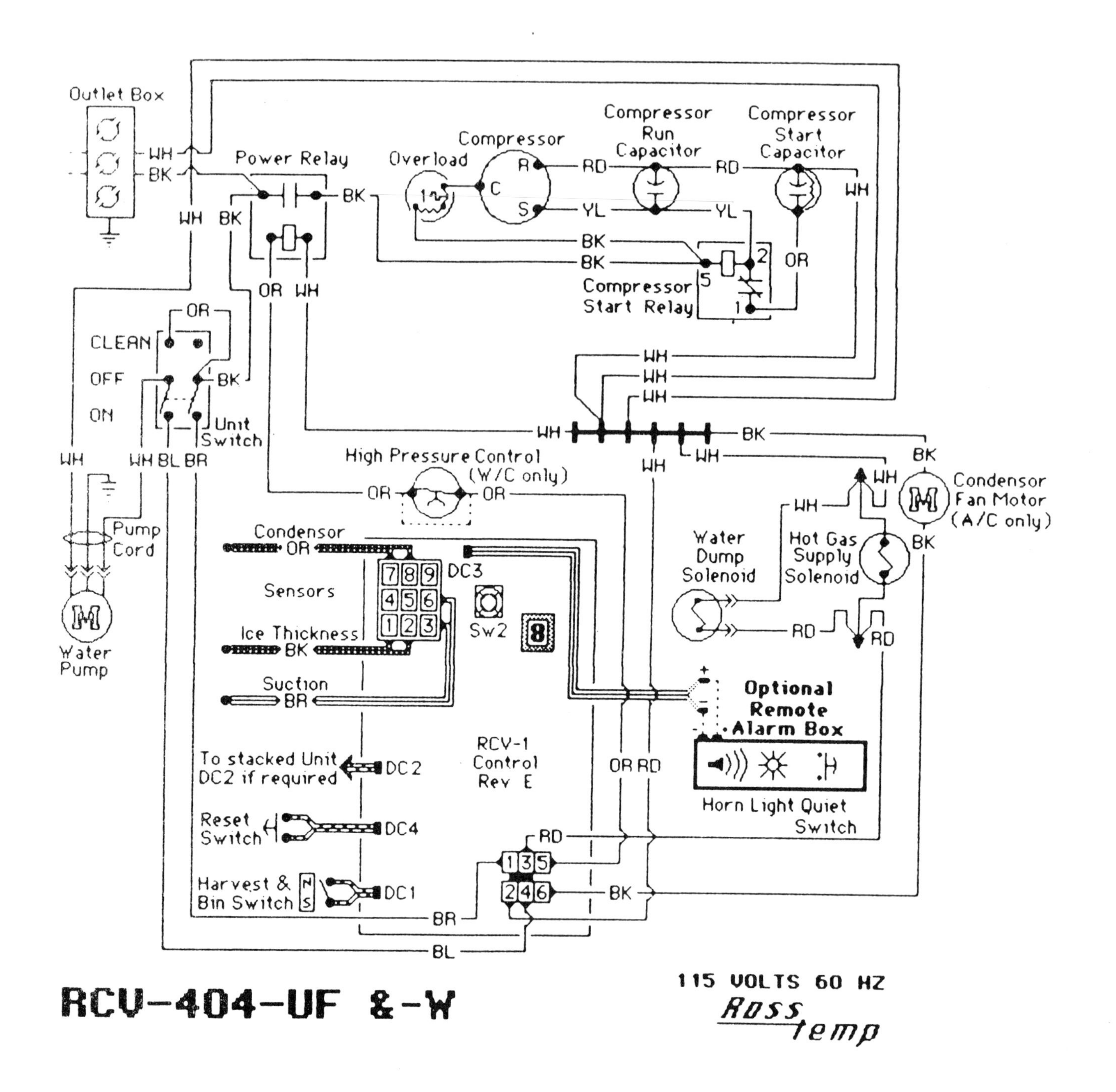

RCU-404-UF &-W

115 VOLTS 60 HZ

Ross temp

Note The solid state control *DELAYS* the start of the water pump until the evaporator temperature reaches 20° F.

IMPORTANT White and black connector blocks are "KEYED" and MUST be inserted correctly on circuit board. DO NOT USE FORCE.

PART NO. 41001
ARTWORK NO. 50551
REV. E

SANITIZING AND CLEANING PROCEDURE

1. Remove front panel to gain access to the on-off-clean switch.

2. Push switch to "clean" and allow the ice on the evaporator to release or melt away.

3. Remove all ice from storage bin.

4. Mix a sanitizing solution of 1/4 oz. "Calgon Ice Machine Sanitizer" to one gallon of water. Using a non-metallic bristle brush scrub the following:

   a. Inside surfaces of ice bin including top and door.
   b. Inside surfaces of the icemaker to include evaporator section in the ice machine including the top, front panel and evaporator splash curtain.
   c. Make sure splash curtain is correctly positioned.

5. Add 2 oz. of "Calgon Nickel-Safe Ice Machine Cleaner" directly into water reservoir. Circulate for approximately 45 minutes.

6. Remove cap from reservoir drain and allow cleaner to drain away. Replace cap.

7. Fill reservoir with clean fresh water, circulate for approximately 1 minute. Remove cap from reservoir drain and allow water to drain away. Repeat three times.

8. Flush all waste water from ice storage bin with clean fresh water.

9. Push switch from "clean" to "on" position.

10. Replace front panel.

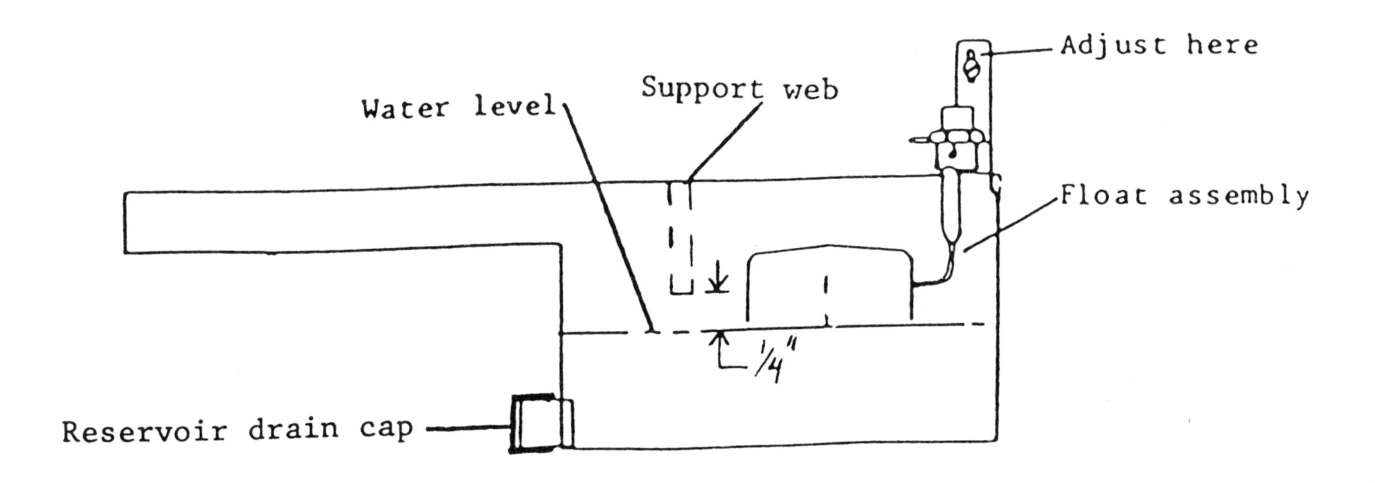

WATER TREATMENT

Automatic ice-making machines can quit working for any number of reasons, mechanical, electrical or faulty refrigeration, but water problems foul them up faster than almost anything else. While ice machines vary in design, you can apply these water treatment tips to all of them.

1. START WITH THE WATER

The mineral content of water varies in different areas and as the chart shows, high hardness and alkalinity counts combine to form insoluble calcium carbonate or lime scale. If this condition is constant, the intake water must be treated constantly to prevent scale formation in the ice machine.

2. PREVENT LIME SCALE FORMATION

We recommend the installation of a Calgon Microment Feeder on the incoming water line. No. X-8B Feeder is recommended for ice machines with a capacity of 400-450 lbs. per day. Fill the feeder with 6R Micromet, the slowly soluble poly-phosphate which lasts six months before renewing the 8-oz. charge.

Constant treatment with 6R Micromet will control lime scale and prevent minerals from sticking to the freezing surfaces in ice machines. Result - smooth movement of ice slabs, good harvest of ice cubes, efficient, automatic production.

3. REMOVE OBJECTIONABLE TASTE OR ODOR

If the bad taste or odor is traceable to the water source, install a Calgon Fine Carbon Filter to the incoming water line. The No. 1-1/2B Fine Carbon Filter is ideal for machines making up to 500 pounds of ice per day and will remove bad taste, odors, and problems caused by chlorine in the water supply. In some instances, slime growths may cause odor problems and these growths can be removed by the use of liquid ice machine cleaner.

4. SERVICE REGULARLY

A service program to clean the ice machine at regular intervals and check on filter and feeder charges is important. In the long run, it will assure adequate water treatment, reduce emergency calls and aid in the trouble-free performance of automatic ice making machines.

WINTER STORAGE

If the unit is to be stored in an area where the temperature will drop below freezing, it is most important that all water lines be drained to prevent them from freezing and possible rupture.

To blow out the water line, disconnect the water supply at the cabinet inlet and use air pressure to force the water into the water reservoir pan. This can then be removed from the water pan.

CLEANING THE CONDENSER

In order to produce at full capacity, the refrigeration condenser must be kept clean. The frequency of cleaning will be determined by surrounding condition. A good maintenance plan calls for an inspection at least every two months.

Remove the lower front panel of the machine. With a vacuum cleaner, remove all accumulated dust and lint that has adhered to the finned condenser.

CAUTION: CONDENSER COOLING FINS ARE SHARP. USE CARE WHEN CLEANING.

1. If the condenser is being cleaned from the back of the machine, remove all accumulated dust, dirt etc., that has adhered to the finned surface with a vacuum cleaner.

2. If the unit is being cleaned from the front, remove lower panel, turn the power switch off and blow through the finned surface of the condenser past the fan blade to remove accumulated dust, etc.

STATUS INDICATOR

STATUS	EXPLANATION	POSSIBLE CAUSE
0	Unit is in freeze cycle, making ice, no problems.	
1	Unit is in harvest cycle, ice should drop shortly, no problems.	
2	Indicates a full bin condition, unit off, water curtain being held open with ice.	If "2" is shown but bin isn't full, check for individual cube holding curtain open. Harvest Bin switch not adjusted properly.
4	Unit OFF due to suction line not pulling down to at least 40°F. Manual reset required.	Low on refrigerant. Defective TXV. Compressor defective or inefficient. Defective power relay, won't close. Defective start relay, won't start compressor. Low voltage to compressor no start. Defective C.P.R. valve. Defective sensor (brown wire). SENSOR NOT INSULATED PROPERLY.
6	Unit is OFF due to condenser temperature climbing too high. Manual reset required.	Dirty condenser. Defective fan motor or blade.* Gross overcharge. Extremely high ambient temperature, above 120°F. Defective sensor (orange wire).**
Decimal Point OFF	Indicates that all sensors, except condenser, are switched off for first six minutes of freeze cycle.	Normal time delay, approximately 6 minutes.
Decimal Point ON	Indicates that evaporator and suction line sensors have switched "ON".	
Decimal Point FLASHING	Indicates evaporator temperature has pulled down and unit will go into harvest after time delay.	Normal time delay of approximately 20 seconds before harvest cycle begins.

FOR MANUAL RESET - PUSH MASTER SWITCH TO "OFF" - WAIT 10 SECONDS - PUSH TO "ON" OR PUSH RESET BUTTON

* Not applicable to Water-Cooled units.

** Not applicable to Remote units.

TROUBLE SHOOTING THE SOLID STATE

CONTROL BOARD

To determine if the circuit board and sensors are functioning correctly under all operating parameters, the adverse conditions must be simulated to check out the digital display status numbers.

PROCEDURE

To check #6 - Block condenser fan blade on start up. Condenser should get hot within two minutes and shut unit off on #6, condenser too hot.

To check #4 - Remove suction line sensor from thermowell anytime during freeze cycle. Machine should shut off on #4, suction line too warm when the evaporator temperature gets low enough to start the harvest cycle.

To check #2 - Hold water curtain open anytime after unit goes into harvest. Machine should shut down within approximately 8 seconds on #2, full bin.

To check #1 - Push defrost button anytime during freeze cycle and unit should go into harvest. #1 indicates a harvest cycle, no problems.

To check #0 - A "0" indicates that the unit is in the freeze cycle and there are no problems.

PLEASE NOTE: In rare cases a "0" can be displayed on the control board and the compressor not running in water cooled and remote air cooled machines. If this occurs, the manual reset high pressure control will be open and must be reset for proper operation. The control is located in the upper rear, right corner of the compressor compartment.

After reset, check out the machine for the possible causes of the problem.

TROUBLESHOOTING THE SENSORS

1. Turn off power to machine.

2. Remove the front panel and electrical box cover of the machine.

3. Cut the suspected sensor wire at least six inches from the thermowell in which it is located.

4. Remove the sensor from the thermowell.

5. Carefully separate the wires and strip the insulation off the end.

6. Pack a glass or container with ice and add some water to make an ice-water solution. Check the temperature of the ice water with an accurate thermometer. Ice water must be 32° F.

7. Insert the sensor into the ice water and soak for a minimum of two minutes.

8. With a zerod ohmmeter measure the resistance across the two wires of the sensor lead. It should read 2815 ohms + or -10% (281 ohms).

NOTE: If the above ohm reading is not within the range stated, the sensor is bad and should be replaced.

RECONNECTION OF A GOOD OR REPLACEMENT SENSOR AFTER TROUBLESHOOTING

1. Carefully separate the wires of the sensor leads coming from the solid state control and strip the insulation off the end of each wire.

2. Reconnect the sensor leads and twist the stripped ends tightly. Secure with the proper sized wire nuts.

3. Tape all the wire nut connections to insulate connections from each other.

REMOVAL OF SOLID STATE CONTROL FROM MACHINE

CAUTION: THE CIRCUIT BOARD IS FRAGILE, HANDLE WITH CARE.

1. Turn off power to machine.
2. Remove front panel.
3. Remove electrical box left front cover.
4. Disconnect the through wire plug connections from circuit board.
5. Carefully lift any corner of the circuit board while pinching closed the top part of the plastic "stand off" support with needle nose pliers. The circuit board has to be gradually worked up over all five of the "stand off" supports. The circuit board will not "pop off" until all supports have been pinched closed and the board is then holding them in that position.

REINSTALLATION OF SOLID STATE CONTROL

1. Align all holes in the circuit board over the plastic stand-off supports.
2. Carefully push downward at all hole locations until board seats on all the stand-off supports. (Sometimes a snap will be heard as this seating takes place.)
3. After the circuit board is seated, carefully connect the three plugs to the circuit board. Note: Plug connects are polarized, make sure the plug is inserted correctly.

PARTS LIST

ILLUS. NO.	DESCRIPTION	PART NO. RCV-404
1	Control, circuit board	40974
2	Switch, on-off-clean	23836
3	Relay, compressor start	21547
4	Relay, power	40980
5	Capacitor, compressor start	21544
6	Capacitor, compressor run	28632
7	Drier	21850
8	N/A	N/A
9	Valve, Schrader	20654
	Core, Schrader valve	21214
	Cap, Schrader valve	23988
10	Compressor	22620
11	Motor, condenser fan	25242
12	Blade, condenser fan	23527
13	N/A	N/A
14	Condenser	25266
15	Sensor, condenser temp.	38703
16	Pump, water	41011
17	Valve, thermostatic expansion	41415
18	Sensor, suction line temp.	38703
19	Valve, hot gas	9214
20	Tube, water pan to pump inlet	38790
21	Cap, reservoir drain	45681
22	Reservoir, water	41448
23	Float and valve	21924
24	Bracket, float & valve	45922
25	Hose, pump	45680
26	Tee	987
27	Hose	43412
28	Orifice, restrictor	29784
29	Evaporator	45905
30	End cap	22279
31	Distributor, water	43056
32	Cover, evaporator	45907
33	Bracket, front cover mount	43530
34	Condenser coil, water cooled	22429
35	Valve, water regulating	1211
36	Bracket, back cover mount	38743
37	Valve, water dump solenoid	42297
38	Switch, harvest-bin proximity	43446
39	N/A	N/A
40	N/A	N/A
41	N/A	N/A
42	N/A	N/A
43	N/A	N/A
44	N/A	N/A
45	N/A	N/A
46	N/A	N/A
47	N/A	N/A
48	N/A	N/A
49	Control, high pressure (water cooled)	7024
50	Sensor, evaporator	38703
51	Switch, reset	42680
52	Splash guard	41464

ILLUSTRATED PARTS BREAKDOWN

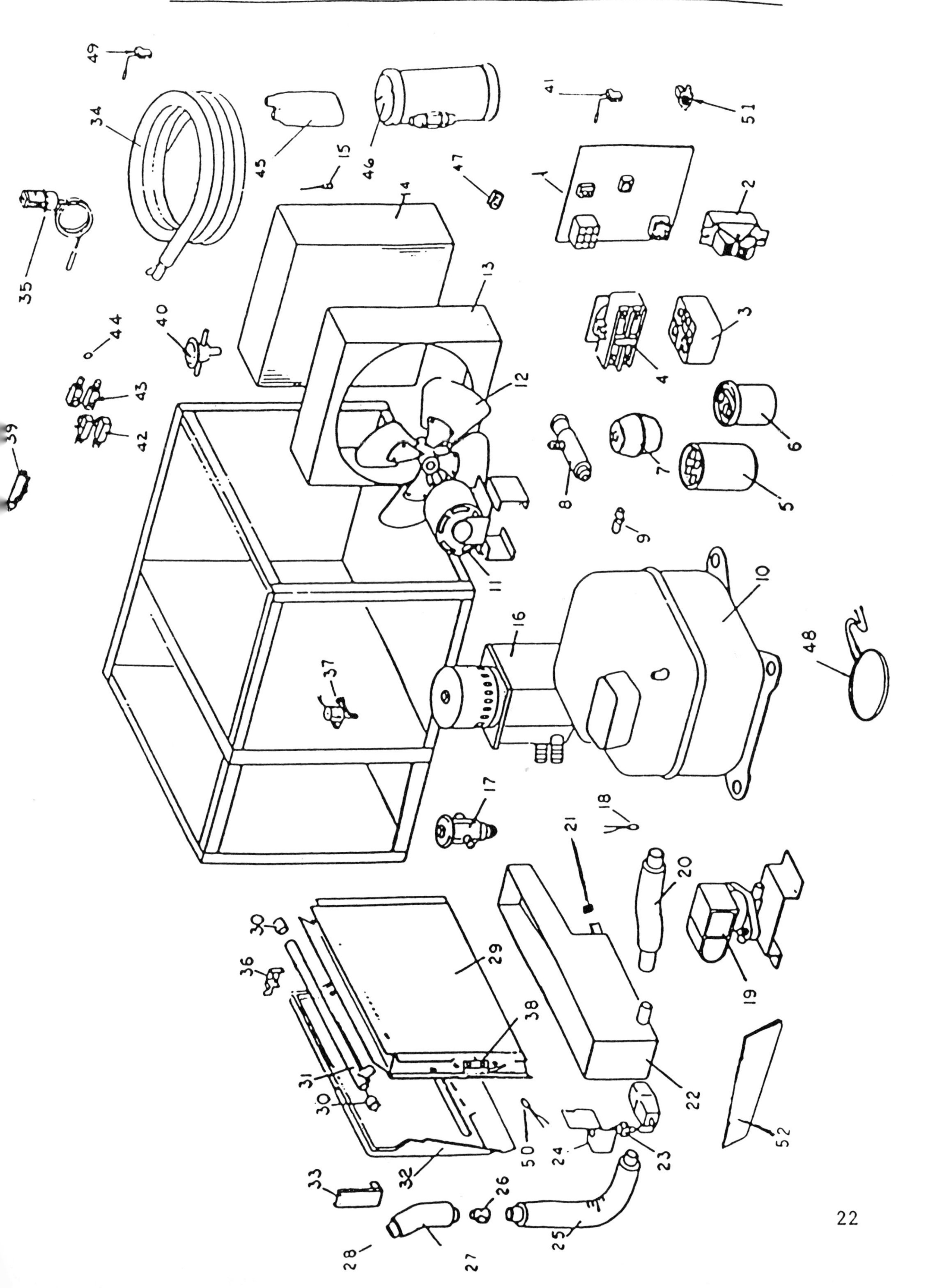

ROSS-TEMP

Service Manual No. 3

MODEL

RF-151
RF-351
RF-452
RF-600
RF-951

SELF-CONTAINED FLAKED ICE MAKER

P/N 36060
4/88

INTRODUCTION

We at ROSS-TEMP have strived to produce a quality product. The design has been kept simple, thus insuring trouble-free operation.

This manual has been prepared to assist servicemen and users with information concerning installation, construction and maintenance of ROSS-TEMP equipment. The problems of the serviceman and user have been given special attention in the development of all ROSS-TEMP Ice Makers.

If you encounter a problem which is not covered in this manual, please feel free to write or call the service department. We will be happy to assist you in any way we can.

When writing, please state the Model and Serial number of the machine.

Address all correspondence to:

ROSS-TEMP Service Department
2421 15th Street S.W.
Mason City, Iowa 50401

Phone: (515) 424-6150

TABLE OF CONTENTS

SPECIFICATIONS

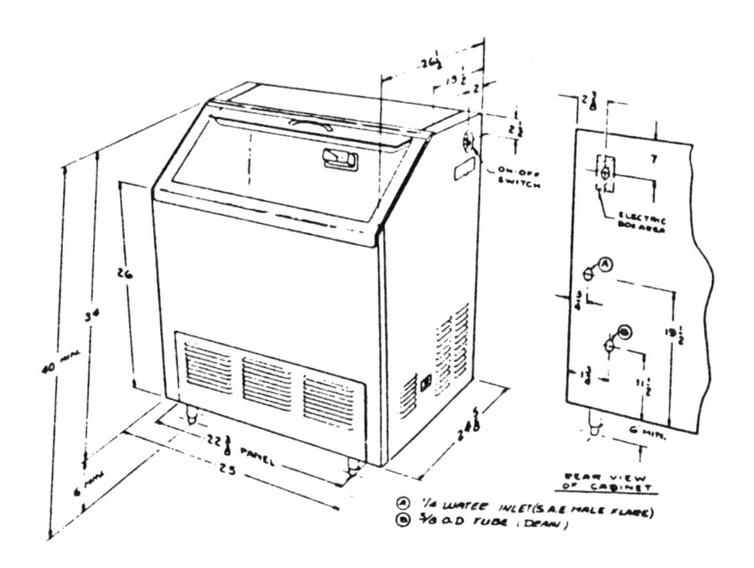

ICE PRODUCTION CAPACITY

RF-151 (AIR COOLED)

Ambient Room Temp.	°F / °C	70 / 21	80 / 27	90 / 32	100 / 38
Inlet Water Temp.	°F / °C	50 / 10	60 / 16	70 / 21	80 / 27
Production per 24 hrs.	Lbs.	200	175	150	100

	RF-151
COMPRESSOR ELECTRICAL RATING	1/4 H.P.
COMPRESSOR MODEL	Tec. AE1410A
CONDENSER	Air cooled
REFRIGERANT CHARGE	R-12 10 oz.
REFRIGERANT CONTROL	Capillary tube
INLET WATER SUPPLY	1/4" SAE male flare
VOLTAGE	115V.
GEARMOTOR ELECTRICAL RATING	1/8 H.P.
GEARMOTOR AMP RATING	2.25 amps
TOTAL AMP DRAW	7.0 amps

SPECIFICATIONS

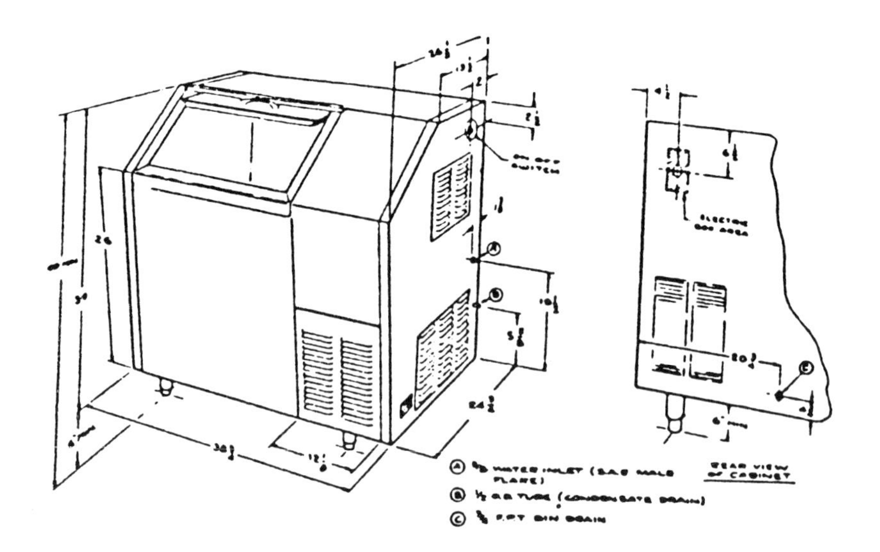

ICE PRODUCTION CAPACITY

RF-351 (AIR COOLED)

Ambient Room Temp.	°F / °C	70 / 21	80 / 27	90 / 32	100 / 38
Inlet Water Temp.	°F / °C	50 / 10	60 / 16	70 / 21	80 / 27
Production per 24 hrs.	Lbs.	350	320	285	240

(WATER COOLED)

Ambient Room Temp.	°F / °C	70 / 21	80 / 27	90 / 32	100 / 38
Inlet Water Temp.	°F / °C	50 / 10	60 / 16	70 / 21	80 / 27
Production per 24 hrs.	Lbs.	350	330	310	285

RF-452 (AIR COOLED)

Ambient Room Temp.	°F / °C	70 / 21	80 / 27	90 / 32	100 / 38
Inlet Water Temp.	°F / °C	50 / 10	60 / 16	70 / 21	80 / 27
Production per 24 hrs.	Lbs.	600	540	465	408

(WATER COOLED)

Ambient Room Temp.	°F / °C	70 / 21	80 / 27	90 / 32	100 / 38
Inlet Water Temp.	°F / °C	50 / 10	60 / 16	70 / 21	80 / 27
Production per 24 hrs.	Lbs.	588	549	471	444

	RF-351	RF-452
COMPRESSOR ELECTRICAL RATING	1/3 H.P.	1/2 H.P.
COMPRESSOR MODEL	Tec. AE-2415A	Tec. AJ7441A
CONDENSER	Air or water cooled	Air or water cooled
REFRIGERANT CHARGE	13 oz. R-12 air cooled 11 oz. R-12 water cooled	14 oz. R-12
REFRIGERANT CONTROL	Capillary tube	Capillary tube
INLET WATER SUPPLY	3/8" SAE male flare	3/8" SAE male flare
VOLTAGE	115 V.	115 V.
GEARMOTOR ELECTRICAL RATING	1/8 H.P.	1/8 H.P.
GEARMOTOR AMP DRAW	2.25 amps	2.25 amps
TOTAL AMP DRAW	7.5 amps	12.0 amps

SPECIFICATIONS

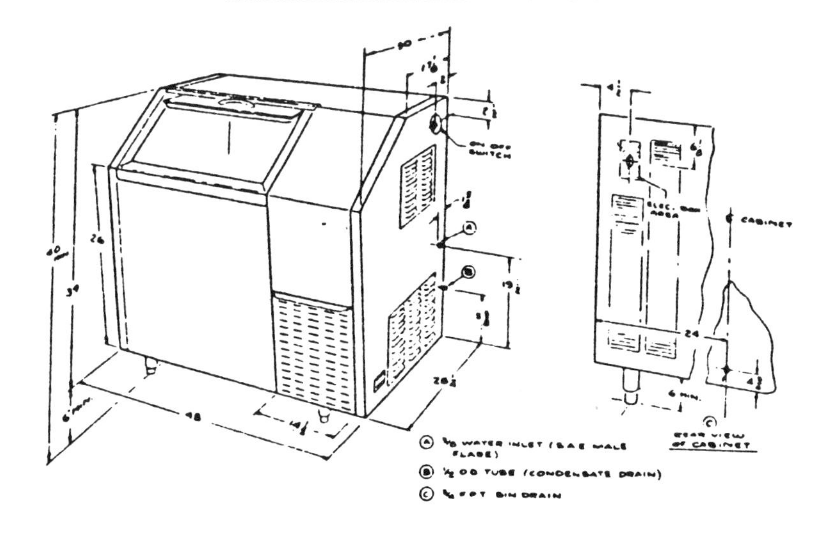

ICE PRODUCTION CAPACITY

RF-600 (AIR COOLED)

Ambient Room Temp.	°F / °C	70 / 21	80 / 27	90 / 32	100 / 38
Inlet Water Temp.	°F / °C	50 / 10	60 / 16	70 / 21	80 / 27
Production per 24 hrs.	Lbs.	690	590	490	390

(WATER COOLED)

Ambient Room Temp.	°F / °C	70 / 21	80 / 27	90 / 32	100 / 38
Inlet Water Temp.	°F / °C	50 / 10	60 / 16	70 / 21	80 / 27
Production per 24 hrs.	Lbs.	756	675	618	480

RF-951 (AIR COOLED)

Ambient Room Temp.	°F / °C	70 / 21	80 / 27	90 / 32	100 / 38
Inlet Water Temp.	°F / °C	50 / 10	60 / 16	70 / 21	80 / 27
Production per 24 hrs.	Lbs.	1020	849	756	642

(WATER COOLED)

Ambient Room Temp.	°F / °C	70 / 21	80 / 27	90 / 32	100 / 38
Inlet Water Temp.	°F / °C	50 / 10	60 / 16	70 / 21	80 / 27
Production per 24 hrs.	Lbs.	1032	910	807	708

	RF-600	RF-951
COMPRESSOR ELECTRICAL RATING	3/4 H.P.	1 H.P.
COMPRESSOR MODEL	Tec. AJ-7465A	Tec. AH-7511A
CONDENSER	Air or water cooled	Air or water cooled
REFRIGERANT CHARGE	19 oz. R-12 air cooled 14 oz. R-12 water cooled	19 oz. R-12
REFRIGERANT CONTROL	AXV	AXV
VOLTAGE	115 V	208/230-115 3 wire
INLET WATER SUPPLY	3/8" SAE male flare	3/8" SAE male flare
GEARMOTOR ELECTRICAL RATING	1/8 H.P.	1/8 H.P.
GEARMOTOR AMP DRAW	2.25	2.25
TOTAL AMP DRAW	14 amps	10.5 amps

INSTALLATION INSTRUCTIONS

You will get better service from the ice machine, longer life and greater convenience if you choose its location with care.

Here are a few points to consider:

1. Select a location as close as possible to where you are going to use the ice.
2. Allow a minimum of 6" space at sides and rear of machine for ventilation.
3. A kitchen installation is not desirable as a rule. If a kitchen installation is necessary, locate the machine as far away from the cooking area as possible. Grease laden air will form a greasy deposit on the condenser. This reduces the ice making efficiency and necessitates thorough cleaning quite often.
4. If you install the unit in a storeroom, be sure the room is well ventilated.

   NOTE: Do not install where the ambient and incoming water temperature will drop below 50 degrees or rise to over 100 degrees.

   NOTE: If water pressure exceeds 50 pounds, a water pressur regulator should be installed in water inlet line between water shut off valve and strainer. Minimum incoming water pressure required is 22 pounds.

A. Uncrate the unit by removing the staples or nails from the bottom of the carton and lift off.

B. Remove the bolts holding the skid to the machine.

C. Set machine in place and level properly with adjustable legs.

D. Provide a cold water supply line to the area selected for the installation of the unit. The incoming water line should have a shut-off valve provided at a convenient location close to the icemaker. The water strainer provided with the unit should be installed on the down-stream side of the shut-off valve. The supply line must be adequately sized to compensate for the lengths of the incoming water run. The machine is equipped with a 3/8" male flare connection for the incoming water line. NOTE: ALWAYS FLUSH OUT WATER LINES BEFORE STARTING UNIT.

   Water cooled units have a water regulating valve that is factory set to operate at 120# refrigerant head pressure. This should be checked at the time the unit is being installed.

   Two water inlet connections are provided on water cooled units, one for the ice making (evaporator) section, the other is for the water cooled condenser. Both connections are 3/8" male flare fittings. Inlet water to the condenser will go to the water regulating valve first, then to the condenser coil and out the drain.

   The reason for separate water inlet connections is that some installations use a water tower for cooling the water used in the water cooled condenser and some installations use treated water for the ice making inlet water.

   A separate drain will be required for the outlet of the water cooled condenser.

INSTALLATION INSTRUCTIONS CON'T.

E. Provide a suitable trapped open drain as close as possible to the area where the ice cube maker is going to be installed. This may be an existing floor or a 1-1/4" trapped open drain. Connect the drain line to the rear of the unit and run it with a good fall to the open drain. All plumbing must be installed according to local codes. The storage bin drains by gravity, and therefore the drain line must maintain a gradual slope to an open drain and should be insulated.

NOTE: IN SOME CASES IT MAY BE NECESSARY TO INSULATE THE WATER SUPPLY LINE AND DRAIN LINE. CONDENSATE DRIPPING TO THE FLOOR CAN CAUSE SERIOUS STAINING OF CARPETS OR HARDWOODS.

F. Connect a drain hose to the condensate drain stub tube.

NOTE: All plumbing must be done in accordance with national and local codes.

G. Connect the electrical supply line to the unit.

NOTE: Make sure the proper voltage and number of wires are provided. See serial plate for this information.

NOTE: All wiring must conform to national and local codes.

H. Turn on water supply and observe the water level in evaporator sections which should be 1/4" below the inclined discharge chute of the shell.

I. Turn machine on and check for proper voltage and amp draw on the entire unit as well as components such as the gearmotor and fan motor.

J. Check refrigerant circuit and all plumbing connections for leaks, etc.

K. Check bin thermostat or mechanical shut-off for proper operations. In the mid-range the bin thermostat will open at 42° and has a 6° differential.

WARNING

WATER LEVEL MUST BE MAINTAINED

AT THE TOP OF THE EVAPORATOR.

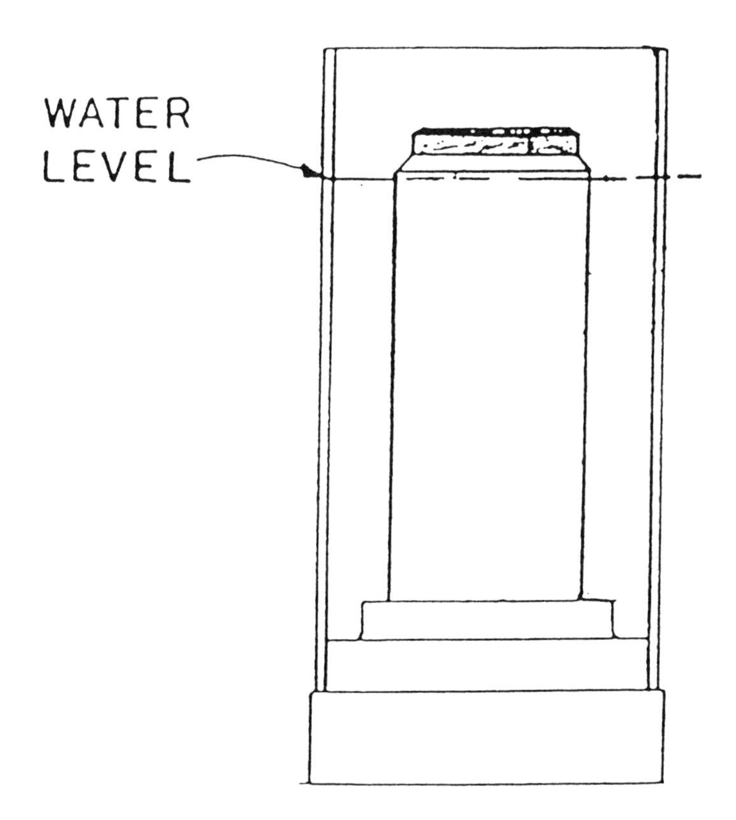

ADJUSTMENT PROCEDURE

1. Remove gearmotor and auger.

2. Adjust float valve to get water level to top seam of the evaporator.

3. Re-install auger. WATER LEVEL WILL RISE WHEN AUGER IS INSERTED BUT WHEN THE MACHINE IS TURNED BACK ON AND ICE STARTS BEING MADE, THE WATER LEVEL WILL GO BACK TO THE ORIGINAL SETTING.

4. Re-install gearmotor assembly and start machine.

TYPICAL WATER CIRCUIT

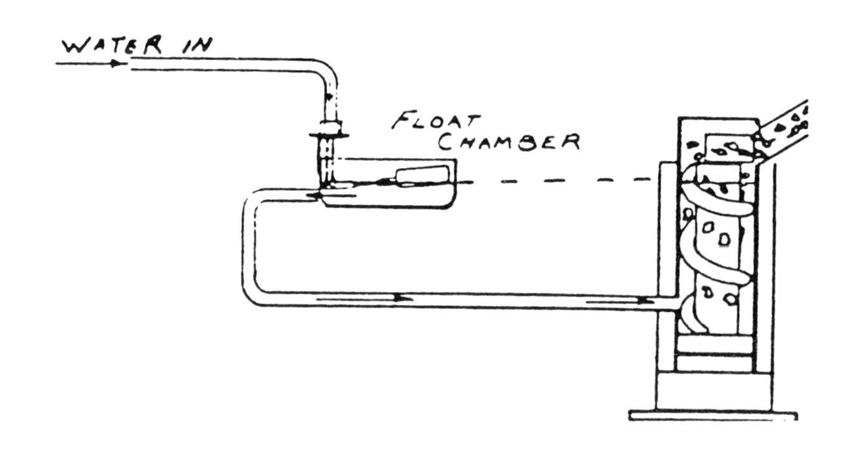

The supply water enters the float chamber through a small orifice. The water level rises and lifts the bouyant float with it. The float attached to the float arm seats a valve to shut off any further water supply. As water leaves the float chamber, the level drops along with the float and arm, causing the valve to open and admit more water. Thus the water level is maintained automatically as the machine operates.

Water now flows through a hose connected to the float chamber and enters the opening of the evaporator shell. The water level in the shell will rise to the same level that is maintained in the float chamber. The water that is in immediate contact with the center post evaporator will be reduced in temperature. As a result, freezing occurs and ice forms on the surface of the evaporator.

As more water is frozen, the thickness of the ice increases until it exceeds the distance allowed between the evaporator and auger. The auger rotates at a slow speed to wipe off the accumulated ice as well as help it to the surface. After the ice reaches the surface it is discharged through the top opening in the shell. An ice chute attached to the shell conveys the ice to the storage bin where it accumulates in the insulated bin until it is used. The ice will pile up to a point where the bin thermostat tubing is located. When the ice touches this brass tubing, the unit will shut off and remain off until enough ice is used or melted to reduce the pile. Any ice that melts will pass through the drain and drain hose to an open drain.

TYPICAL REFRIGERANT CIRCUIT

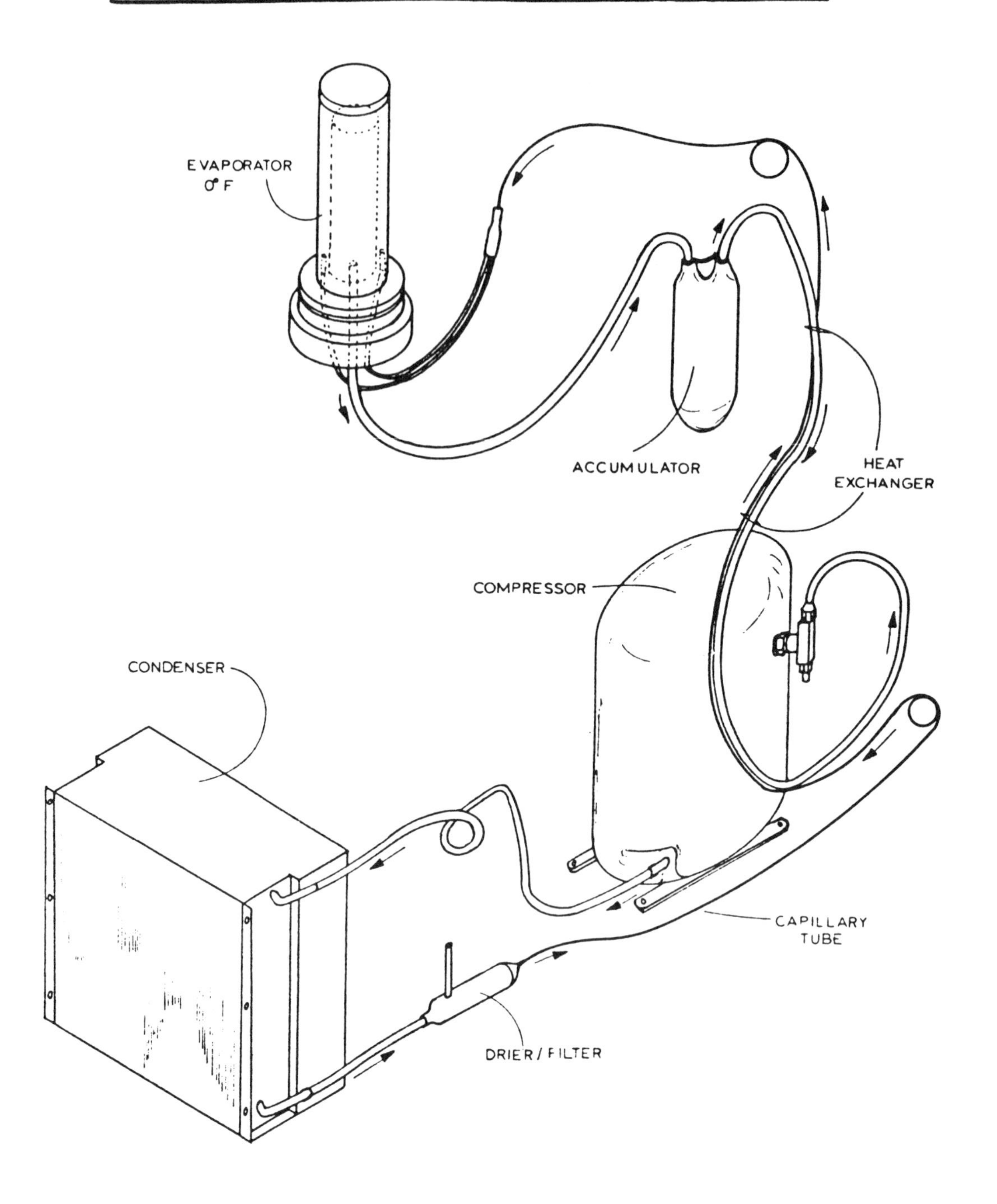

Heat always flows from hot to cold and therefore, the "heat load" supplied to the evaporator section by water gives up its heat to the refrigerant which is at a temperature below the freezing point of water. This refrigerant now passes through the heat exchanger back to the compressor, as a low pressure gas.

This low pressure gas is compressed in the compressor, as it leaves the compressor at a high pressure in vapor form it enters the top of the condenser. The condenser has a rapid flow of cool air across it which removes much of the heat from the hot refrigerant gas. As the gas passes through the condenser and loses heat it reverts back to a liquid since it is still under high pressure and cooler than when it entered the condenser. The liquid refrigerant then passes through the drier/filter still under pressure and goes through the heat exchanger where further cooling takes place. As the refrigerant leaves the capillary tube or automatic expansion valve, the pressure has dropped, causing the refrigerant to vaporize and boil off as it picks up heat in the evaporator and since the pressure is low, the refrigerant will be cold.

ELECTRICAL CIRCUIT

CIRCUIT DESCRIPTION

As the manual on-off-dispenser switch is pushed to "on", an electrical circuit is completed to the evaporator gearmotor via the external gearmotor overload, the N.C. contact of the power relay and the bin thermostat. As the gearmotor start relay comes "in" the gearmotor starts. Also, at this time the condenser fan motor starts as does the compressor via the high pressure control (water units only) and the compressor starting relay.

COMPONENT DESCRIPTION

EXTERNAL GEARMOTOR OVERLOAD - In the event of gearmotor overload for any reason, a heater strip in the external gearmotor overload will warp (due to excessive amps) and close a N.O. contact which in turn allows a circuit to be completed to the coil of the interlock relay that is normally de-energized.

INTERLOCK RELAY - This relay is only energized by the external gearmotor overload. It contains a set of N.C. power contacts that breaks the circuit to the gearmotor, power relay and fan motor. At the same instant, the N. O. contact is closed completing a parallel circuit to its own coil thus "interlocking" the relay. As the external gearmotor overload resets (opens) due to the cooling off of the control, nothing will happen. Only by pushing the on-off switch to the "off" position can the holding circuit be broken. This will allow the interlock relay to return to its normal position.

BIN THERMOSTAT - This is electrically in "series" with the ice making system. When the bin is full, the contact opens, terminating power to the machine.

GEARMOTOR START RELAY - This is a current type relay which means as the gearmotor run winding comes "on" the line, the current draw initially is relatively heavy through the relay coil (coil is in series with run winding). It then acts like a normal relay and the N.O. start contact "makes", completing a circuit through the start capacitor to the start winding. As the gearmotor picks up speed, the amp draw through the relay coil drops off allowing the armature to return to its normal position (start contact "opens"). This action removes the start winding from the circuit.

POWER RELAY - This relay controls the compressor power only. NOTE: This relay is utilized on 1/2 h.p. compressor units and higher.

ON-OFF SWITCH - This switch interrupts power to the entire unit.

LOW AMBIENT SWITCH - (when used) The function of this switch is to maintain condensing pressures at a satisfactory level during low ambient conditions. The switch breaks the circuit to the condenser fan motor at 110 PSI and makes the circuit at 150 PSI.

DELAY THERMOSTAT

This thermostat keeps the gearmotor running until the suction line temperature reaches 45° after the full bin switch terminates power to the power relay.

ELECTRICAL CIRCUIT CON'T.

HIGH PRESSURE CONTROL (Water units only) - The high pressure cut out is electrically in series with the compressor. As the head pressure rises (water shut off?) to a preset level, the contact opens, thus breaking the circuit to the compressor. As the pressure drops the control resets, allowing the compressor to restart.

ON-OFF CARBONATOR SWITCH - (DD Models Only) - This single pole, single throw switch is in the electrical circuit to turn off the drink dispensing components while the ice making components are still functional.

SWITCH, ON-OFF-DISPENSER - (DD Models Only) - This switch is a double pole, double throw switch with three positions. The "on" or up position of this switch supplies power to the ice maker section and the drink dispenser system. The "on-drink dispenser" or "down" position of this switch supplies power to the drink dispenser only. The switch in this position shuts off the ice maker section but still supplies power to the drink dispensing components. The center position of this switch is the OFF position.

COMPRESSOR START RELAY- This is a current type relay and contains a N.O. contact which is connected in series with the start winding of the compressor. The relay coil is electrically in series with the run winding. When power is applied, the compressor draws high current which sets up a magnetic field around the magnet coil which causes the relay to operate, closing the relay contact. As the compressor approaches operating speed the current flowing through the coil decreases, permitting the relay contact to open, thereby opening the starting circuit.

POTENTIAL RELAYS - The potential relay is sometimes used as a compressor starting relay. The contact in the potential relay is N.C. The magnet coil is connected across (parallel the start winding and is affected by induced voltage, generated by the start winding. As the compressor comes up to design speed the voltage across the relay coil increases and at running speed is sometimes as much as 2 1/2 times the supply voltage. This voltag sets up a magnetic field which causes the relay to operate. The starting relay is calibrated to remove the start capacitor (open the starting circuit) at approximately 85% of the motor design speed.

NOTE: BOTH TYPES OF RELAYS ARE DESIGNED TO OPERATE WITHIN VERY NARROW LIMITS OF VOLTAGE AND CURRENT DICTATED BY MOTOR DESIGN, THEREFORE, WHEN MAKING A REPLACEMENT OF A RELAY ALWAYS PROVIDE AN EXACT REPLACEMENT, RECOMMENDED BY THE COMPRESSOR MANUFACTURER.

CAPACITORS - GENERAL - An electrical capacitor is a device which stores up electrical energy. Capacitors are used with single phase motors to provide starting torque and improve running characteristics; by feeding this energy to the start winding in step with the run winding.

Any capacitor has three (3) essential parts, two (2) of which are usually metal plates separated and insulated by the third part called the dialectric.

Two general types of capacitors are used with electric motors. The electrolytic starting capacitor usually uses a very thin film of oxide on the metallic plate as the dialectric. The running capacitor usually is of the oil filled type.

The amount of electricity a capacitor will hold depends on the voltage applied. Doubling the voltage of a given capacitor will double the amount of electrical energy forced upon its plates.

The capacity of a capacitor, usually expressed in microfarads (MFD) is the ratio of the quantity of electricity and the voltage. This is dependent upon the size and spacing of the conducting plates and the type of insulating or dialectric medium between the plates.

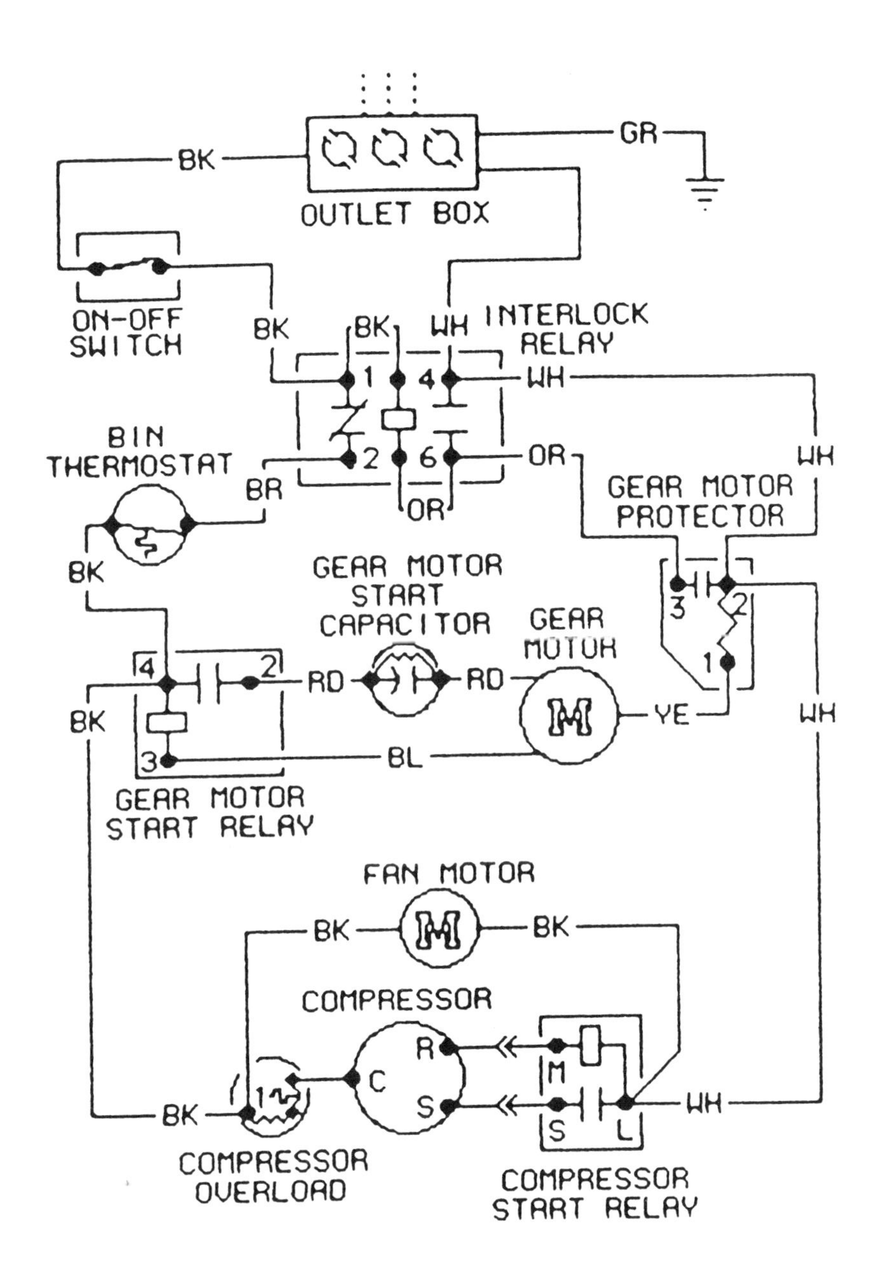

RF-151-SC

120 VOLTS 60 HZ

Ross temp

PART NO. 20541
ARTWORK NO. 50172
REV. B

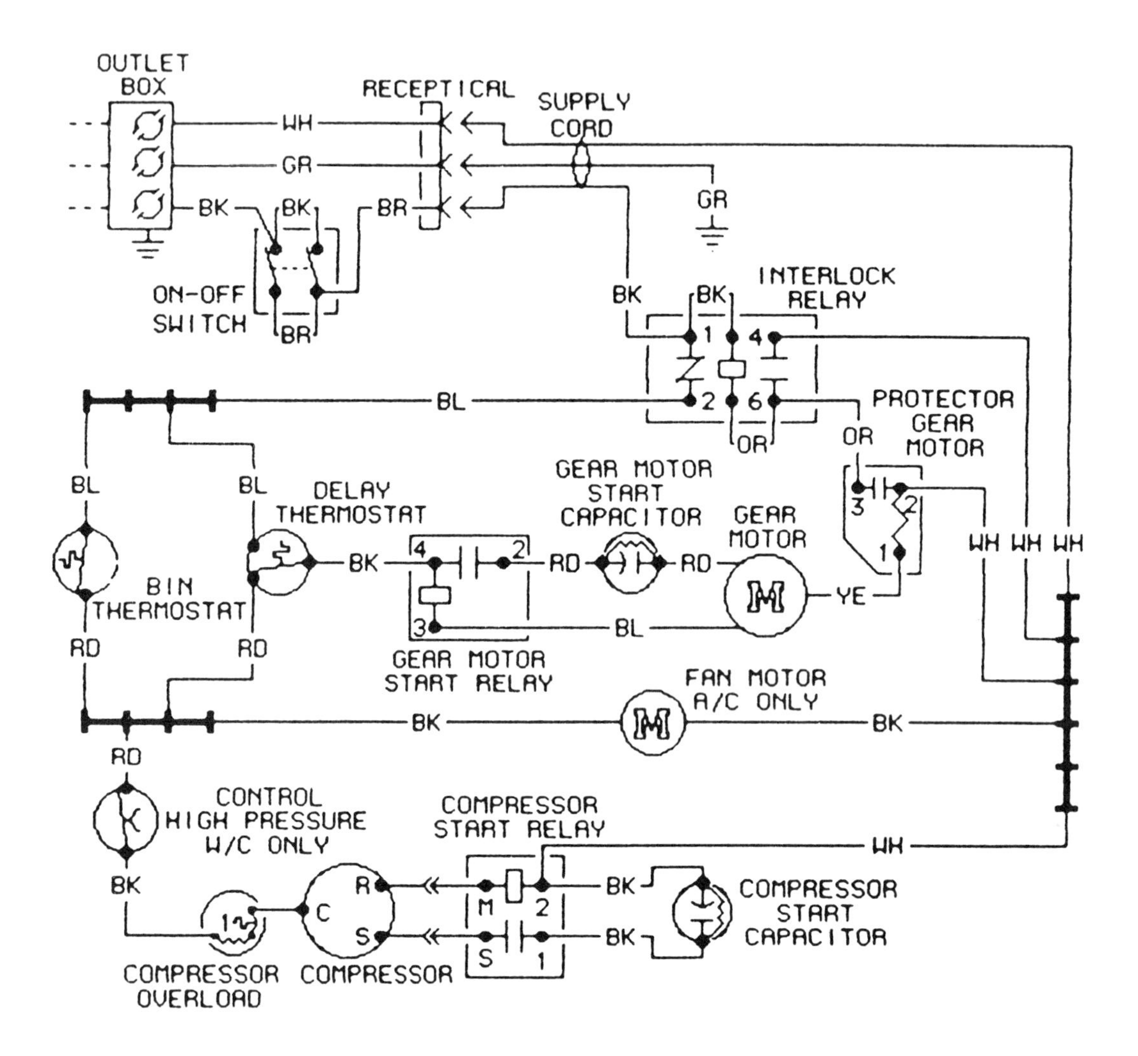

RF-351-SC & -W

115 VOLTS 60 HZ

Ross temp

PART NO. 36084
ARTWORK NO. 50353
REV. B

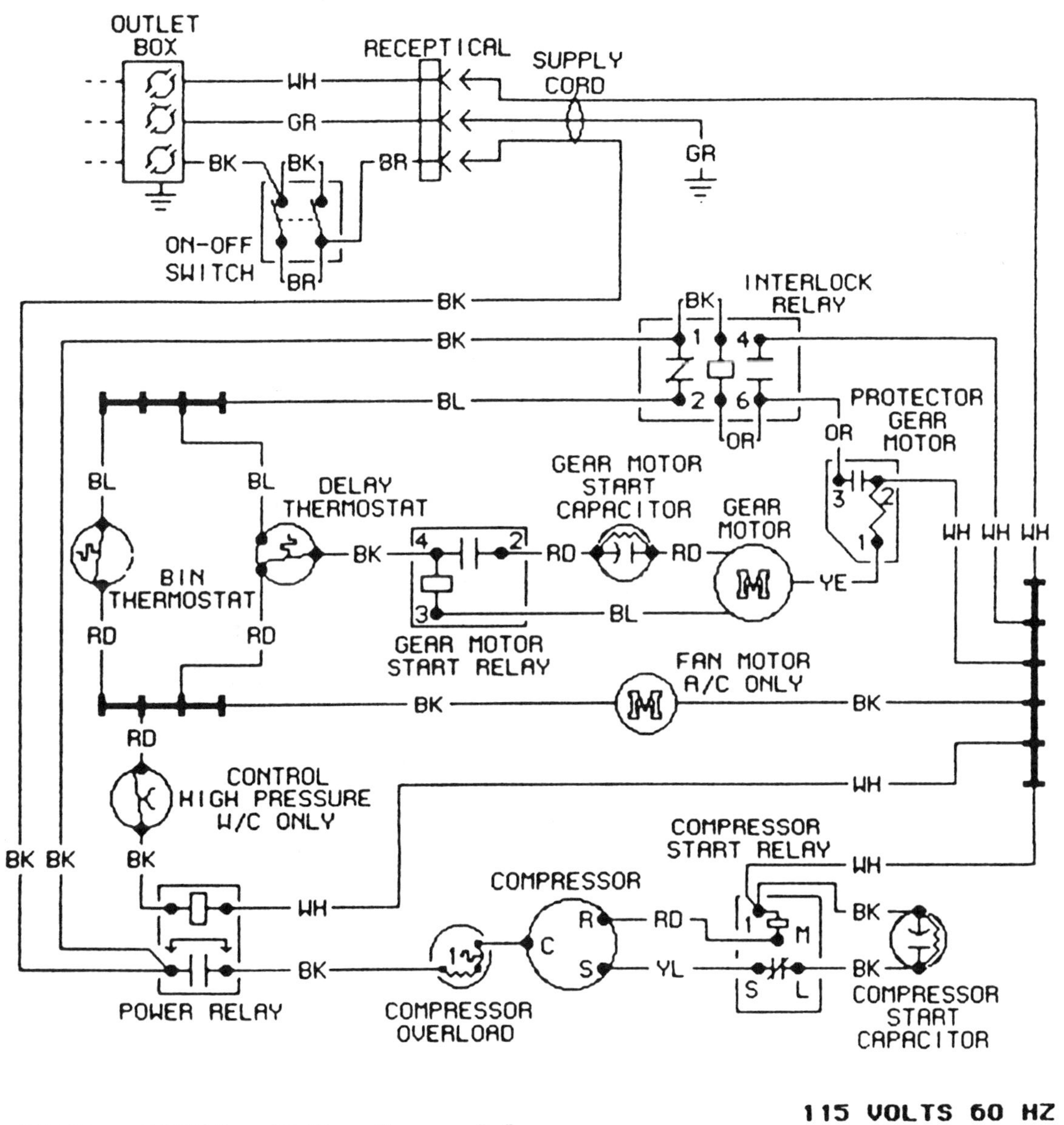

RF-452-SC & -W

115 VOLTS 60 HZ

Ross temp

PART NO. 36127
ARTWORK NO. 50356
REV. A

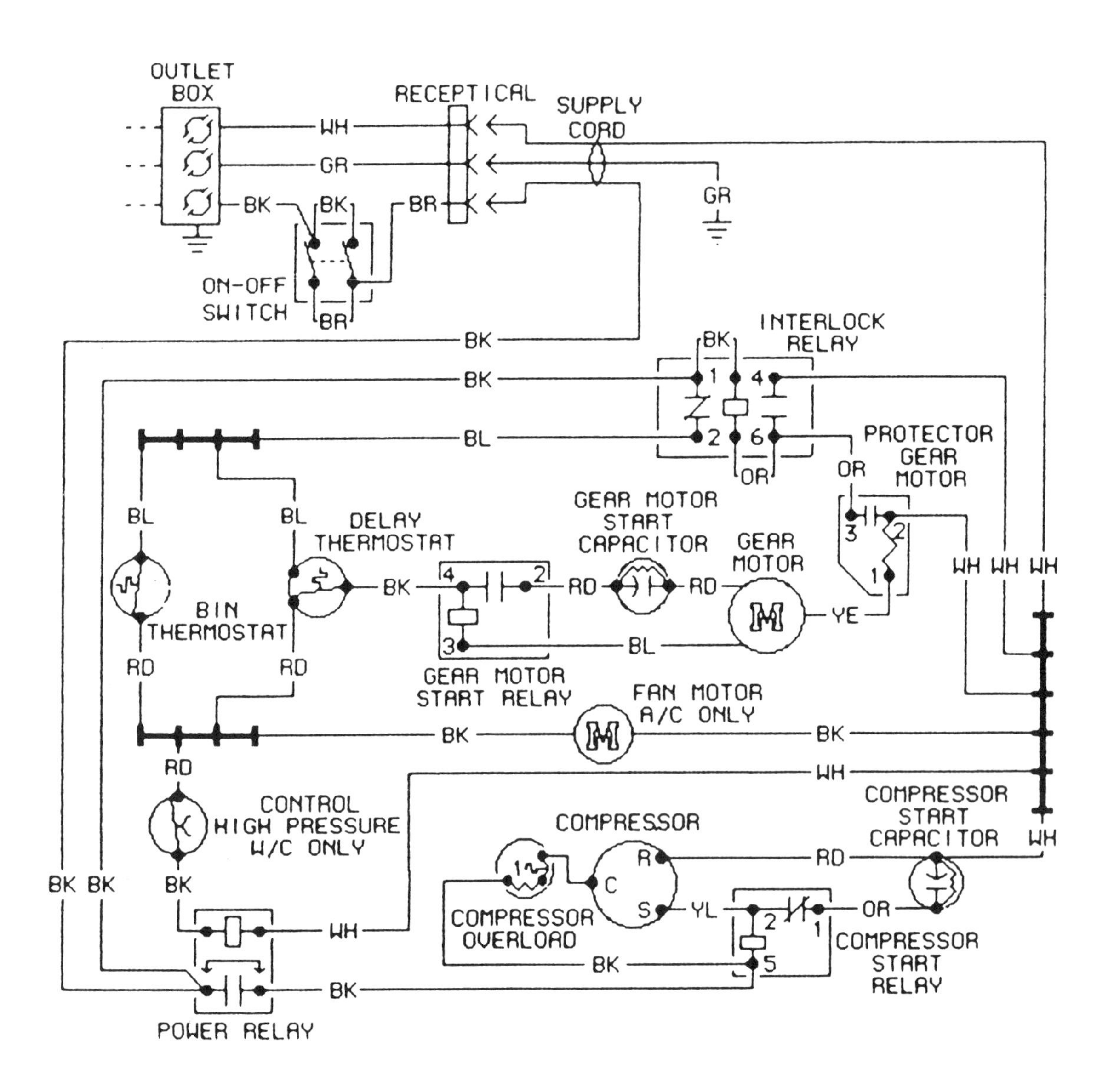

RF-600-SC & -W

115 VOLTS 60 HZ

Ross temp

PART NO. 36129
ARTWORK NO. 50358
REV. B

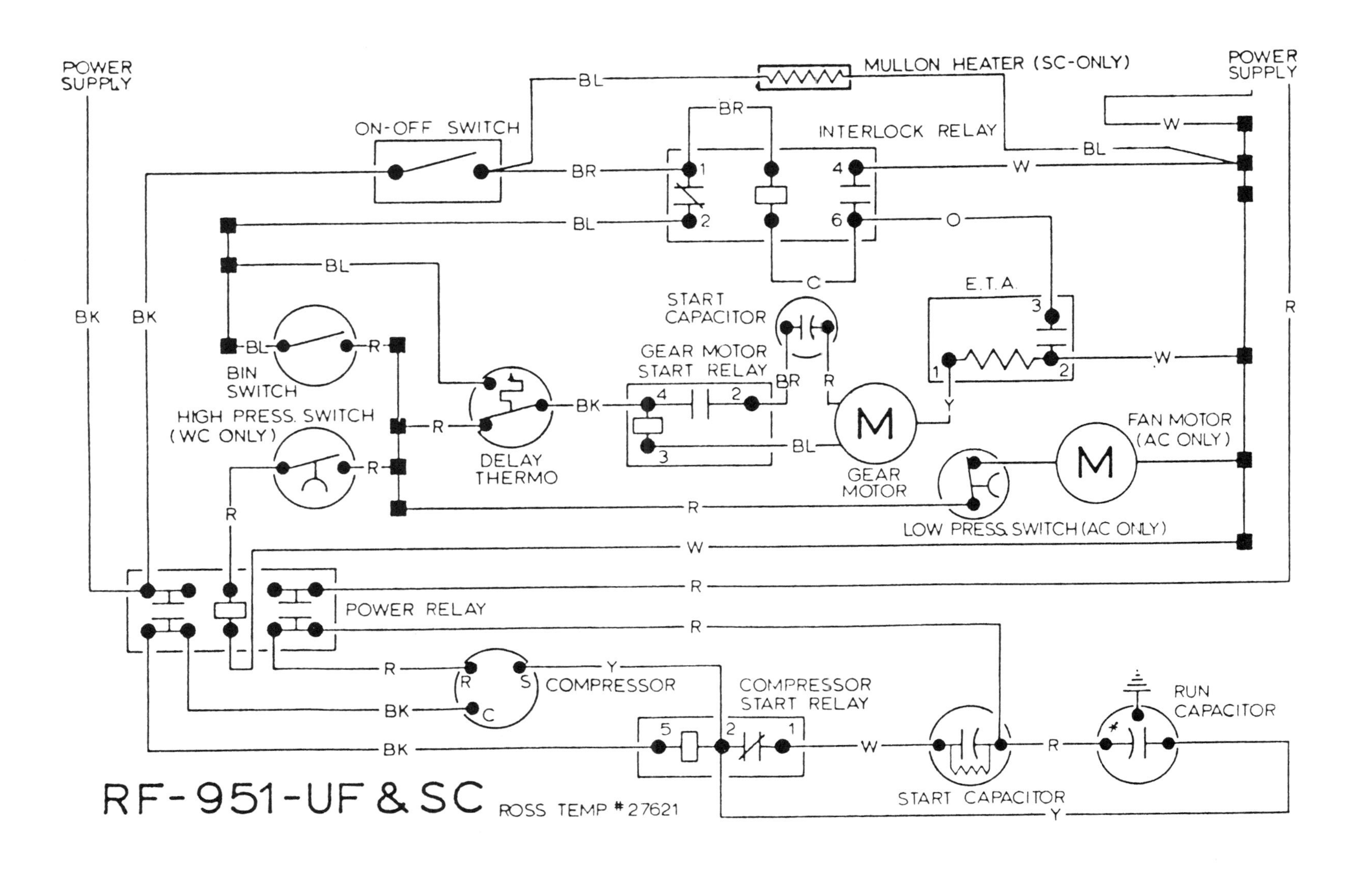
POWER SUPPLY
ON-OFF SWITCH
MULLON HEATER (SC-ONLY)
POWER SUPPLY
INTERLOCK RELAY
E.T.A.
START CAPACITOR
GEAR MOTOR START RELAY
BIN SWITCH
HIGH PRESS. SWITCH (WC ONLY)
DELAY THERMO
GEAR MOTOR
FAN MOTOR (AC ONLY)
LOW PRESS. SWITCH (AC ONLY)
POWER RELAY
COMPRESSOR
COMPRESSOR START RELAY
RUN CAPACITOR
START CAPACITOR
RF-951-UF & SC
ROSS TEMP # 27621

THE FOLLOWING MAINTENANCE SHOULD BE PERFORMED AT LEAST EVERY SIX MONTHS ON FLAKED ICE MACHINES:

1. Check power supply with machine running for proper voltage.

2. Check water level in the float tank reservoir. Water level should be maintained at the top of the evaporator. Adjust if necessary. (See illustration and adjustment procedure)

3. Clean the air-cooled condenser coil with a stiff brush or vacuum cleaner. (see procedure)

   CAUTION: CONDENSER COOLING FINS ARE SHARP, USE CARE WHEN CLEANING.

4. Clean the ice storage bin and flush the bin drain at least once a month.

5. If a water conditioner is installed in the inlet water line, change, replace or clean the filter, strainer or cartridge as required.

6. If heavy mineral deposits on the auger and evaporator shell are encountered due to bad local water conditions, follow sanitizing and cleaning procedure.

7. Loosen hold-down cam locks and remove gearmotor assembly.

8. Check thrust washer;replace if noticeably worn.

9. Lift out auger and examine for wear. The corkscrew auger guide bushing pressed into the drive block should be checked for wear. Replace if loose or if worn flat with auger drive block. If the Helix auger on the corkscrew auger round bar becomes flat on the inside more than 1/8 of an inch over a length of two inches or more it should be replaced.

   NOTE: HELIX AUGERS DO HAVE MACHINED FLAT RELIEF SURFACES. DON'T CONFUSE THEM WITH WORN FLAT AREAS.

   Check the insert in the bottom ring of the Helix auger and replace if excessively worn.

10. Check shell vertical strips for wear. Replace the shell if excessive wear shown.

11. Check O ring, replace if worn or cut.

12. Re-assemble, steps 6 through 10.

    CAUTION: IN RE-ASSEMBLING THE AUGER GEARMOTOR, THE HOLD DOWN CLAMPS MUST BE RIGHT AND SECURE. IN RE-INSTALLING THE EVAPORATOR SHELL, BE ABSOLUTELY SURE THAT THE "O" RING IS NOT PINCHED OFF AS THIS WOULD CAUSE A WATER LEAK AROUND THE BASE OF THE EVAPORATOR. LUBRICATE THE "O" RING WITH FOOD GRADE LUBRICANT BEFORE RE-ASSEMBLING SHELL.

13. Check for alignment of ice chute. Make sure chute gasket is not blocking path of ice flow.

14. Check bin thermostat operation. In the mid-range position the bin thermostat will open at 42° and has a 6° differential.

15. Check operation of mechanical shut-off assembly on RF-1251 units.

SANITIZING AND CLEANING PROCEDURE

1. Turn switch to "OFF" to stop unit.

2. (a) Turn water off and remove water hose from float chamber and proceed to drain the float chamber and evaporator; or

   (b) Remove float chamber cover and while holding float up to prevent more water from entering the float chamber, remove water hose from float chamber and proceed to drain the float chamber and evaporator.

3. With the float still held closed or water still off, restore water hose to float chamber and add 1/4 oz. of "sanitizer" (see note below) to the float chamber. Release float arm so that chamber fills with water. Turn water on, if necessary.

4. Turn switch to "ON" position and make ice for a minimum of 10 minutes.

5. Remove all ice from bin.

6. During the period in which the ice is being made, mix up the "sanitizer" for the bin and use as follows:

   To a gallon of lukewarm water, add 1/2 oz. "sanitizer" and with a non-metallic bristle brush, scrub the interior of the ice bin including the under side of the door, and the outside of the ice chute that protrudes into the ice bin.

7. Rinse interior of bin with clean water. (NOTE: Steps 6 and 7 should take no less than 10 minutes.)

8. Not less than 10 minutes after completing step #4, turn switch to "OFF" and repeat step #2. After the evaporator and float chamber have drained, restore hose to float chamber, release float, replace float cover and turn water on if necessary.

9. Turn switch to "ON". Collect and throw away the first gallon of ice.

   NOTE: APPROVED SANITIZER: "Calgon Ice Machine Sanitizer"

WATER TREATMENT

During the freezing process, as water passes under the freezing plate, the impurities in the water have a tendency to be rejected and the plate will freeze only the pure water.

However, the more dissolved solids in the water, the more troublesome the freezing operation will be. Bicarbonates in the water are the most troublesome of the impurities. These impurities will cause scaling on the evaporator, clogging of the water distributor head, float valve mechanism and other parts in the water system. If the concentration of impurities is high, cloudy cubes or mushy ice may be the result.

Parts of the ice maker, that are in contact with the water or ice, may corrode if the water is high in acidity. In some areas, water may have to be treated in order to overcome some of the problems that arise because of the mineral content.

One of the most economical and successful ways of doing this is whey a polyphosphate feeder which is inserted in the water inlet to the ice cube maker. These feeders correct problems caused by dissolved iron, preventing iron stains and deposits. It also prevents liming and scaling. These feeders must be installed according to the manufacturer's recommendations in order to satisfactorily treat the water.

WINTER STORAGE

If the unit is to be stored in an area where the temperature will drop below freezing, it is most important that all water lines be drained to prevent them from freezing and possible rupture.

To blow out the water line, disconnect the water supply at the cabinet inlet and use air pressure to force the water into the water reservoir pan. This can then be removed from the water pan.

WATER COOLED CONDENSER - To remove water from condenser unhook water supply and attach compressed air hose. Start machine. As head pressure reaches the appropriate level opening the water regulating valve, the compressed air will force the water out. Do not let the machine operate longer than necessary.

CLEANING THE CONDENSER (AIR COOLED)

In order to produce at full capacity, the refrigeration condenser must be kept clean. The frequency of cleaning will be determined by surrounding condition. A good maintenance plan calls for an inspection at least every two months.

Remove the unit compartment grille at the front. With a vacuum cleaner, remove all accumulated dust and lint that has adhered to the finned condenser.

CAUTION: CONDENSER COOLING FINS ARE SHARP. USE CARE WHEN CLEANING.

SERVICE ANALYSIS

1. Unit will not run	On-off switch in "off" position.	Turn switch to "on".
	Defective on-off switch.	Check and replace.
	Blown fuse.	Replace fuse and check for cause of blown fuse.
	Thermostat set too warm for ambient.	Ajust colder.
	Power relay contacts corroded.	Check and clean.
	Defective thermostat.	Check and replace
	Loose electrical connection	Check wiring
	Gearmotor overload protector has cut off machine.	Turn switch to off then to on.
2. Compressor cycles intermittently.	Low voltage	Check line voltage
	Dirty condenser	Clean condenser
	Air circulation restricted	Remove restriction
	Defective condenser fan motor	Check and replace
	Defective relay, overload protector or starting capacitor.	Check and replace
	Loose electrical connection	Check wiring
3. Making wet ice.	Surrounding air temperature too high	Correct or move unit
	High water level in float reservoir	Lower water level, see step 1, page
	Dirty condenser	Clean condenser
	Faulty compressor	Check and replace
	Refrigerant leak	Check and repair
	"O" ring leaking at bottom of evaporator shell	Check and replace
4. Unit runs but makes no ice.	Leak in refrigerant system	Check and repair
	Moisture in system	Check, dehydrate and add drier to system
	No water	Check water supply
	"O" ring leaking at bottom of evaporator shell	Check and replace "O" ring
	Compressor not running	Check and replace "O" ring

SERVICE ANALYSIS CON'T

5. Water leaks	Worn or bad float valve	Check and replace
	Float and arm assembly stuck	Check and replace
	"O" ring leaking at bottom of evaporator shell	Check and replace
	Storage bin drain and tubing	Check and repair
6. Excessive noise or chattering.	Mineral or scale deposits on inside of evaporator shell	Remove and clean inside surfaces by immersing evaporator shell in ice machine cleaner
	Intermitten water supply	Check inlet water line
	Water level in float tank too low	Check and adjust water level
	Auger gearmotor end-play or worn bearings	Repair or replace
	Air lock in gravity water supply line from float tank to evaporator shell	Check and adjust warmer
7. Machine runs with full bin of ice	Storage bin thermostat set too cold	Check and adjust warmer
	Bin thermostat thermowell out of path of ice	Adjust thermowell
8. Unit off on reset	Ice jams up in evaporator shell	Clean inside surface of evaporator shell
	Bin thermostat will not shut off machine. Set too cold	Check and adjust or replace
	Auger motor has worn bearings	Check and replace
9. Unit goes off on reset.	Ice chute out of alignment, restricted ice flow out of evaporator section	Re-align
	Ice chute center separator bent restricting ice flow out of evaporator section	Replace ice chute
	Incoming water temperature too cold	Maintain temperature above 50° F
	Bin thermostat does not shut off when bin is full of ice	Replace bin thermostat if necessary
	Mineral or scale deposits on inside of evaporator shell and evaporator	Inspect and clean
	Strips loose or missing on inside of evaporator shell	Inspect and replace evaporator shell if necessary
	Low ambient temperature in room where unit is located	Maintain temperature above 50 degrees
	Gearmotor sticking which causes it to draw excessive amperage (over 3.5 amps)	Check amp draw of gear-motor with an amprobe (2.25 amps)

9. Unit goes off on reset	Plugged capillary tube or expansion valve, causing low back pressure	Check back pressure, replace cap tube or valve, evacuate and re-charge system, replace drier-strainer
	Slight leak, causing low back pressure	Check backpressure, find gas leak, repair leak, evacuate system, add drier and recharge
	Loose hold-down assy.	Check and tighten or replace
	Auger worn excessively on the inside surfaces causing thicker flaked ice to be made	Replace auger
	Auger out of line causing excessive wear on the lower outside surface where it rubs against evaporator shell liner at the bottom	Replace auger and evaporator shell
	Broken auger	Replace auger
	Evaporator surfaces worn or gouged, causing thicker ice to be made	Inspect and replace evaporator if necessary
	Auger guide bushing worn down	Replace auger guide bushing (corkscrew type augers only)
	Loose gearmotor mounting plate	Check and tighten
	Low water level in float tank reservoir	Adjust float arm to maintain correct water level
	Worn thrust washer	Replace

PARTS LIST

Illus. No.	Description	RF-151	RF-351
1	Compressor	25225	25340
2	Pad, masonite	9115	9115
3	Hold-down, evaporator	2558	2429
4	Evaporator	1501	1085
5	O ring	122	122
6	Auger	4062	6353
7	Shell, evaporator	1508	8712
8	Gasket, evaporator shell	1916	1916
9	Washer, thrust	8043	8043
10	Plate, gearmotor mounting	4060	4060
11	Chute, inclined ice	8821	8821
12	Gearmotor (Spiroid)	21700	21700
	Gearmotor (Von Weiss)	29516	29516
13	Condenser	2908	7463
14	Shroud, condenser	2722	2722
15	Blade, condenser fan	131	131
16	Motor, condenser fan	7470	7470
17	Bracket, condenser fan motor mounting	260	260
18	Coil, condenser W-C	N/A	3844
19	Clamp, coil W-C	N/A	35987
20	Bracket, condenser coil W-C	N/A	36052
21	Valve, water regulating W-C	N/A	1211
22	Drier	20557	20557
23	Blade, compressor cooling fan	N/A	9355
24	Motor, compressor cooling fan	N/A	9361
25	Bracket, compressor cooling fan	N/A	260
26	Thermostat, bin	9570	9570
27	Protector, gearmotor overload (Spiroid gearmotor)	25170	25170
	Protector, gearmotor overload (Von Weiss gearmotor)	35707	35707
28	Relay, gearmotor start (Spiroid gearmotor)	25865	25865
	Relay, gearmotor start (Von Weiss gearmotor)	29520	29520
29	Relay, interlock	27618	27618
30	Control, high pressure	N/A	7024
31	Cord	N/A	36153
32	Float tank, complete	22030	22030
33	Float valve only	21924	21924
34	Thermowell	20773	29541
	Thermowell, drink dispensers only	N/A	7686
35	Relay, compressor start	25447	25425
36	Capacitor, compressor start	N/A	25246
37	Protector, compressor overload	25445	25411
38	Bushing, auger guide	2040	2040
39	Switch, on-off	28513	28513
	Switch, on-off dispense (DD only)	23836	23836
40	Screw, wing	890	890
41	Thermostate, delay (potted)	N/A	25865
42	Clip, delay thermostat	N/A	25871
43	Suction line assembly	25228	36111
44	Accumulator	2798	2798
45	Receptacle	N/A	36130
46	Capacitor, gearmotor start (Von Weiss)	29518	29518

PARTS LIST

Illus. No.	Description	RF-452	RF-600	RF-951
1	Compressor	28007	22620	22966
2	Masonite pad	3182	3182	3182
3	Hold-down, evaporator	3492	3492	3492
4	Evaporator	41100	41100	41100
5	O Ring	3120	3120	3120
6	Auger	3796	3796	3796
7	Shell, evaporator	9182	9182	9182
8	Gasket, evaporator shell	8065	8065	8065
9	Thrust washer	8043	8043	8043
10	Plate, gearmotor mounting	3163	3163	3163
11	Chute, inclined ice	8720	8720	8720
12	Gearmotor (Spiroid)	21700	21700	21700
	Gearmotor (Von Weiss)	29516	29516	29516
13	Condenser	23418	25466	26266
14	Shroud, condenser	36074	36579	23571
15	Blade, condenser fan	2861	25578	25874
16	Motor, condenser fan	23526	25242	25242
17	Bracket, condenser fan motor	36075	36075	8226
18	Condenser coil, W-C	8646	1210	8823
19	Clamp, condenser coil, W-C	35987	1197	8286
20	Bracket, condenser coil mtg., W-C	36052	5011	5011
21	Valve, water regulating, W-C	1211	1211	1211
22	Drier	20556	20556	21850
23	Blade, compressor cooling fan, W-C	9355	9355	9355
24	Motor, compressor cooling fan, W-C	9361	9361	9361
25	Bracket, compressor cooling fan motor	260	260	260
26	Thermostat, bin	9570	9570	9570
27	Protector, gearmotor overload (Spiroid)	25170	25170	25170
	Protector, gearmotor overload (Von Weiss)	35707	35707	35707
28	Relay, gearmotor start (Spiroid)	25865	25865	25865
	Relay, gearmotor start (Von Weiss)	29520	29520	29520
29	Relay, interlock	27618	27618	27618
30	Control, high pressure W-C	7024	7024	7024
	Control, low ambinet fan, A-C	N/A	N/A	23838
31	Cord	36303	36303	36303
32	Float tank (complete)	22030	22030	22030
33	Float valve (only)	21924	21924	21924
34	Thermowell (poly bin liner)	29541	29541	29541
	Thermowell (stainless bin liner)	7686	7686	7686
35	Capacitor, compressor run	N/A	21545	21545
36	Relay, power	27617	27617	23839
37	Thermostat, gearmotor relay (potted)	25865	25865	25865
38	Clip, delay thermostat	25871	25871	25871
39	Valve, automatic expansion	N/A	27653	27653
40	Suction line assembly	36301	36322	26125
41	Accumulator	N/A	3128	3128
42	Relay, compressor start	28008	21547	22968
43	Capacitor, compressor start	28009	21544	22969
44	Capacitor, gearmotor start (Spiroid)	N/A	7013	7013
	Capacitor, gearmotor start (Von Weiss)	29518	29518	29518
45	Protector, compressor overload	28010	21543	N/A
46	Screw, wing	890	890	890
47	Switch, on-off	28513	28513	28513
	Switch, on-off-dispense (DD models only)	23836	23836	23836
48	Receptacle	36130	36130	36130
49	Disc, centering	20956	20956	20956

CABINET PARTS LIST

Illus. No.	Description	RF-151	RF-351	RF-452	RF-600	RF-951
1	Bin door	9120	9120	9120	36555	9123
2	Door track, LH	25007	25007	25007	25007	25007
3	Door track, RH	8824	8824	8824	8824	8824
4	Door trim	9112	9112	9112	8681	8681
5	Back panel	56	N/A	N/A	N/A	N/A
6	Front cover	25845	26856	26856	26858	26858
	Front cover, SS	N/A	26857	26857	26859	26859
7	Side louvered panel	25041	N/A	N/A	N/A	N/A
8	Upper louvered panel	N/A	25041	25041	24153	24153
	Upper louvered panel, SS	N/A	25042	25042	24154	24154
9	Lower louvered panel	N/A	24153	24153	25012	25012
	Lower louvered panel, SS	N/A	24154	24154	25011	25011

ILLUSTRATED PARTS BREAKDOWN

RF-151

RF-351

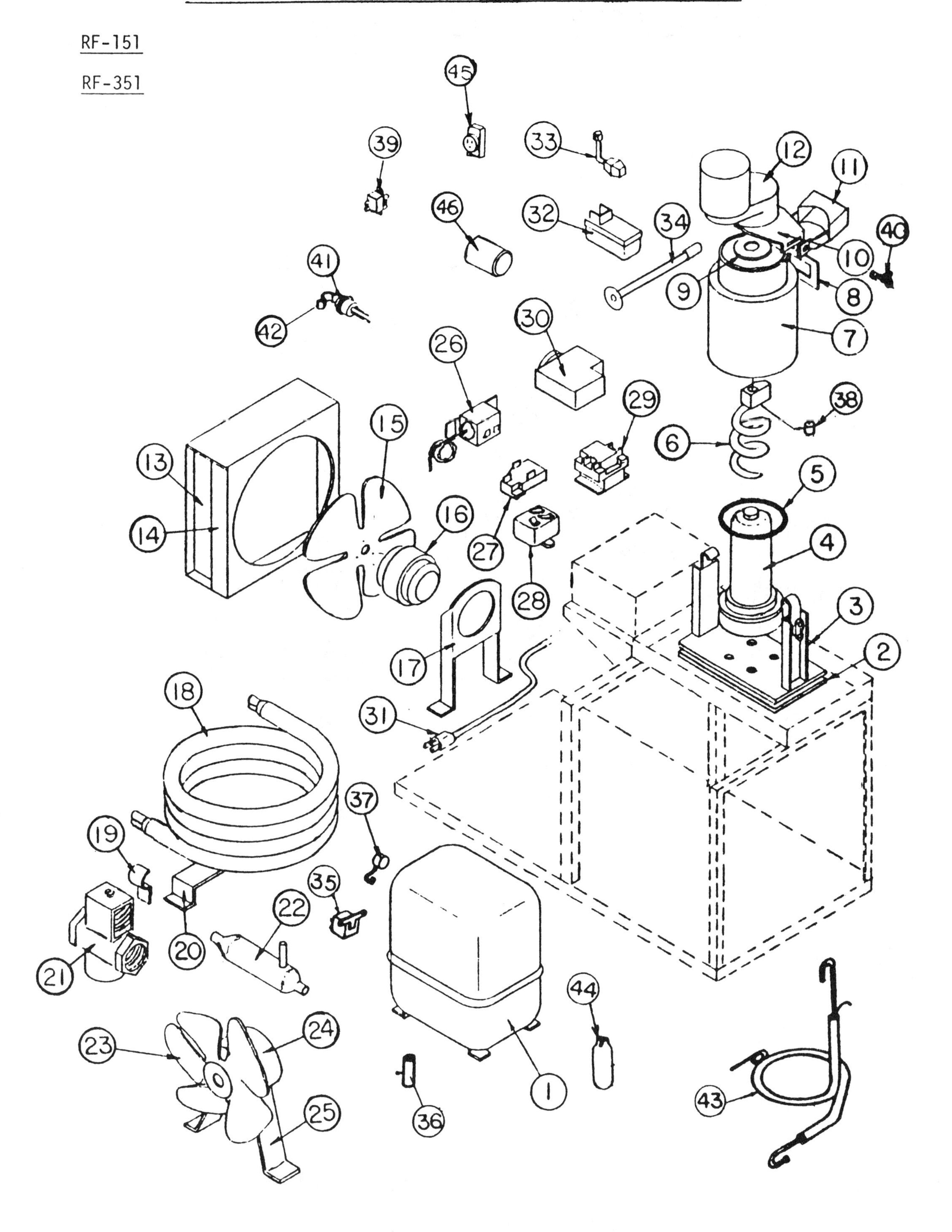

ILLUSTRATED PARTS BREAKDOWN

RF-452

RF-600

RF-951

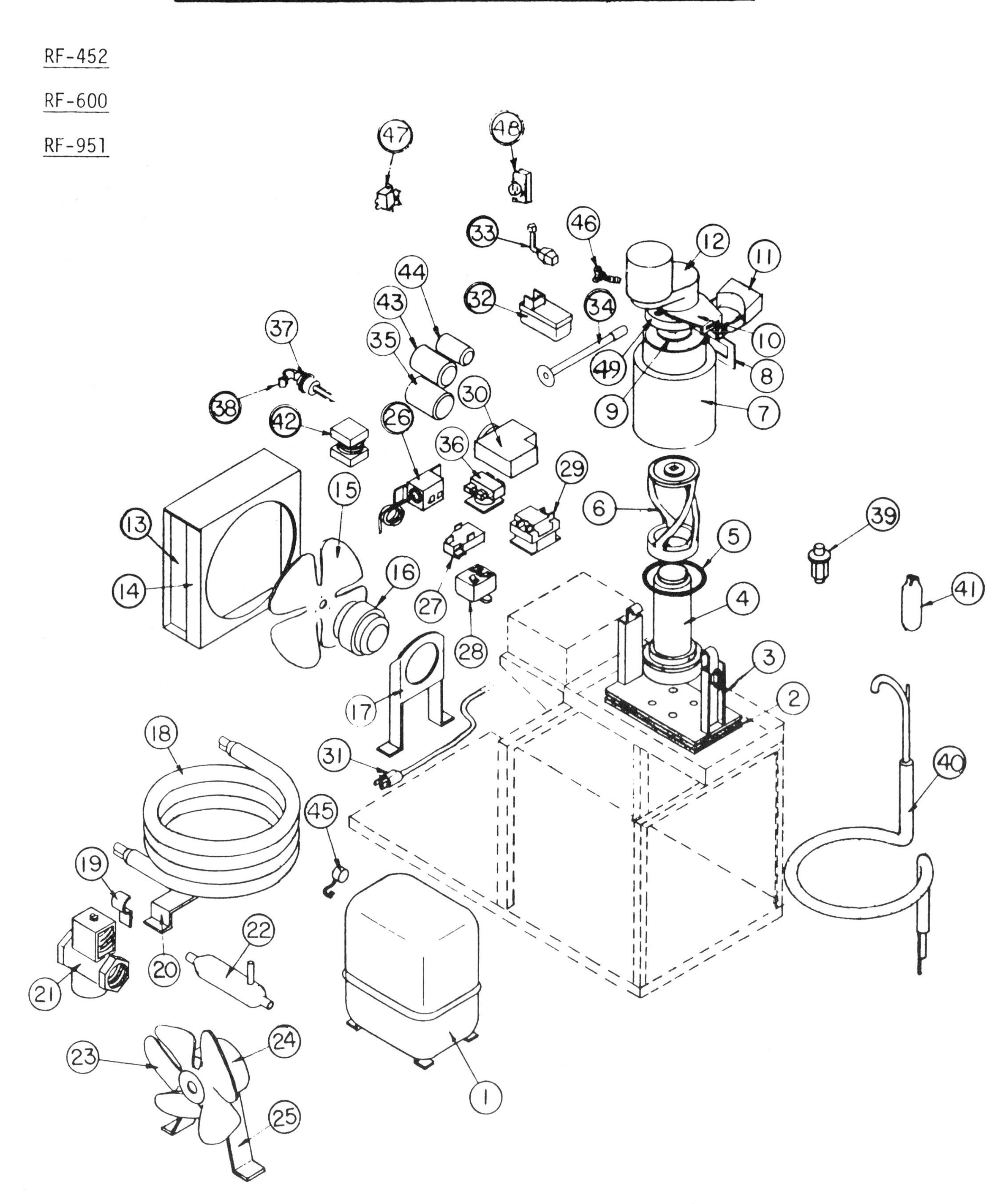

ILLUSTRATED CABINET PARTS BREAKDOWN

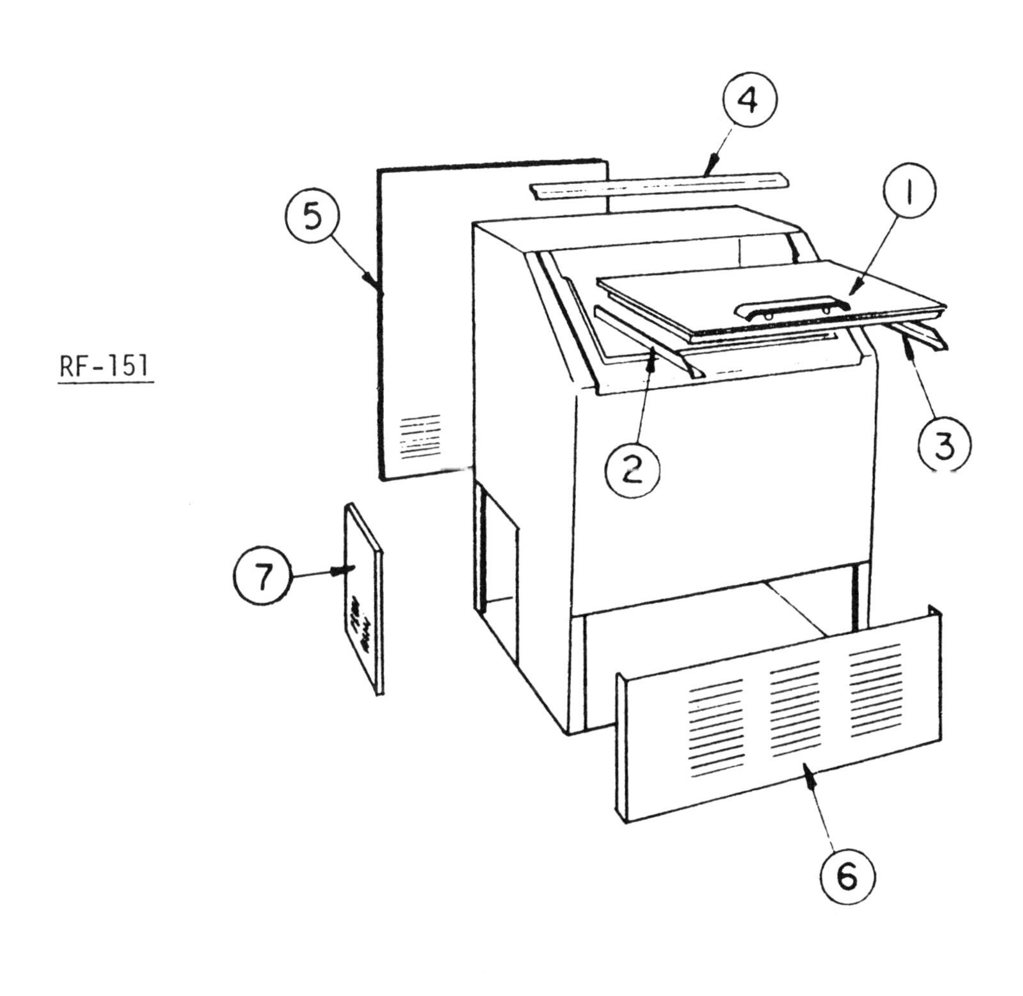

RF-351
RF-452
RF-600
RF-951

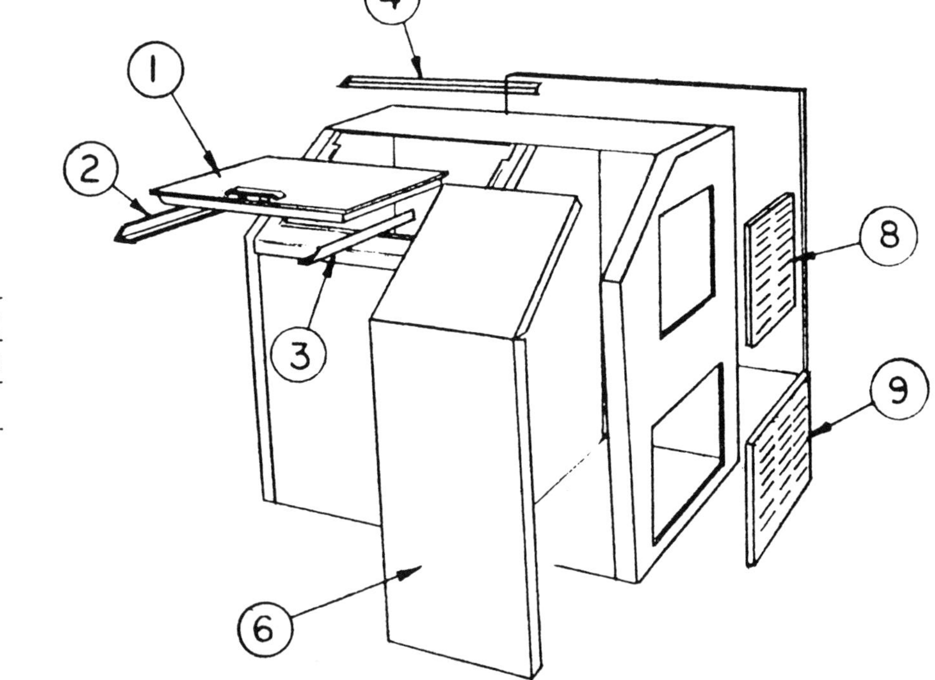

Scotsman

SCOTSMAN®

COMMERCIAL ICE SYSTEMS

Service Manual for **Modular Cuber**

MODEL CM650 E

CM650 TABLE OF CONTENTS

CM650 FOR THE INSTALLER

INTRODUCTION

These instructions provide the specifications and the step-by-step procedures for the installation, start up and operation for the Scotsman Model CM650 Modular Cuber.

The Model CM650 Modular Cubers are quality designed, engineered and constructed, and are thoroughly tested icemaking systems, providing the utmost in flexibility to fit the needs of a particular user.

NAMEPLATE (ON BACK)

SERIAL NUMBER PLATE

NAME PLATE

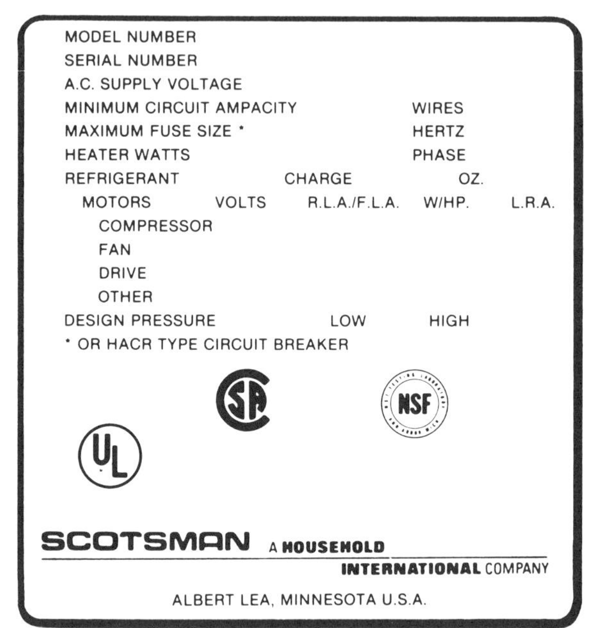
MODEL NUMBER
SERIAL NUMBER
A.C. SUPPLY VOLTAGE
MINIMUM CIRCUIT AMPACITY WIRES
MAXIMUM FUSE SIZE * HERTZ
HEATER WATTS PHASE
REFRIGERANT CHARGE OZ.

MOTORS	VOLTS	R.L.A./F.L.A.	W/HP.	L.R.A.
COMPRESSOR				
FAN				
DRIVE				
OTHER				

DESIGN PRESSURE LOW HIGH
* OR HACR TYPE CIRCUIT BREAKER

SCOTSMAN A HOUSEHOLD INTERNATIONAL COMPANY
ALBERT LEA, MINNESOTA U.S.A.

SERIAL NUMBER PLATE

MODEL NUMBER
SERIAL NUMBER
VOLTS/HERTZ/PHASE
MAXIMUM FUSE SIZE
REFRIGERANT CHARGE OZ.

SCOTSMAN CONTOUR CUBE

This product qualifies for the following listings:

This icemaker has been engineered to our own rigid safety and performance standards. The National Sanitation Foundation (NSF) seal, signifies that it is listed with NSF and that it complies with the materials and construction standards of NSF. In addition, the Underwriters Laboratories, Inc., (UL) Listing Mark and the Canadian Standards Association (CSA) Monogram, both signify that its construction and design have been inspected and tested by them. NSF, UL and CSA inspectors also periodically examine production icemakers at the factory, to assure continued compliance.

To retain the safety and performance built into this icemaker, it is important that installation and maintenance be conducted in the manner outlined in this manual.

CM650 FOR THE INSTALLER

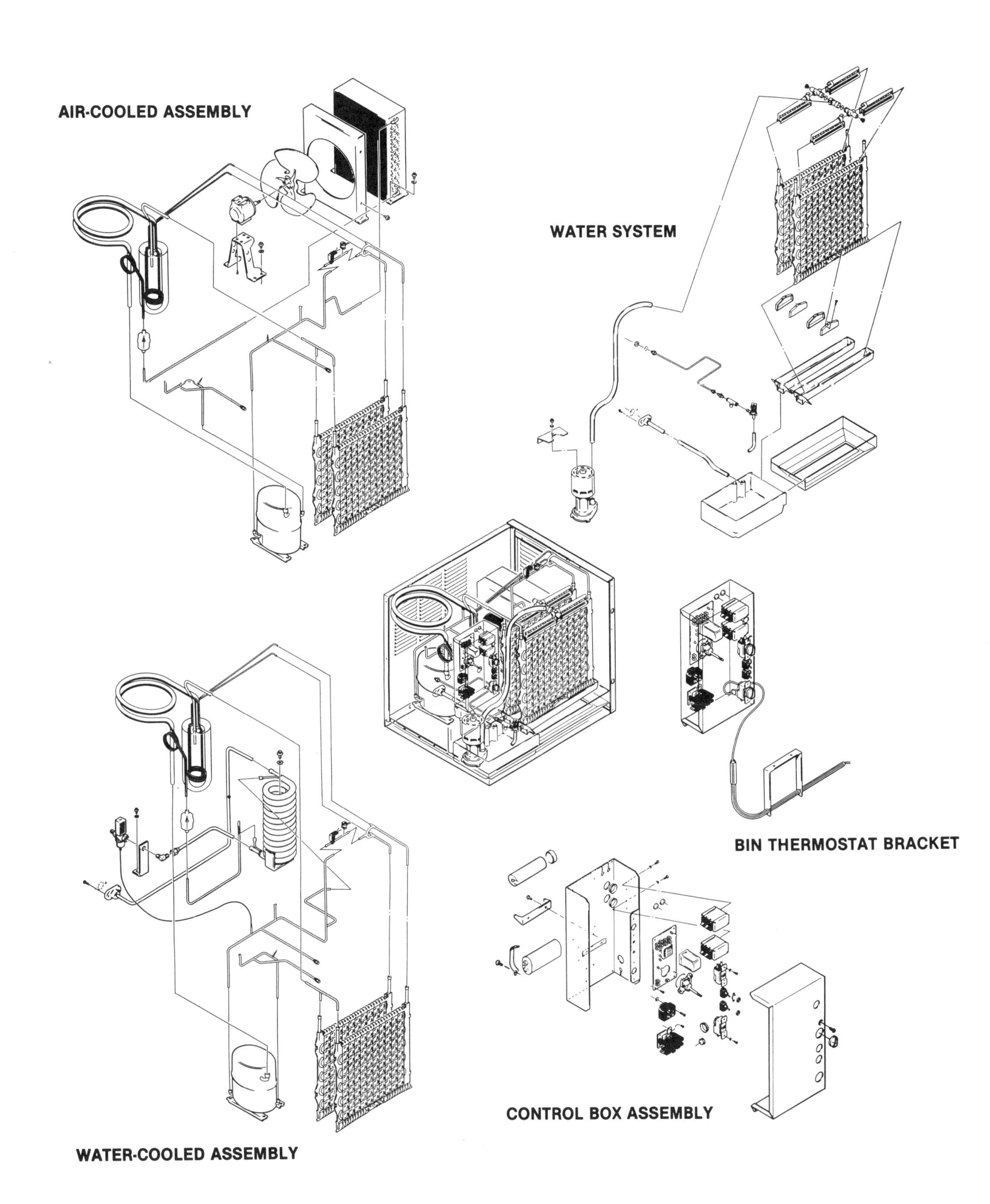

CM650E FOR THE INSTALLER

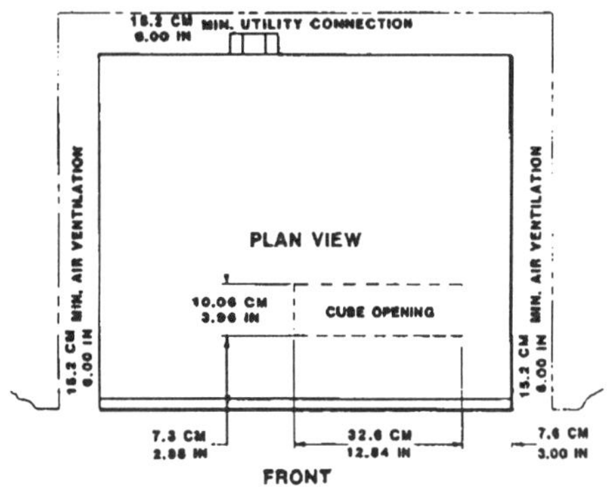

INSTALLATION NOTE: Allow 6" minimum space at sides and back for ventilation and utility connections.

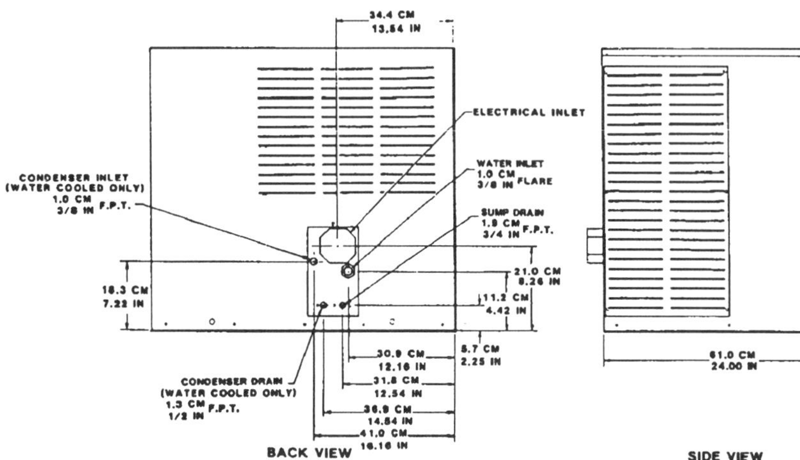

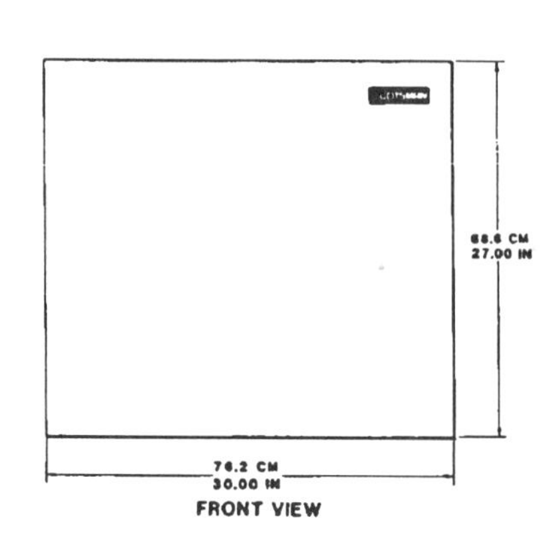

The CM650 is designed to fit the following Scotsman storage bins:

- HTB500 or BH550

When installing a new system, check to be sure that you have everything you need before beginning:

- Correct Bin
- Correct Ice Machine either CM650E-32E or CM650E-3E
- All kits, legs, and information required for the specific job.
- Optional Stainless Steel Panel Kit, SPKCMD1 contains the front, top, 2 ends, and 2 service panels.

Installation Limitations:

This ice system is designed to be installed indoors, in a controlled environment:

	Min ....Max
Air Temperature	50^0F.................100^0F
Water Temperature	40^0F.................100^0F
Water Pressure	20 80
Voltage	-5%...................+10%

(Compared to the nameplate)

Operating the machine outside of the limitations is misuse and can void the warranty.

SPECIFICATIONS

Model Number	(Height - w/0 Bin) Dimensions H" x W" x D"	Ice Type	Condenser Type	Basic Electrical	Comp. H.P.	No. of Wires	Min. Circ. Ampacity+	Max. Fuse Size
CM650AE-32E	27 x 30 x 24	Contour Cube	Air	208-230/60/1	1.5	2	19.8	30
CM650WE-32E	same	same	Water	same	1.5	2	18.6	30
CM650AE-3E	same	same	Air	208-230/60/3	1.5	3	9.2	15
CM650WE-3E	same	same	Water	same	1.5	3	8.5	15

+Use this value to determine wire size and type per National Electric Code. Standard finish is enamel sandalwood.

Scotsman Ice Systems are designed and manufactured with the highest regard for safety and performance. They meet or exceed the standards of UL, NSF, and CSA.

Scotsman assumes no liability or responsibility of any kind for products manufactured by Scotsman that have been altered in any way, including the use of any part and/or other components not specifically approved by Scotsman.

Scotsman reserves the right to make design changes and/or improvements at any time. Specifications and design are subject to change without notice.

CM650 FOR THE INSTALLER

LOCATION & LEVELING

1. Arrange for proper electric, water and drain. See instructions for the plumber and for the electrician.
2. After mounting the legs, position the ice storage bin in the selected location which should have a minimum room temperature of 50-degrees F. and maximum room temperature of 100-degrees F. Level the bin, adjusting the leg levelers in both the front to rear and side to side directions. Select a well-ventilated location for the air-cooled condenser.
3. Inspect the bin top mounting gasket which should be flat, with no wrinkles, to provide a good water seal when the Cuber is installed on top of the bin. Remove baffle from bin to gain access.
4. Install the modular cuber on top of the bin using care to be sure a good seal is made between the two cabinets. Align the holes in the bottom rear of the cabinet to mate with the two mounting straps on the top rear of the bin.
5. Use bolts and straps found in hardware package to secure the ice machine. When alignment and leveling are completed, tighten the bolts to secure the mounting straps.

STACKING INSTRUCTIONS

NOTE: This model can only be stacked onto Scotsman units with the same size cabinet as itself. Earlier models are not compatable.

When stacking two units, remove the top panel from the lower icemaker. (The top removed from the lower icemaker will no longer have any function.)

Carefully lift the uncrated top unit onto the bottom one and align the two cabinets. Use of a mechanical lift is recommended for this step.

At the back of the two icemakers, bolt the upper icemaker cabinet to the lower icemaker cabinet using the mounting straps and bolts from the hardware package.

BIN THERMOSTAT INSTALLATION

1. Remove cap from bin thermostat bracket. Insert bin thermostat bracket up through routing hole in sump.
2. Attach the bin thermostat bracket to the bottom of the cuber base using the thumb screws and threaded holes provided.
3. Carefully uncoil the capillary tube to the icemaker and route the capillary tube through the routing hole provided in the sump and through the full length of the bin thermostat bracket.

NOTE: When stacking, route the bin thermostat control from the upper ice machine through the routing hole provided in the sump of the lower icemaker and through the bin thermostat control bracket.

NOTE: Extend bin thermostat capillary tube through entire length of the tubular section of the bin thermostat bracket. When properly installed, the tip of the capillary tube should be flush with the tip of the bin thermostat bracket.

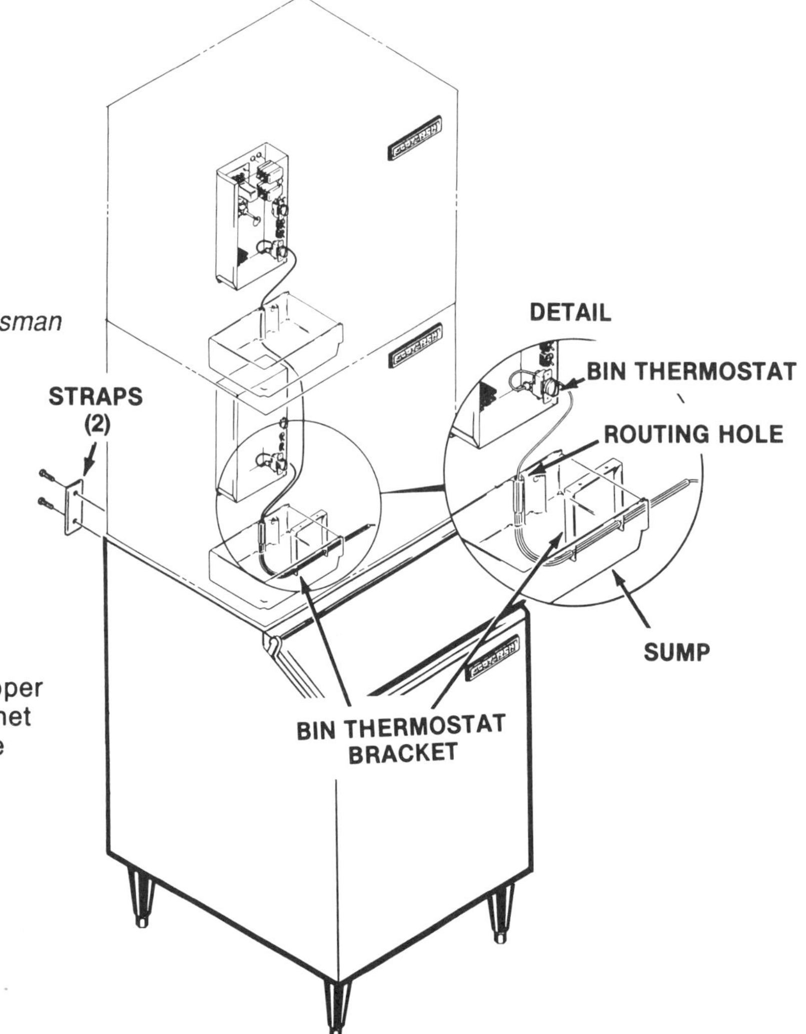

CM650 FOR THE ELECTRICIAN

CONFORM TO ALL APPLICABLE CODES

ELECTRICAL CONNECTIONS

SEE NAMEPLATE for current requirements to determine wire size to be used for electrical hookup. The cuber requires a solid chassis-to-chassis earth ground wire. See Wiring Diagram.

Be certain the cuber is connected to its own electrical circuit and individually fused. Voltage variation should not exceed ten percent of the nameplate rating, even under starting conditions. Low voltages can cause erratic operation and may be responsible for serious damage to the icemaker.

Electrical connections are made at the rear of the icemaker, inside the junction box.

All external wiring should conform to the national, state and local electrical code requirements. Usually an electrical permit and services of a licensed electrician will be required.

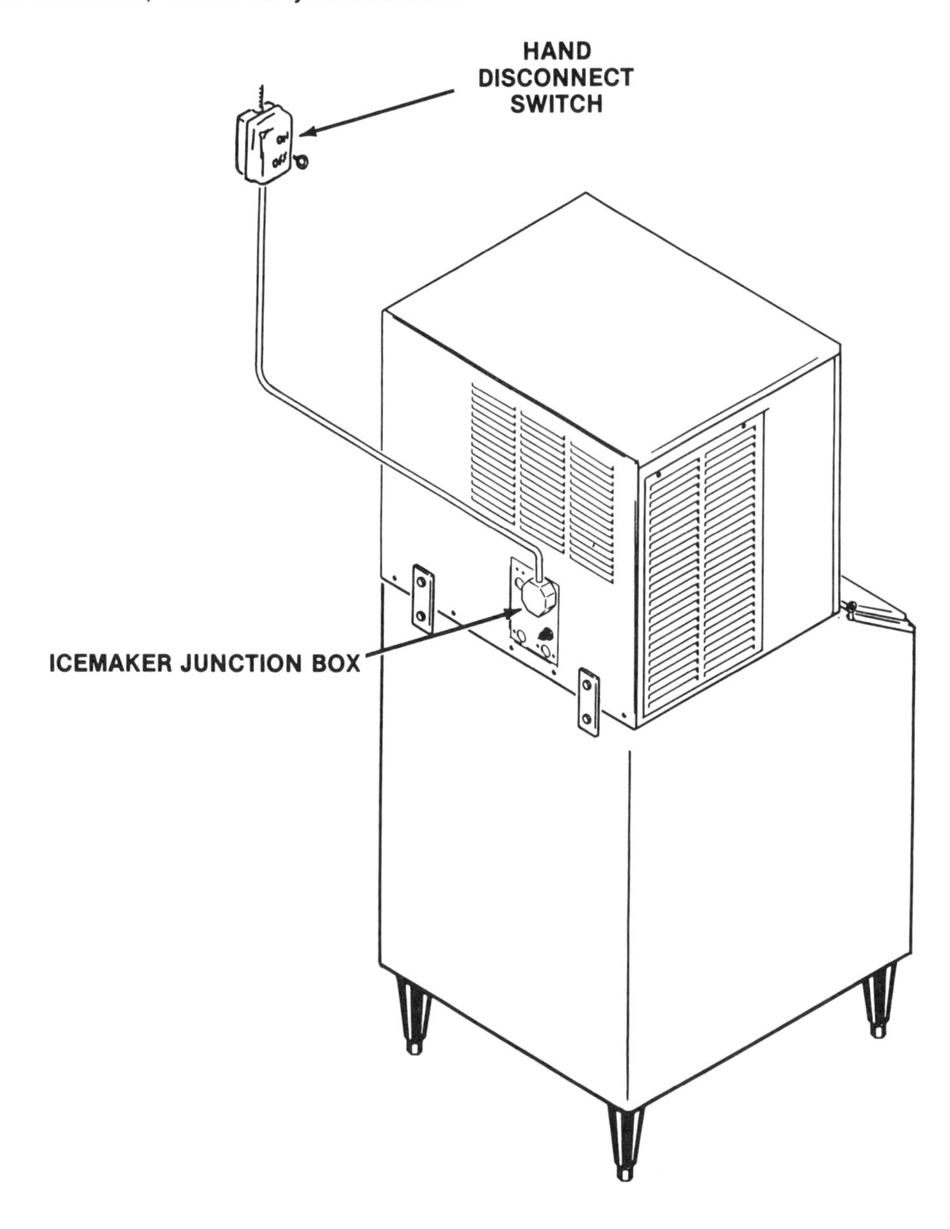

CM650 FOR THE PLUMBER

CONFORM TO ALL APPLICABLE CODES

WATER SUPPLY AND DRAIN CONNECTIONS

The recommended water supply line is a 3/8-inch O.D. copper tubing with a minimum operating pressure of 20 PSIG and a maximum of 120 PSIG. Connect to cold water supply line with standard plumbing fittings, with shutoff valve installed in an accessible place between the water supply and the cuber. In some cases a plumber will be required.

DRAIN CONNECTIONS: All drains are gravity type and must have a minimum of 1/4-inch fall per foot on horizontal runs. The drains to be installed to conform with the local plumbing code. Install a vertical open vent on drain line high point to ensure good draining. The ideal drain receptacle is a trapped and vented floor drain.

Recommended bin drain is 5/8-inch O.D. copper tubing and should be vented and run separately. Insulation for high humidity areas is recommended.

The ice machine sump drain is 3/4″ FPT. There must be a vent at this connection for proper sump drainage.

WATER-COOLED MODELS: On water-cooled models, a separate cold water supply inlet is required, to be connected to a second 3/8-inch female pipe thread (FPT) fitting at the rear of the cabinet. Additional drain lines are required to drain the water-cooled condenser, and they must not interconnect to any other of the units drains.

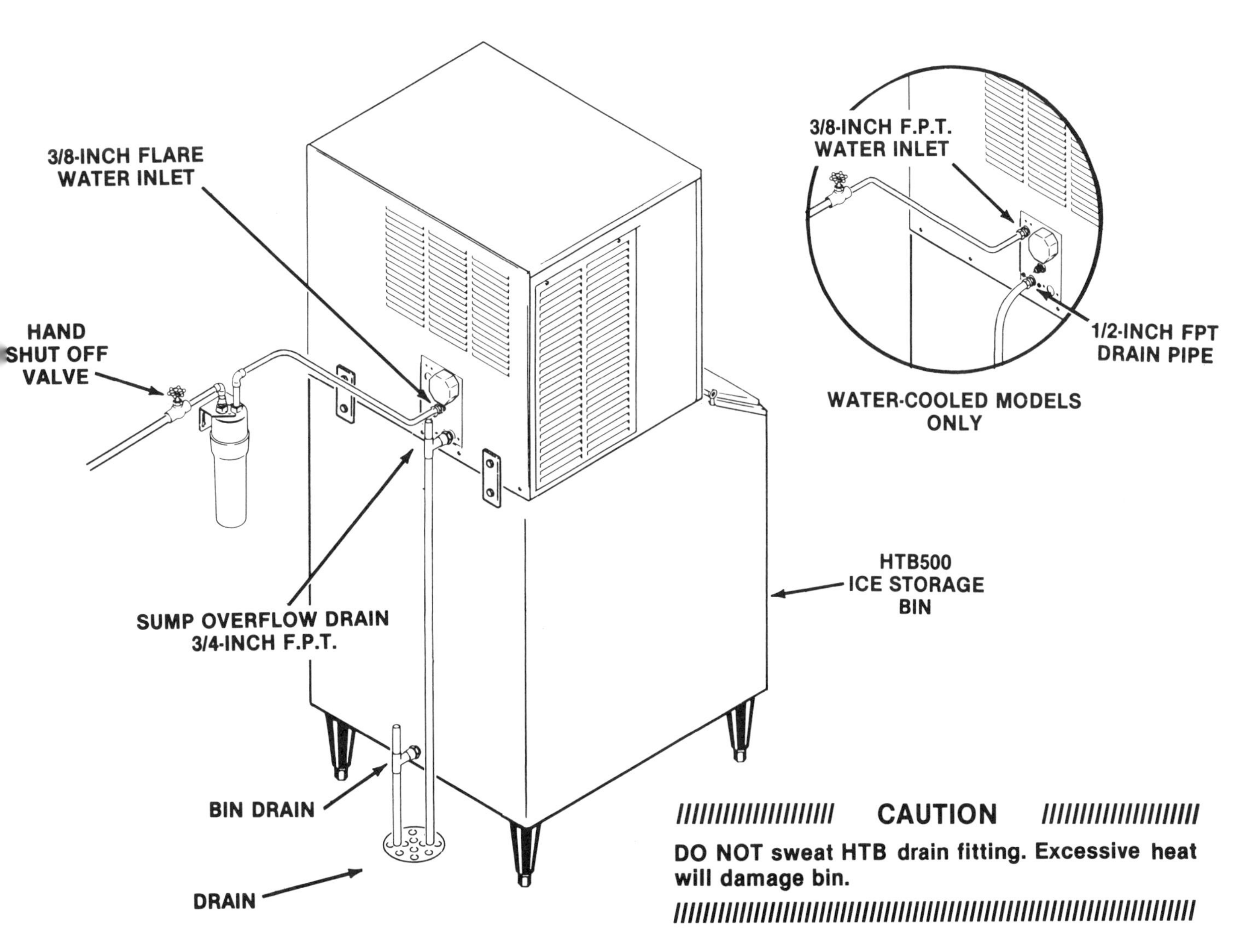

CAUTION

DO NOT sweat HTB drain fitting. Excessive heat will damage bin.

Water Supply and Drain Connection

CM650 FOR THE INSTALLER

FINAL CHECK LIST

1. Is the cabinet in a room where ambient temperatures are within the minimum and maximum temperatures specified?
2. Is there at least six inches clearance at both sides of the cabinet for proper air circulation? (Air-Cooled models)
3. Has water supply pressure been checked to insure a minimum of 20 PSIG and a maximum of 120 PSIG operating pressure?
4. Is the cabinet level?
5. Check that any shipping material has been removed from inside the cabinet.
6. Has the bin thermostat and bracket been properly installed?
7. Check that the drain troughs are properly secured to the bottom of the evaporator plates.
8. Have all electrical and piping connections been made?
9. Is the water supply line shut-off valve installed and electrical wiring properly connected?
10. Check all refrigerant lines and conduit lines, to guard against vibration or rubbing and possible failure.
11. Have the bin and cabinet been wiped clean?
12. Has the Manufacturer's Registration Card been properly filled out? Check for correct model and serial numbers from Serial nameplate, then mail the completed card to the SCOTSMAN factory.
13. Has the owner/user been given the Service Manual and instructed how to operate and maintain the icemaker?
14. Has the owner been given the name and telephone number of the authorized SCOTSMAN Service Agency serving him?

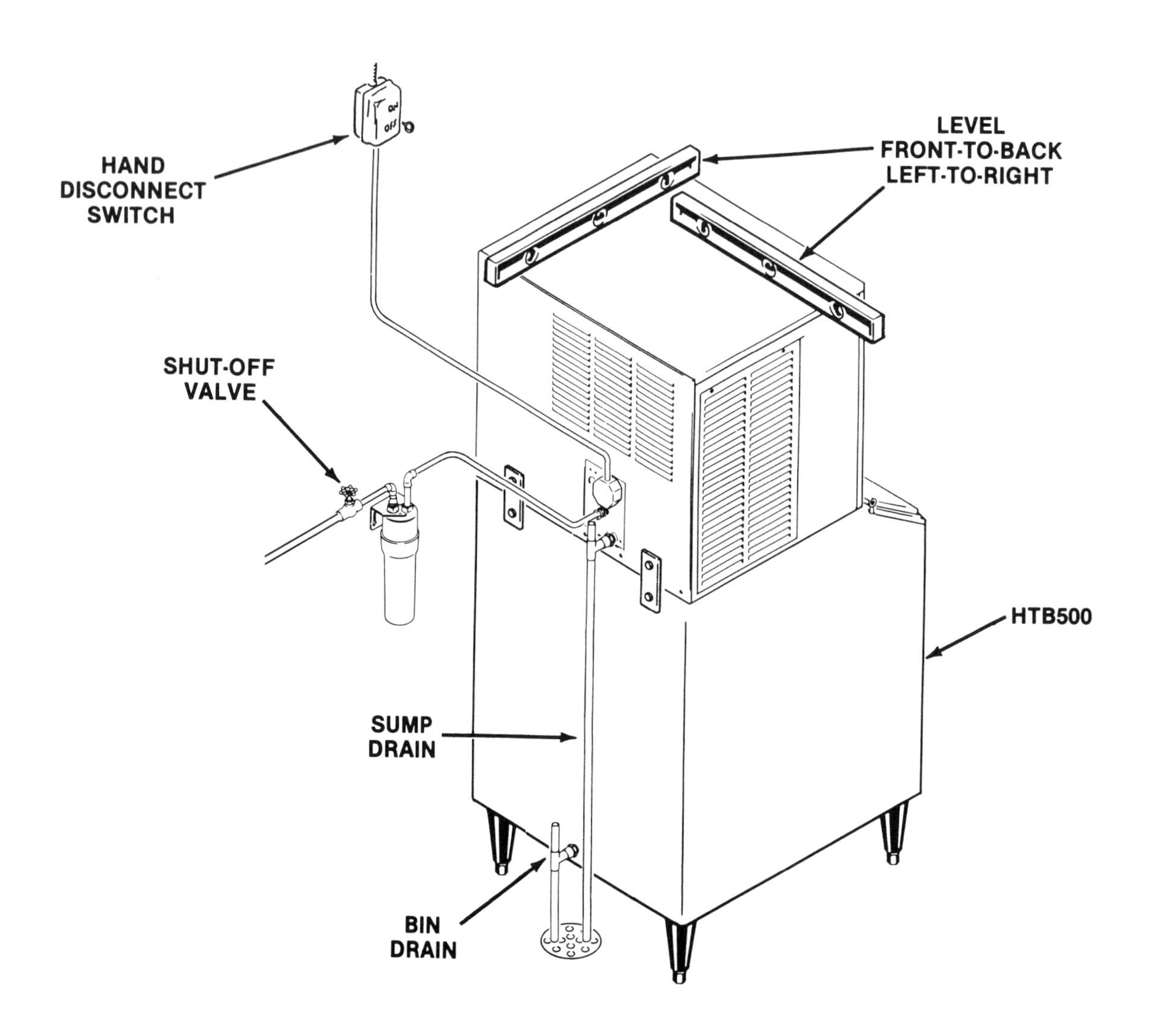

CM650 START-UP

START-UP

1. Remove front panel by pulling out to unsnap.
2. Remove two screws and the control box cover.
3. Remove four thumb screws and the evaporator cover.
4. Check that the two toggle switches, the MASTER ON-OFF toggle switch and the COMPRESSOR ON-OFF toggle switch are in the OFF position, on the control box.
5. OPEN the water supply line shutoff valve.
6. Inside the control box is the shaft of the timer and the switch assembly. Rotate the shaft of the timer and switch assembly clockwise to start the timer. The timer starts when the actuator arm on the microswitch drops off outer cam into cam slot. See Timer Cam Positions. An audible click can be heard, but in a noisy area, look at the cam and switch to observe the event.

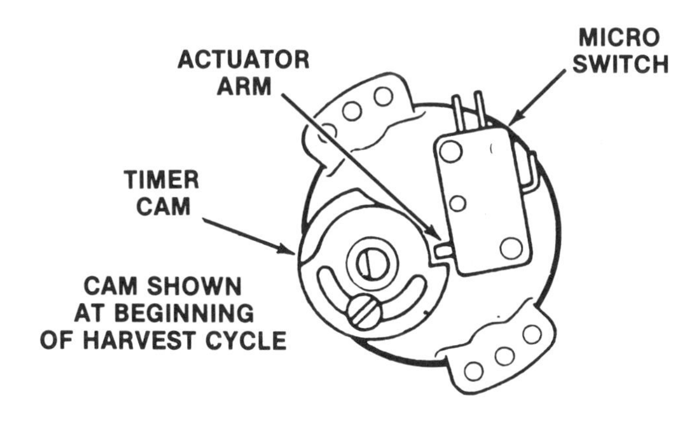

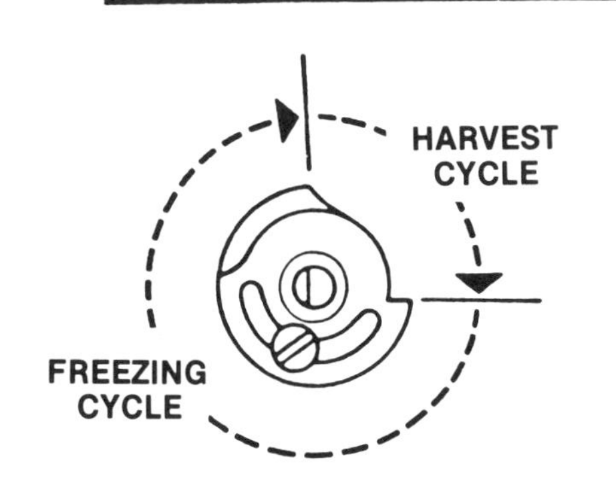

CAM SHOWN DIVIDED INTO TYPICAL FREEZING & HARVEST CYCLES

Timer Cam Positions

7. Move the master ON-OFF toggle switch, the top toggle switch, to the ON position.
8. Observe water fill cycle:

   Water pump operates and water inlet solenoid valve OPENS, incoming water flows from the valve through the tubing, the reservoir fills and excess water is overflowed through the stand pipe. This cycle will take about three minutes. Timer will close the water inlet solenoid valve and the water fill cycle is complete.

   Advancing the shaft of the timer and switch assembly into a new harvest cycle, restarts the timer and allows a check that: Water inlet solenoid valve OPENS and the reservoir overflows through the stand pipe. Water inlet valve CLOSES, stopping water overflow.

   The water pump is operating, as seen by water moving through the tygon tube, up to the water distributor at the top of each evaporator plate, where water is uniformly dispensed and cascades down both sides of the evaporator plate and drains back into the sump assembly for recirculation.
9. Check that the water cascades down over each cube mold and into the sump.
10. When the second cycle is completed, move the compressor ON-OFF toggle switch, to the ON position.
11. Check operation of the freezing cycle:

    The compressor is operating.

    The icemaking process begins; feeling the metal parts of the evaporator plate reveals cold temperature, very shortly ice begins to form. Tubing will become frosted at the top of the evaporator plate.

    Freezing time will range between 15 and 30 minutes. Longer time for temperatures above 70-degrees F. and shorter time required when temperatures are below 70-degrees F. Average complete cycle time is about 18 minutes.

12. Observe second and third cube harvest.

    Check size of SCOTSMAN CONTOUR CUBE

    Unlike other SCOTSMAN cubes which are made in a definite molded shape, contour cubes are produced in indentations and many shapes and sizes of contour cubes may be produced — only ONE size and shape combination is correct.

    An under-charged refrigeration system produces smaller cubes at the top of the evaporator plate and large cubes at the bottom. Charge system per NAMEPLATE specifications.

    Charge Refrigeration System with **REFRIGERANT 502 ONLY.**

    In areas where extreme problem water condition exists, filtering or purifying equipment is recommended. Contact SCOTSMAN ICE SYSTEMS, Service Department, Albert Lea, Minnesota 56007 for further details.

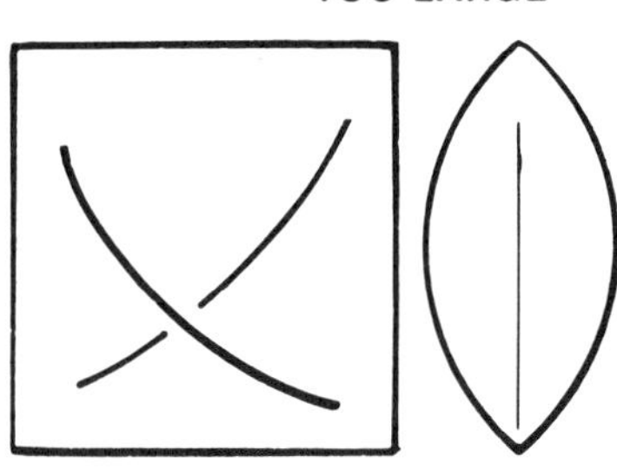

Too LARGE — may cause evaporator freeze ups. Adjust cube size control counter-clockwise to obtain smaller cubes

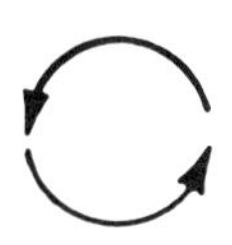

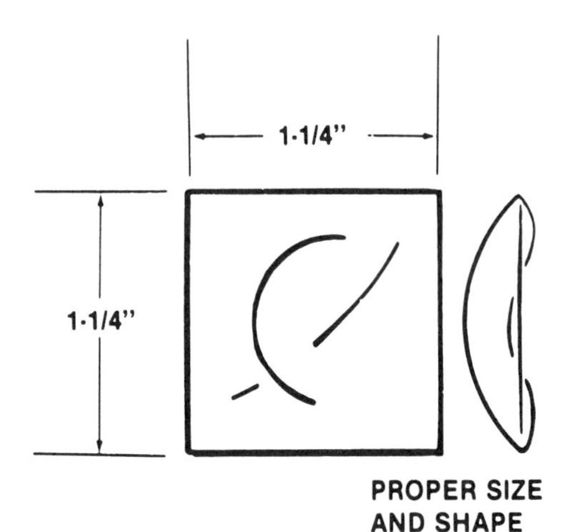

PROPER SIZE AND SHAPE of the contour cube. Icemaker operates at peak efficiency when a cube this size and shape is produced. A finely tuned system produces vertical strips of ice which easily break when they fall.

TOO SMALL

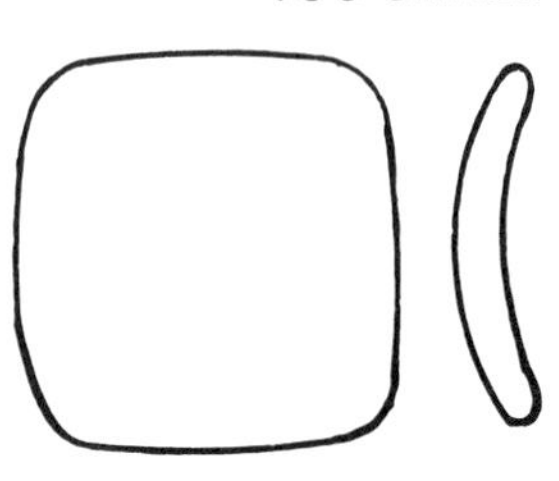

TOO SMALL. To obtain proper size cubes, adjust cube size control clockwise. May cause freeze up problems due to poor harvest.

Ice Cube Size & Shape

ADJUSTMENT OF CUBE SIZE

To produce SMALLER sized ice cubes:

Locate cube size control knob, in the front of the control box.

Rotate the adjusting screw one-eighth of a turn COUNTERCLOCKWISE to WARMER.

Observe size of the ice cubes in the next two ice cube harvests and adjust in one-eighth turn or less increments, until correct ice cube is achieved.

To produce LARGER ice cubes:

Locate cube size control on the front of the control box

Rotate the adjusting screw one-eighth of a turn CLOCKWISE to COLDER.

Observe size of ice cubes in the next two cube harvests and adjust in one-eighth turn or less increments, until correct ice cube size is achieved.

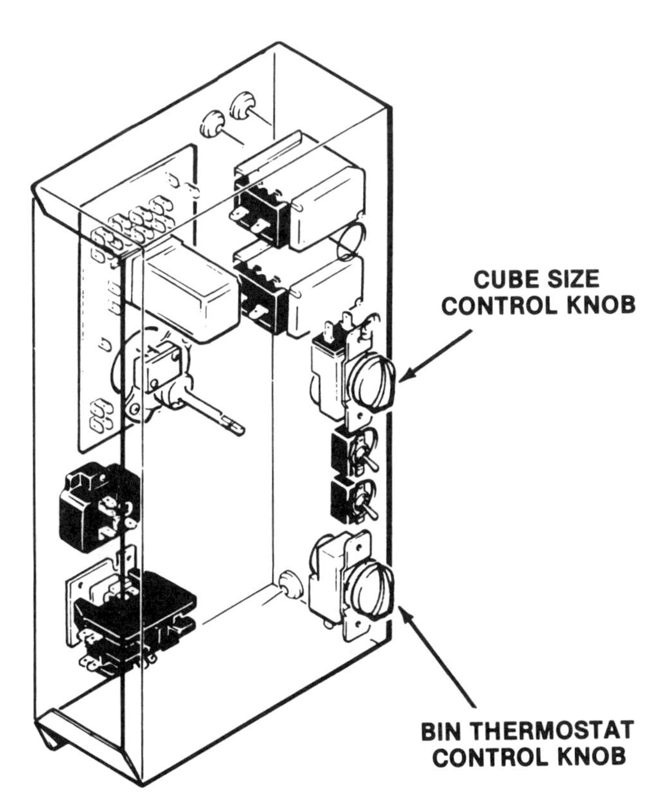

ROTATE ADJUSTMENT KNOB COUNTER-CLOCKWISE TO PRODUCE SMALLER CUBES.

ROTATE ADJUSTMENT KNOB CLOCKWISE TO PRODUCE LARGER CUBES.

BIN THERMOSTAT OPERATION

Check texture of ice cubes; when partially cloudy throughout, suggests icemaker is operating short of water; or, possibly an extreme problem water condition exists, where filtering or purifying equipment is recommended. Contact SCOTSMAN ICE SYSTEMS, Service Department, Albert Lea, Minnesota 56007 for further details. See SERVICE DIGNOSIS chart, for shortage of water symptoms and corrections.

13. With the icemaker in the harvest cycle, hold ice against the bin thermostat control bulb to test shutoff, which should cause the icemaker to shut OFF at the END OF THE HARVEST CYCLE.

    Within minutes after the ice is removed from the sensing bulb, the bulb will warm up and cause the icemaker to restart. This control is factory set and should not be reset until testing is performed.

14. Replace control box cover and all cabinet panels and screws.

15. Thoroughly explain to the owner/user the significant specifications of the icemaker, the start up and operation, going through the procedures in the operating instructions. Answer all questions about the icemaker by the owner; and inform the owner of the name and telephone number of the authorized SCOTSMAN Distributor, or service agency serving him.

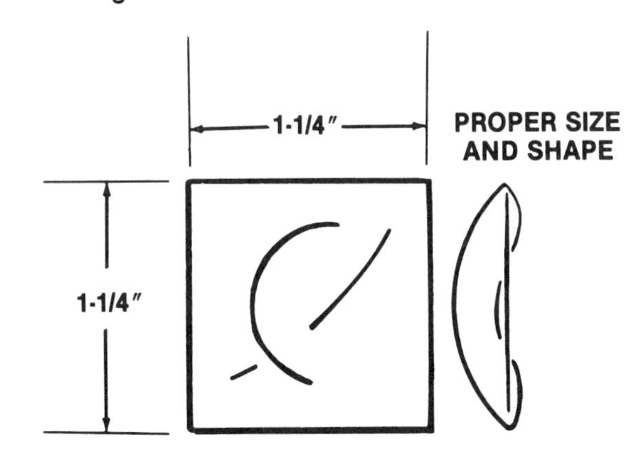

OPERATION

FREEZING CYCLE

Water from the sump assembly is pumped to the water distributor system, at the top of each evaporator plate. From the water distributor the water cascades by gravity over all cells of the plate and to the sump assembly below, for recirculation. At the beginning of the freezing cycle, the electrical circuit is completed to the compressor and the water pump. The water pump operates continuously, through both the freezing cycle and the harvest cycle.

In water-cooled models, water also flows through the condenser and out the drain. While in the condenser, water removes heat from the refrigerant and allows the refrigerant to condense from a gas to a liquid.

In the compressor, gaseous refrigerant is compressed and discharged into the condenser, as a high pressure, high temperature gas. The refrigerant is cooled by either air or water, and condenses into a high pressure, medium temperature liquid. This liquid refrigerant then passes through a small capillary tube, where the temperature and pressure of the liquid refrigerant are lowered and it next enters the evaporator plates. The refrigerant is warmed by the water cascading over the Evaporator plate and begins to boil off and become a gas. The refrigerant next travels through the accumulator and the heat exchange area of the suction line where any remaining liquid refrigerant boils off and returns to the compressor as a low pressure, low temperature gas, and the cycle starts again.

During the freezing cycle, the hot gas solenoid valve is CLOSED and the water inlet solenoid valve is CLOSED.

When the ice cubes are partially formed, the cube size control will sense the temperature at which it is preset to CLOSE. This will complete the electrical circuit to the timer. The timer then controls the remainder of the freezing cycle.

The timer will keep the icemaker operating in the freezing cycle for a selected length of time. This will give the ice cubes time to fully form. after that selected length of time, the timer will switch the icemaker into the harvest cycle, through the contacts of the timer assembly microswitch.

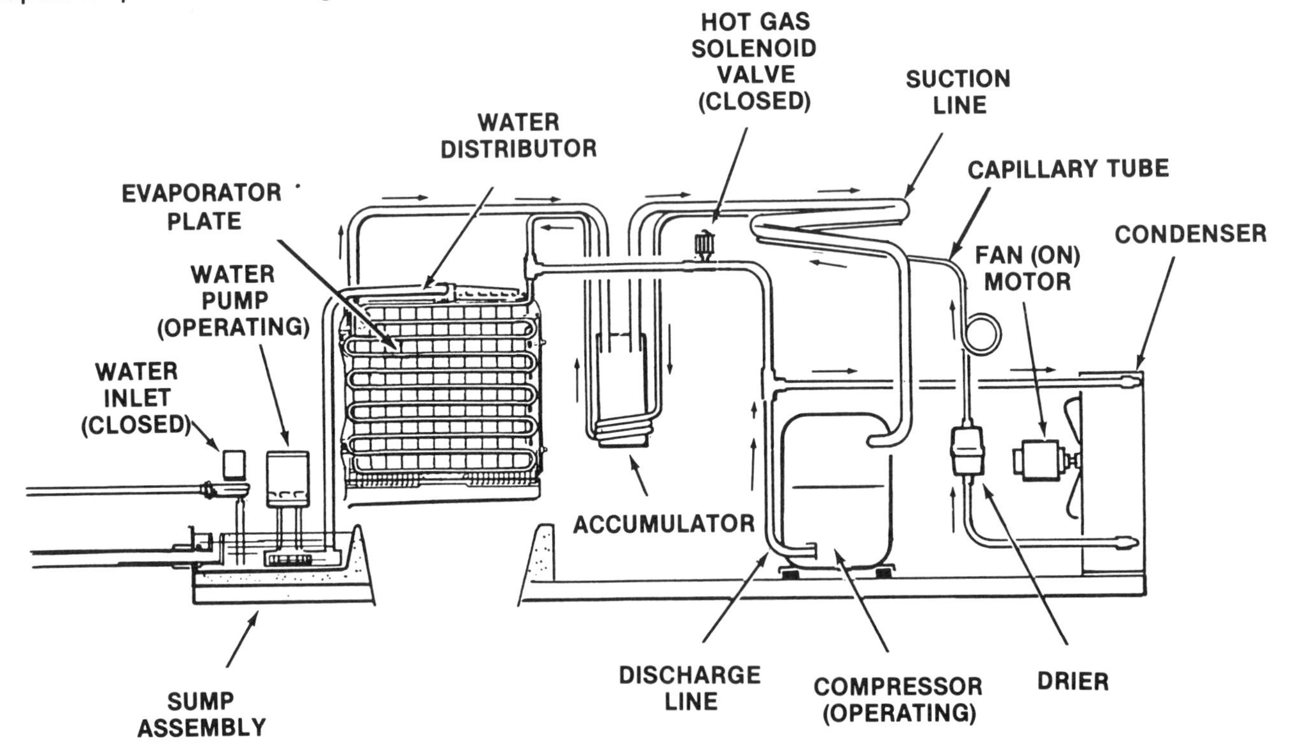

Freezing Cycle

HARVEST CYCLE - HOT GAS BYPASS

When the timer switches the icemaker into the harvest cycle, high pressure, high temperature gas refrigerant being discharged from the compressor is diverted from the condenser through the hot gas solenoid valve into each evaporator plate. During this cycle, the gaseous refrigerant bypasses the condenser and the capillary tube.

In the electrical circuit, both the compressor and the water pump are operating and the hot gas solenoid valve is energized and OPEN and the water inlet solenoid valve is OPEN.

The finished ice cubes are released from the sides of each evaporator plate, by the warming effect of the hot gas condensing in each evaporator plate and the water cascading over the ice cubes. The released ice cubes drop into the ice storage bin below.

At the end of the harvest cycle, the timer cam will push the actuator arm to the microswitch IN. If the bin thermostat is still CLOSED, a whole new cycle will begin. If the bin thermostat is OPEN, the icemaker will shut OFF, at this time.

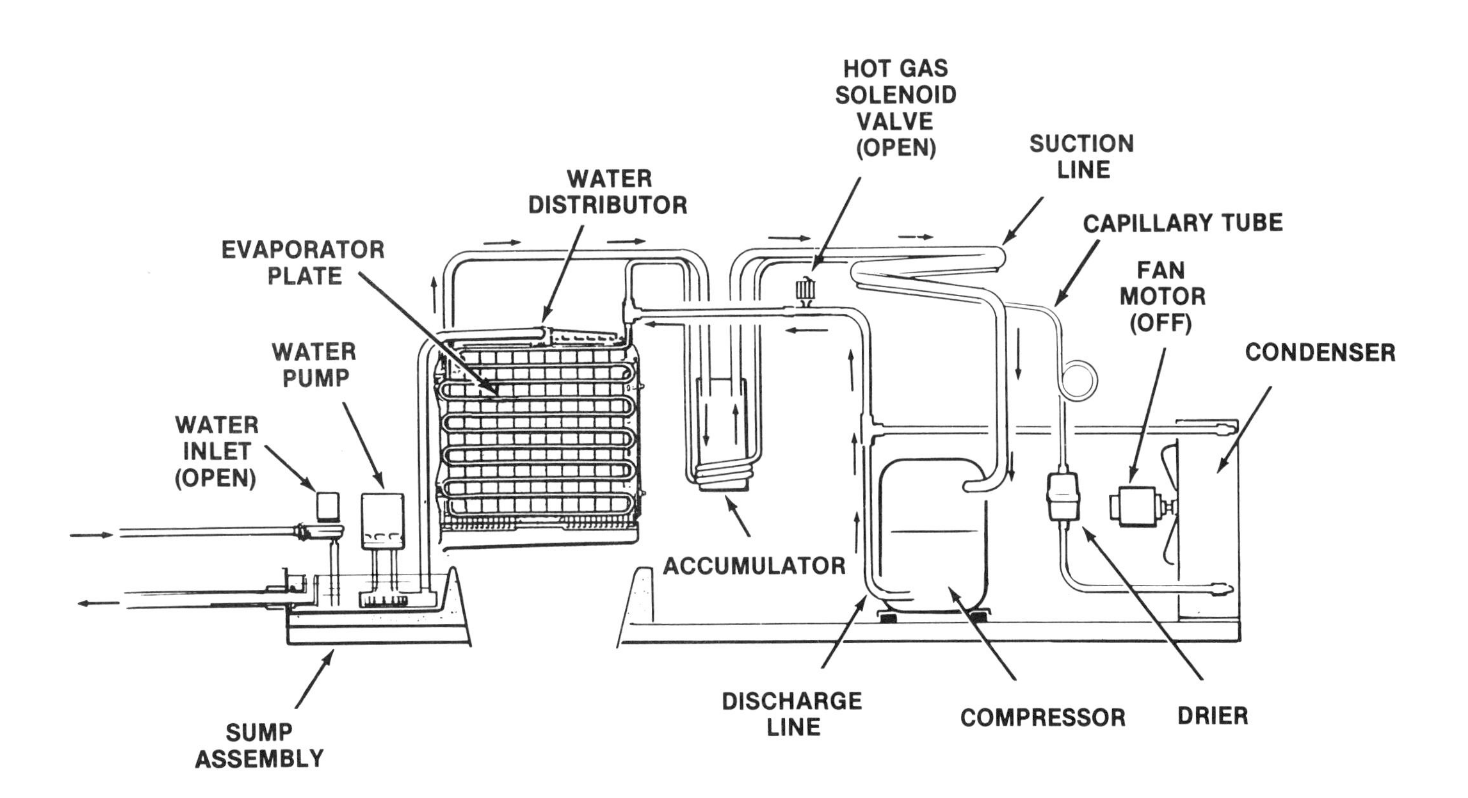

Harvest Cycle

CM650 OPERATION - ELECTRICAL SEQUENCE

The following charts illustrate which switches and which components are ON or OFF during a particular phase of the icemaking cycle.

Refer to the wiring diagram for a reference. Remember, the wiring diagram shows the unit as it is in the Timed Freeze Cycle.

BEGINNING FREEZE

ELECTRICAL COMPONENTS (LOADS)	ON	OFF
Compressor	X	
*Fan Motor (Air-cooled only)	X	
Hot Gas Valve		X
Inlet Water Valve		X
P.C. Board Relay Coil	X	
Timer		X
Water Pump	X	

SWITCHES	ON	OFF
Bin Thermostat	X	
Cube Size Thermostat		X
H.P. Fan Control	X	
Toggle - Master	X	
Toggle - Compressor	X	
H.P. Control	X	
Timer Micro Switch N.C.	X	
Timer Micro Switch N.O.	X	
Hi Temp Switch	X	

*The fan control will open and close with changes in discharge pressure, but will usually be closed during early freeze so the fan will usually be ON.

At the beginning of the freeze cycle the timer is not turning, but the icemaker is refrigerating the water starting to turn the water into ice.

CM650 OPERATION - ELECTRICAL SEQUENCE
TIMED FREEZE

ELECTRICAL COMPONENTS (LOADS)	ON	OFF
Compressor	X	
*Fan Motor (Air-cooled only)	X	X
Hot Gas Valve		X
Inlet Water Valve		X
P.C. Board Relay Coil	X	
Timer	X	
Water Pump	X	

SWITCHES	ON	OFF
Bin Thermostat	X	
Cube Size Thermostat	X	
H.P. Fan Control	X	X
Toggle Switch - Master	X	
Toggle Switch - Compressor	X	
H.P. Control	X	
Timer Micro Switch N.C.	X	
Timer Micro Switch N.O.	X	
Hi Temp Switch	X	

After the icemaker has cooled the water and formed some ice on the evaporator, the evaporator will have gotten cold enough to have the cube size control close. All this does is start and run the timer.

*The fan control will be opening and closing with the changes in discharge pressure, so the fan will be turning ON and OFF.

HARVEST

ELECTRICAL COMPONENTS (LOADS)	ON	OFF
Compressor	X	
*Fan Motor (Air-cooled only)		X
Hot Gas Valve	X	
Inlet Water Valve	X	
P.C. Board Relay Coil		X
Timer	X	
Water Pump	X	

SWITCHES	ON	OFF
Bin Thermostat	X	
Cube Size Thermostat		X
H.P. Fan Control		X
Toggle Switch - Master	X	
Toggle Switch - Compressor	X	
H.P. Control	X	
Timer Micro Switch N.C.	X	
Timer Micro Switch N.O.	X	
Hi Temp Switch	X	

The timer has now turned far enough so that the micro-switch plunger has dropped into the gap in the cam, this breaks the circuit to the relay in the P.C. Board - and that puts the machine into the Harvest cycle, where the hot gas valve and inlet water valve have opened to harvest the ice. When ice is on the bin control, it will open and at the end of the harvest cycle shut off the machine.

*The fan control will keep the fan OFF during the harvest cycle until the discharge pressure builds up to 190 PSIG, probably near the very end of the harvest cycle, then the fan will come ON.

CM650 SERVICE SPECIFICATIONS

In servicing a machine, it is often useful to compare that individual unit's operating characteristics to those of a normally operating machine. The numbers and facts listed on this page are for NEW, CLEAN machines.

Use these numbers as a guideline only.

COMPONENT

Timer: 1 revolution of the cam, in minutes	8	
Harvest Time, preset, in minutes	2-1/4	
Inlet Water Valve, water flow in g.p.m.	3/4	
	Close	Open
Cube Size Thermostat, Reverse Acting, Temperature Range	(+12^{0}F. to -6^{0}F.)	n/a
Bin Thermostat Temperature Range	38.5^0 - 43.5^0	33.5^0 - 38.5^0
High Pressure Safety Switch, Air Cooled PSIG	manual	450
High Pressure Safety Switch, Water Cooled, PSIG	manual	350
Fan Control	210	193

OPERATING CHARACTERISTICS

This model is air cooled or water cooled, and on air cooled models during the freezing cycle, the discharge pressure will slowly decline as the unit forms ice on the evaporators. At the same time the suction pressure is also dropping, reaching it's lowest point at the end of the freeze cycle. Compressor amps experience a similar decline. Water cooled models have a constant discharge pressure, usually about 220 psig.

During the harvest or defrost cycle, the suction pressure goes up dramatically with the opening of the hot gas by pass valve. The discharge pressure falls when this happens. On air cooled the fan may cycle during either cycle. Compressor amps reach their peak during harvest.

Freeze Cycle:

For example, on air cooled with conditions at 70^{0}F. air, and 50^{0}F. water, the discharge pressure will decline from about 250 at the begining to 200 by the end. Suction pressure at the end of freeze will be about 23 PSIG.

Freeze cycle time will be about 12-13 minutes.

Single phase total amps 5 minutes into the freeze cycle will be about 7.

At 90^0/70^0, the discharge pressures will go from about 350 down to 220. Suction pressure will be about 33 PSIG at the end of the 16-17 minute freeze cycle.

Harvest Cycle:

Discharge pressure at 70^0/50^0 will be about 140-150, and suction pressure will be around 95-100 PSIG.

At 90^0/70^0, discharge pressure will be about 160. Suction pressure will be about 110-120 PSIG.

The ice per cycle is 5.5 to 6 pounds.

The values listed are representative of values seen at a wide range of air and water temperatures and are for a normal cube size. When comparing these figures to field data, allow a variation from each end of the range given.

After servicing refrigeration system, always torque the access valve caps to 60-75 inch pounds.

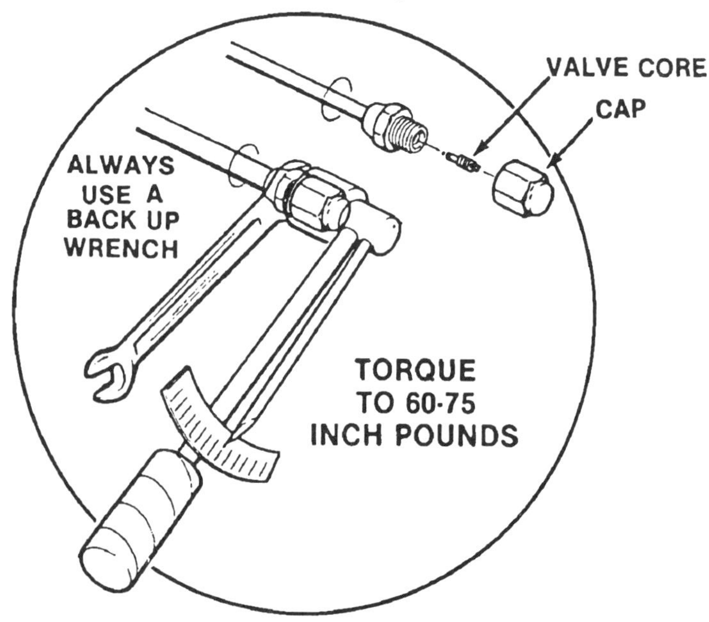

CM650 MAINTENANCE & CLEANING INSTRUCTIONS

A Scotsman Ice System represents a sizable investment of time and money in any company's business. In order to receive the best return for that investment, it MUST receive periodic maintenance.

It is the USER'S RESPONSIBILITY to see that the unit is properly maintained. It is always preferable, and less costly in the long run, to avoid possible down time by keeping it clean; adjusting it as needed; and by replacing worn parts before they can cause failure. The following is a list of recommended maintenance that will help keep the machine running with a minimum of problems.

Maintenance and Cleaning should be scheduled at a minimum of twice per year.

ICEMAKER

THE FOLLOWING MAINTENANCE SHOULD BE SCHEDULED AT LEAST TWO TIMES PER YEAR ON THIS ICEMAKER. CALL YOUR AUTHORIZED SCOTSMAN SERVICE AGENCY.

1. Check and clean or service any optional water treatment devices, if any.
2. Clean the water strainer.
3. On air cooled models, clean the air cooled condenser. Do not use a wire brush. Vacuum or blow out any dust in the fins of the condenser: the condenser is only clean when light can be seen through the fins.
4. Check that the cabinet is level in the side to side and front to back directions.
5. Clean the water system, evaporator plates and sump assembly, using a solution of Scotsman Ice Machine Cleaner. Refer to CLEANING - Icemaker.

   Cleaning requirements vary according to local water conditions and to how much the machine runs. Continuous check of the clarity and shape of the ice cubes, with a visual inspection of the water system parts, evaporator parts and the reservoir will indicate if more frequent cleaning is needed.
6. Check and tighten all bolts and screws.
7. Check for water leaks and make corrections.
8. Check that the bin thermostat operates correctly: holding ice on the thermostat control tube in the bin should cause the icemaker to shut off at the end of the harvest cycle. After the ice is removed, the icemaker should restart within a few minutes.
9. Check cube size, adjust if required.
10. Check harvest time, adjust if required.

ICE STORAGE BIN

The interior liner of the bin is in contact with a **food** product: **ice.** The storage bin must be **cleaned** regularly to maintain a **sanitary** environment. Once a week cleaning with soap and water, a hot water rinse and an air dry is a basic procedure.

Every 90 days, the liner should be sanitized with a commercial ice machine sanitizer, according to the directions of the sanitizer, or with a solution of household bleach and water:

1. Mix the bleach and water using the ratio of two ounces of bleach to two gallons of water.
2. Wipe all interior surfaces of the ice storage bin with the bleach and water.
3. Allow to air dry.

CLEANING: ICEMAKER

1. Remove front panel.
2. Switch the compressor switch to OFF. Switch the master switch to OFF.
3. Remove 4 thumbscrews and the front liner.
4. Twist forward and remove the front drain trough, and switch the master switch to ON, pumping the water from the reservoir into the bin. Replace the drain trough.
5. Mix 8 ounces of Scotsman Ice Machine Cleaner with 1 gallon of warm (95^0F. - 115^0F.) water and pour into the reservoir until full.

WARNING

Scotsman Ice Machine Cleaner contains Phosphoric and Hydroxyacetic acids. These compounds are corrosive and may cause burns. If swallowed, DO NOT induce vomiting. Give large amounts of water or milk. Call Physician Immediately. In case of external contact, flush with water. KEEP OUT OF THE REACH OF CHILDREN.

6. Let the unit operate for 30 minutes with the compressor off, then switch the unit off.

7. Remove the front drain trough again.

8. Switch on the master switch to pump water from the sump into the bin. Continue to add fresh water to flush residual cleaner from the system. Switch the master switch off.

9. Wash the plastic and stainless liners of the freezer section with a solution of household bleach (1 ounce of bleach to 1 gallon of water) and warm (95^0F.-115^0F.) water. Allow to air dry.

10. Replace the drain trough and front liner.

11. Switch the master and compressor switches back on.

12. Replace the front panel

13. Check the next batch of cubes to make sure all of the acid taste is gone.

////////////////////////////////////CAUTION/////////////////////////////

DO NOT use ice cubes produced from the cleaning solution. Be sure none remain in the bin.

//

14. Pour hot water into the storage bin to melt the cubes and also clean out the bin drain.

15. The unit is now ready for continued automatic operation.

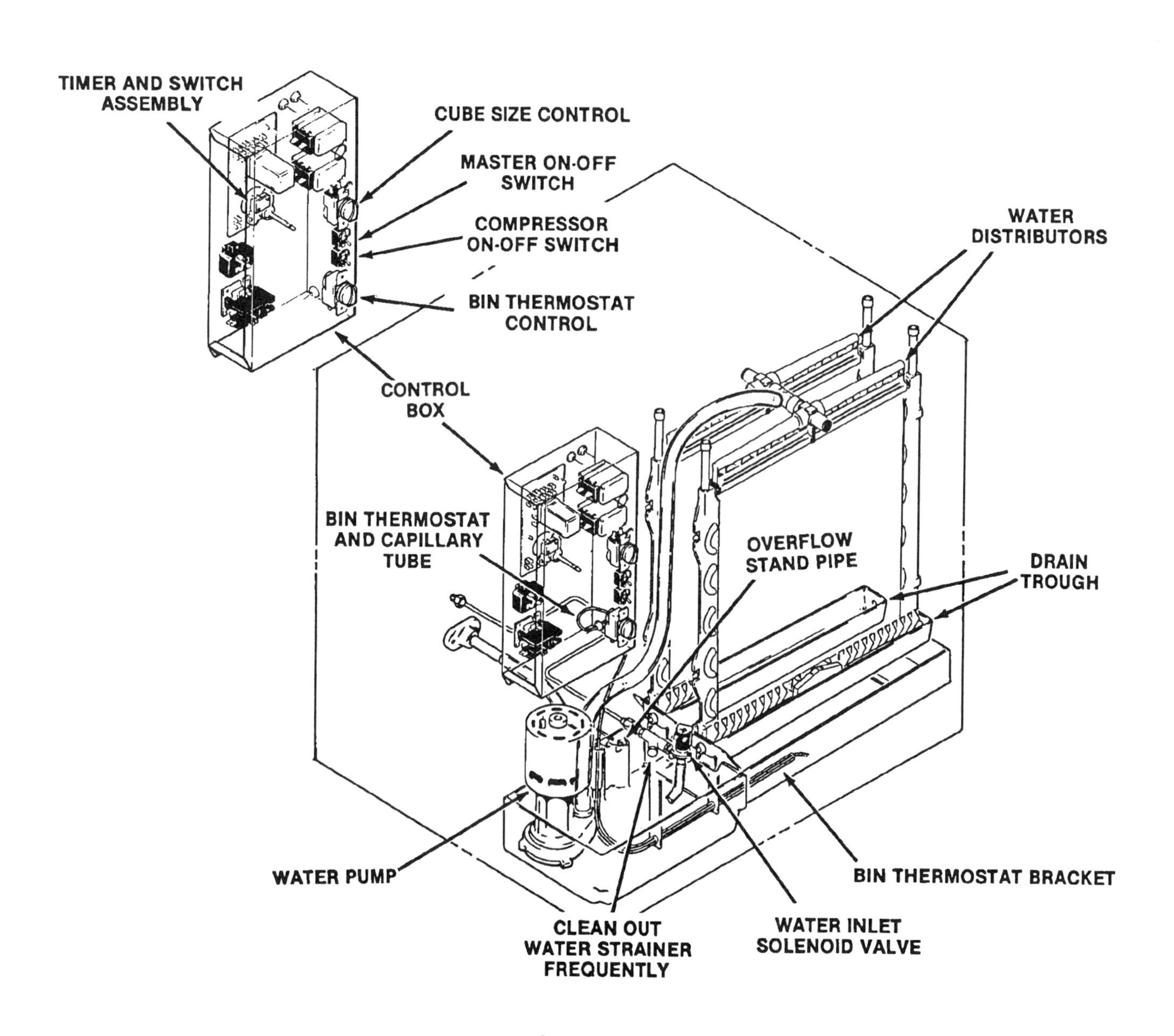

CM650 COMPONENT DESCRIPTION

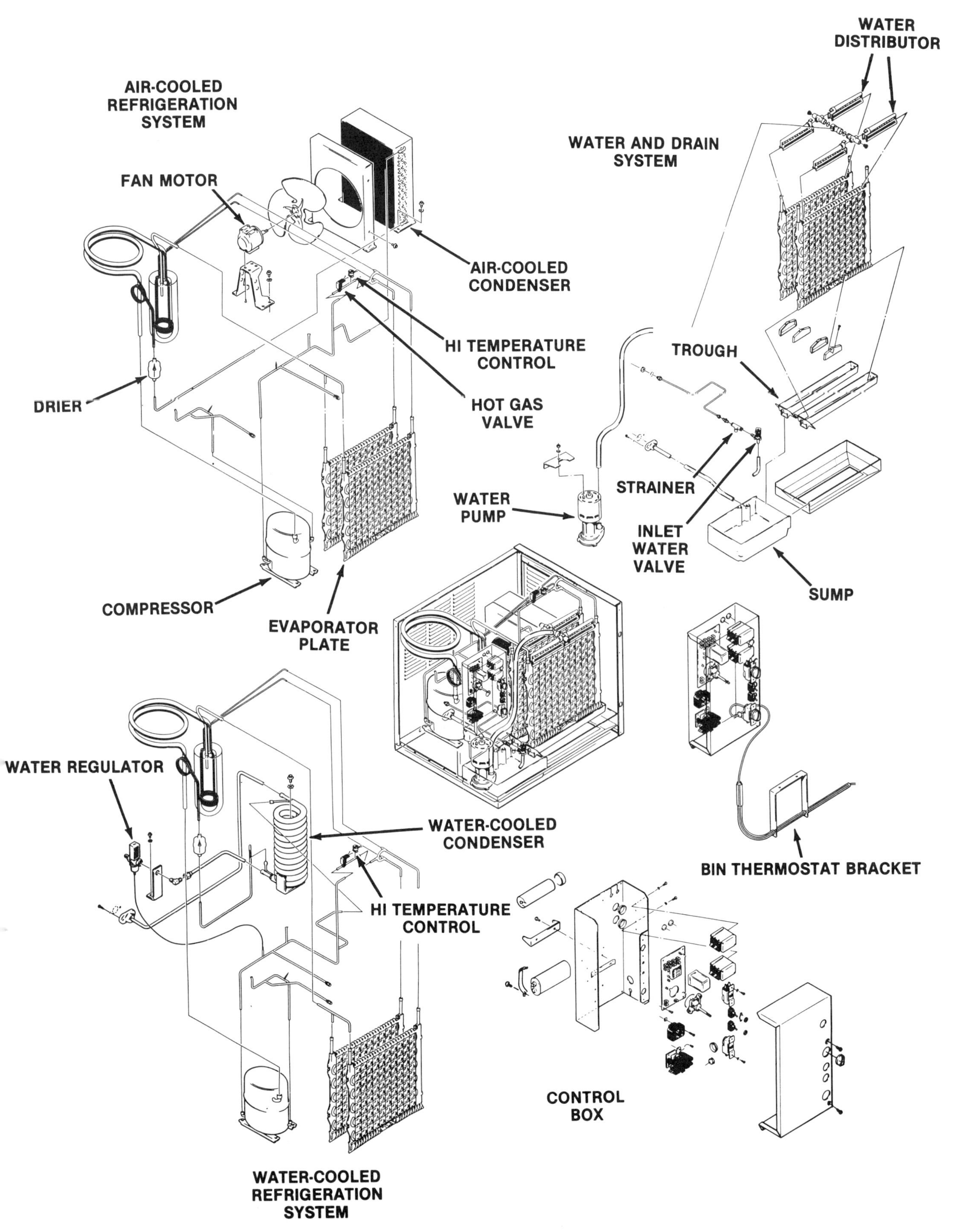

CM650 COMPONENT DESCRIPTION

BIN THERMOSTAT CONTROL

The bin thermostat is located on the front edge of the control box with an accessible knob on the front. The sensing capillary tube is routed from the control out the side of the control box down and across the front edge of the sump and down into the bin thermostat bracket. The bin thermostat control functions to automatically shut OFF the icemaker, when the ice storage bin is filled and ice contacts the capillary tube. It also signals the RESTART of the icemaker, when the capillary tube starts to warm up, after ice has been removed from the bin.

Bin thermostat control adjustment should ONLY be performed on icemakers installed in extreme warm or cold locations and adjust only in increments of one eighth turn at a time.

COMPRESSOR CONTACTOR

The compressor contactor functions to carry the compressor line current. The contactor is wired so any control in the pilot circuit, such as the bin thermostat, and high pressure controls, etc., will cause the contactor holding coil to be de-energized, when the control contact OPENS, thereby breaking the circuit to the compressor.

CUBE SIZE CONTROL

The temperature sensing cube size control affects the length of the freezing cycle prior to initiating the finishing timer. The cube size control closes its contacts when the evaporator reaches a preset temperature, starting the finishing timer. A variation in either ambient air or incoming water temperature will affect the efficiency of the refrigeration system. This will vary the length of time it takes the evaporator to reach the temperature at which the cube size control is preset to CLOSE; which, in turn, will affect the overall cycle time.

See *Cube Size Adjustment* BEFORE attempting to adjust the control.

RELAY

The multi-function, three pole, double-throw, plug-in relay is installed directly into a receptacle on the printed circuit board in the control box. The relay functions in part to by-pass the bin thermostat control to prevent the icemaker from shutting OFF, when a filled-bin condition occurs during the freezing cycle. The by-pass action serves to ensure full-sized ice cubes with each harvest cycle; and, to prevent short cycling on the bin thermostat control.

TIMER — Timer & Switch Assembly

The function of the timer begins when activated by the cube size control. The outer surface, or large diameter lobe of the timer cam, determines the timer cycle for finish freezing of the ice cubes, while the inner surface, or small diameter lobe, determines the time cycle for the harvest sequence. All electrical circuitry is connected through the printed circuit board and the timer and shunted by the single-pole, double-throw microswitch to either the freezing cycle or the harvest cycle. The microswitch is actuated by a cam assembly directly connected to the timer motor. The timer cam can be adjusted to vary the defrost line, as required. One complete rotation of the cam will take eight minutes. Harvest is preset at two and one-fourth minutes.

HIGH PRESSURE SAFETY CONTROL

This is a manual reset control that shuts down the icemaker, should the discharge pressure ever reach 450 PSIG on air-cooled and 350 PSIG on water-cooled.

CM650 COMPONENT DESCRIPTION

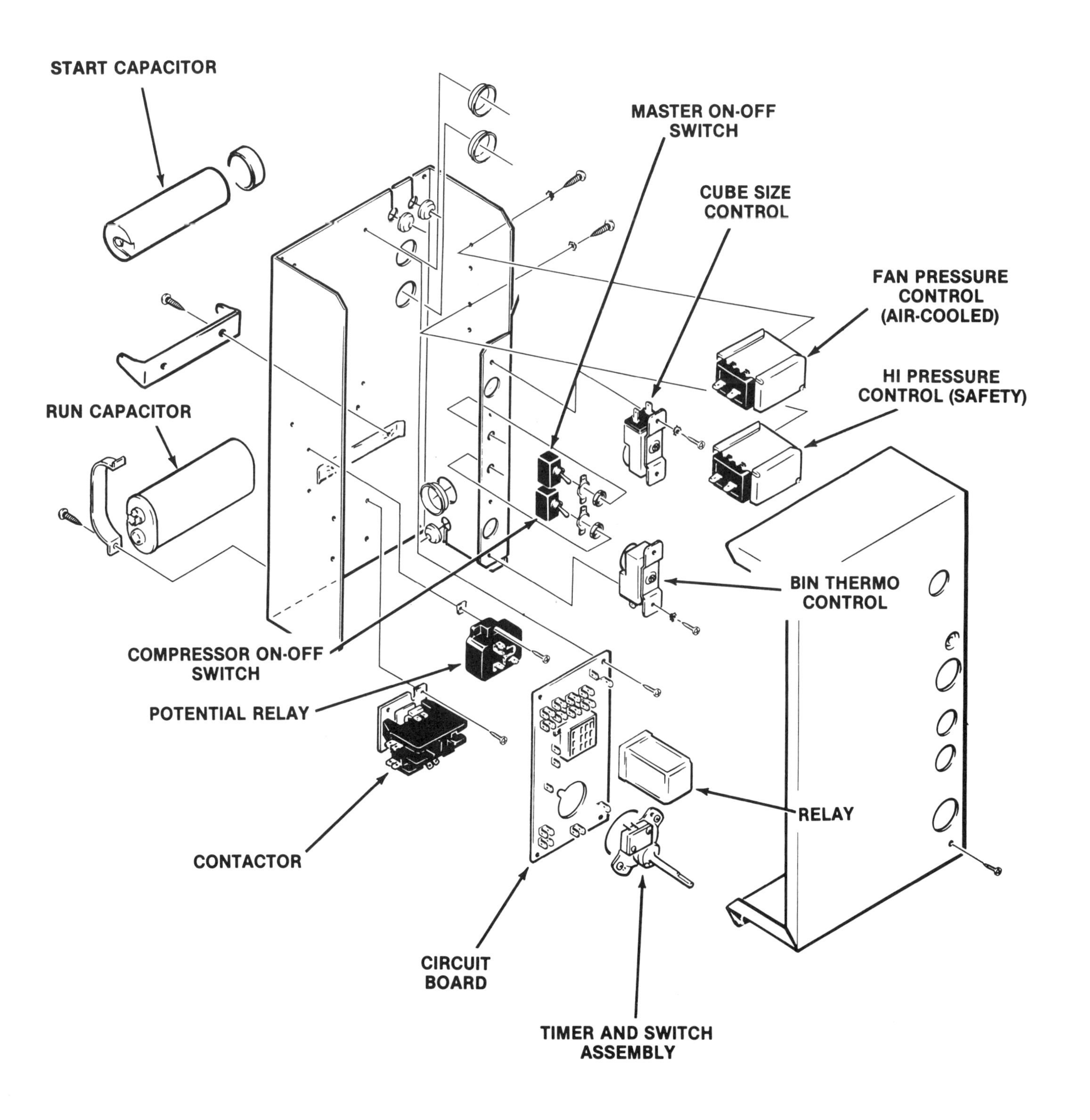

CM650 COMPONENT DESCRIPTION

WATER INLET SOLENOID VALVE

The water inlet solenoid valve functions to fill the sump assembly with water and overflow out the overflow standpipe located at the bottom of the sump. This action fills and rinses the sump during each harvest cycle. The flow rate is .75 g.p.m.

WATER REGULATOR VALVE — Water-Cooled Model

The water regulator valve functions to maintain a constant compressor head pressure, by regulating the amount of inlet water flow through the condenser on water-cooled models. The valve operates through the refrigerant system high side pressure. Rotating the adjusting screw, located on top of the valve, can INCREASE or DECREASE the water flow through the water-cooled condenser, which in turn, will DECREASE or INCREASE the compressor operating head pressure. It is to be set at 220 PSIG.

When installing a replacement water regulator valve, be sure the replacement valve is installed with the arrow positioned in the direction of the water flow.

WATER DISTRIBUTION SYSTEM

The water distribution system functions to evenly supply water to all cells of the evaporator plates. The water pump pumps water from the sump up the vertical tygon tube to the tee. From there water is channeled through the water manifold to the water distributors, above each evaporator plate, and from six holes within each distributor, water flows to the cells of each side of the evaporator plates. Gravity flow returns the unfrozen excess portion of water to the sump reservoir for recirculation.

HOT GAS SOLENOID VALVE

The hot gas solenoid valve functions only during the harvest cycle, to divert the hot discharge gas from the compressor, bypassing the condenser and capillary tube, for direct flow in the evaporator plates to release ice cubes from the ice cube molds. The hot gas solenoid valve is comprised of two parts, the body & plunger and the coil & frame assemblies. Installed in the discharge line of the compressor, the energized solenoid coil lifts the valve stem within the valve body, to cause the hot discharge gas to be diverted when the finishing timer has advanced to the start of the harvest cycle.

FAN PRESSURE CONTROL — Air-Cooled Models only. In both freeze and harvest cycles, the fan pressure control functions to maintain a minimum discharge pressure by cycling the fan on and off. The approximate C.I. is 210 PSIG and C.O. is 193 PSIG.

STRAINER

A water strainer is located in the potable inlet water line before the water inlet solenoid valve. Clean the water strainer frequently.

CM650 COMPONENT DESCRIPTION

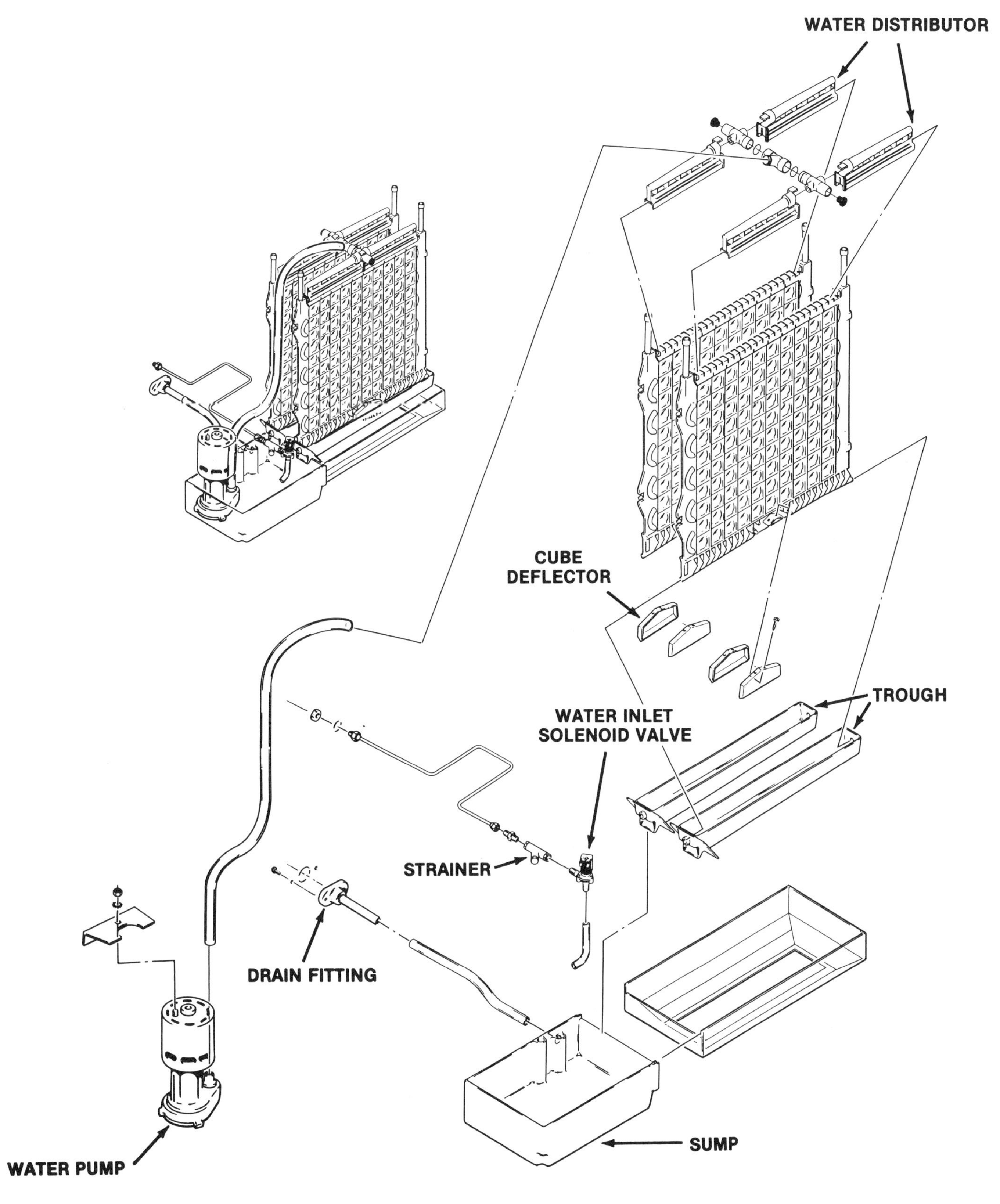

CLEAN OUT
WATER STRAINER
FREQUENTLY

CM650 SERVICE DIAGNOSIS

The service diagnosis section is for use in aiding the serviceman in diagnosing a particular problem for pin-pointing the area in which the problem lies, thus an ever available reference for proper corrective action.

The following chart lists corrective actions for the causes of known symptoms of certain problems that can occur in the icemaking-refrigeration system.

ICEMAKING - REFRIGERATION SYSTEM

SYMPTOM	POSSIBLE CAUSE	CORRECTION
Irregular size cubes some cloudy.	Some distributor holes plugged.	Clean distributor holes.
		Clean water sump.
	Shortage of water.	See shortage of water CORRECTION.
	Unit not level.	Level cabinet, as required.
Cubes too large.	Dirty air-cooled condenser.	Clean condenser.
	Cube size control set too cold.	Rotate cube size control dial toward WARMER.
Cubes too small.	Cube size control set too warm.	Rotate cube size control dial toward COLDER.
	Partially restricted capillary tube. (Kinked, pinched off, etc.)	Replace heat exchange assembly.
	Moisture in refrigeration system.	Blow refrigerant charge; replace drier; evacuate system; add proper refrigerant charge.
	Shortage of water.	See shortage of water CORRECTION.
Cloudy cubes.	Shortage of water.	See shortage of water SYMPTOM.
	Dirty water supply.	Check water quality and install water purification system.
	Accumulated impurities.	Use SCOTSMAN Ice Machine Cleaner and clean icemaker.
Shortage of water.	Short harvest cycle.	Adjust cam of timer and switch assembly.
	Water leak in sump area or off evaporator plate.	Locate Leak and correct condition.
	Partial restriction in water inlet strainer or inlet water valve.	Clean or replace strainer or valve.
	Water pressure too low.	Check for 20 PSI flowing water. Restore pressure.

CM650 SERVICE DIAGNOSIS

SYMPTOM	POSSIBLE CAUSE	CORRECTION
Decreased ice capacity.	High head pressure, result of dirty condenser or faulty fan motor or fan pressure switch or on water-cooled - not enough water through condenser.	Clean condenser. Repair or replace fan motor or switch. Check with regulator valve.
	Non-condensable gas in the system	Purge the system, evacuate and recharge per nameplate requirements.
	Poor air circulation or extreme hot location.	Relocate the cabinet; or provide ventilation.
	Overcharge of refrigerant.	Evacuate and recharge per nameplate.
	Hot gas solenoid valve leaking.	Replace valve.
	Partially restricted capillary tube.	See cubes too small CORRECTION.
	Defective compressor.	Replace compressor.
Poor harvests. Icemaker does not harvest.	Too short defrost time.	Check and adjust harvest cycle.
	Restriction in water inlet line.	Check strainer and inlet water valve.
	Hot gas solenoid does not open. Binds or burned out.	Replace solenoid, coil or valve as applicable.
	Undercharge of refrigerant.	Charge to nameplate requirements.
	Water pressure too low.	Check for 20 PSI flowing water. Restore pressure.
	Cube size too large or too small.	Adjust size with cube size control.
	Water-cooled models: water Regulator valve leaks water thru during harvest.	Replace water regulator valve.
Compressor cycles intermittently.	Low voltage.	Check for circuit overload. Check building supply voltage, if low, contact power company.
	Dirty condenser.	Clean condenser with vacuum cleaner or brush. DO NOT USE A WIRE BRUSH.
	Air circulation blocked.	Locate cabinet with adequate air space for proper air flow.
	Defective fan motor.	Replace fan motor.
	Non-condensable gases in system.	Purge the system and recharge per nameplate requirements.
Icemaker will not operate.	Blown fuse in line.	Replace fuse and check for cause.
	Master switch in OFF position.	Set switch to ON position.
	Faulty master switch.	Replace switch.
	Timer contacts open.	Replace timer microswitch.
	Faulty bin thermo.	Replace.
	Off on high head pressure safety switch.	Water-Cooled - water interruption correct. Air-Cooled - condenser not getting air.

ADJUSTMENT OF THE BIN THERMOSTAT CONTROL

The control for the bin thermostat is the temperature control, located in the front of the control box.

The bin thermostat control requires adjustment only if the icemaker shuts off prematurely. Turn adjusting knob, in 1/8 turn or less increments, in the appropriate direction until the icemaker shuts OFF.

Remove ice from the capillary bulb; then, place warm hand on the capillary bulb to restart the icemaker.

Place handful of ice against the capillary bulb and observe that the icemaker should shut OFF. (Only at the end of harvest).

CAUTION

The adjusting screws on the temperature control device have very sensitive response to adjustment. DO NOT attempt to adjust the screw until after thoroughly reading and understanding the instructions and illustrations. Over-adjusting or erratic guessing, can foul the instrument and cause ultimate delay and part replacement, WHICH COULD HAVE BEEN PREVENTED.

ALTITUDE ADJUSTMENT
over 2,000 feet above sea level only

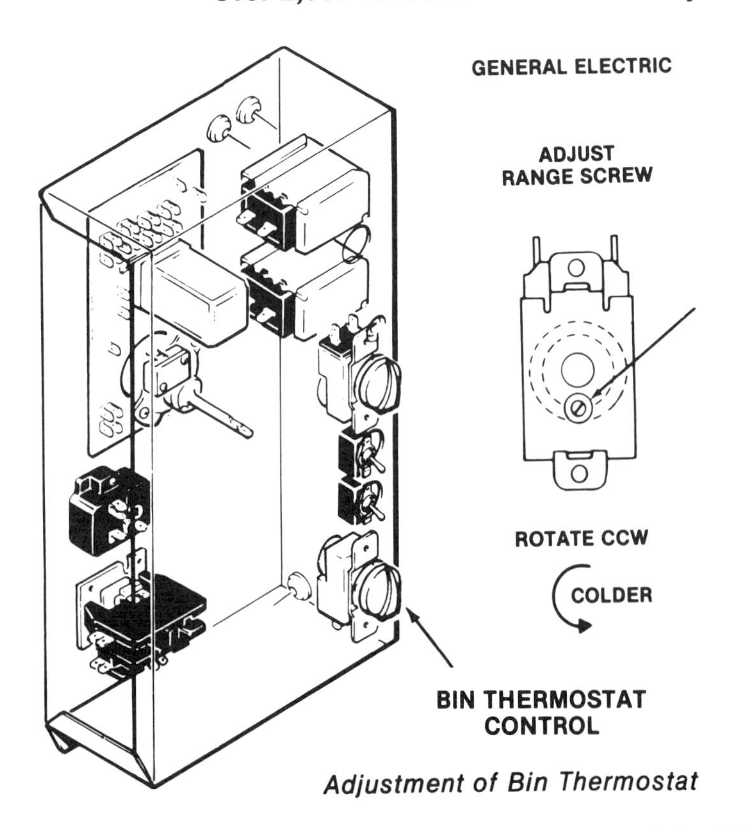

Adjustment of Bin Thermostat

ADJUSTMENT OF THE WATER REGULATOR ASSEMBLY — WATER-COOLED MODELS

The correct compressor head pressure on water-cooled models is 220 PSIG. Adjusting the water regulator valve increases or decreases the rate of flow of water, through the water-cooled condenser; which increases or decreases the affected temperature/pressure of the compressor head pressure, INCREASED water flow, results in DECREASED or LOWER head pressure; while, DECREASED water flow, results in INCREASED or HIGHER head pressure.

To adjust the water regulator assembly:

To INCREASE the head pressure: Rotate the adjusting screw COUNTERCLOCKWISE.

To DECREASE the head pressure: Rotate the adjusting screw CLOCKWISE.

Check change in compressor head pressure, and repeat adjustment as necessary, to achieve desired operating head pressure.

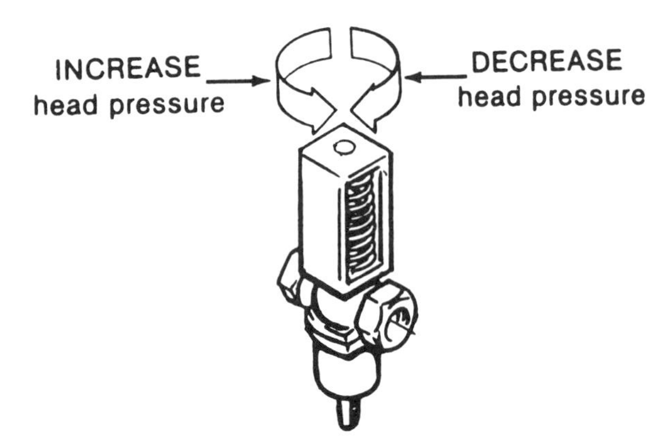

WATER REGULATOR

ADJUSTMENT OF THE CUBE SIZE CONTROL

////////////////// CAUTION //////////////////

BEFORE performing actual adjustments to the cube size control, check other possible causes for cube size problems, refer to SERVICE DIAGNOSIS for problem review and analysis. DO NOT perform adjustment when a new cube size control is installed, until the control bulb has been properly installed in the tube well, on the evaporator outlet tube and the icemaker has progressed through several complete freezing and harvest cycles, to observe size and quality of ice cubes freezing and harvest cycles, and whether or not a cube size problem exists.

//

As a reverse acting temperature control, adjustment on the cube size control is performed to cause either larger sized ice cubes or smaller sized ice cubes to be produced.

To produce LARGER sized ice cubes:

1. Locate the cube size control, on the front of the control box.
2. Rotate the adjusting knob one-eighth of a turn CLOCKWISE toward COLDER.
3. Observe size of ice cubes in next two ice cube harvests and repeat step 2 above, in one-eighth turn increments, until correct ice cube size is achieved.

To produce SMALLER sized ice cubes:

1. Locate the cube size control, on the front of the control box.
2. Rotate the adjusting knob one-eighth of a turn COUNTERCLOCKWISE toward warmer.
3. Observe size of ice cubes in next two ice cube harvests and adjust in one-eighth turn increments, until correct ice cube size is achieved.

ADJUSTMENT OF THE TIMER & SWITCH ASSEMBLY

The timer and switch assembly if factory set, so one complete revolution of the cam on the timer represents eight minutes. Five and one-half minutes comprise the freezing cycle event during cam rotation, and the final two and one-half minutes program the defrost and harvest cycle. Rotating the shaft of the timer cam CLOCKWISE will allow positioning the actuator arm of the microswitch on the cam at the selected start position for either the freezing cycle or harvest cycle, as required in the cleaning instructions.

////////////////// WARNING //////////////////

Disconnect electrical power supply to icemaker whenever adjustment procedures are performed.

//

To adjust the timer & switch assembly:

A. HARVEST CYCLE: Slowly rotate the shaft of the timer and switch assembly, located in the control box, CLOCKWISE, until the actuator arm on the microswitch initiates the harvest cycle. An audible click can be heard, but in a noisy area, look at the cam and switch to observe the event.

B. FREEZING CYCLE: Slowly rotate the shaft of the timer and switch assembly, located in the control box, CLOCKWISE, until the actuator arm on the microswitch initiates the freezing cycle.

C. The length of the harvest cycle can be changed by loosening the adjustment screw on the cam. The normal setting is two and one-quarter minutes, as set at the factory. It is important that the length of the harvest cycle allow enough time for all the ice cubes to fall from the evaporator. Too short of a time will cause the evaporator to freeze up and stop ejecting ice into the bin. Too much time wastes icemaking capacity, energy and water. Adjustment of the harvest cycle may require a corresponding adjustment of the cube size control.

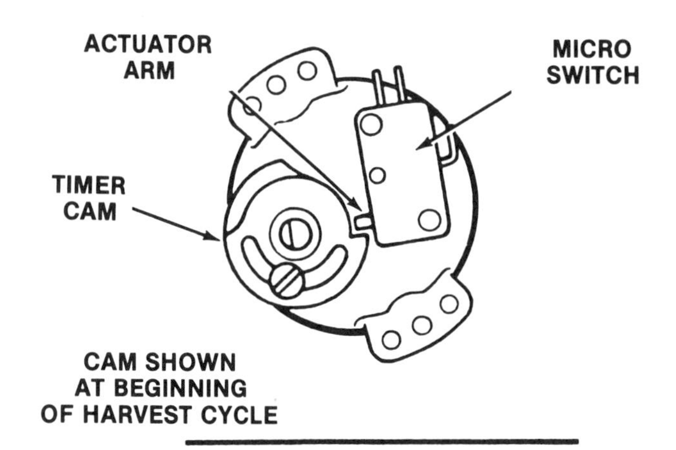

CAM SHOWN AT BEGINNING OF HARVEST CYCLE

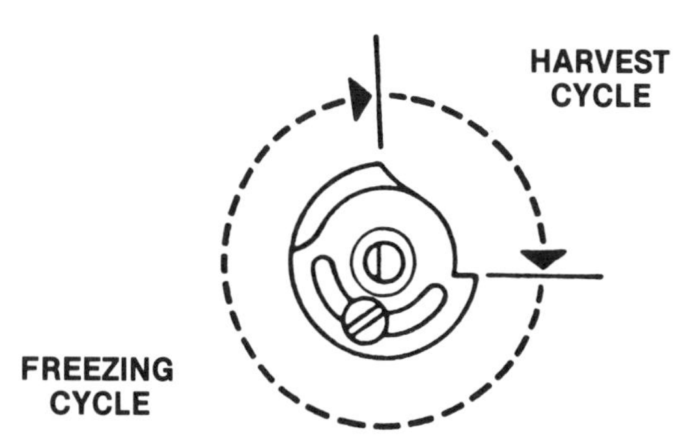

CAM SHOWN DIVIDED INTO TYPICAL FREEZING AND HARVEST CYCLES

CM650 REMOVAL AND REPLACEMENT PROCEDURES

//////////////////// **WARNING** ////////////////////

Be sure the electrical power supply circuit breaker and the inlet water supply are OFF, BEFORE starting any of the following REMOVAL AND REPLACEMENT procedures as a precaution to prevent possible personal injury or damage to equipment.

//

REMOVAL AND REPLACEMENT OF THE DRIER

To remove the drier:

1. Pull out to unsnap catches and remove the front panel.
2. Remove screws and remove left side access panel.
3. Bleed off or blow the refrigerant charge through the Schrader valve.
4. Unsolder refrigeration lines at both ends of the drier, and remove the drier.

To replace the drier:

//////////////////// **CAUTION** ////////////////////

If the factory seal is broken on the replacement drier, exposing it to the atmosphere more than a few minutes, the drier will absorb moisture from the atmosphere and lose substantial ability for moisture removal.

Be sure the replacement drier is installed with the arrow positioned in the direction of the refrigerant flow.

//

1. Remove the factory seals from the replacement drier and install the drier in the refrigerant lines with the arrow positioned in the direction of the refrigerant flow.
2. Solder the drier into the lines, two places, taking precautions to NOT OVERHEAT the drier body, during installation soldering.
3. Purge the system and check for leaks.
4. Thoroughly evacuate the system to remove moisture and non-condensables.
5. Charge the system with refrigerant, by weight. SEE NAMEPLATE.
6. Replace and attach the left side panel.

//////////////////// **NOTE** ////////////////////

Always install a replacement drier, anytime the sealed refrigeration system is opened. Do not replace the drier until all other repair or replacement has been completed.

REMOVAL AND REPLACEMENT OF THE EVAPORATOR PLATE ASSEMBLY

1. Remove the front and top panels.
2. Bleed off or blow the refrigerant charge through the Schrader valve.
3. Disconnect Tygon water inlet tube(s), at the water manifold tee(s), above the evaporator plates.
4. Unsnap the sump assembly from the lower left and right attachment points on each evaporator plate.
5. Slide the water distributor tubes about 1/8-inch along the top of the evaporator plate to be removed, until the left water distributor tube can be lifted upward.
6. Lift the end of the water distributor tube and slide the distributors toward the left along the top of the evaporator plate, until the flexible right notch is cleared.
7. Unsnap and disconnect each left and right water distributor tube from the water manifold section.

//////////////////// **CAUTION** ////////////////////

Use EXTRA PRECAUTION to protect the plastic parts during the next step to unsolder the refrigerant lines, two places, at the top of the evaporator plate. Position wet cloths over top of plates, as well as over the plastic liner at the rear, or sides, to prevent accidental heat damage, or possible fire from torch flame.

//

8. Unsolder and remove the refrigerant lines at the top of the evaporator plate to be replaced.
9. Remove nuts at left of the evaporator then, loosen the braces just enough to remove the evaporator plate. Temporarily replace the braces, to support the remaining evaporator plate.

To replace the evaporator plate, reverse the removal procedures. See Nameplate. Weigh in proper charge of R-502.

REMOVAL AND REPLACEMENT OF THE COMPRESSOR ASSEMBLY

To remove the compressor assembly:

1. Pull out to unsnap catches and remove the front panel.
2. Remove screws and remove cabinet top and left side access panel.
3. Bleed off or blow the refrigerant charge through the Schrader valve.
4. Remove the cover from the terminal box on the compressor; then, remove electrical leads from the compressor.
5. Unsolder suction, discharge and process header from compressor.
6. Remove four bolts and washers which secure the compressor to the chassis mounting base.
7. Remove the compressor from the cabinet.

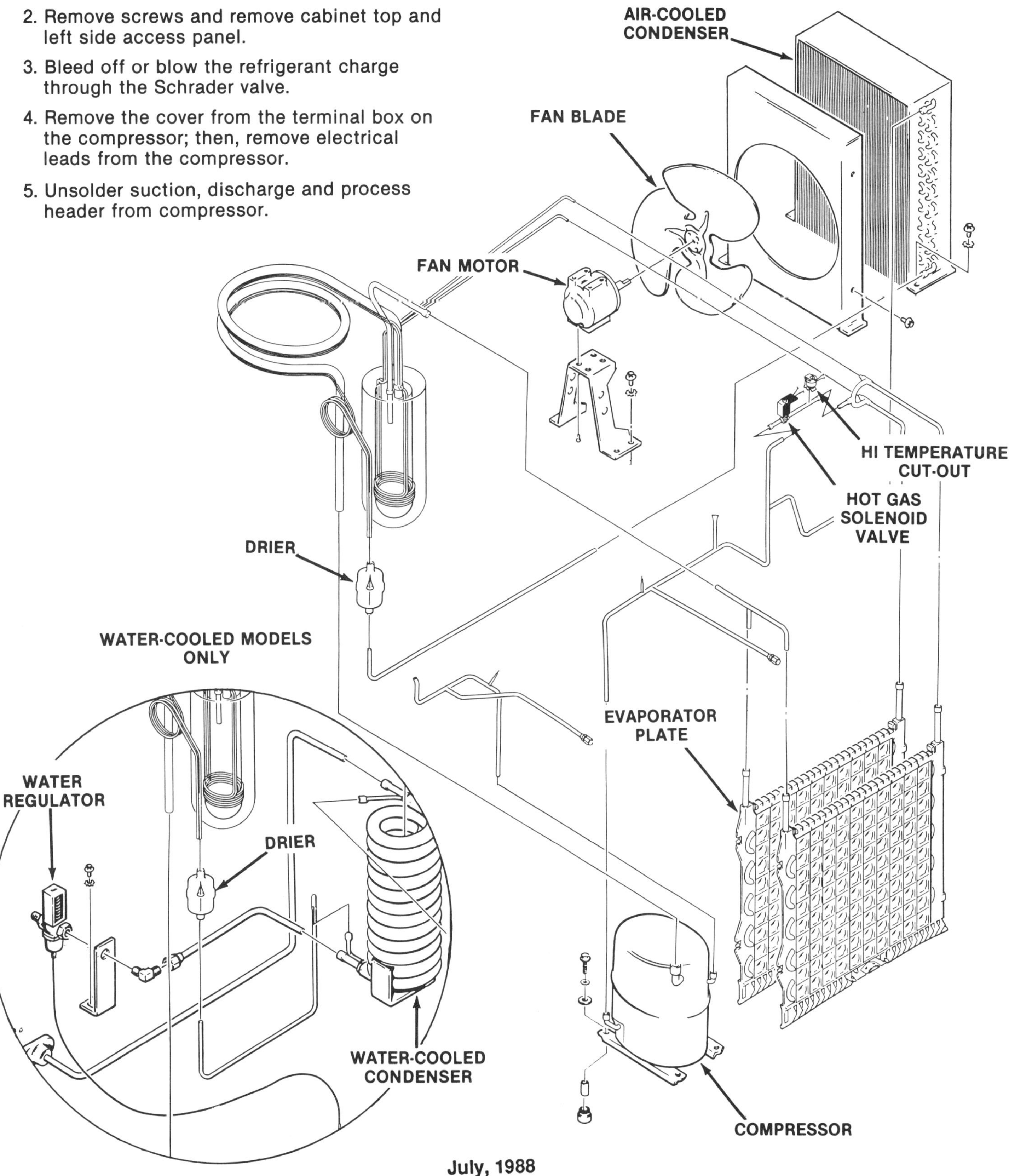

////////////////// **WARNING** //////////////////

Disconnect electrical power supply to icemaker whenever adjustment procedures are performed.

//

REMOVAL AND REPLACEMENT OF THE CONDENSER — AIR-COOLED MODELS

To remove the condenser:

1. Pull out to unsnap catches and remove the front panel.
2. Remove screws and remove cabinet top and right service side panels.
3. Bleed off or blow the refrigerant charge through the Schrader valve.
4. Unsolder and disconnect refrigerant lines from the condenser.
5. Unsolder and remove the drier from the refrigerant lines connecting to the condenser.
6. Remove screws, lockwashers and the condenser from the chassis base.

To replace the air-cooled condenser, reverse the removal procedure.

REMOVAL AND REPLACEMENT OF THE CONDENSER — WATER-COOLED MODELS

To remove the condenser:

1. Remove screws and the right side panel.
2. Bleed off or blow the refrigerant charge through the Schrader valve.
3. Check to be sure building source water inlet supply shutoff valve to rear of chassis is OFF.
4. Disconnect water-cooled condenser inlet water line at the water regulator assembly outlet fitting.
5. Unsolder the refrigerant capillary tube distributor line at the outlet end of the drier, the refrigerant liquid line from the bottom of the condenser, and remove the drier from the lines.
6. Unsolder the compressor discharge line, at the top of the water-cooled condenser.
7. Unsolder the compressor discharge line at the top of the water-cooled condenser and the refrigerant line from the bottom of the water regulator valve.
8. Remove two screws and washers and the water-cooled condenser from the cabinet.

To replace the water-cooled condenser, reverse the removal procedures.

REMOVAL AND REPLACEMENT OF THE FAN MOTOR — AIR-COOLED MODELS

To remove the fan motor assembly:

1. Pull out to unsnap catches and remove the front panel.
2. Remove unit top or left side panel.
3. Disconnect the two electrical leads, from the fan motor, at the control box assembly.

Before next step, measure or make accurate mental note of the distance the fan blades extend beyond the fan shroud, so during replacement the correct distance is maintained.

4. Remove screws and the fan motor and bracket assembly; and remove the fan motor and bracket from the chassis base.
5. Loosen set screws on the fan blade and remove the fan blade from the fan motor. Mark or note position of the blade on the shaft of the fan motor, for replacement.
6. Remove screws and lockwasher and separate the fan motor from the fan motor bracket.

To replace the fan motor assembly, reverse the removal procedure.

Be sure to replace the fan blade with the hub of the fan blade facing the fan motor, to ensure air flow is toward the fan motor; and, is set at marked location on shaft and setscrews tightened on the FLAT part of the shaft of the fan motor.

// **NOTE** //

Always install a replacement drier, anytime the sealed refrigeration system is opened. Do not replace the drier until all other repair or replacement has been completed.

Thoroughly evacuate the system to remove moisture and non-condensables.

CM650 REMOVAL AND REPLACEMENT PROCEDURES

//////////////////// **WARNING** ////////////////////

Disconnect electrical power supply to icemaker whenever adjustment procedures are performed.

//

REMOVAL AND REPLACEMENT OF THE CUBE SIZE CONTROL

To remove the cube size control:

1. Remove front panel.
2. Remove cover from control box.
3. Trace capillary tube, from the cube size control to the refrigerant suction line.
4. Remove the coiled capillary tube bulb from the tube well on the suction line.
5. Remove electrical leads from the cube size control.
6. Remove screws and the cube size control.

To replace the cube size control, reverse the removal procedure.

REMOVAL AND REPLACEMENT OF THE WATER DISTRIBUTOR TUBES AND MANIFOLD TUBES

To remove the water distributor tube and manifold tube:

1. Pull out to unsnap catches and remove the front panel.
2. Remove 4 thumb screws and remove the evaporator cover.
3. Slide the water distributor tube to the left about 1/8-inch along the top of the evaporator plate, until the water distributor tube can be unsnapped from the flexible notch and lifted upward to the right side.
4. Unsnap and disconnect water distributor tubes from the water manifold section.

To replace the water distributor tubes and manifold tubes, reverse the removal procedure. BE SURE the notches in the water manifold tubes properly engage the alignment keys in the tee.

BE SURE the water distributor tube is securely fastened at the notch at both sides of the evaporator plate.

Check identical attachment for the left water distributor tube and notch; also, that the distributor/manifold connections at the top center of each evaporator plate is snug against the top of the plate.

REMOVAL AND REPLACEMENT OF THE BIN THERMOSTAT CONTROL

To remove the bin thermostat control:

1. Remove front panel.
2. Remove screws and the control box cover.
3. Remove wire leads from the bin thermostat control.
4. Unthread the capillary tube and remove from the bin thermostat control bracket at the bottom right side of the evaporator section.
5. Remove the two screws attaching the bin thermostat control to the side of the control box; then, carefully pull the capillary tube out of the evaporator section. Carefully remove the bin thermostat control and capillary tube from the control box.

To replace the bin thermostat control, reverse the removal procedure.

REMOVAL AND REPLACEMENT OF THE INLET WATER SOLENOID VALVE ASSEMBLY

To remove the inlet water solenoid valve assembly:

1. Shut OFF water supply to machine.
2. Remove screws and pull the water solenoid valve out to gain access.
3. Loosen and remove inlet water line fitting from the inlet water solenoid valve assembly.
4. Remove inlet water Tygon tubing from the water solenoid valve.

To replace the inlet water valve assembly, reverse the removal procedures.

//////////////////// **NOTE** ////////////////////

Always install a replacement drier, anytime the sealed refrigeration system is opened. Do not replace the drier until all other repair or replacement has been completed.

CM650 REMOVAL AND REPLACEMENT PROCEDURES

////////////////////// WARNING //////////////////////

Disconnect electrical power supply to icemaker whenever adjustment procedures are performed.

//

REMOVAL AND REPLACEMENT OF THE WATER PUMP ASSEMBLY

To remove the water pump assembly:

1. Pull out to unsnap catches and remove the front panel.
2. Remove the Tygon tube to the pump assembly.
3. Disconnect electrical leads from the water pump assembly.
4. Loosen thumbscrew, loosen Phillips headscrew, lift pump up and remove from unit.
5. Remove one nut and mounting bracket from top of water pump.

To replace the water pump assembly, reverse the removal procedure.

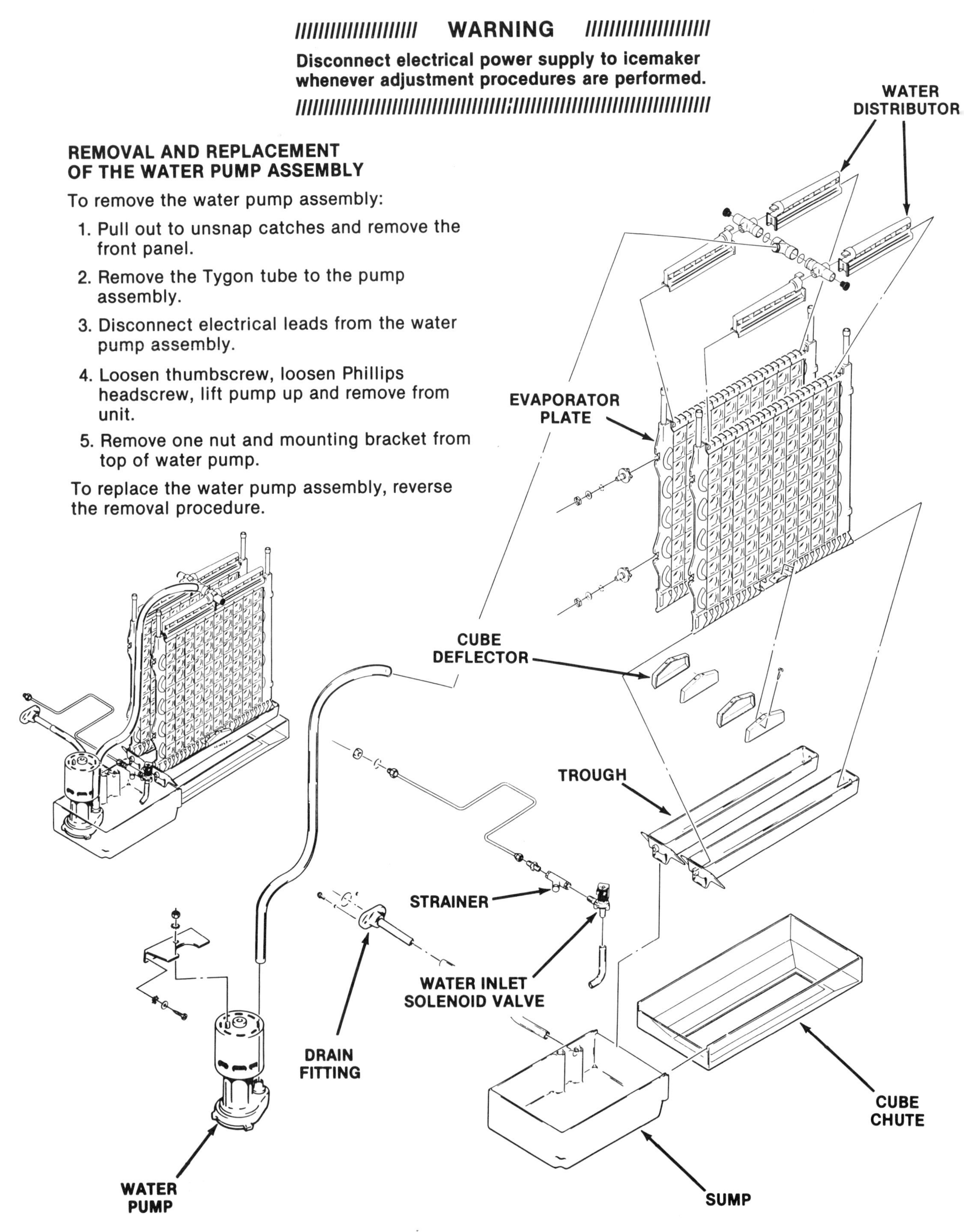

CM650E SERVICE PARTS

This section contains the parts illustrations and parts lists for the CM650E.
A "no number" designation, when found in the part number column, indicates the part in question is not available from Scotsman as an assemby. This designation is used only for clarity in making the list.
When ordering parts, to avoid costly delays and errors, give the part number, the complete description as shown in the list, and the quantity of each part or assembly required.

CM650's have been manufactured in "A", "C", "D" and "E" series (as in CM650AE-32E). This parts list is for the model CM650 "E". Check the complete model number to see that it is an E model.

TABLE OF CONTENTS

CM650E SERVICE PARTS

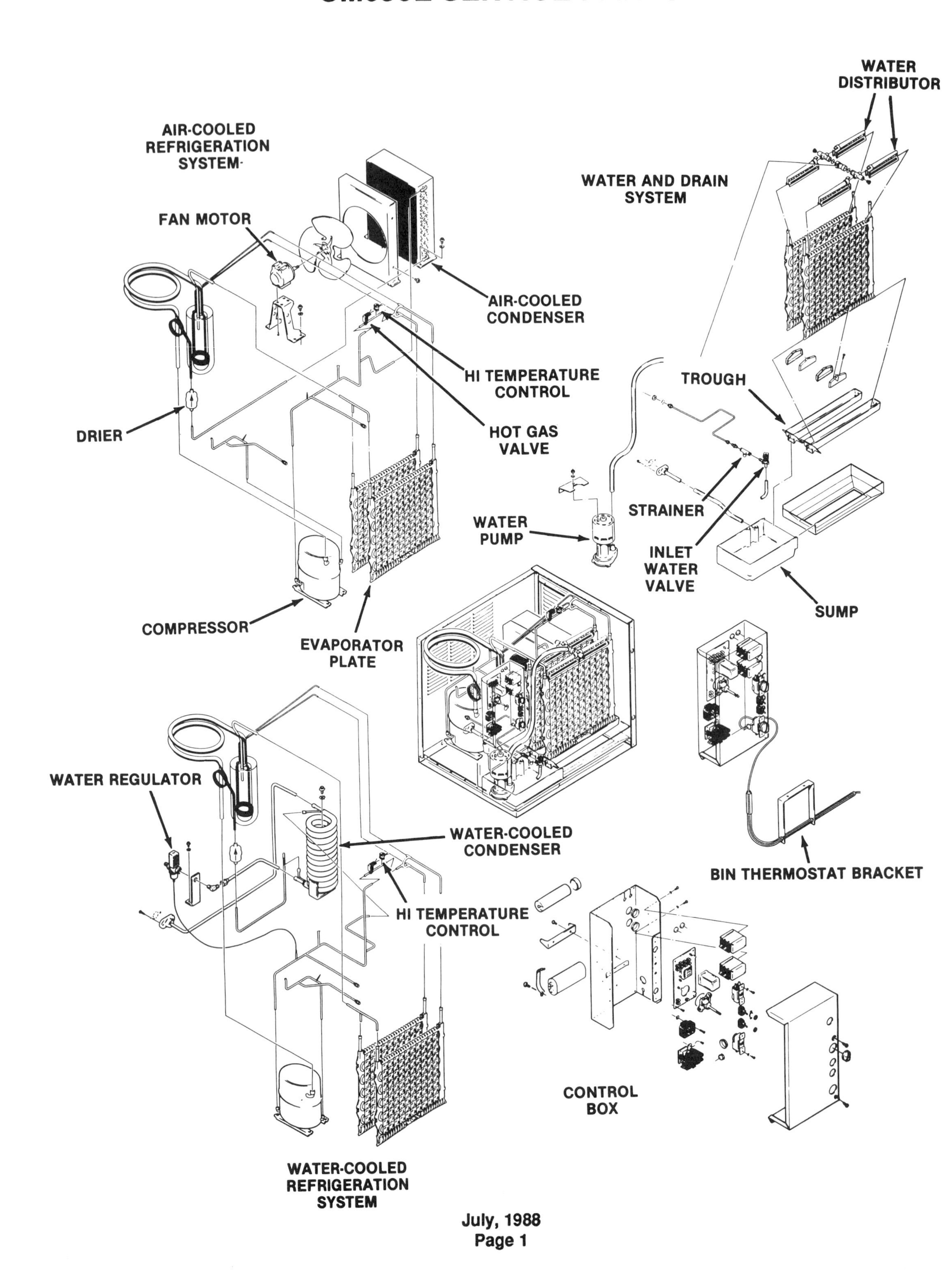

CM650E SERVICE PARTS

CABINET ASSEMBLY

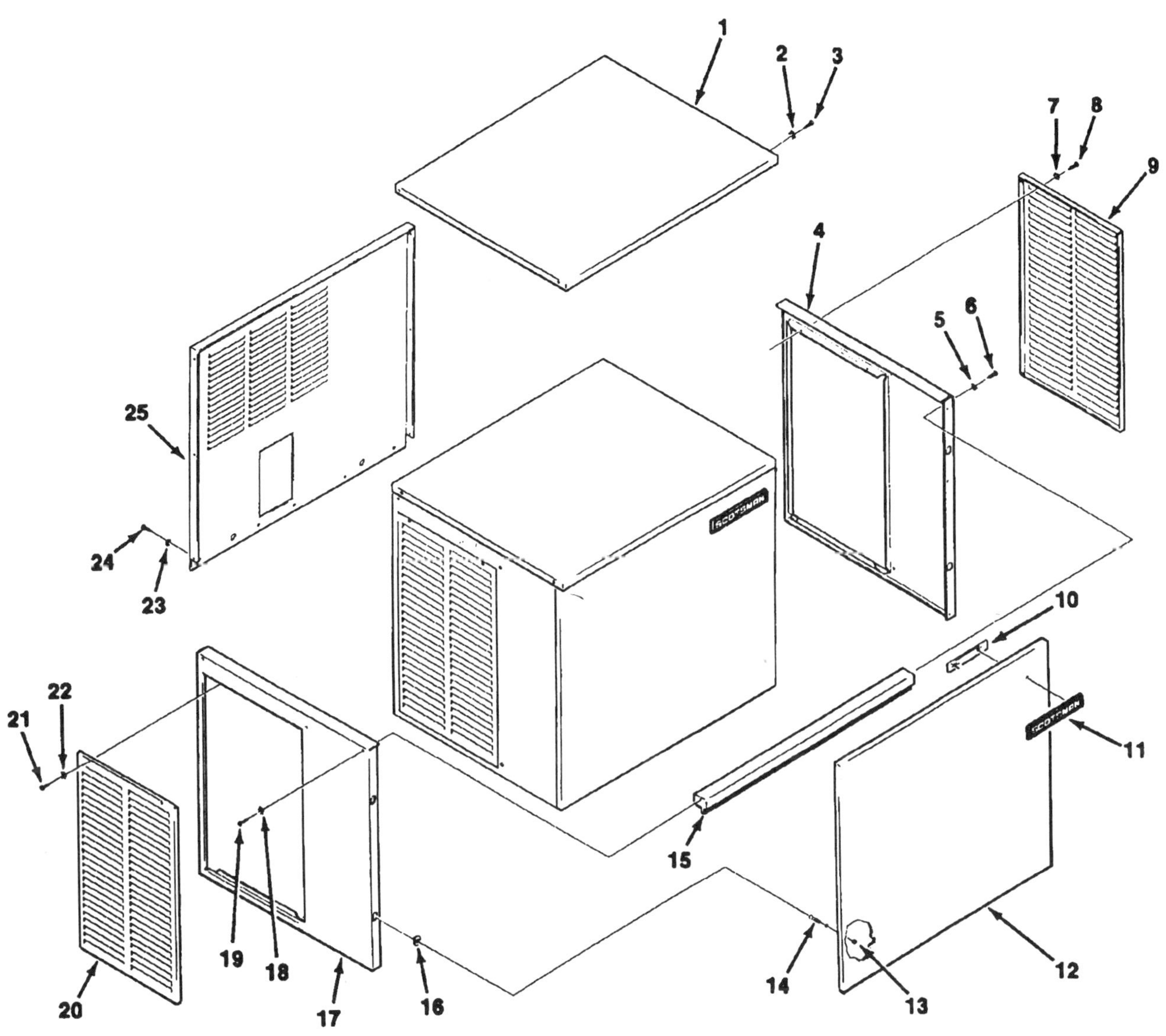

ITEM NUMBER	PART NUMBER	DESCRIPTION
1	A32159-001	Top Panel, Painted
	A32159-002	Top Panel, S. S.
2	03-1417-03	Lockwasher
3	03-1531-01	Screw
4	A32162-001	Right End Panel, Paint
	A32162-002	Right End Panel, S.S.
5.	03-1417-03	Lockwasher
6	03-1531-01	Screw
7	03-1417-03	Lockwasher
8	03-1404-12	Screw
9	A32161-001	Service Panel, Painted
	A32161-002	Service Panel, S.S.
10	03-0271-00	Speed Nut
11	15-0711-01	Emblem
12	A32160-001	Front Panel, Painted
	A32160-002	Front Panel, S.S.

ITEM NUMBER	PART NUMBER	DESCRIPTION
13	03-1406-04	Nut
14	15-0411-00	Strike
15	A32169-001	Front Support
16	02-0836-00	Cabinet Catch
17	A32163-001	Left End Panel, Paint
	A32163-002	Left End Panel, S.S.
18	03-1417-03	Lockwasher
19	03-1531-01	Screw
20	A32161-001	Service Panel, Paint
	A32161-002	Service Panel, S.S.
21	03-1404-12	Screw
22	03-1412-03	Lockwasher
23	03-1417-03	Lockwasher
24	03-1531-01	Screw
25	A32202-001	Back Panel

CM650E SERVICE PARTS

CABINET INTERIOR

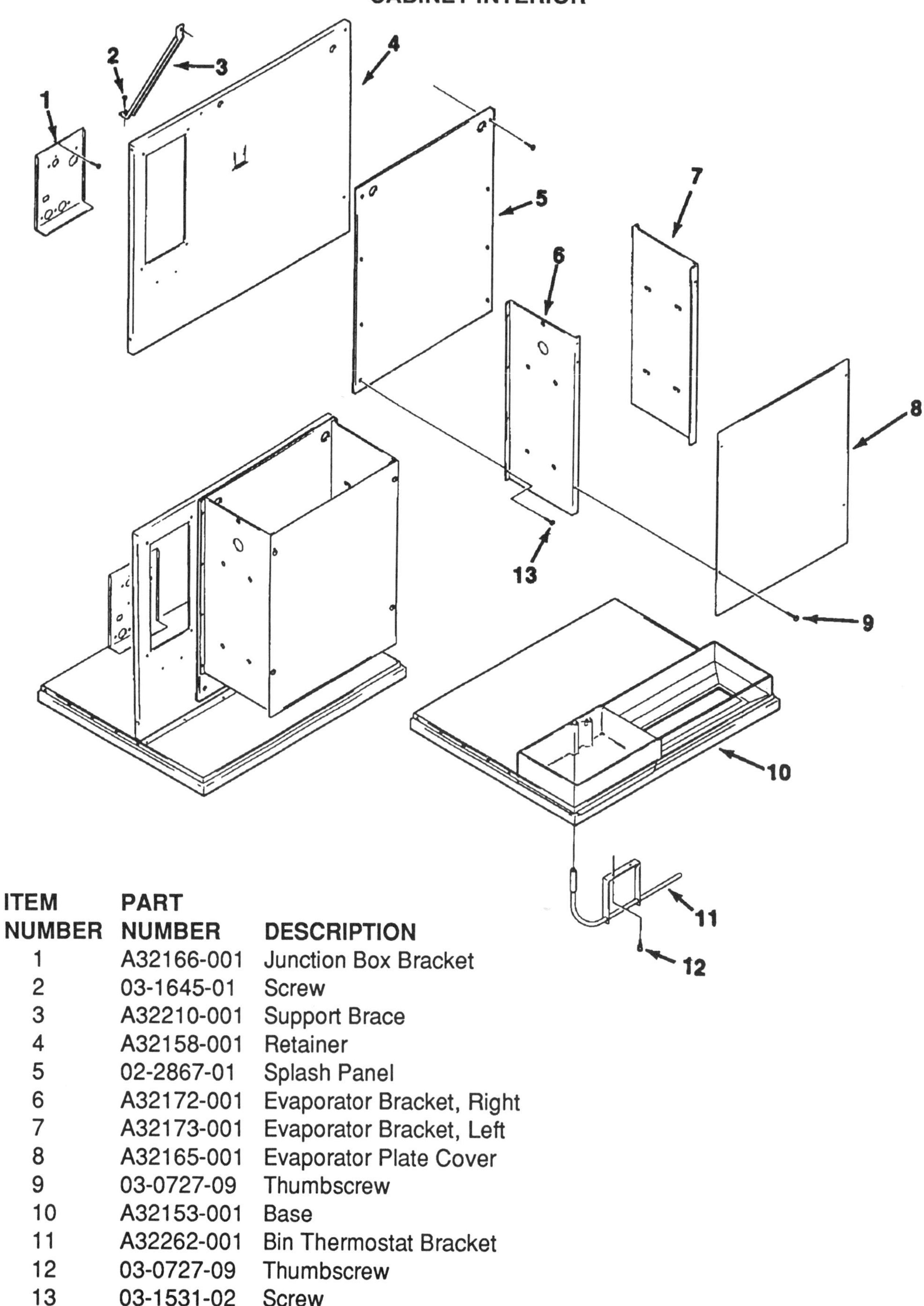

ITEM NUMBER	PART NUMBER	DESCRIPTION
1	A32166-001	Junction Box Bracket
2	03-1645-01	Screw
3	A32210-001	Support Brace
4	A32158-001	Retainer
5	02-2867-01	Splash Panel
6	A32172-001	Evaporator Bracket, Right
7	A32173-001	Evaporator Bracket, Left
8	A32165-001	Evaporator Plate Cover
9	03-0727-09	Thumbscrew
10	A32153-001	Base
11	A32262-001	Bin Thermostat Bracket
12	03-0727-09	Thumbscrew
13	03-1531-02	Screw

CM650E SERVICE PARTS

AIR COOLED REFRIGERATION

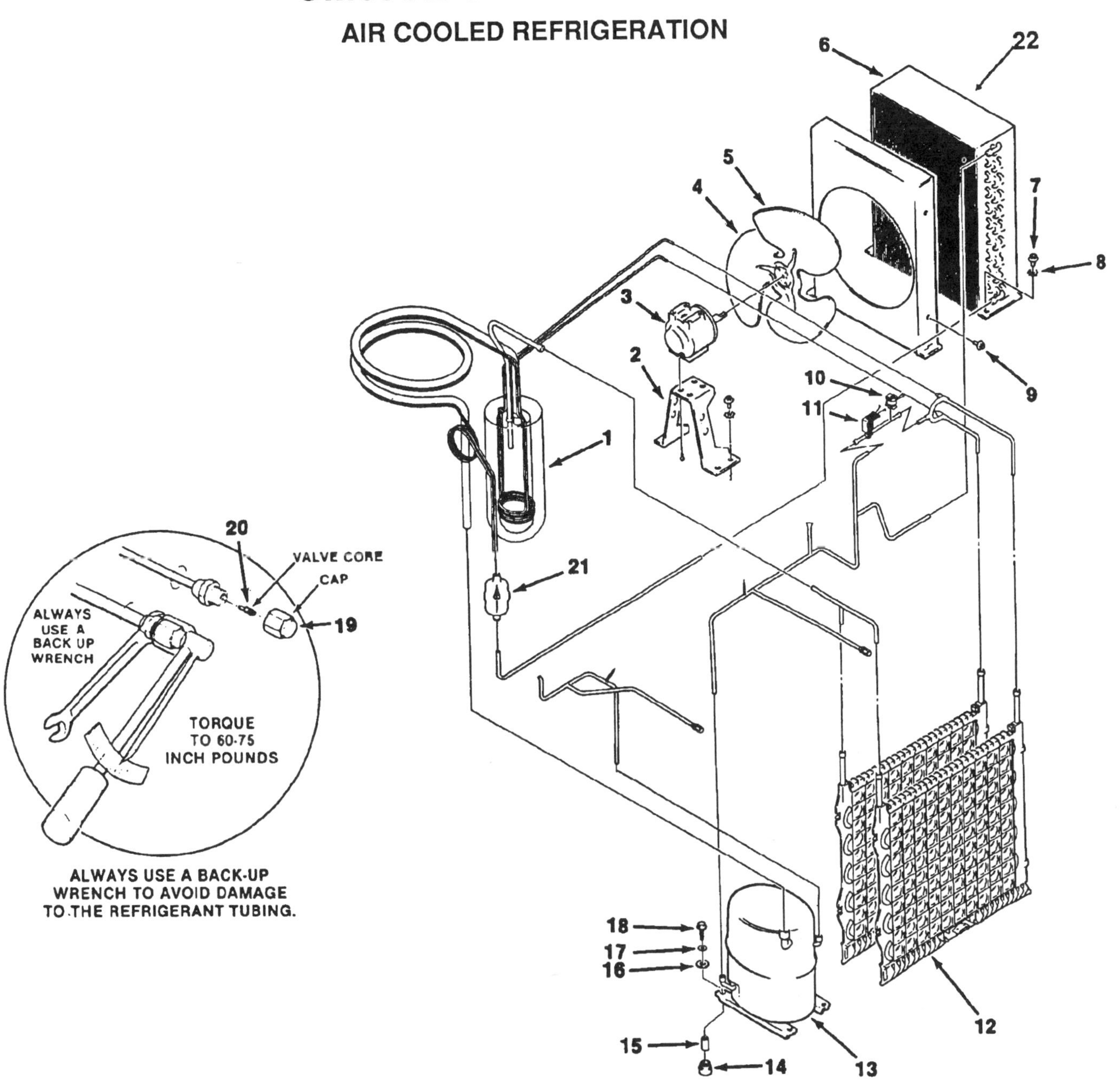

ALWAYS USE A BACK-UP WRENCH TO AVOID DAMAGE TO THE REFRIGERANT TUBING.

ITEM NUMBER	PART NUMBER	DESCRIPTION
1	A32174-020	Heat Exchanger
2	18-3739-01	Fan Motor Bracker
3	18-5105-11	Fan Motor
4	18-3732-01	Fan Blade
5	A32170-001	Fan Shroud
6	18-3709-01	Condenser
7	03-1645-01	Screw
8	03-1410-03	Lockwasher
	03-1407-05	Washer
9	03-1531-01	Screw
10	12-2300-03	Hi Temp Cut Out
11	12-2135-01	Hot Gas Valve
12	A30296-020	Evaporator
13	18-8000-04	Compressor (203-230/60/1)
	18-8000-05	Compressor (208-230/60/3)
	(Both have an internal overload)	
14	18-6800-01	Grommet
15	18-2300-26	Sleeve
16	03-1407-07	Washer
17	no number	washer
18	03-1405-20	Screw
19	16-0563-00	Cap
20	16-0560-00	Core
21	02-2426-02	Drier
22	A32975-001	Foam Filter

CM650E SERVICE PARTS

WATER COOLED REFRIGERATION

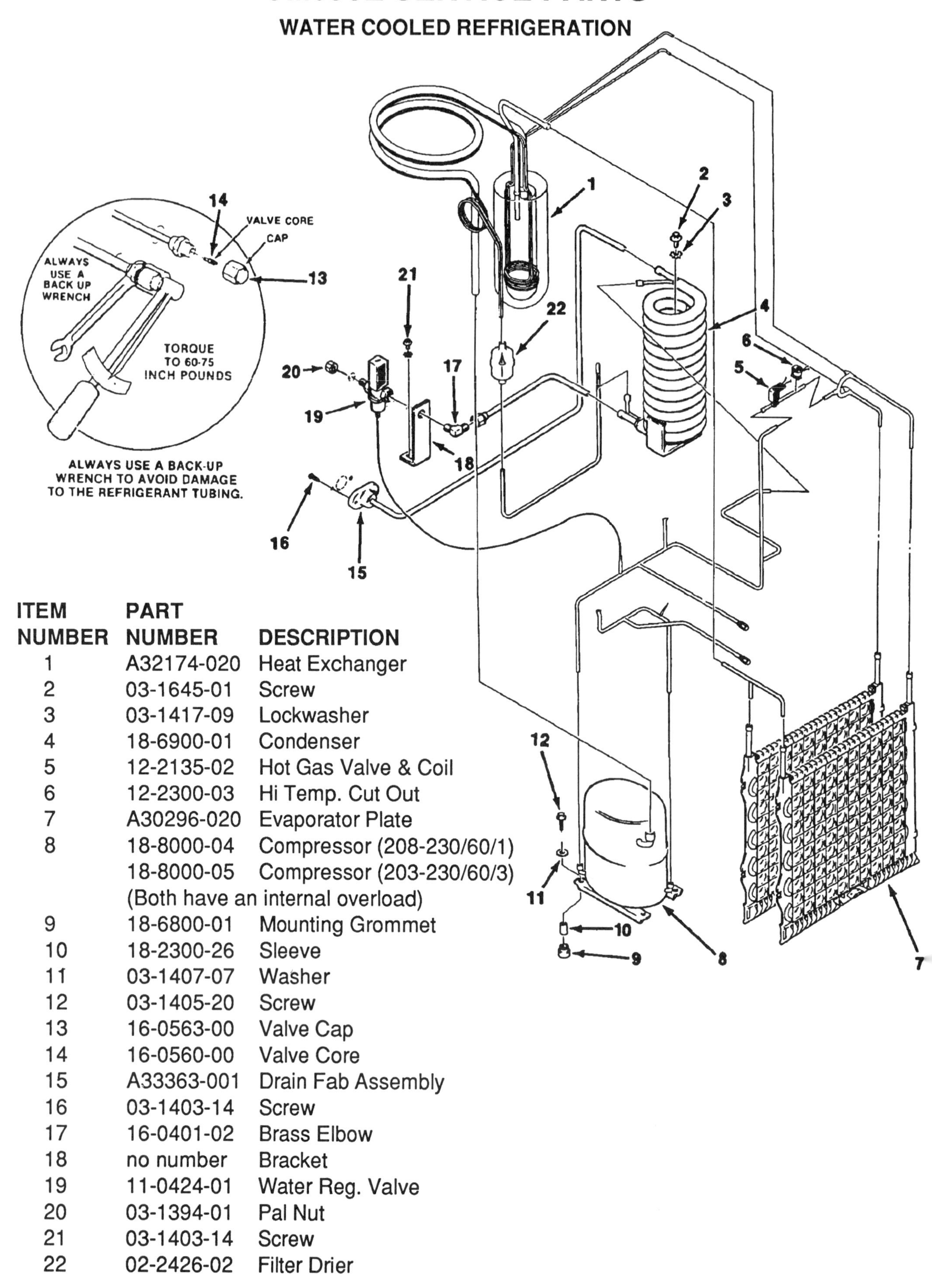

ITEM NUMBER	PART NUMBER	DESCRIPTION
1	A32174-020	Heat Exchanger
2	03-1645-01	Screw
3	03-1417-09	Lockwasher
4	18-6900-01	Condenser
5	12-2135-02	Hot Gas Valve & Coil
6	12-2300-03	Hi Temp. Cut Out
7	A30296-020	Evaporator Plate
8	18-8000-04	Compressor (208-230/60/1)
	18-8000-05	Compressor (203-230/60/3)
	(Both have an internal overload)	
9	18-6800-01	Mounting Grommet
10	18-2300-26	Sleeve
11	03-1407-07	Washer
12	03-1405-20	Screw
13	16-0563-00	Valve Cap
14	16-0560-00	Valve Core
15	A33363-001	Drain Fab Assembly
16	03-1403-14	Screw
17	16-0401-02	Brass Elbow
18	no number	Bracket
19	11-0424-01	Water Reg. Valve
20	03-1394-01	Pal Nut
21	03-1403-14	Screw
22	02-2426-02	Filter Drier

CM650E SERVICE PARTS

WATER SYSTEM

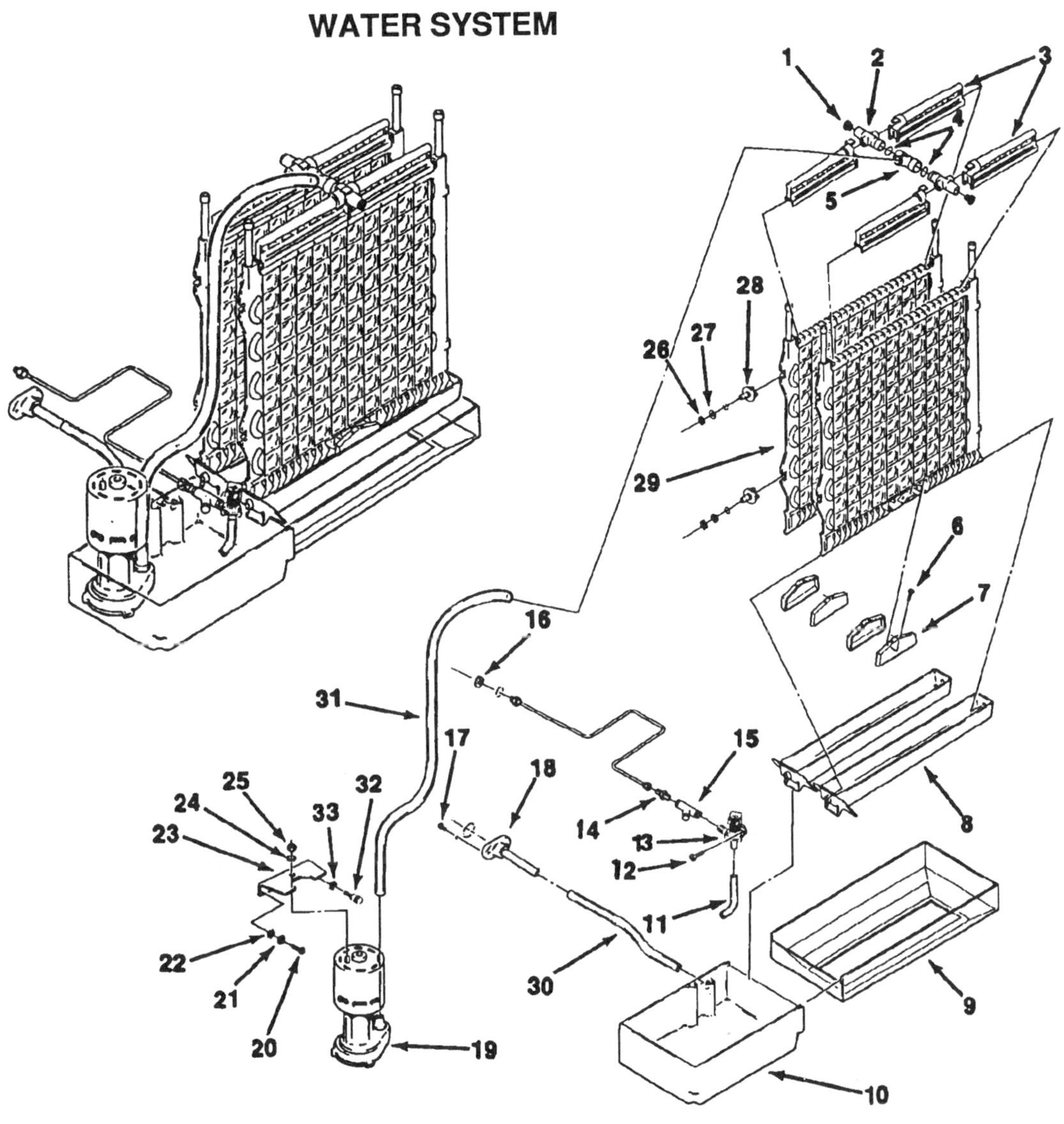

ITEM NUMBER	PART NUMBER	DESCRIPTION
1	13-0840-01	Manifold Plug
2	A29703-001	Manifold Water Tube
3	02-2527-01	Water Distrib. Tube
4	13-0617-48	O-Ring (2)
5	02-2519-01	Plastic Tee
6	03-1404-10	Screw
7	02-2878-01	Sump End Cover
8	02-2868-01	Drain Trough
9	no number	Cube Chute
10	02-2869-01	Sump
11	A30918-001	Sump Inlet Tube
12	03-1531-01	Screw
13	12-2313-03	Solenoid Valve
14	16-0791-01	Half Union
15	16-0162-00	Stainer
16	03-1394-01	Pal Nut
17	03-1645-01	Screw

ITEM NUMBER	PART NUMBER	DESCRIPTION
18	A32282-003	Drain Ass. (Air Cooled)
19	12-2265-22	Water Pump Assembly
20	03-1403-17	Screw
21	03-1407-03	Washer
22	03-1408-31	Washer
23	A30914-002	Pump Mounting Bracket
24	03-1417-05	Lockwasher
25	03-1406-05	Nut
26	03-1394-03	Pal Nut
27	03-1408-34	Special Washer
28	02-2642-01	Plate Mount
29	A30296-020	Evaporator
30	13-0674-07	Drain Hose, 8"
31	13-0674-07	Pump Hose, 33"
32	03-0727-09	Thumbscrew
33	03-1417-03	Lockwasher

CM650E SERVICE PARTS

CONTROL BOX

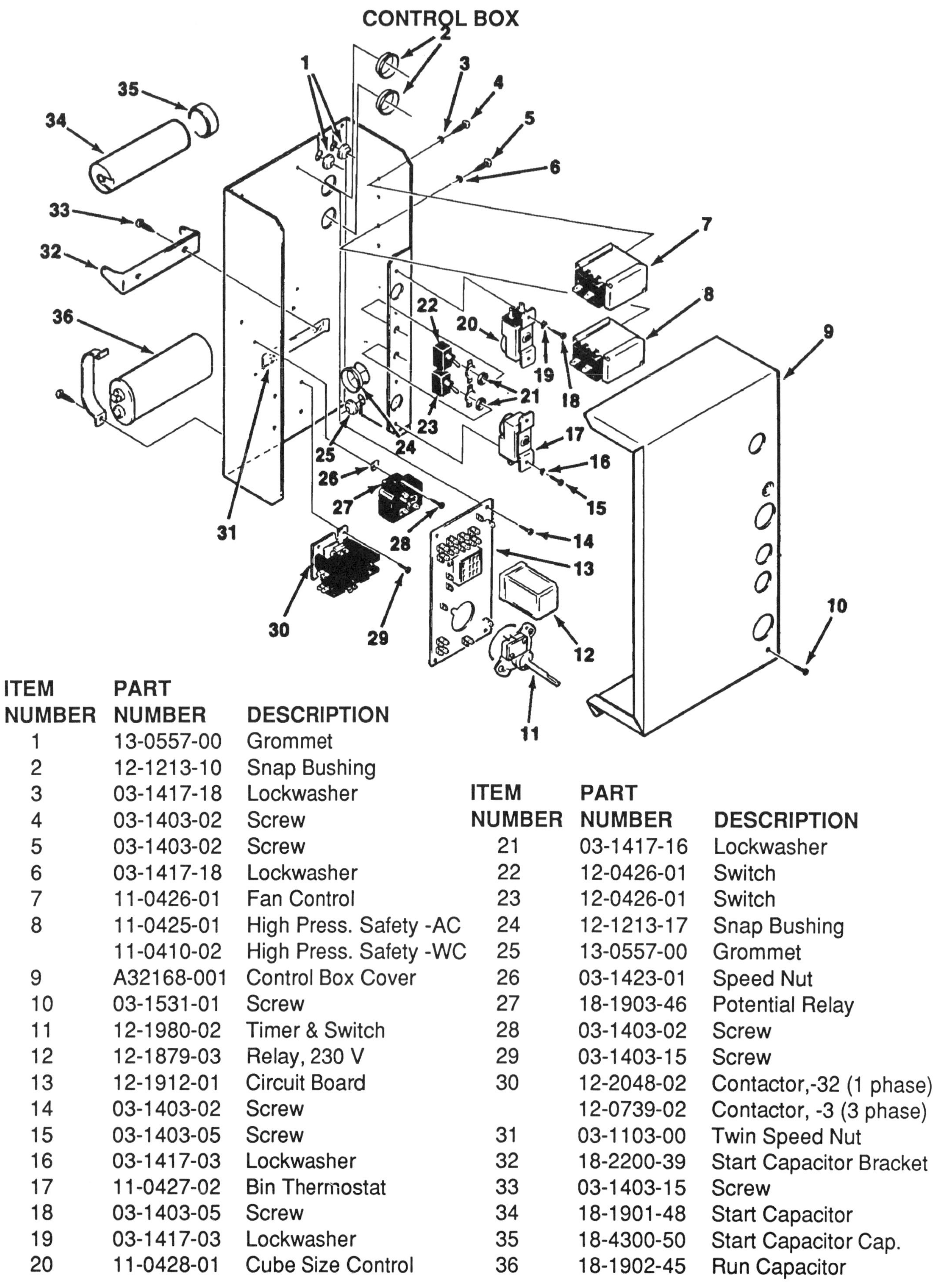

ITEM NUMBER	PART NUMBER	DESCRIPTION
1	13-0557-00	Grommet
2	12-1213-10	Snap Bushing
3	03-1417-18	Lockwasher
4	03-1403-02	Screw
5	03-1403-02	Screw
6	03-1417-18	Lockwasher
7	11-0426-01	Fan Control
8	11-0425-01	High Press. Safety -AC
	11-0410-02	High Press. Safety -WC
9	A32168-001	Control Box Cover
10	03-1531-01	Screw
11	12-1980-02	Timer & Switch
12	12-1879-03	Relay, 230 V
13	12-1912-01	Circuit Board
14	03-1403-02	Screw
15	03-1403-05	Screw
16	03-1417-03	Lockwasher
17	11-0427-02	Bin Thermostat
18	03-1403-05	Screw
19	03-1417-03	Lockwasher
20	11-0428-01	Cube Size Control

ITEM NUMBER	PART NUMBER	DESCRIPTION
21	03-1417-16	Lockwasher
22	12-0426-01	Switch
23	12-0426-01	Switch
24	12-1213-17	Snap Bushing
25	13-0557-00	Grommet
26	03-1423-01	Speed Nut
27	18-1903-46	Potential Relay
28	03-1403-02	Screw
29	03-1403-15	Screw
30	12-2048-02	Contactor,-32 (1 phase)
	12-0739-02	Contactor, -3 (3 phase)
31	03-1103-00	Twin Speed Nut
32	18-2200-39	Start Capacitor Bracket
33	03-1403-15	Screw
34	18-1901-48	Start Capacitor
35	18-4300-50	Start Capacitor Cap.
36	18-1902-45	Run Capacitor

CM650E WIRING DIAGRAMS

Component Layout Diagram for CM650E-32E (208-230/60/1) Air or Water Cooled. (Water Cooled has no Fan or Fan Control)

A34311-032

Caution: More than one disconnect means may be required to disconnect all power to this unit.

USE COPPER CONDUCTORS ONLY

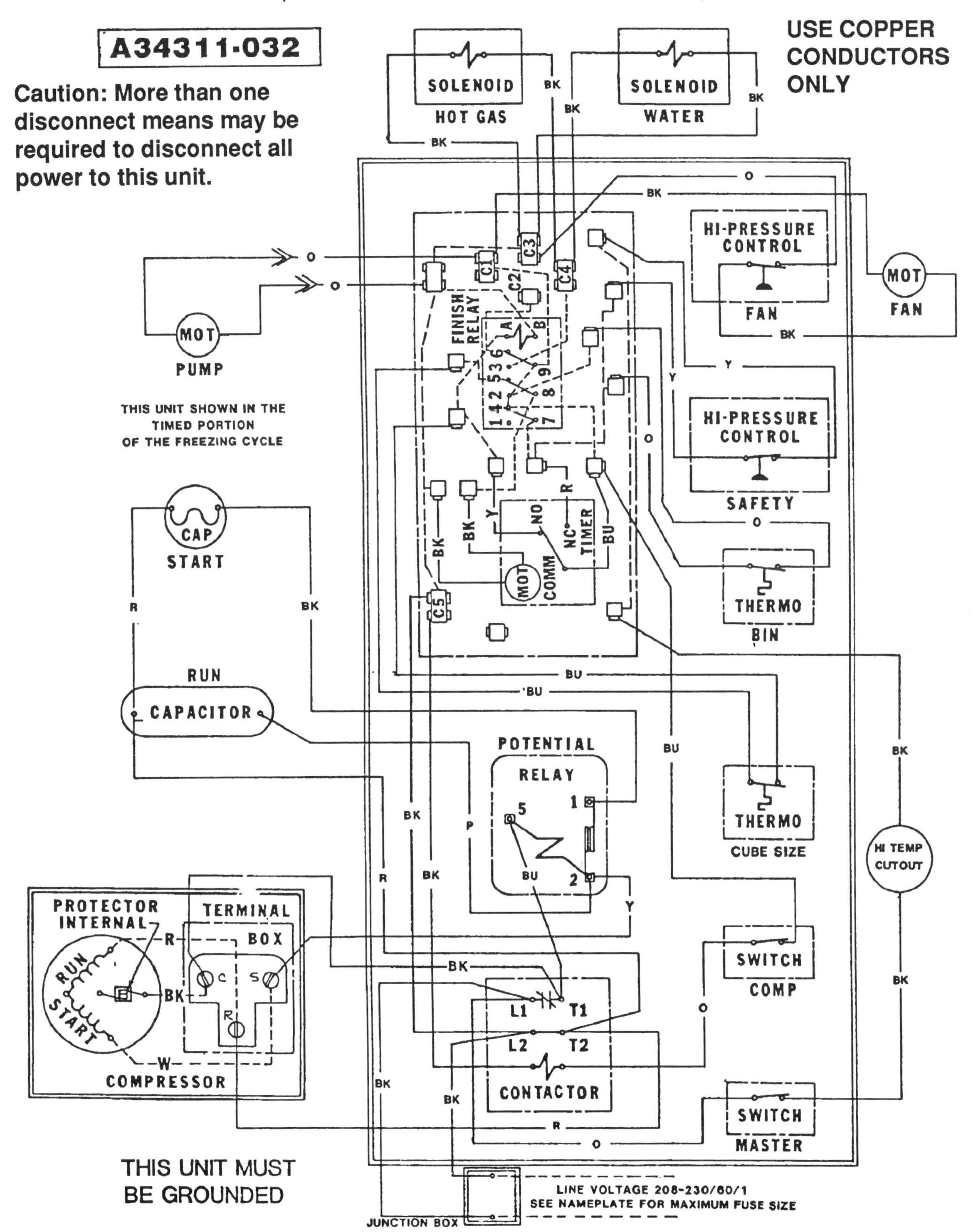

THIS UNIT MUST BE GROUNDED

CM650E WIRING DIAGRAMS

Schematic Diagram for CM650E-32E (208-230/60/1) Air or Water Cooled.

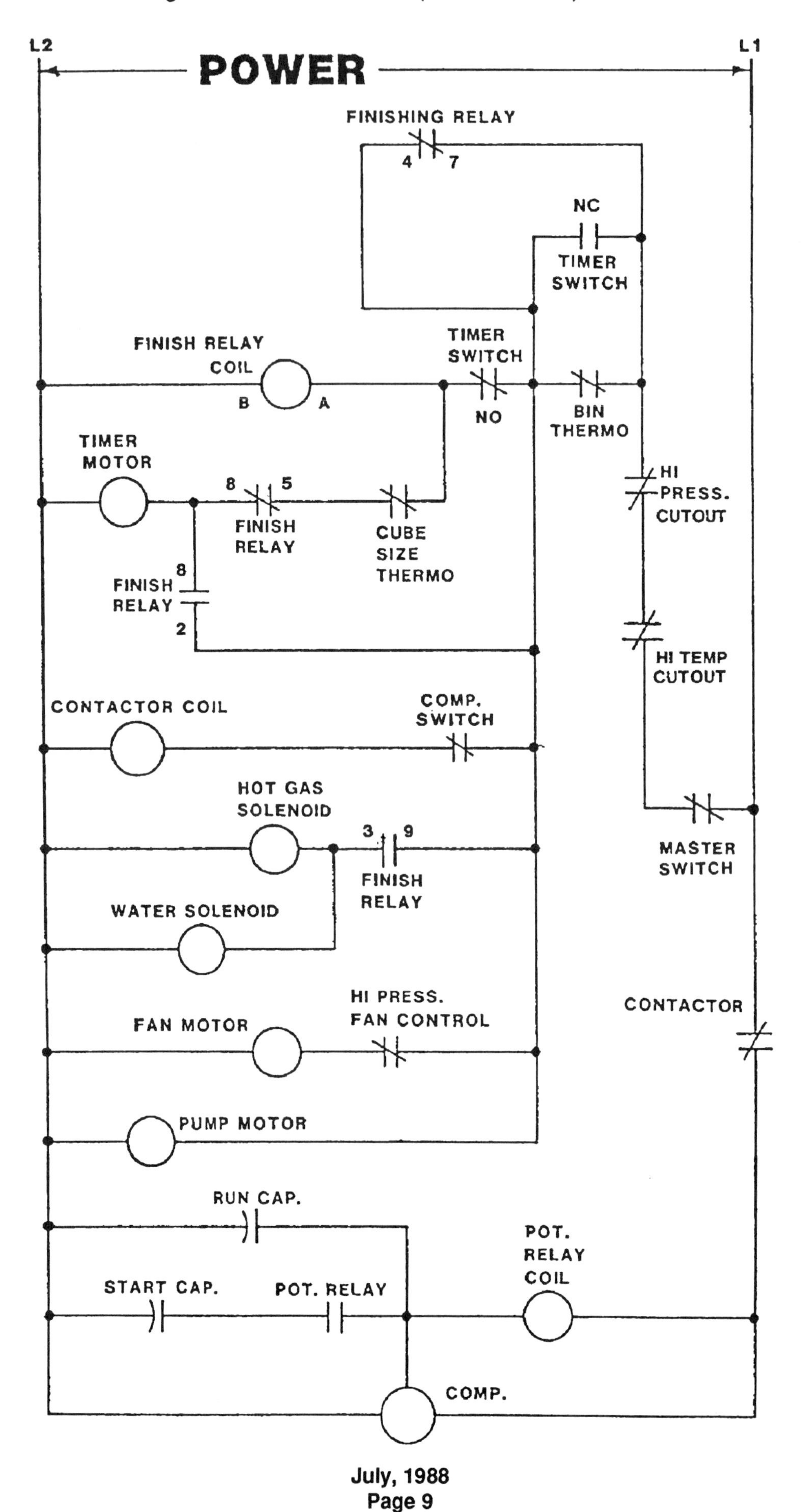

CM650E WIRING DIAGRAMS

Component Layout Diagram for CM650E-3E (208-230/60/3) Air and Water Cooled.

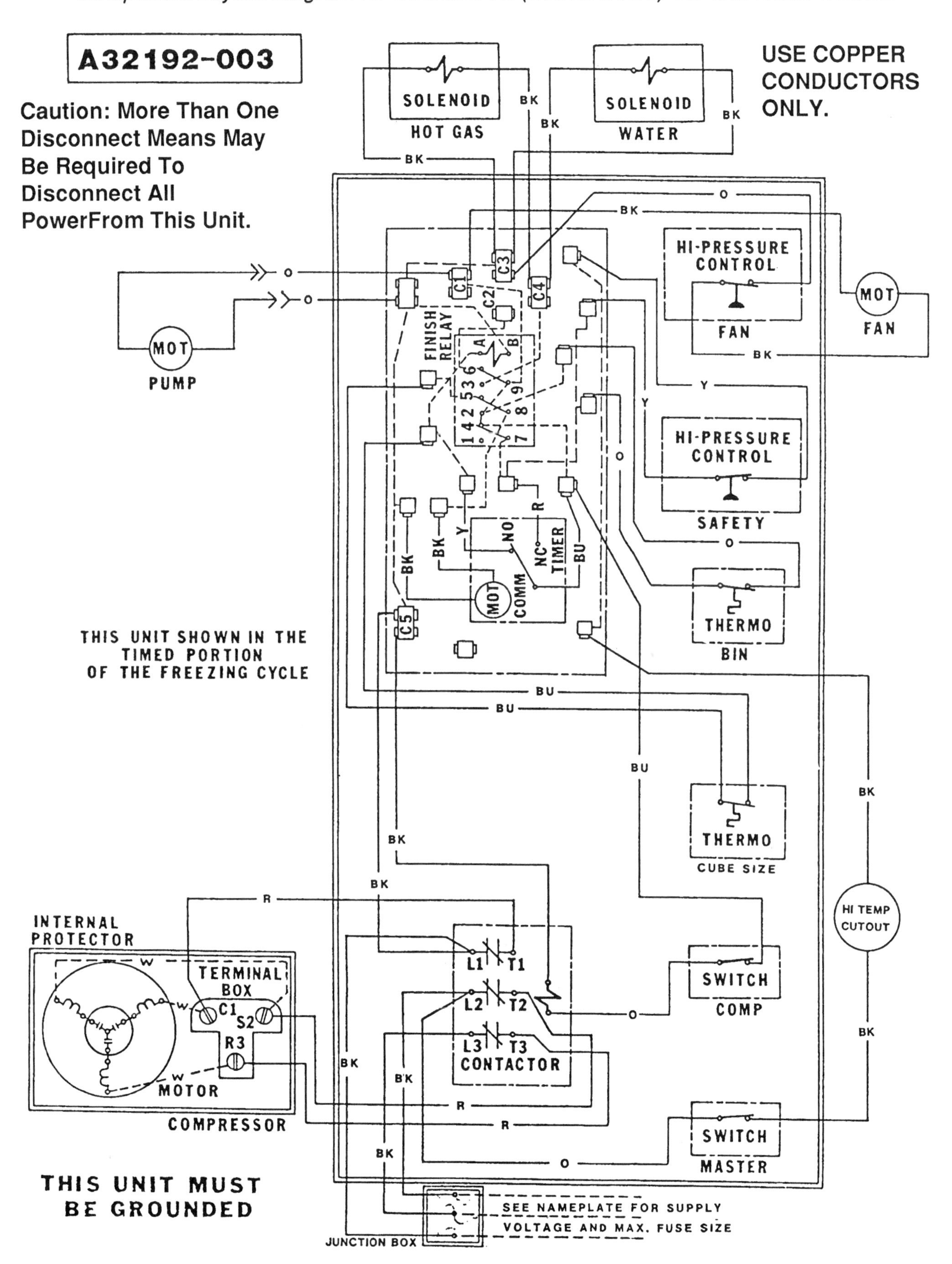

CM650E WIRING DIAGRAMS

Schematic Diagram for CM650E-3E, (208-230/60/3) Air or Water Cooled.

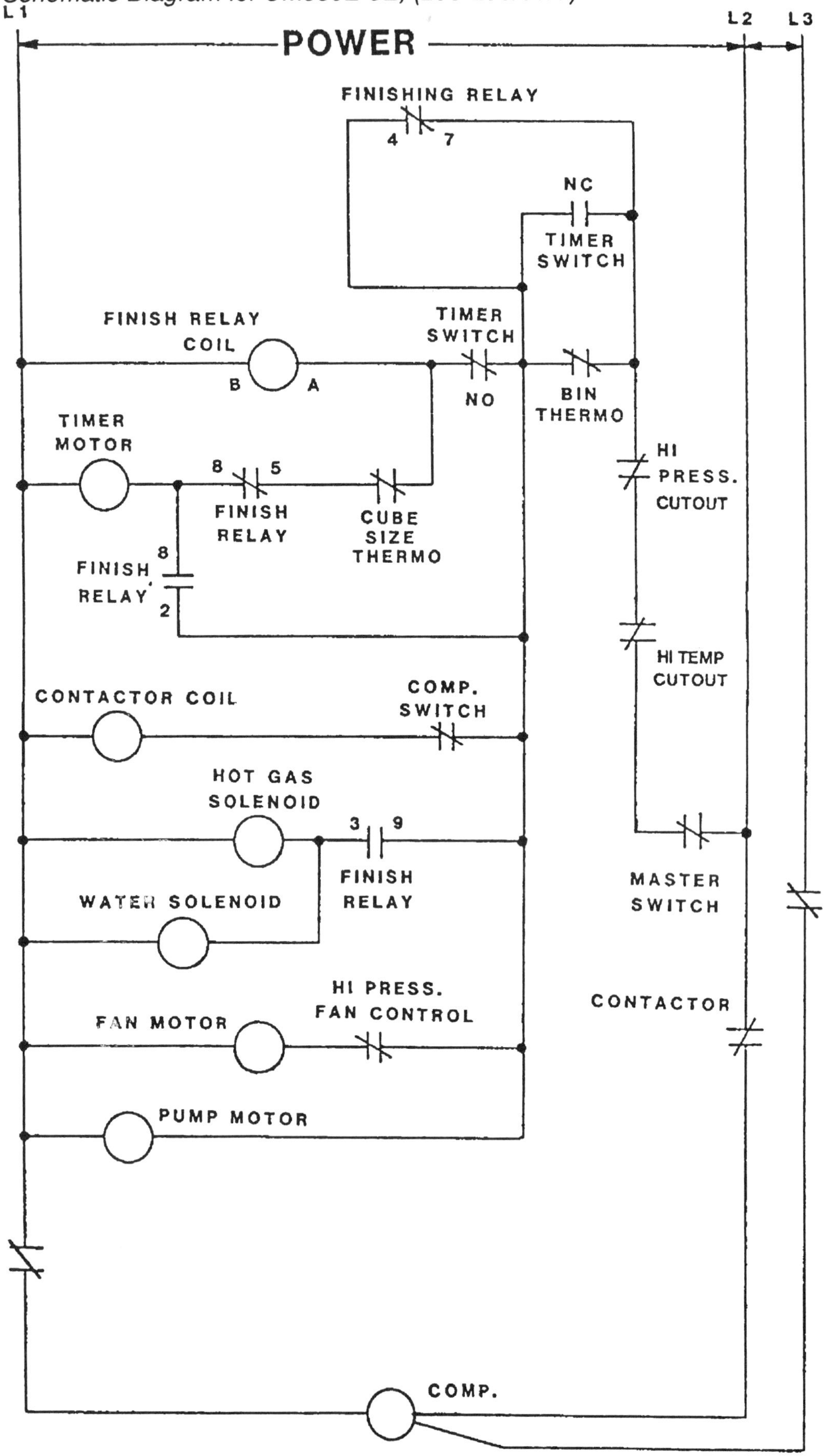

L1
L2
L3
POWER
FINISHING RELAY
4
7
NC
TIMER SWITCH
FINISH RELAY COIL
B
A
TIMER SWITCH
NO
BIN THERMO
TIMER MOTOR
8
5
FINISH RELAY
CUBE SIZE THERMO
HI PRESS. CUTOUT
8
FINISH RELAY
2
HI TEMP CUTOUT
CONTACTOR COIL
COMP. SWITCH
HOT GAS SOLENOID
3
9
FINISH RELAY
MASTER SWITCH
WATER SOLENOID
HI PRESS. FAN CONTROL
FAN MOTOR
CONTACTOR
PUMP MOTOR
COMP.

July, 1988

SCOTSMAN®

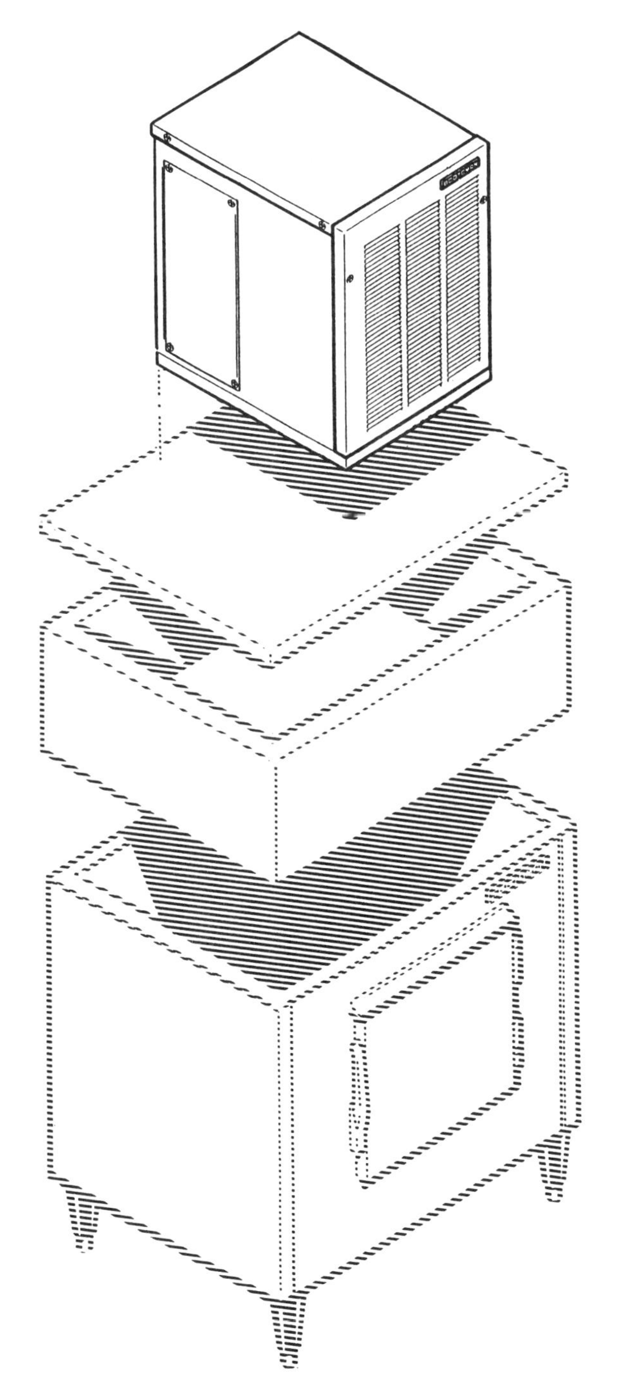

Service Manual for

Modular Nugget Ice Maker

MODEL NM650

INTRODUCTION

To the owner or user: The service manual you are reading is intended to provide you, and the maintenance or service technician, with the information needed to install, start up, clean, maintain, and service this ice system.

The NM650 is a modular ice system that fits a variety of Scotsman ice storage bins.

It features: front service for the freezer, gearmotor, control box, water reservoir, and bin control; an electronic circuit for monitoring ice and water level; a thermostatic expansion valve; and R-502 as the refrigerant.

TABLE OF CONTENTS

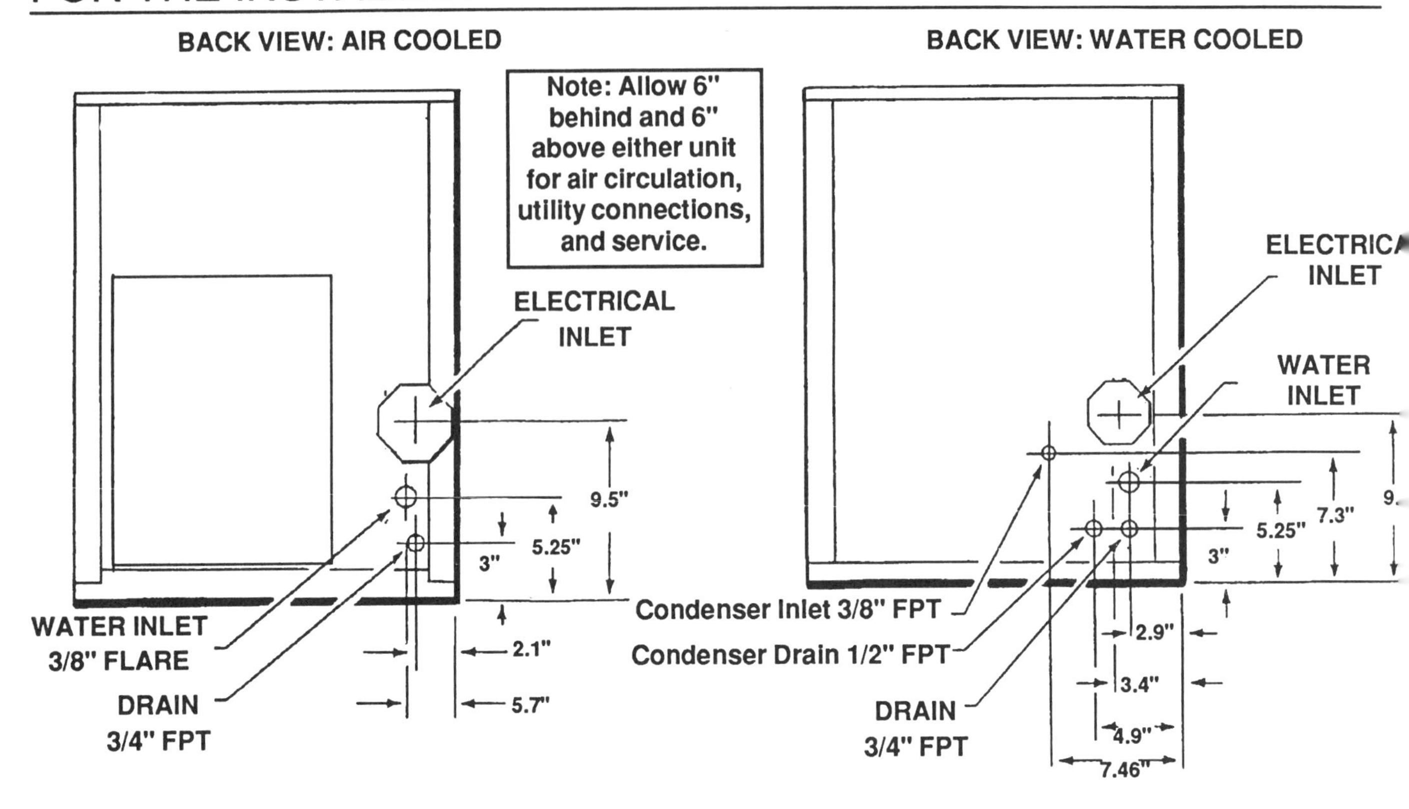

The NM650 is designed to fit the following Scotsman storage bins:

- B590 and extensions (with bin top KBT18)
- BH800 using bin top KBT15 (one unit).
- BH800 (two units, no bin top required).
- BH1000 using bin top KBT16.

When installing a new system, check to be sure that you have everything you need before beginning:

- Correct Bin
- Correct Ice Machine
- Correct Bin Top (if required)
- All kits, legs, and information required for the specific job.

Installation Limitations:

This ice system is designed to be installed indoors, in a controlled environment:

	Min	Max
Air Temperature	50^0F	100^0F
Water Temperature	40^0F	100^0F
Water Pressure	20 PSI	80 PSI
Voltage	-5%	+10%

(Compared to the nameplate)

Operating the machine outside of the limitations is misuse and can void the warranty.

The normal finish for the ice machine is enamel-sandalwood. A stainless steel panel kit, SPKFM21 may be field installed to convert the unit to a stainless steel finish.

SPECIFICATIONS: ICE MAKER

Model Number	Diminsions (w/o bin) H X W X D	Basic Electrical	Ice Type	Condenser Type	Minimum Circuit Ampacity	Max Fuse Size	Comp. H.P.
NM650AE-1A	27" x 21" x 24"	115/60/1	NUGGET	Air	20.7	35	3/4
NM650WE-1A	same	same	same	Water	16.8	30	3/4
NM650AE-32A	same	208-230/60/1	same	Air	11.4	20	3/4
NM650WE-32A	same	same	same	Water	10.2	20	3/4

Note: Minimum Circuit Ampacity is used to determine wire size and type per national electric code.

FOR THE INSTALLER

Water Limitations:

An ice machine is a food manufacturing plant: it takes a raw material, water, and transforms it into a food product, ice. The purity of the water is very important in obtaining pure ice and in maximizing product life. This section is not intended as a complete resource for water related questions, but it does offer these general recommendations:

1. Check with a water treatment specialist for a water test, and recommendations regarding filters and treatment.

2. In most cases, the water used to make ice should be filtered or treated, depending upon the water. *There is no one type of water filter that is effective in all situations.* That is why a water test is important.

Note:

Scotsman Ice Systems are designed and manufactured with the highest regard for safety and performance. They meet or exceed the standards of UL, NSF, and CSA.

Scotsman assumes no liability or responsibility of any kind for products manufactured by Scotsman that have been altered in any way, including the use of any part and/or other components not specifically approved by Scotsman.

Scotsman reserves the right to make design changes and/or improvements at any time. Specifications and design are subject to change without notice.

Typical Storage Bin - B590

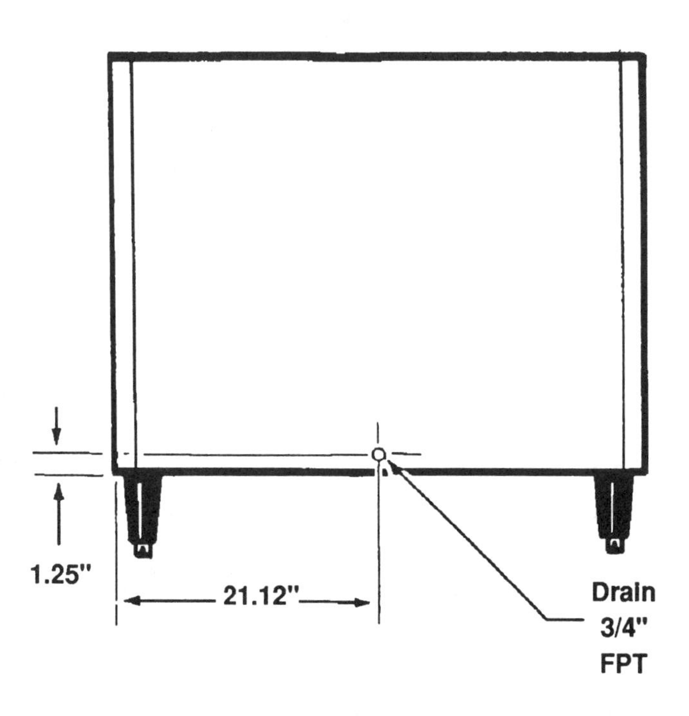

Typical Storage Bin with Extension and Bin Top

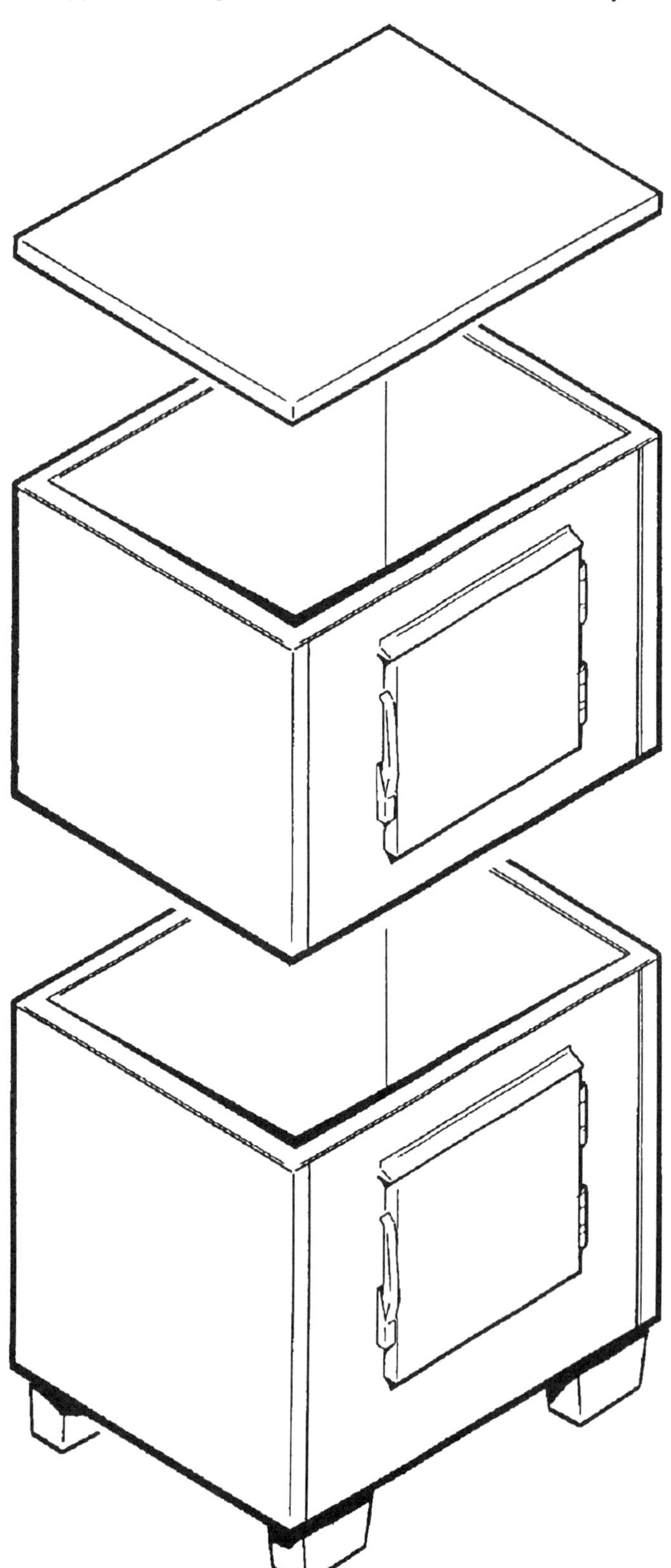

Location:

After uncrating and inspection, the unit is ready for installation. It is important that the machine be installed in a location where it has enough space around it to be accessible for service, and minimum of 6" be allowed at the back for air circulation on air cooled models. Try to avoid hot, dirty and crowded locations. Be sure that the location for the machine is within the environmental limitations.

Storage Bin:

Tip the storage bin on its back, using parts of the carton to protect the exterior finish. Install the legs into the threaded holes in the bottom of the bin. Turn the leg levelers all the way in preparation for leveling later. Return the bin to the upright position, remove paper covering the bin gasket.

Note: Do not push bin into position, but lift it there. Pushing a bin, especially one with ice in it, can cause damage to the legs and the leg mounts.

Install the appropriate bin top on the bin, according to the instructions for the bin top.

Ice Maker:

The machine is heavy, so the use of a mechanical lift is recommended for lifting the machine high enough to install on top of the bin. After the unit is placed on the bin, line it up so it is even with the back side. Secure the machine to the bin with the hardware provided with the machine.

Remove the front panel and remove any shipping blocks.

FOR THE INSTALLER: Location

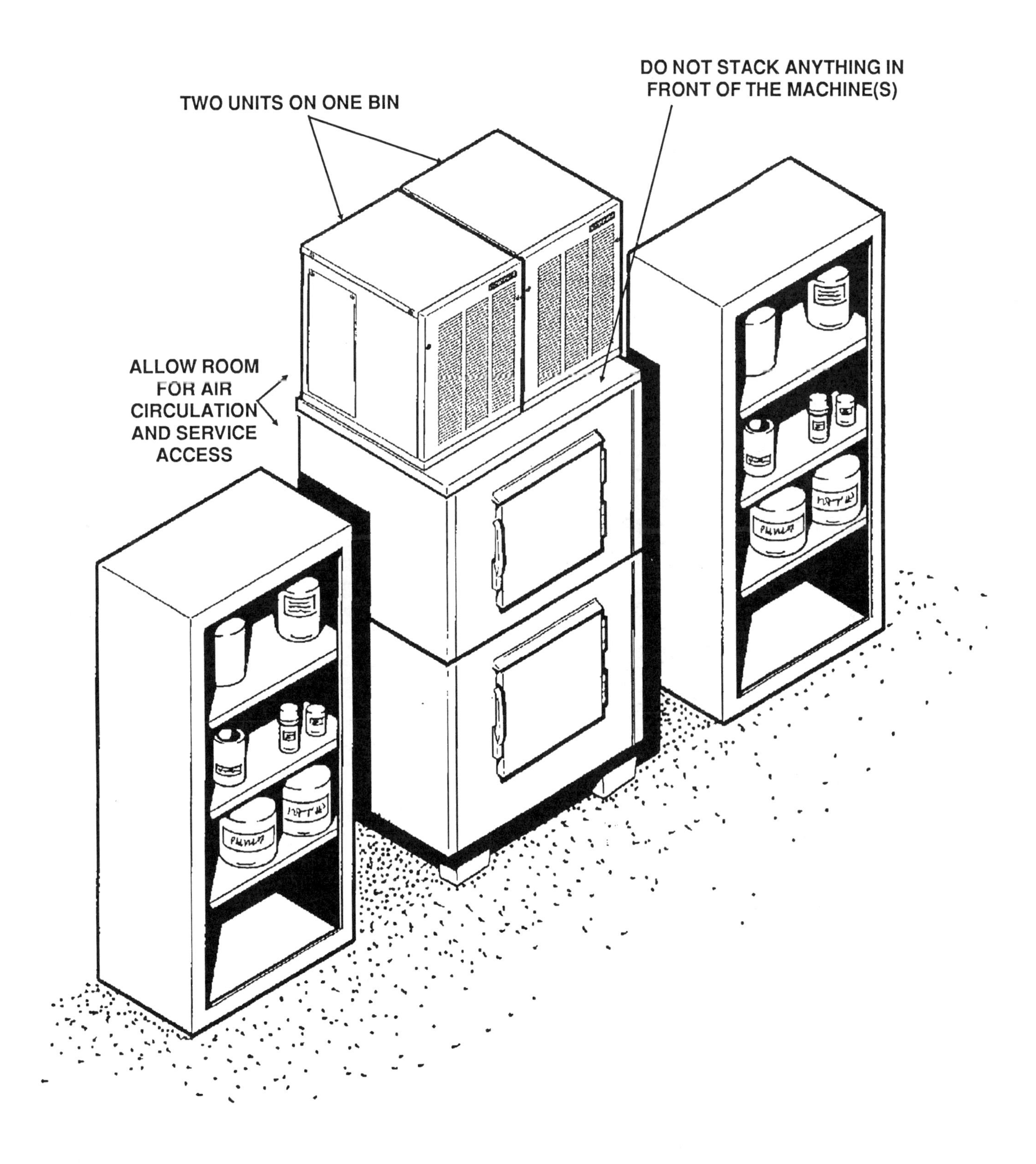

CONFORM TO ALL APPLICABLE CODES

Water Inlet

Air Cooled Models: The recommended water supply is clean, cold water. Use 3/8" O.D. copper tubing, connect to the 3/8" male flare at the back of the cabinet. Install a hand valve near the machine to control the water supply.

Water Treatment: In most areas, a water filter of some type will be useful. In areas where the water is highly concentrated with minerals the water should be tested by a water treatment specialist, and the recommendations of the specialist regarding filtration and/or treatment should be followed.

Water Cooled Models: A separate 3/8" O.D. copper line is recommended, with a separate hand valve to control it. It is connected to a 3/8" FPT condenser inlet at the back of the cabinet. The water pressure to all lines must always be above 20 psig, and below 120 psig.

Drains

Air Cooled Models: There is one 3/4" FPT drain at the back of the cabinet, the drain line is of the gravity type, and 1/4 inch per foot fall is an acceptable pitch for the drain tubing. There should be a vent at the highest point of the drain line, and the ideal drain receptacle would be a trapped and vented floor drain. Use only 3/4" rigid tubing.

Water Cooled Models: In addition to the above mentioned drain, a separate condenser drain must be installed. Connect it to the 1/2" condenser drain connection at the back of the cabinet.

Storage Bin: A separate gravity type drain needs to be run, similar to the air cooled drain. Insulation of this drain line is recommended.

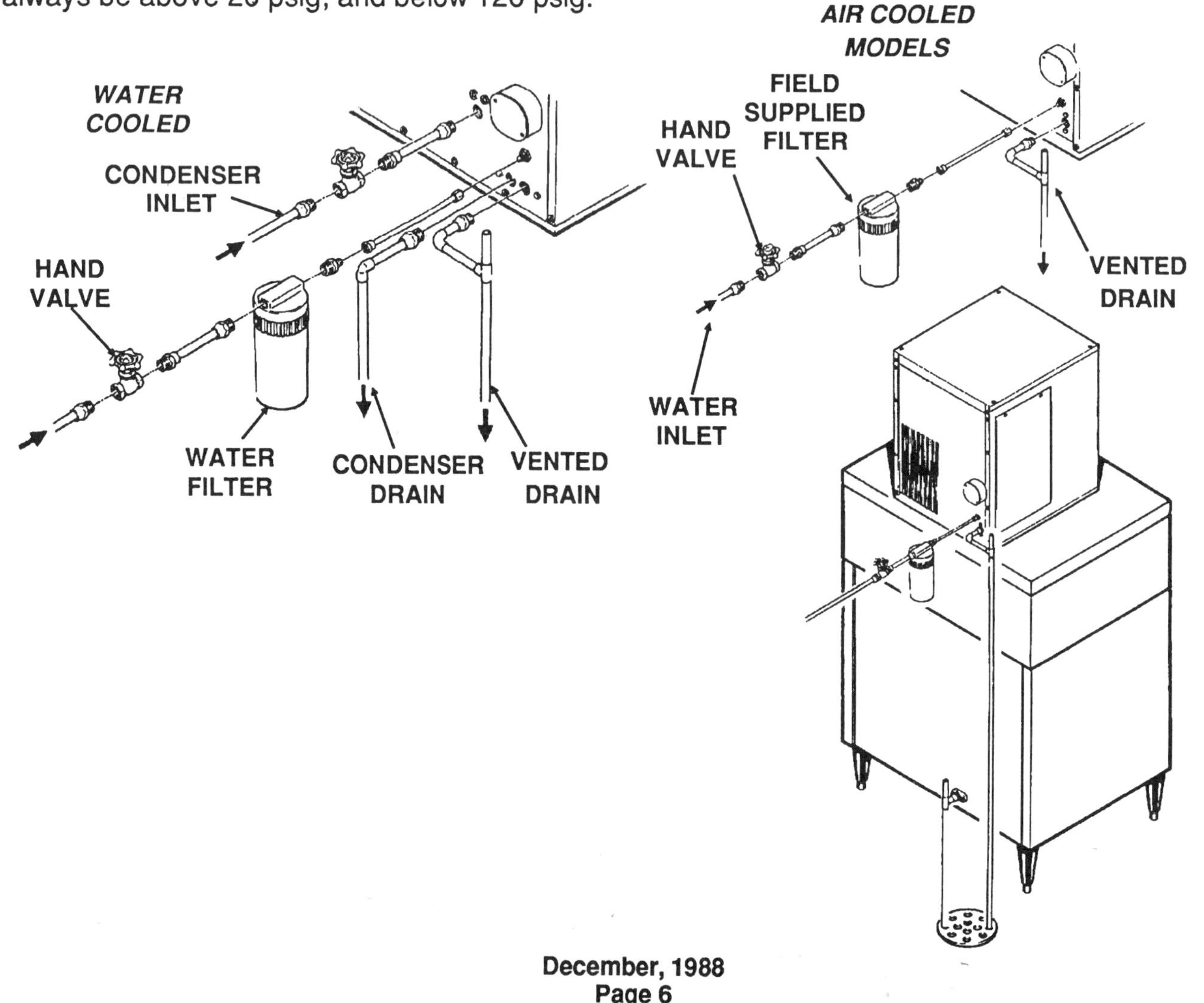

FOR THE ELECTRICIAN

CONFORM TO ALL APPLICABLE CODES

The electrical power to the unit is supplied through the junction box at the rear of the machine.

Check the nameplate (located on the back panel) for the voltage requirements, and for the minimum circuit ampacity. The machine requires a solid chassis to earth ground wire.

The ice maker should be connected to its own electrical circuit so it would be individually fused. Voltage variation must remain within design limitations, even under starting conditions.

All external wiring must conform to national, state, and local electrical codes. The use of a licensed electrician is required to perform the electrical installation.

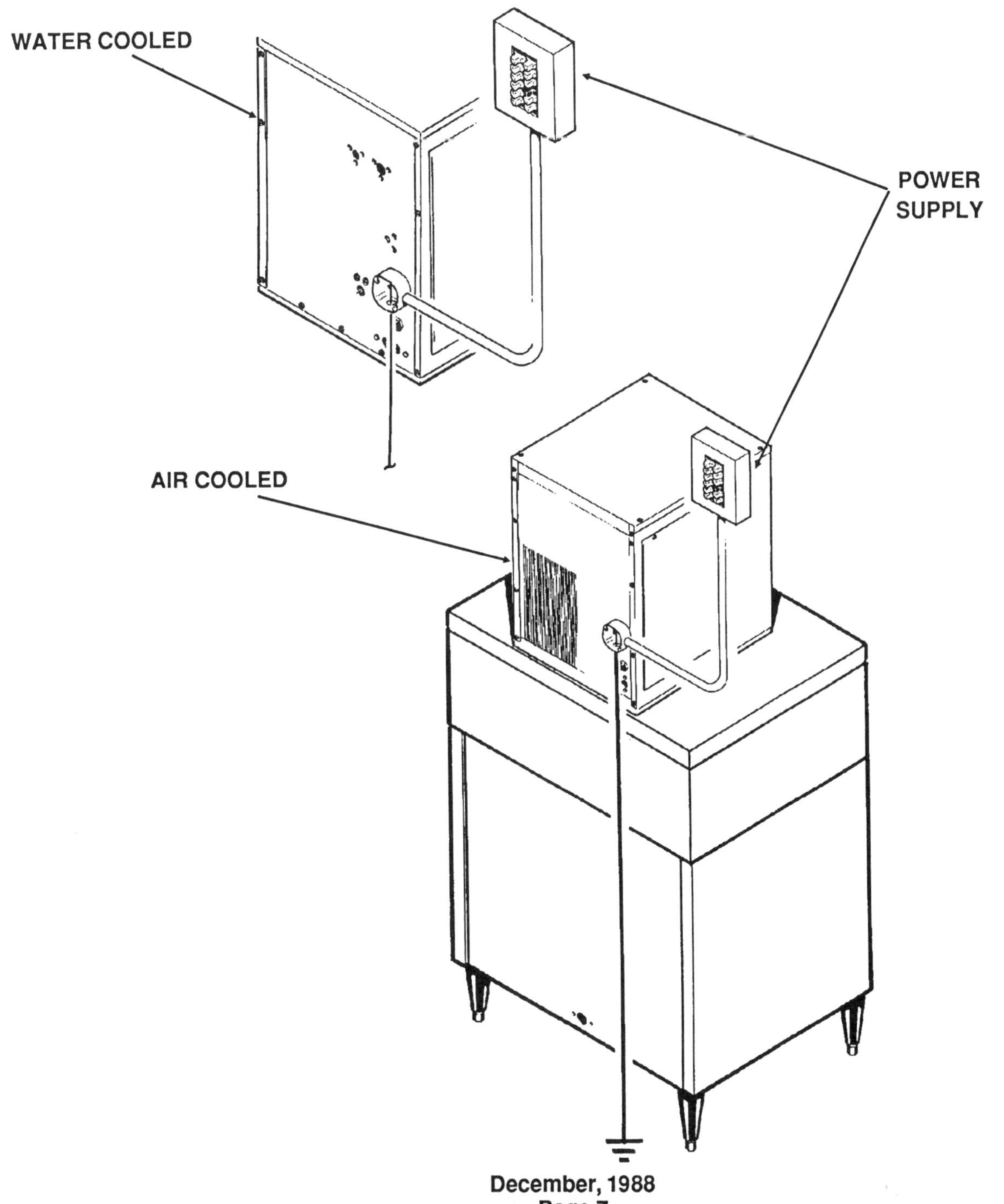

Final Check List

1. Is the ice system installed indoors in a location where the air and water temperatures are controlled, and where they do not exceed the design limitations?

2. Is there an electrical service disconnect within sight of the installed machine? Has the voltage been checked, and compared to nameplate requirements?

3. Have all the plumbing connections been made and checked for leaks?

4. Has the machine and bin been leveled?

5. Is there a minimum of 6" clearance at the back of the machine for proper service access and air circulation?

6. Is the water pressure a minimum of 20 psig?

7. Has the machine been secured to the bin?

8. Is there clearance over the top of the machine for service access?

9. Is there a water shut off valve installed near the machine?

10. Have all of the shipping blocks been removed?

START UP

Pre-Start Inspection

1. Remove the front and side service panels.

2. Check that the styrofoam shipping blocks have been removed.

3. Inspect the interior of the machine for loose screws or wires. Check that no refrigerant lines are rubbing each other. Check that the fan blade turns freely (air cooled).

4. Check that the unit is installed correctly according to the final check list (page 8).

Start Up

1. Go through the prestart inspection.

2. Open the hand valve, observe that water enters the water reservoir, fills the tube from the reservoir to the evaporator, and then shuts off. Check for leaks.

3. Switch the master switch on.

The electrical start up sequence is now on automatic.

A. There should be a short (15 second) delay before the gearmotor starts.

B. After the gearmotor starts, the compressor will start.

4. On air cooled models, the condenser will begin to discharge warm air, on water cooled models, the water regulating valve will open, and warm water will be discharged into the drain.

5. The unit should soon be making ice, if desired, the low side pressure can be checked: it should be 30 psig + or - 4 psig.

The suction line temperature at the compressor is normally very cold, nearly to the point of frost up to the compressor body, but not on it.

The air cooled discharge pressure will depend upon air and water temperatures, but should be between 200 psig and 280 psig.

The water cooled discharge pressure should be constant at about 220 psig.

The above numbers are for new, clean machines, you can expect to see some values higher, and some lower between different units.

6. **THERE ARE NO ADJUSTMENTS TO MAKE**, so replace the panels.

7. Clean and/or sanitize the storage bin interior, wipe off the exterior with a clean, damp cloth.

8. Give the owner/user the service manual, instruct him/her in the operation of the unit, and make sure they know who to call for service.

9. Fill out the manufacturers registration card, and mail it to the Scotsman Factory.

10. Fill out the field quality audit form, and mail it to the Scotsman factory.

NM650
COMPONENT DESCRIPTION

Control Box: Contains the electrical controls that operate the machine.

High Pressure Cut Out Switch: A manual reset switch sensing the high side refrigeration pressure. It is set to shut the machine off, and illuminate the reset switch light if the discharge pressure should ever exceed 450 psig.

Compressor: The refrigerant vapor pump.

Reservoir: Float operated, it maintains the water level in the evaporator at a constant level, it also contains the water level sensor.

Water Level Sensor: Senses if there is water in the reservoir to make ice out of. Will shut the machine off it there is none.

Ice Discharge Chute: Directs the ice produced by the evaporator into the storage bin.

Ice Level Sensor: An electronic "eye", it senses the presence of ice in the bottom of the ice discharge chute. Operates to turn the ice machine on and off automatically as the level of ice in the bin changes.

Gear Motor: An oil filled, speed reduction gearbox, driving the auger.

Condenser: Air or water cooled, where the heat removed in ice making is discharged.

Expansion valve: The refrigerant metering device.

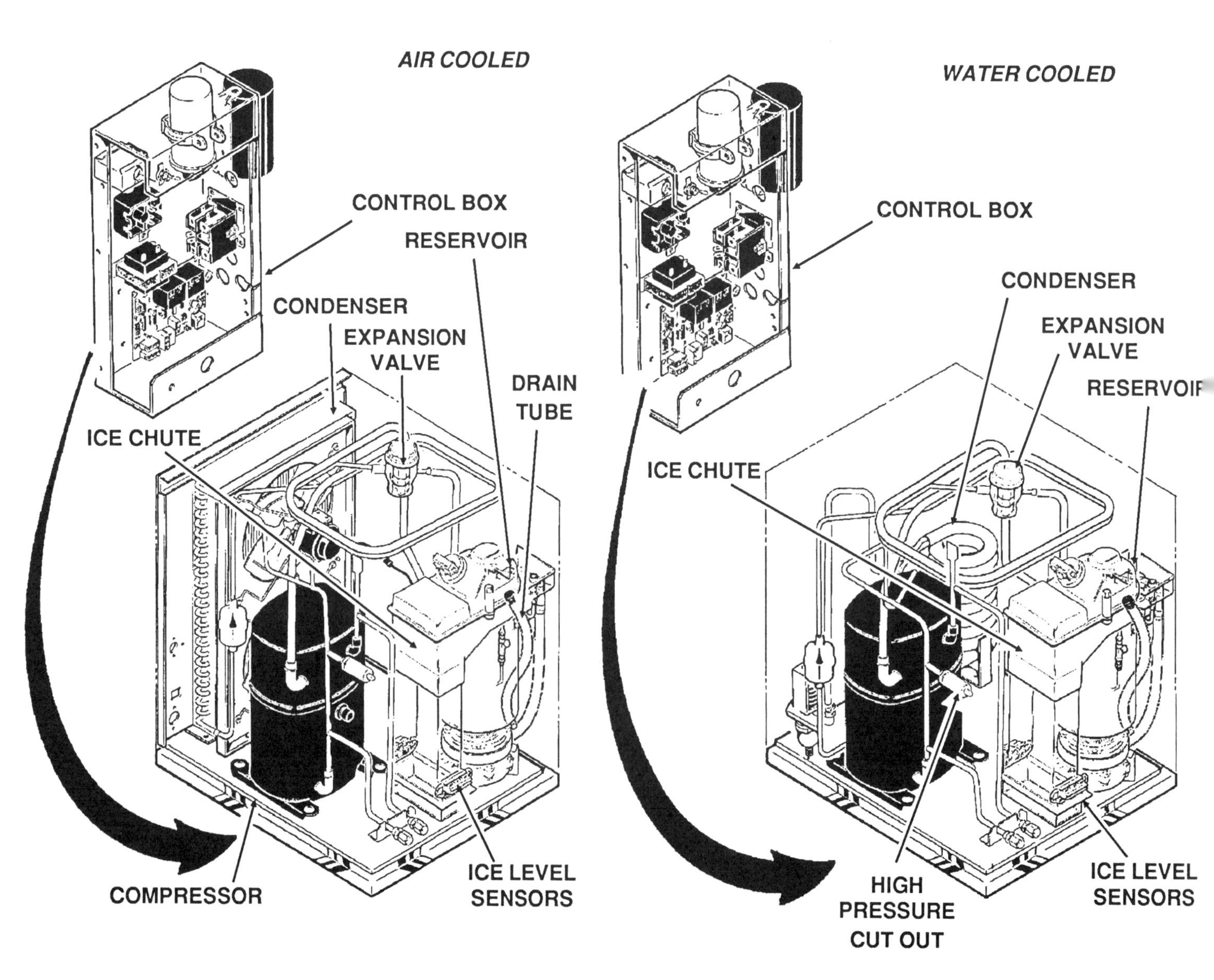

COMPONENT DESCRIPTION: Control Box

Contactor: A definite purpose contactor connecting the compressor and the remote condenser fan motor to the power supply.

Circuit Board: Controlling the ice machine through sensors and relays. The sensors are for ice level and water level. The relays are for the gear motor (with a built in time delay to clear the evaporator of ice when the unit turns off) and for the liquid line valve. The reset switch is mounted on the circuit board.

Transformer: Supplies low voltage to the circuit board.

Low Pressure Cut Out Switch: A manual reset control that shuts off the ice machine when the low side pressure drops below a preset point, 0-4 psig.

Potential Relay: The compressor start relay.

On/Off Switch: Manual control for the machine.

Reset Switch: Part of Circuit Board, manual reset. Lights up when unit shuts off from: ice discharge chute being overfilled (opening the microswitch at the top of the chute); low or high pressure switches opening.

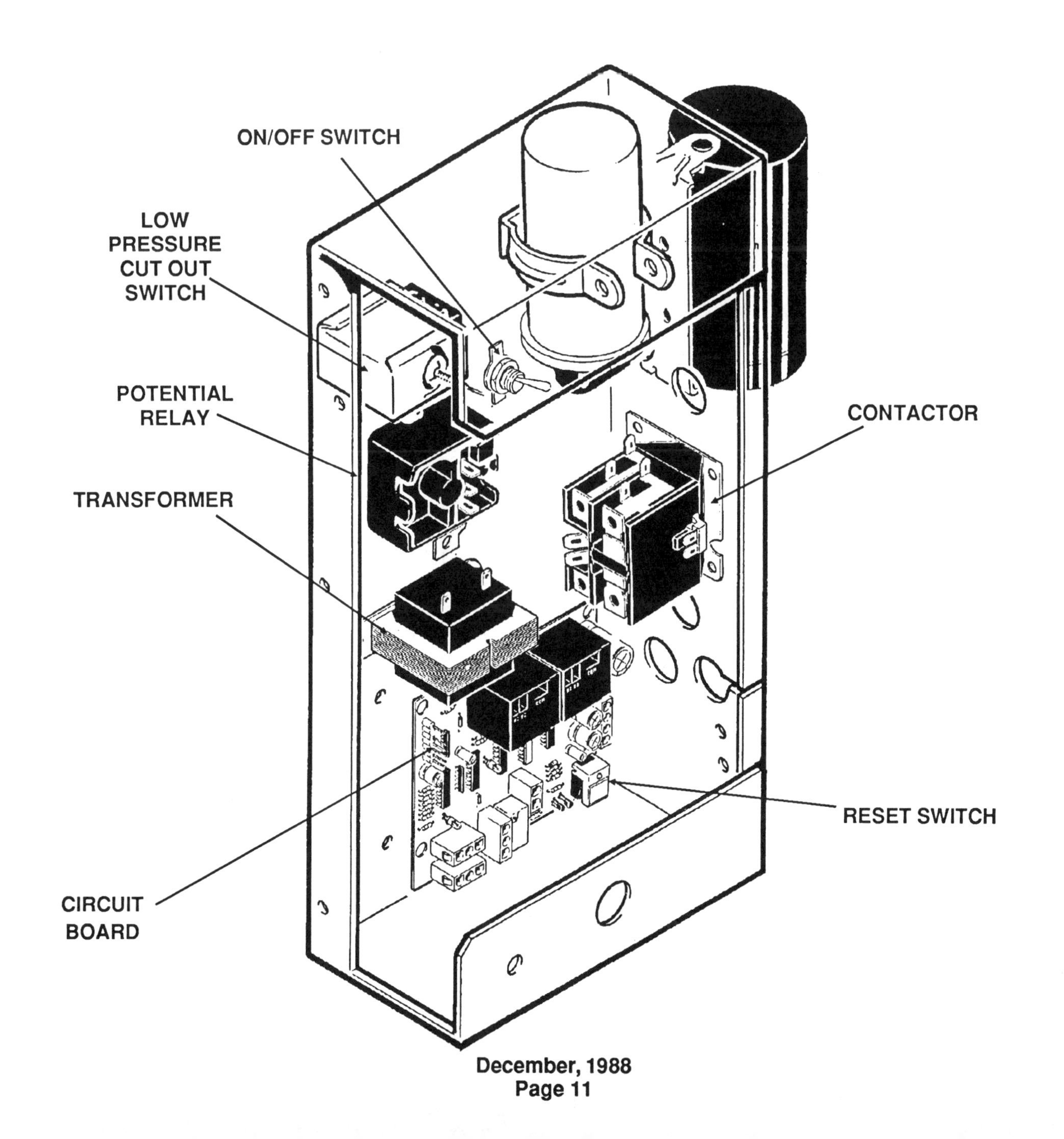

COMPONENT DESCRIPTION

Evaporator: A refrigerated vertical tube filled with water and containing a water seal and auger.

Auger: A solid stainless steel double spiral auger, it pushes the ice crystals up to the top of the evaporator.

Water Seal: A two part "face" seal, the top half rotating with the auger, the bottom half stationary, the sealing action being where the two seal "faces" meet.

Ice Sweep: A plastic cap with "fingers". It revolves with the auger to "sweep" the ice into the ice chute.

Breaker (Divider): Where the ice is compressed and much of the extra water is squeezed out of it before it is discharged into the bin.

Motor: A permanent split capacitor motor that drives the gear reducer.

Thrust Bearing: As the ice is pushed up the evaporator, the auger is thrust down, and pressure from the auger thrust is taken up by this bearing.

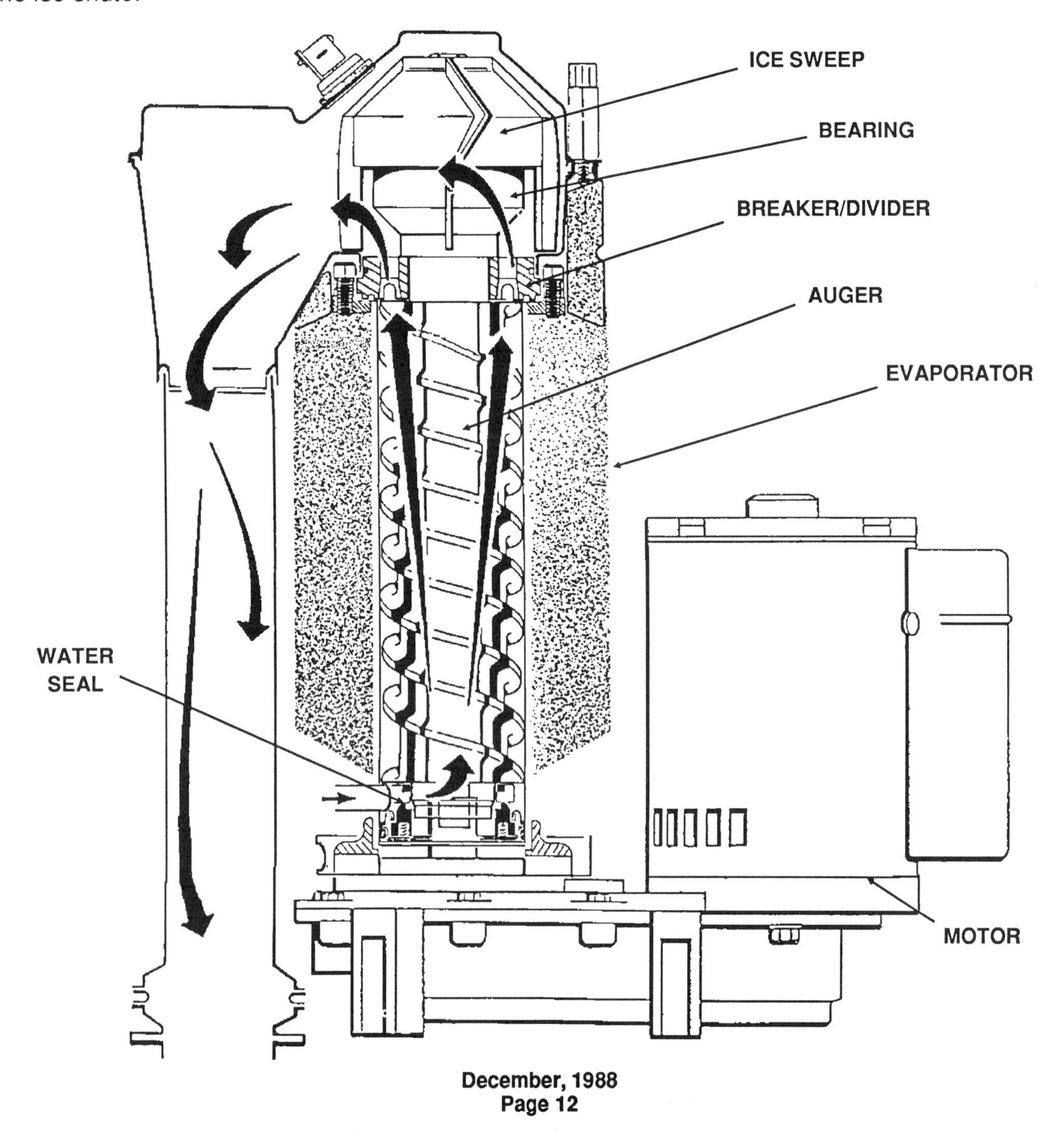

ELECTRICAL SEQUENCE

Refer the wiring diagram as needed.

If the machine is switched off at the master switch, but is otherwise ready to go, switching the master switch to on does the following:

- The bin empty light on the circuit board goes on
- There is a 15 second delay
- If there is enough water in the reservoir, the circuit board will allow the machine to start up.

Start up consists of:

- The compressor relay and auger motor relay become energized, connecting power to the windings of the auger motor.
- The auger motor starts, and the centrifugal switch closes, connecting power to the compressor contactor coil.
- The contactor is energized, connecting power to the compressor, and the compressor starts.
- As ice goes past the ice level sensors, the bin empty light will stay on, and the machine will continue to run, unless the ice stays between the sensors for more than 15 seconds (bin full). At that point, the bin empty light goes out, and the machine shuts down.

Shut Down consists of:

- The compressor relay opens.
- The compressor contactor opens
- The compressor stops
- The auger motor is run by the circuit board for 2.5 more minutes, clearing out ice in the evaporator, and then
- The auger motor relay opens, and the auger motor stops.

If the ice level sensor is clear (bin empty) for more than 15 seconds, the machine will start up again.

Another purpose of the circuit board is to turn the machine off if there is not enough water in the machine.

- When the water level in the reservoir falls below the water level sensor, the machine will "shut down"
- When the water refills the reservoir, the machine will start up again.

Separate from the circuit board:

- If the high pressure control (cut out switch) opens, the machine will stop immediately (through the relays on the circuit board) and cause the reset switch on the circuit board to light up. It must be manually reset at the control and at the reset switch on the circuit board.
- If the low pressure control (cut out switch) opens, the machine will stop immediately (through the relays on the circuit board) and cause the reset switch on the circuit board to light up. It must be manually reset at the control and at the reset switch on the circuit board.
- If the spout switch opens, the machine will stop immediately (through the relays on the circuit board) and cause the reset switch on the circuit board to light up. After it recloses the reset switch on the circuit board must be manually reset.
- The master switch is the manual control for the complete machine, but it is not a service disconnect.

Water enters the machine through the 3/8" male flare at the rear of the cabinet, goes to a strainer and then to the water reservoir which it enters through the float valve. The water then goes out the bottom of the reservoir tank to the bottom of the evaporator.

Reservoir overflow or evaporator condensation is routed to the drain. Water cooled models have a separate water circuit for the cooling water: it enters the fitting at the rear, goes to the water regulating valve, then to the water cooled condenser and down the drain.

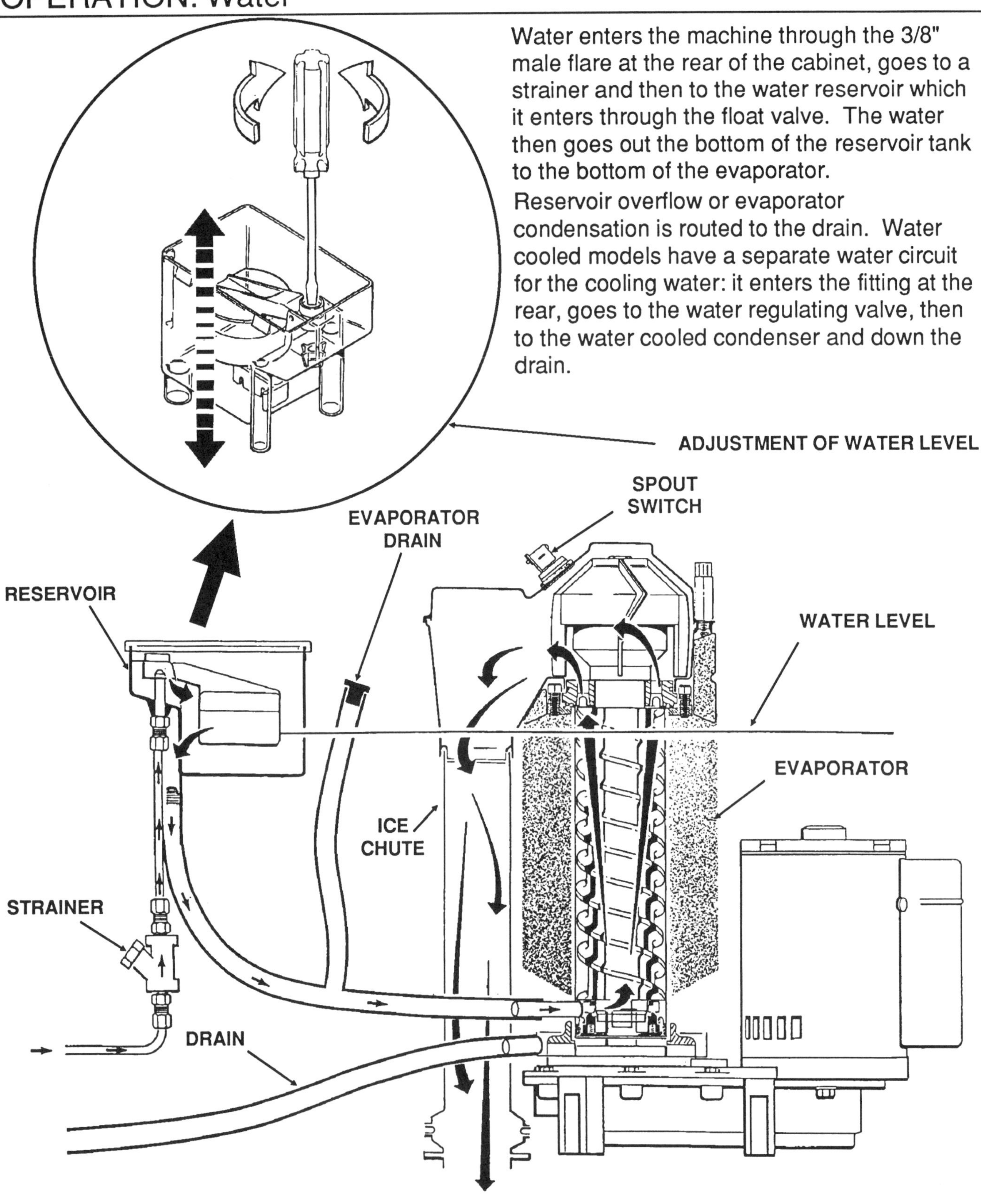

WATER SCHEMATIC

OPERATION: Refrigeration

Beginning at the compressor, the refrigerant 502 is compressed into a high temperature gas. The discharge line directs this gas to the condenser. At the condenser (air or water cooled) the gas is cooled by either air or water and it then condenses into a liquid. This high pressure liquid then goes through the liquid line to the expansion valve. The thermostatic expansion valve meters liquid refrigerant into the evaporator, the volume of liquid refrigerant depending upon the temperature of the evaporator; warmer evaporators get more refrigerant and colder evaporators get less. At the evaporator, the refrigerant enters an area of relatively low pressure, where it can easily "boil off" or evaporate. As it evaporates, it absorbs heat from the evaporator and whatever is in contact with it (such as the water inside it). After the evaporator, the refrigerant, now a low pressure vapor, goes through the suction line back to compressor, where the cycle is repeated.

Refrigeration Schematic

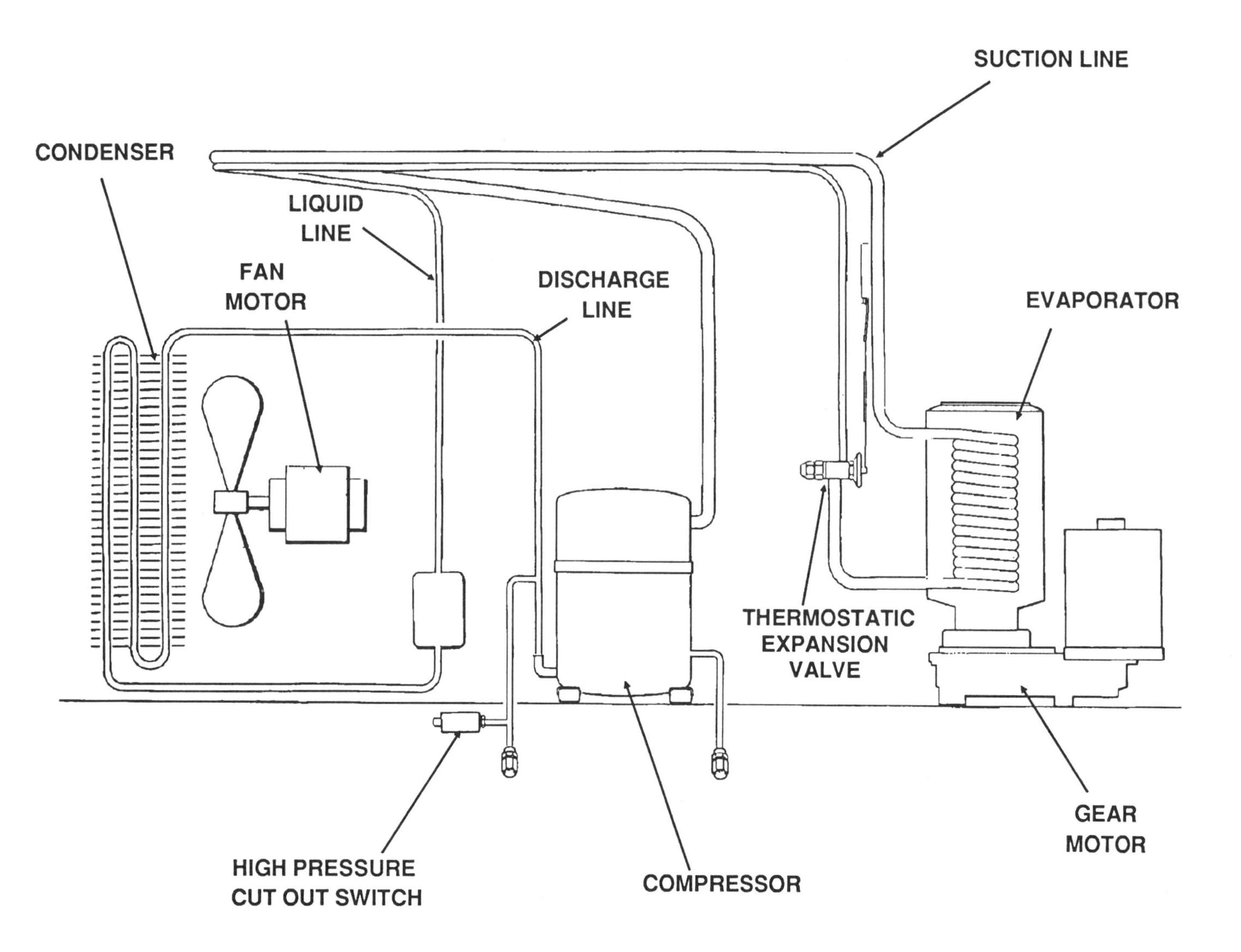

MAINTENANCE AND CLEANING

//

A Scotsman Ice System represents a sizable investment of time and money in any company's business. In order to receive the best return for that investment, it MUST receive periodic maintenance.

It is the USER'S RESPONSIBILITY to see that the unit is properly maintained. It is always preferable, and less costly in the long run, to avoid possible down time by keeping it clean; adjusting it as needed; and by replacing worn parts before they can cause failure. The following is a list of recommended maintenance that will help keep the machine running with a minimum of problems.

Maintenance and Cleaning should be scheduled at a minimum of twice per year.

//

//////////////////////////WARNING//////////////////////////

Electrical power will be ON when doing in place cleaning. Switch it OFF before completing the cleaning procedures.

//

ICEMAKING SYSTEM: In place cleaning

1. Check and clean any water treatment devices, if any are installed.
2. Remove screws and the front and top panels.
3. Move the ON-OFF switch to OFF.
4. Remove all the ice from the storage bin.
5. Remove the cover to the water reservoir and block the float up.
6. Drain the water reservoir and freezer assembly using the drain tube attached to the freezer water inlet. Return the drain tube to its normal upright position and replace the end cap.

//////////////////////////WARNING//////////////////////////

Scotsman Ice Machine Cleaner contains Phosphoric and Hydroxyacetic acids. These compounds are corrosive and may cause burns. If swallowed, DO NOT induce vomiting. Give large amounts of water or milk. Call Physician immediately. In case of external contact, flush with water. KEEP OUT OF THE REACH OF CHILDREN.

//

7. Prepare the cleaning solution: Mix eight ounces of Scotsman Ice Machine Cleaner with three quarts of hot water. The water should be between 90-115 degrees F.
8. Slowly pour the cleaning solution into the water reservoir until it is full. Wait 15 minutes, then switch the master switch to ON.
9. As the ice maker begins to use water from the reservoir, continue to add more cleaning solution to maintain a full reservoir.
10. After all of the cleaning solution has been added to the reservoir, and the reservoir is nearly empty, switch the master switch to OFF.
11. After draining the reservoir, as in step 6, wash and rinse the water reservoir.
12. Remove the block from the float in the water reservoir.
13. Switch the master switch to ON
14. Continue ice making for at least 15 minutes, to flush out any cleaning solution. Check ice for acid taste - continue icemaking until ice tastes sweet.

//////////////////////////WARNING//////////////////////////

DO NOT USE any ice produced from the cleaning solution.

Be sure no ice remains in the bin.

//

15. Remove all ice from the storage bin.
16. Add warm water to the ice storage bin and thoroughly wash and rinse all surfaces within the bin.
17. Sanitize the bin interior with an approved sanitizer using the directions for that sanitizer.
18. Replace the panels.

//////////////////////////////**WARNING**//////////////////////////////

Disconnect electrical power before beginning.

//

1. The bin control uses devices that sense light, therefore they must be kept clean enough so that they can "see". At least twice a year, remove the bin control sensors from the base of the ice chute, and wipe the inside clean, as illustrated.

2. The ice machine senses water level by a probe located in the water reservoir. At least twice a year, the probe should be removed from the reservoir, and the tip wiped clean of mineral buildup.

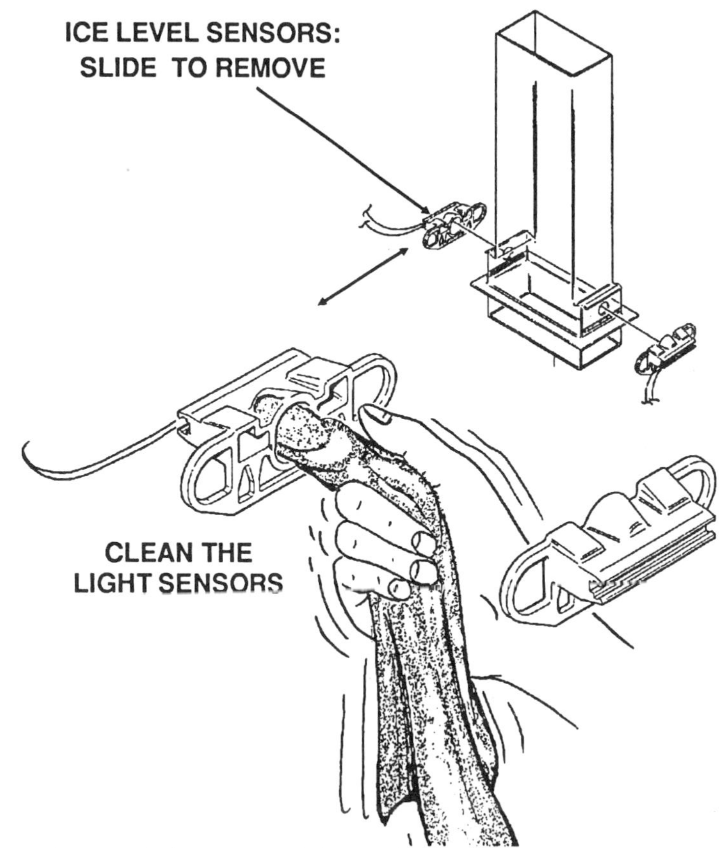

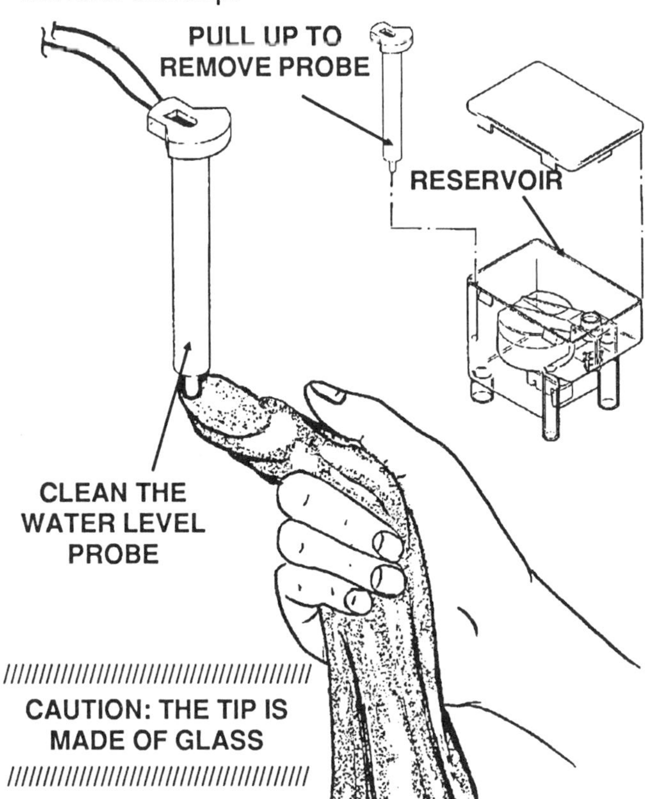

//

CAUTION: THE TIP IS MADE OF GLASS

//

3. The bearing in the breaker should also be checked at least **two times per year**.

A. Check the bearing by:

- removing the ice chute cover
- unscrewing the ice sweep
- removing the water shed
- using a spanner wrench and unscrewing the breaker cover.
- unscrewing the auger stud

Inspect the assembly, looking for wear. See Removal and Replacement to replace bearing or seals. Reverse to reassemble.

4. Check and tighten all bolts and screws.

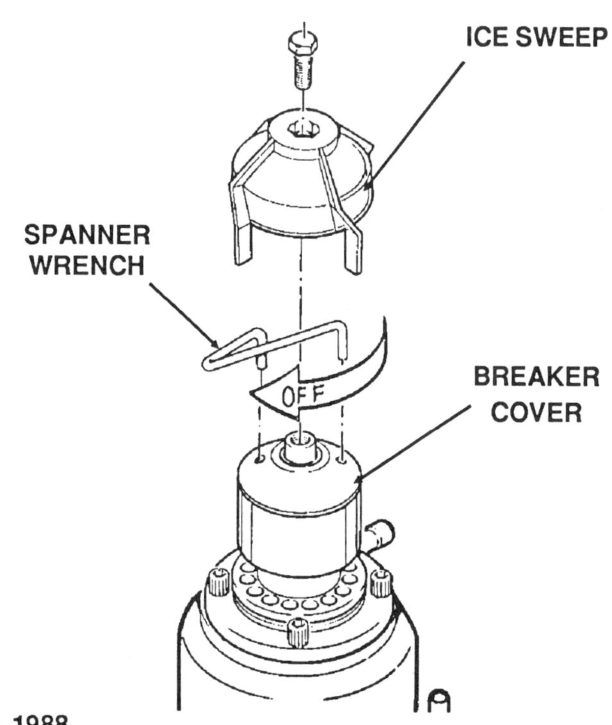

//////////////////////////WARNING////////////////////////////

Disconnect electrical power before beginning.

//

5. Clean the air cooled condenser.

The air flow on this model is from front to back, so the inside of the machine will have to be available to clean the air cooled condenser. Use a vacuum cleaner or coil cleaner if needed. Do NOT use a wire brush.

A. Disconnect electrical power, and remove the filter. The filter may be cleaned or replaced.

B. Clean the condenser: the condenser may appear to be clean on the surface, but it can still be clogged internally. Check with a flash light from the front to see if light can be seen though the condenser fins. Reverse to reasemble.

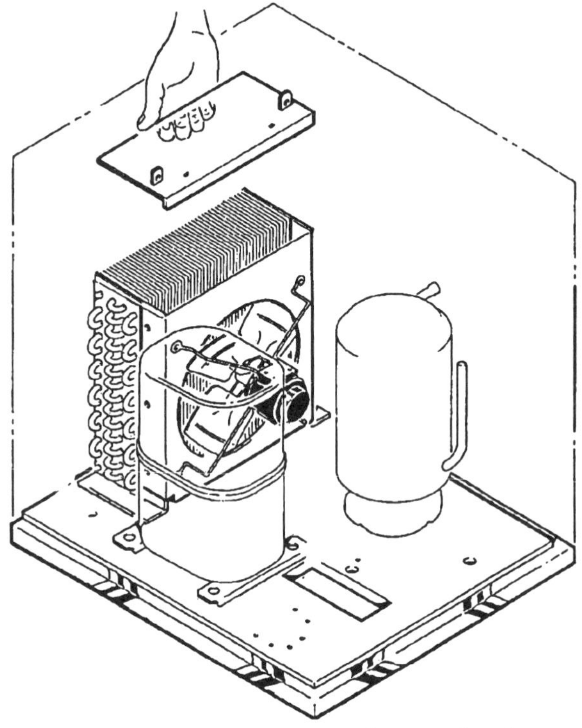

Step 2: Remove the top portion of the fan shroud.

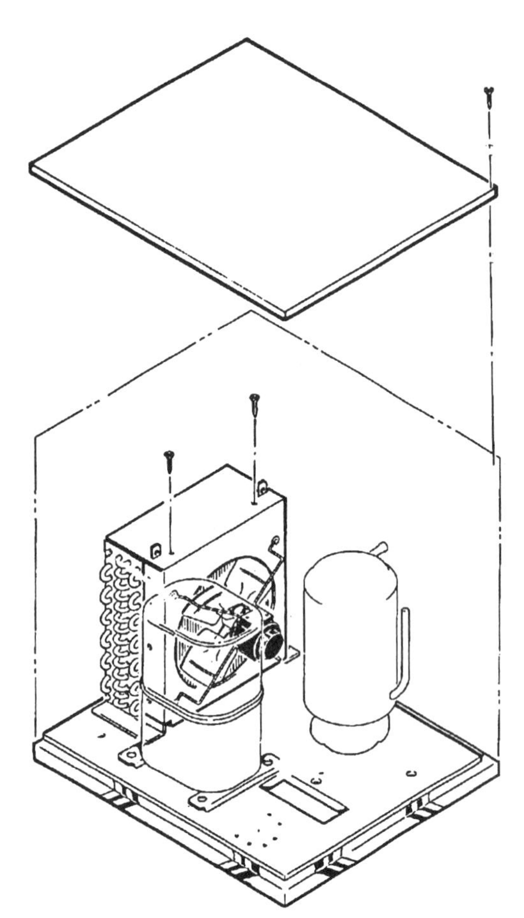

Step 1: Remove the top panel.

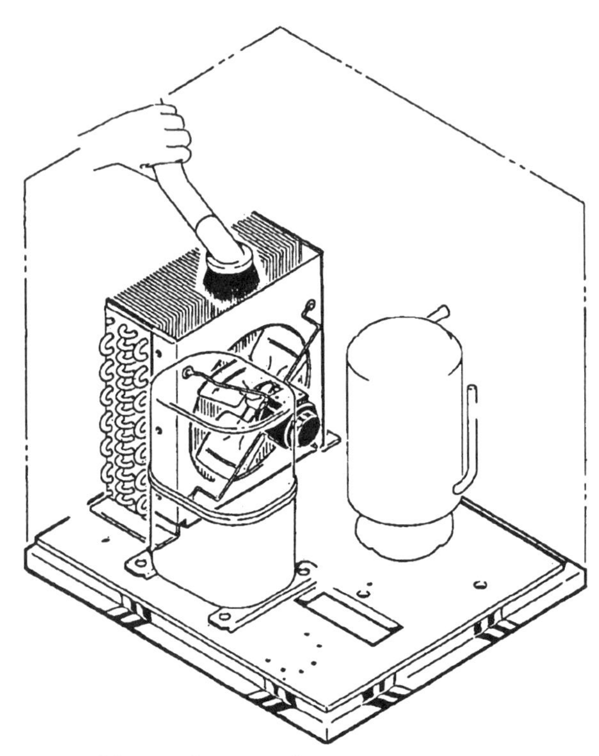

Step 3: Clean the condenser.

MAINTENANCE AND CLEANING: Auger

In some areas, the water supply to the ice maker will contain a high concentration of minerals, and that will result in an evaporator and auger becoming coated with these minerals, requiring a more frequent removal than twice per year. If in doubt about the condition of the evaporator and auger, the auger can be removed so the parts can be inspected.

Note: Water filters can filter out suspended solids, but not dissolved solids. "Soft" water may not be the complete answer. Check with a water treatment specialist regarding water treatment.

For more information on removal of these parts, see REMOVAL AND REPLACEMENT.

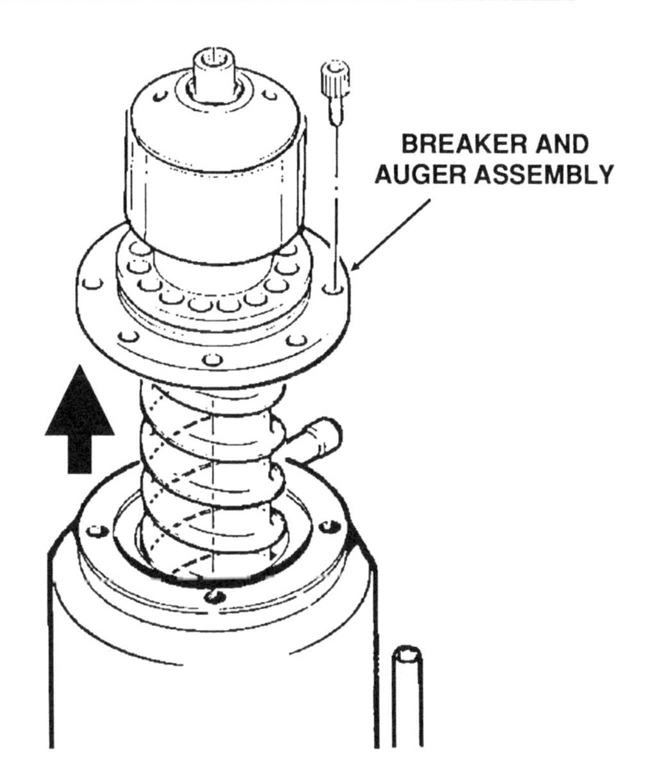

///////////////////////////**WARNING**/////////////////////////

Disconnect electrical power, and shut off the water supply.

Use care when removing the auger, it has sharp edges.

///

1. To remove the auger, remove the front and top panels.
2. Remove 3 hex studs holding ice chute cover to ice chute, and remove cover.
3. Unscrew and remove ice sweep.
4. Loosen band clamp under ice chute, and remove ice chute from evaporator.
5. Remove 4 allen screws holding breaker to evaporator.
6. Pull up to remove auger.

After the auger has been removed, allow the auger to dry: if the auger is not bright and shiny, it must be cleaned.

Clean the auger and evaporator as required. DO NOT HONE THE EVAPORATOR.

7. Replace the water seal.
8. Reverse to reassemble.

SERVICE DIAGNOSIS: Condition - No Ice Being Produced

STATUS: *NOTHING OPERATES*

A. Check: **Voltage** to the unit, restore it if there is none. Compare to the nameplate.

B. Check: The **master switch**, switch ON if off.

C. Check: The **3 reset switches**, (circuit board, high and low pressure): depress and release each switch. If the still does not start, check: the spout switch; the high and the low side pressures.

D. Check the **low pressure cut out,** if closed, go to **E;** if open, it could be due to:

- Low refrigerant charge
- The auger not turning
- Restricted system
- TXV not opening

1. Check the low side pressure, the low pressure cut out opens at pressure below 4 psig. If open, reset and:

    a. Check if the **auger** is turning, if it is not, remove the **gearbox** and:

        Check for internal damage, repair and replace in the machine.

    b. Check for low charge, add some refrigerant, if the unit will operate,(normal low side pressure being about 30 psig) stop and look for a leak, repair, replace the drier, evacuate, and weigh in the nameplate charge. If, with added charge, the unit does **not** operate:

        Check for a restricted system, replace the drier, evacuate, and weigh in a nameplate charge.

        Check for a **Thermostatic Expansion Valve** that does not open, if defective, replace it. Replace the drier, evacuate, and weigh in the nameplate charge.

E. Check the **high pressure cut out,** if closed go to **F**; if open check:

1. The pressure control opens at 450 psig. **Check** the high side pressure, reset the control, and observe: on water cooled, that water soon begins to flow from the condenser drain; or, on air cooled, that the fan is forcing air through the condenser. If the unit trips out on pressures below 450 psig, replace the control. If the pressures rise above the trip out point, and the unit shuts down:

    a. Check for adequate water flow on water cooled, if adequate, clean the interior of the condenser. If the pressures are still too high replace the water regulating valve.

    b. Check for adequate air flow on air cooled. Clean the condenser and (if used) the filter. If the air flow is poor because of the installation, advise the user that the unit should be moved, or the air around it kept cooler.

        Check the fan motor for tight bearings and proper rotation.

        Check that the fan blades are clean, and the fan secure to the fan motor shaft.

F. Check the **spout switch**. It opens from excess pressure of ice inside the ice chute: this should only happen when the machine does not shut off when the ice storage bin is full. This switch will reset when the ice melts, but the machine will not resart until the reset switch on the circuit board is pressed.

G. Check the **water level** in the reservoir. The machine will not run if there is not enough water in the reservoir.

1. Restore/adjust water level. See the next step.

SERVICE DIAGNOSIS: Condition - No Ice Being Produced

STATUS: *NOTHING OPERATES*

H. Check: The gear **motor**, if it will not run, the compressor will not run. If no power to it:
Check: The **indicator lights** on the circuit board, the **bin empty** light should be ON, the **no water** light should be OFF .

1. If the **bin empty** and **no water** lights are off, check the **transformer**.
   a. Transformer "load" side should have 12 to 15 volts. If not, check the "line" side. The line side should have between 208-230 volts. If the line side has the correct voltage and the load side does not, replace the transformer.
2. If the transformer is good, and the **bin empty** light is OFF, check the **ice level sensors**.
   a. Remove sensors by sliding them sideways out of the ice chute. Visually inspect them, clean if needed.
   b. Look through the ice chute "eye" hole for something blocking the ice chute.
   c. If the unit still does not run, replace the ice level sensors.
   d. If the bin empty light is still OFF, check the **circuit board**.
      1. Unplug "opto trans" and "LED" connectors from the circuit board.
      2. Plug "opto trans" and "LED" connectors from the Scotsman Electronic Control Testor Model NM1 into the circiut board.
         a.Move the "bin full" switch on the tester to the full position. The bin full light on the tester should be ON, if not, replace the circuit board.
         If the bin full light on the tester is ON, move the tester switch to "bin empty" the light on the tester should go OFF and the bin empty light on the circuit board should go ON. If not, replace the circuit board. If it does as above, and the machine still does not run, replace the ice level sensors.
3. If the transformer is fine, and the "no water" light is ON, check the **water level sensor.**
   a. Check the water level in the **reservoir,** restore if low. If the water level is ok:
   b. Remove the water level sensor from the reservoir and clean the tip if dirty.
   CAUTION: THE TIP IS MADE OF GLASS
   c. Replace the water level sensor. If the no water light is still on, check that the "water sen" plug is firmly plugged into the circuit board.
   d. If the no water light is still on,
      1. Unplug the "water sen" connector from the circuit board.
      2. Plug "water sen" connector from the Scotsman Electronic Control tester into the circuit board.
         a. Move the water switch on the tester to "no water" and the no water light on the circuit board should go on. If not, replace the board.
         b. Move the water switch to the "water" position, the no water light should go off, if not, replace the circuit board.
         c. If after the above, the machine still will not run, replace the water level sensor

MORE INFORMATION ON THE TESTER CAN BE FOUND ON THE LAST PAGES OF THE MANUAL.

SERVICE DIAGNOSIS: Condition - No Ice Produced

STATUS: *GEARMOTOR OPERATES, COMPRESSOR DOES NOT*

A. Check the **compressor relay.**

The relay is on the circuit board, if it does not supply power to the contactor coil, the compressor will not run.

1. Check for power at the contactor coil, if none:

   a. Check for power at the compressor relay at the circuit board.

   If there is power at the relay, but none at the contactor coil,

   Check for an open wire between the relay and the contactor.

2. Check the contactor coil. If the coil is open, replace the contactor.

3. Check the auger drive motor centrifugal switch. If, when the drive motor is running, contact 4 (black wire removed) has no power, and all of the above switches have been checked, replace the centrifugal switch, or the drive motor.

4. If the compressor relay on the circuit board has power on the NO contact, but not on the COM contact, replace the circuit board.

B. Check the compressor

1. Check the compressor start relay.
2. Check the start capacitor.
3. Check the windings of the compressor for open windings or shorts to ground.

Replace those items found defective.

SERVICE DIAGNOSIS: Condition - Low Ice Production

STATUS: *EVERYTHING IS OPERATING*

A. Check the air cooled condenser for dirt. Clean as required. Check the head pressure on water cooled. Adjust as required. If the head pressure is very high:

1. Air cooled. Check for high air temperatures, or restrictive air flow. Correct as needed.
2. Water cooled. Check for high water temperatures, or low water pressure. Correct as needed.
3. The refrigerant may contain non condensable gases, purge, evacuate, and recharge per nameplate.

B. Check the evaporator

1. Clean the evaporator, the mineral build up will adversely affect the ice machines production.
2. Check the evaporator for water leaks, replace the water seal if found to be leaking.
3. Check the low side pressure; normal is about 30 psig. If low, assume a refrigerant leak, locate, repair and recharge.

   If no leak, the TXV may be restricted, defective or not adjusted properly. If needed, replace the TXV, evacuate, and recharge per nameplate.
4. Check the insulation on the evaporator. It should be dry, with no wet spots or frost. If the insulation has failed: repalace the evaporator or add extra insulation in the form of foam tape to the evaporator.

C. Check the compressor.

1. The compressor may be inefficient.
   - a. Check the amp draw, if low change the compressor.
   - b. if the amp draw is normal, pinch off the suction line to check the pull down capability of the compressor. The compressor should pull down to 25 inches of vacuum and hold there for three to five minutes.

REMOVAL AND REPLACEMENT: Water Reservoir & Bin Controls

WATER RESERVOIR

1. Shut off the water supply to the icemaker.
2. Remove front panel and reservoir cover.
3. To remove float only, pry the mounting flanges apart enough to lift one float pivot pin out of the flange hole, and pull float up and out of the reservoir.
4. To remove reservoir, disconnect water inlet compression fitting at reservoir inlet.
5. Remove drain hose from reservoir.
6. Remove evaporator inlet hose from reservoir.
7. Remove mounting screws from reservoir bracket, and remove reservoir from icemaker.
8. Reverse to reassemble.

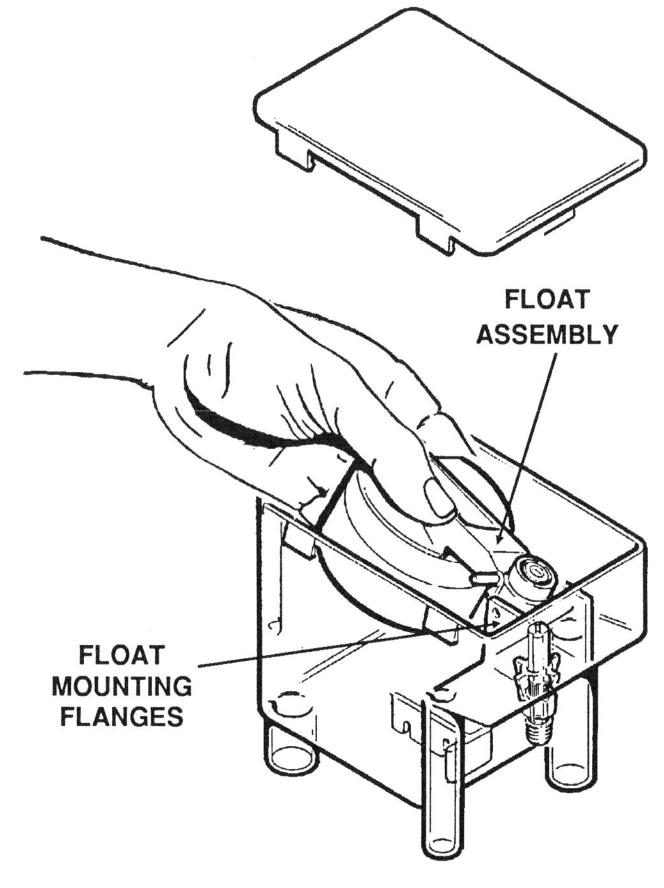

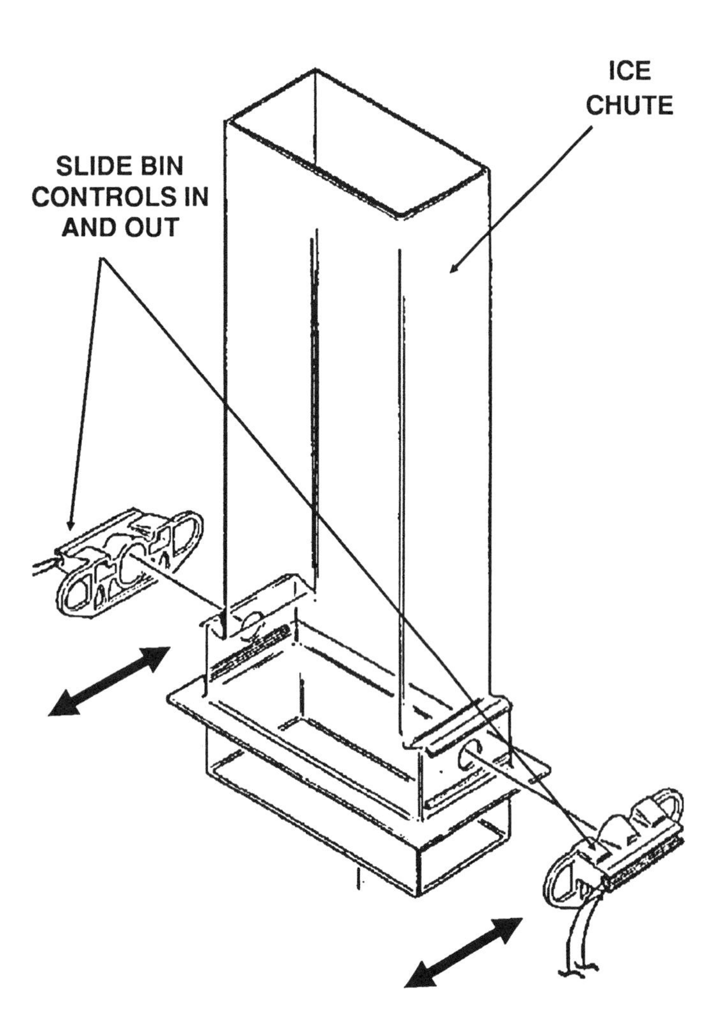

BIN CONTROLS (Ice Level Sensors)

1. Disconnect electrical power.
2. Remove front panel.
3. Remove control box cover.
4. Locate ice chute, at the base of the chute, in front of and behind it are two plastic bin control mounts.
5. Slide each bin control to the left, and in the control box, disconnect the electrical leads connecting the bin control to the circuit board.
6. Reverse to reassemble, be certain that the bin controls are aligned so that the ice level sensors are visible (centered) through the holes in the ice chute.

REMOVAL AND REPLACEMENT: Bearing And Breaker

Note: Removal of the auger, water seal, evaporator and gearmotor must begin at the top of the assembly.

To Remove the Breaker Bearing Assembly:

///////////////////////////////**WARNING**///////////////////////////

Disconnect the electrical power to the machine at the building source BEFORE proceeding with any repair.

//

1. Remove panels and disconnect electrical power.
2. Unscrew three studs and remove ice chute cover.
3. Unscrew and remove ice sweep.
4. Remove insulation halves from outside of ice chute, loosen band clamp under ice chute, lift up and remove ice chute.
5. The breaker may be removed from the auger and evaporator without disturbing the auger.

 a. Use spanner wrench and unscrew breaker cover from breaker (left hand threads)

 b. Unscrew auger stud from top of auger.

 c. Unscrew 4 allen head cap screws holding breaker to evaporator.

 d. Lift up, and remove breaker/bearing assembly from auger & evaporator.

6. Service the bearing. Check for rust, rough spots and damage.

 a. The bearing is pressed into the breaker, to remove the bearing and replace it an arbor press is needed.

 b. Replace lower seals before installing new bearing in breaker.

Note: seals must be pressed in with a tool pushing against the outer edge only, they will not install by hand.

Replace parts as required. Re-grease bearing with Scotsman part no. 19-0609-01 bearing grease. Replace top seal, and check the o-rings, replace if cut or torn.

7. Reverse to reassemble: specific tools and materials are required to install properly.

a. Add food grade grease such as Scotsman part number 19-0569-01 to the seal area before installing on the auger.

b. Check the seal to shaft areas for cuts, or rough spots: none are permitted.

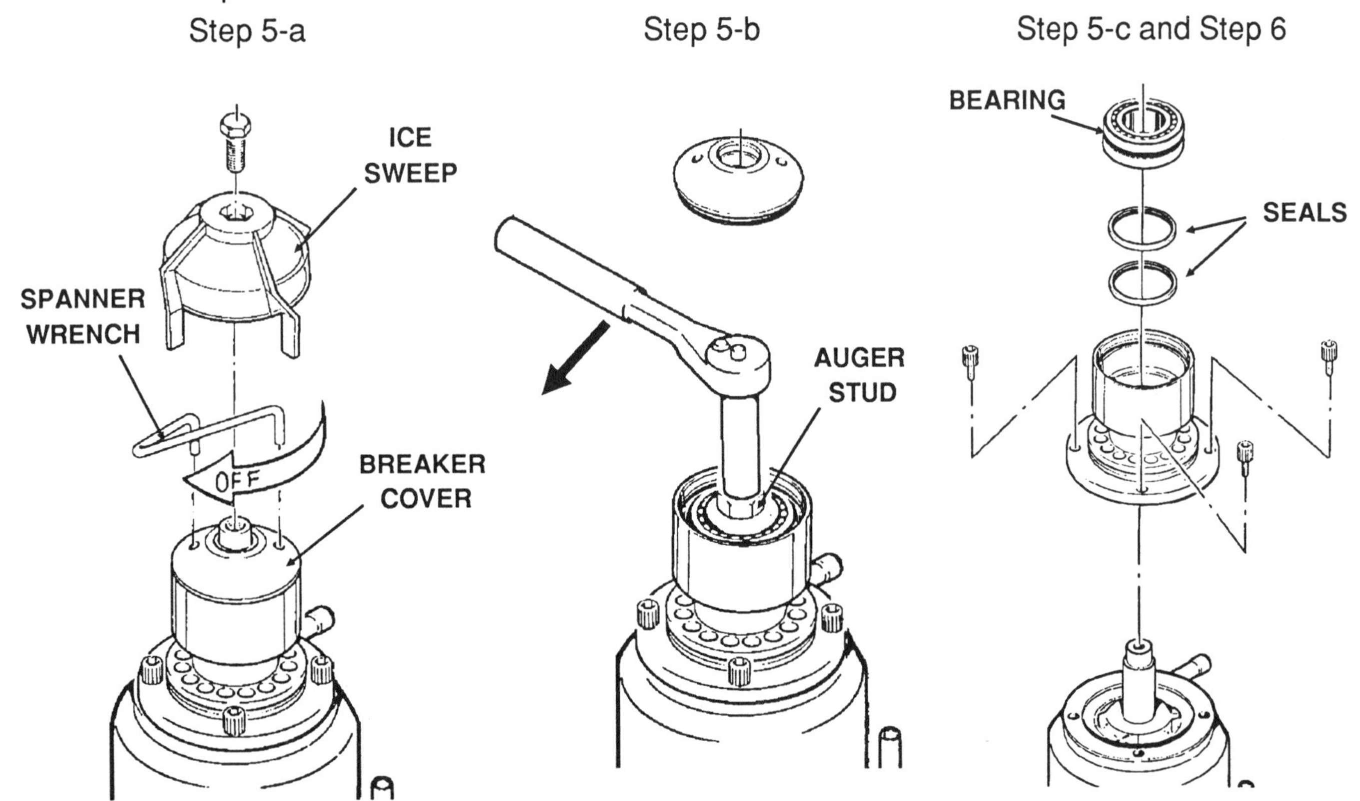

NM650
REMOVAL AND REPLACEMENT: Auger

//////////////////////////////**WARNING**//////////////////////////

Disconnect electrical power.

Use care when handling auger, it has sharp edges.

//

To Remove the Auger:

Turn off the water to the machine, and unclip the evaporator drain hose, pull it down and drain the evaporator into the bin or a container.

1. The top panel must be removed.
2. Remove ice chute cover.
3. Unscrew ice sweep.
4. Loosen band clamp and remove ice chute body.
5. The auger and breaker/bearing may now be removed as an assembly.

a. Unscrew 4 allen head cap screws holding breaker to evaporator.

b. Lift up on breaker and remove auger from evaporator.

BREAKER AND AUGER ASSEMBLY

Note: If the auger is stuck, the breaker must be removed from the auger.

The breaker may be removed from the auger and evaporator without disturbing the auger.

a. Use spanner wrench and unscrew stainless breaker cover from breaker (left hand threads)

b. Unscrew auger stud from top of auger.

c. Unscrew 4 allen head cap screws holding breaker to evaporator.

d. Lift up & remove breaker from evaporator.

e. If the auger is stuck use a slide hammer type puller to pull on the auger at the threaded hole. The size of that hole is 5/8"-18.

Inspect the auger, the critical areas of the auger are:

1. The auger body. It should be clean and shining. Sometimes an auger will appear clean when wet, but after it is dry it will be seen to be stained. Scrub the auger with ice machine cleaner and hot water.

//////////////////////////WARNING////////////////////////////////

Ice machine cleaner is an acid. Handle it with extreme care, keep out of the reach of children.

//

2. The water seal area. Because the auger has been removed, the water seal will have to be replaced. Remove the water seal top half from the auger, and inspect the auger for minerals clean as required.

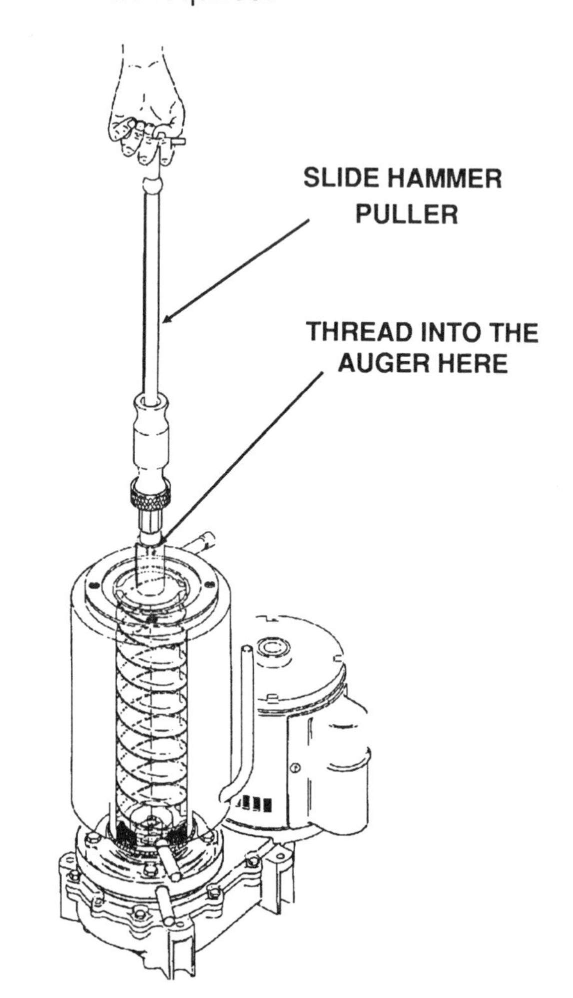

REMOVAL AND REPLACEMENT: Water Seal

To Remove the Water Seal:

(Assuming all steps to remove the auger have been performed.)

1. The gearmotor/evaporator assembly will have to be exposed.

2. Remove the 4 hex head cap screws holding the evaporator to the gearmotor assembly. Lift the evaporator up and off of the gearmotor.

3. Remove the snap ring or wire retainer from the grove under the water seal.

REMOVAL OF THE WATER SEAL

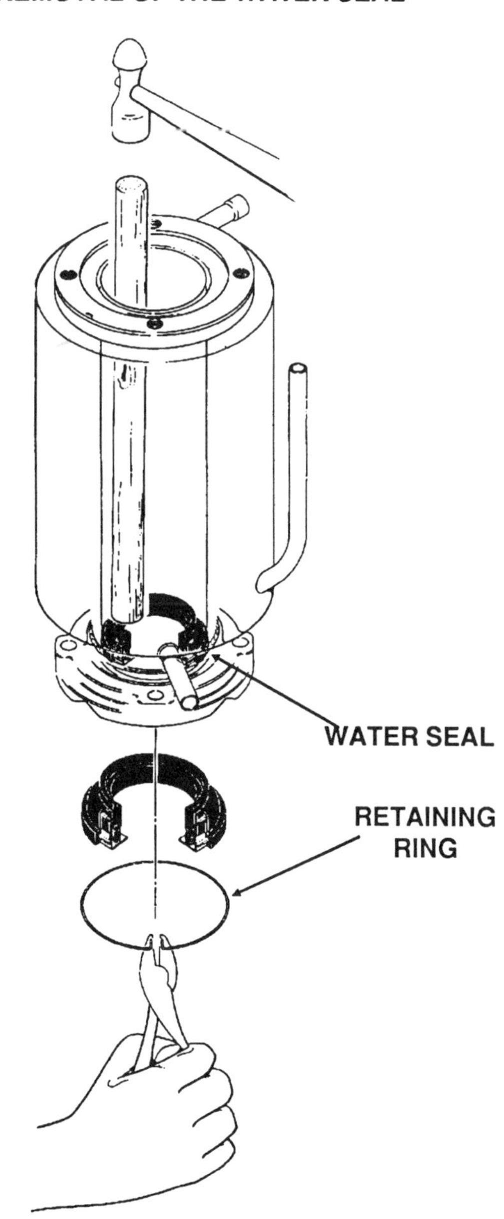

4. Pull or drive out the lower half of the water seal.

To Replace the Water Seal:

1. Lubricate the water seal with water, and push the water seal into the bottom of the evaporator slightly past the grove for the snap ring.

2. Replace the snap ring and pull the water seal down against it.

3. The part of the water seal that rotates with the auger must also be replaced. Remove the old part from the auger and clean the mounting area.

REPLACING THE WATER SEAL

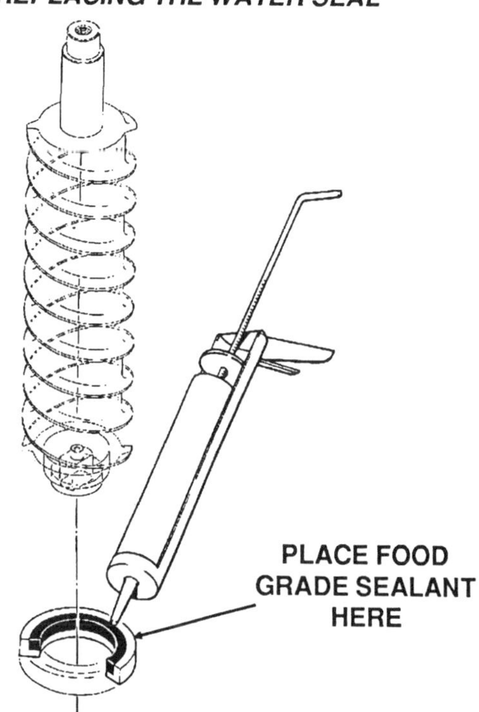

4. Place a small bead of food grade silastic sealant (such as 732 RTV or Scotsman part number 19-0529-01) on the area of the auger where the water seal is to be mounted.

5. Carefully push the water seal (rubber side against the auger shoulder and the silastic.)

//////////////////////////////CAUTION////////////////////////////

Do not get any silastic onto the face of the seal.

//

6. Allow the auger and seal to air dry until the silastic is dry on the surface.

7. If the original water seal was leaking, it would be a good idea to inspect the interior of the gearmotor.

//WARNING///
Disconnect electrical power before begining removal procedures.
///

To Replace the Evaporator:

(Assuming all the steps for removal of the thrust bearing, breaker, auger, and water seal have been performed.)

1. Discharge the refrigerant from the ice maker.

2. Unsweat the refrigerant connections:

a) At the thermostatic expansion valve outlet.

//////////////////////////////CAUTION//////////////////////////

Heat sink the TXV body when unsweating or resweating the adjacent tubing.

//

b) At the suction line at the joint about 3" from the evaporator.

3. Remove the evaporator.

4. Unsweat the drier from the liquid line.

5. After installing a new water seal in the new evaporator (see "To Replace the Water Seal") sweat in the new evaporator at the old tubing connections.

6. Install an new drier in the liquid line.

7. Evacuate the system until dehydrated, then weigh in the nameplate charge. Check for leaks.

8. Install auger, breaker, breaker bearing assembly, and ice discharge chute in reverse order of disassembly. See "To Reassemble Evaporator and Auger"

To Reassemble the Evaporator and Auger

1. After the gearmotor has been inspected, fasten the evaporator to the gear motor, be sure that the number of shims indicated on the gear case cover is in place between the gearcase cover and the drip pan gasket. Torque the bolts to 110 inch pounds.

2. Lower the auger into the evaporator barrel, slightly turning it to match up with the drive end. Do Not Drop Into the Evaporator.

3. Complete the reassembly by reversing the disassembly for the breaker & thrust bearing assembly.

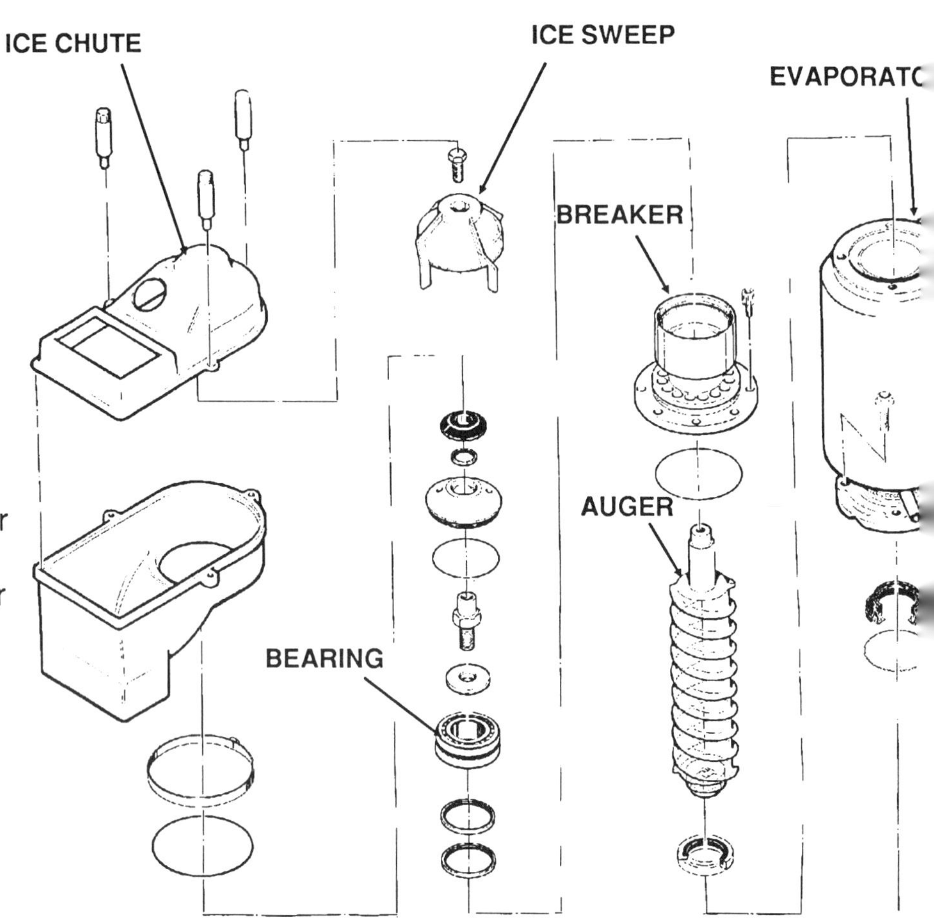

REMOVAL AND REPLACEMENT: Gearmotor

//////////////////////////////**WARNING**//////////////////////////////

Disconnect Electrical Power Before Begining Removal Procedures.

//

To Remove and Repair the Gearmotor Assembly:

(Assuming that the procedures through removal of the water seal have been performed.)

1. Remove the electrical wires from the gear drive motor.
2. Unscrew the 4 cap screws holding the gearmotor to the base of the machine.
3. Remove the gearmotor from the icemaker.

Bench test the gearmotor, check for oil leaks, noise, and amp draw.

To Inspect the gearmotor.

A) Remove the cap screws holding the gearmotor case halves together and pry the two cases apart.

B) To lift off the cover, lift up until you can feel internal contact, then pull the cover towards the output gear end, and then lift the cover (with drive motor attached) up and away from the gear motor case.

Note: The gearcase cover, output gear, bearings and output shaft are a pressed together assembly. Replace as a unit.

C) Inspect the oil, gears, and bearings. If the oil level and condition is acceptable, quickly check the gears and bearings. They are likely to be fine if the oil is. If there is evidence of water in the oil (rusty bearings and gears; the oil having a creamy white appearance; oil level too high) carefully inspect the bearings and gears. If in doubt about the condition of a part, replace it. The oil quantity is 16 fluid ounces, do not overfill.

Note: The gears and bearings are available only as pressed together sets.

D) After replacing parts as required, (if any) reassemble the gearcase. The two smaller gears and the oil should be in the lower case, the output gear will be with the cover. As you lower the cover onto the lower case, the cover will have to be moved closer to the second gear after the output gear has cleared the second gear top bearing.

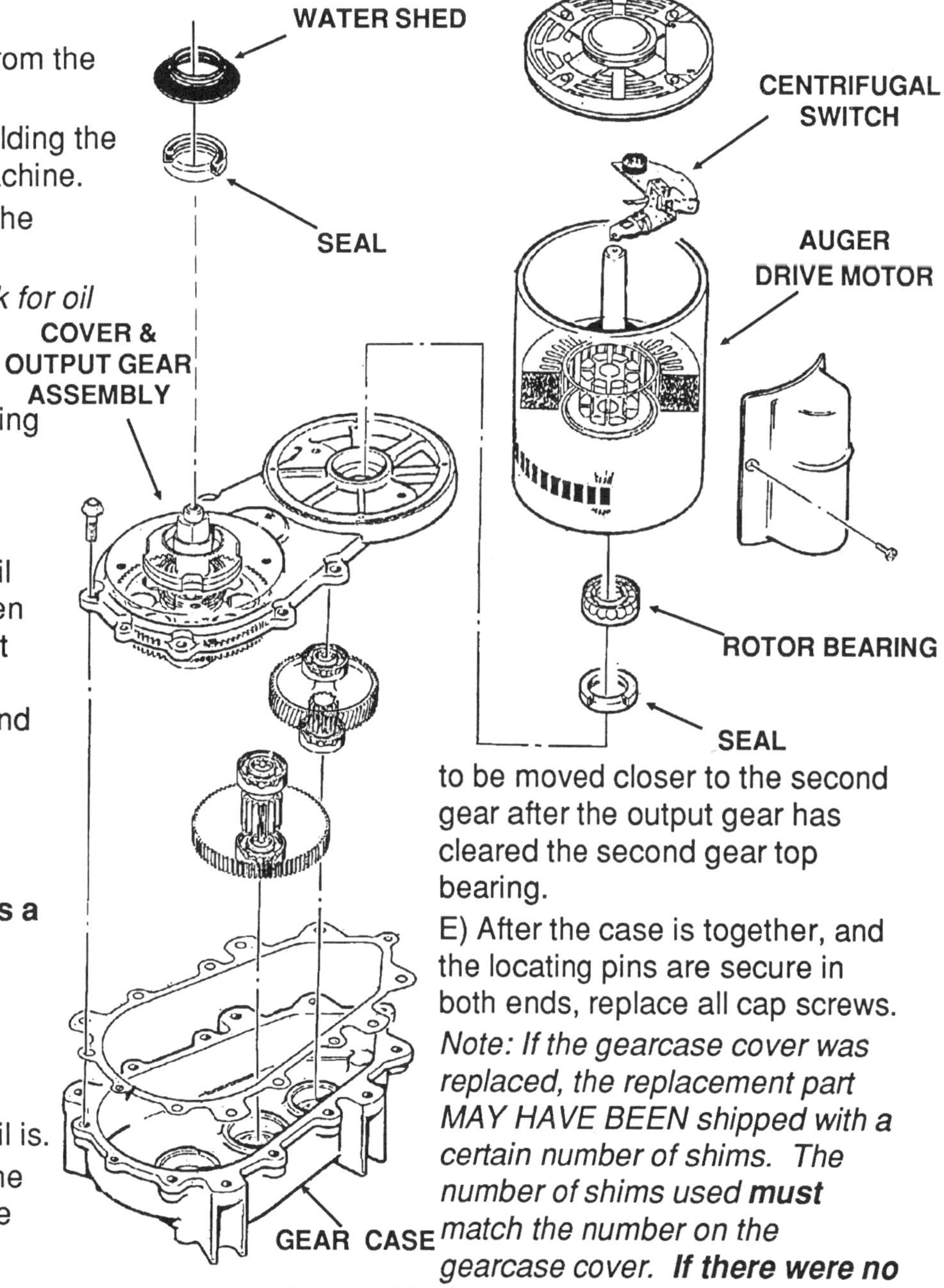

E) After the case is together, and the locating pins are secure in both ends, replace all cap screws.

*Note: If the gearcase cover was replaced, the replacement part MAY HAVE BEEN shipped with a certain number of shims. The number of shims used **must** match the number on the gearcase cover.* ***If there were no shims with the new parts, do not use any shims.*** *Do not use the old shims.*

Bench test the gearmotor, check for oil leaks, noise, and amp draw.

WATER SHED
CENTRIFUGAL SWITCH
SEAL
AUGER DRIVE MOTOR
COVER & OUTPUT GEAR ASSEMBLY
ROTOR BEARING
SEAL
GEAR CASE

//**WARNING**//

These procedures require the machine to be connected to the power supply. The voltages of the electronic circuit are very low, but HIGHER VOLTAGES ARE PRESENT IN THE UNIT. Do not touch anything but the tester while the unit is being checked out. Make all connections to the circuit board with the ELECTRICAL POWER OFF.

//

INSTRUCTIONS FOR USING TESTER, model FC1 (Optional, order part no. A33942-001)

*(These instructions assume that the unit **will not run,** and prior investigation of electric power, controls, and mechanical parts indicates that the electronic circuit may be at fault.)*

If the "Reset" indicator (located in the "reset" switch) is off and the "NO WATER" indicator is lit, but inspection shows that the water level in the reservoir is above the top of the water level sensor, OR the "BIN EMPTY" indicator is off while inspection shows that the ice level sensors are properly aligned, clean and not obstructed, use the tester as follows:

Bin Control *Note: All testing is done with the electrical power on, the master switch on, and all reset switches "reset".*

1. Unplug "photo trans" and "LED" connectors from the circuit board.

2. Plug "photo trans" and "LED" connectors from the tester into the circuit board.

a. Move the "bin full" switch on the tester to Full. The light on the tester should be ON.

If the light on the tester is not on, the circuit board should be replaced.

b. If the light on the tester IS on, move the "bin full" switch to Bin Empty. The light on the tester should go OFF, and the Bin Empty light on the circuit board should go ON.

If the Bin Empty light is ON, wait 10-20 seconds for the machine to start, if the machine starts, replace the ice level sensors. If the Bin Empty light does not come ON, the circuit board should be replaced.

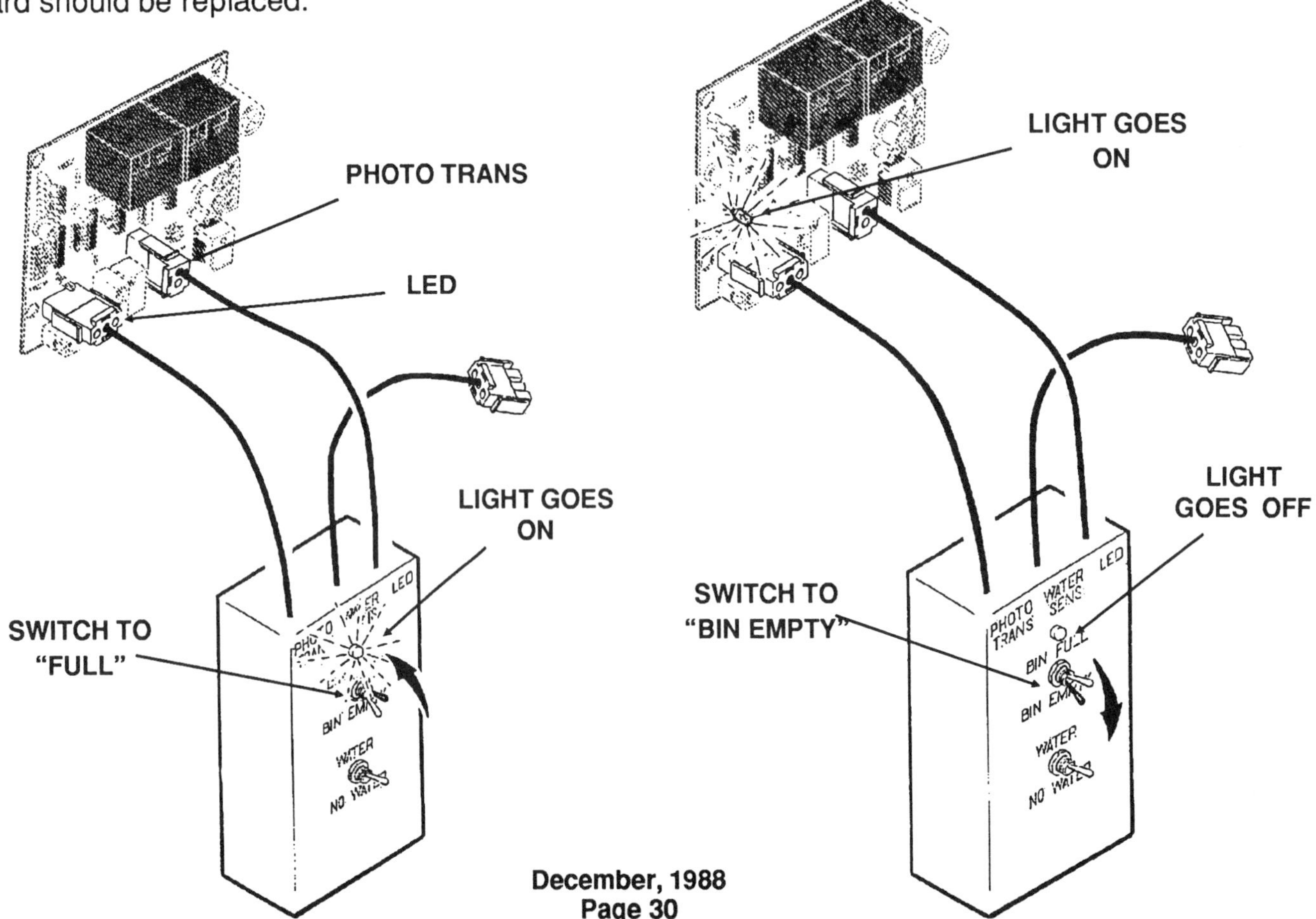

CIRCUIT BOARD TESTING

Water Level

1. Unplug "water sen" connector from control board.
2. Plug "water sen" connector from Scotsman tester into circuit board.

a. Move "water" switch on tester to No Water position. The No Water light on the circuit board should go ON. If not, replace the circuit board.

b. Move the "water" switch on the tester to the Water position. The No Water light on the board should go OFF. If not replace the circuit board. If the light does go off, replace the water level sensor.

If the Bin Empty light is ON, wait 10-20 seconds for the machine to start. The machine should start.

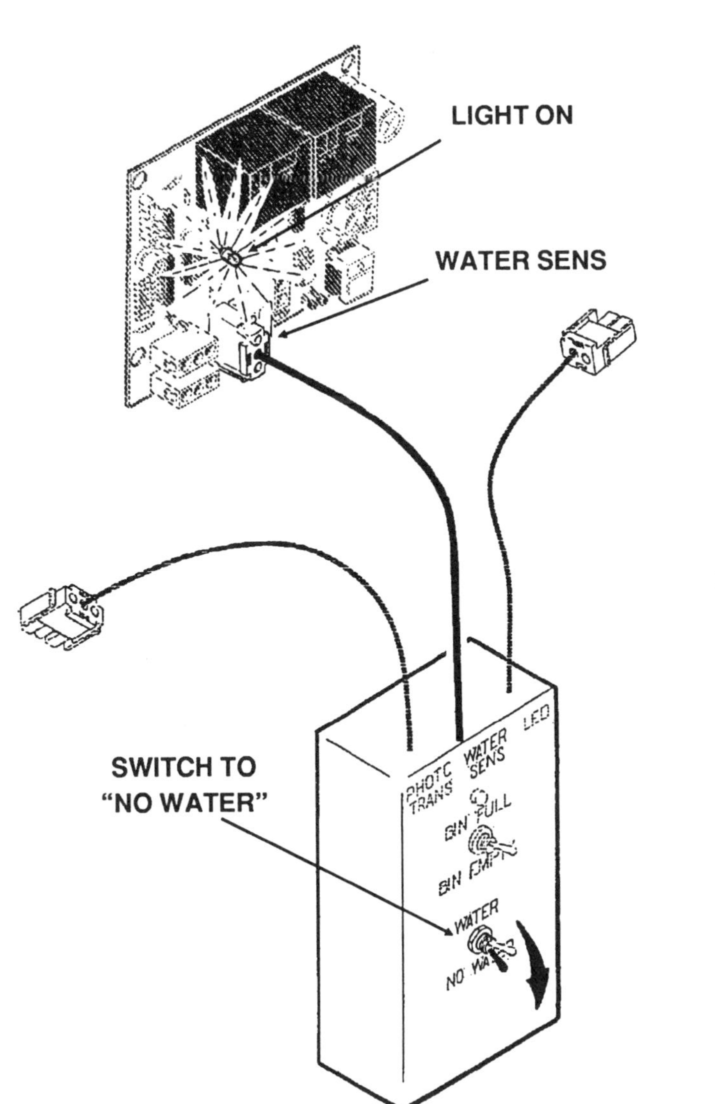

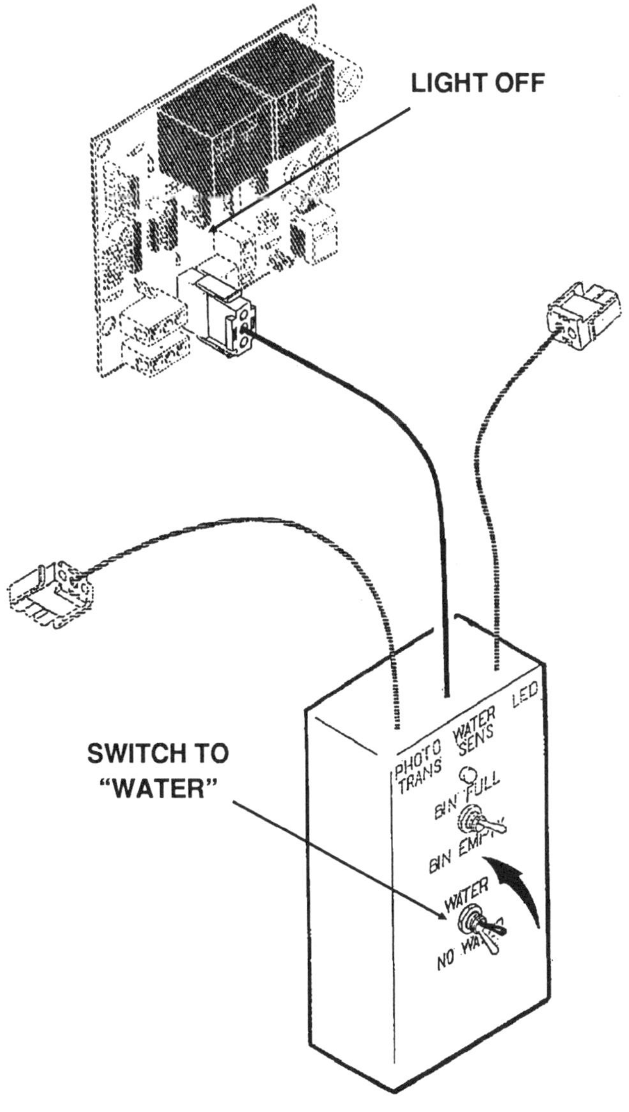

NM650 SERVICE PARTS

This parts list contains the service parts and wiring diagrams for this model. Check the model number in question to be sure that it is applicable to this parts list.
Caution: This model ice maker has been manufactured in two voltages, for electrical components, check to be certain that the part is the correct voltage for the machine.

TABLE OF CONTENTS

NM650 SERVICE PARTS

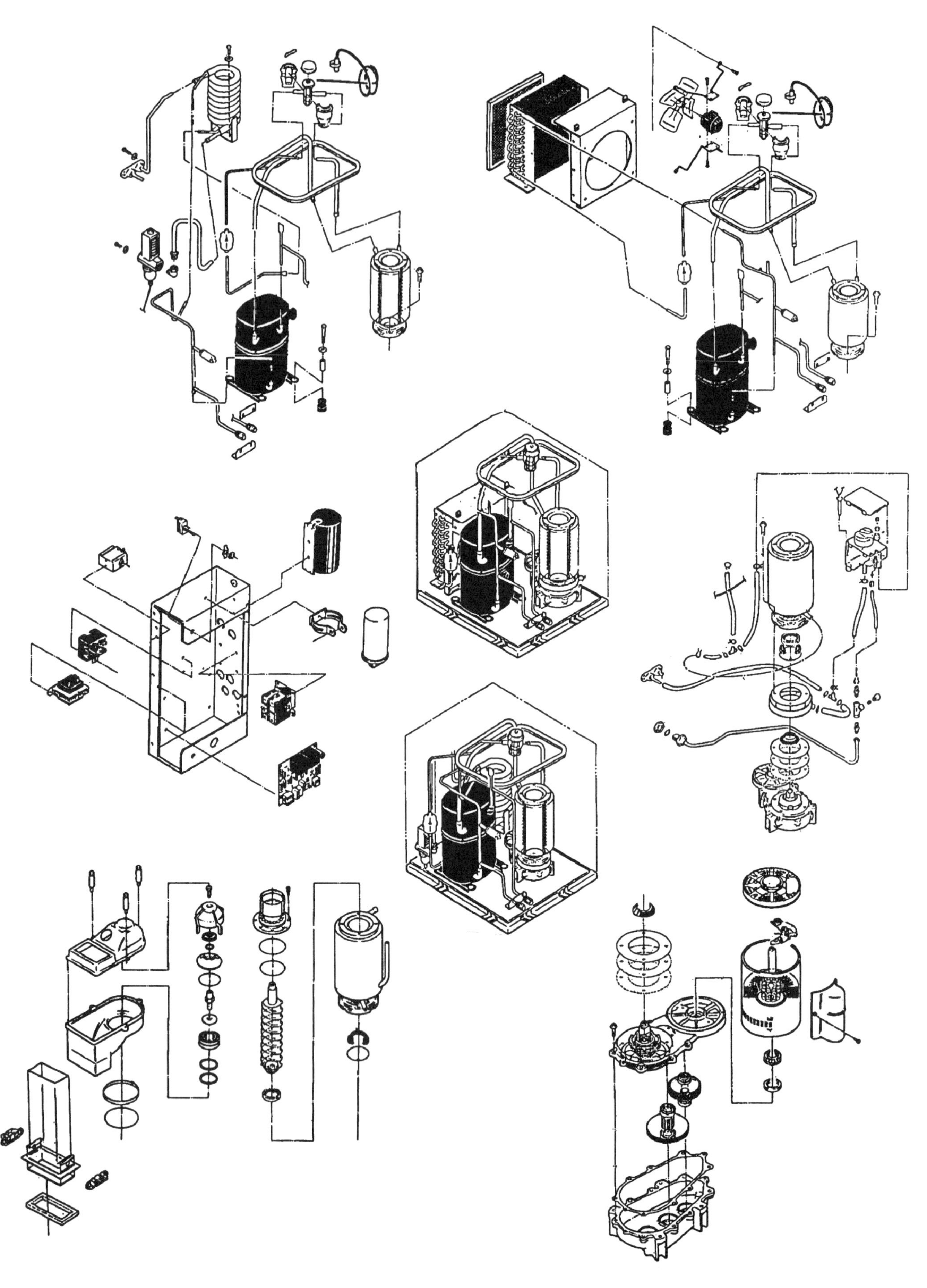

NM650 SERVICE PARTS
Cabinet Assembly

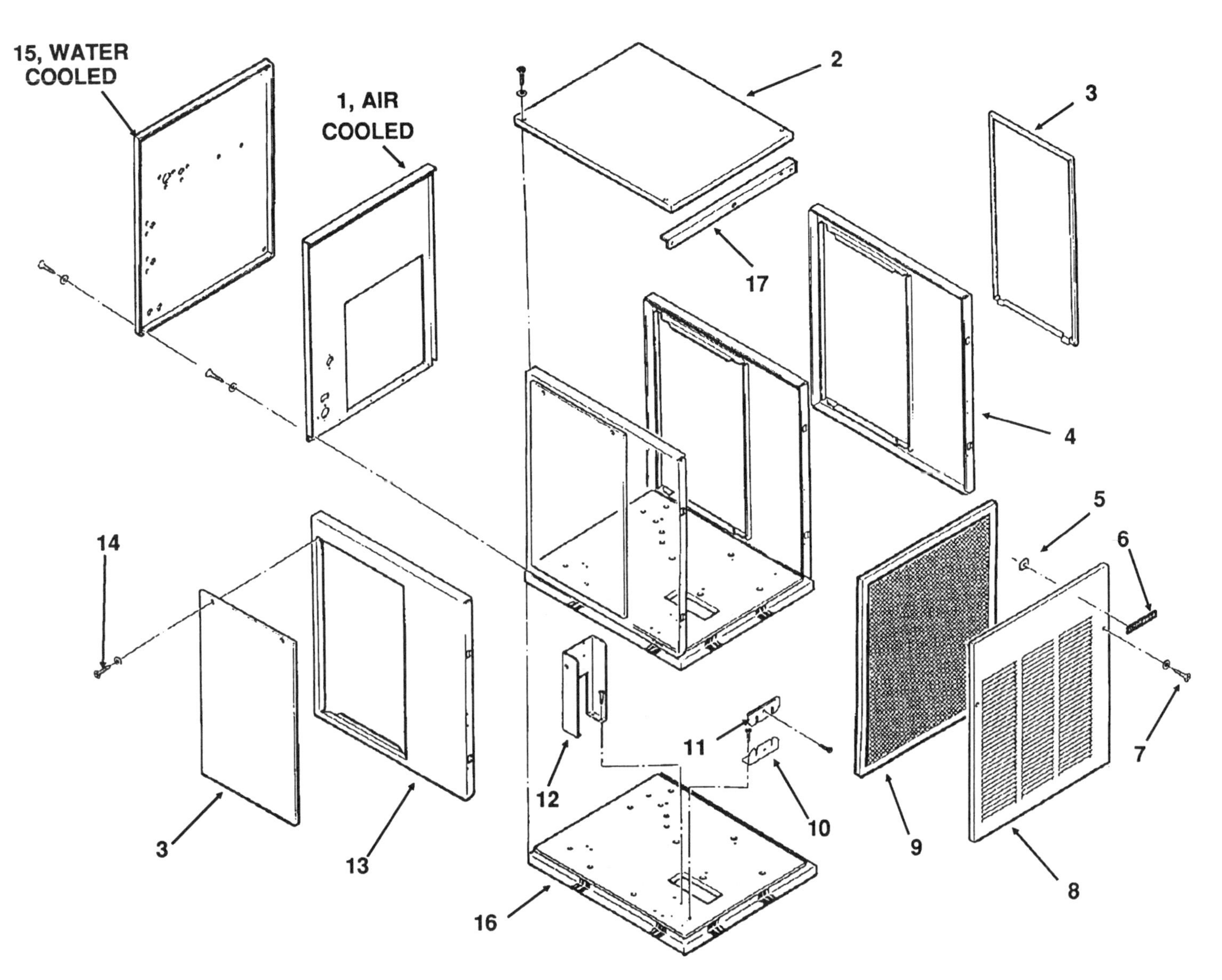

ITEM NUMBER	PART NUMBER	DESCRIPTION	ITEM NUMBER	PART NUMBER	DESCRIPTION
1	A34544-001	Back Panel, Air Cooled	9	02-2796-01	Filter Assembly
2	A33256-001	Top Panel, Painted	10	A31462-001	Tube Bracket
	A33256-002	Top Panel, S.S.	11	A31461-001	Tube Bracket
3	A34038-001	Service Panel, Painted	12	A34039-001	Control Box Support
	A34038-002	Service Panel, S.S.	13	A32435-001	Left Side Panel, Painted
4	A33292-001	Right Side Panel, Painted		A32435-002	Left Side Panel, S.S.
	A33292-002	Right Side Panel, S.S.	14	03-1404-12	Screw
5	03-0271-00	Speed Clip	15	A33252-001	Rear Panel, Water Cooled Painted Only
6	15-0711-01	Emblem			
7	03-1419-08	Screw			
8	A33255-001	Front Panel, Painted	16.	A34042-001	Base
	A33255-002	Front Panel, S.S.	17	A34348-001	Front Brace, Air Cooled

NM650 SERVICE PARTS
Air Cooled Refrigeration

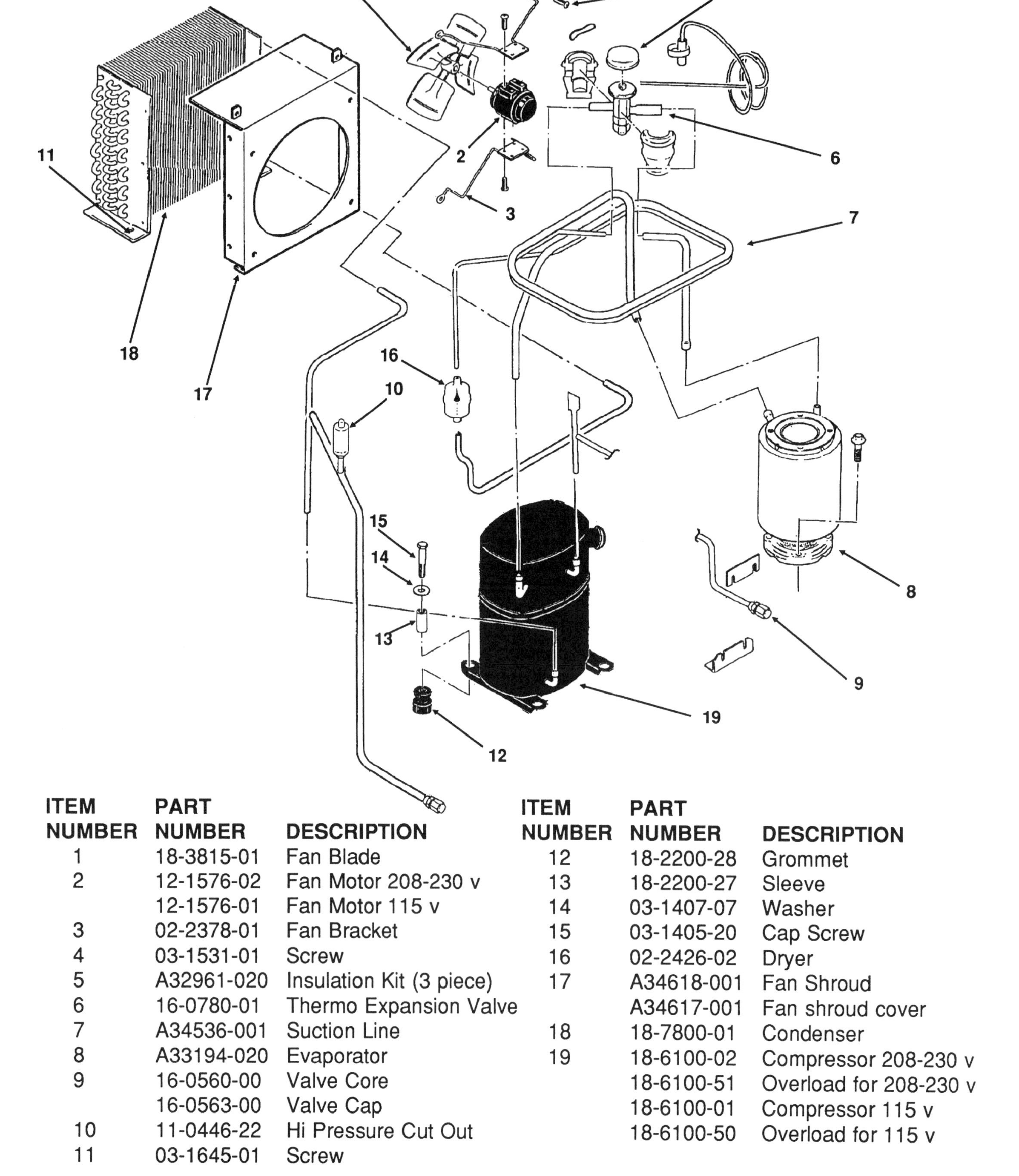

ITEM NUMBER	PART NUMBER	DESCRIPTION
1	18-3815-01	Fan Blade
2	12-1576-02	Fan Motor 208-230 v
	12-1576-01	Fan Motor 115 v
3	02-2378-01	Fan Bracket
4	03-1531-01	Screw
5	A32961-020	Insulation Kit (3 piece)
6	16-0780-01	Thermo Expansion Valve
7	A34536-001	Suction Line
8	A33194-020	Evaporator
9	16-0560-00	Valve Core
	16-0563-00	Valve Cap
10	11-0446-22	Hi Pressure Cut Out
11	03-1645-01	Screw

ITEM NUMBER	PART NUMBER	DESCRIPTION
12	18-2200-28	Grommet
13	18-2200-27	Sleeve
14	03-1407-07	Washer
15	03-1405-20	Cap Screw
16	02-2426-02	Dryer
17	A34618-001	Fan Shroud
	A34617-001	Fan shroud cover
18	18-7800-01	Condenser
19	18-6100-02	Compressor 208-230 v
	18-6100-51	Overload for 208-230 v
	18-6100-01	Compressor 115 v
	18-6100-50	Overload for 115 v

NM650 SERVICE PARTS
Water Cooled Refrigeration

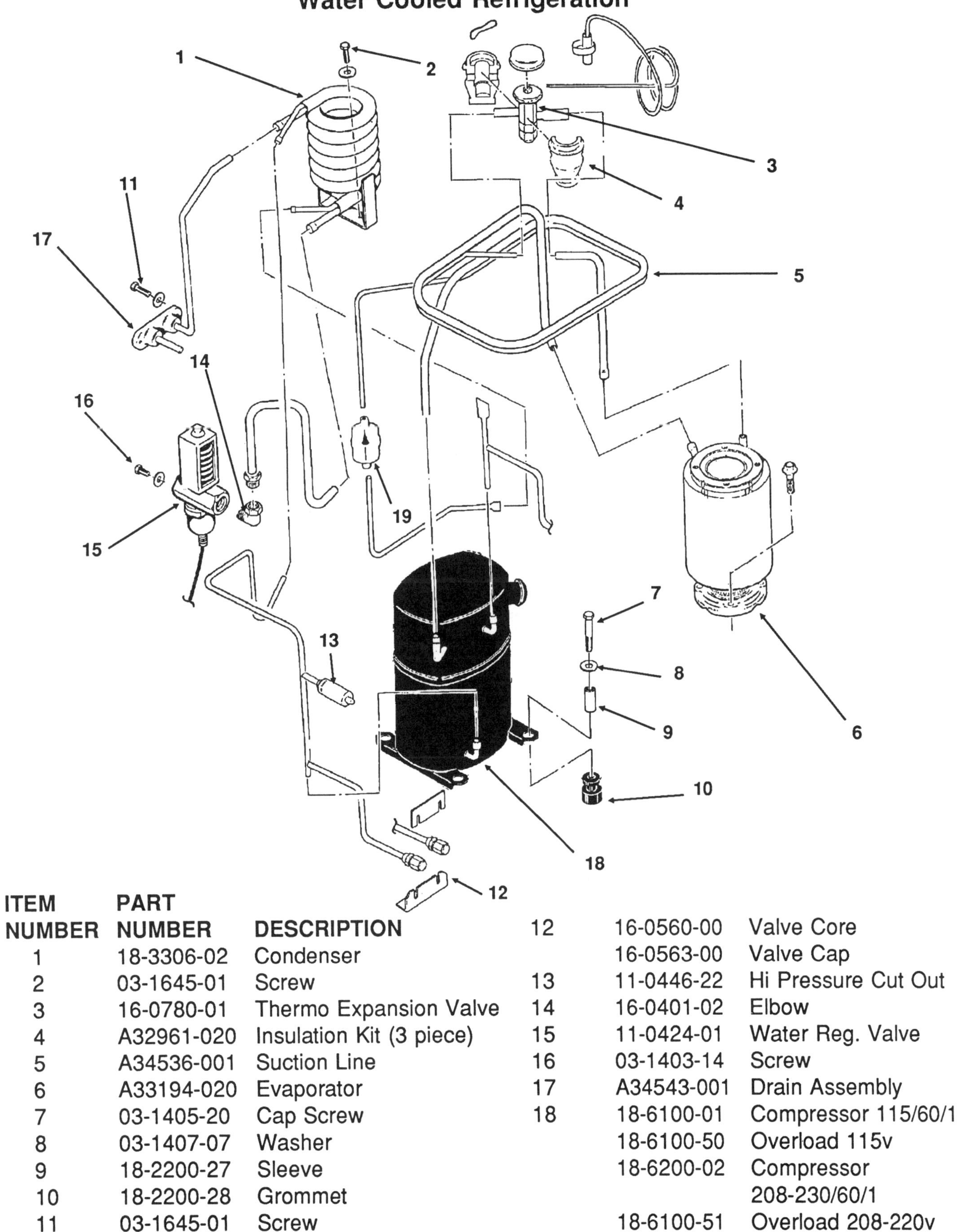

ITEM NUMBER	PART NUMBER	DESCRIPTION
1	18-3306-02	Condenser
2	03-1645-01	Screw
3	16-0780-01	Thermo Expansion Valve
4	A32961-020	Insulation Kit (3 piece)
5	A34536-001	Suction Line
6	A33194-020	Evaporator
7	03-1405-20	Cap Screw
8	03-1407-07	Washer
9	18-2200-27	Sleeve
10	18-2200-28	Grommet
11	03-1645-01	Screw
12	16-0560-00	Valve Core
	16-0563-00	Valve Cap
13	11-0446-22	Hi Pressure Cut Out
14	16-0401-02	Elbow
15	11-0424-01	Water Reg. Valve
16	03-1403-14	Screw
17	A34543-001	Drain Assembly
18	18-6100-01	Compressor 115/60/1
	18-6100-50	Overload 115v
	18-6200-02	Compressor 208-230/60/1
	18-6100-51	Overload 208-220v
19	02-2426-02	Dryer

NM650 SERVICE PARTS
Water System

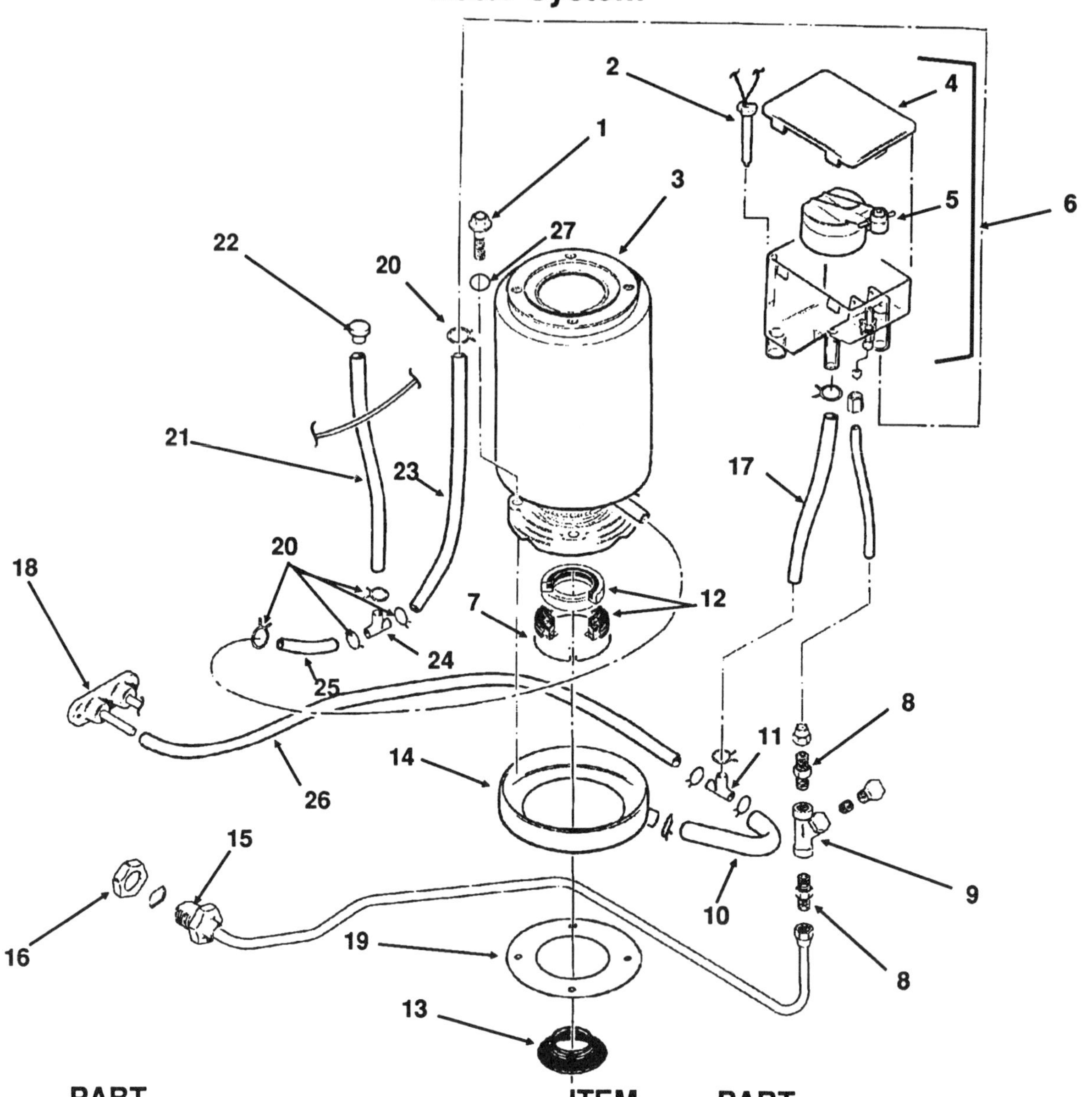

ITEM NUMBER	PART NUMBER	DESCRIPTION
1	03-1420-03	Cap Screw
2	A33101-021	Water Level Sensor
3	A33194-020	Evaporator
4	02-2936-01	Res. Cover
5	A32922-020	Float Assembly with pin housing seal, rubber seal, insert & float
6	A32929-020	Reservoir Assy Complete
7	A32777-001	Retaining Ring for seal
8	16-0791-01	Half Union
9	16-0162-00	Strainer
10	13-0079-03	Tube, 3.2" req.
11	16-0670-02	Tee
12	02-0929-02	Water Seal
13	13-0868-01	Water Shed
14	A32050-001	Drip Pan
15	A27318-001	Water Inlet Fitting
16	03-1394-01	Pal Nut
17	13-0079-03	Res. Overflow, 8" req.
18	A32282-004	Drain Casting, air cooled
19	13-0704-00	Gasket
20	02-2814-08	Clamps
21	13-0674-06	Evap. Drain, 14" req.
22	A33205-001	Plug
23	13-0674-06	Evap. Inlet Tube, 5" req.
24	16-0670-01	Tee
25	A33203-001	Tube, Preformed
26	13-0079-03	Drain Tube, 24" req.
27.	03-1417-13	Lockwasher

NM650 SERVICE PARTS

Gearmotor

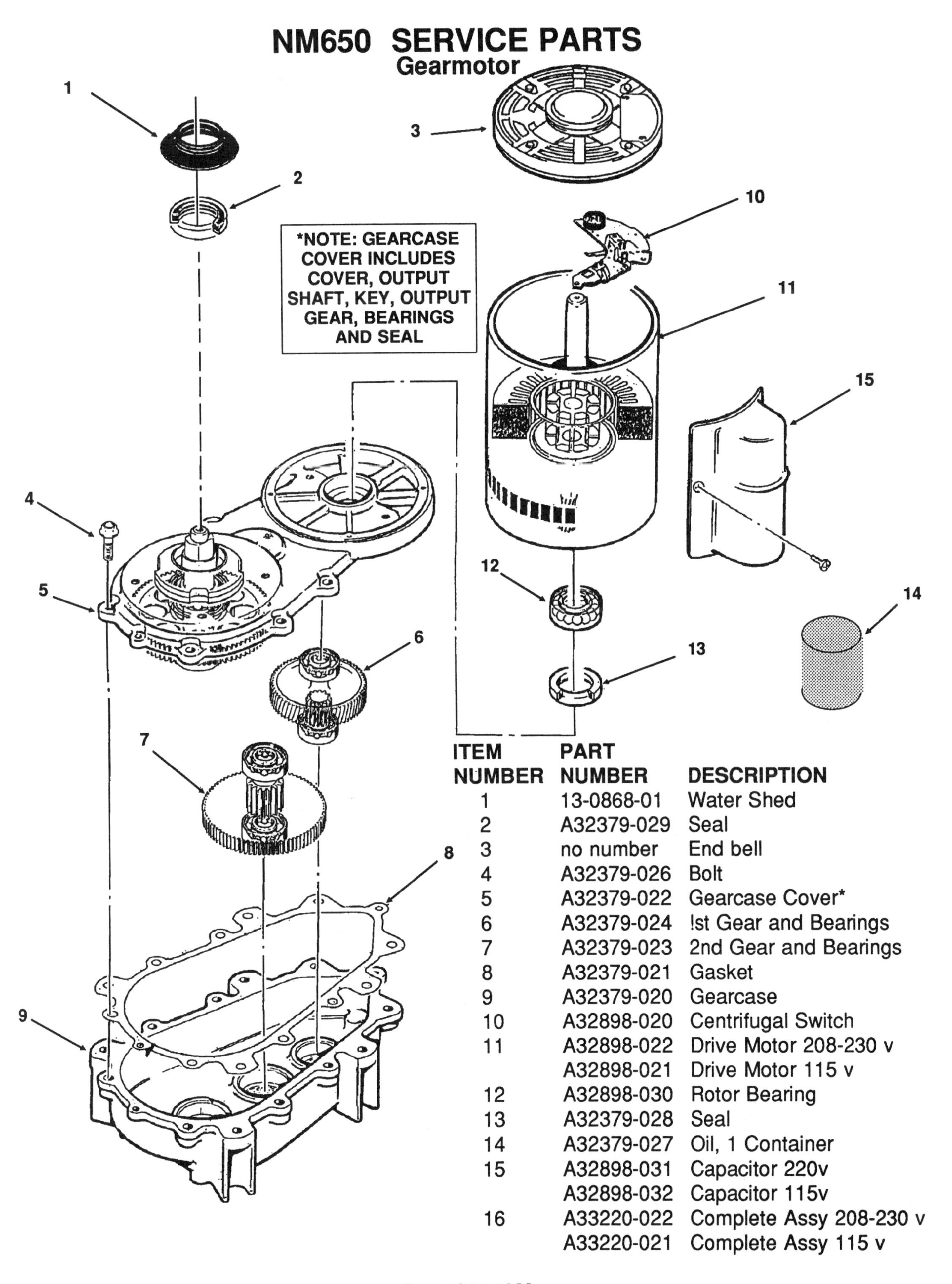

ITEM NUMBER	PART NUMBER	DESCRIPTION
1	13-0868-01	Water Shed
2	A32379-029	Seal
3	no number	End bell
4	A32379-026	Bolt
5	A32379-022	Gearcase Cover*
6	A32379-024	!st Gear and Bearings
7	A32379-023	2nd Gear and Bearings
8	A32379-021	Gasket
9	A32379-020	Gearcase
10	A32898-020	Centrifugal Switch
11	A32898-022	Drive Motor 208-230 v
	A32898-021	Drive Motor 115 v
12	A32898-030	Rotor Bearing
13	A32379-028	Seal
14	A32379-027	Oil, 1 Container
15	A32898-031	Capacitor 220v
	A32898-032	Capacitor 115v
16	A33220-022	Complete Assy 208-230 v
	A33220-021	Complete Assy 115 v

NM650 SERVICE PARTS

Evaporator

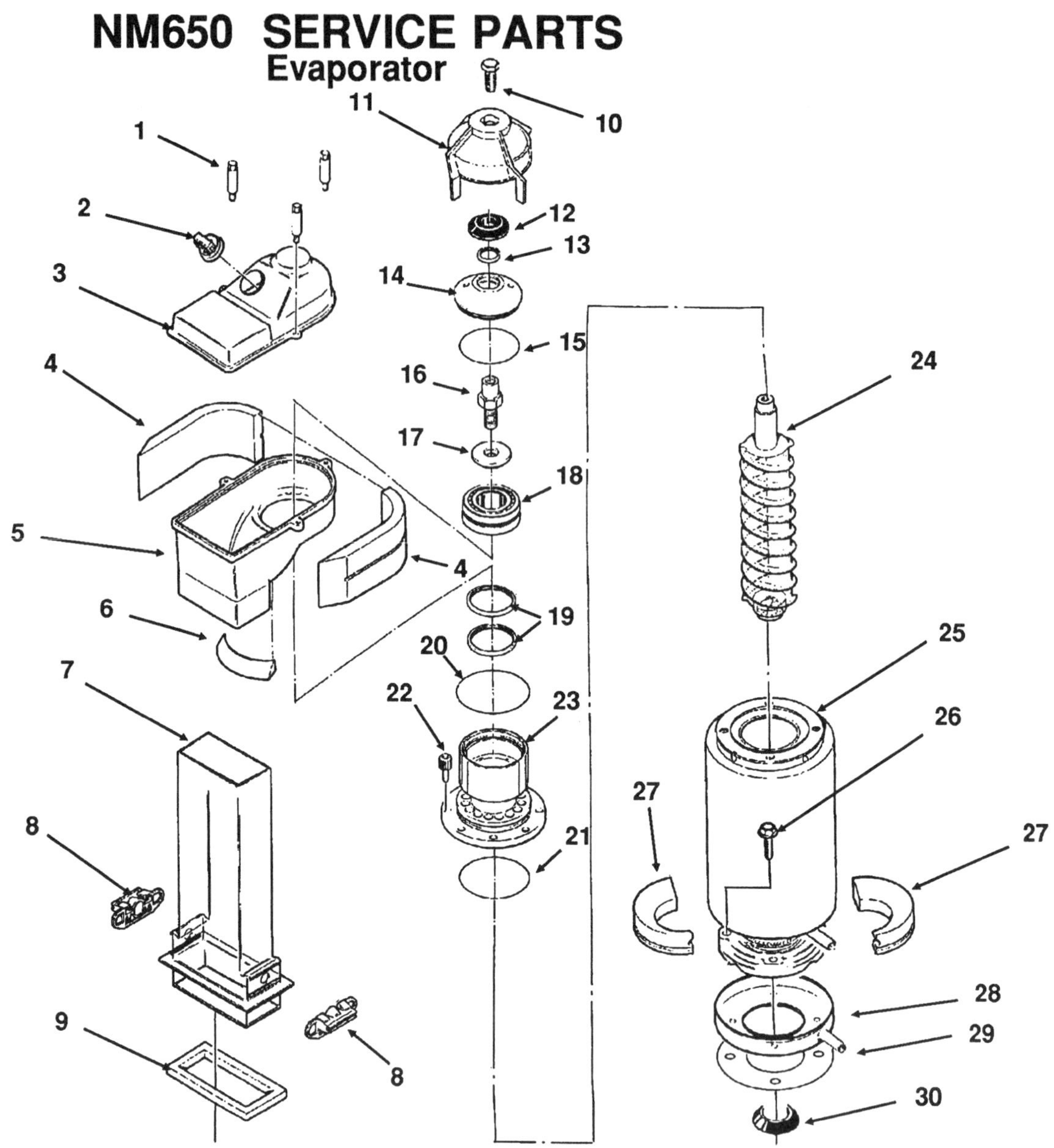

ITEM NUMBER	PART NUMBER	DESCRIPTION
1.	02-2933-01	Hex Stud
2.	A32891-001	Switch & housing assy
3.	02-2930-01	Ice Chute Cover
4.	A32963-001	Insulation Top (half)
5.	02-2926-01	Ice Chute Body
6.	A33102-001	Insulation Collar Inside
	A33210-001	Rubber Band
7.	02-2929-02	Ice Chute Lower
8.	A33117-020	Bin Controls Assy. (set of 2)
9.	A32930-001	Chute Gasket
10.	03-1405-52	Hex Cap Screw
11.	02-3001-01	Ice Sweep
12.	13-0871-01	Water Shed
13.	02-2978-01	Lip Seal
14.	08-0650-01	Breaker Cover
15.	13-0617-54	O-Ring
16.	08-0653-01	Auger Stud
17.	08-0654-01	Thrust Washer
18.	A34559-020	Bearing
19.	02-2977-01	Lip Seal
20.	13-0617-52	O-Ring
21.	13-0617-45	O-Ring
22.	03-1544-08	Soc. Head Screw
23.	A32900-020	Breaker (includes 18 & 19)
24.	02-3002-01	Auger
25.	A33194-020	Evaporator
26.	03-1420-03	Cap Screw
27	A32962-001	Insulation Bot. (half)
	A32962-002	Insulation half
28	A32050-001	Drip Pan
29.	13-0704-00	Drip Pan Gasket
30.	13-0868-01	Water Shed

NM650 SERVICE PARTS

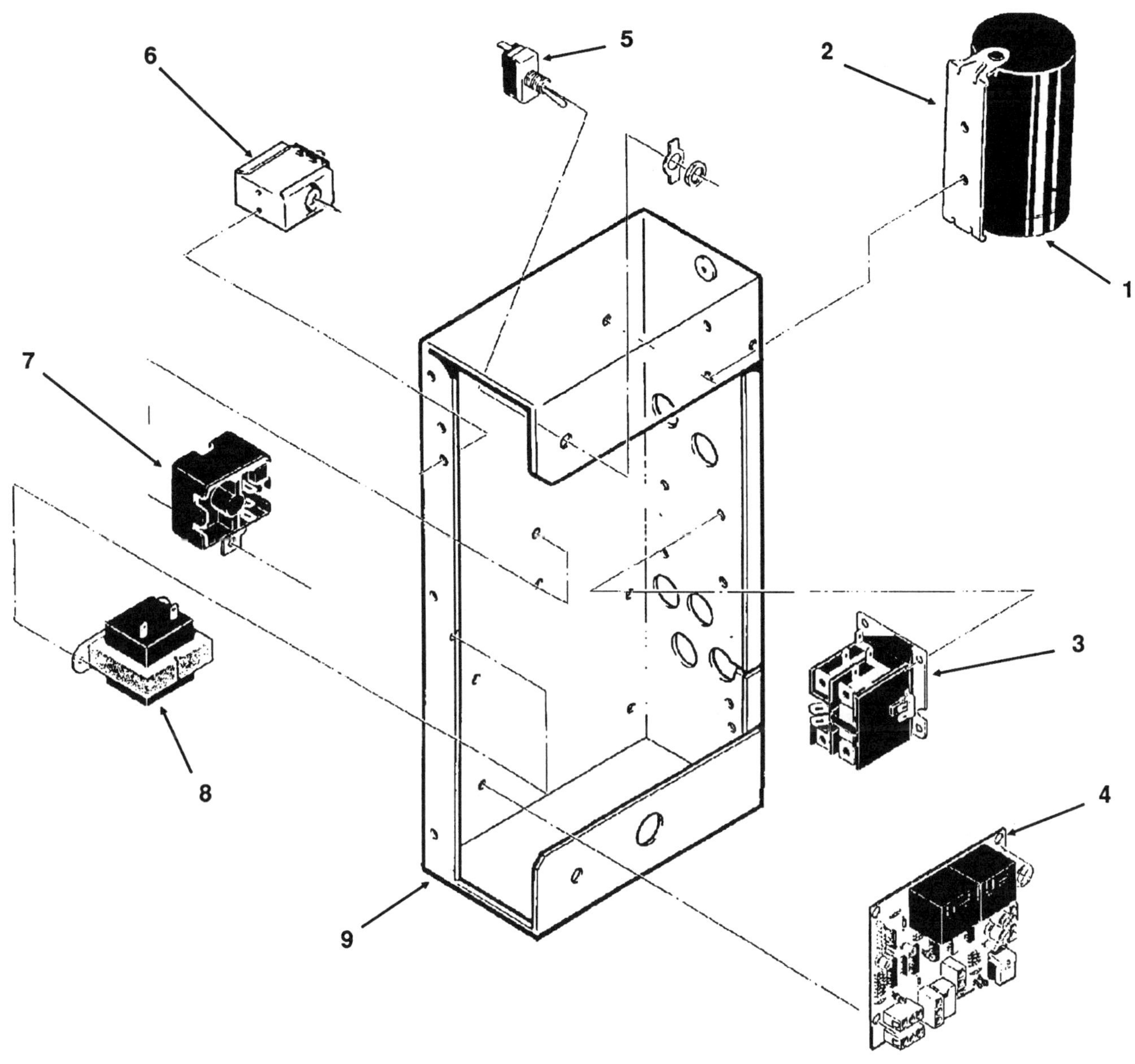

ITEM NUMBER	PART NUMBER	DESCRIPTION
1	18-1901-42	Start Cap 208-230 v
	18-1901-40	Start Capacitor 115 v
	18-4300-50	Start Capacitor Cap.
2.	18-2200-39	Start Capacitor Bracket
	03-1638-04	Screw
	03-1417-15	Lockwasher
3.	12-2048-02	Contactor 208-230 v
	12-2048-01	Contactor 115 v
4	A32976-020	Circuit Board
5.	12-0426-01	Master Switch
6.	11-0447-20	Low Pressure Control
7	18-1903-44	Potential Relay 208-230 volt
	18-1903-33	Potential Relay 115 volt
8.	12-2285-21	Transformer 115v
	12-2285-22	Transformer 220v
9	no number	Control Box
NOT ILLUSTRATED		
10.	12-2337-02	3 Pin Plug Assy (Safety Circuit to Circuit Board)
11.	12-2340-01	4 Pin Plug Assy (Transformer to Circuit Board)
12.	A34055-001	Control Box Cover

NM650 WIRING DIAGRAMS

Component Layout Diagram

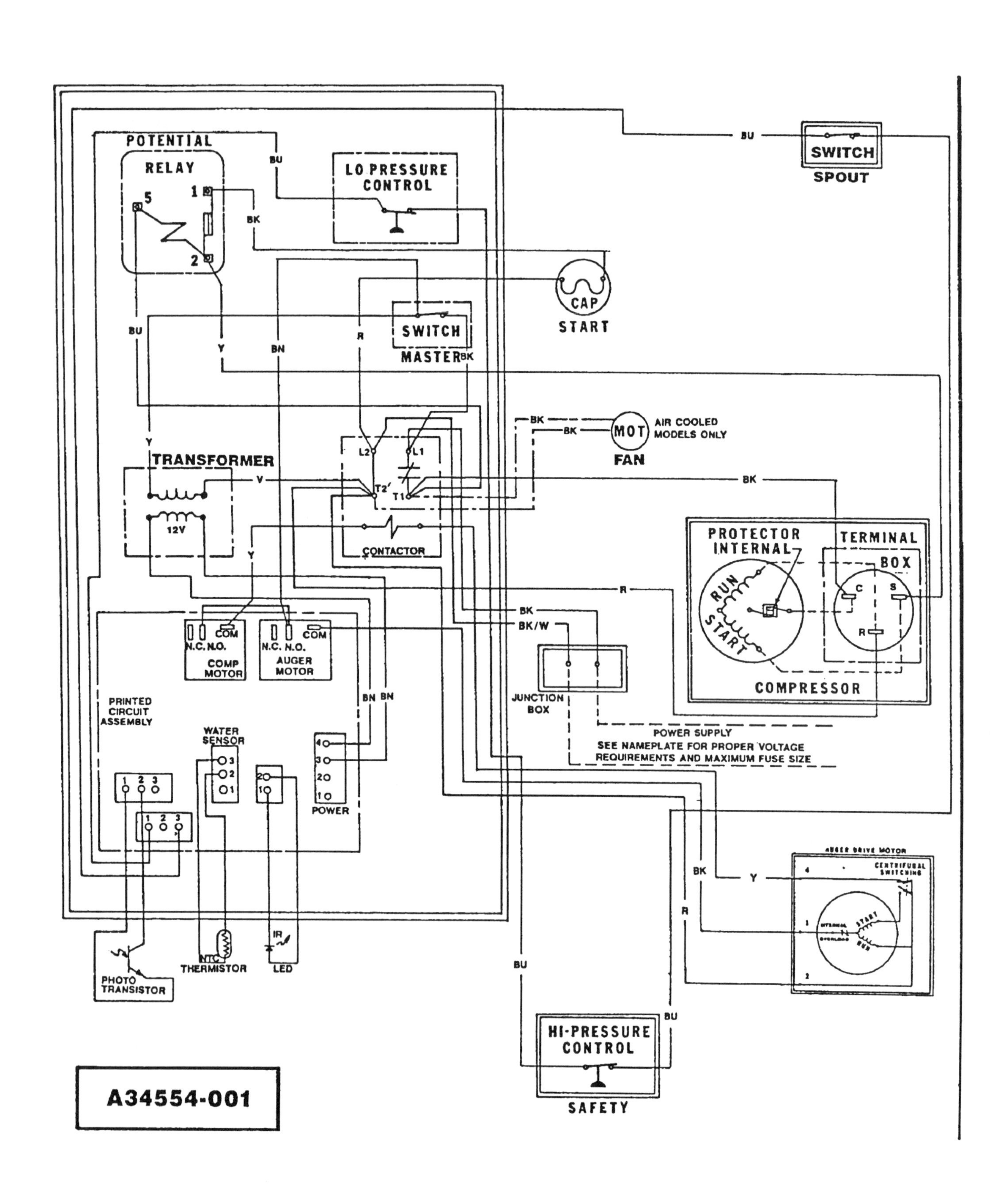

A34554-001

NM650 WIRING DIAGRAMS

Schematic Diagram

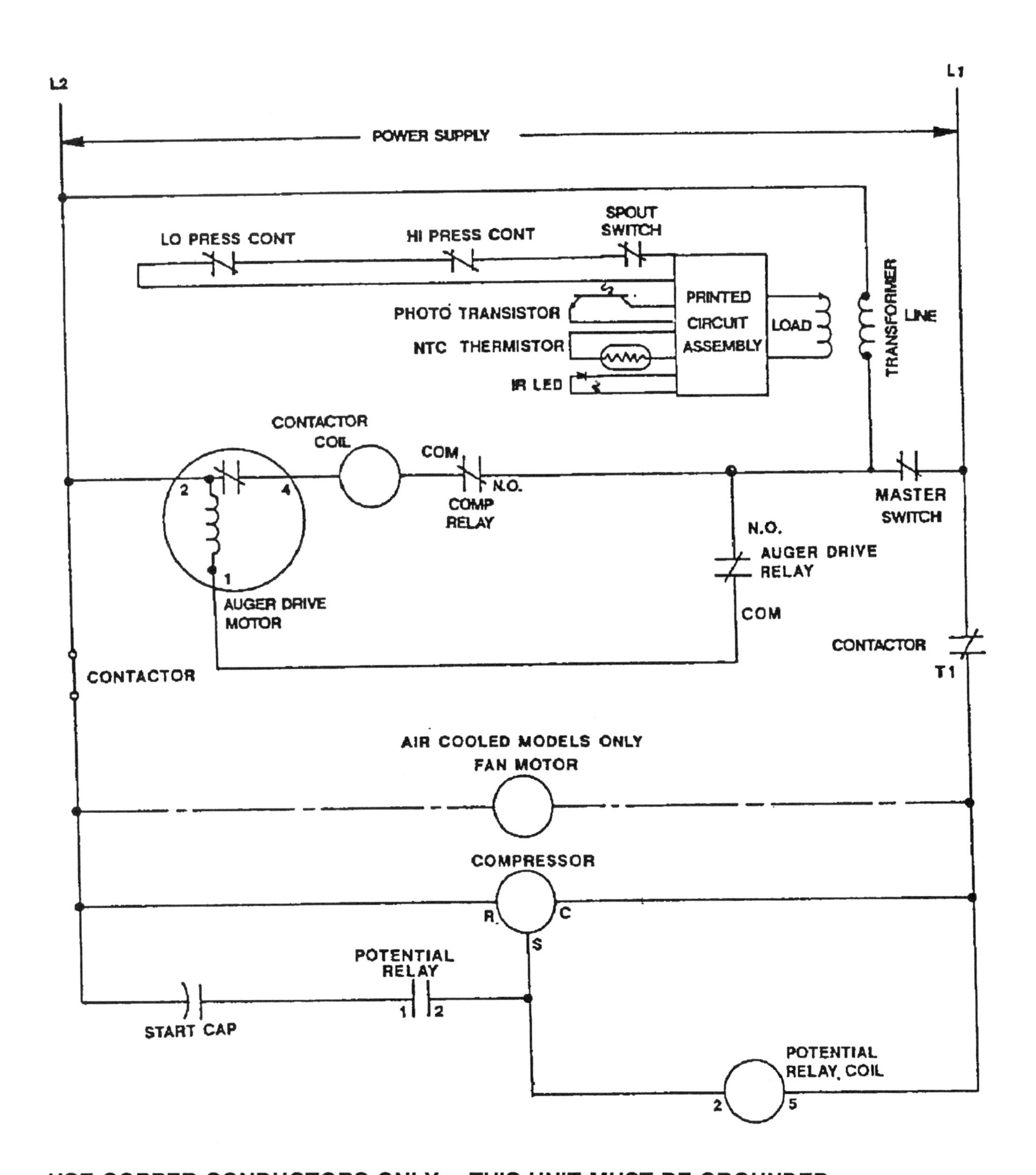

USE COPPER CONDUCTORS ONLY THIS UNIT MUST BE GROUNDED

All Controls Shown in Normal Ice Making Mode.

Note: Dashed lines indicate field wiring which must be installed in accordance with the National Electric Code Using a Minimum of 14 ga. awg. wire.

SCOTSMAN
775 CORPORATE WOODS PARKWAY
VERNON HILLS, ILLINOIS 60061
U. S. A.

312/634-3505

17-1927-20

Hoshizaki

HOSHIZAKI

MODULAR CRESCENT CUBER

MODEL KM-450MAB
KM-450MWB
KM-450MRB

SERVICE MANUAL

HOSHIZAKI AMERICA, INC.

618 Highway 74 South, Peachtree City, GA 30269

Phone: (404) 487-2331

NO.	U1HJ - 178
ISSUED:	FEB. 27, 1990
REVISED:	————

SERVICE MANUAL

MODULAR CRESCENT CUBER

KM - 450MAB
KM - 450MWB
KM - 450MRB

HOSHIZAKI AMERICA, INC.

FOREWORD

This Service Manual contains the specifications and information in regard to transporting, unpacking, installing, operating and servicing the machine. You are encouraged to read it thoroughly in order to obtain maximum performance. You will find details on the construction, installation and maintenance.

If you encounter any problem not covered in this Service Manual, feel free to contact Hoshizaki America, Inc. We will be happy to provide whatever assistance is necessary.

Keep this Service Manual handy, and read it again when questions arise.

HOSHIZAKI AMERICA, INC.

618 HIGHWAY 74 SOUTH, PEACHTREE CITY,
GEORGIA 30269 U.S.A.
PHONE: 404-487-2331

U1HJ1789002

CONTENTS

U1HJ1789002

I. SPECIFICATIONS

1. NAMEPLATE RATING

[a] KM-450MAB (Air-cooled)

HOSHIZAKI ICE MAKER

MODEL NUMBER	KM-450MAB
AC SUPPLY VOLTAGE	115-120/60/1
AMPERES	12 AMPS
MAXIMUM FUSE SIZE	20 AMPS
MINIMUM CIRCUIT AMPACITY	20 AMPS
DESIGN PRESSURE	HI-380 PSI LO-206 PSI
REFRIGERANT 502	1 lb. 10 oz.
MOTOR-COMPRESSOR THERMALLY PROTECTED	
SERIAL NUMBER	

HOSHIZAKI AMERICA, INC.
Peachtree City, GA

See NAMEPLATE for electrical and refrigeration specifications. This nameplate is located on the upper right hand side of rear panel.

Since this nameplate is located on the rear panel of the icemaker, it cannot be read when the back of the icemaker is against a wall or against another piece of kitchen equipment. Therefore, the necessary electrical and refrigeration information is now also on the rating label, which can be easily seen by removing only the front panel of the icemaker.

We reserve the right to make changes in specifications and design without prior notice.

[b] KM-450MWB (Water-cooled)

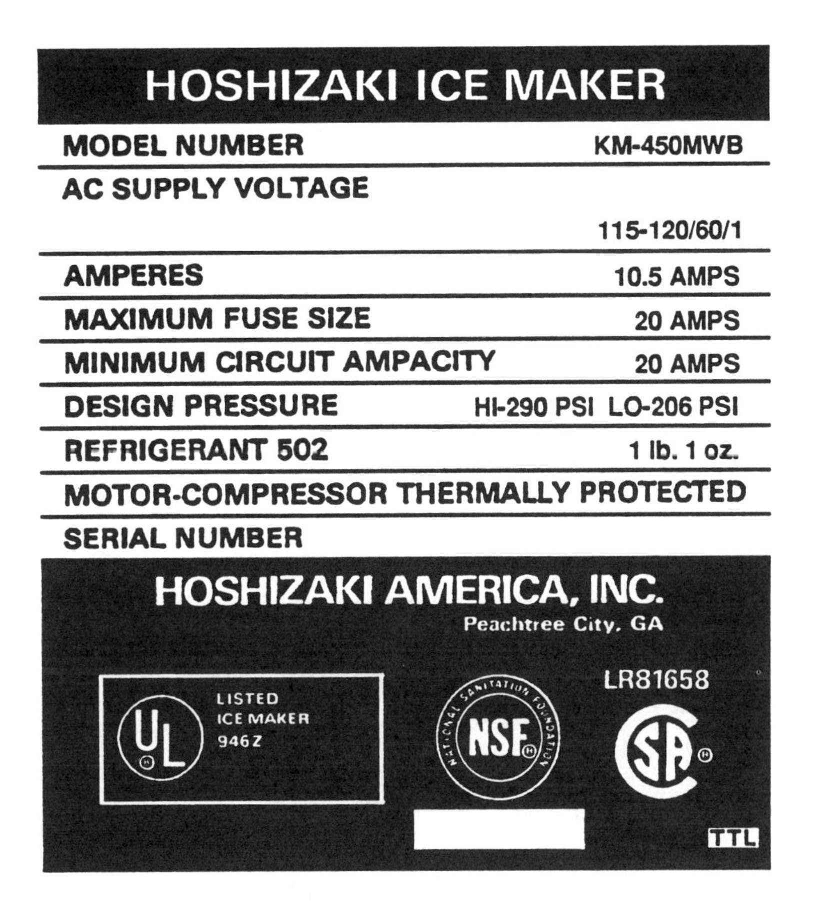

See NAMEPLATE for electrical and refrigeration specifications. This nameplate is located on the upper right hand side of rear panel.

Since this nameplate is located on the rear panel of the icemaker, it cannot be read when the back of the icemaker is against a wall or against another piece of kitchen equipment. Therefore, the necessary electrical and refrigeration information is now also on the rating label, which can be easily seen by removing only the front panel of the icemaker.

We reserve the right to make changes in specifications and design without prior notice.

[c] KM-450MRB (Remote Air-cooled)

HOSHIZAKI ICE MAKER

MODEL NUMBER	KM-450MRB
AC SUPPLY VOLTAGE	115-120/60/1
AMPERES	12 AMPS
MAXIMUM FUSE SIZE	20 AMPS
MINIMUM CIRCUIT AMPACITY	20 AMPS
DESIGN PRESSURE	HI-330 PSI LO-206 PSI
REFRIGERANT 502	
MOTOR-COMPRESSOR THERMALLY PROTECTED	
SERIAL NUMBER	

HOSHIZAKI AMERICA, INC.
Peachtree City, GA

UL LISTED ICE MAKER 946Z
NSF
LR81658
SA
TTL

See NAMEPLATE for electrical and refrigeration specifications. This nameplate is located on the upper right hand side of rear panel.

Since this nameplate is located on the rear panel of the icemaker, it cannot be read when the back of the icemaker is against a wall or against another piece of kitchen equipment. Therefore, the necessary electrical and refrigeration information is now also on the rating label, which can be easily seen by removing only the front panel of the icemaker.

We reserve the right to make changes in specifications and design without prior notice.

2. DIMENSIONS/CONNECTIONS

[a] KM-450MAB

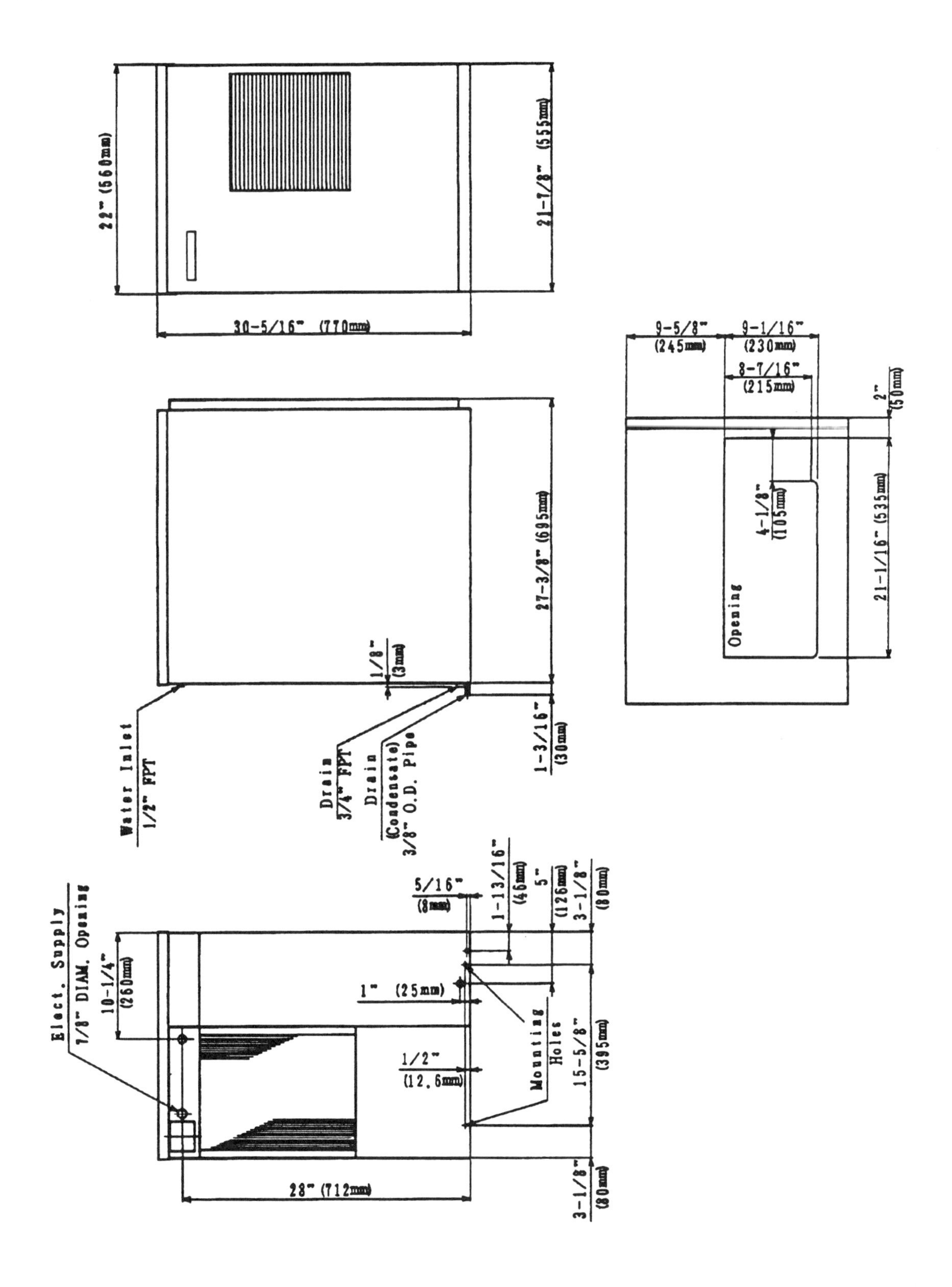

[b] KM-450MWB

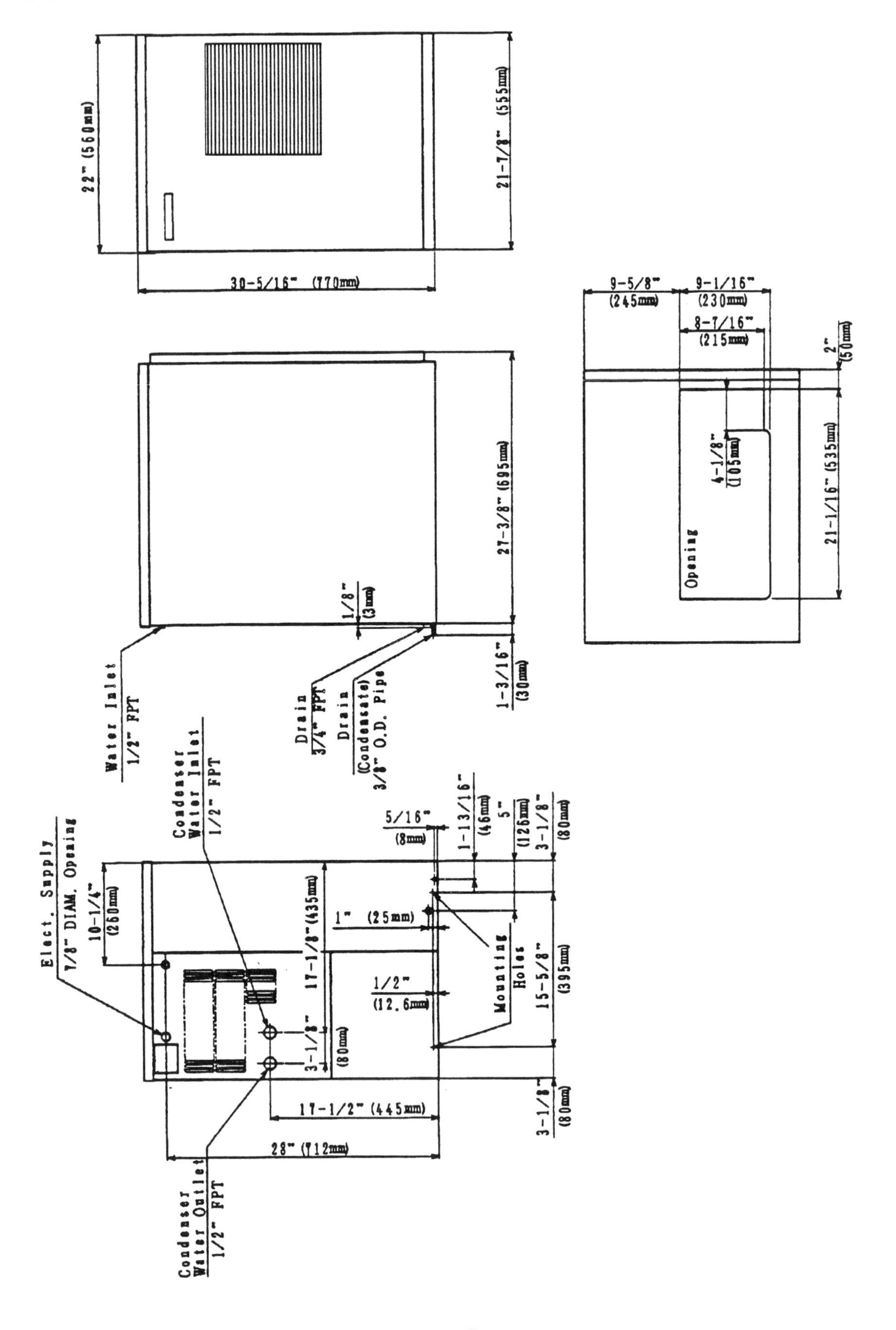

22" (560mm)
21-7/8" (555mm)
30-5/16" (770mm)
27-3/8" (695mm)
1/8" (3mm)
1-3/16" (30mm)
Water Inlet
1/2" FPT
Drain
3/4" FPT
Drain
(Condensate)
3/8" O.D. Pipe
Condenser
Water Inlet
1/2" FPT
Elect. Supply
7/8" DIAM. Opening
10-1/4" (260mm)
17-1/8" (435mm)
3-1/8" (80mm)
1" (25mm)
1/2" (12.6mm)
5/16" (8mm)
1-13/16" (46mm)
5" (126mm)
Mounting
Holes
15-5/8" (395mm)
17-1/2" (445mm)
28" (712mm)
Condenser
Water Outlet
1/2" FPT
9-5/8" (245mm)
9-1/16" (230mm)
8-7/16" (215mm)
2" (50mm)
4-1/8" (105mm)
21-1/16" (535mm)
Opening

[c] KM-450MRB

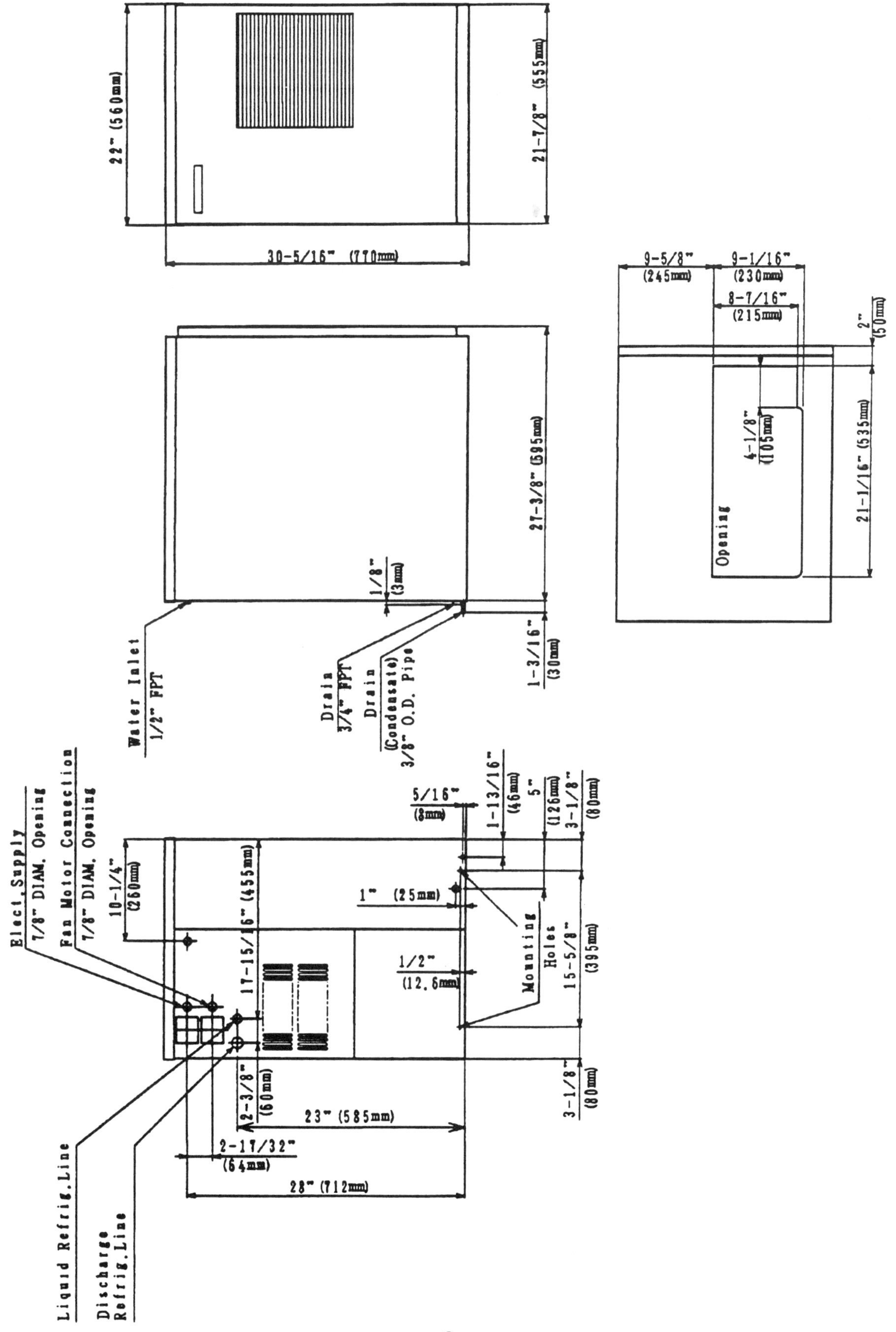

22" (560mm)
21-7/8" (555mm)
30-5/16" (770mm)
9-5/8" (245mm)
9-1/16" (230mm)
8-7/16" (215mm)
2" (50mm)
4-1/8" (105mm)
21-1/16" (535mm)
Opening
27-3/8" (595mm)
1/8" (3mm)
1-3/16" (30mm)
Water Inlet 1/2" FPT
Drain 3/4" FPT
Drain (Condensate) 3/8" O.D. Pipe
Elect. Supply 7/8" DIAM. Opening
Fan Motor Connection 7/8" DIAM. Opening
10-1/4" (260mm)
17-15/16" (455mm)
5/16" (8mm)
1-13/16" (46mm)
5" (126mm)
3-1/8" (80mm)
1" (25mm)
1/2" (12.6mm)
Mounting Holes
15-5/8" (395mm)
3-1/8" (80mm)
2-3/8" (60mm)
23" (585mm)
2-17/32" (64mm)
28" (712mm)
Liquid Refrig. Line
Discharge Refrig. Line

3. SPECIFICATIONS

[a] KM-450MAB

AC SUPPLY VOLTAGE	115-120/60/1
AMPERAGE	12 A (5 Min. Freeze AT 104 °F/ WT 80 °F)
MINIMUM CIRCUIT AMPACITY	20 A
MAXIMUM FUSE SIZE	20 A
APPROXIMATE ICE PRODUCTION PER 24 HR. (MAX. CUBE SIZE) lbs./day (kg/day) Reference without *marks	(see table below)

Ambient Temp. (°F)	Water Temp. (°F) 50	70	90
70	*460(209)	410(186)	350(159)
80	420(190)	370(168)	330(150)
90	375(170)	*340(154)	300(136)
100	340(154)	295(133)	260(118)

SHAPE OF ICE	Crescent Cube
ICE PRODUCTION PER CYCLE	10.4 lbs. (4.7 kg) 480 pcs.
APPROXIMATE STORAGE CAPACITY	N/A
PERFORMANCE	90 °F/ 70 °F 70 °F/ 50 °F See details of
ELECTRIC CONSUMPTION	1150 W 1050 W PERFORMANCE DATA
WATER CONSUMPTION PER 24 HR.	114 gal. 198 gal. NO.89194
EXTERIOR DIMENSIONS (WxDxH)	22" x 27-3/8" x 30-5/16" (560 x 695 x 770 mm)
EXTERIOR FINISH	Stainless Steel, Galvanized Steel (Rear)
WEIGHT	Net 150 lbs.(68 kg), Shipping 170 lbs.(77 kg)
CONNECTIONS - ELECTRIC	Permanent Connection
- WATER SUPPLY	Inlet 1/2" FPT
- DRAIN	Outlet 3/4" FPT 3/8" OD Pipe
CUBE CONTROL SYSTEM	Float Switch, Adjustable Cube Size
HARVESTING CONTROL SYSTEM	Hot Gas and Water, Thermistor and Timer
ICE MAKING WATER CONTROL	Timer Controlled, Overflow Pipe
COOLING WATER CONTROL	N/A
BIN CONTROL SYSTEM	Thermostat
COMPRESSOR	Hermetic, Model JSL5-0075-CAA
CONDENSER	Air-cooled, Fin and Tube type
EVAPORATOR	Vertical type, Stainless Steel and Copper
REFRIGERANT CONTROL	Thermostatic Expansion Valve
REFRIGERANT CHARGE	R502, 1 lb. 10 oz. (750 g)
DESIGN PRESSURE	High 380 PSIG, Low 206 PSIG
P.C.BOARD CIRCUIT PROTECTION	High Voltage Cut-out Relay
COMPRESSOR PROTECTION	Auto-reset Overload Protector (External)
REFRIGERANT CIRCUIT PROTECTION	Auto-reset High Pressure Control Switch
LOW WATER PROTECTION	Float Switch
ACCESSORIES - SUPPLIED	Ice Scoop
- REQUIRED	Ice Storage Bin
OPERATION CONDITIONS	VOLTAGE RANGE 104 - 132 V
	AMBIENT TEMP. 45 - 100 °F
	WATER SUPPLY TEMP. 45 - 90 °F
	WATER SUPPLY PRESS. 7 - 113 PSIG
DRAWING NO. (DIMENSIONS)	437538

* We reserve the right to make changes in specifications and design without prior notice.

[b] KM-450MWB

AC SUPPLY VOLTAGE	115-120/60/1
AMPERAGE	10.5 A (5 Min. Freeze AT 104 °F/ WT 80 °F)
MINIMUM CIRCUIT AMPACITY	20 A
MAXIMUM FUSE SIZE	20 A

APPROXIMATE ICE PRODUCTION PER 24 HR. (MAX. CUBE SIZE) lbs./day (kg/day) Reference without *marks

Ambient Temp. (°F)	Water Temp. (°F) 50	70	90
70	*490 (222)	425 (193)	360 (163)
80	485 (220)	420 (190)	350 (159)
90	475 (215)	*420 (190)	330 (150)
100	460 (209)	395 (179)	330 (150)

SHAPE OF ICE	Crescent Cube
ICE PRODUCTION PER CYCLE	10.4 lbs. (4.7 kg) 480 pcs.
APPROXIMATE STORAGE CAPACITY	N/A
PERFORMANCE	90 °F/ 70 °F 70 °F/ 50 °F See details of
ELECTRIC CONSUMPTION	1020 W 980 W PERFORMANCE DATA
WATER CONSUMPTION PER 24 HR.	1062 gal. 816 gal. NO.89195
EXTERIOR DIMENSIONS (WxDxH)	22" x 27-3/8" x 30-5/16" (560 x 695 x 770 mm)
EXTERIOR FINISH	Stainless Steel, Galvanized Steel (Rear)
WEIGHT	Net 146 lbs.(66 kg), Shipping 170 lbs.(77 kg)
CONNECTIONS - ELECTRIC	Permanent Connection
- WATER SUPPLY	Inlet 1/2" FPT Cond. Inlet 1/2" FPT
- DRAIN	Outlet 3/4" FPT 3/8" OD Pipe Cond. Outlet 1/2" FPT
CUBE CONTROL SYSTEM	Float Switch, Adjustable Cube Size
HARVESTING CONTROL SYSTEM	Hot Gas and Water, Thermistor and Timer
ICE MAKING WATER CONTROL	Timer Controlled, Overflow Pipe
COOLING WATER CONTROL	N/A
BIN CONTROL SYSTEM	Thermostat
COMPRESSOR	Hermetic, Model JSL5-0075-CAA
CONDENSER	Water-cooled, Tube in Tube type
EVAPORATOR	Vertical type, Stainless Steel and Copper
REFRIGERANT CONTROL	Thermostatic Expansion Valve
REFRIGERANT CHARGE	R502, 1 lb. 1 oz. (470 g)
DESIGN PRESSURE	High 290 PSIG, Low 206 PSIG
P.C.BOARD CIRCUIT PROTECTION	High Voltage Cut-out Relay
COMPRESSOR PROTECTION	Auto-reset Overload Protector (External)
REFRIGERANT CIRCUIT PROTECTION	Auto-reset High Pressure Control Switch
LOW WATER PROTECTION	Float Switch
ACCESSORIES - SUPPLIED	Ice Scoop
- REQUIRED	Ice Storage Bin
OPERATION CONDITIONS	VOLTAGE RANGE 104 - 132 V AMBIENT TEMP. 45 - 100 °F WATER SUPPLY TEMP. 45 - 90 °F WATER SUPPLY PRESS. 7 - 113 PSIG
DRAWING NO. (DIMENSIONS)	437539

* We reserve the right to make changes in specifications and design without prior notice.

U1HJ1789002

[c] KM-450MRB

AC SUPPLY VOLTAGE	115-120/60/1
AMPERAGE	12 A (5 Min. Freeze AT 104 °F/ WT 80 °F)
MINIMUM CIRCUIT AMPACITY	20 A
MAXIMUM FUSE SIZE	20 A
APPROXIMATE ICE PRODUCTION PER 24 HR. (MAX. CUBE SIZE) lbs./day (kg/day) Reference without *marks	see table below

Ambient Temp.(°F)	Water Temp. (°F) 50	70	90
70	*470(213)	430(195)	380(172)
80	430(195)	400(181)	360(163)
90	400(181)	*370(168)	330(150)
100	360(163)	330(150)	290(132)

SHAPE OF ICE	Crescent Cube
ICE PRODUCTION PER CYCLE	10.4 lbs. (4.7 kg) 480 pcs.
APPROXIMATE STORAGE CAPACITY	N/A
PERFORMANCE	90 °F/ 70 °F 70 °F/ 50 °F See details of
ELECTRIC CONSUMPTION	1140 W 1080 W PERFORMANCE DATA
WATER CONSUMPTION PER 24 HR.	141 gal. 210 gal. NO.89196
EXTERIOR DIMENSIONS (WxDxH)	22" x 27-3/8" x 30-5/16" (560 x 695 x 770 mm)
EXTERIOR FINISH	Stainless Steel, Galvanized Steel (Rear)
WEIGHT	Net 143 lbs.(65 kg), Shipping 170 lbs.(77 kg)
CONNECTIONS - ELECTRIC	Permanent Connection
- WATER SUPPLY	Inlet 1/2" FPT
- DRAIN	Outlet 3/4" FPT 3/8" OD Pipe
CUBE CONTROL SYSTEM	Float Switch, Adjustable Cube Size
HARVESTING CONTROL SYSTEM	Hot Gas and Water, Thermistor and Timer
ICE MAKING WATER CONTROL	Timer Controlled, Overflow Pipe
COOLING WATER CONTROL	N/A
BIN CONTROL SYSTEM	Thermostat
COMPRESSOR	Hermetic, Model JSL5-0075-CAA
CONDENSER	Air-cooled remote, Condenser Unit URC-4A recommended
EVAPORATOR	Vertical type, Stainless Steel and Copper
REFRIGERANT CONTROL	Thermostatic Expansion Valve
REFRIGERANT CHARGE	R502, 5 lbs. 5 oz. (2400 g) (Ice maker 3 lbs. 8 oz. Cond. unit 1 lb. 13 oz.) High 330 PSIG, Low 206 PSIG
P.C.BOARD CIRCUIT PROTECTION	High Voltage Cut-out Relay
COMPRESSOR PROTECTION	Auto-reset Overload Protector (External)
REFRIGERANT CIRCUIT PROTECTION	Auto-reset High Pressure Control Switch
LOW WATER PROTECTION	Float Switch
ACCESSORIES - SUPPLIED	Ice Scoop
- REQUIRED	Ice Storage Bin
OPERATION CONDITIONS	VOLTAGE RANGE 104 - 132 V AMBIENT TEMP. 45 - 100 °F WATER SUPPLY TEMP. 45 - 90 °F WATER SUPPLY PRESS. 7 - 113 PSIG
DRAWING NO. (DIMENSIONS)	437540

* We reserve the right to make changes in specifications and design without prior notice.

4. CONDENSER UNIT MODEL URC-4A

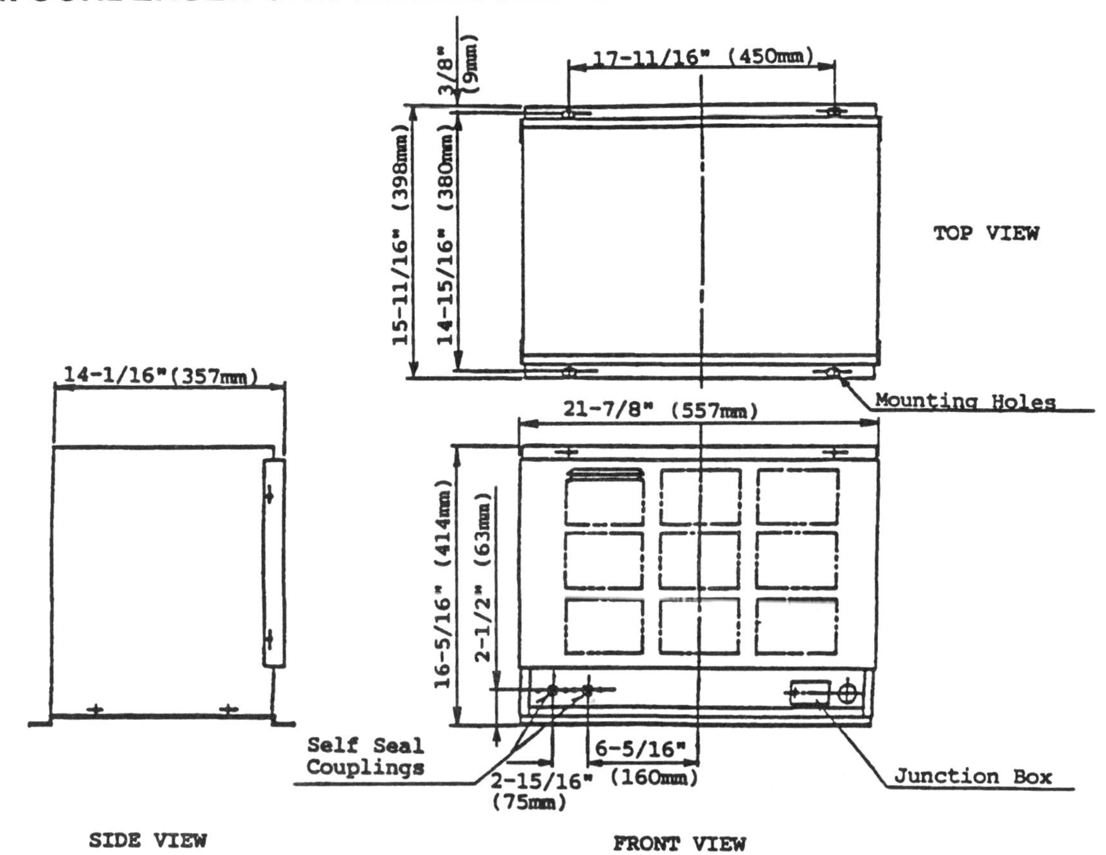

SPECIFICATIONS

MODEL: CONDENSER UNIT
URC-4A

EXTERIOR	Galvanized Steel
DIMENSIONS (H x D x W)	16-5/16" x 15-11/16" x 21-7/8" (414mm x 398 x 557)
REFRIGERANT CHARGE	R502 1 lb. 12 oz. (800 g)
WEIGHT	Net 37 lbs. (17 kg) Shipping 44 lbs. (20 kg)
CONNECTIONS	
REFRIGERATION	Self Seal Couplings (Aeroquip)
ELECTRICAL	Junction Box - Fan Motor
HEAD PRESSURE CONTROL	Condensing Pressure Regulator
AMBIENT CONDITION	Min. -20°F - Max. +122°F (-29°C to +50°C) Full weather

II. GENERAL INFORMATION

1. CONSTRUCTION

[a] KM-450MAB

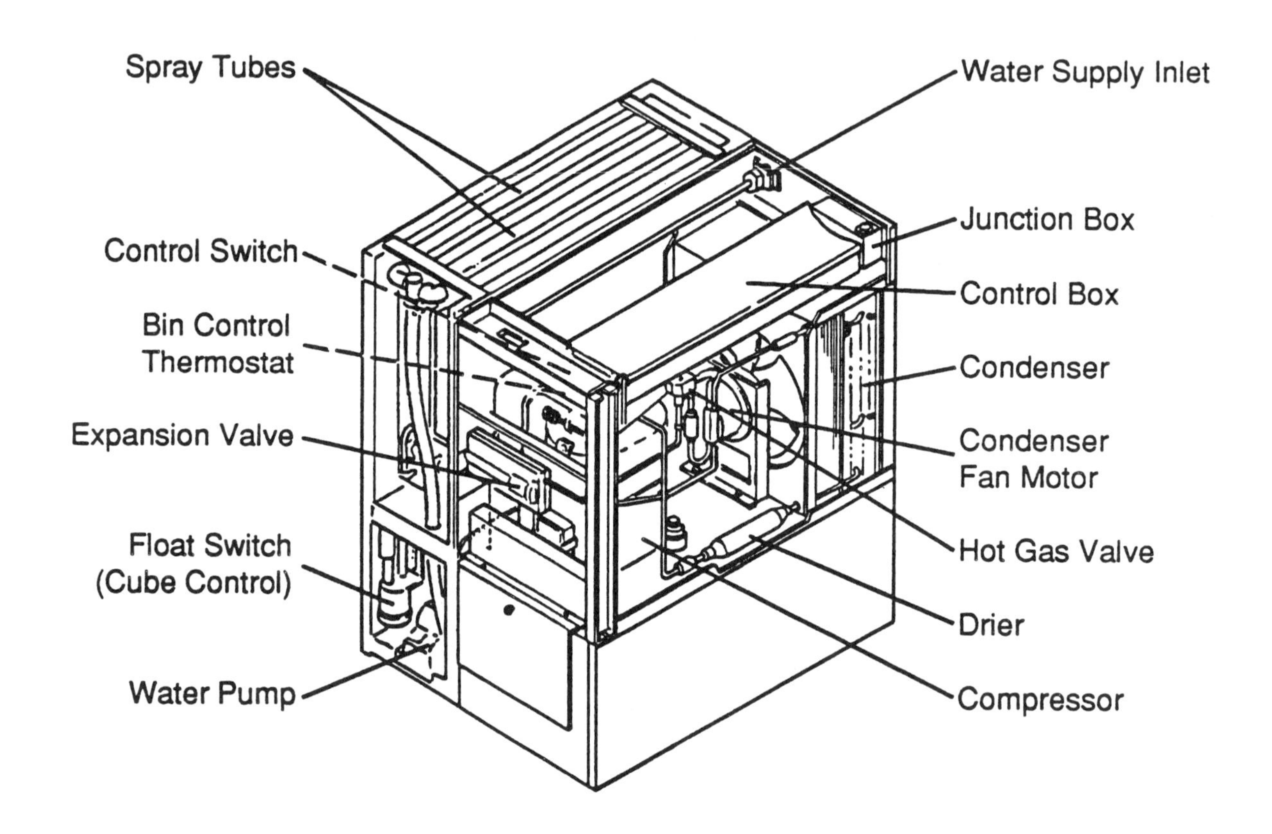

[b] KM-450MWB

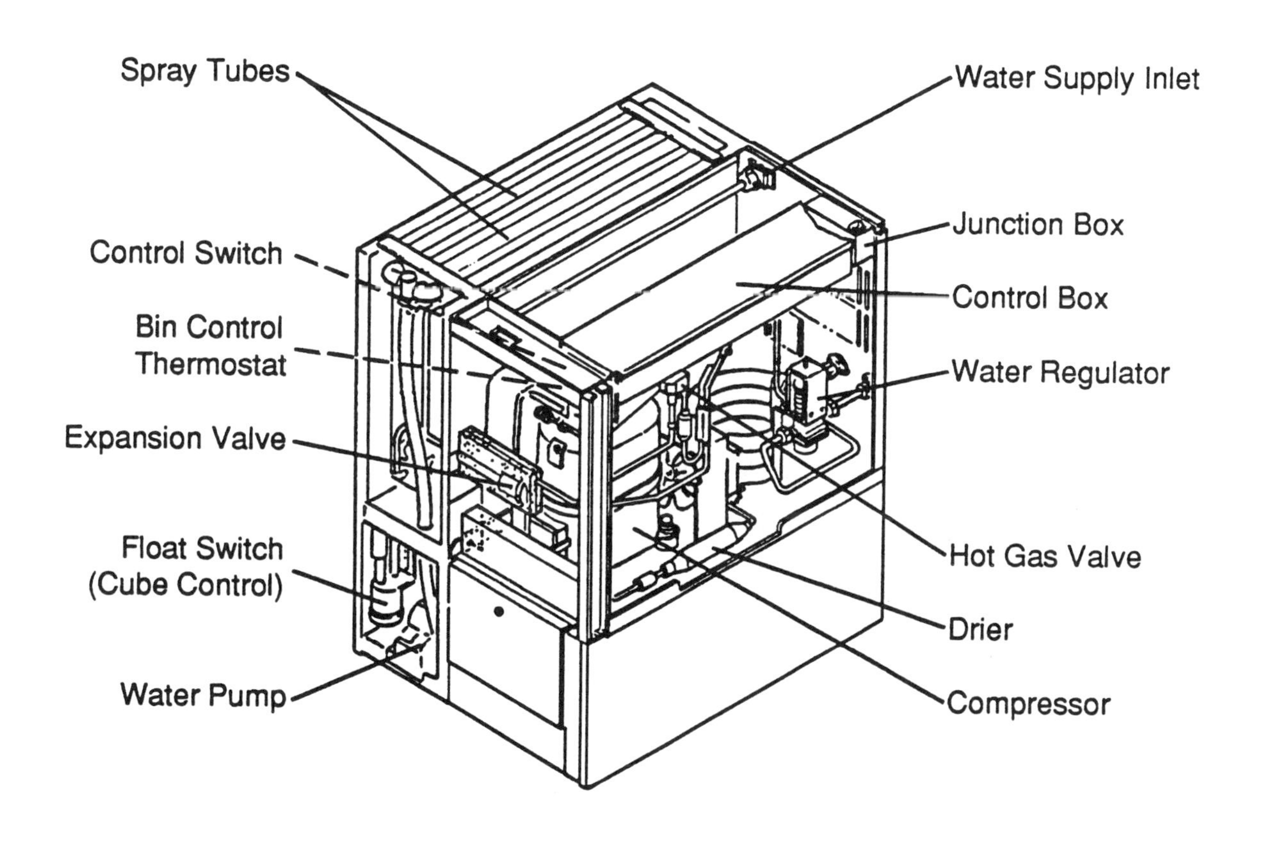
Spray Tubes
Control Switch
Bin Control
Thermostat
Expansion Valve
Float Switch
(Cube Control)
Water Pump
Water Supply Inlet
Junction Box
Control Box
Water Regulator
Hot Gas Valve
Drier
Compressor

[c] KM-450MRB

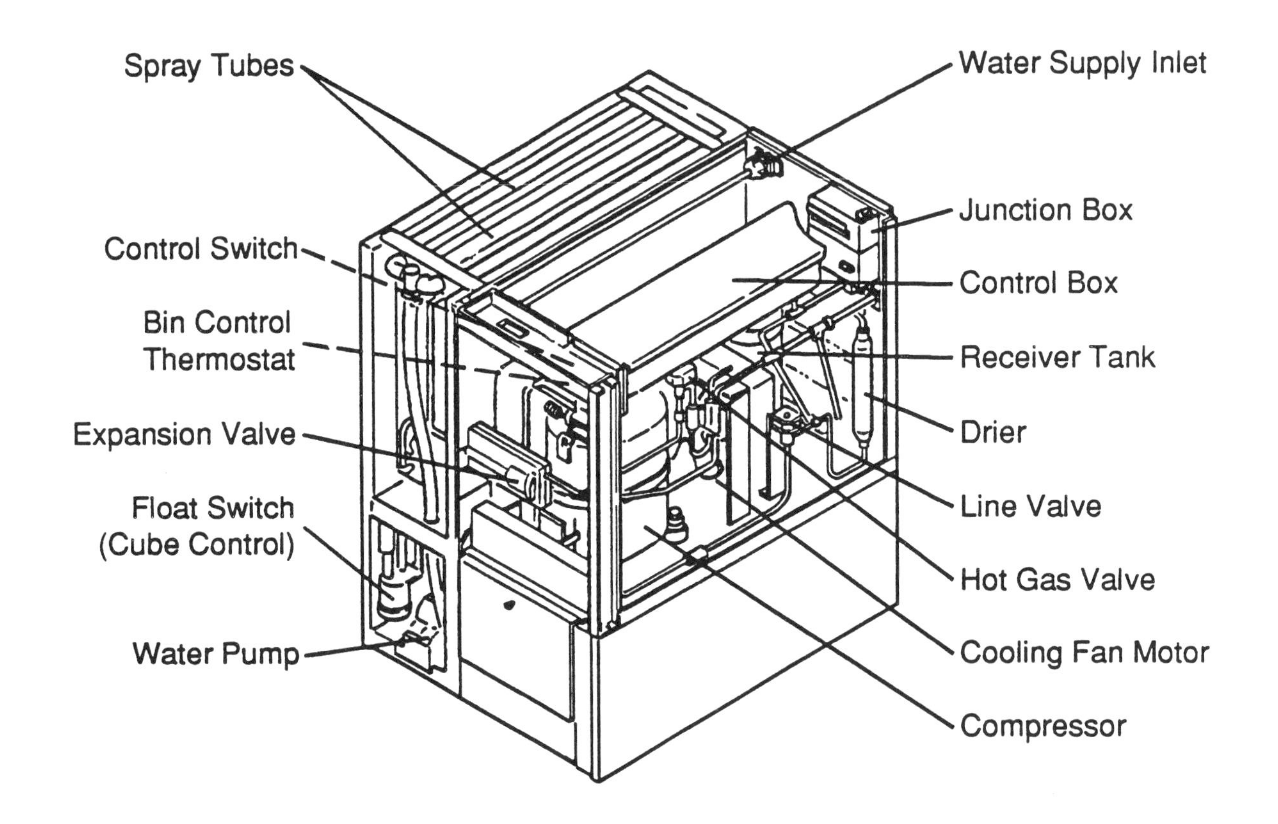

Spray Tubes
Control Switch
Bin Control Thermostat
Expansion Valve
Float Switch (Cube Control)
Water Pump
Water Supply Inlet
Junction Box
Control Box
Receiver Tank
Drier
Line Valve
Hot Gas Valve
Cooling Fan Motor
Compressor

2. CONTROLLER BOARD

[a] SOLID-STATE CONTROL

1) A HOSHIZAKI EXCLUSIVE solid-state control is employed in KM-450M series, modular crescent cubers.

2) A Printed Circuit Board (hereafter called "Controller Board") includes a stable and high quality control system.

3) All models are pretested and factory-adjusted.

[b] CONTROLLER BOARD

CAUTION

1. FRAGILE, handle very carefully.
2. A controller board contains integrated circuits, which are SUSCEPTIBLE TO FAILURE DUE TO STATIC DISCHARGE. It is especially important to touch the metal part of the unit when handling or replacing the board.
3. Do not touch the electronic devices on the board or at the back of the board to prevent damage to the board.
4. Do not change wiring and connections. Especially, never misconnect K3, K5 and K6, because the same connector is used for the Thermistor, Float Switch and Jumper.
5. Do not fix the electronic devices or parts on the board in the field. Always replace the whole board assembly when it goes bad.
6. Do not short out power supply to test for voltage.

A controller board, Part Code 2U0103-02, is used for KM-450M series. See "[c] SEQUENCE."

Note: (1) Maximum Water Supply Period - 6 minutes

Water solenoid valve opening, in the Defrost (Harvest) Cycle, is limited by maximum period. Water valve cannot be opened exceeding the maximum period. And water valve closes even within the maximum period, when the defrost cycle is completed.

(2) High Temperature Safety - 127 ± 7°F

The temperature of suction line in the refrigerant circuit is limited by High Temperature Safety.
Thermistor detects to start the Defrost Timer. After the defrost starts, the temperature rises, and the ice making starts working in normal operation. If the connection of thermistor is loosen, thermistor lead is broken, or any other trouble happens, thermistor does not detect the temperature properly. Therefore, the defrost cycle continues, and the temperature gets higher.
But after High Temperature Safety operates, the circuit shuts down and the icemaker automatically stops. To reset the safety, turn the power off and back on again.
Thus, the temperature of the suction line is prevented from rising excessively.

(3) Low Water Safety

If the Pump Motor is operated without water, the mechanical seal can fail. Therefore if there is no water in the sump when the defrost cycle finishes, the unit does not start the freeze cycle but continues the initial water supply cycle.

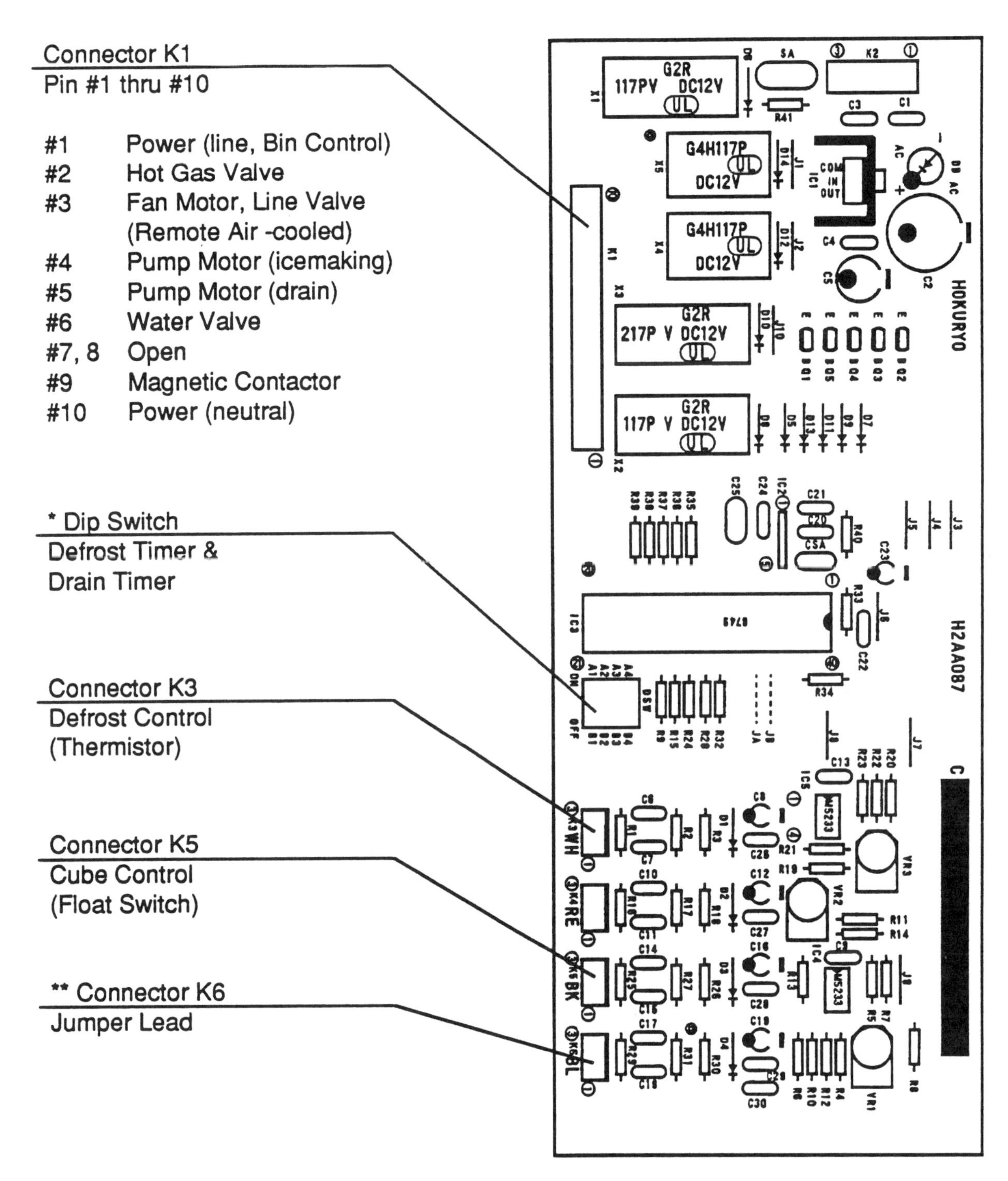

Note: * Defrost Timer (Dip Switch No. 1 & 2) is adjustable, Min. 1 minute to Max. 3 minutes.
Drain Timer (Dip Switch No. 3 & 4) is adjustable, Min. zero to Max. 20 seconds. Do not adjust it.

** On all models, Jumper Lead is required. See "[e] CHECKING CONTROLLER BOARD."

Fig. 1 Controller Board

[c] SEQUENCE

1st Cycle

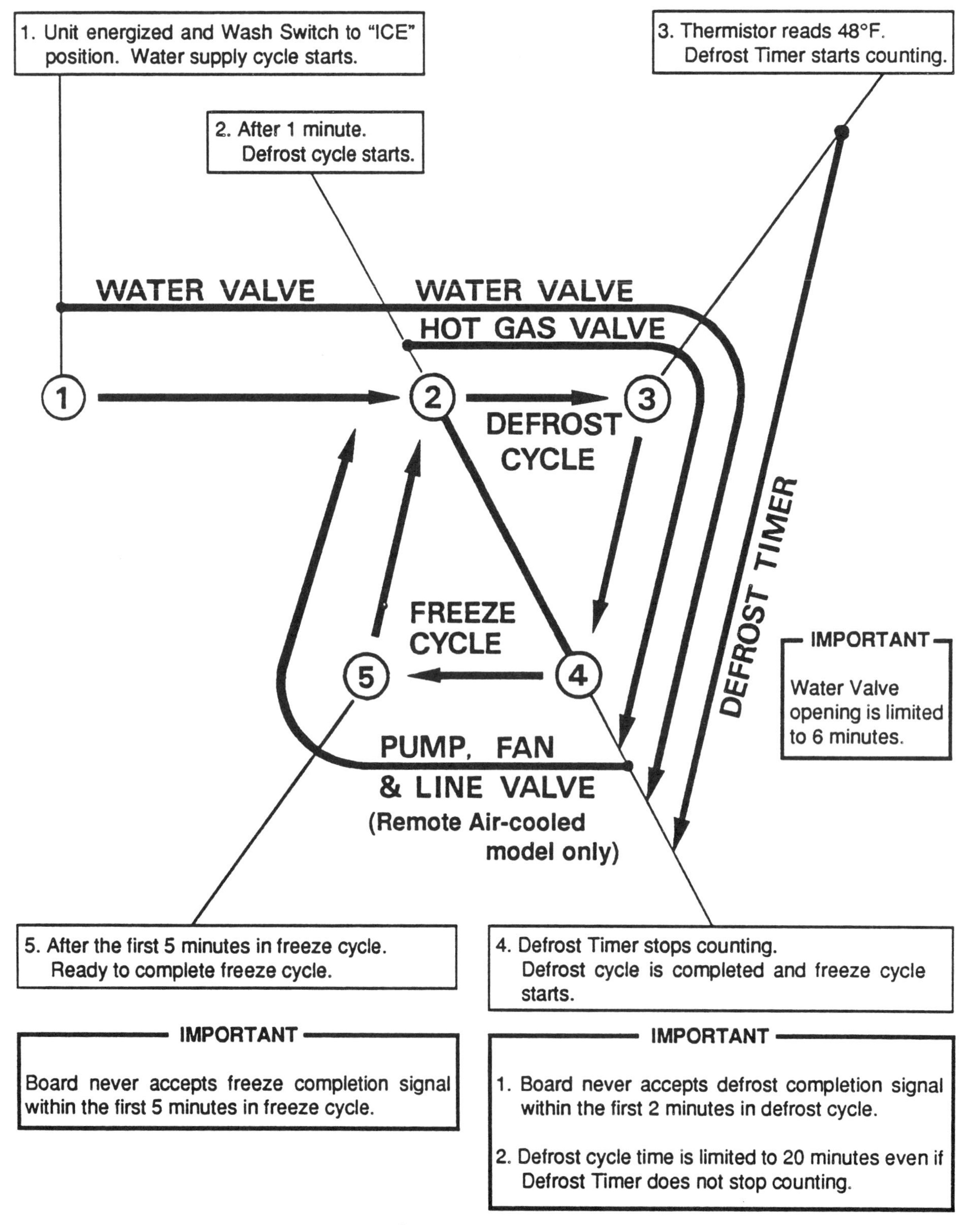

IMPORTANT

Board never accepts freeze completion signal within the first 5 minutes in freeze cycle.

IMPORTANT

1. Board never accepts defrost completion signal within the first 2 minutes in defrost cycle.
2. Defrost cycle time is limited to 20 minutes even if Defrost Timer does not stop counting.

Fig. 2 Sequence - 1st Cycle

2nd Cycle and after with pump drain

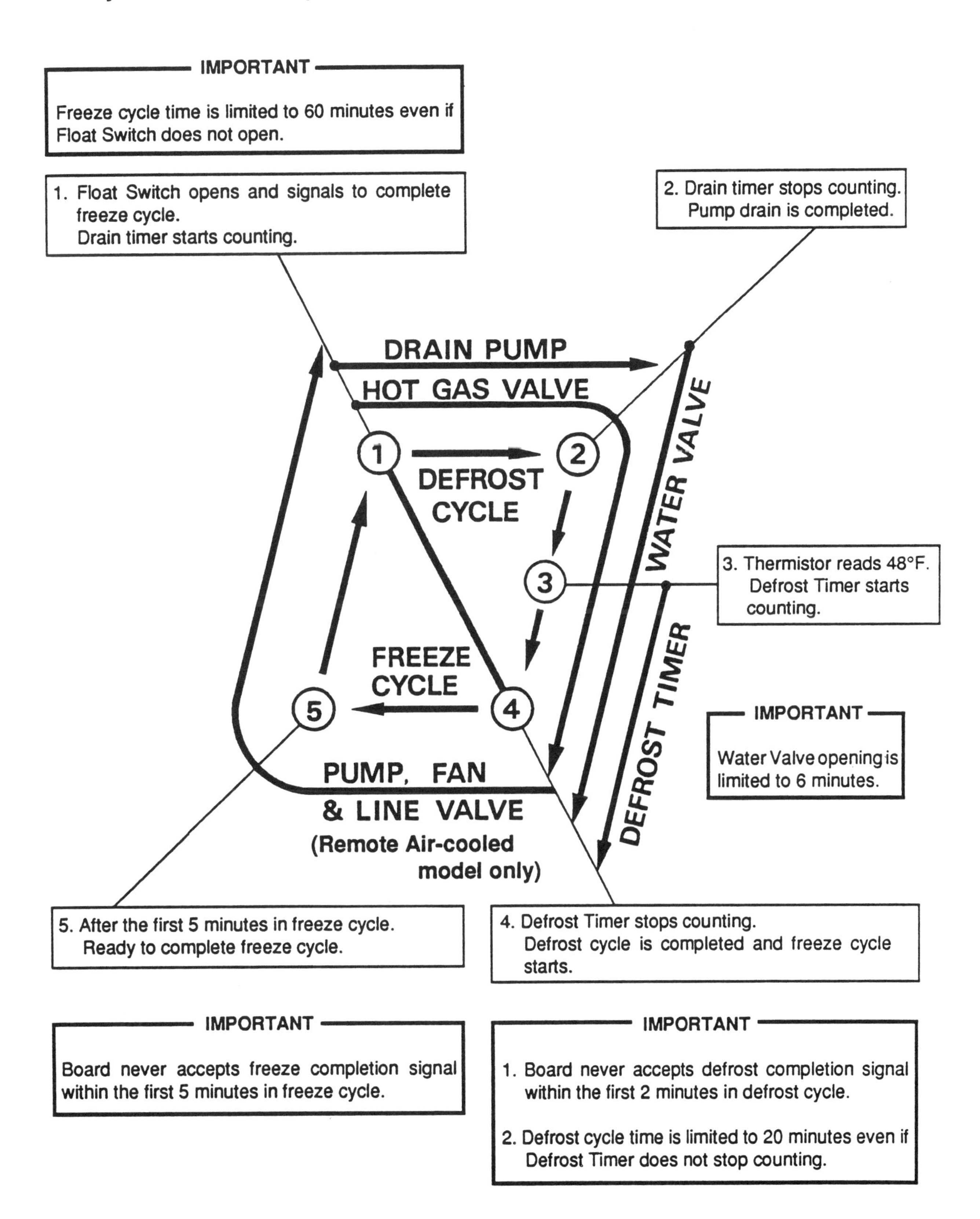

Fig. 3 Sequence - 2nd Cycle and after with pump drain

2nd Cycle and after with no pump drain

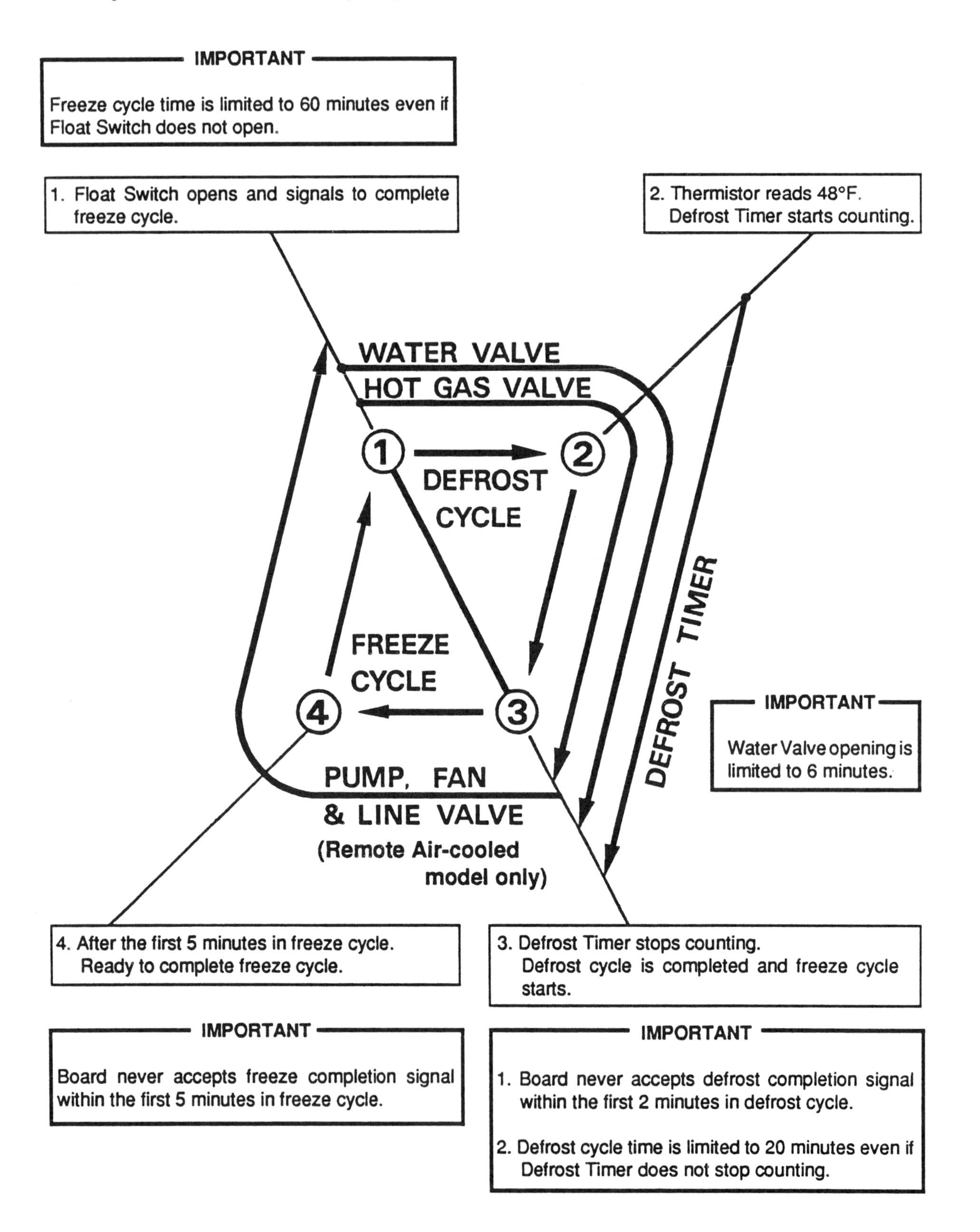

Fig. 4 Sequence - 2nd Cycle and after with no pump drain

[d] CONTROLS AND ADJUSTMENT

1) Cube Control

A float switch, located next to the Pump Motor, signals to complete the freeze cycle. This switch is factory-adjusted to the lowest position, or Max. Cube Size position. Setting to upper position makes cube size smaller.

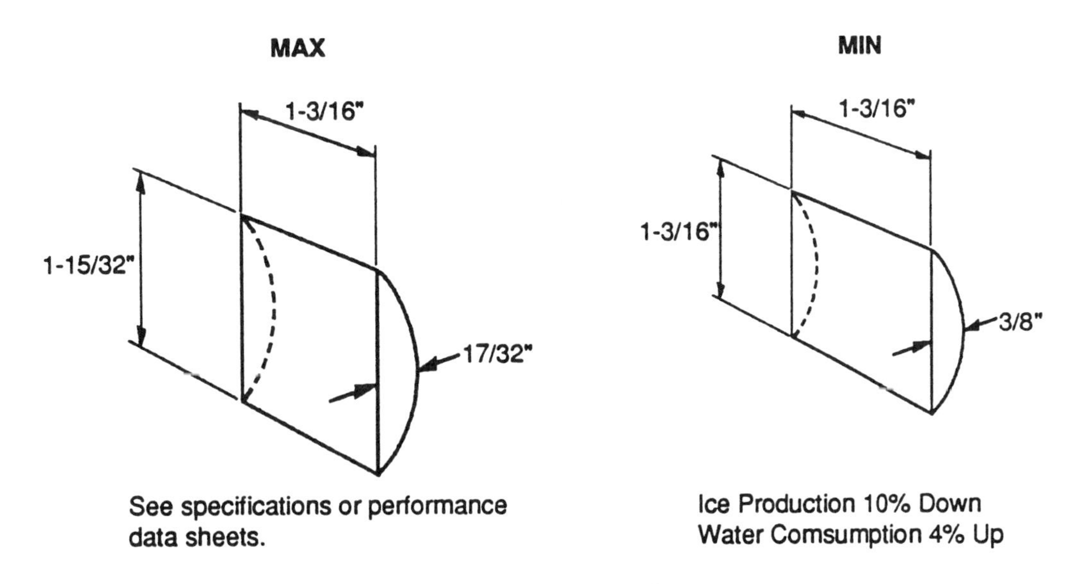

See specifications or performance data sheets.

Ice Production 10% Down
Water Comsumption 4% Up

Fig. 5 Cube Size

2) Defrost Control

A thermistor (Semiconductor) is used for a defrost control sensor, whose resistance varies depending on the Suction Line temperatures. The Thermistor detects the temperature of the Evaporator outlet to start the Defrost Timer. No adjustment is required. If necessary, check for resistance between Thermistor leads, and visually check the Thermistor mounting, located on the Suction Line next to the Evaporator outlet.

Temperature (°F)	Resistance (kΩ)
0	14.401
10	10.613
32	6.000
50	3.871
70	2.474
90	1.633

Check a thermistor for resistance by using the following procedures.

(i) Disconnect the connector K3 on the board.

(ii) Remove the Thermistor. See "VII. 10. REMOVAL AND REPLACEMENT OF THERMISTOR."

(iii) Immerse the Thermistor sensor portion in a glass containing ice and water for 2 or 3 minutes.

(iv) Check for a resistance between Thermistor leads.
Normal reading is within 3.5 to 7 kΩ. Replace the Thermistor if it exceeds the normal reading.

3) Defrost Timer

No adjustment is required under normal use, as the Defrost Timer is adjusted to the suitable position. However, if necessary when all the ice formed on the Evaporator does not fall into the bin in the harvest cycle, adjust the Defrost Timer to longer position by setting the Dip Switch (No. 1 & 2) on the Controller Board.

SETTING		TIME
Dip Switch No. 1	Dip Switch No. 2	
OFF	OFF	60 second
ON	OFF	90 second
OFF	ON	120 second
ON	ON	180 second

Note: When shipped, the Defrost Timer is adjusted to 60 second.

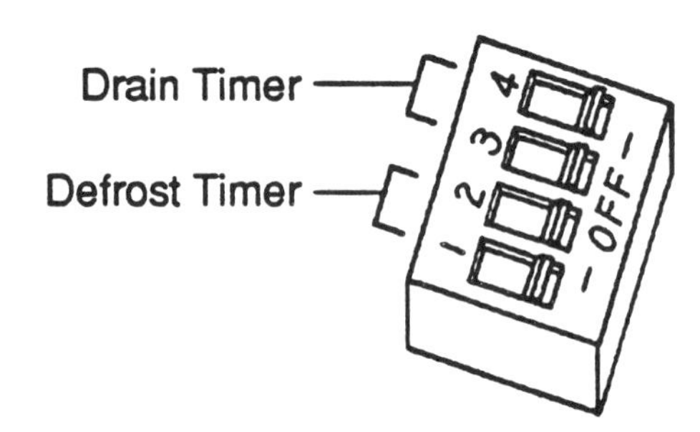

Fig. 6 Dip Switch for Defrost and Drain Timer

4) Drain Timer

The Drain Timer is factory-adjusted to 10 second position, and no adjustment is required.

SETTING		TIME	FLOAT SWITCH POSITION
Dip Switch No. 3	Dip Switch No. 4		
OFF	OFF	10 second	ANY

5) Bin Control

CAUTION

When the ambient temperature is below 45°F, the Bin Control Thermostat operates to stop the icemaker even if the Ice Storage Bin is empty.

No adjustment is required under normal use, as the Bin Control is factory-adjusted. Adjust it, if necessary, so that the icemaker stops automatically in approximately 6 to 10 seconds after ice contacts the Bin Control Thermostat Bulb, which is attached to a bracket on the left side wall near the bin opening.

[e] CHECKING CONTROLLER BOARD

1) Visually check the sequence with the icemaker operating.

2) Visually check the Controller Board by using the following procedures.

(i) Adjust the Defrost Timer to minimum position.
Disconnect the Thermistor from the Controller Board.
Connect a 1.5 kΩ - 3.5 kΩ resistor to the Connector K3 (pins #1 and #2), and energize the unit.

After the 1 minute ± 5 second water supply cycle and the 2 minute ± 10 second defrost cycle, the unit should start the freeze cycle.

(ii) After the above step (i), disconnect the Float Switch leads from the Controller Board within the first 5 minutes of the freeze cycle.

The unit should go into the defrost cycle after the first 5 minutes ± 20 seconds of the freeze cycle.

(iii) After the above step (i), disconnect the Float Switch leads from the Controller Board after the first 5 minutes of the freeze cycle.

At this point, the unit should start the defrost cycle.

(iv) Adjust the Defrost Timer to minimum position. Disconnect the Thermistor from the Controller Board, and energize the unit.
After the 1 minute water supply cycle, the defrost cycle starts.
Connect a 1.5 kΩ - 3.5 kΩ resistor to the Connector K3 (pins #1 and #2) after the first 2 minutes of the defrost cycle.

The unit should start the freeze cycle after 1 minute ± 5 second from the resistor connection.

3) Check the Controller Board by using test program of the Controller Board.

(i) Disconnect the Connectors K1 and K6 from the Controller Board. Set the Dip Switch No. 1 and 2 on the Controller Board to the "OFF" position, and energize the unit.

(ii) The current draws to each Relay (from X1 to X4) one after another for 5 seconds, and then the contacts close. See the following table, and check "OPEN" and "CLOSE" of Pins of the Connector K1 at each step.

(iii) If the checks are completed, turn off the icemaker, plug the Connectors K1 and K6 into the Controller Board as before, and set the Dip Switch as before.

* TEST PROGRAM OF CONTROLLER BOARD

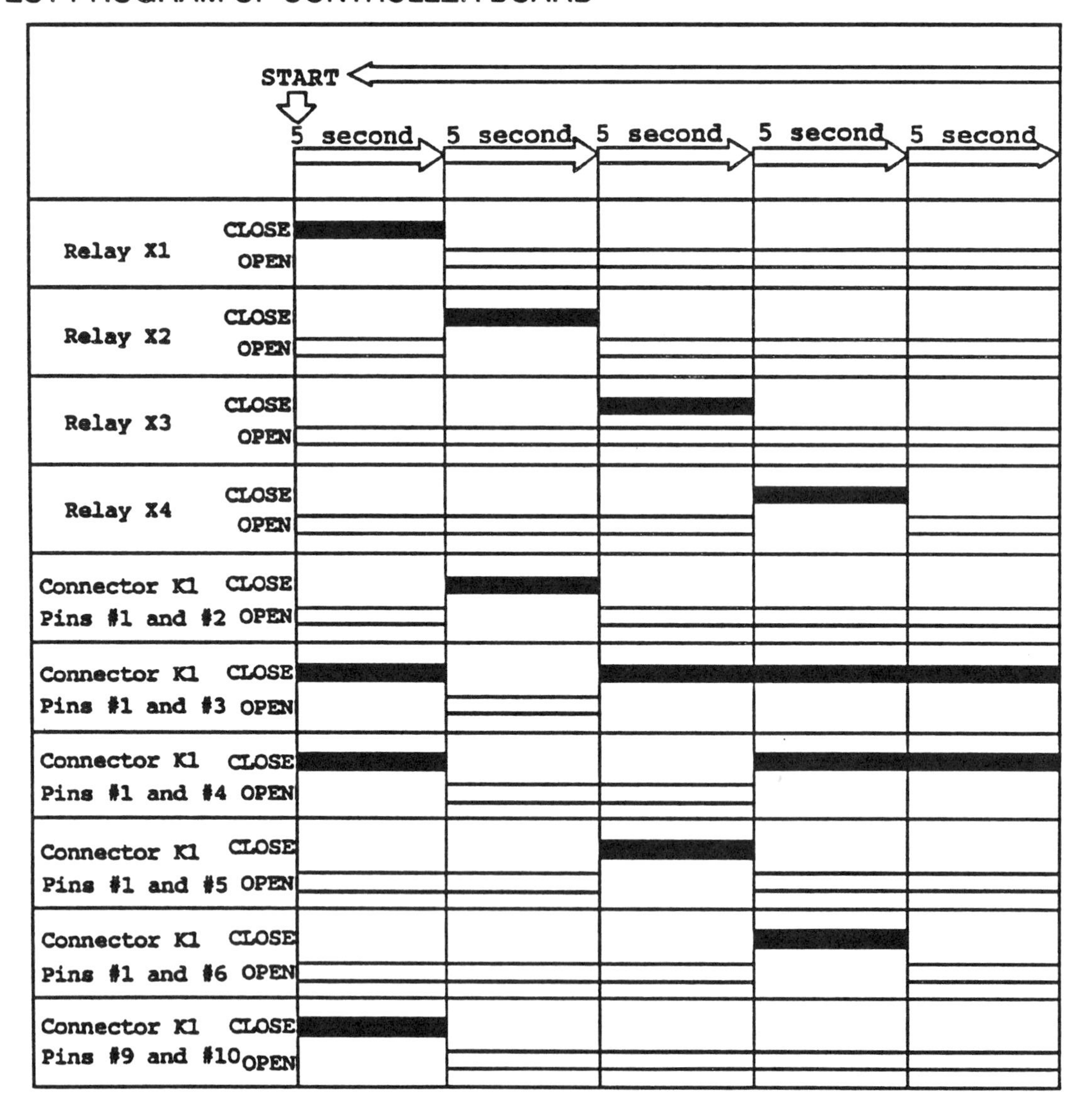

III. INSTALLATION AND OPERATING INSTRUCTION

1. CHECKS BEFORE INSTALLATION

WARNING

Remove shipping carton, tapes and packing. If packing material is left in the icemaker, it will not work properly.

1) Remove the panels to prevent damage when installing the icemaker. See "2. HOW TO REMOVE PANELS."

2) Remove the package containing the accessories.

3) Remove the protective plastic film from the panels. If the icemaker is exposed to the sun or to heat, remove the film after the icemaker cools.

4) Check that the refrigerant lines do not rub or touch lines or other surfaces, and that the fan blade moves freely.

5) Check that the Compressor is snug on all mounting pads.

6) See NAMEPLATE on the Rear Panel, and check that your voltage supplied corresponds with the voltage specified on the Nameplate.

7) On remote air-cooled models, a remote condenser unit is needed. The recommended remote condenser unit is HOSHIZAKI CONDENSER UNIT, MODEL URC-4A.

2. HOW TO REMOVE PANELS - See Fig. 7

a) Front Panel .... Remove the screw.
Lift up and pull toward you.

b) Top Panel .... Lift up slightly its front, push away, and then lift off.

c) Side Panel (R)...... Remove the screw.
Pull slightly toward you, and lift off.

d) Insulation Panel ... Lift up slightly, and pull toward you.

e) Base Cover .... Lift up slightly, and pull toward you.

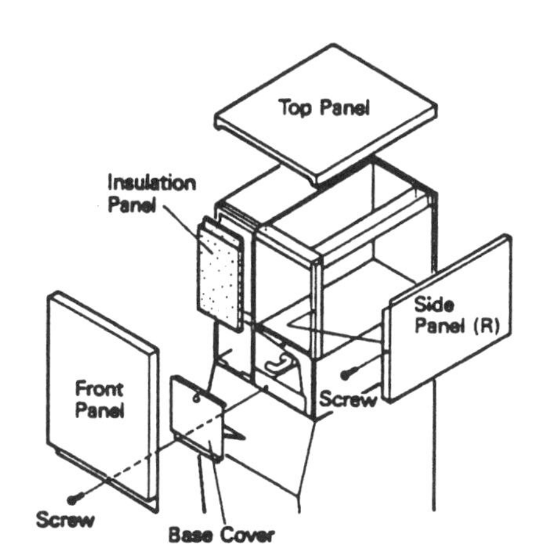

Fig. 7

3. LOCATION

WARNING

This icemaker is not designed for unsheltered outdoor installations. Normal operating ambient temperature should be within +45°F to +100°F; Normal operating water temperature should be within +45°F to +90°F. Operation of the icemaker, for extended periods, outside of these normal temperature ranges may affect production capacity

For best operating results:

* Icemaker should not be located next to ovens, grills or other high heat producing equipment.

* Location should provide a firm and level foundation for the equipment.

* Allow 6" clearance at rear, sides and top for proper air circulation and ease of maintenance and/or service should they be required.

4. SET UP

1) Unpack the storage Bin, and attach the four adjustable legs provided (bin accessory) to the bottom of the storage bin.

2) Position the Storage Bin in the selected permanent position.

3) Place the icemaker on the top of the Storage Bin.

4) Secure the icemaker to the Storage Bin, by using the two mounting bracket and the four bolts provided. See Fig. 8

5) Level the icmaker/Storage Bin, in the both the left-to-right and front-to-rear directions. Adjust the Ice Bin Legs to make the icemaker level.

6) Replace the panels in their correct position.

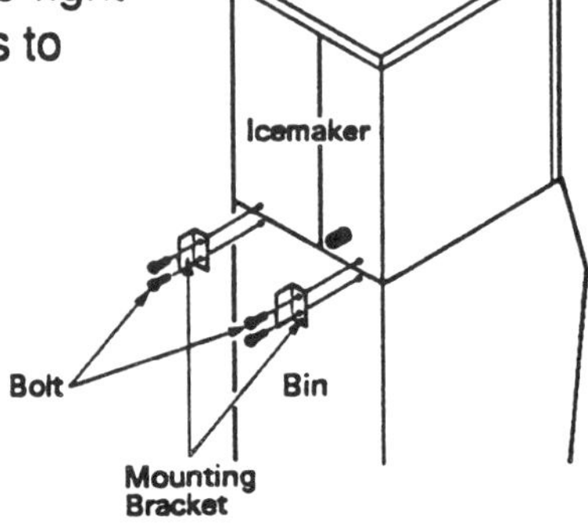

Fig. 8

5. ELECTRICAL CONNECTION

WARNING

1. Electrical connection must be made in accordance with the installations on a "WARNING" tag, provided with the pig tail leads in the Junction Box.

2. This icemaker requires a ground that meets the national and local electrical code requirements. To prevent possible electrical shock to individuals or extensive damage to the equipment, install a proper ground wire to the icemaker.

* A WHITE lead must be connected to the neutral conductor of the power source. Miswiring results in severe damage to the icemaker. See Fig. 9.

* This icemaker must have a separate power supply or receptacle of proper capacity. See NAMEPLATE.

* Usually an electrical permit and services of a licensed electrician are required.

WARNING

ELECTRICAL CONNECTION

A white lead must be connected to the neutral conductor of the power source. Miswiring results in severe damage to the icemaker. (See below Fig.)

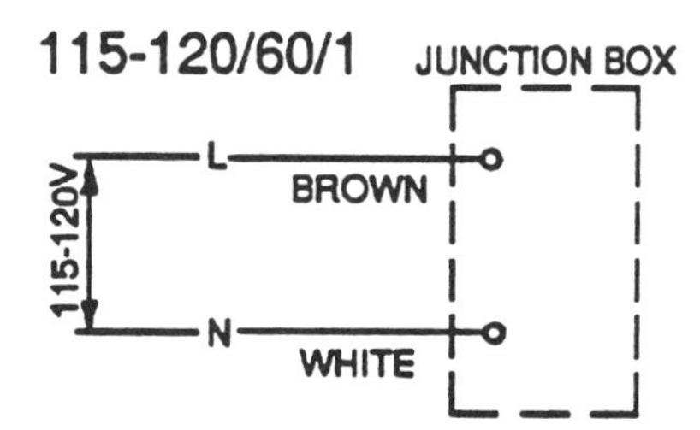

Fig. 9

6. INSTALLTION OF REMOTE CONDENSER UNIT

[a] CHECKS BEFORE INSTALLATION

1) Unpack and remove shipping carton, tapes and packing.

2) Check that the refrigerant lines do not rub or touch lines or other surfaces, and that the fan blade moves freely.

[b] LOCATION

The condenser unit must be positioned in a permanent site under the following guidelines.

* A firm and flat site. Use of a roof curb is preferred.

* A dry and well ventilated area with 24" clearance on both front and rear for ease of maintenance and service should they be required.

* Normal condenser ambient temperature: -20°F to +122°F. Temperatures not within this operating range may affect the production capacity of the icemaker.

* The normal refrigerant line length is 66 ft. Should an installation require a longer line length, please call 1-800-233-1940 for recommendations.

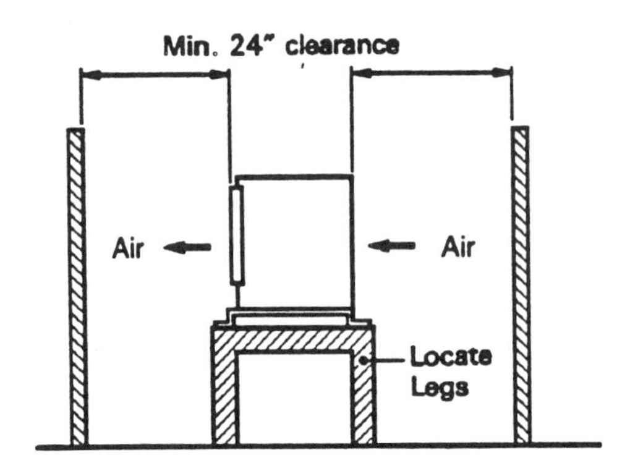

Fig. 10

Note: If the recommended guidelines of the installation are exceeded, the icemaker performance may be reduced.

[c] SET UP

1) Secure the Legs to the condenser unit with eight M8 x 16 mm Hexagon Bolts and M8 Nuts as shown in the illustration. (See Fig. 11.)

   Note: Locate the Legs symmetrically.

2) The Legs have eight mounting holes. Secure the Legs with eight bolts (not included).

3) Install enough length of two copper tubings provided with Aeroquip couplings between the icemaker and the condenser unit. The two copper tubings should be insulated separately. (See Fig. 12.)

   * Precharged tubing kits, available as optional equipment from HOSHIZAKI AMERICA are recommended.

4) Line sets fabricated in the field should be evacuated through the charging ports on the Aeroquip couplings and charged with R-502 refrigerant vapor to a pressure of 15 - 30 PSIG.

   Note: Factory fabricated tubing kits are precharged and do not need to be evacuated.

5) Remove the plastic caps protecting the couplings. Attach the two refrigerant lines to the male couplings on the icemaker and the remote condenser unit. Each refrigerant line must be connected as follows:

   Icemaker discharge refrigerant line - 1/2" OD tubing to "DIS" of condenser unit

   Icemaker liquid refrigerant line - 3/8" OD tubing to "LIQ" of condenser unit

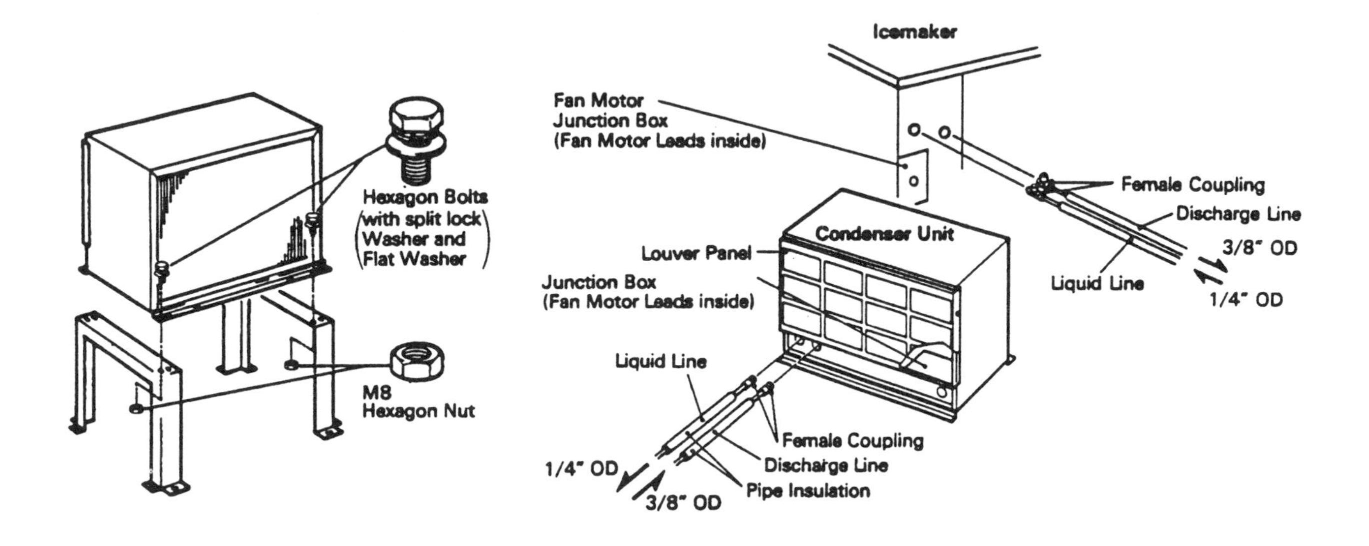

Fig. 11

Fig. 12

[d] ELECTRICAL CONNECTION

WARNING

This remote condenser unit requires a ground that meets the national and local electrical code requirements. To prevent possible electrical shock to individuals or extensive damage to equipment, install a proper ground wire to the condenser unit.

1. This condenser unit must be connected to the Fan Motor Junction Box on the icemaker.

2. Usually an electrical permit and services of a licensed electrician are required.

1) Remove the Louver Panel.

2) Remove the Junction Box Cover.

3) Connect the Fan Motor leads in the Junction Box of the remote condenser unit to the Fan Motor leads in the Junction Box of the HOSHIZAKI remote air-cooled icemaker.

4) Install a ground wire from the icemaker to the remote condenser unit.

5) Replace the Junction Box Cover and the Louver Panel in their correct position.

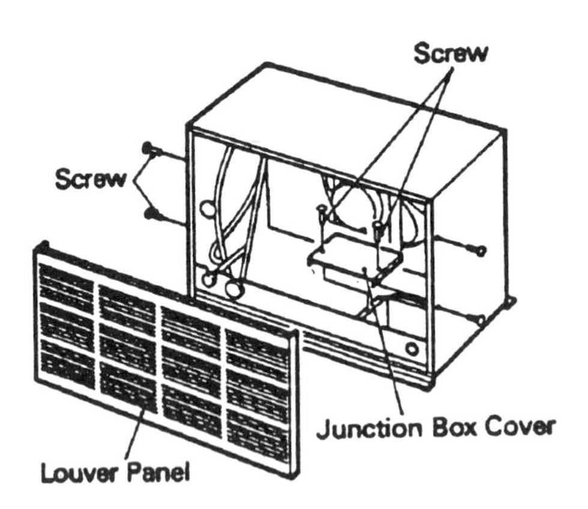

Fig. 13

[e] STACKING CONDENSER UNIT

1) Secure the lower condenser unit to the Legs with eight bolts (not included).

2) Attach the upper condenser unit on the top of the lower.

3) Secure the upper condenser unit with the four screws provided.

4) Install refrigerant lines, and make electrical connection for each Fan Motor as shown in Items [c] and [d].

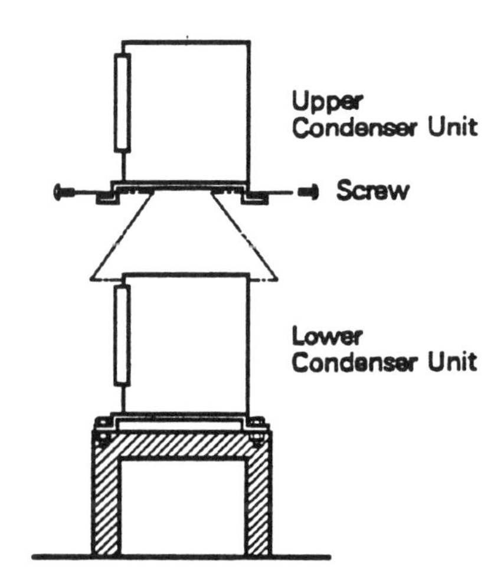

Fig. 14

7. WATER SUPPLY AND DRAIN CONNECTIONS - See Fig. 15

To prevent damage to the pump assembly, do not operate the icemaker when the water supply is OFF, or if the pressure is below 7 PSIG. Do not run the icemaker until the proper water pressure is reached.

* Water supply inlet is 1/2" female pipe thread (FPT).

  Note: On water-cooled model, two water supply inlets are provided. One is for the ice making water inlet, and the other is for the cooling water inlet.

* A water supply line shut-off valve and drain valve should be installed. A minimum of 3/8" OD copper tubing is recommended for the water supply lines.

  Note: An optional strainer should be installed next to the water supply inlet in the water supply line. HOSHIZAKI recommended optional strainer Part Code 311166A01.

* Water supply pressure should be a minimum of 7 PSIG and a maximum of 113 PSIG. If the pressure exceeds 113 PSIG, the use of a pressure reducing valve is required.

* Drain outlet for icemaker dump is 3/4" FPT. The icemaker drain and the condenser drain piping connections must be made separately from the bin drain.

  Note: On water-cooled model, a 1/2" FPT is provided for the condenser drain outlet.

* The drains must be 1/4" fall per foot on horizontal runs to get a good flow.

* A plumbing permit and services of a licensed plumber may be required in some areas .

KM-450MAB
KM-450MRB

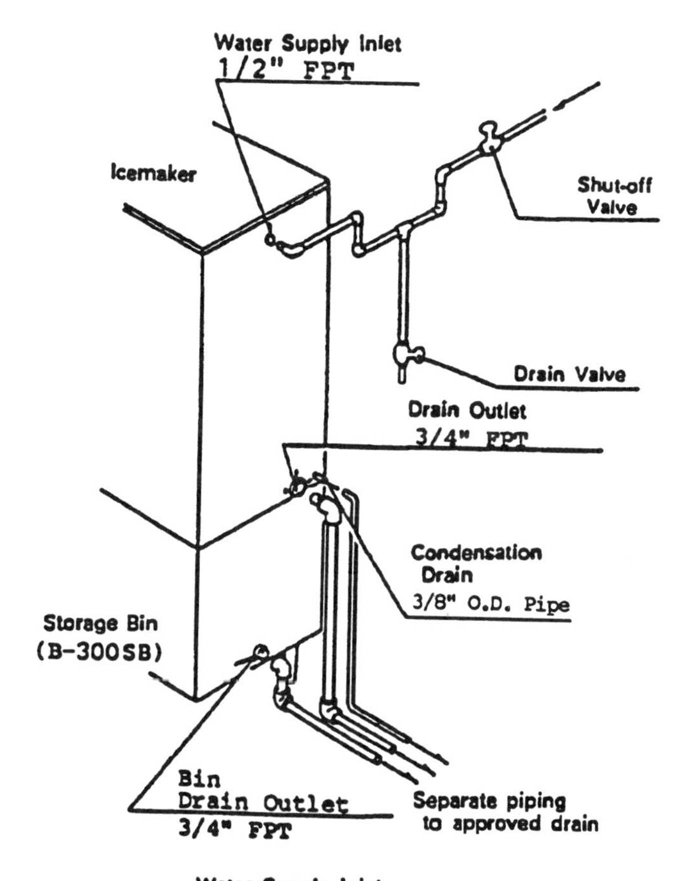

KM-450MWB

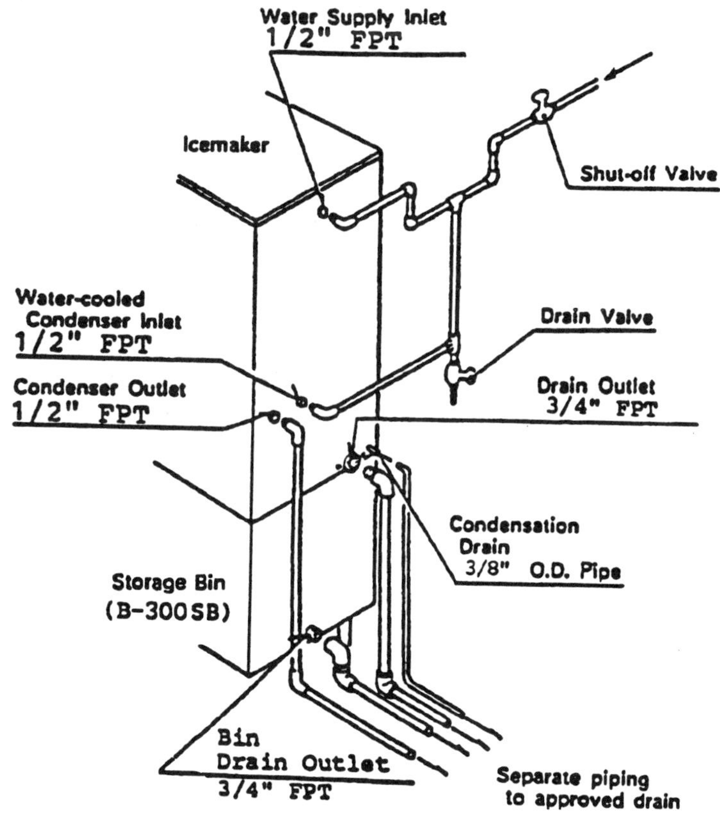

Fig. 15

8. FINAL CHECK LIST

1) Is the icemaker level?

2) Is the icemaker in a site where the ambient temperature is within +45°F to +100°F and the water temperature within +45°F to +90°F all year around?

3) Is there at least 6" clearance at sides, rear at top of the icemaker for maintenance or service?

4) Have all shipping carton, tapes and packing been removed from the icemaker?

5) Have all electrical and piping connections been made?

6) Has the power supply voltage been checked or tested against the nameplate rating? And has a proper ground been installed in the icemaker?

7) Are the Water Supply Line Shut-off Valve and Drain Valve installed? And has the water supply pressure been checked to ensure a minimum of 7 PSIG and a maximum of 113 PSIG?

   Note: The icemaker stops running when the water supply is OFF, or if the pressure is below 7 PSIG. When the proper water pressure is reached, the icemaker automatically starts running again.

8) Have the compressor hold-down bolts and refrigerant lines been checked against vibration and possible failure?

9) Has the Bin Control Switch been checked for correct operation? When the icemaker is running, hold an ice cube in contact with the Bulb. The icemaker should stop within 10 seconds.

10) Has the end user been given the instruction manual, and instructed on how to operate the icemaker and the importance of the recommended periodic maintenance?

11) Has the end user been given the name and telephone number of an authorized service agent?

9. START UP

WARNING

1. All parts are factory-adjusted. Improper adjustments may result in failure.
2. If the unit is turned off, wait for at least 3 minutes before restarting the icemaker to prevent damage to the Compressor.

1) Open the Water Supply Line Shut-off Valve.

2) Remove the Front Panel.

3) Move the Toggle WASH-OFF-ICE Control Switch, on the Control Box, to the "ICE" position.

4) Replace the Front Panel in its correct position.

5) Turn on the power supply, and start the washing process for 10 minutes.

6) Turn off the power supply, and remove the Front Panel and the Base Cover.

7) Remove one end of the Pump Tubing, and drain the Water Tank.

8) Replace the Pump Tubing in its correct position.

9) Clean the Storage Bin.

10) Replace the Base Cover and the Front Panel in their correct position.

11) Turn on the power supply, and start the automatic icemaking process.

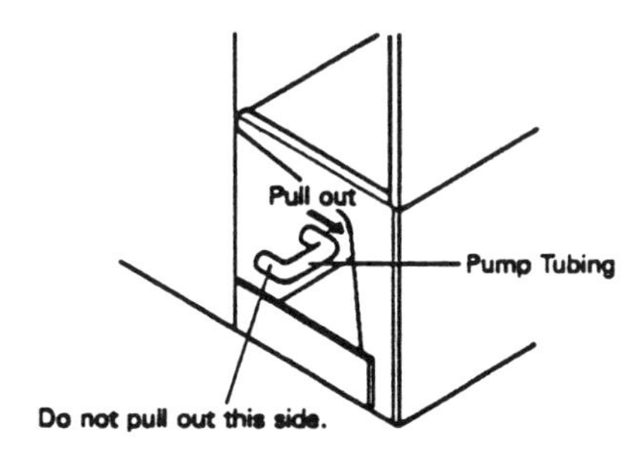

Fig. 16

10. PREPARING THE ICEMAKER FOR LONG STORAGE

WARNING

When shutting off the icemaker for an extended time, drain out all water from the water line and remove the ice from the Storage Bin. The Storage Bin should be cleaned and dried. Drain the icemaker to prevent damage to the water supply line at sub-freezing temperatures, using a foot or hand pump. Shut off the icemaker until the proper ambient temperature is resumed.

1) Turn off the power supply.

2) Remove the Front Panel.

3) Move the Toggle WASH-OFF-ICE Control Switch, on the Control Box, to the "OFF" position.

4) Close the Water Supply Line Shut-off Valve, and remove the Base Cover.

5) Remove one end of the Pump Tubing, and drain the Water Tank.

6) Remove the Pump Tubing and the Base Cover in their correct position.

7) Remove all ice from the Storage Bin, and clean the Storage Bin.

8) Replace the Front Panel in its correct position.

Note: (1) When shutting off the icemaker at sub-freezing temperatures, run the icemaker, with the Water Supply Line Shut-off Valve closed and the Drain Valve opened, and blow out the water inlet line, by using air pressure. See Fig. 17.

(2) When the icemaker is not used for two or three days, just move the Toggle WASH-OFF-ICE Control Switch to the "OFF" position.

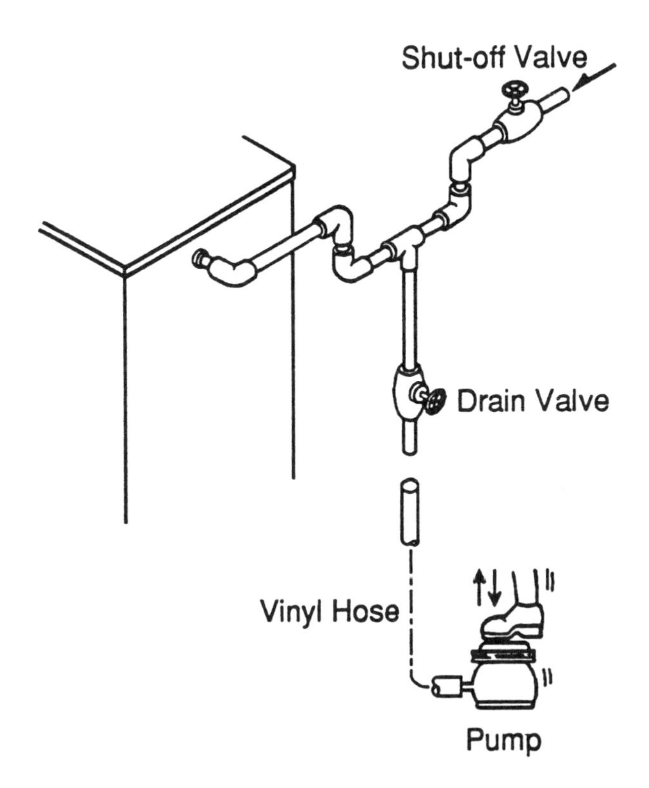

Fig. 17

IV. MAINTENANCE AND CLEANING INSTRUCTIONS

1. CLEANING INSTRUCTIONS

WARNING

1. HOSHIZAKI recommends cleaning this unit at least once a year. More frequent cleaning, however, may be required in some existing water conditions.

2. To prevent injury to individuals and damage to the icemaker, do not use ammonia type cleaners.

3. Always wear liquid-proof gloves for safe handling of the cleaning and sanitizing solution. This will prevent irritation in case the solution contacts with skin.

[a] CLEANING PROCEDURE

* STANDARD

1) Dilute approximately 16 fl. oz. of recommended cleaner ("LIME-A-WAY," Economics Laboratory, Inc.) with 3 gal. of water.

2) Turn off the power supply and close the Water Supply Line Shut-off Valve.

3) Remove the Front Panel, the Insulation Panel and the Base Cover.

4) Remove all ice from the icemaker and the Storage Bin.

5) Remove one end of the Pump Tubing, and drain the Water Tank. See Fig. 16.

6) Replace the Pump Tubing in its correct position.

7) Pour the cleaning solution into the Water Tank.

8) Move the Toggle WASH-OFF-ICE Control Switch, on the Control Box, to the "WASH" position.

9) Replace the Insulation Panel and the Front Panel in their correct position.

10) Turn on the power supply, and start the washing process.

11) Turn off the power supply in 30 minutes.

12) Remove the Front Panel.

13) Remove one end of the Pump Tubing, and drain the Water Tank. Replace the Pump Tubing in its correct position after the icemaker has been drained.

14) Open the Water Supply Line Shut-off Valve.

15) Move the Toggle WASH-OFF-ICE Control Switch to the "ICE" position.

16) Replace the Front Panel in its correct position.

17) Turn on the power supply, and start the rinse process.

18) Turn off the power supply in 10 minutes.

19) Remove the Front Panel.

20) Remove one end of the Pump Tubing, and drain the Water Tank. Replace the Pump Tubing in its correct position after the icemaker has been drained. Replace the Front Panel in its correct position.

21) Repeat the above steps 17) through 20) three more times to rinse thoroughly.

Note: Replace the Base Cover when you don't proceed to [b] SANITIZING PROCEDURE.

* ALTERNATE

1) Turn off the power supply, and close the Water Supply Line Shut-off Valve.

2) Dilute approximately 16 fl. oz. of cleaner ("LIME-A-WAY", Economics Laboratory, Inc. recommended) with 3 gal. of water.

3) Remove the Front Panel, the Insulation Panel and the Base Cover.

4) Remove the Water Pump Discharge Tubing at tee. Prepare and use a bucket to receive water through the Pump Tubing.
Move the Toggle WASH-OFF-ICE Control Switch, on the front of the Control Box, to the "WASH" position, and then turn on the power supply. In about 2 minutes, the Pump Motor starts to rotate, discharging water in the Water

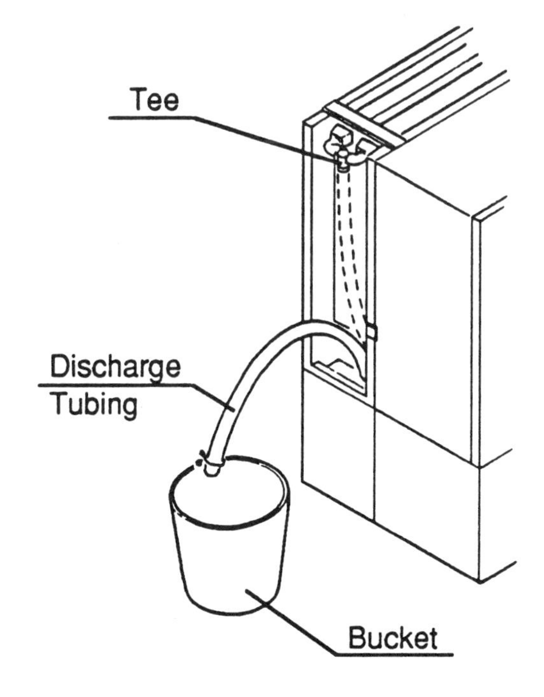

Fig. 18

Tank. And then move the Toggle WASH-OFF-ICE Control Switch to the "OFF" position, and place the Pump Discharge Tubing in position.

5) Pour the cleaning solution into the Water Tank.

6) Place the Insulation Panel in position.

7) Move the Toggle WASH-OFF-ICE Control Switch, on the front of the Control Box, to the "WASH" position.

8) Turn off the icemaker in 30 minutes.

9) Drain the Water Tank using step 4).

10) Open the Water Supply Line Shut-off Valve.

11) Move the Toggle WASH-OFF-ICE Control Switch to the "ICE" position.

12) Turn on the icemaker, and start the rince process. Turn off the icemaker in 10 minutes, and drain the Water Tank using step 4).

13) Repeat the above step 12) three more times to rinse thoroughly.

[b] SANITIZING PROCEDURE - Following Cleaning Procedure

1) Dilute approximately 24 fl. oz. of a 5.25 % Sodium Hypochlorite solution with 3 gal. of water.

2) Turn off the power supply and close the Water Supply Line Shut-off Valve.

3) Remove the Front Panel and the Insulation Panel.

4) Pour the sanitizing solution into the Water Tank.

5) Move the Toggle WASH-OFF-ICE Control Switch to the "WASH" position.

6) Replace the Insulation Panel in its correct position.

7) Turn on the power supply, and start the sanitizing process.

8) Turn off the power supply in 15 minutes.

9) Remove one end of the Pump Tubing, and drain the Water Tank. Replace the Pump Tubing in its correct position after the icemaker has been drained.

10) Open the Water Supply Line Shut-off Valve.

11) Move the Toggle WASH-OFF-ICE Control Switch to the "ICE" position.

12) Replace the Front Panel in its correct position.

13) Repeat the above steps 17) through 20) in "[a] CLEANING PROCEDURE" two more times to rinse thoroughly.

Note: Place the Base Cover in its correct position before replacing the Front Panel.

14) Clean the Storage Bin by using clear water.

15) Turn on the power supply, and start the automatic icemaking process.

2. MAINTENANCE

IMPORTANT

This icemaker must be maintained individually, referring to the instruction manual and labels provided with the icemaker.

1) Stainless Steel Exterior

To keep the exterior from corrosion, wipe occasionally with a clean and soft cloth. Use a damp cloth containing a neutral cleaner to wipe off oil or dirt build up.

2) Storage Bin and Scoop

* Wash your hands before removing ice. Use the plastic scoop provided (Accessory).

* The Storage Bin is for ice use only. Do not store anything else in the bin.

* Keep the scoop clean. Clean it by using a neutral cleaner and rinse thoroughly.

* Clean the bin liner by using a neutral cleaner. Rinse thoroughly after cleaning.

3) Air Filter (Air-cooled model only)

A plastic mesh air filter removes dirt or dust from the air, and keeps the Condenser from getting clogged. As the filter gets clogged, the icemaker's performance will be reduced. Check the filter at least twice a month. When clogged, use warm water and a neutral cleaner to wash the filter. Handle it with care to prevent damage to the mesh.

4) Condenser (Air-cooled model only)

Check the Condenser once a year, and clean if required by using a brush or vacuum cleaner. More frequent cleaning may be required depending on the location of the icemaker.

V. TECHNICAL INFORMATION

1. WATER CIRCUIT AND REFRIGERANT CIRCUIT

[a] KM-450MAB (Air-cooled)

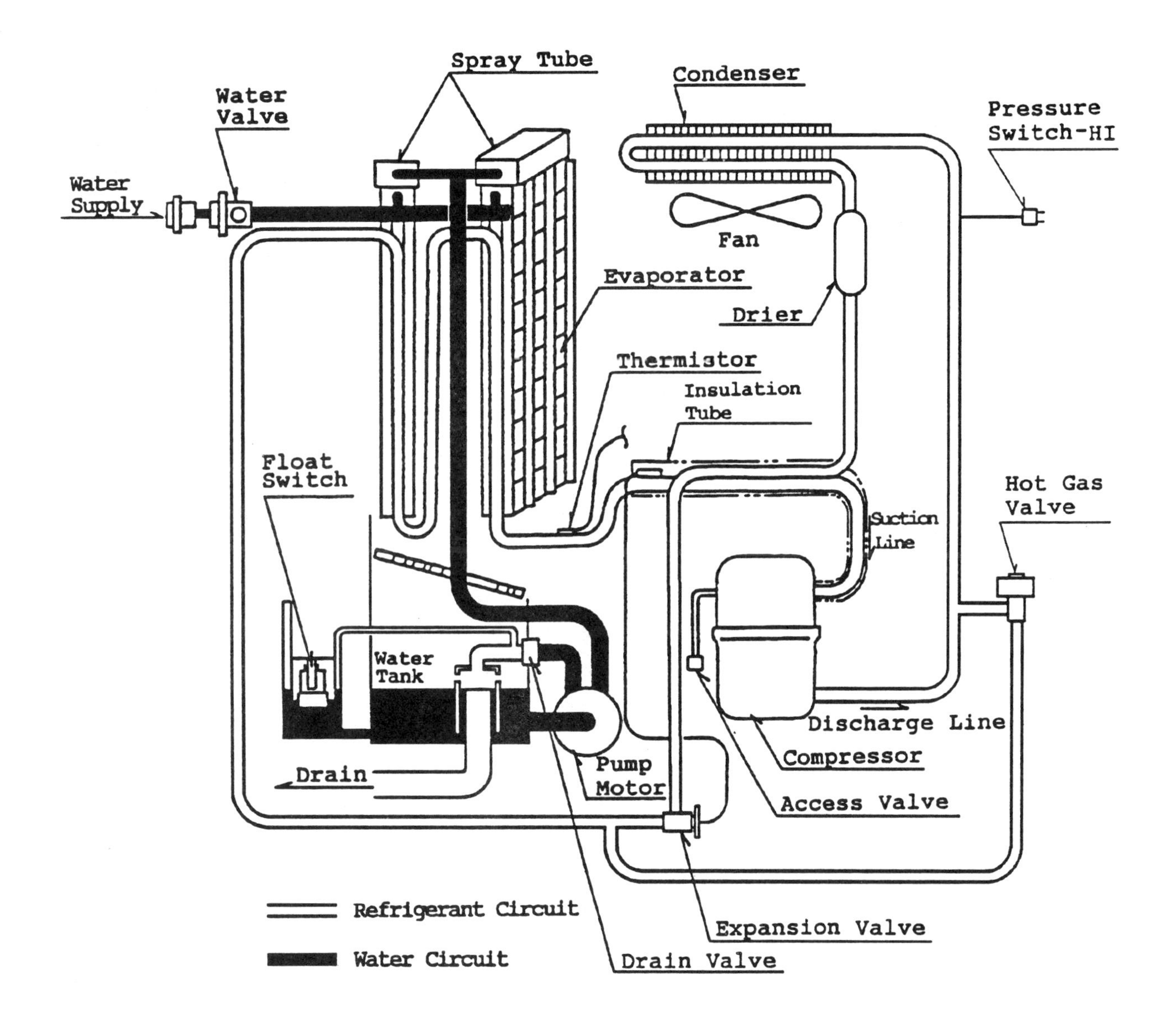

Fig. 19 Water and Refrigerant Circuit - KM-450MAB

U1HJ1789002

[b] KM-450MWB (Water-cooled)

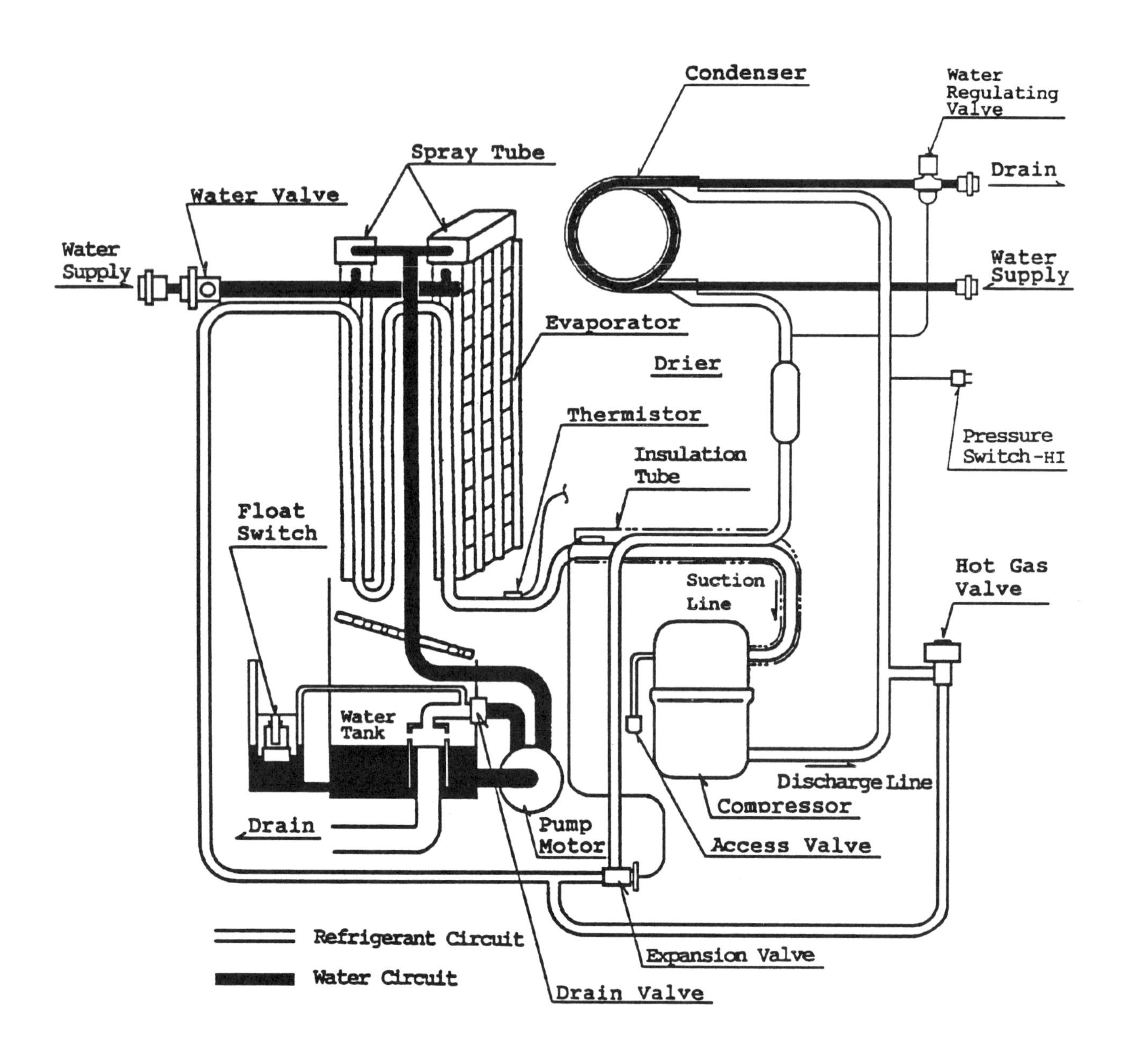

Fig. 20 Water and Refrigerant Circuit - KM-450MWB

[c] KM-450MRB (Remote Air-cooled)

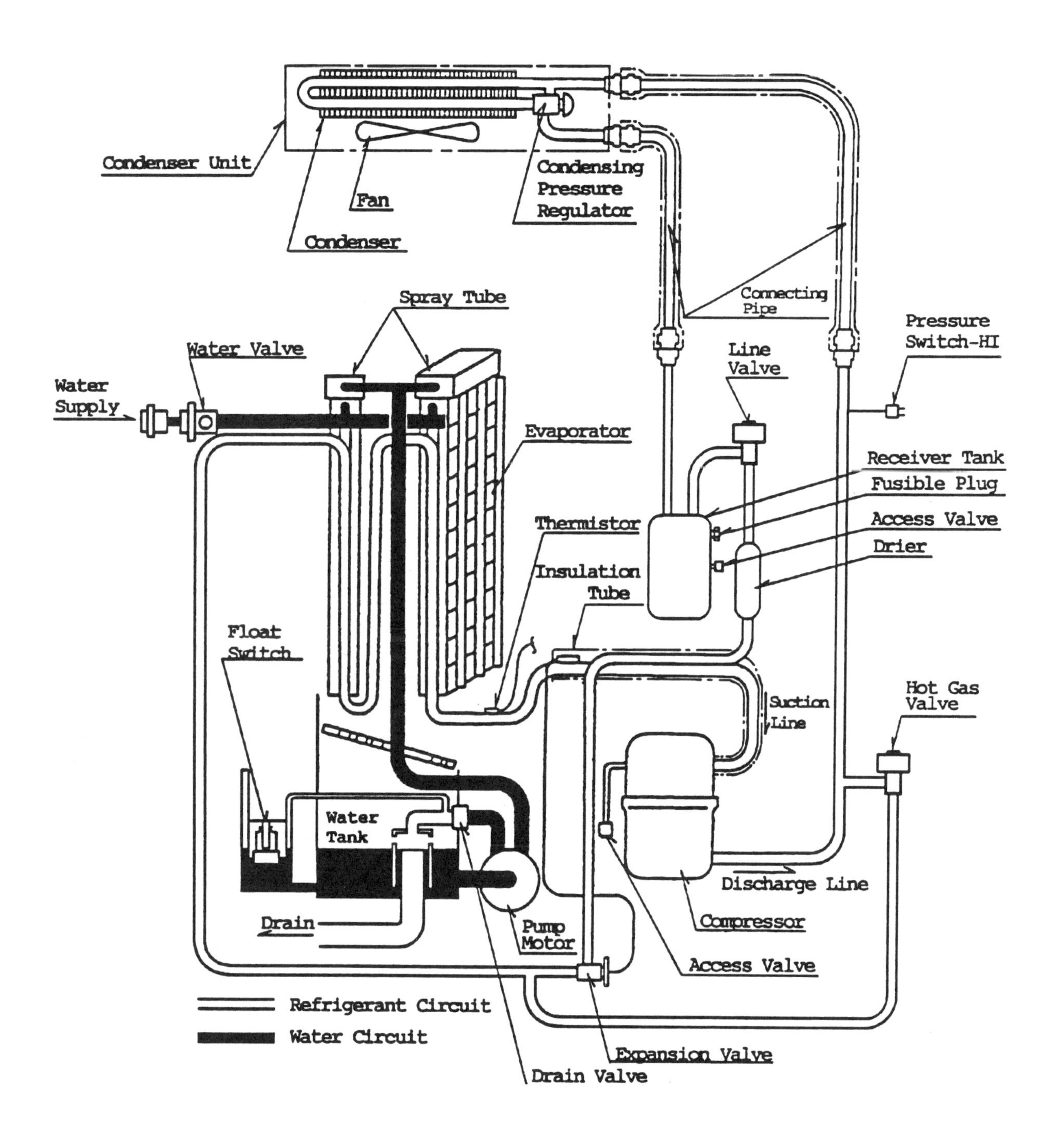

Fig. 21 Water and Refrigerant Circuit - KM-450MRB

2. WIRING DIAGRAM

[a] KM-450MAB, MWB

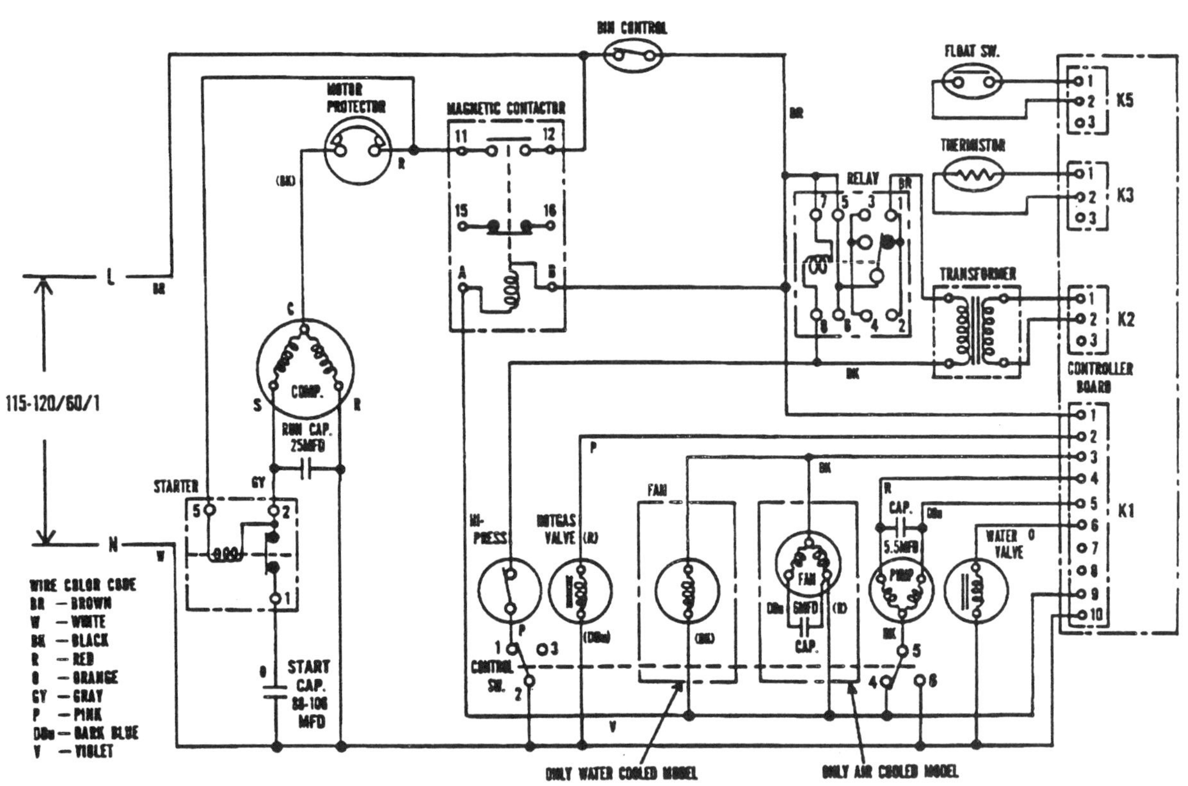

Fig. 22 Wiring Diagram - KM-450MAB, MWB

Note: Pressure Switch - HI

	KM-450MAB		KM-450MWB	
Cutout	384.0	PSIG	355.6	PSIG
Cutin	284.5±21.3	PSIG	256.0±21.3	PSIG

[b] KM-450MRB

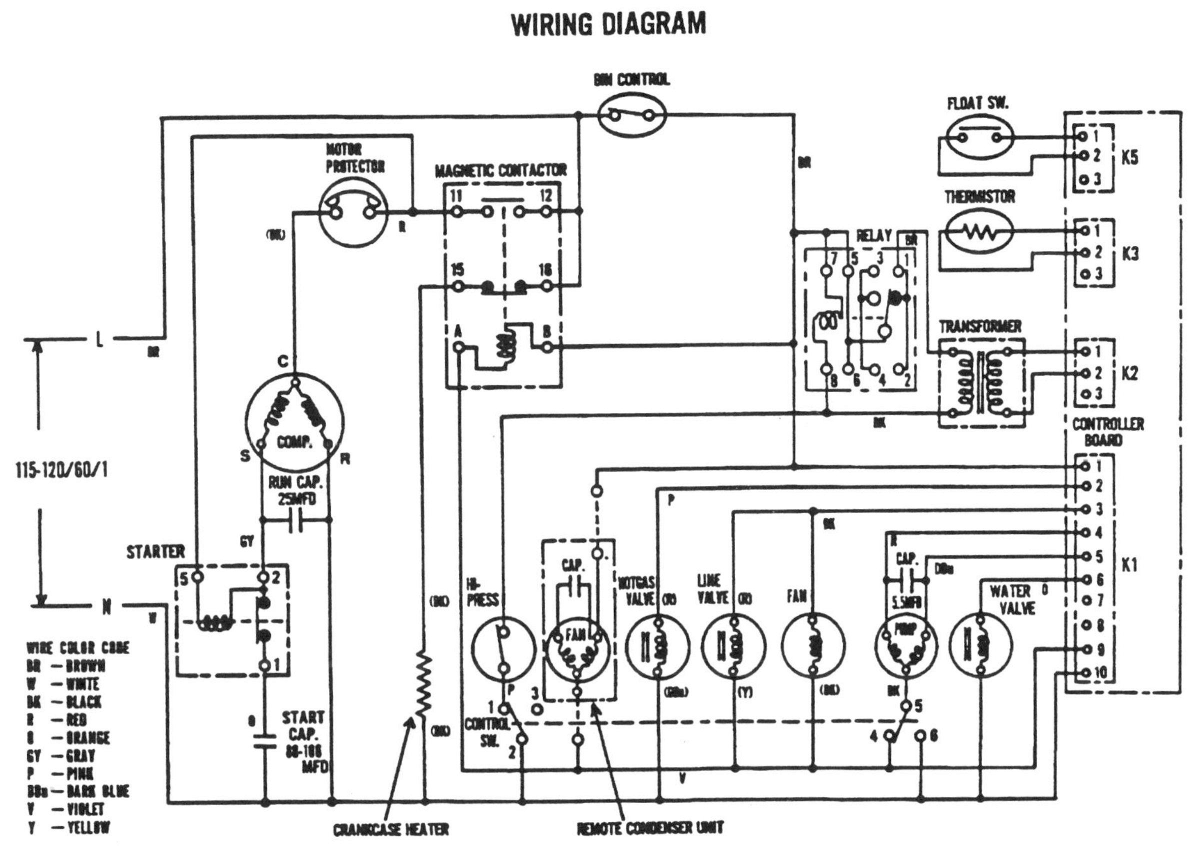

WIRING DIAGRAM

BIN CONTROL
FLOAT SW.
MOTOR PROTECTOR
MAGNETIC CONTACTOR
11 12 15 16 A B
RELAY
7 5 3 1 8 6 4 2
THERMISTOR
TRANSFORMER
K5 K3 K2 K1
CONTROLLER BOARD
L
N
115-120/60/1
C S R
COMP.
RUN CAP. 25MFD
STARTER
START CAP. 88-106 MFD
HI-PRESS
CAP.
FAN
HOTGAS VALVE
LINE VALVE
FAN
CAP. 5.5MFD
PUMP
WATER VALVE
CONTROL SW.
CRANKCASE HEATER
REMOTE CONDENSER UNIT

WIRE COLOR CODE
BR — BROWN
W — WHITE
BK — BLACK
R — RED
O — ORANGE
GY — GRAY
P — PINK
DBu — DARK BLUE
V — VIOLET
Y — YELLOW

Fig. 23 Wiring Diagram - KM-450MRB

Note: Pressure Switch - HI

Cutout	384.0	PSIG
Cutin	284.5±21.3	PSIG

3. TIMING CHART

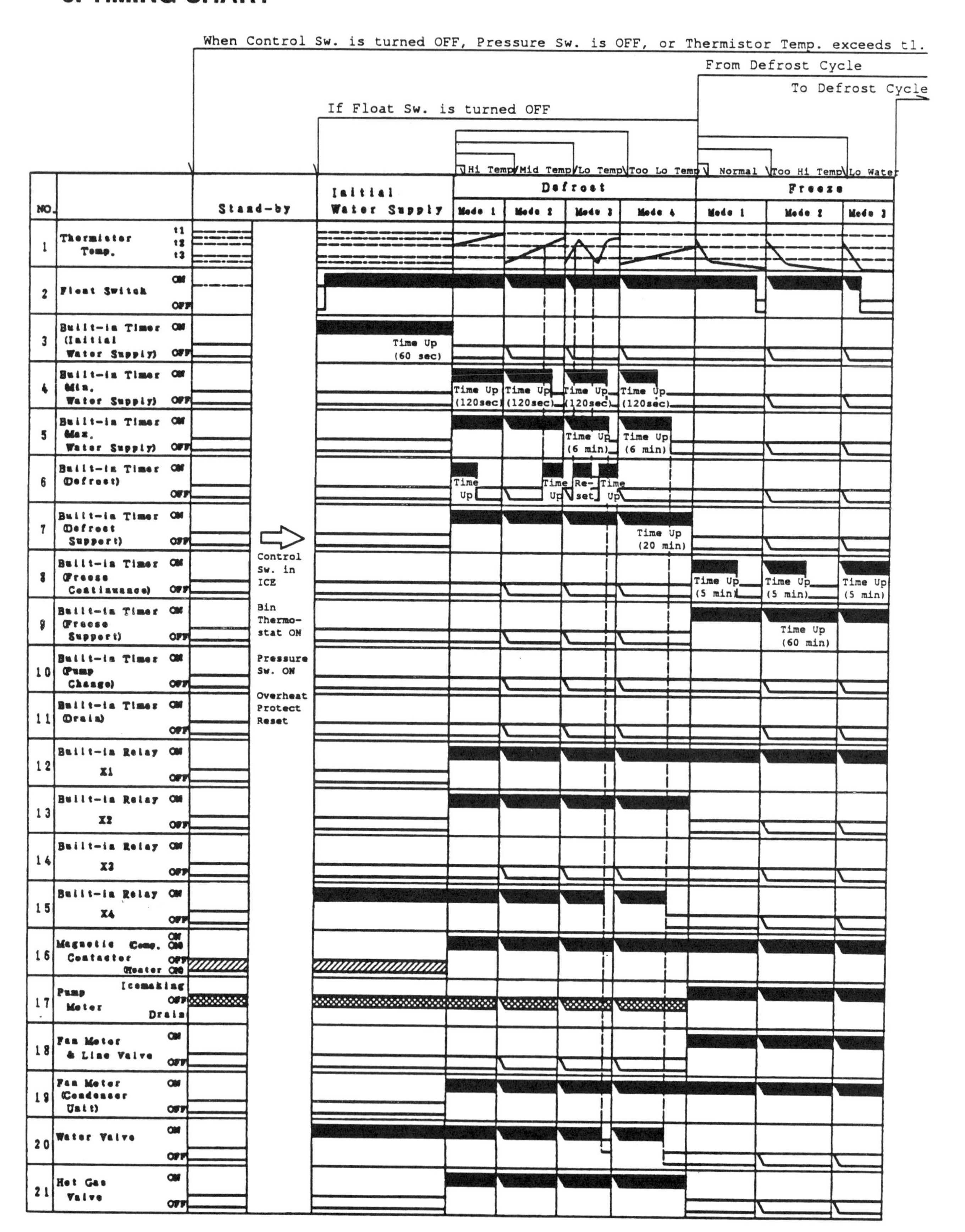

U1HJ1789002

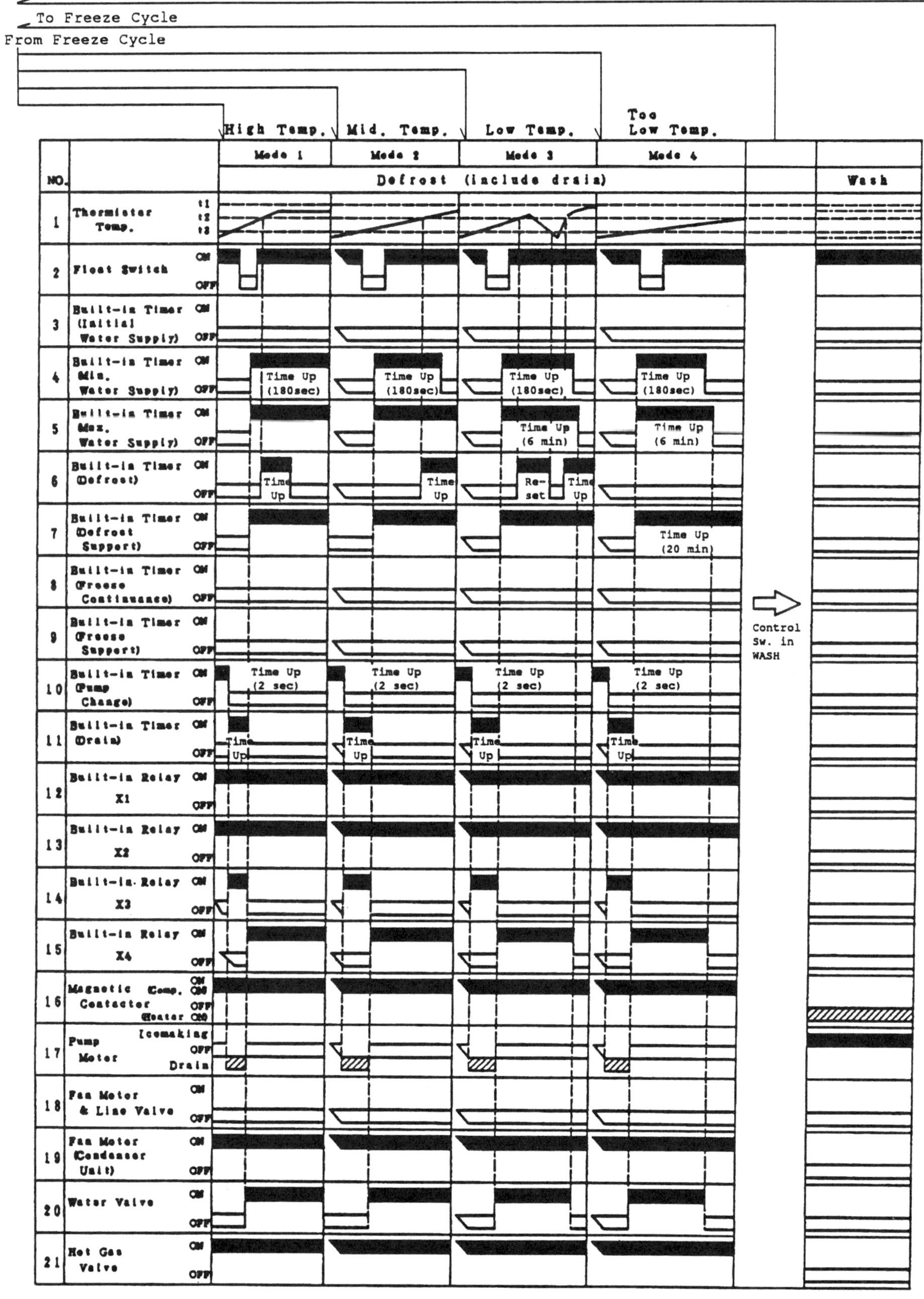
To Stand-by Cycle (When Control Sw. is turned OFF, Pressure Sw. is OFF, or Thermistor Temp. exceeds t1.)
To Freeze Cycle
From Freeze Cycle
High Temp.
Mid. Temp.
Low Temp.
Too Low Temp.
Mode 1
Mode 2
Mode 3
Mode 4
NO.
Defrost (include drain)
Wash
1 Thermistor Temp.
t1
t2
t3
2 Float Switch
3 Built-in Timer (Initial Water Supply)
4 Built-in Timer Min. Water Supply
Time Up (180sec)
5 Built-in Timer Max. Water Supply
Time Up (6 min)
6 Built-in Timer (Defrost)
Time Up
Re-set
7 Built-in Timer (Defrost Support)
Time Up (20 min)
8 Built-in Timer (Freeze Continuance)
9 Built-in Timer (Freeze Support)
Control Sw. in WASH
10 Built-in Timer (Pump Change)
Time Up (2 sec)
11 Built-in Timer (Drain)
12 Built-in Relay X1
13 Built-in Relay X2
14 Built-in Relay X3
15 Built-in Relay X4
16 Magnetic Contactor
Comp.
Heater
17 Pump Motor
Icemaking
Drain
18 Fan Motor & Line Valve
19 Fan Motor (Condenser Unit)
20 Water Valve
21 Hot Gas Valve
ON
OFF

4. PERFORMANCE DATA

[a] KM-450MAB

	Ambient Temp. (°F)	Water Temp. (°F) 50	70	90
APPROXIMATE ICE PRODUCTION PER 24 HR. (MAX. CUBE SIZE) lbs./day (kg/day)	70	*460 (209)	410 (186)	350 (159)
	80	420 (190)	370 (168)	330 (150)
	90	375 (170)	*340 (154)	300 (136)
	100	340 (154)	295 (133)	260 (118)
APPROXIMATE ELECTRIC CONSUMPTION watts	70	*1050	1080	1080
	80	1080	1120	1120
	90	1110	*1150	1160
	100	1130	1160	1180
APPROXIMATE WATER CONSUMPTION PER 24 HR. gal./day (m³/day)	70	198 (0.75)	145 (0.55)	132 (0.50)
	80	185 (0.70)	127 (0.48)	116 (0.44)
	90	153 (0.58)	114 (0.43)	106 (0.40)
	100	119 (0.45)	100 (0.38)	92 (0.35)
FREEZING CYCLE TIME min.	70	30	33	41
	80	35	37	40
	90	38	42	46
	100	41	49	53
HARVEST CYCLE TIME min.	70	4	3	3
	80	4	3	3
	90	4	3	3
	100	3	3	3
HEAD PRESSURE PSIG (kg/cm²G)	70	216 (15.2)	219 (15.4)	220 (15.5)
	80	242 (17.0)	249 (17.5)	253 (17.8)
	90	270 (19.0)	270 (19.0)	276 (19.4)
	100	306 (21.5)	306 (21.5)	309 (21.7)
SUCTION PRESSURE PSIG (kg/cm²G)	70	43 (3.0)	43 (3.0)	43 (3.0)
	80	43 (3.0)	43 (3.0)	44 (3.1)
	90	44 (3.1)	44 (3.1)	47 (3.3)
	100	46 (3.2)	46 (3.2)	47 (3.3)
TOTAL HEAT OF REJECTION	9250 BTU/h (AT 90°/WT 70°F)			

Note: Pressure data is recorded first 5 minutes in freezing cycle.
The data without *marks should be used for reference.

U1HJ1789002

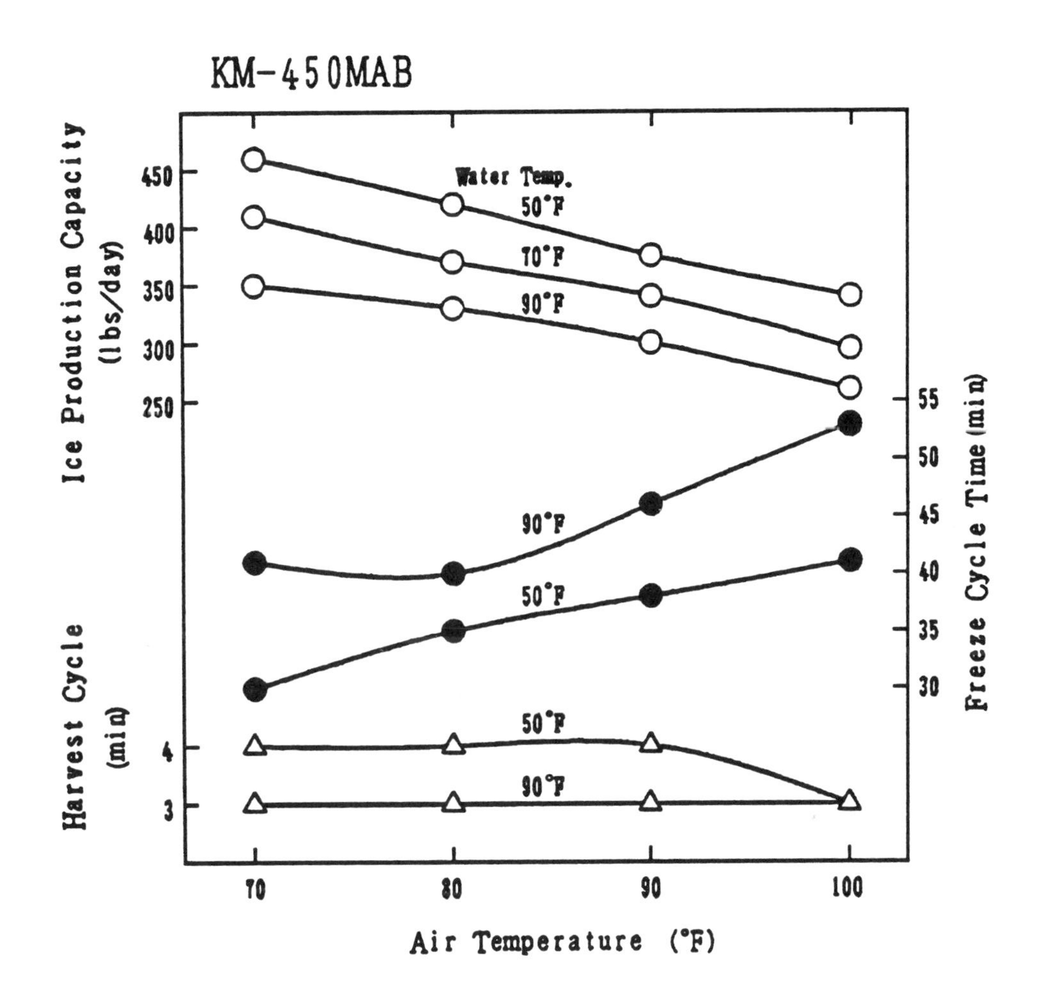

Fig. 24 Performance Data (1) - KM-450MAB

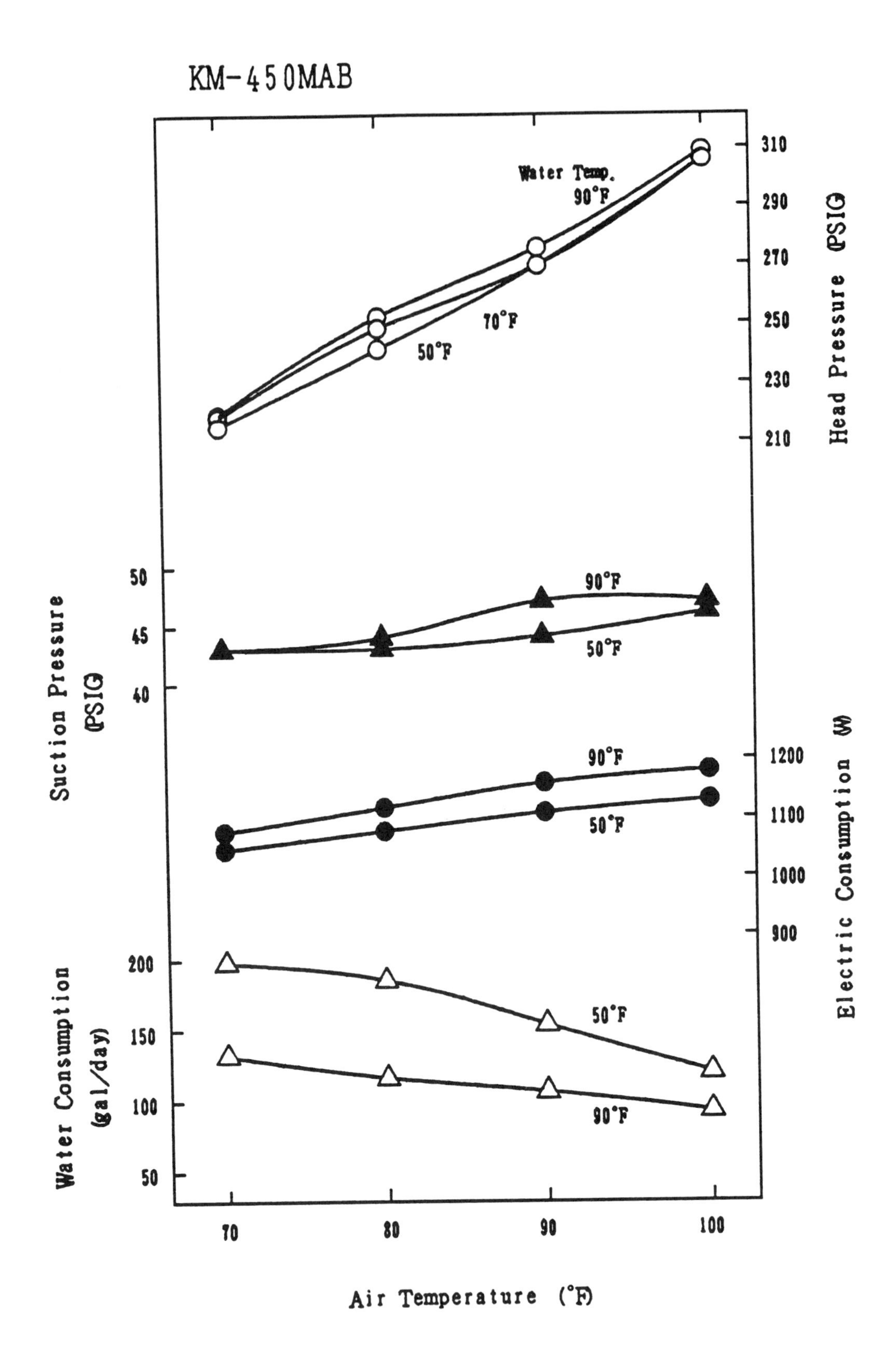

Fig. 25 Performance Data (2) - KM-450MAB

[b] KM-450MWB

	Ambient Temp. (°F)	Water Temp. (°F) 50	70	90
APPROXIMATE ICE PRODUCTION PER 24 HR. (MAX. CUBE SIZE) lbs./day (kg/day)	70	*490 (222)	425 (193)	360 (163)
	80	485 (220)	420 (190)	350 (159)
	90	475 (215)	*420 (190)	330 (150)
	100	460 (209)	395 (179)	330 (150)
APPROXIMATE ELECTRIC CONSUMPTION watts	70	* 980	980	1020
	80	1000	1000	1020
	90	1000	*1020	1020
	100	1000	1030	1030
APPROXIMATE WATER CONSUMPTION PER 24 HR. gal./day (m^3/day)	70	816 (3.09)	1041 (3.94)	1981 (7.50)
	80	830 (3.14)	1051 (3.98)	1992 (7.54)
	90	845 (3.20)	1062 (4.02)	2002 (7.58)
	100	859 (3.25)	1075 (4.07)	2016 (7.63)
FREEZING CYCLE TIME min.	70	29	34	39
	80	29	34	40
	90	30	35	41
	100	33	36	41
HARVEST CYCLE TIME min.	70	4	3	3
	80	4	3	3
	90	4	3	3
	100	4	3	3
HEAD PRESSURE PSIG (kg/cm²G)	70	230 (16.2)	232 (16.3)	249 (17.5)
	80	232 (16.3)	235 (16.5)	249 (17.5)
	90	232 (16.3)	235 (16.5)	249 (17.5)
	100	232 (16.3)	235 (16.5)	250 (17.6)
SUCTION PRESSURE PSIG (kg/cm²G)	70	43 (3.0)	43 (3.0)	43 (3.0)
	80	43 (3.0)	43 (3.0)	44 (3.1)
	90	44 (3.1)	46 (3.2)	47 (3.3)
	100	46 (3.2)	46 (3.3)	47 (3.3)

HEAT OF REJECTION FROM CONDENSER	9160 BTU/h (AT 90°/WT 70°F)
HEAT OF REJECTION FROM COMPRESSOR	1450 BTU/h (AT 90°/WT 70°F)
WATER FLOW FOR CONDENSER	70 gal./h (AT 100°/WT 90°F)
PRESSURE DROP OF COOLING WATER LINE	less than 7 PSIG

Note: Pressure data is recorded first 5 minutes in freezing cycle.
The data without *marks should be used for reference.

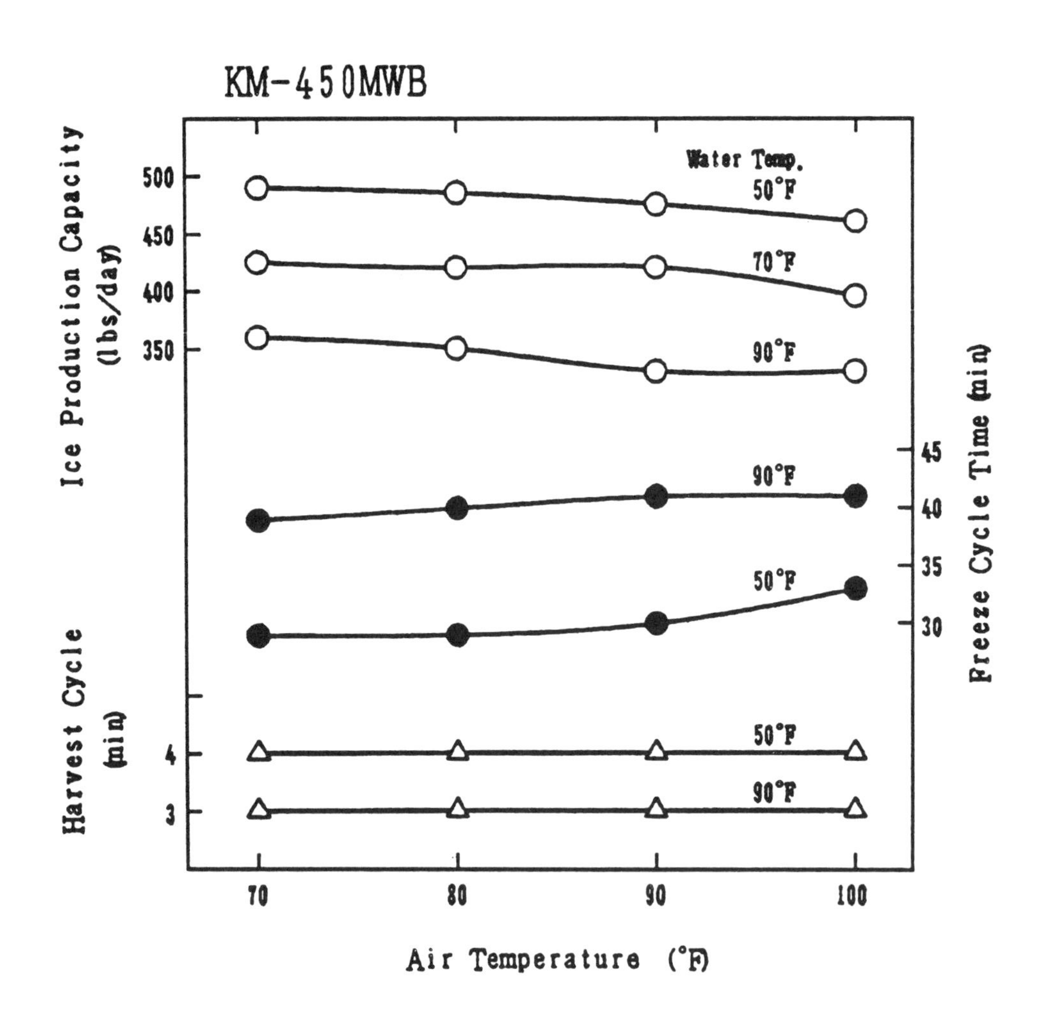

Fig. 26 Performance Data (1) - KM-450MWB

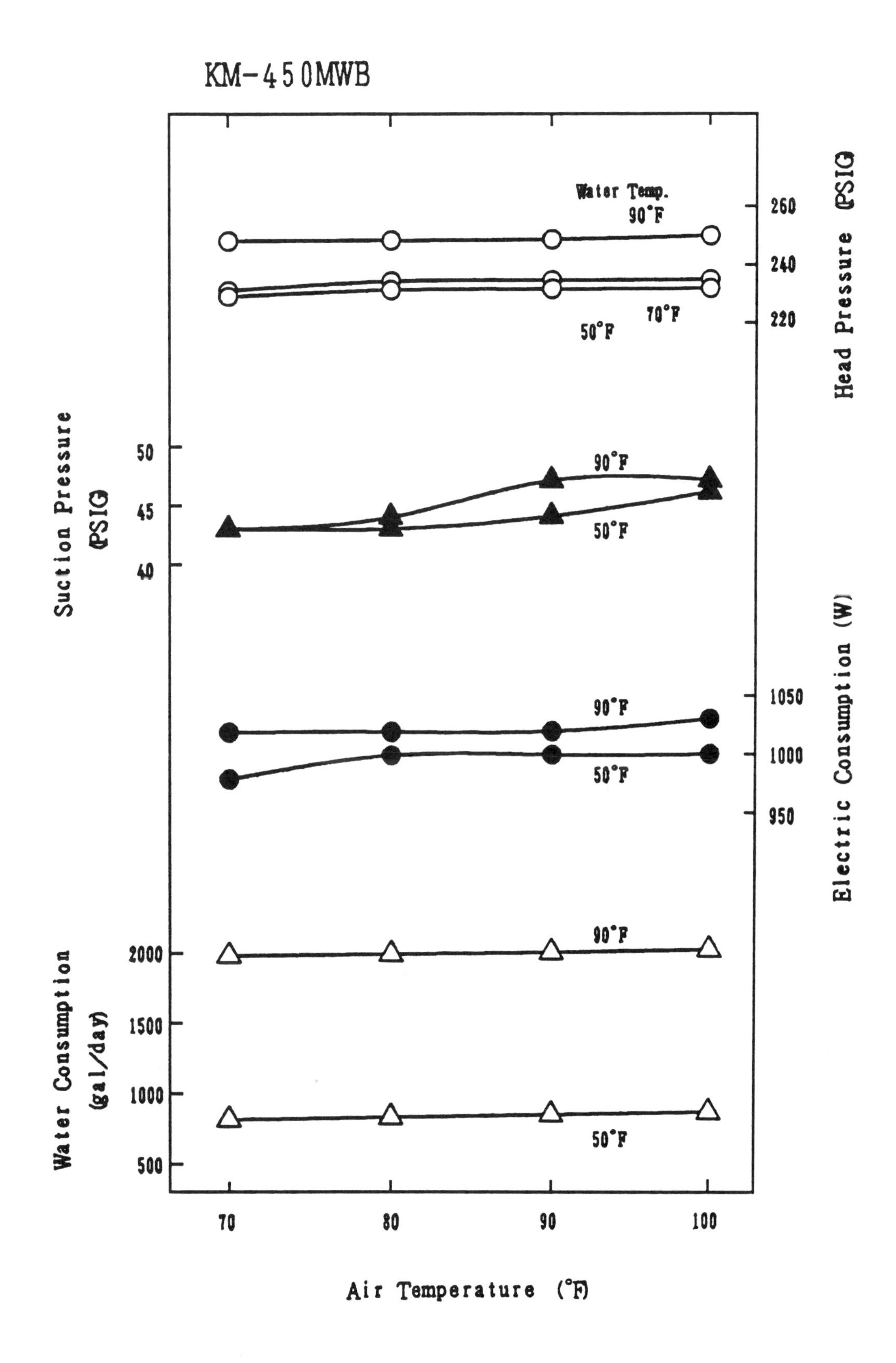

Fig. 27 Performance Data (2) - KM-450MWB

[c] KM-450MRB

	Ambient Temp. (°F)	Water Temp. (°F) 50	70	90
APPROXIMATE ICE PRODUCTION PER 24 HR.(MAX. CUBE SIZE) lbs./day (kg/day)	70	*470(213)	430(195)	380(172)
	80	430(195)	400(181)	360(163)
	90	400(181)	*370(168)	330(150)
	100	360(163)	330(150)	290(132)
APPROXIMATE ELECTRIC CONSUMPTION watts	70	*1080	1080	1090
	80	1090	1100	1130
	90	1120	*1140	1160
	100	1160	1160	1200
APPROXIMATE WATER CONSUMPTION PER 24 HR. gal./day (m^3/day)	70	210(0.82)	169(0.64)	148(0.57)
	80	198(0.77)	153(0.59)	141(0.55)
	90	159(0.62)	141(0.55)	131(0.51)
	100	137(0.53)	127(0.49)	116(0.45)
FREEZING CYCLE TIME min.	70	28	29	33
	80	30	32	35
	90	32	35	38
	100	36	39	43
HARVEST CYCLE TIME min.	70	4	3	3
	80	4	3	3
	90	3.5	3	3
	100	3	3	3
HEAD PRESSURE PSIG (kg/cm²G)	70	199(14.0)	199(14.0)	199(14.0)
	80	218(15.3)	222(15.6)	223(15.7)
	90	242(17.0)	245(17.2)	249(17.5)
	100	280(19.7)	272(19.1)	284(20.0)
SUCTION PRESSURE PSIG (kg/cm²G)	70	43(3.0)	43(3.0)	43(3.0)
	80	43(3.0)	43(3.0)	44(3.1)
	90	44(3.1)	44(3.1)	47(3.3)
	100	46(3.2)	46(3.2)	47(3.3)

HEAT OF REJECTION FROM CONDENSER	8800 BTU/h (AT 90°/WT 70°F, URC-4A)
HEAT OF REJECTION FROM COMPRESSOR	1450 BTU/h (AT 90°/WT 70°F)
CONDENSER VOLUME	54.3 cu.in.(URC-4A)

Note: Pressure data is recorded first 5 minutes in freezing cycle.
The data without *marks should be used for reference.

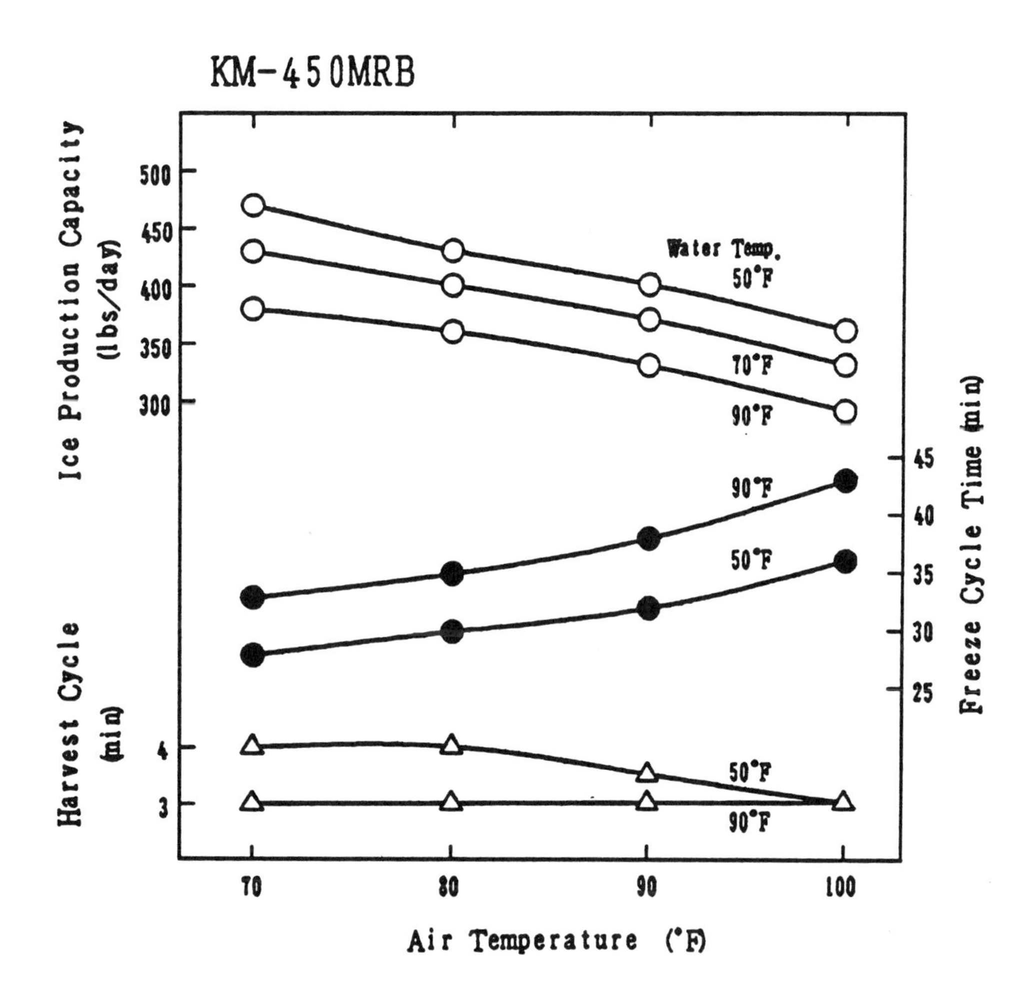

Fig. 28 Performance Data - KM-450MRB

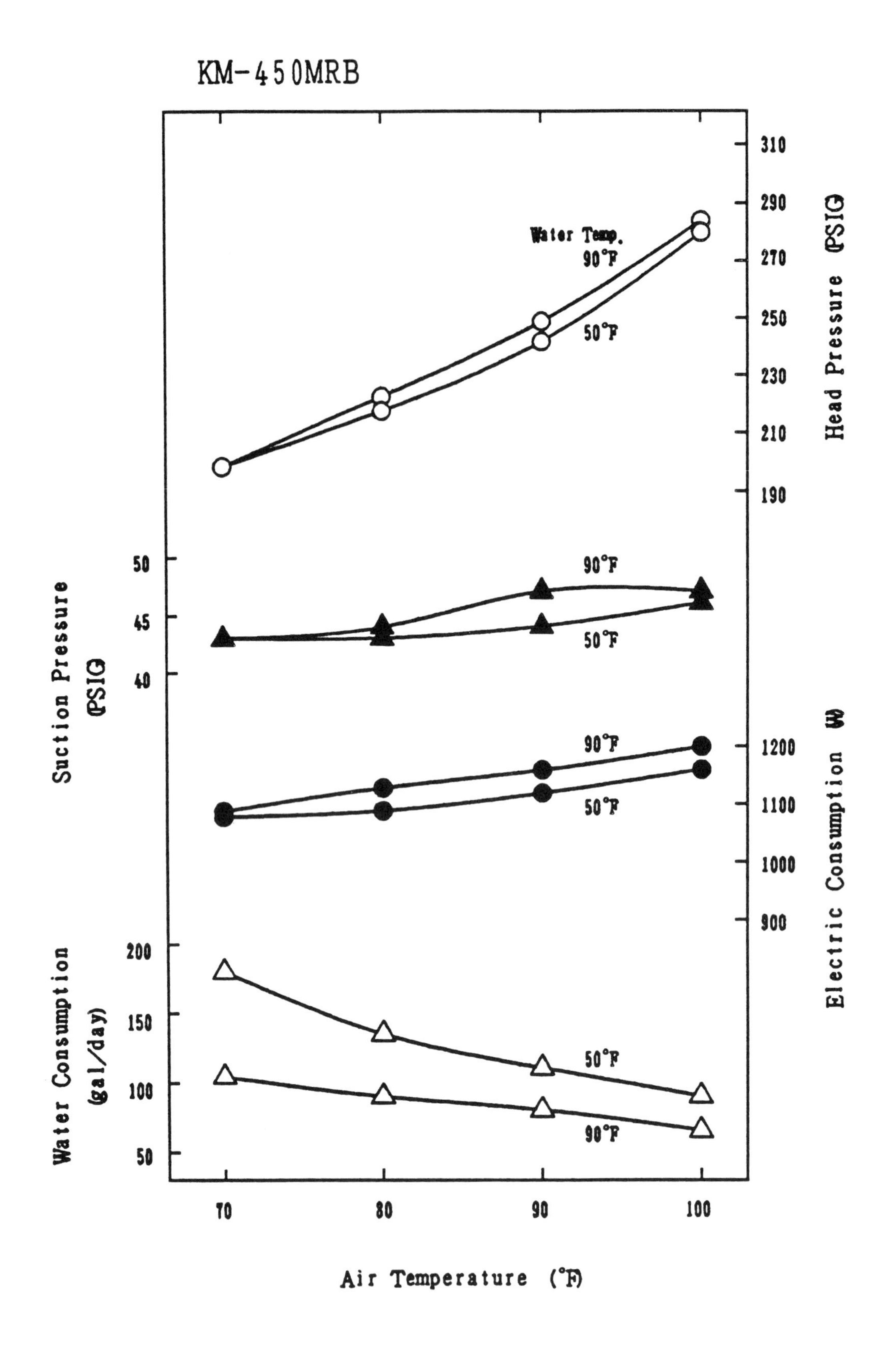

Fig. 29 Performance Data (2) - KM-450MRB

VI. SERVICE DIAGNOSIS

1. NO ICE PRODUCTION

PROBLEM	POSSIBLE CAUSE		REMEDY
[1] The icemaker will not start.	a) Power Source	1. OFF position.	1. Move to ON position.
		2. Loose connections.	2. Tighten.
		3. Bad contacts.	3. Check for continuity and replace.
	b) Fuse (Inside Fused Disconnect, if any)	1. Blownout.	1. Check for short-circuit and replace.
	c) Control Switch	1. OFF position.	1. Move to ICE position.
		2. Bad contacts.	2. Check for continuity and replace.
	d) Bin Control Thermostat	1. Tripped with bin filled with ice.	1. Remove ice.
		2. Ambient temperature too cool.	2. Get warmer.
		3. Set too warm.	3. See "II. 2. [d] CONTROLS AND ADJUSTMENT, 5) Bin Control."
		4. Bulb out of position.	4. Place in position.
		5. Bad contacts or leaks from bulb.	5. Check for continuity and replace.
	e) High Pressure Control	1. Bad contacts.	1. Check for continuity and replace.
	f) Transformer	1. Thermal fuse blowout or coil winding opened.	1. Replace.
	g) Wiring to Controller Board	1. Loose connections or open.	1. Check for continuity and replace.
	h) Thermistor	1. Leads short-circuit or open and High Temperature Safety operates.	1. See "II. 2. [d] CONTROLS AND ADJUSTMENT, 2) Defrost Control."
		2. Connector is disconnected and High Temperature Safety operates.	2. Place in position.
	i) Hot Gas Solenoid Valve	1. Continues to open in freeze cycle and High Temperature Safety operates.	1. Check for power OFF in freeze cycle and replace.
	j) Water Supply Line	1. Water supply OFF and water supply cycle does not finish.	1. Check and get recommended pressure.
		2. Condenser water pressure too low or OFF and High Temperature Safety operates.	2. Check and get recommended pressure.

	k) Water Solenoid Valve	1. Mesh filter or orifice get clogged and water supply cycle does not finish.	1. Clean.
		2. Coil winding opened.	2. Replace.
		3. Wiring to water valve.	3. Check for loose connection or open, and replace.
	l) Controller Board	1. Defective.	1. See "II. 2. [e] CHECKING CONTROLLER BOARD."
[2] Water continues to be supplied, and the icemaker will not start.	a) Cube Control Float Switch	1. Connector disconnected.	1. Place in position.
		2. Leads opened or defective switch.	2. Check and replace.
		3. Float does not move freely.	3. Clean or replace.
	b) Controller Board	1. Defective.	1. Replace.
[3] Compressor will not start, or operates intermittently.	a) Wash Switch	1. WASH position.	1. Move to ICE position.
		2. Bad contacts.	2. Check and replace.
	b) High Pressure Control	1. Dirty air filter or condenser.	1. Clean.
		2. Ambient or condenser water temperature too warm.	2. Get cooler.
		3. Refrigerant overcharged.	3. Recharge.
		4. Condenser water pressure too low or off.	4. Check and get recommended pressure.
		5. Fan not operating.	5. See chart 1 - [6].
		6. Refrigerant line or components plugged.	6. Clean and replace drier.
	c) Water Regulator	1. Set too high.	1. Adjust lower.
	d) Overload Protector	1. Bad contacts.	1. Check for continuity and replace.
		2. Voltage too low.	2. Get higher.
		3. Refrigerant overcharged or undercharged.	3. Recharge.
		4. Line valve continues to close in freeze cycle and Overload Protector operates.	4. Check line valve's operation in freeze cycle and replace.
	e) Starter	1. Bad contacts.	1. Check and replace.
		2. Coil winding opened.	2. Replace.
	f) Start Capacitor or Run Capacitor	1. Defective.	1. Replace.
	g) Magnetic Contactor	1. Bad contacts.	1. Check for continuity and replace.
		2. Coil winding opened.	2. Replace.

	h) Compressor	1. Wiring to compressor	1. Check for loose connection or open, and replace.
		2. Defective.	2. Replace.
[4] Water continues to be supplied in freeze cycle.	a) Water Solenoid Valve	1. Diaphragm does not close.	1. Check for water leaks with icemaker OFF.
	b) Controller Board	1. Defective.	1. See "II. 2. [e] CHECKING CONTROLLER BOARD."
[5] No water comes from Spray Tubes, water pump will not start, or freeze cycle time is too short.	a) Water Supply Line	1. Water pressure too low and water level in water tank too low.	1. Check and get recommended pressure.
	b) Water Solenoid Valve	1. Dirty mesh filter or orifice and water level in water tank too low.	1. Clean.
	c) Water System	1. Water leaks.	1. Check connections for water leaks, and replace.
		2. Clogged.	2. Clean.
	d) Pump Motor	1. Motor winding opened.	1. Replace.
		2. Bearing worn out.	2. Replace.
		3. Wiring to pump motor.	3. Check for loose connection or open, and replace.
		4. Defective capacitor.	4. Replace.
		5. Defective or bound impeller.	5. Replace or clean.
		6. Mechanical Seal worn out.	6. Check and replace.
	e) Controller Board	1. Defective.	1. See "II. 2. [e] CHECKING CONTROLLER BOARD."
[6] Fan motor will not start, or is not operating.	a) Fan Motor	1. Motor winding opened.	1. Replace.
		2. Bearing worn out.	2. Replace.
		3. Wiring to Fan Motor.	3. Check for loose connection or open, and replace.
		4. Defective capacitor	4. Replace.
		5. Fan blade bound	5. Check and replace.
	b) Controller Board	1. Defective.	1. See "II. 2. [e] CHECKING CONTROLLER BOARD."
[7] All components run, but no ice is produced.	a) Refrigerant	1. Undercharged.	1. Check for leaks and recharge.
		2. Air or moisture trapped.	2. Replace drier, and recharge.
	b) Compressor	1. Defective valve	1. Replace.
	c) Hot Gas Solenoid Valve	1. Continues to open in freeze cycle.	1. Check and replace.

	d) Line Valve (Remote Air-cooled model)	1. Continues to close in freeze cycle.	1. Check and replace.

2. EVAPORATOR IS FROZEN UP

PROBLEM	POSSIBLE CAUSE		REMEDY
[1] Freeze cycle time is too long.	a) Cube Control Float Switch	1. Leads short-circuit or defective switch.	1. Check and replace.
		2. Float does not move freely.	2. Clean or replace.
	b) Water Solenoid Valve	1. Diaphragm does not close.	1. Check for water leaks with icemaker OFF.
	c) Controller Board	1. Defective.	1. See "II. 2. [e] CHECKING CONTROLLER BOARD."
[2] All ice formed on evaporator does not fall into bin in harvest cycle.	a) Controller Board	1. Defrost Timer is set too short.	1. Adjust longer, referring to "II. 2. [d] CONTROLS AND ADJUSTMENT, 3) Defrost Timer."
		2. Defective.	2. See "II. 2. [e] CHECKING CONTROLLER BOARD."
	b) Thermistor	1. Out of position or loose attachment.	1. See "VII. 10. REMOVAL AND REPLACEMENT OF THERMISTOR."
	c) Evaporator	1. Deformed.	1. Replace.
	d) Ambient and /or water temperature	1. Too cool.	1. Get warmer.
	e) Water Supply Line	1. Water pressure too low.	1. Check and get recommended pressure.
	f) Water Solenoid Valve	1. Dirty mesh filter or orifice.	1. Clean.
		2. Diaphragm does not close.	2. Check for water leaks with icemaker OFF.
	g) Line Valve (Remote Air-cooled model)	1. Continues to open in harvest cycle.	1. Check operation in harvest cycle and replace.
[3] Others	a) Spray Tubes	1. Clogged.	1. Clean.
		2. Out of position.	2. Place in position.
	b) Water System	1. Dirty.	1. Clean.
	c) Refrigerant	1. Undercharged.	1. Check for leaks and recharge.
	d) Expansion Valve	1. Bulb out of position or loose attachment.	1. Place in position.
		2. Defective.	2. Replace.
	e) Hot Gas Solenoid Valve	1. Coil winding opened.	1. Replace.

		2. Plunger does not move.	2. Replace.
		3. Wiring to hot gas valve.	3. Check for loose connection or open, and replace.

3. LOW ICE PRODUCTION

PROBLEM	POSSIBLE CAUSE		REMEDY
[1] Freeze cycle time is long.	a) See chart 1 - [3], and check dirty air filter or condenser, ambient or water temperature, water pressure, water regulator or refrigerant charge.		
	b) See chart 2 - [1], and check cube control float switch, water solenoid valve or controller board.		
[2] Harvest cycle time is long.	a) See chart 2 - [2], and check controller board, thermistor, evaporator, ambient and/or water temperature, water supply line, water solenoid valve or line valve.		

4. ABNORMAL ICE

PROBLEM	POSSIBLE CAUSE		REMEDY
[1] Small cube	a) Cube Control Float Switch	1. Upper position.	1. Move to lower position.
	b) Ice Cube Guide	1. Out of position. Circulated water falls into bin.	1. Place in position.
	c) See chart 1 - [5], and check water supply line, water solenoid valve, water system, pump motor or controller board.		
	d) Drain Valve	1. Dirty.	1. Clean.
[2] Cloudy or irregular cube	a) See chart 2 - [1] and - [3], and check cube control float switch, water solenoid valve, controller board, spray tubes, water system, refrigerant charge or expansion valve.		
	b) Spray Guide	1. Dirty.	1. Clean.
	c) Water Quality	1. High hardness or contains impurities.	1. Install a water filter or softener.

5. OTHERS

PROBLEM	POSSIBLE CAUSE		REMEDY
[1] Icemaker will not stop when bin is filled with ice.	a) Bin Control Thermostat	1. Set too cold.	1. Adjust warmer.
		2. Defective.	2. Replace.
[2] Abnormal noise	a) Pump Motor	1. Bearings worn out.	1. Replace.
	b) Fan Motor	1. Bearings worn out.	1. Replace.
		2. Fan blade deformed.	2. Replace fan blade.
		3. Fan blade does not move freely.	3. Replace.
	c) Compressor	1. Bearings worn out, or cylinder valve broken.	1. Replace.

		2. Mounting pad out of position.	2. Reinstall.
	d) Refrigerant Lines	1. Rub or touch lines or other surfaces.	1. Replace.
[3] Ice in storage bin often melts.	a) Bin Drain	1. Plugged.	1. Clean.

VII. REMOVAL AND REPLACEMENT OF COMPONENTS

1. SERVICE FOR REFRIGERANT LINES

[a] REFRIGERANT DISCHARGE

The all models are provided with a Refrigerant Access Valve on the LOW-SIDE line. For the air-cooled and water-cooled models, install a proper fitting on the HIGH-SIDE line, if necessary, to check for gauge pressure. Only the remote air-cooled model is provided with a Refrigerant Access Valve on the Receiver Tank.

[b] EVACUATION AND RECHARGE

1) Attach Charging Hoses, a Service Manifold and a Vacuum Pump to the system. For the remote air-cooled model, be sure to connect Charging Hoses to both HIGH-SIDE and LOW-SIDE lines to evacuate the system.

2) Turn on the Vacuum Pump.

3) Allow the Vacuum Pump to pull down to a 29.9" Hg vacuum. Evacuating period depends on pump capacity.

4) Close the Low-side Valve and High-side Valve on the Service Manifold.

5) Disconnect the Vacuum Pump, and attach a Refrigerant Service Cylinder.to the Low-side line. Remember to loosen the connection, and purge the air from the Hose. For the air-cooled and water-cooled models, see NAMEPLATE for required refrigerant charge. For the remote air-cooled model, see the Charge Label on the left side wall of the machine compartment.

6) Open the Low-side Valve. Do not invert the Service Cylinder. A LIQUID CHARGE will damage the Compressor.

7) Turn on the icemaker when charging speed gets slow. Turn off the icemaker when the Low-side Gauge shows approximately 0 PSIG. DO NOT run the icemaker at negative pressures. Close the Low-side Valve when the Service Cylinder gets empty.

8) Repeat the above steps 4) through 7), if necessary, until a required amount of refrigerant has entered the system.

9) Close the one or two Refrigerant Access Valve(s), and disconnect the Hoses, Service Manifold and so on.

10) Cap the Access Valve(s) to prevent possible leak.

2. BRAZING

DANGER

1. Refrigerant R502 itself is not flammable, explosive and poisonous. However, when exposed to an open flame, R502 creates Phosgene gas, hazardous in large amounts.

2. Always purge the refrigeration system through hose vented to the outside, because it is dangerous for the room to be filled with R502 and short of oxygen.

3. Do not use silver alloy or copper alloy containing Arsenic.

Note: All brazing-connections in the Evaporator Case are clear-paint coated. Sandpaper the brazing-connections before unbrazing the components. Use a good abrasive cloth to remove paint.

3. REMOVAL AND REPLACEMENT OF COMPRESSOR

IMPORTANT

Always install a new Drier every time the sealed refrigeration system is opened. Do not replace the Drier until after all other repairs or replacement have been made.

1) Turn off the power supply.

2) Remove the Front Panel, the Top Panel and the Right Side Panel.

3) Discharge the refrigerant.

4) Remove the Barrier and the Terminal Cover on the Compressor, and disconnect the Compressor Wiring.

5) Remove the Discharge and Suction Pipes using a brazing equipment.

6) Remove the Hold-down Bolts, Washers and Rubber Grommets.

7) Slide and remove the Compressor. Unpack a new Compressor package. Install a new Compressor.

8) Attach the Rubber Grommets of the prior Compressor.

9) Sandpaper the Suction, Discharge and Process Pipes.

10) Place the Compressor in position, and secure it using the Bolts and Washers.

11) Remove plugs from the Suction, Discharge and Process Pipes.

12) Braze the Process, Suction and Discharge lines (DO NOT change this order), with nitrogen gas flowing at the pressure 3 - 4 PSIG.

13) Replace the Drier.

14) Check for leaks using nitrogen gas (140 PSIG) and soap bubble.

15) Evacuate the system, and charge it with refrigerant. For the air-cooled and water-cooled models, see NAMEPLATE for required refrigerant charge. For the remote air-cooled model, see the Charge Label on the left side wall of the machine compartment.

16) Connect the Terminals, and place the Terminal Cover in position. Attach the Barrier with screws.

17) Open the Water Supply Line Shut-off Valve.

18) Turn on the power supply.

4. REMOVAL AND REPLACEMENT OF DRIER

IMPORTANT

Always install a new Drier every time the sealed refrigeration system is opened. Do not replace the Drier until after all others repairs or replacements have been made.

1) Turn off the power supply.

2) Remove the Front Panel, Top Panel and Right Side Panel.

3) Discharge the refrigerant.

4) Remove the Drier.

5) Install a new Drier, with the ARROW on the Drier, in the DIRECTION of the REFRIGERANT FLOW. Use nitrogen gas at the pressure of 3 - 4 PSIG when brazing the tubings.

6) Check for leaks using nitrogen gas (140 PSIG) and soap bubble.

7) Evacuate the system, and charge it with refrigerant. For the air-cooled and water-cooled models, see NAMEPLATE for required refrigerant charge. For the remote air-cooled model, see the Charge Label on the left side wall of the machine compartment.

8) Reassemble the body.

9) Open the Water Supply Line Shut-off Valve.

10) Turn on the power supply.

5. REMOVAL AND REPLACEMENT OF EXPANSION VALVE

IMPORTANT

Sometimes moisture in the refrigerant circuit exceeds the Drier capacity and freezes up at the Expansion Valve. Always install a new Drier every time the sealed refrigeration system is opened. Do not replace the Drier until after all other repairs or replacement have been made.

1) Turn off the power supply.

2) Remove the Top Panel, the Upper and Lower Rear Panels and the Louver.

3) Discharge the refrigerant.

4) Remove the Expansion Valve Bulb at the Evaporator outlet.

5) Remove the Expansion Valve Cover, and disconnect the Expansion Valve using a brazing equipment.

6) Braze a new Expansion Valve, with nitrogen gas flowing at the pressure of 3 - 4 PSIG.

WARNING

Always PROTECT the valve body using a DAMP CLOTH to prevent the valve from overheat. DO NOT braze with the valve body exceeding 250°F.

7) Check for leaks using nitrogen gas (140 PSIG) and soap bubble.

8) Evacuate the system, and charge it with refrigerant. For the air-cooled and water-cooled models, see NAMEPLATE for required refrigerant charge. For the remote air-cooled model, see the Charge Label on the left side wall of the machine compartment.

9) Attach the Bulb to the suction line in position. Be sure to secure it with clamps and to insulate it.

10) Place a new set of Expansion Valve Covers in position.

11) Place the panels and the Louver in position.

12) Turn on the power supply.

6. REMOVAL AND REPLACEMENT OF HOT GAS VALVE AND/OR LINE VALVE

CAUTION

Always use a copper tube of the same diameter and length when replacing the hot gas lines; otherwise the performance may be reduced.

IMPORTANT

Always install a new Drier every time the sealed refrigeration system is opened. Do not replace the Drier until after all other repairs or replacement have been made.

1) Turn off the power supply.

2) Remove the Front Panel, Top Panel and Right Side Panel.

3) Discharge the Refrigerant.

4) Remove the screw and the Solenoid.

5) Disconnect the Hot Gas Valve and/or the Line Valve using a brazing equipment.

6) Install a new valve.

WARNING

Always PROTECT the valve body using a DAMP CLOTH to prevent the valve from overheat. DO NOT braze with the valve body exceeding 250°F.

7) Replace the Drier.

8) Check for leaks using nitrogen gas (140 PSIG) and soap bubble.

9) Evacuate the system, and charge it with refrigerant. For the air-cooled and water-cooled models, see NAMEPLATE for required refrigerant charge. For the remote air-cooled model, see the Charge Label on the left side wall of the machine compartment.

10) Cut the leads of the Solenoid at the Connector, and connect the new Solenoid.

11) Attach the Solenoid to the valve body, and secure it with a screw.

12) Connect the Connector.

13) Open the Water Supply Line Shut-off Valve.

14) Turn on the power supply.

7. REMOVAL AND REPLACEMENT OF WATER REGULATING VALVE - WATER-COOLED MODEL ONLY

IMPORTANT

It is better to install a new Drier every time the sealed refrigeration system is opened. Do not replace the Drier until after all other repairs or replacement have been made.

1) Turn off the power supply.

2) Close the Water Supply Line Shut-off Valve.

3) Remove the Front Panel, the Top panel and the Right Side Panel.

4) Discharge the refrigerant.

5) Disconnect the Capillary Tube at the Condenser outlet using a brazing equipment.

6) Disconnect the Flare-connections of the valve.

7) Remove the screws and the valve from the Bracket.

8) Install a new valve, and braze the Capillary Tube.

9) Check for leaks using nitrogen gas (140 PSIG) and soap bubble.

10) Evacuate the system, and charge it with refrigerant. For the air-cooled and water-cooled models, see NAMEPLATE for required refrigerant charge. For the remote air-cooled model, see the Charge Label on the left side wall of the machine compartment.

11) Connect the Flare-connections.

12) Open the Water Supply Line Shut-off Valve.

13) Check for water leaks.

14) Place the panels in position.

15) Turn on the power supply.

8. ADJUSTMENT OF WATER REGULATING VALVE - WATER-COOLED MODEL ONLY

The Water Regulating Valve or also called "WATER REGULATOR" is factory-adjusted. No adjustment is required under normal use. Adjust the Water Regulator , if necessary, using the following procedures.

1) Attach a pressure gauge to the high-side line of the system. Or prepare a thermometer to check for the condenser drain temperature.

2) Rotate the adjustment screw by using a flat blade screwdriver, so that the pressure gauge shows 235 PSIG, or the thermometer reads 108 - 113°F, in 5 minutes after a freeze cycle or icemaking process starts. When the pressure exceeds 235 PSIG, or the condenser drain temperature exceeds 113°F, rotate the adjustment screw counterclockwise.

3) Check that the pressure or the condenser drain temperature holds a stable setting.

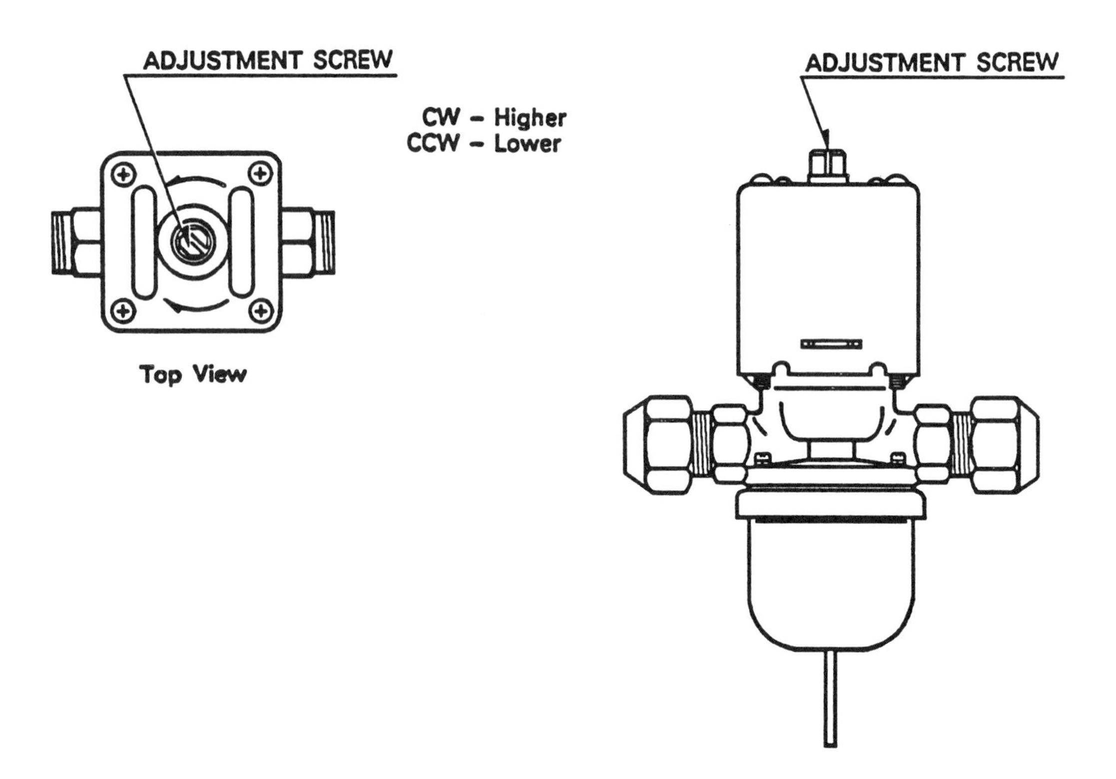

Fig. 30 Water Regulating Valve

9. REMOVAL AND REPLACEMENT OF CONDENSING PRESSURE REGULATOR (C.P.R.) - REMOTE AIR-COOLED MODEL ONLY

IMPORTANT

Always install a new Drier every time the sealed refrigeration system is opened. Do not replace the Drier until after all other repairs or replacements have been made.

1) Turn off the power supply.

2) Remove the panels from the remote condenser unit.

3) Discharge the refrigerant.

4) Remove the C.P.R. using a brazing equipment.

5) Install a new C.P.R. Always PROTECT the body using a DAMP CLOTH to prevent damage to valve against overheat. DO NOT braze with the body exceeding 250°F. Use nitrogen gas at the pressure of 3 - 4 PSIG when brazing the C.P.R..

6) Replace the Drier.

7) Check for leaks using nitrogen gas (140 PSIG) and soap bubble.

8) Evacuate the system and charge it with refrigerant. See the Charge Label on the left side wall of the machine compartment.

9) Place the panels in position.

10) Turn on the power supply.

10. REMOVAL AND REPLACEMENT OF THERMISTOR

CAUTION

1. FRAGILE, handle very carefully.
2. Always use a recommended sealant (High Thermal Conductive Type), Model KE4560RTV manufactured by SHINETSU SILICONE, Part Code 60Y000-11, or equivalent.
3. Always use a recommended foam insulation (Non-absorbent Type) or equivalent.

1) Turn off the power supply.

2) Remove the Front Panel.

3) Remove the Control Box Cover.

4) Disconnect the Thermistor leads from the K3 Connector on the Controller Board.

5) Remove the screw, Strap, Cable Tie, Foam Insulation, Thermistor Holder and Thermistor.

6) Scrape away the old sealant on the Thermistor Holder and the Suction Pipe.

7) Smoothly apply recommended sealant (KE4560RTV, Part Code 60Y000-11) to the Thermistor Holder concave.

8) Wipe off moisture or condensate on the Suction Pipe.

9) Attach a new Thermistor to the Suction Pipe very carefully to prevent damage to the leads. And secure it using the Thermistor Holder and recommended foam insulation.

10) Secure the Insulation using plastic cable ties.

11) Secure the Thermistor attachment portion using the Strap and the screw.

12) Connect the Thermistor leads through the bushing of the Control Box to the K3 Connector on the Controller Board.

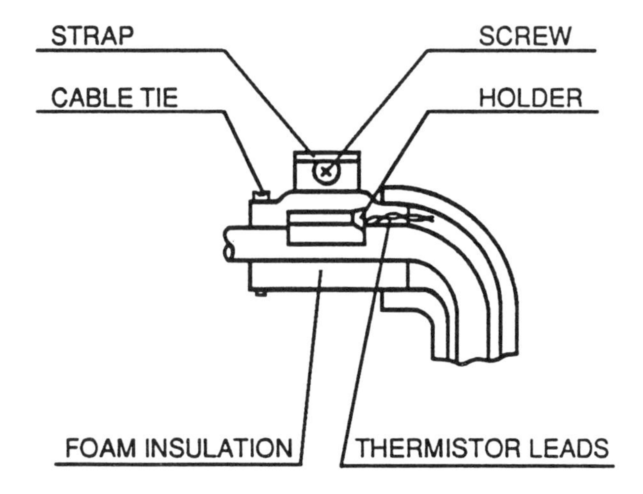

Fig. 31 Thermistor Attachment

13) Place the Control Box Cover and the panels in position.

14) Turn on the power supply.

11. REMOVAL AND REPLACEMENT OF FAN MOTOR

1) Turn off the power supply.

2) Remove the Front Panel, the Top Panel and the Right Side Panel.

3) Remove the closed end connectors from the Fan Motor leads.

4) Remove the Fan Motor Bracket and Fan Motor.

5) Install a new Fan Motor in reverse order.

6) Place the panels in position.

7) Turn on the power supply.

12. REMOVAL AND REPLACEMENT OF WATER VALVE

1) Turn off the power supply.

2) Close the Water Supply Line Shut-off Valve.

3) Remove the Valve Outlet Tubing by releasing the Clamp.

4) Remove the Bracket from the unit.

5) Remove the Fitting Nut and Water Vlave.

6) Disconnect the Terminals from the Water Valve.

7) Install a new Water Valve in reverse order.

8) Open the Water Supply Line Shut-off Valve.

9) Turn on the power supply.

10) Check for leaks.

11) Place the panels in position.

13. REMOVAL AND REPLACEMENT OF PUMP MOTOR

1) Turn off the power supply.

2) Remove the Front panel, the Top Panel and the Base Cover.

3) Remove one end of the Pump Tubing, and drain the Water Tank.

4) Disconnect the Pump Suction and Discharge Hoses.

5) Remove the screw and the Pump Motor Bracket.

6) Remove the closed end connectors from the Pump Motor leads.

7) Remove the Pump Motor from the Pump Motor Bracket.

8) Remove the Pump Housing, and check the Impeller.

9) If the Impeller is defective, install a new Pump Motor.

10) Install a new motor or new parts in reverse order.

11) Turn on the power supply, and check for leaks.

12) Place the panels in position.

14. REMOVAL AND REPLACEMENT OF SPRAY TUBES

1) Turn off the power supply.

2) Close the Water Supply Line Shut-off Valve.

3) Remove the Front Panel and the Front Insulation.

4) Remove the Rubber Hoses from the Spray Tubes (Water Supply Pipe).

5) Release the Clamps, and disconnect the Rubber Hoses.

6) Remove the Spray Tubes by clipping the side tabs.

7) Install new Spray Tubes in reverse order.

VIII. PROCEDURES FOR CHECKING INSIDE THE CONTROL BOX

1) Remove the two screws fixing the front of the Control Box to the Front Frame.

2) Slightly lift the rear of the Control Box to remove it from the Hook of the Junction Box, and then pull the Control Box out.

3) Check the inside of the Control Box.

Replace the Control Box in the reverse order of steps 1) and 2).
At this time, take sufficient care so that the rear of the Control Box goes inside the guides of the Top Frame and also that the power leads are not pinched between the Top Frame and the Control Box.

HOSHIZAKI

HOSHIZAKI AMERICA, INC.

618 HIGHWAY 74 SOUTH, PEACHTREE CITY,
GEORGIA 30269 U.S.A.
PHONE: 404-487-2331

HOSHIZAKI
MODULAR FLAKER

MODEL F-1000MAB
F-1000MWB
F-1000MRB

SERVICE MANUAL

HOSHIZAKI AMERICA, INC.
618 Hwy. 74 South Peachtree City, GA 30269
Telephone: (404) 487-2331

NO. U2AC-194
ISSUED: JAN. 29, 1991
REVISED: ———

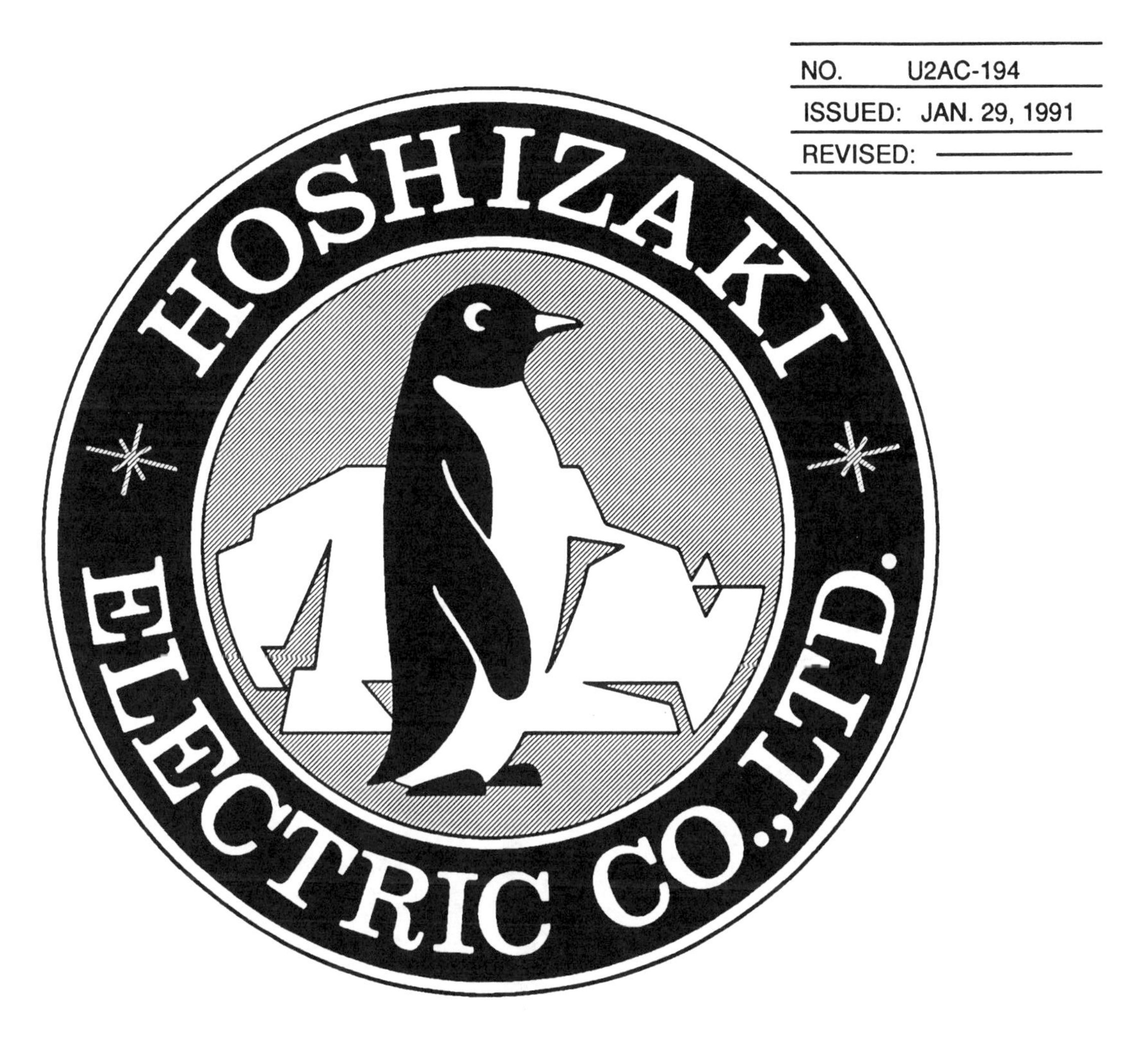

HOSHIZAKI
MODULAR FLAKER

MODEL F-1000MAB
F-1000MWB
F-1000MRB

SERVICE MANUAL

FOREWORD

This Service Manual contains the specifications and information in regard to transporting, unpacking, installing, operating and servicing the machine. You are encouraged to read it thoroughly in order to obtain maximum performance. You will find details on the construction, installation and maintenance.

If you encounter any problem not covered in this Service Manual, feel free to contact Hoshizaki America, Inc. We will be happy to provide whatever assistance is necessary.

Keep this Service Manual handy, and read it again when questions arise.

HOSHIZAKI AMERICA, INC.

618 HIGHWAY 74 SOUTH, PEACHTREE CITY,
GEORGIA 30269 U.S.A.
PHONE: 404-487-2331

HOSHIZAKI ELECTRIC CO., LTD.

TOYOAKE, AICHI, JAPAN
PHONE: 0562-97-2111
TELEX-NO: 04486-514 HOSHI J

U2AC1949101

CONTENTS

PAGE

I. SPECIFICATIONS

1. NAMEPLATE RATING

[a] F-1000MAB (Air-cooled)

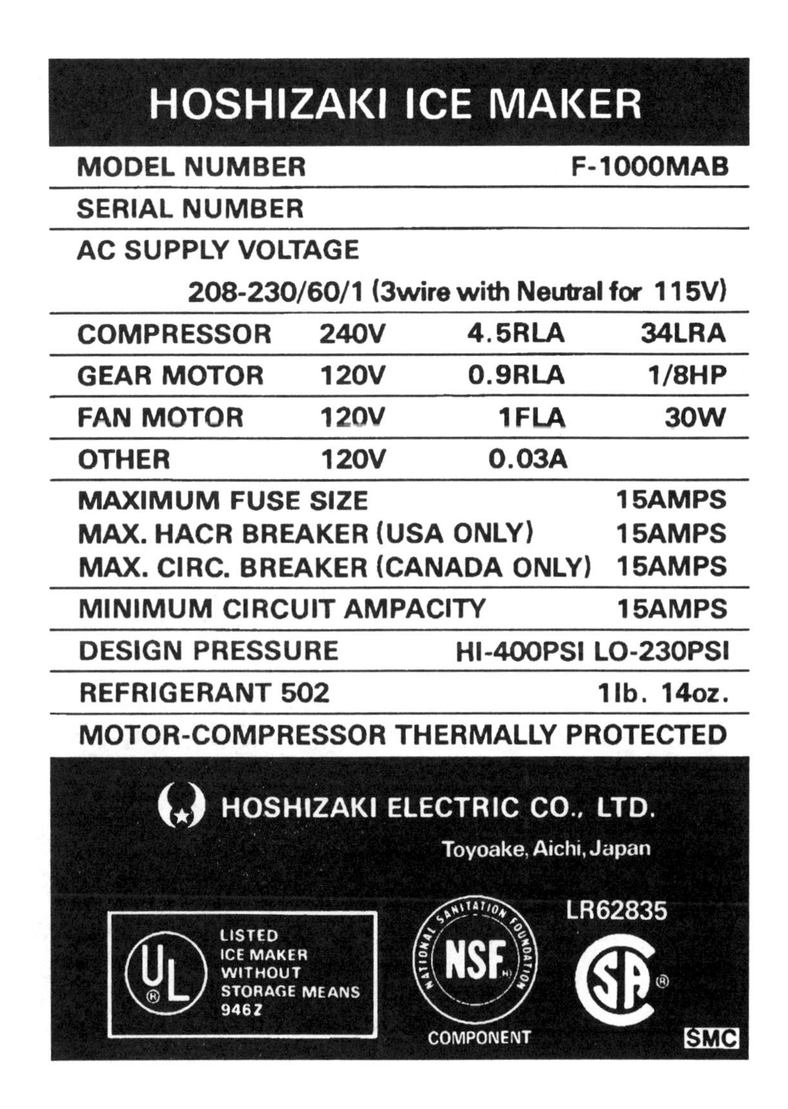

See the Nameplate for electrical and refrigeration specifications. This Nameplate is located on the upper right hand side of rear panel.

We reserve the right to make changes in specifications and design without prior notice.

[b] F-1000MWB (Water-cooled)

HOSHIZAKI ICE MAKER			
MODEL NUMBER			F-1000MWB
SERIAL NUMBER			
AC SUPPLY VOLTAGE 208-230/60/1 (3wire with Neutral for 115V)			
COMPRESSOR	240V	4.3RLA	34LRA
GEAR MOTOR	120V	0.9RLA	1/8HP
OTHER	120V	0.03A	
MAXIMUM FUSE SIZE			15AMPS
MAX. HACR BREAKER (USA ONLY)			15AMPS
MAX. CIRC. BREAKER (CANADA ONLY)			15AMPS
MINIMUM CIRCUIT AMPACITY			15AMPS
DESIGN PRESSURE			HI-400PSI LO-230PSI
REFRIGERANT 502			1lb.
MOTOR-COMPRESSOR THERMALLY PROTECTED			

HOSHIZAKI ELECTRIC CO., LTD.
Toyoake, Aichi, Japan

LISTED ICE MAKER WITHOUT STORAGE MEANS 946Z

NATIONAL SANITATION FOUNDATION NSF COMPONENT

LR62835

SMC

See the Nameplate for electrical and refrigeration specifications. This Nameplate is located on the upper right hand side of rear panel.

We reserve the right to make changes in specifications and design without prior notice.

[c] F-1000MRB (Remote Air-cooled)

HOSHIZAKI ICE MAKER			
MODEL NUMBER			F-1000MRB
SERIAL NUMBER			
AC SUPPLY VOLTAGE 208-230/60/1 (3wire with Neutral for 115V)			
COMPRESSOR	240V	4.5RLA	34LRA
GEAR MOTOR	120V	0.9RLA	1/8HP
FAN MOTOR	120V	0.5FLA	8W
FAN MOTOR	REMOTE	120V	3A MAX
OTHER	120V	0.03A	
MAXIMUM FUSE SIZE			15AMPS
MAX. HACR BREAKER (USA ONLY)			15AMPS
MAX. CIRC. BREAKER (CANADA ONLY)			15AMPS
MINIMUM CIRCUIT AMPACITY			15AMPS
DESIGN PRESSURE			HI-400PSI LO-230PSI
REFRIGERANT 502			
MOTOR-COMPRESSOR THERMALLY PROTECTED			

HOSHIZAKI ELECTRIC CO., LTD.
Toyoake, Aichi, Japan

LISTED ICE MAKER WITHOUT STORAGE MEANS 946Z

NATIONAL SANITATION FOUNDATION NSF COMPONENT

LR62835

SMC

See the Nameplate for electrical and refrigeration specifications. This Nameplate is located on the upper right hand side of rear panel.

We reserve the right to make changes in specifications and design without prior notice.

2. DIMENSIONS/CONNECTIONS

[a] F-1000MAB

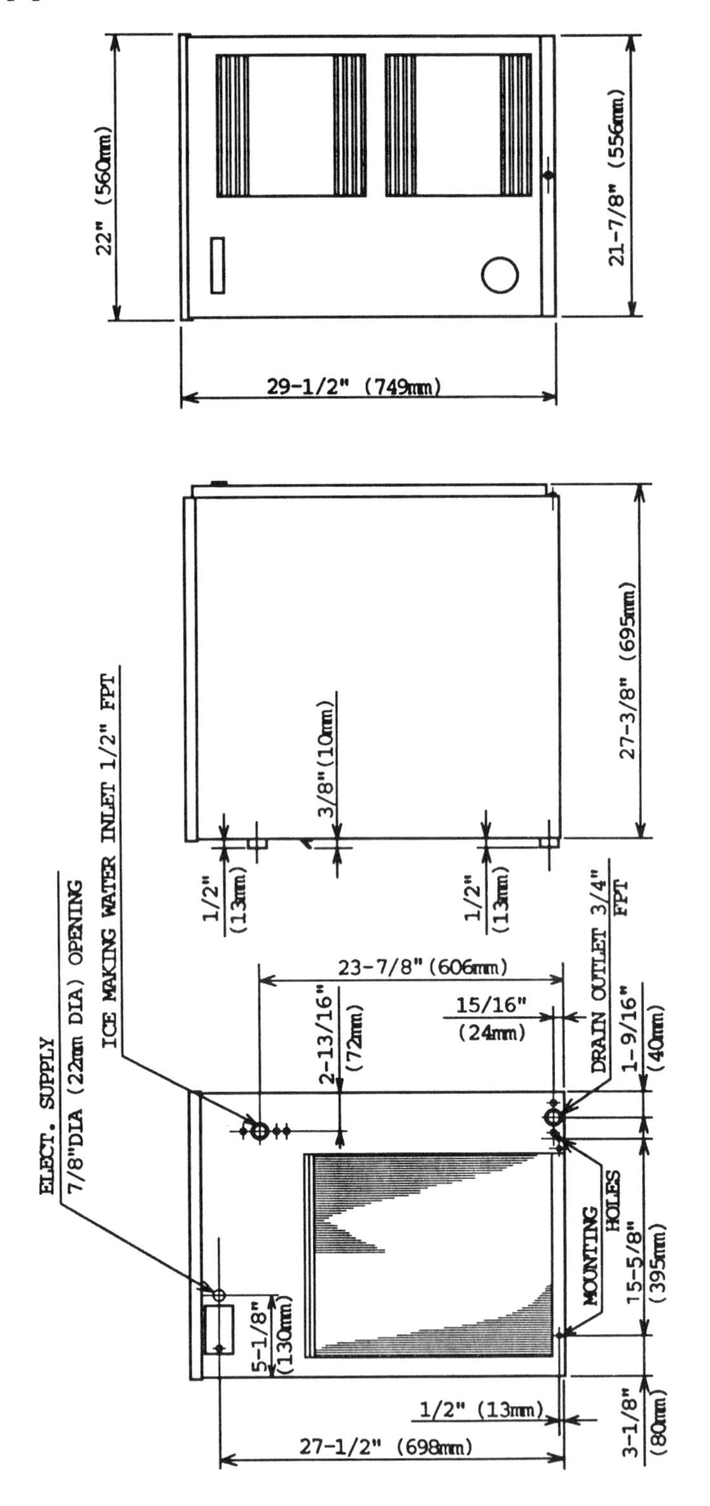

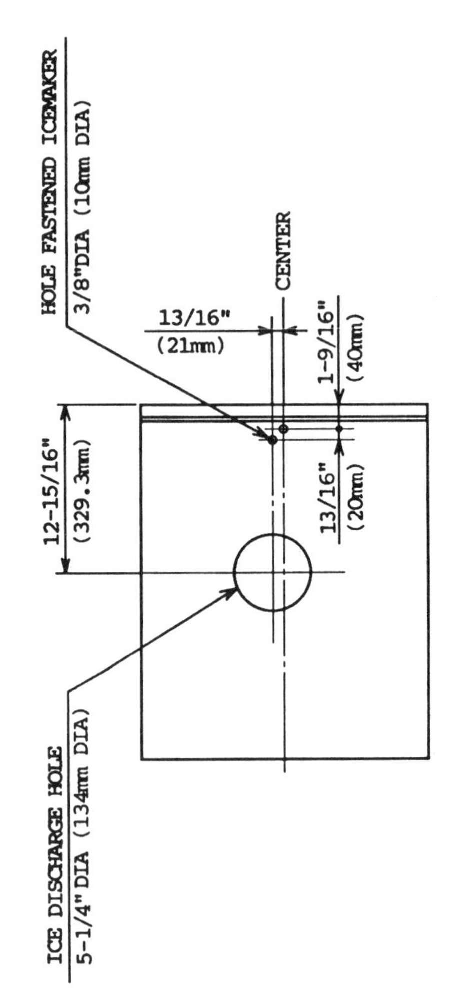

U2AC1949101

[b] F-1000MWB

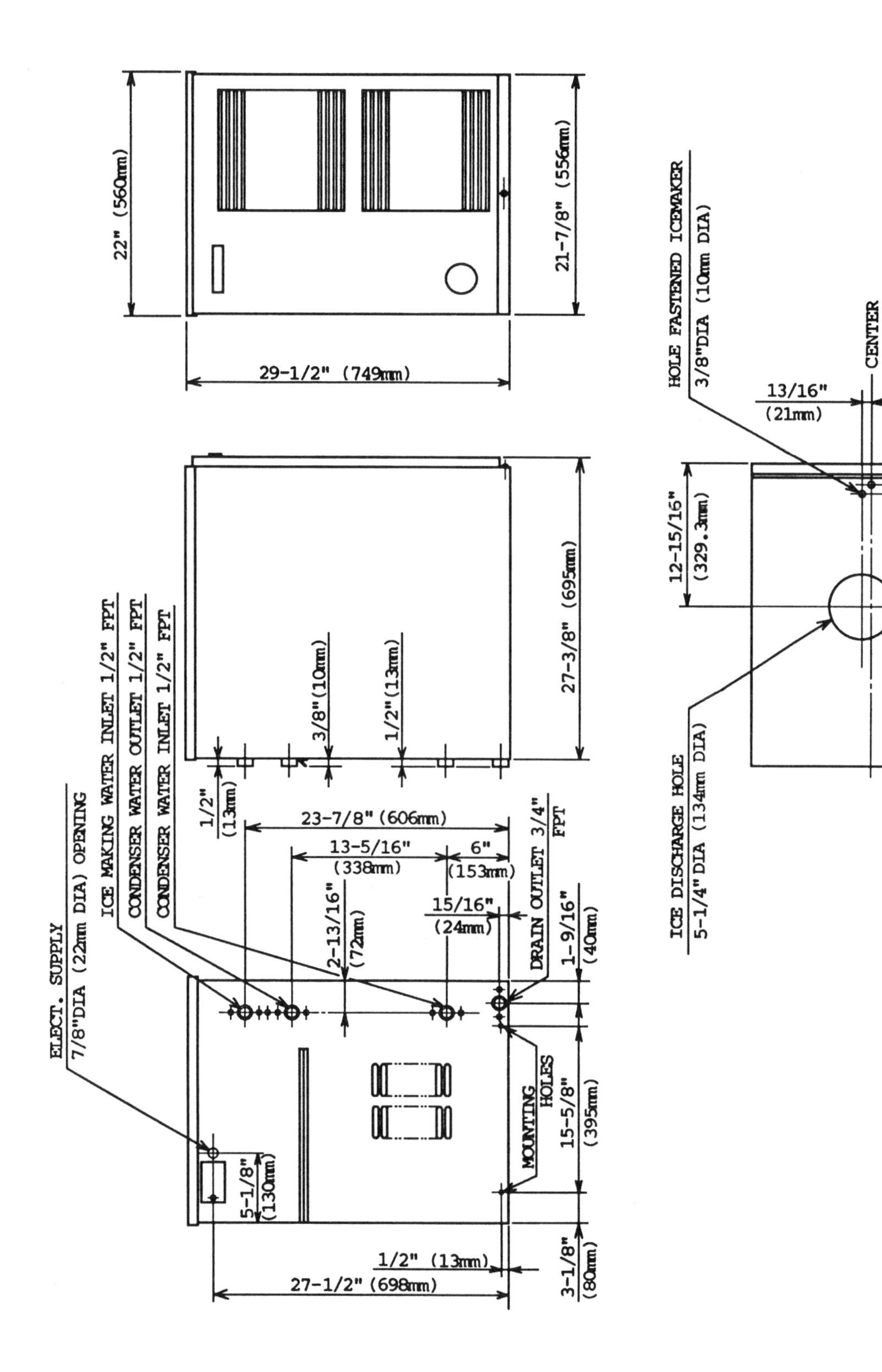

22" (560mm)
21-7/8" (556mm)
29-1/2" (749mm)
27-3/8" (695mm)
3/8"(10mm)
1/2"(13mm)
ELECT. SUPPLY
7/8"DIA (22mm DIA) OPENING
ICE MAKING WATER INLET 1/2" FPT
CONDENSER WATER OUTLET 1/2" FPT
CONDENSER WATER INLET 1/2" FPT
1/2" (13mm)
23-7/8" (606mm)
13-5/16" (338mm)
6" (153mm)
2-13/16" (72mm)
15/16" (24mm)
DRAIN OUTLET 3/4" FPT
1-9/16" (40mm)
MOUNTING HOLES
15-5/8" (395mm)
5-1/8" (130mm)
1/2" (13mm)
27-1/2" (698mm)
3-1/8" (80mm)
HOLE FASTENED ICEMAKER
3/8"DIA (10mm DIA)
CENTER
13/16" (21mm)
1-9/16" (40mm)
12-15/16" (329.3mm)
13/16" (20mm)
ICE DISCHARGE HOLE
5-1/4"DIA (134mm DIA)

[c] F-1000MRB

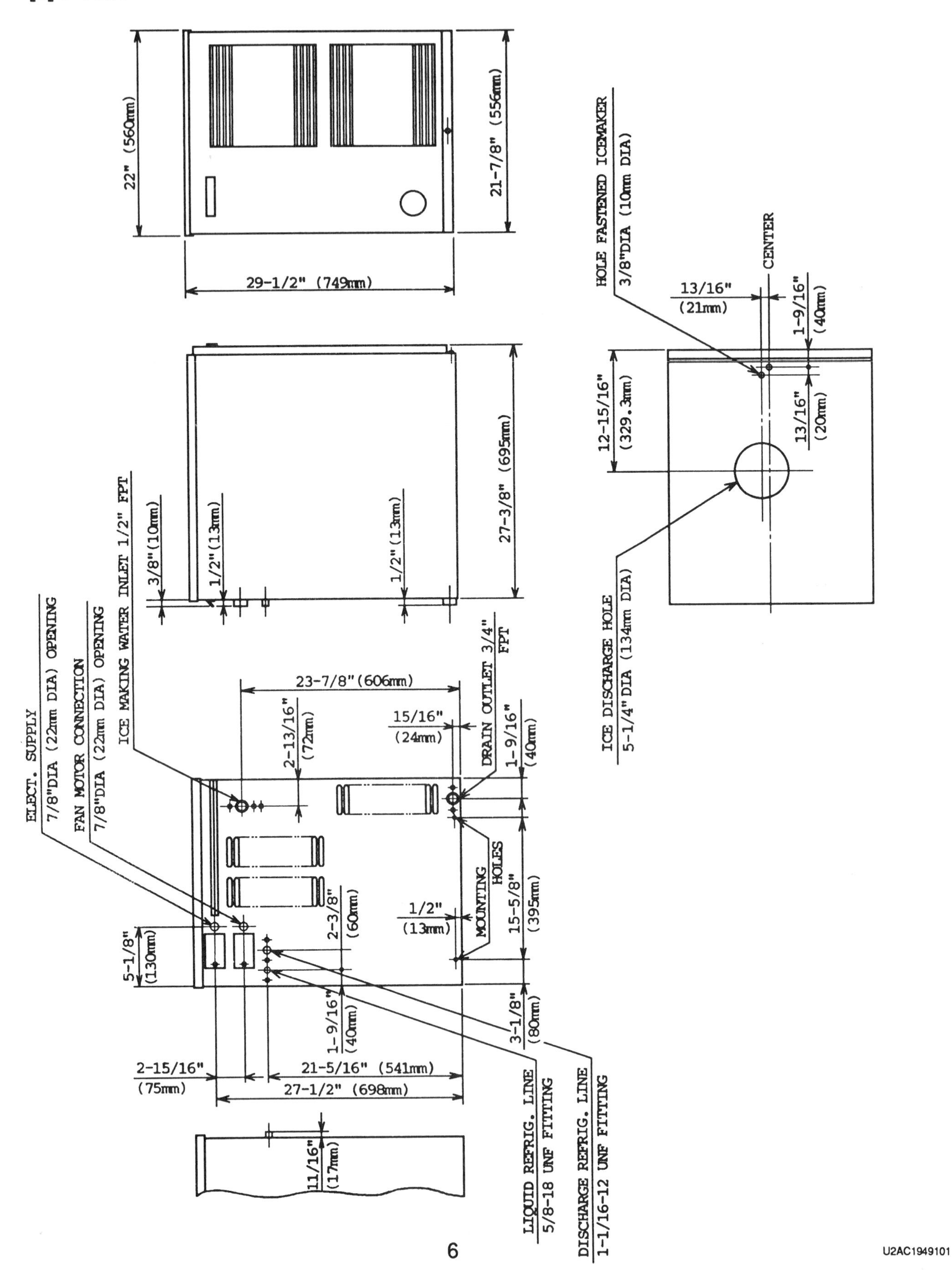

U2AC1949101

3. SPECIFICATIONS

[a] F-1000MAB

AC SUPPLY VOLTAGE	208-230/60/1 (3 wire with neutral for 115 V)
COMPRESSOR	240V 4.5RLA 34LRA
GEAR MOTOR	120V 0.9RLA 1/8HP
FAN MOTOR	120V 1FLA 30W
OTHER	120V 0.03A
MINIMUM CIRCUIT AMPACITY	15 A
MAXIMUM FUSE SIZE	15 A

APPROXIMATE ICE PRODUCTION PER 24 HR. lbs./day (kg/day) Reference without *marks

Ambient Temp. (° F)	Water Temp. (° F) 50	70	90
70	1020(463)	920(417)	840(381)
80	950(431)	865(392)	790(358)
90	880(399)	800(363)	740(336)
100	810(367)	730(331)	680(308)

SHAPE OF ICE	Flake
ICE QUALITY	Approx. 70 %, Ice (90° F/ 70° F Conductivity 200 μs/cm)
APPROXIMATE STORAGE CAPACITY	N/A
ELECTRIC & WATER CONSUMPTION	90° F/ 70° F 70° F/ 50° F
ELECTRIC W (kWH/100 lbs.)	1250 (3.8) 1200 (2.8)
POTABLE WATER gal./24HR (gal./100 lbs.)	96 (12.0) 122 (12.0)
EXTERIOR DIMENSIONS (WxDxH)	22" x 27-3/8" x 29-1/2" (560 x 695 x 749 mm)
EXTERIOR FINISH	Stainless Steel, Galvanized Steel (Rear)
WEIGHT	Net 177 lbs. (80 kg), Shipping 220 lbs. (100 kg)
CONNECTIONS - ELECTRIC	Permanent Connection
- WATER SUPPLY	Inlet 1/2" FPT
- DRAIN	Outlet 3/4" FPT
ICE MAKING SYSTEM	Auger type
HARVESTING SYSTEM	Direct driven Auger (100 W Gear Motor)
ICE MAKING WATER CONTROL	Float Switch
COOLING WATER CONTROL	N/A
BIN CONTROL SYSTEM	Mechanical Bin Control (Proximity Sw.)
COMPRESSOR	Hermetic 860W, Model REK3-0125-PFV
CONDENSER	Air-cooled, Fin and Tube type
EVAPORATOR	Copper Tube on Cylinder
REFRIGERANT CONTROL	Thermostatic Expansion Valve
REFRIGERANT CHARGE	R502, 1 lb. 14 oz (850 g)
DESIGN PRESSURE	High 400 PSIG, Low 230 PSIG
P.C.BOARD CIRCUIT PROTECTION	High Voltage Cut-out Relay
COMPRESSOR PROTECTION	Internal Protector
GEAR MOTOR PROTECTION	Manual reset Circuit Breaker and Thermal Protector
REFRIGERANT CIRCUIT PROTECTION	Auto reset High Pressure Control Switch
LOW WATER PROTECTION	Float Switch and Timer
ACCESSORIES - SUPPLIED	Ice Scoop, Spare Fuse
- REQUIRED	Ice Storage Bin
OPERATION CONDITIONS	VOLTAGE RANGE 187 - 264 V 104-132V
	AMBIENT TEMP. 45 - 100° F
	WATER SUPPLY TEMP. 45 - 90° F
	WATER SUPPLY PRESS. 7 - 113 PSIG

* We reserve the right to make changes in specifications and design without prior notice.

[b] F-1000MWB

AC SUPPLY VOLTAGE	208-230/60/1 (3 wire with neutral for 115 V)
COMPRESSOR	240V 4.3RLA 34LRA
GEAR MOTOR	120V 0.9RLA 1/8HP
OTHER	120V 0.03A
MINIMUM CIRCUIT AMPACITY	15 A
MAXIMUM FUSE SIZE	15 A

APPROXIMATE ICE PRODUCTION PER 24 HR. lbs./day (kg/day) Reference without *marks

Ambient Temp. (° F)	Water Temp. (° F) 50	70	90
70	*1010(458)	890(404)	750(340)
80	1005(456)	880(399)	745(338)
90	995(451)	*865(392)	740(336)
100	980(460)	845(383)	730(331)

SHAPE OF ICE	Flake
ICE QUALITY	Approx. 70 % Ice (90° F/ 70° F, Conductivity 200 μs/cm)
APPROXIMATE STORAGE CAPACITY	N/A
ELECTRIC & WATER CONSUMPTION	90° F/ 70° F, 70° F/ 50° F,
ELECTRIC W (kWH/100 lbs.)	1060 (2.9) 1060 (2.5)
POTABLE WATER	104 (12.0) 121 (12.0)
WATER-COOLED CONDENSER gal./24HR (gal./100 lbs.)	811 (94) 524 (52)
EXTERIOR DIMENSIONS (WxDxH)	22" x 27-3/8" x 29-1/2" (560 x 695 x 749 mm)
EXTERIOR FINISH	Stainless Steel, Galvanized Steel(Rear)
WEIGHT	Net 177 lbs. (80 kg), Shipping 220 lbs. (100 kg)
CONNECTIONS - ELECTRIC	Permanent Connection
- WATER SUPPLY	Inlet 1/2" FPT Cond. Inlet 1/2" FPT
- DRAIN	Outlet 3/4" FPT Cond. Outlet 1/2" FPT
ICE MAKING SYSTEM	Auger type
HARVESTING SYSTEM	Direct driven Auger (100 W Gear Motor)
ICE MAKING WATER CONTROL	Float Switch
COOLING WATER CONTROL	Automatic Water Regulator
BIN CONTROL SYSTEM	Mechanical Bin Control (Proximity Sw.)
COMPRESSOR	Hermetic 860W, Model REK3-0125-PFV
CONDENSER	Water-cooled, Tube in Tube type
EVAPORATOR	Copper Tube on Cylinder
REFRIGERANT CONTROL	Thermostatic Expansion Valve
REFRIGERANT CHARGE	R502, 1 lb. (450 g)
DESIGN PRESSURE	High 400 PSIG, Low 230 PSIG
P.C.BOARD CIRCUIT PROTECTION	High Voltage Cut-out Relay
COMPRESSOR PROTECTION	Internal Protector
GEAR MOTOR PROTECTION	Manual reset Circuit Breaker and Thermal Protector
REFRIGERANT CIRCUIT PROTECTION	Auto reset High Pressure Control Switch
LOW WATER PROTECTION	Float Switch and Timer
ACCESSORIES - SUPPLIED	Ice Scoop, Spare Fuse
- REQUIRED	Ice Storage Bin
OPERATION CONDITIONS	VOLTAGE RANGE 187 - 264 V 104 - 132V
	AMBIENT TEMP. 45 - 100° F
	WATER SUPPLY TEMP. 45 - 90° F
	WATER SUPPLY PRESS. 7 - 113 PSIG

* We reserve the right to make changes in specifications and design without prior notice.

[c] F-1000MRB

AC SUPPLY VOLTAGE	208-230/60/1 (3 wire with neutral for 115 V)
COMPRESSOR	240V 4.5RLA 34RLA
GEAR MOTOR	120V 0.9RLA 1/8HP
FAN MOTOR	120V 0.5FLA 8W
FAN MOTOR REMOTE	120V 3A MAX
OTHER	120V 0.03A
MINIMUM CIRCUIT AMPACITY	15 A
MAXIMUM FUSE SIZE	15 A

APPROXIMATE ICE PRODUCTION PER 24 HR.
lbs./day (kg/day)
Reference without *marks

Ambient Temp. (° F)	Water Temp. (° F) 50	70	90
70	1020(463)	920(417)	840(381)
80	950(431)	865(392)	790(358)
90	880(399)	800(363)	740(336)
100	810(367)	730(331)	680(308)

SHAPE OF ICE	Flake
ICE QUALITY	Approx. 70% Ice (90° F/ 70° F, Conductivity 200 μs/cm)
APPROXIMATE STORAGE CAPACITY	N/A
ELECTRIC & WATER CONSUMPTION	90° F/ 70° F, 70° F/ 50° F
ELECTRIC W (kWH/100 lbs.)	1300 (3.9) 1250 (2.9)
POTABLE WATER gal./24HR (gal./100 lbs.)	96 (12.0) 122 (12.0)
EXTERIOR DIMENSIONS (WxDxH)	22" x 27-3/8" x 29-1/2" (560 x 695 x 749 mm)
EXTERIOR FINISH	Stainless Steel, Galvanized Steel (Rear)
WEIGHT	Net 177 lbs. (80 kg), Shipping 220 lbs. (100 kg)
CONNECTIONS - ELECTRIC	Permanent Connection
- WATER SUPPLY	Inlet 1/2" FPT
- DRAIN	Outlet 3/4" FPT
- REFRIGERATION CIRCUIT	Discharge Line 1-1/16-12 UNF Fitting Liquid Line 5/8-18 UNF Fitting
ICE MAKING SYSTEM	Auger type
HARVESTING SYSTEM	Direct driven Auger (100 W Gear Motor)
ICE MAKING WATER CONTROL	Float Switch
COOLING WATER CONTROL	N/A
BIN CONTROL SYSTEM	Mechanical Bin Control (Proximity Sw.)
COMPRESSOR	Hermetic 860W, Model REK3-0125-PFV
CONDENSER	Air-cooled Remote Condenser Unit URC-6A Recommended
EVAPORATOR	Copper Tube on Cylinder
REFRIGERANT CONTROL	Thermostatic Expansion Valve Condensing Pressure regulator on URC-6A
REFRIGERANT CHARGE	R502, 6 lbs. 10 oz. (3000 g) (Icemaker 4 lbs. 7 oz. Cond. unit 2 lbs. 3 oz.)
DESIGN PRESSURE	High 400 PSIG, Low 230 PSIG
P.C. BOARD CIRCUIT PROTECTION	High Voltage Cut-out Relay
COMPRESSOR PROTECTION	Internal Protector
GEAR MOTOR PROTECTION	Manual reset Circuit Breaker and Thermal Protector
REFRIGERANT CIRCUIT PROTECTION	Auto reset High Pressure Control Switch
LOW WATER PROTECTION	Float Switch and Timer
ACCESSORIES - SUPPLIED	Ice Scoop, Spare Fuse
- REQUIRED	Ice Storage Bin
OPERATION CONDITIONS	VOLTAGE RANGE 187 - 264 V 104 - 132 V AMBIENT TEMP. 45 - 100° F WATER SUPPLY TEMP. 45 - 90° F WATER SUPPLY PRESS. 7 - 113 PSIG

* We reserve the right to make changes in specifications and design without prior notice.

4. CONDENSER UNIT MODEL URC-6A

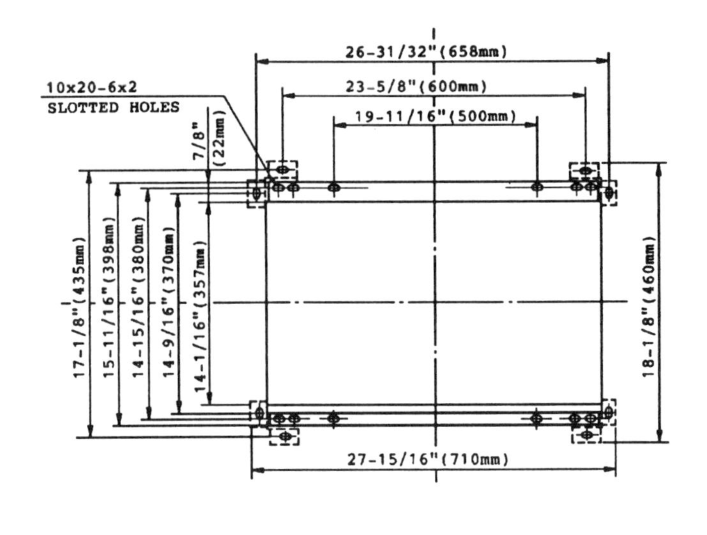

TOP VIEW

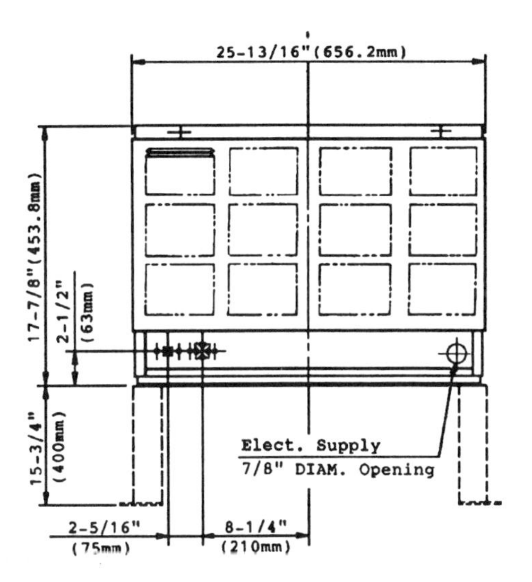

FRONT VIEW

SPECIFICATIONS

MODEL: CONDENSER UNIT URC-6A	
EXTERIOR	Galvanized Steel
DIMENSIONS (W x D x H)	25-13/16" x 15-11/16" x 17-7/8" (656 x 398 x 454 mm)
REFRIGERANT CHARGE	R502 2 lbs. 3 oz. (1000 g)
WEIGHT	Net 46 lbs. (21 kg) Shipping 53 lbs. (24 kg)
CONNECTIONS REFRIGERATION ELECTRICAL	 One Shot Coupling (Aeroquip) Permanent Connection
CONDENSER	Air-cooled
HEAD PRESSURE CONTROL	Condensing Pressure Regulator
AMBIENT CONDITION	Min. -20°F - Max. +122°F (-29°C to +50°C) Outdoor use

II. GENERAL INFORMATION

1. CONSTRUCTION

[a] F-1000MAB

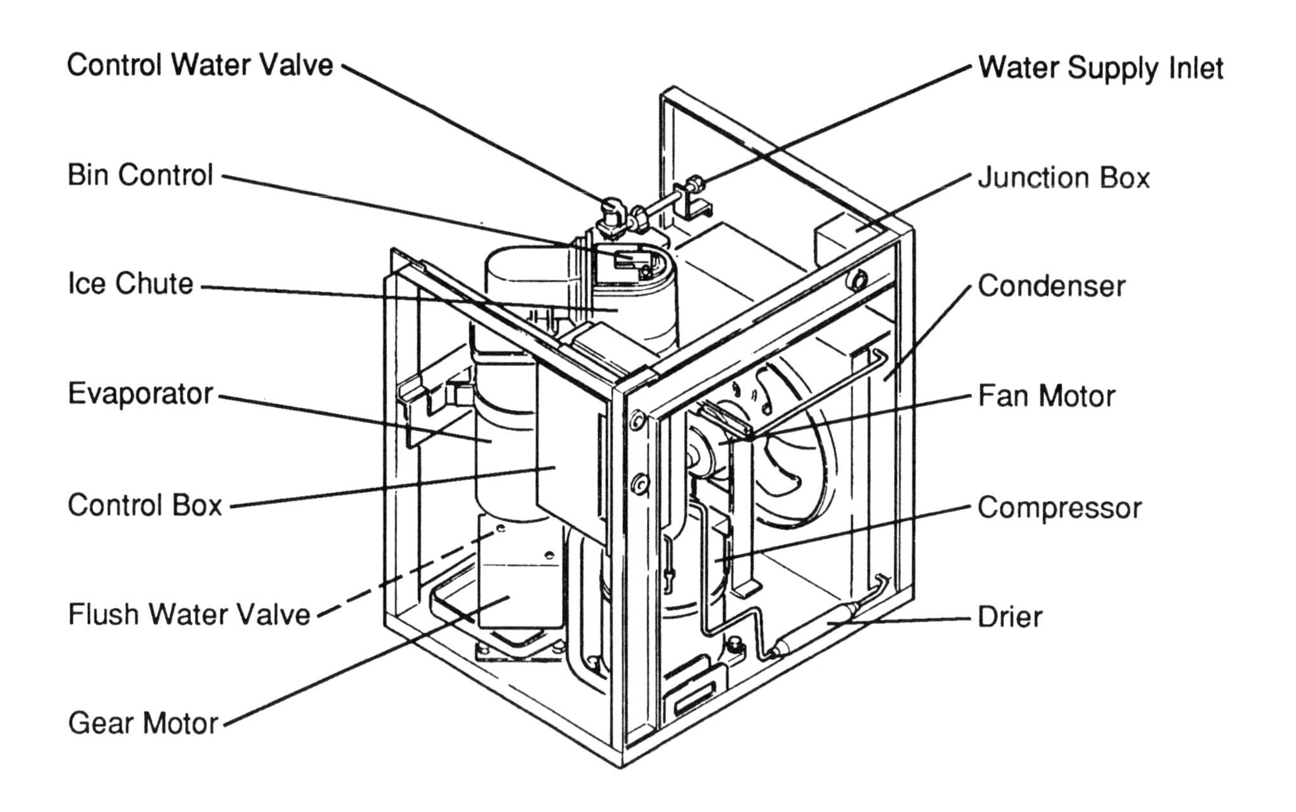

[b] F-1000MWB

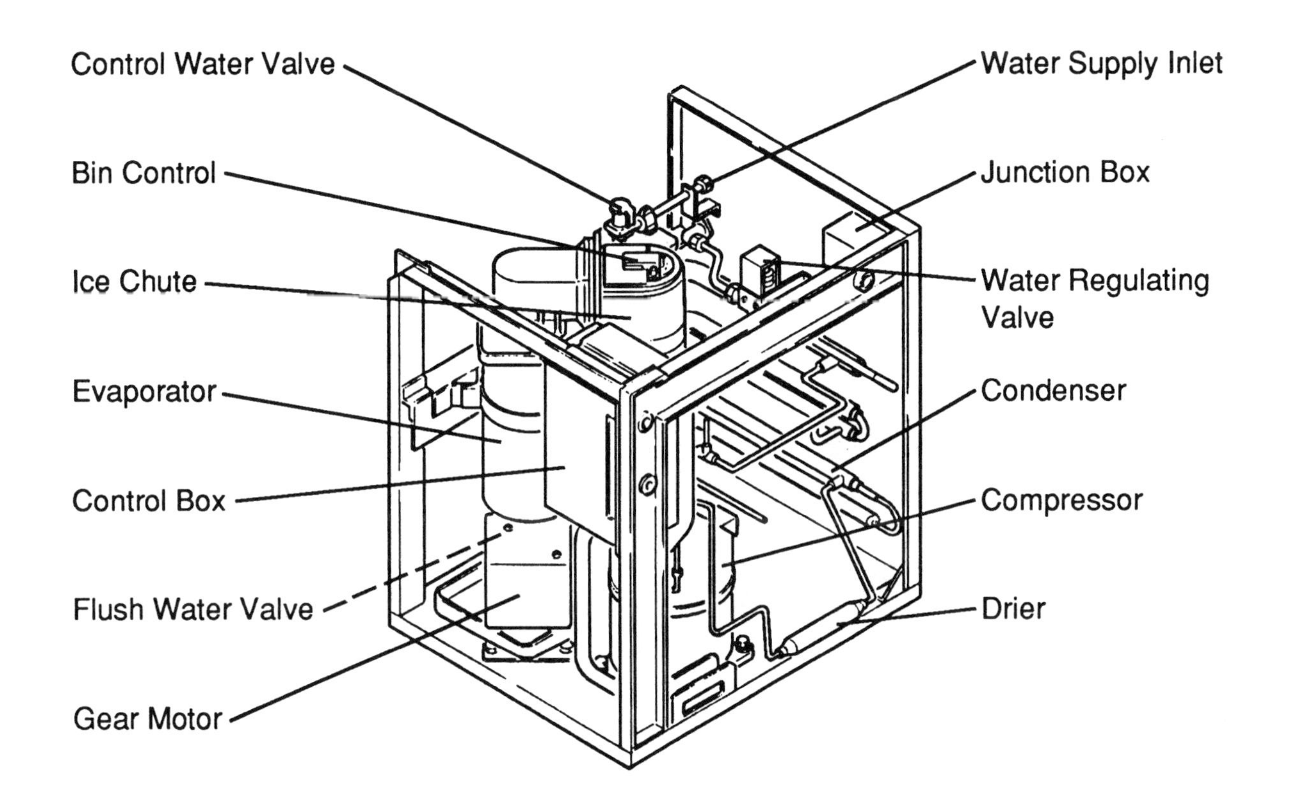

Control Water Valve
Bin Control
Ice Chute
Evaporator
Control Box
Flush Water Valve
Gear Motor
Water Supply Inlet
Junction Box
Water Regulating Valve
Condenser
Compressor
Drier

[c] F-1000MRB

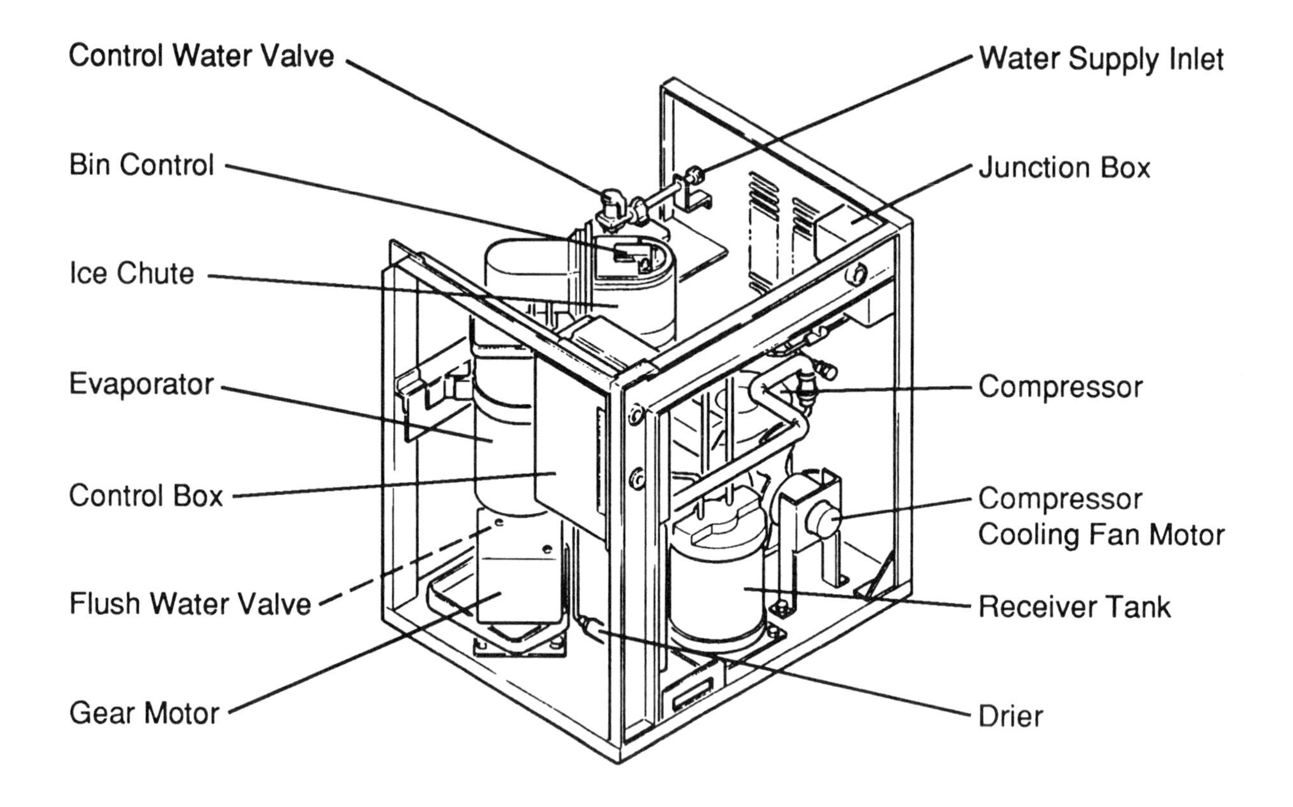
Control Water Valve
Bin Control
Ice Chute
Evaporator
Control Box
Flush Water Valve
Gear Motor
Water Supply Inlet
Junction Box
Compressor
Compressor
Cooling Fan Motor
Receiver Tank
Drier

2. CONTROL BOX LAYOUT

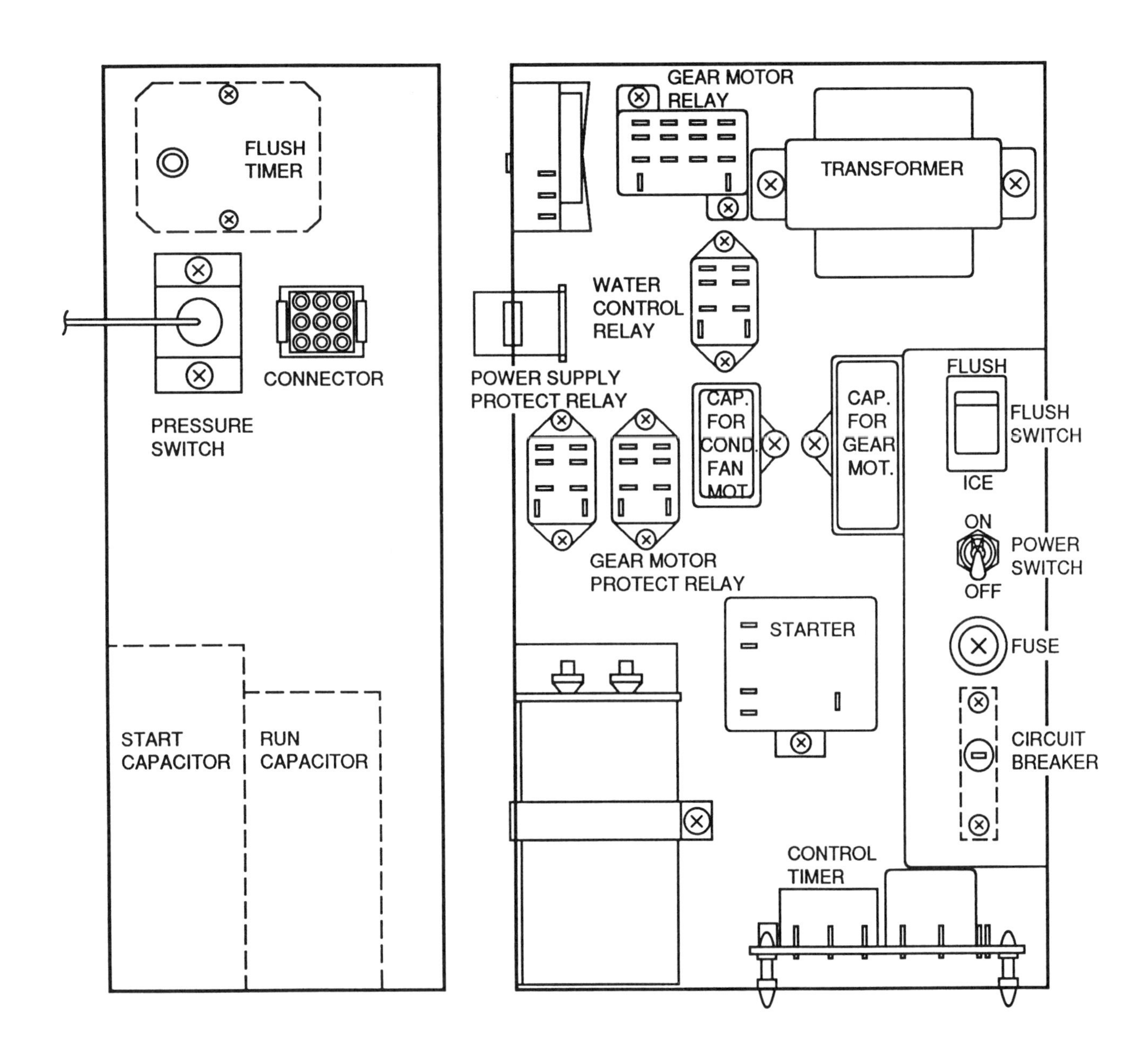

Note: The above component names are identical with the Wiring Label, but not with the Parts List.

III. INSTALLATION AND OPERATING INSTRUCTIONS

1. CHECKS BEFORE INSTALLATION

WARNING

Remove shipping carton, tape(s) and packing. If packing material is left in the icemaker, it will not work properly.

1) Remove the Front, Top and Side Panels to prevent damage when installing the icemaker. (See "2. HOW TO REMOVE PANELS.")

2) Remove the package containing the accessories from inside the icemaker.

3) Remove the protective plastic film from the panels. If the icemaker is exposed to the sun or to heat, remove the film after the icemaker cools.

4) Check that the refrigerant lines do not rub or touch lines or other surfaces, and that the fan blade turns freely.

5) Check that the Compressor is snug on all mounting pads.

6) See the Nameplate on the Rear Panel, and check that your voltage supplied corresponds with the voltage specified on the Nameplate.

7) This icemaker needs a storage bin. The recommended storage bin is HOSHIZAKI ICE STORAGE BIN, Model B-300 series.

8) On remote air-cooled models, a remote condenser unit is needed. The recommended remote condenser unit is HOSHIZAKI CONDENSER UNIT, Model URC-6A.

2. HOW TO REMOVE PANELS - See Fig. 1

a) Front Panel .... Remove the screw. Lift up and pull toward you.

b) Top Panel .... Lift up front, push away, and then lift off.

c) Side Panel .... Remove the screw. Pull slightly toward you, and lift off.

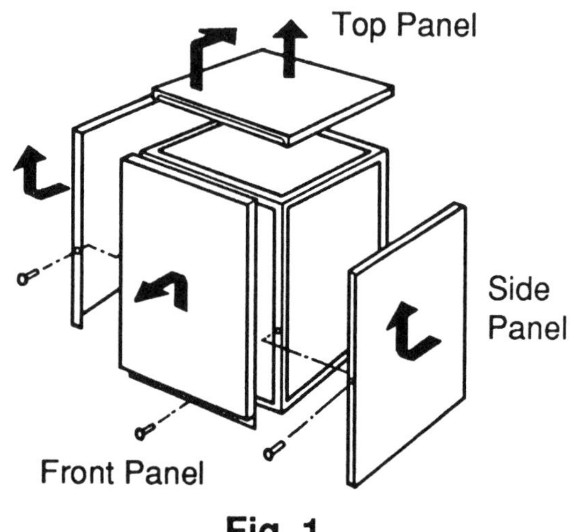

Fig. 1

3. LOCATION

WARNING

This icemaker is not designed for unsheltered outdoor installations. Normal operating ambient temperature should be within +45°F to +100°F; Normal operating water temperature should be within +45°F to +90°F. Operation of the icemaker, for extended periods, outside of these normal temperature ranges may affect production capacity

For best operating results:

* Icemaker should not be located next to ovens, grills or other high heat producing equipment.

* Location should provide a firm and level foundation for the equipment.

* Allow 6" clearance at rear and sides for proper air circulation and ease of maintenance and /or service should they be required. Allow 24" clearance at top to allow for removal of the Auger.

4. SET UP

1) Unpack the Storage Bin, and attach the four adjustable legs provided (bin accessory) to the bottom of the Storage Bin.

2) Position the Storage Bin in the selected permanent position.

3) Place the icemaker on the top of the Storage Bin.

4) Secure the icemaker to the Storage Bin, by using the two mounting brackets and four bolts provided. See Fig. 2.

5) Seal the seam between the icemaker and the Storage Bin.

6) Level the icemaker/Storage Bin in both the left-to-right and front-to-rear directions. Adjust the Ice Bin Legs to make the icemaker level.

7) Replace the Top Panel and the Front Panel in their correct position, and secure the Front Panel by a screw.

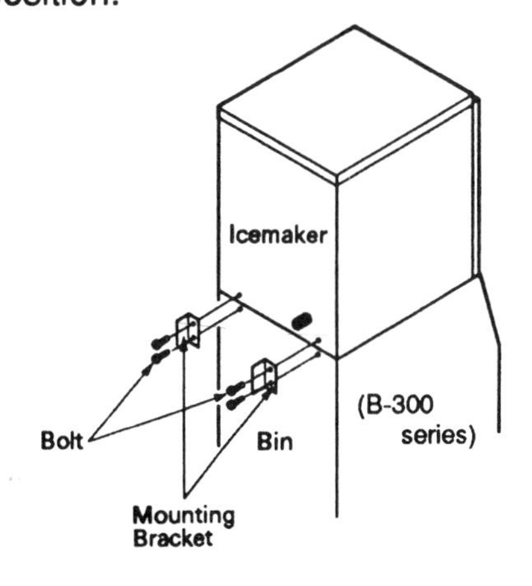

Fig. 2

5. ELECTRICAL CONNECTION

WARNING

1. Electrical connection must be made in accordance with the instructions on a "WARNING" tag, provided with the pig tail leads in the Junction Box.

2. This icemaker requires a ground that meets the national and local electrical code requirements. To prevent possible electrical shock to individuals or extensive damage to the equipment, install a proper ground wire to the icemaker.

* The white lead must be connected to the neutral conductor of the power source. Miswiring results in severe damage to the icemaker. See Fig. 3.

* This icemaker must have a separate power supply or receptacle of proper capacity. See the Nameplate.

* The opening for the power supply connection is 7/8" DIA to fit a 3/4" trade size conduit.

* Usually an electrical permit and services of a licensed electrician are required.

WARNING

ELECTRICAL CONNECTION

The white lead must be connected to the neutral conductor of the power source. Miswiring results in severe damage to the icemaker. (See below Fig.)

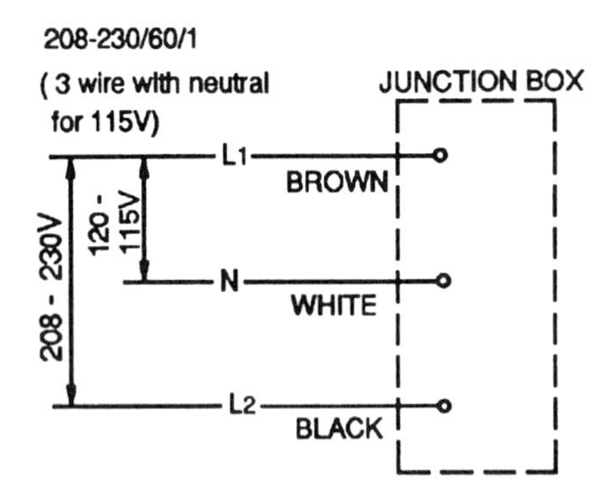

Fig. 3

6. INSTALLATION OF REMOTE CONDENSER UNIT

[a] CHECKS BEFORE INSTALLATION

1) Unpack and remove shipping carton, tape(s) and packing.

2) Check that the refrigerant lines do not rub or touch lines or other surfaces, and that the fan blade turns freely.

[b] LOCATION

The condenser unit must be positioned in a permanent site under the following guidelines.

* A firm and flat site.

* A dry and well ventilated area with 24" clearance on both front and rear for ease of maintenance and service should they be required.

* Normal condenser ambient temperature: -20°F to +122°F. Temperatures not within this operating range may affect the production capacity of the icemaker.

* The normal refrigerant line length is 66 ft. Should an installation require a longer line length, please call 1-800-233-1940 for recommendations.

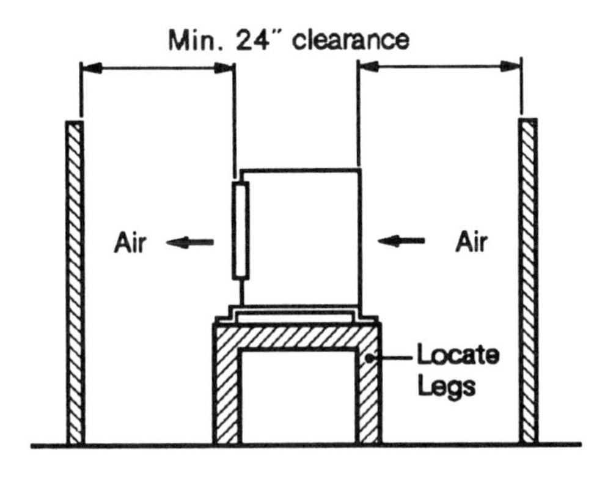

Fig. 4

Note: If the recommended guidelines of the installation are exceeded, the icemaker performance may be reduced.

[c] SET UP

1) Secure the Legs to the condenser unit with eight M8 x 16 mm Hexagon Bolts and M8 Nuts as shown in the illustration. See Fig. 5.

   Note: Locate the Legs symmetrically.

2) The Legs have eight mounting holes. Secure the Legs with eight bolts (not included).

3) Install enough length of two copper tubings provided with Aeroquip couplings between the icemaker and the condenser unit. The two copper tubings should be insulated separately. See Fig. 6.

   * Precharged tubing kits, available as optional equipment from HOSHIZAKI AMERICA are recommended.

4) Line sets fabricated in the field should be evacuated through the charging ports on the Aeroquip couplings and charged with R-502 refrigerant vapor to a pressure of 15 - 30 PSIG.

   Note: Factory fabricated tubing kits are precharged and do not need to be evacuated.

5) Remove the plastic caps protecting the couplings. Attach the two refrigerant lines to the male couplings on the icemaker and the remote condenser unit. Each refrigerant line must be connected as follows:

   Icemaker discharge refrigerant line - 3/8" OD tubing to "DIS" of condenser unit

   Icemaker liquid refrigerant line - 1/4" OD tubing to "LIQ" of condenser unit

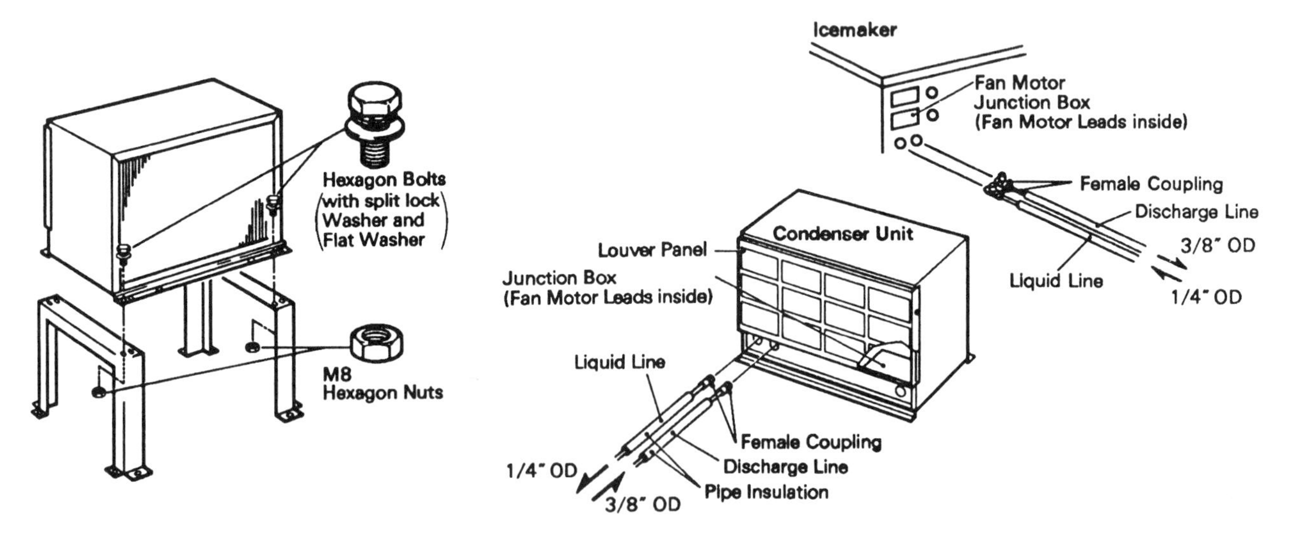

Fig. 5

Fig. 6

[d] ELECTRICAL CONNECTION

WARNING

This remote condenser unit requires a ground that meets the national and local electrical code requirements. To prevent possible electrical shock to individuals or extensive damage to equipment, install a proper ground wire to the condenser unit.

* This condenser unit must be connected to the Fan Motor Junction Box on the icemaker.

* The opening for the power supply connection is 7/8" DIA to fit a 1/2" trade size conduit.

* Usually an electrical permit and services of a licensed electrician are required.

1) Remove the Louver Panel.

2) Remove the Junction Box Cover.

3) Connect the Fan Motor leads in the Junction Box of the remote condenser unit to the Fan Motor leads in the Junction Box of the HOSHIZAKI remote air-cooled icemaker.

4) Install a ground wire from the icemaker to the remote condenser unit.

5) Replace the Junction Box Cover and the Louver Panel in their correct position.

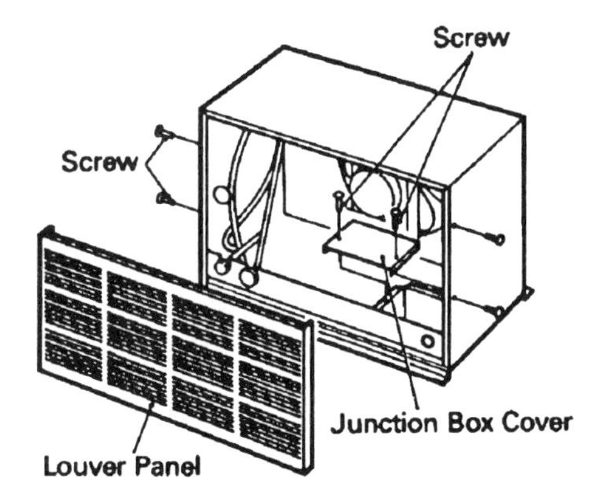

Fig. 7

[e] STACKING CONDENSER UNIT

1) Secure the lower condenser unit to the Legs with eight bolts (not included).

2) Attach the upper condenser unit on the top of the lower.

3) Secure the upper condenser unit with the four screws provided.

4) Install refrigerant lines, and make electrical connection for each Fan Motor as shown in Items [c] and [d].

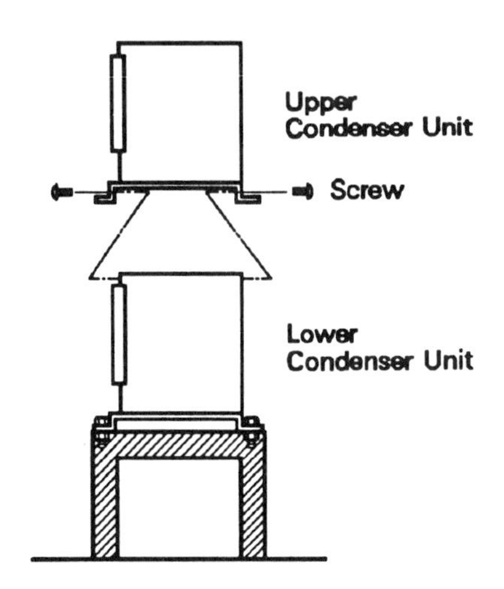

Fig. 8

7. WATER SUPPLY AND DRAIN CONNECTIONS - See Fig. 9

* External filters, strainers or softeners may be required depending on the water quality.

* Water supply inlet is 1/2" female pipe thread (FPT).

  Note: On water-cooled model, two water supply inlets are provided. One is for the ice making water inlet, and the other is for the cooling water inlet.

* A water supply line shut-off valve and drain valve should be installed. A minimum of 3/8" OD copper tubing is recommended for the water supply lines.

* Water supply pressure should be a minimum of 7 PSIG and a maximum of 113 PSIG. If the pressure exceeds 113 PSIG, the use of a pressure reducing valve is required.

* Drain outlet for icemaking is 3/4" FPT. The icemaker drain and the condenser drain piping connections must be made separately from the bin drain.

  Note: On water-cooled model, a 1/2" FPT is provided for the Cooling water outlet.

* The drains must have 1/4" fall per foot on horizontal runs to get a good flow.

* A plumbing permit and services of a licensed plumber may be required in some areas .

* A back flow preventer may be required in some areas.

F-1000MAB
F-1000MRB

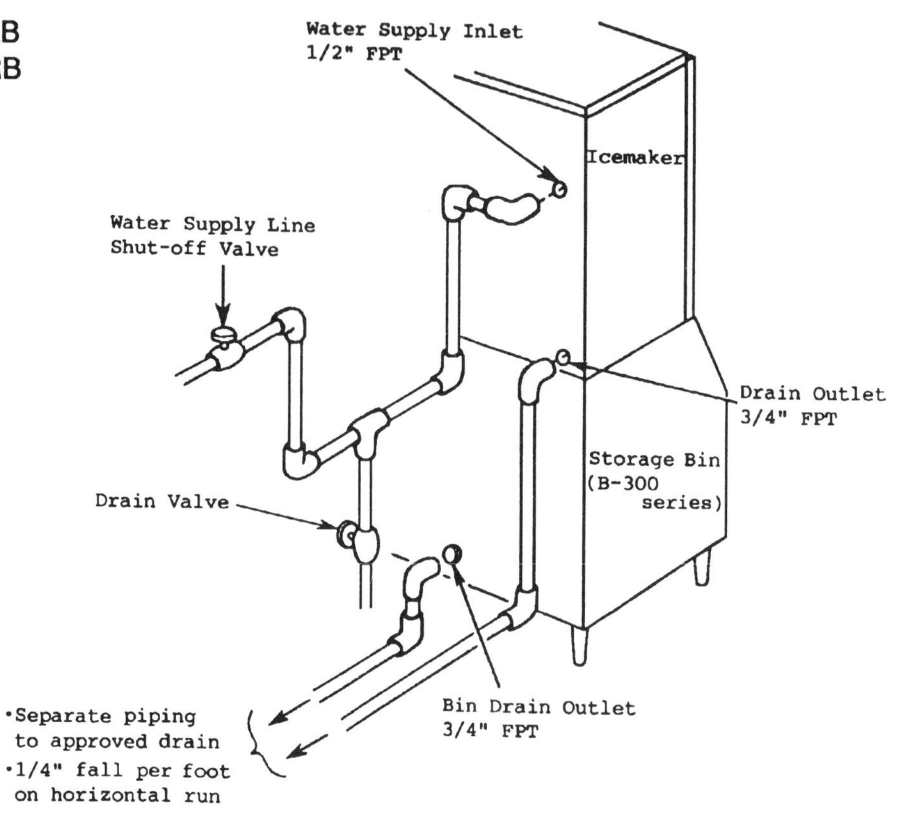

F-1000MWB

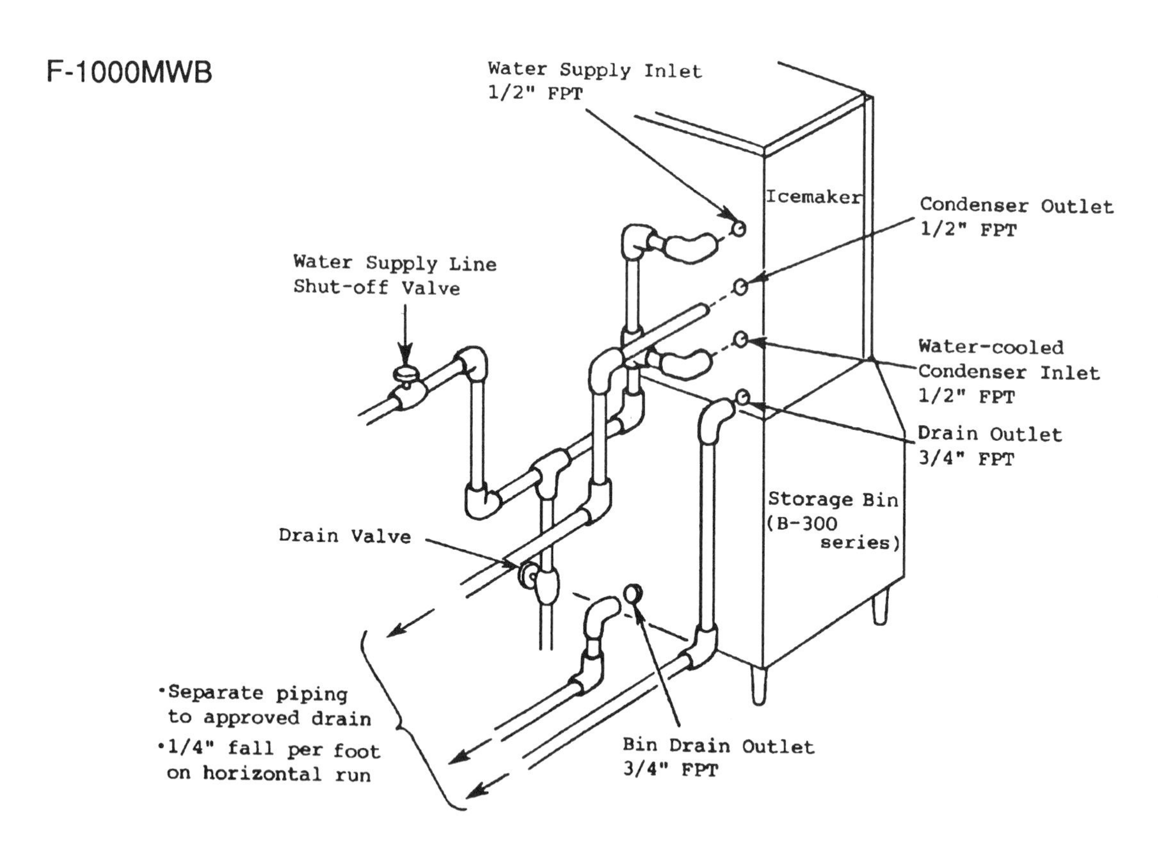

Fig. 9

8. FINAL CHECK LIST

1) Is the icemaker level?

2) Is the icemaker in a site where the ambient temperature is within +45°F to +100°F and the water temperature within +45°F to +90°F all year around?

3) Is there at least 6" clearance at rear and sides and 24" at top for maintenance or service?

4) Have all shipping carton, tape(s) and packing been removed from the icemaker?

5) Have all electrical and piping connections been made?

6) Has the power supply voltage been checked or tested against the nameplate rating? And has a proper ground been installed in the icemaker?

7) Are the Water Supply Line Shut-off Valve and Drain Valve installed? Has the water supply pressure been checked to ensure a minimum of 7 PSIG and a maximum of 113 PSIG?

   Note: The icemaker may stop running when the water supply is OFF, or if the pressure is below 7 PSIG. When the proper water pressure is reached, the icemaker automatically starts running again.

8) Have the compressor hold-down bolts and refrigerant lines been checked against vibration and possible failure?

9) Has the Bin Control Switch been checked for correct operation? Move the Activator located on the top of the Chute. The Compressor should stop in 5 seconds, and the Gear Motor in 5 seconds.

10) Has the end user been given the instruction manual, and instructed on how to operate the icemaker and the importance of the recommended periodic maintenance?

11) Has the end user been given the name and telephone number of an authorized service agent?

12) Has the warranty tag been filled out and forwarded to the factory for warranty registration?

9. START UP

WARNING

1. All parts are factory-adjusted. Improper adjustments may result in failure.
2. If the unit is turned off, wait for at least 3 minutes before restarting the icemaker to prevent damage to the Compressor.

1) Clean the Storage Bin. (See "IV. 2. MAINTENANCE.")

2) Open the Water Supply Line Shut-off Valve.

3) Move the Flush Switch on the Control Box to the "ICE" position.

4) Turn on the Power Switch on the Control Box.

5) Replace the Side, Top and Front Panels in their correct position.

6) Turn on the power supply.

10. PREPARING THE ICEMAKER FOR LONG STORAGE - See Fig. 10

WARNING

When shutting off the icemaker for an extended time, drain out all water from the water line and remove the ice from the Storage Bin. The Storage Bin should be cleaned and dried. Drain the icemaker to prevent damage to the water supply line at sub-freezing temperatures, using air or carbon dioxide. Shut off the icemaker until the proper ambient temperature is resumed.

* Air-cooled and Remote Air-cooled Models

1) Run the icemaker with the Water Supply Line Shut-off Valve closed.

2) Open the Drain Valve and blow out the water inlet line by using air pressure.

3) Turn off the power supply.

4) Remove the Front Panel.

5) Move the Flush Switch on the Control Box to the "FLUSH" position.

6) Turn on the power supply, and then drain out all water from the water line.

7) Turn off the power supply.

8) Turn off the Power Switch on the Control Box.

9) Replace the Front Panel in its correct position.

10) Close the Drain Valve.

11) Remove all ice from the Storage Bin, and clean the bin.

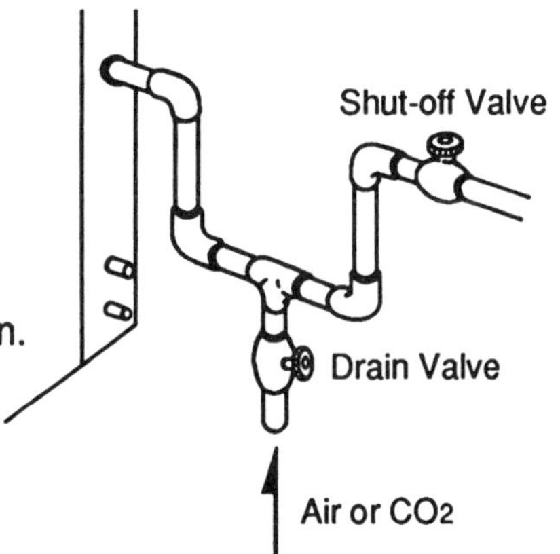

Fig. 10

* Water-cooled Model

1) Turn off the power supply and wait for 3 minutes.

2) Turn on the power supply and wait for 20 seconds.

3) Close the Water Supply Line Shut-off Valve.

4) Open the Drain Valve and quickly blow the water supply line from the Drain Valve to drain water in the Condenser.

5) Follow the above steps 3) through 11) in "* Air-cooled and Remote Air-cooled Models."

IV. MAINTENANCE AND CLEANING INSTRUCTIONS

1. CLEANING INSTRUCTIONS

WARNING

1. HOSHIZAKI recommends cleaning this unit at least once a year. More frequent cleaning, however, may be required in some existing water conditions.

2. To prevent injury to individuals and damage to the icemaker, do not use ammonia type cleaners.

3. Always wear liquid-proof gloves for safe handling of the cleaning and sanitizing solution. This will prevent irritation in case the solution contacts with skin.

<STEP 1>

Dilute the solution with water as follows.

Cleaning solution: 4.8 fl. oz. of recommended cleaner ("LIME-A-WAY," Economics Laboratory, Inc.) with 0.8 gallon of water. This is a minimum amount. Make more solution, if necessary.

Sanitizing solution: 2.5 fl. oz. of 5.25 % sodium hypochlorite solution with 5 gallons of water.

IMPORTANT

For safety and maximum effectiveness, use the solution immediately after dilution.

<STEP 2>

Use the cleaning solution to remove lime deposits in the water system.

1) Turn off the power supply.

2) Close the Water Supply Line Shut-off Valve.

3) Remove all ice from the Storage Bin.

4) Remove the Front Panel and the Top Panel.

5) Move the Flush Switch to the "FLUSH" position.

6) Turn on the power supply and drain out all water from the water line.

7) Turn off the power supply.

8) Remove the Water Supply Valve by releasing the Fitting Nut. Do not lose the Packing.

9) Remove the Cover of the Reservoir.

10) Fill the Reservoir with the cleaning solution.

11) Replace the Cover of the Reservoir and the Water Supply Valve in their correct position.

Note: This unit is designed to start operating when the Reservoir is filled with water.

12) Move the Flush Switch to the "ICE" position.

13) Replace the Top Panel and the Front Panel in their correct position.

14) Allow the icemaker to set for about 10 minutes before the operation. Then, turn on the power supply, and make ice using the solution until the icemaker stops icemaking.

15) Remove the Front Panel.

16) Move the Flush Switch to the "FLUSH" position to drain the cleaning solution.

17) Move the Flush Switch to the "ICE" position.

18) Replace the Front Panel in its correct position.

19) Open the Water Supply Line Shut-off Valve, and supply water to the Reservoir.

20) Turn off the power supply when the Gear Motor starts.

21) Drain out all water from the water line. See 4) through 7).

<STEP 3>

Use 3/4 gallon of the sanitizing solution to sanitize the icemaker.

1) Close the Water Supply Line Shut-off Valve.

2) Remove the Water Supply Valve by releasing the Fitting Nut.

3) Remove the Cover of the Reservoir.

4) Fill the Reservoir with the sanitizing solution.

5) Replace the Cover of the Reservoir and the Water Supply Valve in their correct position.

6) Move the Flush Switch to the "ICE" position.

7) Replace the Top Panel and the Front Panel in their correct position.

8) Allow the icemaker to set for about 10 minutes before the operation. Then, turn on the power supply, and make ice using the solution until the icemaker stops icemaking.

9) Remove the Front Panel.

10) Move the Flush Switch to the "FLUSH" position to drain the sanitizing solution.

11) Move the Flush Switch to the "ICE" position.

12) Replace the Front Panel in its correct position.

13) Open the Water Supply Line Shut-off Valve, and supply water to the Reservoir.

14) Turn off the power supply when the Gear Motor starts.

15) Drain out all water from the water line. See 4) through 7) in STEP 2.

16) Move the Flush Switch to the "ICE" position.

<STEP 4>

Use the sanitizing solution to sanitize removed parts.

1) Remove the Thumbscrew securing the Bin Control Switch on the Chute Assembly.

2) Remove the Band connecting the Chute Head with the Chute Assembly, and take out the Chute Assembly from the icemaker.

3) Remove the Gasket at the bottom of the Ice Chute and another at the Chute Head.

4) Remove the three Ties and the Insulation of the Chute.

5) Remove the six Wing Nuts and two Baffles.

IMPORTANT

When installing the Baffles, make sure that the bent surface (the one without the studs) faces the Activator so that the bent surface can guide the ice to the center of the Activator.

6) Remove the two Thumbscrews and the Plate, and then remove the Bin Control Assembly by sliding it slightly toward the Chute Head and lifting it off.

7) Disassemble the Bin Control Assembly by removing the two Snap Pins, Axle and Activator.

8) Remove the three Thumbscrews and the Chute Head.

9) Remove the Rubber O-ring and Nylon O-ring at the top of the Cylinder.

10) Soak or wipe the removed parts.

11) Rinse these parts thoroughly.

IMPORTANT

If the solution is left on these parts, they will rust.

12) Replace the removed parts and the panels.

13) Turn on the power supply and run the icemaker.

14) Turn off the power supply after 30 minutes.

15) Pour warm water into the Storage Bin to melt all ice, and then clean the Bin Liner with the solution.

16) Flush out any solution from the Storage Bin.

17) Turn on the power supply and start the automatic icemaking process.

IMPORTANT

1. After cleaning, do not use ice made from the sanitizing solution. Be careful not to leave any solution in the Storage Bin.
2. Follow carefully any instructions provided with the bottles of cleaning or sanitizing solution.
3. Never run the icemaker when the Reservoir is empty.

2. MAINTENANCE

IMPORTANT

This icemaker must be maintained individually, referring to the instruction manual and labels provided with the icemaker.

1) Stainless Steel Exterior

To prevent corrosion, wipe the exterior occasionally with a clean and soft cloth. Use a damp cloth containing a neutral cleaner to wipe off oil or dirt build up.

2) Storage Bin and Scoop

* Wash your hands before removing ice. Use the plastic scoop provided (Accessory).

* The Storage Bin is for ice use only. Do not store anything else in the bin.

* Keep the scoop clean. Clean using a neutral cleaner and rinse thoroughly.

* Clean the bin liner using a neutral cleaner. Rinse thoroughly after cleaning.

3) Air Filter (Air-cooled model only)

A plastic mesh air filter removes dirt or dust from the air, and keeps the Condenser from getting clogged. As the filter gets clogged, the icemaker's performance will be reduced. Check the filter at least twice a month. When clogged, use warm water and a neutral cleaner to wash the filter.

4) Condenser (Air-cooled model only)

Check the Condenser once a year, and clean if required by using a brush or vacuum cleaner. More frequent cleaning may be required depending on the location of the icemaker.

V. TECHNICAL INFORMATION

1. WATER CIRCUIT AND REFRIGERANT CIRCUIT

[a] F-1000MAB (Air-cooled)

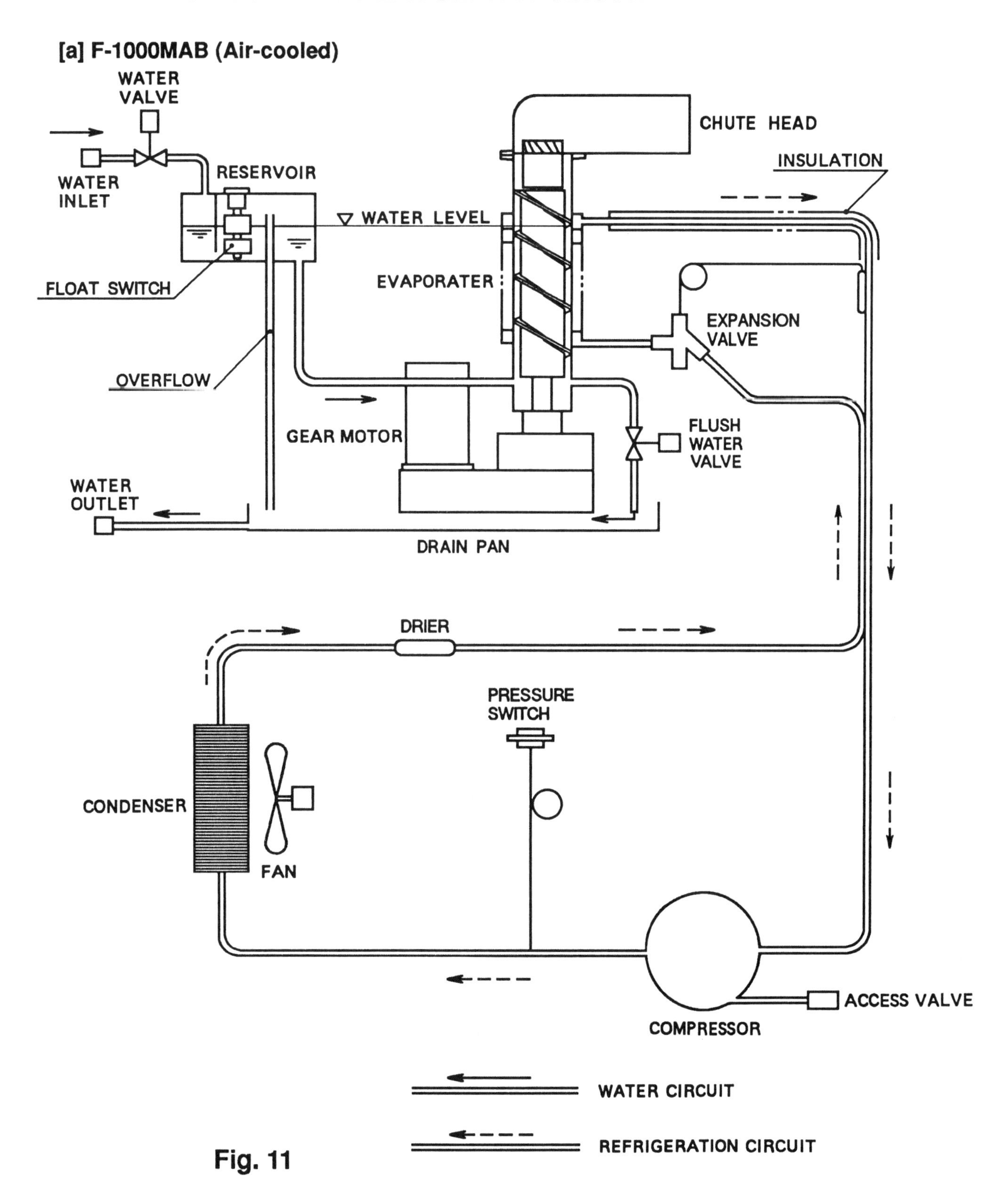

Fig. 11

[b] F-1000MWB (Water-cooled)

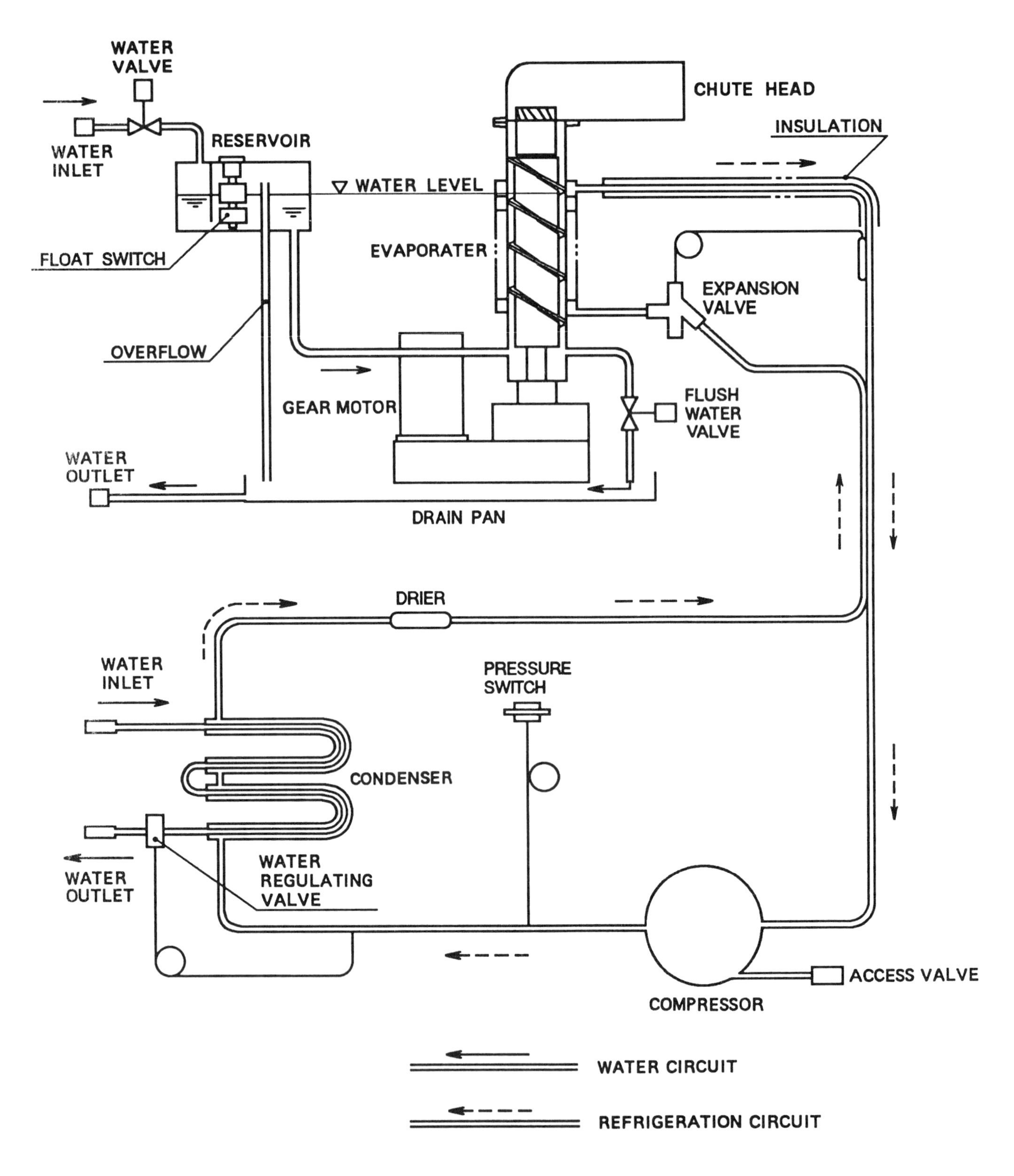

Fig. 12

[c] F-1000MRB (Remote Air-cooled)

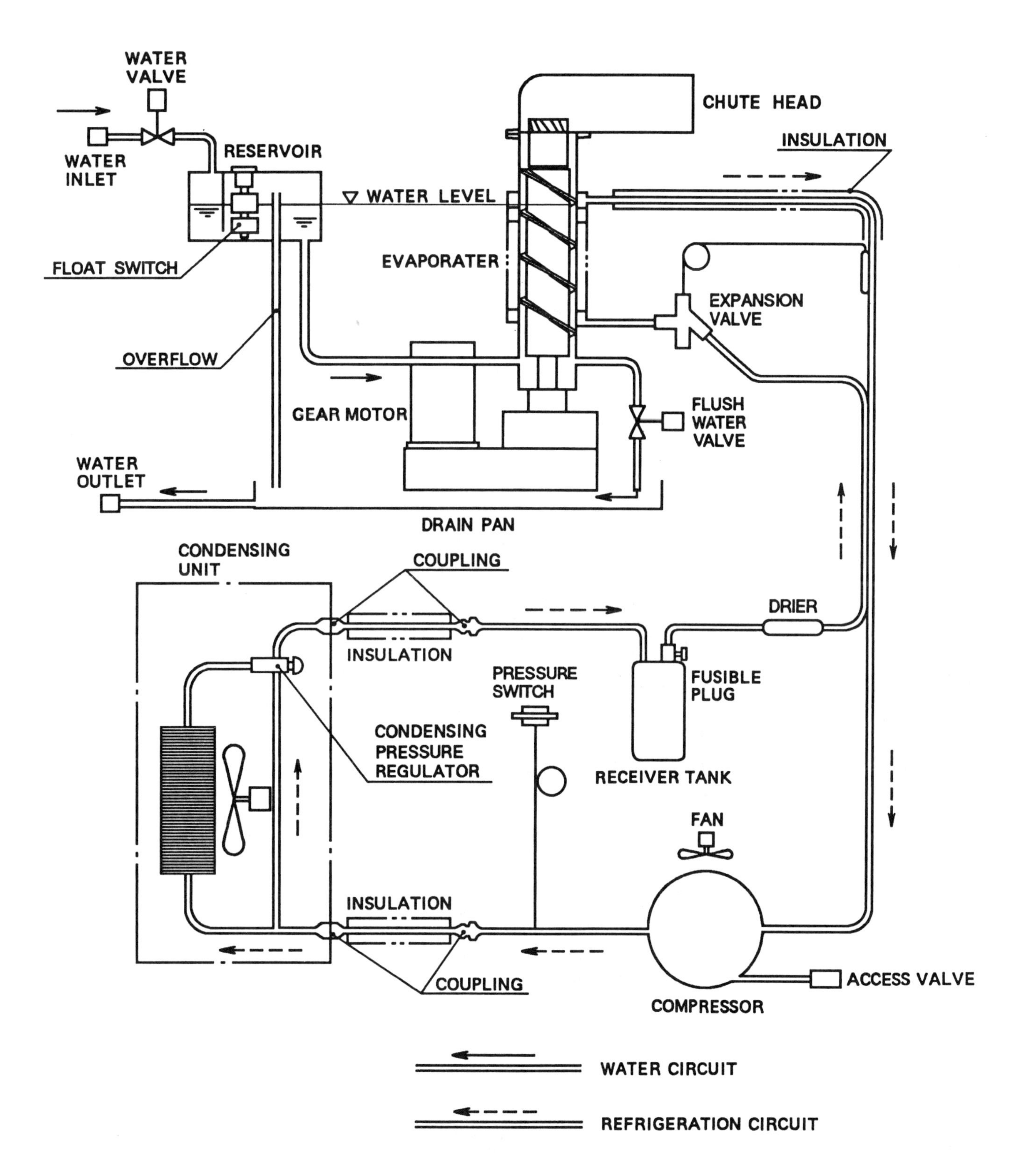

Fig. 13

2. WIRING DIAGRAM

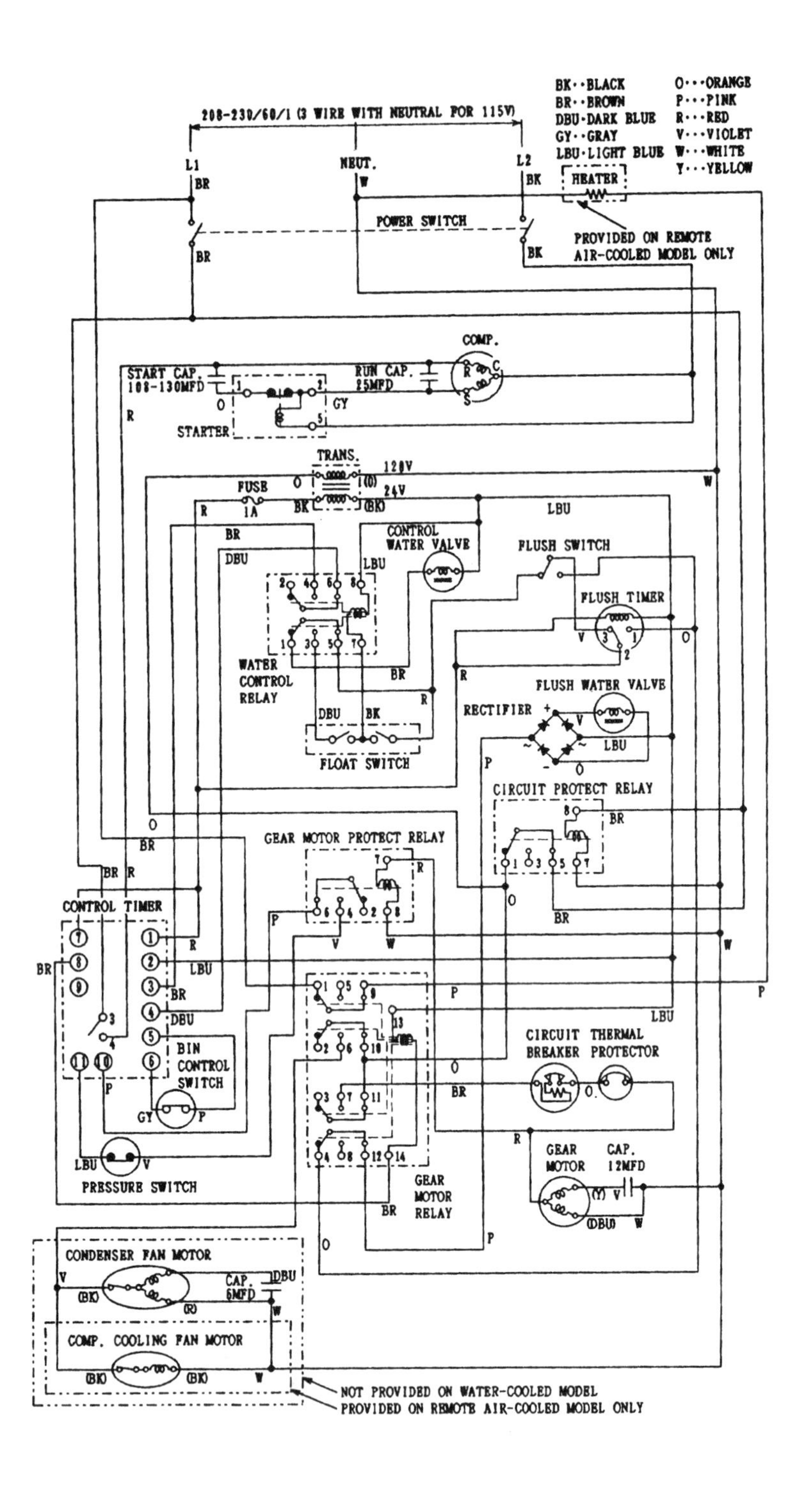

Fig. 14

3. SEQUENCE OF ELECTRICAL CIRCUIT

[a] When the Power Switch is moved to the "ON" position and the Flush Switch to the "ICE" position, water starts to be supplied to the Reservoir.

Note: When 208-230V supplied to the Circuit Protect Relay, it operates and protects the circuit from miswiring. If the power supply is properly connected, the contact of the Circuit Protect Relay does not move even when the coil is energized.

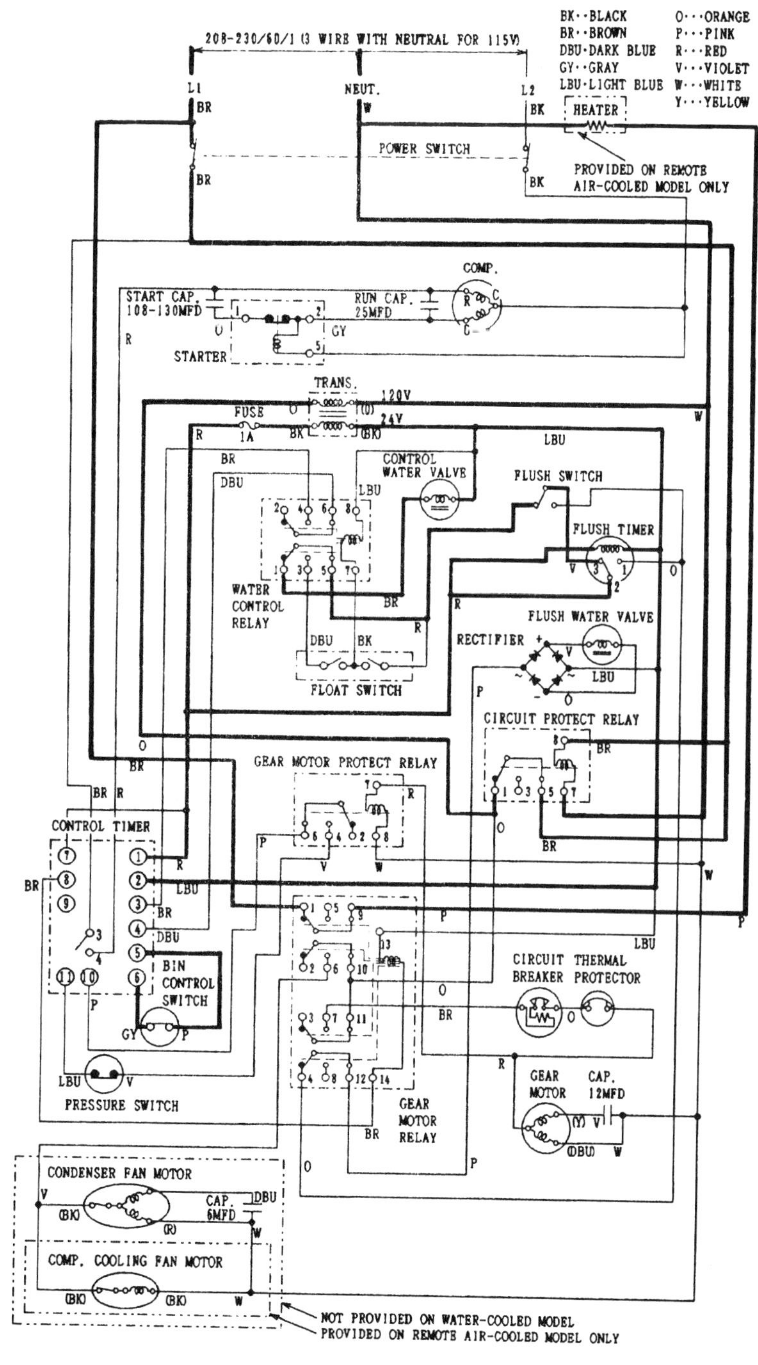

Fig. 15

[b] When the Reservoir is filled up, the Gear Motor starts immediately.

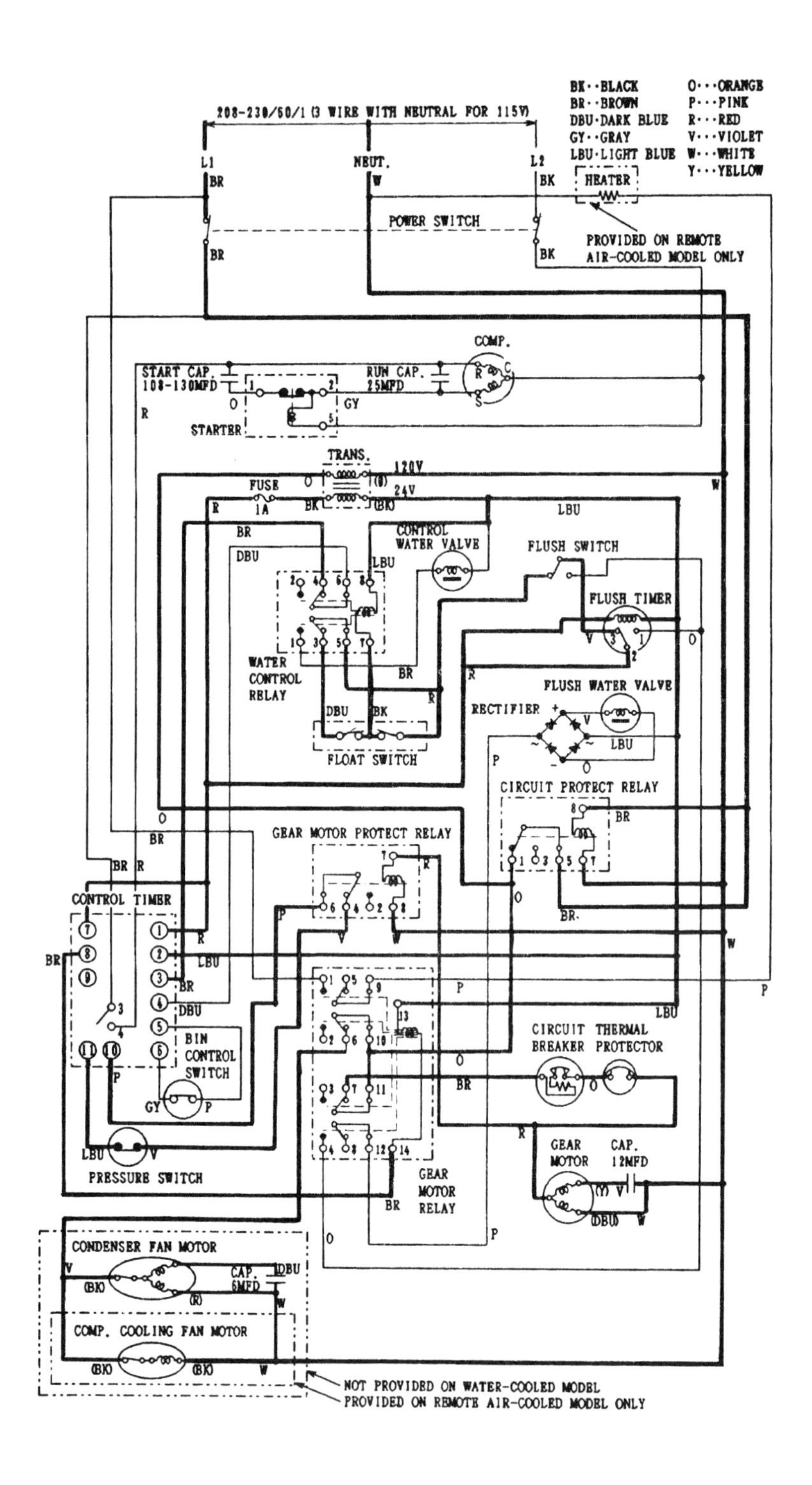

Fig. 16

[c] The Compressor starts about 60 sec. after the Gear Motor starts.

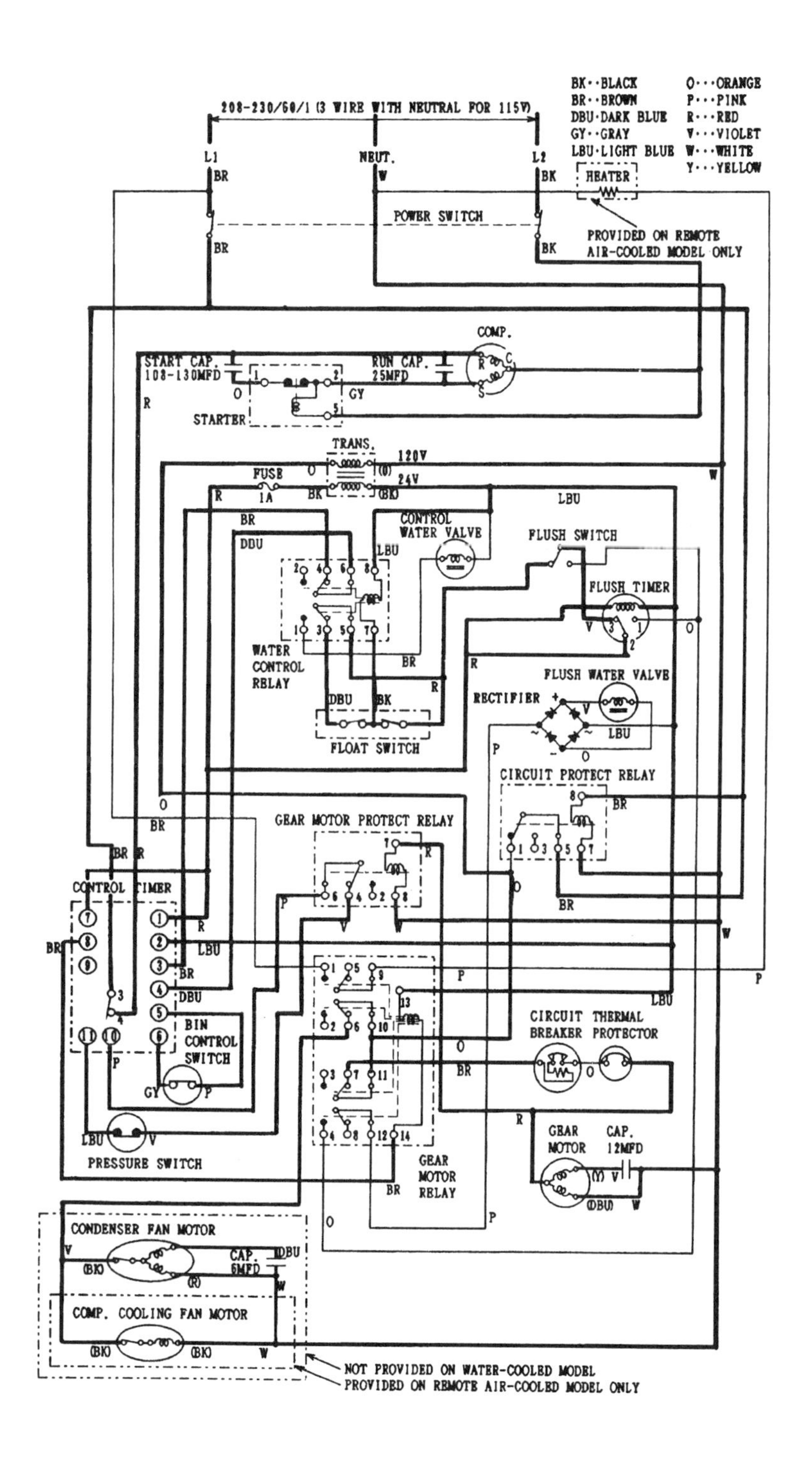

Fig. 17

[d] The Bin Control operates, and about 6 sec. later the Compressor and the Gear Motor stop simultaneously.

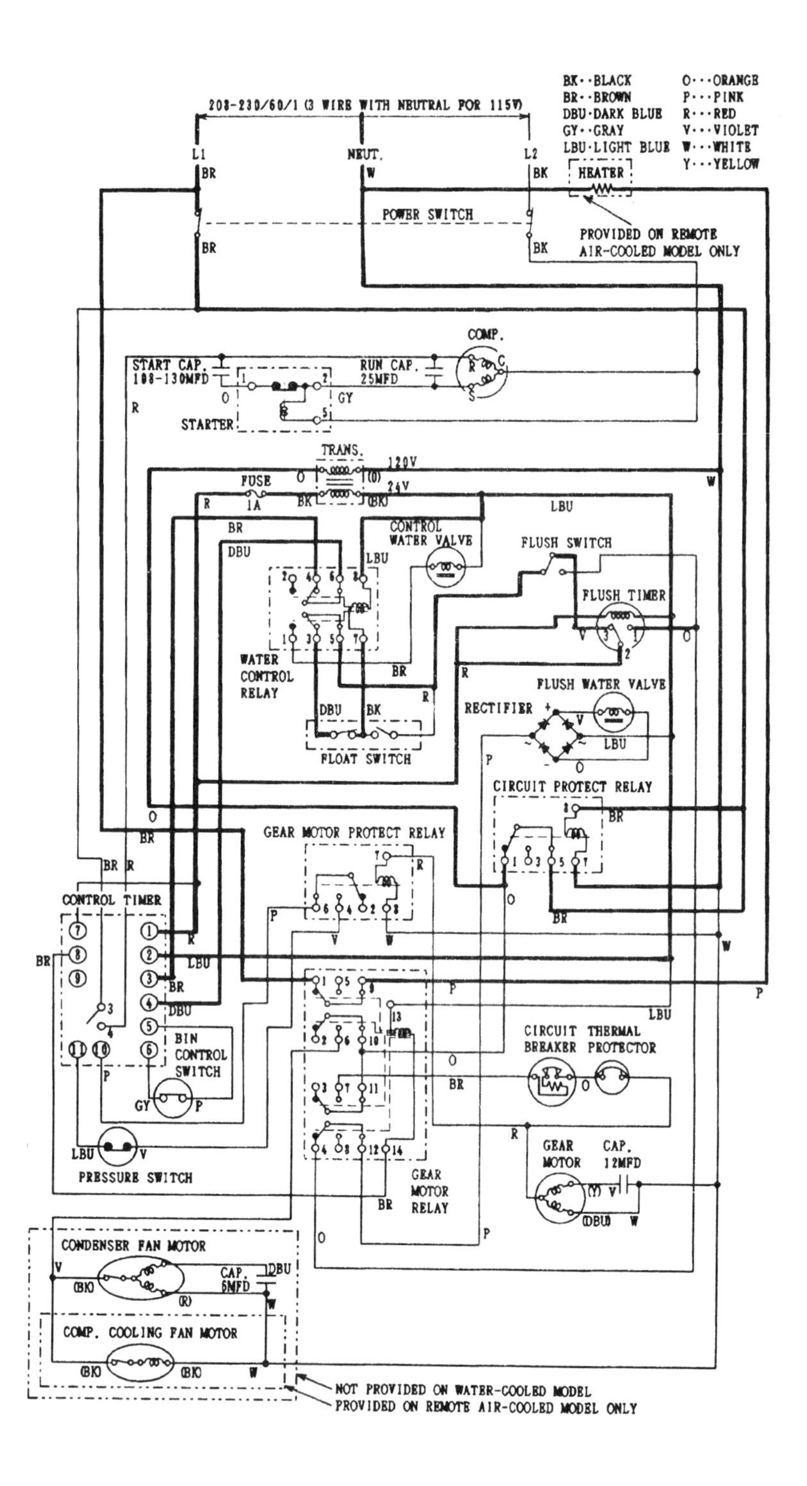

Fig. 18

[e] Low Water (Air-cooled and Remote Air-cooled Models)

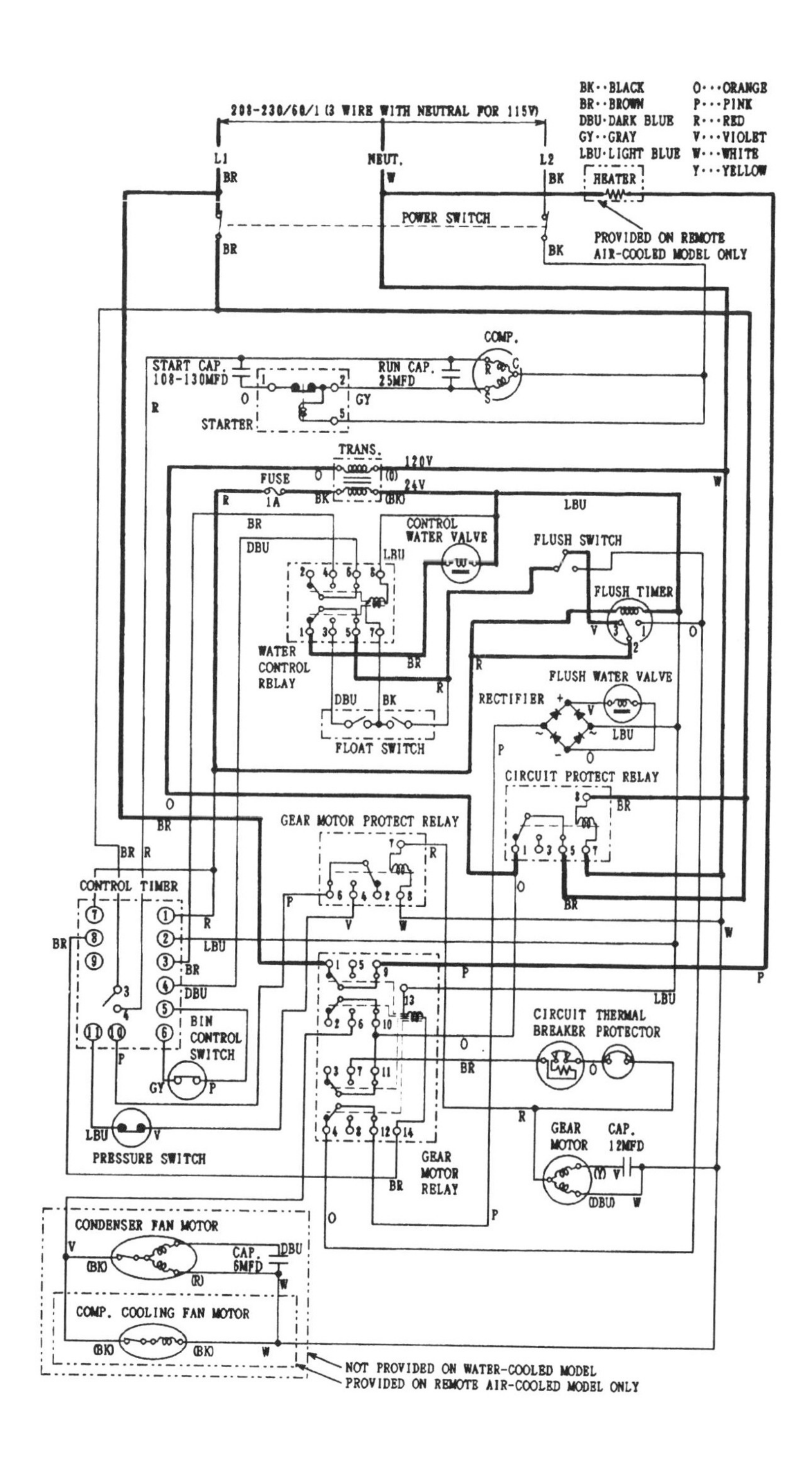

Fig. 19

[f] Low Water (Water-cooled Model)
The Compressor operates intermittently.

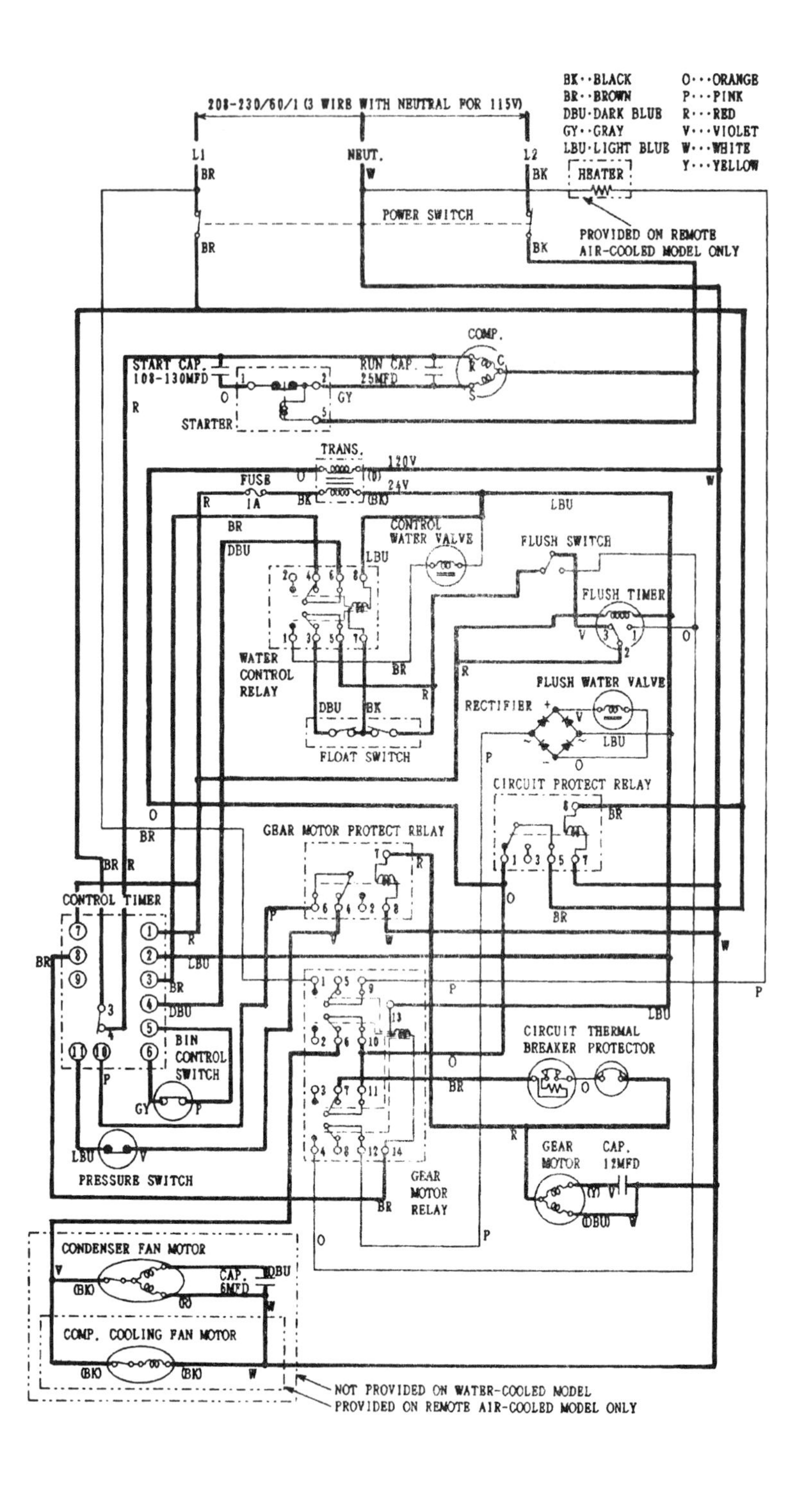

Fig. 20

[g] **When the Flush Timer operates (for 15 min. every 12 hours) or the Flush Switch is moved to the "FLUSH" position, the Flush Water Valve opens and flushes out water in the Reservoir and the Evaporator.**

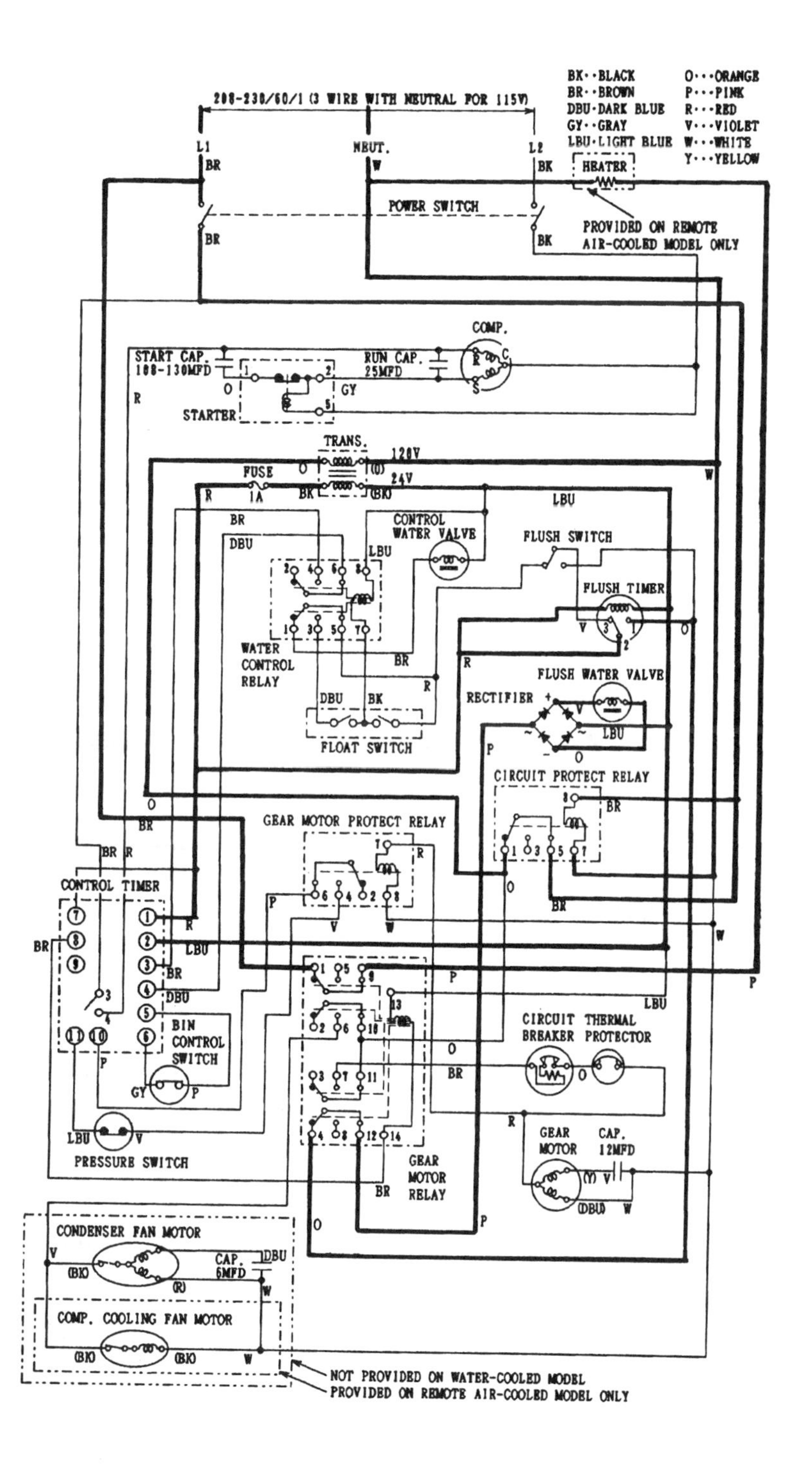

Fig. 21

4. TIMING CHART

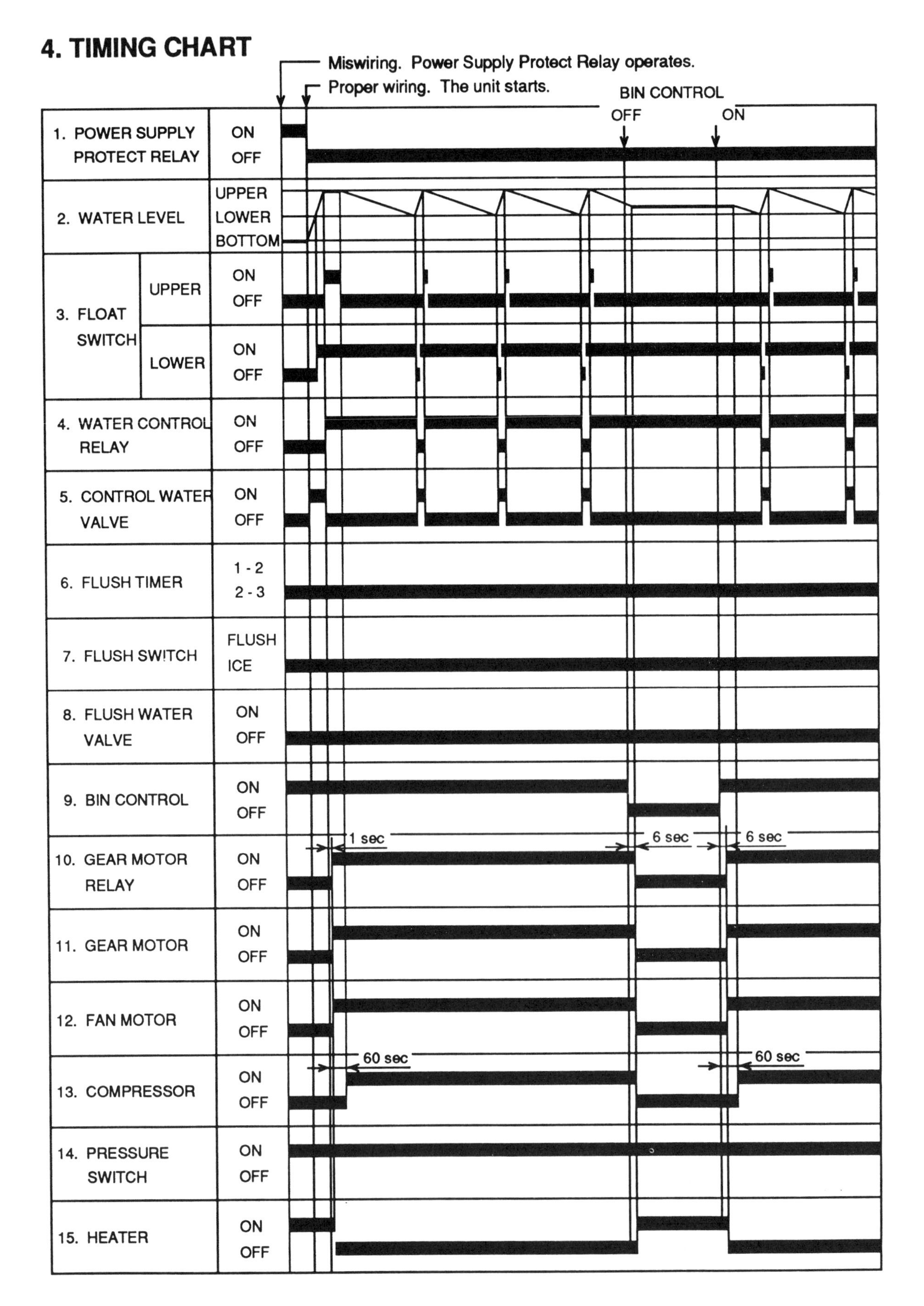

U2AC1949101

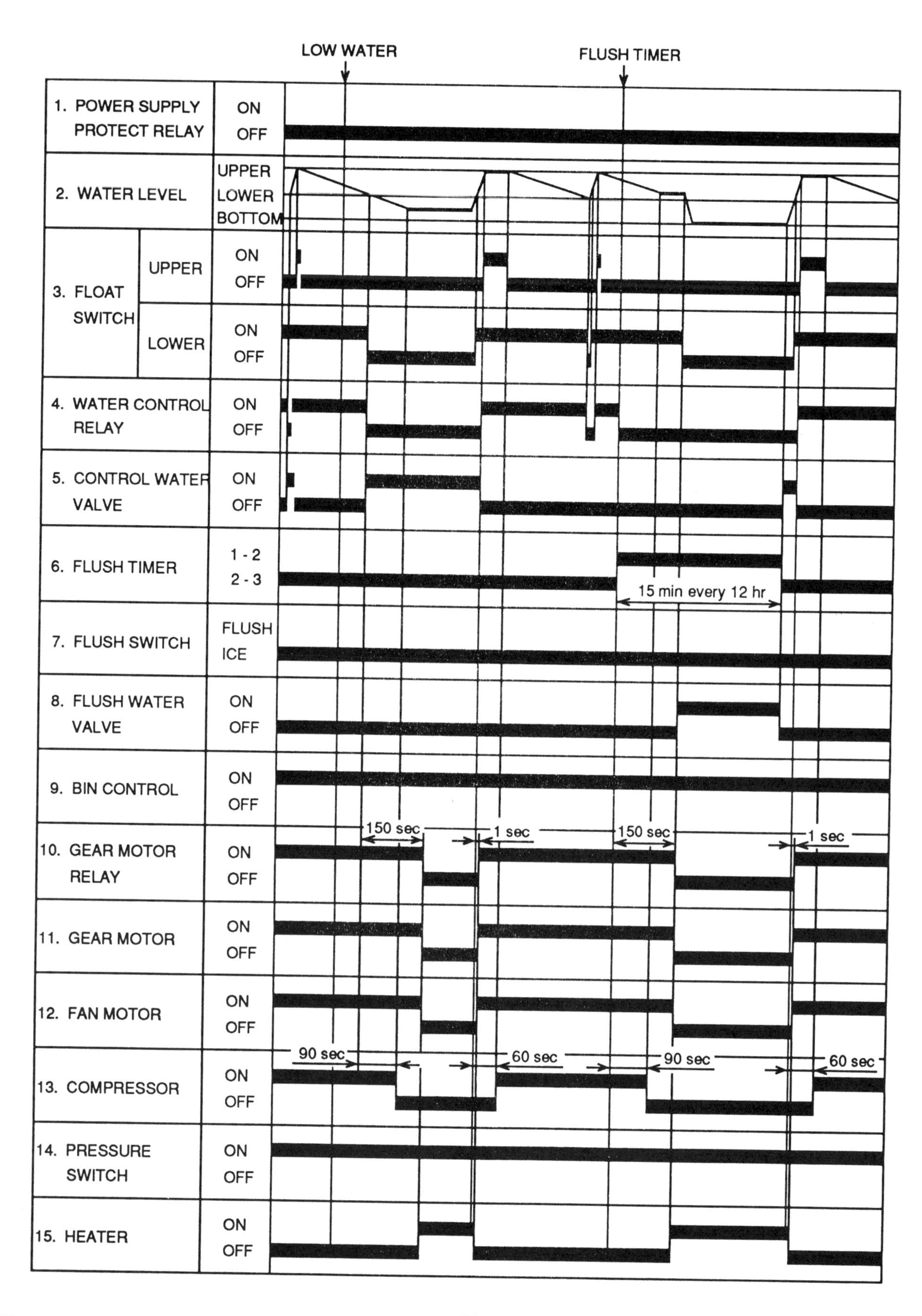

U2AC1949101

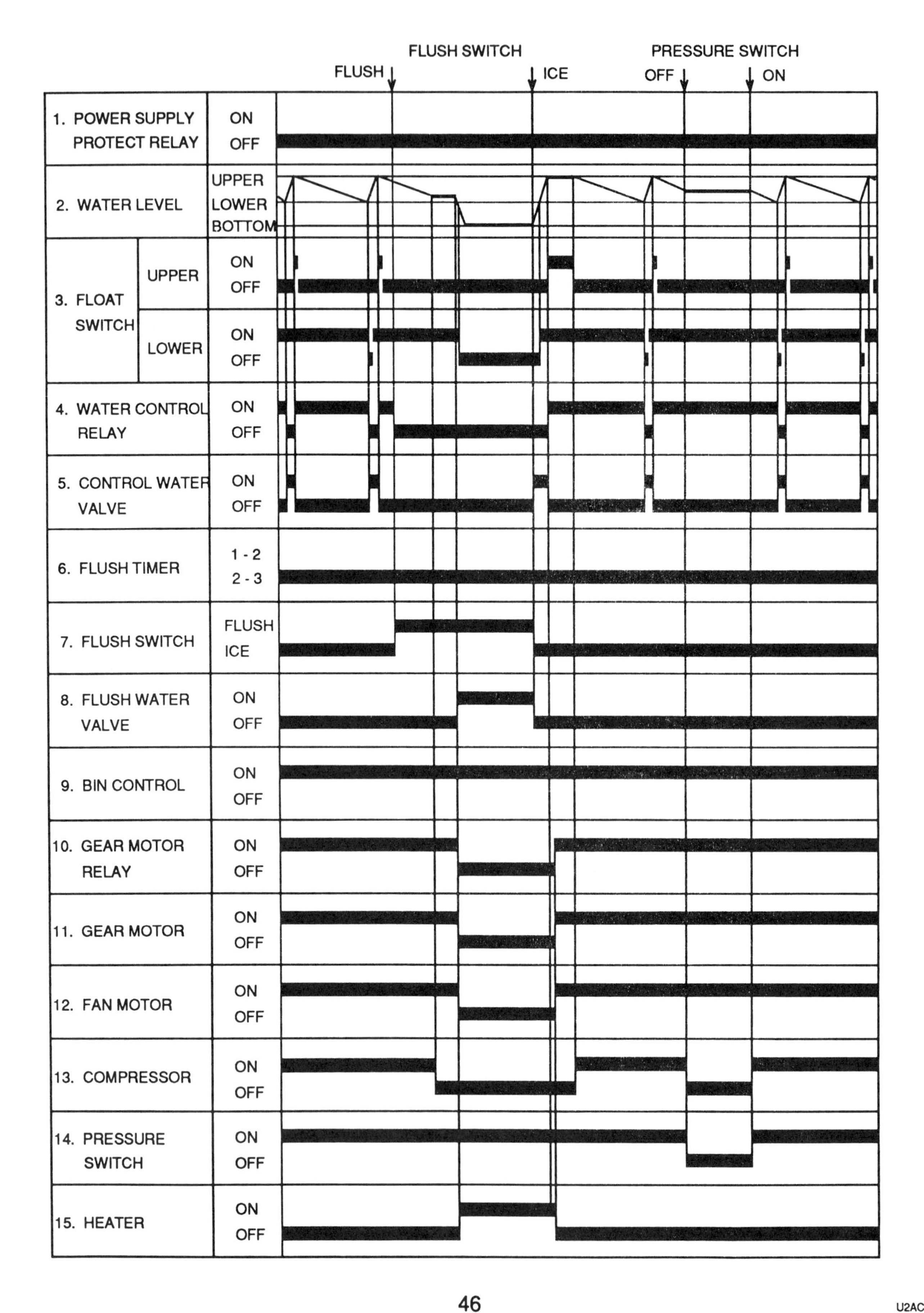
FLUSH SWITCH
FLUSH
ICE
PRESSURE SWITCH
OFF
ON
1. POWER SUPPLY PROTECT RELAY
ON
OFF
2. WATER LEVEL
UPPER
LOWER
BOTTOM
3. FLOAT SWITCH
UPPER
ON
OFF
LOWER
ON
OFF
4. WATER CONTROL RELAY
ON
OFF
5. CONTROL WATER VALVE
ON
OFF
6. FLUSH TIMER
1 - 2
2 - 3
7. FLUSH SWITCH
FLUSH
ICE
8. FLUSH WATER VALVE
ON
OFF
9. BIN CONTROL
ON
OFF
10. GEAR MOTOR RELAY
ON
OFF
11. GEAR MOTOR
ON
OFF
12. FAN MOTOR
ON
OFF
13. COMPRESSOR
ON
OFF
14. PRESSURE SWITCH
ON
OFF
15. HEATER
ON
OFF

5. PERFORMANCE DATA

[a] F-1000MAB

	Ambient Temp. (°F)	Water Temp. (°F) 50	70	90
APPROXIMATE ICE PRODUCTION PER 24 HR. lbs./day (kg/day)	70	*1020(463)	920(417)	840(381)
	80	950(431)	865(392)	790(358)
	90	880(399)	* 800(363)	740(336)
	100	810(367)	730(331)	680(308)
APPROXIMATE ELECTRIC CONSUMPTION watts (115/230V)	70	*1200	1200	1200
	80	1220	1220	1220
	90	1250	*1250	1250
	100	1280	1280	1280
APPROXIMATE WATER CONSUMPTION PER 24 HR. gal./day (ℓ/day)	70	122(463)	110(417)	101(381)
	80	114(431)	104(392)	95(358)
	90	105(399)	96(363)	89(336)
	100	97(367)	87(331)	81(308)
EVAPORATOR OUTLET TEMP. °F (°C)	70	21(-5)	21(-5)	21(-5)
	80	21(-5)	21(-5)	21(-5)
	90	25(-4)	21(-5)	21(-5)
	100	27(-3)	27(-3)	27(-3)
HEAD PRESSURE PSIG (kg/cm²G)	70	195(13.7)	200(14.0)	200(14.0)
	80	223(15.6)	223(15.6)	223(15.6)
	90	250(17.5)	250(17.5)	250(17.5)
	100	285(20.0)	285(20.0)	285(20.0)
SUCTION PRESSURE PSIG (kg/cm²G)	70	29(2.0)	30(2.1)	30(2.1)
	80	33(2.3)	32(2.2)	32(2.2)
	90	34(2.4)	34(2.4)	36(2.5)
	100	37(2.6)	37(2.6)	37(2.6)
TOTAL HEAT OF REJECTION	8480 BTU/h (AT 90°F/ WT 70°F)			

Note: The data without *marks should be used for reference.

F-1000MAB

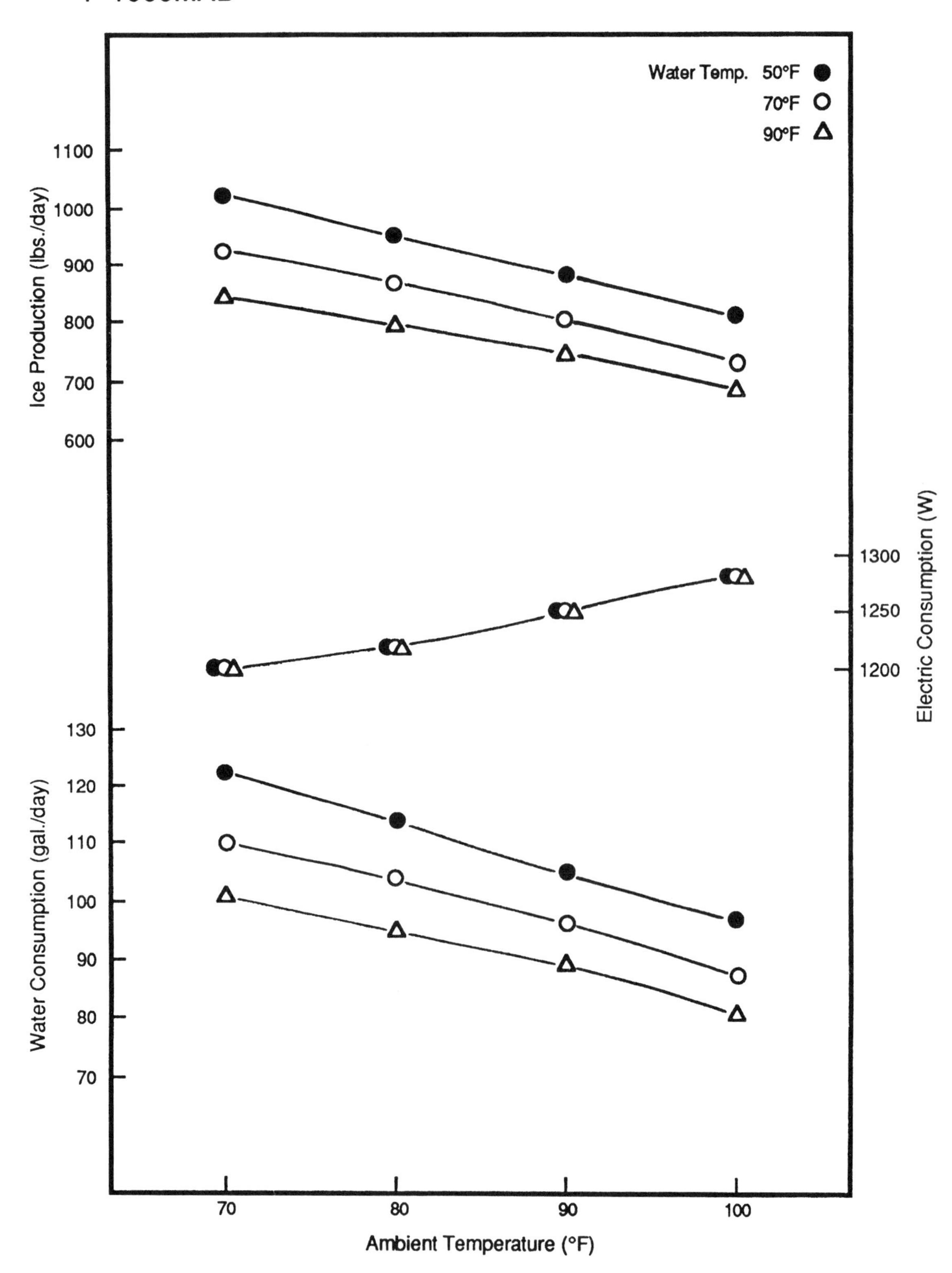

Fig. 22

F-1000MAB

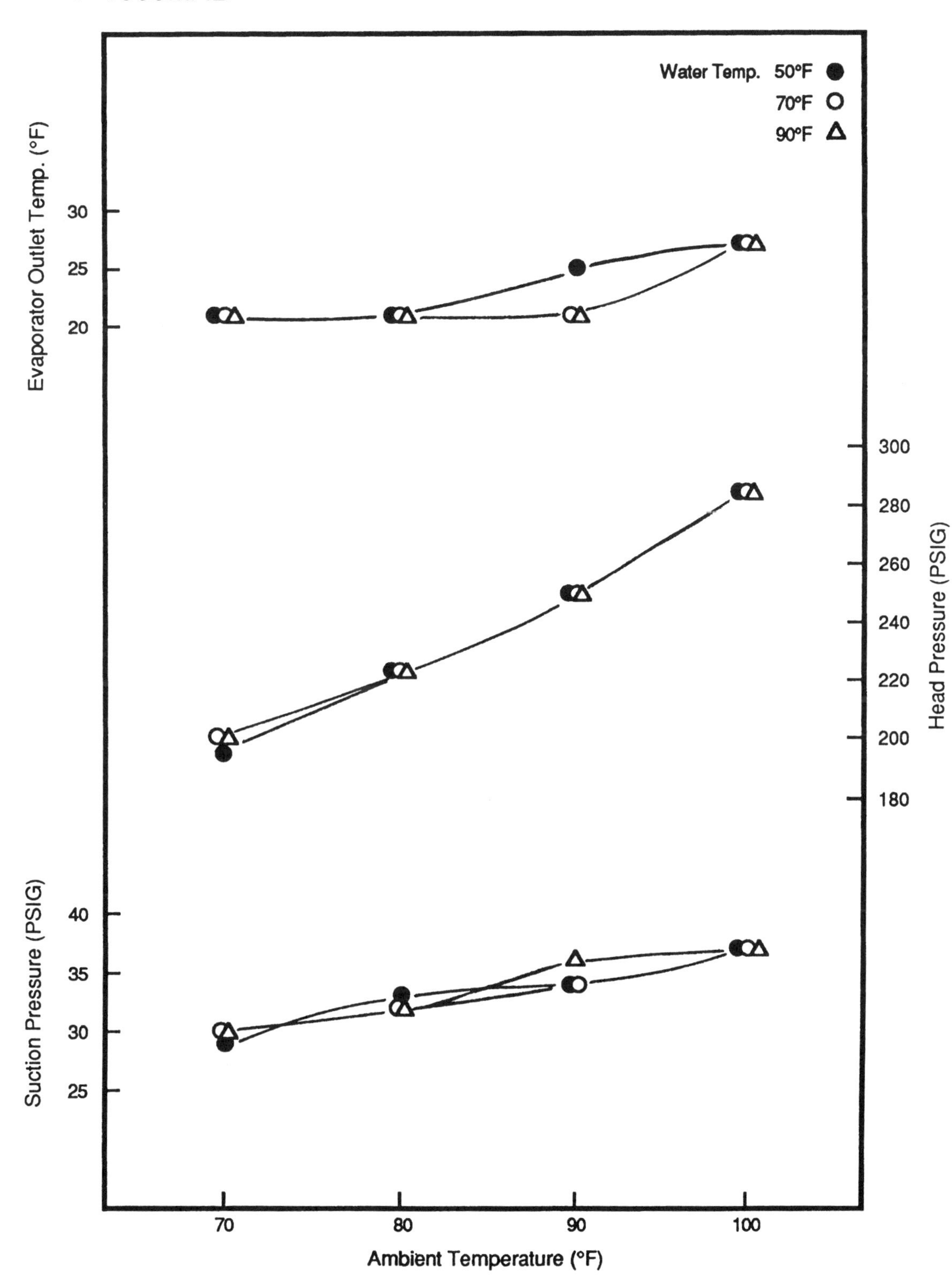

Fig. 23

[b] F-1000MWB

	Ambient Temp. (°F)	Water Temp. (°F) 50	70	90
APPROXIMATE ICE PRODUCTION PER 24 HR. lbs./day (kg/day)	70	*1010(458)	890(404)	750(340)
	80	1005(456)	880(399)	745(338)
	90	995(451)	*865(392)	740(336)
	100	980(445)	845(383)	730(331)
APPROXIMATE ELECTRIC CONSUMPTION watts (230V)	70	*1060	1060	1100
	80	1060	1060	1100
	90	1060	*1060	1100
	100	1100	1100	1100
APPROXIMATE WATER CONSUMPTION PER 24 HR. gal./day (m³/day)	70	645(2.44)	900(3.41)	1790(6.78)
	80	655(2.48)	910(3.44)	1835(6.95)
	90	655(2.48)	915(3.46)	1950(7.38)
	100	685(2.59)	945(3.58)	2100(7.95)
EVAPORATOR OUTLET TEMP. °F (°C)	70	21(-6)	21(-6)	21(-5)
	80	21(-6)	21(-6)	21(-5)
	90	21(-6)	21(-6)	21(-5)
	100	21(-6)	21(-5)	21(-5)
HEAD PRESSURE PSIG (kg/cm²G)	70	215(15.1)	215(15.1)	225(15.8)
	80	215(15.1)	215(15.1)	225(15.8)
	90	215(15.1)	215(15.1)	225(15.8)
	100	215(15.1)	220(15.4)	230(16.1)
SUCTION PRESSURE PSIG (kg/cm²G)	70	30(2.1)	30(2.1)	32(2.2)
	80	30(2.1)	30(2.1)	32(2.2)
	90	30(2.1)	30(2.1)	34(2.4)
	100	32(2.2)	33(2.3)	34(2.4)

WATER FLOW FOR CONDENSER	84 gal/h (AT 100°F/ WT 90°F)
PRESSURE DROP OF COOLING WATER LINE	Less than 7 PSIG
HEAT OF REJECTION FROM CONDENSER	6700 BTU/h (AT 90°F/ WT 70°F)
HEAT OR REJECTION FROM COMPRESSOR	1260 BTU/h (AT 90°F/ WT 70°F)

Note: The data without *marks should be used for reference.

F-1000MWB

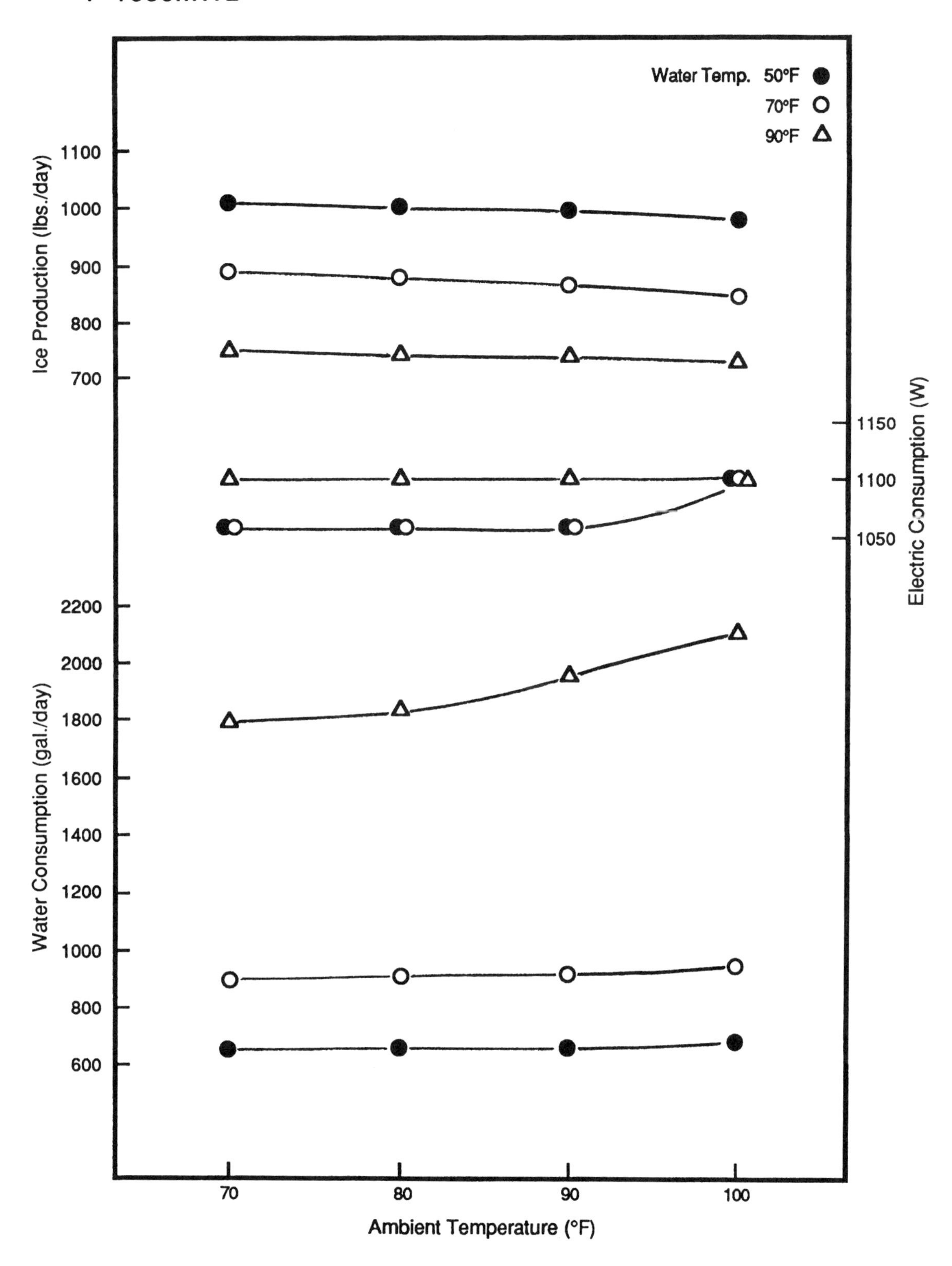

Fig. 24

F-1000MWB

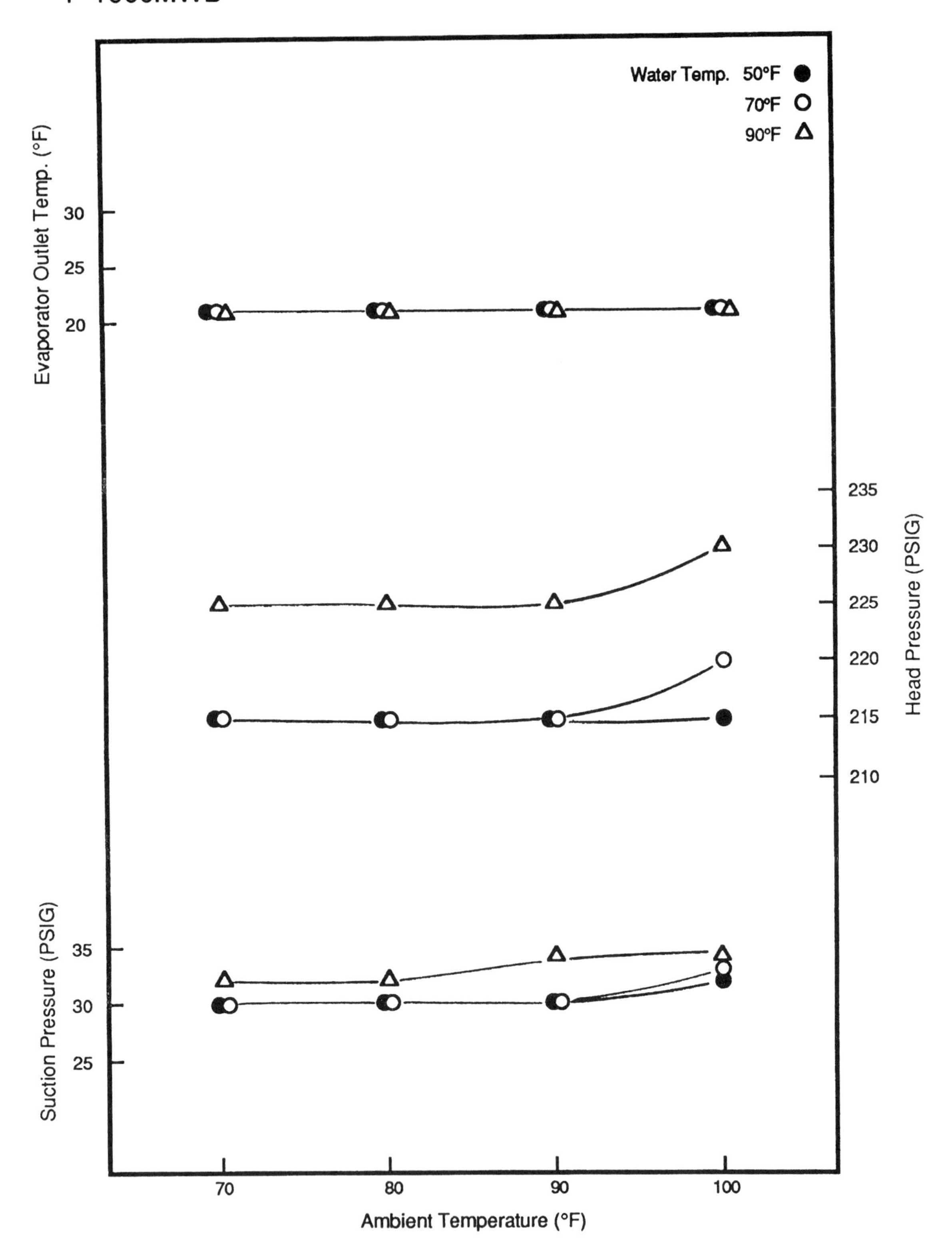

Fig. 25

[c] F-1000MRB

	Ambient Temp. (°F)	Water Temp. (°F)		
		50	70	90
APPROXIMATE ICE PRODUCTION PER 24 HR. lbs./day (kg/day)	70 80 90 100	*1020(463) 950(431) 880(399) 810(367)	920(417) 865(392) *800(363) 730(331)	840(381) 790(358) 740(336) 680(308)
APPROXIMATE ELECTRIC CONSUMPTION watts (115/230V)	70 80 90 100	*1250 1270 1300 1300	1250 1270 *1300 1300	1250 1270 1300 1300
APPROXIMATE WATER CONSUMPTION PER 24 HR. gal./day (ℓ/day)	70 80 90 100	122(463) 114(431) 105(399) 97(367)	110(417) 104(392) 96(363) 87(331)	101(381) 95(358) 89(336) 81(308)
EVAPORATOR OUTLET TEMP. °F (°C)	70 80 90 100	21(-6) 21(-6) 21(-6) 21(-5)	21(-6) 21(-6) 21(-6) 21(-5)	21(-6) 21(-6) 21(-6) 21(-5)
HEAD PRESSURE PSIG (kg/cm²G)	70 80 90 100	190(13.3) 207(14.5) 233(16.3) 265(18.6)	190(13.3) 207(14.5) 233(16.3) 265(18.6)	190(13.3) 207(14.5) 233(16.3) 265(18.6)
SUCTION PRESSURE PSIG (kg/cm²G)	70 80 90 100	36(2.5) 36(2.5) 37(2.6) 39(2.7)	36(2.5) 36(2.5) 37(2.6) 39(2.7)	36(2.5) 36(2.5) 37(2.6) 39(2.7)

HEAT OF REJECTION FROM CONDENSER	8800 BTU/h (AT 90°F/WT 70°F)
HEAT OR REJECTION FROM COMPRESSOR	1260 BTU/h (AT 90°F/WT 70°F)
CONDENSER VOLUME	93 cu in

Note: The data without *marks should be used for reference.

F-1000MRB

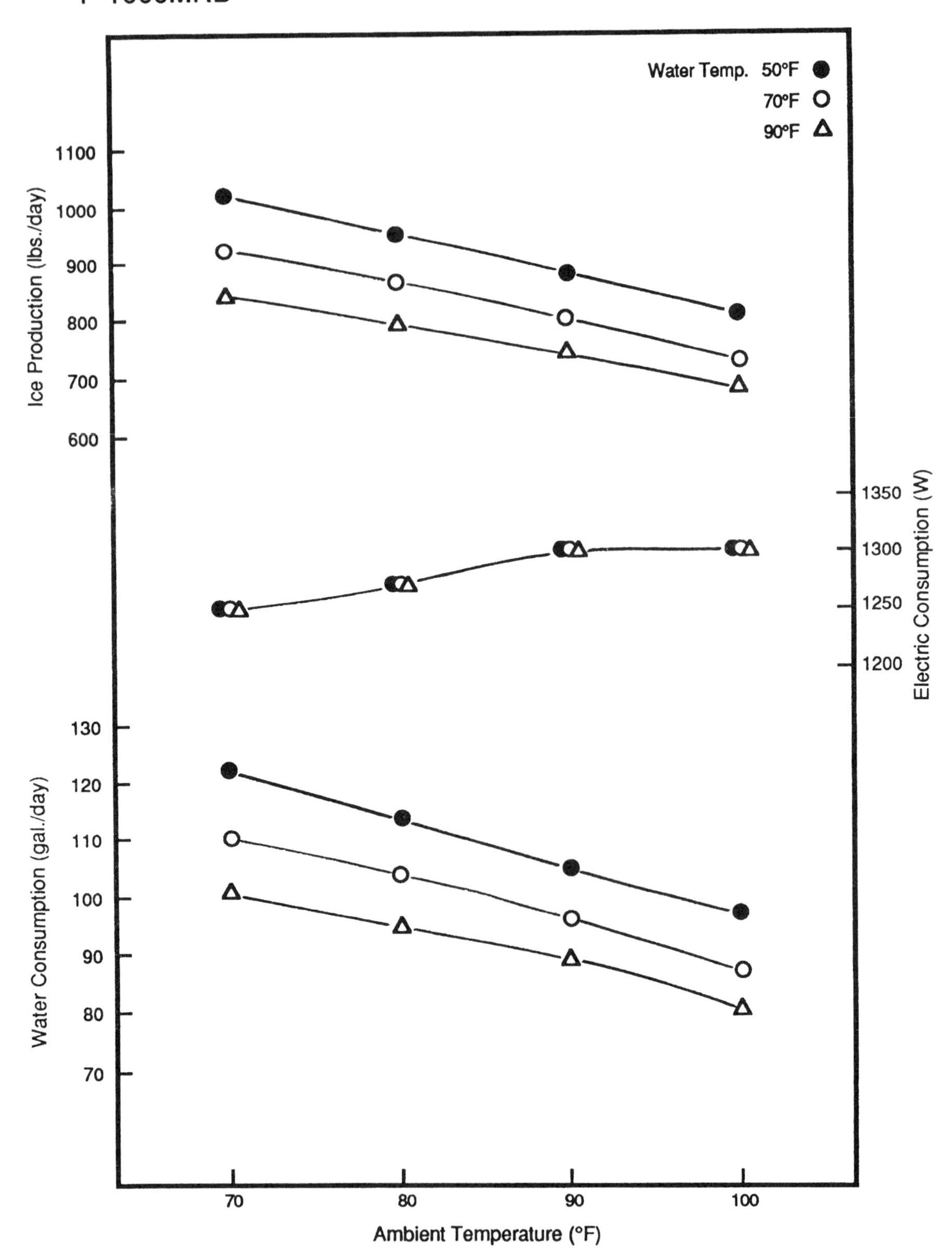

Fig. 26

F-1000MRB

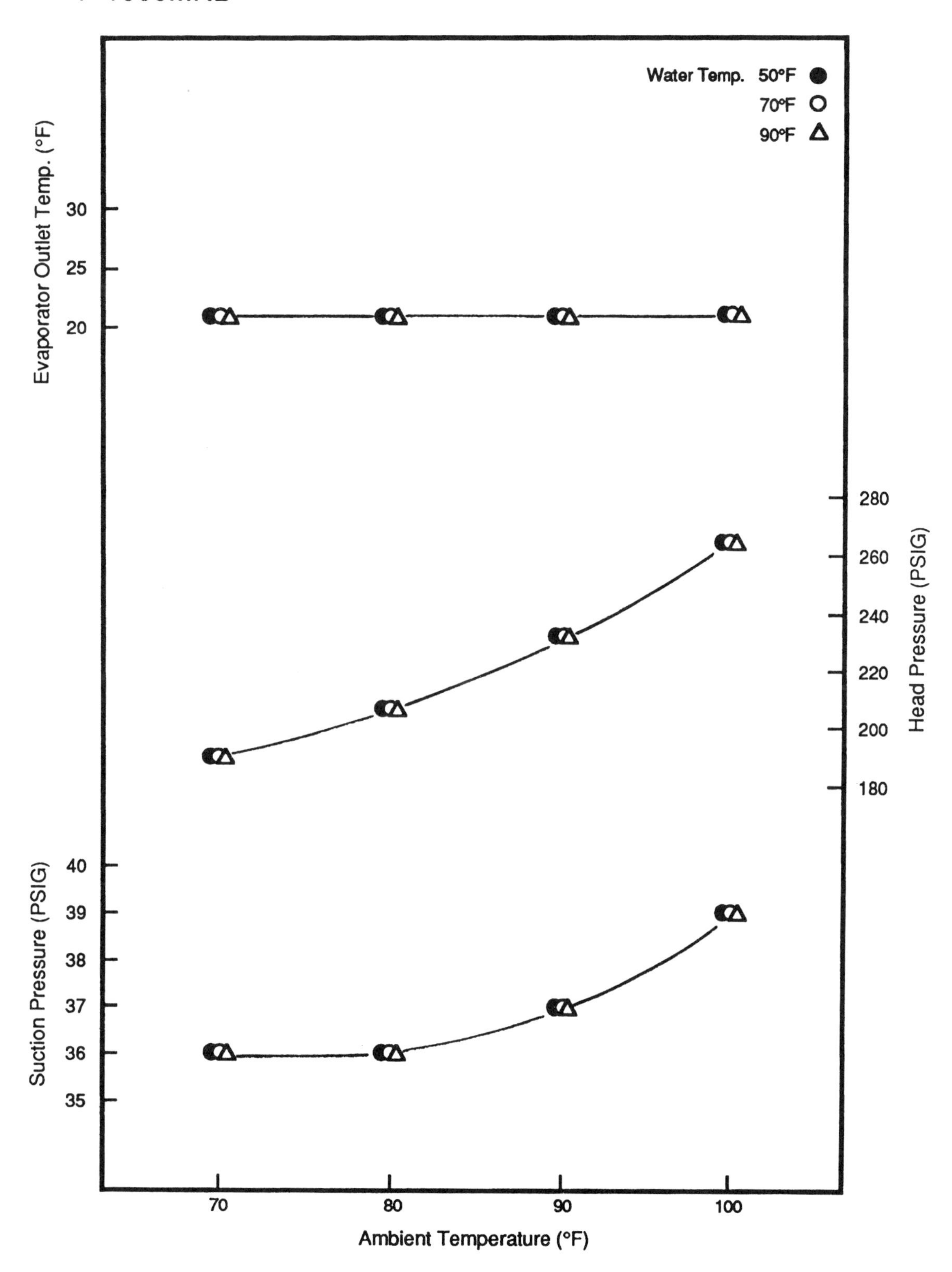

Fig. 27

VI. ADJUSTMENT OF COMPONENTS

1. ADJUSTMENT OF WATER REGULATING VALVE - WATER-COOLED MODEL ONLY

The Water Regulating Valve or also called "WATER REGULATOR" is factory-adjusted. No adjustment is required under normal use. Adjust the Water Regulator , if necessary, using the following procedures.

1) Attach a pressure gauge to the high-side line of the system. Or prepare a thermometer to check the condenser drain temperature.

2) Rotate the adjustment screw by using a flat blade screwdriver, so that the pressure gauge shows 230 PSIG, or the thermometer reads 100 - 104°F, in 5 minutes after the icemaking process starts. When the pressure exceeds 230 PSIG, or the condenser drain temperature exceeds 104°F, rotate the adjustment screw counterclockwise.

3) Check that the pressure or the condenser drain temperature holds a stable setting.

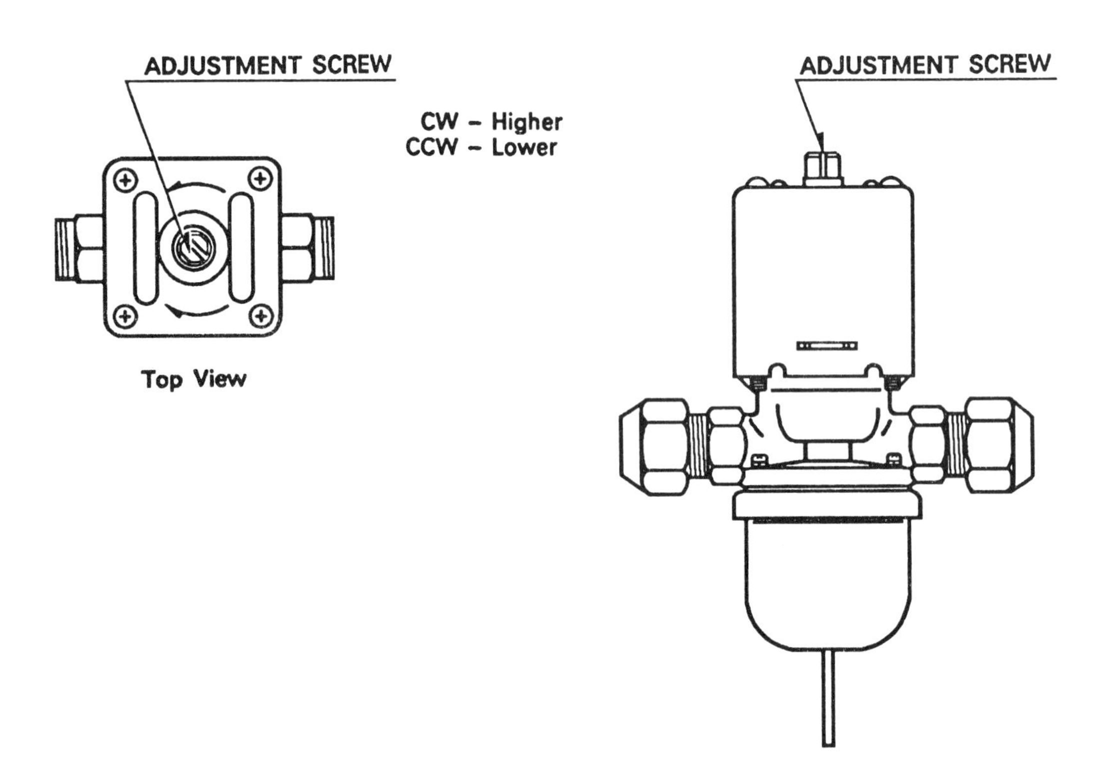

Fig. 28

2. ADJUSTMENT OF EXPANSION VALVE

The icemaker unit is provided with a Thermal Expansion Valve.
This valve controls the temperature at the bulb to be approx. 9 - 18 °F higher than the evaporation temperature in the Expansion Valve Outlet Pipe (Evaporator Inlet Pipe). The Expansion Valve is already adjusted to get the maximum ice production capacity. Readjust the valve only when the temperature at the bulb is too high (Compressor overheated) or too low (too much liquid refrigerant flowing back to Compressor).

1) Remove the Cap Nut on the end to give access to the adjusting screw.

   (a) Turn the screw clockwise to decrease the refrigerant flow.

   (b) Turn the screw counterclockwise to increase the refrigerant flow.

CAUTION

This is a sensitive device. Check conditions after each 1/4 to 1/2 turn.

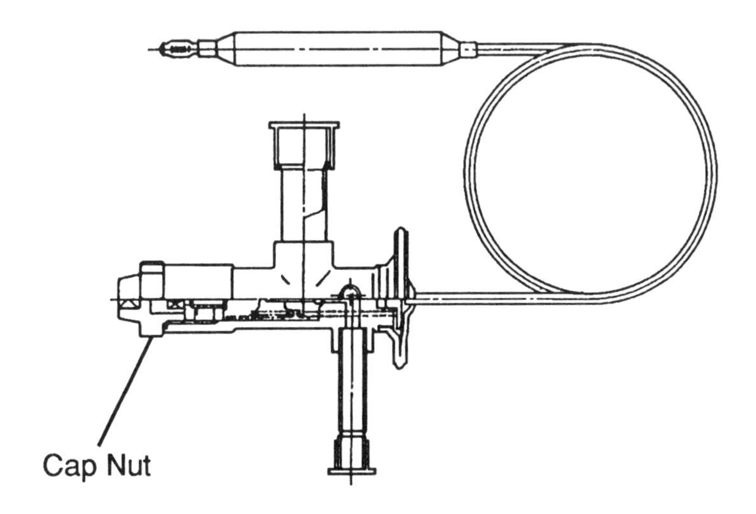

Fig. 29

3. ADJUSTMENT OF CUTTER

To adjust the ice size, change the direction of the Cutter on the top of the Auger, according to the following procedure.

1) Remove the Bolt and the Cutter.

2) The Cutter is marked with "L", "M" and "S" and adjusted in "M" before shipping. Adjust it in "L" or "S", if necessary.

SETTING	TYPE OF ICE
Large	Coarse
Medium	Flake
Small	Fine

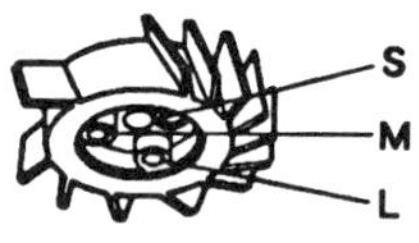

S
M
L

Fig. 30

3) Place the Cutter so that the pin on the top of the Auger fits into the hole next to the marking "L", "M" or "S".

4) Secure the Cutter by the Bolt.

VII. SERVICE DIAGNOSIS

1. NO ICE PRODUCTION

PROBLEM	POSSIBLE CAUSE		REMEDY
[1] The icemaker will not start.	a) Power Supply	1. OFF position.	1. Move to ON position.
		2. Loose connections.	2. Tighten.
		3. Bad contacts.	3. Check for continuity and replace.
		4. Blown fuse.	4. Replace.
	b) Power Switch (Control Box)	1. OFF position.	1. Move to ON position.
		2. Bad contacts.	2. Check for continuity and replace.
	c) Fuse (Control Box)	1. Blown out.	1. Check for short-circuit and replace.
	d) Power Supply Protect Relay	1. Miswiring.	1. Check power supply voltage and wire properly.
	e) Flush Timer	1. Flushing out.	1. Wait for 15 minutes.
		2. Bad contacts.	2. Check for continuity and replace.
	f) Flush Switch	1. FLUSH position.	1. Move to ICE position.
		2. Bad contacts.	2. Check for continuity and replace.
	g) Transformer	1. Coil winding opened.	1. Replace.
	h) Water Valve	1. Coil winding opened.	1. Replace.
	i) Shut-off Valve	1. Closed.	1. Open.
		2. Water failure.	2. Wait till water is supplied.
	j) Plug and Receptacle (Control Box)	1. Disconnected.	1. Connect.
		2. Terminal out of Plug or Receptacle.	2. Insert Terminal back in position.
[2] Water does not stop, and the icemaker will not start.	a) Water Control Relay	1. Contacts fused.	1. Replace.
		2. Coil winding opened.	2. Replace.
	b) Float Switch	1. Bad contacts.	1. Check for continuity and replace.
		2. Float does not move freely.	2. Clean or replace.
	c) Flush Water Valve	1. Valve sheet clogged and water leaking.	1. Clean or replace.
	d) Hoses	1. Disconnected.	1. Connect.
[3] Water has been supplied, but the icemaker will not start.	a) Water Control Relay	1. Bad contacts.	1. Check for continuity and replace.
	b) Bin Control	1. Bad contacts.	1. Check for continuity and replace.
		2. Activator does not move freely.	2. Clean Axle and its corresponding holes or replace Bin Control.
	c) Gear Motor Protector (Circuit Breaker)	1. Tripped.	1. Find out the cause, get rid of it, and press Reset Button on Circuit Breaker.

PROBLEM	POSSIBLE CAUSE		REMEDY
	d) Gear Motor Relay	1. Coil winding opened.	1. Replace.
		2. Bad contacts.	2. Check for continuity and replace.
	e) Control Timer (Printed Circuit Board)	1. Broken.	1. Replace.
	f) Gear Motor Protect Relay	1. Coil winding opened.	1. Replace.
		2. Bad contacts.	2. Check for continuity and replace.
[4] Gear Motor starts, but Compressor will not start or operates intermittently.	a) Pressure Switch	1. Dirty Air Filter or Condenser.	1. Clean.
		2. Ambient or condenser water temperature too warm.	2. Get cooler.
		3. Condenser water pressure too low or off.	3. Check and get recommended pressure.
		4. Water Regulating Valve set too high.	4. Adjust it lower.
		5. Fan not rotating.	5. See "3. [1] a) Fan Motor."
		6. Refrigerant overcharged.	6. Recharge.
		7. Refrigerant line or components plugged.	7. Clean and replace drier.
		8. Bad contacts.	8. Check for continuity and replace.
		9. Loose connections.	9. Tighten.
	b) X2 Relay on Control Timer	1. Bad contacts.	1. Check for continuity and replace Timer.
		2. Coil winding opened.	2. Replace Timer.
	c) Starter	1. Bad contacts.	1. Check for continuity and replace.
		2. Coil winding opened.	2. Replace.
		3. Loose connections.	3. Tighten.
	d) Start Capacitor or Run Capacitor	1. Defective.	1. Replace.
	e) Compressor	1. Loose connections.	1. Tighten.
		2. Motor winding opened or grounded.	2. Replace.
		3. Motor Protector tripped.	3. Find out the cause of overheat or overcurrent.
	f) Power Supply	1. Circuit Ampacity too low	1. Install a larger-sized conductor.

PROBLEM	POSSIBLE CAUSE		REMEDY
[5] Gear Motor and Compressor start, but no ice is produced.	a) Refrigerant Line	1. Gas leaks	1. Check for leaks with a leak detector. Reweld leak, replace drier and charge with refrigerant. The amount of refrigerant is marked on Nameplate or Label.
		2. Refrigerant line clogged	2. Replace the clogged component.

2. LOW ICE PRODUCTION

PROBLEM	POSSIBLE CAUSE		REMEDY
[1] Low ice production	a) Refrigerant Line	1. Gas leaks	1. See "1. [5] a) Refrigerant Line."
		2. Refrigerant line clogged	2. Replace the clogged component.
		3. Overcharged	3. Recharge.
	b) High-side Pressure Too High	1. Dirty Air Filter or Condenser	1. Clean.
		2. Ambient or condenser water temperature too warm.	2. Get cooler.
		3. Condenser water pressure too low or off.	3. Check and get recommended pressure.
		4. Fan rotating too slow.	4. See "3. [1] a) Fan Motor."
		5. Water Regulator clogged.	5. Clean.
	c) Expansion Valve (not adjustable)	1. Low-side pressure exceeding the limit	1. Replace.

3. OTHERS

PROBLEM	POSSIBLE CAUSE		REMEDY
[1] Abnormal noise	a) Fan Motor	1. Bearings worn out.	1. Replace.
		2. Fan blade deformed.	2. Replace fan blade.
		3. Fan blade does not move freely.	3. Replace.
	b) Compressor	1. Bearings worn out, or cylinder valve broken.	1. Replace.
		2. Mounting pad out of position.	2. Reinstall.
	c) Refrigerant Lines	1. Rub or touch lines or other surfaces.	1. Replace.
	d) Gear Motor (Ice Making)	1. Bearing or Gear wear/damage	1. Replace.

PROBLEM	POSSIBLE CAUSE		REMEDY
	e) Evaporator	1. Low-side pressure too low.	1. See if Expansion Valve bulb is mounted properly, and replace the valve if necessary.
		2. Scale on inside wall of Freezing Cylinder.	2. Remove Auger. Use "LIME-A-WAY" solution to clean periodically. If water is found to be hard by testing, install a softener.
[2] Overflow from Reservoir (Water does not stop.)	a) Water Supply	1. Water pressure too high.	1. Install a Pressure Reducing Valve.
	b) Water Valve	1. Diaphragm does not close.	1. Clean or replace.
	c) Float Switch	1. Bad contacts.	1. Check for continuity and replace.
[3] Gear Motor Protector operates frequently.	a) Power Supply Voltage	1. Too high or too low.	1. Connect the unit to a power supply of proper voltage.
	b) Evaporator Assy	1. Bearings or Auger worn out.	1. Replace Bearing or Auger.

VIII. REMOVAL AND REPLACEMENT OF COMPONENTS

1. SERVICE FOR REFRIGERANT LINES

[a] REFRIGERANT DISCHARGE

The icemaker unit is provided with a Refrigerant Access Valve on the low-side line. Install a proper fitting on the high-side line, if necessary, to check the gauge pressure.

[b] EVACUATION AND RECHARGE

1) Attach Charging Hoses, a Service Manifold and a Vacuum Pump to the system.

2) Turn on the Vacuum Pump.

3) Allow the Vacuum Pump to pull down to a 29.9" Hg vacuum. Evacuating period depends on pump capacity.

4) Close the Low-side Valve on the Service Manifold.

5) Disconnect the Vacuum Pump, and attach a Refrigerant Service Cylinder to the low-side line. Remember to loosen the connection, and purge the air from the Hose. See the Nameplate for required refrigerant charge. (For the Remote Air-cooled model, see the label on the Control Box.)

6) Open the Low-side Valve. Do not invert the Service Cylinder. A liquid charge will damage the Compressor.

7) Turn on the icemaker when charging speed gets slow. Turn off the icemaker when the Low-side Gauge shows approximately 0 PSIG. Do not run the icemaker at negative pressures. Close the Low-side Valve when the Service Cylinder gets empty.

8) Repeat the above steps 4) through 7), if necessary, until a required amount of refrigerant has entered the system.

9) Close the Refrigerant Access Valve, and disconnect the Hoses and Service Manifold.

10) Cap the Access Valve to prevent possible leak.

2. BRAZING

DANGER

1. Refrigerant R502 itself is not flammable, explosive and poisonous. However, when exposed to an open flame, R502 creates Phosgene gas, hazardous in large amounts.

2. Always purge the refrigeration system through hose vented to the outside, because it is dangerous for the room to be filled with R502 which displaces oxygen.

3. Do not use silver alloy or copper alloy containing Arsenic.

4. In its liquid state, the refrigerant can cause frostbite because of the low temperature.

3. REMOVAL AND REPLACEMENT OF COMPRESSOR

IMPORTANT

Always install a new Drier every time the sealed refrigeration system is opened. Do not replace the Drier until after all other repairs or replacement have been made.

1) Turn off the power supply.

2) Remove the panels.

3) Remove the Terminal Cover on the Compressor, and disconnect the Compressor Wiring.

4) Discharge the refrigerant.

5) Remove the Discharge, Suction and Access Pipes form the Compressor using brazing equipment.

WARNING

When repairing a refrigerant system, be careful not to let the burner flame contact the lead wires or insulation.

6) Remove the Bolts and Rubber Grommets.

7) Slide and remove the Compressor. Unpack a new Compressor package. Install a new Compressor.

8) Attach the Rubber Grommets of the prior Compressor.

9) Sandpaper the Discharge, Suction and Access Pipes.

10) Place the Compressor in position, and secure it using the Bolts.

11) Remove plugs from the Discharge, Suction and Access Pipes.

12) Braze the Access, Suction and Discharge lines (Do not change this order), while purging with nitrogen gas flowing at the pressure of 3 - 4 PSIG.

WARNING

Always protect the valve body using a damp cloth to prevent the valve from overheat. Do not braze with the valve body exceeding 250°F.

13) Install a new Drier.

14) Check for leaks using nitrogen gas (140 PSIG) and soap bubble.

15) Evacuate the system, and charge it with refrigerant. See the Nameplate for required refrigerant charge. (For the Remote Air-cooled model, see the label on the Control Box.)

16) Connect the Terminals to the Compressor, and place the Terminal Cover in position.

17) Place the panels in position.

18) Turn on the power supply.

4. REMOVAL AND REPLACEMENT OF DRIER

IMPORTANT

Always install a new Drier every time the sealed refrigeration system is opened. Do not replace the Drier until after all other repairs or replacement have been made.

1) Turn off the power supply.

2) Remove the panels.

3) Discharge the refrigerant.

4) Remove the Drier using brazing equipment.

5) Install a new Drier with the arrow on the Drier in the direction of the refrigerant flow. Use nitrogen gas at the pressure of 3 - 4 PSIG when brazing the tubings.

6) Check for leaks using nitrogen gas (140 PSIG) and soap bubble.

7) Evacuate the system, and charge it with refrigerant. See the Nameplate for required refrigerant charge. (For the Remote Air-cooled model, see the label on the Control Box.)

8) Place the panels in position.

9) Turn on the power supply.

5. REMOVAL AND REPLACEMENT OF EXPANSION VALVE

IMPORTANT

Sometimes moisture in the refrigerant circuit exceeds the Drier capacity and freezes up at the Expansion Valve. Always install a new Drier every time the sealed refrigeration system is opened. Do not replace the Drier until after all other repairs or replacement have been made.

1) Turn off the power supply.

2) Remove the Front Panel.

3) Discharge the refrigerant.

4) Remove the Expansion Valve Bulb at the Evaporator outlet.

5) Remove the Expansion Valve Cover, and disconnect the Expansion Valve using brazing equipment.

5) Braze a new Expansion Valve with nitrogen gas flowing at the pressure of 3 - 4 PSIG.

WARNING

Always protect the valve body using a damp cloth to prevent the valve from overheat. Do not braze with the valve body exceeding 250°F.

6) Install a new Drier.

7) Check for leaks using nitrogen gas (140 PSIG) and soap bubble.

8) Evacuate the system, and charge it with refrigerant. See the Nameplate for required refrigerant charge. (For the Remote Air-cooled model, see the label on the Control Box.)

9) Attach the Bulb to the suction line. Be sure to secure the Bulb using a band and to insulate it.

10) Place a new set of Expansion Valve Covers in position.

11) Replace the Front Panel in position.

12) Turn on the power supply.

6. REMOVAL AND REPLACEMENT OF WATER REGULATING VALVE - WATER-COOLED MODEL ONLY

IMPORTANT

Always install a new Drier every time the sealed refrigeration system is opened. Do not replace the Drier until after all other repairs or replacement have been made.

1) Turn off the power supply.

2) Close the Water Supply Line Shut-off Valve.

3) Remove the panels.

4) Discharge the refrigerant.

5) Disconnect the Capillary Tube using brazing equipment.

6) Disconnect the Flare-connections of the valve.

7) Remove the screws and the valve from the Bracket.

8) Install a new valve, and braze the Capillary Tube.

9) Install a new Drier.

10) Check for leaks using nitrogen gas (140 PSIG) and soap bubble.

11) Connect the Flare-connections.

12) Open the Water Supply Line Shut-off Valve.

13) Turn on the power supply.

14) Check for water leaks.

15) Evacuate the system, and charge it with refrigerant. See the Nameplate for required refrigerant charge. (For the Remote Air-cooled model, see the label on the Control Box.)

16) See "VI. 1. ADJUSTMENT OF WATER REGULATING VALVE." If necessary, adjust the valve.

17) Place the panels in position.

7. REMOVAL AND REPLACEMENT OF CONDENSING PRESSURE REGULATOR (C.P.R.) - REMOTE AIR-COOLED MODEL ONLY

IMPORTANT

Always install a new Drier every time the sealed refrigeration system is opened. Do not replace the Drier until after all other repairs or replacement have been made.

1) Turn off the power supply.

2) Remove the panels from the remote condenser unit.

3) Discharge the refrigerant. See "1. [a] REFRIGERANT DISCHARGE."

4) Remove the C.P.R. using brazing equipment.

5) Braze a new C.P.R. with nitrogen gas flowing at the pressure of 3 - 4 PSIG.

WARNING

Always protect the valve body using a damp cloth to prevent the valve from overheat. Do not braze with the valve body exceeding 250°F.

6) Install a new Drier in the icemaker.

7) Check for leaks using nitrogen gas (140 PSIG) and soap bubble.

8) Evacuate the system and charge it with refrigerant. See the Label on the Control Box in the icemaker.

9) Place the panels in position.

10) Turn on the power supply.

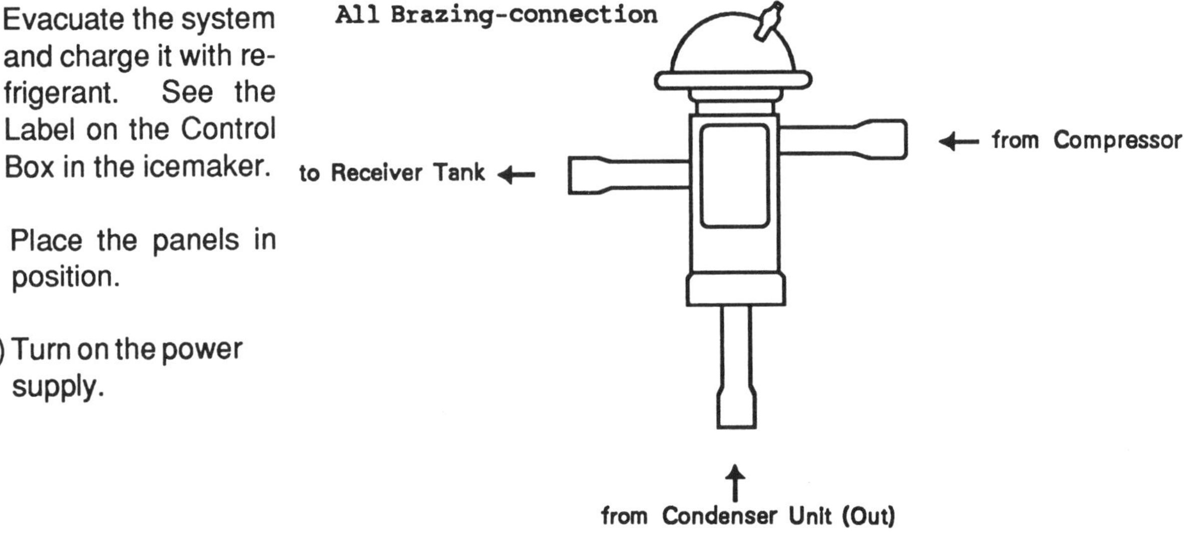

Fig. 31

8. REMOVAL AND REPLACEMENT OF EVAPORATOR ASSEMBLY - See Fig. 32

1) Turn off the power supply.

2) Remove the panels.

3) Move the Flush Switch to the "FLUSH" position to drain the water in the Evaporator.

4) Remove the Band connecting the Chute Head with the Chute Assembly.

5) Remove the three Thumbscrews and take off the Chute Head from the Evaporator.

CUTTER

6) Remove the Bolt and lift off the Cutter.

7) Remove the Rubber O-ring and the Nylon O-ring at the top of the Evaporator.

EXTRUDING HEAD

8) Remove the three Socket Head Cap Screws and lift off the Extruding Head.

9) Replace the Bearing inside the Extruding Head, if it is worn or scratched.

Note: Replacing the Bearing needs a fitting tool. If it is not available, replace the whole Extruding Head.

AUGER

10) Lift off the Auger. If the Bearing is worn or the Blade scratched, replace the Auger.

EVAPORATOR

Note: Skip the following steps 11) through 13) when the Evaporator does not need replacement.

11) Discharge the refrigerant.

IMPORTANT

Always install a new Drier every time the sealed refrigeration system is opened. Do not replace the Drier until after all other repairs or replacement have been made.

12) Remove the Bulb of the Expansion Valve.

13) Disconnect the brazing-connections of the Expansion Valve and the Copper Tube - Low Side from the Evaporator, using brazing equipment.

14) Remove the two Truss Head Machine Screws and the Strap securing the Evaporator.

15) Disconnect the three Hoses from the Evaporator.

16) Remove the four Hexagon Socket Cap Bolts securing the Evaporator with the Bearing-Lower.

17) Lift off the Evaporator.

BEARING-LOWER AND MECHANICAL SEAL

18) The Mechanical Seal consists of two parts. One moves along with the Auger, and the other is fixed on the Bearing-Lower. If the contact surfaces of these two parts are worn or scratched, the Mechanical Seal may cause water leaks and should be replaced.

19) Remove the O-ring on the Bearing-Lower.

20) Remove the six Bolts and the Bearing-Lower from the Gear Motor. Replace the Bearing inside the Bearing-Lower using a fitting tool, if it is worn or scratched.

GEAR MOTOR

21) Remove the Coupling-Spline on the Gear Motor Shaft.

22) Remove the Barrier on the top of the Gear Motor.

23) Remove the Terminal Cover of the Gear Motor and cut the Connectors.

24) Remove the three Hexagon Socket Cap Bolts securing the Gear Motor.

25) Assemble the removed parts in the reverse order of the above procedure.

WARNING

Be careful not to scratch the surface of the O-ring, or it may cause water leaks. Handle the Mechanical Seal with care not to scratch nor to contaminate its contact surface.

26) When replacing the Evaporator;

(a) Braze a new Evaporator with nitrogen gas flowing at the pressure of 3 - 4 PSIG.

(b) Replace the Drier.

(c) Check for leaks using nitrogen gas (140 PSIG) and soap bubble.

(d) Evacuate the system, and charge it with refrigerant. See the Nameplate for required refrigerant charge. (For the Remote Air-cooled model, see the label on the Control Box.)

27) Turn on the power supply.

28) Move the Flush Switch to the "ICE" position.

29) Place the panels in position.

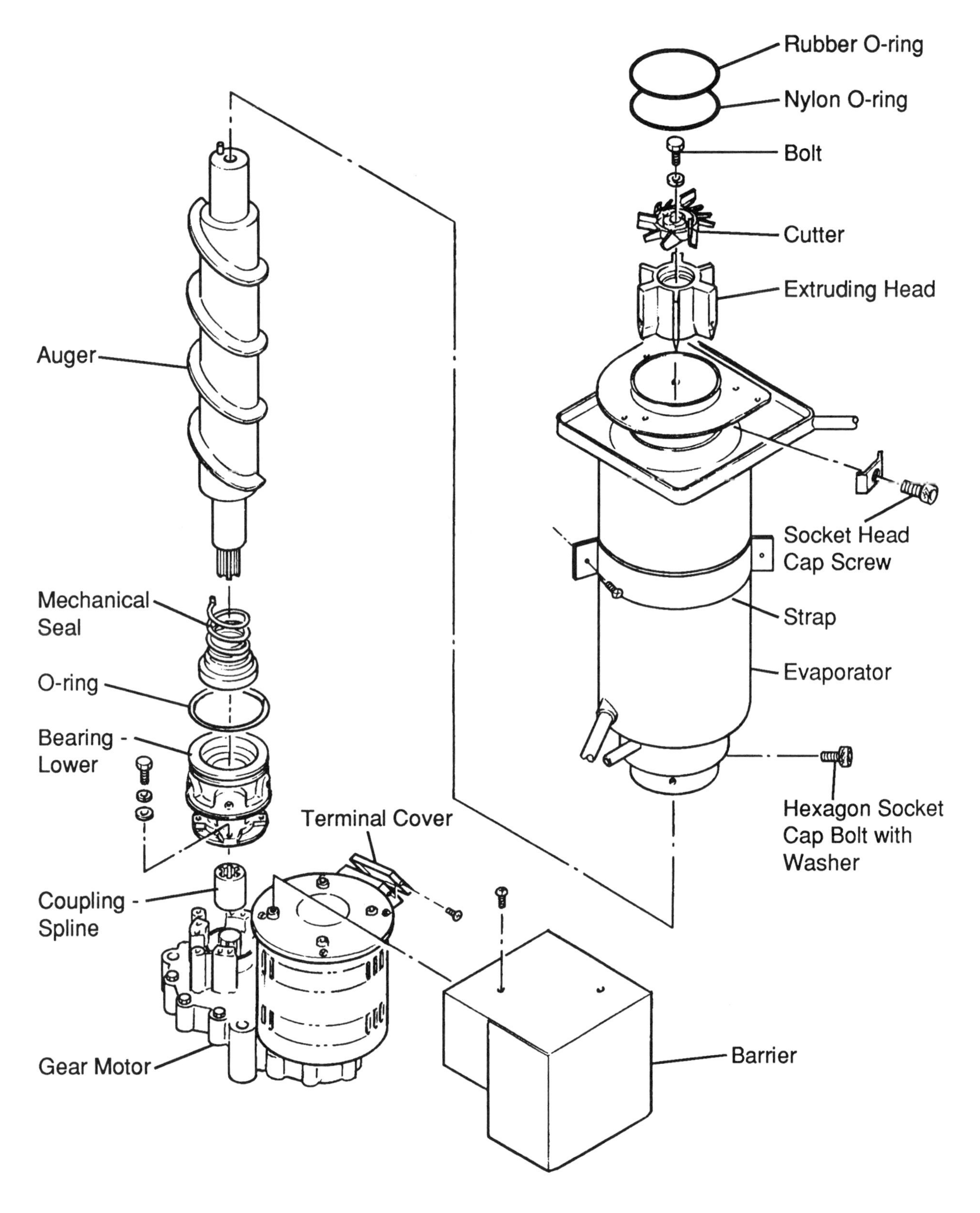
Rubber O-ring
Nylon O-ring
Bolt
Cutter
Extruding Head
Auger
Socket Head
Cap Screw
Strap
Evaporator
Mechanical
Seal
O-ring
Bearing -
Lower
Hexagon Socket
Cap Bolt with
Washer
Terminal Cover
Coupling -
Spline
Gear Motor
Barrier

Fig. 32

9. REMOVAL AND REPLACEMENT OF FAN MOTOR

1) Turn off the power supply.

2) Remove the panels.

3) Remove the wire connectors from the Fan Motor leads.

4) Remove the Fan Motor Bracket and Fan Motor.

5) Install a new Fan Motor.

6) Replace the Fan Motor Bracket and the wire connectors.

7) Place the panels in position.

8) Turn on the power supply.

10. REMOVAL AND REPLACEMENT OF CONTROL WATER VALVE

1) Turn off the power supply.

2) Close the Water Supply Line Shut-off Valve.

3) Remove the panels.

4) Disconnect the Terminals from the Control Water Valve.

5) Remove the Cover-Reservoir Inlet from the Control Water Valve.

6) Loosen the Fitting Nut on the Control Water Valve Inlets, and remove the Control Water Valve. Do not lose the Packings inside the Fitting Nut.

7) Install a new Control Water Valve.

8) Assemble the removed parts in the reverse order of the above procedure.

9) Open the Water Supply Line Shut-off Valve.

10) Turn on the power supply.

11) Check for water leaks.

12) Place the panels in position.

11. REMOVAL AND REPLACEMENT OF FLUSH WATER VALVE

1) Turn off the power supply.

2) Close the Water Supply Line Shut-off Valve.

3) Remove the panels.

4) Remove the Clamp and disconnect the Hose from the Flush Water Valve.

   Note: Water may still remain inside the Evaporator. Be sure to drain the water into the Drain Pan.

5) Disconnect the Terminals from the Flush Water Valve.

6) Remove the Flush Water Valve from the Bracket.

7) Remove the Drain Pipe from the Flush Water Valve.

8) Connect the Drain Pipe to a new Flush Water Valve, and place the valve in position.

9) Connect the Hose to the Flush Water Valve and secure it with the Clamp.

10) Open the Water Supply Line Shut-off Valve.

11) Turn on the power supply.

12) Move the Flush Switch to the "ICE" position and the Power Switch to the "ON" position.

13) Check for water leaks.

14) Move the Flush Switch to the "FLUSH" position, and make sure water is flushing.

15) Move the Flush Switch to the "ICE" position.

16) Place the panels in position.

SerVend

ICE MAKER/BIN

SERVICE MANUAL

(Note: Service manual reference of Ice King is for SerVend ice machines).

SerVend International, Inc.
2100 Future Drive
Sellersburg, Indiana 47172-1868
812-246-6363
FAX: 812-246-9922

SERVICE MANUAL TABLE OF CONTENTS

Installation Instructions

Freight Claim Loss or Damage:

1. The delivering freight company, distributor or dealer is responsible for loss or damage to your merchandise. All claims must be made with the party that delivers your merchandise.
2. Check the number of containers delivered against the number shown on your receipt. If the total is not correct, have the driver note the shortage on your receipt.
3. Check all cartons for visible damage, open and check the contents of any carton in question before the driver leaves. Be sure the driver notes the type and degree of damage on your receipt. As damaged merchandise must be inspected within 15 days of delivery, notify your carrier immediately.
4. If concealed damage is found when merchandise is unpacked, place the packing material with merchandise and request an inspection from the delivering carrier.
5. File your claim for loss or damage at once. Delays in filing will only hinder achieving a satisfactory resolution to your claim.

Installation:

We recommend that installation and start-up be performed by the Dealer "Professionals" where your ice maker was purchased.

Location:

For best performance, select a location away from all heat sources, such as radiators, ovens, refrigeration equipment, direct sunlight, etc.

Avoid placing air cooled models in kitchens whenever possible as grease, flour or other air-borne particles will collect on the condenser and fan blade, requiring increased preventative maintenance and will reduce efficiency.

Discuss the best location with your Dealer "Professional". Allow a minimum of 6" clearance around the ice maker for air circulation (both sides, top and back). Restricted air circulation will affect the maintenance-free life of your ice maker and its efficiency.

Your ice maker will perform at optimum efficiency in an approximate 70°F room with 50°F water. Increased air or water temperatures will decrease performance. Never operate your machine in rooms with temperatures below 40°F or above 100°F.

If the ice maker is located in an unheated area, it must be protected from freezing temperatures or shut down and winterized.

Set-Up of the Storage Bin and Cuber:

1. Remove the top cap and sleeve from the storage bin. Lay the bin on its back on the corrugated carton to prevent scratching. Screw the legs into the bin bottom.
2. Set the bin on its legs. Move the bin to its final location, level by screwing the feet either up or down.
3. Remove the carton from ice maker and place the ice maker on the bin, align it with the bin back and sides.
4. Remove the ice maker internal packing, compressor, shipping blocks, etc. Connect the siphon hose to the water pan. Connect siphon clip and pull hose from back of unit.

Plumbing Lines and Connections:

All plumbing (water and drain) connections must conform to local and national codes.

To prevent condensation, insulate all water and drain lines. Insulating water lines will increase efficiency as well.

Shut-off valves should be located in water inlet lines, both ice make-up and condenser (water cooled units).

The ice make-up water supply should be connected to a water treatment system. See your Hussmann / Ice King Dealer / Distributor for type and size system required. Water treatment will pay for itself through decreased maintenance, higher efficiency and quality of product.

Note: *When installed without mechanical water treatment, a 100 mesh strainer with clean out plug must be installed in ice make-up water line.*

Incoming water pressure, both ice make-up and condenser must not exceed 120 PSI and not be less than 20 PSI.

Note: *A minimum of 10 PSI pressure drop must be maintained across the condenser when water models are connected to recirculation systems, cooling towers, etc.*

Drains:

To ensure trouble-free drainage, vent cuber and bin drains to the atmosphere at the cabinet (see sketch).

Drain lines from the cabinet require 1-1/2" drop per five feet of run. Lines should terminate over an open trapped and vented floor drain.

Electrical:

All supply wiring and connections must conform to national and local electrical codes. Size wiring and fuse per name-plate specification. Connect the cuber to a separately fused circuit. Conduit must be connected to the electrical control box inside the cuber, not the rear panel. Place a manual disconnect in a convenient location between the cuber and fuse box. The cuber must be grounded by the control box ground screw or conduit connection.

SERVEND

Installation Check List:

1. Has cuber been levelled?
2. Are electrical connections complete per nameplate specification? Is manual disconnect "ON"?
3. Check water and drain connections:
   a) Has water inlet pressure been checked?
   b) Are shut-off valves open?
   c) Is water filter installed?
   d) Is bin drain tube assembled to bin?
   e) When used, is the water cooled condenser drained separately?
   f) Are drain lines vented?
   g) Are all lines insulated and sloped to floor drain?
4. Is there a 6" clearance at cuber sides and back for air circulation?
5. Is the cuber installed where ambient temperatures will not be below 40°F or above 100°F? Will incoming water temperature be maintained between 35°F and 90°F? If not, are you prepared to winterize?
6. Does water curtain move freely?
7. Check float valve.
8. Has storage bin and cuber been sanitized?

Start-Up Procedure:

1. Place the On/Off switch in the water pump position. Only the water pump will start. Permit the sump to fill to float shut-off. Check for continuous and even water flow over top plastic extrusion and evaporator. Is water flowing from all water distribution tube holes? Check water level in siphon tube.
2. Place switch in "off" position. Water should fill the sump drain tube starting a siphoning action. The siphon should drain the sump and stop flowing in 1 to 2 minutes. If not, lower the water level to approximately 3/8" below the drain tube overflow point to stop the flow.
3. Place switch in the ice position, the compressor, condenser fan motor and water pump will operate. Depress the manual harvest switch, the water pump and fan motor will stop and the harvest solenoid will open. Swing the water curtain open and hold for 10 seconds, the compressor will stop. Release the water curtain and the compressor, fan motor and water pump will start, initiating a new ice making cycle.
4. For maximum ice production and cube separation, the ice bridge (link) between the cubes should be 1/8" thick at the center of the ice sheet. The ice bridging will vary slightly from top to bottom and side to side. Permit the ice maker to complete two full cycles before checking bridging.

Owner/Operator:

Before leaving, be sure the owner/operator understands the ice maker operation and the value of preventative maintenance.

Do they know:

1. Location of the electrical disconnect switch and water shut-off valves?
2. How to start and shut down unit, adjust bridging, clean and sanitize unit?
3. Bin fill operation, function of high pressure cut-out (water models) and replacing circuit board fuse?
4. Proper method of cleaning condenser and fan blades?
5. To inspect distribution tubes and evaporator for mineral deposits?
6. How to identify when filter needs replacing?
7. Who to contact for product information or service? We suggest placing the organization's name and phone number inside the front panel.

Circuit Board Installation

This board is designed for universal usage.

For 115 volt (C2 & C4 series)
White on #1 and #3
Black on #2 and #4

For 208/240 volt (C7 & C9 series)
Black on #2 and #3
White on #1 and N/C

Installation

1. Transfer connections one at a time, terminal locations are the same.
2. Set potentiometer dials to the same corresponding position as the removed board.
3. If ice size requires adjustment, use fine adjustment only.

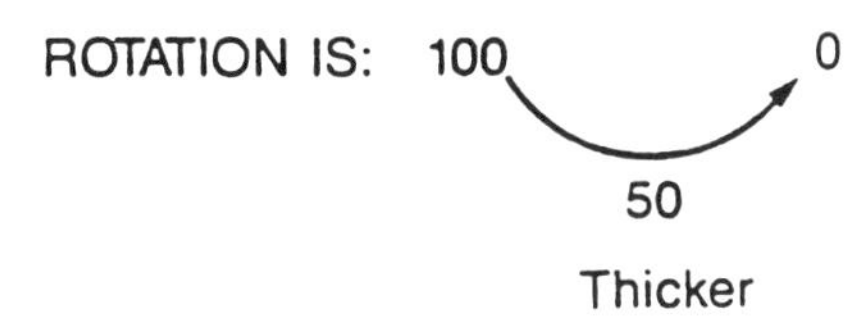

4. If board is completely out of adjustment,
   a. Center "Fine" Pot to "50".
   b. Rotate "Coarse" Pot to "0".
   c. Build ice until slab appears to be correct thickness.
   d. With curtain in place, slowly advance "Coarse" adjustment until harvest begins.
   e. Make small "Fine" Pot adjustments, ½" to 1" division at a time, until ⅛" to $3/16$" bridge is achieved.

SERVEND

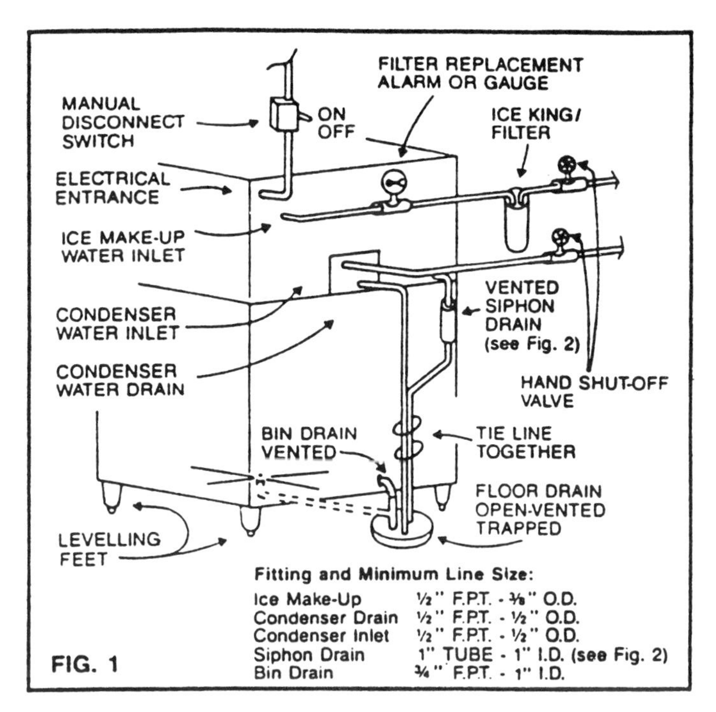

FIG. 1

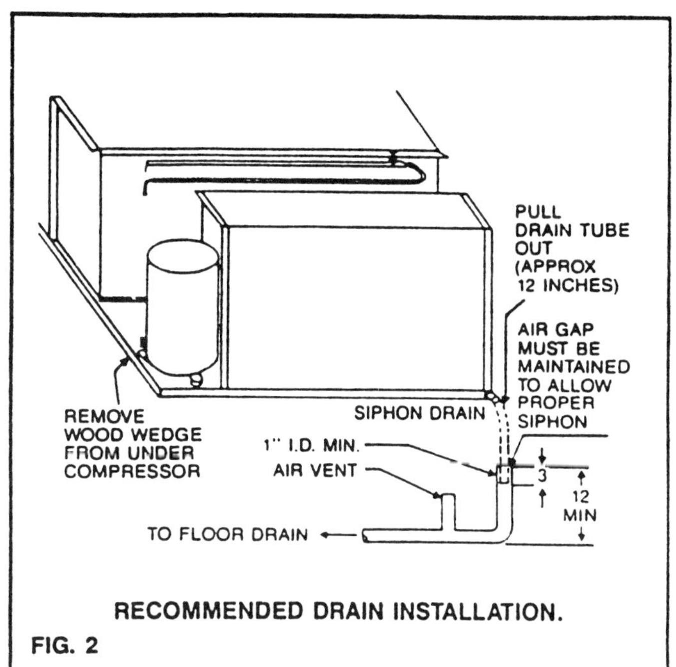

RECOMMENDED DRAIN INSTALLATION.

FIG. 2

Replacement Parts List

DESCRIPTION	PART NO.	C2F	C2M	C4F	C4M	C7F	C7M	C9F	C9M
Evap Assy. Full Cube	F010-295984	1							
Evap Assy. Full Cube	F010-295986			1					
Evap Assy. Full Cube	F010-295988					1			
Evap Assy. Full Cube	F010-295990							1	
Clip Dist Tube	P207-294940	2	2	2	2	2	2	2	2
Pin Tube Rotation	P014-297639	1	1	1	1	1	1	1	1
Evap Assy. Mini Cube	F010-295983		1						
Evap Assy. Mini Cube	F010-295985				1				
Evap Assy. Mini Cube	F010-295987						1		
Evap Assy. Mini Cube	F010-295989								1

DESCRIPTION	PART NO.	C2A	C2W	C4A	C4W	C7A	C7W	C9A	C9W
Fan Motor	P209-295993	1							
Fan Motor	P209-295994			1					
Fan Motor	P209-294922					1			
Fan Motor	P209-295995							1	
Blade Fan	P211-292270	1							
Blade Fan	P211-292271			1					
Blade Fan	P211-292272					1			
Blade Fan	P211-292273							1	
Condenser Air	P421-294971	1							
Condenser Air	P421-291186			1					
Condenser Air	P421-291187					1			
Condenser Air	P421-291188							1	
Condenser Water	P421-296452		1						
Condenser Water	P321-296453				1				
Condenser Water	P321-296454						1		
Condenser Water	P321-296455								1
Valve Water Regulator	P309-048125		1		1		1		1
Valve Access	P209-295943	2	2	2	2	2	2	2	2
Compressor W/Grommets Sleeves, S. Relay S. Capacitor	P016-294977	1	1						
Relay Start	P016-368707	1	1						
Capacitor Start	P016-368706	1	1						
Comp W/Grommet & Sleeve	P016-296415			1	1				
Comp W/Grommet & Sleeve	P016-206416					1	1		
Comp W/Grommet & Sleeve	P016-294918							1	1
Overload Protector	P016-368708	1	1						
Overload Protector	P016-368709			1	1				
Overload Protector	P016-368710					1	1		
Overload Protector	P016-368711							1	1
Relay Comp Start	P011-295941			1	1				
Relay Comp Start	P011-296421					1	1	1	1
Capacitor Start	P016-368726			1	1				
Capacitor Start	P011-295933					1	1	1	1
Capacitor Run	P011-295932					1	1		
Capacitor Run	P011-296422							1	1
Thermostat Safety	P211-269839			1	1	1	1	1	1
Thermostat Safety	P211-296838	1	1						
Cut Out High Pressure	P011-297676		1		1		1		1

DESCRIPTION	PART NO.	C2A	C2W	C4A	C4W	C7A	C7W	C9A	C9W
Fuse Circuit Board	P011-368712	1	1	1	1	1	1	1	1
Pin Curtain	P012-295931	2	2	2	2	2	2	2	2
Plate Curtain	F222-297693	2	2	2	2	2	2	2	2
Block Curtain	P214-297614	2	2	2	2	2	2	2	2
Nut Speed Pin Curtain	P012-297698	2	2	2	2	2	2	2	2
Screw Pin Plate Curtain	P012-297696	4	4	4	4	4	4	4	4
Nut Pin Plate Curtain	P012-297697	4	4	4	4	4	4	4	4
Front Panel Assy.	M300-297627	1	1						
Front Panel Assy.	M300-297629			1	1	1	1		
Front Panel Assy.	M300-297628							1	1
Clips Front Panel	P012-163873	2	2	2	2	2	2	2	2
Screws W/Washer Front Panel	P212-298871	2	2	2	2	2	2	2	2

DESCRIPTION	PART NO.	C2A	C2W	C4A	C4W	C7A	C7W	C9A	C9W
Drier	P209-295938	1	1	1	1	1	1		
Drier	P209-296420							1	1
Expansion Valve	P209-296861	1	1						
Expansion Valve	P209-296862			1	1	1	1		
Expansion Valve	P209-296863							1	1
Clamp Exp Valve Bulb	P009-365148	2	2	2	2	2	2	2	2
Coil - Solenoid Valve	P211-0295934	1	1	1	1				
Coil - Solenoid Valve	P211-296419					1	1	1	1
Valve - Solenoid	P211-295936	1	1	1	1	1	1		
Valve - Solenoid	P211-296418							1	1
WATER SYSTEM									
Pump Water	P011-297640	1	1	1	1				
Pump Water	P011-297641					1	1	1	1
Impeller/Magnet Assy. & 'O' Ring	P111-368718	1	1	1	1	1	1	1	1
Float Valve and Flow Restrictor Assy.	M200-368771	1	1	1	1	1	1	1	1
Flow Restrictor	P007-297695	1	1	1	1	1	1	1	1
Dist Tube Assy.	M300-294940	1	1						
Dist Tube Assy.	M300-294941			1	1				
Dist Tube Assy.	M300-294942					1	1	1	1
Curtain, Target	M100-368772	1	1						
Curtain, Target & Stop	M100-368773			1	1	1	1		
Curtain, Target & Stop	M100-368774							1	1
Target	F202-298836	1	1	1	1	1	1	1	1
Pan - Water	F311-294915	1	1	1	1	1	1	1	1
Fitting Reducer Water Inlet	P007-298833	1	1	1	1	1	1	1	1
Fitting Water Inlet Assy.	P209-297642	1	1	1	1	1	1	1	1
Circuit Board	P011-298804	1	1	1	1	1	1	1	1
Proximity Switch	P211-297675	1	1	1	1	1	1	1	1
Thermistor Evaporator W/Cup	M100-368775	1	1	1	1	1	1	1	1
Thermistor Ambient	P011-365147	1	1	1	1				
Thermistor Ambient	P011-365146					1	1	1	1
Switch On-Off	P011-917953	1	1	1	1	1	1	1	1
Contactor	P011-295942	1	1	1	1				
Contactor	P011-295947					1	1	1	1

EVAPORATOR ASSEMBLY

ITEM NUMBER 1,2,3,4,26,27,28 OR 29

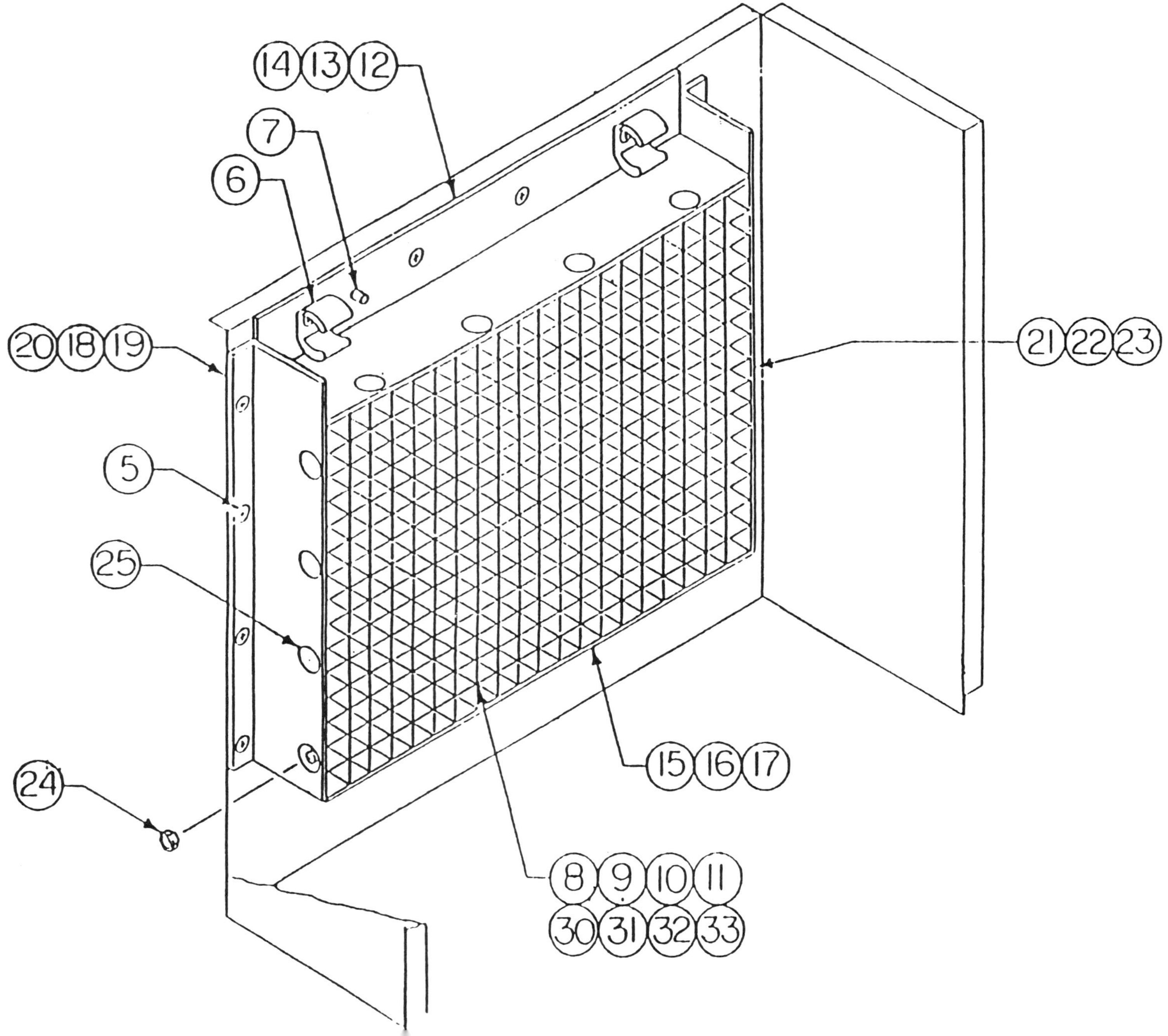

ITEM NO.	DESCRIPTION EVAPORATOR	PART NO.	QTY/UNIT C2-*F*	C2-*M*	C4-*F*	C4-*M*	C7-*F*	C7-*M*	C9-*F*	C9-*M*	SERIAL NO. START	END
1	Evaporator Assy.	F010-295984	1									
2	Evaporator Assy.	F010-295986			1							
3	Evaporator Assy.	F010-295988					1					
4	Evaporator Assy.	F010-295990							1			
5	Screws assy. to liner	P012-295927	9	9	12	12	12	12	12	12		
6	Clip-Distribution Tube	P207-294948	2	2	2	2	2	2	2	2		
7	Pin - Tube Rotation	P014-297639	1	1	1	1	1	1	1	1		
8	Evaporator Plated	F302-294407	1									
9	Evaporator Plated	F302-294412			1							
10	Evaporator Plated	F302-294413					1					
11	Evaporator Plated	F302-294414							1			
12	Extrusion Top	P314-292802	1	1								
13	Extrusion Top	P314-292803			1	1						
14	Extrusion Top	P314-292804					1	1	1	1		
15	Extrusion Bt'm	P314-292805	1	1								
16	Extrusion Bt'm	P314-292806			1	1						
17	Extrusion Bt'm	P314-292807					1	1	1	1		
18	Extrusion L.H.	P314-294927	1	1								
19	Extrusion L.H.	P314-294928			1	1	-	1				
20	Extrusion L.H.	P314-293011							1	1		
21	Extrusion R.H.	P314-292808	1	1								
22	Extrusion R.H.	P314-292809			1	1	1	1				
23	Extrusion R.H.	P314-292810							1	1		
24	Nuts - Mount Extrusion	P012-430490	12	12	16	16	16	16	16	16		
25	Sealent - Mounting Nuts	P014-294153										
26	Evaporator Assy.	F010-295983		1								
27	Mini Cube	F010-295985				1						
28	"	F010-295987						1				
29	"	F010-295989								1		
30	Evaporator Plated	F302-294935		1								
31	Evaporator Plated	F302-294408				1						
32	Evaporator Plated	F302-294409						1				
33	Evaporator Plated	F302-294410								1		

CONDENSER
COMPARTMENT

REPLACEMENT PARTS LIST

DATE February 8, 1989

ITEM NO.	DESCRIPTION CONDENSER COMPARTMENT	PART NO.	QTY/UN..								SERIAL NO.	
			C2-A**	C2-W**	C4-A**	C4-W**	C7-A**	C7-W**	C9-A**	C9-W**	START	END
1.	Fan Motor W/Mtg.Sc.& Bld. Nut	P209-295993	1									
2.	Fan Motor " " "	P209-295994			1							
3.	Fan Motor " " "	P209-294922					1					
4.	Fan Motor " " "	P209-295995							1			
5.	Blade Fan	P211-292270	1									
6.	Blade Fan	P211-292271			1							
7.	Blade Fan	P211-292272					1					
8.	Blade Fan	P211-292273							1			
9.	Bracket - Fan Motor	F302-296461	1		1							
10.	Bracket - Fan Motor	F302-292895					1		1			
11.	Well Nut - Motor Brkt	P012-294923	4		4		4		4			
12.	Screw - Brkt to shroud	P012-129595	4		4		4		4			
13.	Fan Shroud	F302-296406	1									
14.	Fan Shroud	P302-292931			1							
15.	Fan Shroud	F302-292932					1					
16.	Fan Shroud	F302-292933							1			
17.	Screws - Shroud to Cond.	P012-000460	6		6		6		6			
18.	Nuts - Shroud to base	P012-130490	2		2		2		2			
19.	Condenser air	P421-294971	1									
20.	Condenser air	P421-291186			1							
21.	Condenser air	P421-291187					1					
22.	Condenser air	P421-291188							1			
23.	Condenser water	P421-296452		1								
24.	Condenser water	P321-296453				1						
25.	Condenser water	P321-296454						1				
26.	Condenser water	P321-296455								1		
27.	Valve - Water regulator	P309-048125		1		1		1		1		
28.	Washer - Cond.	P012-167743		1		1		1		1		
29.	Nuts - Cond. bracket	P012-115197		1		1		1		1		
30.	Fitting - Cond. Water inlet	P209-297642		2		2		2		2		
31.	Screws - Water fitting	P012-297696		4		4		4		4		
32.	Locknut - Wat. Fitting	P012-297697		4		4		4		4		

COMPRESSOR COMPARTMENT

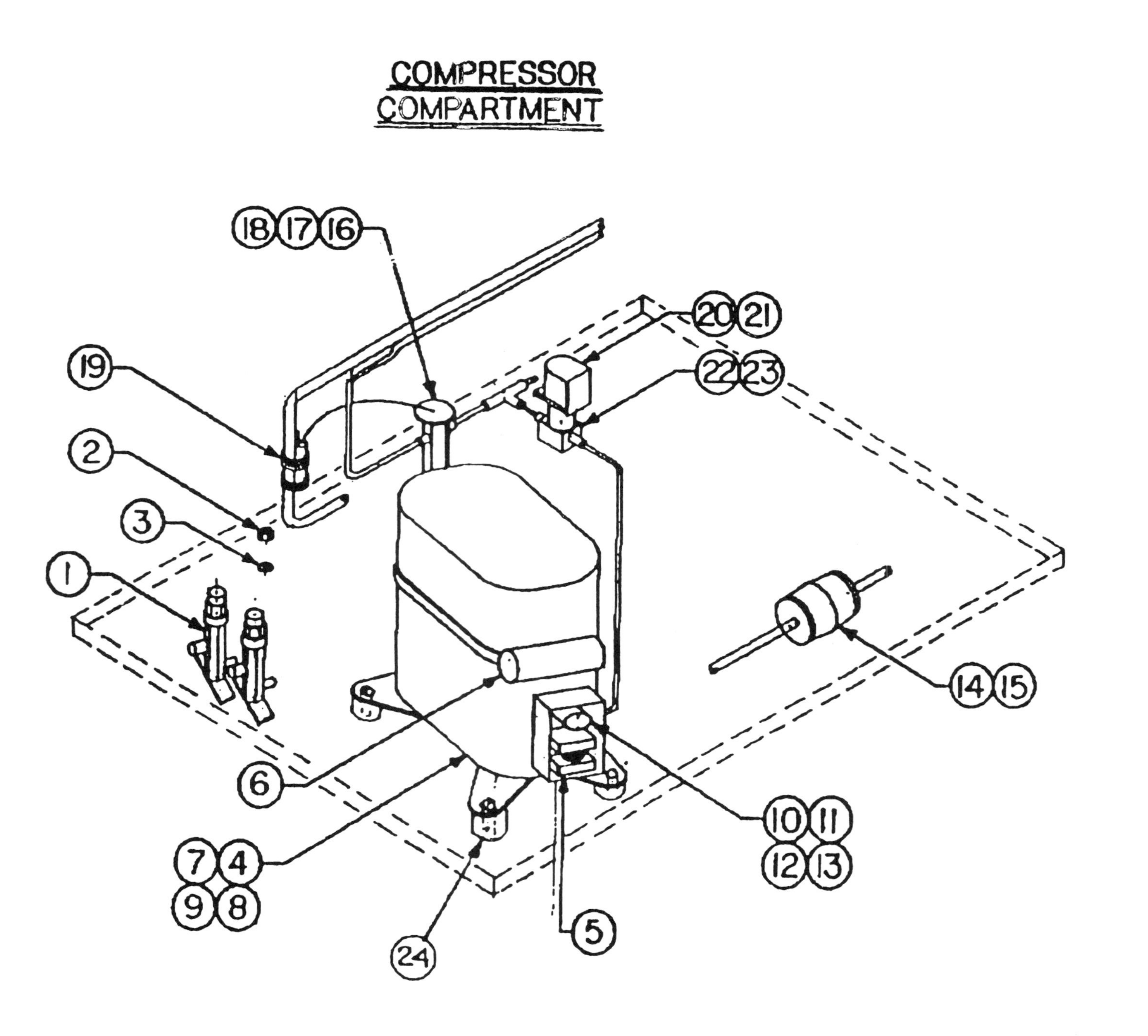

REPLACEMENT PART LIST

DATE February 8, 1989

ITEM NO.	DESCRIPTION COMPRESSOR COMPARTMENT	PART NO	QTY/UNIT C2-A**	C2-K**	C4-A**	C4-K**	C7-A**	C7-K**	C9-A**	C9-K**	SERIAL NO. START	END
1.	Valve Access	P209-295943	2	2	2	2	2	2	2	2		
2	Nut - Mount Access Valve	P012-130490	4	4	4	4	4	4	4	4		
3.	Lockwasher - Mount acc. valve	P012-296474	4	4	4	4	4	4	4	4		
4.	Compressor - W/grommets Sleeves, S. relay, S.Capacitor	P016-294977	1	1								
5.	Relay - start	P016-368707	1	1								
6.	Capacitor start	P016-368706	1	1								
7.	Compressor- W/grommets Sleeves, S.relay & Capacitors	P016-0368710			1	1						
8.	Compressor- W/grommets Sleeves, S.relay & Capacitors	P016-0368711					1	1				
9.	Compressor- W/grommets Sleeves, S.relay & Capacitors	P016-0368781							1	1		
10.	Overload - Protector	P016-0368708	1	1								
11.	Overload - Protector	P016-0368709			1	1						
14.	Drier	P209-295938	1	1	1	1	1	1				
15.	Drier	P209-296420							1	1		
16.	Expansion Valve	P209-296861	1	1								
17.	Expansion Valve	P209-296862			1	1	1	1				
18.	Expansion Valve	P209-296863							1	1		
19.	Clamp worm - Exp. valve Bulb	P009-365148	2	2	2	2	2	2	2	2		
20.	Coil - Solenoid Valve	P211-295934	1	1	1	1						
21.	Coil - Solenoid Valve	P211-296419					1	1	1	1		
22.	Valve - Solenoid	P211-295936	1	1	1	1	1	1				
23.	Valve - Solenoid	P211-296418							1	1		
24.	Grommet & Sleeve	P016-368776	1	1	1	1	1	1	1	1		

WATER SYSTEM

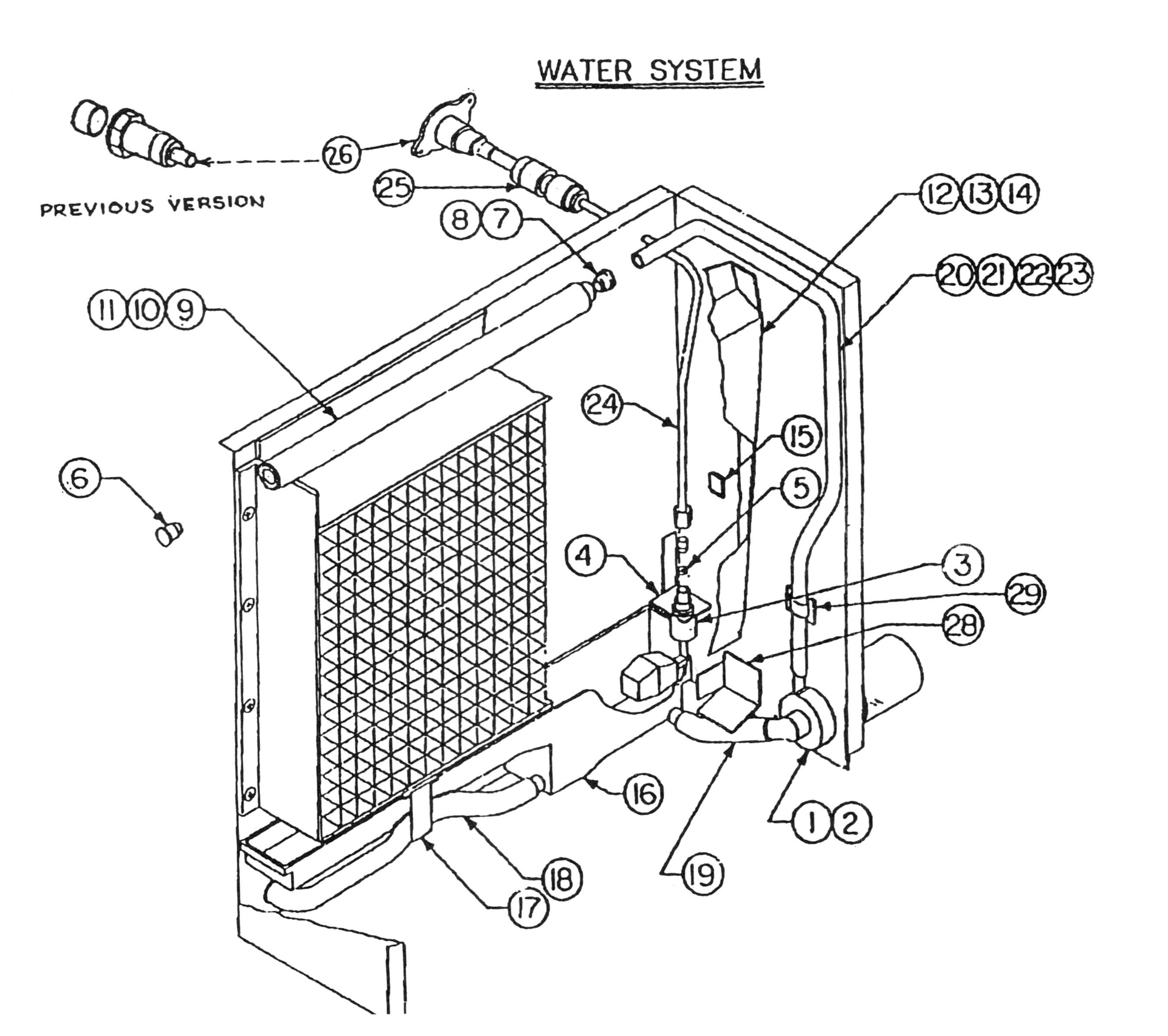

REPLACEMENT PARTS LIST

DATE February 8, 1989

ITEM NO.	DESCRIPTION WATER SYSTEM	PART NO.	QTY/UNIT C2-A**	C2-W**	C4-A**	C4-K**	C7-A**	C7-K**	C9-A**	C9-W**	SERIAL NO. START	END
1.	Pump Water	P011-297640	1	1	1	1						
2.	Pump Water	P011-297641					1	1	1	1		
3.	Float Valve Assy. W/FLOW RESTR.	P209-294949	1	1	1	1	1	1	1	1		
4.	Bracket Float Valve	F322-297690	1	1	1	1	1	1	1	1		
5.	Flow Control Washer	P007-297695	1	1	1	1	1	1	1	1		
6.	End plug - Distribution tube	P214-292901	1	1	1	1	1	1	1	1		
7.	Restrictor Plug Distributor	P214-299711	1	1								
8.	Restrictor Plug Distributor	P214-299712			1	1	1	1	1	1		
9.	Distributor tube assy.	M300-294940	1	1								
10.	Distributor tube assy.	M300-294941			1	1						
11.	Distributor tube assy.	M300-294942					1	1	1	1		
12.	Curtain Assy. W/TARGET	M100-0368772	1	1								
13.	Curtain Assy. W/TARGET & STOP	M100-0368773			1	1	1	1				
14.	Curtain Assy. W/TARGET & STOP	M100-0368774							1	1		
15.	Target	F202-298836	1	1	1	1	1	1	1	1		
16.	Pan - Water	P311-294915	1	1	1	1	1	1	1	1		
17.	Clip - Siphon hose	F222-294991	1	1	1	1	1	1	1	1		
18.	Hose - Siphon	P212-297674	1	1	1	1	1	1	1	1		
19.	Hose - Pan to pump	P212-297674	1	1	1	1	1	1	1	1		
20.	Hose - pump to distributor	P312-296426	1	1								
21.	Hose - pump to distributor	P212-296427			1	1						
22.	Hose - pump to distributor	P212-294993					1	1				
23.	Hose - pump to distributor	P212-294964							1	1		
24.	Tube - Water inlet	P214-294990	1	1	1	1	1	1	1	1		
25.	Fitting Reducer	P007-298833	1	1	1	1	1	1	1	1		
26.	Fitting Water Inlet	P209-297642	1	1	1	1	1	1	1	1		
26.	Fitting Water Inlet	P007 298831	1	1	1	1	1	1	1	1		
28.	Bracket - Water pan	F322-297689	1	1	1	1	1	1	1	1		
29.	Hose clamp - Liner wall	P014-295924	2	2	2	2	2	2	2	2		

CONTROL CIRCUIT

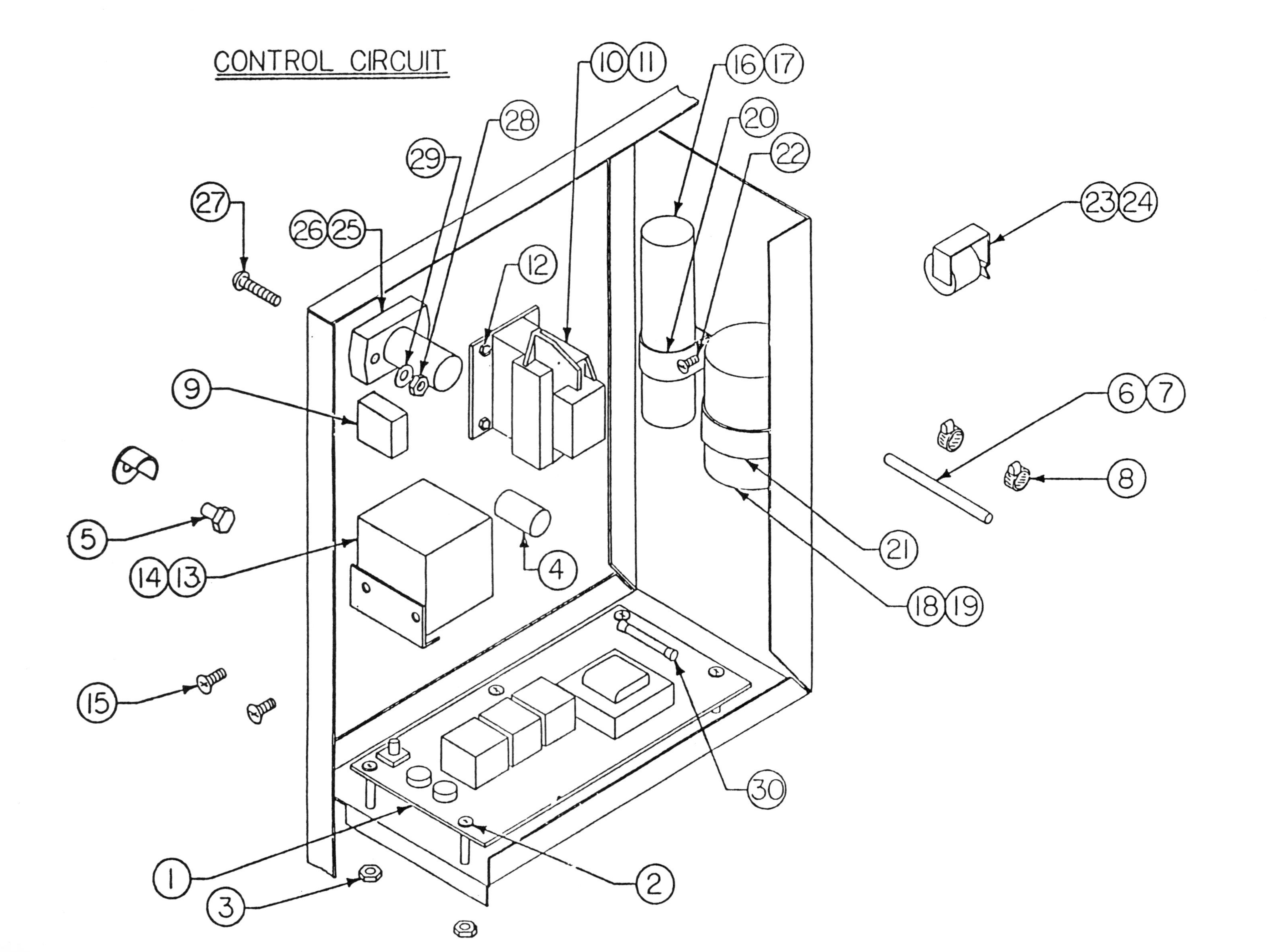

ITEM 19	DESCRIPTION CONTROL CIRCUIT	PART NO.	C2-A**	C2-W*	C4-A*	C4-W**	C7-A**	C7-W**	C9-A**	C9-W**	SERIAL NO. START	END
	Circuit Board	P011-298804	1	1	1	1	1	1	1	1		
2.	Screw to mount board	P012-070090	5	5	5	5	5	5	5	5		
3.	Nuts - to mount board	P012-130490	5	5	5	5	5	5	5	5		
4.	Proximity switch	P211-297675	1	1	1	1	1	1	1	1		
5.	Thermistor evaporator (w/cap)	M100-368775	1	1	1	1	1	1	1	1		
6.	Thermistor Ambient	P011-365147	1	1	1	1						
7.	Thermistor Ambient	P011-365146					1	1	1	1		
8.	Clamp Horm-Amb. therm.	P009-365188	2	2	2	2	2	2	2	2		
9.	Switch - ON-OFF	P011-917953	1	1	1	1	1	1	1	1		
10.	Contactor	P011-295942	1	1	1	1						
11.	Contactor	P011-295947					1	1	1	1		
12.	Nuts - Contactor mount	P012-296474	3	3	3	3	3	3	3	3		
13.	Relay - Comp start	P011-295941			1	1						
14.	Relay - Comp start	P011-296421					1	1	1	1		
15.	Screws - Relay mount	P012-296483			2	2	2	2	2	2		
16.	Capacitor start	P016-368726			1	1						
17.	Capacitor - start	P011-295933					1	1	1	1		
18.	Capacitor - run	P011-295932					1	1				
19.	Capacitor - run	P011-296422							1	1		
20.	Strap - start capacitor	P202-297633			1	1	1	1	1	1		
21.	Strap - run capacitor	F202-297632					1	1	1	1		
22.	Screw - capacitor straps	P012-297643			1	1	2	2	2	2		
23.	Thermostat - safety	P211-296839			1	1	1	1	1	1		
24.	Thermostat - safety	P211-296838	1	1								
25.	Cut out - High pressure	P011-297676		1		1	1			1	RETROFIT W/P011-0295937	
27.	Screws - to mount Hi out	P012-70090		2		2	2			2		
29.	Washer - Lock - Cutout	P012-000195		2		2	2			2		
30.	Fuse - Circuit board	P011-368712	1	1	1	1	1	1	1	1		

WATER SYSTEM

ITEM NO.	DESCRIPTION WATER SYSTEM	PART NO.	QTY/UNIT C2-A**	C2-W**	C4-A**	C4-K**	C7-A**	C7-W**	C9-A**	C9-W**	SERIAL NO. START	END
1.	Pump Motor	P011-368720	1	1	1	1						
2.	Pump Motor	P011-368721					1	1	1	1		
3.	Impeller - Magnet Assy & 'O' Ring - Viton	P011-368718	1	1	1	1	1	1	1	1		
			1	1	1	1	1	1	1	1		
			1	1	1	1	1	1	1	1		
			1	1	1	1	1	1	1	1		
			6	6	6	6	6	6	6	6		
			1	1	1	1	1	1	1	1		
			1	1	1	1	1	1	1	1		
			1	1	1	1	1	1	1	1		
			1	1	1	1	1	1	1	1		
			4	4	4	4	4	4	4	4		
			4	4	4	4	4	4	4	4		
14	Pin - Curtain	P012-295931	2	2	2	2	2	2	2	2		
15.	Plate - Curtain pin	P222-297693	2	2	2	2	2	2	2	2		
16.	Block - Curtain pin	P214-297614	2	2	2	2	2	2	2	2		
17.	Nut - Speed pin	P012-297698	2	2	2	2	2	2	2	2		
18.	Screw - Pin plate	P012-297696	4	4	4	4	4	4	4	4		
19.	Nut - Pin plate	P012-297697	4	4	4	4	4	4	4	4		
20.	Outer tube	P314-292888	1	1								
21.	Outer tube	P314-292907			1	1						
22.	Outer tube	P314-292912					1	1	1	1		
23.	Inner tube	P214-292887	1	1								
24.	Inner tube	P214-292906			1	1						
25.	Inner tube	P214-292911					1	1	1	1		
26.	Grommet	P214-292900	2	2	2	2	2	2	2	2		
27.	Block - L.H. Water pan	P214-294989	1	1	1	1	1	1	1	1		
28.	Block - R.H. Water pan	P214-296858	1	1	1	1	1	1	1	1		
29.	Screws - Pan block	P012-297643	4	4	4	4	4	4	4	4		

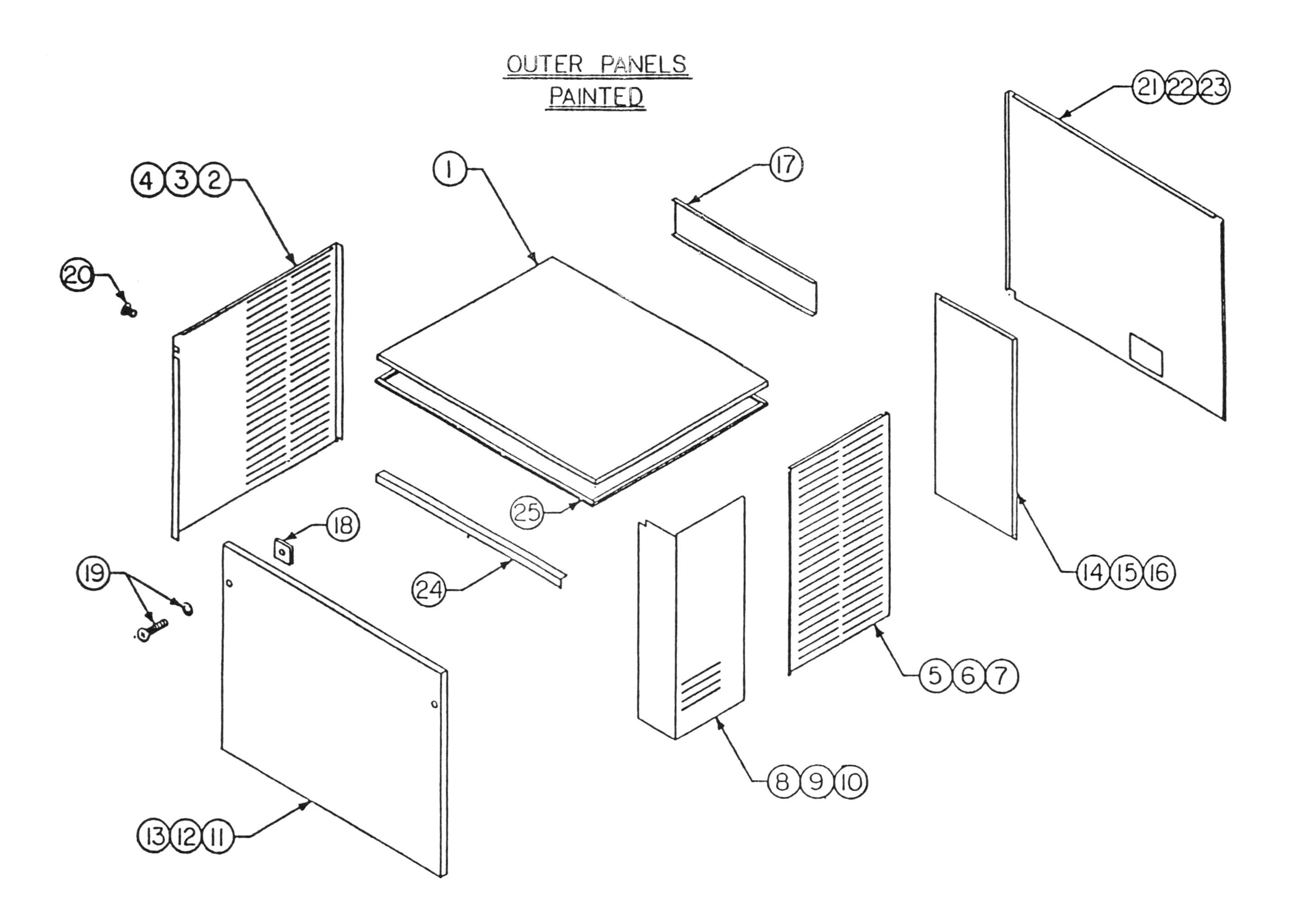
OUTER PANELS
PAINTED
21 22 23
4 3 2
1
17
20
25
18
19
24
14 15 16
5 6 7
8 9 10
13 12 11

ITEM NO.	DESCRIPTION	PART NO.	C2-A**	C2-W**	C4-A**	C4-W**	C7-A**	C7-W**	C9-A**	C9-W**	SERIAL NO. START	SERIAL NO. END
	OUTER PANELS											
	BEIGE POWDER											
1.	Cover top	F302-292264	1	1	1	1	1	1	1	1		
2.	Side panel L.H.	F302-295901	1	1								
3.	Side Panel L.H.	F302-292267			1	1	1	1				
4.	Side Panel L.H.	F302-296413							1	1		
5.	Side Panel R.H.	F302-295902	1	1								
6.	Side Panel R.H.	F302-294970			1	1	1	1				
7.	Side Panel R.H.	F302-292969							1	1		
8.	Cover - Control Box	F202-296408	1	1								
9.	Cover - Control Box	F202-294992			1	1	1	1				
10.	Cover - Control Box	F202-295904							1	1		
11.	Front Panel - Assy.	M300-297627	1	1								
12.	Front Panel - Assy.	M300-297629			1	1	1	1				
13.	Front Panel - Assy.	M300-297628							1	1		
14.	Rear Panel - Air	F302-296403	1									
15.	Rear Panel - Air	F302-292965			1		1					
16.	Rear Panel - Air	F302-292977							1			
17.	Filler - Back Panel	F202-294947							1			
18.	Clips - Front Panel	P012-286079	2	2	2	2	2	2	2	2	*incorrect (should Be 163873)*	
19.	Screws w/washer front panel	P212-298871	2	2	2	2	2	2	2	2		
20.	Screws - Outer panels	P012-297687	26	26	26	26	26	26	29	29		
21.	Rear Panel - Water	F302-298806		1								
22.	Rear Panel - Water	F302-298805				1		1				
23.	Rear Panel - Water	F302-298807								1		
24.	Brace - Liner	F202-294994	1	1	1	1	1	1	1	1		
25.	Foam - Insulation Tape (Inch)	P004-048701	75	75	75	75	75	75	75	75		

OUTER PANELS
STAINLESS STEEL

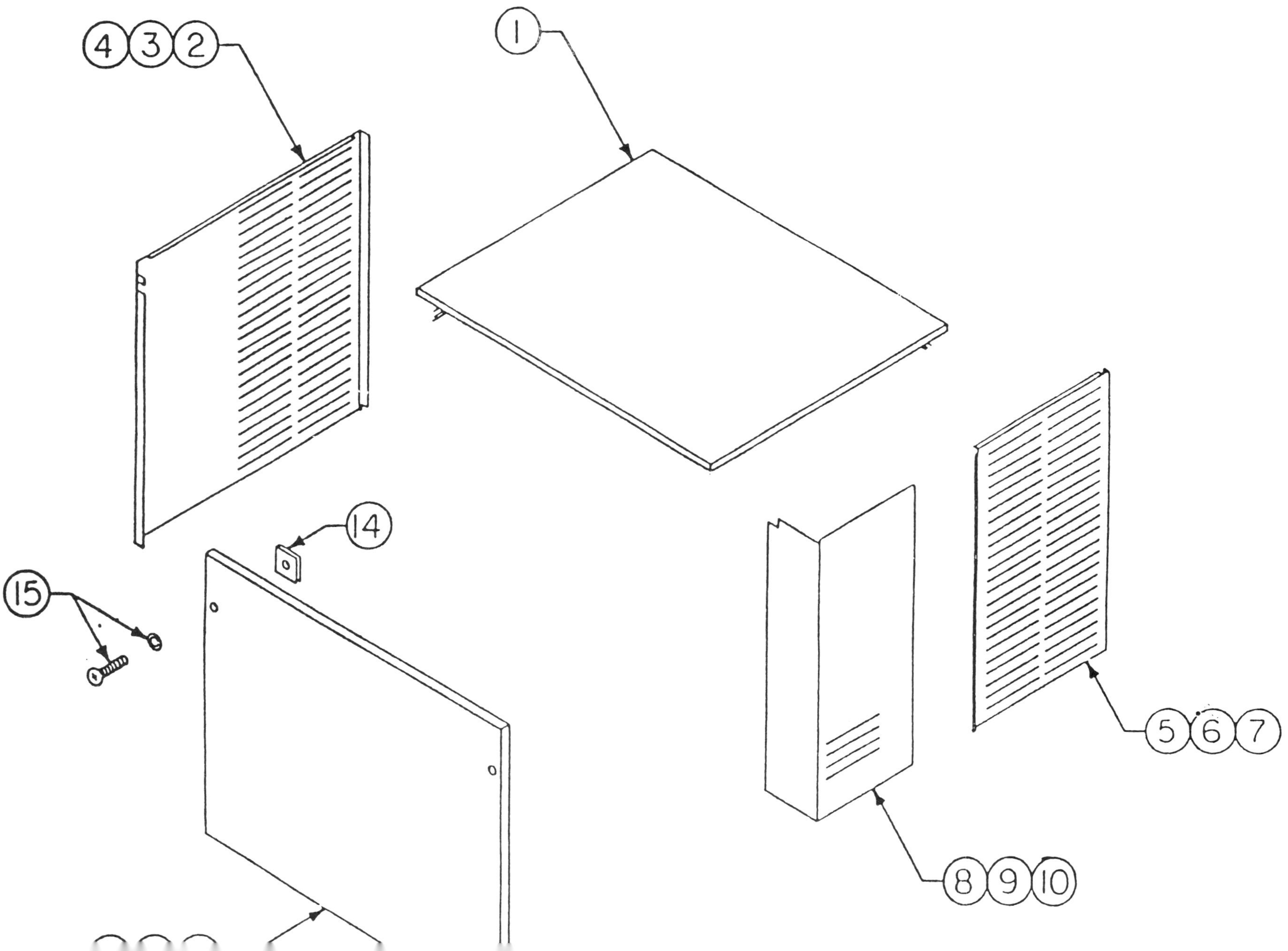

ITEM NO.	DESCRIPTION	PART NO.	UNIT C2-A**	C2-W**	C4-A**	C4-W**	C7-A**	C7-W**	C9-A**	C9-W**	SERIAL NO. START	END	
	OUTER PANELS												
	STAINLESS STEEL												
1.	Cover top	F322-296432	1	1	1	1	1	1	1	1			
2.	Side Panel L.H.	F322-296429	1	1									
3.	Side Panel L.H.	F322-296435			1	1	1	1					
4.	Side Panel L.H.	F322-296440							1	1			
5.	Side Panel R.H.	F322-298862	1	1									
6.	Side Panel R.H.	F322-298861			1	1	1	1					
7.	Side Panel R.H.	F322-298863							1	1			
8.	Contr. Box Cover Assy.	M300-365112	1	1									
9.	Contr. Box Cover Assy.	M300-365113			1	1	1	1					
10.	Contr. Box Cover Assy.	M300-365111							1	1			
11.	Front Panel Assy.	F022-297624	1	1									
12.	Front Panel Assy.	F022-297621			1	1	1	1					
13.	Front Panel Assy.	F022-297617							1	1			
14.	Clips - Front Panel	P012-286079	2	2	2	2	2	2	2	2			
15.	Screw W/Washer Front Panel	P012-130485	2	2	2	2	2	2	2	2			

CARTON ASSY

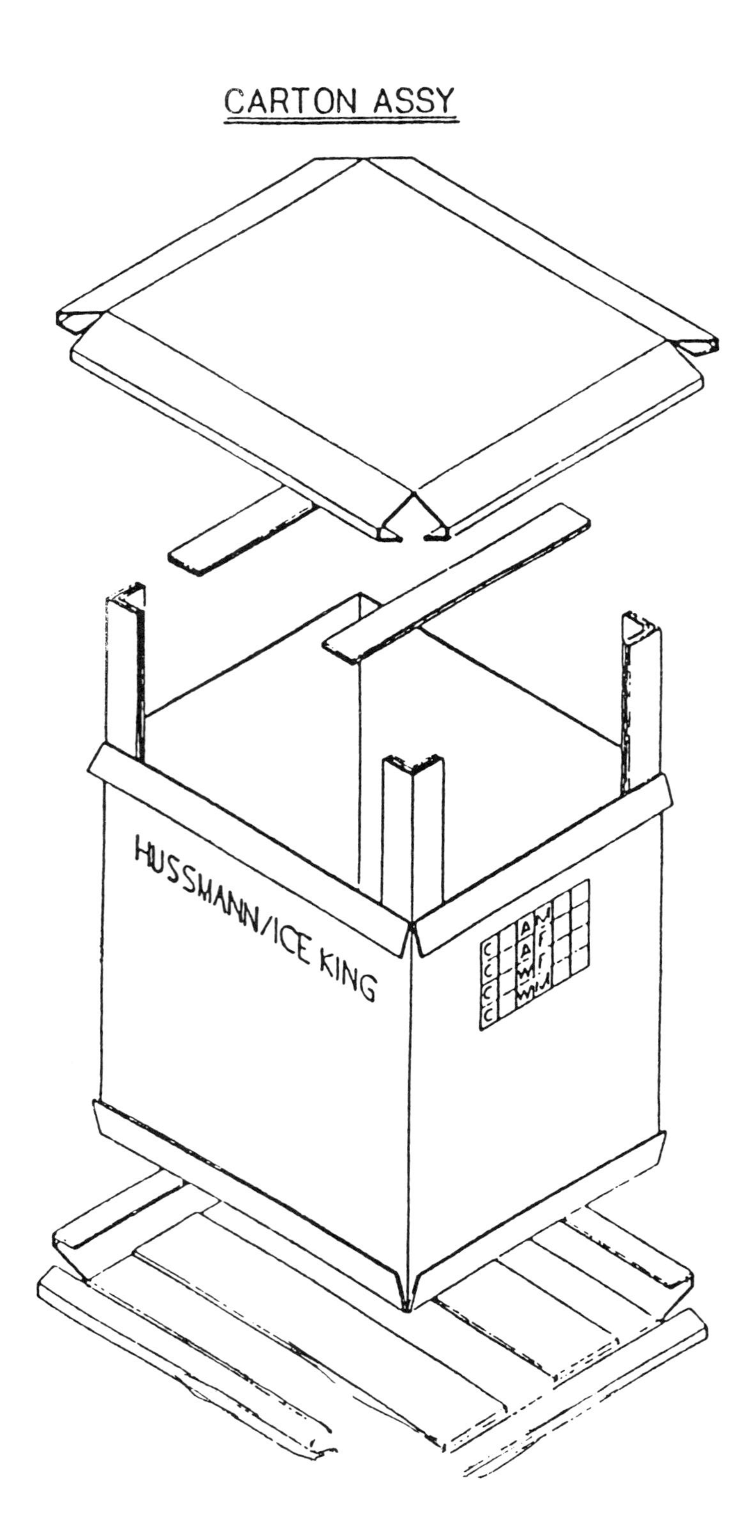

ITEM NO.	DESCRIPTION CARTON ASSY. AND MISCELLANEOUS PARTS	PART NO.	C2-A**	C2-W**	C4-A**	C4-W**	C7-A**	C7-W**	C9-A**	C9-W**	SERIAL NO. START	SERIAL NO. END
1.	Carton Assy.	P213-365140	1	1								
2.	Carton Assy.	P213-365141			1	1	1	1				
3.	Carton Assy.	P213-365142							1	1		
4.	Wiring Diagram	P115-296814	1	1								
5.	Wiring Diagram	P115-296811			1	1						
6.	Wiring Diagram	P115-296812					1	1	1	1		
7.	Curtain Release Label	P215-299710	1	1	1	1	1	1	1	1		
8.	Low Side/High Side Label	P215-296806	1	1	1	1	1	1	1	1		
9.	Elect. Inlet Label	P215-296804	1	1	1	1	1	1	1	1		
10.	Moving Parts Label	P215-296803	1		1		1		1			
11.	On/Off Label	P215-296802	1	1	1	1	1	1	1	1		
12.	Water Inlet Label	P215-296801	1	1	1	1	1	1	1	1		
13.	Shock Hazard Label	P215-296800	1	1	1	1	1	1	1	1		
14.	Operating Instruction Label	P115-296840	1	1	1	1	1	1	1	1		
15.	Potentiometer Direction Label	P215-298867	1	1	1	1	1	1	1	1		
16.	Cuber Drain Installation Label	P214-296816	1	1	1	1	1	1	1	1		
17.	Harvest Termination Label	P315-298822	1	1	1	1	1	1	1	1		
18.	High Pressure Reset Label	P215-296805		1		1		1		1		
19.	Condenser Water Inlet Label	P215-296807		1		1		1		1		
20.	Condenser Drain	P215-296808		1		1		1		1		

ITEM NO.	DESCRIPTION BIN ASSY	PART NO.	QTY/UNIT B2	B4	B6	B10					SERIAL NO. START	END	
1.	Door Assy	P314-295917	1										
2.	Door Assy	P314-294987		1									
3.	Door Assy	P314-365100			1								
4.	Door Assy	P314-295974				1							
5.	Hinge Assy (L.H.)	F202-292875	1	1	1	1							
6.	Hinge Assy (R.H.)	F202-292876	1	1	1	1							
7.	Hinge Door	F222-296469	2	2	2	2							
8.	Screw (Door Hinge)	PO12-296483	4	4	4	4							
9.	Screw (Hinge Assy)	PO12-295927	4	4	4	4							
10.	Lockwasher (Door Hinge)	PO12-034073	4	4	4	4							
11.	Drain (Bottom)	P014-228830	1	1		1							
12.	washer	PO14-228831	1	1		1							
13.	Locknut	P007-289398	1	1		1							
14.	Scoop-Ice	P014-294924	1	1	1	1							
15.	Leg-Bin	P012-295922	4	4	4	4							
16.	Carton Assy 2	P313-295920	1										
17.	Carton Assy 4	P313-295921		1									
18.	Carton Assy 6	PO13-365109			1								
19	Carton Assy 10	P313-295978				1							
20.	Gasket (Roll 50 Ft)	P004-285602	1	1	1	1							
21.	Panel Header (2)	F322-292874	1										
22.	Panel Header (4)	F322-292257		1									
23.	Panel Header (6)	F322-299799			1								
24.	Panel Header (10)	F322-295956				1							

SYSTEM OPERATION

WATER SYSTEM

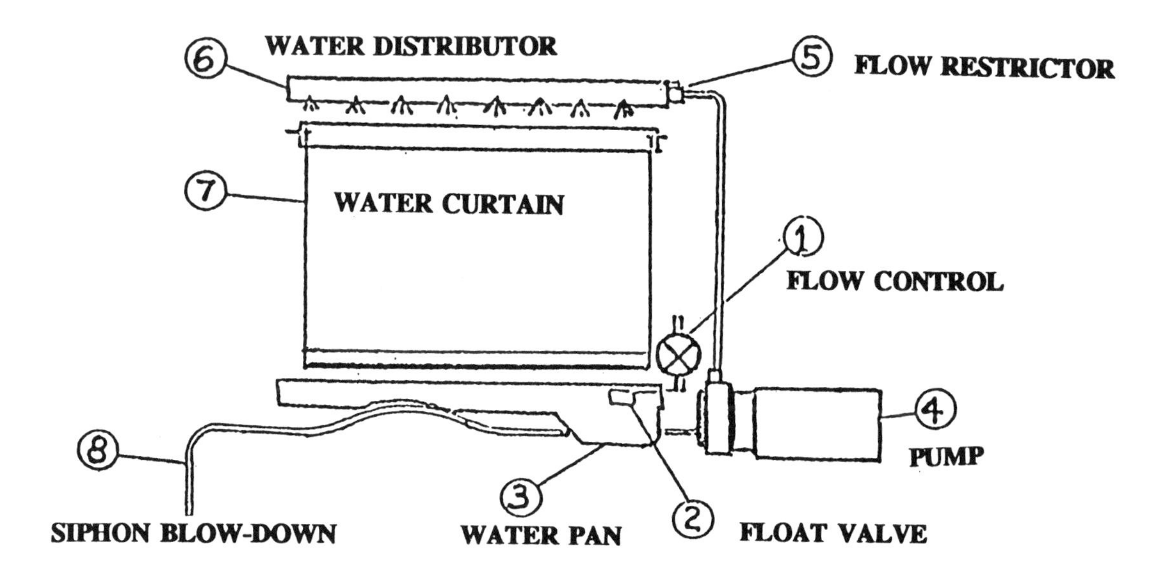

WATER SYSTEM OPERATION

1. FLOW CONTROL
   - A. Limits volume or maximum water flow through the float valve (control up to 125 lbs. of inlet pressure).
   - 2. Stops float valve chatter caused by over pressure.

2. FLOAT VALVE
   - A. Modulates - opens and closes metering water flow into pan.
   - B. Sets height of water in pan.

3. WATER PAN
   - A. Reservoir supplies water pump. At the end of the freeze cycle, the water pump stops, water rises in the water pan flooding the hose, activating the siphon.

4. WATER PUMP
   - A. Lifts water to the distribution tube.

5. FLOW RESTRICTOR
   - A. Limits water volume into the distribution tube.

6. DISTRIBUTION TUBE
   - A. Distributes water evenly over the evaporator.

7. WATER CURTAIN
   - A. Direct excess water back to the water pan for recirculation.

8. SIPHON
   - A. Removes mineral saturated water from the pan. For the siphon to break,the volume of water leaving the pan must be greater than the amount of water entering through the float. The siphon must be vented so air can be drawn through the siphon hose.

REFRIGERATION SYSTEM

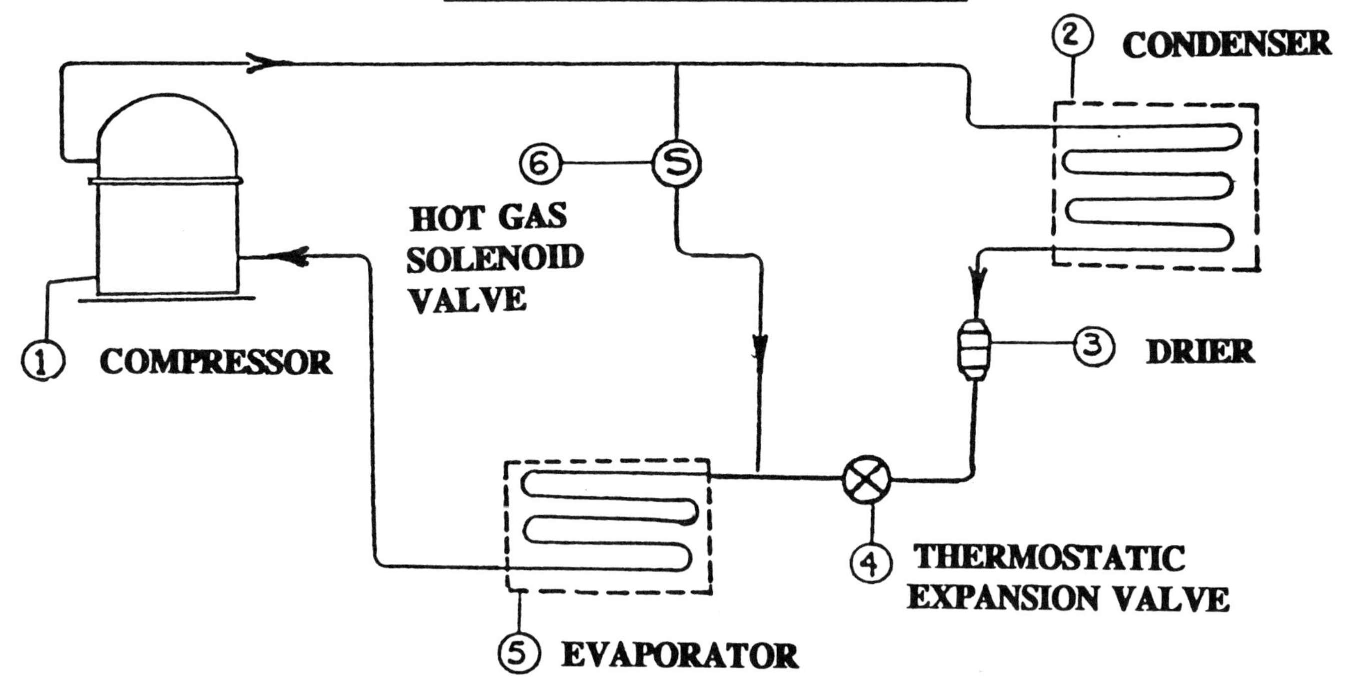

6 COMPONENTS

1) COMPRESSOR
   TO PUMP HEATED GAS

2) CONDENSER
   TO COOL HOT GAS TO LIQUID

3) DRIER
   TO REMOVE MOISTURE AND ACID

4) EXPANSION VALVE
   TO METER LIQUID REFRIGERANT

5) EVAPORATOR
   FREEZING PLATE

6) SOLENOID
   TO PERMIT HOT GAS TO FLOW TO THE FREEZING PLATE

EVAPORATOR THERMISTOR

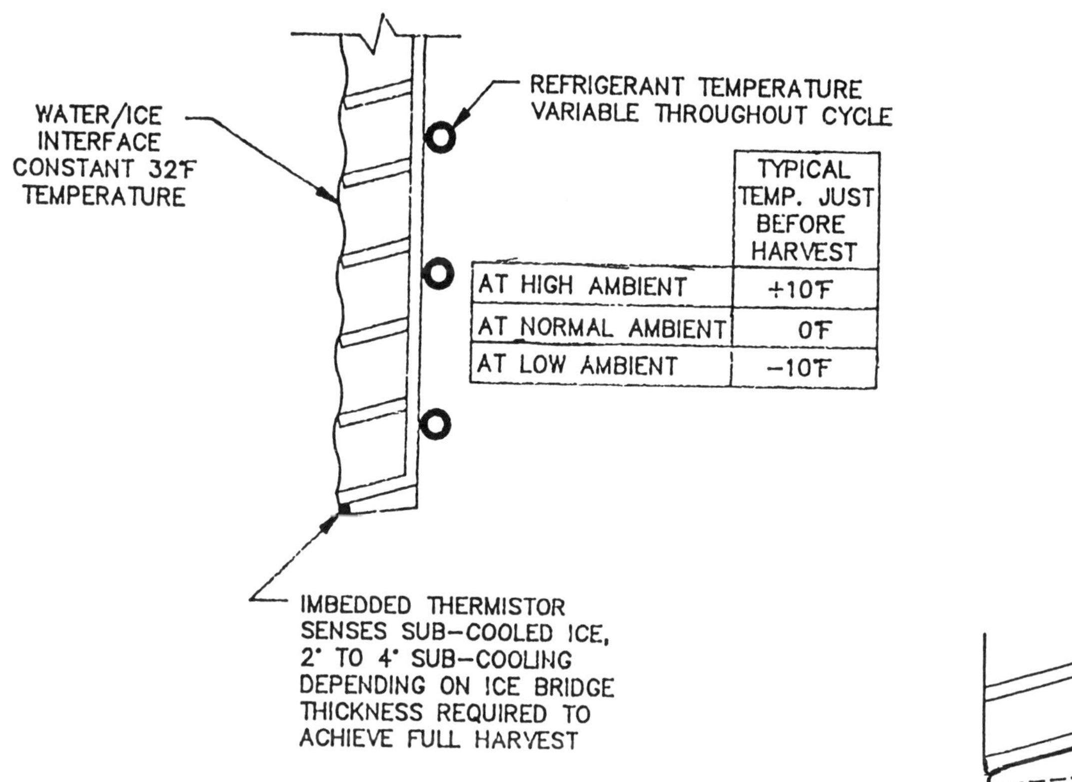

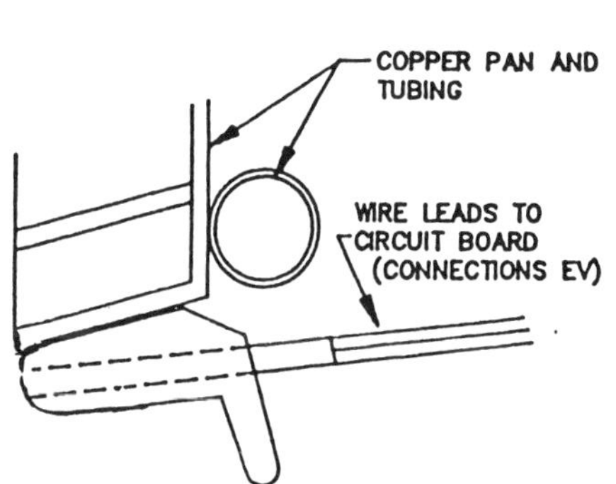

If the thermistor is "open", the cuber will start in harvest:

A. Adjusting the pots will not put the unit into freeze.

B. Bin switch will trip harvest light off and on.

If the thermistor is "shorted", the cuber will remain in the freeze cycle:

A. Manual harvest switch will not initiate harvest.

B. Adjusting the pots will not put the unit into harvest.

EVAPORATOR THERMISTOR TEMPERATURE RESISTANCE READINGS

TEMP F	TEMP C	RESISTANCE
-0.4	-18	86494
32	0	32650
75.2	24	10450

THERMISTOR (NEW STYLE) AFTER 8/90

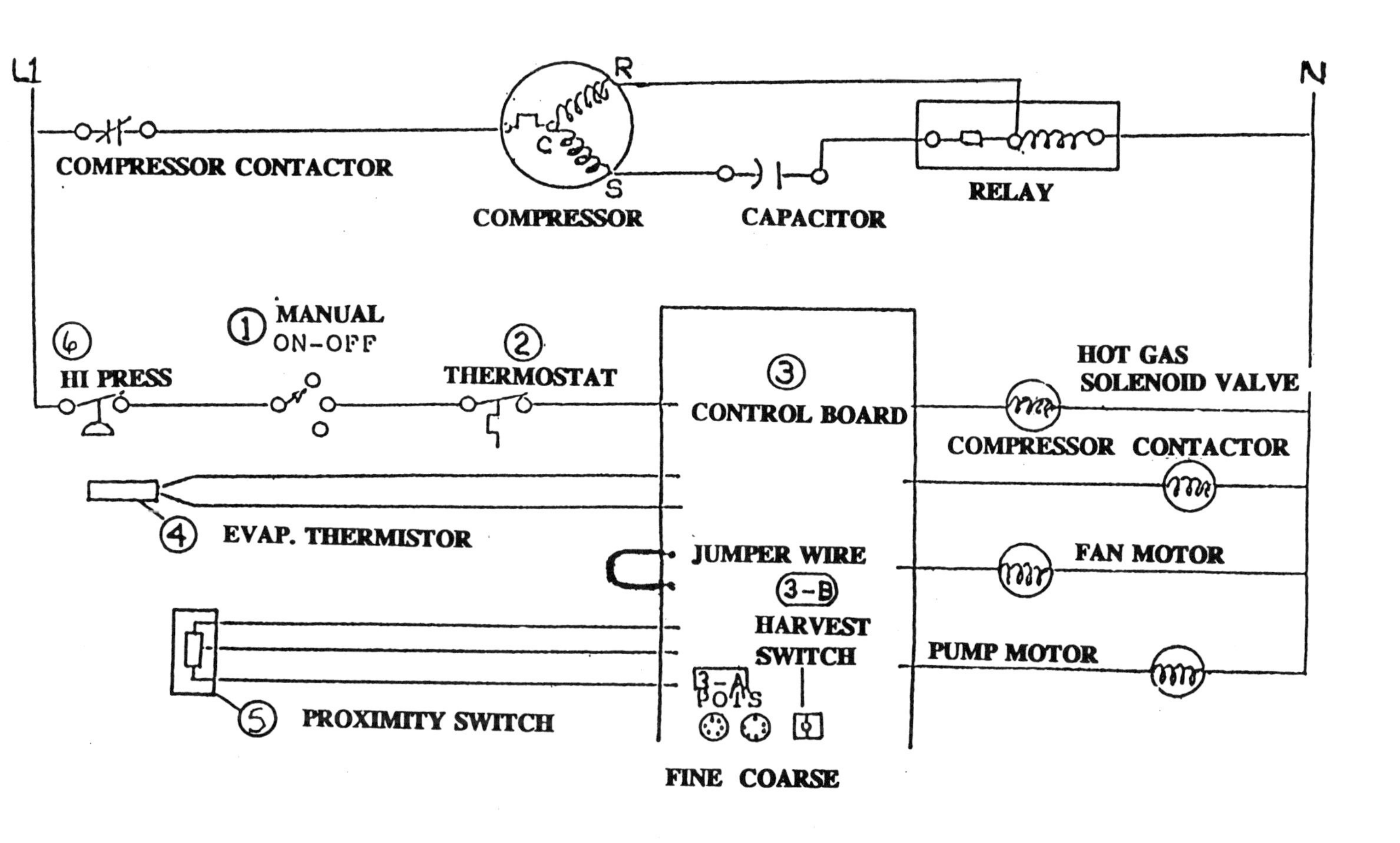

CONTROL CIRCUIT - 6 PARTS

1. **3 POSITION ON/OFF SWITCH:**

   ON - POWERS CONTROL BOARD
   OFF - SHUTS DOWN THE CUBER
   PUMP - CLEANING CYCLE

2. **THERMOSTAT - SAFETY THERMODISC:**

   PROTECTS UNIT IF IT FAILS TO COME OUT OF HARVEST:

   TWO LOCATIONS:
   WATER COOLED - INLET OF EVAPORATOR
   AIR COOLED - SUCTION LINE

3. **CONTROL BOARD**

   A. **FREEZE CYCLE:** RELAYS #2 & #3
   COMPRESSOR CONTACTOR
   FAN MOTOR/WATER PUMP

   B. **HARVEST CYCLE:**
   HOT GAS SOLENOID VALVE - RELAY #1
   COMPRESSOR CONTACTOR - RELAY #2

   C. FINE & COARSE POTENTIOMETERS:
   ADJUST THE BRIDGE THICKNESS

   D. MANUAL HARVEST SWITCH:
   BYPASSES FREEZE CYCLE

4. **EVAPORATOR THERMISTOR:**

   MONITORS ICE TEMPERATURE

5. **PROXIMITY SWITCH:**

   A. TERMINATES HARVEST CYCLE
   B. BIN CONTROL

6. **HIGH PRESSURE SAFETY CONTROL:**

   PROTECTS UNIT IF THE OPERATION PRESSURE IS TOO HIGH

SERVICE BULLETIN

BOTTOM MOUNT EVAPORATOR THERMISTOR

1. Shut off power and water supply to ice cuber. Remove siphon clip from water pan and slide the left side of the pan forward and downward to permit room to reach under the bottom molding of the evaporator.

2. Drill 1/8 inch diameter hole through the bottom plastic molding approximately 3.5 inches from the right side of the evaporator. The hole is to be located directly under the vertical divider. See illustration.

3. Disconnect the evaporator thermistor wires from the "EV" terminals and the liquid line thermistor wires from the "CD" (screw) terminals on the circuit board. See illustration. Clip the thermistor wires from the evaporator and liquid line. (It is not necessary to remove disconnected thermistors from the evaporator and liquid line.)

4. Attach the jumper wire across the liquid line thermistor terminals "CD" on the circuit board. Note: The liquid line thermistor is used on air cooled models only. See illustration.

5. Connect the bottom mount thermistor to terminals "EV" on circuit board. Pass thermistor through plastic bushing in control box side, down and under evaporator side molding and insert in back of the hole drilled in the bottom of the molding. See illustration. The tip of the thermistor should be flush to the slightly exposed molding surface.

6. Seal thermistor to molding at both entry and exit points with G.E. silicone adhesive. Allow 10 minutes for silicone to set up. (Silicone not included.)

7. Remove plastic drill chips from water pan, re-assemble siphon and pan. Place the ON-OFF switch in wash position, turn on water and power, clean evaporator with LIME-A-WAY. Flush and sanitize unit.

8. Place switch in "ice" position. Adjust ice bridge thickness per the following "Bottom Mount Thermistor Bridge Adjustment Instructions".

BOTTOM MOUNT THERMISTOR BRIDGE ADJUSTMENT INSTRUCTIONS:

POTENTIOMETER ROTATION IS : 100 → 0

THINNER BRIDGE 50 THICKER BRIDGE

A. Center "Fine" pot to "50".
B. Center "Coarse" pot to "50".
C. Build ice until slab appears to be correct thickness.
D. With curtain in place, slowly advance "Course" adjustment until harvest begins.
E. Make small "Fine" pot adjustments, 1/2 to 1 division at a time, until 1/8 to 3/16 bridge is achieved.

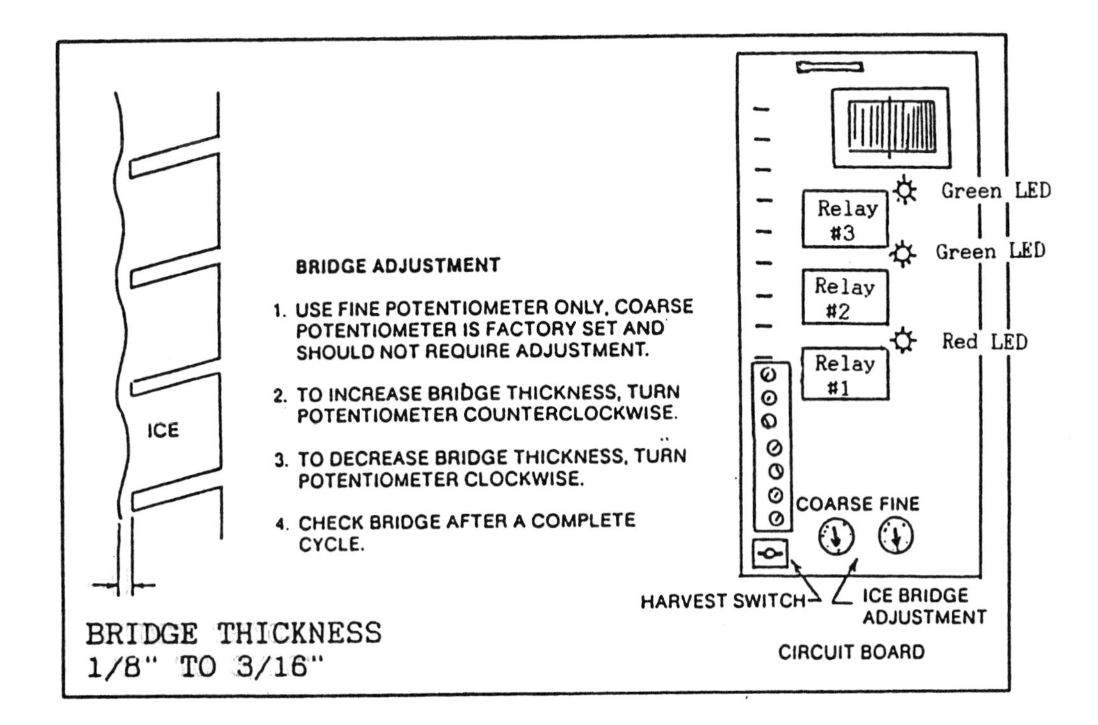

THE LARGER THE POTS NUMBER, THE LIGHTER THE ICE BRIDGE.
THE SMALLER THE POTS NUMBER, THE THICKER THE ICE BRIDGE.

2100 Future Drive
Sellersburg Indiana 47172-1868

1-800-423-3878 or (812) 246-5452

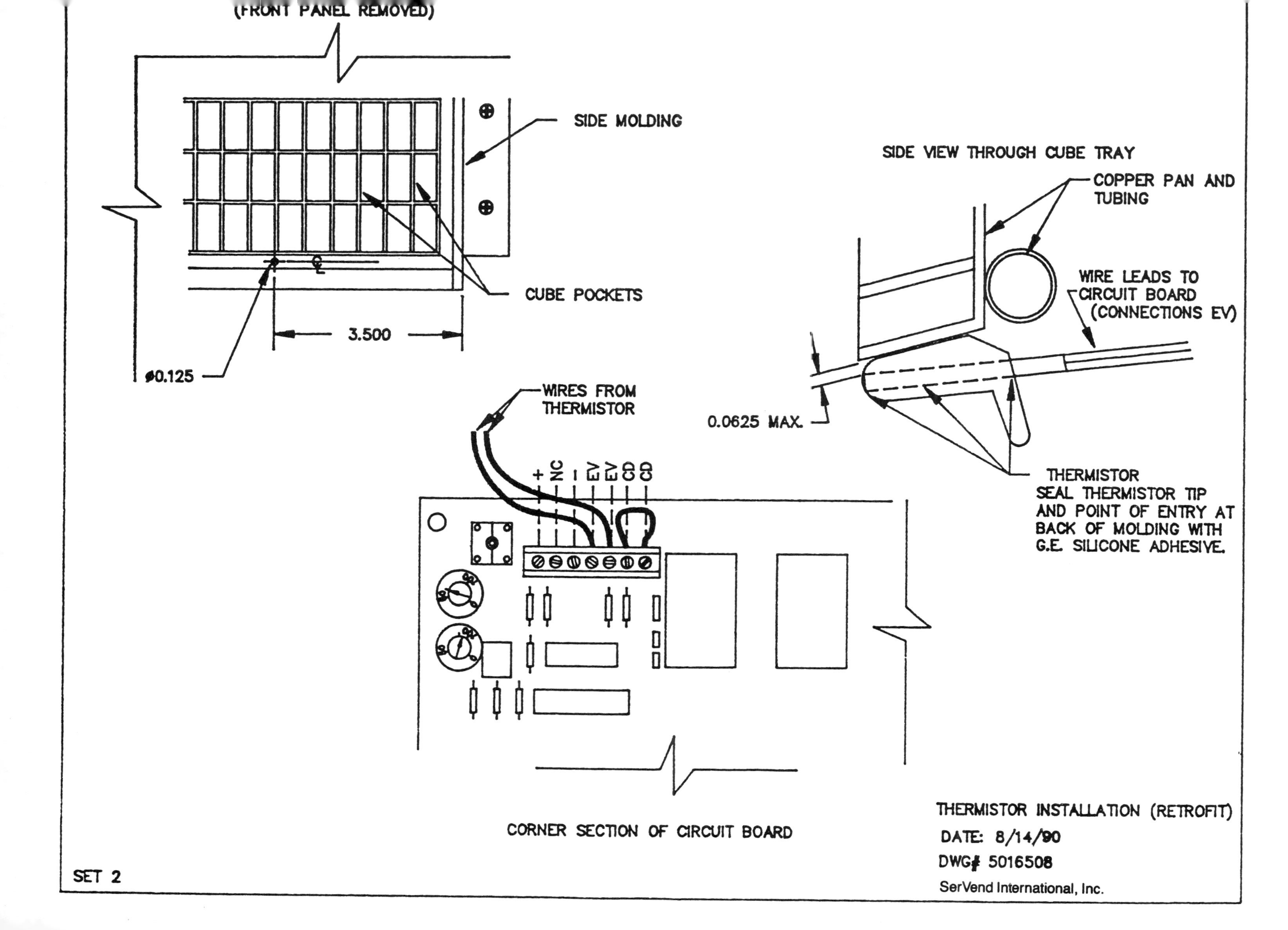
(FRONT PANEL REMOVED)
SIDE MOLDING
CUBE POCKETS
3.500
ø0.125
SIDE VIEW THROUGH CUBE TRAY
COPPER PAN AND TUBING
WIRE LEADS TO CIRCUIT BOARD (CONNECTIONS EV)
0.0625 MAX.
THERMISTOR
SEAL THERMISTOR TIP AND POINT OF ENTRY AT BACK OF MOLDING WITH G.E. SILICONE ADHESIVE.
WIRES FROM THERMISTOR
+ NC - EV EV CD CD
CORNER SECTION OF CIRCUIT BOARD
THERMISTOR INSTALLATION (RETROFIT)
DATE: 8/14/90
DWG# 5016508
SerVend International, Inc.
SET 2

PROXIMITY SWITCH CHECK PROCEDURE

To check proximity switch, switch must be connected to circuit board.

NO TARGET:

A. Cuber will not start (except first cycle warm up after long off period.

B. Voltage between

+ and N/C	13 to 24 VDC
+ and -	13 to 24 VDC
N/C and -	approx. 61 MVDC

WITH TARGET:

A. Voltage between

N/C and -	13 to 24 VDC
+ and -	13 to 24 VDC
+ and N/C erratic	

Voltage above 24 VDC will damage switch, replace circuit board and switch.

Voltage below 13 VDC switch will not operate, replace circuit board.

With voltage between 13 and 24 VDC, if LED on Switch is out or does not blink with target movement, replace switch.

If the jumper wire is loose (or open), unit will start in freeze, pull down and remain in harvest until thermodisc opens.

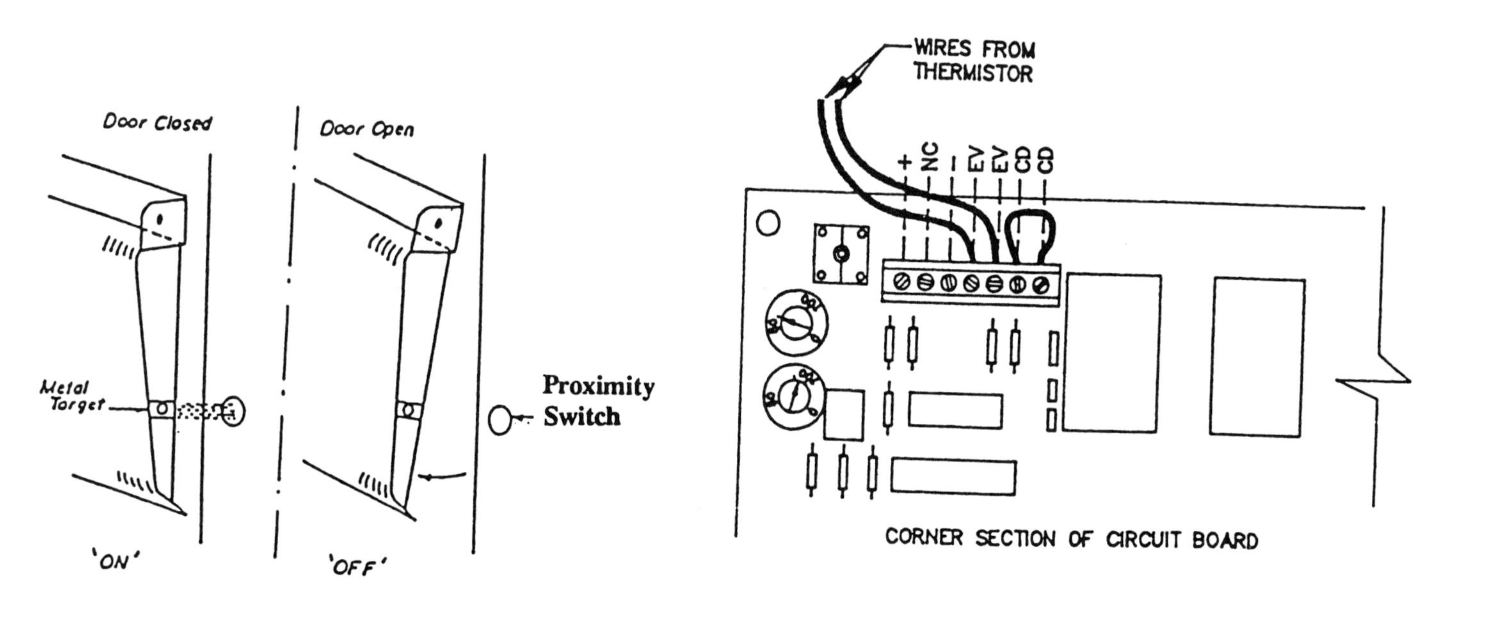

CIRCUIT BOARD INSTALLATION

A. This board designed for universal usage:

for 115 volt **(C2 & C4 series)**

White on #1 and #3
Black on #2 and #4

for 208/240 volt **(C7 & C9 series)**

Black on #2 and #3
White on #1 and N/C

B. Installation:

1. Transfer connections one at a time, terminal locations are the same.

2. Set potentiometer dials to the same corresponding position as the removed board.

3. If ice size requires adjustment, use fine adjustment only.

4. If board is completely out of adjustment:

   a. Center "Fine" Pot to "50"

   b. Rotate "Coarse" Pot to "50"

   c. Build ice until slab appears to be correct thickness.

   d. With curtain in place, slowly advance the "Fine" Pot adjustments 1/2 to 1 division at a time, until 1/8" to 3/16" bridge is achieved.

START UP
TROUBLE SHOOTING

WATER LEVEL AND SIPHON ADJUSTMENT

1. WITH THE WATER PUMP OFF, ADJUST THE WATER LEVEL TO 3/8" (+/-) 1/8" BELOW THE OVERFLOW LEVEL.

2. START THE PUMP AND PERMIT THE WELL TO REFILL UNTIL THE FLOAT VALVE IS SATISFIED.

3. STOP PUMP AND CHECK SIPHON FOR OPERATION. SIPHON MUST ACTIVATE DRAIN WATER WELL AND THEN STOP COMPLETELY.

WATER RISES TO ACTIVATE THE SIPHON WHEN THE PUMP STOPS.

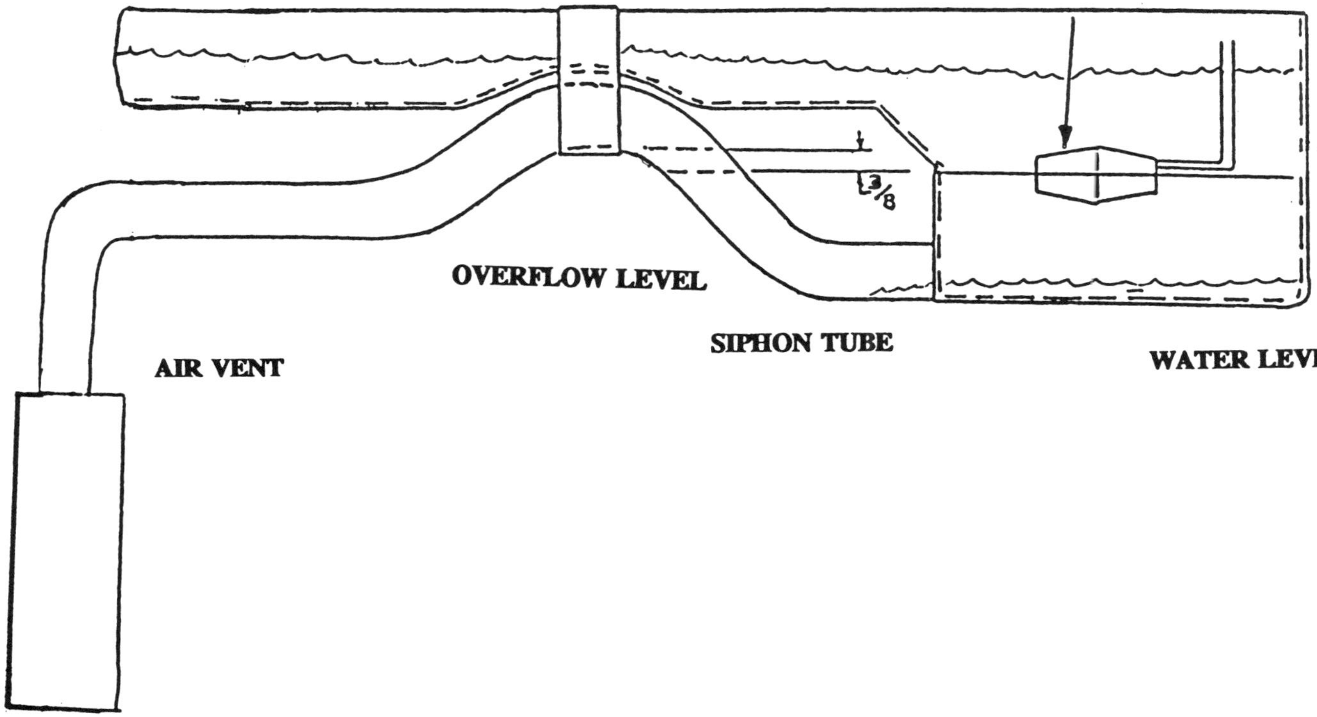

START UP OF HARVEST

WATER RESERVOIR

FLOAT VALVE

WATER LEVEL
DURING FREEZE CYCLE

3/8

OVERFLOW LEVEL

SIPHON TUBE

AIR VENT

WATER LEVEL AT END OF SIPHON ACTION

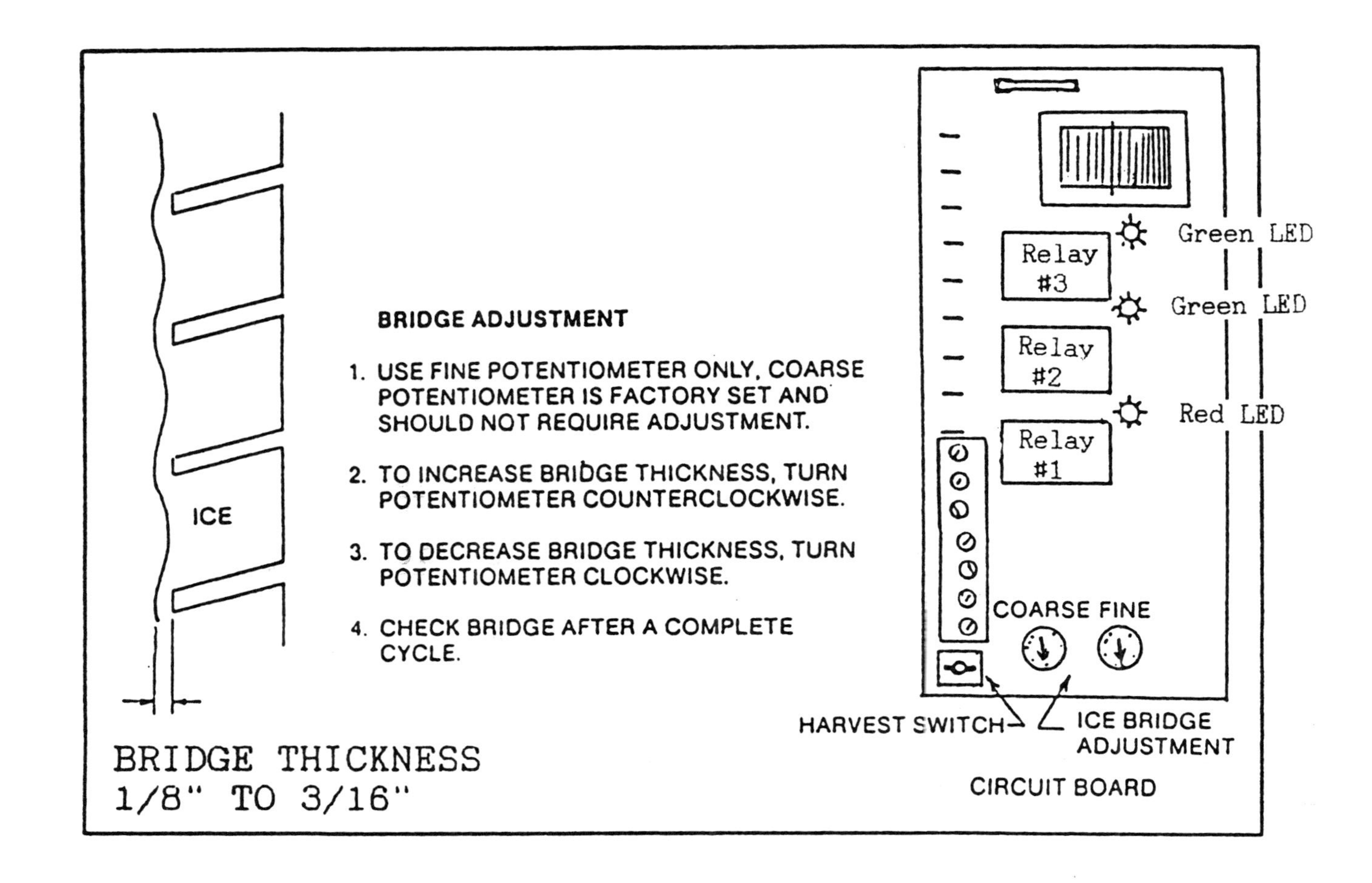

Relay #1 - Red LED - Hot Gas Solenoid
Relay #2 - Green LED - Compressor contactor
Relay #3 - Green LED - Waterpump, Condenser fan
Relay #2 & #3 (both Green LED) - Freeze Cycle

WATER CURTAIN ADJUSTMENT

ITEM #	DESCRIPTION
14	PIN - CURTAIN
15	PLATE - CURTAIN PIN
16	BLOCK - CURTAIN PIT
17	NUT - SPEED PIN

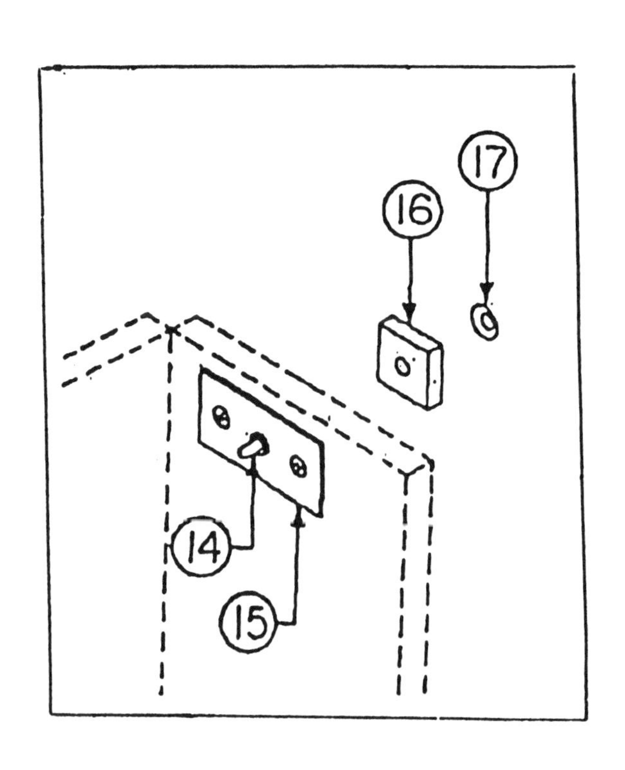

CURTAIN SLIDE OF DRIFT -

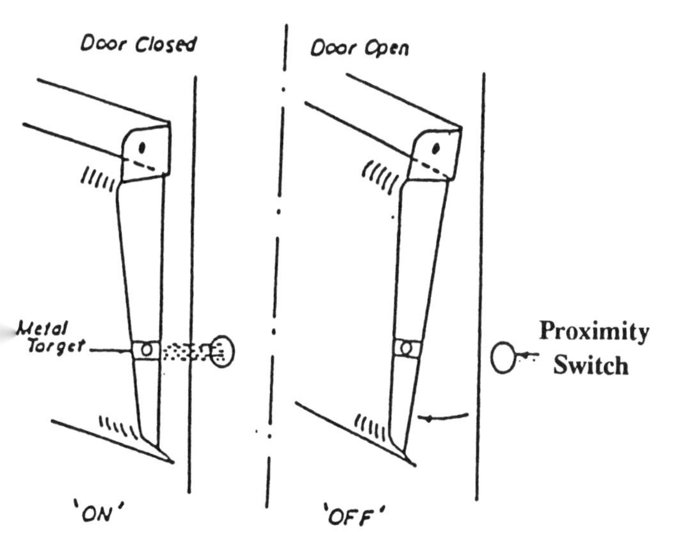

STEP 1: HORIZONTAL MOVEMENT (LEFT OR RIGHT) MAY PERMIT THE CURTAIN TO HANG UP ON THE PROXIMITY SWITCH OR MOVE OUT OF RANGE OF THE MAGNETIC FIELD.

TO CHECK PROPER CURTAIN PLACEMENT, PLACE THE UNIT IN HARVEST AND MOVE THE CURTAIN TO THE LEFT. IF HARVEST LIGHT (RED) GOES OUT, SHIM THE CURTAIN.

STEP 2: IF THE HARVEST LIGHT (RED) STAYS ON, MOVE THE CURTAIN TO THE RIGHT. IF THE CURTAIN TOUCHES THE PROXIMITY SWITCH, ADJUST THE SWITCH AND REPEAT STEP 1.

TO INSTALL THE SHIM: LOOSEN THE TWO SCREWS ON THE RIGHT CURTAIN PLATE AND SLIDE THE SHIM BEHIND IT. TIGHTEN SCREWS AND TEST CURTAIN OPERATION.

START UP TROUBLE SHOOTING

PROBLEMS WITH AN UNLEVEL CUBER - IT IS VERY IMPORTANT TO MAKE SURE THE CUBER IS LEVEL DURING INSTALLATION TO ENSURE PROPER HARVEST AND ICE PRODUCTION.

IF THE EVAPORATOR IS TIPPED TO FAR BACKWARDS (TOWARDS THE BACK OF THE UNIT) THE SLOPE IS LOST AND ICE WILL NOT HARVEST.

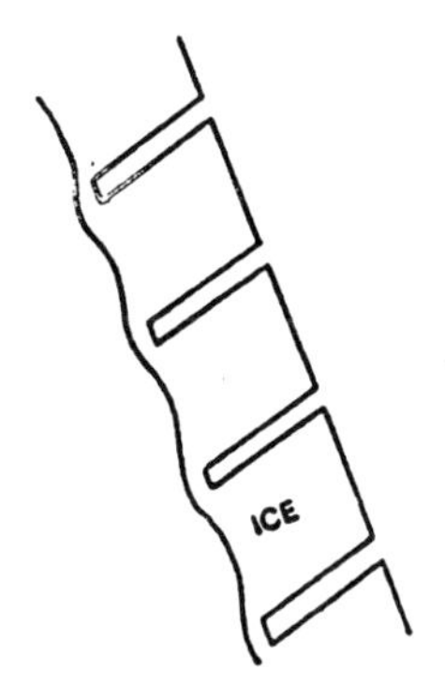

IF THE EVAPORATOR IS TIPPED TO FAR FORWARD, (TOWARDS THE FRONT OF THE UNIT) THE WATER CANNOT MIGRATE TO THE BACK OF THE CUBE POCKET, WHICH WILL CAUSE POORLY FORMED CUBES.

PROBLEM ANALYSIS GUIDE

ROBLEM	CAUSE	CORRECTIVE ACTION
uber not operating.	Toggle switch in center "off" position.	Place switch in "on" position
. Lights "off" no power to circuit board.	Toggle switch	See check out - bad, replace.
	Thermodisc open	See check out - replace if bad, reset automatically, determine reason for open switch.
	High pressure cut-out open (water models only)	Depress manual reset, determine cause: Water supply shut-off, Water pressure too low, Water valve defective or out of adjustment, Water condenser dirty or corroded, Unit over-charged, Water inlet pressure too high.
. Lights "off" power to circuit board.	No target on curtain.	Replace target.
	Target not in proximity switch field.	Water curtain drifting out of switch range, reduce clearance between curtain and pivot pin bracket
		No curtain movement, adjust proximity switch.
	Fuse blown on circuit board (8 amp).	Replace fuse.
	Faulty circuit board.	See check out, replace.
	Faulty proximity switch.	See check out, replace.
	Unit off-on bin full.	Remove ice from door, eliminate curtain hang up.

PROBLEM ANALYSIS GUIDE

...hts "on" on ...cuit board.	Compressor does not run.	
	a. Check circuit board.	Replace.
	b. Check contactor.	Replace.
	c. Compressor overload open.	Permit overload to cool and reset.
	d. Check compressor and start components.	Replace.
	Compressor run does not cool.	
	a. Internal by-pass open, compressor noisy.	Permit pressures to equalize.
	b. Hot gas-solenoid leaking.	Replace.
	c. Defective expansion valve.	Replace.
	d. Inefficient compressor	Replace.
	e. Low charge.	Leak check. Recharge.
...r remains in ...ze cycle.	Liquid line thermister open.	
	a. Check wire connection at circuit board.	Tighten, re-attach.
	b. Check thermister.	Replace.
	Evaporator thermister shorted.	
	a. Check wire connections.	Separate.
	b. Check thermister.	Replace.
	Ice bridge setting, set too low.	Adjust.
	Expansion valve failure (will not pull down).	Tighten bulb, replace, see check out.

ber remains in arvest cycle.	Evaporator thermister open (starts in harvest).	
	a. Loose connection at circuit board.	Tighten, re-attach wire.
	b. Check thermister.	Replace.
	Proximity switch light "out".	
	a. Loose wire connection at circuit board.	Tighten, re-attach wire.
	b. Proximity switch bad.	Replace, see check out.
	Proximity switch light "on".	
	a. Circuit board failure.	Replace, see check out.
	b. Ice weight too light.	Adjust bridge.
	c. Potentiometer set too high.	Adjust bridge.
	d. Water curtain stuck.	Check, curtain frozen to plate. Curtain hung on water pan, proximity switch, water hose.
ong freeze cycle.	Siphon not breaking.	
	a. Siphon not vented.	Correct.
	b. Float set too high.	Adjust.
	c. Float stuck.	Clean or replace.
	d. Flow control washer missing.	Replace.
	Water leaking around pan or curtain.	Correct.
	Dirty condenser or fan blade.	Clean.

PROBLEM ANALYSIS GUIDE

	Louvers at condenser obstructed.	Remove obstruction.
	Ambient or water temperature too high.	Reduce.
	Low charge.	Check leak evacuate. Recharge.
	Solenoid valve leaking (not seated).	Replace.
	Water regulator valve set too high or stuck (water units only).	Adjust, clean or replace.
Long harvest cycles.	Ice weight set too light.	Adjust bridge.
	Water curtain restricted or frozen in.	Remove restriction.
	Low head pressure	
	a. Ambient too low.	
	b. Water valve set too low (water units only) or leaking during harvest.	Adjust water regulator valve or replace.
	Scale build up on evaporator.	Clean. See instructions.
	Solenoid valve not opening. Slow rise of low side pressure.	See pressure charts. replace valve.
	Expansion valve leaking.	Replace valve.
	Unit not level.	Level.
Ice weight light at top of plate and heavy at bottom.	Siphon operating intermittently or continuously leaking.	
	a. Water level set too high.	Adjust float.
	b. Float stuck open.	Replace.

	c. Flow control washer missing from float.	Replace.
	d. Siphon not vented.	Correct.
	Water leaking round curtain and pan.	Correct.
	a. Curtain frozen in	
	Expansion valve flooding or starving.	Tighten and seal bulb. Replace valve.
	Low charge.	Leak check, evacuate and charge.
	Solenoid valve leaking.	Replace.
Soft white ice or water pump not pumping.	Distribution tube or water pump scaled.	Disassemble Pump impeller and shaft clean, clean distribution system.
	a. Water volume insufficient.	
	Inadequate water supply.	
	a. Water pressure too low.	Correct.
	b. Float plugges or damaged.	Replace.
	c. Water filter plugged.	Replace.
	d. Float set too low.	Adjust.
	Water pump drawing air.	
	a. Float set too high causing unit to siphon.	Adjust float.
	b. Air leaking by pump o'ring.	Tighten screws on face plate.
	c. Pump sucking air through hose at pump inlet or pan.	Replace hose.
	d. Hose too high at distribution tube inlet causing air lock.	Shorten hose.

PROBLEM ANALYSIS GUIDE

Unit turns off before bin is full.	Curtain target moved out of range of proximity switch.	
	a. Clearance too great at curtain ends, drifts away from proximity switch.	Add spacer, adjust switch.
	b. Curtain mechanically hang up.	Correct obstruction.
	c. Ice weight too light, shell cubes holding curtain open.	Adjust ice bridge.
Ice weight light difficult to adjust after normal operation.	Water pump and distribution system scaled. A. Water volume insufficient.	Clean distribution system, Disassemble and clean pump impeller and shaft.

OPERATING
CHARACTERISTICS

AIR-COOLED MODELS AND W[illegible]ER-COOLED MODELS

SERIES MODEL	C2-A	C2-W	C4-A	C4-W	C7-A	C7-W	C9-A	C9-W
UNIT								
VOLTS	115	115	115	115	208/230	208/230	208/230	208/230
PHASE	1	1	1	1	1	1	1	1
HERTZ	60	60	60	60	60	60	60	
MIN CIRCUIT AMPS	12.5	11.7	15.4	14.3	12.0	11.3	14.6	13.8
MAX FUSE SIZE	15A	15A	20A	20A	20A	20A	25A	25A
COMPRESSOR								
VOLTS	115	115	115	115	208/230	208/230	208/230	208/230
PHASE	1	1	1	1	1	1	1	1
HERTZ	60	60	60	60	60	60	60	60
LRA	42.0	42.0	51.0	51.0	53.0	53.0	64.0	64.0
RLA	7.4	7.4	9.5	9.5	8.1	8.1	10.0	10.0
CONDENSER FAN MOTOR								
VOLTS	115		115		230		230	
PHASE/HERTZ	1/60		1/60		1/60		1/60	
AMPS	0.82		1.1		0.7		0.85	
WATTS	9		25		35		50	
WATER PUMP								
VOLTS	115	115	115	115	230	230	230	230
PHASE/HERTZ	1/60	1/60	1/60	1/60	1/60	1/60	1/60	1/60
AMPS	1.05	1.05	1.05	1.05	0.55	0.55	0.55	.55
REFRIGERANT								
TYPE	R-502	R-502	R-502	R-502	R-502	R-502	R-502	R-502
WEIGHT	20 oz.	14 oz.	24 oz.	16 oz.	30 oz.	17 oz.	38 oz.	18 oz.

MODEL : C2-A

OPERATING CHARACTERISTICS

TEMP. IN DEGREES F AIR/WATER	* FREEZE CYCLE			HARVEST CYCLE			DESIGNED ICE WT. RANGE	
	HEAD PRESSURE MAX/MIN	SUCTION PRESSURE MAX/MIN	CYCLE TIME MINUTES	HEAD PRESSURE MAX/MIN	SUCTION PRESSURE MAX/MIN	CYCLE TIME MINUTES	LOW	HIGH
40/40	/	/	-	/	/	-		
60/50	165 / 145	39 / 22	10.5 - 11.5	125 / 115	95 / 85	1.0 - 1.75	2.2	2.5
70/50	185 / 165	41 / 23	11.0 - 13.0	140 / 130	120 / 100	1.0 - 1.75	2.1	2.4
80/70	215 / 190	45 / 24	12.5 - 14.5	160 / 150	135 / 125	1.0 - 1.5	2.1	2.4
90/70	250 / 220	47 / 24	14.0 - 17.0	180 / 170	155 / 145	0.5 - 1.5	2.2	2.5
100/70	285 / 250	49 / 25	20.0 - 22.0	210 / 200	160 / 150	0.5 - 1.5	2.25	2.55
110/70	320 / 280	54 / 25	26.0 - 29.0	235 / 225	165 / 155	0.5 - 1.5	2.5	2.8

MODEL : C2-W

AMBIENT TEMP. IN DEGREES F	* FREEZE CYCLE			HARVEST CYCLE			DESIGNED ICE WT. RANGE	
	HEAD PRESSURE MAX/MIN	SUCTION PRESSURE MAX/MIN	CYCLE TIME MINUTES	HEAD PRESSURE MAX/MIN	SUCTION PRESSURE MAX/MIN	CYCLE TIME MINUTES	LOW	HIGH
40/40	230 / 227	44 / 32	10.0 - 11.0	130 / 120	100 / 90	0.5 - 1.5	1.75	2.05
70/50	230 / 226	45 / 29	12.5 - 13.5	140 / 130	115 / 105	0.5 - 1.5	1.9	2.2
70/70	230 / 226	45 / 29	13.5 - 14.5	140 / 130	115 / 105	0.5 - 1.5	1.9	2.2
90/70	230 / 226	47 / 29	17.5 - 18.5	140 / 130	120 / 110	0.5 - 1.5	2.1	2.3
90/90	234 / 224	47 / 29	17.5 - 18.5	150 / 140	130 / 120	0.5 - 1.5	2.1	2.3

NOTE: MAXIMUM PRESSURES TAKEN TWO MINUTES INTO CYCLE
MINIMUM PRESSURES TAKEN ONE MINUTE BEFORE END OF CYCLE

MODEL : C4-A

OPERATING CHARACTERISTICS

TEMP. IN DEGREES F AIR/WATER	* FREEZE CYCLE			HARVEST CYCLE			DESIGNED ICE WT. RANGE	
	HEAD PRESSURE MAX/MIN	SUCTION PRESSURE MAX/MIN	CYCLE TIME MINUTES	HEAD PRESSURE MAX/MIN	SUCTION PRESSURE MAX/MIN	CYCLE TIME MINUTES	LOW	HIGH
40/40	140 / 125	40 / 24	11.5 - 12.5	105 / 95	70 / 60	3.5 - 4.5	4.9	5.2
60/50	195 / 170	43 / 29	12.0 - 13.0	135 / 125	90 / 80	1.0 - 1.75	4.4	4.7
70/50	210 / 190	43 / 30	13.5 - 14.5	140 / 130	100 / 90	1.0 - 1.75	4.35	4.65
80/70	220 / 210	45 / 31	14.5 - 16.0	155 / 145	115 / 105	1.0 - 1.75	4.3	4.6
90/70	260 / 245	48 / 32	19.5 - 21.0	180 / 170	135 / 125	0.5 - 1.5	4.2	4.5
100/70	295 / 270	54 / 31	25.0 - 26.0	190 / 180	140 / 130	0.5 - 1.5	4.6	4.9
110/70	325 / 300	57 / 32	30.5 - 32.0	205 / 195	140 / 130	0.5 - 1.5	4.8	5.1

MODEL : C4-W

AMBIENT TEMP. IN DEGREES F	* FREEZE CYCLE			HARVEST CYCLE			DESIGNED ICE WT. RANGE	
	HEAD PRESSURE MAX/MIN	SUCTION PRESSURE MAX/MIN	CYCLE TIME MINUTES	HEAD PRESSURE MAX/MIN	SUCTION PRESSURE MAX/MIN	CYCLE TIME MINUTES	LOW	HIGH
40/40	226 / 224	45 / 36	10.5 - 11.5	130 / 120	90 / 80	1.0 - 2.0	3.75	4.15
70/50	230 / 227	47 / 36	13.5 - 14.5	130 / 120	95 / 85	0.5 - 1.75	4.2	4.5
70/70	230 / 226	47 / 36	16.5 - 17.5	135 / 125	95 / 85	0.5 - 1.5	4.2	4.5
90/70	230 / 226	48 / 36	17.5 - 19.0	140 / 130	105 / 95	0.5 - 1.5	4.2	4.5
90/90	234 / 226	56 / 36	19.0 - 21.0	145 / 135	110 / 100	0.5 - 1.5	4.2	4.5

*NOTE: MAXIMUM PRESSURES TAKEN TWO MINUTES INTO CYCLE
MINIMUM PRESSURES TAKEN ONE MINUTE BEFORE END OF CYCLE

MODEL : C7-A

OPERATING CHARACTERISTICS

TEMP. IN DEGREES F AIR/WATER	*FREEZE CYCLE			HARVEST CYCLE			DESIGNED ICE WT. RANGE	
	HEAD PRESSURE MAX/MIN	SUCTION PRESSURE MAX/MIN	CYCLE TIME MINUTES	HEAD PRESSURE MAX/MIN	SUCTION PRESSURE MAX/MIN	CYCLE TIME MINUTES	LOW	HIGH
40/40	165 / 150	34 / 23	7.5 - 8.5	130 / 120	60 / 50	3.8 - 4.5	4.8	5.1
60/50	205 / 180	35 / 26	8.5 - 9.5	155 / 145	70 / 60	1.0 - 1.75	4.9	5.2
70/50	215 / 200	35 / 27	9.5 - 10.5	160 / 150	80 / 70	.75 - 1.5	5.1	5.4
80/70	245 / 215	40 / 26	12.5 - 13.0	175 / 160	90 / 80	.5 - 1.5	5.4	5.6
90/70	280 / 240	40 / 26	15.0 - 16.0	195 / 185	100 / 90	.5 - 1.5	5.5	5.8
100/70	320 / 250	40 / 26	17.5 - 18.5	220 / 210	120 / 110	.5 - 1.5	5.6	5.9
110/70	350 / 310	43 / 29	20.5 - 21.5	240 / 230	120 / 110	.5 - 1.5	5.7	6.0

MODEL : C7-W

AMBIENT TEMP. IN DEGREES F	*FREEZE CYCLE			HARVEST CYCLE			DESIGNED ICE WT. RANGE	
	HEAD PRESSURE MAX/MIN	SUCTION PRESSURE MAX/MIN	CYCLE TIME MINUTES	HEAD PRESSURE MAX/MIN	SUCTION PRESSURE MAX/MIN	CYCLE TIME MINUTES	LOW	HIGH
40/40	226 / 222	36 / 29	8.5 - 9.5	155 / 145	80 / 70	1.0 - 1.75	4.6	4.9
70/50	228 / 224	36 / 27	9.5 - 10.5	160 / 150	85 / 75	1.0 - 1.75	4.9	5.2
70/70	230 / 224	36 / 26	10.5 - 11.5	160 / 150	85 / 75	1.0 - 1.75	4.9	5.2
90/70	230 / 224	37 / 26	13.0 - 14.0	180 / 170	95 / 85	0.5 - 1.5	5.1	5.4
90/90	235 / 228	39 / 26	15.5 - 16.5	180 / 170	100 / 90	0.5 - 1.5	5.3	5.5

*NOTE: MAXIMUM PRESSURES TAKEN TWO MINUTES INTO CYCLE
MINIMUM PRESSURES TAKEN ONE MINUTE BEFORE END OF CYCLE

MODEL : C9-A

OPERATING CHARACTERISTICS

TEMP. IN DEGREES F AIR/WATER	*FREEZE CYCLE			HARVEST CYCLE			DESIGNED ICE WT. RANGE	
	HEAD PRESSURE MAX/MIN	SUCTION PRESSURE MAX/MIN	CYCLE TIME MINUTES	HEAD PRESSURE MAX/MIN	SUCTION PRESSURE MAX/MIN	CYCLE TIME MINUTES	LOW	HIGH
40/40	160 / 140	33 / 20	6.5 - 7.5	120 / 110	60 / 50	4.0 - 6.0	5.9	6.2
60/50	195 / 175	35 / 23	7.0 - 8.0	145 / 135	80 / 70	1.0 - 2.0	5.8	6.1
70/50	220 / 200	37 / 23	8.0 - 9.0	155 / 145	90 / 80	1.0 - 1.75	6.2	6.5
80/70	250 / 225	40 / 25	10.0 - 11.0	170 / 160	95 / 85	1.0 - 1.75	6.2	6.5
90/70	285 / 250	43 / 25	12.5 - 13.5	190 / 180	105 / 95	0.5 - 1.5	6.1	6.4
100/70	325 / 280	45 / 25	15.0 - 16.0	215 / 210	110 / 105	0.5 - 1.5	6.2	6.5
110/70	350 / 310	47 / 32	16.0 - 18.0	230 / 220	120 / 110	0.5 - 1.5	6.4	6.7

MODEL : C9-W

AMBIENT TEMP. IN DEGREES F	*FREEZE CYCLE			HARVEST CYCLE			DESIGNED ICE WT. RANGE	
	HEAD PRESSURE MAX/MIN	SUCTION PRESSURE MAX/MIN	CYCLE TIME MINUTES	HEAD PRESSURE MAX/MIN	SUCTION PRESSURE MAX/MIN	CYCLE TIME MINUTES	LOW	HIGH
40/40	230 / 226	40 / 27	8.5 - 9.5	140 / 130	75 / 65	1.0 - 2.0	6.2	6.5
70/50	232 / 228	40 / 27	10.0 - 11.0	140 / 130	80 / 70	1.0 - 1.75	6.4	6.7
70/70	232 / 228	40 / 27	11.0 - 12.5	145 / 135	80 / 70	1.0 - 1.75	6.5	6.8
90/70	232 / 228	40 / 27	12.5 - 13.5	155 / 145	90 / 80	1.0 - 1.75	6.6	6.9
90/90	235 / 230	41 / 28	14.5 - 15.5	160 / 150	90 / 80	1.0 - 1.75	6.7	7.0

*NOTE: MAXIMUM PRESSURES TAKEN TWO MINUTES INTO CYCLE
MINIMUM PRESSURES TAKEN ONE MINUTE BEFORE END OF CYCLE

ELECTRICAL DIAGRAMS

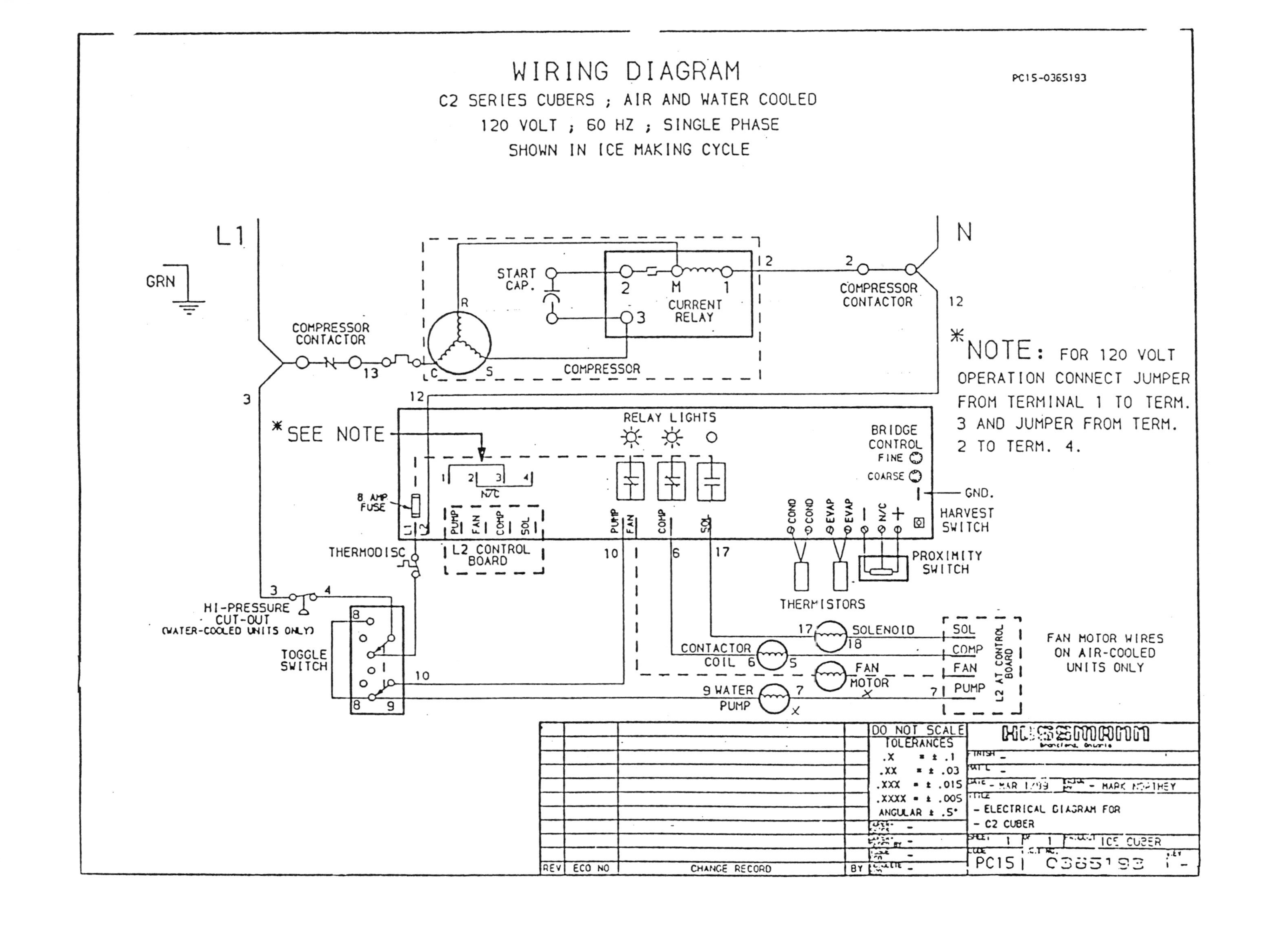
WIRING DIAGRAM
C2 SERIES CUBERS ; AIR AND WATER COOLED
120 VOLT ; 60 HZ ; SINGLE PHASE
SHOWN IN ICE MAKING CYCLE
PC15-0365193
L1
N
GRN
COMPRESSOR CONTACTOR
START CAP.
CURRENT RELAY
COMPRESSOR
*NOTE: FOR 120 VOLT OPERATION CONNECT JUMPER FROM TERMINAL 1 TO TERM. 3 AND JUMPER FROM TERM. 2 TO TERM. 4.
*SEE NOTE
RELAY LIGHTS
BRIDGE CONTROL
FINE
COARSE
GND.
HARVEST SWITCH
8 AMP FUSE
N/C
PUMP
FAN
COMP
SOL
L2 CONTROL BOARD
COND
EVAP
PROXIMITY SWITCH
THERMODISC
THERMISTORS
HI-PRESSURE CUT-OUT
(WATER-COOLED UNITS ONLY)
TOGGLE SWITCH
SOLENOID
CONTACTOR COIL
FAN MOTOR
WATER PUMP
L2 AT CONTROL BOARD
FAN MOTOR WIRES ON AIR-COOLED UNITS ONLY
DO NOT SCALE
TOLERANCES
.X = ± .1
.XX = ± .03
.XXX = ± .015
.XXXX = ± .005
ANGULAR ± .5°
- ELECTRICAL DIAGRAM FOR
- C2 CUBER
ICE CUBER
PC151
REV
ECO NO
CHANGE RECORD
BY

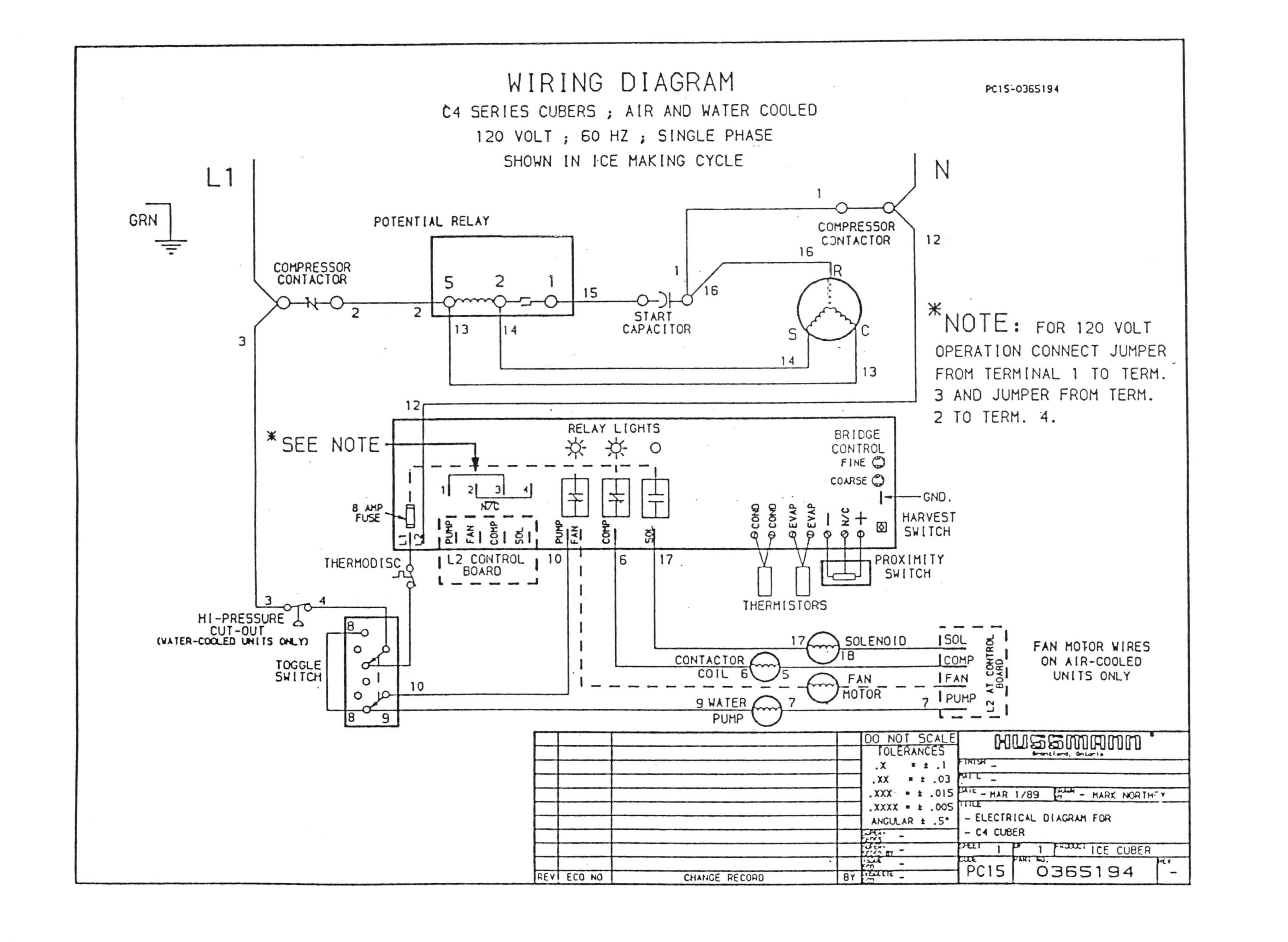

WIRING DIAGRAM
C4 SERIES CUBERS ; AIR AND WATER COOLED
120 VOLT ; 60 HZ ; SINGLE PHASE
SHOWN IN ICE MAKING CYCLE
PC15-0365194
L1
N
GRN
POTENTIAL RELAY
COMPRESSOR CONTACTOR
COMPRESSOR CONTACTOR
START CAPACITOR
*NOTE: FOR 120 VOLT OPERATION CONNECT JUMPER FROM TERMINAL 1 TO TERM. 3 AND JUMPER FROM TERM. 2 TO TERM. 4.
*SEE NOTE
RELAY LIGHTS
BRIDGE CONTROL
FINE
COARSE
GND.
HARVEST SWITCH
PROXIMITY SWITCH
THERMISTORS
8 AMP FUSE
THERMODISC
L2 CONTROL BOARD
HI-PRESSURE CUT-OUT
(WATER-COOLED UNITS ONLY)
TOGGLE SWITCH
SOLENOID
CONTACTOR COIL
FAN MOTOR
WATER PUMP
L2 AT CONTROL BOARD
FAN MOTOR WIRES ON AIR-COOLED UNITS ONLY
DO NOT SCALE
TOLERANCES
.X = ± .1
.XX = ± .03
.XXX = ± .015
.XXXX = ± .005
ANGULAR ± .5°
HUSSMANN
- MAR 1/89
- MARK NORTHEY
- ELECTRICAL DIAGRAM FOR
- C4 CUBER
ICE CUBER
PC15
0365194
REV
ECO NO
CHANGE RECORD
BY

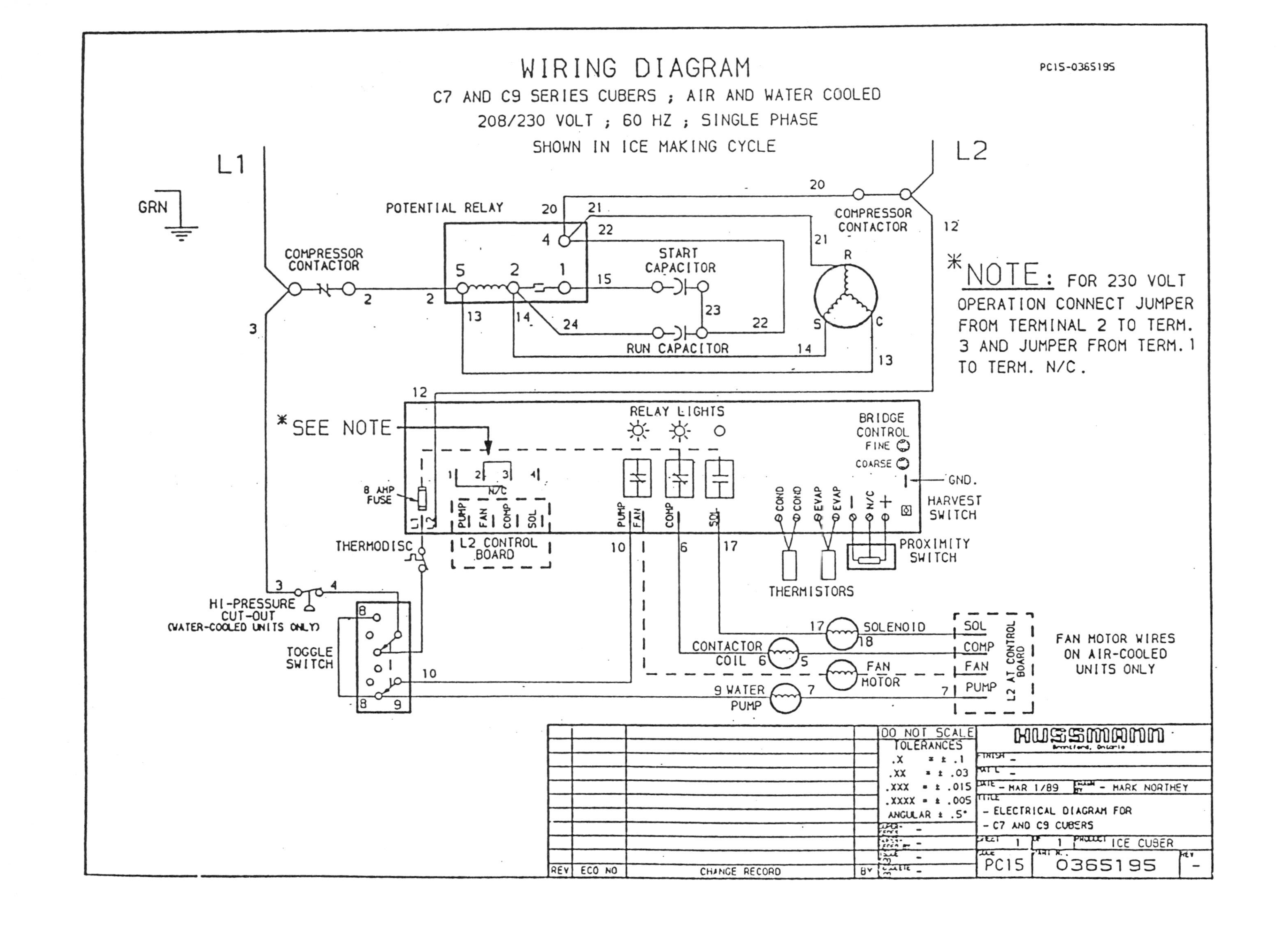
WIRING DIAGRAM
C7 AND C9 SERIES CUBERS ; AIR AND WATER COOLED
208/230 VOLT ; 60 HZ ; SINGLE PHASE
SHOWN IN ICE MAKING CYCLE
PC15-0365195
L1
L2
GRN
POTENTIAL RELAY
COMPRESSOR CONTACTOR
COMPRESSOR CONTACTOR
START CAPACITOR
RUN CAPACITOR
R
S
C
*NOTE: FOR 230 VOLT OPERATION CONNECT JUMPER FROM TERMINAL 2 TO TERM. 3 AND JUMPER FROM TERM. 1 TO TERM. N/C.
*SEE NOTE
RELAY LIGHTS
BRIDGE CONTROL
FINE
COARSE
GND.
HARVEST SWITCH
8 AMP FUSE
N/C
L1
L2
PUMP
FAN
COMP
SOL
L2 CONTROL BOARD
THERMODISC
COND
COND
EVAP
EVAP
PROXIMITY SWITCH
THERMISTORS
HI-PRESSURE CUT-OUT
(WATER-COOLED UNITS ONLY)
TOGGLE SWITCH
SOLENOID
CONTACTOR COIL
FAN MOTOR
WATER PUMP
L2 AT CONTROL BOARD
FAN MOTOR WIRES ON AIR-COOLED UNITS ONLY
DO NOT SCALE
TOLERANCES
.X = ± .1
.XX = ± .03
.XXX = ± .015
.XXXX = ± .005
ANGULAR ± .5°
HUSSMANN
Brantford, Ontario
DATE - MAR 1/89
MARK NORTHEY
TITLE
- ELECTRICAL DIAGRAM FOR
- C7 AND C9 CUBERS
SHEET 1 OF 1
PRODUCT ICE CUBER
PC15
0365195
REV
ECO NO
CHANGE RECORD
BY

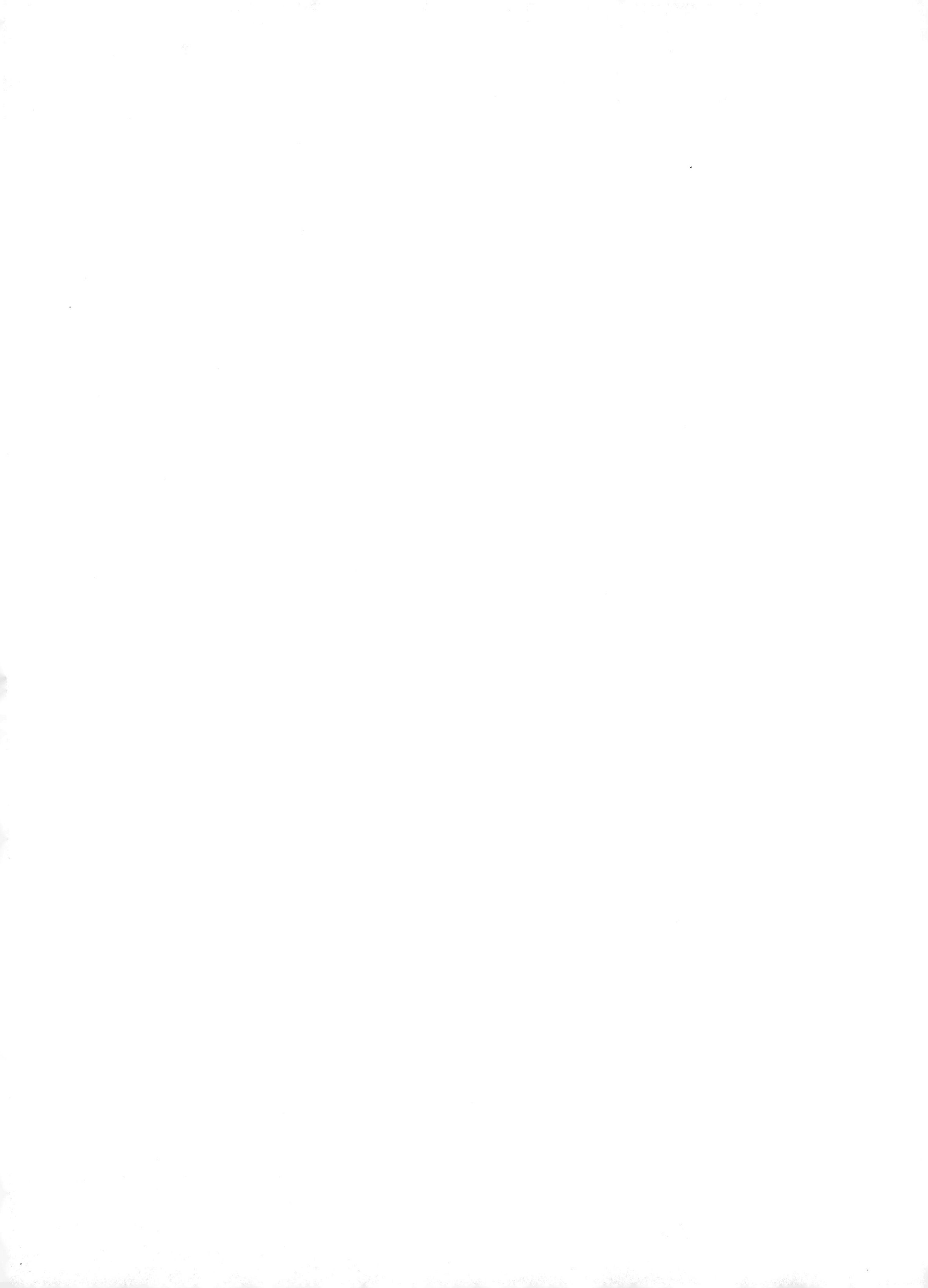

NOW AVAILABLE FROM BUSINESS NEWS PUBLISHING COMPANY

Troubleshooting & Servicing Air Conditioning Equipment

Don Swenson

This practical, concise book presents an in-depth study of troubleshooting and repairing air conditioning and heat pump equipment. The material is organized to progress from general principles of safety, tools, chart interpretation, instrumentation and troubleshooting, to descriptions of specific systems and components. The rest of the text is then devoted to applying this general information to commonly occurring problems in each of the component parts and subsystems of air conditioning, heat pump and refrigeration units.

Special features of this text include:

- **A focused approach:** The book clearly and concisely covers essentials and procedures.
- **Concept review:** Early chapters provide a review of system operation and provide basic but essential information for diagnosing and servicing.
- **Tool and safety information:** Excellent coverage is provided concerning the hazards and safety procedures required when working with air conditioning and heat pump equipment.
- **Troubleshooting charts:** Extensive troubleshooting and servicing flow-charts provide concise summaries of procedures described in the text and serve as quick diagnosing reference for job-site situations.
- **Systematic coverage:** A general systematic method of troubleshooting air conditioning and heat pump equipment is presented.
- **Review problems:** A number of problems and questions are included at the end of each chapter to aid in understanding the material presented when used in a training environment.

Troubleshooting and Servicing Air Conditioning Equipment is intended for use by both service professionals and technicians in training. A great addition to any field service professional's troubleshooting arsenal.

Table of Contents

Chapter 1: **Troubleshooting and Service: Basic Concept**
Chapter 2: **Tools, Instruments and Supplies**
Chapter 3: **Hazards and Safety**
Chapter 4: **Troubleshooting and Service**
Chapter 5: **Refrigeration Systems**
Chapter 6: **Air Conditioning System**
Chapter 7: **Troubleshooting Charts**
Chapter 8: **Service Charts**
Chapter 9: **Electric Power Supply System**
Chapter 10: **Control System**
Chapter 11: **Electric Motors**
Chapter 12: **Servicing Electric Motors and Controls**
Chapter 13: **Servicing Procedures Common to All Refrigeration Systems**
Chapter 14: **Compressor and High-Pressure Side of the System**
Chapter 15: **Servicing the High Pressure Side of the System**
Chapter 16: **Expansion Device and Low Pressure Side of the System**
Chapter 17: **Servicing the Low-Pressure Side of the Systems**
Chapter 18: **Air Distribution System**
Chapter 19: **Heat Pump System**
Chapter 20: **Service Procedures Common to All Heat Pump Systems**
Chapter 21: **Heat Pump Service Charts**
Chapter 22: **Checking Specific Parts of Heat Pump Systems**
Chapter 23: **Maintenance**

Technical Book Division • Business News Publishing Co. • P. O. Box 2600 • Troy • MI • 48007 • (313)362-3700

THE BEST HEAT PUMP REFERENCE AVAILABLE

Finally, all the information you need to install, repair and maintain heat pumps has been compiled into one up-to-date, comprehensive guide. *Heat Pump Systems & Service* is the first in a new series of professional service guides from the BNP Technical Book Division — leading hvac/r publisher.

The text contains 8 comprehensive sections describing heat pump origination, operation, and components, all supplemented with illustrations, charts, troubleshooting tips, and repair strategies. The appendix contains manufacturers' data for selected models. Includes instructions for—

- **unit placement**
- **installation & charging**
- **service & maintenance**
- **electrical connections & terminal ID**
- **accessories & replacement parts**

Knowing how specific units are configured makes service easier and more accurate, and improves your reputation as a trustworthy serviceman.

The entire book format makes understanding and accessing the information you need even easier. Written in a clear, straightforward style, this 8 1/2 x 11, 380 page volume is bound so it can be laid flat, folded over, or propped up for immediate reference when working in the field.

Includes select manufacturer's data covering the most popular models from:

• Carrier • Coleman • Comfortmaker • Day & Night, Payne • Goettl •
• Heil-Quaker • Lennox • Rheem • Sun Dial • Weathertron • York •

Never before has so much valuable, manufacturer specific heat pump information been collected in one handy source. Every service professional or tech-in-training will find this book absolutely indispensable.

PUT THIS VERSATILE BOOK TO WORK FOR YOU TODAY!

Technical Book Division • Business News Publishing Company
P. O. Box 2600 • Troy • Michigan • 48007

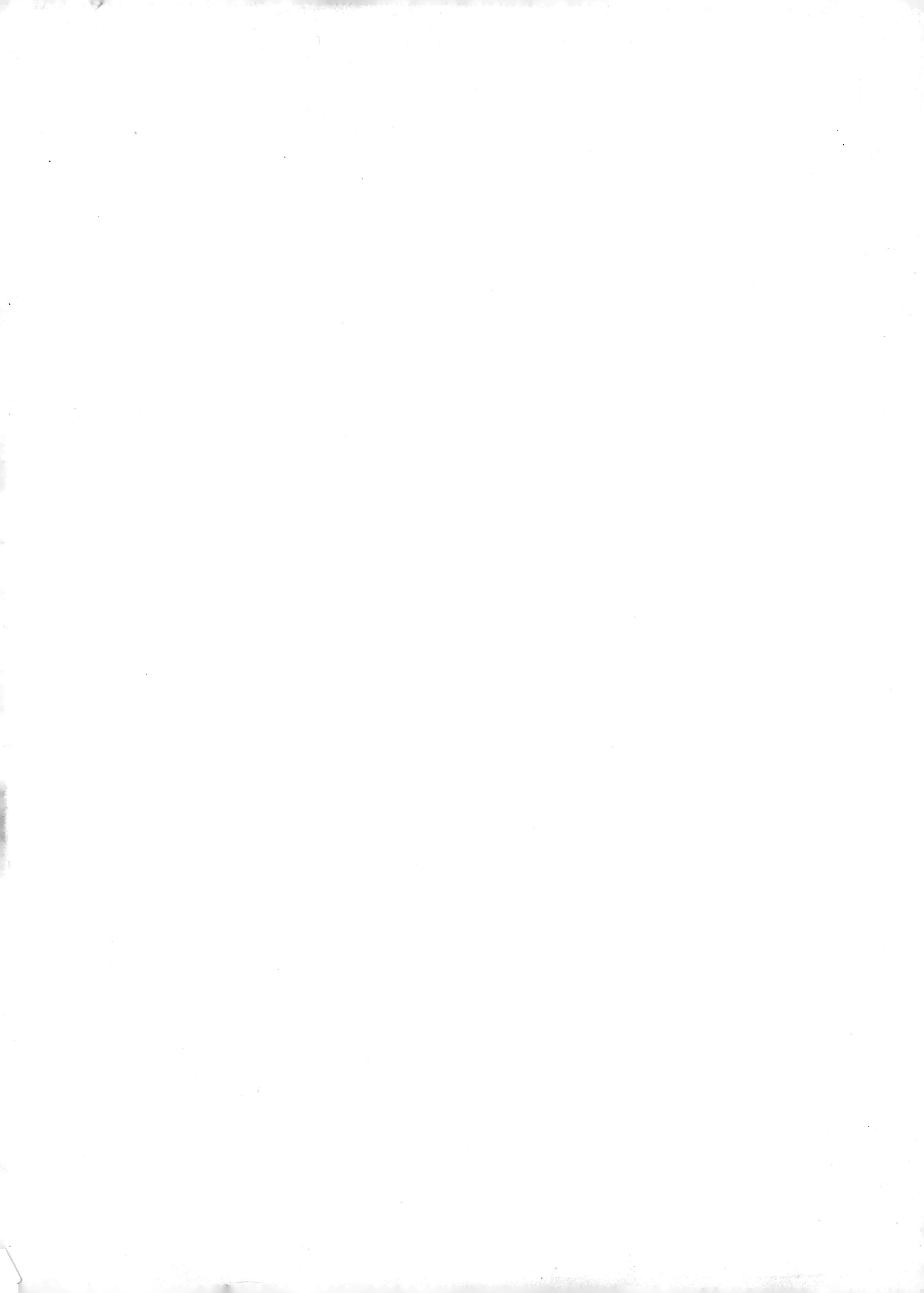